W9-BAI-874

IAASTD

International Assessment of Agricultural Knowledge, Science and Technology for Development

Global Report

IAASTD

International Assessment of Agricultural Knowledge, Science and Technology for Development

UN
DP

UNEP

THE WORLD BANK

WHO

GLOBAL
ENVIRONMENT
FACILITY

IAASTD

International Assessment of Agricultural Knowledge, Science and Technology for Development

Global Report

Edited by

Beverly D. McIntyre
IAASTD Secretariat

Hans R. Herren
Millennium Institute

Judi Wakhungu
African Centre for
Technology Studies

Robert T. Watson
University of East Anglia

Copyright © 2009 IAASTD. All rights reserved. Permission to reproduce and disseminate portions of the work for no cost will be granted free of charge by Island Press upon request: Island Press, 1718 Connecticut Avenue, NW, Suite 300, Washington, DC 20009.

Island Press is a trademark of The Center for Resource Economics.

Library of Congress Cataloging-in-Publication data.

International assessment of agricultural knowledge, science and technology for development (IAASTD) : global report / edited by Beverly D. McIntyre . . . [et al.].
    p.   cm.
  Includes bibliographical references and index.
  ISBN 978-1-59726-538-6 (cloth : alk. paper) —
ISBN 978-1-59726-539-3 (pbk. : alk. paper)
  1. Agriculture—International cooperation. 2. Sustainable development. I. McIntyre, Beverly D. II. Title: Global report.
  HD1428.I544 2008
  338.9′27—dc22   2008046043

British Cataloguing-in-Publication data available.

Printed on recycled, acid-free paper

Interior and cover designs by Linda McKnight, McKnight Design, LLC.

Manufactured in the United States of America

10 9 8 7 6 5 4 3 2 1

Contents

Statement by Governments

All countries present at the final intergovernmental plenary session held in Johannesburg, South Africa in April 2008 welcome the work of the IAASTD and the uniqueness of this independent multistakeholder and multidisciplinary process, and the scale of the challenge of covering a broad range of complex issues. The Governments present recognize that the Global and sub-Global Reports are the conclusions of studies by a wide range of scientific authors, experts and development specialists and while presenting an overall consensus on the importance of agricultural knowledge, science and technology for development they also provide a diversity of views on some issues.

All countries see these Reports as a valuable and important contribution to our understanding on agricultural knowledge, science and technology for development recognizing the need to further deepen our understanding of the challenges ahead. This Assessment is a constructive initiative and important contribution that all governments need to take forward to ensure that agricultural knowledge, science and technology fulfills its potential to meet the development and sustainability goals of the reduction of hunger and poverty, the improvement of rural livelihoods and human health, and facilitating equitable, socially, environmentally and economically sustainable development.

In accordance with the above statement, the following governments accept the Global Report.

Armenia, Azerbaijan, Bahrain, Bangladesh, Belize, Benin, Bhutan, Botswana, Brazil, Cameroon, China (People's Republic of), Costa Rica, Cuba, Democratic Republic of Congo, Dominican Republic, El Salvador, Ethiopia, Finland, France, Gambia, Ghana, Honduras, India, Iran, Ireland, Kenya, Kyrgyzstan, Lao People's Democratic Republic, Lebanon, Libyan Arab Jamahiriya, Maldives, Republic of Moldova, Mozambique, Namibia, Nigeria, Pakistan, Panama, Paraguay, Philippines, Poland, Republic of Palau, Romania, Saudi Arabia, Senegal, Solomon Islands, Swaziland, Sweden, Switzerland, United Republic of Tanzania, Timor-Leste, Togo, Tunisia, Turkey, Uganda, United Kingdom of Great Britain, Uruguay, Viet Nam, Zambia (58 countries)

While approving the above statement the following governments did not fully approve the Global Report and their reservations are entered in Annex G.

Australia, Canada, and United States of America (3 countries)

Foreword

The objective of the International Assessment of Agricultural Knowledge, Science and Technology for Development (IAASTD) was to assess the impacts of past, present and future agricultural knowledge, science and technology on the
- reduction of hunger and poverty,
- improvement of rural livelihoods and human health, and
- equitable, socially, environmentally and economically sustainable development.

The IAASTD was initiated in 2002 by the World Bank and the Food and Agriculture Organization of the United Nations (FAO) as a global consultative process to determine whether an international assessment of agricultural knowledge, science and technology was needed. Mr. Klaus Töepfer, Executive Director of the United Nations Environment Programme (UNEP) opened the first Intergovernmental Plenary (30 August – 3 September 2004) in Nairobi, Kenya, during which participants initiated a detailed scoping, preparation, drafting and peer review process.

The outputs from this assessment are a global and five subglobal reports; a global and five subglobal Summaries for Decision Makers; and a cross-cutting Synthesis Report with an Executive Summary. The Summaries for Decision Makers and the Synthesis Report specifically provide options for action to governments, international agencies, academia, research organizations and other decision makers around the world.

The reports draw on the work of hundreds of experts from all regions of the world who have participated in the preparation and peer review process. As has been customary in many such global assessments, success depended first and foremost on the dedication, enthusiasm and cooperation of these experts in many different but related disciplines. It is the synergy of these interrelated disciplines that permitted IAASTD to create a unique, interdisciplinary regional and global process.

We take this opportunity to express our deep gratitude to the authors and reviewers of all of the reports—their dedication and tireless efforts made the process a success. We thank the Steering Committee for distilling the outputs of the consultative process into recommendations to the Plenary, the IAASTD Bureau for their advisory role during the assessment and the work of those in the extended Secretariat. We would specifically like to thank the cosponsoring organizations of the Global Environment Facility (GEF) and the World Bank for their financial contributions as well as the FAO, UNEP, and the United Nations Educational, Scientific and Cultural Organization (UNESCO) for their continued support of this process through allocation of staff resources.

We acknowledge with gratitude the governments and organizations that contributed to the Multidonor Trust Fund (Australia, Canada, the European Commission, France, Ireland, Sweden, Switzerland, and the United Kingdom) and the United States Trust Fund. We also thank the governments who provided support to Bureau members, authors and reviewers in other ways. In addition, Finland provided direct support to the secretariat. The IAASTD was especially successful in engaging a large number of experts from developing countries and countries with economies in transition in its work; the Trust Funds enabled financial assistance for their travel to the IAASTD meetings.

We would also like to make special mention of the Regional Organizations who hosted the regional coordinators and staff and provided assistance in management and time to ensure success of this enterprise: the African Center for Technology Studies (ACTS) in Kenya, the Inter-American Institute for Cooperation on Agriculture (IICA) in Costa Rica, the International Center for Agricultural Research in the Dry Areas (ICARDA) in Syria, and the WorldFish Center in Malaysia.

The final Intergovernmental Plenary in Johannesburg, South Africa was opened on 7 April 2008 by Achim Steiner, Executive Director of UNEP. This Plenary saw the acceptance of the Reports and the approval of the Summaries for Decision Makers and the Executive Summary of the Synthesis Report by an overwhelming majority of governments.

Signed:

Co-chairs
Hans H. Herren,
Judi Wakhungu

Director
Robert T. Watson

Preface

In August 2002, the World Bank and the Food and Agriculture Organization (FAO) of the United Nations initiated a global consultative process to determine whether an international assessment of agricultural knowledge, science and technology (AKST) was needed. This was stimulated by discussions at the World Bank with the private sector and nongovernmental organizations (NGOs) on the state of scientific understanding of biotechnology and more specifically transgenics. During 2003, eleven consultations were held, overseen by an international multistakeholder steering committee and involving over 800 participants from all relevant stakeholder groups, e.g. governments, the private sector and civil society. Based on these consultations the steering committee recommended to an Intergovernmental Plenary meeting in Nairobi in September 2004 that an international assessment of the role of agricultural knowledge, science and technology (AKST) in reducing hunger and poverty, improving rural livelihoods and facilitating environmentally, socially and economically sustainable development was needed. The concept of an International Assessment of Agricultural Knowledge, Science and Technology for Development (IAASTD) was endorsed as a multi-thematic, multi-spatial, multi-temporal intergovernmental process with a multistakeholder Bureau cosponsored by the Food and Agricultural Organization of the United Nations (FAO), the Global Environment Facility (GEF), United Nations Development Programme (UNDP), United Nations Environment Programme (UNEP), United Nations Educational, Scientific and Cultural Organization (UNESCO), the World Bank and World Health Organization (WHO).

The IAASTD's governance structure is a unique hybrid of the Intergovernmental Panel on Climate Change (IPCC) and the nongovernmental Millennium Ecosystem Assessment (MA). The stakeholder composition of the Bureau was agreed at the Intergovernmental Plenary meeting in Nairobi; it is geographically balanced and multistakeholder with 30 government and 30 civil society representatives (NGOs, producer and consumer groups, private sector entities and international organizations) in order to ensure ownership of the process and findings by a range of stakeholders.

About 400 of the world's experts were selected by the Bureau, following nominations by stakeholder groups, to prepare the IAASTD Report (comprised of a Global and 5 sub-Global assessments). These experts worked in their own capacity and did not represent any particular stakeholder group. Additional individuals, organizations and governments were involved in the peer review process.

The IAASTD development and sustainability goals were endorsed at the first Intergovernmental Plenary and are consistent with a subset of the UN Millennium Development Goals (MDGs): the reduction of hunger and poverty, the improvement of rural livelihoods and human health, and facilitating equitable, socially, environmentally and economically sustainable development. Realizing these goals requires acknowledging the multifunctionality of agriculture: the challenge is to simultaneously meet development and sustainability goals while increasing agricultural production.

Meeting these goals has to be placed in the context of a rapidly changing world of urbanization, growing inequities, human migration, globalization, changing dietary preferences, climate change, environmental degradation, a trend toward biofuels and an increasing population. These conditions are affecting local and global food security and putting pressure on productive capacity and ecosystems. Hence there are unprecedented challenges ahead in providing food within a global trading system where there are other competing uses for agricultural and other natural resources. AKST alone cannot solve these problems, which are caused by complex political and social dynamics, but it can make a major contribution to meeting development and sustainability goals. Never before has it been more important for the world to generate and use AKST.

Given the focus on hunger, poverty and livelihoods, the IAASTD pays special attention to the current situation, issues and potential opportunities to redirect the current AKST system to improve the situation for poor rural people, especially small-scale farmers, rural laborers and others with limited resources. It addresses issues critical to formulating policy and provides information for decision makers confronting conflicting views on contentious issues such as the environmental consequences of productivity increases, environmental and human health impacts of transgenic crops, the consequences of bioenergy development on the environment and on the long-term availability and price of food, and the implications of climate change on agricultural production. The Bureau agreed that the scope of the assessment needed to go beyond the narrow confines of S&T and should encompass other types of relevant knowledge (e.g., knowledge held by agricultural producers, consumers and end users) and that it should also assess the role of institutions, organizations, governance, markets and trade.

The IAASTD is a multidisciplinary and multistakeholder enterprise requiring the use and integration of information, tools and models from different knowledge paradigms including local and traditional knowledge. The IAASTD does not advocate specific policies or practices; it assesses the major issues facing AKST and points towards a range of AKST options for action that meet development and sustainability goals. It is policy relevant, but not policy prescriptive. It integrates scientific information on a range of topics that are critically interlinked, but often addressed independently, i.e., agriculture, poverty, hunger, human health, natural resources, environment, development and innovation. It will enable decision makers to bring a richer base of knowledge to bear on policy and management decisions on issues previously viewed in isolation. Knowledge gained from historical analysis (typically the past 50 years) and an analysis of some future development alternatives to 2050 form the basis for assessing options for action on science and technology, capacity development, institutions and policies, and investments.

The IAASTD is conducted according to an open, transparent, representative and legitimate process; is evidence-based; presents options rather than recommendations; assesses different local, regional and global perspectives; presents different views, acknowledging that there can be more than one interpretation of the same evidence based on different worldviews; and identifies the key scientific uncertainties and areas on which research could be focused to advance development and sustainability goals.

The IAASTD is composed of a Global assessment and five sub-Global assessments: Central and West Asia and North Africa—CWANA; East and South Asia and the Pacific—ESAP; Latin America and the Caribbean—LAC; North America and Europe—NAE; Sub-Saharan Africa—SSA. It (1) assesses the generation, access, dissemination and use of public and private sector AKST in relation to the goals, using local, traditional and formal knowledge; (2) analyzes existing and emerging technologies, practices, policies and institutions and their impact on the goals; (3) provides information for decision makers in different civil society, private and public organizations on options for improving policies, practices, institutional and organizational arrangements to enable AKST to meet the goals; (4) brings together a range of stakeholders (consumers, governments, international agencies and research organizations, NGOs, private sector, producers, the scientific community) involved in the agricultural sector and rural development to share their experiences, views, understanding and vision for the future; and (5) identifies options for future public and private investments in AKST. In addition, the IAASTD will enhance local and regional capacity to design, implement and utilize similar assessments.

In this assessment, agriculture is used in the widest sense to include production of food, feed, fuel, fiber and other products and to include all sectors from production of inputs (e.g., seeds and fertilizer) to consumption of products. However, as in all assessments, some topics were covered less extensively than others (e.g., livestock, forestry, fisheries and the agricultural sector of small island countries, and agricultural engineering), largely due to the expertise of the selected authors. Originally the Bureau approved a chapter on plausible futures (a visioning exercise), but later there was agreement to delete this chapter in favor of a more simple set of model projections. Similarly the Bureau approved a chapter on capacity development, but this chapter was dropped and key messages integrated into other chapters.

The IAASTD draft Report was subjected to two rounds of peer review by governments, organizations and individuals. These drafts were placed on an open access Web site and open to comments by anyone. The authors revised the drafts based on numerous peer review comments, with the assistance of review editors who were responsible for ensuring the comments were appropriately taken into account. One of the most difficult issues authors had to address was criticisms that the report was too negative. In a scientific review based on empirical evidence, this is always a difficult comment to handle, as criteria are needed in order to say whether something is negative or positive. Another difficulty was responding to the conflicting views expressed by reviewers. The difference in views was not surprising given the range of stakeholder interests and perspectives. Thus one of the key findings of the IAASTD is that there are diverse and conflicting interpretations of past and current events, which need to be acknowledged and respected.

The Global and sub-Global Summaries for Decision Makers and the Executive Summary of the Synthesis Report were approved at an Intergovernmental Plenary in April 2008. The Synthesis Report integrates the key findings from the Global and sub-Global assessments, and focuses on eight Bureau-approved topics: bioenergy; biotechnology; climate change; human health; natural resource management; traditional knowledge and community based innovation; trade and markets; and women in agriculture.

The IAASTD builds on and adds value to a number of recent assessments and reports that have provided valuable information relevant to the agricultural sector, but have not specifically focused on the future role of AKST, the institutional dimensions and the multifunctionality of agriculture. These include FAO State of Food Insecurity in the World (yearly); InterAcademy Council Report: Realizing the Promise and Potential of African Agriculture (2004); UN Millennium Project Task Force on Hunger (2005); Millennium Ecosystem Assessment (2005); CGIAR Science Council Strategy and Priority Setting Exercise (2006); Comprehensive Assessment of Water Management in Agriculture: Guiding Policy Investments in Water, Food, Livelihoods and Environment (2007); Intergovernmental Panel on Climate Change Reports (2001 and 2007); UNEP Fourth Global Environmental Outlook (2007); World Bank World Development Report: Agriculture for Development (2007); IFPRI Global Hunger Indices (yearly); and World Bank Internal Report of Investments in SSA (2007).

Financial support was provided to the IAASTD by the cosponsoring agencies, the governments of Australia, Canada, Finland, France, Ireland, Sweden, Switzerland, US and UK, and the European Commission. In addition, many organizations have provided in-kind support. The authors and review editors have given freely of their time, largely without compensation.

The Global and sub-Global Summaries for Decision Makers and the Synthesis Report are written for a range of stakeholders, i.e., government policy makers, private sector, NGOs, producer and consumer groups, international organizations and the scientific community. There are no recommendations, only options for action. The options for action are not prioritized because different options are actionable by different stakeholders, each of whom have a different set of priorities and responsibilities and operate in different socio-economic-political circumstances.

1

Context, Conceptual Framework and Sustainability Indicators

Coordinating Lead Authors
Hans Hurni (Switzerland) and Balgis Osman-Elasha (Sudan)

Lead Authors
Audia Barnett (Jamaica), Ann Herbert (USA), Anita Idel (Germany),
Moses Kairo (Kenya), Dely Pascual-Gapasin (Philippines), Juerg
Schneider (Switzerland), Keith Wiebe (USA)

Contributing Authors
Guéladio Cissé (Cote d'Ivoire), Norman Clark (UK), Manuel
de la Fuente (Bolivia), Berhanu Debele (Ethiopia), Markus
Giger (Switzerland), Udo Hoeggel (Switzerland), Ulan Kasimov
(Kyrgyzstan), Boniface Kiteme (Kenya), Andreas Klaey (Switzerland),
Thammarat Koottatep (Thailand), Janice Jiggins (Netherlands),
Ian Maudlin (UK), David Molden (USA), Cordula Ott (Switzerland),
Marian Perez Gutierrez (Costa Rica), Brigitte Portner (Switzerland),
Riikka Rajalahti (Finland), Stephan Rist (Switzerland), Gete Zeleke
(Ethiopia)

Review Editors
Judith Francis (Trinidad and Tobago) and JoAnn Jaffe (Canada)

Key Messages

Key Messages

1. Agriculture is multifunctional. It provides food, feed, fiber, fuel and other goods. It also has a major influence on other essential ecosystem services such as water supply and carbon sequestration or release. Agriculture plays an important social role, providing employment and a way of life. Both agriculture and its products are a medium of cultural transmission and cultural practices worldwide. Agriculturally based communities provide a foundation for local economies and are an important means for countries to secure their territories. Agriculture accounts for a major part of the livelihood of 40% of the world's population and occupies 40% of total land area; 90% of farms worldwide have a size of less than 2 hectares. Agriculture includes crop-, animal-, forestry- and fishery-based systems or mixed farming, including new emerging systems such as organic, precision and peri-urban agriculture. Although agricultural inputs and outputs constitute the bulk of world trade, most food is consumed domestically, i.e., where it is produced.

2. Agricultural systems range across the globe from intensive highly commercialized large-scale systems to small-scale and subsistence systems. All of these systems are potentially either highly vulnerable or sustainable. This variability is rooted in the global agrifood system, which has led to regional and functional differences around the world—the social, economic and ecological effects of which have not yet been assessed and compared. The global agricultural system faces great challenges today, as it has to confront climate change, loss of biological and agrobiological diversity, loss of soil fertility, water shortage and loss of water quality, and population growth. Sustainable agricultural production is dependent on effective management of a range of interdependent physical and natural resources—land, water, energy, capital and so on—as well as on full internalization of currently externalized costs. The sustainability of production also depends on the continuing availability of and generalized access to public goods. Finding ways of dealing with these challenges is a highly contested matter: strategies differ because they are based on different visions of agriculture, different interests and diverging values. However, while agriculture is a strong contributor to the most critical problems we face today; it can also play a major role in their resolution.

3. Agricultural Knowledge, Science and Technology (AKST) can address the multifunctionality of agriculture. It plays a key role in shaping the quality and quantity of natural, human and other resources as well as access to them. AKST is also crucial in supporting the efforts of actors at different levels—from household to national, sub-global and global—to reduce poverty and hunger, as well as improve rural livelihoods and the environment in order to ensure equitable and environmentally, socially and economically sustainable development. On the one hand, tacit and locally-based agricultural knowledge has been, and continues to be, the most important type of knowledge particularly for small-scale farming, forestry and fishery activities. On the other hand, the development of formal agricultural knowledge has been enormously successful particularly since the 1950s, and forms a dominating part of agricultural knowledge today. Challenges ahead include the development and use of transgenic plants, animals and microorganisms for increased productivity and other purposes; access to and use of agrochemicals; the emerging challenges of biofuel and bioenergy development, and in a broader sense, the political, social and economic organization of agriculture as a component of rural development. All these challenges have implications (both positive and negative) on the environment, human health, social well-being and economic performance of rural areas in all countries. The combination of community-based innovation and local knowledge with science-based approaches in AKST holds the promise of best addressing the problems, needs and opportunities of the rural poor.

4. The majority of the world's poorest and hungry live in rural settings and depend directly on agriculture. Over 70% of the world's poor live in rural areas. These 2.1 billion people live on less than US$2 a day. This is not inevitable, and an improved economic environment and greater social equity at local, national, and global scales have the potential to ensure that agriculture is able to provide improved livelihoods. Inextricably linked to poverty are vulnerabilities relating to production and consumption shocks, poor sanitation, and lack of access to health care and deficient nutrient intake, placing many in agrarian societies at risk. AKST may help mitigate these negative effects by supporting appropriate interventions, but it may also increase the vulnerability of poor farmers if no attention is paid to the risks and uncertainties to which these farmers are exposed. The livelihoods of many poor farmers are oriented towards meeting basic needs, particularly food. With insufficient income, households have little money to invest in increasing the productivity or sustainability of their production systems. The global trend has been towards a decapitalization of poor farmers and their resources (as well as rural areas), as they experience declining terms of trade and competition with low-cost producers. AKST offers opportunities to contribute to recapitalization of such farming households.

5. A vicious circle of poor health, reduced working capacity, low productivity and short life expectancy is typical, particularly for the most vulnerable groups working in agriculture. All persons have a right to sufficient, safe, nutritious and culturally acceptable food. Good nutrition is a prerequisite for health. Although global production of food calories is sufficient to feed the world's population, millions die or are debilitated every year by hunger and malnutrition which makes them vulnerable to infectious diseases (e.g., HIV/AIDS, malaria and tuberculosis). In many developing countries hunger and health risks are exacerbated by extreme poverty and poor and dangerous working conditions. In contrast, in industrialized countries, overnutrition and food safety issues, including food-borne illnesses affecting human health as well as diseases associated with agricultural production systems, are predominant concerns. Notwithstanding, in industrialized countries there is also a significant incidence of undernutrition among the poor, and a higher burden of both infectious and noncommunicable diseases associated with metabolic syndromes. AKST has an important role to play in both moving towards food security and food sovereignty, and breaking the malnutrition–poor health–low productivity cycle.

6. A range of fundamental natural resources (e.g., land, water, air, biological diversity including forests, fish) provide the indispensable base for the production of essential goods and services upon which human survival depends, including those related to agricultural ecosystems. During the last 50 years, the physical and functional availability of natural resources has shrunk faster than at any other time in history due to increased demand and/or degradation at the global level. This has been compounded by a range of factors including human population growth, and impacts have comprised unprecedented loss of biodiversity, deforestation, loss of soil health, and water and air quality. In many cases, such negative impacts can be mitigated; and in some cases, they are. Given the multifunctional nature of agriculture, it is critical to consider links across ecosystems in which agricultural systems are embedded, as these have important implications for the resilience or vulnerability of these systems. Linkages between natural resource use and the social and physical environment across space and time are an important issue for AKST, with significant implications for sustainable development and the mitigation of adverse impacts.

7. Social equity issues, including gender, are major concerns in agriculture, as they relate to poverty, hunger, nutrition, health, natural resource management and environment, which are affected by various factors resulting in greater or lesser degrees of equity. As a majority of the world's poor and hungry live in rural settings and are directly dependent on agriculture for their livelihoods, political, economic, cultural and technological factors contribute to mitigating or reinforcing inequality. Women and men have differing roles and responsibilities in productive households, and they can derive varying degrees of benefits from AKST and innovations. Gender-based patterns are context-specific, but a persistent feature is that women have a key role in agricultural activities, yet they have limited access to, and control of, productive resources such as land, labor, credit and capital. Agricultural development sometimes strengthens patterns that are unfavorable to women, such as male bias of the agricultural extension system in many countries. Societies can develop governance institutions, legal systems, social policy tools, and social/gender sensitive methods (e.g., gender analysis) that seek to minimize disparities and even opportunities out among women and men.

8. Agriculture today is faced with several emerging challenges and opportunities; the evaluation of those relating to climate change, land degradation, reduced access to natural resources (including genetic resources), bioenergy demands, transgenics and trade require special efforts and investments in AKST. About 30% of global emissions leading to climate change are attributed to agricultural activities. Climate change in turn affects all types of agricultural production systems, from farming to forestry, livestock production and fishery; it particularly affects resource-poor agriculture. Current as well as future damage due to temperature increases and more extreme weather events and their consequences on the hydrology of watersheds and groundwater resources are yet to be detected in detail. Agricultural households and enterprises need to adapt to climate change but they do not yet have the experience in and knowledge of handling these processes, including increased pressure due to biofuel production. Bioenergy is seen as a potential to mitigate the impact of using fossil fuels as a source of energy, thereby mitigating the impact on climate. While on-farm bioenergy production is emerging as a possibility to make better use of farm residues and excrements, the substitution of fossil fuel through biofuel plantation for transport and mobility is under contention and thus a matter of concern for AKST. The development and use of transgenics is seen very differently by the different stakeholders, ranging from a purely positivist view of genetically modified organisms (GMOs) as the solution to the problems of agriculture, to a purely negativist view that considers GMOs to be uncontrollable and life threatening. Finally, agricultural trading conditions, rules and standards are changing; together with emerging alternatives, they offer challenges and opportunities.

1.1 Setting the Scene

1.1.1 The IAASTD

IAASTD, *the International Assessment of Agricultural Knowledge, Science and Technology for Development,* comes at a time of rapid change that is affecting both rural and urban areas, as well as the climate and other natural resources—in ways that present unprecedented threats. However these changes also provide opportunities for sustainable development and poverty alleviation, and require increased knowledge, science and technology in conjunction with appropriate policies, institutions and investments.

The main goal of IAASTD is to provide decision makers with the information they need to reduce hunger and poverty, improve rural livelihoods, and facilitate equitable, environmentally, socially and economically sustainable development through the generation of, access to and use of agricultural knowledge, science and technology (AKST). IAASTD uses a conceptual framework that enables systematic analysis and appraisal of the above challenges based on common concepts and terminology.

The development and sustainability goals of the IAASTD are to:
(1) reduce hunger and poverty,
(2) improve rural livelihoods, human health and nutrition, and
(3) promote equitable and socially, environmentally and economically sustainable development.

Sustainable development is crucial to meet the needs of the present without compromising the ability of future generations to meet their own needs (see WCED, 1987). Using AKST to achieve development and sustainability goals will depend on the choices of different actors related to AKST development and application.

Agriculture plays a prominent role for human well-being on Earth; the IAASTD concentrates on how knowledge, science and technology can contribute to agricultural development. This assessment is a specific step among several global efforts to achieve sustainable development that have emerged in follow-up processes and policies of the World Conference in Rio de Janeiro in 1992. AKST will contribute to the achievement of these goals. Specifically, the IAASTD will contribute to knowledge-based decision making for future sustainable development by assessing: (1) those interrelations within AKST relevant to sustainable development; (2) knowledge and scientific development, technology diffusion, innovation, and adaptation of ecosystem management; and (3) the integration of AKST within international, regional, national and local development policies and strategies.

What is an assessment?

International assessments are very useful when they address complex issues of supranational interest and dimensions. A number of assessments have been undertaken by many organizations and individuals in the past two decades: the Global Biodiversity Assessment (GBA), the Ozone Assessment, the Intergovernmental Panel on Climate Change (IPCC) reports, the Millennium Ecosystem Assessment (MA), the Comprehensive Assessment of Water Management in Agriculture (CA), the Global Environment Outlook (GEO), and now, the International Assessment of Agriculture, Knowledge, Science and Technology for Development (IAASTD).

The evidence-based analyses that underpin the outcomes of the various assessments have common characteristics. A key point is that an assessment is not simply a review of the relevant literature; it can be based, in part, on a literature review, but also needs to provide an assessment of the veracity and applicability of the information and the uncertainty of outcomes in relation to the context of the identified questions or issues within a specified authorizing environment (Table 1-1).

To be effective and legitimate, the assessment process was designed to be open, transparent, reviewed, and widely representative of stakeholders and relevant experts, and the resulting documents to be broadly reviewed by independent experts from governments, private and nongovernmental organizations, as well as by representatives of the participating governments. Obtaining a balance of opinions in a global assessment based on a literature review and relevant expertise is an ongoing and iterative challenge to ensure that it encompasses a broad range of disciplinary and geographical experience and different knowledge systems. The IAASTD has been designed in a way that attempts to ensure effectiveness and legitimacy.

The role of Agricultural Knowledge, Science and Technology (AKST). Agricultural knowledge, science and technology are seen as key factors and instruments for future adjustment of indirect and direct drivers of agricultural outputs, as well as of ecosystem services. Assessing AKST sets the stage for an informed choice by decision-makers among various options for development. It indicates how policy and institutional frameworks at all organizational levels might affect sustainability goals. Specifically, it provides the basis for designing

AKST in a way that mitigates detrimental development dynamics such as growing disparities, the decreasing share of agricultural value-added and the degradation of ecosystems. In other words, the assessment draws lessons about what conditions have led AKST to have an impact on development that has been positive for human and ecosystem well-being, and where, when and why impacts have been negative. Moreover, it explores the demands that are likely to be made on agricultural systems (crops, livestock and pastoralism, fisheries, forestry and agroforestry, biomass, commodities and ecosystem services) in the future, asking what agricultural goods and services society will need under different plausible future scenarios in order to achieve the goals related to hunger, nutrition, human health, poverty, equity, livelihoods, and environmental and social sustainability, and whether and how access to these goods and services is hindered. The result is an evidence-based guide for policy and decision-making.

IAASTD commitment to sustainable development. IAASTD sees the assessment of AKST and its implications for agriculture as a prerequisite for knowledge-based decision-making for future sustainable development portfolios. Specifically, IAASTD aims to contribute to knowledge-based, decision-making for future sustainable development by:
1. Identifying interrelations between agricultural knowledge, science and technology in view of sustainable development;
2. Exploring knowledge and scientific development, technology diffusion, innovations and adaptations of ecosystem management;
3. Supporting the integration of agricultural knowledge, science and technology (AKST) within international and national development policies and strategies.

IAASTD's relationship to the Millennium Development Goals (MDGs) and the Millennium Ecosystem Assessment (MA). The MDGs and the MA are cornerstones for development policy and serve as major references for the IAASTD. In addition to these frameworks, the IAASTD assesses AKST in relation to the objective of meeting broader development and sustainability goals. It is generally assumed that AKST can play a major role in efforts to achieve the MDGs, particularly that of eradicating extreme poverty and hunger (MDG 1) by improving the productivity of agriculture in general and the competitiveness of smallholders and marginalized groups in the expanding global, national and local markets in particular, as well as by creating employment among poor rural people and making food available to consumers everywhere. AKST can also contribute directly or indirectly to improving primary education and social and gender equity, reducing child mortality, improving maternal health, combating HIV/AIDS, malaria and other diseases (MDG 2-6), and ensuring environmental sustainability (MDG 7) by delivering a variety of supporting, regulating and cultural services (MDG 8). The IAASTD assessment enables a more adequate consideration of the linkage between poverty reduction and environmental change.

Key questions for the IAASTD. The major question for this assessment is: "How can we reduce hunger and

Table 1-1. Differences between a review and an assessment.

	Scientific Reviews	Assessment
Audience	Undertaken for scientists	Undertaken for decision-makers from a specified authorizing environment
Conducted by	One or a few scientists	A larger and varied group based on relevant geographic and disciplinary representation
Issues/Topics	Often deal with a single topic	Generally a broader and complex issue
Identifies gaps in	Research issues generally driven by scientific curiosity	Knowledge for implementation of outcomes; problem-driven
Uncertainty statements	Not always required	Essential
Judgment	Hidden; a more objective analysis	Required and clearly flagged
Synthesis	Not required, but sometimes important	Essential to reduce complexity
Coverage	Exhaustive, historical	Sufficient to deal with main range of uncertainty associated with the identified issues

Source: Watson and Gitay, 2004.

poverty, improve rural livelihoods, and facilitate equitable, environmentally, socially and economically sustainable development through the generation of, access to, and use of AKST?" Three questions recur throughout the global and sub-global assessments of IAASTD. They concern:
1. *Social disparities:* How have changing markets and changing access to markets affected development and sustainability goals, and how has this been influenced by AKST? How and by what have cultural values, traditions and social equity (including gender equity) been influenced? What are projected implications of market changes in the future, and how can AKST contribute to informed decision-making?
2. *Ecology:* How has availability of, access to and management of natural resources (particularly water and soil resources, as well as plant, animal, genetic and other resources) affected the development and sustainability goals of IAASTD? How can AKST enhance knowledge of natural resource management?
3. *AKST:* What have been, and what are projected to be, the implications of institutional and policy changes and funding (e.g., private versus public investment; intellectual property rights [IPR]; legislative frameworks) on access to AKST, on innovation systems and on ownership of knowledge? How will AKST influence social, environmental and economic outcomes of agricultural and food systems?

Other central issues relating to hunger, nutrition, human health, poverty, livelihoods and the economy, as well as productivity and technologies are part of the sustainability goals and thus further emphasized in the document.

Diversity of views and value systems represented in the IAASTD
AKST is not an entity; it is a diverse field of knowledge and values. Achieving development and sustainability goals requires probing and experimentation, negotiation, and learning among diverse voices and interpretations, as well

as taking into account place-based and context-relevant factors and voices to address the multiple functions of agriculture. The IAASTD has made clear how contested AKST are among the hundreds of professionals involved, especially formal AKST. Conflicting perspectives on AKST have led to different options for policy-making, and understanding the competing interpretations of AKST does not guarantee a consensual outcome. IAASTD focuses on AKST issues most relevant to development and sustainability goals.

1.1.2 Agriculture and its global context
Importance of agriculture. Agriculture as the source of human food, animal feed, fiber and fuel plays a key role in efforts to achieve global sustainable development. It is a major occupational sector in developing countries, with the poorest countries being those with predominantly agricultural economies and societies (FAO, 2000). Approximately 2.6 billion people—men, women and children —rely on agricultural production systems, be it farming, livestock production, forestry or fishery. Food security for a growing world population is positioned to remain a challenge in the next few decades. Most food is produced in Asia and other densely populated poor regions, and most of that food is consumed domestically. Because of the high diversity of agricultural systems across the world IAASTD decided to carry out five sub-global assessments in addition to the global one, in order to adequately address issues in the major agricultural regions of the world. These regions have developed to their current state for a variety of reasons, and a more specific reorientation of AKST is likely to be more effective if it addresses region-specific issues in agriculture, development and sustainability. The IAASTD has put particular emphasis on addressing issues relevant to tackling poverty reduction, which is central to the Millennium Development Goals to be achieved by 2015, though these issues are also expected to remain important long beyond that date.

In parallel with the spread and growth of human population, particularly during the last 300 years, but at a particularly impressive rate since 1950, the transformation of natural ecosystems into agriculturally used and managed land has accelerated, which coincides with the time when formal AKST began to have a significant impact. The world population grew from about 2.5 billion people in 1950 to 6.5 billion in 2005, i.e., by a factor of 2.6; in most countries, growth rates have just recently begun to decrease. Trends indicate that the global population will reach between 7.5 and 11 billion people by 2050, depending on the expected average number of children per women (Figure 1-1).

World agricultural output, or more specifically, food output as measured in cereal and meat production, in turn, increased even more during the same period, due to large increases in fertilizer use, herbicides, plant and animal breeding, and extension of irrigated area (Figure 1-2). The total cultivated area increased much less, i.e., from 1.4 to 1.5 million ha between 1950 and 2005 (Wood et al., 2000, based on FAO data), although fallow systems were greatly reduced.

For similar figures indicating equally moderate growth of crop area see also the Millennium Ecosystem Assessment (MA, 2005a). However, more land was converted to cropland in the 30 years after 1950 than in the 150 years between 1700 and 1850 (MA, 2005a). More than half of all the synthetic nitrogen fertilizer ever applied on the planet has been used since 1985, and phosphorus use tripled between 1960 and 1990 (MA, 2005b). Globally, agricultural output has been growing at about 2% per year since 1960, with higher rates in developing countries because area productivity, particularly in sub-Saharan Africa and Latin America, is still much lower than in industrialized countries and in Asia (FAO, 2006a). Along with an increase in agricultural output, water use in agriculture has increased to 7,130 cubic kilometers today and is expected to double by 2050 (CA, 2007). Another form of competition has recently been observed between the use of crops for food and feed and the use of the same crops (e.g., maize) for biofuels; moreover, a competition at the world level is rising for the supply of protein-rich animal feeds.

Today's land use patterns in general reveal the importance of agriculture as a major land management system transforming and making use of natural ecosystems. Given a global land surface (without Antarctica) of 13,430 million ha (FAOSTAT, 2006), there are still about 30% forest ecosystems (nearly 4,000 million ha), part of which are the least converted in a biological sense. About a further 26% (3,400 million ha) are pastureland (FAOSTAT, 2006), of which about half was converted from natural grassland and the rest from forestland or woodland. About 11.5% are cropland (1,500 million ha) (FAOSTAT, 2006), most of which was also converted from forestland. The remaining share of the global land surface are deserts, shrubland and tundra (about 25%), inland water surfaces and wetlands (about 4%), and built up land for human settlements and other infrastructure (about 5%). In sum, more than half of the earth's land surface is intensively used for agricultural purposes such as cultivation, grazing, plantation forestry and aquaculture; and since 1950 one third of the soil has been profoundly altered from its natural ecosystem state because of moderate to severe soil degradation (Oldeman et al., 1990).

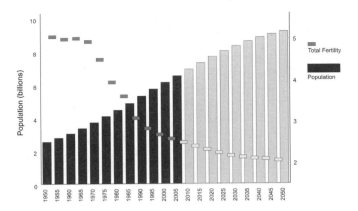

Figure 1-1. *Total world population 1950-2050 and average number of children per woman (total fertility).* Source: UNFPA, 2007

Multifunctionality of agriculture. As an activity, agriculture has multiple outputs and contributes to several ends at the same time (Abler, 2004). Agricultural resource management thus involves more than maintaining production systems. Services such as mitigating climate change, regulating water, controlling erosion and support services such as soil formation, providing habitats for wildlife, as well as contributions to cultural activities such as use and preservation of landscapes and spiritual sites are some of the positive functions that agriculture provides. The OECD identifies two key elements of multifunctionality: externalities and jointness (OECD, 2005). Agriculture uses public goods—natural resources (landscapes, plants, animals, soils, minerals, water and atmospheric N and C) for the production of public services, common goods, and private goods (food, feed, fiber, fuel). These natural resources are controlled and distributed partly through public entities and partly via privately producing and marketing entities; hence the issue of externality of costs are borne by the public. Agriculture is embedded in local and regional contexts and is always bound to particular, socially defined relationships and interdependencies between the production of private goods and the use and production of multifunctional public goods, which leads to the issue of jointness (Abler, 2004).

Globalization in agriculture

Globalization in agriculture, aided by information and communication technologies (ICT), has resulted in economic opportunities as well as challenges, particularly in developing countries. Globalization is typified by the increased interlinkage and concentration at almost all stages of the production and marketing chain, with functional and regional differentiations, and includes transnational corporations that are vertically and horizontally integrated in globalization and their increasing power over consumers and agricultural producers. Globalization is also characterized by growing investments in agriculture, food processing and marketing, and increasing international trade in food facilitated by reduced trade barriers (FAO, 2003). The creation of intellectual property rights has become an increasingly important source of competitive advantage and accumula-

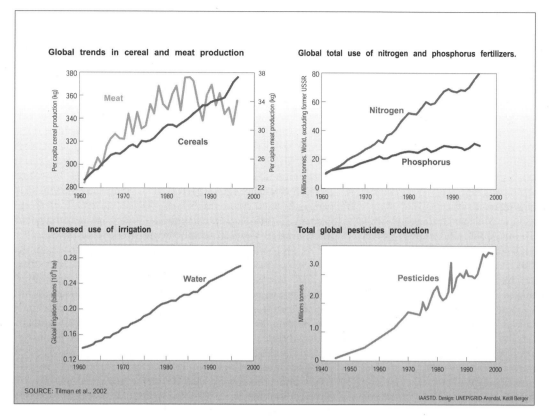

Figure 1-2. *Global trends in cereal and meat production; nitrogen and phosphorus fertilizer use; irrigation, and pesticide production.*

tion in the production and trade of agricultural goods. Globalization has resulted in national and local governments and economies ceding some sovereignty as agricultural production has become increasingly subject to international agreements, such as the World Trade Organization's Agreement on Agriculture (WTO, 1995).

The progressive expansion of commercial-industrial relations in agriculture has put further strain on many small-scale farmers in developing countries who must also contend with direct competition from production systems that are highly subsidized and capital intensive, and thus able to produce commodities that can be sold more cheaply. Newly industrialized countries like India have increasingly subsidized inputs in agriculture since the early 1980s (IFPRI, 2005).

Competition, however, does not only originate in subsidies from agricultural policies of richer countries; it may also derive from large entrepreneurial holdings that have low production costs, which are primarily but not exclusively found in developing countries. Three phenomena related to globalization are specific for a number of countries: the growing impact of supermarkets and wholesalers, of grades and standards, and of export horticulture, have substantially favored large farms (Reardon et al., 2001, 2002, 2003) except when small farmers get special assistance through subsidies, micro-contracts or phytosanitary programs (cf. Minten et al., 2006), for example. A steady erosion of local food production systems and eating patterns

has accompanied the net flow of food from poorer to richer countries (Kent, 2003).

While average farm sizes in Europe and North America have increased substantially, in Asia, Latin America, and in some highly densely populated countries in Africa, average farm sizes have decreased significantly in the late 20th century, although they were already very small by the 1950s (Eastwood et al., 2004; Anriquez and Bonomi, 2007). These averages conceal vast and still growing inequalities in the scale of production units in all regions, with larger industrialized production systems becoming more dominant particularly for livestock, grains, oil crops, sugar and horticulture and small, labor-intensive household production systems generally becoming more marginalized and disadvantaged with respect to resources and market participation. In industrialized countries, farmers now represent a small percentage of the population and have experienced a loss of political and economic influence, although in many countries they still exercise much more power than their numbers would suggest. In developing countries agricultural populations are also declining, at least in relative terms, with many countries falling below 50% (FAO, 2006a). Although, there are still a number of poor countries with 60-85% of the population working in small-scale agricultural systems. The regional distribution of the economically active population in agriculture is dominated by Asia, which accounts for almost 80% of the world's total active population, followed by Africa with 14%. Although the overall number of women in agriculture is falling, the relative share is rising, i.e., women make up an increasing fraction of the labor force in agriculture, especially in sub-Saharan Africa where hoe agriculture is practiced extensively (Spieldoch, 2007). While the agrifood sector *in toto* may still account for a large portion of national economies, with the production of inputs, industrial transformation and marketing of food, and transport becoming more important in terms of value and employment, agricultural production itself accounts for a diminishing share of the economy in many countries while the other sectors are expanding. Average farm sizes vary greatly by region (see Table 1-2).

Trade and the agricultural sector

International trade and economic policies can have positive and negative effects on different development and sustainability goals. In addition, AKST can have substantial roles in the formation of better policies. Poverty-affected agricultural producers in particular have been poorly served by trade; unless they have better access to efficient and equitable market systems, they cannot easily benefit from AKST initiatives (IFAD, 2003). Trade policies and market dynamics are thus key determinants of whether and how AKST systems can effectively address poverty, hunger, rural livelihoods and environmental sustainability. Although most agricultural production is not traded internationally, national agricultural planning and AKST investment is increasingly oriented towards export markets and designed to comply with international trade rules. Agricultural trade has been increasing in developing country regions particularly since the 1970s (FAO, 2005a).

The focus on export has left many small-scale producers, i.e., the majority of the rural poor, vulnerable to volatile international market conditions and international competition, often from subsidized producers in the North. The globalization of agriculture has been accompanied by concentration of market power away from producers into the hands of a limited number of large-scale trade and retail agribusiness companies. Corporate concentration along the agrifood value chain can have a significant impact on international commodity prices, which have recently risen but have generally been low relative to industrial and manufactured goods (FAO, 2005a). In addition, increased international trade in agricultural commodities has often led to overexploitation of natural resources, and increased energy use and greenhouse gas (GHG) emissions. Overall the impact of trade liberalization has been uneven in industrialized and developing countries.

Table 1-2. **Approximate farm size by world region.**

World region	Average farm size, ha
Africa	1.6
Asia	1.6
Latin America and Caribbean	67.0
Europe*	27.0
North America	121.0

Source: Nagayets, 2005; von Braun, 2005.

*data includes Western Europe only.

Small-scale farming as a particular challenge for agriculture

Despite the crucial role that agriculture has for rural populations in transition and developing countries, agriculture-based livelihoods and rural communities are endangered by poverty worldwide. Based on FAO census data, it has been estimated that about 525 million farms exist worldwide, providing a livelihood for about 40% of the world's population. Nearly 90% of these are small farms defined as having less than two hectares of land (see e.g., Nagayets, 2005). Small farms occupy about 60% of the arable land worldwide and contribute substantially to global farm production (Figure 1-3). In Africa, 90% of agricultural production is derived from small farms (Spencer, 2002). If a high percentage of a country's population is engaged in agriculture and derives its livelihood from small-scale farming, the whole sector is predominantly subsistence-oriented, which makes livelihoods extremely vulnerable to changes in direct drivers such as diseases, pests, or climate, even though its sensitivity to indirect drivers such as markets, infrastructure and external inputs is less pronounced. Not surprisingly, subsistence farmers tend to be very aware of their multiple vulnerabilities and therefore adopt diverse risk-minimizing and mitigating strategies.

Poorly developed market infrastructure such as rural roads and postharvest facilities are among the factors that have limited market access for outputs and inputs (e.g., fertilizer) for the majority of small-scale farmers (FAO, 2005a) (Figure 1-4).

Growing disparities have developed over the last 50 years between small-scale farming that follows local practices and industrial agricultural systems that have incorporated formal AKST. A key factor is the tremendous increase in labor productivity in industrialized agriculture and the stagnating labor productivity in most small-scale systems

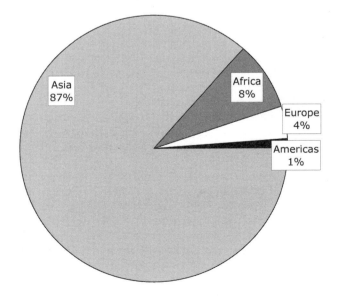

Figure 1-3. *Regional distribution of small-scale farms.* Source: Nagayets, 2005 based on FAO 2001c, 2004c and national statistical agencies.

Note: Small-scale farms are defined as those of less than 2 hectares. The total number of small-scale farms is 404 million.

Figure 1-4. *Small-scale farmer heterogeneity; access and market gap.* Source: Huvio et al., 2004

in developing countries (see Mazoyer and Roudard, 1997; see Figure 1-5). In parallel, work incomes increased most in industrialized countries (Mazoyer, 2001) and prices of industrial manufactured goods generally increased relative to agricultural goods, adding to disparities due to differences between productivity levels.

Many small-scale systems have not been able to compete with industrialized production systems for a number of reasons, including subsidies given to farmers in industrialized countries, cheap fossil energy in mechanized systems compared to metabolic energy in small-scale systems, stabilized market prices in industrialized countries as opposed to completely liberalized prices in developing countries, and the inability to access inputs on favorable terms as compared to large-scale systems. Countries and communities based mainly on small-scale economies are the poorest in the world today, as well as the most threatened by ecosystem degradation (UNEP, 2002). Most small farms with a size of less than two hectares are in Asia (87%), followed by Africa (8%), Europe (4%) and America (1%) (Nagayets, 2005). While the trend in industrial countries has been an increase in average farm size (from about ten to more than 100 ha), it has been the opposite in densely populated developing countries (from about 2 to <1 ha). In some contexts small farm size may be a barrier to investment, however, small farms are often among the most productive in terms of output per unit of land and energy. As yet they are often ignored by formal AKST.

Historical trends suggest that small-scale farms will continue to dominate the agricultural landscape in the developing world, especially in Asia and Africa, at least for the coming two to three decades (Nagayets, 2005). The absolute number of small farms is still increasing in a number of countries on these continents, due to further subdivision of landholdings and expansion of agricultural land. This is also reflected in the labor force differences between countries (see Figure 1-6).

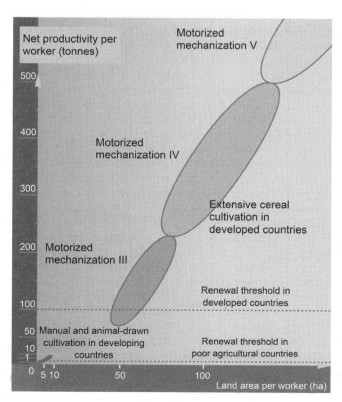

Figure 1-5. *Productivity in developing country cereal systems using motorized mechanization and chemicals and in those using manual or animal-drawn cultivation.* Source: Mazoyer, 2001.

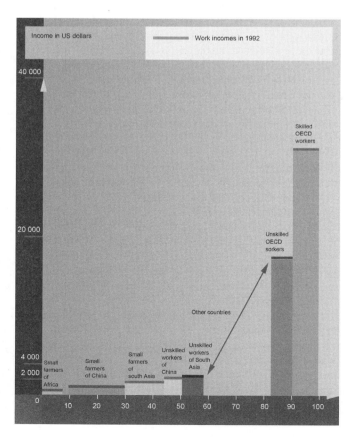

Figure 1-6. *Labor force diversity and income circa 1992.* Source: Mazoyer, 2001.

Ecological changes induced by all types of agriculture

Agricultural activities require change of the natural ecosystem to an agricultural system that is oriented towards human use. This concerns local agricultural practices as well as industrial models and all forms in between. Deforestation was, and still is, the first major step to convert primary tree vegetation into cropland or grazing land, thereby reducing biological diversity in most instances. Other environmental impacts relate to soil, physical, biological and chemical degradation and problems of water quality and quantity.

On the one hand, even in traditional agricultural systems cropping involves tillage operations that may cause accelerated soil erosion. Soil degradation is highest on cropland, but it also affects grazing land and even forest plantations and other agricultural activities (Hurni et al., 1996). Small-scale farming can damage the environment, particularly when practiced under increasing population pressure and with scarce suitable land, involving shortened fallow periods and expansion of cropland areas into unsuitable environmental situations such as steep slopes. This process was particularly accelerated during the past 100 years due to the expansion of farming, despite the emergence of agroecological practices and widespread efforts to introduce sustainable land management technologies on small farms (Liniger and Critchley, 2007).

On the other hand, the advancement of industrial models in agriculture has promoted the simplification of agroecosystems, with reductions in the number of and variability within species. Increased specialization at the field, farm, and landscape levels produces monocultures that potentially increase environmental risks because they reduce biodiversity, ecosystem functions and ecological resilience, and they may be highly vulnerable to climate change. These systems have both benefited and endangered human health and the environment in many industrial countries. While industrial production systems yield large volumes of agricultural commodities with relatively small amounts of labor, they are often costly in terms of human health (Wesseling et al., 1997; Antle et al., 1998; Cole et al., 2002), have additional negative environmental impacts, and are frequently inefficient in terms of energy use. Runoff and seepage of synthetic fertilizers and concentrated sources of livestock waste damage aquifers, rivers, lakes, and even oceans—with costly effects on drinking water quality, fish habitat, safety of aquatic food, and recreational amenities (FAO, 1996a; WWAP, 2003; FAO, 2006b; CA, 2007). This is occurring particularly rapidly in some emerging industrialized countries. However, in countries with increasing industrial production one may also observe more effective food regulation and safety protocols, providing enhanced health protection against foodborne illness. Commercial pesticides often affect non-target organisms and their habitats, and especially when used without strict attention to recommended usage and safety protocols, can negatively affect the health of farm workers (WWAP, 2003). The international transportation of crops, livestock and food products has promoted the global spread of agricultural pests and disease organisms. Many recent significant disease outbreaks have been due to informal, unregulated trade, smuggling, or the industrial restructuring of food systems. The global atmospheric transport of agricultural pollutants, including pesticides, the breakdown products of other agrichemicals, and greenhouse gases, means that environmental costs are also borne by populations far removed from sites of production (Commoner, 1990; UNEP, 2005).

Food security and food sovereignty

Improvement of rural livelihoods, human health and nutrition. Livelihoods are a way of characterizing the resources and strategies individuals and households use to meet their needs and accomplish their goals. Livelihoods are often described in terms of people, their capabilities and their means of living (Chambers and Conway, 1991). Livelihoods encompass income as well as the tangible and intangible resources used by the household to generate income. Livelihoods are basically about choices regarding how, given their natural and institutional environments, households combine resources in different production and exchange activities, generate income, meet various needs and goals, and adjust resource endowments to repeat the process.

Food security exists when all people of a given spatial unit, at all times, have physical and economic access to safe and nutritious food that is sufficient to meet their dietary needs and food preferences for an active and healthy life, and is obtained in a socially acceptable and ecologically sustainable manner (WFS, 1996). *Food sovereignty* is defined as the right of peoples and sovereign states to democratically determine their own agricultural and food policies.

Food sovereignty, the right to food, equitable distribution of food, and the building of sufficient reserves to ensure food security for unexpected events of unpredictable duration and extent (such as hurricanes or droughts), have so far been strategies at the national and international levels with obvious advantages (Sen and Drèze, 1990, 1991). Assumptions that national average food production figures can indicate food security are belied by internal distribution constraints, political limitations on access, inabilities to purchase available food, overconsumption in segments of a population, policies which encourage farmers to shift from family food production to cash crops, crop failure, storage losses, and a range of other factors. Unless all persons feel food secure and are confident in their knowledge of the quality, quantity, and reliability of their food supply, global food security averages cannot be extrapolated to specific cases. The ability to access adequate food covers industrial and cash-cropping farmers, subsistence farmers during crop failures, and non-agriculturists. Access can be limited by local storage failures, low purchasing power, and corrupt or inefficient distribution mechanisms, among other factors. Quality of food, in terms of its nutritional value, is determined by freshness or processing and handling techniques, variety, and chemical composition. A new component in the food security debate is increasing malnutrition in agricultural areas where cash crops, including biofuel crops, replace local food crops.

Food insecurity has been defined in terms of availability, access, stability and utilization. Food insecurity occurs when there is insufficient food over a limited period of time, such as a "hungry season" prior to harvest, or for extended or recurring periods. Food insecurity may affect individuals, households, specific population groups or a wider population. The basic unit for food security within a poor community is the family. Parental sacrifices for children's welfare are demonstrated daily under conditions of scarcity. Families are affected by certain policies, which, perhaps as externalities, create unemployment, inconsistent agricultural prices, and credit-based farming and lifestyles; this is why they are the logical focus for definitions of food security. A family's food supply must be secure "at all times", not simply on average, thereby implying that local storage facilities must be effective, that staples are available out of season, and that distribution systems are uninterrupted by adverse weather, political or budgetary cycles. Food insecurity can be limited to small pockets or affect entire regions. Famine, in contrast, is used to define chronic hunger affecting entire populations over an extended period of time in a famine-affected area, potentially leading to the death of part of the population. Famine may have multiple causes, from political and institutional ones to social, ecological and climatic causes (WFS, 1996).

Temporary food insecurity may be overcome when a harvest comes or when conditions such as weather, wages or employment opportunities improve; it may require action before, during, and even after the period of food insecurity. Household livelihood strategies reflect this. For example, a household that anticipates an upcoming "hungry season" may seek to accumulate savings in advance in the form of cash, grain, or livestock, or it may diversify its economic activities by sending a household member away to seek employment elsewhere. A household experiencing a hungry season may draw on those savings or receive remittances from household members working elsewhere. In more severe cases, a household may borrow money, draw on informal social networks, seek food aid, or even be forced to sell assets (decapitalization)—perhaps achieving temporary food security only at the expense of the ability to generate income in subsequent periods. Other strategies include post-harvest technologies, which may improve storage of products and hence increase both the quantity and quality of available food.

In seeking to meet current needs, some households may be forced to deplete their resources to the point that they remain food insecure for extended periods of time or for recurring periods over many years. In extreme cases, households may have depleted their reserves, exhausted other assets, and be reduced to destitution—with their labor being their only remaining asset. The worst off may, in addition, be burdened with debt and poor health, further limiting their ability to meet current needs, let alone begin rebuilding their capacity to face future challenges.

Whether addressing temporary or chronic food insecurity, it is clear that the challenge goes well beyond ensuring sufficient food in any given period of time. Rather, understanding and meeting the challenge requires a broader perspective on the full range of needs and choices faced by households, the resources and external conditions that influence those choices, and the livelihood strategies that could enable families to meet their food needs over time.

Availability of and access to animal genetic resources can be a problem for pastoralists and poor households. An emerging problem is management of epidemics, as currently illustrated in Asia and increasingly in other parts of the world, where thousands of animals are killed prophylactically because of avian influenza (GTZ, 2006).

1.1.3 Emerging issues

What can agriculture offer globally to meet emerging global demands, such as mitigating the impacts of climate change, dealing with competition over (dwindling) resources? Projections of the global food system indicate a tightening of world food markets, with increasing scarcity of natural and physical resources, adversely affecting poor consumers. Improved AKST in recent years has helped to reduce the inevitable negative environmental impacts of trade-offs between agricultural growth and environmental sustainability at the global scale. Growing pressure on food supply and natural resources require new investment and policies for AKST and rural development in land-based cropping systems.

AKST is well placed to contribute to emerging technologies influencing global change, such as adaptations to climate change, bioenergy, biotechnologies, nanotechnology, precision agriculture, and information and communication technologies (ICT). These technologies present both opportunities and challenges, and AKST can play a central role in accessing the benefits while managing the potential risks involved.

About 30% of *global emissions* leading to climate change are attributed to agricultural activities, including land use changes such as deforestation. Additionally, environmental variations resulting from climate change have also adversely affected agriculture. In extreme cases, severe droughts and floods attributed to climate change make millions of people in resource poor areas particularly vulnerable when they depend on agriculture for their livelihoods and food. AKST can provide feasible options for production systems, manufacturing and associated activities which will reduce the dependence on depleting fossil fuels for energy. Similarly, AKST can provide information about the consequences of agricultural production on the hydrology of watersheds and groundwater resources. AKST can also be harnessed to reduce greenhouse gas (GHG) emissions from agriculture, as well as increase carbon sinks and enhance adaptation of agricultural systems to climate change impacts (Chapter 6).

Continuing structural changes in the *livestock* sector, driven mainly by rapid growth in demand for livestock products, bring about profound changes in livestock production systems. Growing water constraints are a major driver of future AKST. Soil degradation continues to pose a considerable threat to sustainable growth of agricultural production and calls for increased action at multiple levels; this can be strongly supported by AKST. Forestry systems will remain under growing pressure, as land use systems and urbanization continue to spread particularly into these ecologically favorable areas. Biodiversity is in danger as a result of some agricultural practices. Finally, there is significant scope for AKST and supporting policies to contribute to more sustainable fisheries and aquaculture, leading to a reduction of overfishing in many of the world's oceans.

Bioenergy is being promoted in several countries as a

lucrative option to reduce GHG emissions from fossil fuels; however, controversy is increasing on the economic, social and ecological cost/benefit ratio of this option. On-farm bioenergy production utilizing farm residues has potential. However, studies have revealed that bioenergy demand is sensitive not only to biomass supply, but also to total energy demand and competitiveness of alternative energy supply options (Berndes et al., 2003). Additionally, the environmental consequences and social sustainability aspects of the processing of crops and feedstocks as biofuels have not yet been thoroughly assessed.

Biotechnology has for millennia contributed to mankind's well-being through the provision of value-added foods and medicines. It has deep roots in local and traditional knowledge and farmer selection and breeding of crops and animals, which continues to the present day. Micropropagation of plants by tissue culture is now a common technique used to produce disease-free plants for both the agricultural and ornamental industries. Recent advances in the area of genomics, including the ability to insert genes across species, have distinguished "modern biotechnology" from traditional methods. Resulting transgenic crops, forestry products, livestock and fish have potentially favorable qualities such as pest and disease resistance, however with possible risks to biodiversity and human health. Other apprehensions relate to the privatization of the plant breeding system and the concentration of market power in input companies. Such issues have underpinned widespread public concern regarding transgenic crops. Less contentious biotechnological applications relate to bioremediation of soils and the preparation of genetically engineered insulin. Commercial transgenic agricultural crops are typically temperate varieties such as corn, soya and canola, which have been engineered to be herbicide resistant or to contain the biological agent Bt (bacillus thuringiensis), traits that are not yet widely available for tropical crops important to developing countries. Transgenic crops have spread globally since 1996, more in industrialized than in developing countries, covering about 4% of the global cropland area in 2004 (CGIAR Science Council, 2005).

Current trends indicate that transgenic crop production is increasing in developing countries at a faster rate than in industrialized nations (Brookes and Barfoot, 2006). This is occurring against a background of escalating concerns in the world's poorest and most vulnerable regions regarding environmental shocks that result from droughts, floods, marginal soils, and depleting nutrient bases, leading to low productivity. Plant breeding is fundamental to developing crops better adapted to these conditions. The effectiveness of biotechnologies will be augmented, however, by integrating local and tacit knowledge and by taking into account the wider infrastructural and social equity context. Taking advantage of provisions under the international protocol on biosafety (Cartagena Protocol on Biosafety) as well as establishing national and regional regulatory regimes are essential elements for using AKST in this domain.

1.2 Conceptual Framework of the IAASTD

1.2.1 Framework for analysis—centrality of knowledge
Conceptual framework of the AKST assessment (Figure 1-7). There is huge diversity and dynamics in agricultural production systems, which depend on agroecosystems and are embedded in diverse political, economic, social and cultural contexts. Knowledge about these systems is complex. The AKST assessment considers that knowledge is coproduced by researchers, agriculturalists (farmers, forest users, fishers, herders and pastoralists), civil society organizations and public administration. The kind of relationship within and between these key actors of the AKST system defines to what degree certain actors benefit from, are affected by or excluded from access to, control over and distribution of knowledge, technologies, and financial and other resources required for agricultural production and livelihoods. This puts policies relating to science, research, higher education, extension and vocational training, innovation, technology, intellectual property rights (IPR), credits and environmental impacts at the forefront of shaping AKST systems.

Knowledge, innovation and learning play a key role in the inner dynamics of AKST. But it is important to note that these inner dynamics depend on how the actors involved respect, reject or re-create the values, rules and norms implied in the networks through which they interrelate. The IAASTD considers that its own dynamics strongly depend on related development goals and expected outputs and services, as well as on indirect and direct drivers mainly at the macro level, e.g., patterns of consumption or policies.

The AKST model emphasizes the centrality of knowledge. It is therefore useful to clarify the differences between "information" and "knowledge". Knowledge—in whatever field—empowers those who create and possess it with the capacity for intellectual or physical action (ICSU, 2003). Knowledge is fundamentally a matter of cognitive capability, skills, training and learning. Information, on the other hand, takes the shape of structures and formatted data that remain passive and inert until used by those with the knowledge needed to interpret and process them (ICSU, 2003). Information only takes on value when it is communicated and there is a deep and shared understanding of what that information means—thus becoming knowledge—both to the sender and the recipient.

Such an approach has direct implications for the understanding of science and technology. The *conventional* distinction between science and technology is that science is concerned with searching for and validating knowledge, while technology concerns the application of such knowledge in economic production (defined broadly to include social welfare goals). In most countries institutional and organizational arrangements are founded on this distinction.

However, this distinction is now widely criticized in contemporary science and development literature, both from a conceptual point of view and in terms of practical impacts. Gibbons and colleagues are a good example of this critical debate: they distinguish between "mode 1" and "mode 2" styles of knowledge development (Gibbons et al., 1994; Nowotny et al., 2003). In very simple terms, the distinction is that "mode 1" approaches (the traditional view) argue for a complete organizational separation between scientific research on the one hand and its practical applications for economic and social welfare on the other. Conversely "mode 2" approaches argue for institutional arrangements that build science policy concerns directly into the conduct

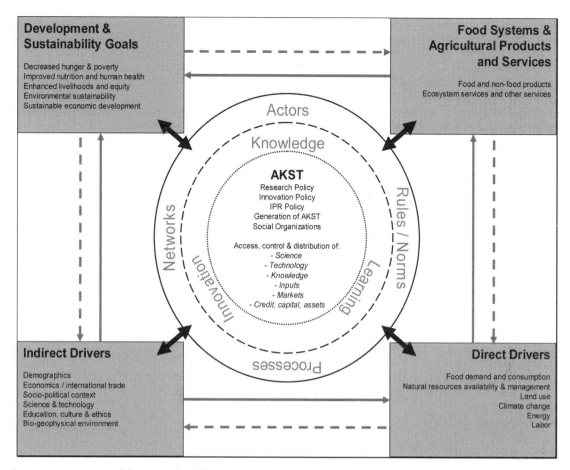

Figure 1-7. *Conceptual framework of the IAASTD.*

of research and development (R&D). As a practical contemporary example, this debate is very much at the heart of current discussions about how agricultural research should be conducted in all countries.

Innovation and innovation systems. Scientific and technological knowledge and information can (1) add value to resources, skills, knowledge, and processes, and (2) create entirely novel strategies, processes, and products (World Bank, 2006a). An innovation system may be defined as the network of agents, usually organized in an interdisciplinary and transdisciplinary manner, with interactions that determine the innovative impact of knowledge interventions, including those associated with scientific research. The concept is now used as a kind of shorthand for the network of interorganizational linkages that apparently successful countries have developed as a support system for economic production. In this sense it has been explicitly recognized that economic creativity actually relies on the quality of "technology linkages" and "knowledge flows" amongst and between economic agents. Where interactions are dynamic and progressive, great innovative strides are often made. Conversely, where systemic components are compartmentalized and isolated from each other, the result is often that relevant research bodies are not innovative. Some approaches suggest that innovation systems cannot be separated from the social, political and cultural context from

which they emerge (Engel, 1997), and this context therefore has to be included in the analysis of AKST. This implies a need to focus on those factors that enable the emergence of "innovative potential," rather than on factors related directly to specific innovations.

Collaborative learning processes. The creation of favorable conditions making it possible for different actors to engage in collaborative learning processes—i.e., the increase in space and capacity for innovativeness—has thus gained prime importance. Approaches based on linear understandings of research-to-extension-to-application are being replaced by approaches focusing on processes of communication, mutual deliberation, and iterative collective learning and action (van de Fliert, 2003). More concretely, this implies that sustainable use of natural resources requires a shift from a focus on technological and organizational innovation to a focus on the norms, rules and values under which such innovation takes place (Rist et al., 2006). The enhanced AKST model considers that values, rules and norms that are relevant to the promotion of agricultural development are constantly produced and reproduced by social actors who are embedded in the social networks and organizations to which they belong. Social networks are important spaces where the actors involved in the coproduction of knowledge share, exchange, compare and eventually socialize their individually realized perceptions of what is important, good, and bad, and enable the visions they have for their own families, communities and wider social categories to which they belong.

AKST-related policies. For the IAASTD model of AKST, policy referring to AKST must be understood in a broad sense. Policy can be thought of as a course or principle of action designed to achieve particular goals or targets. The idea of policy is usually associated with government bodies, but other types of organization also formulate policies—for example a local NGO may establish a policy about who is eligible for its programs (DFID, 2001). "Policy analysis" is the process through which the interactions at and between these various levels are explored and articulated. Policy relating to the AKST model is thus understood as the attempt to systematically intervene in the process of shaping and reshaping the interrelationships between the different actors, networks and organizations involved in the processes of coproduction of knowledge for more sustainable and pro-poor agriculture and food production.

1.2.2 Development and sustainability goals

Reduction of poverty and hunger. Poverty can be defined in different ways, each requiring its own measurement. Poverty can be measured in terms of access to the basic needs of life, such as nutrition, clean water and sanitation, education, housing and health care. An income level of US$1 per day is widely accepted as a rough indicator of poverty although there is general agreement that the multidimensional nature of poverty cannot be captured with this measure. Worldwide, about 1,200 million people live on less than US$1 per day; in percentage terms this is expected to drop from 19% of the world population in 2002 to 10% by 2015 (World Bank, 2006b), although in absolute numbers the difference will be smaller because by then the total population will be larger by about 800 million people. Moreover, many countries, particularly in Africa and South Asia, are not on track regarding achievement of the Millennium Development Goals (Global Monitoring Report, 2006) (Figure 1-8). Furthermore, these numbers should be interpreted with caution. Any change from the nonmonetary provision of goods and services to the cash market, such as a shift from subsistence to commercial crops, will appear as an increase in income whether or not there has been a concomitant improvement in standard of living or reduction in poverty. This indicator focuses our attention exclusively on income derived from market transactions and ignores other components of livelihood.

Approximately 852 million people are unable to obtain enough food to live healthy and productive lives (FAO, 2004a). Hunger is discussed here in the wider sense of encompassing both food and nutritional insecurity (UN Millennium Project, 2005). An estimated 800 million persons, i.e., more than half of the people living in extreme poverty, are occupied in the agricultural sector (CGIAR Science Council, 2005). Their livelihoods are usually derived from small-scale farming. In 1996, around 2.6 billion people, or 44% of the total world population were living in agriculture-dependent households, mostly in Asia and Africa (Wood et al., 2000). Poverty is thus disproportionately rural (poor farmers and landless people) despite ongoing migra-

tion from rural to urban areas. Among other factors such as civil wars and diseases, migration has led to an increase in female-headed households and intensified the already heavy workload of rural women (García, 2005).

Decapitalization (e.g., through sale of livestock and equipment), deterioration of infrastructure and natural capital (e.g., soils), and the general impoverishment of peasant communities in large areas in developing countries (for Africa, see Haggblade et al., 2004) remains a serious threat to livelihoods and food security. The loss or degradation of production assets is linked to the overexploitation of scarce resources (land, water, labor), markets that are inequitable (IFAD, 2003) and difficult to access, competition from neighboring farms, and in some instances the combined effects of competition from the industrialized sector (leading to low prices), and the direct and indirect taxation of agriculture. It may also be a consequence of the barriers to capital accumulation and investment associated with the realities faced by some small-scale farmers (Mazoyer and Roudard, 1997). On the other hand, agricultural growth can, despite this difficult context, lead to important benefits for poverty alleviation (Byerlee et al., 2005). In some cases the beneficiaries are people remaining in small-scale agriculture but there may also be important opportunities for those who work, for example, in agriculture-related product processing activities.

Improvement of livelihoods, human health and nutrition. Even though a large number of people depend entirely on agriculture, off-farm income is important for many households that depend on agriculture for their livelihoods. The resulting variety of livelihood strategies can be thought of in terms of adjustments in the quantity and composition of an individual's or household's resource endowment. Different resource endowments and different goals imply different incentives, choices, and livelihood strategies.

Health is fundamental to live a productive life, meet basic needs and contribute to community life. Good health offers individuals wider choices regarding how to live their lives. It is an enabling condition for the development of human potential. The components of health are multiple and their interactions complex. The health of an individual is strongly influenced by genetic makeup, nutritional status, access to health care, socioeconomic status, relationships with family members, participation in community life, personal habits and lifestyle choices. The environment—natural, climatic, physical, social or workplace—can also play a major role in determining the health of individuals. For example, in most societies, biomass fuel collection is a woman's task. Women often spend hours collecting and carrying fuelwood back home over long distances. Poor women are among the more than two billion people who are unable to obtain clean, safe fuels and have to rely on burning biomass fuels such as wood, dung or crop residues. The time and labor spent in this way limits their ability to engage in other productive activities; and their health suffers from hauling heavy loads and from cooking over smoky fires (Lambrou and Piana, 2006). On the other hand about 50% of the health burden of malnutrition is attributable to poor water, sanitation and hygiene (Prüss-Üstün and Corvalán, 2006). For example, some long-standing problems such as mycotoxins continue

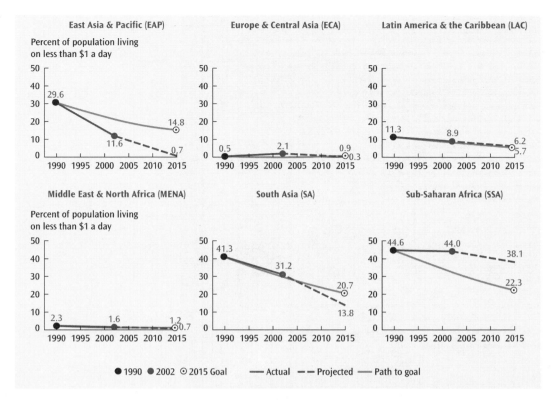

Figure 1-8. *Poverty by region, 1990-2002, and forecasts to 2015.* Source: Global Monitoring Report, 2006.

to significantly add to the health burden, especially of infants (Gong et al., 2004), and cause widespread problems with basic foodstuffs (Strosnider et al., 2006). This has become an issue for formal AKST, as is water quality, which is linked to improving rural livelihoods, human health and nutrition and to the covering of protein requirements, particularly in the case of children. Human health is also linked with animal health: numerous examples of zoonoses are reported, including avian influenza, hoof and mouth disease, and brucellosis.

In 50% of cases undernutrition is due to poor sanitation and diseases (Prüss-Üstün and Corvalán, 2006). This fundamental issue is reflected throughout the AKST context and emphasizes traditional food safety, including hygiene issues related to animal husbandry and phytosanitary protection, food storage in homes and food handling in developing countries. Furthermore, in developing countries such phytosanitary issues as Claviceps purpurea or ergotism, (which are no longer problems in the North because of highly protected industrial food production), are significantly adding to the health burden, especially of infants (Gong et al., 2004) and cause hygienic problems amongst billions with basic foodstuffs (Strosnider et al., 2006). Poverty and undernourishment are intimately linked. The MDG targets for 2015 are expected to be met by most regions except for sub-Saharan Africa in particular and South Asia (see Figure 1-9).

A direct consequence of poverty is undernourishment, which is an issue not only for the urban poor and for landless persons, but particularly for the underprivileged such as women and children. Undernourishment also affects rural people producing agricultural goods and services on farms that are too small, not productive enough, or too degraded to produce sufficient outputs for a decent living. Good nutrition has thus much to contribute to poverty reduction. It is intrinsic to the accumulation of human capital, since sound nutrition provides the basis of good physical and mental health, and thus of intellectual and social development and a productive life. If global poverty is to be reduced, agricultural development will have to pay particular attention to the problems faced by deprived small-scale producers and their families. Science and technology are expected to contribute to the achievement of this goal.

Promotion of socially equitable and environmentally and economically sustainable development. Sustainable development is about meeting current needs without compromising the ability of future generations to meet their own needs. Within this context, sustainability is envisaged within three key dimensions: social, environmental and economic, all three of which have direct and indirect linkages to agriculture. In the context of the IAASTD, the term "agriculture" encompasses crop cultivation, livestock production, forestry and fishery. This broader definition provides future opportunities for maximizing synergies in achieving development and sustainability goals. It serves the primary goal of providing sufficient and nutritious food for humankind, in the present and in the future. It is indisputable that agriculture as a sector cannot meet this goal on its own. Agriculture, however, fulfills a series of additional goals besides food production. Last but by no means least, agriculture ensures the delivery of a range of ecosystem services. In view of a globally sustainable form of development, the importance

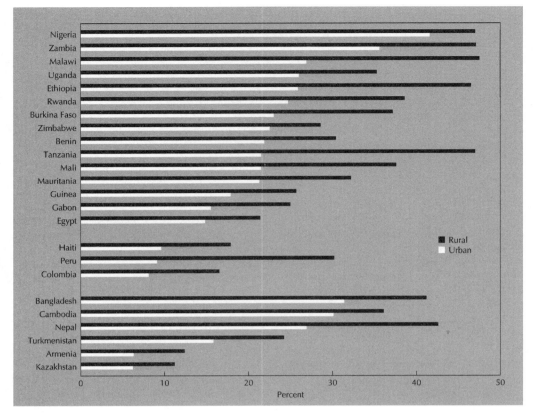

Figure 1-9. *Child malnutrition (low height for age) among preschool children in surveys since 1999.*
Source: Rosegrant et al., 2006

of this role may increase and become central for human survival on this planet.

1.2.3 Agricultural Knowledge, Science and Technology (AKST)

A challenge for formal AKST is the great imbalance in numbers of researchers per million inhabitants: this number is 65 times smaller in Africa than in industrialized countries (Hurni et al., 2001). Nearly half of public agricultural research expenditures, amounting to US$23.0 billion in 2000, are spent in developed countries, i.e., benefiting only a few million, though highly productive, farmers (Pardey et al., 2006). While private agricultural research spending is somewhat higher than public spending in developed countries, private spending in developing countries is very low, accounting for only 8% of total public and private investments in AKST (see chapter 8 for details of AKST investment levels).

Public agricultural research in industrial countries also benefits farmers in other countries, since much public agricultural research is basic research that may later be applied to a variety of agricultural settings through technology transfer, and public research often leads to publicly available crop varieties that are widely distributed. Traditional experimental systems and many emerging farmers' programs—some initiated by international institutions such as FAO but most from farmers' organizations and social movements—are also considered as a component of agricultural research.

Regional shares in public agricultural research expendi-

tures have been changing in the past 40 years (Pardey et al., 2006). While overall investments nearly doubled, industrialized countries, which had 55% of all investments in 1981, received a smaller share—44% in 2000, while in China and other Asian states investments increased manyfold. In general, research and development (R&D) investments have so far generated high returns (Byerlee and Alex, 2003; Chapter 8.2), however at a high ecological cost. For example, trends in cereal production since 1960 show that area productivity increased by a factor of 2.5 in industrialized countries, from 2.1 to 4.9 tonnes ha^{-1} on average on a total of 140 million hectares. In developing countries, the factor was even higher, i.e., 2.8, and the increase was from 1 to 2.8 tonnes ha^{-1} on a total cropped area of 440 million ha (Cassman, 2003). It must be noted, however, that stagnation in land productivity increase has been observed in many areas since about 1985 (Cassman, 2003).

Some recent changes in thinking have raised a number of cognate issues in formal AKST systems. The policy agenda has evolved from a formal "science push" approach to one that places more emphasis on participatory, multi-stakeholder, inter- and transdisciplinary, and client-driven research agendas. Donors, supranational structures, regional organizations, and governments are looking for stronger interinstitutional support for development projects in order to attract private sector investments. Largely, this has been driven by changing contexts and circumstances since the days of the Green Revolution. Perhaps the biggest challenge is to fill the gap in research and technology that is relevant to the poorest. In particular much private and public R&D is spent on corn, wheat, maize, and rice, while very little is devoted to cassava, millet, sorghum and potatoes. However, it has not proved easy for research and extension organizations to adapt their established practices (Graham et al., 2001) to the new way of understanding rural development as part of an AKST system based on the idea that knowledge is coproduced by all actors involved. The most important of these issues are summarized in the present subchapter. Thinking on rural development has shifted from the 1960s to the 1990s and has reached a balanced state between the productive and social sectors, and between state and market interventions.

Effectiveness of formal AKST organizations. It is well known that many public research and development (R&D) bodies of national agricultural research systems (NARS) are finding it difficult to deal with poor farmer- and peasant economy-based issues in many countries. The problems range from resource constraints on the one hand to rigid, disciplinary-bound research planning on the other (IAC, 2004). Often there is a lack of engagement with client sectors and unwillingness to exchange and co-generate knowledge with other research bodies in the sector. This is also related to the process of identifying research problems, which is often based solely on perceptions of disciplinary-based researchers with incentive systems usually grounded mainly on the number of publications. The inevitable result is that all too often resource allocation to the NARS does not pay off in terms of economic, social and environmental development possibilities for poor farmers. While a number of countries have initiated some remedial policies for these issues, the relevant literature shows that there is still some way to go. The difficulties of more equality-based engagement with farmers, peasants, or "clients" has also to do with an understanding of the reasons guiding rural actors' decisions, actions and livelihoods that is too narrow (see Yapa, 1993 for Asia; Wiesmann, 1998 for Africa; Trawick, 2003 for Latin America).

Promotion of other stakeholders' AKST. Traditionally, the passing on of results of agricultural research to users was handled by state-funded extension services. Not only have these suffered through structural adjustment measures, but an increasing number of questions have also been raised by the extension systems themselves as operational organizational mechanisms (Farrington et al., 2002; IAC, 2004). There is also evidence of an increased need to engage in partnerships in order to reconceptualize (in theory and practice) the delivery of technology in the context of an AKST system that is based on the paradigms of knowledge coproduced by scientists, policy makers and client groups. These partners include private sector organizations, but they also involve NGOs, community-based organizations (CBOs) and social movements that are able to bring skills and knowledge to bear simply due to the close relationships they have established with specific communities. Today's challenges in community development in developing countries make it more compelling for higher education to reach effective changes of vision and prepare professionals to lead innovative rural development processes. Training, capability building, and reinforcement of small-scale farmers' skills to enable them to participate in the agriculture supply chain are urgent tasks.

Coproduction of agricultural knowledge

The combination of various forms of exogenous scientific knowledge, e.g., from the natural, agronomic, economic or social sciences, with the many and highly diverse forms of so-called "local", traditional or endogenous knowledge is a basic challenge. These different forms of knowledge are represented by different local (farmers, traders, craftsmen, etc.) and external actor groups (civil servants, extensionists, researchers, service providers, etc.). One can therefore call them "knowledge systems". Combining endogenous and exogenous knowledge is achieved by increased participation of "end users"—including marginalized and poor actors—in the different forms of research and development. While the initial focus of combining knowledge was on increasing participation at local levels, today emphasis is shifting towards upscaling participatory processes into the meso- and macro-levels of social organization (Gaventa, 1998) resulting in multilevel and multistakeholder approaches.

When taking into account the centrality and value of endogenous, traditional or local forms of knowledge related to agricultural development—e.g., through ethnological approaches in sciences studying agricultural soils, plants and animals (Nazarea, 1999; Winklerprins, 1999)—it is necessary to reflect on the ethical and epistemological implications related to the integration of different knowledge systems (Dove and Kammen, 1997; Olesen et al., 2000; Rist and Dahdouh-Guebas, 2006). Integration of, and cooperation between, different knowledge systems is often hampered by interaction that does not take into account the need for the process of communication to move beyond the practical and generally tangible technological economic, ecological and social effects of innovations. In the long run, innovation can only be successful if it "makes sense" to all those involved, i.e., it needs to be integrated into (and by) the different knowledge systems involved. This is also particularly important for innovations in rural development (Dove and Kammen, 1997; Olesen et al., 2000).

There is also growing consensus among researchers concerned with sustainable agriculture that no single group of actors should appropriate the right to define what type of combination should exist between scientific and "local" forms of knowledge (Röling and Wagemakers, 2000; Rist and Dahdouh-Guebas, 2006). As a consequence, participatory forms of coproduction of knowledge, based on social learning among actors involved, have become a key feature of sustainable agriculture and resource management (Wollenberg et al., 2001; Rist et al., 2003; Pahl-Wostl and Hare, 2004). This means that the role of science within a process of participatory knowledge production must be redefined. Instead of striving to find and voice the ultimate instance of "truth", the scientific community must complement conventional and generally discipline-based knowledge production with inter- and transdisciplinary approaches. The particularity of a transdisciplinary approach is that it implies examining "real-world problems" from a perspective that (1) goes beyond specific disciplines by combining natural, technical, economic and social sciences, and (2) is based on broad participation, characterized by systematic cooperation with those concerned (Hurni and Wiesmann, 2004). A major task of sciences relating to society in a transdisciplinary perspective is to assure that the diversity of actors, interests, complexity and dynamics of the processes involved are given adequate consideration. More concretely this means bringing three basic and interrelated questions into societal debates on sustainable agriculture: (1) How do processes constitute a problem field, and where is the need for change? (2) What are more sustainable practices? (3) How can existing practices be transformed (Hirsch Hadorn et al., 2006).

Engagement with agribusiness opportunities. Agricultural research partly faces the agenda of an agricultural research system which is frequently inappropriate for the emerging realities of the often poverty-affected agricultural sector in developing countries. While production, sale and consumption of major food crops remains important, a number of niche sectors with impressive growth rates are emerging, and this is coupled with fundamental changes in the nature of the sector as a whole. New and rapidly growing markets are emerging, e.g., for livestock, horticulture and cut flowers, pharmaceutical and nutriceutical crops, natural beauty products, and industrial use products such as biofuels and starch. The role of the private sector is increasing, and with it new issues arise, such as corporatization of craft-based industries, the exposure of producers and firms to competition, changing international trade rules and regulations such as sanitary and phytosanitary standards, intellectual property rights (IPR, see below), the knowledge-intensive nature of these niche sectors, and the importance of innovation as a source of competitive advantage under rapidly evolving market and technology conditions.

Transfer and use of imported AKST. The recent report of Task Force 10 on Science, Technology and Innovation (UN Millennium Project, 2005) emphasizes the general importance for all actors involved in agricultural production and marketing of acquiring knowledge in a globalized world. A key change is the emergence of private sector research. This is partly a result of strengthened intellectual property protection regimes and technical advances in biotechnology. Also significant are the opportunities that economic and trade liberalization and globalization are now offering for private investments in agroindustries such as seed production. The net result is that on the one hand, public agricultural research systems have to consider more complex agendas including for example how to appropriately acquire genetic resources and how to establish equitable benefit-sharing regimes for those societies and communities from whose livelihood sphere the primary ingredients for corporate patents often originate. On the other hand, this also implies that research and development centers have to learn how to better respond to sociopolitical debates that can shape and define the societal preconditions that influence the amounts, use and allocation of financial and human resources available for research and development in rural areas. Technocratic, hierarchical and disciplinary-based definitions of research and development policies are no longer adequate in the context of civil society organizations' growing participation in defining policies related to research and technology development. Against this background, an especially important issue is related to local knowledge, which was perceived as an "obstacle" for development, and is now considered an important resource that contributes to better targeted development efforts (Scoones and Thompson, 1994; Blaikie et al., 1997).

International agreements and implications for AKST. A related issue is that of the growing number of relevant international agreements that many developing countries have signed and ratified. One good example is the Convention on Biological Diversity (CBD), with a number of articles on opportunities for sustainable agricultural development. For example, Article 15 on access to genetic resources enjoins members to rationalize the use of biological resources in ways that promote exploitation of such resources for socioeconomic purposes. Many countries are aware that there are significant opportunities here for the acquisition of significant off-farm income generation that could go some way towards alleviating poverty, but there is often a severe shortage of technological capacity to realize these opportunities (Glowka et al., 1994). The key point is that such agreements imply a need for developing countries to increase AKST capacity relevant to the new contexts.

Management of relevant "intellectual property rights" (IPR). Management (and protection) of intellectual property (IP) in agriculture is now recognized as a fundamental task of knowledge-based development. But while large international companies have moved forward in this respect, many developing countries still have great difficulties ensuring that their creativity can achieve similar protection. Part of the problem is clearly institutional. Scientists find it difficult to understand that their research will often give rise to significant IP and that they have additional responsibilities in this respect, if only to protect the novel public goods that they have helped to create. Similarly the organizations in which they work are often trapped in a "mode 1" world (Gibbons et al., 1994) and see their responsibilities as ending with the publication of scientific papers in refereed journals. Moreover, patents on life forms create broad controversies, especially those connected with a ban on using harvested grain as seed. Patent claims for animals currently regard whole breeds.

Therefore, questions that arise in this context have to do with the creation of capacity and related initiatives which ensure that knowledge coproduction and technology development in developing countries are as fully informed as possible in these respects. However, it remains open whether the global tendency to protect IP rights is realistic, considering the fact that numerous instances of intellectual property are based on societies' centuries-old intellectual and empirical inputs. In such situations, the quest for equitable benefit sharing may seem impossible, thus calling into question the entire discussion about IPR. The patenting case of Neem extracts (*Azadirachta indica*) may be quoted as an example. By challenging the patent on a Neem product, the Indian Government was able to prove that the same Neem product was industrialized and has been used in India for several millennia (Sheridan, 2005).

Access to and reform of AKST education

A broader set of issues concerns the formal training of scientists and related workforce. As the MDG Task Force 10 has emphasized, higher education is increasingly being recognized as a critical aspect of the development process; at the same time, however, most universities are ill-equipped to meet the challenge. Outdated curricula, under-motivated faculties, poor management and a continuous struggle for funds have undermined the capacity of universities to play their roles as engines of community or regional development (UN Millennium Project, 2005).

A report by the InterAcademy Council (IAC, 2004) recently underlined the relative decline of the agricultural research and education system in Africa in the past decades. Among the reasons discussed in the report are the relative weakness of science education in African schools, low investment in research in general, and the growth of student numbers (by 8% per year), with funding falling short of this increase and funding decline accentuated by structural adjustment. The report also notes an unexpected renewal phase initiated by a half dozen African universities in the recent past.

Some MSc and PhD programs in industrialized countries do not always suit the needs of less industrialized countries. The implications both for curriculum revision and access are therefore considerable from an AKST standpoint and will be covered at various points in this report. A positive example is the higher education system in Costa Rica, which is making significant efforts to focus agricultural development on knowledge and technological innovation. It is also important to take into account the gender disparity in training as well as the lack of focus on gender analysis in the curricula of agricultural universities in developing and—most often also in industrialized—countries.

Besides overcoming shortcomings with regard to quantitative aspects of human and financial resources, it will also be of paramount importance to combine an increase in resource allocation and further capacity development of actors involved in research and extension aimed at a qualitative shift towards more societal modes of knowledge production emphasizing inter- and transdisciplinary approaches (Hurni and Wiesmann, 2004).

Capacity development is broadly defined here and includes developing (1) common understandings of problems, solutions and ways to approach them, using a variety of interpersonal and intra-social processes; (2) social and cultural resources, not just human resources; (3) multiple, strategic skills across a range of areas to intervene and advocate, not just passive receipt of programs and policies, and (4) institutional and organizational bases of power. If policies for organizational reforms are introduced, medium- to high-level scientific resources are made available for formal higher and tertiary education systems, and organizational change is initiated in the structure of relevant governance procedures, such as those concerned with the management of extension services, funding of R&D, mobilizing of informal inputs from NGO and related bodies, optimizing the use of for-

eign technology, and providing procedures for a balanced use of the private sector, deployment of AKST will become far more effective. Indeed, such changes will enable more adequate analysis of agroecosystem services, which is usually not included in production-oriented AKST, and the finding of strategies to mitigate negative impacts ("damages") caused by agricultural practices to such services. Further improvements can be achieved by promoting knowledge of interventions that are environmentally and socially sustainable, including measures to empower women to a much greater degree than has been the case in the past.

Measurement of "knowledge" categories

There is a large gap in research intensity (measured as public R&D investment as a ratio of agricultural GDP) between developing and industrialized countries. In 2000, the intensity ratio for the developing countries as a total averaged 0.53%, compared to 2.36% for the developed countries as a group (Pardey et al., 2006). This intensity gap has increased over the past decades as a result of a much higher growth in agricultural output in developing countries as group than in the developed countries.

One of the problems in dealing with AKST policy (indeed, KST of all types) is that of measurement—both for "inputs", i.e., investment in AKST, and "outputs", i.e., indicators of resultant knowledge impacts. In the case of the former, a range of proxies are used, the most common being agricultural R&D expenditures in the public sector. Another is the number of persons with PhDs currently working in agricultural R&D organizations. Both are unsatisfactory for the obvious reason that they probably give a distorted picture of knowledge investment. For example, they do not account for external inputs from overseas, which may be higher than the internal inputs. A similar problem exists on the output side since outputs can also take a variety of forms, for instance number of patents, number of new plant varieties registered or number of relevant scientific papers published in refereed journals. Again, all kinds of problems involved in the interpretation of these data are due to paucity of information, lack of disaggregation, variations in national practices, and of course the fact that they often do not pick up on several types of tacit knowledge. It is therefore worth noting that attempts to be quantitative in this area need to be treated with great care.

Giving local knowledge due recognition means to specifically monitor its integration into the processes of knowledge production at the interface of research and practice. The above indicators must be differentiated more accurately, taking into account the share of research and development expenditures per sector, number of PhDs, and scientific publications, explicitly in relation to the search for new modes of knowledge production that focus on the integration of local forms of knowledge. Indicators must not only allow quantification of resources allocated to local and traditional components of AKST systems. They must also make visible to what degree the resources allocated to these components of an AKST system reflect the overall relationship that local or traditional knowledge and external knowledge actually have in ensuring the livelihood systems of rural people in general and of poor and marginalized people in particular.

Multifunctionality and sustainability would require indicators of both local and scientific knowledge.

1.2.4 Agrifood systems, agricultural products and services

Agricultural systems, outputs and services. The major outputs generated by the multiple agricultural systems worldwide may be referred to as "provisioning services" (MA, 2003):

- Food consisting of a vast range of food products derived from plants, animals, and microbes for human consumption;
- Feed products for animals such as livestock or fish, consisting of grass, herbs, cereals or coarse grains and other crops;
- Fiber such as wood, jute, hemp, silk, and other products;
- Fuel such as wood, dung, biofuel plants and other biological materials as sources of energy;
- Genetic resources including genes and genetic information used for animal and plant breeding, and for biotechnology;
- Biochemicals, natural medicines, and pharmaceuticals including medicines, biocides, food additives, and biological materials;
- Ornamental resources including animal products such as skins and shells, and ornamental plants and lawn grass; and
- Freshwater from springs and other sources, as an example of the linkage between provisioning and regulating services.

Agricultural systems are highly complex, embracing economic, biophysical, sociocultural and other parameters. They are based on fragile and interdependent natural systems and social constructions. Agriculture has a potential to play positive roles at different scales and in different spheres (Table 1-3).

Diversity of agricultural systems

Globally, agricultural systems have been changing over time in terms of intensity and diversity, as agriculture undergoes transition driven by complex and interacting factors related to production, consumption, trade and political concerns. There are a multitude of agricultural systems worldwide. They range from small subsistence farms to small-scale and large commercial operations across a variety of ecosystems and encompassing very diverse production patterns. These can include polycultures or monocultures, mixed crop and livestock systems, extensive or intensive livestock systems, aquaculture systems, agroforestry systems, and others in various combinations. In Africa alone, there are at least 20 major farming systems combining a variety of agricultural approaches, be they small- or large-scale, irrigated or non-irrigated, crop- or tuber-based, hoe- or plough-based, in highland or lowland situations (Spencer et al., 2003).

Agricultural systems are embedded in a multiplicity of different economic, political and social contexts worldwide. The importance of the agricultural sector in these economies, or the type of agricultural policy enforced will therefore depend on the national economies. It is thus crucial

to gain a clear knowledge of the state of agriculture in the different ecological and socioeconomic contexts to be able to assess the potential for further development of this sector in relation to development and sustainability goals. The different contexts have led to economic disparities within and among regions, countries and especially between industrial and small-scale farmers (FAO, 2000). Apart from differences in labor productivity, examples of disparities are average farm sizes (121 ha in North America vs. 1.6 ha in Asia and Africa, see von Braun, 2005; 100,000 ha in Russia, Ukraine and Kazakhstan, see Serova, 2007) and the crop yield gap between high- and low-income countries.

The last 50 years have seen a tremendous increase in agricultural food production, at a rate more rapid than human population growth. This was mainly due to the increase in area productivity, which differed between the regions of the world, while cereal-harvested area stagnated almost everywhere (Cassman, 2003).

In all regions of the world, however, a decrease in the economic importance of the agricultural sector at different stages of economic development can be observed. But there is insufficient recognition of the fact that, in a monetized economy, the central functions of agriculture support the performance of other sectors. The regulating and supporting functions of global ecosystems are insufficiently understood. The findings of the Millennium Ecosystem Assessment (MA, 2005b) show the key role of agriculture not only in productive and social aspects but also in preserving or endangering ecosystem functions.

The crops component of agriculture

World crop and livestock output growth fell in 2005 to the lowest annual rate since the early 1970s, and well below the rates reached in 2003 and 2004, with a strong decline in industrialized countries as a group and negative 1.6% growth in 2004 (FAO, 2006a). This was mainly due to a decrease in output growth in the crops sector from 12% in 2004 to negative 4% in 2005 in industrialized countries. But with growing resource scarcity, future food production depends more than ever on increasing crop yields and livestock productivity (FAO, 2006a). The positive and negative effects of technological progress have raised uncertainties. Two groups of crops are cited here as examples.

Cereal crops. World cereal production, after several years of stagnation, increased sharply in 2004/2005, reaching 2,065 million tonnes, a 9% increase from the previous year, and global utilization continued an upward trend (FAO, 2006a). However, cereal yields in East Asia rose by an impressive 2.8% a year in 1961–2004, much higher than the 1.8% growth in industrialized countries, mainly due to widespread use of irrigation, improved varieties, and fertilizer (Evenson and Gollin, 2003).

The green revolution doubled cereal production in Asia between 1970 and 1995, yet the total land area cultivated with cereals increased by only 4% (Rosegrant and Hazell, 2001) while in sub-Saharan Africa it changed little in the same period.

Slowing down expansion of cultivated areas through intensification benefited the environment by preserving the forests, wetlands and biodiversity. But there are negative

Table 1-3. **Positive functions of agriculture.**

	Environmental	Social	Food Security	Economic	Cultural
Global	Ecosystem resilience Mitigation of climatic change (carbon sequestration, land cover) Biodiversity	Social stability Poverty alleviation	Food security/ food for all	Growth, international trade	Cultural diversity
Regional/National	Ecosystem resilience Soil conservation (erosion, siltation, salinization) Water retention/availability (flood and landslide prevention) Biodiversity (agricultural and wildlife) Pollution abatement	Balanced migration Social stability (and sheltering effects during crisis) Unemployment prevention Poverty alleviation	Access to food National security Food safety	Economic stability Employment Foreign exchange Tourism	Landscapes Cultural heritage Cultural identity Social capital
Local	Ecosystem resilience Soil conservation Water retention Biodiversity Pollution abatement	Social stability (employment, family) Livelihoods Balanced gender relations	Local and household food security	Employment effects on secondary and tertiary sectors	Landscapes Indigenous, local knowledge Traditional technologies Cultural identity

environmental impacts such as excessive use of agrochemicals (fertilizers and pesticides) resulting in water pollution, which affects human and animal health and indirectly damages ecosystems. An example is the intensive and continuous monoculture of rice-wheat systems in the Indo-Gangetic Plain of India and Pakistan, which led to soil and water degradation that has canceled the gains from the green revolution (Ali and Byerlee, 2002). In all regions, especially in the heterogeneous and risky rainfed systems of sub-Saharan Africa, there is a need for sustainable technologies that increase the productivity, stability and resilience of production systems (Conway, 1999). It is important to note that most rice-producing areas such as China, India, Japan and Indonesia have experienced stagnation in rates of productivity increase as of 1985-2000 (Cassman, 2003).

The fishery component of agriculture

Fisheries play a very important role in agriculture and the world economy. Rapid population growth in developing countries, changing consumer preferences and increased disposable income have increased global demand for fishery products. About 200 million people worldwide, most of them in developing countries, live on fishing and aquaculture, and fish provides an important source of food, cash income for many poor households, and is a widely traded food commodity (Kurien, 2006; WorldFish Center, 2006). Over a billion people worldwide rely on fish as their main source of protein or their most inexpensive source of animal protein. In 2004, aquaculture production accounted for 43% of fish consumption (FAO, 2006b). Fish contributes to national food sufficiency through direct consumption and through trade and exports.

Total fishery production in 2004 was 150.5 million tonnes, of which 45.5 million tonnes were from aquaculture, of which 40% that entered international trade reached a value of US$71.5 billion (FAO, 2006c). While capture increased moderately from 1970 to 1998, aquaculture multiplied by a factor of 15 in the same period, from about 2 million tonnes in 1970 to about 30 tonnes in 1998 (Delgado et al., 2003; Figure 1-10). Fishery exports have become a significant foreign currency earner for many developing countries, contributing slightly less than 50% of such exports. The export value of world trade in fish, US$58 billion in 2002, is more than the combined value of net exports of rice, coffee, sugar and tea (World Bank, 2004a). Demand for fish products is increasing rapidly as income levels rise in Asia and the population grows in Africa. Led by Asia, developing nations produce nearly three times as much fish as industrialized countries (Delgado et al., 2003).

World capture fishery production was from 90 to 100 million tonnes in 2005, an increase of about 5% from 2003 (FAO, 2006c). Aquaculture may substitute for wild catch but can create environmental problems, especially when practiced intensively, such as in large-scale, intensive operations, most of which (with the exception of shrimp farming) are found in temperate countries.

The forestry component of agriculture

Forests are intensively linked to agriculture, providing products (i.e., wood, fuelwood, food, medicines), inputs for crop and livestock production (fodder, soil nutrients, pollination, etc.), and services (i.e., watershed protection, climate regulation, carbon storage, biodiversity conservation). World roundwood production in 2004 reached an estimated 3,418

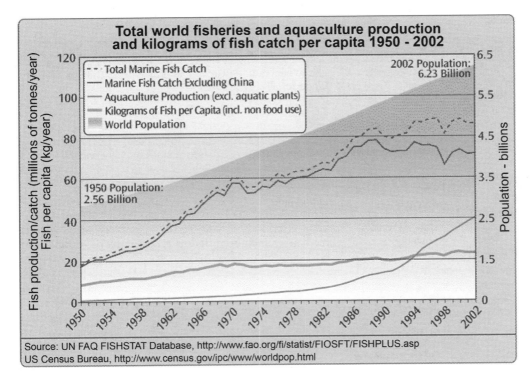

Figure 1-10. *Global capture fisheries and aquaculture production, 1950-2002.* Source: FAO, 2007b; US Census Bureau, 2007.

million cubic meters, of which 60% was produced in developing countries where wood is the most important source of energy (FAO, 2006d).

Forests cover 31% of global land surface (FAO, 2007a) and have potential to provide products and services, hence, could contribute to meeting development and sustainability goals. All types of forests contribute to agriculture in two main ways: (1) the world's forests act as a buffer against climate change, storing 50% carbon in their biomass, deadwood, litter and soil, i.e., more than the amount of carbon dioxide in the atmosphere alone; and (2) they are a principal source of biodiversity. Forests also play a key role in agriculture as the source of much of the land and soils for agriculture. "Slash and burn" agriculture is dependent on forest ecosystems for regeneration of soils, and forests are the source of many types of fruit, meat, timber, fuelwood, medicine, etc. for rural people.

Almost a quarter of a billion people live in or near tropical forests, and their well-being depends on them (CIFOR, 2006). Two billion people, a third of the world's population, use fuelwood and charcoal, most of which is harvested in the forest; and two billion people rely on traditional medicines, much of which depends on forest products (CIFOR, 2006). The rapid development of agriculture has proceeded through conversion of natural forests, mainly due to rapid population growth, and the higher food production and cash income that can be obtained from farming rather than from forestry. Deforestation, mainly due to conversion of forests to agricultural land, continues at the rate of 13 million ha per year (FAO, 2005b). The net global change in forest area in 2000-2005 is estimated at -7.3 million ha per year, down from -8.9 million ha per year in 1990-2000 (FAO, 2005b).

The deforestation trend is increasingly being reversed as forest goods and services are becoming scarce. Changes in cropland show that most of this deforestation has not been for conversion of cropland. Eighty percent of incremental crop production in developing countries by 2030 will come from intensification and only 20% from area expansion (FAO, 2003).

The livestock component of agriculture
Global livestock production continues to grow more rapidly than crop agriculture, with growth rates of 5% in the 1990s, but has slowed down since 2004 (FAO, 2006a). The volume of livestock production in developing countries has steadily increased since the early 1980s, both for internal consumption and for export (COAG, 2005), driven by rising demand for poultry, pork and eggs as income rises. Livestock production accounts for 40% of the agricultural GDP (FAO, 2006a), produces about one-third of humanity's protein intake, employs 1.3 billion people and creates livelihoods for one billion of the world's poor (Steinfeld et al., 2006). The social and energy benefits of livestock production have long been recognized, as well as its economic contribution outside the formal market system. Women play a key role in small-scale livestock production, and in processing and marketing animal products.

Outbreaks of animal diseases, in particular avian influenza, and subsequent consumer fears, trade bans and declines in poultry prices have caused slow growth rates. Livestock production systems also cause environmental problems, with negative impacts on land, climate, water quality and quantity, and biodiversity (FAO, 2006a). As poverty declines, there is predicted to be increased demand for animal protein in diets, exacerbating already fragile environmental conditions in developing countries and causing further loss of biodiversity.

Much of livestock production is on small farms, where it is an integrated component of the farming system, often with multipurpose uses (Dolberg, 2001; LivestockNet, 2006). However, there are also nomadic systems, particularly in Africa, in extreme northern Asia, Europe and America (in the tundra) where livestock continues to be the primary source of livelihoods.

Some emerging agricultural systems

Organic agriculture. In the past few years, organic agriculture has developed rapidly with more than 31 million ha in at least 623,174 farms worldwide in 120 countries (Willer and Yussefi, 2006).

Global sales of organic food and drink increased by about 9% to US$27.8 billion in 2004, with the highest growth in North America, where organic product sales are expanding by over US$1.5 billion per year, with the United States accounting for US$14.5 billion sales in 2005 (Willer and Yussefi, 2006).

Organic agriculture is a holistic production management system that promotes and enhances agroecosystems health including biodiversity, biological cycles, and soil biological activity (Codex Alementarius Commission, 2001). It emphasizes the use of management practices in preference to the use of off-farm inputs, taking into account that regional conditions require locally adapted systems. This can be accomplished by using, wherever possible, cultural, biological, and mechanical methods instead of using synthetic materials, to fulfill any specific function within the system. Organic agriculture can contribute to socially, economically and ecologically sustainable development, firstly, because organic practices use local resources (local seed varieties, dung, etc.) and secondly, because the market for organic products has high potential and offers opportunities for increasing farmers' income and improving their livelihood. It also contributes to *in situ* conservation and sustainable use of genetic resources. But organic agriculture also has negative environmental impacts such as overuse of animal manure, which can lead to nitrite pollution of water supplies; on the other hand, insufficient application of organic manure can lead to soil mining and long-term productivity declines (World Bank, 2004a).

The sustainability of organic agriculture is often debated, with divergent views regarding its feasibility and productivity potential in resource-poor areas. Most information is from temperate countries and the technological needs in low-potential areas are not addressed. Organic production requires a high level of managerial knowledge, the ability to protect crops from pests and diseases, and compliance with production process requirements. Certification is one of the most important cost items. Reliable and independent accreditation and control systems are essential to enforce organic standards and regulations and to meet phytosanitary standards and general quality requirements.

Urban and peri-urban agriculture refers to growing plants and raising animals for food and other uses within and around cities and towns, and related activities such as the production and delivery of inputs and the processing and marketing of products (van Veenhuizen, 2006). It has received increasing attention from development organizations and national and local authorities, and is likely to do so in future as well, as migration of poor people from rural to urban areas will continue to be a major trend in developing countries. This results in shifting poverty from rural areas to urban slums and increasing the importance of urban and peri-urban agriculture, as it contributes to reliable food supply and provides employment for a large number of urban poor, especially women (World Bank, 2004a). It is an integral part of the urban economic, social and ecological system (Mougeot, 2000).

Urban and peri-urban agriculture includes a range of production systems from subsistence production and processing at household level to fully commercialized agriculture. It may include different types of crops (grains, root crops, vegetables, mushrooms, fruit) or animals (poultry, rabbits, goats, sheep, cattle, pigs, guinea pigs, fish) or combinations of these (ETC-Netherlands, 2003). Non-food products include aromatic and medicinal herbs, ornamental plants, tree products (seed, wood, fuel), and tree seedlings. For example in Hanoi, Vietnam, urban and peri-urban agriculture supplies about one-half of the food demand and engages more than 10% of the urban labor force in processing, marketing, retailing, input supply, and seed and seedling production (Anh et al., 2004)

Urban and peri-urban agriculture is characterized by closeness to markets, high competition for land, limited space, use of urban resources such as organic solid wastes and wastewater, a low degree of farmer organization, mainly perishable products, and a high degree of specialization (van Veenhuizen, 2006). Some critical issues include the use of pesticides; use of urban waste in agricultural production; environmental pollution caused by agricultural activities in densely populated areas; conflicts over land and water between agricultural, industrial, and housing uses; unhygienic food marketing; and inability of producers, wholesalers, retailers and other agents engaged in food processing and marketing to adapt to coordinated food chains (World Bank, 2004a). Urban planning will need to take into account the potential environmental impacts of urban and peri-urban agriculture.

Conservation agriculture. Conservation or zero-tillage agriculture is one of the most important technological innovations in developing countries, as part of Sustainable Land Management approaches. It is a holistic agricultural system that incorporates crop rotations, use of cover crops, and maintenance of plant cover throughout the year, with positive economic, environmental and social impacts (Pieri, et al., 2002). It consists of four broad intertwined management practices: (1) minimal soil disturbance (no plowing and harrowing); (2) maintenance of permanent vegetative soil cover; (3) direct sowing; and (4) sound crop rotation.

The United States has the longest experience in conservation agriculture approaches, which were first implemented in large and medium-sized farms. Conservation agriculture then began to be widely used in diverse farming systems in Brazil and adapted to small farms in the southern part of the country. It is rapidly being adapted to irrigated rice-wheat systems in the Indo-Gangetic Plains, especially in India, where 0.8 million hectares were planted in 2004 using this system (Malik, Yadav and Singh, 2005).

Broader adoption of conservation agriculture practices would result in numerous environmental benefits such as decreased soil erosion and water loss due to runoff, decreased carbon dioxide emissions and higher carbon sequestration, reduced fuel consumption, increased water productivity, less flooding, and recharging of underground aquifers (World Bank, 2004a).

Agriculture, agrifood systems and value chains

Agrifood systems are described as including a range of activities involved at every step of the food supply chain from producing food to consuming it, the actors that both participate in and benefit from these activities, and the set of food security, environmental and social welfare outcomes to which food system activities contribute (Ericksen, 2006). They include the primary agriculture sector and related service industries (i.e., veterinary and crop dusting services); the food and beverage, tobacco and non-food processing sectors; the distribution sector (wholesale and retail); and the food service sector. Value chains are multinational enterprises or systems of governance that link firms together in a variety of sourcing and contracting arrangements for global trade. Lead firms, predominantly located in industrialized countries and comprising multinational manufacturers, large retailers and brand-name firms, construct these chains and specify all stages of product production and supply (Gereffi et al., 2001). The value chain perspective shifts the focus of agriculture from production alone to a whole range of activities from designing to marketing and consumption.

Agrifood systems range from traditional systems that are localized where food, fuel and fiber are consumed close to the production areas using local resources, to large agrifood industries that are globalized and linked to integrated value chains. Traditional systems may include hunter-gathering and peasant agriculture that meet the needs of the community from local resources. The major traditional agrifood systems comprise small family farms that supply products to the local markets but are continuously being transformed in response to market signals. At the other end, there are large agrifood industries consisting of international or transnational companies that are globalized and integrated into complete value chains. These systems are continuously being transformed by market and consumer demands, with new agrifood systems emerging that consider social and environmental aspects and use technological innovations. Organic agriculture is an example, which showed rapid growth in the 1990s in Europe, where 4% of EU agricultural land area is now organic, compared with only 0.3% in North America (Willer and Yussefi, 2006).

Agrifood systems have a strong influence on culture, politics, societies, economics and the environment, and their interactions affect food system activities. Agrifood system activities can be grouped accordingly: producing, processing and packaging, distributing and retailing, and consuming (Zurek, 2006). As the agrifood systems become more sophisticated and globalized, they have to adhere to regulations and standards to meet product safety and quality, and consumers' specific needs in order to survive. New and more innovative technology in food production, post-harvest treatment, processing, packaging and sanitary treatment are now playing a more important role.

Agriculture and the environment

Land cover and biodiversity changes. Beyond its primary function of supplying food, fiber, feed and fuel, agricultural activity can have negative effects such as leading to pollution of water, degradation of soils, acceleration of climate change, and loss of biodiversity. Conversion of land for production of food, timber, fiber, feed and fuel is a main driver of biodiversity loss (MA, 2005b). Many agricultural production systems worldwide have not sufficiently adapted to the local/regional ecosystems, which has led to disturbances of ecosystem services that are vital for agricultural production. Requirements for cropland are expected to increase until 2050 by nearly 50% in a maximum scenario, but much less in other, more optimistic scenarios (CA, 2007; see Figure 1-11).

Soil degradation has direct impacts on soil biodiversity, on the physical basis of plant growth and on soil and water quality. Processes of water and wind erosion, and of physical, chemical and biological degradation are difficult to reverse and costly to control once they have progressed. The Global Assessment of Human-induced Soil Degradation (GLASOD) showed that soil degradation in one form or another occurs in virtually all countries of the world. About 2,000 million hectares are affected by soil degradation. Water and wind erosion accounted for 84% of these damages, most of which were the result of inappropriate land management in various agricultural systems, both subsistence and mechanized (Oldeman et al., 1990).

Water quality and quantity changes. Access to enough, safe and reliable water is crucial for food production and poverty reduction. Most people without access to an improved water source are in Asia, but their number has been rapidly decreasing since 1995, which is less the case in sub-Saharan Africa, Latin America, West Asia and Northern Africa (see Figure 1-12).

However, putting more water into agricultural services threatens environmental sustainability. Water management in agriculture thus has to overcome this dilemma (CA, 2007). Intensive livestock production is probably the largest sectoral source of water pollution and is a key player in increasing water use, accounting for over 8% of global human water use (Steinfeld et al., 2006). Excessive use of agrochemicals (pesticides and fertilizers) contaminates waterways. Better management of human and animal wastes will improve water quality. Agriculture uses 85% of freshwater withdrawals in developing countries, mainly for use in irrigation, and water scarcity is becoming an acute problem, limiting the future expansion of irrigation (CA, 2007). Water conservation and harvesting also have an important potential for rainfed farming (Liniger and Critchley, 2007) as water scarcity is widespread.

Climate change: Climate change influences and is influenced by agricultural systems. The impact of climate change on agriculture is due to changes in mean temperature and to seasonal variability and extreme events. Global mean temperature is very likely to rise by 2-3°C over the next

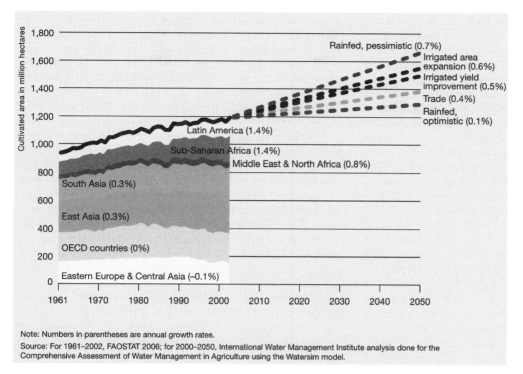

Note: Numbers in parentheses are annual growth rates.
Source: For 1961–2002, FAOSTAT 2006; for 2000–2050, International Water Management Institute analysis done for the Comprehensive Assessment of Water Management in Agriculture using the Watersim model.

Figure 1-11. *Scenarios of land requirements by regions from 2000 to 2050.*

50 years, with implications for rainfall and the frequency and intensity of extreme weather events (Stern, 2006). The outcomes of this change will vary heavily by region. Crop-climate models predict an increase in crop production in slight to medium warming scenarios of less than 3°C (Parry et al., 2007). Livestock production is one of the major contributors to climate change within agriculture (Steinfeld et al., 2006).

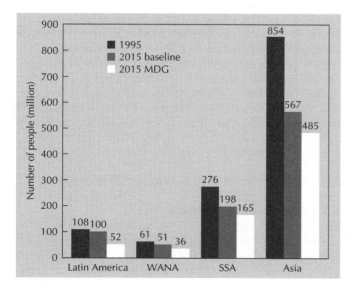

Figure 1-12. *Number of people in 1995 without access to an improved water source, MDG goal and projection to 2015.*
Source: Rosegrant et al., 2006.

1.2.5 Direct and indirect drivers

Direct drivers of change

Changes in human well-being, as characterized by the development and sustainability goals of the IAASTD, come about as the result of a multitude of factors at a variety of scales. For example, change for a particular household may occur as the direct effect of a better harvest due to use of an improved technology. The improved technology itself may have been developed as a result of investment in agricultural research, science and technology and its adoption may have been facilitated by changes in prices or improvements in education and market infrastructure. Effective policy measures depend on a careful distinction between direct and indirect drivers of change.

Following the framework, direct drivers of change include food demand and consumption patterns, land use change, the availability and management of natural resources, climate and climate change, energy and labor, as well as the development and use of AKST.

Relevant natural resources include land resources—i.e., soil, water, flora and fauna—and climate. Growing demand for food, feed, fiber and fuel drives the pace of changes in land use. These changes may include clearing or planting of forests, drainage of wetlands, shifts between pasture and cropland, and conversion to urban uses. Climate change has the potential to change patterns of temperature and precipitation as well as the distribution of pests and diseases. Other natural, physical, and biological drivers include evolution, earthquakes, and epidemics, the use of labor, energy, inputs such as chemical fertilizers, pesticides, and irrigation, and the use of new plant and animal species or varieties. Finally, direct drivers include AKST development and use, including new tools and new techniques such as soil and water conservation or biotechnology. This may also comprise aspects of access to, control over and distribution of AKST, such as extension and dissemination efforts, credit markets and capital assets, and markets for information and knowledge. Species introduction or removal may be intentional or unintentional. Epidemics are increasing the vulnerability of plant and animal production in a globalized economy and are therefore also considered to be direct drivers.

These changes may enhance the well-being of some people and diminish that of others; they may have beneficial effects in the short term but adverse effects over time (or the reverse), and they may have beneficial effects locally but adverse effects at larger scales (or vice versa).

Indirect drivers of change

Many indirect drivers result in turn from a variety of other indirect drivers. Demographic factors include total population and its composition and spatial distribution in terms of age, gender, urbanization, and labor, as well as pressure on land resources within a farm or between farms. Economic factors include prices and other market characteristics, globalization, trade, land tenure and access regulations, agribusiness, credits, markets, and technology. Sociopolitical factors include governance, formal and informal institutions, legal frameworks such as international dispute mechanisms, kinship networks, social and ethnic identity, and political stability. Indirect drivers also include infrastructure such as transportation, communication, utilities, and irrigation. Indirect drivers of science and technology include institutions and policy, funding for R&D, knowledge and innovations systems, advances and discoveries in biotechnology, intellectual property rights, communication systems and information technology, harnessing and adapting local knowledge, and local and institutional generation of AKST. Education, culture and ethics (e.g., in cultural and religious developments or choices individuals make about what and how much to produce and consume and what they value) may also influence decisions regarding direct drivers. Whether direct or indirect, some drivers may have cumulative effects that are felt only when a critical threshold level is reached, as for example when rising pollutant levels exceed a watershed's natural filtration capacity.

Finally, improvements in AKST are driven both by factors that help generate new AKST as well as factors that encourage its adoption and use. Factors that help generate AKST include research policy and funding, intellectual property rights, and farmers' innovation capacity. Factors that affect adoption and use of AKST include extension services, education, and access to natural, physical, and financial resources. These will be explored fully in the chapters to follow.

Conditions determined by political, economic, social and cultural contexts

Agriculture and AKST are strongly bound to the human context in which they are embedded. For example, in the context of Switzerland, where the agricultural sector constitutes merely 3% of the tax-paying workforce, small-scale farmers with an average farm size of 16 ha which they may use for livestock breeding, will not generate sufficient income for the family for a decent livelihood. Because of the importance of agriculture for nonproductive services such as cultural landscape preservation, recreation forests, and water management, Swiss farmers are paid by society for their environmental and social services, up to a total of over 50% of their income, thus reaching the minimum national income standard of about US$35,000 in 2005 (BFS, 2006).

A farming household in Ethiopia, by contrast, typically survives on one hectare of cultivated land and some communal pastureland for livestock rearing. This family produces about one tonne of cereals and pulses per year, of which about 10-20% is marketed and the rest is used for home consumption. Such a household has to pay head taxes but only very marginally profits from investment programs by government or foreign aid. There are millions of farming households all over the world in the same situation, which have an average annual per capita GNP of less than US$200.

Any assessment of the potential of AKST to contribute to more equitable development will thus have to take into account the political, economic, social and cultural contexts in which agricultural land users operate. Additionally, AKST assessments are inherently inter- or multidisciplinary and generate knowledge through transdisciplinary approaches.

Conditions determined by ecosystems, agricultural systems and production systems

The concept of ecosystems provides a valuable framework for analyzing and acting on the linkages between people and the environment (MA, 2005a). An ecosystem is defined as a dynamic complex of plant, animal and microorganism communities and their nonliving environment, interacting as a functional unit (UN, 1992). The AKST conceptual framework uses ecosystems as the broadest context within which agricultural production/farming systems are analyzed.

The predominance of the "cultivated" ecosystem category for agriculture is immediately apparent in the table, followed by mountain ecosystems, which constitute 26% of the Earth's land surface, followed by forestland, covering about 30% of the land surface, as well as drylands, which constitute about one third of all land area worldwide. Together these land cover areas provide about 93% of agricultural products. It should be noted, however, that other services provided by agroecosystems will have a considerably different balance. An example is forests, which provide clean water, reduce flooding, offer biodiversity protection and recreational and spiritual value, which adds to the importance of the forests' production value.

1.3 Development and Sustainability Issues

1.3.1 Poverty and livelihoods

Eradication of extreme poverty and hunger is a key goal of the assessment. Progress has been particularly striking in Asia, but the proportion of people in sub-Saharan Africa who live in extreme poverty has changed little since 1990. Hunger is inextricably linked to poverty, and here again progress is evident but uneven, with reductions in Asia and Latin America partly offset by increases in Africa and the Middle East. Poverty and hunger arise out of the interaction between economic, environmental, and social conditions and the choices people make. Livelihoods depend not only on current incomes but on how individuals, households, and nations use resources over the long term. Physical and financial capital is critical and relatively easily measured. Equally important but less easily measured are sustainable use of natural capital and investment in human and social capital.

Poverty and hunger

Extreme poverty (crudely measured by the percentage of people living on less than US$1 per day) in developing countries decreased from 28% in 1990 to 19% in 2002, and is projected to fall further to 10% by 2015 (World Bank, 2006c). Progress has been particularly striking in East Asia and the Pacific, where the target of the MDGs has already been achieved, and in South Asia, where progress is on track. But the proportion of people in sub-Saharan Africa who live in extreme poverty has changed little since 1990, and remains at about 44% (World Bank, 2006c). The prevalence of undernourishment has fallen from 20% of the population of developing countries to 16% over the past decade, with reductions in Asia and Latin America partly offset by increases in Africa and the Middle East (World Bank, 2006c). Poverty is most pronounced in Africa and South Asia.

In the simplest terms, hunger can be thought of as the situation that occurs when consumption falls short of some level necessary to satisfy nutritional requirements. Similarly, poverty can be thought of as the situation that occurs when income falls short of some level defined by society, usually in terms of the ability to afford sufficient food and other basic needs. These definitions provide a starting point, but simple definitions mask more complex relationships. In fact, income and consumption fluctuate in response both to changing conditions and to choices made by farmers and others. This challenges us to consider more carefully how hunger and poverty arise out of the interaction between economic, environmental, and social conditions and the choices people make.

Hunger is still the result of insufficient consumption, but insufficient consumption may itself arise for several reasons. For example, household income may be insufficient to acquire sufficient food to meet the nutritional requirements of its members, or food may be inequitably distributed within the household. Alternatively, income may allow a household to acquire sufficient food, but doing so may leave insufficient income to meet other needs, such as paying costs associated with schooling—forcing the household to choose between competing priorities. Similarly, poverty is still the result of insufficient income, but insufficient income may itself arise for a variety of reasons. For example, drought or illness might reduce the amount of crops or labor a household has to sell, while low wages or prices may reduce its value (Sen, 1981). Alternatively, income may be low (or high) in part because of choices a household made earlier in the season, such as which crops to plant, or how much fertilizer to apply, or whether to migrate in search of employment. These choices in turn depend on the resources available to the household. Resources may include natural resources such as land and water as well as the household's

labor power, tools and financial resources. Resources also include the household's social and institutional settings, which shape property rights and access to infrastructure and social support services.

To complete the cycle, the quality and quantity of the household's resources in turn depend, at least in part, on the consumption and investment choices the household made previously. Given its income last week (or last year), for example, a household will make decisions about how much to spend on food, health care or education (each of which affects the quality of its labor resources), how much to spend on seeds, fertilizer and other agricultural inputs, and how much to save or invest in other ways. Once we recognize the dynamic interaction between household resources, choices, and outcomes, it becomes clear that a more complete understanding of hunger and poverty requires not only a broader understanding of the factors that affect them, but also a longer-term perspective on how they interact over time.

Livelihoods

Livelihoods are a way of characterizing the resources and strategies individuals and households use to meet their needs and accomplish their goals or in other words: "people, their capabilities and their means of living" (Chambers and Conway, 1991). Livelihoods encompass income as well as the tangible and intangible resources used by people to generate income and their entitlements to them. In 2003 about 2.6 billion people, or 41% of the world's population, depended on agriculture, forestry, fishing or hunting for their livelihoods (FAOSTAT, 2006), even while agriculture (including forestry and fishing) represented only 12% of GDP in developing countries in 2004, and 4% for the world as a whole (World Bank, 2006c).

Diversification of livelihoods, both within agriculture and beyond, i.e., focusing on other sectors of the economy, is particularly important for countries where the proportion of people engaged in the primary sector is above 40% of all employment (ILO, 2004). This concerns about half of all countries worldwide. The share of households with wage-specialized earnings appears to considerably contribute to an increase of household GDP per capita (Hertel, 2004).

Migration is another livelihood strategy pursued by nearly 200 million people. Reasons for migration are manifold; they range from labor seeking, economic interest and family reunification to displacement due to natural or cultural disasters. Temporary migration and commuting to national and international, rural and urban destinations are now a routine part of the livelihood strategies of many households, including farm households, both in industrialized and developing countries. The effects of migration on agriculture are highly diverse—migration can be a negative phenomenon that creates labor shortage in rural areas, leaving the land abandoned; or it can mitigate population pressure and resource use, and the remittances from family members can boost agricultural development (IOM, 2007).

Income

Economic well-being is most commonly thought of in terms of income (measured as a flow over a particular period of time). For a farm household, for example, this may be in kind (such as food crops produced on the farm) as well as in cash, and may come from both on-farm and off-farm sources. Gross national income per capita averaged US$1,502 in developing countries in 2004, or about US$4 per day (World Bank, 2006c); half the people in developing countries live on less that US$2 per day, and 19% live on less than US$1 per day (World Bank, 2006c). By contrast, income per capita in high-income countries averaged US$32,112 in 2004, or about US$88 per day. Generally, there is a strong correlation between the average income per capita and the share agriculture takes in GDP. The lower this share is, the higher the income (see Figure 1-13).

A simple measure of economic well-being can be derived by comparing an individual's or household's income over a given period of time with their needs or wants over that same period of time. The disadvantage of such a simple measure is that it could indicate that a household was well-off at present even if it was increasing its income in the short term by depleting its resources in a way that is unsustainable over the long term. Thus a more complete measure of economic well-being requires knowledge about the resources from which an individual or household derives its income.

Resources. Control of resources shapes income-generating opportunities, and determines how resilient households are when incomes fluctuate in response to changing economic conditions or natural disasters. Resources can be grouped in various ways, e.g., natural, human and social capital and wealth (or man-made capital) (Serageldin, 1996). Wealth can be further divided into physical and financial forms (Chambers and Conway, 1991). Access to different forms of capital varies widely across and within regions, affecting the choices that households make in combining resources in their diverse livelihood strategies, and also affecting the types of AKST investments that are most relevant in any particular context.

An important aspect of resources is the discussion of labor productivity versus land productivity. These are often compared to show the differences between the achievements of formal AKST versus local knowledge. The value of production in industrialized countries is much higher than in developing countries (Wood et al., 2000), simply because the energy balance is hardly taken into account when comparing mechanized with manual agricultural systems. It is noteworthy, however, that there are groups of countries where labor productivity made particular progress (probably through mechanization), while others made most progress in land productivity, and sub-Saharan countries had only little advances over the past 30 years, although with probably the best energy balance (Byerlee et al., 2005). Latin America has the highest levels of labor productivity in the developing world, followed by the Middle East and North Africa, and transition economies. South-East Asia and sub-Saharan Africa have considerably lower labor productivity levels although, in terms of growth, China is leading (ILO, 2004; see Figure 1-14).

Education. Reported gross primary school enrolment rates are near universal in developing countries already, but completion rates are lower, and more than 100 million children of primary school age remain out of school. Gross enrolment rates drop to 61% for secondary education and 17% for tertiary education (World Bank, 2006c). Primary and secondary education are near-universal in high-income countries, and drop to 67% for tertiary education. Adult literacy rates in developing countries are 86 and 74% for men and women, respectively (World Bank, 2006c).

Research. Expenditures for public agricultural research and development (R&D) averaged 0.5% of agricultural GDP in high-income countries (Pardey et al., 2006), which in view of the high disparity between the GDPs themselves must be seen as a potentially greatly underrated difference. Moreover, about five times as many scientific and technical journal articles were published by authors from high-income

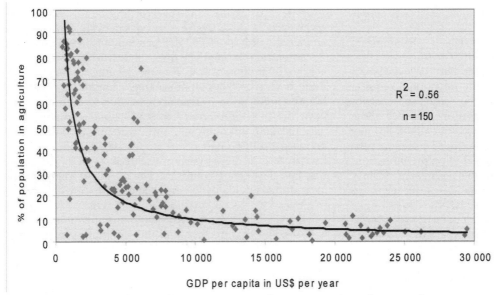

Figure 1-13. *World distribution of GDP per capita and percentage of population working in agriculture (Average of years 1990-2002).* Source: Based on Hurni et al., 1996, with data from World Bank, 2006c; ILO, 2007.

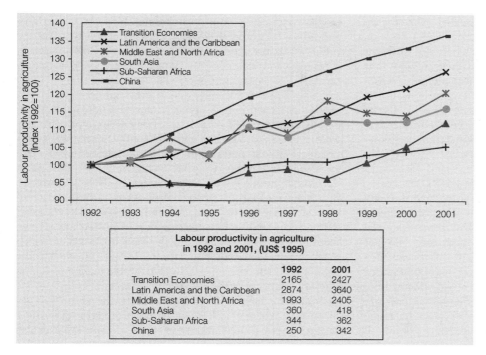

Figure 1-14. *Labor productivity in agriculture by region (1992-2001) and labor productivity levels in 1992 and 2001.* Source: ILO, 2004

countries as were published by authors from developing countries in 2001 (World Bank, 2006c).

Measurement of the different forms of capital poses many challenges, particularly for those forms that are non-marketed. In an effort to better understand the importance of different types of capital, the World Bank (1997) undertook to estimate the value of human resources, produced assets, and natural capital. They noted that human resources include both raw labor power and the embodied knowledge that comes from education, training and experience. Monetary values are admittedly imprecise, but what was striking about their results was the uniform dominance of human resources, which accounted for 60-80% of total wealth in all regions except for the Middle East, where natural capital, in the form of energy reserves, accounted for an unusually high proportion.

Livelihoods, resilience, and coping strategies. Even though a large number of people depend entirely on agriculture, off-farm income is important for the livelihoods of many farming households. Agriculture's share of GDP was declining in both developing and high-income countries, while the share accounted for by the service sector was increasing—to 52% in developing countries and 72% in high-income countries (World Bank, 2006c). Data are scarce, but in many developing countries the informal sector accounts for a large (and in some cases rising) share of urban employment (World Bank, 2006c). Remittances from workers abroad form an increasing share of income in most developing regions, totaling US$161 billion in 2004 and accounting for more than 3% of GDP in South Asia (World Bank, 2006c). A household may be able to avoid hunger and maintain its human capital during a drought by depleting its financial, physical or natural capital (for example, by drawing on its savings or selling its livestock or failing to maintain the fertility of its soils). But this may threaten its ability to survive over the longer term. Alternatively, a household may accept severe cuts in consumption in the short term, with consequences for health and strength, precisely in order to protect its endowment of other resources and its ability to recover in future.

Different resource endowments and different goals imply different incentives, choices, and livelihood strategies. For example, two households that have the same endowments of land, labor, and materials may choose different cropping strategies if one household does not have access to savings, credit or insurance and the other one does. In this case the first household may choose to plant a safe but low-yielding crop variety while the second household will plant a riskier variety—expecting higher yields while at the same time knowing that additional financial capital could help sustain income (and consumption levels) even if it were to suffer a poor harvest.

Likewise different livelihood strategies and different weather and market conditions imply different outcomes, which in turn imply different endowments. In the example just mentioned, the first household may suffer smaller losses in a drought year, but also smaller gains in average and good years. Even when both households suffer losses, their coping strategies might differ. The first, in order to meet consumption needs, might be forced to sell assets. If many other households are in a similar position, asset prices might fall, making it even more difficult to exchange them for sufficient food. Households with sufficient food or financial reserves, by contrast, may be in a position to buy assets at discounted prices, increasing not only their own ability to survive future droughts but also the degree of inequality in the region (Basu, 1986).

These sometimes desperate tradeoffs between different components of the resource endowment illustrate why simple or short-term definitions of poverty, hunger and food security provide an incomplete understanding of household's livelihood strategies. They have important implications for economic sustainability, which we will explore in the next subchapter. They also have important implications for environmental sustainability and social equity.

Economic dimensions of sustainability

Sustainability, like food security, has been defined in many ways. The Brundtland Commission (WCED, 1987) defined sustainable development as "development that meets the needs of the present without compromising the ability of future generations to meet their own needs." But even such an intuitively appealing definition raises difficult operational questions regarding both needs and ability (Serageldin, 1996). Abilities depend on the resources that individuals and households have at their disposal, and the ways in which they can be combined and exchanged to produce goods and services that they desire.

Sustainability can, in turn, be understood in terms of maintaining or increasing a household's ability to produce desired goods and services—which may or may not involve maintaining or increasing the level of each particular component of the household's resource endowment. A very narrow interpretation of sustainability involves maintaining each component of the resource endowment at its current level or higher. In its strictest sense this would mean that non-renewable resources could not be used at all, and that renewable resources could be used only at rates less than or equal to their growth rates. Such a requirement would preclude extraction of oil to improve human capital, for example by investing in education for girls (Serageldin, 1996). A broader interpretation of sustainability by contrast, involves maintaining the total stock of capital at its present level or higher, regardless of the mix of different types of capital. This would require the unrealistic assumption that different types of capital can be substituted completely for one another, and that complete depletion of one type is acceptable as long as it is offset by a sufficient increase in another. An intermediate alternative involves maintaining the total stock of capital, but recognizing that there may be critical levels of different types of capital, below which society's (or an individual's, or a household's) ability to produce desired goods and services is threatened.

Measuring the different forms of capital poses considerable challenges, and these in turn complicate assessments of sustainability. In an effort to improve such assessments, the World Bank (1997) sought to adjust national accounts and savings rates for investment in and depletion of natural and other forms of capital not traditionally included in those accounts. Accounting for changes in natural capital and human resources, they found that high-income OECD countries have had "genuine savings rates" of around 10% per year over the past several decades—less than traditional measures of investment, but still positive (and thus sustainable, at least in the broad sense). Asia and Latin America have also had positive genuine savings rates, most notably

in East Asia (with rates approaching 20% per year). Sub-Saharan Africa and the Middle East/North Africa, on the other hand, have consistently had negative genuine savings rates of -5 to -10% per year (World Bank, 1997). Such patterns and concerns continue today.

The World Bank's measure of adjusted net savings currently begins with gross savings, adds expenditures on education, and subtracts measures of consumption or depletion of fixed (i.e., produced) capital, energy, minerals, forest products and damages from carbon dioxide and particulate emissions. In contrast to gross savings of 27.5% of GNI in developing countries and 19.4% in high-income countries in 2004, adjusted net savings after accounting for selected changes in human, physical, and natural capital were 9.4 and 8.7% in the two regions, respectively. Adjusted net savings were highest in East Asia and the Pacific (23.9% of GNI) and lowest in sub-Saharan Africa (-2.0%) and the Middle East and North Africa (-6.2%) (World Bank, 2006c). These findings reinforce concerns about sustainability by any of the measures described above. Similarly, the recent growth in crops, livestock, and aquaculture production has come at the expense of declines in the status of most other provisioning, regulating and cultural services of ecosystems (MA, 2005a).

1.3.2 Hunger, nutrition and human health

Some key characteristics of hunger, nutrition and human health are related to working conditions in agriculture and the effects of HIV/AIDS on rural livelihoods. Health is fundamental to live a productive life, to meet basic needs and to contribute to community life. Good health offers individuals wider choices in how to live their lives. It is an enabling condition for the development of human potential. Societies at different stages of development exhibit distinct epidemiological profiles. Poverty, malnutrition and infectious disease take a terrible toll among the most vulnerable members of society. Good nutrition, as a major component of health, has much to contribute to poverty reduction and improved livelihoods.

Health

Health has been defined as "a state of complete physical, mental and social well-being and not merely the absence of disease or infirmity" (WHO, 1946). It is an enabling condition for the development of human potential. The components of health are multiple and their interactions complex. The health of an individual is strongly influenced by genetic makeup, nutritional status, access to health care, socioeconomic status, relationships with family members, participation in community life, personal habits and lifestyle choices. The environment—whether natural, climatic, physical, social or at the workplace—can also play a major role in determining the health of individuals.

The health profile of a society can be framed in terms of both measurable aspects—for example, access to clean water, safe and nutritious food, improved sanitation, basic health care, and education; mortality and morbidity rates for various segments of the population; the incidence of disease and disability; the distribution of wealth across the population—as well as factors that are less easily quantifiable. Among these are issues of equity or discrimination as evidenced in a society's treatment of minority groups, such as indigenous peoples, immigrants and migrant workers, and of vulnerable groups, such as women, children, the elderly and the infirm. These factors influence not only the general sense of social well-being but also the health of individuals and groups. Multiple measurement approaches can maximize data accuracy; however, the cost of such measurements must be taken into account.

Societies at different stages of development exhibit distinct epidemiological profiles. The prevalence of various causes of death, average life expectancy, disability-adjusted life years, infant and under-five mortality rates and maternal mortality rates all fluctuate in discernible patterns as the economic underpinnings of society change. For example, societies that depend on hunting and gathering typically have short average life expectancies and deaths due to accident or injury are more prevalent. Agrarian societies show a greater prevalence of death from infectious disease as the major cause of death, particularly among children. In industrial societies, death from cardiovascular disease is predominant, whereas in a service-based post industrial society, the major cause of death is cancer. In the societal form now emerging, it is expected that the predominant cause of death will be senescence—age-related disorders (Horiuchi, 1999).

Such a typology is useful as a rough guide when examining the health statistics or "health profiles" of countries at different stages of development. They demonstrate the linkages between socioeconomic development and human health: the heavy burden of infectious disease in poor, predominantly agrarian countries; the double burden of both infectious and noncommunicable diseases in middle-income developing countries where basic sanitation, clean water and health care systems have already considerably reduced under-five and maternal mortality rates and thereby lengthened average life spans. However, great differences still exist in the health status of rural and urban population groups; and advanced industrialized economies, with aging populations and a predominance of "lifestyle" diseases often related to excessive consumption, inadequate physical activity and the use of tobacco.

Health gains in recent decades are nowhere more evident than in the extension of life expectancy at birth from a global average of 46 years in 1950-55 to 65.4 years a half century later. This progression is expected to continue, reaching an estimated global average life expectancy of 75.1 years in the period 2045-2050 (UN, 2005a). These positive gains are also witnessed in the speed with which developing countries have narrowed the gap in life expectancy between more industrialized and less developed regions of the world, from a difference of 25 years in the period 1950-1955 to slightly over 12 years in 2000-2005 (UN, 2005b). This rapid improvement is due principally to greater access to clean water, sanitation, immunization, basic health services and education: all factors that have transformed the health profile of populations.

While these average figures demonstrate considerable global progress, they also mask wide disparities at the local, national and regional levels. For example, for the past decade, largely due to the ravages of AIDS, life expectancy in Africa has been declining, reaching the current level of

45 years, more than 20 years lower than the global average. The gap in life expectancy between sub-Saharan Africa and the industrialized economies of Europe and North America in 2000 was wider than at any time since 1950 (World Bank, 2006a).

Quality of life questions gain in importance as average life expectancy grows, and here too the gaps between richer and poorer countries and regions are evident. People living in developing countries not only have lower average life expectancies, but also spend a greater proportion of their lives in poor health, than do those in industrialized countries. More than 80% of the global years lived with disability occur in developing countries, and almost half occur in high-mortality developing countries. Healthy life expectancy, that is, total life expectancy reduced by the time spent in less than full health due to disease or injury, ranges from a low of 41 years in sub-Saharan Africa to 71.4 years in Western Europe, with the proportion of lost healthy years ranging from 9% in Europe and the Western Pacific to 15% in Africa (WHO, 2005).

Infectious disease has ceded its place to noncommunicable illnesses, such as heart disease, cancer and degenerative conditions, as the primary cause of mortality worldwide. Noncommunicable diseases accounted for about 60% of all deaths and 47% of the global burden of disease in 2002, and figures are expected to rise to 73% and 60% by 2020 (WHO, 2003b). Yet, once again, sub-Saharan Africa is the striking exception to the rule, since more than 60% of deaths in that region are attributable to infectious disease, with HIV/AIDS as the number one killer of adults aged 15-59 (WHO, 2003b). The resurgence of infectious disease, whether due to the growth of drug-resistant germs, as in tuberculosis, or the transmission to humans of viral pathogens of animal origin continue to pose health threats worldwide.

Poverty, malnutrition and infectious disease take a terrible toll among the most vulnerable members of society. Of the 57 million deaths worldwide in 2002, 10.5 million were among children less than five years of age. More than 98% of those childhood deaths occurred in developing countries. The principal causes were peri-natal conditions, lower respiratory tract infections, diarrhea-related disease and malaria, with malnutrition contributing to all (WHO, 2003b). Infections and parasitic diseases accounted for 60% of the total (WHO, 2003b). The prevalence of malnutrition and infectious disease among the young has important implications for the health and well-being of the population as a whole, since the functional consequences of ill health in early childhood are likely to be felt throughout life, affecting the individual's physical and mental development, susceptibility to disease and capacity for work. In rural areas, in particular, where much work requires sustained physical effort, lack of strength and endurance can lower labor capacity, productivity and earnings. Much of the burden of death as a result of malnutrition is attributable to moderate, rather than severe undernutrition (Caulfield et al., 2004). Young children with mild to moderate malnutrition had 2.2 times the risk of dying compared to their better nourished counterparts, and for those who were severely malnourished the risk of death was 6.8 times greater (Schroeder and Brown, 1994). Children from poor households had a significantly higher risk of dying than those from richer households (WHO, 2003b).

Hunger

At the turn of the millennium, the world produced sufficient food calories to feed everyone, mainly because of increased efficiency brought about by the evolving plant science industry and innovative agricultural methods, including pesticides. The dietary energy supply for the global population was estimated to be 2803 kcal per person per day, comfortably within the range of average energy intake considered adequate for healthy living. Yet close to 800 million people were undernourished. Uneven distribution and consumption patterns across regions and among population groups, however, meant that the average actual food supply ranged from 3273 kcal per capita per diem in industrialized countries to 2677 in developing countries. Even these averages mask tremendous disparities. Dietary energy supply per capita per diem in Afghanistan, Burundi, the Democratic Republic of Congo and Eritrea was less than half that in Austria, Greece, Portugal and the United States (FAO, 2004a).

While global production of food calories has outpaced population growth, thanks to improved farming methods and advances in plant and animal sciences, the number of people potentially supported by the world's food supply depends heavily on the kind of diet people consumed. There are vast regional differences in the prevalence of undernourishment (see Figure 1-15), which is increasing the vulnerability to hunger and famine.

It has been calculated that the global food supply in 1993 was adequate to feed 112% of the world population on a near vegetarian diet, but only 74% of the population on a diet composed of 15% animal foods and just 56% of the population on a diet in which 25% of calories were derived from animal products (Uvin, 1995; DeRose et al., 1998). By the early 1990s, roughly 40% of the world's grain

supply was consumed in animal feed, with grain-to-livestock ratios conservatively estimated at two kilos of grain to produce one kilo of chicken, four kilos for one kilo of pork and seven kilos for one kilo of beef (Messer and DeRose, 1998). Demand for meat is increasing in many parts of the world and feedlot livestock production will cause ever heavier demands on food resources as the proportion of industrially produced animal products increases.

Almost 60% of the world's undernourished people live in South Asia, whereas the highest incidence of undernourishment is in sub-Saharan Africa, where approximately one-third of the population is underfed and hunger is on the increase (FAO, 2006a). In sub-Saharan Africa, food production per capita has not grown in the past three decades. Indeed, it declined during the 1970s and has remained stagnant ever since (FAO, 2006a).

Poor households spend a proportionately larger share of their income on food than do wealthier households, and this budget share tends to decline as income rises. It is not unexpected, therefore, that per capita GDP is correlated with underweight of children under 5 (Haddad, 2000; see Figure 1-16).

In low-income countries, average expenditure on food, beverages and tobacco represented 53% of household spending, compared to 35% in middle-income and 17% in high income countries. The budget share ranged from 73% of total household budget in Tanzania to less than 10% in the United States. The composition of the foodstuffs purchased varied according to income levels as well, with households in low-income countries spending significant portions (over one-third) of their budget on cereals, and fruit and vegetables, including roots and tubers, whereas meat, dairy and tobacco took up higher shares in high-income countries. Low value staple foods accounted for more than a quarter of consumers' total food budget in low-income countries, compared to less than one-eighth in wealthier countries

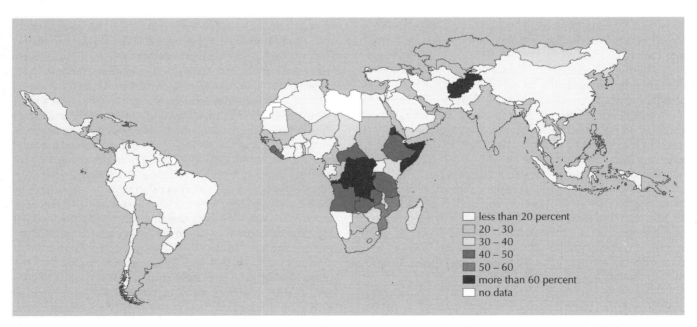

Figure 1-15. *Proportion of the population unable to acquire sufficient calories to meet their daily caloric requirements, 2003 estimates.*
Source: Rosegrant et al., 2006.

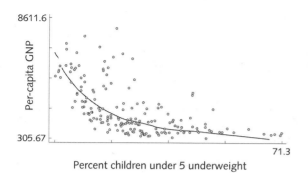

Figure 1-16. *Relationship between per capita gross national product (GNP) and nutrition.* Source: Haddad, 2000.

(Seale et al., 2003). Per capita meat consumption in high-income countries was more than 11 times higher than that in low-income countries in 2002 (WRI, 2006).

Poor rural infrastructure contributes to high food costs for rural poor people. For example, in the upper east region of Ghana, expenditure on food averages over 75% of the household budget. Farmers who lack storage facilities or access to credit are forced by necessity to sell their crops soon after harvest when prices are at their lowest. During the dry season, they buy food when prices are at their highest. In many cases it is women who spend the greatest effort in ensuring food security for the family, cultivating garden plots, carrying out income generating activities and spending the largest portion of their income on food, followed by health. In some cases, women's enterprises pay their women employees in food, in order to ensure that the household benefits directly from the woman's work and that cash earnings are not diverted to other purposes (IFAD, 1998).

Nutrition

Nutrition is one of the major components of health. A healthy diet is typically seen as one which provides sufficient calories to meet the individual's energy needs, as well as adequate protein, vitamins, minerals, essential fatty acids and trace elements to ensure growth and maintenance of life. While the volume of food intake is important, an adequate intake of calories does not in itself ensure that the need for micronutrients has been met. Good nutrition is based on principles of variety, proportion and balance in the choice of foods. Good nutrition has much to contribute to poverty reduction. It is intrinsic to the accumulation of human capital, since sound nutrition provides the basis for good physical and mental health, and is thus a foundation for intellectual and social development and a productive life.

Malnutrition is often linked to poverty and disease, for each one lays the groundwork for the others and contributes to its perpetuation. In developing countries where nutrient deficiencies are most prevalent, malnutrition in children is the result of a range of factors including insufficient food, poor food quality, and severe and repeated infectious disease. It is a contributing factor to childhood death from diarrhea, acute respiratory illness and to a lesser extent, malaria, all among the leading causes of under-five mortality. Even children with mild to moderate malnutrition are at an increased risk of dying (Rice et al., 2000; Caulfield

et al., 2004). Improving nutritional status, particularly of biologically vulnerable groups such as infants, children, and pregnant and lactating women, weakens the transmission of poverty from one generation to the next. AKST has a role to play in developing food crops of high nutritional value that can be produced at affordable prices.

More than 50 nutrients are needed to maintain good health, but the scope and global impact of inadequate nutrition have been studied for only a few critical nutrients, such as iron, iodine, vitamin A and protein. Of these, iron deficiency anemia is the most prevalent nutritional deficiency worldwide and is associated with parasitic infestation, chronic infection as well as other micronutrient deficiencies. It impairs physical and cognitive development in children and leads to reduced capacity for work and lower productivity in adults. In pregnant women, iron deficiency anemia contributes to maternal morbidity and mortality and increases the risk of fetal morbidity, mortality and low birth weight (UNSCN, 2004). Inadequate iodine in the diet affects nearly two billion people, approximately 23% of the global population, and is the primary cause of preventable mental retardation in children (UNSCN, 2004). Vitamin A deficiency, which affects an estimated 140 million preschool children and seven million pregnant women every year, can lead to night blindness, anemia, growth retardation and increased vulnerability to infectious disease and death (UNSCN, 2004).

Malnutrition can result from either excessive or inadequate intake of nutrients. Protein-energy malnutrition, for example, results from an imbalance between the intake of protein and carbohydrates and the body's actual need for them. Inadequate intake leads to malnutrition in the form of wasting, stunting and low weight; excessive intake leads to excess weight and obesity. Child malnutrition is particularly serious and more prevalent in rural than in urban areas.

A healthy diet is often pictured as a pyramid of food groups, with cereals and other staples at the base and progressively smaller layers of fruits and vegetables, followed by meat, poultry, fish, eggs and dairy products, and finally culminating in small amounts of fats and sugar at the peak. A balanced diet would draw on a variety of foods from each of the main groups, respecting the proportions assigned to each. Current patterns of food consumption involving over-consumption of fat, sugar and salt coupled with inadequate intake of whole grains, fruits and vegetables as well as the trend towards excess weight and obesity in many countries demonstrate how far from the ideal the modern diet has become. As the global burden of disease shifts to chronic illnesses, such as diabetes, cardiovascular disease, hypertension and cancer, there is a growing recognition of the impact of dietary habits, environmental hygiene and lifestyle choices on health outcomes (WHO, 2003a).

In recent years, efforts have been directed to analyzing the nutritional content of traditional, locally produced foods, taking into account food availability and eating patterns, in order to draw up dietary guidelines that are culturally meaningful and easily applicable in local conditions. Such food-based guidelines go beyond nutrients and food groups to a more holistic vision of nutrition based on how foods are produced, prepared, processed and developed. The health implications of agricultural practices, production and distribution of food products, sanitary standards and common culinary practices are all considered. The guidelines encourage the consumption of locally available foods and healthy traditional dishes and suggest an increase in food variety based on healthy alternatives (WHO, 1999). "Eat local" campaigns geared towards supporting local agriculture have engendered awareness of the benefits of fresh foods, as well as renewed social interactions, contributing to overall community health.

Food safety

Food-borne disease is estimated to affect 30% of the population in industrialized countries and to account for an estimated 2.1 million deaths in developing countries annually (Heymann, 2002). Globally, the proportion of the population at high risk of illness or death from food-borne pathogens is rising in many countries due to factors such as age, chronic diseases, immunosuppressive conditions and pregnancy. Well-publicized incidences of bovine spongiform encephalopathy (BSE), hoof-and-mouth disease, avian influenza and the mass culling resulting from these outbreaks have raised public concerns with regard to intensified food production, particularly of meat. The reemergence of bovine tuberculosis and brucellosis as well as the outbreaks of illness due to food-borne pathogens, such as salmonella, e. coli, and listeria, that may contaminate fruit, vegetables, poultry, beef or dairy products, have pointed to the need for strict food safety standards "from the farm to the fork", and raised awareness of the fact that the distances from the point of production to the point of consumption continue to grow. As the general public has become increasingly interested in the linkages between agricultural production systems and human health, the list of food-related health concerns has continued to grow. It includes uncertainty with regard to the effects of GMOs on human health, fear of pesticide residues on foodstuffs, recognition of the role that widespread use of antimicrobial agents have had in the emergence of infectious pathogens resistant to antibiotics, and concern with the impact of intensive, industrial-style poultry production on animal health and welfare. Such public concerns have all begun to affect food purchasing decisions in many countries (FAO, 2001a).

Both industrialized and developing countries have made efforts to improve surveillance and investigative capabilities regarding food-borne disease outbreaks over the past two decades. The experience acquired so far, together with molecular biology techniques, ICT, as well as new risk assessment and mitigation methodologies have improved prospects for targeted interventions to control and prevent disease. Safety assurance systems, which provide complete traceability from food production units through to the ultimate consumer, are being put in place in many countries. Such upstream and downstream management systems augment food inspection systems, which have proven unable to cope with the rapidly expanding trade in food products.

Working conditions in agriculture

Much agricultural work is arduous by nature. It is physically demanding, involving long periods of standing, stooping, bending, and carrying out repetitive movements. Poor tool design, difficult terrain and exposure to heat, cold, wind and rain lead to fatigue and raise the risk of accidents. New

technology has brought about a reduction in the physical drudgery of much agricultural work, but has also introduced new risks, notably associated with the use of machinery and the intensive use of chemicals without appropriate information, safety training or protective equipment. The level of accidents and illness is high in some countries and the fatal accident rate in agriculture is twice the average for other industries. Worldwide, agriculture accounts for some 170,000 occupational deaths each year. Machinery and equipment, such as tractors and harvesters, account for the highest rates of injury and death (ILO, 2000).

Exposure to pesticides and other agrochemicals constitutes one of the principal occupational hazards, with poisoning leading to illness or death. The WHO has estimated that between two and five million cases of pesticide poisoning occur each year and result in approximately 40,000 fatalities. Pesticide sales and use continue to rise around the world. In developing countries, the risks of serious accident is compounded by the use of toxic chemicals banned or restricted in other countries, unsafe application techniques, the absence or poor maintenance of equipment, lack of information available to the end-user on the precautions necessary for safe use and inadequate storage practices, and handling and disposal practices (ILO, 1999). The health risks associated with pesticides have spurred efforts to reduce or eliminate their use, for example, through the development of integrated pest management (IPM) and the increase in organic agriculture.

Farmers, agricultural workers and their families live on the land. Their living and working conditions are interwoven, raising the threat of environmental spillover from the occupational risks mentioned above. Wider community exposure to pesticides may come in the form of contamination of foodstuffs, the reuse of containers for food or water storage, the diversion of chemically-treated seeds for human consumption, and the contamination of ground water with chemical wastes. Extensive public education efforts are needed to raise awareness of the dangers involved in the improper handling, storage and disposal of agrochemicals as well as of safe work practices that can prevent accidents and reduce exposure. National systems of chemical safety management can help to ensure that agrochemicals are properly packaged and labeled throughout the distribution chain so that end users in rural communities have the information they need to handle these substances with the necessary precaution.

Animal handling and contact with dangerous plants and biological agents give rise to allergies, respiratory disorders, zoonotic infections and parasitic diseases. In developing countries, in particular, a number of well-known and preventable animal diseases, such as brucellosis, leishmaniasis and echinococcosis, are transmitted to those working closely with animals, affecting millions each year. New threats to human health are posed by pathogens originating in animals and animal products. Indeed, three-quarters of the new diseases that have emerged over the past decade have arisen from this source (WHO-VPH, 2007). Yet, many countries lack effective veterinary and public health systems, let alone the multisectoral environmental health practices, required to prevent the spread of disease.

The interaction between poor living and working conditions determines a distinctive morbidity-mortality pattern among agricultural workers. A large number of rural workers live in extremely primitive conditions, often without adequate food, water supply or sanitation or access to health care. Poor diet combined with diseases prevalent among the rural population (such as malaria, tuberculosis, gastrointestinal disorders, anemia, etc.), occupational disorders, and complications arising from undiagnosed or untreated diseases can be deadly and is certainly debilitating. A vicious circle of poor health, reduced working capacity, low productivity and shortened life expectancy is a typical outcome, particularly for the most vulnerable groups, such as those working in subsistence agriculture (i.e., wage workers in plantations, landless daily paid laborers, temporary and migrant workers and child laborers).

While difficult to quantify, child labor in agriculture is known to be widely prevalent. It is estimated that of the 250 million working children in the world, roughly 70% are active in agriculture. Many of these children work directly for a wage or as part of a family group, exposed to the same work hazards as adults; they endure long daily and weekly hours of work under strenuous conditions. Exposure to agrochemicals, injuries due to machinery or tools, and the repeated shouldering of heavy loads have a negative impact on their health and development with life-long consequences. Conditions of poverty, including poor housing, an inadequate diet and lack of sanitation, little access to health care and loss of educational opportunity, compound these health problems and mortgage their future (ILO, 2006).

HIV/AIDS and its effects on rural livelihoods

The HIV/AIDS epidemic provides a compelling example of the linkages among poverty, illness, food insecurity and loss of productive capacity as well as the differentiated effects on sufferers, caregivers, other family members and the wider community. An estimated 40.3 million people were living with HIV in 2005, two-thirds of whom were in sub-Saharan Africa, where agriculture is the mainstay of most economies and women comprise the backbone of the agricultural labor force. In that region, 57% of adults (15-49) living with HIV were women (UNAIDS and WHO, 2005).

While the epidemic affects people of all ages and in all walks of life, the disease cuts to the heart of the rural economy, afflicting adults in the prime of life, reducing their capacity to earn a living and provide for their families, whether from off-farm activities or from cultivation of the land. Women and girls, who already carry out the bulk of the work in small-scale, labor-intensive agriculture, split their waking hours between care for the sick and the orphaned, their traditional productive work and additional tasks taken on to compensate for the lost labor of family members struck down by the disease (UNAIDS and WHO, 2005).

The viability of rural households is undermined by the loss of family labor and the increased cash requirements to meet medical costs and eventually funeral expenses, which can trigger sales of crops, livestock, farm tools and other assets. The death of a male head of household can lead to destitution for wives and children in societies where customary law prevents women from inheriting property, or where "widow inheritance" transfers a surviving wife to another male family member. Stigmatization further marginalizes surviving family members from the community (UNAIDS, 2005).

HIV/AIDS has become a major factor in the pervasiveness of food insecurity, as it undermines farm families' ability to cultivate adequate food for their members. Irregular and poor quality nutrition, in turn, hastens the onset of AIDS in those weakened by HIV and increases vulnerability to opportunistic infections.

The global labor force had lost 28 million economically active people to AIDS by 2005, a figure which is expected to rise to 48 million by 2010 and 74 million by 2015. Two-thirds of these labor losses will be in Africa, where four countries are expected to lose over 30% of their workforce by 2015 (ILO, 2005). Fewer workers mean more families left without providers, more children left without parents, and the loss of transmission of knowledge, skills and values from one generation to the next. Orphans are left in the care of the elderly or to fend for themselves in poverty and without access to education.

Agriculture and health are interlinked in complex ways. Agriculture produces the products on which humanity depends for its health—food—and yet, most of the poverty and malnutrition in the world is found in rural areas among those who work in agriculture. AKST has an important role to play in ensuring that future food supplies are available to meet growing demand for nutritious, safe and health-giving foods so that these can be made available at affordable prices to those who need them most.

1.3.3 Environment and natural resources

Natural resource issues

Natural resources are an indispensable basis for agriculture. A range of ecosystems produce the wide range of goods and services on which human survival depends. Production of these goods and services, including those related to agriculture such as food, is supported by a range of basic natural resources including soil, water and air. The demand for food will continue to rise as the human population increases, and while in the short-to-medium term production is expected to rise to meet this demand, there is growing concern about the vulnerability of the productive capacity of many agroecosystems to stress imposed by intensification, e.g., water scarcity and soil degradation (Thrupp, 1998; Conway, 1999; MA, 2005c; CA, 2007). Thus for instance, loss of biodiversity through simplification of habitats when monocultures are established in large areas is a major concern (Ormerod et al., 2003). The negative impact of increased soil erosion on downstream aquatic ecosystems and other activities such as fisheries can also be discerned. The positive and negative impacts of chemical inputs, particularly inorganic fertilizers and pesticides, are also well documented.

Sustainable use of natural resources is critical for sustainable livelihoods, and it has a direct impact on the improvement of natural capital. Both the poor and the rich impact the environment. Where access is easy and extraction is not capital-intensive, poor people may overuse natural resources; the poor also tend to be the most vulnerable to the effects of environmental degradation. By contrast, where extraction is highly capital intensive—such as in the case of deep groundwater extraction—the rich tend to have the biggest impact (Watson et al., 1998).

Agriculture is sustainable if the productive resource base is maintained at a level that can sustain the benefits obtained from it. These benefits are physical, economic and social. Ecological sustainability thus needs to be defined in relation to the sustainable use of natural resources, i.e., maintaining the productive capacity of an ecosystem.

Pressures on ecosystems have important consequences for agricultural production. In turn, agriculture has ecological impacts on ecosystems, and on the services provided by ecosystems.

The IAASTD recognizes that in agriculture, there is most often a continuum between a farming system and a natural ecosystem, as the term agroecosystem indicates. Farmers have a pivotal role as managers of these systems, and as stewards of their resource base. Their role includes for example the conservation of soil properties and water availability, the development and maintenance of crop species and the pursuit of multipurpose production objectives. Issues relating to NRM management are often framed as specific problems such as soil degradation, water pollution, biodiversity loss. We should also frame agriculture's contribution to NRM positively: farmers create and enhance resources such as arable soil, agrobiodiversity, productive forest stands. Working with the natural resource base, they often enrich and enhance it.

Drivers of natural resource degradation and depletion. As with other ecosystems, a range of direct and indirect drivers influence changes in natural resources in agricultural ecosystems. These drivers can act directly or indirectly to cause change. They may range from well defined drivers to those involving complex interactions. Among the key drivers assessed here is the role of decision makers and identification of those drivers that influence their decisions. Also important are the specific temporal, spatial and organizational scale dependencies as well as linkages and interactions between these drivers. The approach adopted also assumes that decisions are made at local, regional and international levels. Many globally recognized drivers are likely to influence natural resources in the context of agriculture, including demographic, economic, sociopolitical, science and technology, cultural and religious, and physical, biological and chemical drivers (see Figure 1-7).

Definition of natural resources
No unanimously accepted definition of natural resources exists. Natural resources can be defined as "factors of production provided by nature. This includes land suitable for agriculture, mineral deposits, and water resources useful for power generation, transport and irrigation. It also includes sea resources, including fish and offshore minerals" (Black, 2003). Natural resources may also be more broadly referred to as resources that "include all functions of nature that are directly or indirectly significant to humankind, i.e., economic functions as well as cultural and ecological functions that are not taken into account in economic models or which are not entirely known" (CDE, 2002). Climate can also be considered as a natural resource.

In these broader definitions, resources such as timber or fish are part of ecosystems that are living environments containing forests, rivers, wetlands and drylands as well agroecosystems embedded in broader ecosystems that make use of selected resources within the ecosystem (WRI, 2005). From here, it is a short step to integrating natural resources in the "ecosystem services" concept (MA, 2005a), i.e., to describe natural resources as system elements that ensure human well-being through a range of interdependent regulating, supporting, provisioning and sociocultural functions.

Availability of natural resources. The Millennium Ecosystem Assessment concluded that the global availability of natural resources is shrinking. "Over the past 50 years, humans have changed ecosystems more rapidly than in any comparable period of time in human history, largely to meet rapidly growing demand for [natural resources]. This has resulted in a substantial and largely irreversible loss in the diversity of life on earth" (MA, 2005a). Ecosystem change means that availability of natural resources should not be expressed exclusively in terms of physical availability. Their functional availability needs to be indicated as well.

Natural resource dynamics. As a result of intensifying global interactions, spatial and temporal effects become more interlinked and these are related to the weak recognition of the multifunctional nature of agroecosystems at all hierarchical levels. Resource degradation in one location may lead to pollution in another location. High discount rates for agricultural investments, in particular in developing countries, have been an incentive for short-term decision making, with the effect that farmers undervalue both future benefits and the costs of their present resource use. However, hunger may influence a household's view of the agricultural discount rate. Thus, while many households are aware that their decision-making is short term, the severe cost of hunger makes long-term considerations of benefits of natural resources irrelevant to them. Both poverty-induced expansion of agricultural activities into fragile and vulnerable lands (Bonfiglioli, 2004), and capital-intensive extraction of resources such as groundwater can contribute to increased vulnerability of natural resources.

The functionality of ecosystems and the temporal effects of system alterations are insufficiently understood. For example, understanding and using ecosystem functions in agriculture could result in enormous ecological savings while at the same time contributing to sustainable production of food (e.g., Costanza et al., 1997). There is an increased risk of non-linear changes as a result of system alteration (MA, 2005a). Therefore, the understanding of spatial and temporal effects of natural resource use for agricultural production is an increasingly important issue for science and technology in agricultural development.

Vulnerability and resilience of natural resources. The loss of ecosystems such as wetlands and mangroves has reduced natural protection of resources by destroying all or part of the inherent system functionality (MA, 2005a). The differences between damage caused by the December 2004 tsunami on shores protected by functional coral reefs and shores where reefs had been degraded exemplifies the increase of vulnerability as a result of unsustainable human activity (IUCN, 2005).

Natural ecosystems often have had to bear the brunt of intensification in agriculture. The degradation of forests, grasslands, coastal ecosystems and inland waters threatens their services to, and thus the long-term productive capacity of, agroecosystems. It is known that in many cases agricultural activities have depleted natural resources (forests, soil, water) to an extent that has resulted in net productivity losses; these developments are caused by a wide range of drivers. In other cases (e.g., rainfed agriculture or sustainable soil conservation) agricultural practices have been operated by generations of successive farmers in a sustainable way.

Natural resources and their management

Forestry. Agriculture has had an intimate and productive relation with forests: many historical and contemporary farming systems are built partly on that relationship. Swidden agriculture in tropical areas, for example, uses forests as a means of soil and nutrient restoration.

Agroforestry and home garden systems are ways of combining trees and other species with crop production or animal husbandry. Up to the present, forests and agroforests have played an important role in contributing to the food security of a large part of the world's food insecure people. They provide products (timber, fuelwood, food, and medicines), inputs for crop and livestock production (fodder, soil nutrients, and pollination) and services (watershed protection, climate regulation, carbon storage, and biodiversity conservation) (FAO, 2006a).

Some 350 million of the world's poorest people are considered to be largely dependent on forests for their living, including for food production (WCSFD, 1999). A majority of farmers manage some trees on their land, or benefit from forests adjacent to their land, often for environmental services (e.g., to shelter or shade homes, crops and livestock, or for soil conservation), as well as for diverse products (such as fuelwood and fruit) (Scherr et al., 2004; Molnar et al., 2005). Approximately 1.5 billion people use products from trees as key elements of their livelihoods (Leakey and Sanchez, 1997).

Deforestation has been identified as a major problem facing forest resources. The expansion of agriculture in its many forms at the expense of forestland is one of the factors contributing to deforestation, though not the only one. The conversion of forest to another land use or the long-term reduction of the tree canopy cover below the minimum 10% threshold is one definition of forestry. The rate of deforestation is proceeding at 13 million ha per year (FAO, 2007a).

Recent estimates show that forests cover about 31% of global land surface (FAO, 2007a). Since pre-agricultural times, forests have been reduced by 20 to 50% (Matthews et al., 2000). Patterns of forest management and use vary across the globe. Thus, for instance, while the last two and a half decades have seen an increase in forest area in industrial countries, developing countries have on average witnessed a decline of about 10% (FAO, 2007a; Figure 1-17). An increasing trend is also the rapid expansion of mixed forest/

agriculture zones encroaching on formerly intact forest areas. 80% of the fiber and fuelwood production is derived from primary and secondary growth forests and therein lies the importance of management of this important resource. In addition to fiber and fuel, forests provide a range of ecosystems services. Forests make up two thirds of the more than 200 ecoregions identified by WWF as outstanding representatives of the worlds' ecosystems that include important endemic bird areas and more than three quarters of the centers of plant biodiversity (Olson and Dinerstein, 1998). Forest soils and vegetation store about 40% of all carbon in the terrestrial biosphere. However, due to deforestation rates that exceed growth, forests are currently a net source of atmospheric carbon. Loss of forest cover in watersheds has secondary effects on water resources through increased erosion, and alteration of water quantity and possibly floods. It has been estimated that roughly 0.75 ha of forest is now needed to supply each person on the planet with shelter and fuel (Lund and Iremonger, 1998).

Biological corridors play an important role in mitigating incidental or secondary effects. Thus, in some regions in Central America, using local and foreign funds, international organizations, governing institutions and rural committees are working to connect natural reserves by planting native tree species in deforested areas. These new green spots will open routes for the safe migration and mating of wild animals, as well as preserve the wild and native flora.

Grasslands

Grasslands are mostly associated with drylands where plant production is limited by water availability—the dominant users are large mammals, herbivores including livestock, and cultivation. Drylands include cultivated lands, scrublands, shrublands, grasslands, semideserts, and true deserts (MA, 2005c). They are, as their name implies, natural landscapes where the dominant vegetation is grass. Grasslands usually receive more water than deserts, but less than forested regions. Worldwide, these ecosystems provide livelihoods for nearly 800 million people. Grasslands are also a source of

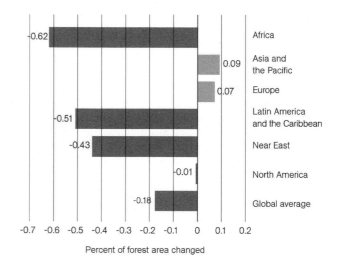

Figure 1-17. *Annual net change in forest area, 2000-2005.* Source: FAO, 2007a.

forage for livestock, wildlife habitat, and a host of other resources (White et al., 2000).

Grasslands provide feed for livestock farming across the globe as well as a wide range of ecosystem services. For instance, grasslands provide part of the cover to some of the world's major watersheds. Most of the world's meat comes from animals that forage on grasslands. World meat production has doubled since 1975, from 116 million to 233 million tonnes in 2000 (UNEP, 2002). Grasslands are also a major component of important areas of bird endemism and wildlife sanctuaries, and store approximately 34% of the global stock of carbon in terrestrial ecosystems.

Nearly 49% of grasslands are lightly to moderately degraded and at least 5% are considered strongly to extremely degraded (White et al., 2000). The degree of degradation is dependent on geographical location and management practices as well as on characteristics of the soil, vegetation, and grazing patterns. Cultivation and urbanization of grasslands, and other modifications can be a significant source of carbon to the atmosphere. For instance, biomass burning, especially on tropical savannas, contributes over 40% of gross global carbon dioxide emissions (Baumert et al., 2005).

Fisheries

Fish play a key role as an economic commodity of significance to a great number of farming households and rural poor people. Inland fisheries and aquaculture—for example in irrigated rice agroecosystems—are not only important as a direct food source: fish are also a high value commodity that can be traded for cash, for other needs and cheaper foods, by small-scale farmers and the poor, and provide a source of direct employment for 38 million and indirect employment for about 160 million people (FAO, 2004b; ICTSD, 2006). The highest share of fish workers (fishers and aquaculture workers) is in Asia (87%), followed by Africa (7%), Europe, North and Central America and South America (about 2% each) and Oceania (0.2%) (FAO, 2004b).

In 2002, about 76% (100.7 million tonnes) of estimated world fisheries production was used for direct human consumption. The remaining 24% (32.2 million tonnes) was destined for non-food products, mainly the manufacture of fishmeal and oil, slightly (0.4%) above levels in 1999 but 5.8% below levels in 2000 (FAO, 2004b). In 2002, total capture fisheries production amounted to 93.2 million tonnes. Marine capture fisheries production contributed 84.5 million tonnes. Between 2000 and 2003, the reported landings of marine capture fisheries have fluctuated between 80 and 86 million tonnes: a slight increase over the preceding decade (mean = 77 million tonnes). Production from different capture and culture systems varies greatly (CA, 2007).

At the global level, inland capture fisheries have been increasing since 1984. In 1997, inland fisheries accounted for 7.7 million tonnes, or almost 12% of total capture available for human consumption, a level estimated to be at or above maximum sustainable yields (Revenga et al., 2000). In 2000-2002, inland capture fisheries were estimated at around 8.7 million tonnes. However, there is still a lack of reliable data on global inland fisheries production, which are therefore estimated to be underreported by two or three times (FAO, 2004b).

In 2004, aquaculture accounted for 43% of the world's food fish production and is perceived as having the greatest potential to meet the growing demand for aquatic food (FAO, 2006c). World aquaculture has grown at an average annual rate of 8.8% from 1950 to 2004. In recent years, Asia and Africa have shown the highest growth with Latin America displaying only moderate growth. Production in North America, Europe and the former Soviet states has however declined. The average growth rate for the Asia and the Pacific region was 9.8%, while production in China, considered separately, has grown at a rate of 12.4% per year (FAO, 2006c).

In 2004, freshwater aquaculture was the predominant form of aquaculture, accounting for 56% of the total production while mariculture contributed 36% and brackish-water aquaculture 7.4% (FAO, 2006c). During the last decade, inland capture production has remained relatively stagnant. For instance, during the period 2000-2005, production ranged between 8.8-9.6 million tonnes. During the same period, aquaculture grew from 21.2 to 28.9 million tonnes. Similar trends have been observed in marine environments. Thus overall, the total aquaculture production grew from 35.5 to 47.8 million tonnes. Despite this increase in landings, maintained in many regions by fishery enhancements such as stocking and fish introductions, the greatest overall threat for the long-term sustainability of inland fishery resources is the loss of fishery habitat and the degradation of the terrestrial and aquatic environment.

About 40% of the world's population lives within 100 km of a coast. Because of the current pressures on coastal ecosystems, and the immense value of the goods and services derived from them, there is an increasing need to evaluate trade-offs between different activities that may be proposed for a particular coastal area. This important habitat is increasingly becoming disturbed due to human activity. Many coastal habitats such as mangroves, wetlands, sea-grasses, and coral reefs, which are important as nurseries, are disappearing at a fast pace. About 75% of all fish stocks for which information is available are in urgent need of better management (Burke et al., 2001; FAO, 2004b).

A recent assessment of fish stocks by the FAO indicates that only 20% of fish species is moderately exploited and only 3% is underexploited. Of the remaining 76%, 52% of stocks is fully exploited, 17% is overexploited and 7% is depleted (FAO, 2004b).

Depletion of marine resources is so severe that some commercial fish species, such as the Atlantic Cod, five species of tuna, and haddock are now threatened globally, as are several species of whales, seals, and sea turtles. The scale of the global fishing enterprise has grown rapidly and exploitation of fish stocks has followed a predictable pattern, progressing from region to region across the world's oceans. As each area in turn reaches its maximum production level, it then begins to decline (Grainger and Garcia, 1996).

Apart from being an important food source, fish can also be a source of contamination. In heavily polluted areas, in waters that have insufficient exchange with the world's oceans, e.g., the Baltic Sea and the Mediterranean Sea, in estuaries, rivers and especially in locations that are close to industrial sites, concentrations of contaminants that exceed natural load can be found. These increasing amounts may also be found in predatory species as a result of biomagnifications, which is the concentration of contaminants in higher levels of the food chain, posing a risk for human health (FAO, 2004b).

Water resources

In the hydrological cycle water resources can be divided into "blue" and "green" water. The main source of water is rain falling on the earth's land surfaces (110,000km³) (CA, 2007). Blue water refers to the water flowing or stored in rivers, lakes, reservoirs, ponds and aquifers (Rockström, 1999). Globally, about 39% of rain (43,500 km³) contributes to blue water sources, important for supporting biodiversity, fisheries and aquatic ecosystems. Blue water withdrawals are about 9% of total blue water sources (3,800 km³), with 70% of withdrawals going to irrigation (2,700 km³). The concept of green water (Falkenmark, 1995) is now used to refer to water that is stored in unsaturated soil and is used as evapotranspiration (Savenije and van der Zaag, 2000). Green water is the water source of rainfed agriculture. Total evapotranspiration by irrigated agriculture is about 2,200 cubic kilometers (2% of rain), of which 650 cubic kilometers are directly from rain (green water) and the remainder from irrigation water (blue water). To date, sub-Saharan Africa has the smallest ration of irrigated to rainfed water and more than half of irrigated land is in Asia (HDR, 2006; see Figure 1-18).

Technological advancements, especially in the construction of dams, have markedly increased the volume and availability of blue water for consumption and irrigation purposes. Similarly, improvements in pumping have motivated farmers to extract more and more groundwater. Moreover, the demand for water has increased at more than double the rate of population increase, leading to serious depletion of surface water resources (Penning De Vries et al., 2003; Smakhtin et al., 2004). Seventy percent of blue water abstraction is for irrigation; given increasing competition from other users water productivity is a priority concern. Furthermore, much of water used in irrigation is lost to less-than-optimal evaporation, not profiting plant growth.

On the other hand, half of the world's wetlands are estimated to have been lost during the last century, as land was converted to agriculture and urban use, or filled to combat diseases, such as malaria. Yet these freshwater wetlands provide a range of services including flood control, storage and purification of water as well as being an important habitat for biodiversity. Worldwide water quality conditions appear to have been degraded in almost all regions with intensive agriculture and other developments (Molden and de Fraiture, 2004). Pollution is a growing problem in most inland water systems around the world while waterborne diseases from fecal contamination of surface waters continue to be a serious problem in developing countries (Revenga et al., 2000).

There is no agriculture without water. Agriculture's sustainability agenda as regards water is twofold: access to clean water for the poor on the one hand, improvements in water productivity and institutional arrangements on the other (CA, 2007).

Half of the world's 854 million malnourished people are small-scale farmers who depend on access to secure water supplies for food production, health, income and employment. Improving their access to clean water potentially has an enormous impact on their livelihoods and productive strategies by reducing poverty and vulnerability (HDR, 2006). With scarcity and competing demands for water increasingly becoming evident, growing more food with less water is a high priority. There is much scope for better water productivity both in low-production rainfed areas and in irrigated systems (CA, 2007). Blue water used in irrigation has a particularly important role, as 40% of global crop production is produced on irrigated soils (WWAP, 2003). In addition, irrigation often depends on dams that impact the environment in various ways, leading to disturbance or destruction of habitats and fisheries (WCD, 2000). To mitigate these impacts, water use efficiency is also paramount. Responses by AKST aiming at improving water use effectiveness include developing micro-irrigation systems (Postel, 1999) and more precise management techniques generally, but also breeding of drought-tolerant crop varieties such as in maize (Edmeades et al., 1999).

Soils

Soil is the source of nutrients required for plant growth and itself the result of organic processes of living organisms. It is therefore the primary environmental stock that supports agriculture. Soil condition varies widely but global estimates suggest that 23% of all used land is degraded to some degree, which is a cause of serious concern (Oldeman, 1994; Wood et al., 2000). The key soil degradation processes include: erosion, salinization and water logging, compaction and hard setting, acidification, loss of soil organic matter, soil nutrient depletion, biological degradation, and soil pollution. Agricultural activities influence all these processes (Scherr, 1999).

In crop cultivation, the resilience of arable soils is an issue of great concern. Different soil types have very different erodibility characteristics, i.e., their ability to resist soil erosion caused by water, wind, or plowing varies a great deal. Some soils will hardly recover once eroded, while others may regenerate within a relatively short time. There are two dimensions to the degradation of soils: first their sensitivity to factors causing degradation, and second their resilience to degradation, which is their ability to recover their original

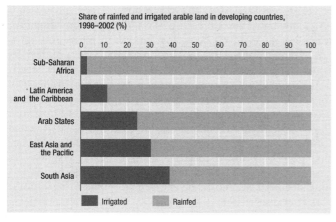

Figure 1-18. *Rainfed and irrigated arable land in developing countries, 1998–2002.* Source: HDR, 2006.

properties after degradation has occurred. Sensitivity and resilience depend on climate and the biophysical structures of the soil, and whether degradation has exceeded a threshold of resilience (such as loss of all organic matter or severe compaction) beyond which recovery is not possible without active intervention (Blaikie and Brookfield, 1987).

Soil, just like water, is a key resource for agricultural production. Sometimes erroneously subsumed under "land" issues, the availability of soils for growing crops often seem to be taken for granted. Yet in both the developing and the industrialized world, the loss of productive agricultural soils to urban development is enormous. In addition, according to an estimate by the Global Assessment of Human-induced Soil Degradation (GLASOD), degradation had affected 38% of the world's cropland, to some extent as a result of human activity (Oldeman et al., 1991). However, GLASOD did not estimate productivity losses associated with land degradation. In the absence of data on the productivity impacts of land degradation, estimates based on different methods vary widely (Wiebe, 2003).

The direct influence of agricultural practices cannot be neglected: they account for about a quarter of total soil degradation (GACGC, 1994). AKST is, and always has been, crucial to address these problems both through more classical approaches (e.g., proposing mechanical protection such as bunds and terraces to control surface runoff) and through more comprehensive frameworks aiming at greater integration of water conservation and soil protection and the use of biological methods (Shaxson et al., 1989; Sanders et al., 1999; WOCAT, 2006).

The impact of nitrates from fertilizers and livestock production on soil and water resources is a related issue. This impact can be described in general terms as the nitrification of the global ecosystem from inorganic fertilizers and alteration of the global nitrogen cycle. Eutrophication as a consequence of nutrient runoff from agriculture poses problems both for human health and the environment. Impacts of eutrophication have been easily discernible in some areas such as the Mediterranean Sea and northwestern Gulf of Mexico (Wood et al., 2000).

Some agricultural activities have led to a reduction of system productivity. For instance, irrigated agriculture has contributed to water logging and salinization, as well as depletion and chemical contamination of surface and groundwater supplies (Revenga et al., 2000; Wood et al., 2000; CA, 2007). Manure from intensive livestock production has exacerbated the problem of water contamination. Misuse of pesticides has led to contamination of land and water, to negative impacts on non-target species, and to the emergence of pesticide-resistant pests. These problems compound to reduce system productivity (Thrupp, 1998; Conway, 1999). The capacity of coastal and marine ecosystems to produce fish for human harvest is highly degraded by overfishing, destructive trawling techniques, and loss of coastal nursery areas. This is exacerbated by the decline of mangroves, coastal wetlands, and seagrasses with resultant loss of pollutant filtering capacity of coastal habitats.

Biodiversity
Biodiversity underpins agriculture by providing the genetic material for crop and livestock breeding, raw materials for industry, chemicals for medicine as well as other services that are vital for the success of agriculture, such as pollination. The last century has seen the greatest loss of biodiversity through habitat destruction, for instance through conversion of diverse ecosystems to agriculture. Other factors such as the growing threat from introduction of invasive alien species, fostered by globalization of trade and transport, have further exacerbated the situation. On small islands, introduction of invasive alien species, many through agriculture-related activities, is the main threat to biodiversity. In freshwater systems, an estimated 20% of fish species have become extinct (Wood et al., 2000). Globally, the cost of damage caused by invasive species is estimated to run to hundreds of billions of dollars per year (Pimentel et al., 2001). In developing countries, where agriculture, forestry and fishing account for a high proportion of GDP, the negative impact of invasive species is particularly acute. Globalization and economic development through increasing trade, tourism, travel and transport also increase the numbers of intentionally or accidentally introduced species (McNeely et al., 2001). It is widely predicted that climate change will further increase these threats, favoring species migration and causing ecosystems to become more vulnerable to invasion.

While agriculture is based on the domestication and use of crop and livestock species, the continuum between (wild) biodiversity and agrobiodiversity has been recognized both in research on plant genetic resources and in conservation efforts for many decades—starting with the hypothesis of "centers of diversity" of crop species proposed by Vavilov in the 1920s. More recently an emphasis on the provisioning services of biodiversity has been added: "Biodiversity, including the number, abundance, and composition of genotypes, populations, species, functional types, communities, and landscape units, strongly influences the provision of ecosystem services and therefore human well-being. Processes frequently affected by changes in biodiversity include pollination, seed dispersal, climate regulation, carbon sequestration, agricultural pest and disease control, and human health regulation. Also, by affecting ecosystem processes such as primary production, nutrient and water cycling, and soil formation and retention, biodiversity indirectly supports the production of food, fiber, potable water, shelter, and medicines" (MA, 2005c).

Agrobiodiversity is the very stuff of food production and an essential resource for plant and animal breeding. Yet it is a resource that is being lost *in situ*: in farms and agroecosystems (FAO, 1996b; Thrupp, 1998; CBD, 2006). Its conservation is somewhat framed by a paradox: new breeds have boosted agricultural productivity, but simultaneously they displaced traditional cultivars. In response, gene or seed banks have been created to fulfill a double function: to resource plant breeders with the agrobiodiversity needed for further crop development, and to conserve crop diversity that may have disappeared from agricultural systems. *Ex situ* conservation in seed repositories and gene banks has long been considered to be the central pillar of agrobiodiversity conservation.

To be effective, agrobiodiversity management needs to operate at several levels: local, national, and international. Against the overall trend of declining diversity in agricultural systems, crop diversity is still being created and preserved locally, and the importance of local *in situ* conservation efforts has more recently been acknowledged under Article 8 of the CBD. *In situ* conservation of crops and seeds on the farm or community level operates under a number of constraints, partly organizational, partly economic. These constraints can more easily be overcome if biodiversity management is part of an integrated approach—such as sustainable land management.

It is notable that plant varieties and animal breeds —very much like farming systems—are intricately linked to languages, environmental knowledge, farming systems, and the evolution of human societies. They embody history, both in their form which is a result of selection and adaptation to human needs, and through the knowledge that is associated with them. In participatory research and selection, such knowledge has increasingly been validated and valued.

In the contemporary context of rapid land use change, the complex coevolution of agrobiodiversity, ecosystems and human societies needs to be documented, analyzed and validated. An appropriate level for this task is the landscape. Cultural landscapes are complex but spatially bounded expressions of ecosystems that have evolved under the influence of biophysical factors as well as of human societies. They provide the context to understand how management practices have shaped the productive and characteristic landscapes of cultivated systems, and how crop knowledge fits into these patterns (Brookfield et al., 2003).

Agriculture and climate change

Agriculture contributes to climate change through the release of greenhouse gases in its production processes. It is a significant emitter of CH_4 (50% of global emissions) and N_2O (70%) (Bathia et al., 2004). The levels of its emissions are determined by various aspects of agricultural production: frequency of cultivation, presence of irrigation, the size of livestock production, the burning of crop residues and cleared areas. In many cases, emissions are difficult to mitigate because they are linked to the very nature of production; in a number of cases, however, technical measures can be adopted to mitigate emissions from specific sources.

Agricultural activities account for 15% of global greenhouse gas (methane, nitrous oxide and carbon dioxide) emissions (Baumert et al., 2005). Two-fifths of these emissions are a result of land use or soil management practices. Methane emissions from cattle and other livestock account for just over a quarter of the emissions. Wetland rice production and manure management also contribute a substantial amount of methane. Land clearing and burning of biomass also contributes to carbon dioxide production.

Changes in land use, especially those associated with agriculture, have negatively affected the net ability of ecosystems to sequester carbon. For instance the carbon rich grasslands and forests in temperate zones have been replaced by crops with much lower capacity to sequester carbon. By storing up to 40% of terrestrial carbon, forests play a key role, and despite a slow increase in forests in the northern hemisphere, the benefits are lost due to increased deforestation in the tropics (Matthews et al., 2000).

There is considerable potential in agriculture for mitigating climate change impacts. Changing crop regimes and modifying crop rotations, reducing tillage, returning crop residues into the soil and increasing the production of renewable energy are just a few options for reducing emissions (Wassmann and Vlek, 2004).

Climate change poses the question of risks for food security both globally and for marginal or vulnerable agroecological zones. People's livelihoods are threatened, as we know, if they lack resilience and the purchasing power to bridge production losses on their farms. The magnitude of the threat to the agricultural sector, and to small-scale farmers in particular, is thus also dependent on the performance of the non-agricultural sectors of developing economies, and on the opportunities they provide. Adaptation to climate change is therefore an important topic for AKST. The need and the capacity to adapt vary considerably from region to region, and from farmer to farmer (Smit, 1993; McCarthy et al., 2001).

Change in water runoff by 2050 is expected to be considerable (Figure 1-19). Some regions will have up to 20% less runoff, while others will experience increases of the same order, and only few countries will have similar conditions as at present (HDR, 2006). Improving water use efficiency, adapting to the risks related to topography, and changing the timing of farming operations are some examples of adaptation that will be required.

Adaptation has a cost and often requires investments in infrastructure. Therefore, where resource endowments are already thin, adverse impacts may be multiplied by the lack of resources to respond. Farmers are masters in adapting to changing environmental conditions because this has been their business for thousands of years. This is a knowledge base farmers will need to maintain and improve, even if climate change may pose challenges that go beyond problems tackled in the past.

Sustainability implications of AKST

A key objective of agricultural policies since the 1950s, both in industrialized and in developing countries, has been to increase crop production. In its production focus, these policies have often failed to recognize the links between agricultural production and the ecosystems in which it is embedded. By maximizing provisioning services, crop production has often affected the functioning of the supporting ecosystem services.

In the 1960s and 1970s, for instance, irrigated agriculture was intensified in Asia and elsewhere to boost production of one major food crop: rice. The effort was underpinned with massive public investments in crop research, infrastructure and extension systems. While successful in terms of production and low commodity prices, this Green Revolution led in some cases to environmentally harmful practices such as excessive use of fertilizers or pesticides. As evidence of negative impacts on the environment—particularly on soil and water—emerged, a number of corrective measures were envisaged.

In Indonesia, for example, a major effort was undertaken in the 1980s to introduce integrated pest management (IPM) in intensive rice production (Röling and van de Vliert, 1994). This required that farmers have better knowledge of pests and their predators—knowledge that could be used to reestablish pest-predator balances in rice agroecosystems,

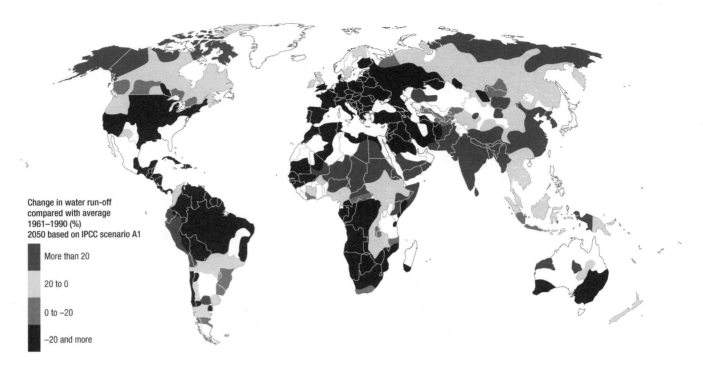

Figure 1-19. *Climate change and water run-off.* Source: HDR, 2006.

and to avoid the harmful use of pesticides. The successful, practical application of IPM is an example of the ecological services provided by agroecosystems, and the monetary, health and environmental benefits they provide.

In the 1990s, management has become a key term in most debates on natural resources, agriculture included. The multifunctional character of agriculture implies a serious consideration of the links with the ecosystems in which agricultural systems are embedded, beyond measures and policies addressing specific resources such as water and soil. This is a very complex challenge concerning a multitude of actors.

AKST and natural resource management (NRM)
There is now a strategic understanding that "the management of natural resources clearly has social and behavioral components, the understanding of which is indispensable for orienting biophysical research to these resources. Behavioral and sociocultural variables of resource management are no less important for resource sustainability than physical parameters" (CGIAR, 2000).

Practitioners of NRM research in agricultural development have adjusted their research agendas to address this problem, often under the headings "policies", "institutions", and "processes". This allows them to frame the debate on how access to resources should be regulated, and what types of institutional regimes are needed to ensure environmental sustainability of resource use in agriculture. Management of natural resources is articulated on at least two levels: the household and its livelihood, and the larger resource regimes on the community, the national and the international levels. For this aspect, AKST has benefited from research that

deals with common property and common pool resources (Ostrom, 1990). A balanced research agenda focusing both on institutional aspects of resource management and on biophysical parameters of the systems is key for managing the multifunctional base and effects of agricultural production. AKST has also benefited from research on traditional agricultural systems and their knowledge base. While local knowledge forms are rarely equipped to respond to all the changes in contemporary agricultural systems, participatory research in AKST has demonstrated its value for grounded and adapted solutions.

While national policies are evidently key in these areas, some approaches have become agreed notions in multilateral processes, like Agenda 21. Sustainable Land Management (SLM), for example, is defined as "the use of land resources, including soils, water, animals and plants, for the production of goods to meet changing human needs, while ensuring the long-term productive potential of these resources and the maintenance of their environmental functions" (UN, 1993a). This is a pertinent and comprehensive definition. However, its impact on the promotion of innovative management strategies and on national and international policies is scarcely visible to date. We may also note that efforts are devoted on the one hand to soil and water conservation, and on the other to conservation of biotic resources (agrobiodiversity), with little interlinkages between the two.

In sum, a shift towards the integrated analysis of natural resource management has begun to transform the agricultural research agenda and AKST. However further progress in integrating biophysical with sociocultural and behavioral variables, and the recognition—in practice—of the multifunctional nature of agriculture may be needed. In addition to techniques aiming at specific resources, the overall management of natural resources has become a concern in agricultural development.

1.3.4 Social equity

The sense of justice and injustice is a universal feature of human society; yet complexity, stratification and inequality are enduring hallmarks of social organization. Nowhere is this more evident than in agriculture, where patterns of land ownership, land tenure, social status, employment and division of labor have evolved in highly diverse ecological, social and cultural contexts.

Social equity is intimately linked to a sense of justice both in terms of processes and outcomes. In its ideal form, it incorporates notions of equality, as in equal rights under the law, and of equivalence as in differentiated treatment that produces outcomes of comparable value or significance for beneficiaries in disparate circumstances. In legal terms, equity originated as a system of jurisprudence developed to correct injustices caused by inflexibility in the law. It was based on the principle of natural justice. In this sense, equity serves to bridge the gap between legality and legitimacy of outcomes, for example, when equal treatment would result in the perpetuation of injustice.

Political, economic and cultural factors contribute to greater or lesser degrees of equity in society, sometimes mitigating, sometimes reinforcing inequality. Many sources of inequality are determined by the circumstances of birth. Sex, ethnicity, the wealth or poverty of parents, their educational status, birth in a rural or urban setting are among these. Other sources of inequality are cultural constructs. These include gender roles in the world of work; the rights and duties of family members as defined by age, sex or birth order; parental expectations of sons and daughters; the loci of decision-making power within households and in the wider community; and the formal and informal rules that determine access to land, water and other resources. Whether determined by birth or culture, these sources of inequality tend to widen or narrow the opportunities that individuals have to develop their inherent talents and their productive potential. Combating corruption can help improve equity, as corruption is undermining justice in many parts of the world. Corruption is the abuse of entrusted power for private gain; this may include material and non-material gain from political interference to bribery (TI, 2007). This will occur unless society develops institutions of governance, legal systems and social policy tools that tend to lessen disparities and equalize opportunities. With improved women's economic and social rights corruption is generally reduced.

While economic forces tend to favor some to the detriment of others, it is common for social policy instruments to attempt to redress the balance in some measure by promoting equality of opportunity, ensuring that basic services are available to all and assisting vulnerable groups in meeting their needs. Equity concerns underpin efforts to eliminate discrimination, widen opportunities for social and economic advancement, increase access to public goods and services, such as education and health care, provide fairer access to resources and promote empowerment through participation in decision-making (ILO, 1962). All of these are critical to reducing poverty and building a just society based on rights for all.

Rights-based approach. Since the adoption of the Universal Declaration of Human Rights in 1948, there has been a growing worldwide consensus that abject poverty, hunger, and deprivation are an affront to human dignity and that conditions must be created whereby all persons may enjoy basic human rights (UNICCPR, 1966; UNICESCR, 1966). Whether these rights are of a civil, political, economic, social or cultural nature, they are considered to be "universal, indivisible and interdependent and inter-related" (UN, 1993b).

Civil and political rights—such as political voice and representation, freedom of association, and equal protection under the law—are important in themselves, but also in their function as enabling rights. Such rights enable individuals and groups to participate in public debate, influence the decisions that affect the life of their communities, defend their common interests, build more responsive economic and social institutions, and manage conflicts through peaceful, democratic means. Economic, social and cultural rights—such as the right to education, health care, food and an adequate standard of living—help to create the conditions under which civil and political rights can be freely exercised.

Social equity concerns and agriculture. Social equity concerns are gaining in importance in countries where large numbers of people are engaged in agricultural production and where productivity improvements are needed to keep pace with or exceed population growth, in other words, in most developing countries. Globalization has placed the agricultural sector in many countries under tremendous pressure as generally declining commodity prices, rising input costs, low levels of investment and lack of credit take their toll, particularly on small-scale farmers, their families and agricultural workers. Loss of status, uncertainty of income, indebtedness, unfulfilled needs and the deterioration in their economic and social condition are among the factors that have spurred able-bodied men and youth to leave rural areas in search of opportunities elsewhere. Many swell the ranks of the urban unemployed, lacking the skill sets needed to prosper in the new environment, subsisting through informal activities. Those remaining in agriculture—particularly, ethnic minorities, women, the elderly, children and youth—find themselves increasingly on the margins of economic, social, and political life. They form the majority of the world's poor.

Potential beneficiaries of AKST are a heterogeneous group living in highly diverse social, economic and environmental contexts. Research, development and dissemination efforts need to take their capacities and constraints into account in order to ensure that innovations are practical, affordable and offer real benefits to the poor among them. Social equity concerns challenge policy-makers, researchers, practitioners and donors to work together across their respective disciplines to provide not only the technological means, but also the social support needed to encourage and enable uptake of new techniques by those who may not previously have had access to skills training, extension services or credit facilities.

A major social equity issue in agriculture is the perpetuation of poverty from one generation to the next due to the high incidence of child labor. Approximately 70% of all child labor is found in agriculture. Unpaid work on the family farm may or may not have an incidence on the child's school attendance and performance, depending on the hours and conditions of work. However, time lost to education, particularly if low achievement levels lead to early drop-out, has lifelong consequences on earnings. Much child labor in commercial agriculture is invisible and unacknowledged, although it may account for a considerable portion of family earnings (WDR, 2007).

Social equity issues, such as child labor, must be addressed if broad-based agricultural development is to contribute positively to both economic growth and poverty reduction. The principal challenges are twofold: raising the living standards of those working in agriculture, particularly the poorest among them, and lessening the demographic burden on agriculture by providing opportunities for more diversified and rewarding economic activity outside the sector. Educating rural children and preparing them for a productive future addresses both those concerns and AKST can be instrumental in achieving this in a number of ways. For example, well targeted AKST can enable poor farmers to increase their earnings sufficiently to keep their children in school, rather than at work. The adoption by parents of innovative farming practices can teach children the experience of lifelong learning, openness to technological change and the benefits of applying knowledge to production. Incorporating AKST into rural school programs could provide young people with practical skill sets to help them make the transition to more productive work in agriculture or in rural support services, or could inspire them to pursue other science based studies.

The labor requirements of various crops or cultivation methods are an important variable that needs to be considered. AKST is not employment-neutral, nor can it be if it is to improve the livelihoods of the rural poor. In some poor communities and households, the greatest challenge is to generate productive employment for able-bodied workers. In such circumstances, the development of high-value, highly nutritious, labor-intensive crops may offer opportunities for improving livelihoods and well-being. In other cases, labor-saving crops and techniques may offer better outcomes, for example, for labor-poor female headed households, or rural communities suffering from a high incidence of HIV/AIDS or other debilitating illnesses.

Many observers note a dichotomy between small-scale agriculture and industrialized agriculture. Indeed, the uneven competition that has emerged between small- and large-scale production systems raises serious social equity issues within the agricultural sector as a whole. The two systems differ greatly in terms of resource consumption, capital intensity, access to markets and employment opportunities. The economic and political power of agribusiness enterprises and their relative importance in national economies enable them to influence decisions regarding domestic support packages, infrastructure investment, the direction of agricultural research and development and the setting of international trade rules in ways that small-scale farmers cannot. Another major difference lies in their capacity to provide employment. Large-scale production systems are often in a position to offer better terms of employment, but they tend to shed labor as productivity gains are realized through technology and more efficient work organization. Although the number of persons working in small-scale agriculture has decreased as a percentage of the global population in recent decades, it has steadily increased in absolute numbers and is estimated to include approximately 2.6 billion people or 40% of the world's population (Dixon et al., 2001).

While the notion of dichotomy may be useful in drawing out such contrasts, it tends to mask the wide range of ownership patterns, relationships to the land, forms of labor force participation and employment relationships that generate profound social equity issues. It is instructive to consider how just one set of rights—property rights—affects the livelihoods of various stake-holders in the agriculture sector: plantation owners, medium to small-scale owner-cultivators, tenant farmers, share-croppers, squatters, landless laborers, bonded laborers, migrant workers, or members of an indigenous community sharing common lands. These categories are not discrete; indeed, there is frequent overlap among them, and cutting across all these categories are issues of gender, which further define or delimit rights of ownership, access, use and inheritance of the land.

Choices to be made: agricultural productivity and poverty reduction

Most discussions of broad-based agricultural development focus on the interaction of five main factors—innovation, inputs, infrastructure, institutions and incentives (Hazell, 1999). Equity issues are inherent, though they may not be explicitly evoked, in the policy decisions that guide the investment of resources in these areas. For example, agricultural research and development is needed to generate productivity-enhancing technologies, but choices must be made as to the orientation of research efforts. The improvement of local food crops to better satisfy nutritional needs, the development of drought-resistant breeds to provide a more reliable harvest to those living on marginal lands, or the development of horticultural produce suitable for export may all be worthy goals in themselves, but have very different potential beneficiaries. Whether or not these activities lead to improved livelihoods for the poor depends on many factors, not least among them being the social characteristics of particular rural communities and the convergence of innovation with other productivity factors. Ownership or control of land and other assets, knowledge and skill levels, roles and responsibilities with regard to production, access to affordable credit, and rights with regard to distribution of services vary considerably across and within social groups. Ethnicity, class, sex and age all affect the capacity of those who work the land to access and use new technologies effectively and profitably, but take-up can be modified with well-targeted interventions. Productivity enhancement is not so much a technical issue, as one of political, economic and social choices and constraints, hence an issue of equity (HDR, 2006).

This is well illustrated by a number of "equity modifiers" that have been suggested as a means to reduce poverty and contribute to growth through broad-based agricultural development. These include targeting small and medium-sized family farms as priority beneficiaries for publicly funded agricultural research and extension, marketing, credit and input supplies; undertaking land reform, where needed; investing in human capital to raise labor productivity and increase opportunities for employment; ensuring that agricultural extension, education, credit and small business assistance programs reach rural women; setting public investment priorities through participatory processes; and actively encouraging the rural non-farm economy (Hazell, 1999). It is noteworthy that all six modifiers imply some form of human capital enhancement.

Adoption and implementation of such transformational policies would require political will and political power, but the potential beneficiaries, indeed, the major actors, are largely absent from the decision-making process. The geographical locus of decision-making tends to be in the country's capital or major commercial centers and competition for government resources tends to be heavily weighted in favor of urban areas, where populations are concentrated, vocal and potentially active. Rural poor people in general and rural women in particular tend to be "invisible" to policy makers and service providers, and are without voice or representation in political decision-making.

Perhaps as a result of this, the rural sector has suffered years of neglect, notably during the course of structural adjustment. Lack of investment in roads, water systems, education and health services, and the dismantling of public extension systems have all left their mark on rural areas and on the people who live there. Rural poverty rates consistently exceed those in urban areas. In all 62 countries for which data sets were available, a greater percentage of rural people were living below the national poverty line compared to their urban counterparts. In several cases, the rural-urban poverty gap was more than 30% (World Bank, 2006b). If it were measurable, the urban-rural disparity in political power would most likely be greater. The male-female power disparity certainly is.

Government ministries dealing with agriculture and rural development have a minority of women among their professional and technical staff, and only a small percentage at decision-making levels. For example, a 1993 study of women in decision-making positions found that overall, women held 6% of decision-making positions in ministries and government bodies in Egypt. Cooperative agricultural societies had an almost exclusively male membership, agrarian reform societies were entirely within male hands, and land reclamation societies had no women members. In Benin, women held only 2.5% of high-level decision-making positions in government, and comprised only 7.3% of the decision-making and technical staff at the Ministry of Rural Development (FAO-CDP, 2007).

Local government might appear to provide opportunities for greater involvement of women in political life, yet proportional representation is nowhere the rule. In many countries, patriarchal social systems, cultural prejudices, financial dependence and lack of exposure to political processes have made it difficult for women to participate in public life. The maleness of political institutions and the high cost of campaigning prevent many women from en-

tering electoral politics. When they do so, however, many see themselves as role models whose political actions should have a positive impact on people's lives. A survey of women in local government in 13 Asian and Pacific countries found that women also brought a more transformational political agenda to the fore, one more attuned to social concerns, such as employment, care of the elderly, poverty alleviation, education, health care and sanitation—all subjects of critical importance to rural people. Women in politics understood the positive impact that female decision makers had on women's participation generally (UNESCAP, 2001).

Gender

Gender is a key category for understanding agrarian societies, as anthropological and historical research has consistently shown (Boserup, 1965; Linares, 1985; McC Netting, 1993). The category refers not, as is often assumed, to the role of women as such, but to the specific social ascription of roles and functions according to gender. In agrarian societies, these roles and responsibilities have been, in most cases, clearly and specifically assigned to either men or women in productive households. In addition, not only work, but also assets are as a rule accessed and controlled according to gender-based patterns. These patterns vary with time and place; a persistent feature is that women have a key role in agricultural work, yet they have often limited access to, or control over, the resource base such as land.

Hence, the management of resources in agriculture is related to gender. What does this imply for sustainability? It certainly means that research needs to closely look at existing gender-related patterns of resource access and control, to arrive at meaningful conclusions (Linares, 1985). While sustainability has to be a target of farm operations, there may be differential factors at work here.

Agricultural development has sometimes strengthened patterns that do not favor women. Two factors are considered in this context. First, the double male bias of agricultural extension systems: it is mainly men who represent the state and its agencies, so men control information and communications; and it is men who are considered to represent the community or farming household, so they are the ones addressed. Second, as agricultural industrialization often implies a need for investments, market integration—handling larger sums of money—has favored men in many contexts, as women are usually not considered eligible for credit.

With growing awareness of this imbalance, the international agricultural research community has developed research to address the issues of women and discriminating gender roles in agriculture. This has often implied establishing a participatory research agenda (Lilja et al., 2000), such as in the CGIAR Systemwide Program on Participatory Research and Gender Analysis (CGIAR, 2005). While this is a welcome trend towards research products that have been developed with a greater involvement of women, it is not a sufficient condition to change a social fabric that discriminates against women.

Gender and other identity issues in natural resource management

The status and development potential of an individual depend on many social factors. In particular, they depend on a person's assigned gender, defined as the economic, social, political and cultural attributes and opportunities associated with being male and female (OECD, 1998). Other aspects of social identity such as caste, ethnicity, age and religion are just as influential with regard to an individual's status and development potential, and therefore need to be taken into account in much the same way as outlined below in the case of gender.

As a result of the gender division of labor, women and men relate to different economic spheres. In addition, they do not have the same stake in natural resources, social institutions and decision-making processes in the household and society. Nor do women and men have the same power to act and make decisions. Women and men are therefore affected differently by development. The dichotomy between men's and women's spheres is, on the one hand, a social challenge, but on the other hand it is an opportunity to make resource management truly stakeholder-oriented. Hence, for the assessment it is necessary to differentiate between male and female spheres by integrating disaggregated data.

In many instances and for a number of reasons women's access to natural resources is limited and their power to make decisions regarding natural resource management is socially restricted (Worldwatch Institute, 2003). Yet the majority of women in developing countries live and work in close association with natural resources (UNDP, 2005) and are particularly affected by ecosystem changes (MA, 2005a). Therefore, demands for a gender focus in natural resource management range from "experimentation with institutional forms that are more hospitable to women and marginalized groups" (Colfer, 2005), to demands calling for increased emphasis on the needs of women when addressing aspects of natural resource sustainability (Müller, 2006) and calls for a strategy for making women's as well as men's concerns and experiences an integral dimension of the policies and programs in all political, economic and societal spheres so that women and men benefit equally, and inequality is not perpetuated (UN, 1997).

Much has been written in recent years regarding the feminization of agriculture. As men have migrated to urban areas to seek better livelihoods, small-scale farming has been gradually feminized, with a larger percentage of women acting as head of household in rural areas, although their percentage in relation to all economically active women has been dropping since 1980 worldwide, in developing countries as well as in low-income food-deficit countries (FAO, 2001b; Figure 1-20). Feminization does not represent an equalization of opportunities, but rather a further marginalization of small-scale farms, since many female heads of household are younger and less educated than male heads of household, have less land, less capital and less access to credit. Fewer than 10% of women farmers in India, Nepal and Thailand own land and credit schemes in five African countries award women less than 10% of the credit awarded to male small-scale farmers (FAO-Gender, 2007) In most countries, the proportion of female-headed households is far less than 50% of the total.

A lack of sex-disaggregated data means that women's roles in agriculture and their specific needs are still poorly understood. It is noteworthy that about one-fifth of farms are headed by women. It is clear, however, that rural women are not a homogeneous group. Gender roles and the gender division of labor are highly specific to location, farming systems and peoples, but they are not fixed. Men and women constantly renegotiate their roles and relationships as circumstances change, both within the household and in the wider community. Their relative bargaining power can be influenced by many factors, their economic importance within the household, kinship relations, cultural norms of behavior, not to mention their individual character. Women as well as men have the capacity to exercise agency, that is, to make choices and decisions that can alter outcomes in their lives. In many countries, however, institutions of governance, legal systems and social policies have not equalized opportunities between men and women or created greater social equity between urban and rural dwellers, but have reinforced disparities instead.

A growing body of evidence suggests that economic efficiency gains can be realized through more widespread enjoyment of rights and more just distribution of opportunity. Conversely, persistent inequality is increasingly seen to limit the rate and quality of economic growth, threaten national unity and fuel social conflict (WDR, 2007). The challenge facing policymakers and practitioners is to mediate the modernization of agriculture in such a way that it leads to improved social and economic outcomes for those working in the sector, while supporting the transition to more value-adding activities for others. Investing in people will be the key to achieving these goals.

1.4 Sustainability Indicators

1.4.1 Indicators for the IAASTD
Indicators are part of what we observe in the world around us as we attempt to detect patterns and extract information meaningful for directing action. Indicators are quantitative and qualitative variables that provide a simple and reliable means to track achievement, reflect changes connected to an intervention or trend, or help assess the performance of an organization, an economic sector, or a policy measure with respect to set targets and goals.

In science, state variables of high precision and generality tend to be favored as indicators. In everyday life, there is a strong preference for trend indicators. An indicator, however, does not exist independently of the observer. Once an indicator is established, there still remain multiple issues of interpretation and meaning. Experts use indicators to inform policy and to increase their own scientific understanding (Table 1-4).

On a methodological level, an assessment is not simply a review of relevant literature; it can be based, in part, on a literature review, but also needs to provide an assessment of the veracity and applicability of the information and the uncertainty of outcomes within the context of the identified questions or issues within a specified environment. To be effective and legitimate, an assessment process should be open, transparent, reviewed, and include a broad representation of stakeholders and relevant experts.

Additional methodological elements include the selection of units of analysis, integrating biophysical and human systems as the context of agricultural practice, temporal and spatial scales of assessments from regional to global, issues

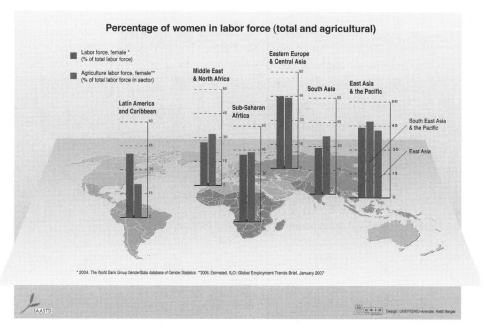

Figure 1-20. *Percentage of women in labor force (total and agricultural).* Source: World Bank, 2004b; ILO, 2007.

of values and valuation, dealing with uncertainty, dealing with different knowledge systems, as well as modeling issues and developing scenarios.

1.4.2 Working with indicators

What are indicators for? Indicators are used both for specialist purposes and in everyday life. In specialist applications the purposes are defined within the domain of expertise. In everyday life, they form part of the repertoire of heuristics—simple rules for making decisions when time is pressing, information limited or partial, and deep reflection a luxury (Gigerenzer et al., 1999). Indicators become part of what we observe in the world around us as we attempt to detect patterns and extract information relevant to effective action. In this everyday sense, they can be accurate and powerful (Gigerenzer et al., 1999) but also, if wrongly observed or interpreted, contribute to systemic failures (Dörner, 1996).

Referents and contexts. All indicators require a referent measurement situation. To allow meaningful interpretation of indicators and utilization that will appropriately inform policy processes, there is also a need for awareness of the context of use. Strictly speaking, indicators require application in a controlled environment (with/without, before/after). Rarely, however, is such a design possible in reality, for obvious practical and ethical reasons. Thus the present assessment has to accept that information is not perfect. One approach to handle uncertainty is through scenarios that are built on available indicators and assumptions.

State variables and trend indicators. The IAASTD uses two kinds of indicators, describing either state or trends. State variables, of high precision and generality, tend to be favored in science, as they represent the current state of an object or process and are thus measurable. In everyday life,

there is a strong preference for accurate trend indicators. Especially at policy level, information is required on whether situations are improving or worsening, and whether policy objectives are getting closer to their goals or farther away. Trend indicators tend to focus more on identifying thresholds that might indicate an imminent change of state, and less on constant values—the more favored emphasis of many sciences. In many usages trend indicators are also used as learning devices, leading to reestimation of achievement and redefinition of goals as trend data move through time.

Precision, accuracy, and generality. There is agreement in the philosophy of logic and statistics that precision, accuracy, and generality cannot be simultaneously optimized. Any pair of the three may be. The construction and choice of indicator thus has to take into consideration which combination is the most pertinent to the problem or situation for which the indicator might be used. There is a need to identify appropriate indicators and the relationships of these when used at various spatial and temporal hierarchical levels. This is partly a matter of scale and structure of systems hierarchies, and partly a matter of whether it is the state variables or dynamics that the user considers important to observe and monitor.

The dilemmas of interpretation and meaning

An indicator does not exist independently of the observer: as mentioned above, a range of pre-analytic choices are made before an indicator is constructed or brought into use. These choices are inevitably value-laden, and enriched with meaning that the indicator itself does not possess. Take, for example, poverty indicators: one can construct income-based, nutrition-based, gender-based (etc.) indicators. Each type of indicator both reveals what is important for the user's purpose but also conceals what is not considered pre-analytically to be of importance.

Table 1-4. Overview of issues addressed by indicators in the IAASTD framework.

IAASTD framework components	Issues addressed by indicators
Development and sustainability goals	• Decreased hunger and poverty • Improved nutrition and human health • Sustainable economic development • Enhanced livelihoods and equity • Environmental sustainability
AKST systems	• Research/Innovation policies • Local and institutional setting of AKST • Social organization • Generation, dissemination, access to, adoption and use of AKST • Agricultural markets
Agricultural outputs and services	• Biomass, livestock, fish, crop production • Forestry for food • Fiber • Carbon sequestration • Energy • Ecosystem services
Indirect drivers	• Economic • Demographic • Sociopolitical
Direct drivers	• Economic • Demographic • Availability and management of natural resources

Once an indicator is established, multiple issues of interpretation and meaning remain to be solved. Is an increasing mechanization in agriculture that contributes to increased area productivity on the one hand, yet increases externalities of various kinds on the other, an indicator of agricultural modernization or an indicator of the increasing lack of sustainability of that particular food system? Available indicators for agricultural mechanization in most cases provide inadequate information. Only if indicators are placed in a context of meaning determined by prior adoption of frameworks that incorporate value systems and perceptions, can indicators be used for decision making. Unfortunately, frameworks are rarely articulated explicitly, thereby greatly decreasing the utility of indicators.

The conceptual framework of IAASTD does indeed provide tools to interpret indicators for agricultural mechanization, for example. While on the one hand, an increase in mechanization could contribute to food production in the component "Development and Sustainability Goals" and "Food System and Agricultural Products and Services", on the other hand, such an increase generates a number of negative externalities in the component "Direct / Indirect Drivers". The four components of the IAASTD conceptual framework, in turn, influence rules, norms and processes where actors are involved. This, i.e., the outer ring of the

AKST component in the conceptual framework, is exactly the level at which the implications of a given indicator need to be negotiated, agreed upon and fed into the policy process.

Similarly, an indicator on female employment in agriculture needs to be interpreted in terms of the components of the conceptual framework. An increased employment rate could have a positive impact on family nutrition, but might be negatively interpreted in terms of an increased workload for women. Therefore, an interpretation of the meaning of an indicator as suggested by the outer ring of the conceptual framework needs to take place in order to equip the indicator with context and meaning.

Expert-based versus participatory indicator construction and use. Experts use indicators all the time to inform policy and to increase their scientific understanding. These are legitimate and powerful usages. Problems arise, however, when assumptions are made about indicators as information tools, and as motivators of the actions of others, because indicators rapidly lose their originally intended meaning when they are moved to other domains. A further implication of the IAASTD conceptual framework is that indicators are powerful in developing our understanding and in motivating reflection and action when they are constructed with, rather than extended to, other actors.

1.4.3 Indicators in the IAASTD

The scope of the AKST assessment includes the relevance of agricultural systems and encompasses major aspects of human well-being and environmental sustainability. This extended view of agricultural development is in line with the major international initiatives addressing sustainable development, such as the MDGs and the Millennium Ecosystem Assessment (MA). The assessment thus suggests indicators that assist in observing critical changes in the area of human development, the environment, agriculture, and AKST. The particular challenge for indicators is that they must be able to link AKST with these three areas of sustainable development in a meaningful way.

This broad, sustainable development-oriented view of the process of agricultural development has also been adopted by major international actors in development for the past two decades, e.g., the Agenda 21 of the UN Conference on Environment and Development (UNCED) in 1992 and the World Summit on Sustainable Development (WSSD) in 2002. The indication of effects of agricultural development on the broader aspects of human development and the environment poses major challenges to the identification of impact and process indicators.

Identification of indicators for the AKST assessment

This global assessment occasionally uses some key indicators to show how different global and sub-global trends and drivers—including effectiveness of investments in AKST systems—affect the main agricultural outcomes and services, and more importantly, how they impact on the global population and their well-being, and on the ecological systems used and/or affected. A global assessment like IAASTD gains in efficiency and effectiveness if it focuses on a limited number of representative indicators. Indicators are quantitative and qualitative variables that provide a simple and reliable means to track achievement, reflect changes connected to an intervention or trend, or help assess the performance of an organization, an economic sector, or a policy measure against set targets and goals. Tracking changes over time relative to a reference point ("baseline") using indicators, can provide useful feedback and help improve data availability and thus support decision-making at all levels.

For the purpose of the assessment, two main types of indicators have been considered:

Impact indicators show impacts of AKST on society and the environment in terms of poverty, livelihoods, equity, or hunger. These impacts are influenced by various technical, environmental and socioeconomic drivers and pressures, e.g., immediate outcomes of AKST investments. The targets and goals used in this assessment are closely linked to the internationally agreed MDGs.

Process/performance indicators show the influence of key drivers on AKST, on AKST and main agricultural outputs/services, and on AKST and human well-being as defined in the MDGs.

Because of their considerable policy relevance and practical use, the selection and presentation of the indicators is of critical importance in the assessment. However, most of the underlying data that is needed to derive the desired indica-

tors is either organized along individual sectors (agriculture, health, and environment), or highly aggregated into indexes like the Human Development Index (HDI) or the Gender Empowerment Measure (GEM). Therefore, the challenge is to identify indicators which clearly describe the relationship between agricultural science and technology and sustainable development in the various aspects described above.

Indicator characteristics. As indicators are used for various purposes, it is necessary to define general criteria for selecting indicators and validating their choice. Indicators (Hardi and Zdan, 1997; Prescott-Allen, 2001) can be characterized by their:

Relevance to measure change: for an indicator to be relevant, it must cover the most important aspects of the topic "human capacity for AKST". It must also be a sign of the degree to which an objective is met.

Reliability from well-established data sources: an indicator is likely to be reliable if it is well founded, accurate, and measured in a standardized way using an established or peer-reviewed method, and sound and consistent sampling procedures.

Feasibility: an indicator is feasible if it depends on data that are readily available or obtainable at reasonable cost.

To be consistent, an indicator must illustrate trends over time, as well as differences between places and groups of people. The usefulness of indicators depends on how well they meet the above criteria. When no direct indicators can be found that adequately meet these criteria, then indirect indicators or "proxies" and/or a combination of indicators or aggregate indices can be used. The selection of variables and indicators, together with underlying methodologies and data sets, must also be clearly documented and referenced. The more rigorous and systematic the choice of indicators and indices, the more transparent and consistent an assessment will be. And the more involved decision makers and other stakeholders are in the selection process, the higher the chance of acceptance of assessment results.

However, three potential problems need to be noted here:

1. Not all potential indicators are practical: data may not be available; and data may be either too difficult or too expensive to collect. For this reason, more distant (proxy) indicators need to be selected. These may not be the most appropriate and reliable indicators, but they can be interpreted to reflect the issue being monitored. For example, if one is comparing innovation levels in different countries, the proxy indicator of the number of patents issued per million people per year may be used to save time and resources, making use of existing reliable data sources in order to give an approximate idea of different innovation levels in different countries.

2. Experience with indicator identification for this assessment shows that one cannot expect to find clear and concise indicators for many of the critical IAASTD areas such as (1) AKST and sustainable development in general, exemplified through the MDGs; (2) AKST and human health; (3) AKST and social equity, etc. Therefore, indicators selected for this assessment will often

need to compromise between being "exactly wrong or approximately right".

3. The time and technical skills required for selecting indicators might make it difficult for decision makers and stakeholders to participate fully in the selection of indicators. At the same time, experts carrying out the assessment have the responsibility of ensuring that the selection of indicators and the assessment as a whole are technically and scientifically sound.

Hence, in the area of indicators, a way must be found to maximize both the technical excellence of the assessment and the commitment of participants from government, civil society, and business.

The focus of this assessment on poverty, sustainable livelihoods and sustainable ecosystems marks a clear trend that future agricultural development is moving away from the exclusive production focus of the past. However, indicators available today can support assessment of these broadened goals of agricultural development only partially: more efforts are needed to develop sufficiently appropriate indicators.

Units of analysis and reporting. The IAASTD uses indicators which measure at several scales, from individual to farm, nation, region and global levels. Numeric indicators use metric units while qualitative indicators are descriptive. Information from smaller units will be aggregated up to sub-global and global assessment levels. The results will thus be generic but presented in such a way that it makes sense to other units of analysis.

Dealing with systems

The IAASTD basically deals with two different sets of systems, a biophysical and a socioeconomic set. On the one hand, there is the biophysical set with the underlying ecosystem in which the agricultural system and the unit-based production system is established. Primary ecosystems have been altered to a greater or lesser extent by agricultural production systems that define themselves according to economic criteria of efficiency as opposed to the multifunctional character of ecosystems. Usually, forest ecosystems are converted into grassland for livestock rearing, or a system with bare soils for cultivation. Depending on the capacity and suitability of this new agricultural land, production takes place over shorter or longer periods of time, from a single or a few years to decades and even centuries on the most suitable land. Assessing the future of these production systems requires taking into account their current suitability, including the degradation of ecosystems or parts thereof which has taken place, and the potential of these land areas to support agricultural production of goods. In addition, the multifunctional character of ecosystems has to be considered as a crucial aspect important to societies and the global community.

On the other hand, political, economic, social and cultural sets of systems shape human livelihoods and agricultural production systems in the different contexts in which the latter operate. A large disparity exists between these contexts. A majority of agricultural workers are poor small-scale farmers in developing countries, with a high degree of dependence on subsistence systems, i.e., production by households for their own consumption, and a high degree of dependence on both the biophysical and socioeconomic systems. A minority of agricultural workers live on larger production units and in industrialized nations, profiting from wealthy economies and a variety of subsidies to maintain their production and/or production systems. Assessing the future of agricultural systems will require thorough analysis and evaluation of these different contexts and the livelihoods derived from them through agricultural activities.

Many of these contexts and systems are evolutionary; shifts in parameters must be expected, and the state of natural and human environments will continuously change, be it through factors such as opportunity (e.g., new business options or access to new resources) or constraints (such as further decapitalization of small-scale farmers). The degrees of uncertainty are rather great and difficult to foresee.

Dealing with scales (spatial and temporal)

Assessments need to be conducted at spatial and temporal scales appropriate to the process or phenomenon being examined. Analysis of issues must take place across several spatial scales simultaneously because an analysis at a single scale will miss important interactions. For example, national policies embedded in a global system have an impact on local decisions regarding AKST. Moreover, vulnerabilities are related to various scales. A comparison of a larger scale poultry production system with a decentralized backyard poultry system reveals different scales. While an infection of the former system is relatively easy to prevent, a possible outbreak would be catastrophic. In the latter system an infection of the flock is harder to prevent while an outbreak would affect a smaller number of poultry. Most of the analysis in the IAASTD is carried out at national and regional levels, but informed by experience from ground realities.

The IAASTD is structured as a multiscale assessment in order to enable its findings to be of greater use at the many levels of decision-making. A global assessment cannot meet the needs of local farmers, nor can a local assessment meet the collective needs of parties to a global convention. A multiscale assessment can also help remedy the biases that are inevitably introduced when an evaluation is done at a single geographic scale. For example, while a national AKST assessment might identify substantial national benefits from a particular policy change, a local assessment would be more likely to identify whether that particular community might be a winner or loser as a result of the policy change. For example, in contrast to privately funded research, where the donor derives benefits, benefits derived from public goods research does not go to the funding agency itself, rather to other members of society, and there is no direct incentive to do more (CGIAR Science Council, 2005).

Dealing with values and valuation

The IAASTD deals with two valuation paradigms at the same time. The utilitarian paradigm is based on the principle of human preference for satisfaction (welfare). AKST systems provide value to human societies because people derive utility from their use, either directly or indirectly. Within this utilitarian concept of value, people also give value to AKST aspects that they are not currently using (non-use values), for example people value education systems even though they themselves have completed their school education. Non-use values often rely on deeply held historical, national, ethical, religious, and spiritual values. A different, non-utilitarian value paradigm holds that something can have intrinsic value; that is, it can be of value in and for itself, irrespective of its utility for someone else. For example, birds are valuable, regardless of what people think about them. The utilitarian and non-utilitarian value paradigms overlap and interact in many ways, but they use different metrics, with no common denominator, and cannot usually be aggregated, although both value paradigms are used in decision-making processes.

How decisions are made will depend on the value systems endorsed in each society, the conceptual tools and methods at their disposal, and the information available. Making the appropriate choices requires, among other things, reliable information on current conditions and trends of ecosystems and on the economic, political, social, and cultural consequences of alternative courses of action. Assessments strive to be value free, using evidence-driven results. But in fact, all people involved in assessments come with value systems and need to explicitly state these values wherever they are at work. Another way to take advantage of different ways of thinking is to create diversity in the assessment in terms of background, region, gender, and experience in order to balance views.

Dealing with different knowledge systems

The IAASTD aims to incorporate both formal scientific information and traditional or local knowledge. Traditional societies have nurtured and refined systems of knowledge of direct value to those societies and their production systems, but also of considerable value to assessments undertaken at regional and global scales. To be credible and useful to decision makers, all sources of information, whether from scientific, local, or practitioner knowledge, must be critically assessed and validated as part of the assessment process through procedures relevant to the specific form of knowledge.

Substantial knowledge concerning both AKST and policy interventions is held within the private (and public) sector by "practitioners" of AKST, yet only a small proportion of this information is ever published in scientific literature, and much is kept in less accessible gray literature. Again, broad participation can help include as many sources of knowledge as possible.

Effective incorporation of different types of knowledge in an assessment can both improve the findings and help to increase their adoption by stakeholders if the latter believe that their information has contributed to those findings. At the same time, no matter what sources of knowledge are incorporated in an assessment, effective mechanisms must be established to judge whether the information provides a sound basis for decisions.

Modeling issues

Models are used in the IAASTD to analyze interactions between processes, fill data gaps, identify regions for data collection priority, or synthesize existing observations into appropriate indicators of ecosystem services. Models also provide the foundations for elaborating scenarios. As a result, models will play a synthesizing and integrative role in the IAASTD, complementing data collection and analytical efforts.

It is relevant to note that all models have built-in uncertainties linked to inaccurate or missing input data, weaknesses in driving forces, uncertain parameter values, simplified model structure, and other intrinsic model properties. One way of dealing with this uncertainty in the IAASTD is to encourage the use of alternative models for computing the same ecosystem services and then compare the results of these models. Having at least two independent sets of calculations can add confidence to the robustness of model calculations, although it will not eliminate uncertainty.

It should be stressed that the majority of "human system models" focus on economic efficiency and the economically optimal use of natural resources. Thus the broader issues of human well-being, including such factors as freedom of choice, security, equity and health, will require a generation of new models. To deal with these issues IAASTD must rely on qualitative analysis.

References

Abler, D. 2004. Multifunctionality, agricultural policy, and environmental policy. Agric. Resour. Econ. Rev. 33(1):8-17.

Ali, M., and D. Byerlee. 2002. Productivity growth and resource degradation in Pakistan's Punjab: A decomposition analysis. Econ. Dev. Cult. Change 50(4):839-863.

Anh, M.T.P., M. Ali, H.L. Ahn, and T.T.T. Hua. 2004. Urban and peri-urban agriculture in Hanoi: Resources and opportunities for food production. Tech. Bull. 26. AVRDC, Tainan.

Anriquez, G., and G. Bonomi. 2007. Long-term farming trends: An inquiry using agricultural censuses: ESA Working Pap. No. 07-20. Agric. Dev. Econ. Div., FAO, Rome.

Antle, J.M., D.C. Cole, and C.C. Crissman. 1998. Further evidence on pesticides, productivity and farmer health: Potato production in Ecuador. Agric. Econ. 2(18):199-208.

Basu, K. 1986. The market for land: An analysis for interim transactions. J. Dev. Econ. 20(1):163-177.

Bathia, A., H. Pathak, and P.K. Aggarwal. 2004. Inventory of methane and nitrous oxide emissions from agricultural soils of India and their global warming potential. Curr. Sci. 87(3):317-324.

Baumert, K.A., T. Herzog, and J. Pershing. 2005. Navigating the numbers: Greenhouse gas data and international climate policy. World Resour. Inst., Washington DC.

Berndes, G., M. Hoogwijk, and R. van den Broek. 2003. The contribution of biomass in the future global energy supply: A review of 17 studies. Biomass Bioenergy 25:1-28.

BFS. 2006. Statistical data on Farming and Forestry. (In German or French). Available at http://www.bfs.admin.ch/bfs/portal/de/index/themen/07.html. Swiss Fed. Stat. Off., Neuchâtel.

Black, J. 2003. A dictionary of economics. Available at http://www.oxfordreference.com/views/ENTRY.html?entry=t19.e2083&srn=2&ssid=205945865#FIRSTHIT. Oxford Ref. Online, Oxford Univ. Press, UK.

Blaikie P., K. Brown, M. Stocking, L. Tang, P. Dixon, and P. Sillitoe. 1997. Knowledge in action: Local knowledge as a development resource and barriers to its incorporation in natural resource research and development. Agric. Syst. 55:217-237.

Blaikie, P., and H. Brookfield. 1987. Land degradation and society. Methuen, London.

Bonfiglioli, A. 2004. Land of the poor. Local environmental governance and the decentralized management of natural resources. UN Capital Dev. Fund, NY.

Boserup, E. 1965. The conditions of agricultural growth: The economics of agrarian change under population pressure. George Allen and Unwin, London.

Bravo-Ortega, C., and D. Ledermann. 2005. Agriculture and national welfare around the world: Causality and international heterogeneity since 1960. World Bank, Washington DC.

Brookes, G., and P. Barfoot. 2006. GM crops: The first ten years: global, socio-economic and environmental impacts. Brief No. 36. ISAAA, Ithaca.

Brookfield H., H. Parsons, and M. Brookfield (ed) 2003. Agrodiversity: Learning from farmers across the world. UN Univ. Press, Tokyo.

Burke, L., Y. Kura, K. Kassem, M. Spalding, C. Revenga, and D. McAllister. 2001. Pilot analysis of global ecosystems: Coastal ecosystems. World Resour. Inst., Washington DC.

Byerlee, D., and G. Alex. 2003. Designing investments in agricultural research for enhanced poverty impacts. ARD Working Pap. No. 6. World Bank, Washington DC.

Byerlee, D., X. Diao, and Ch. Jackson. 2005. Agriculture, rural development, and pro-poor growth: Country experiences in the post-reform era. ARD Disc. Pap. 21. World Bank, Washington DC.

CA. 2007. Water for food, water for life: A comprehensive assessment of water management in agriculture. Earthscan and IWMI, London and Colombo.

Cassman, K.G., A. Dobermann, D.T. Walters, and H. Yang. 2003. Meeting cereal demand while protecting natural resources and improving environmental quality. Ann. Rev. Environ. Resour. 28:315-358.

Caulfield, L.E., M. de Onis, M. Blössner, and R.E. Black. 2004. Undernutrition as an underlying cause of child deaths associated with diarrhea, pneumonia, malaria and measles. Am. J. Clin. Nutr. 80:193-8.

CBD. 2006. Global Biodiversity Outlook 2. Convention on Biological Diversity, Montreal.

CDE. 2002. NRM - Natural resource management. Available at http://www.cde.unibe.ch/Themes/NRM_Th.asp. Centre for Dev. Environ., Berne.

CGIAR. 2005. Systemwide program on participatory research and gender analysis. Available at http://www.prgaprogram.org/. CGIAR, Rome.

CGIAR Science Council. 2005. Science for agricultural development: Changing contexts, new opportunities. Science Council Secretariat, Rome.

Chambers, R., and G. Conway. 1991. Sustainable rural livelihoods: Practical concepts for the 21st century. IDS Disc. Pap. 296. Inst. Dev. Studies, Brighton.

CIFOR. 2006. Earning a living from the forest. Available at http://www.cifor.cgiar.org/Publications/Corporate/FactSheet/livelihood.htm. CIFOR, Bogor Barat.

COAG. 2005. The globalizing livestock sector: Impact of changing markets. Comm. Agric., FAO, Rome.

Codex Alementarius Commission. 2001. Organically produced foods. FAO and WHO, Rome.

Cole, D.C., S. Sherwood, C. Crissman, V. Barrera, and P. Espinosa. 2002. Pesticides and health in highland Ecuadorian potato production: Assessing impacts and developing responses. Int J. Occup Med. Environ. Health 8(3):182-190.

Colfer, C.J.P. 2005. The complex forest: Communities, uncertainties, and adaptive collaborative management. RFF Press, Washington DC.

Commission for Africa. 2005. Our common interest. Rep. Commission for Africa. DFID, London.

Commoner, B. 1990. Making peace with the planet. Pantheon, NY.

Conway, G. 1999. The doubly green revolution: Food for all in the 21st century. Cornell Univ. Press, Ithaca.

Costanza, R., R. d'Arge, R. de Groot, S. Farber, M. Grasso, B. Hannon et al. 1997. The value of the world's ecosystem services and natural capital. Nature 387:253-260.

Delgado, C.L., N. Nikolas, M.W. Rosegrant, S. Meijer, and A. Mahfuzuddin. 2003. Fish to 2020: Supply and demand in changing global markets. IFPRI and WorldFish Center, Washington DC and Penang.

DeRose, L., E. Messer, and S. Millman. 1998. Who's hungry and how do we know? Food shortage, poverty and deprivation. UN Univ. Press, Tokyo.

DFID. 2001. Sustainable livelihoods guidance sheets. Sec. 8, Glossary. DFID, London.

Dixon, J., A. Gulliver, and D. Gibbon. 2001. Farming systems and poverty: Improving farmers' livelihoods in a changing world. FAO and World Bank, Rome and Washington DC.

Dolberg, F. 2001. A livestock development approach that contributes to poverty alleviation and widespread improvement of nutrition among the poor. Livestock Res. Rural Dev. 13(5).

Dörner, D. 1996. The logic of failure. Recognising and avoiding error in complex situations. Metropolitan Books, NY.

Dove, M. R., and D.M. Kammen. 1997. The epistemology of sustainable resource use: Managing forest products, swiddens, and high-yielding variety crops. Human Organ. 56:91-101.

Eastwood, R., M. Lipton, and A. Newell. 2004. Farm size. Available at http://www.sussex.ac.uk/Units/PRU/farm_size.pdf. Univ. Sussex, UK.

Edmeades, G.O., J. Bolanos, S.C. Chapman, H.R. Lafitte, and M. Bänziger. 1999.

Selection improves drought tolerance in tropical maize populations. Crop Sci. 39(5):1306-1315.

Engel, P.G.H. 1997. The social organization of innovation: A focus on stakeholder interaction. Roy. Trop. Inst., KIT Press, Amsterdam.

Ericksen, P.J. 2006. Conceptualizing food systems for global environmental change (GEC) research. GECAFS Working Pap. 2. Global Environ. Change Food Syst., Environ. Change Inst., Oxford.

ETC-Netherlands. 2003. Annotated Bibliography on Urban Agriculture. ETC-Netherlands Urban Agric. Prog., Leusden.

Evenson, R., and D. Gollin. 2003. Assessing the impact of the green revolution, 1960-2000. Science 300(5620):758-762.

Ezzati, M., A. Rodgers, A.D. Lopez, S.V. Hoorn, and C.J.L. Murray. 2004. Mortality and burden of disease attributable to individual risk factors. In Comparative quantification of health risks: Global and regional burden of disease attributable to selected major risk factors, Vol. 2, Chap. 26. WHO, Geneva.

Falkenmark, M. 1995. Coping with water scarcity under rapid population growth. SADC Ministers Conf., Pretoria, 23-24 Nov 1995.

FAO. 1996a. Control of water pollution from agriculture: FAO Irrig. Drainage Pap. 55. FAO, Rome.

FAO. 1996b. Global plan of action for the conservation and sustainable utilization of plant genetic resources for food and agriculture. Available at http://www.fao.org/ag/AGP/AGPS/GpaEN/gpatoc.htm. FAO, Rome.

FAO. 2000. The state of food and agriculture: Lessons from the past 50 years. FAO, Rome.

FAO. 2001a. Genetically modified organisms, consumers, food safety and the environment. FAO Ethics Series. FAO, Rome.

FAO. 2001b. Women, agriculture and food security. FAO, Rome.

FAO. 2001c. Supplement to the report on the 1990 World Census of Agriculture. FAO, Rome.

FAO. 2003. Report of the panel of eminent experts on ethics in food and agriculture, Second Session, 18-20 March 2002. FAO, Rome.

FAO. 2004a. The state of food and agriculture 2003: Agricultural biotechnology: Meeting the needs of the poor? FAO, Rome.

FAO. 2004b. The state of world fisheries and aquaculture. FAO, Rome.

FAO. 2004c. World census on agriculture, 1980, 1990, 2000 rounds. Available at http://www.fao.org/es/ess/census/wcares/default.asp. FAO Statistics Division, Rome.

FAO. 2005a. The state of food and agriculture. Agricultural trade and poverty: Can trade work for the poor? FAO, Rome.

FAO. 2005b. Global forest resource assessment 2005: Progress towards sustainable forest management. FAO, Rome.

FAO. 2006a. The state of food and agriculture 2006: Food aid for food security? FAO. Rome.

FAO. 2006b. The State of world fisheries and aquaculture. FAO, Rome.

FAO. 2006c. State of world aquaculture 2006: Fish. Tech. Pap. 500. FAO, Rome.

FAO. 2006d. Global forest resources assessment 2005: Progress towards sustainable forest management: FAO For. Pap. 147. FAO, Rome.

FAO. 2007a. State of the world's forests 2007. FAO, Rome.

FAO. 2007b. Yearbook of fishery statistics : Summary tables. Available at ftp://ftp.fao.org/fi/stat/summary/default.htm. FAO, Rome.

FAO-CDP. 2007. FAO Corporate document repository: Fact sheet Egypt and Fact sheet Benin. Available at www.fao.org/docrep/V9648e/v9648e01.htm and www.fao.org/docrep/V7948e/v7948e.02.htm. FAO, Rome.

FAO-Gender. 2007. FAO Gender and food security in agriculture. Available at www.fao.org/GENDER7en/agrib4-e.htm. FAO, Rome.

FAOSTAT. 2006. FAO statistical databases of the FAO. Available at http://faostat.fao.org/. FAO, Rome.

Farrington, J., I. Christoplos, A.D. Kidd, and M. Beckman. 2002. Extension, poverty and vulnerability: The scope for policy reform: Final report of a study for the Neuchâtel Initiative. Overseas Dev. Inst., London.

GACGC. 1994. World in transition: The threat to soils: Annual report. German Advisory Council Glob. Change, Bonn.

García, Z. 2005. Impact of agricultural trade on gender equity and rural women's position in developing countries. FAO, Rome.

Gaventa, J. 1998. The scaling up and institutionalization of PRA: Lessons and challenges. p. 153-166. In P. Blackburn, and J. Holland (ed) Who changes? Institutionalizing participation in development. Intermediate Tech. Publ., London.

Gereffi, G., J. Humphrey, R. Kaplisky, and T.J. Sturgeon. 2001. Introduction: Globalisation, value chains and development: IDS Bulletin 32(3). Inst. Dev. Studies, Univ. Sussex, UK.

Gibbons, M., C. Limoges, H. Nowotny, S. Schwartzman, P. Scott, and M. Trow. 1994. The new production of knowledge. Sage Publ., London.

Gigerenzer, G., P.M. Todd, and the ABC Research Group. 1999. Simple heuristics that make us smart. Oxford Univ. Press, Oxford.

Gitay, H. 2005. Assessment process for Comprehensive Assessment of Water Management in Agriculture: Meeting of the working group on water pricing in agriculture. Montpellier, 3-4 June. Available at http://www.iwmi.cgiar.org/assessment/files/montpellier/assessment/process/203.1.ppt. CA Secretariat, Colombo.

Global Monitoring Report. 2006. Millennium development goals: Strengthening mutual accountability, aid, trade, and governance. World Bank, Washington DC.

Glowka, L., F. Burhenne-Guilmin, H. Synge (with J.A. McNeely and L. Gundling). 1994. A guide to the Convention on Biological Diversity: Environmental Policy Law Pap. No. 30. The World Conserv. Union, Gland.

Gong, Y., A. Hounsa, S. Egal, P.C. Turner, A.E. Sutcliffe, A.J. Hall et al. 2004. Postweaning exposure to aflatoxin results in impaired child growth: A longitudinal study in Benin, West Africa. Environ. Health Perspect. 112:1334-38.

Graham, T., E. van de Fliert, and D. Campilan. 2001. What happened to participatory research at the International Potato Center? Agric. Human Values 18:429-446.

Grainger, R.J.R., and S.M. Garcia. 1996. Chronicles of marine fishery landings (1950-94): Trend analysis and fisheries potential. FAO Fish. Tech. Pap. No. 359. FAO, Rome.

GTZ. 2006. Policies against hunger v. food security and poultry production: How to cope with Avian Influenza. Int. Workshop. 19-20 October 2006, Berlin. GTZ, Berlin.

Haddad, L.J. 2000. A conceptual framework for assessing agriculture-nutrition linkages. Food Nutr. Bull. 21(4):367-373.

Haggblade, S., P. Hazell, I. Kirsten, and R. Mkandawire. 2004. African agriculture: Past performance, future imperatives. In S. Haggblade (ed) Building on successes in African agriculture. 2020 Focus No. 12. IFPRI, Washington DC.

Hardi, P., and T. Zdan (ed) 1997. Assessing sustainable development: Principles in practice. Int. Inst. Sustain. Dev., Winnipeg.

Hazell, P. 1999. Agricultural growth, poverty alleviation, and environmental sustainability: Having it all. 2020 Brief No. 59, March 1999. IFPRI, Washington DC.

HDR. 2006. Human development report. Beyond scarcity: Power, poverty and the global water crisis. UNDP, New York.

Hertel, Th.W., M. Ivanic, P.V. Preckel, and J.A.L. Cranfield. 2004. The earnings effects of multilateral trade liberalization: Implications for poverty. World Bank Econ. Rev. 18(2):205-236.

Heymann, D.L. 2002. Food safety, an essential public health priority. Available at http://www.fao.org/docrep/meeting/004/Y3680E/Y3680E05.htm. FAO, Rome.

Hirsch Hadorn, G., D. Bradley, C. Pohl, S. Rist, and U. Wiesmann. 2006. Implications of transdisciplinarity for sustainability research. Ecol. Econ. 60(1):119-128.

Horiuchi, S. 1999. Epidemiological transitions in developed countries: Past, present and future. p. 54-71. In J. Chamie, and R. Cliquet (ed) Health and mortality: Issues of global concern. CBGS, Flemish Sci. Inst., NY and Leuven.

Hurni, H. (with assistance from contributors). 1996. Precious earth: From soil and water conservation to sustainable land management. Int. Soil Conserv. Org. Centre Dev. Environ., Wageningen and Berne.

Hurni, H., U. Christ, J.A. Lys, D. Maselli, and U. Wiesmann. 2001. The challenges in enhancing research capacity. p. 50-58. In KFPE. Enhancing research capacity in developing and transition countries: Experience, discussions, strategies and tools for building research capacity and strengthening institutions in view of promoting research for sustainable development. Geo. Bernensia, Berne.

Hurni, H., and U. Wiesmann. 2004. Towards transdisciplinarity in sustainability-oriented research for development. p. 31-42. In H. Hurni et al. (ed) Research for mitigating syndromes of global change: Perspectives of the Swiss NCCR North-South, Vol.1. Univ. Berne, Geo. Bernensia, Berne.

Huvio, T., J. Kola, and T. Lundström (ed) 2005. Proc. Seminar. Haikko, Finland, 18-19 Oct. 2004. Agric. Policy Publ. No. 38. Dep. Econ. Manage., Univ. Helsinki.

IAC. 2004. Realizing the promise and potential of African agriculture: Science and technology strategies for improving agricultural productivity and food security in Africa. InterAcademy Council Secretariat, Amsterdam.

ICSU. 2003. Optimizing knowledge in the information society. Int. Council Science, Paris.

ICTSD. 2006. Fisheries, international trade and sustainable development: Policy Disc. Pap. Int. Centre Trade Sustain. Dev., Geneva.

IFAD. 1998. Ghana — Women's contribution to household food security. Available at http://www.ifad.org/hfs/learning/46.htm. IFAD, Rome.

IFAD. 2003. Promoting market access for the rural poor in order to achieve the millennium development goals: Disc. Pap. Available at http://www.ifad.org/gbdocs/gc/26/e/markets.pdf. IFAD, Rome.

IFPRI. 2005. South Asia. Agricultural and rural development. Proc. Seminars. New Delhi, Lahore, Chennai, Dhaka. IFPRI, Washington DC.

ILO. 1962. Convention concerning basic aims and standards of social policy: Convention No. 117. ILO, Geneva.

ILO. 1999. Safety and health in agriculture: Int. Labour Conf., 88th Session, Geneva. 30 May-15 June, 2000. Rep. VI (1). ILO, Geneva.

ILO. 2000. Occupational safety and health in agriculture. Available at http://www.ilo.org/public/english/protection/safework/agriculture/intro.htm. ILO, Geneva.

ILO. 2004. World Employment Report: Employment Productivity and Poverty Reduction. ILO, Geneva.

ILO. 2005. HIV/AIDS and employment. ILOAIDS brief, September. ILO, Geneva.

ILO. 2006. The end of child labour within reach. Int. Labour Conf., 95th Session, 2006, Rep. 1(B). ILO, Geneva.

ILO. 2007. Global employment trends. Available at http://www.ilo.org/public/english/employment/strat/global.htm. ILO, Geneva.

IOM. 2007. Migration initiatives appeal 2007. Int. Org. Migration, Geneva.

IUCN. 2005. Healthy corals fared best against tsunami. Press release. Available at http://www.iucn.org/en/news/archive/2005/12/cordio_iucn_report_2005.pdf. The World Conservation Union, Gland.

Kent, G. 2003. Principles for food trade. SCN News (26):43-44.

Kurien, J. 2006. Achieving a sustainable global fish trade. Id21 insights 65:4.

Lambrou, Y., and G. Piana. 2006. Energy and gender issues in rural sustainable development. FAO, Rome.

Leakey, R.R.B., and P.A. Sanchez. 1997. How many people use agroforestry products? Agrofor. Today 9:4-5.

Lilja, N., J.A. Ashby, and L. Sperling (ed) 2000. Proceedings of the seminar on assessing the impact of participatory research and gender analysis Sept 1998, Quito. CGIAR, Cali.

Linares, O.F. 1985. Cash crops and gender constructs: The Jola (Diola) of Senegal. Ethnology 24(2):83-93.

Liniger, H.P., and W. Critchley. 2007. Where the land is greener: Case studies and analysis of soil and water conservation initiatives worldwide. Centre Dev. Environ., Berne.

LivestockNet. 2006. Livestock production and the Millennium Development Goals: The role of livestock for pro-poor growth. Available at http://www.livestocknet.ch/ne_news.htm. Swiss College Agriculture/InfoAgrar, Zollikofen.

Lund, H.G. and S. Iremonger. 1998. Omissions, commissions, and decisions: The need for integrated resource assessments. p. 182-189. In Environmental Research Institute of Michigan (ed) Proc. Int. Conf. Geospatial Information in Agriculture and Forestry. Decision Support, Technology and Applications. Lake Buena Vista, Florida, 1-3 June. 1998. ERIM Int., Ann Arbor.

MA (Millennium Ecosystem Assessment). 2003. Ecosystems and human well-being: A framework for assessment. Island Press, Washington DC.

MA (Millennium Ecosystem Assessment). 2005a. Ecosystems and human well-being: Synthesis. Island Press, Washington DC.

MA (Millennium Ecosystem Assessment). 2005b. Our human planet: Summary for decision makers. Island Press, Washington DC.

MA (Millennium Ecosystem Assessment). 2005c. Ecosystems and human well-being: Current states and trends. Vol. 1. Island Press, Washington DC.

Malik, R.K., A. Yadav, and S. Singh. 2005. Resource conservation technologies in rice-wheat cropping systems in Indo-Gangetic Plains. In I.P. Abrol et al. (ed), Conservation agriculture: Status and prospects. Center Advance. Sustainable Agric., New Delhi.

Matthews, E., R. Payne, M. Rohweder, and S. Murray. 2000. Pilot analysis of global ecosystems: Forest ecosystems. World Resourc. Inst., Washington DC.

Mazoyer, M. 2001. Protecting small farmers and the rural poor in the context of globalization. FAO, Rome.

Mazoyer, M., and L. Roudard. 1997. History of the world's agriculture. (In French). Editions du Seuil, Paris.

McC Netting, R. 1993. Smallholders, householders: Farm families and the ecology of intensive, sustainable agriculture. Stanford Univ. Press, CA.

McCarthy, J.J., O.F. Canziani, N.A. Leary, D.J. Dokken, and K.S. White (ed) 2001. Climate change 2001: Impacts, adaptation and vulnerability. Third Assessment Report IPCC. Cambridge Univ. Press, UK.

McNeely, J.A., H.A. Mooney, L.E. Neville, P.J. Schei, and J.K. Waage (ed) 2001. Global strategy on invasive alien species. The World Conservation Union and Global Invasive Species Prog., Cambridge.

Messer, E., and L.F DeRose. 1998. Food shortage. p. 53-91. In L. DeRose et al. (ed) Who's hungry and how do we know? Food shortage, poverty and deprivation. UN Univ. Press, Tokyo.

Minten, B., L. Randrianarison, and J.F.M. Swinnen. 2006. Global retail chains and poor farmers: Evidence from Madagaskar. LICOS Disc. Pap. 164. Centre for Transition Economies, Katholieke Universiteit, Leuven.

Molden, D., and C. de Fraiture. 2004. Investing in water for food, ecosystems and livelihoods: Blue Pap. CA, IWMI, Colombo.

Molnar, A., S.J. Scherr, and A. Khare. 2005. Who conserves the world's forests? A new assessment of conservation and investment trends. Forest Trends and Ecoagriculture Partners, Washington DC.

Moss, R.H., and S.H. Schneider. 2000. Uncertainties in the IPCC TAR: Recommendations to lead authors for more consistent assessment and reporting. p. 33-51. In R. Pachauri et al. (ed) Guidance papers on the cross-cutting issues of the third assessment report of the IPCC. WMO, Geneva.

Mougeot, L. 2000. Urban agriculture: Definition, presence, potentials and risks. In N. Bakker et al. (ed) Growing cities, growing food. Deutsche Stiftung für Entwicklung, Feldafing.

Müller, C. 2006. Synthesis and conclusion. p. 331-352. In S. Premchander, and C. Müller (ed) Gender and sustainable development: Case studies from NCCR North-South. Perspectives of the Swiss NCCR North-South, Univ. Bern, Vol.2. Geo. Bernensia, Berne.

Nagayets, O. 2005. Small farms: Current status and key trends. Information brief. Future of Small Farms Research Workshop. Wye, 26-29 June. IFPRI, Washington DC.

Nazarea, V.D. (ed) 1999. Ethnoecology: Situated knowledge/located lives. Univ. Arizona Press, Tucson.

Nowotny, H., P. Scott, and M. Gibbons. 2003. "Mode 2" revisited: The new production of knowledge. Minerva 41:179-194.

ODI. 2002. Rethinking rural development. Briefing Pap. March. Overseas Dev. Inst., London.

OECD. 1998. Reaching the goals in the S-21: Gender equality and the environment.

Working Party on Gender Equality. OECD, Paris.

OECD. 2005. Multifunctionality in agriculture: What role for private initiatives? OECD, Paris.

Oldeman, L.R. 1994. The global extent of soil degradation. p. 99-118. In D.J. Greenland and I. Szabolcs (ed) Soil resilience and sustainable land use. CABI, Wallingford.

Oldeman, L.R., R.T.A. Hakkeling, and W.G. Sombroek. 1991. World map of the status of human-induced soil degradation: An explanatory note. Int. Soil Reference and Information Centre and UNEP, Wageningen and Nairobi.

Olesen, I., A.F. Groen, and B. Gjerde. 2000. Definition of animal breeding goals for sustainable production systems. J. Anim. Sci. 78:570-582.

Olson, D.M., and E. Dinerstein. 1998. The global 200: A representation approach to conserving the earth's most biologically valuable ecoregions. Conserv. Biol. 12(3):502-515.

Ormerod, S.J., E. J. P. Marshall, G. Kerby and S. P. Rushton. 2003. Meeting the ecological challenges of agricultural change: Editors' introduction. J. Appl. Ecol. 40, 939-946.

Ostrom, E. 1990. Governing the commons: The evolution of institutions for collective action. Cambridge Univ. Press, UK.

Pahl-Wostl, C., and M. Hare. 2004. Processes of social learning in integrated resources management. J. Community Appl. Soc. Psychol. 14:193-206.

Pardey, P.G., N.M. Bientema, S. Dehmer, and S. Wood. 2006. Agricultural research: A growing global divide? IFPRI Food Policy Rep. IFPRI, Washington, D.C.

Parry, M., C. Rosenzweig, and M. Livermore. 2007. Climate change, global food supply and risks of hunger. Philos. Trans. R. Soc. Lond., Ser. B. 360(1463):2125-2136.

Penning de Vries, F.W.T., H. Acquay, D. Molden, S.J. Scherr, C. Valentin, and O. Cofie. 2003. Integrated land and water management for food and environmental security: CA Res. Rep. 1. Comprehensive Assessment Secretariat, Colombo.

Pieri, C., G. Evers, J. Landers, P. O'Connell, and E. Terry. 2002. No-till farming for sustainable rural development. ARD Working Pap. World Bank, Washington DC.

Pimentel, D., S. McNair, J. Janecka, J. Wightman, C. Simmonds, C. O'Connell et al. 2001. Economic and environmental threats of alien plant, animal and microbe invasions. Agric. Ecosyst. Environ. 84:1-20.

Postel, S. 1999. Pillar of sand: Can the irrigation miracle last? Worldwatch and W.W. Norton, NY.

Prescott-Allen, R. 2001. The wellbeing of nations: A country by country index of quality of life and the environment. Island Press, Washington DC.

Prüss-Üstün, A., and C. Corvalán. 2006. Preventing disease through healthy environments. Towards an estimate of the environmental burden of disease. WHO, Geneva.

Reardon, T., J. Berdegué, and J. Farrington. 2002. Supermarkets and farming in Latin America: Pointing directions for elsewhere? Perspectives 81 (Dec). DFID, London.

Reardon, T., J.-M. Codron, L. Busch, J. Bingen, and C. Harris. 2001. Global change in agrifood grades and standards: Agribusiness strategic responses in developing countries. Int. Food Agrib. Manage. Rev. 2(3).

Reardon, T., C.P. Timmer, C.B. Barrett, and J. Berdegué. 2003. The rise of supermarkets in Africa, Asia, and Latin America. Am. J. Agric. Econ. 85(5):1140-1146.

Revenga, C., J. Brunner, N. Henninger, K. Kassem, and R. Payne. 2000. Pilot analysis of global ecosystems: Freshwater systems. World Resourc. Inst., Washington DC.

Rice, A.L., L. Sacco, A. Hyder, and R.E. Black. 2000. Malnutrition as an underlying cause of childhood deaths associated with infectious diseases in developing countries. Bull. WHO 78(10):1207-1221.

Rist, S., M. Chiddambaranathan, C. Escobar, U. Wiesmann, and A. Zimmermann. 2006. Moving from sustainable management to sustainable governance of natural resources: The role of social learning processes in rural India, Bolivia and Mali. J. Rural Stud. 23(1):23-37.

Rist, S., and F. Dahdouh-Guebas. 2006. Ethnosciences: A step towards the integration of scientific and non-scientific forms of knowledge in the management of natural resources for the future. Environ. Dev. Sust. 8(4):467-493.

Rist, S., F. Delgado, and U. Wiesmann. 2003. The role of social learning processes in the emergence and development of Aymara land use systems. Mountain Res. Dev. 23:263-270.

Rockström, J. 1999. On-farm green water estimates as a tool for increased food production in water scarce regions. Phys. Chem. Earth 24(4):375-383.

Röling, N., and A. Wagemakers (ed) 2000. Facilitating sustainable agriculture: Participatory learning and adaptive management in times of environmental uncertainty. Cambridge Univ. Press, New York.

Röling, N., and E. Van de Fliert. 1994. Transforming extension for sustainable agriculture: The case of integrated pest management in rice in Indonesia. Agric. Human Values 11(2- 3):96-108.

Rosegrant, M.W., and P.B.R. Hazell. 2001. Transforming the rural Asian economy: The unfinished revolution: 2020 Vision Brief 69. IFPRI, Washington DC.

Rosegrant, M.W., C. Ringler, T. Benson, X. Diao, D. Resnick, J. Thurlow et al. 2006. Agriculture and achieving the Millennium Development Goals. World Bank, Washington DC.

Sanders, D.W., P.C. Huszar, S. Sombatpanit, and T. Enters. 1999. Incentives in soil conservation. Science Publ., Enfield.

Savenije, H.H.G., and P. van der Zaag. 2000. Conceptual framework for the management of shared river basins. Water Policy 2:9-45.

Scherr, S., A. White, A. Khare, M. Inbar, and A. Molnar. 2004. For services rendered: The current status and future potential of markets for the ecosystem services provided by forests. Tech. Ser. No. 21. Int. Trop. Timber Org., Yokohama.

Scherr, S.J. 1999. Soil degradation: A threat to developing-country food security? 2020 Vision Food Agric. Environ. Disc. Pap. No. 27. IFPRI, Washington DC.

Schroeder, D.G., and K.H. Brown. 1994. Nutritional status as a predictor of child survival: Summarizing the association and quantifying its global impact. Bull. WHO 72(4):569-79.

Scoones, I., and J. Thompson. 1994. Knowledge, power and agriculture: Towards a theoretical understanding. p. 16-31. In I. Scoones and J. Thompson (ed) Beyond farmers first: Rural people's knowledge, agricultural research and extension practice. Intermediate Tech. Publ., IIED, London.

Seale, Jr., J., A. Regmi, and J. Bernsteing. 2003. International evidence on food consumption patterns: Tech. Bull. No.1904. Available at www.ers.usda.gov/publications/tb1904. ERS, USDA, Washington DC.

Sen, A. 1981. Poverty and famines: An essay on entitlement and deprivation. Clarendon Press, Oxford.

Sen, A., and J. Drèze (ed) 1990, 1991. The political economy of hunger in 3 volumes. Clarendon Press, Oxford.

Serageldin, I. 1996. Sustainability and the wealth of nations: First steps in an ongoing journey: ESDS and Mono. Ser. No. 5. World Bank, Washington DC.

Serova, E. 2007. Vertical integration in Russian agriculture. In J. Swinnen (ed) Global supply chains, standards and the poor: How the globalization of food systems and standards affects rural development and poverty. CABI, Oxfordshire.

Shaxson, T.F., N.W. Hudson, D.W. Sanders, E. Roose, and W.C. Moldenhauer. 1989. Land husbandry: A framework for soil and water conservation. Soil Water Conserv. Soc., Ankeny IA.

Sheridan, C. 2005. EPO neem patent revocation revives biopiracy debate. Nature Biotechnol. 23:511-512.

Smakhtin, V., C. Revenga, and P. Döll. 2004. Taking into account environmental water requirements in global-scale water resources assessments: CA Res. Rep. 2. Comprehensive Assessment Secretariat, Colombo.

Smit, B. (ed) 1993. Adaptation to climatic variability and change: Report of the Task Force on Climate Adaptation. Canadian Climate Prog., Univ. Guelph, Ontario.

Spencer, D. 2002. The future of agriculture in sub-Saharan Africa and South Asia: W(h)ither the small farm? In Sustainable food security for all by 2020: Proc. Int. Conf.

Bonn. 4-6 Sept. 2001. IFPRI, Washington DC.

Spencer, D.S.C., P.J. Matlon, and H. Löffler. 2003. African agricultural production and productivity in perspective. InterAcademy Council, Amsterdam.

Spieldoch, A. 2007. A row to hoe: The gender impact of trade liberalization on our food system, agricultural markets and women's human rights. Friedrich-Ebert-Stiftung, Geneva.

Steinfeld, H., P. Gerber, T. Wassenaar, M. Rosales, and C. de Haan. 2006. Livestock's long shadow: Environmental issues and options. Livestock, Environment and Development. FAO, Rome.

Stern, N. 2006. Stern review: Economics of climate change. United Kingdom's Treasury. London.

Strosnider, H., E. Azziz-Baumgartner, M. Banziger, R.V. Bhat, R. Breiman, M. Brune, et al. 2006. Workgroup report: Public health strategies for reducing Aflatoxin exposure in developing countries. J. Environ. Health Perspect. 114:1898-1903.

Thrupp, L.A. 1998. Cultivating diversity: Agrobiodiversity and food security. World Resourc. Inst., Washington DC.

Tillman, D., K.G. Cassman, P.A. Matsons, R. Naylor, and S. Polasky. 2002. Agricultural sustainability and intensive production practices. Nature 418:671-677.

Trawick, P. 2003. Against the privatization of water: An indigenous model for improving existing laws and successfully governing the commons. World Dev. 31:977-996.

UN Millennium Project. 2005. Innovation: Applying knowledge in development: Task Force 10 on Sci. Tech. Innovation. Earthscan, Washington DC.

UN. 1992. Convention on Biological Diversity. Vol. 1760, I-30619. United Nations, N.Y.

UN. 1993a. Vienna Declaration and Programme of Action. World Conf. Human Rights. Vienna, 14-15 June. United Nations, N.Y.

UN. 1993b. Agenda 21: Earth Summit: The United Nations Programme of Action from Rio. UN Publ., NY.

UN. 1997. Coordination of the policies and activities of the specialized agencies and other bodies of the United Nations System: Mainstreaming the gender perspective into all policies and programmes in the United Nations system: E/1997/100. UN Econ. Soc. Council, Geneva.

UN. 2005a. World population prospects: The 2004 revision. Vol. 1. Dep. Econ. Soc. Affairs. United Nations, NY.

UN. 2005b. World population prospects: The 2004 revision. Vol. III. Dep. Econ. Soc. Affairs. United Nations, NY.

UNAIDS, WHO. 2005. AIDS epidemic update: Special report on HIV prevention. UN Prog. HIV/AIDS and WHO, New York and Geneva.

UNAIDS. 2005. Resource pack on gender and HIV/AIDS: Interagency task team on gender and HIV/AIDS. Royal Trop. Inst. (KIT) Publ., Amsterdam.

UNDP. 2005. Women's empowerment: Biodiversity for development: The gender dimension. Available at http://www.undp.org/women/mainstream/index.shtml. UNDP, NY.

UNEP. 2002. Global Environment Outlook 3: Past, present and future perspectives. Earthscan, London and Sterling VA.

UNEP. 2005. GEO Year Book 2004/5: An overview of our changing environment. UNEP, Nairobi.

UNESCAP. 2001. Women in local government in Asia and Pacific: A comparative analysis of thirteen countries. UN Econ. Soc. Comm. Asia and the Pacific, Bangkok.

UNFPA. 2007. State of world population. Available at http://www.unfpa.org/swp/2004/english/ch1/page7.htm. UNFPA, New York.

UNICCPR. 1966. International covenant on civil and political rights. UN, NY.

UNICESCR. 1966. International covenant on economic, social and cultural rights. UN, NY.

UNSCN. 2004. 5th Report on the world nutrition situation: Nutrition for improved development outcomes. UN Standing Comm. Nutrition, Geneva.

US Census Bureau. 2007. World population information. Available at http://www.census.gov/ipc/www/idb/worldpopinfo.html. US Census Bureau, Washington DC.

Uvin, P. 1995. The state of world hunger. p. 1-18. In E. Messer and P. Uvin (ed) The hunger report: 1995. Gordon and Breach Sci. Publ., Amsterdam.

van de Fliert, E. 2003. Recognizing a climate for sustainability: Extension beyond transfer of technology. Aust. J. Exp. Agric. 43:29-36.

van Veenhuizen, R. (ed) 2006. Cities farming for the future: Urban agriculture for green and productive cities. Resource Centres on Urban Agric. Food Security Foundation, Leusden.

von Braun, J. 2005. Small-scale farmers in a liberalized trade environment. In Small-scale farmers in a liberalized trade environment. Available at http://honeybee.helsinki.fi/mmtal/abs/Pub38.pdf. Dep. Econ. Manage., Univ. Helsinki.

Wassmann, R. and P.L.G Vlek. 2004. Mitigating greenhouse gas emissions from tropical agriculture: Scope and research priorities. Environ. Dev. Sustain. 6(1-20):1-9.

Watson, R., and H. Gitay. 2004. Mobilization, diffusion and use of scientific expertise. Institut du développement durable et des relations internationales, Paris.

Watson, R.T., J.A. Dixon, S.P. Hamburg, A.C. Janetos, and R.H. Moss. 1998. Protecting our planet, securing our future: Linkages among global environmental issues and human needs. UNEP, NASA, and World Bank, Washington DC.

WCD. 2000. Dams and development: A new framework for decision-making. Earthscan, London.

WCED. 1987. Our common future. World Commission on Environment. Oxford Univ. Press, NY.

WCSFD. 1999. Our forests, our future. World Commission on Forests and Sustainable Development. Cambridge Univ. Press, Cambridge, New York, and Melbourne.

WDR. 2007. World development report 2007: Development and the next generation. World Bank and Oxford Univ. Press, Washington DC and Oxford.

Wesseling, C., R. McConnell, T. Partanen, and C. Hogstedt. 1997. Agricultural pesticides use in developing countries: Health effects and research needs. Int. J. Health Serv. 27(2):273-308.

WFS. 1996. World Food Summit. FAO, Rome.

White, R., S. Murray, and M. Rohweder. 2000. Pilot analysis of global ecosystems: Grassland ecosystems. World Resourc. Inst., Washington DC.

WHO. 1946. Preamble to the Constitution of the World Health Organization as adopted by the International Health Conference, New York, 19-22 June, 1946. WHO, Geneva.

WHO. 1999. Development of food-based dietary guidelines for the Western Pacific Region. WHO, Reg. Off. Western Pacific, Manila.

WHO. 2003a. Diet, nutrition and the prevention of chronic diseases: WHO Tech, Rep. Ser. No. 916. Geneva, 2003.

WHO. 2003b. The world health report 2003: Shaping the future. WHO, Geneva.

WHO. 2005. The world health report 2005: Make every mother and child count. WHO,Geneva.

WHO. 2006. Preventing disease through healthy environments. WHO, Geneva.

WHO-VPH. 2007. Veterinary public health (VPH) webpage. Available at www.who.int.zoonoses/vph/en. WHO, Geneva.

Wiebe, K. (ed) 2003. Land quality, agricultural productivity, and food security: Biophysical processes and economic choices at local, regional, and global scales. Edward Elgar Publ., Cheltenham.

Wiesmann, U. 1998. Sustainable regional development in rural Africa: Conceptual framework and case studies from Kenya. Geographica Bernensia, Berne.

Willer, H., and M. Yussefi. 2006. The world of organic agriculture: Statistics and emerging trends 2006. International Federation of Organic Agriculture Movements and Res. Inst. Organic Agriculture (FiBL), Bonn, Germany and Frick, Switzerland.

Winklerprins, A.W.G.A. 1999. Insights and applications of local soil knowledge: A tool

for sustainable land management. Soc. Nat. Resour. 12(2):151-161.

WOCAT. 2006. World overview of conservation approaches and technologies. Available at http://www.wocat.net/. Centre Dev. Environ., Berne.

Wollenberg, E., D. Edmunds, L. Buck, J. Fox, and S. Brodt. 2001. Social learning in community forests. CIFOR and the East-West Centre, Bogar Barat.

Wood, S., K. Sebastian, and S.J. Scherr. 2000. Pilot analysis of global ecosystems: Agroecosystems. IFPRI and World Resourc. Inst., Washington DC.

World Bank. 1997. Expanding the measure of wealth: Indicators of environmentally sustainable development: ESDS Mono. Ser. No. 17. World Bank, Washington DC.

World Bank. 2001. Engendering development: Through gender equality in rights, resources, and voice: Policy Res. Rep. 21776. World Bank and Oxford Univ. Press, UK.

World Bank. 2004a. Agriculture investment sourcebook. The World Bank, Washington, D.C.

World Bank. 2004b. Genderstats database. Available at http://genderstats.worldbank.org/home.asp. World Bank, Washington DC.

World Bank. 2006a. World Development Report 2006: Equity and development. Oxford Univ. Press, NY.

World Bank. 2006b. Enhancing agricultural innovation: How to go beyond the strengthening of research systems? World Bank, Washington DC.

World Bank. 2006c. 2006 World development indicators. World Bank, Washington DC.

WorldFish Center. 2006. The threat to fisheries and aquaculture from climate change: Policy Brief. WorldFish Center, Penang.

Worldwatch Institute. 2003. State of the world 2003. W.W. Norton, NY.

WRI. 2005. World resources 2005: The wealth of the poor: Managing ecosystems to fight poverty. World Resourc. Inst., Washington DC.

WRI. 2006. EarthTrends. Environmental information. Available at http://earthtrends.wri.org. World Resourc. Inst., Washington DC.

WTO. 1995. Agreement on agriculture. WTO, Geneva.

WWAP. 2003. Water for people, water for life: World water development report. UNESCO and Berghahn Books, Paris.

Yapa, L. 1993. What are improved seeds? An epistemology of the green revolution. Econ. Geogr. 69:254-273.

Zurek, M.B. 2006. A short review of global scenarios for food systems analysis: GECAFS Working Pap. 1. Global Environmental Change and Food Systems, Environ. Change Inst., Oxford Univ. Centre Environ., Oxford.

2

Historical Analysis of the Effectiveness of AKST Systems in Promoting Innovation

Coordinating Lead Authors
Fabrice Dreyfus (France), Cristina Plencovich (Argentina), Michel Petit (France)

Lead Authors
Hasan Akca (Turkey), Salwa Dogheim (Egypt), Marcia Ishii-Eiteman (USA), Janice Jiggins (UK), Toby Kiers (USA), Rose Kingamkono (Tanzania)

Contributing Authors
Emily Adams (USA), Medha Chandra (India), Sachin Chaturvedi (India), Chris Garforth (UK), Michael Halewood (Canada), Andy Hall (UK), Niels Louwaars (Netherlands), Jesus Moncada (Mexico), Cameron Pittelkow (USA), Jeremy Schwartzbord (USA), Matthew Spurlock (USA), Jeff Waage (UK)

Review Editors
Stephen Biggs (UK) and Gina Castillo (Ecuador)

Key Messages

Key Messages

1. Acknowledging and learning from competing and well evidenced historical narratives of knowledge, science and technology processes and understanding the flaws in past and existing institutional arrangements and maintaining the space for diverse voices and interpretations is crucial for designing policies that are effective in reaching the integrated goals of productivity, environmental sustainability, social equity and inclusion. Agricultural Knowledge, Science and Technology (AKST) encompass diverse agricultural practices, interventions, institutional arrangements and knowledge processes. Different and often conflicting interpretations of the contributions of AKST to productivity, environmental and social sustainability and equity exist side-by-side but are not equally heard or recognized. Political power and economic influence have tended to privilege some types of AKST over others. Dominant institutional arrangements have established the privileged interpretations of the day and set the agenda for searching for and implementing solutions. The narrative used to explain past events and AKST choices has important implications for setting future priorities and projecting the future design of AKST.

2. In the prevailing AKST arrangements of the past, key actors have been excluded or marginalized. Preference has been given to short-term goals vs. longer-term agroecosystem sustainability and social equity and to powerful voices over the unorganized and voiceless. Development of appropriate forms of partnerships can help bring in the excluded and marginalized and open AKST to a larger set of policy goals. Many effective participatory approaches exist that facilitate the establishment and operation of such partnerships. Targeted public support can help address the biases in the dominant arrangements.

3. The Transfer of Technology (ToT) model has been the most dominant model used in operational arrangements and in policy. However, the TOT model has not been the most effective in meeting a broader range of development goals that address the multiple functions and roles of farm enterprises and diverse agroecosystems. In this model, science and technology are mobilized under the control of experts in the definition of problems and the design of solutions, problem setting and solving. Other types of knowledge have sometimes been tapped, although mainly for local adaptation purposes. Where the TOT model has been applied appropriately with the conditions necessary for achieving impact, it has been successful in driving yield and production gains. These conditions include properly functioning producer and service organizations, the social and biophysical suitability of technologies transferred in specific environments and proper management of those technologies at plot, farm and landscape levels.

4. Successful education and extension programs have built on local and traditional knowledge and innovation systems, often through participatory and experiential learning processes and multi-organizational partnerships that integrate formal and informal AKST. Basic and occupational education empowers individuals to innovate in farming and agroenterprises, adapt to new job opportunities and be better prepared for migration. Attention to overcoming race, ethnic and gender biases that hamper the participation of marginalized community members, diverse ethnic groups and women, is essential. Education and training of government policymakers and public agency personnel, particularly in decentralized participatory planning and decision-making, and in understanding and working effectively with rural communities and other diverse stakeholders has also proven effective. Effective options include but are not limited to experiential learning groups, farmer field schools, farmer research circles, Participatory Plant Breeding, social forestry and related community-based forest landscape management, study clubs and community interaction with school-based curriculum development.

5. Investment in farmers and other rural actors' learning and capacity to critically assess, define and engage in locally-directed development processes has yielded positive results. Modern ICTs are beginning to open up new and potentially powerful new opportunities for extending the reach and scope of educational and interactive learning. Extension and advisory services complement but do not substitute for rural education. The development and implementation of successful learning and innovation programs require skills in facilitating processes of interaction among partners, interdisciplinary science and working with all partners' experience and knowledge processes. Active development of additional options are needed to extend these arrangements and practices to include more marginalized peoples and areas and in ways that respect and uphold their roles, rights and practices.

6. Innovation is a multisource process and always and necessarily involves a mix of stakeholders, organizations and types of knowledge systems. Innovative combinations of technology and knowledge generated by past and present arrangements and actors have led to more sustainable practices. These include for example, integrated pest management, precision farming, and local innovations in crop management (e.g., push-pull in Africa). Further experimentation with facilitated innovation is needed to capitalize on new opportunities for innovation under market-oriented development.

7. Partnerships in agricultural and social science research and education offer potential to advance public interest science and increase its relevance to development goals. Industry, NGOs, social movements and farmer organizations have contributed useful innovations in ecologically and socially sustainable approaches to food and agriculture. Increased private sector funding of universities and research institutes has helped fill the gap created by declining public sector funds but has mixed implications for these institutions' independence and future research directions. Effective codes of conduct can strengthen multi-stakeholder partnerships and preserve public institutions' capacity to perform public good research.

8. Public policy, regulatory frameworks, and international agreements informed by scientific evidence and public participation have enabled decisive and effective global transitions towards more sustainable practice. New national, regional and international agreements will be needed to support further shifts towards ethical, equitable and sustainable food and agriculture systems in response to the urgent challenges posed by declining availability of clean water, climate change, and insupportable labor conditions.

9. Awareness of the importance of ensuring full and meaningful participation of multiple stakeholders in international and public sector AKST policy formation has increased. For example, in some countries, pesticide policies today are developed by diverse group of actors including civil society and private sector actors, informed by science and empirical evidence and inclusive of public interest concerns. These policies have focused on the multifunctionality of agriculture.

10. The number and diversity of actors engaged in the management of agricultural resources such as germplasm has declined over time. This trend reduces options for responding to uncertainties of the future. It increases asymmetries in access to germplasm and increases the vulnerabilities of the poor. Participatory plant breeding provides strong evidence that diverse actors can be engaged in an effective practice for achieving and sustaining broader goals of sustainability and development by bringing together the skills and techniques of advanced and conventional breeding and farmers' preferences and germplasm management capacities and skills, including seed production for sale. Further development and expansion would require adjustment of varietal release protocols and appropriate policy recognition under the International Union for the Protection of New Varieties of Plants (UPOV).

11. The debates surrounding the use of synthetic pesticides have led to new arrangements that have increased awareness, availability and effectiveness of the range of options for pest management. Institutional responses have included the strengthening of regulatory controls over synthetic chemical pesticides at global and national levels, growing consumer and retail markets for pesticide-free and organic products, removal of highly toxic products from sale, development of less acutely toxic products and more precise means of delivery and education of users in safe and sustainable practices. What constitutes safe and sustainable practice has been defined in widely varying ways by different actors reflecting different conditions of use as well as different assessments of acceptable tradeoffs. The availability of and capacity to assess, compare and choose from a wide range of options in pest management is critical to strengthening farmers' ability to incorporate effective strategies that are safe, sustainable and effective in actual conditions of use.

12. Integrated Pest Management exemplifies a flexible and wide-reaching arrangement of actors, institutions and practices that better address the needs of diverse farmers. Although definitions, interpretations and outcomes of IPM programs vary widely among actors, IPM typically incorporates KST from a broad range of sciences, including social sciences, and the experience and knowledge of a diverse set of actors. IPM has become more common in high value production systems and has been adopted by an increasing number of important commercial actors in food processing and retailing. Successful approaches to introducing IPM to small-scale producers in the tropics include farmer field schools, push-pull approaches, advisory services provided under contractual arrangements for supply to central processing facilities and creative use of communication tools such as farmer-to-farmer videos and focused-message information campaigns. A combination of such approaches, backed by strong policy reform to restrict the sale of outdated and highly toxic synthetic controls, will be needed to meet future development goals. Further experimentation with and operational fine-tuning of the institutional arrangements for IPM in the field in different settings is also needed to ensure optimal efficacy. These can be evaluated by comparative assessment using a combination of social, environmental and economic measures that include positive and negative externalities.

13. Local food systems, known to sustain livelihoods at micro level, are currently challenged by globalized food systems. This trend brings opportunities but also threatens livelihoods and sovereignties of marginalized communities and indigenous peoples. In some countries, social, ethical and cultural values have been successfully integrated in commercial mechanisms. Fair trade and ethnic labeling are examples of institutional options that can be considered by those who wish to promote effective measures to protect the interests of the marginalized and revitalize rural livelihoods and food cultures. The addition of a geographic indication can promote local knowledge and open opportunities for other agroenterprises such as tourism and specialty product development, as well as collaboration with utilities such as water companies. Production systems dominated by export markets are weakened by erratic changes in international markets and have sparked growing concerns about the sustainability of long-distance food shipping and the ecological footprint and social impacts of international trade practices. Local consumption and domestic outlets for farmers' products can alleviate the risks inherent in international trade.

2.1. Science, Knowledge, Technology and Innovation in Agriculture

The Asian AgriHistory Foundation translates historical writings that remind us that formal processes for generating technology-led innovation were in place in some countries more than 3000 years ago. This subchapter focuses on AKST processes and institutional arrangements, how these have been brought to bear on agricultural problems and combined to bring about innovation in agricultural systems when mobilized for different policy purposes. Subchapter 2.2 assesses the roles that various knowledge actors have played in different contexts, noting changes over time from different perspectives so as to minimize the risk that past actions are judged by current values or by those of only one set of actors. The drivers are assessed at three levels—local, regional, global. The assessments are further elaborated (2.3) in order to provide depth and detail in terms of three thematic narratives—(1) genetic resources management; (2) pest management; (3) food system management.

2.1.1 The specificity of agriculture as an activity

At the beginning of the period under assessment, policy makers and other knowledge actors around the world had vividly in mind the fact that food is a basic necessity of life and that its supply and distribution is vulnerable to a range of disruptions that cannot always be well controlled. Only for those for whom food is reliably abundant can food be treated as an industrial good subject to the laws of elasticity of price. The special characteristics of farming as a human activity for supplying a basic necessity of life and as the cultural context of existence for a still large if declining proportion of the world's people are central to meaningful historical assessment of AKST.

2.1.1.1 The characteristics of agriculture as a multidimensional activity

Agriculture is based on local management decisions made in interaction with the biophysical, ecological and social context, this context to a large extent itself evolving independently of agriculture. It follows that AKST includes both a set of activities that happen to deal with the particular domain of agriculture and activities that necessarily coevolve with numerous other changes in a society. AKST thus involves many types of knowledge and many suppliers of that knowledge acting in relation to vast numbers of (semi) autonomous enterprises and decision makers. This characteristic has provided special challenges but also opportunities in the design of institutional arrangements for AKST (Yunus and Islam, 1975; Yunus, 1977; Izuno, 1979; Symes and Jansen, 1994; Scoones et al., 1996; Buck et al., 1998; Stroosnijder and Rheenen, 2001; Edgerton, 2007).

A place-based activity. Agriculture as a place-based activity relies on unique combinations of bioclimatic conditions and local resources in their natural, socioeconomic and cultural dimensions. Agricultural practices depend on and also influence these conditions and resources (Herdt and Mellor, 1964). Specific knowledge of the locality is an asset decisive for the outcomes actually achieved through application of any technology (Loomis and Beagle, 1950; Hill, 1982; Giller, 2002; Tittonell et al., 2005, 2007; Vanlauwe et al., 2006; Wopereis et al., 2006; Zingore et al., 2007) yet a dominant trend over the period is the evolution of agricultures driven by nonlocal changes and by the introduction of technologies designed by actors and in places far removed from their site of application (Merton, 1957; Biggs, 1978; Anderson et al., 1991; Seur, 1992; Matson et al., 1997; Harilal et al., 2006; Leach and Scoones, 2006). This trend has been tightly associated with the adoption of a science-based approach to the industrialization of farming. It has allowed greater control by farmers of production factors and the simplification and homogenization of production situations particularly for internationally-traded commodities and

high-value crops (Allaire, 1996). This has enabled large surpluses of a narrow range of basic grains and protein foods to be generated, traded and also moved relatively quickly to meet emergency and humanitarian needs. It has eased hunger and reduced poverty as well as kept food prices stable and low relative to other prices and allowed investment in other economic sectors (FAO, 2004). However, the ecological and cultural context of farming is always and necessarily "situated" and cannot—unlike functions such as water use or carbon trading—be physically exchanged (Berkes and Folke, 1998; Hubert et al., 2000; Steffen et al., 2004; Lal et al., 2005; Pretty, 2005). Advances especially in the ecological sciences and socioeconomic research as well as drivers originating in civil society movements (2.2, 2.3) have mobilized science, knowledge and technology in support of approaches appreciative of place-specific, multidimensional and multifunctional opportunities (Agarwal et al., 1979; Byerlee, 1992; Symes and Jansen, 1994; Gilbert, 1995; de Boef, 2000; Fresco, 2002). Examples include (Cohn et al., 2006), trading arrangements connecting those willing to pay for specific ecological values and those who manage the resources that are valued (Knight, 2007), urban councils using rate levies to pay farmers for the maintenance of surrounding recreational green space or for ecosystem services such as spreading flood water on their fields; hydroelectric companies such as Brazil-Iguacú paying farmers to practice conservation tillage to avoid silting behind the dams and improve communal water supplies; farmers' markets; and community-supported agriculture.

An embedded activity. The resulting flows of products and services are embedded in a web of institutional arrangements and relationships at varying scales, such as farmers' organizations, industrial districts, commodity chains, *terroirs*, production areas, natural resource management areas, ethnic territories, administrative divisions, nations and global trading networks. Farmers are simultaneously members of a variety of institutions and relationships that frame their opportunities and constraints, offering incentives and penalties that are sometimes contradictory; farmers require strategic ability to select and interpret the relevant information constituted in these institutions and relationships (Chiffoleau and Dreyfus, 2004). The various ways of organizing science, knowledge and technology over the last sixty years have taken different approaches to farmers' strategic roles (see 2.1.2).

A collective activity. Farmers are not wholly independent entrepreneurs; their livelihoods critically depend on relationships that govern access to resources. With asymmetrical social relations, access is not equitably or evenly distributed. Individuals, groups and communities attempt to cope with inequalities by developing relational skills and capacity for collective action that help them to protect or enhance their access to and use of resources (Barbier and Lémery, 2000); the form that collective action takes changes over time and place and between genders. As commercial actors such as supermarkets have become dominant in food and farming systems, many farmers have transformed their production-oriented organizations into market-oriented organizations.

A disadvantaged activity. Agriculture is disadvantaged as an economic sector in the sense that the majority of small-scale producers and farm workers even today, in developing countries particularly, suffer from restricted access to formal education and opportunities to learn more about science and technology. Women and indigenous communities in particular tend to be more disadvantaged than others in this respect (Moock, 1976; Muntemba and Chimedza, 1995; ISNAR, 2002; IFAD, 2003; FAO, 2004; UNRISD, 2006). Investment in educating farmers in their principal occupation has been low compared to need throughout the period in most contexts. Master farmer classes, farmer field schools, study clubs, land care groups and interactive rural school curricula are among the options that have been developed in part simply to fill the gaps; few assessments exist of their comparative cost-effectiveness as educational investments. The potential of AKST to stimulate economic growth is affected in multiple ways by educational opportunity although these effects have not been well quantified (Coulombe et al., 2004; FAO, 2004). Overcoming educational disadvantages by contracting out extension to private suppliers as in Uganda poses new challenges (Ekwamu and Brown, 2005; Ellis and Freeman, 2006).

Wherever the structural and systemic disadvantages have been coupled to a lack of effective economic demand among cash-poor households, farmers in most parts of the developing world have been excluded also from formal decision making in agriculture and food policy and from priority setting in agricultural research unless special arrangements have been made to include them, such as the PRODUCE foundations in Mexico (Paredes and Moncado, 2000; Ekboir et al., 2006). Even under these arrangements it is the better educated and socially advantaged who participate; the inclusion of poor farmers, women, and laborers in research agenda-setting typically requires additional effort, for example by use of Citizen Juries (Pimbert and Wakeford, 2002). Given poor farmers' relative lack of education they also have been and remain vulnerable to exploitation in commercial relations (Newell and Wheeler, 2006), a growing problem as competitive markets penetrate deeper into rural areas. Market-oriented small-scale agriculture in developing countries is disadvantaged also by the huge and growing gap in the average productivity of labor between small-scale producers relying mainly on hand tools and the labor efficiency of farmers in areas that contribute the largest share of international market deliveries (Mazoyer, 2005; Mazoyer and Roudard, 2005).

2.1.1.2 The controversy on multifunctionality

How AKST should or could address multifunctionality is controversial; while some have sought to balance the multiple functions of agriculture, others have made tradeoffs among them, creating large variation in outcomes at different times and in changing contexts. The concept of multifunctionality itself has been challenged (Barnett, 2004). In general (Figure 2-1) it refers to agriculture as a multi-output activity producing not only commodities (food, fodder, fibers, biofuel and recently pharmaceuticals) but also non-commodity outputs such as environmental benefits, landscape amenities and cultural heritages that are not traded in organized markets (Blandford and Boisvert, 2002). The

frequently cited working definition proposed by OECD in turn associates multifunctionality with particular characteristics of the agricultural production process and its outputs: (1) the existence of multiple commodity and non-commodity outputs that are jointly produced by agriculture; and that (2) some of the non-commodity outputs may exhibit the characteristics of externalities or public goods, such that markets for these goods function poorly or are nonexistent (OECD, 2001).

A multi-country FAO study (2008), Roles of Agriculture, identified the multifunctional roles of agriculture at different scales (Table 2-1). The project's country case studies underlined the many cross-sector links through which agricultural growth can support overall economic growth and highlighted the importance to sustainable farming of balancing the interests of rural and urban populations; social stability, integration, and identities; food safety and food cultures and the interests of nonhuman species and agroecological functioning.

In the early years under review the multifunctionality of agriculture was undervalued in the tradeoffs made in technology choices and in formal AKS arrangements that were responding to urgent needs to increase edible grain output and high protein foods such as meat or fish. The success in meeting this essential but somewhat narrow goal tended to lock AKST into a particular pathway that perpetuated the initial post World War II focus. The political environment evolved in a direction that gave further stimulus to the organization of AKST devoted to the production of internationally traded goods (as advocated, for example, by the Cairns group of nations) rather than to sustaining multidimensional, place-based functionality in both its biophysical and sociocultural dimensions. This suited the circumstances of countries with large agricultural trade surpluses and relatively few small-scale producers in the areas where the surpluses were grown (Brouwer, 2004). For the majority of nations, agriculture throughout the period has

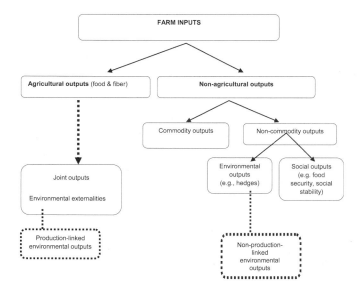

Figure 2-1. *Multiple outputs produced from farm inputs.* Source: Adapted from OECD, 2001; Verhaegen et al., 2002; Wustenberghs et al., 2004, 2005.

Table 2-1. **Roles of agriculture.**

Role	Environmental	Social	Food Security	Economic	Cultural
Global	Ecosystem resilience Mitigation of climatic change (carbon sequestration, land cover) Biodiversity	Social stability Poverty alleviation	Food security	Growth	Cultural diversity
Regional/National	Ecosystem resilience Soil conservation (erosion, siltation, salinization) Water retention (flood and landslide prevention) Biodiversity (agricultural, wild life) Pollution abatement/generation	Balanced migration Social stability (and sheltering effects during crisis) Unemployment prevention Poverty alleviation Gender relations	Access to food National security Food safety	Economic stability Employment Foreign exchange Tourism	Landscape Cultural heritage Cultural identity Social capital
Local	Ecosystem resilience Soils conservation Water retention Biodiversity Pollution abatement/generation	Social stability (employment, family)	Local and household food safety	Employment effects on secondary and tertiary sectors	Landscape Indigenous and local knowledge Traditional technologies Cultural identity

Source: Adapted from FAO, 2008

remained a domestic issue, based in part on large numbers of small-scale producers who still need to ensure basic food security and here a different calculus of interests (Conway, 1994). Countries such as Japan, Switzerland, Norway and the European Union opted for redirecting AKST toward maintaining the multifunctional capacity of agriculture once food surplus was assured (De Vries, 2000; Huylenbroeck and Durand, 2003; Sakamoto et al., 2007). In recent decades, changes in consumer demand and renewed emphasis by citizens on food quality, ethical issues, rural community livelihoods as well as changes in policy concerns (including resource conservation, tourism, biomass energy production and environmental sustainability) have led to expectations in many countries that agriculture will be able to play a balanced and sustainable role in meeting multifunctional goals (Cahill, 2001; Hediger and Lehmann, 2003; Rickert, 2004; Paxson, 2007).

Debates about multifunctionality were taken up by the OECD and FAO leading to a clarification of the policy implications and a broader recognition among trading partners that agriculture does play multiple roles and that AKST arrangements can and do have a part. The additional broad benefits potentially associated with multifunctional agriculture, including conservation of biodiversity, animal welfare, cultural and historical heritage values and the liability and viability of rural communities (Northwest Area Foundation, 1994; de Haan and Long, 1997; Cahill, 2001; Hediger and Lehmann, 2003) were in many countries returned to core AKST agendas. A growing body of evidence concerning the social and environmental costs of past and current tradeoffs among functions also began to be systematically quantified (Pimentel et al., 1992, 1993; Pretty and Waibel, 2005; Pretty, 2005a; Stern, 2006) as well as the benefits of reintroducing multifunctionality to industrial agricultural environments (NRC, 1989; Northwest Area Foundation, 1995; Winter, 1996; Buck et al., 1998). The role of local knowledge and technology processes also became more widely recognized

and formed the basis of AKST arrangements that sought to offer rural youth a motivation and realistic opportunities to stay in farming and develop agroenterprises (Breusers, 1998; FAO, 2004; Richards, 2005).

At some scales the multifunctional roles and functions that different agricultural systems actually play today are well described for many contexts and are noncontroversial. However, many of the variables are difficult to assess and are recognized as requiring the development of new knowledge routines if they are to be addressed adequately (Raedeke and Rikoon, 1997). In particular, some of the ecological and social goods, services and amenities that are not subject to commercial transactions have proven difficult to measure and hence in recent years greater reliance has been placed on developing alternatives. These include the use of relevant and efficient proxy indicators (Akca et al., 2005; Mukherjee and Kathuria, 2006), "water footprint" estimations (Chapagain and Hoekstra, 2003; Hoekstra and Chapagain, 2007) that show the extent to which farming systems, production practices, consumption patterns and the composition of agricultural trade affect net water balances at national levels (Chapagain and Hoekstra, 2003) and environmentally adjusted macroeconomic indicators for national economies (O'Connor, 2006). The experience has been mixed of applying these to actual decision-making. Developing and using computer-simulated modeling of multifunctionality (McCown et al., 2002) at field-scale (e.g., McCown et al., 1996) or farm-to-landscape scale (e.g., Parker et al., 2002) has led to robust applications in support of interactive learning among diverse users (Walker, 2002; van Ittersum et al., 2004; Nidumolu et al., 2007) seeking to balance interests in processes of adaptive management (Buck et al., 2001).

2.1.2 Knowledge processes

Knowledge processes refer to the collective processes of creating, transforming, storing and communicating about knowledge (Beal et al., 1986). The organization of knowledge processes in agricultural development has been subsumed in powerful mental models of how science, knowledge and technology "get agriculture moving" (Mosher, 1966; Borlaug and Dowswell, 1995). Each of the main models (Albrecht et al., 1989, 1990) has its own logic and fitness for purpose. They and their variants are discussed and compared; in each case for the sake of clarity they are first presented as commonly accepted abstractions followed by assessment of the dynamic ways in which the model has been applied within specific institutional arrangements in particular contexts. Institutional arrangements are important to the assessment because they provide different ways of distributing power and influence among sources of knowledge and hence are consequential for understanding the kinds of impact that can be expected and were in fact realized.

2.1.2.1 Transfer of Technology as a model for organizing knowledge and diffusion processes

One model in particular has dominated as a guide to the organization of knowledge processes in the public sector in developing countries, the Transfer of Technology (ToT) model. It was formally elaborated as a practical model for guiding action and investment in specific AKST arrangements on the basis of empirical studies of knowledge management and diffusion processes in the midwest of America (Lionberger, 1960; Havelock, 1969). Science is positioned in this model as a privileged problem-defining and knowledge generating activity carried out mainly by universities and research stations whose knowledge, embedded in technologies, messages, and practices is transferred by extension agents to farmers. The model assumes a linear flow of technological products and information. Each of the entities described in the model is treated more or less as a "black box." Although in practice much local level interaction takes place between extension agents, farmers and research specialists, the underlying assumption of the model is that farmers are relatively passive cognitive agents whose own knowledge is to be replaced and improved as a result of receiving messages and technologies designed by others and communicated to them by experts (Röling, 1988; Compton, 1989; Eastman and Grieshop, 1989; Lionberger and Gwin, 1991; Blackburn, 1994; Röling and Wagemakers, 1998).

The model mirrored the prevailing AKST organizational arrangements of states gaining their independence in the 1950s and 60s. Many explicitly favored centrally-planned economic development and most relied heavily on state organizations as the catalyst of agricultural development and commodity marketing (Hunter, 1969, 1970; Dayal et al., 1976). Extension field staff were positioned on the lowest rung in a hierarchy of relationships under the direction of departments of agriculture and publicly funded research stations and universities (Maunder, 1972; Peterson et al., 1989). Social, educational and political biases reinforced the idea that lack of access to "modern knowledge" was a constraint to production (Mook, 1974; Morss, et al., 1976). District development plans and projects to develop cooperatives, farmer service societies and the like received considerable attention (Halse, 1966; Lele, 1975; Hunter et al., 1976).

The ToT model assumes that wide impact is achieved on the basis of autonomous diffusion processes; this indeed can be so (Rogers, 1962). The classic study of *diffusion of innovations* was published in 1943 based on the rapid autonomous spread of hybrid maize among farmers in Iowa (Ryan and Gross, 1943). The diffusion of innovations became a popular subject for empirical social science research, generating well over 2000 studies and much was learned that was helpful concerning the conditions in which rapid and widespread diffusion can occur, what helps and hinders such processes and the limitations of diffusion for achieving impact. Diffusion research has continued even after the late Everett Rogers (well-known for his classic decadal overviews of research on the diffusion of innovations) (Rogers, 1962, 1983, 1995, 2003) himself spoke of the "passing of a dominant paradigm" (Rogers, 1976). The role of autonomous diffusion among farmers persists as one of the pillars of the common understanding of the pathways of science impact. The history of the rapid spread in Africa of exotic crops such as cassava, maize, beans and cocoa is added testimony to the power of diffusion processes to change the face of agriculture even without the kinds of scientific involvement of more recent years.

The positive impact of the ToT model. The ToT model gained credibility from the rapid and widespread adoption of the first products of the Green Revolution (GR) emerging from basic and strategic research (Jones and Rolls, 1982; Evenson, 1986; Jones, 1986; Evenson and Gollin, 2003a). For example, in the poor, populous, irrigated areas of Asia the GR allowed Bangladesh to move in 25 years from a net importer of rice to self sufficiency while its population grew from 53 million to 115 million (Gill, 1995) and India, Indonesia, Vietnam, and Pakistan to avert major famine and keep pace with population growth (Repetto, 1994). In China, wheat imports dropped from 7.2 tonnes in 1994 to 1.9 tonnes in 1997 and by 1997 net rice exports had risen to 1.1 tonnes. The Green Revolution not only increased the supply of locally available staples but also the demand for farm labor, increasing wage rates and thus the work-based income of the "dollar-poor" (Lipton, 2005). National food security in food staples in the high population areas of developing countries throughout the world was achieved except in sub-Saharan Africa. The diet of many households changed as more milk, fish and meat became available (Fan et al., 1998). Investment in industrialized food processing and in agricultural engineering, often stimulated by heavy government subsidies, in turn began to transform subsistence farming into a business enterprise and created new employment opportunities in postharvest operations i.e., storage, milling, marketing and transportation (Sharma and Poleman, 1993). The ToT model clearly proved fit for the overall purposes of disseminating improved seed, training farmers in simple practices and input use and disseminating simple messages within the intensive, high external input production systems characterizing the relatively homogeneous irrigated wheat and rice environments of South and Southeast Asia. Positive impacts were recorded also in parts of sub-Saharan Africa (Moris, 1981, 1989; Carr, 1989).

The ToT model's drawbacks with respect to development and sustainability goals. Criticism of the ToT model began to emerge strongly in the late 1970s as evidence of negative socioeconomic and environmental impacts of the GR accumulated (UNRISD, 1975; Freebairn, 1995) leading to sharp controversies that are still alive today (Collinson, 2000). Sometimes a technology itself was implicated; in other cases the institutional and economic conditions for using a new technology effectively and safely were not in place or the services needed for small-scale producers to gain access to or realize the benefits were inadequate, especially for the resource-poor, the indigent and the marginalized and women (Hunter, 1970; Roling et al., 1976; Ladejinsky, 1977; Swanson, 1984; Jiggins, 1986). The loss of entitlements to subsistence brought about by changes in the agricultural sector itself and in societies as a whole; weather-related disasters; civil unrest; and war also left many millions still vulnerable to malnutrition, hunger, and starvation (Sen, 1981; Johnson, 1996). The evidence highlighted three areas of concern:

Empirical: The ToT model was shown to be unfit for organizing knowledge processes capable of impacting heterogeneous environments and farming populations (Hill, 1982) and did not serve the interests of resource-poor farmers in risky, diverse, drought prone environments (Chambers, 1983). In the absence of measures to address women's technology needs and social condition, technologies transferred through male-dominated extension services largely bypassed women farmers and women in farm and laboring households (Hanger and Moris, 1973; Leonard, 1977; Harriss, 1978; Buvinic and Youssef, 1978; Fortmann, 1979; Bettles, 1980; Dauber and Cain, 1981; Evans, 1981; Deere and de Leal, 1982; Safilios-Rothschild, 1982; Mungate, 1983; Carloni, 1983; IRRI, 1985; Gallin and Spring, 1985; Muzale with Leonard, 1985; Nash and Safa, 1985; Staudt, 1985; Gallin et al., 1989; Gallin and Ferguson, 1991; Samanta, 1995). In addition, the improved seeds rapidly displaced much of the genetic diversity in farmers' fields that sustained local (food) cultures (Howard, 2003) and which had allowed farmers to manage place-dependent risks (Richards, 1985); the higher use of pest control chemicals in irrigated rice in the tropics had detrimental effects on beneficial insects, soils and water (Kenmore et al., 1984; Georghiou, 1986; Gallagher, 1988; Litsinger, 1989) as well as on human health (Whorton et al., 1977; Barsky, 1984). The evidence of negative effects on equity was claimed by some to be a first generation effect. Analysis of data from the Northern Arcot region of Tamil Nadu, India, indicated that the differences in yield found between large and small-scale producers in the 1970s had disappeared by the 1980s (Hazell and Ramaswamy, 1991) but further empirical studies failed to resolve the extent to which the second generation effects were the result of "catch up" by later adopters or the result of smaller farmers having lost their land or migrated out of farming (Niazi, 2004).

Theoretical: A basic assumption of the ToT model that "knowledge" can be transferred was shown to be wrong. It is information and communications about others' knowledge and the products of knowledge that can be shared (Beal et al., 1986). No one is merely a passive "receiver" of information and technology since every one engages in the full range of knowledge processes as a condition of human survival (Seligman and Hagar, 1972; Maturana and Varela, 1992; Varela et al., 1993). Information about people's existing knowledge, attitudes and practices was found to be a poor predictor of their response to new ideas, messages, or technologies because knowledge processes and behaviors interact with the dynamic of people's immediate environment (Fishbein and Ajzen, 1975). The organization of processes for generating knowledge that is effective in action (Cook and Brown, 1999; Hatchuel, 2000; Snowden, 2005) was shown to take many forms. Where the rights of individuals and communities to be agents in their own development and considerations of equity, human health, and environmental sustainability were important policy goals, the comparative advantages of the ToT model also appeared less compelling (Jones and Rolls, 1982; de Janvry and Dethier, 1985; Swanson, 1984; Jones, 1986).

Practical: The mix of organizational support and services needed to gain maximum impact from the ToT model often were inadequate, imposed high transaction costs or were not accessible to the poor and to women (Howell, 1982; Korten and Alfonso, 1983; Ahmed and Ruttan, 1988; Jiggins, 1989). The positive role of local organizations as intermediaries in rural development was demonstrated but also the tendency for agricultural services organized along ToT lines to bypass these (Esman and Uphoff, 1984). The credit markets introduced to support technology adoption for instance typically were selective and biased in favor of resource rich regions and individuals (Howell, 1980; Freebairn, 1995) although pioneering initiatives such as the Grameen Bank in Bangladesh demonstrated that alternative approaches to the provision of microcredit to poor producers, women and farm laborers were possible (Yunus, 1982). Institutional analyses demonstrated how and why ToT arrangements that worked well in one context might fail to perform as well when introduced into other contexts. A recent authoritative assessment concludes that after "twenty-five years in which agricultural extension received the highest level of attention it ever attracted on the rural development agenda" political support for ToT in the form of "relatively uniform packages of investments and extension practices in large state and national programs" had disappeared (Anderson et al., 2006).

2.1.2.2 Other models of knowledge generation and diffusion processes

By the early 1970s, empirical studies and better theoretical understanding indicated that better mental models of knowledge processes were needed to guide practice if broader development goals were to be reached (Hunter, 1970). The first wave of institutional innovation in the organization of knowledge processes in noncommunist states sought to make more effective the process of moving science "down the pipeline" and technologies "off the shelf" by creating mechanisms and incentives for obtaining feedback from producers so that their local knowledge and priorities could be taken into account in targeting the specific needs of different categories of farmers. *The Training and Visit (T&V) approach* is a particularly well known example of this effort (Benor et al., 1984). It was heavily supported by the World Bank and became standard practice in the majority of noncommunist developing countries. Among other aims it sought to strengthen the management of diffusion processes by selection of "contact" or "leading farmers" and in some cases also contact groups. Extension agents report back "up the line" the problems and priorities of the farmer and farmer groups that they trained during their fortnightly field visits (Benor et al., 1984). The T&V approach was criticized almost from its inception as an inadequate response to the widespread evidence of the limitations of ToT approaches (Rivera and Schram, 1987; Howell, 1988; Gentil, 1989; Roberts, 1989). Little remains today of national T&V investments and service structures (Anderson et al., 2006).

Farming systems research and extension (FSRE) is another well-known response. In this model, feedback came directly through diagnostic surveys carried out by multidisciplinary teams, by farm level interactions between researchers and farmers in the course of technology design, testing and adaptation and by the organization of farmer visits to research stations (Rhoades and Booth, 1982; Bawden, 1995; Collinson, 2000). Wide impact in this case was sought by the designation of farming systems within agroecological "recommendation domains" for which a specific technology or practice was designed to be effective and profitable. FSRE practitioners explicitly took into account the contextual conditions that might compromise the effectiveness or profitability of a problem-solution as well as sociocultural factors such as women's roles in farming. How well they managed to do so was disputed (Russell et al., 1989). FSRE produced interesting results but failed to have wide impact. Although largely abandoned as an institutional arrangement, its influence lived on (Dent and McGregor, 1994) through methodological innovations addressing the highly differentiated livelihood needs of the rural poor (Dixon and Gibbon, 2001), the stimulus it gave to revaluation of the multifunctionality of farming (Pearson and Ison, 1997) and the ways in which it forged connections across scientific disciplines that endure within the organizational arrangements of numerous research communities (Engel, 1990).

Neither T&V nor FSRE addressed the institutional challenge of creating "the mix" of support services necessary for articulating innovation along the chain from producer to consumer (Lionberger, 1986). In the private commercial sector the production of tea, coffee, palm oil, rubber, pineapples and similar commodities in the small-scale sector typically used the *core-estate-with-out-growers model* to address the challenge (Chambers, 1974; Hunter et al., 1976; Compton, 1989), positioning producers under contract to supply outputs to a processing facility that provided inputs and services. The company assumed responsibility for assembling the scientific and market knowledge required as well as the technology and infrastructure for securing company profits, drawing largely on knowledge resources in the home country or from within the company's international operations. The approach provided reliable income to producers, employees and companies and through commodity taxes or export levies to governments. It was criticized for locking small-scale producers into low income contracts. It also proved open to corruption when applied through government owned Commodity Boards, with profits siphoned off to intermediaries and elites (Chambers and Howe, 1979; Sinzogan et al., 2007).

The challenge was addressed in Communist states by state seizure of the means of production and by state control of the provision of inputs and services and the distribution of the product. The scientific knowledge base to support such a high degree of planning was strong. However, the means chosen within the prevailing ideology to translate knowledge generated at the scientific level into knowledge that was effective for practice was based on *command and control*. Support to the knowledge processes and experiential capacity of those actually working the land—albeit under direction of others—was not encouraged. In the exceptional historical experiences of states such as Cuba (Carney, 1993; Wright, 2005) or Vietnam state-directed knowledge processes contributed to basic food security but in general the command and control approach did not prove efficient in generating surplus nor a continuing stream of innovation in agriculture and became a source of vulnerability for state survival (Gao and Li, 2006). Since the fall of the Berlin Wall in 1989, the command and control model has been largely abandoned.

A parallel wave of innovation in the organizational design of knowledge processes was centered in producers' own capacity to engage in "knowledge work" and on the role of local organizations in meeting development and sustainability goals (Chambers and Howes, 1979; Chambers, 1981). Models for what became known as Farmer Participatory Research and Extension (FPRE) were elaborated in practice by drawing on local traditions of association, knowledge generation and communication. Experience generated under labels such as Participatory Learning and Action Research, Farmer Research Circles, Community Forestry, Participatory Technology Development and FAO's People's Participation Program (Haverkort et al., 1991; Scoones and Thompson, 1994; Ashby, 2003; Coutts et al., 2005; IIRR, 2005) showed that if time is taken to create effective and honest partnerships in FPRE the results are significant and can offer new opportunities to socially marginalized communities and those excluded under other knowledge arrangements. They share a number of generic features *viz.* learner-centered, place dependent, ecologically informed and use of interactive communication and of facilitation rather than extension skills (Chambers and Ghildyal, 1985; Ashby, 1986; Farrington and Martin, 1987; Gamser, 1988; Biggs, 1989; Haverkort et al., 1991; Ashby, 2003). Science and off-the-shelf technologies are positioned as stores of knowledge and as specialized problem-solving capacities that can be called upon as needed. An FPRE approach has been used for example in the development and promotion of on-farm multipurpose tree species in Kenya (Buck, 1990) that had wide-scale impact and complemented the mobilization of women in tree-planting under the Green Belt movement (Budd et al., 1990). The development and promotion through farmer-to-farmer communication and training of a range of soil fertility and erosion control techniques in Central America similarly was based on an FPRE approach

(Bunch and Lopez, 1994; Hocdé et al., 2000; Hocdé et al., 2002) as were integrated rice-duck farming in Bangladesh (Khan et al., 2005) and the testing and adaptation of agricultural engineering prototypes by farmer members of the Kondomin Group network in Australia. Nongovernment organizations (NGOs), community-based organizations (CBOs), universities and the Consultative Group on International Agriculture Research (CGIAR) played key roles in elaborating effective practice and supporting local FPRE initiatives (Lumbreras, 1992; Dolberg and Petersen, 1997; IIRR, 1996, 2005).

Participatory Plant Breeding (PPB) is a particular adaptation of FPRE: its client-oriented interactive approach to demand-driven research has been shown to be particularly effective for grains, beans and roots (de Boef et al., 1993; Sperling et al., 1993; Farrington and Witcombe, 1998; CIAT, 2001; Fukuda and Saad, 2001; Chiwona-Karltun, 2001; Mkumbira, 2002; Ceccarelli et al., 2002; Witcombe et al., 2003; Virk et al., 2005). It is a flexible strategy for generating populations, pure lines and mixes of pure lines in self-pollinated crops as well as hybrids, populations, and synthetics in cross-pollinated crops. Biodiversity is maintained or enhanced because different varieties are selected at different locations (Joshi et al., 2001; Ceccarelli et al., 2001ab). Recent assessments of over 250 participatory plant breeding projects in over 50 countries in Latin America, Europe, south and southeast Asia and sub-Saharan Africa led by farmers, NGOs or by national or international researchers or some mix of these actors (Atlin et al., 2001; Joshi et al., 2001; Cleveland and Soleri, 2002; Ashby and Lilja, 2004; Almekinders and Hardon, 2006; Mangione, 2006; Ceccarelli and Grando, 2007; Joshi et al., 2007) demonstrate that PPB is a cost-effective practice that is best viewed along a continuum of plant breeding effort. French researchers, e.g., are working with marker-assisted selection to develop virus resistant rice varieties for Central America and the Cameroon in the context of PPB activities (www .ird.fr/actualites/2006/fas247.pdf). GIS and satellite-based imaging are adding additional value to PPB activities.

While over 8000 improved varieties of food grains with wide adaptability have been released over a 40-yr period by the CGIAR institutes (Evenson and Gollin, 2003b), PPB has shown capacity to generate multiples of this output for target environments, specific problems and the needs of farmers overlooked by conventional breeding efforts. The three major differences of PPB compared to conventional breeding are that testing and selection take place on the farm instead of on-station; the key decisions are taken jointly by farmers and breeders; the process can be independently implemented in a large number of locations. The activity also incorporates seed production with farmers multiplying promising breeding material in village-based seed production systems. The assessments also highlights the improved research efficiencies and program effectiveness gained by faster progress toward seed release and the focus on the multiplication of varieties known to be farmer-acceptable. Decentralized selection in target environments for specific adaptations allows women's seed preferences to be addressed (Sperling et al., 1993; Ashby and Lilja, 2004; Almekinders and Hardon, 2006). Sustained PPB activity has the additional advantage of bringing about the progressive empowerment of individual farmers and farmer communities (Almekinders and Hardon, 2006; Cecccarelli and Grando, 2007). However, the tightening of UPOV regulations and the increasing trend toward seed patenting and IPR over genetic material has given rise to concern (Walker, 2007) that despite PPB's demonstrated advantages in a wide variety of contexts and for multiple purposes the space for PPB may be closing.

As the case of PPB shows, wider scale impact in the case of FPRE relies on the replication of numerous initiatives in response to specific markets and non-market demands rather than on supply-push and diffusion of messages or technologies, although diffusion processes can and do amplify the outcomes of FPRE. The process of replication can be strengthened through investment in farmer-to-farmer networking (Van Mele and Salahuddin, 2005), support to farmer driven chain development (as in poultry or dairy chains serving local markets) and in the creation of "learning alliances" among support organizations that aim to promote shared learning at societal scales (Pretty, 1994; Lightfoot et al., 2002). FPRE has proved to be cost-effective and fit for the purposes of meeting integrated development and sustainability goals (Bunch, 1982; Hyman, 1992) and for natural resource management (NRM) in agrarian landscapes (Campbell, 1992, 1994; Hilhorst and Muchena, 2000; CGIAR, 2000; Stroosnijder and van Rheenen, 2001; Borrini-Feyerabend et al., 2004). However, it has been criticized for failing in specific cases to take advantage of the "best" science and technology available, as self-indulgent by supporting farm systems that some consider insufficiently productive to provide surplus to feed the world's growing urban populations; as sometimes misreading the gender power dynamics of local communities (Guijt and Shah, 1998) and as incapable of involving a sufficient number of small-scale producers (Biggs, 1995; Richards, 1995; Cooke and Kothari, 2001). NGOs and community-based organizations have raised issues of equity. It has also been criticized as too locally focused (see critiques of Australia's Landcare experience in Lockie and Vanclay, 1997; Woodhill, 1999) and thus unable to address higher level economic and governance constraints and tradeoffs. This criticism has prompted recent institutional experimentation with applying FPRE under catchment scale regional development authorities (Australia) and in sustainable water development (South Africa and Europe) (Blackmore et al., 2007) within normative policy frameworks that explicitly seek the sustainability of both human activity and agroecologies.

Innovations in the organization of knowledge processes also occurred in relation to farmer-developed traditions of agroecological farming (e.g., Fukuoka, 1978; Dupré, 1991; Gonzales, 1999; Furuno, 2001), gathering and domestication of wild foods and non-timber forest products (Scoones et al., 1992; Martin, 1995) and landscape management (Fairhead and Leach, 1996). For example migrants from the Susu community first encountered the rice-growing ethnic Balantes in Guinea Bissau around 1920; later on, the Susu (and the related Baga peoples) hired migrant Balantes to carry out rice cultivation in the brackish waters of coastal Guinea Conakry where the skills are now recognized as traditional knowledge (Sow, 1992; Penot, 1994).

Indigenous long-standing technologies include the use of Golden Weaver ants as a biocontrol in citrus and mango orchards (Bhutan, Vietnam and more recently, with WAR-DA's assistance, introduced to West Africa); stone lines and planting pits for water harvesting and conservation of soil moisture (West African savannah belt); *qanats* and similar underground water storage and irrigation techniques (Iran, Afghanistan and other arid areas); tank irrigation (India, Sri Lanka); and many aspects of agroforestry, e.g., rubber, cinnamon, and damar agroforests in Indonesia. Over the years they have supported wildlife and biodiversity and rich cultural developments.

It is this continuing *indigenous capacity for place-based innovation* that has been almost entirely responsible for the initial bringing together of the science, knowledge and technology arrangements for what have become over time certified systems of agroecological farming such as organic farming, confusingly known also as biological or ecoagriculture; (Badgely et al., 2007) and variants such as permaculture (Mollison, 1988; Holmgren, 2002). Systems such as these are knowledge-intensive, tend to use less or no externally supplied synthetic inputs and seek to generate healthy soils and crops through sustainable management of agroecological cycles within the farm or by exchange among neighboring farms. Although there is considerable variation in the extent to which the actors in diverse settings initially drew on formal science and knowledge, as the products have moved onto local, national, and international markets under various certification schemes the relationships between formal AKST actors and producer organizations have become stronger along the entire chain from seed production to marketing (Badgely et al., 2007). A distinctive feature in these arrangements is the role of specialist farmers in producing certified seed on behalf of or as members of producer organizations.

The relative lack of firm evidence of the sustainability and productivity of these kinds of certified systems in different settings and the variability of findings from different contexts allows proponents and critics to hold entrenched positions about their present and potential value (Bindraban and Rabbinge, 2005; Tripp, 2005; Tripp, 2006a). However, recent comprehensive assessments conclude that although these systems have limitations, better use of local resources in small scale agriculture can improve productivity and generate worthwhile innovations (Tripp, 2006b) and agroecological/organic farming can achieve high production efficiencies on a per area basis and high energy use efficiencies and that on both these criteria they may outperform conventional industrial farming (Pimentel et al., 2005; Sligh and Christman, 2007; Badgely et al., 2007). Despite having lower labor efficiencies than (highly mechanized) industrial farming and experiencing variable economic efficiency, latest calculations indicate a capability of producing enough food on a per capita basis to provide between 2,640 to 4,380 kilocalories/per person/per day (depending on the model used) to the current world population (Badgely et al., 2007). Their higher labor demand compared to conventional farming can be considered an advantage where few alternative employment opportunities exist. Organic agriculture as a certified system by 2006 was in commercial practice on 31 million ha in 120 countries and generating US$40 billion per year.

Innovations with comparable goals but originating in private commercial experience (Unilever, 2005) or in the context of partnerships among a range of farmers' organizations, public and private commercial enterprises by the mid 1990s were reported with increasing frequency (Grimble and Wellard, 1996). The Northwest Area Foundation experience in the USA (Northwest Area Foundation, 1995), the New Zealand dairy industry (Paine et al., 2000) or farming and wildlife advisory groups in the UK are among the numerous compelling examples of an emerging practice. They indicate a convergence of experience toward a range of options for bringing multifunctional agriculture into widespread practice in diverse settings by working with farmer-participatory approaches in combination with advanced science solutions (Zoundi et al., 2001; Rickert, 2004).

The *continuing role of traditional and local knowledge in AKST* for most of the world's small-scale producers in generating innovations that sustain individuals and communities also merits highlighting. Indigenous knowledge (IK) is a term without exact meaning but it is commonly taken to refer to locally bound knowledge that is indigenous to a specific area and embedded in the culture, cosmology and activities of particular peoples. Indigenous knowledge processes tend to be nonformal (even if systematic and rigorous), dynamic and adaptive. Information about such knowledge is usually orally transmitted but also codified in elaborate written and visual materials or artifacts and relates closely to the rhythms of life and institutional arrangements that govern local survival and wellbeing (Warren and Rajasekaran, 1993; Darré, 1999; Hounkonnou, 2001). Indigenous and local knowledge actors are not necessarily isolated in their experience but actively seek out and incorporate information about the knowledge and technology of others (van Veldhuizen et al., 1997). Sixty years ago such knowledge processes were neglected except by a handful of scholars. From the 1970s onward a range of international foundations, NGOs, national NGOs and CBOs began working locally to support IK processes and harness these in the cause of sustainable agricultural modernization, social justice and the livelihoods of the marginalized (IIRR, 1996; Boven and Mordhashi, 2002). Much more is known today about the institutional arrangements that govern the production of IK in farming (Colchester, 1994; Howard, 2003; Balasubramanian and Nirmala Devi, 2006). Poverty and hunger persist at local levels and among indigenous peoples and this indeed may arise from inadequacies in the knowledge capacity of rural people or the technology available, but field studies of knowledge processes of indigenous peoples, their empirical traditions of enquiry and technology generation capabilities (Gonzales, 1999) establish that that these also can be highly effective at both farm (Brouwers, 1993; Song, 1998; Hounkounou, 2001) and landscape scales (Tiffen et al., 1994; Darré, 1995). IK related to agriculture and natural resource management is assessed today as a valuable individual and social asset that contributes to the larger public interest (Reij et al., 1996; Reij and Waters-Bayer, 2001; World Bank, 2006) and likely to be even more needed under mitigation of and adaptation to climate change effects.

However, empirical research shows how economic drivers originating in larger systems of interest tend to undermine the autarchic gains made at local levels or to block further development and upscaling (Stoop, 2002; Unver, 2005). A major challenge to IK and more broadly to FPRE over the last few decades has been the emergence of IPR regimes (see 2.3.1) that so far do not adequately protect or recognize individual farmers' and communities' ongoing and historic contributions to knowledge creation and technology development or their rights to the products and germplasm created and sustained under their management. Even so, innovative ways forward can be found: formal breeders and commercial organizations in the globally important Dutch potato industry cooperate with Dutch potato hobby specialists in breeding and varietal selection; farmers negotiate formal contracts which give them recognition and reward for their intellectual contribution in all varieties brought to market.

The inequities in access and benefit sharing under the various protocols and conventions negotiated at international levels have given rise to a strong civil society response (2.2.1; 2.2.3) reflected in the Declaration on Indigenous Peoples' rights to genetic resources and IK—a collective statement on an international regime on access and benefit sharing issued by the indigenous peoples and organizations meeting at the Sixth Session of the United Nations Permanent Forum on Indigenous Issues, in New York on 14-25 May, 2007 (ICPB-Net Indigenous Peoples' Council on Biocolonialism, http://lists.ipcb.org/listinfo.cgi/ipcb-net-ipcb.org). Recent experience with the development of enforceable rights for collective innovations (Salazar et al., 2007) offers ground for evolution of currently dominant IPRs. There are new concerns that clean development mechanisms (CDMs), international payments for environmental services or payments for avoided deforestation and/or degradation will override the rights of indigenous peoples and local communities.

The final model considered here is by far the most dominant model of knowledge processes associated with commercial innovation in the private sector, *the chain-linked model* (Kline and Rosenberg, 1986). A distinctive feature is the effort made throughout every stage of product development to obtain feedback from markets and end users (Blokker et al., 1990); it is demand-driven rather than supply-push. It has given significant impulse to the development of market economies wherever the enabling conditions exist but has had little to offer where science organizations have remained weak and consumer markets are unable to articulate monetary demand—as in fact has been the case for much of the period among the rural and urban poor and especially among women and other marginalized peoples.

The recent emphasis among policy makers on developing market-oriented and market-led opportunities along entire value chains for small-scale producers and other rural people (DFID 2002, 2005; NEPAD, 2002; IAC, 2004; FAO, 2005c; UN Millennium Project, 2005; World Bank, 2005c; OECD, 2006a) has created wider interest in the model as a platform where diverse actors in public-private partnerships can find each other and organize their respective roles. Today it is being extended with varying energy mainly in the "new consumer economies" i.e., countries with populations over 20 million (Argentina, Brazil, China, Colombia, India, Indonesia, Iran, Malaysia, Mexico, Pakistan, Philippines, Poland, Russia, Saudi Arabia, South Africa, South Korea, Thailand, Turkey, Ukraine, Venezuela). However, evidence of the extent to which small-scale producers can participate effectively, if at all, in these arrangements in the absence of

strong producers' organizations (Reardon et al., 2003) and of the impact on knowledge management (Spielman and Grebner, 2004; Glasbergen et al., 2007) has shown that the interests of private research and public-private partners may diverge from the combined public interest goals of equity, sustainability and productivity. Holding on to benefits may be difficult for employees and national research systems in globalizing markets as the recent rapid switch of a number of commercial cut flower operations from Kenya to Ethiopia illustrates, while global retailers' ability to determine price, quality, delivery and indirectly also labor conditions for suppliers and producers in the chain means that the burdens of competition may be transferred to those least able to sustain them (Harilal et al., 2006).

2.1.2.3 New challenges and opportunities
Transfer of technology has become important in recent years as a means of shifting technological opportunity and knowledge among private commercial actors located in different parts of the world and through science networks that stretch across geographic boundaries. It continues to guide practice as a means of promoting farm level change in what are still large public sector systems in countries such as China (Samanta and Arora, 1997). However, increasingly ToT has to find its place in an organizationally fragmented and complex context that emphasizes demand-driven rather than supply-push arrangements (Rivera, 1996; Leeuwis and van den Ban, 2004; Ekwamu and Brown, 2005). The shift toward contracting or other forms of privatization of research, extension and advisory services in an increasing number of countries (Rivera and Gustafson, 1991; Byerlee and Echevveria, 2002; Rivera and Zijp, 2002; van den Ban and Samantha, 2006) is an effort to reorganize the division of power among different players in AKST. In the process the central state is losing much of its ability to direct technological choice and the organization of knowledge processes. The effects and the desirability in different contexts of altering the balance between public and private arrangements remain under debate as the expanding diversity of financing and organizational arrangements has not yet been fully assessed (Allegri, 2002; Heemskerk and Wennink, 2005; Pardey et al., 2006a).

Decentralization and devolution of development-related governance powers from central to more local levels in an increasing number of developing countries has opened the space for many more instances of FPR&E in an increasingly diverse array of partnerships that are not easy to classify and demand new frames of understanding (Dorward et al., 1998; AJEA, 2000). At the same time, the push for export-oriented agriculture and in an increasing number of countries, the strong growth in domestic consumer demand has opened the space for the chain-linked model to be expressed more widely and with deeper penetration into small-scale farming communities. In addition, the "core estate-without-growers" model has taken on new life as international food processors and retailers contract organized producer associations to produce to specification. The partnership between IFAD and the Kenya Tea Development Authority to introduce sustainable production techniques to small-scale outgrowers by means of Field Schools is a strong example of how changing values in consuming countries can have positive knock-on effects for the poor. Some models are more fit than others for meeting development and sustainability goals (Table 2-2).

The growing recognition of the complexity of knowledge processes and relations among a multiplicity of diverse actors has led to renewed attention to the role of *information and communication processes* (Rogers and Kincaid, 1981). All parties in communication play roles of "senders" and "receivers," "encoders" and "decoders," of information but communication typically is neither neutral nor symmetric: empirical studies demonstrate the extent to which social, cultural and political factors determine whose voices are heard and listened to (Holland and Blackburn, 1998). The history of the last sixty years may be read in part as a history of struggle to get the voices of the poor, of women and other marginalized people heard in the arenas where science and technology decisions are made (Leach et al., 2005; IDS, 2006).

By the 1980s the technologies of the digital age began to revolutionize the ability to obtain and disseminate information. Computer communication technologies and mobile telephony are becoming available to populations in developing countries (ITU, 2006). Mobile telephony by end 2006 had become a US$25 billion industry across Africa and the Middle East and Indian operators were signing up 6.6 million new subscribers a month. In the last five years low cost mobile telephony has begun to overtake computer-based technology as the platform for information-sharing

and communication. For the first time, poor producers in remote places no longer have to remain isolated from market actors or to rely on bureaucrats or commercial middlemen for timely market information (Lio and Meng-Chun, 2006). Initiatives such as TradeNet (Ghana) connect buyers and sellers across more than ten countries in Africa and Trade at Hand provides daily price information to vegetable and fruit exporters in Burkina Faso and Senegal.

The new ICTs are also opening up formal education opportunities, ranging from basic literacy and numeracy courses to advanced academic, vocational and professional training. Free online libraries (e.g., IDRIS) and new institutional arrangements offer potential for further innovation in knowledge processes. For instance, the Digital Doorway, a robust portable computer platform with free software for downloading information, is being initiated at schools and community forums throughout southern Africa by Syngenta and the University of Pretoria to support locally adapted curricula for Schools in the Field covering a range of crops, animals, poultry, small rural agroenterprises and soil and water management. Insufficient information is available as yet to make robust assessments of these trends but the early evidence is that their impact may be at least as important as technologies originating within AKST development. Nonetheless, the rate of expansion of access to modern ICTs continues to be much greater in developed than developing countries and among urban more than rural populations, raising concerns about how to avoid ICTs reinforcing

Table 2-2. Characteristics of models of knowledge processes in relation to fitness for purpose.

Model	Model Characteristics	Fit for Purpose
ToT	Science as the source of innovation; linear communication flows through hierarchically organized linkages; farmers as passive cognitive agents serving public interests	Productivity increase on the basis of substitutable technologies, simple messages, simple practices; catalyzing Cochrane's "treadmill" (1958) i.e., forcing farmers to adopt the latest price-cutting, yield increasing measures in order to stay competitive in the market. Not fit for promoting complicated technologies & management practices, complex behavior change, and landscape scale innovations
Farmer-Scientist Collaboration	Innovations as place dependent and multi-sourced, based on widely distributed experimental capacity; communication flows multi-sided, through networked social and organizational linkages among autonomous actors serving their own interests	Socially equitable, environmentally sustainable livelihood development at local levels, multi-stakeholder landscape management, and empowerment of self-organizing producers and groups. Not fit for rapid dissemination of simple messages, substitutable technologies, simple practices
Contractual Arrangements	Science as an on-demand service to support production to specification; communication flows framed by processors' and retailers' need to supply to known market requirements; farmers as tied agents serving company interests	Sustains yield and profit in company interests; can be environmentally sustainable but not necessarily so. Contractual arrangements can trap poor farmers in dependent, unequal relationships with the company. Crop focused, thus not fit for promoting whole system development or landscape scale innovations
Chain-linked	Science as a store of knowledge and a specialized problem-solving capacity; structured communication among product/technology development team around iterative prototyping, continuously informed by market information; farmers sometimes as team members but primarily as market actors serving private interests	Motor of innovation in the private commercial sector in the presence of monetized markets, consumers able to articulate demand, and adequate science capacity. Increasingly, practitioners have begun to internalize within company R&D practices a range of environmental and sustainable livelihood concerns—the "triple bottom line"—under pressure from citizens and regulation

existing patterns of inequality (Gao and Li, 2006).The history of broadcast radio suggests that over time the "digital divide" may become narrower. Issues of the quality and relevance of the information available are likely to become more important than those of access and ability to use the technology.

2.1.3 Science processes
Science processes are those involved in the creation and dissemination of scientific knowledge; including processes within the scientific community and interactions between scientific communities and other actors. Members of a scientific community are defined here as those who are principally involved as professional actors in such activities as pre-analytic theorizing, problem identification, hypothesis formulation and testing through various designs and procedures (such as mathematical modeling, experimentation or field study), data collection, analysis and data processing and critical validation through peer review and publication, i.e., activities commonly viewed as core practices of scientists.

Intellectual investment in these activities by individual scientists is driven in part by human motivations such as curiosity and the pleasure of puzzle-solving but also by the structure of professional incentives that encourages—even demands—that scientists pay closer attention to obtaining the recognition of their work by peers in the scientific community rather than by other segments of society. However, scientific institutions cannot ignore the preoccupations and knowledge wielded by other actors (Girard and Navarette, 2005) and in other societal forums. This is particularly obvious in the case of agriculture; no matter the science involved in the origin and initial development of an idea, to be effective it has to become an applied science with potential for wide impact whose results are visible to all in the form of changes in agricultural landscapes. Thus it is unsurprising that opinions and drivers outside the domain of science itself condition science for agriculture. This tension between the incentives faced by individual scientists and the societal demands placed on scientific institutions in agriculture has been growing in recent decades, posing a strong challenge for the governance of scientific institutions (Lubchenco, 1998).

2.1.3.1 Cultures of science
Agricultural science processes in our period have been associated with the cultures of thought distinguished by two intellectual domains known respectively as "positivist realism" and "constructivism". The positivist realist understanding of modern science as a neutral, universal, and value-free explanatory system has dominated the processes of scientific inquiry in agriculture for the period under review. The basic assumptions are that reality exists independently of the human observer (realism), and can be described and explained in its basic constitution (positivism).This mind set is legitimate for the work that professional scientists do and enables transparent and rigorous tests of truth to guide their work. However, others (Kuhn, 1970; Prigogine and Stengers, 1979; Bookchin, 1990; Latour, 2004) have found this scheme problematic for explaining causality in their own disciplines for a number of reasons: it appears to exclude the qualitative (even if quantitative) ambiguous and highly contextualized interpretations that human subjects

give to the meaning of reality and it does not allow sufficiently for the unpredictability of the social effects of any intervention nor for the reflexive nature of social interactions (the object of enquiry never stabilizes; learning that something has happened changes decisions about what actions to take, in an unending dance of co-causality). This difference in legitimate perspective provides a partial explanation of why "the history" of the last sixty or so years cannot stabilize around a single authoritative causal interpretation of what has happened.

For scientists working within positive realist traditions the locus of scientific knowledge generation is largely confined to public and private universities, independent science institutions and laboratories and to an increasing extent corporate research and development (R&D) facilities. These offer the conditions for highly specialized expertise to be applied to study of immutable laws governing phenomena that allow for prediction and control. Technology is conceived in this logic as applied science, i.e., as a design solution developed by experts removed from the site of application. The main task of the agricultural sciences in this perspective thus becomes that of developing the best technical solutions to carefully described problems (Gibbons et al., 1994; Röling, 2004). The problem description can and often does include scientists' understanding of environmental and social dimensions.

The paradigm of positive realism has attracted large-scale support for public and private science institutions as a way of thinking about and organizing innovation in tropical agriculture. It was harnessed to the expectation of maximizing yields and compensating for shortfalls in the quantity or quality of the biotic and abiotic factors of production by the provision of supplementary inputs, such as fertilizers and services to improve the productivity of labor and land. As such this paradigm lies at the heart of what is often called "productivism", a doctrine of agricultural modernization giving primary emphasis to increased productivity rather than the multifunctionality of agriculture or to the role of agriculture in rural development. It has constituted for much of the period under review a primary justification for science investments for development (Evenson et al., 1979).

The dominance of this paradigm has had notable institutional consequences. University agricultural faculties progressively became divided into highly specialized departments. This split created "knowledge silos" that reflected the increasing specialization of scientific disciplines that reduced agriculture as an integrated practice into smaller and smaller fractions that largely excluded the human manager. This reductionism made it harder to mobilize multidisciplinary teams to address more complex problems (Bentley, 1994) and was consistent with the increasing specialization in modern farm sectors, developing countries and the social sciences.

More inclusive and integrated science practices began to emerge from the 1970s onwards (Werge, 1978; Agarwal, 1979; Izuno, 1979; Biggs, 1980, 1982; Rhoades, 1982; Biggs, 1983). The drivers for this included the emergence of gender studies and women in agricultural development projects (Jiggins, 1984; Appleton, 1995; Doss and McDonald, 1999); the impact studies, analyses, and evaluations commissioned through the reporting cycles of the UN Human

Development agency and the FAO's Food and Hunger reports that showed the persistence of widespread hunger, rural unemployment and food insecurity for vulnerable populations; and studies of the land degradation, water pollution, and loss of flora or fauna species associated with narrow technological interventions. (Repetto, 1985; Loevinsohn, 1987; Repetto et al., 1989; Repetto, 1990). The growing experience of alternative ways of mobilizing science capacity (noted in 2.1.2-2.1.4) complemented these efforts and stimulated a more critical reflection within scientific communities (ODI, 1994) on the governance of agricultural science and the accountability of science as a source of innovation not only for "success" but also for "failure" in agricultural development. Institutional responses included the creation around 1995 of a system-wide program on gender analysis and participatory research within the CGIAR and the beginning of the sustained long term research that fed into the Millennium Ecosystem Assessment (MA, 2005).The ethical and political questions posed by scientific and technological choices stimulated the spread and more rigorous use of ethics committees to address a broader range of societal considerations and renewed efforts to bring together the natural, technical and social sciences. This often involved the creation of specialist cross-disciplinary organizational units charged with the task of integration around selected themes and of new knowledge networks.

Scientists trained to specialize often struggled to understand their role in these arrangements. A different paradigm, constructivism, offered a sound epistemological base for the kinds of interactive and integrative work that challenged scientists as professionals to think about themselves and their work in new ways. The epistemological position of constructivism is that reality and knowledge are actively created through social relationships and through interactions between people and their environment. These relationships and interactions are seen as affecting the ways in which scientific knowledge is produced, organized and validated (Schütz, 1964; Berger and Luckmann, 1966). An authoritative overview of empirical research studies (Biggs and Farrington, 1991) robustly demonstrated the ways in which institutional and political factors affected both the conduct of agricultural science and the translation of research results into farming practices. An important distinction became more widely understood: i.e., between knowledge as a lived experience of inquiry and hence transient and continuously re-created and knowledge products that can be stabilized (e.g., in journal articles, technologies, artifacts and in the norms of organizational behavior) and shared and under the right conditions, will diffuse. It opened the door to science not only as a *source* of innovation but as potentially a *co-creator of knowledge* in processes of enquiry shared with other actors (Borrini-Feyerabend, et al., 2004).

Collaboration among science disciplines tended to assume one of three forms: combining multiple disciplines in a single study; to a variable extent dissolving disciplinary boundaries in purposive learning from each others' disciplines and non-science actors; and transdisciplinary effort that actively sought to build new frames of meaning and understanding (Figure 2-2). The founding precepts of General Systems Theory, introduced by the biologist von Bertalanffy in 1950 informed these efforts, especially from the 1970s

onwards (Spedding, 1975; Cox and Atkins, 1979; Altieri, 1987). Strong interdisciplinary collaboration in developing systemic approaches to agroecology occurred throughout the world in the 1980s, often led by NGOs. The boundaries expanded to include on-farm fisheries, the role of wild and semi-domesticated foods and medicines (Scoones et al., 1992), forests and non-timber forest products (Ball, Carl, and Del Lungo, 2005). The agricultural sciences were newly positioned at the interface of two complex and complementary systems: natural and social systems. Translation of this understanding into practice nonetheless faced strong barriers within the scientific community and from market specialization and the dominance of economic drivers over social and ecological sustainability concerns.

2.1.3.2 A changing contract between science and society

In the immediate post World War II period in what later became grouped as OECD countries there was a tacit understanding between science and society that what was good for science was good for humanity and that science would deliver solutions to societal problems. The output response in OECD agricultures and under the Green Revolution's early successes in Asia and then Latin America consolidated this view and led over time to significantly higher national investments in AKST and science in general. The less strong impacts experienced in sub-Saharan Africa (Beintema and Stads, 2006) reflected both the weakness of the scientific infrastructures and personnel around the time of independence and the overall economic and social conditions of the time, leading to a prolonged period of donor investment to strengthen capacity (see Chapter 8). Although a few "islands of success" were created, the lack of sustained national investments meant that the capacity for science and technology development at the university, research institute or enterprise level in most of sub-Saharan Africa by the 1990s had fallen to an exceptionally low level (Eisemon, 1986; Eisemon and Davis, 1992; Gaillard and Waast, 1992). Recent renewed efforts by African leaders to build a stronger contract between their societies and science have not yet translated into adequate national investments in their own science base.

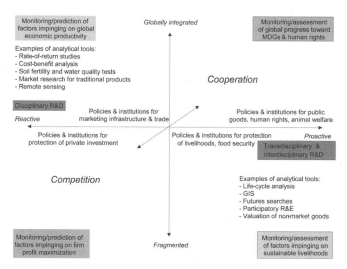

Figure 2-2. *Modes of science.*

Over time science as a human activity began to be viewed more critically as the increasing reliance on science and technology to drive national economic growth progressively revealed also the technical risks of scientific development. This view resulted in a growing public mistrust in some countries concerning the effectiveness of science as the unqualified promoter of the public good, (Nelkin, 1975; Calvora, 1988; Gieryn, 1995; BAAS, 1999) although in others, such as Sweden, public confidence in science has remained high. For example, public concerns, themselves informed by science, surfaced for instance concerning the impact of synthetic chemicals on other species, human health and the environment. As these issues began to figure more strongly in agricultural and food science research priorities (Byerlee and Alex, 1998) science began to occupy an ambiguous position as a supplier of the objective knowledge needed to generate new kinds of formal knowledge and technology as well as that needed to identify and measure risks and the evidence of harm that applications of knowledge and technology might cause in particular conditions of use; science as a human activity thus became implicated in societal controversies (Nash, 1989; Brimblecombe and Pfister, 1993; Gottlieb, 1993; Sale, 1993; Shiva, 2000; Maathai, 2003). It experienced both optimistic support from the public about its potential social utility and loss of credibility when it was found in specific instances to have produced unintended or undesirable results. At the same time, the lines between public good science, not-for-profit science and science carried out for commercial gain began to blur as the public sector in many countries began to yield its role as a direct supplier and the private commercial sector emerged as a major source of funding for agricultural science and technology development.

The imbalance between science investments, infrastructures and staffing in OECD countries compared to tropical countries (UNESCO, 1993; Annerstedt, 1994) for much of the period meant that "science's contract with society" for the goals of international agricultural development and sustainability had to be mobilized with the support of OECD country electorates. That is, the resources had to be mobilized by appeals to values and interests of people distanced from those experiencing the effects. This process stimulated the growth of civil society and NGOs working on international development and the introduction of the broader concerns of citizens into the science agenda. As science institutions by the 1990s in the poorest developing countries became heavily dependent on foreign funding and foreign training opportunities the concerns of donors tended to drive their agendas. Other countries such as Brazil, South Africa, China and India identified S&T as key drivers of their own economic development while giving relatively lower attention specifically to the agricultural sciences. Private commercial investment in science tended to concentrate on technologies such as food preservation and processing, pest control technologies, feed stuffs, veterinary products and more recently also on transgenic crops for which profits could be more easily captured; under competitive commercial pressures the concerns of better-off consumers and urban residents also began to influence the AKST agenda.

As a consequence of these complex interweaving trends, public support for international agricultural development and sustainability was and remains peculiarly susceptible to crises (EC, 2001; 2005). These include crises in intensive agricultures, in the public mind in Europe associated with "the silent spring" (Carson, 1962) or diseases such as BSE (bovine spongiform encephalopathy—"mad cow disease") and more recently the risks of the spread of avian flu to the intensive poultry industries of Europe and beyond. The actual or potential human health consequences provided an extra emotional dimension. Environmental crises, such as the drying of Lake Aral through diversion of its waters to feed the Soviet Union's cotton farming or the unsustainable use of surface and groundwater in irrigated farming in the southwest of the United States or in the Punjab or crises of acute hunger and starvation, drought or flooding similarly brought the agricultural sciences into question. Fear of the unknown and suspicion of the concentration of ownership in commodity trading, food industries and input supply (Tallontire and Vorley, 2005) and increasing private control over new opportunities in agriculture arising from advances within science (WRI/UNEP/WBCSD, 2002) also fed into public concerns. The first generation technologies resulting from genomics e.g., raised concerns about the risks of increased spread of known allergens, toxins or other harmful compounds, and horizontal gene transfer particularly of antibiotic-resistant genes and unintended effects (Ruan and Sonnino, 2006). An important consequence is that demand has grown for stronger accountability, stricter regulation and publicly funded evaluation systems to determine objectively the benefits of new sciences and technologies.

Today in many industrialized countries an increasing percentage of the funding for university science comes from private commercial sources. It tends to be concentrated in areas of commercial interest or in advanced sciences such as satellite imaging, nanotechnologies and genomics rather than in applications deeply informed by knowledge of farming practice and ecological contexts. License agreements with universities may include a benefit sharing mechanism that releases funds for public interest research, but product development, especially the trials needed to satisfy regulatory authorities, is expensive and companies (as well as universities) need to recover costs. Hence a condition of funding is that the source of funds often determines who is assigned first patent rights on faculty research results. In some cases the right to publication and the uninhibited exchange of information among scholars are also restricted. The assumption under these arrangements that scientific knowledge is a private good changes radically the relationships within the scientific community and between that community and its diverse partners

2.1.4 Technology and innovation processes

The relationship between technology and innovation has remained a matter of debate throughout the period under review. The analysis by scholars around the world of literally thousands of empirical studies of the processes that have led to changes in practice and technology (not only in agriculture but in related sectors such as health) over time has forced acceptance that innovation requires much more than a new technology, practice, or idea and that not all change is innovation. Innovation processes have been driven mainly by for-profit drivers but there has been also an as yet incomplete convergence toward AKST relationships, arrangements and processes that foster innovations supportive of socially inclusive and ecologically sustainable and productive agricultures.

2.1.4.1 Changes in perspective: from technologies to innovations

The proposition that technical change could be a major engine of economic growth was demonstrated in the 1950s (Solow, 1957). Later analysis of empirical evidence showed that small-scale producers, although handicapped by severe constraints, made rational adaptations over time in their practices and technologies in response to those constraints. Insofar as externally introduced technology released some of the constraints, technology could become a driver of significant change (Schultz, 1964). The Green Revolution subsequently appeared to vindicate the analysis and it quickly became dominant in the agricultural economics profession (Mosher, 1966). The model that this analysis pointed toward is the dominant way of organizing knowledge and diffusion processes, i.e., "the transfer of technology model" (2.1.2) (e.g., Chambers and Jiggins, 1986). It is known also as a policy model, variously as "the agricultural treadmill" (e.g., Cochrane, 1958) and "the linear model" (e.g., Kline and Rosenberg, 1986); and its role in policy is assessed here. In its simplest form it recommends *technology supply-push*, i.e., developing productivity enhancing component technologies through research for delivery, transfer, or release to farmers, the "ultimate users".

The model emerged in a specific historical context, the American Midwest in the decennia after WWII (Van den Ban, 1963); similar models were elaborated from empirical findings in other economic sectors. Although these mechanisms driving the model's impact are familiar to economists, they are not necessarily as familiar to others so the persistence of technology supply push as the dominant policy model for stimulating technology change in agriculture warrants a full explanation of the mechanisms. In the case of agriculture the empirical data robustly confirm the following features:

1. *Diffusion of innovations.* Some technologies diffuse quite rapidly in the farming community after their initial release, typically following the S-curve pattern of a slow start, rapid expansion and tapering off when all farmers for whom the innovation is relevant or feasible have adopted. The classic case is hybrid maize in Iowa (Ryan and Gross, 1943). Diffusion multiplies the impact of agricultural research and extension effort "for free". But diffusion is mainly observed ex-post: it is difficult to predict (or ensure) that it will take place (Rogers, 2003).

2. *Agricultural treadmill.* The treadmill refers to the same phenomenon but it focuses on the economics (Cochrane, 1958). Farmers who adopt early use of a technology that is more productive or less costly than the prevailing state-of-the-art technology, i.e., when prices have not as yet decreased as a result of increased efficiency, capture a windfall profit. When others begin to use the new technology, total production increases and prices start to fall. Farmers who have not yet adopted the technology or practice experience a price squeeze: their incomes decrease even if they work as hard as before. Thus they must change; the treadmill refers to the fact that the market propels diffusion: it provides incentives for early adoption and disincentives for being late.

3. *Terms of trade.* A key underlying aspect of the treadmill is that farmers cannot retain the rewards of technical innovation. Because none of the thousands of small firms who produce a commodity can control the price, all try to produce as much as possible against the going price. Given the low elasticity of demand of agricultural products, prices are under constant downward pressure. During the last decennia, the price of food has continuously declined both in real and relative terms (World Bank, 2008). The farm subsidies in the US and Europe can be seen as a necessary cost for societal benefit without rural impoverishment.

4. *Scale enlargement.* In the tail of the diffusion process, farmers who are too poor, too small, too old, too stupid or too ill to adopt eventually drop out. Their resources are taken over by those who remain and who usually capture the windfall profits. This shakeout leads to economies of scale in the sector as a whole.

5. *Internal rate of return.* Investing in agricultural research and extension to feed the treadmill has a high internal rate of return (Evenson et al., 1979). The macro effects of relatively minor expenditures on technology development and delivery are major in terms of (1) reallocating labor from agriculture to other pursuits as agriculture becomes more efficient, (2) improving the competitive position of a country's agricultural exports on the world market, and (3) reducing the cost of food. An advantage is that farmers do not complain. Their representatives in the farmers' unions are among those who capture windfall profits and benefit from the process, even though in the end the process leads to loss of farmers' political power as their numbers dwindle to a few percent of the population. The treadmill encourages farmers to externalize social and environmental costs, which tend to be difficult to calculate and hence usually are not taken into account. One may note here that this process, first described at the national level in the case of the USA, also explains the growing gap in the productivity of agricultural labor between industrialized and developing countries and that it leads to overall efficiencies in production and reduced prices for consumers, outcomes that have favored its persistence as a dominant policy model.

However, other business analysts and social scientists throughout the period under review have stressed the concept of innovation rather than mere technical change as a measure of development. The evidence that technical change itself requires numerous, often subtle, but decisive steps before an adoption decision is made reinforced this view (Rogers, 1983). Others pointed to biophysical, sociocultural, institutional and organizational factors such that when the same technology is brought into use in different contexts the effects vary (Dixon et al., 2001). Recently more emphasis has been given to development of "best fit" technology options for a given situation, reflecting further discoveries of institutional and sociological factors that shape technical opportunities (Herdt, 2006; Ojiem et al., 2006). This understanding has deep roots in extension research (e.g., Loomis and Beagle, 1950; Ascroft et al., 1973; Röling et al., 1976), farming systems research (Collinson, 2000), 1980s gender research (e.g., Staudt and Col, 1991; Sachs, 1996), and 1990s policy research (e.g., Jiggins, 1989; Christopolos et al., 2000). However, the reasons that thinking about policy began to change likely had little to do with this research and more to do with the realization that technology supply-push could fuel massive social problems wherever there were no alternative opportunities for those who could not survive in farming. This lack of survival contributed to the growth of megacity slums (UN Habitat, 2007), the ease with which displaced youngsters eagerly turned toward civil disorder and even civil war (Richards, 2002; UNHCR, 2007) and the growing numbers of internal and transboundary migrants (UNHCR, 2007). Supply-push arrangements were shown to produce agricultures accounting for 85% of the world's water withdrawals and 21% (rising to 35%) of gaseous emissions contributing to climate change; and to the declining material condition of natural resources and biophysical functioning (MA, 2005; UNEP, 2005). The cumulative evidence indicated a policy change was overdue.

The concept of innovation systems offered itself as a policy model for sustaining agricultures to meet ecological and social needs. Effective innovation systems were shown to need systemic engagement among a diversity of actors (Havelock; 1986; Swanson and Peterson, 1989; Röling and Engel, 1991; Bawden and Packham, 1993; Engel and Salomon, 1997; Röling and Wagemakers, 1998; Chema et al., 2003; Hall et al., 2003, 2006). However, people and organizations interact in diverse ways for the purposes of creating innovation for sustainable development; the range of actors needed to develop a specific innovation opportunity is potentially large and thus becomes increasingly difficult to classify (Figure 2-3) (see 2.3). The "innovation systems" concept, widely used in other industries, usefully captures the complexity (Hall et al., 2006) by drawing attention to the totality of actors needed for innovation and growth; consolidating the role of the private sector and the importance of interactions within a sector; and emphasizing the outcomes of technology and knowledge generation and adoption rather than the strengthening of research systems and their outputs.

Empirical studies emphasize that the dominant activity in the process is working with and reworking the stock of knowledge (Arnold and Bell, 2001) in a social process that is realized in collaborative effort to generate individual and collective learning in support of an explicit goal. Innovation processes focus on the creation of products and technologies through ad hoc transformations in locally specific individual or collective knowledge processes. As such innovation is neither science nor technology but the emergent property of an action system (Crozier and Friedberg, 1980) in which knowledge actors are entangled. The design of the action system thus is a determinant of the extent to which an innovation meets sustainability and development goals.

2.1.4.2 Market-led innovation

From about the 1990s onwards innovation processes in agriculture principally have been driven by a rise in market-led development. Typical responses to market pressures in North America and Europe in terms of the way in which technical requirements, market actors, and market institutions interact can provide an understanding of the "innovation space" for socially and ecologically sustainable agriculture (NAE Chapter 1; Figure 2-3).

2.1.4.3 Technological risks and costs in a globalizing world

The risk outlook fifty years ago could be described in general terms as high local output instability, relative autonomy of food systems and highly diverse local technology options: an agricultural technology that failed in one part of the world had few consequences for health, hunger or poverty in other regions. The increase in aggregate food output and the trend toward liberalizing markets and globalizing trade has smoothed out much of the instability; it has integrated food systems (mostly to the benefit of poor consumers) and it has spread generic technologies throughout the world for local adaptation. The mechanisms of food aid, local seed banks and other institutional innovations have been put in place to cope with catastrophic loss of entitlements to food or localized production shortfalls. Yet the world faces technological risks in food and agriculture that have potential for widespread harm and whose management requires the mobilization of worldwide effort (Beck et al., 1994; Stiglitz, 2006). A robust conclusion is that human beings are not very good at managing complex systemic interactions (Dörner, 1996).

Immediate costs of risks that cause harm typically are carried by the poor, the excluded and the environment, for instance with regard to choices of irrigation technologies (Thomas, 1975; Biggs, 1978; Repetto, 1986); crop management (Repetto, 1985; Kenmore, 1987; Loevinsohn, 1987); and natural resource and forestry management (Repetto et al., 1989; Repetto, 1990; Repetto, 1992; Hobden, 1995). The weight of the evidence is that power relations and pre-analytic assumptions about how institutions and organizations actually work in a given context shape how scientific information and technologies are developed and used in practice, producing necessarily variable and sometimes damaging effects (Hobart, 1994; Alex and Byerlee, 2001). Recent assessments for instance of the "long shadow" of livestock farming systems (Steinfeld et al., 2006) and of agricultural use of water (Chapagrain and Hoekstra, 2003) lead to a well-founded conclusion that estimations of agricultural technologies' benefits, risks and costs have been in the past too narrowly defined. The mounting scale of risk exposure in agriculture is delineated in the Millennium Ecosystem Assessment (2005), Global Water Assessment (2007), and IPCC reports (2007).The accumulating weight of evidence that past technology choices in agriculture have given rise to unsustainable risks has led to efforts to develop more appropriate technological risk assessment methods (Graham and Wiener, 1995; Jakobson and Dragun, 1996; NRC, 1996) and to take on differing perspectives on what levels of harm are acceptable and for whom (Krimsky and Golding, 1992; Funtowicz and Ravetz, 1993; Funtowicz et al., 1998; Scanlon, 1998; Stagl et al., 2004). Important experience has been gained in working with civil society on technological risk assessments and sustainability appraisals, sometimes involving large numbers of citizens (Pimbert and Wakeford, 2002; IIED, 2006; Pimbert et al., 2006).

2.2 Key Actors, Institutional Arrangements and Drivers

Actors and institutions have power to set policy agendas and influence how research and development investments are made. All knowledge actors develop processes for generating AKST and innovation that evolve within their own IAs and culture of understanding. These processes can generate stress when key actors are excluded or marginalized by new or old arrangements (Table 2-3).

The main actors considered here are in the vast majority *farmers and farm laborers*, many of whom are poor, with limited access to external resources and formal education, but rich in traditional and local knowledge and increasingly organized and adept at sharing knowledge and innovating. *Additional domestic actors* affecting the development and innovation of AKST include local, provincial and national governments, and the agencies, departments and ministries devoted to agriculture, environment, education, health, trade, finance, etc. Still *other actors* with direct impacts on AKST include regional consortia and international institutions, FAO, the Global Integrated Pest Management (IPM) Facility, the World Bank, CGIAR, private foundations, and others. Each organization develops and brings its own sets of priorities, perspectives and agendas to the business of

AKST. Private sector actors who have played increasingly important roles are commercial and corporate players and civil society organizations (CSOs), including farmer and consumer organizations, foundations and those working for nonhuman species and the environment, as well as a range of development and relief NGOs.

The currently dominant AKST systems are the product of a long history of attempts by diverse combinations of these actors, under numerous institutional arrangements (IAs), to meet the needs and challenges of agriculture in different contexts, as well as the actors' own individual or institutional needs. Their histories are made up of successes, but also failures and frustrations, often leading to new attempts at meeting both local and global challenges. In many instances, crises have led to the emergence of new actors and the reshuffling of roles and relationships. Institutional arrangements formally or informally coordinate the work of knowledge producers and engage them in distinctive knowledge processes, thus favoring the emergence of different kinds of innovation. Some become long-standing permanent arrangements; others are *ad hoc* initiatives or of more recent origin.

2.2.1 Farmer and community-based arrangements
The emergence of major producer organizations representing their members' interests and rights at district, national, regional and international levels may be seen as an increasingly strong driver of change over the last decades. Most of them are actively engaged in the provision of technology and information services and have entered into partnerships

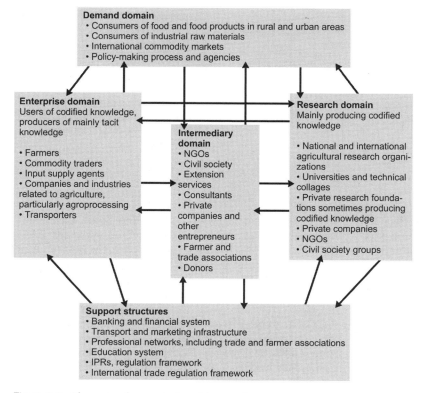

Figure 2-3. *Elements of an agricultural innovation system.* Source: Adapted from Arnold and Bell (2001) in Hall et al. 2006.

Table 2-3 **Analytic map of the main features of AKST for development paradigms.**

Label	Features of Production System	Features of AKST	Direct Drivers	Indirect Drivers
Pre-modern/Traditional	Diverse products locally; "natural" systems; small-scale units; local/recycled inputs	Local knowledge generation and repositories	*Biophysical:* soils, local climate *Resources:* labor availability *Social factors:* mutual help, social capital *Economic:* local economy/food need	*Policy and economic:* tax systems, access to markets *Social:* cultural practices related to farming *Cognitive:* focus on meeting local needs
Industrial agriculture in capitalist contexts	Mechanization; less diverse products—greater specialization; larger scale units external inputs; private sector production	Formal R&D (public and private); dissemination of knowledge	*Cognitive:* profit and yield maximization through science *Policy:* subsidy for production goals *Economic:* agribusiness corporations *Institutional:* formal research institutions	*Social and economic:* consumer demand *Trade:* international trade agreements *Economic:* cheap energy; externalization of health and environmental costs
Industrial agriculture in socialist contexts	Mechanization; larger scale units; external inputs; collective ownership of resources (labor, land); central planning	Public sector R&D, dissemination by state institutions	*Policy:* national food self-sufficiency *Institutional:* funding for research/extension	
High external input intensive agriculture in south (e.g., Green Revolution; some plantation systems)	HYVs; package of external inputs; pest management and nutrient management through chemical inputs	National agric. universities and research stations; CGIAR; global transfer through aid agencies/projects; local knowledge has little influence	*Cognitive:* increase production to keep up with population; science provides solutions *Policy:* state support/subsidy *Institutional:* research community *Technological:* growth of new technologies *Trade:* focus on export-led growth	*Economic and policy:* post-colonial drive for food self-sufficiency *Cognitive:* faith in rational science & expert advice *Globalization and trade:* multinational agribusiness and agrochemical corporations; aid conditionalities *Social:* loss of local knowledge; perceived inefficiencies in previous production systems
Low external input agriculture in South (not necessarily sustainable)	Marginal land resources; low yields; low priority crops (national and trade perspective); prone to natural shocks; minimal use of synthetic inputs	Little attention from formal R&D; reliance on local knowledge and innovation	*Institutional and policy:* low provision of credit and technical assistance	*Institutional and policy:* high potential lands have been prioritized *Trade:* low value of output means little attention from input manufacturers and agribusiness

continued next page

with R&D providers. Many now have websites that act as an information umbrella for and communication link to thousands of affiliated farmers' groups organized at local levels. Examples include the Network of Farmers' Organizations and Agricultural Producers from western Africa (http://www.roppa-ao.org); the International Land Coalition (www.landcoalition.org/partners/partact.htm); the International Federation of Agricultural Producers (www.ifap.org); and Peasants Worldwide (www.agroinfo.nl/scripts/website.asp).

The focus on local mobilization masks the wide scale of effort and impact (Boven and Mordhashi, 2002). For example, in 2004 Catholic Relief Services was working directly with 120,000 poor producers in community-based seed system development (www.crs.org) and South East Asian Regional Initiatives for Community Empowerment (SEA-RICE) (see 2.2.3). The local seeds movement pioneered by such organizations has given rise to information exchange networks that assert individual and community rights to "first publication" so as to safeguard native IPR and germplasm. Over time, such organizations have strengthened their own R&D networks by commissioning research and through organizing national and international technical conferences, such as the International Farmers' Technical Conference held in conjunction with the 2005 Convention on Biodiversity meeting.

Farmer research partnerships typically bring together farmers, professors, scientists and researchers to compose a

Table 2-3 *continued*

Label	Features of Production System	Features of AKST	Direct Drivers	Indirect Drivers
Organic/ Low impact/ Sustainable farming in South and North	Low use of external inputs; crop nutrition and pest management; based on natural systems; focus on maintaining/building quality of soil and water resources	South Local learning, e.g., through Farmer Field Schools; documentation and dissemination of local knowledge; Cuba's model of centers to reproduce biological pest control agents North producers' organizations; independent R&D institutions networking among producers; government funding for research on organic and sustainable farming	South *Social:* social capital, collective effort *Economic:* high cost of external inputs; negative impact on yields of high input agriculture. *Policy:* sustainability *Cognitive:* farmer concern with resource/ecosystem damage *Trade:* high demand for organic/niche products in northern markets *Institutional:* emergence of local NGOs for dissemination of sustainable practices; increase in aid for low input agriculture North (EU) *Cognitive:* idea of "natural" and ecological farming popularized *Policy:* funding, subsidy and support for conversion *Economic and social:* public awareness of organic products *Institutional:* good support structure of organizations and extension services	South *Globalization and investments:* international organizations (IFOAM) *Cognitive:* farmer and researcher recognition of externalities of high external input agriculture North *Cognitive and social:* recognition of negative environmental effects of high input ag., and problems faced by family farms *Globalization and trade:* disease outbreaks leading to trade restrictions *Institutional:* rise of Green movements and political parties

technical pool of expertise dedicated to collaboration with farmers in research and development. These IA's emphasize the centrality of primary producers, food processors and laborers in agricultural and food systems. In general, they initially capitalize volunteerism and fund-raising activities to implement farmer-led projects, but often move on to a holistic approach to development of livelihoods and welfare, community empowerment and measures to extend farmer control over agricultural biodiversity. For instance, MASIPAG (Farmer-Scientist Partnership for Development, Inc.) was established in the Philippines in 1987, after more than five years' collaboration between farmers concerned about the negative impacts of high-yield rice and associated technologies on their livelihoods, local genetic resources, and environment, and a few progressive scientists. It then rapidly developed into a large farmer-led network of people's organizations, NGOs and scientists, promoting the sustainable use and management of biodiversity through farmers' control of genetic and biological resources, agricultural production and associated knowledge based on a strategy of placing command of the skills and knowledge of the agronomic sci-

ences in the hands of small-scale producers. By 2004, MASIPAG was working with four national/regional civil society networks and organizations, seven Philippino universities and research centers and seven local government authorities and line agencies. MASIPAG's network of trial and research farms included 72 in 16 provinces in the island of Luzon, 60 in 10 provinces in Visayas and 140 in 14 provinces in Mindanao. MASIPAG today is recognized worldwide as a leading example of highly effective farmer-led and largely farmer-funded and farmer-managed, R&D and extension that is building small-scale farm modernization, resource conservation and food sector development on ecological principles (Salazar, 1992; Araya, 2000). At the other end of the spectrum, systematic testing has been carried out of user involvement in the barley breeding cycle in Syria (Ceccarelli et al., 2000). The researchers initially designed four types of trials: by farmers in their fields, with farmers on-station, by breeders in farmers' fields and by breeders on-station. Their experience of the rigor, reliability, and comparative costs and benefits of the four led them to concentrate on testing and selection by farmers in their own fields, complemented by seed multiplication on station. Similar achievements have been recorded in southwest China for maize (Vernooy and Song, 2004).

Local research and innovation: the contribution of occupational education. Local level innovation can be promoted if appropriate investments are made in educating farmers but this has been a relatively neglected area. One of the major breakthroughs has been the development and spread of farmer field schools (FFS) (Braun et al., 2006). Based on adult education principles, the schools take groups of farmers through field-based facilitated learning curricula organized in cycles of observation, experimentation, measurement, analysis, peer review and informed decision-making. FFS are making in aggregate a significant and influential contribution to sustainable and more equitable small farm modernization, particularly in the rain fed areas where two-thirds of the world's poor farm households live. Kenya, Tanzania and Uganda have included the approach in national research and extension strategies, as has India. Systematic review of available impact data (Braun et al., 2006; van den Berg and Jiggins, 2007) and area-based impact studies (Braun et al., 2006; Pontius et al., 2002; Bunyatta et al., 2005; Mancini, 2006) demonstrate positive to strongly positive achievements. Contributing effectively to farmer empowerment also contributes to the strengthening of civil society and self-directed development (Mancini et al., 2007). Others have criticized their cost in relation to the scale of impact (Quizon et al., 2000; Feder et al., 2004ab), noted the weak diffusion of specific technologies, lack of diffusion of informed understanding (Rola et al., 2002) and failure in some instances to develop enduring farmer organizations (Bingen, 2003; Tripp et al., 2005). Further experimentation is warranted to test if combining farmer education such as FFS with complementary extension efforts will overcome the perceived shortcomings (Van Mele and Salahuddin, 2005).

World Learning for International Development, the Alaska Rural Systemic Initiative project and the Global Fund for Children similarly have documented gains (World Bank, 2005a) in the effectiveness and efficiency of local research, school-based science education and the development of agroecological literacy at the grass roots brought about by investing in farmers' occupational education (Coutts et al., 2005).

Farmer-funded R&D and extension. Innumerable examples exist of effective technological advances pioneered by farmers themselves; e.g., grafting against pests, biological control agents such as the golden ant in citrus in Bhutan (Van Schoubroeck, 1999) and soil management and farming system development in the Adja Plateau, Benin (Brouwers, 1993). Yet the economic value of local and traditional innovations has not been much researched. One study in Nigeria in the early 1990s estimated the contribution of the informal agricultural sector where farmers are using mostly indigenous innovations at about US$12 billion per year, providing income for an estimated 81 million people (ECA, 1992). This estimate, however, does not include the cost of opportunities foregone or traditional practices that do not work. Recent literature begins to sketch out the strengths

and weaknesses that might be taken into account if a more comprehensive cost-benefit analysis were to be attempted (Almekinders and Louwaars, 1999).

2.2.2 Producers of AKST at national level

Countries have developed a complex array of public institutions, IAs and actors responsible for planning, funding, implementing, assessing, and disseminating public interest agricultural research. They include national, regional/municipal agricultural research institutions, universities and other higher education institutes and extension services. Most of these arrangements historically have been publicly financed because agricultural research investments involve externalities, and are subject to long gestation periods (cfr. Chapter 8, Table 8.1; Lele and Goldsmith, 1989; Beintema and Stads, 2006; Pardey et al., 2006ab).

In the 1960s and 70s National Agricultural Research Systems (NARS) in developing countries (see 2.1), especially agricultural research institutes (ARIs), received strong financial support from governments and international donors to launch agricultural modernization through the dissemination of Green Revolution technologies (Chema et al., 2003). In the 1980s, as a result of budgetary crises and adjustment programs, public funds for agricultural research failed to keep up with expanding demand. Public expenditure declined as proportion of total research and development spending; expenditure per researcher declined much more because staffing continued to expand faster than budgets. From the 1980s onwards, the main drivers of institutional development of the NARS were structural reforms in national economies and adjustment policies, global political changes; ideological demands for reduced public sector involvement and intervention; a greater private sector role and significant biotechnological breakthroughs (Byerlee and Alex, 1998; Iowa State Univ., 2007). These events have given rise to a diverse institutional landscape responding to both domestic and global priorities and opportunities. Brazil's EMBRAPA, for instance, has become an exporter of capacity, in 2007 opening liaison offices in West and East Africa whereas NARS in sub-Saharan Africa continue to face many constraints (Jones, 2004).

Sub-Saharan Africa's National Research Systems. Overall budget constraints throughout the period have weakened public sector NARS in most African states. The general panorama today is of deep attrition of human resources, equipment facilities, capital funding and revenue, despite islands of promise such as the revitalization of capacity in Uganda under vigorous decentralization policies, in Ghana in relation to agroindustrial developments and in post-apartheid South Africa. Nongovernmental organizations, the CGIAR, private commercial actors and recently the establishment of sub-regional bodies (Central African Council for Agricultural Research and Development, CORAF), (Association for Strengthening of Agricultural Research in Eastern and Central Africa, ASARECA) and similar arrangements for southern Africa supported by the Forum for Agricultural Research in Africa (FARA), have filled the gaps only in part. An alliance largely funded by a US-based philanthropic trust recently has been established to transfer germplasm and advanced biotechnology skills to African NARS to catalyze

Africa's "rainbow revolution". Agricultural research trust funds set up to lever matching research contracts from commercial enterprises, donors and government organizations, have not succeeded; although farmer-managed funds are meeting with some modest success.

The Agricultural Research Council (ARC) model. Some large countries with complex research systems have established agricultural research councils to coordinate the work carried out at research institutes. The ARC typically is a public body which has—*inter alia*—the functions of managing, coordinating or funding research programs. Management of the councils has proved effective because they are both autonomous and accountable to users and donors for planning and executing research. In India, the Indian Council of Agricultural Research (ICAR) has coordinated the higher agricultural education system since the 1950s and in 1996 established an agricultural education accreditation board (http://www.icar.org.in/aeac/ednac.htm). In Africa, the role of ARCs has varied widely as some have moved beyond a policy and coordinating role to undertake research themselves (Bingen and Brinkerhoff, 2000). However, the councils that have proliferated have failed to live up to expectations, become bureaucratized (Chema et al., 2003) and been unable to influence national research budgets or coordinate agricultural research among institutions to reach out to small-scale farmers (Byerlee, 1998; Rukuni et al., 1998; Bingen and Brinkerhoff, 2000).

The National Agricultural Research Institute (NARI) model. This model is common in Latin American countries, where agricultural research has been conducted primarily at the national level. They control, direct and manage all publicly funded agricultural research; they may be autonomous or semiautonomous in budgetary support, scientist recruitment, financial norms and disciplines with experiment stations as the basis for research organization. Their creation in the 1950s and early 1960s was driven mainly by the recognition of the leading role of technological change in the modernization of agriculture. In the late 1990s, rural development and poverty alleviation efforts became differentiated from research and technology development, accompanied by the increasing participation by private sector entities in financing and implementing R&D activities. These shifts were driven by changes in the wider socioeconomic and political context within which the NARIs operated (i.e., state reform, deregulation, economic liberalization), and changes in the scientific processes underlying agricultural research (i.e., privatization of knowledge, plant breeders' rights, patent protection for R&D results. In Latin America, two important constraints have limited the role of the NARIs: the decline in government funding and the weak incentives for coordination and cooperation among research system components within each country. In two cases the NARIs also had responsibility for extension: the National Institute of Agriculture (INTA), Argentina, and the National Institute of Agriculture (INIA), Chile. In 2005 INTA created a Center for Research and Technological Development for small-scale family agriculture (CIPAF), with three regional institutes. This signaled a decisive transition from the supply-push Transfer of Technology approach that hitherto characterized the NARI model throughout Latin America, to a client-oriented demand-pull approach based on participatory action-research (http://www.inta.gov.ar/cipaf/cipaf.htm). Since 2003 Brazil has promoted biotechnology as a national policy priority for the Brazilian Agriculture and Livestock Research Company (EMBRAPA) in order to boost productivity in both family farms and large scale agroenterprises. EMBRAPA is collaborating in the federal government's Fome Zero (Zero Hunger) program (http://www.fomezero .gov.br), taking a lead role in the global Cassava Biotechnology Net (CBN) through the Biotechnology Research Unit of Mandioca e Fruticultura (http://www.cnpmf.embrapa .br) and in Participatory Plant Breeding, principally through EMBRAPA-CNPMF, Cruz das Almas, Bahia, together with the Bahian Company of Agricultural Development (http:// www.ebda.ba.gov.br), Caetité, southeast Bahia and farmer communities also located at Caetité.

The Ministry of Agriculture (MOA) model. This model was dominant in communist countries and in the immediate postcolonial era and still prevails in countries where there is less agricultural research capacity. It is characterized by centralized governance and bureaucratic practice. However, in recent years new organizational patterns have begun to emerge that provide greater flexibility. Collectivization and nationalization resulted in significant and often irrational concentration of agricultural production in state or quasi cooperatives managed as industrial enterprises, affecting the whole social and economic life of villages and rural areas in countries such as Tanzania and in the former soviet bloc countries (Swinnen and Vranken, 2006). Adjustment to new economic and political conditions has demanded significant AKST role changes (Petrick and Weingarten, 2004) including redefinition of the role of government in agricultural research; separation of research funding, priority setting and implementation; decentralization of agricultural research both geographically and in terms of decision making; strengthening of system linkages among multiple innovation partners including CSOs, traders, input and processing industries (Swinnen and Vranken, 2006; Petrick and Weingarten, 2004).

Universities and other higher education models. Universities are institutions placed amidst three coordinating forces: the academic oligarchy, the state and the market (Clark, 1983). These three forces are seldom in balance; they act in a continuous and dynamic tension, which often brings about intellectual, practical and organizational conflicts and ruptures (Bourdieu, 1988) often leading to diffuse and contradictory missions (Weick, 1976; Busch et al., 2004). In agricultural universities (schools/colleges or faculties) there are many such divides between purpose and mission; social and scientific power, among managers, teachers, researchers and extensionists; between the established canonical agricultural disciplines and disciplines, such as sociology, ethics and public administration (Readings, 1996; Delanty, 2001). Urgent societal demands, such as those posed by hunger, poverty, inequality, exclusion and solitude, and more recently also natural resource degradation and climate change have had to find their place against the background noises of collaboration and dissent. Universities, nonetheless, are widely identified as key actors in national research systems (Castells, 1993; Clark 1995; Edquist, 1997; Mowery and Sampat, 2004), but their contribution to agricultural research, real or potential, often has been neglected in cost-benefit analyses. Yet they have been and remain the major educators of agricultural scientists, professionals and technicians, a voice of reason (and sometimes partiality) in controversial debates about bioethics, transgenic seeds, IPR, food quality and safety issues, etc., and a source of factual information (Atchoarena and Gasperini, 2002). Robust indicators do not exist for the comparative assessment of the efficiency and effectiveness of universities in generating knowledge, science and technologies for sustainability and development. For example, in a survey of Argentine agricultural scientists (1996 to 1998), the number of journal publications was a proxy measure (Oesterheld et al., 2002), despite known limitations (Biggs, 1990; Gómez and Bordons, 1996; Garfield, 1998; Amin and Mabe, 2000; Bordons and Gómez, 2002). Output was found to be highly variable and on average, low but higher than in other institutions such as the National Institute of Agricultural Technology (INTA), and the National Council for Science and Technology (CONICET) (Oesterheld et al., 2002).

Higher-level agricultural education institutions can be subdivided into (1) agricultural colleges embedded in a comprehensive university, (2) land grant universities, patterned after the US land grant universities, and (3) tertiary level agrotechnological institutes that are not part of a university and depend on a ministry of education or of agriculture. They all have similar constraints to achieving the diversity of their roles and purposes (Table 2-4).

(1) *Agricultural schools or college/faculties model embedded in a comprehensive university*. This model is shaped after the German Humboldt tradition and has teaching, research and extension as central functions. It has diffused to other European countries as well as to other parts of the world, mainly the Americas.

Until recently in many countries research universities were autonomous, with public funds provided as block grants by the Treasury to the Ministry of Education, which transferred them to the central university governing body; the agricultural colleges then had to compete against other interests. In Latin American countries, research budgets are often less that 0.5% of the total university budget (Gentili, 2001) and little of this has reached the agricultural departments, colleges and schools. However, in the last decades research has been financed by the use of competitive funds open to all public research institutions and in some cases to private universities. International donors, philanthropic foundations and increasingly also commercial enterprises also contribute to financing (Echeverría et al., 1996; Kampen, 1997; Gill and Carney, 1999). Their main asset is research and their internal system of reward and promotion is designed to protect standards in this core activity. The pressure to "publish or perish" favors acceptance of actors and types of AKST that is produced in conditions that support such performance and thus tends to increase the gap between developed and developing countries' national academic and research systems. It also further marginalizes scientists and academics in the latter countries where funds

for research, in particular for basic research, are scarce. The incentive system legitimated the dominant position of universities in colonial and later in OECD countries as the centers of basic and strategic research in a hierarchy of AKST providers. Students as well as trained agricultural scientists and professionals continue to leave employment in tropical countries wherever national governments have failed to invest in "catch up" institutional development at tertiary levels.

In the United States policies were important in assisting the commercialization of research products and services. The US Bayh-Dole Act passed in 1980 gave universities and corporations the right to patent federally funded research and was buttressed by the Federal Technology Transfer Act of 1986 (Kennedy, 2001; Bok, 2003). These acts succeeded in their primary purpose but widened existing gaps with most developing countries. Incentive systems designed for the commercialization by universities of private good research appear to perform less well in promoting public goods research and its application in agriculture and food industries (Byerlee and Alex, 1998; Berdahl, 2000; Bok, 2003; Washburn, 2005).

The most immediate challenges tertiary institutions face is how to respond to the often divergent interests of private and public actors, consumers and citizens as AKST systems become more demand-driven and hence also develop or strengthen their capacity to become engaged in problem solving in specific settings, and continue to provide generic potential for sustainable development.

(2) *Land-grant colleges and state universities.* These have been patterned after the land-grant model originating in the 19th century in the United States. Key components are the agricultural experiment station program (Hatch Act 1887) (Kerr, 1987; Mayberry, 1991; Christy and Williamson, 1992; BOA, 1995), and the link via extension programs to farmer advisory, leadership development and training activities in the community. The grant of land to finance research and education ensured in the original conception a high degree of accountability to the application of science to local, practical problem solving and entrepreneurship. These distinctive features tended to attenuate over time or progressively decline as the model spread to and then merged into different contexts. After World War II, the Rockefeller and Ford Foundations and the United States Agency for International Development (USAID) played leading roles in the establishment of state agricultural universities in India modeled on the US land-grant universities. State agricultural universities of Pakistan and the Philippines also adopted the model as their guide. In sub-Saharan Africa, the research and extension missions of the land-grant model generally introduced under Ministries of Higher Education came into conflict with research and extension departments in ministries of agriculture. By the 1980s most of the land-grant universities in SSA had become comprehensive universities emphasizing training. Nevertheless, the model proved powerful; land-grant universities in the USA throughout the 20th century have been central to North America's farm modernization, dominance in commodity trade and preeminence in global food industries (Ferleger and Lazonick, 1994; Slaybaugh, 1996; Fitzgerald, 2003). The land-grant construct explicitly

Table 2-4. **Constraints of university arrangements.**

Funds	Universities have to share budgetary allocations with other public sector agencies for agricultural research. In Latin America, e.g., expenditure per researcher diminished strongly in the 1980s, and then recovered in the nineties but without reaching the previous position.
Scientific culture	Different knowledge paradigms and scientific culture pervade teaching, research and extension activities addressing societal problems. Most public concerns or problems are multidisciplinary, while most university departments are disciplinary. Research, especially in the agricultural colleges—produces fundamental knowledge under standards of rigor focused on "manageable" (well defined) or "technical problems," not always pertinent to social needs. Teaching follows the same disciplinary pattern, moving from simple units to complex ones in five to six or more year programs. There is little latitude for interdisciplinary or multidisciplinary work, though professional practice deals with ill-defined, complex and practical problems of agriculture that are "incapable of technical solution" and are intertwined with social and cultural patterns and ethical issues. Needs for synthesis of diverse elements, and interdisciplinary approaches. Outreach requires a different epistemology of science, because it faces real, synthetic and complex problems, and needs training in communicative competences and participatory approaches.
Promotion and reward	Academic staff are usually promoted and rewarded on the peer review system. Although this system has served certain fields of agricultural science well, it does not allow much credit for societal value or social pertinence of research contributions, and gives less value to teaching and extension. It also emphasizes the big gap between basic and applied research and between wealthy and developing countries´ academic and research systems and also marginalizes basic research in industrialized countries.
Curriculum policies	In many universities, curricula were broadened to encompass environmental sustainability, poverty alleviation, hunger elimination and gender issues. But this trend has not always been followed by specific fund allocation to programs oriented to these goals, nor have interdisciplinary courses and social sciences–sociology of organizations, cultural anthropology, IP issues, food security, and some crosscutting subjects, such as Ethics, always been included. Change is sometimes cosmetic.
Enrollment and graduation rates	Enrollment of agricultural students is today very low compared to total university enrollment. This is a generalized trend even in countries with a high share of agricultural GDP in total GDP and a high ratio of rural to urban population, mostly in non-industrialized countries. Likewise, graduates in agricultural programs (agriculture, forestry and fishery and veterinary) have a very low percent of total graduates. In many countries where agriculture is a major source of income, employment and export earnings, and thus critical to alleviating rural poverty and safeguarding natural resources, the percent of graduates is low (UNESCO, 2005).
Gender issues	Despite their key role in agricultural and food production and security, agricultural information and education is not reaching women and girls. Greater awareness of women's contributions to agriculture and changing discriminatory practices and attitudes are needed to foster their participation in agricultural education and extension. Not many women professionals are trained in agriculture due to factors rooted in the gendered nature of culture and society. Women's participation in higher education in agriculture is increasing, but is still lower than that of men, even in the developed countries and in Latin America and the Caribbean, where women participate in higher education in nearly equal numbers with men (UNESCO, 2005).

rests on concern for both agriculture and rural communities; enterprise development, revenue and welfare; education and research as a privileged knowledge and information activity for faculty and students and as a service to meet citizens' needs. The task of forming, educating and empowering farmers and young farm leaders has been a key strategic objective, resting on tripartite funding contributions from education, agriculture and state agencies at various levels. Farmers have opportunities as well as a right to participate in forming and assessing university research priorities and outputs. Outreach and service count in professional advancement; and the universities' own institutional advancement—even survival—rests on accountability to the broad constituency it serves.

On the other hand, in industrialized countries, particularly in the US, universities have emerged as the nation's main source for the three key ingredients to continued growth and prosperity: highly trained specialists, expert knowledge and scientific advances. There is some evidence that more recent shifts in the balance of public and private funding is affecting the type of research and teaching and hence narrowing the range of available AKST systems. One paradigmatic and controversial case was the agreement between the Novartis Agricultural Discovery Institute and the Department of Plant and Microbial Biology at the University of Berkeley. Under this agreement Novartis provided $5 million per year in support of basic research at the department and in return was given the right to license patents held by the University for up to one-third of the patentable intellectual property developed by the department, with the University retaining the patent rights and earning royalties from the patents. Participating faculty, in turn, received access to proprietary databases held by Novartis (Berdahl, 2000; Busch et al., 2004). The Novartis agreement disquieted those who believed it indicated a transition toward the privatization of public universities; critics argued that by allowing Novartis to participate even as a minority vote on the funding committee, the University was allowing a private company to chart the course of research at the University (Berdahl, 2000). Others pointed out that faculty members applying for research support from the federal government possibly also tailored their applications to increase their chances of support. This situation illustrates the need for codes of conduct in universities to guide their interactions with industry (Washburn, 2005) in order to preserve independence and capacity to deliver disinterested public goods and maintain public trust (Vilella et al., 2002). More public-private partnerships without ensuring such codes may reduce the space for public interest science (Washburn, 2005), although under certain conditions, university partnerships with private actors may contribute to equitable and sustainable development. For instance, the Seed Nursery at the Faculty of Agronomy, Buenos Aires University (www.agro.uba.ar) and the Argentine Agrarian Federation (www.faa.com.ar) developed high-yielding non-Bt corn hybrids (FAUBA 207, 209 and 3760), which are locally adapted and affordable by small-scale farmers and were released to market at less than half the price of the main competitors (Vilella et al., 2003; Federacion Agraria Argentina, 2005; http://www.todoagro .com.ar/todoagro2/nota.asp?id=6542).

(3) Agrotechnical institutes. Postsecondary institutes that are not part of the university system usually depend on public funding from Ministries of Education or Agriculture. They mostly train technicians in agricultural competences related to local labor demand in order to bridge the gap between untrained farmers, semi-skilled technicians and university graduates. However, many developing countries have given little attention to the training demanded by agricultural service agencies and agroindustries. Other countries, such as India or Brazil, invested heavily in such training. In Brazil, the Federal Centers of Technological Education (CEFETs) originated in agrotechnical or technical schools that were upgraded to tertiary-level institutes in the mid-1990s. They have developed good links with the private sector and sometimes share resource training activities through "sandwich courses." They have become drivers for the application of technology, but an extension worker with certificate-level training and field experience can seldom bridge to a degree program (Atchoarena and Gasperini, 2002; Plencovich and Costantini, 2003). The Sasakawa Africa Fund for Extension Education (SAFE) specifically addresses this need in SSA. Other countries have chosen an alternative agricultural school system shaped after the Maisons Familiale Rurale (rural family house) (Granereau, 1969; Forni et al., 1998; García-Marirrodriga and Puig Calvó, 2007). Today there are more than 1,300 such schools in forty countries, alternating residential training and experience on the family farm. In Argentina, a large group of secondary public

schools managed privately by NGOs, foundations, and other private actors and federated under the umbrella of an apex organization (FEDIAP; http://www.fediap.com.ar/) manages 3,000 teachers and about 15,000 students, taking occupational education deep into marginal and vulnerable areas (Plencovich and Costantini, 2006).

2.2.3 Producers of AKST at regional and international levels

The institutional arrangements for development-oriented AKST at international levels have evolved from rather simple relationships organized largely by and in support of colonial interests, through focused support organized through the CGIAR system largely under the guidance of multilateral and bilateral development organizations, to arrangements that are rapidly diversifying under market pressures. The increasing attention to environmental issues, especially from the early 1980s onwards, also gave rise to arrangements that made effective use of collective capacity to address shared practical and policy problems related to such issues as watershed management, vector-borne diseases and biodiversity conservation efforts. Examples include CSOs, such as the South East Asian Regional Initiatives for Community Empowerment (SEARICE) in the Philippines (cf. http://www .searice.org.ph/), which serves as the secretariat for region-wide advocacy, lobbying and action among networks of CSOs to promote and protect farmers' seed exchanges and sales and to ensure legal recognition of farmer-bred varieties and of community registries of local plants, animals, birds, trees, and microorganisms. SEARICE has become a major actor in the establishment of community-based native seeds research centers, such as CONSERVE in the Philippines and farmer field schools for plant genetic resource conservation development and use in Laos, Bhutan, Vietnam and community biodiversity conservation efforts in Vietnam, Thailand, and the Philippines. SEARICE today is recognized as an effective and legitimate partner in sustainable and equitable development. The Mekong River Commission (MRC) offers a different kind of arrangement; founded in 1995 by the Agreement on the Cooperation for the Sustainable Development of the Mekong River Basin (http://www.mrcmekong .org/). It is funded by contributions from the downstream member countries (Cambodia, Laos, Thailand, Viet Nam) and donors and is considered an important institutional innovation that is successfully bringing together cross-sectoral knowledge and helping actors to learn from policy experiments. However economic drivers within the member states resulting in upstream development of irrigation and hydroelectric power in China are undermining local efforts to forge more sustainable development pathways (Jensen, 2000; MRC, 2007). In SSA regional AKST arrangements have emerged and today their NARIs also act as regional service centers. ASARECA and CORAF were established in the late 1990s in eastern and western Africa respectively to fill gaps and build on strengths but no assessment can yet be made of their effectiveness. In southern Africa the formalization of interstate collaboration in AKST has not yet occurred. The South African Agricultural Research Council and universities continue to provide a regional backup service and various R&D networks seek to fill some of the severe gaps in public and private capacity.

The tropical AKST institutions established by the colonial powers, such as the Royal Tropical Institute (Netherlands) or the Institut de Recherche pour le Développement (formerly ORSTOM) (France) and their supporting university networks similarly have surrendered their dominance and undergone major transformations over the period (Jiggins and Poulter, 2007), yet they remain collectively the largest source of knowledge on the diversity of ecological and ethnic situations in the tropics. These institutional arrangements were generally effective for their initial purpose, but they badly neglected the food crops consumed by indigenous populations, with the exception of a few such as the federal research station for French West Africa created in 1935 to increase food production (Benoît-Cattin, 1991). The International Agricultural Research Centers (IARCs), subsequently grouped under the CGIAR umbrella, was in part a response to this gap.

CGIAR. Assessing the role of the CGIAR is fraught with difficulties, mainly because of the controversies raised by this important actor, since its inception. Several external reviews of the CGIAR took place in the 1990s (World Bank, 2003), most of them organized by the CGIAR itself, indicating a willingness to change and adapt but also some uneasiness about the way the CGIAR worked, chose its priorities and was governed. However, the reviews did not fully address some of the more fundamental questions raised by the critics. There is insufficient space here to do justice to all these debates.

Creation of the CGIAR. The role of the two US-based foundations, Rockefeller and Ford, in the creation of the first international centers has been well-documented (Baum, 1986). The first international research center, the International Maize and Wheat Improvement Center—in Spanish, CIMMYT—was devoted to wheat and maize, the second one—the International Rice Research Institute (IRRI) established in the Philippines in 1960—to rice. This early emphasis on cereals, i.e., on staple food crops, was a direct reaction, befitting the philanthropic nature of the two foundations, to the emphasis on plantation crops during the colonial era.

The emergence of this new type of institutional configuration had a profound impact on the IAs for agricultural research in developing countries. In this respect, the rapid evolution of the role of IRRI is exemplary. The first high-yielding (HY) rice cultivar released by IRRI (IR8) grew out of a dwarf gene which originated in Japan. Soon, however, its limitations became obvious. The new variety was sensitive to multiple pests and did not have the taste desired by many in Asia. The second generation of HY cultivars released by IRRI grew out of elaborate collaborations among many national research institutions in Asia, permitting a quantum jump in the exchange of genetic material and the coordinated testing of new genetic material in multiple locations. These new kinds of IA's, based on networking among public research institutions, with the hub located at an international center, set a pattern for the future. The role of the international centers in the development of new and more productive staple crop varieties has been well documented (Dalrymple, 1986) and is not by itself a controversial issue.

Early criticisms. But early on, extensive criticisms were expressed; in particular, it was pointed out that a technological change, however rapid and even if called a revolution as in the expression "green revolution", could fall short of the radical changes in agrarian structures which many felt were necessary to tackle the most glaring inequalities associated with unequal access to productive resources, land in particular (Griffin, 1979; Griffin and Khan, 1998). One must recall in this respect that the Green Revolution (GR) came after many attempts at promoting land reforms or agrarian reforms. Many of the reform attempts were made in a climate of bitter social struggles, often violent. In this context, the promotion of an international consensus in support of a technology-led green revolution could be seen as an alignment with conservative forces, nationally and internationally (Frankel, 1971). A similar criticism saw the GR, the CGIAR and their promoters such as the World Bank, which indeed had played a crucial role in the formal creation of the CGIAR, patterned on other consultative groups sponsored by the Bank, as an attempt at "liquidating peasantries" in developing countries (Feder, 1976). These criticisms prompted a large body of empirical research and interpretative analyses to evaluate the impact of the GR on poverty and the survival of small-scale producers. The assessment of the merits and limitations of the transfer of technology (ToT) model draws on that literature (2.1) (Harris, 1977; Lipton and Longhurst, 1989; Biggs and Farrington, 1991; Hazell and Ramaswamy, 1991; Lipton, 2005). One important lesson was that the social impacts of the technological changes associated with the GR varied greatly in space and in time. This should not come as a surprise since we know that technological change is only one component in the complex evolution of social realities yet the implications for how AKST were conducted within the CGIAR and with the CG's partners did not immediately sink in. The controversies themselves also reflect the fact that many views expressed in the controversies were oversimplifications drawn from limited empirical data, giving privileged attention to some aspects of the complex phenomena involved.

Similarly, the debates on the role of the CGIAR in the impact of the GR on the environment have been heated (2.1). Those who defend the GR and the CGIAR emphasize the millions of hectares of primary forests and other lands saved from destruction through the intensification of existing cropland that the GR permitted (see Borlaug's numerous public speeches on the topic). There is no doubt, however, that negative environmental effects, ranging from pollution to degradation of land and water resources also have been significant (Byerlee, 1992). Another environmental consequence, the increase in the uniformity of crop germplasm, with all the risks that the corresponding loss of biodiversity entails, roused similar controversies (Hogg, 2000; Falcon and Fowler, 2002).

Subsequent evolution of the CGIAR. These debates and the recognition that many issues were not well addressed led to changes within the CGIAR. For instance, it was recognized that the focus on individual crops had serious limitations. Mixed farming—the basis of many small-scale farming systems—agroecosystem sustainability and the management of natural resources had not been addressed systematically.

The two livestock-focused centers had not achieved impacts comparable to the crop-focused centers. These concerns led to the creation in the 1970s and 1980s of a further wave of international agricultural research centers that were initially outside the ambit of the CGIAR (e.g., IWMI: water and irrigation, IBSRAM: soils, ICRAF: agroforestry, ICLARM: aquatic resources, and INIBAP: plantains and bananas). Generally speaking, the newer institutions developed more extensive networks of partnerships with a wider range of civil society and public agencies than the original crop research centers. In the early 1990s, some of the new centers were brought into the CGIAR ambit, after much discussion and resistance by those who feared that the expansion of the CGIAR would entail a reduction in funding for the original centers. Two major concerns drove this expansion: the perceived need to widen the research agenda to include a systematic focus on natural resource management, and a broad recognition of the need for CGIAR centers to diversify their partnerships and networking capacity. The international centers were thus driven by a growing pressure to address new issues, mainly related to natural resource management, and to address more directly than before the needs of the poorest producers and of undervalued clients, such as women (Jiggins, 1986; Gurung and Menter, 2004).

In response to donor calls for more efficient, collaborative and cost-effective approaches, the CG centers opened up to new modes of collaboration, including "system-wide programs" that draw together expertise from across the range of centers and other AKST actors in order to focus on specific themes and the development of "partnerships for innovation". The increasing focus on innovation in turn required the centers to pay more attention to institutional issues and the contexts in which a technology is inserted and to seek a wider range of partners in recognition of the emerging global architecture for AKST (Petit et al., 1996). However, the rate of change within the CG was considered by its funders to be too slow and indecisive. One of the solutions was the introduction of well-resourced, multistakeholder, regionally focused "Challenge Programs" (CGIAR, 2001), often including a competitive research grant component. Their emphasis on multiple partnerships is a potentially significant institutional development for the CGIAR system. As yet however, there is insufficient evidence to assess their contribution to sustainable development or to driving change within the CG. The Global Forum on Agricultural Research (GFAR) was established in 1996 as a complementary initiative to promote global leadership in AKST for shared public interest goals; currently there is insufficient data for an assessment of GFAR's effectiveness.

Current debates. In spite of the changes briefly sketched above, the debates and controversies about the CGIAR have not disappeared. For some, "the CGIAR and the GR that it created have arguably been the most successful investments in development ever made" (Falcon and Naylor, 2005). Yet criticisms abound. The old fundamental questions regarding the insufficient inclusion of the poor and marginalized and the consequences on the environment, particularly the loss of biodiversity, have not been resolved in the eyes of many. Another criticism, often heard but seldom formalized, is that the CGIAR is very much part of the "establishment"

and not sufficiently receptive enough to new ideas. An illustration of this resistance to change is the assessment by social scientists (other than economists) that their expertise has not been used as effectively as possible (a few have now been integrated into some CG centers) (Rhoades, 2006; Cernea and Kassam, 2006). Another frequent criticism, often heard in donor circles but not often openly expressed, is that many centers are not open enough to broad partnerships with multiple and diverse actors. Others continue to fear a dilution of the main mission and unique role of the CGIAR, lest it drift more and more towards becoming a broad based development agency. Thus, some convincingly argue for a stronger CGIAR focus on international public goods through its attention higher productivity, particularly for orphan crops (Falcon and Naylor, 2005). One lesson to draw from this debate may be the relevance of, but also the difficulties associated with, the use of the concept of global public goods (Dalrymple, 2006; Unnevehr, 2004).

Food and Agriculture Organization of the United Nations (FAO) was founded in October 1945 under the United Nations as a key pillar of the post WWII reconstruction, with a mandate to raise levels of nutrition and standards of living, to improve agricultural productivity and the condition of rural populations. From 1994 onwards, it has undergone significant restructuring in an effort to increase the voice of tropical countries in its governance and priority setting and in response to advances within AKST and the changing architecture of public and private provision. Although remaining heavily male-dominated in its staffing and leadership, it has been a significant global actor in creating awareness of gender issues, stimulating growth with equity and in linking nutrition, food security and health issues.

It has played a leading role in organizing disinterested technical advice in the international response to the health and environmental concerns associated with synthetic chemical pesticides (see 2.3.2), leading among other important outcomes to the International Code of Conduct on the Distribution and Use of Pesticides and efforts to remove stockpiles of obsolete pesticides. This code has encouraged many countries to adopt pesticide legislation and regulations although governments may experience difficulty in implementing and managing pesticide regulations in the face of competing interests (Dinham, 1995). The FAO similarly has played a critical role also in international efforts to protect crop genetic diversity through the International Treaty on Plant Genetic Resources for Food and Agriculture. One of the important spin-offs so far is the Global Crop Diversity Trust hosted jointly by FAO and IPGRI (http://www.fao.org/ag/cgrfa/itpgr.htm).

The World Bank. The World Bank Group was established as another of the key pillars of post World War II reconstruction. It consists of the International Bank of Rural Development (IBRD), the International Development Agency (IDA), International Finance Corporation (IFC) and Multilateral Investment Agency (MIGA). The Group has been and remains a leading global player in development policy, funding and advisory efforts. It has invested heavily in economic and service infrastructures in rural areas; it was an early backer and consistent supporter of the emerging CG system and particularly through the 1980s dominated investments in agricultural extension and advisory systems in developing countries. The World Bank directly shapes the development path of many borrower countries through its research and through structural adjustment programs that restructure national economies or specific sectors (including agriculture). Yet Bank agricultural lending has decreased steadily over the past 60 years; currently it constitutes less than 10% of IBRD and IDA lending. The very mixed effects of these trends and shifts in financing on AKST and on innovation in the agricultural sector have been assessed in the 2008 World Development Report on Agriculture (World Bank, 2008). Internal as well as external analysts over the last 15 years have recommended that the trend be reversed.

Like other development actors, the World Bank has evolved over the decades in response to different drivers, external pressures and internal experiences (Stone and Wright, 2007). According to one narrative, the Bank initially perceived its central role as the transfer of capital from rich countries to poor ones. The bulk of its portfolio lay in infrastructure projects developed by engineers. In the 1970s, Bank management concluded that infrastructure development alone was not sufficient to eliminate poverty and so Bank agricultural economists focused on "poverty alleviation." In the 1980s, macroeconomists, who played a leading role in designing investment projects at the Bank, viewed the debt crisis as evidence that sectoral development efforts could not succeed in the presence of major macroeconomic imbalances. Powerful interests in industrialized countries (where commercial banks feared that the loans they had made to developing country governments were at risk), pressured their government representatives in the Bank, and in the IMF to intervene. Accordingly, Bank management promoted structural adjustment programs as a condition of its lending. In the 1990s, the Asian economic crisis demonstrated that a narrow macroeconomic perspective was not appropriate for the pursuit of a sustainable development agenda, and the role of social sciences was gradually recognized. Changes in the hierarchy of professional disciplines within the Bank did not come about smoothly. Struggles eventually led to greater inclusion of the social sciences (Kardan, 1993); Ismail Serageldin, a Bank vice-president, spelled out why non-economic social scientists were not listened to earlier and delineated key intellectual challenges that remained to be faced (Cernea, 1994).

Political economic, anthropological and ethnographic analyses have also assessed the role of the Bank (Wade, 1996, 1997, 2001, 2004; Ferguson, 1990; Harris, 2001; Mosse, 2004a; Broad, 2006; Bebbington et al., 2007). Simple causal linkages between external event, internal analysis, policy formation and subsequent implementation have been questioned (Mosse, 2004b). Evidence suggests that the Bank through its principal research unit has constructed, defended, maintained and promoted a neoliberal paradigm, despite changing contexts and emerging empirical evidence that challenge this paradigm (Broad, 2006). Organizational dynamics and international political economy have consistently shaped policy statements produced by the Bank over the period, while organizational culture—the everyday imperative to disburse loans and move projects through the pipeline, the internal incentive structure, hiring, staffing and subcontracting decisions and, importantly, power relationships within the Bank and between it and other actors—have been more decisive determinants of the outcomes of Bank interventions than its policy statements (Liebenthal, 2002; Mosse, 2004a; Bebbington et al., 2007). The empirical evidence indicates a need for more political economy and social science-based analyses of the World Bank's institutional behavior, culture, internal and external power relations and dynamics, and outcomes in terms of equitable and sustainable development.

The positive role of the Bank in the provision of financial resources to AKST includes loans to many governments in support of public agricultural research and extension institutions. Such support usually accompanied commitment to institutional reforms of these institutions. For instance, in Mali a Bank loan permitted the creation of research user committees at the level of regional research centers. These committees were designed to give a voice to farmers in the selection of research topics and in the evaluation of the usefulness to them of research results. The initiative gave some space for educated farmers with more resources to participate in research agenda-setting. In India, a large loan made in the late 1990s promoted significant reforms in the large Indian public sector agricultural research establishment, which had become quite bureaucratized. In Brazil, the volume of the loan made in the late 1990s was relatively modest; it was used by the then new leaders of EMBRAPA, the national research institute, to facilitate institutional reforms. The impact of Bank supported projects have been assessed and documented by the Bank's own Operation Evaluation Department (OED), a quasi-independent body which, while providing a degree of critical analysis, admittedly often reflects the dominant ideology in the Bank. Accordingly, the final and critical evaluation by OED of the T&V agricultural extension system, long promoted by the Bank particularly in Africa, was published only in 2003 (Anderson et al., 2006) i.e., long after the shortcomings of T&V had been emphasized by its critics; thus internal institutional learning and reform has been slow.

In 1992, the Bank joined 178 governments in committing itself to Agenda 21, a global effort to articulate the link between environment and development issues. An internal World Bank review of progress towards environmentally sustainable development found that it had failed to integrate environmental sustainability into its core objectives or to forge effective cross-sectoral linkages between environmental and other development goals (Liebenthal, 2002). External assessments similarly found that the Bank had not lived up to the expectations of its Agenda 21 commitment (FOE/Halifax Initiative, 2002). A four-year Structural Adjustment Participatory Review Initiative, in which the World Bank participated, reported that the effects of the Bank's structural adjustment programs on the rural poor over the past 20 years had been largely negative (SAPRIN, 2002). External analyses likewise found that these programs tended to drive the evolution of AKST towards high external input models of production, while the pressure of debt repayment schedules in turn prevented governments from investing in poverty-oriented multi-sector sustainable development programs at home (Hammond and McGowan, 1992; Danaher, 1994; Korten, 1995; Oxfam America, 1995; Clapp, 1997; McGowan, 1997; Hellinger, 1999; SAPRIN, 2002).

An important lesson is that both because of its size and its role as a financial institution, the World Bank has not been deft in its interventions in countries' institutional arrangements, particularly at the local level. This is a damaging limitation because as other subchapters demonstrate, appropriate institutional arrangements, particularly at the local level, are critical to the effectiveness of AKST in terms of the Assessment's criteria of equitable and sustainable development. The Bank also has faced numerous demands in the area of AKST from other development funding agencies that are willing to fund initiatives through "trust fund arrangements." The danger that the Bank could be drifting too far from its primary role as a financial institution has been keenly felt by some senior managers; as a result, the Bank has at times taken up and then dropped AKST initiatives that may have been worthwhile in advancing broader development goals. The consequences of its brief attention to these issues have not been well assessed. The other regional development banks have not held the central and symbolic role of the World Bank. But they also have played an important role in their region and have sent powerful signals regarding their AKST priorities to client governments. More in-depth social scientific analyses of the nature of the banks' interactions with other AKST actors and their contributions to equitable and sustainable development is warranted.

2.2.4 Public-private and private sectoral arrangements

Public-private arrangements. A number of countries have relied on multi-organizational partnerships to carry AKST to small-scale producers. For instance, the Foundation for the Participatory and Sustainable Development of the Small-scale producers of Colombia (Spanish acronym, PBA) brings together members of the Ministry of Agriculture and Rural Development, the Ministry of the Environment and the DNP (National Planning Department); international research centers, such as CIAT (International Center for Tropical Agriculture); research agencies such as CORPOICA (Colombian Corporation of Agricultural Research) and CONIF (National Agency for Forestry Research); national and regional universities and local farmers' organizations. It is responsible for bringing together at local levels the expertise and support required for small-scale producers and rural entrepreneurs in research, technology generation, and extension and agroenterprise development. The Andean consortium, established in the early 1990s on the initiative of the PBA, brings together five Andean countries (Venezuela, Colombia, Ecuador, Peru and Bolivia) under a regional project in order to strengthen participative exchange of research and technology with small-scale producers, as well as mobilize international cooperation in AKST and funding. The project has significantly advanced understanding of the small farm economy, established a strong nucleus of expertise in participatory research, developed the scientific, adaptive and applied research infrastructure and established key agroenterprises for the production of clean seed and bioinputs, and initiated links with private commercial actors in the development of value-adding chains in export-oriented markets, e.g., cut flowers, tropical fruits and counter-season vegetable supply.

Organizations such as *Solidaridad* have extended the concept and practices of public-private partnerships by linking fair trade to high return markets, such as the fashion industry and more recently, by moving an increasing amount of fair trade product into mass marketing. This effort is being guided by the multistakeholder negotiation of Codes of Conduct. For instance, the Common Code for the Coffee Industry was introduced in September 2004. It is currently operating in Vietnam and Uganda, with major expansion from 2006 onwards under the sponsorship of the German Ministry for Development Cooperation, the German Coffee Association, producer associations and major coffee processors, such as Nestlé, Tchibo, Kraft and Sara Lee, and international organizations such as Consumers International.

Private sector arrangements for profit. The last sixty years have witnessed a rapid increase in the concentration of commercial control by a handful of companies over the sale of planting seed for the world's major traded crops—by 1999, seven companies controlled a high percentage of global seed sales and the concentration has since increased through takeovers and company mergers. The budgets of the leading six agrochemical companies in 2001-2002 combined equaled US$3.2 billion—compared to a total CGIAR budget in 2003 of US$330 million, an order of magnitude less (Dinham, 2005). At the same time, national small and medium-sized seed companies have emerged, playing an important role for small-scale producers and niche markets. They may result in improved market access by small-scale farmers to locally adapted and affordable seed but this remains to be proven. Interesting innovations include the following three examples. The Seeds of Development Program (SODP) is a capacity development and network initiative that seeks to alleviate rural poverty through improved access to appropriate seed varieties. It offers an innovative program for small and medium sized indigenous seed companies in Africa. The network currently includes 25 seed companies in eight African countries. The SODP has been developed by Market Matters, Inc., a US-based organization working in collaboration with Cornell University. Private seed companies operating in India for many years relied on ICRISAT-bred hybrid parents and while gradually developing their own research and development capabilities; over time they became a major channel for large-scale farm level adoption of hybrids derived from ICRISAT-bred hybrid parents or their derivatives. ICRISAT realized that such partners have better integrated perceptions of farmers' preferences and this triggered the initiation in 2000 of the ICRISAT-Private Sector Hybrid Parent Research Consortia for sorghum and pearl millet. The consortia expanded to include pigeon pea in 2004. Small and medium sized manufactures of agricultural machinery and equipment, specialized for conservation agricultural equipment (e.g., no-till seeders and planters), especially in Brazil, provide agronomic assistance to farmers and advice on conservation agriculture, which simultaneously increases their own market.

2.2.5 NGOs and other civil society networks

Nongovernmental organizations (NGOs) are the so-called "third sector" of development, which is different from but interacts with both the state (public) and the for-profit private sector in AKST relationships ranging from complementarily to challenge (Farrington and Lewis, 1993; Farrington et al., 1993). The NGO sector developed in response to the actual and perceived failures or shortcomings of the state, a desire to examine developmental questions from motives other than those of profit and to question and analyze interests, priorities and the conditionalities imposed by donor agencies and other organizational actors. A fundamental basis of NGO activity is voluntarism (Uphoff, 1993) and this conditions NGO perspectives and scope of action and imposes a degree of similarity on what is an otherwise diverse domain. The diversity in the domain may be usefully classified by the origin of the NGO (Southern, Northern, Northern with activities in the South, etc.); the nature of the work—grassroots organizations (such as communities, cooperatives, neighborhood communities, etc.), organizations that give support to the grassroots, and those that (whether in addition to other activities or solely focused on this) are engaged in networking and lobbying activities; their funding; relationships with the state and private sector; their membership base; their size, staffing and relationships with their constituencies (which could be as diverse as rural farmers, urban slum dwellers, indigenous tribes), and the mechanisms and procedures in place for accountability (Farrington and Lewis, 1993; Farrington et al., 1993). In the case of the agricultural sector, the main types of NGOs encountered are those working directly with farmers with close involvement in dissemination of farming techniques and processes, provision of agricultural inputs, technologies, access to markets and implements (i.e., developmental NGOs); NGOs that are engaged in conducting research on agricultural crops, processes and products (research NGOs); NGOs that lobby for specific issues related to agriculture ranging from farm-worker health, to gender empowerment among farming communities, to advocating for specific regional, national and international agriculture and trade policies (advocacy NGOs); NGOs focusing on activities such as microcredit for farmers and agricultural communities (support NGOs).

The nature of activities that NGOs undertake, their relationship with the state and the private sector, their core constituency and nature of their involvement with it, their own organizational character and staff profile determine the attitude of an NGO towards the kinds of knowledge it considers valid and consequently the nature of knowledge processes it engages with and utilizes in its interactions with its constituency (Pretty, 1994). The processes of engagement range from the commissioning of research providers to inform NGO action, top-down dissemination of knowledge through NGO community trainers to engagement with farming communities in research and enquiry through user groups and participatory committees and direct involvement of farming communities in research agenda setting and knowledge selection. NGOs have become significant players in AKST. One of the largest member-based NGOs, BirdLife International has become a significant player in organizing civil society-based collection of data that informs local, national and global environmental policy and conservation effort. Local groups affiliated to this NGO and to WWF

(World Wildlife Fund) were instrumental in ensuring that attention was paid to the impacts on native biodiversity in UK trials of GM oilseed rape and other selected field crops. Collaboration among three Indian NGOs (Deccan Development Society, Andhra Pradesh Coalition in Defence of Diversity and Permaculture Association of India) supported the first thorough assessment of Bt Cotton from farmers' perspectives (Qayum and Sakkhari, 2005).

2.3 AKST Evolutions over Time: Thematic Narratives

The implementation and evolution of different IAs (see 2.2), have been causes as well as consequences of the main changes in AKST. Although it now appears that AKST presents itself as a whole, or at least as a tightly intertwined ensemble of domains, it has not always been the case. Progressively, over centuries, a hierarchy has developed between scientific knowledge, technological knowledge and agricultural production, the latter being progressively limited to the execution of external recipes. Paralleling this hierarchy, science itself established a hierarchy between emerging and evolving disciplines: chemistry, biology, genetics, botany, entomology, economy, sociology, and anthropology are permanently struggling for recognition, status and resources, and scientists engage in alliances with other actors in this purpose. Science allied with technology branched out in different domains of application that resulted in new professions related to various aspects of agricultural production, its products and impacts. Hence, in modern times is that the role of scientific research in maximizing agricultural productivity has increased exponentially (Cernea and Kassam, 2006). However, through the last decades, a reverse movement has occurred and the division between the different branches of AKST have been blurred, the great divide of labor between science and technology is currently challenged, the hierarchy among disciplines reveals its shortcomings and the role of public and private actors has changed.

The following narratives are illustrative of how AKST contributed and shaped (as well as resulted from) the management of three major elements: seeds, pests, and food. These narratives identify trends, turns, and bifurcations in each domain and look at the major actors who managed them, in response to drivers relevant for them.

2.3.1 Historical trends in germplasm management and their implications for the future

2.3.1.1 Summary of major trends in the history of global germplasm management

Genetic resource management over the past 150 years has been marked by an institutional narrowing with the number and diversity of actors engaged in germplasm management declining. Breeding has largely become an isolated activity, increasingly separated from agricultural and cultural systems from which it evolved (Box 2-1).

This narrowing is illustrated in history by four major trends: (1) a movement from public to private ownership of germplasm; (2) unprecedented concentration of agrochemical, seed corporations, and commodity traders; (3) tensions between civil society, seed corporations, breeders and farmers in the drafting of IPR; (4) stagnation in funding for common goods germplasm. These trends have reduced options for using germplasm to respond to the uncertainties of the future. They have also increased asymmetries in access to germplasm and benefit sharing and increases vulnerabilities of the poor.

For example, farmers have received no direct compensation for formerly held public accessions that have been sold on to the private sector but have generally benefited from public breeding arrangements. It remains a question if farmers now have to pay for accessing seed stock and germplasm that contain lines and traits that originally were bred by them and originated in their own farming systems. Meanwhile, decreases in funding for public breeding has stagnated research innovations for the public good (e.g., lack of research on orphan crops). New ownership and IPR regimes have restricted movement and made development of noncommercial (public) good constructs more expensive. These changes have limited those actors that do not have legal, commercial and financial power.

2.3.1.2 Genetic resources as a common heritage
Farmers as managers of genetic resources. Historically, farmers have been the principal generators and stewards of crop genetic resources (e.g., Simmonds, 1979). This means that genetic resources have been viewed as a common heritage to be shared and exchanged. The concept places farmers at the center of control of their own food security. The planting of genetically diverse, geographically localized landraces by farmers can be conceptualized as a decentralized management regime with significant biological (Brush, 1991; Tripp, 1997; Almekinders and Louwaars, 1999) and political (e.g., Ellen et al., 2000; Stone, 2007) implications. Studies of traditional farming systems suggest that farmers in Africa (Mulatu and Zelleke, 2002; van Leur and Gebre, 2003) the Americas (Quiros et al., 1992; Bellon et al., 1997, 2003; Perales et al., 2003) and Asia (Trinh et al., 2003; Jaradat et al., 2004;) managed and continue to manage existing varieties and innovate new ones through a variety of techniques including hybridization with wild species, regulation of cross-pollination, and directional selection (Bellon et al., 1997). In many parts of the world, it is women's knowledge systems that select and shape crop genetic resources (Tsegaye, 1997; Howard, 2003; Mkumbira et al., 2003). The fear is that erosion of crop diversity is commonly paralleled by erosion of the farmer's skills and farmer empowerment (Bellon et al., 1997; Brown, 2000; Mkumbira et al., 2003; Gepts, 2004). This loss of farmer's skills (i.e., agricultural deskilling; see Stone, 2007) means a loss of community sovereignty as less of the population is able to cultivate and control their own food (see 2.3.3).

Development of public and private sector. The public sector emerged to catalyze formal crop improvement, focusing on yield with high input requirements and wide adaptability (Tripp 1997; Almekinders and Louwaars, 1999). Major benefits arose from breeding with large, diverse germplasm populations. These advancements had both negative and positive impacts on farming communities as more uniform crops replaced locally adapted crops. Meanwhile, expedi-

tions to collect global germplasm were underway by several nations and gene banks were established for the conservation of germplasm for use in research and breeding.

Public sector institutions were the dominant distributors of improved varieties in first half of the 20th century, aiming to reach as large a constituency possible. Where different forms of mass selection formed the main breeding method in the 19th century, the rediscovery of Mendel's laws of heredity (1900) and the discovery of heterosis (1908) spurred the growth of the commercial industry, most notably with the founding of Pioneer Hi-Bred in 1919 (Crow, 1998; Reeves and Cassaday, 2002). Throughout the 20th century, universities and research institutes gradually specialized in basic research while the private sector increased its capacity in practical breeding. The public sector assumed primary responsibility for pre-breeding, managing genetic resources and creating scientific networks that acted as conduits of information and technology flow (Pingali and Traxler, 2002), and creating regulatory bodies for variety testing, official release, and seed certification.

The first institutional arrangements exported to developing countries. The education, research and extension system triangle commonly found in industrial countries was exported to developing countries to help foster agricultural development and food security, mainly through the development of broadly adapted germplasm. With the aid of the Rockefeller Foundation (and later the Ford Foundation), a collaborative research program on maize, wheat and beans in Mexico was founded in 1943. This laid the foundation for the first international research centers of the CGIAR, with the initial focus to improve globally important staple crops (see 2.2.4).

The formation of the CGIAR centers laid the groundwork for the emergence of the technologies of the Green Revolution. Borrowing from breeding work in developed countries, high yielding varieties (HYV) of rice, wheat, and maize were developed in 1960s and 70s. By the year 2000, 8000 modern varieties had been released by more than 400 public breeding programs in over 100 countries. The FAO launched a significant program to establish formal seed production capacities and so-called "lateral spread" systems in developing countries to make the new varieties available to as many farmers as possible. These public seed projects, financed by UNDP, World Bank and bilateral donors were subsequently commercialized, often as parastatal companies, before national or multinational seed companies were established in these developing countries (World Bank, 1995; Morris 1998; Morris and Ekesingh, 2001).

The FAO has estimated the economic and social consequences of crop genetic improvement gains emanating from the international agricultural research centers using the IFPRI based model "IMPACT" (Evenson and Gollin, 2003b). Without CGIAR input, it is estimated that world food and feed grain prices would have been 18-20% higher: world food production 4-5% lower, and imports of food in developing countries about 5% higher. Debates continue as to whether increases in food production, such as those of the Green Revolution, necessarily lead to increases in food security (IFPRI, 2002; Box 2-2; see 2.3.3).

Box 2-1. Timeline of genetic resource management.

10,000 years of agricultural history. Farmers as the generators & stewards of crop genetic resources (e.g conservation, selection, and management of open pollinated varieties)

1800s. Agricultural genetic resources—apart from plantation crops—not a policy issue, and valued and managed by farmers as a common good; First commercial seed companies (e.g., Sweden) and agricultural experiment stations in Germany and England; National school of agriculture founded in Mexico (1850s); Discoveries of Darwin and Mendel (rediscovered and applied in 1900 only). 1883 Paris Convention on patents (not applied to plants for a full century).

1910s. George Shull produces first hybrids (1916); Wheat rust resistance breeding program in India

1920s. First maize hybrids available; Vavilov collects crop genetic resources systematically and develops the concept of Centers of Diversity.

1930s. 1930 Plant Patent Act (USA) to cover plants that are reproduced asexually (e.g., apples and roses), excluding bacteria and edible roots and tubers (potato).

1940s. Bengal Famine 1943-1944; International Agricultural Research is conceived and funded; Rockefeller Foundation sets up research program on maize, wheat and beans with Mexican government. Breeder's rights laws develop in Europe.

1950s. Ford and Rockefeller Foundations place agricultural staff in developing countries. Mexico becomes self-sufficient in wheat as a result of plant breeding efforts. Watson and Crick describe the double helix structure of DNA and Coenberg discovers and isolates DNA polymerase which became the first enzyme used to make DNA in a test tube; Reinart regenerates plants from carrot callus culture—important techniques for genetic engineering. The National Seed Storage Laboratory (NSSI) was opened in USA.

1960s. South Asian subcontinent on the brink of famine. High Yielding Varieties (HYV) introduced. International Convention for the Protection of New Varieties of Plants (UPOV, 1961) providing a sui generis protection to crop varieties with important exemptions for farmers and breeders. Establishment of IRRI, CIMMYT, IITA, CIAT. Crop Research and Introduction Center established by the FAO in Turkey for the study of regional germplasm.

1970s. Public inbred lines of maize disappear from USA. European Patent Convention states that plants and animals are not patentable. Further development of international agricultural research centers under the auspices of the CGIAR; IR8 (high-yielding semi-dwarf rice) grown throughout Asia. Hybrid rice introduced in China. First recombinant DNA organism by gene splicing. Genentech Inc founded and dedicated to products based on recombinant DNA technology. First international NGOs focus on the seed sector (FAFI). Technical meetings on genetic resources organized by FAO.

1980s. First patents granted to living organisms by US courts. Large scale mergers in the seed sector. International funding for R&D begins to decline. Methods developed for Participatory Variety Selection and Plant Breeding as new institutional arrangement for breeding for development. (1985). Establishment of the FAO Commission on Plant Genetic Resources for Food and Agriculture (CPGRFA) and the FAO-International Undertaking (IU-PGRFA): Legally non-binding undertaking that confirms a "heritage of mankind" principle over plant genetic resources and recognises Farmers' Rights. US EPA approved the release of the first GE tobacco plants.

1990s. Agrochemical, pharmaceutical, and seed companies merge into "life science" companies; Major technological advances (e.g., marker assisted breeding, gene shuffling, genetic engineering, rDNA Technology, and Apomixis); Share of HYV increases to 70% for wheat and rice in selected developing countries. Acceleration towards consolidation of seed industry with agrochemical companies as main investors. Introduction of first commercial transgenic crops (e.g., Calgene's "Flavr-Savr" tomato and herbicide and insect-tolerant crops); Gradual change in CIMMYT approach from selection in high input environments to include drought and nitrogen stress. Rate of funding of CGIAR stagnant—more NRM-focused centers established. Regions where agricultural R&D relies on donors are particularly hard-hit. IU-PGRFA recognizes national sovereignty over PGRFA in the wake of CBD. CBD as legally binding agreement among all countries (except USA and some tiny states in Europe) lays the foundation for bilateral negotiations over access and benefit sharing to genetic resources, including PGRFA. Cartagena Protocol seeks to regulate international movement of transgenics. Agreement on Trade Related Aspects of Intellectual Property Rights (TRIPS) spurs a debate on plants and varieties in developing countries; European Patent Office moves to grant patents on plants (1999). UPOV 1978 treaty closed to new accessions. Latest UPOV Act prohibits farmers from sharing seed of protected varieties. Campaigns against strong IPRs in medical and agricultural research grow, notably against "terminator technology".

2000s. International Treaty on Plant Genetic Resources for Food and Agriculture (IT-PGRFA) facilitating access and benefit sharing and defining Farmers' Rights; World Intellectual Property Organization member states set up an Intergovernmental Committee on Intellectual Property and Genetic Resources, Traditional Knowledge and Folklore. Developing countries join UPOV or develop their own *sui generis* protection (e.g., India, Thailand). Free Trade Agreements put pressure on developing countries for stronger than TRIPs protection. Over 180 transgenic crop events, involving 15 traits deregulated or approved in at least one of 27 countries. Top 10 companies control half of the world's commercial seed sales; however farmer-seed systems remain key source of seed. Nanotechnologies enter agricultural sciences.

Box 2-2. Historical limitations of CGIAR arrangements.

Formal on-station breeding programmers have historically resulted in homogeneous varieties that favor uniform conditions, such as obtained with high inputs, rather than the low-input heterogeneous ecological clines that characterize the majority of small farmer's fields. The prevalence of pests, disease, and variability of climate and land requires a wide range of locally adapted heterogeneous varieties (Brush, 1991; Wolfe, 1992; Lenne and Smithson, 1994; Brouwer et al., 1993). In many cases, small farmers have been economically constrained from using high-input varieties. For instance, in Zimbabwe, drought in the 1990s affected poorer farmers, who had adopted hybrid maize, whereas richer farmers who had benefited from an early adoption of the varieties had diversified into cattle, leaving them better protected from drought shock. Weak performance of the hybrid maize under drought conditions left poor farmers poorer. Following early lessons, the CIMMYT program began to develop varieties in sub-Saharan Africa under conditions of low nitrogen input and drought (CIMMYT, 2002). Gender played a role in the adoption of new varieties, with women preferring open-pollinated traditional varieties disseminated by social networks, while the men preferred the improved varieties. Networks and social relationships have both facilitated and constrained technology dissemination (Meinzen-Dick et al., 2004).

Sharing of genetic resources as historical norm. Until the 1970s, there were very few national and international laws creating proprietary rights or other forms of explicit restriction of access to plant genetic resources. The common heritage concept of genetic resources as belonging to the public domain had been the foundation of farming communities for millennia where seed was exchanged and invention was collective (Brush, 2003). Farmers and professional breeding have relied on genetic resources, in the public domain or in the market, to be freely available for use in research and breeding. The public-sector research "culture" is based on this tradition of open-sharing of resources and research findings (Gepts, 2004) although this is changing (see below), with serious social and political implications. Indeed, the global collaboration required for the development of the HYVs of the Green Revolution demonstrated the effectiveness of an international approach to sharing of germplasm. The International Undertaking on Plant Genetic Resources, 1983, encapsulated this spirit citing the "universally accepted principle that plant genetic resources are a heritage of mankind and consequently should be available without restriction." Since that time, in many ways, the common heritage principle has been turned on its head, with the gradual encroachment of claims for control over access to and use of genetic resources grounded in IP laws, assertions of national sovereignty (Safrin, 2004) and or the intentional use of technologies that cannot be reused by farmers.

The common heritage or public goods approach to the use of Plant Genetic Resources for Food and Agriculture (PGRFA)

has not been entirely eclipsed. It is worth noting in this regard that the International Union for the Protection of New Varieties of Plants (UPOV) Conventions through their several revisions to further strengthen breeders' rights have consistently maintained a "breeders' exemption" which allows researchers/breeders to use protected materials in the development of new varieties without the permission of the owners (as long as the new varieties are not "essentially derived" from the protected varieties). Furthermore, in what might be considered a surprise development in the context of the overall shift in the genetic rights paradigm, the International Treaty on PGRFA creates an international research and breeding commons within which individuals and organizations in member states, and international organizations that sign special agreements, enjoy facilitated access (and benefit sharing) on preset, minimal transaction costs. Farmers and other target groups of this assessment have been inadvertently, and largely negatively, affected by the battles over genetic rights.

2.3.1.3 Major changes in germplasm management
The development of IPR in breeding. The business environment and size of the market are important factors for investment. Intellectual property rights (IPR) provides a level of protection. With the introduction of IPR, the private seed industry has benefited from the ability to appropriate profits to recoup investments and foster further research, organizational capability and growth (Heisey et al., 2001). The stakes are high; IPR regimes have transformed the US$21 billion dollar global seed market and contribute to the restructuring of the seed industry (ETC, 2005).

The increasingly international character of IPR regimes is a reflection of widespread and integrated trade in germplasm resources as well as global trends toward liberalization of markets and trade, privatization, and structural adjustment that reduce the role of the public sector (Tripp and Byerlee, 2000).

An evolution towards stronger IPR protection. Germplasm protections have been both biological; (e.g., hybrid maize) and legal. Initially plants were excluded from patentability for moral, technical and political reasons. For example, special, so-called *sui generis* protection was developed for asexually reproduced plants (US Plant Act 1930). In Europe protection for all varieties in the 1940s was harmonized through the International Union for the Protection of New Varieties of Plants (UPOV) (1961). This Plant Variety Protection (PVP) system recognized farmers and breeders exemptions. While PVP offers protection to private seed producers by prohibiting others from producing and selling the protected variety commercially, it does not restrict anyone from using a protected variety as parental material in future breeding. This is known as "farmer's privilege" and responds to the traditional seed handling mechanisms which allows farmers to save and exchange seed (1978 Act), a provision which was interpreted very widely in the USA, leading to large scale "brown bagging".

Utility Patents on a bacterium in 1980 signaled the advent of an era of strong IPR (Falcon and Fowler, 2002), marking the end of "farmer's privilege", which was restricted in the latest revision in UPOV (1991 Act). This loss of privilege generated heated debate among ratifying countries, especially developing nations, because it limits the rights of farmers to freely save, exchange, reuse and sell agricultural seeds (Tansey and Rajotte, 2008).

Patents entered plant breeding initially through court decisions in the USA in the 1980s via association with biotechnology. They were subsequently granted in other OECD countries, and offered greater protection to a wider array of products and processes, such as genes, traits, molecular constructs, and enabling technologies (Lesser and Mutschler, 2002). However, varieties are excluded from patentability in most countries. The EU introduced a breeder's exemption into its patent law, and some EU countries have introduced a farmer's privilege to avoid the pitfalls of excessively strong protection (World Bank, 2006).

IPR limitations. Even though IPR may be important for private seed sector development, some sectors have been successful in developing countries without IP protection. For example, the private seed sector in India has grown and diversified without the benefit of IPRs but in the context of liberal seed laws and in many cases through the use of hybrids as a means of appropriation (Louwaars et al., 2005).

Some indicators suggest that the IPR in developing countries may have occurred primarily as costs, as many patents are thought to slow down research. This problem is described as "the problem of the anti-commons" (Heller and Eisenberg, 1998) or "patent thickets" (Shapiro, 2001; Pray et al., 2005). Consider the example of Veery wheat, which is the product of 3170 different crosses involving 51 parents from 26 countries that were globally, publicly released. The development cycle of Veery would have been very difficult if, for each parent and each cross, it was necessary to negotiate a separate agreement (SGRP, 2006). Even though IPR tends to be territorial, i.e., granted at the national level, trade agreements have led to greater "harmonization" of IPR regimes (Falcon and Fowler, 2002) with countries adopting laws and rules that may not benefit seed-saving farmers (Box 2-3).

In many developing countries, institutional infrastructure required for implementation and enforcement of IPR regimes is still lacking. Opposition against TRIPS and the IP-clauses of free trade agreements concentrates on the lack of incentives for development of the seed industry in developing countries due to the harmonization approach. However, in agricultural biotechnology development, which is highly concentrated, the IPR issues precipitate more in the form of licensing practices and policies, shaping the impact of patent systems to a large extent. Consequently, there has been a misconception that existing problems can be best solved through reshaping patent regulations and laws alone. There is a related need to examine how licensing agreements contribute to many problems at the intersection of IP and agricultural biotechnology (CIPP, 2004).

Sharing of genetic resources; challenge and necessity. A reaction to IPR: national sovereignty and equity issues. The lack of explicit rules governing germplasm rights was the historical standard in agriculture until the 1990s. As pressure to protect IPR in improved varieties and "inventions" increased, the atmosphere concerning access to and use of genetic resources became increasingly politicized. This was

Box 2-3. Emergence of TRIPs-Plus.

International IPR regimes under the TRIPS agreements of the WTO allow for flexibilities for plant varieties, which may be exempted from patentability under the condition that an effective *sui generis* protection is provided for. This flexibility has been introduced by UPOV member countries, and creates a broad option for developing countries to develop their own systems, often balancing the rights of breeders with those of farmers. However, bilateral and multilateral trade agreements with IPR components dubbed "TRIPS-plus" often go far beyond the baseline of TRIPS standards, eclipsing the relative flexibility that was offered in TRIPS in favor of "harmonisation" at a more stringent, developed country IPR, level. For instance, TRIPS-plus regimes may force countries to join UPOV under the strict Act of 1991 or to allow patent protection on varieties. TRIPS-plus type regimes may take many forms and raise concerns about bypassing appropriate democratic decision making based on the interest of the national seed systems. Such Free Trade Agreements may be bilateral between regional blocks, such as in the EU or the Andean Community. In addition, the WIPO (World Intellectual Property Organization) is working to harmonise (i.e., strengthen) IPR globally, through the Substantive Patent Law Treaty (SPLT), raising concerns about development or conservation objectives.

augmented with concern, particularly among developing countries, that inequitable global patterns were established in the distribution of benefits associated with the use of genetic resources. Concurrently, there was growing concern that genetic diversity and local knowledge related to the use of those resources continued to be eroded under the pressures of modernization (Gepts, 2004).

In response, the international community attempted to address these tensions and create a new regime for access to genetic resources and the sharing of benefits associated with their use. One of the most significant outcomes was the Convention on Biological Diversity (CBD, 1994) (Box 2-4 and Chapter 7), which came into force in 1993. The CBD emphasized states' sovereign rights over their natural resources and their "authority to determine access to genetic resources, subject to national legislation." The Convention also addresses rights of local and indigenous communities in this respect. Over 160 countries have ratified the CBD; the US is not among them. Most countries have interpreted the access and benefit sharing provisions of the CBD as the basis for establishing much tighter procedural and substantive restrictions to gaining access to genetic resources within their borders. To this end, they have developed, or are developing, bilaterally oriented access laws that require case-by-case negotiations to establish legal conditions for obtaining and using materials from a country although they are not binding, and few countries have reported implementing them. Nonetheless, they are a good indicator that most countries think of the CBD's access and benefit sharing provisions as requiring, or justifying, a bilateral and restrictive approach to regulating access. Very different approaches were taken by individual countries to implement their sovereignty rights. Noticeably, the African Union and some countries in Asia (notably India and Thailand) have developed an approach that combines aspects of access and benefit sharing and breeder's rights in one regulatory framework, thereby clearly indicating the connection between the two issues.

While a restrictive bilateral approach to implementing the CBD may be appropriate for wild endemic species of flora or fauna, it is not well suited to plant genetic resources for food and agriculture (Box 2-4). All domesticated crops are the end result of contributions of farmers from numerous countries or continents over extremely long periods of time.

The CBD explicitly closed the concept of "heritage of mankind" that had been expressed in the 1980s. The nonbinding International Undertaking (Box 2-5) has re-established a commons for the crops and forages included in its Annex 1. CIP and IRRI have reported that since the CBD came into force, movement of plant varieties from and to their gene bank collections have been noticeably reduced and regulation of biological materials has resulted in increased bureaucracy and expense. Very few cases of effective (even non-monetary) benefit sharing as a result of CBD-based regulation during the first decade of the Convention (Visser et al., 2005). The key message is that promoting fair and equitable sharing of the benefits arising from the use of genetic

---

Box 2-4. Convention on Biological Diversity

Goals
1. Conservation of biological diversity
2. Sustainable use of its components
3. Fair and equitable sharing of benefits arising from genetic resources

The CBD asserts sovereignty rights to regulate access to genetic resources. It recognizes, and is to be interpreted consistent with, intellectual property over genetic resources. The sovereignty principal was to be implemented through prior informed consent and mutually agreed terms for access to genetic resources.

The Nairobi Final Act, 1993, resolution 3, signed by the signatories to the CBD acknowledged that the access and benefit sharing framework established by the CBD did not sufficiently address the situation of existing *ex situ* collections of PGRFA held around the world. It further states that it was important to promote cooperation between the CBD and the Global System of Sustainable Use of Plant Genetic Resources for Food and Agriculture as supported by FAO. This resolution set the stage for the further investigation into appropriate access and benefit sharing regime or regimes for PGRFA. This led indirectly to the seven years of negotiations of the International Treaty on Plant Genetic Resources for Food and Agriculture.

Positive Outcomes
- Heightened awareness globally of the inequitable distribution of benefits associated with the use of genetic resources
- Heightened awareness globally of the need to value, use and conserve indigenous and local knowledge, and to promote *in situ* conservation.
- Created a framework for the development of a plan of coordinated work on Agricultural Biodiversity
- Created a framework for funding for *in situ* conservation promotion projects through the Convention's funding mechanism: Global Environmental Facility

Problems
- The CBD does not distinguish between domesticated agricultural resources, collected in the form of ascensions of given crop (intra-species), and other biological resources, such as

wild plants collected for pharmaceutical applications. In fact, the convention seems to have been drafted more with the latter in mind (bio-prospecting).
- The CBD links benefit sharing to being able to identify the country of origin of a resource. The CBD defines the "country of origin of genetic resources" as "the country which possesses those genetic resources in *in situ* conditions." In turn, it defines "*in situ* conditions" as those "conditions where genetic resources exist within ecosystems and natural habitats and, in the case of domesticated or cultivated species, in the surroundings where they have developed their distinctive properties." Pursuant to this definition, the CBD requires more than simply identifying the country of origin of a crop—it requires the identification of the country of origin of the distinctive properties of the crop. Because of the international nature of the development and use of PGRFA, the CBD's method of linking the "origin" of traits to benefit sharing is impractical and often impossible to make work.
- The CBD has contributed to and reinforced exaggerated expectations about the commercial market value for local crop and forages varieties, leading countries to take measures to restrict access to those resources as a means of eventually capturing their market value (through use licenses) rather than sharing them in cooperative research projects that would likely result in significantly higher overall public benefit.

As a result of these factors, some critics feel the convention is inappropriate for the agricultural genetic resources, while allowing that it may still have potential for redistributing benefits associated with the use of other forms of genetic resources.

In the field of agriculture, the CBD was a groundbreaking assertion of national sovereignty over genetic resources. The sovereignty principal was to be implemented through prior informed consent and mutually agreed terms for access to genetic resources. Its implementation is through bilateral agreements between provider country and user.

Adopted at the Earth Summit in Rio de Janeiro, Brazil 1992, came into force 29 December 1993.

resources remains a major goal. Defining a monetary value to estimate the historic or current contribution of farmers' varieties remains elusive (Mendelsohn, 2000). Identifying the actual genetic resource property attributable to specific farming communities or even nations is "problematic" (Peeters and Williams, 1984; Visser et al., 2000). Some proponents have argued that benefit sharing would be more successful in the form of transfer of international capital, e.g., through development assistance to improve rural incomes in genetically diverse farming systems (Brush, 2005). Another approach could be to reduce structural adjustment policies that link agricultural credit to the planting of modern homogeneous varieties, and other crop and technology choices (Morales, 1991; Foko, 1999; Amalu, 2002).

The question of facilitated access. To match the principle of national sovereignty with the needs of sustainable agriculture and food security, an International Treaty for Plant Genetic Resources for Food and Agriculture concluded in 2001 and entered force in June 2004 (Box 2-5 and Chapter 7).

With roughly the same objectives as the CBD, it translates its conservation and sustainable use goals to agriculture, including both *in situ*, on farm and *ex situ* conservation strategies, and various aspects of crop improvement by both farmers and specialized plant breeders in implementing "sustainable use".

The main novelties in the International Treaty are (1) the creation of a Multilateral System for Access and Benefit Sharing for most important food crops and pasture species and (2) the definition of the concept of Farmers' Rights. Farmers' Rights include the right of benefit sharing, of protection of traditional knowledge and of farmers' involvement in relevant policy making. The objective is to have no restrictions on the ability of farmers to save, use, exchange and sell seed. However, signatory countries have freedom in specifying the Farmers' Rights as "subject to national law and as appropriate." The formulation was chosen to avoid conflict with existing and future IPR laws. Some claim that this formulation has thus far prevented an international acceptance of an inclusive Farmers' Rights concept (Brush, 2005).

Box 2-5. International Treaty on Plant Genetic Resources for Food and Agriculture (adopted November 2001, came into force June 2004).

1. Ensure access to and conservation of plant genetic resources.
2. Equitable sharing of benefit arising from agricultural genetic resources.

The treaty is a legally binding mechanism specifically tailored to agricultural crops, in harmony with the CBD. Creates multilateral system (MLS) for access to genetic resources and benefit sharing, which is designed to lower transaction costs of exchanges of materials to be used for research, conservation and training. The International Treaty links benefit sharing to access from the MLS as a whole. A proportion of monetary benefits arising from commercialization of new PGRFA developed using material from the MLS (when others are restricted from using the new PGRFA even for research) will be paid into an international fund, ultimately controlled by the Governing Body of the Treaty. Funds will be used for programs such as conservation and research, particularly in developing countries. The monetary benefit sharing provisions are not triggered when new PGRFA are made freely available for research and breeding. 64 major food crops and forages are included within the MLS. The list could be expanded in the future, by consensus of the Governing Body.

Positive Results
- It appears to be well on its way to becoming a truly global Treaty, with an increasing number of countries ratifying or acceding to it.
- Specifically tailored for agricultural genetic resources.
- Regularizes access to genetic resources under a single uniform multilateral regime using a single fixed legal instrument for all transfers.
- Includes a benefit sharing clauses, triggered through commer-

cialization of new PGRFA products that incorporated materials accessed from the MLS when those new products are not made available for further research.
- Provides a permanent legal status for the *ex situ* collections of PGRFA hosted by the CGIAR Centres, placing the Centres Annex 1 holding within the MLS (and making the Centres' non-Annex 1 holdings available on very similar terms).
- Recognizes the principal of Farmers Rights, and creates some momentum for countries to implement national laws to advance Farmers' rights.

Problems
- Significant crops are excluded from the Treaty, (including soybeans, groundnuts, tomatoes, tropical forages, onions, sugarcane, melons, grapes, cocoa, coffee). The rules applying to those crops is therefore uncertain, falling by default under whatever systems countries put in place to implement the CBD. Of course, additional species or genera can be included within the MLS with the consensus of the Governing Body.
- While a number of major industrial countries have ratified the Treaty, the USA still has not, and it is not clear if or when it will do so.
- The SMTA adopted by the Governing Body in June 2006 is relatively long and relatively complex. It will take some time before the global community fully understands what it says and becomes comfortable using it. In the meantime, ancillary efforts will be necessary, probably lead by organizations that are going to be participants in the MLS and consequently, users of the SMTA, to raise awareness about the MLS, assist countries in developing legal and administrative frameworks to implement the Treaty, and build organizations' capacity and comfort level in participating in the MLS and using the SMTA.

2.3.1.4 Increasing consolidation of the private sector

The changing face of the seed industry. In the context of newly emerging IPR regimes and the development of biotechnology (e.g., identification, cloning and transferring of individual genes), a major theme of consolidiation in the agricultural plant biotechnology and seed industries has emerged (Pingali and Traxler, 2002; Pray et al., 2005). This consolidation significantly altered the course of germplasm management and marked a major shift in the relationship between the public and private sector.

Consolidation of the industry began with mergers of family-owned seed companies by multinational chemical firms to capitalize on synergies between seeds and chemical inputs (Thayer, 2001; Falcon and Fowler, 2002). Consolidation in the seed industry had been ongoing since the 1970s, but the unprecedented concentration in the 1990s resulted in an extreme vertical integration of the seed and biotechnology industries (Hayenga, 1998). This was followed by a horizontal integration of agriculture and pharmaceuticals into life sciences companies.

The first trend was driven by (1) the stagnation of the agrochemical sector; (2) the changing knowledge base and innovations in chemistry and biotechnology; and (3) the policy environment, such as the increased burden of regulations (Hayenga, 1998; Falcon and Fowler, 2002). Between 1995 and 1998, in the US alone, approximately 68 seed companies either were acquired by or entered into joint ventures with the top six multinational corporations (King, 2001). An analysis for thirty UPOV member-countries identified a high degree of concentration in the ownership of plant variety rights for six major crops at the national level in the developed world (Srinivasan, 2004). The area with the greatest concentration intensity in the past decades has been genetic transformation (Pray et al., 2005; Box 2-6). Liberalized foreign investment policies and multinational structure have allowed agribusiness companies to provide upstream research, with the local seed companies providing the crop varieties developed for specific geographical markets. For developing countries, this concentration has implications for (1) the structure and autonomy of their domestic seed industries; (2) their access to protected varieties; and (3) the use of important breeding technologies (Srinivasan, 2004).

Recent research demonstrates that the effects of the increasing concentration of control over agricultural biotechnology has had mixed yield, economic, social and environmental effects in the United States, Argentina, South Africa, India and China (Fukuda-Parr, 2007), with the differences caused in part by differences in technology adopted, the structure of farming, the organization of seed markets and in the regulatory and institutional contexts. For instance, Emergent, the third largest cotton seed company in India was recently acquired by the US based Monsanto (ETC, 2005), yet India maintains substantial domestic seed company interests in GM technologies (Ramaswami and Pray, 2007). Agricultural liberalization in East Africa has led to an increase in the number of seed companies and varieties on the market but this has not led to an increase of maize yields or production per capita since the mid-1980s (De Groote et al., 2005).

Today, the top 10 agribusiness companies (all based in Europe, the US or Japan) represent half of the world's commercial seed sales (ETC, 2005). These ten firms increased their control of biotechnology patents to over 50% in 2000 (Pray et al., 2005); indicating that instead of negotiating for the rights to a competitor's technology, it might be simpler, cheaper, or more advantageous to acquire the competitor outright. Currently, patents on the foundational transformation technologies for grains are held by only three firms: DuPont, Monsanto, and Syngenta (Brennan et al., 2005).

Implications of concentration. A relatively stable market share may encourage corporations to invest in R&D, both in terms of current profitability and a reasonable expectation of future profitability. However, recent analysis suggests that we are seeing the beginning of negative impacts on innovation and competition through increased concentration within the private sector (Brennan et al., 2005). The major concerns are (1) industrial concentration reduces the amount and the productivity of research because R&D expenditures are consolidated and narrowly focused; (2) concentrated markets create barriers to new firms and quell creative startups; (3) concentration allows large firms to gain substantial monopolistic power over the food industry, making food supply chains vulnerable to market maneuvers (see 2.3.3; Pray et al., 2005). For instance, a recent USDA study suggests that consolidation in the private seed industry over the past decade dampened the intensity of research undertaken on maize, cotton, and soybeans crop biotechnology (Fernandez-Cornejo and Schimmelpfennig, 2004). This raises concerns that decreasing levels of research activity would stunt agricultural innovations, and brings into question whether large biotech firms can be relied on to conduct research with an eye on the public good as well as their own profit margins (Pray et al., 2005). There is additional concern that the anticompetitive impacts of concentration have led to higher seed prices. USDA data suggest that cotton seed prices in the US have increased 3-4 times since the introduction of GM cotton and that GM fees have substantially raised the price of cotton seed in developing nations, such as India (Iowa State Univ., 2007).

The dilemma of the public sector. The establishment and strengthening of IPR in agriculture has contributed to a shift in emphasis from public to private breeding (Moschini and Lapan, 1997; Gray et al., 1999). The public research sector is increasingly restricted because fragmented ownership of IPR creates a situation wherein no comprehensive set of IPR rights can be amassed for particular crops. In 2003, the Public-Sector Intellectual Property Resource for Agriculture (PIPRA) regime was introduced by several US universities in collaboration with Rockefeller and McKnight Foundations with the goal of creating a collective public IP asset database. This collective management regime would allow public sector institutions to retain rights to use the newest and best technologies of agricultural biotechnology for the public good when they issue commercial licenses (Atkinson et al., 2003).

These creative IPR management regimes are needed for the public sector because many public breeding programs

Box 2-6. Emergence of genetic engineering.

Genetic engineering (GE) or genetic modification of crops (GM) has emerged as a major agricultural technology over the past decade, mainly in North America, China and Argentina. Soybeans, maize, cotton and canola constitute 99 percent of the world's acreage of GE crops (James, 2004). Although GE traits encompass several categories (pest and disease resistance, abiotic stress tolerance, yield, nutrition and vaccines), herbicide tolerance and insect resistance dominate the market. A controversial dialogue has emerged as to the role of GE technology in addressing agricultural problems. Whether farmers have realized benefits from GE crops is a matter of debate. GE technology is seen as not being scale neutral by some (Benbrook, 2004; Pemsl et al., 2005; Rosset, 2005), and in certain instances, GE crops have been shown to increase income distribution differentials within the agriculture sector, favoring the establishment of large holdings and increased farm size (see Santaniello, 2003; Pengue, 2005), However, there is also evidence that GE has benefited farmers (Huang et al., 2001; Ismael, 2001; Traxler et al., 2001; Huang et al., 2002a; Cattaneo et al., 2006). The impacts on pesticide use are debated, with some studies indicating reduced use of insecticides (Huang et al., 2003) and others indicating significant rise in herbicide use (USDA, 2000; Benbrook, 2004). New evidence of high insecticide use by Chinese growers of GE insecticidal crops (*Bt cotton*) has demonstrated that farmers do not necessarily reduce their insecticide use even when using a technology *designed for that purpose* (Pemsl et al., 2005). This illustrates the frequently documented gap between the reality of how a technology is used (taken up in a given social context) and its "in the box" design.

Globally, agricultural producers are reported as receiving 13% of the benefits of GE soya. In Argentina, soya producers received 90% of the benefits of GE soya, partly owing to weak IP protection (Qaim and Traxler, 2005), hence greatly favoring the expansion of the technology in Argentina. However, this increasing reliance on a single technology in Argentina is causing ecological and social concerns (Benbrook 2004, Pengue 2005). Similarly, social, economic, political and cultural concerns have been raised in Asia, Africa and Latin America, as GMOs have been assessed for their impacts on poverty reduction, equity, food sovereignty (de Grassi, 2003; FOE, 2005, 2006). Meanwhile, the roles and contributions of public institutions, scientists, governments, industry and civil society are now beginning to be closely analyzed (de Grassi, 2003).

GE risk analysis has historically acknowledged the possibility of negative ecological effects from the deliberate or inadvertent releases of transgenes into the environment through pollen mediated gene transfer to weedy relatives of GM crops (Haygood et al., 2003) and horizontal gene transfer. For most crops grown under regulatory approval such as maize in the USA, the likelihood is negligible (Conner et al., 2003). In other cases, such as canola in Canada, low levels of levels of transgenic DNA have entered non-GM seed supplies (Friesen et al., 2003; Mellon and Rissler, 2004). There have also been cases of contamination of food supply chain with possible litigation against farmers for the unintentional presence of transgenic DNA in their crops. This is likely to emerge as an even larger issue as pharmaceuticals are introduced into crops (Nature Biotechnology, 2004; Snow et al., 2005). Despite technical solutions to prevent such gene movement (e.g., controversial "terminator technology" and limitation of transgenes to the chloroplast genome not carried in pollen) and traditional plant variety purity protocols no method is likely to be completely effective in preventing movement of transgenes (NRC, 2004).

GE R&D in developing countries is behind that of the developed world for a number of factors including: (1) private sector in the developed world holds much of the IPR; (2) weak patent protection resulting in low investment by the private sector; (3) consumer resistance and governmental regulations affecting international trade in GM products and flow of germplasm; (4) and rising costs of development that inhibit the private research (Huang et al., 2002b). The costs of regulatory compliance have been cited as the largest obstacle to release of commercial GE crops in many developing countries (Atanassov, 2004; Cohen, 2005) and even developed countries. In developed countries like the UK, where public opinion has been exposed to food safety crises like BSE, studies highlight the mixed feelings about GMOs. More broadly, citizens are concerned about the integrity and adequacy of present patterns of government regulation, and in particular about official "scientific" assurances of safety. Better science is necessary but may never resolve the uncertainties about the effects of new technologies (ESRC, 1999).

Crops derived from GE technologies have faced a myriad of challenges stemming from technical, political, environmental, intellectual-property, biosafety, and trade-related controversies, none of which are likely to disappear in the near future. Advocates cite potential yield increases, sustainability through reductions in pesticide applications, use in no-till agriculture, wider crop adaptability, and improved nutrition (Huang et al., 2002b; Christou and Twyman, 2004). Critics cite environmental risks and the widening social, technological and economic disparities as significant drawbacks (Pengue, 2005). Concerns include gene flow beyond the crop, reduction in crop diversity, increases in herbicide use, herbicide resistance (increased weediness), loss of farmer's sovereignty over seed, ethical concerns on origin of transgenes, lack of access to IPR held by the private sector, and loss of markets owing to moratoriums on GMOs, among others. Finally, because new genetic technologies are not the only hurdle between resource-poor farmers and secure livelihoods (Tripp, 2000), GM technology can be only one component of a wider strategy including conventional breeding and other forms of agricultural research to provide a series of structural, regulatory, and economic evaluations that relate economic, political, and scientific context of GE crops to their region of adoption.

have been unsure of whether to complement or compete with the private sector; confusion has arisen as to how to take advantage of IPR to control the use of public material (Reeves and Cassady, 2002) or to capture royalties for bigger budgets (Fischer and Byerlee, 2002). These trends have triggered concerns that the lure of potential royalty revenue has distorted research priorities in public institutions away from poverty alleviation and sustainability, as has been suggested by research managers in Uganda (Louwaars et al., 2005) and the emergence of the so called "University-Industrial Complex" in which universities are redirecting their research to meet the needs of sponsoring corporations (Press and Washburn, 2000). Historically, public sector institutions have been the dominant distributor and pre-breeder of germplasm (Morris and Ekasingh, 2001). In contrast, the growing private sector has focused on widely commercialized, competitive crops that are well protected by legal or technical IPR (Fernandez-Cornejo and Schimmelpfennig, 2004). This has meant that tropical crops, crops for marginal areas (and other public goods attributes, such as safety, health, and environmental protection), and "orphan crops" have remained outside the orbit of private investment (Naylor et al., 2004; Fernandez-Cornejo and Schimmelpfennig, 2004). This will remain a problem until an incentive is created for private firms to work on marginal crops or funding for these important crops is increased in public institutions.

2.3.1.5 Farmers, public and private sector: roles and relations
Changes in funding and investments and the strengthening of the private sector vis à vis the public sector. While global agricultural research investment has grown dramatically since the 1960s (more than doubling between 1976 and 1995), recent trends indicate a shift from public to private sector dominated research. The top ten multinational bioscience companies spend $3 billion annually on agricultural research while the global CGIAR system will spend just over $500 million in 2007 (see Chapter 8). The system has seen its funding decline over the last 15 years compared to the widening of its mandate to include NRM issues (Pardey and Beintema, 2001). Lack of funding for the CGIAR is expected to have negative consequences for NARS plant breeding, particularly in Africa as more than one-third of the approximately 8,000 NARS released crop varieties were based on IARC germplasm. Additionally, structural adjustment programs have severely affected the ability of developing countries to support their own public R&D budget (Kumar and Sidharthan, 1997; CIPR, 2002; Chaturvedi, 2008). A continued decline in public sector breeding (see Chapter 8), coupled with increased private sector growth will only increase the growing gap in research intensity between rich and poor countries.

Emergence of new institutional arrangements. Public-private partnerships to reach development and sustainability goals. The changing character of the seed industry has highlighted public/private partnerships as potential generators of valuable synergies (Table 2-5). Examples of partnerships that have positively affected small-scale farmers include hybrid rice development in India, insect resistant maize in Kenya, industry led associations to improve seed policy in Kenya

and collaborative efforts to promote biosafety regulation in India (IFPRI, 2005).

Some public-private partnerships have a strong charitable character; others include a clear, but often long term, commercial benefit to the private partner. However, to date few success stories that are pro-poor have emerged, and even fewer examples have surfaced where partnerships have contributed to food security, poverty reduction and economic growth. Major constraints have been identified, including (1) fundamentally different incentive structures between collaborating organizations; (2) insufficient minimization of costs and risks of collaboration; (3) limited use of creative organizational mechanisms that reduce competition over key assets and resources; and (4) insufficient access to information on successful partnership models (see Spielman and von Grebmer, 2004). Creative IPR strategies may help in the establishment of public-private partnerships. Licensing of IP rights by private to public sector actors for humanitarian uses has facilitated technology transfer, e.g., rice rich in pro-vitamin A and Ringspot Virus Resistance for papaya Asia (Al-Babili and Beyer, 2005; Brewster et al., 2005). Partnerships can be successful as in the case of the Daimler Chrysler collaboration with Poverty and the Environment in Amazonia (POEMA) to use coconut fibers and natural latex rubber (Zahn, 2001; Laird, 2002). Additionally, a recent initiative, the Science and Knowledge Exchange Program, to exchange staff between the public and private sectors may effectively develop productive pro-poor partnerships in food and agriculture. In Africa, schemes have been put forward to promote the acquisition of private sector innovations by the public sector at a price based on their estimated value to society (Kremer, 2003; Master, 2003). Private companies would contribute to crop improvement through partnerships that use local varieties and provide source material and information for improved regulatory passage (Keese et al., 2002; Cohen, 2005). However for complicated genetic transformations, dozens of patents are involved in a single transformation (Guerinot, 2000). In this case, all public and private IPR holders must grant licenses to all IP involved in the final product (Al-Babili and Beyer, 2005). Experience suggests that the public sector must take the lead in such initiatives on crops that are essential for food security, but have marginal profitability.

Renewed involvement of farmers in genetic resource management: Participatory Plant Breeding as a new arrangement. Today, farmers remain indispensable actors in any regime that seeks to conserve, improve, and disseminate genetic diversity. It is estimated that 1.4 billion farmers save seed from year to year (Pimentel et al., 1992; Cleveland et al., 1994; Bellon, 1996). There are many advantages of *in situ* conservation, in particular the relationship between diversity and yield stability (Amanor, de Boef, and Bebbington, 1993; Trinh et al., 2003; Abidin et al., 2005). Participatory plant breeding and *in situ* management relies on the collaboration between farmer-breeders and corporate plant breeders (Lipton and Longhurst, 1989; Sthapit et al., 1996; Kerr and Kolavalli, 1999; Almekinders and Elings, 2001; Witcombe et al., 2005). Traditionally, these projects are judged on their ability to produce adapted crop material at

Table 2-5. **Public-private partnerships in the CGIAR.**

Partnership approach research topic	CGIAR center	Private sector partners	Other partners
Collaborative Research – Global Programs			
Apomixis	CIMMYT	Pioneer Hi-bred (US) Syngenta (Switzerland) Limagrain (France)	L'Institut de Recherche pour le Développement (France)
Golden Rice Humanitarium	IRRI	Syngenta	Rockefeller Foundation (US), Swiss Federal Institute of Technology, and others
HarvestPlus	CIAT, IFPRI	Monsanto (US)	
Wheat improvement[e]	CIMMYT		Grains Research & Development Corp. (Aus)
Collaborative Research – Local/Regional Programs			
Sorghum and millet research[e]	ICRISAT	Consortium of private seed companies incl. Monsanto (India), others	
Forage seed improvement	CIAT	Grupo Papalotla (Mexico)	
Insect resistant maize for Africa[e]	CIMMYT		Kenyan Agricultural Research Institute, Syngenta Foundation (Switzerland)
Technology Transfers			
Potato/Sweet potato transformation	CIP	Plant Genetic Systems[a] (US), Axis Genetics[b] (UK), Monsanto	
Genomics for livestock vaccine research[e]	ILRI	The Institute for Genomic Research (US)	
Bt genes for rice transformation	IRRI	(Switzerland), Plantech[d] (Japan)	Consortium of other public research institutions
Positive selection technology for cassava transformation	CIAT	Novartis[c]	

Source: Spielman and von Grebmer, 2004.

[a]Now Bayer CropScience

[b]Insolvent as of 1999

[c]Now Syngenta

[d]subsidiary of Mitsubishi

[e]The definition of a public-private partnership is extended here to include a collaboration between a CGIAR center and a philanthropic organization established by a commercial entity, or an organization established to represent industry interests, on the other.

lower costs than conventional programs and on their ability to produce higher genetic gains per year (e.g., Ceccarelli et al., 2001a, 2003; Smith et al., 2001; Witcombe et al., 2001; Virk et al., 2003, 2005). However, participatory research projects (composed of both formal and informal actors) have also led to the spread of socially responsible, technical innovations and important policy changes (Joshi et al., 2007). These innovations have been shown to improve the welfare of the poor and socially excluded. One of the best examples is a 1997 client-oriented participatory crop improvement (PCI) project in Nepal in which there was formal recognition that informal R&D processes were taking place,

and a move to encourage those processes (Biggs, 2006). This led to changes in National Varietal Release Procedures and to more effective collaboration between different actors. Informal developments were essentially legitimized and supported. Nevertheless, the benefits of farmer participation may not be universal, and adoption of participatory methods has not been as high as expected, notably because of methodological limitations to upscaling (Witcombe et al., 2005).

The quality issue. In developed countries, changes in the consumers' preferences have pushed the labeling of the geographical origin of products, along with the notion of "terroir", with the result that farmers and specialized breeders are reviving old crop varieties (Bérard and Marchenay, 1995; Bonneuil and Demeulenaere, 2007). The development of organic and sustainable food production systems has created additional challenges, e.g., organic production must use seeds that have been produced in organic conditions. Instead of working on larger domains of breeding for conventional agriculture, breeders select for specific adaptability to specific environments and practices. All these trends challenge the classical ways of evaluating varieties. Since the multifactor and multisite experimentation, backed by statistical analysis is more difficult to perform, new ways of assessing varieties and seeds are needed, e.g., simulation modeling (Barbottin et al., 2006). The key conclusion is that knowledge must be shared among different actors, including farmers, users and consumers. The overall globalization of markets is increasingly pushing this issue in developing countries that seek to cater to the needs of specific market niches in industrialized countries.

2.3.1.6 The need for a renewed design with distribution of diverse roles

Germplasm management over the last 150 years has been characterized by standardization and scale of economies. This has been paramount to the rapid spread and success of widely adapted germplasm. It resulted in seed management becoming largely separate from agricultural and cultural systems, with a decline in the number and diversity of actors actively engaged in seed systems. Moreover, the tightening of IPR, access and benefit sharing laws and other forms of controls over genetic resources weakened exchange of genetic resources among breeders. Industrial strategies have been based on strengthened IP arrangements; attempts to balance IPRs with farmers, industry and the public sector has added to hyper-ownership issues. Consolidation of the seed industry has facilitated the spread of rapid technological advances, but not always to the benefit of the poor. The history of germplasm management has revealed shortcomings, specifically in social and ecological arenas.

Asymmetries in access to germplasm and benefit sharing have increased vulnerabilities of the poor. The International Treaty on Plant Genetic Resources is the first major international policy that attempts to proactively address the situation by creating a form of international germplasm exchange and research commons. Other initiatives such as Public-Sector Intellectual Property Resource for Agriculture (PIPRA) aim to create a collective public IPR asset database to allow the public sector to continue to develop public good germplasm. Public-private partnerships could lead to pro-poor advances if current challenges, such as minimization of risks of collaboration, are tackled. This assessment questions the current separation between researchers and farmers and calls for an increased role of user's knowledge, as exemplified in participatory plant breeding. Local and diverse arrangements have been successful at meeting development and sustainability goals for germplasm management. These arrangements will be important for using germplasm to respond to the uncertainties of the future.

2.3.2 Pest management

Multiple approaches to pest management have emerged in different places during different periods in history. Each has been upheld by distinctive organizational arrangements reflecting cultural values, societal norms and political and economic priorities of their time and place. Widely differing interpretations exist that make competing claims regarding the advantages and disadvantages of the range of options; other narratives may describe differently the identification and implementation of sustainable solutions in pest management. The following narrative emerged from analysis of publications of UN agencies, the World Bank, the CGIAR, universities, national IPM programs in numerous countries, and the work of physical and social scientists, researchers, private sector actors including agrochemical companies, and NGOs actively involved on the ground in pest and pesticide management programs.

2.3.2.1 Chemical control

Emergence of chemical control. Chemical control had its roots in US and German chemical research before and after both World Wars and was driven by formal interagency collaboration between military and public sector chemists and entomologists (Russell, 2001). The emphasis on crop protection and risk minimization supported pest control, rather than management and pest eradication using synthetic chemicals (Perkins, 1982; Russell, 2001). The approach underpinned the priorities of industrial countries: maximizing food and fiber production, increasing efficiency and releasing labor to other economic sectors. Research and extension efforts directed at biological, cultural and mechanical management of risk dropped sharply at this time (Perkins, 1982; Lighthall, 1995; Shennan et al., 2005). The pesticide industry grew rapidly, initially financed through government contracts and then loans, a practice that necessitated constant product innovation and marketing to repay debts (Perkins, 1982). Significant concentration has occurred (DFID, 2004; UNCTAD, 2006); by 2005, the top six multinational pesticide corporations accounted for 75% of the US$29,566 million global pesticide market (Agrow World Crop Protection News, 2005; ETC, 2005).

National and global concerns over food security drove the further intensification of agricultural production and adoption of synthetic chemical pesticides across much of Asia and Latin America (Rosset, 2000). The CGIAR played a pivotal role in the Green Revolution that carried synthetic chemicals into widespread use in irrigated systems (see 2.1). Multilateral and bilateral donor and development agencies such as the World Bank, USAID and JICA provided direct or subsidized supplies of synthetic pesticides, sometimes tying agricultural credit to adoption of input packages inclusive of these chemicals (Holl et al., 1990; Hammond and McGowan, 1992; Jain, 1992; Korten, 1995; Clapp, 1997; Ishii-Eiteman and Ardhianie, 2002; USAID, 2004). Direct state intervention in some cases enforced pest control through calendar spraying regimes or established pesticide distribution systems to ensure product use (Meir and Williamson, 2005). Farmers received pest control advice from pesticide sellers and extension agents operating under T&V and similar state-directed systems. In some cases, government extension personnel served also as pesticide distributors (Pemsl et al., 2005; Williamson, 2005) to supplement low government wages. Smaller pesticide production and distribution companies grew rapidly in developing countries such as Argentina, India, China and South Africa, often producing cheaper but more hazardous pesticides than their multinational counterparts (Pawar, 2002; Bruinsma, 2003).

Impacts of the chemical pest control approach. The significant yield gains and achievements in food security obtained in many countries in the 1950s and 60s have been closely linked to the use of hybrid seeds, synthetic fertilizers and other inputs including pesticides and to high levels of political and institutional investment in public sector research and extension (Bhowmik, 1999; Evenson and Gollin, 2003a; Lipton, 2005). Yield losses owing to disease and weed infestations have been reduced through chemical pest control (Bridges, 1992; CropLife, 2005ab); animal health has improved where insect-vectored diseases have been successfully controlled (Singh, 1983; Windsor, 1992; Kamuanga, 2001) and soil resources have been conserved through no-till practices, which sometimes rely on herbicide use (Lal, 1989; Holland, 2004). Some have speculated that widespread famines and devastation of crops from outbreaks of disease and pests have been prevented (Kassa and Beyene, 2001); from a historical evidence-based approach it is difficult to assess the validity of these claims. As early as 1950, evidence of pest resistance to pesticides, resurgence where natural enemy populations had been destroyed and secondary pest outbreaks began to accumulate (Stern et al., 1959; Smith and van den Bosch, 1967; van Emden, 1974). Pesticide resistance (including cross-resistance to new products) became extensive and has been thoroughly documented in the scientific literature (MSU, 2000; Bills et al., 2003).

By the 1960s the adverse environmental and human health effects of pesticide exposure had become known. The impacts, widely documented in the scientific and medical literature and popularized (e.g., Carson, 1962), affected not only pesticide applicators but entire rural communities and diverse biota in aquatic and terrestrial ecosystems and watersheds (reviewed in Wesseling et al., 1997, 2005; Hayes, 2004; Kishi, 2005; Pretty and Hine, 2005; Relyea, 2005; USGS, 2006; Desneux et al., 2007). Acute poisonings by pesticide residues have had immediate adverse effects, including death (Chaudhry et al., 1998; Rosenthal, 2003; Neri, 2005). Social and environmental justice cases have been documented regarding the inequitable distribution of the benefits of chemical control (largely accruing to better resourced farmers and manufacturers) and the harms in actual conditions of use that are experienced disproportionately by the poor and disadvantaged and the "ecological commons" (Wesseling et al., 2001; Reeves et al., 2002; Jacobs and Dinham, 2003; Reeves and Schafer, 2003; Harrison, 2004; Qayum and Sakkhari, 2005). A significant portion of the chemicals applied has proved to be excessive, uneconomic or unnecessary in both industrialized (Pavely et al., 1994; Yudelman et al., 1998; Reitz et al., 1999; Prokopy, 2003; Pimentel, 2005) and developing countries (Ekesi, 1999; Adipala et al., 2000; Jungbluth, 2000; Sibanda et al., 2000;

Asante and Tamo, 2001; Dinham, 2003; Nathaniels et al., 2003). Pesticide reliance has also been linked to agricultural deskilling (Vandeman, 1995; Stone, 2007), evidenced by subsequent erosion of farmers' knowledge of crop-insect ecology and reduced ability to interpret and innovate in response to environmental cues at field level (Thrupp, 1990; Pemsl et al., 2005).

Chemical control remains the cornerstone of pest management in many parts of the world, sustained by its immediate results, the technology treadmill (see 2.1) and path-dependency (wherein a farmer's accumulation of equipment, knowledge and skills over time conditions her potential to change direction). It is also upheld by the professional cultures and training of most advisory and extension programs (Mboob, 1994; Sissoko, 1994; Agunga, 1995; FAWG, 2001; Sherwood et al, 2005; Touni et al., 2007); the dominance of institutions promoting technology-driven intensification of agriculture; product innovations and marketing by the agrochemical industries (FAO/WHO, 2001; Macha et al., 2001; Kroma and Flora, 2003; Touni et al., 2007); and direct and indirect policy supports such as tax or duty exemptions for pesticides (Mudimu et al., 1995; Jungbluth, 1996; Gerken et al., 2000; Williamson, 2005). In recent years, leading agrochemical companies have integrated seed ventures and biotechnology firms, enabling them to establish synergies among key segments of the agricultural market. This trend is expected to continue and lead to increasing convergence between the segments, with possible inhibition of public sector research and of start-up firms (UNCTAD, 2006). The history of chemical control illustrates a phenomenon in agricultural science and technology development, in which early success of a technical innovation (often measured by a single agronomic metric such as productivity gains), when accompanied by significant private sector investment in advertising and public relations (Perkins, 1982) and by direct and indirect policy supports from dominant institutional arrangements (Murray, 1994), translates into narrowing of organizational research and extension objectives, widespread if uncritical grower adoption and delayed recognition of the constraints and adverse effects of the technology (e.g., resistance, health hazards, etc.).

2.3.2.2 Integrated Pest Management (IPM)

Integrated Pest Management (Box 2-7) in its modern form was developed in the 1950s in direct response to the problems caused by use of synthetic insecticides in actual conditions of use (Perkins, 1982). IPM took many forms but in general emphasized cultural and biological controls (Box 2-8) and selective application of chemicals that do not harm populations of pest predators or parasitoids (Stern et al., 1959), based on scientific understanding of agroecosystems described as complex webs of interacting species that can be influenced to achieve crop protection. IPM adoption in industrialized countries was stimulated by growing concern for human health and the environment, consumer desire for low or no pesticide residues in food (Williamson and Buffin, 2005) and public sector recognition that regulatory interventions were needed to remove the most harmful chemicals from sale. The spread of IPM in the South was driven by the high incidence of involuntary pesticide poisonings among farmers and farm workers through occupational exposure (Holl et al., 1990; Wesseling et al., 1993, 1997, 2002; Antle et al., 1998; Cole et al., 2002). Other drivers were state authorities' recognition of the high cost of pesticide purchase for poorer farmers and resulting problems of indebtedness (Van Huis and Meerman, 1997); the potential of new markets spurred by consumer demand for pesticide-free produce both in the North (IFOAM, 2003; Ton, 2003; Martinez-Torres, 2006) and in countries with growing middle class populations (e.g., Thailand, China, India); export requirements of Maximum Residue Limits; and international attention to issues such as pollution of drinking water, human rights to a safe home and workplace and biodiversity loss.

Impacts of IPM paradigm. IPM can deliver effective crop protection and pesticide reduction without yield loss (Heong and Escalada, 1998; Mangan and Mangan, 1998; Barzman and Desilles, 2002; Eveleens, 2004). The yield advantages of IPM have been particularly strong in the South and thus have significant policy implications for food security in developing countries (Pretty, 1999; Pretty, 2002, Pretty et al., 2003). The community-wide economic, social, health and environmental benefits of farmer-participatory ecologically-based IPM have been widely documented (Dilts, 1999; Pontius et al., 2002; Pontius, 2003; Braun, 2006; Braun et al., 2006; Mancini, 2006; Mancini et al., 2007; van den Berg and Jiggins, 2007), including measurable improvements in neurobehavioural status as a result of reduced pesticide exposure (Cole et al., 2007). Large-scale impacts on social equity have not yet been assessed but higher household income, reduced poverty levels and significant reduction in use of WHO Class 1 highly toxic compounds have been shown in some cases (FAO, 2005a).

Difficulties in measuring the cost-effectiveness of large scale farmer-participatory IPM has impeded wider adoption (Kelly, 2005) and raised questions about its fiscal sustainability as a national extension approach (Quizon et al., 2000; Feder, 2004ab). As acknowledged by the authors, these studies did not calculate the economic savings from reduced poisoning and pollution nor attempt to quantify non-economic benefits. An evaluation of IPM research in the CGIAR system points to the need for more comprehensive economic impact analyses that include these variables (CGIAR TAC, 2000). A recent meta-review of 35 published data sets on costs and benefits of IPM farmer field schools has meanwhile substantiated their effectiveness as an educational investment in reducing pesticide use and enabling farmers to make informed judgments about agroecosystem management (van den Berg and Jiggins, 2007).

More widespread adoption of IPM as defined in the FAO Code of Conduct has been constrained by political, structural and institutional factors, principally

- limited capacity of extension services in both industrialized and developing countries in providing adaptive, place-based, knowledge-intensive ecological education and technical support in IPM (Blobaum, 1983; Anderson, 1990; Holl et al., 1990; Agunga, 1995; Paulson, 1995; Altieri, 1999; Norton et al., 2005; Rodriguez and Niemeyer, 2005; Touni et al., 2007);

- inadequate public sector and donor investment in IPM research and extension and poor coordination between relevant agencies (Mboob, 1994; ter Weel and van der Wulp, 1999; Touni et al., 2007);
- insufficient private sector interest in natural controls (Ehler, 2006) and widespread promotion of synthetic chemical controls by pesticide suppliers and distributors (Kroma and Flora, 2003; Touni et al., 2007);
- shifts in funding and research interests in agricultural colleges away from basic biology, entomology and taxonomy and limited resources for ecological investigations (Jennings, 1997; Pennisi, 2003; Herren et al., 2005); an incentives system that discourages multidisciplinary collaboration in pest management (Ehler, 2006); and a growing tendency, e.g., in the United States, to encourage research likely to return financial benefits to the university rather than broader benefits to the public or ecological commons (Kennedy, 2001; Berdahl, 2000; Bok, 2003; Washburn, 2005) while offering private sector partners such as the agrochemical/biotechnology industry a wider role in shaping university research and teaching priorities (Krimsky, 1999; Busch et al., 2004);
- vertical integration of ownership (FAO, 2003b) and concentration in private sector control (Vorley, 2003; DFID, 2004; Dinham, 2005) over chemical, food and agricultural systems, processes that tend to favor larger scale, input-intensive monoculture production over the biodiverse agroecosystems necessary to sustain effective performance by natural enemies; and
- inequitable distribution of risks and costs: in the absence of public sector support, farmers typically bear the upfront transaction costs and risks of conversion to pest management practices that serve the public good (Brewer et al., 2004; Ehler, 2006).

2.3.2.3 Institutional innovations and responses in pest management

Institutional innovations. FAO's paradigm-shifting work in Asia provided (1) the scientific evidence that pesticide-induced pest outbreaks could contribute to crop failures while reduction of pesticide use could improve system stability and yields (Kenmore et al., 1984); (2) empirical evidence of the positive social impacts of field-based experiential learning processes (Matteson et al., 1984; Mangan and Mangan, 1998; Ooi, 1998); and (3) the policy insight that a number of directives (e.g., ban of selected pesticides, removal of pesticide subsidies and national support for IPM) could transform the situation on the ground, as in Indonesia (Kenmore et al., 1984; Settle et al., 1996; Gallagher, 1999). Building on FAO's farmer field school methodology (http://www.farmerfieldschool.info/), participatory field-based educational processes in IPM gained strength in the 1980s (Röling and Wagemakers, 1998). These innovations in knowledge, science, technology and policy subsequently led to an institutional innovation, the establishment of the Global IPM Facility (see 2.2) and the implementation of farmer-participatory IPM across Asia, Latin America, Africa and Central and Eastern Europe (UPWARD, 2002; Jiggins et al., 2005; Luther et al., 2005; Braun et al., 2006). Plant Health Clinics (piloted in

Box 2-7. Integrated Pest Management.

There are many diverse definitions and interpretations of Integrated Pest Management (IPM). The internationally accepted FAO definition is "the careful consideration of a number of pest control techniques that discourage the development of pest populations and keep pesticides and other interventions to levels that are economically justified and safe for human health and the environment. IPM emphasizes the growth of a healthy crop with the least possible disruption of agroecosystems, thereby encouraging natural pest control mechanisms" (FAO, 2002b, 2005b). Additional endorsement of the revised Code is reflected in the European Commission's recent decision to include it in the forthcoming revision of the EU pesticides authorization directive 91/414, and to use it as the basis for proposing mandatory IPM for EU farmers by 2014.

Contrasting interpretations of IPM have emerged over the period, each with different emphases.

- Toolbox IPM combines two or more tactics from an array of tools and is utilized primarily to optimize crop productivity (OTA, 1979; Cate and Hinkle, 1994). IPM is presented as a continuum of practices, with choices ranging from reliance mainly on prophylactic controls and pesticides to more biologically-intensive methods (USDA, 1993). The approach emphasizes a diversity of technical options, rather than the integration of multiple tactics under a broader ecological framework and does not necessarily require monitoring or conservation of natural enemies (Ehler and Bottrell, 2000; Ehler, 2006; Gray and Steffey, 2007).

- Pesticide-based IPM focuses primarily on the discriminate use of pesticides and improving the efficacy of pesticide applications (Ehler, 2006). The approach emphasizes pest monitoring and the use of less hazardous, lower dose and more selective pesticides, improved formulations, new application technologies, and resistance management strategies (CropLife, 2003; Syngenta, 2006). Some pesticide industry IPM programs may also feature use of the manufacturers' chemical products (Sagenmuller, 1999; Dollacker, 2000). Nonchemical approaches such as biocontrol are mentioned in some industry publications, but presented as "generally too often unreliable or not efficient enough to be commercially used on their own" (CropLife, 2003).

- Biointensive IPM, also sometimes described as Preventative IPM (Pedigo, 1989, 1992; Higley and Pedigo 1993) and Ecological Pest Management (Altieri, 1987; Altieri and Nicholls, 2004), emphasize the ecological relationships among species in the agroecosystem (Shennan et al., 2005) and the availability of options to redesign the landscape and ecosystem to support natural controls (Dufour, 2001). Biological and ecological pest management offer robust possibilities to significantly and sustainably reduce pesticide use without affecting production (van Lenteren, 1992; Badgley et al., 2007; Scialabba, 2007). Implementation remains limited globally as it often requires structural changes in production systems (Lewis et al., 1997) and redirection of market, research, policy and institutional support to favor ecosystem-oriented approaches.

- Indigenous pest management, based on detailed Indigenous technical knowledge (ethnoscience) of pest ecology, local biodiversity and traditional management practices, focuses on achieving moderate to high productivity using local resources and skills, while conserving the natural resource base (Altieri, 1993). Weeds, insect pests and crop pathogens are at times tolerated and provide important foods, medicines, ceremonial materials and soil improvers (Bye, 1981; Chacon and Gliessman, 1982; Brown and Marten, 1986). Control methods rely on a wide range of cultural, biological, physical and mechanical practices, water and germplasm management and manipulation of crop diversity (Altieri and Letourneau, 1982; Matteson et al., 1984; Altieri, 1985) and are supported by knowledge of the local agroecosystem and surroundings (Atteh, 1984; Richards, 1985). In Africa, farmers traditionally practice intercropping with various crops, which can drastically reduce pest densities, especially if the associated crop is a non-host of the target pest species (Khan et al., 1997; Schulthess et al., 2004; Chabi-Olaye et al., 2005; Wale et al., 2006), although farmers are not always aware of the beneficial effect that mixed cropping has on pest infestations (Nwanze and Mueller, 1989). Partnerships between formally trained scientists and farmers skilled in ethnoscience show promise for strengthening agroecological approaches (Altieri, 1993).

Nicaragua, currently in use in 16 other countries), the combination of mass media campaigns, and farmer-to farmer extension and education (Brazil, Ecuador, Peru, Vietnam, Bangladesh) similarly have proven effective in promoting IPM. In Africa and Latin America, communities are exploring economic innovations in self-financing mechanisms for IPM field schools (Okoth et al., 2003).

Innovative agroenvironmental partnerships between growers, extensionists and IPM scientists have implemented integrated farming and alternative pest management strategies to reduce organophosphate insecticide use in major commodity crops across California (Warner, 2006ab) and implement resource-conserving IPM in Michigan (Brewer et al., 2004; Hoard and Brewer, 2006). Their success derives from collaborative partnership structures that emphasize co-learning models, social networks of innovation (through informal grower networks and supported by statewide commodity boards) and building capacity in flexible place-based decision-making rather than conventional transfer of technology (Mitchell et al., 2001; Getz and Warner, 2006; Warner, 2006ab).

Policy responses. Governments have responded to the scientific evidence of adverse environmental and health effects of pesticides with legislation, regulatory frameworks and policy initiatives. A growing number of Southern governments have national IPM extension and education programs (Box 2-9), and several countries (Costa Rica, Ecuador, Paraguay,

Box 2-8. Biological control

Biological control refers to the use of natural enemies of pests (i.e., their predators, parasitoids and pathogens) as pest control agents. Globally, the annual economic contribution of natural enemies has been estimated in the hundreds of billions of dollars worldwide, much of this due to indigenous species (Costanza et al., 1997; Naylor and Ehrlich, 1997; Pimentel, 1997; Pimentel et al., 1997; Gurr and Wratten, 2000; Alene et al., 2005; Losey and Vaughan, 2006). Conservation biocontrol supports the activity of indigenous natural enemies by providing them with suitable habitats and resources (Doutt and Nakata, 1973; Jervis et al., 1993; Kalkoven, 1993; Idris and Grafius, 1995; Murphy et al., 1998; Ricketts, 2001; Gurr et al., 2006) and by limiting the use of disruptive pesticides. Since conservation approaches are locally adapted, they rarely produce products that can be widely marketed and have attracted little interest from the private sector. Yet they constitute one of the most economically important types of biocontrol and form the cornerstone of much ecological pest management (Altieri and Nicholls, 2004). Farmers and public sector scientists have demonstrated practical applications in, e.g., the Biologically Integrated Orchard Systems (BIOS) of California (Thrupp, 1996), vineyard habitat management (Murphy et al., 1998), and rice ecosystem conservation (Settle et al., 1996).

The importance of natural enemies is highlighted by the often explosive outbreaks of pests introduced into regions where they lack specific natural enemies. Classical biological control restores natural pest management by the identification and introduction of specific and effective natural enemies from the pest's home region (DeBach, 1964, 1974). Dramatic early successes in the late 19th century (cottony cushion scale in citrus, Caltagirone and Doutt, 1989) spurred classical biocontrol efforts around the world, but these methods were later displaced by the widespread adoption of cheaper and fast-acting synthetic pesticides. Under pressure to deliver fast results, entomologists economized on ecological studies and began releasing potential biocontrol agents prematurely with less success (Greathead, 2003). Confidence in biocontrol declined, until problems arising from pesticide use rekindled interest (Perkins, 1982). With better institutional support and funding, the success rate improved (Greathead, 2003). Initially, work in developing countries focused on large scale commercial, industrial and export tree crops with less direct impact on small-scale farmers (Altieri, et al., 1997). Subsequent programs focused on staple food crops and on building indigenous capacity in biocontrol (Thrupp, 1996).

Institutional arrangements fostering collaboration enabled the scientific and technological processes associated with classical biocontrol in subsistence crops in Africa to provide a range of social, environmental, economic and cultural benefits (Norgaard, 1988; Zeddies et al., 2001; Bokonon-Ganta et al., 2002; de Groote et al., 2003; Neuenschwander et al., 2003; Moore, 2004; Macharia et al., 2005; Maredia and Raitzer, 2006; Omwega et al., 2006; ICIPE, 2006; Kipkoech et al., 2006; Macharia et al., 2007; Löhr et al., 2007). A noteworthy example is the control of cassava mealybug (Herren and Neuenschwander, 1991; Gutierrez et al., 1988; Neuenschwander, 2001, 2004). Follow-on effects included exten-

sive training of African scientists in biocontrol and the establishment of national programmes targeting invasive insect and weed pests across the region (Herren and Neuenschwander, 1991; Neuenschwander et al., 2003). Technical and administrative staff played a key role in designing and maintaining complex networks of collaboration (Wodageneh, 1989; Herren, 1990; Neuenschwander, 1993; Neuenschwander et al., 2003).

Ecologists have raised concerns regarding potential impacts on non-target organisms of introduced biocontrol agents (Howarth, 1990; Simberloff and Stiling, 1996; Strong, 1997). However, after several early failures due to vertebrate and mollusc predator introductions in the late 19th-early 20th century (Greathead, 1971), the safety record of invertebrate biocontrol has become well established (Samways, 1997; McFadyen, 1998; Wilson and McFadyen, 2000; Wajnberg et al., 2001; Hokkanen and Hajek, 2003; van Lenteren et al., 2003). A substantial body of research has investigated nontarget effects of classical biological control (Boettner et al., 2000; Follett and Duan, 2000) and rigorous screening protocols and methodologies for environmental risk assessment of biocontrol agents now exist (Hopper, 2001; Strong and Pemberton, 2001; Bigler et al., 2006). FAO, CABI BioScience and the International Organization of Biological Control have developed a Code of Conduct for the Import and Release of Biological Control Agents to facilitate their safe import and release (Waage, 1996; IPPC, 2005).

In contrast to conservation and classical biocontrol, augmentation involves mass production of naturally-occurring biocontrol agents to reduce pest pressure (DeBach, 1974; Bellows and Fisher, 1999). The decentralized artisanal biocontrol centers of Cuba offer one model of low-cost production for local use (Rosset and Benjamin, 1994; Altieri et al., 1997; Pretty, 2002). Augmentative control in Latin American field crops (van Lenteren and Bueno, 2003) and throughout the European glasshouse system (Enkegaard and Brodsgaard, 2006) offer others. Growing consumer interest in pesticide-free produce has helped establish a small but thriving biocontrol industry (van Lenteren, 2006), mostly in industrialized countries (Dent, 2005), with some uses in developing countries where pesticide use is difficult or prone to trigger pest outbreaks (i.e., sugarcane, cotton and fruit trees). The costs of production, storage and distribution of living organisms have made these products less attractive to the private sector than chemical pesticides; currently they comprise only 1-2% of global chemical sales (Gelertner, 2005). Their relatively limited use also reflects chronic underinvestment in public sector research and development of biological products and a regulatory system that disadvantages biological alternatives to chemicals (Waage, 1997). Biological pesticides, on the other hand, have been more successful because they fit into existing systems for pesticide development and delivery. Nevertheless, the growth of the global market for biocontrol products, recently at 10-20% per annum, is expected to continue (Guillon, 2004), and is most likely to play a key role in crop systems where pesticide alternatives are required.

Box 2-8. continued

Opportunities and constraints

Successful biocontrol systems have required public sector investment, political commitment to maintain and adequately finance research, breeding and release programs, close collaboration between technical and regulatory agencies and donors at national and regional levels, and minimal pesticide use to create a safe environment for biocontrol agents (Neuenschwander, 1993, Neuenschwander et al., 2003; Maredia and Raitzer, 2006; Omwega et al., 2006). Where such commitments have existed (Western Europe; Kazakhstan, post-Soviet Cuba, many countries throughout Africa), biocontrol programs have been important contributors to agricultural production and national food security (Greathead, 1976; van Lenteren et al., 1992; Rosset and Benjamin, 1994; Pretty, 1995; Neuenschwander, 2001; Omwega et al., 2006; Sigsgaard, 2006; van Lenteren, 2006).

Biological control has provided effective control of pests in many cropping systems, while maintaining high agricultural production (DeBach 1964; DeBach and Rosen, 1991; Bellows and Fisher, 1999; Gurr and Wratten, 2000). Yet public sector investments, institutional support for research and practical applications have been uneven over the period, reflecting shifting priorities of dominant institutional arrangements (NRC, 1989; Cate and Hinkle, 1994; Jennings, 1997; Greathead, 2003; Hammerschlag, 2007).

Substantial taxonomic, biological and ecological knowledge is crucial to support successful biocontrol (Pennisi, 2003; Herren et al., 2005), but these fields have been neglected in many research institutions (Jennings, 1997; Kairo, 2005). Greater public and private sector investment in institutional capacity could increase the ability of farmers, extension staff, scientists, policy makers and the food sector to capitalize on opportunities afforded by biocontrol (Neuenschwander, 1993; Waage, 1996; Williamson, 2001; van Lenteren 2006; Hammerschlag, 2007).

Global challenges for biocontrol include a possible growth in exotic pest problems due to globalization and climate change and the threat posed by degraded agricultural and natural ecosystems to maintaining natural enemy communities. The Convention on Biological Diversity raises important conceptual and practical issues for biocontrol: how to develop capacity and ensure safe and equitable sharing of resources, research and benefits among actors and countries (Waage, 1996). Natural enemies have previously demonstrated capacity to adapt to changing climates encountered in expanding their geographic range (Tribe, 2003) and to control invasive species (van Driesche and Hoddle, 2000; Greathead, 2003) in a safe and sustainable manner. These attributes, along with the imperative to reduce pesticide contamination of drinking water supplies, suggest that biological control will play an increasingly important role in future pest management practices.

China, Thailand and Vietnam) have taken the lead in banning WHO Class 1a and 1b pesticides (FAO, 2006a). Various European countries have implemented Pesticide Use Reduction programs with explicit benchmarks for pesticide reduction (Box 2-9) and Organic Transition Payment programs (Blobaum, 1997). Domestic US programs emphasized IPM in the 1970s and 1990s but shifts in political priorities have led to uneven national support and a more narrow interpretation emphasizing pollution mitigation strategies over preventative approaches to ensuring crop health (Cate and Hinkle, 1994; GAO, 2001; USDA/NRCS, 2001; Brewer et al., 2004; Hammerschlag, 2007; see Hoard and Brewer, 2006 and Getz and Warner, 2006 for state-level innovations in IPM). The CGIAR has established an inter-institutional partnership to promote participatory IPM (http://www.spipm.cgiar.org). Bilateral donor agencies have also prioritized biocontrol or IPM in their development aid, e.g., Germany, the Netherlands, Sweden, IPM Europe and the United States (ter Weel and van der Wulp, 1999; SIDA, 1999; Dreyer et al., 2005; USAID, 2007). Maximum Residue Levels (MRL) regulations for pesticides in food have been established at national and international levels (see 2.3.3.). These and other international and national standards continue to undergo revisions in light of emerging scientific findings on possible and actual effects of low dose and chronic exposure to pesticide residues (NRC, 1993; Aranjo and Telles, 1999; Baker et al., 2002; Thapinta and Hudak, 2000; Kumari and Kumar, 2003; Pennycook et al., 2004).

The UN FAO Code on the Distribution and Use of Pesticides (Box 2-9) focuses not only on minimizing hazards

associated with pesticide use but also on promoting IPM. It indicates that "prohibition of the importation, sale and purchase of highly toxic products [such as] WHO Class I a and I b pesticides may be desirable" and recommends that pesticides requiring use of personal protective equipment (e.g., WHO Class II pesticides) should be avoided where such equipment is uncomfortable, expensive or not readily available (e.g., in most developing countries). In 2007, the 131st Session of the FAO Council mandated FAO to pursue a "progressive ban on highly toxic pesticides" (FAO, 2007). FAO has urged chemical companies to withdraw these products from developing country markets and is calling on all governments to follow the example of countries that have already banned WHO Class Ia and Ib pesticides (FAO, 2006a). Also in 2007, FAO hosted an international conference that highlighted organic farming's capacity to meet food security goals without reliance on chemical pesticides (Scialabba, 2007; Sligh and Christman, 2007). The FAO conference confirmed similar findings from numerous recent studies on organic agriculture (Parrott and Marsden, 2002; Pimentel et al., 2005; Badgley, et al., 2007; Halberg, et al., 2007; Kilcher, 2007).

The World Bank revised its pest management policy in 1998, in response to internal impact assessments (Schillhorn van Veen et al., 1997), public pressure (Aslam, 1996; Ishii-Eiteman and Ardhianie, 2002) and donor government concerns (e.g., Denmark, Germany, the Netherlands, Norway, Switzerland, United States). The policy now emphasizes "reducing reliance on chemical pesticides" and promoting "farmer-driven ecologically-based pest control" (World

Box 2-9. Policy instruments affecting pest management

Many national, regional and international policies and agreements have focused on phasing out the most toxic pesticides, increasing public availability of information on pesticide bans and restrictions, and promotion of least toxic sustainable alternatives such as IPM. They include:

National regulatory instruments, policies and programs:
- Pesticide registration legislation, pesticide subsidies, use taxes and import duties; establishment of Maximum Residue Levels (MRLs)
- Pesticide use, residue and poisoning databases; Pesticide Use Reduction programs and Organic Transition Payments (Baerselman, 1992; Imbroglini, 1992; Blobaum, 1997; Reus and Leendertse, 2000; Jensen and Petersen, 2001; Chunyanuwat, 2005)
- National IPM extension programs (Briolini, 1992; Huus-Bruun, 1992; van Lenteren, 1992; FAO, 2005b)

Regional initiatives and frameworks (some examples):
- OECD/DAC Guidelines on Pest and Pesticide Management prioritize IPM and improved pesticide management, with formats for industry data submission and governmental pesticide evaluation reports (OECD, 1995). The OECD has also initiated a Risk Reduction project (OECD, 2006b).
- The European Commission's "thematic strategy" provides a policy framework to minimize hazards and risks of pesticide use (EC, 2006) filling a regulatory gap in the pesticide cycle between the before-use (product approval) and after-use (impact) stages.
- North American Commission on Environmental Co-operation (NACEC) of NAFTA has established a Sound Management of Chemicals Working Group which has developed action plans to reduce use of specific pesticides (http://www.cec.org/).
- Permanent Inter-State Committee for Drought in the Sahel (CILSS) regional convention to support collaborative management and regulation of pesticides (http://80.88.83.202/dbin-sah/index.cfm?lng=en&sect1=avant1&id=28)

International agreements and treaties:
- *The UN Food and Agriculture Organization (FAO) International Code of Conduct on the Distribution and Use of Pesticides* (agreed in 1985 and revised in 2002) sets voluntary standards for the management and use of pesticides and provides guidance for the development of national pesticide legislation (FAO, 2005a; http://www.fao.org/ag/AGP/AGPP/Pesticid/Code/PM_Code.htm). FAO is updating its guidelines on pesticide labelling (FAO, 1995b) to include the UN Globally Harmonized System of chemical classification and labelling (FAO, 2006b) and is working with governments and commercial actors to phase out highly toxic pesticides (FAO, 2006ab, 2007).
- *The Rotterdam Convention on Prior Informed Consent (PIC) Procedure for Certain Hazardous Chemicals and Pesticides in International Trade* (1998) requires that exporting countries provide notification to importing countries of bans and restrictions on listed pesticides (http://www.pops.int/). By 2006, 107 countries had ratified PIC.
- *The Stockholm Convention on Persistent Organic Pollutants (POPs)*, signed in 2001, provides phaseout plans for an initial twelve pollutants—nine of them pesticides—and defines a process for adding new chemicals such as endosulfan, lindance and chlordecone to the list (http://www.pops.int/). By 2006, 126 countries had ratified the POPs treaty. The nongovernmental International POPs Elimination Network (IPEN) works alongside the POPs treaty process.
- *The Montréal Protocol (1987)* mandates the phasing out of the ozone-depleting pesticide, methyl bromide (http://ozone.unep.org/). The Methyl Bromide Action Network, a coalition of environmental, agriculture and labor organizations, was established in 1993 to assist governments in the transition to affordable, environmentally sound alternatives.
- *Intergovernmental Forum on Chemical Safety (1994)* is a WHO sponsored mechanism to develop and promote strategies and partnerships on chemical safety among national governments and intergovernmental and nongovernmental organizations (http://www.who.int/ifcs/en/). The *Inter-Organization Programme for the Sound Management of Chemicals (IOMC) and International Programme on Chemical Safety (IPCS)* are two other international coordinating organizations relating to chemicals. The IFCS sponsors a *Working Group on Acutely Toxic Pesticides*, which maintains a CD-ROM database on acute pesticide poisonings worldwide.
- *UNEP's Strategic Approach to International Chemicals Management* (2006) articulates global commitments, strategies and tools for managing chemicals more safely around the world (http://www.chem.unep.ch/saicm/). The agreement emphasizes principles of prevention, polluter pays, substitution for less harmful substances, public participation, precaution, and the public's right to know.
- *The Basel Convention on the Control of Transboundary Movements of Hazardous Wastes and Their Disposal (1992)* focuses on controlling the movement of hazardous wastes, ensuring their environmentally sound management and disposal, and preventing illegal waste trafficking (http://ozone.unep.org/). Now ratified by 149 countries including 32 of the 53 African countries, the convention explicitly includes obsolete pesticide stockpiles.
- *The Africa Stockpiles Project* brings together diverse stakeholders to clean up and safely dispose of obsolete pesticide stocks from Africa and establish preventive measures to avoid future accumulation (http://www.africastockpiles.org/). Initiated by Pesticide Action Network UK and WorldWide Fund for Nature (WWF) in 2000, the project is led by FAO (technical assistance on elimination and prevention), PAN and WWF (capacity-building, communication and outreach), CropLife International (financial support and management) and the World Bank (administration of funds).

Bank, 1998a). Subsequent external and internal reviews of World Bank lending and project monitoring noted weak implementation of the Bank's IPM policy (Tozun, 2001; Ishii-Eiteman and Ardhianie, 2002; Hamburger and Ishii-Eiteman, 2003; Sorby et al., 2003; Karel, 2004) hampered by lack of trained staff and an organizational culture and incentive system favoring loan approval over project quality (Liebenthal, 2002). Recent analyses of written policy and project design documents suggest compliance may be improving (Karel, 2004; World Bank, 2005d) and a detailed guidebook to support implementation of the Bank's IPM policy has been produced.

Significant international treaties (Box 2-9) are now in force that seek to minimize and eliminate hazards associated with pesticide use. Multistakeholder initiatives such as the Africa Stockpile Program have harnessed the energies of diverse stakeholders in reducing the hazards and risks of pesticides. Together these policy responses and international agreements, informed by scientific evidence and public participation, have enabled decisive and effective transitions towards more sustainable practice.

Civil society responses. Civil society has emerged as a powerful force in the movement towards ecological pest management, in Northern as well as Southern countries (e.g., India, Thailand, Ecuador, Philippines and Brazil). CSOs and independent researchers (as well as FAO, ILO, WHO and some governments) have called for a rights-based approach to agricultural development, that explicitly recognizes agricultural workers' and rural communities' rights to good health and clean environments (Robinson, 2002; Fabra, 2002; Reeves and Schafer, 2003). NGOs working with social justice, environmental and health causes have contributed to national and international treaties and agreements on chemicals management, sustainable agriculture and food safety. Development NGOs (Thrupp, 1996), social movements such as the Brazilian Landless Workers' Movement (Boyce, et al., 2005) and farmer-NGO-scientist partnerships such as MASIPAG in the Philippines, CLADES in Latin America (Chaplowe, 1997a) and the Latin American Scientific Society of Agroecology (Sociedad Científica Latino Americana de Agroecologia or SOCLA) are implementing ecological pest management as a means towards achieving sustainable development goals. Like other development actors, NGOs have limitations in terms of impact, resources, capacity and performance; and accountability mechanisms have been weak (Chaplowe, 1997b). Nevertheless, important contributions to ecological pest management have resulted from NGO efforts (Altieri and Masera, 1993; UNDP, 1995; Chaplowe, 1997b; Altieri, 1999), although scaling up to achieve widespread impact, in the absence of broader policy reforms, remains difficult (Bebbington and Thiele, 1993; Farrington and Lewis, 1993; Farrington et al., 1993).

Market responses. There has been a notable rise in certification and labeling regimes to meet consumers' demand for information about the origins of foods and methods of production. Food retailers are responding by insisting on observance of legal MRL requirements and using pesticide residue data as marketing material. Food industry actors have focused on minimizing or eliminating pesticide use to meet consumer preferences and regulatory requirements and reduce business costs. Some agrifood companies and the US$30 billion food service company Sysco (Hammerschlag, 2007), food processors (e.g., tomato paste, coffee, cacao/chocolate) and some food retailers (Williamson and Buffin, 2005; EurepGap, 2007) have taken steps to source produce from suppliers—including thousands of small-scale producers—using IPM and organic methods. Labels identifying organic or low-pesticide production methods and other successful market-oriented collaborations (IATP, 1998) have encouraged growers to adopt these practices. Local food systems also offer a small but growing alternative to conventional crop production and distribution (Williamson and Buffin, 2005) (see 2.3.3).

Response from pesticide manufacturers. The multinational agrichemical industry has responded to global concerns about pesticides by developing less hazardous, lower dose and more selective pesticides, improved formulations, new application technologies and resistance management strategies (CropLife, 2003; Latorse and Kuck, 2006; Syngenta, 2006). These efforts can significantly reduce pesticide pressure on the environment, particularly in larger farm operations that can afford specialized equipment. Some pesticide manufacturers have formed Resistance Action Committees to assist advisors and growers in implementing pesticide resistance management practices (Jutsum et al., 1998). The Danish chemical company, Cheminova, submitted plans to FAO in 2006 to voluntarily phase out highly hazardous WHO Class I pesticides from developing countries by 2010 (FAO, 2006a). At the same time, public health specialists and development NGOs have criticized multinational pesticide companies for lobbying against stronger public health regulations, for failing to comply with national laws and the FAO Code of Conduct on the Distribution and Use of Pesticides (Dinham, 2007), and in some cases for refusing to voluntarily withdraw recognized highly hazardous active ingredients—including WHO Class 1 pesticides and acutely toxic organophosphate pesticides—in developing countries (Rosenthal, 2003, 2005; Sherwood et al., 2005; Wesseling et al., 2005). Competitive pressure from local generic pesticide manufacturers that continue to produce off-patent pesticides can be a factor (EJF, 2002; Pawar, 2002).

Industry actors have developed their own IPM programs (Dollacker, 2000; CropLife, 2006). Many of these are built around continued or relatively small reductions in use of a company's pesticide products (Sagenmuller, 1999; Dollacker, 2000; Ellis, 2000; CropLife, 2003, 2005ab). One explanation for this is that a company's need to maintain economic returns on its investments renders them less likely to encourage substantial shifts towards pest management strategies that would significantly reduce reliance on their products (CGIAR TAC, 2000; FAO, 2001a; Murray and Taylor, 2001; Sherwood et al., 2005).

Some newer products developed by private firms show potential to strengthen IPM efforts (for instance, synthetic pheromone products to be tried in the context of "push-pull" strategies in Europe). Other programs describe the integration of crop productivity and biodiversity conservation efforts (Dollacker and Rhodes, 2007). Independent assessments of their effects in actual use, particularly in small scale farming conditions in the tropics, have not been made.

The multinational agrichemical industry has also launched "safe use" programs to train farmers in the use and handling of pesticides and to ensure that products are used in a manner consistent with national regulatory frameworks (Syngenta, 2003; CropLife, 2005b). The efficacy of these pesticide use training programs is disputed, with some sources reporting considerable success (Tobin, 1996; Grimaldi, 1998; Syngenta, 2006) and others finding no reduction in poisoning incidence among participating farmers (McConnell and Hruska, 1993; Murray, 1994; Kishi et al., 1995; Murray and Taylor, 2000). "Safe use" measures are often not affordable or feasible in tropical climates and under actual conditions of use in poor countries (Dinham, 1993, 2007; Cole et al., 2000; FAO, 2007). Even when pesticides are used according to label specifications, adverse health effects have been documented (Nurminene et al., 1995; Garry et al., 1996; Wargo, 1996; Schettler et al., 1996; Reeves et al., 2002). The industry's overall contribution to broader equitable and sustainable development goals, particularly in developing countries, has not as yet been clearly demonstrated.

2.3.2.4 Overall assessment of trends and challenges in pest management

Despite the tightening national and international regulatory environment around synthetic pesticides and notwithstanding the documented success of ecological pest management in most crops and a fast-growing market for organic products, sales and use of synthetic pesticides is still growing, especially in developing countries. These trends continue to result in pesticide-induced pest outbreaks (Yudelman et al., 1998) and an unacceptably high level of unintentional pesticide poisonings under conditions of actual use, mostly but not solely in the developing world (Wesseling et al., 1993; Kishi, 2005, London et al., 2005). Public sector commitment to pesticide reduction efforts and investments in IPM and other ecological approaches has not been consistent over time (Cate and Hinkle, 1994). The prevalence of the use of synthetic pesticides today reflects their immediate results, path dependency at farm and institutional support levels, and the significant political and economic influence of agribusiness interests, trade associations and lobbying groups in the regulatory and policy arena (Ferrara, 1998; FAWG, 2001; Irwin and Rothstein, 2003; CAP/OMB Watch, 2004; Mattera, 2004; UCS, 2004; Dinham, 2005; Wesseling et al., 2005; Shulman, 2006; Hardell et al., 2007). This influence has sometimes downplayed research findings on harmful effects and weakened regulatory assessment of risks (Castleman and Lemen, 1998; Watterson, 2001; Hayes, 2004).

Scientific and technological progress has not been linear; successful pathways (e.g., in biocontrol) have gained and lost popularity according to the economic and political priorities of dominant institutional arrangements. Advances in ecological sciences (e.g., population, community, landscape ecology) have contributed to development of pest management options, but have been underutilized by most conventional extension systems. Genetically-engineered crops were expected by many to reduce the need for and therefore use

of synthetic insecticides. However, their impact on both insecticide and herbicide use has been mixed, in some cases leading to increased recourse to synthetic controls. Their cultivation is perceived by some scientists and critics as potentially introducing new environmental hazards (Wolfenbarger and Phifer, 2000; CEC, 2004; Donald, 2004; Snow et al., 2005), reducing efficacy of biocontrol measures (Obryki et al., 2002) or leading to adverse social impacts (de Grassi, 2003; Pengue, 2005; FOE, 2006) and health risks (Ewen and Pusztai, 1999; Prescott et al., 2005), constraining their adoption in sustainable development initiatives.

The central technical issue facing pest management today is no longer yield maximization, but long-term stabilization and resilience in the face of unknown and changing stresses (Reganold et al., 2001). New directions in science and technology can strengthen IPM efforts if the latter have a strong foundation in basic biology (entomology, botany, plant pathology, taxonomy, ecology), economics and the social sciences (CGIAR TAC, 2000). Agroenvironmental partnerships among farmers, extension agents and researchers that balance social and environmental learning (Warner, 2006b; Stone, 2007) and strengthen ecologically-informed decision-making capacities (Röling and Wagemakers, 1998; Getz and Warner, 2006; Warner, 2006a; Mancini et al., 2007; van den Berg and Jiggins, 2007) offer robust possibilities for meeting technical, social and institutional challenges in sustainable pest management.

Policy decisions in pest management knowledge, science and technology often have been implicitly or explicitly based upon perceptions of tradeoffs. The uneven distribution of gains and losses from these decisions reflect power asymmetries between competing actors (Krimsky, 1999; Kleinman and Vallas, 2001). They have fuelled social and political tensions; in some cases, these have contributed to the development of new institutional arrangements such as international treaties and conventions to manage pesticide problems. Dominant approaches to pest control have in many cases failed to ensure the now-recognized human right to a safe home and working environment (Fabra, 2002; Robinson, 2002; Reeves and Schafer, 2003). The evidence shows that if crop production is assessed solely by a simple economic metric, then other societal goals will not be properly valued. Informed decision-making in pest management requires integration of ecological and social equity metrics as well.

The policy and investment choices regarding pest management have significant implications for how successfully societies will respond to major global challenges ahead (associated with, e.g., clean water, climate change, biodiversity, etc). Responses are varied, reflecting the complex and sometimes competing interests of diverse actors. UN agencies such as the FAO, national governments, public health professionals, labor groups, NGOs, development experts and some private firms are working to eliminate WHO Class I and phase out WHO Class II pesticides. Some pesticide manufacturers are developing new less toxic products and improved delivery systems, although many continue to sell and promote highly hazardous pesticides at the same time. Market leaders and innovators in the food industry are moving towards sourcing organic, fairly traded products. Governments, international commissions and initiatives such as UNCED

(UNCED, 1993), and the UN IFCS (Box 2-9) use the precautionary and polluter pays principles in designing policy approaches to chemical use and distribution (EEA, 2001; City and County of San Francisco, 2007; Fisher, 2007). Scientists and researchers in the fields of public health, medicine, ecology and participatory development and extension call for greater public sector investment in agroecological research and education, and establishment of better institutional linkages among farmers, extension agents and physical and social scientists (UNDP, 1995; Wesseling et al., 1997; Röling and Wagemakers, 1998; SIDA, 1999; Sorby et al., 2003; Norton et al., 2005; Warner, 2006a; Cole et al., 2007).

The weight of the evidence points towards the need for more determined institutional and policy support for participatory ecologically-based decision making by farmers; agroenvironmental partnerships to foster social and environmental collaborative learning; stronger and enforceable policy and regulatory frameworks; and investments by public sector, donor and commercial agencies in sustainable and agroecological research, extension, education, product innovation and marketing. More experimentation is needed to develop and test institutional innovations that are likely to enable further societal shifts towards sustainable pest management.

2.3.3 Food systems management

Satisfaction of social needs and desires, and hunger, more than nutritional needs, govern the selection and consumption of foods). Different food systems differently affect food security, safety and sovereignty. Food systems (Figure 2-4, 2-5) include the complex interactive and interrelated processes involved in keeping a community fed and nourished (Ericksen, 2006ab). At the core are food system activities that include production, processing, distribution, consumption and their outcomes: social welfare; food security and environmental welfare. A sustainable food system would incorporate social justice into a more localized system; alleviate constraints on people's access to adequate, nutritious food; develop economic capacity to purchase local food; train people to grow, process, and distribute food; maintain adequate land to produce a high proportion of locally required food; educate people removed from food production, to participate in, and respect, its generation; and integrate environmental stewardship into process (Koc et al., 1999). Food systems are assessed at the local and global level here for the sake of simplicity, although more complex variations (e.g., regional systems) exist and much interaction actually occurs among all the levels.

2.3.3.1 Local food systems activities

At the eve of World War II, local food systems (LFS) prevailed throughout the world. These predominantly fallow/rotational systems used manual labor (Mazoyer and Roudard, 2005), were family-owned, small and highly diversified crop-animal systems with varying productivity (Fogel, 2004; Mazoyer and Roudard, 2005). Food processing in many parts of the world relied on local knowledge of preservation and packaging techniques, such as salting, curing, curding, sun drying, smoking and fermentation (Johnson, 2000). Surplus produce was sold at the farm gate or in local market places directly to consumers

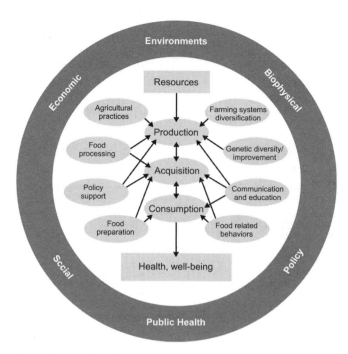

Figure 2-4. *Holistic food system model.* Source: Combs et al., 1996.

or to intermediary traders (Amilien, 2005). LFS directly contributed to the incomes of small-scale farmers, providing fresh and culturally acceptable food to consumers, and allowing direct interaction between consumers and food producers. However, farmers and local processors often experienced high transaction costs, seasonal price highs and lows and flooded markets, while consumers often lacked choice and quality foodstuffs or encountered contaminated or unsafe products (Crosson and Anderson, 2002). Rural households primarily acquired food from their own production (from local markets, relatives and friends; or from gathering, hunting or fishing). LFS sustain livelihoods of a significant number throughout the world, particularly in the southern hemisphere.

2.3.3.2 Global food systems activities

Over the past 50 years there has been a dramatic change in food systems particularly in developed countries (Knudsen et al., 2005; LaBelle, 2005) from local to global, traditional to an industrial, and from state regulated to a market- or transnational corporations-dominated system monopolized by relatively few companies from production to retail (Hendrickson and Heffernan, 2005). LFS production has changed for many into mechanized high-input specialized commodity farming, employing fewer people (Lyson, 2005; Dimitri et al., 2005; Knudsen et al., 2005). This transformation resulted in farm output growing dramatically, except in Africa (Knudsen et al., 2005) and a dramatic rise in GDP (Crosson and Anderson, 2002); spurring rapid growth in average farm size accompanied by an similar rapid decline in the number of farms and rural populations (Lyson, 2005; Knudsen et al., 2005).

Prior to the 20th century, increases in food production were obtained largely by bringing new land into production, With the exception of a few limited areas of East Asia, in

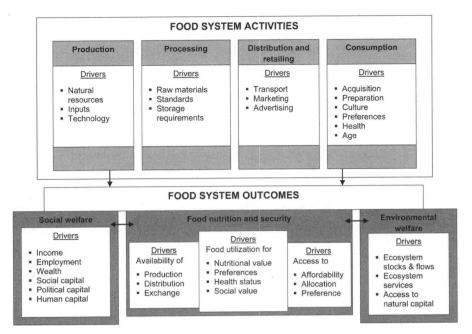

Figure 2-5. *Food system activities and outcomes.* Source: Adapted from GECAFS, 2008.

the Middle East, and in Western Europe (Welch and Graham, 1999; Stringer, 2000; Knudsen et al., 2005). Technology advancements and domestic food and agricultural production policies allowed consumers to spend a smaller portion of their income on food by the end of the 20th century (Johnson, 2000; Khush, 2001) allowed consumers to spend a smaller portion of their income on food (Knudsen et al., 2005). Institutional factors like efficient marketing systems, dynamic production technology and higher education played equally important roles in generating long-term growth in agricultural output per hectare and person employed (Mellor, 1966; Hayami and Ruttan, 1985; Eicher and Staatz, 1998). Food processing and preservation involving new technologies such as cold storage; irradiation; high temperature treatments; chemical additives; canning; milling, labeling and sophisticated computer based controlled systems emerged, both creating and taking advantage of new mass markets. The advantages of pre-prepared time-saving food to rapidly urbanizing populations drove further innovations in food preservation. In OECD countries, a few international food processing giants controlled a wide range of well-known food brands, coexisting with a wide array of small local or national food processing companies. Globalized food trade was originally confined to commodities and non-perishables such as wine, salt, spices and dried fish but expanded to include a wide range of perishable foods transported, sold and consumed at long distances away from their production and processing locality (Young, 2004; Knudsen et al., 2005). Even consumers in rural areas became less dependent on food supplies from local farms and markets (Roth, 1999). Meanwhile, small food retail groceries merged or were swallowed by other emerging and increasingly powerful stores, chains and supermarkets (Smith and Sparks, 1993; Roth, 1999). In the USA for example, from 1990-2000, the market share of the meat industry held by

the nation's top four retailers rose from 17 to 34%. Institutional linkages within local food systems (Lyson, 2005) were thus broken and economies of scale increased by means of new institutional arrangements (Ericksen, 2006ab). Vertical integration in ownership of food supply chains (FAO, 2005c) and increasing concentration in private sector control over food systems (DFID, 2005) has been documented.

2.3.3.3 Food systems outcome trends
The globalized food system (GFS) is considered by some to be economically efficient and productive (Welch and Graham, 1999; LaBelle, 2005) and draws on a range of science, knowledge and technology that extends beyond the agricultural sector. The GFS however hides disparities among agricultural and food systems both in developed and developing countries (Knudsen et al., 2005; LaBelle, 2005). Concerns revolve around social welfare; food and nutritional security; food sovereignty, food safety and environmental welfare (Knudsen et al., 2005; Lyson, 2005) (Figure 2-6).

Social welfare. The GFS widened the gap between the most productive and least productive systems: it increased 20-fold over the last 50 years, particularly between industrialized and developing countries[1] (Knudsen et al., 2005; Mazoyer, 2005). Characterized by capital intensive AKST and seed/animal breeds that required high inputs and favorable agronomic conditions, the GFS favored farming populations with more resources (Knudsen et al., 2005; Lyson, 2005). There is some evidence that the Green Revolution, e.g., in

---

[1] With the exception of some portions of Latin America, North Africa, South Africa and Asia where it has been adopted by large national or foreign farms that have the necessary capital (Knudsen et al., 2005). Africa has the lowest production per unit area of land in the world (Paarlberg, 2002).

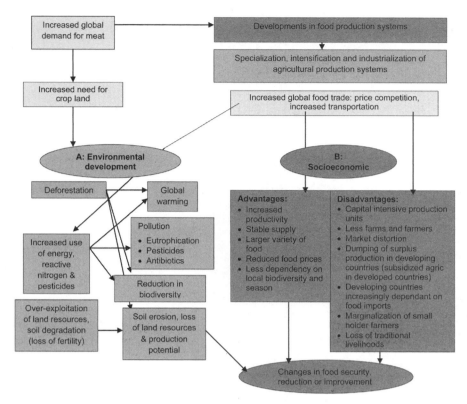

Figure 2-6. *Potentially problematic social and environmental aspects of global food systems sustainability.* Source: adapted from Knudsen et al., 2005.

Bangladesh, benefited the poor and the landless as well as those with resources and that small-scale farmers adopted faster than large scale farmers (Crosson and Anderson, 2002), but in many countries evidence demonstrates that better resourced individuals and firms benefited, sometimes at the expense of the poor and landless (see 2.2).

Food trade. The Uruguay Round of Trade Negotiations saw agriculture and food issues placed firmly within the WTO although some countries and organizations argued against their inclusion, maintaining that countries should have the right to determine their own policies on such an important issue as food security, i.e., they adopted a "food sovereignty" position (FOEI, 2001). Nonetheless, the 1994 WTO Agricultural Agreement adopted minimum import requirements and tariffs and producer subsidies that were accessible to transnational corporations both in USA and Europe (McMichael, 2001; Lyson, 2005), allowing them to operate economies of scale that lowered agricultural product prices all over the world (Welch and Graham, 1999; Wilson, 2005). Consumers and national economies benefited substantially from this agreement. These trends also opened up agricultural and food markets for the northern hemisphere commodities, with USA becoming the major exporter of cereals (with surplus being disposed of as food aid; Johnson, 2000) and Australia and New Zealand of dairy products. This development negatively affected local producers in developing countries; many countries, particularly in sub-Saharan Africa, increasingly became food importers (FAO, 2004). In developed countries, control of the food

system became vertically integrated from seeds; production inputs; processing; transportation and marketing, forming food chain clusters (LaBelle, 2005; Lyson, 2005) and consequently, many small-scale producers lost their livelihoods (Watkins, 1996; Welch and Graham, 1999; Robinson and Sutherland, 2002; Wilson, 2005), migrating to towns where they faced new livelihood challenges and opportunities.

Food security (Box 2-10 and Figures 2-7 and 2-8) greatly improved over the last few decades as a result pf the increase in global food production (Johnson, 2000; Crosson and Anderson, 2002) and the global grain trade (Johnson, 2000). Although increases in global food production (Paarlberg, 2002; Knudsen et al., 2005) surpassed population growth (Crosson and Anderson, 2002; Bruinsma, 2003; Knudsen et al., 2005), and was accompanied by an increase in the poorer country's average food consumption, (Garrett, 1997; Stringer, 2000; Johnson, 2000), food and nutritional insecurity persisted throughout the world even in countries which achieved *national* food security (Mellor, 1990; Stringer, 2000; LaBelle, 2005), particularly in sub-Saharan Africa (Wafula and Ndiritu, 1996; Knudsen et al., 2005).

Protein energy malnutrition in developing countries declined from as high as 46.5% in the early 1960s to as low as 17% in the late 1990s (Khush, 2001; Young, 2004), with Africa contributing about a quarter (24%) of the total undernourished population globally (Young, 2004). This phenomenon corresponds with the proportion of those with prolonged deficits in required energy intake as chronic food shortages fell in Asia and Latin America except sub-Saharan

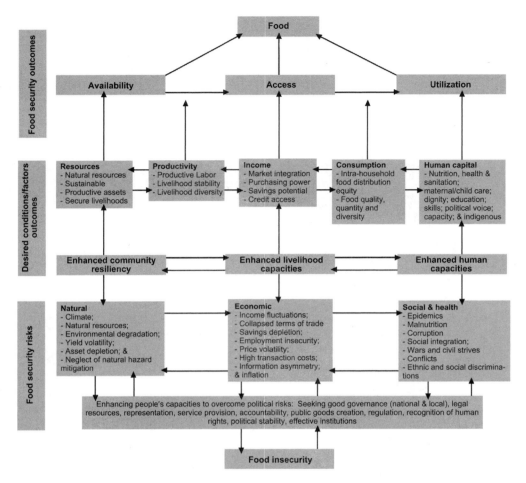

Figure 2-7. *A framework for understanding food security.* Source: Webb and Rogers, 2003.

Africa (FAO, 2001b; Lipton, 2001). In addition to other drivers (Johnson, 2000; Chopra, 2004), the failure of a Green Revolution in Africa (Crosson and Anderson, 2002) may partially be explained by the lack of improvement or worsening of the situation in Africa. Based on the Global Hunger Index (GHI)[2] (Weismann, 2006), 97 developing and 27 transitional countries exhibit poor GHI trends; the malnutrition hot spots are in South Asia and sub-Saharan Africa, where wars and HIV/AIDS exacerbate the situation.

The commoditized monocropping characteristic of the globalized food system (GFS) has resulted in a narrower genetic base for plant[3] and animal production (Knudsen et al., 2005; Lyson, 2005; Wilson, 2005) and in declining nutritional value (Welch and Graham, 1999; Kataki et al., 2001) and has negatively affected micronutrient reserves in the soil (Bell, 2004). A Mexican study, however, suggested that adoption of some improved varieties of maize had enhanced maize genetic diversity (Brush et al., 1988). Increasing and widespread micronutrient malnutrition has developed,

affecting millions of people in industrialized and developing countries alike (Welch and Graham, 1999; Khush, 2001) with quantifiable costs through compromised health resulting from reduced productivity and impaired cognition (Welch and Graham, 1999). However, recent improvements are noted in some parts of the developing world (Mason et al., 2005). Meanwhile, elements of the GFS, for example, subsidies of commodity crops such as corn in the US (Fields, 2004), have contributed to often radical and rapid changes in dietary patterns characterized by an excess of highly refined carbohydrates, sucrose, glucose and syrups (ingredients in fast foods) and animal fats, with a parallel decline in intake of complex carbohydrates (Tee, 1999; Fields, 2004; Young, 2004). These changes, combined with a decline in energy expenditure associated with sedentary lifestyles, motorized transport and household domestic and work place labor-serving devices (Young, 2004) have resulted in the emergence of obesity and other dietary-related chronic diseases afflicting both the affluent as well as the low income population in industrialized and developing countries (Tee, 1999; Fields, 2004; Young, 2004). This paradoxically coexists with undernutrition (Young, 2004), signifying growing imbalances and inequities in food systems.

In the 1980s, food production shifted toward products that were convenient and served ethnic and health-based preferences. This shift has changed the structure of

[2] GHI captures three equally weighted indicators of hunger: insufficient availability of food (the proportion of people who are food energy deficient); prevalence of underweight in children <5 years old; and child mortality (<5 years old mortality rate).

[3] Wheat, rice and maize account for the majority of calories in human diets.

Box 2-10. Evolution of the term *food security*.

Food security [is] a situation that exists when all people, at all times, have physical, social and economic access to sufficient, safe and nutritious food that meets their dietary needs and food preferences for an active and healthy life. (FAO, The State of Food Insecurity 2001)

Food sovereignty is defined as the right of peoples and sovereign states to democratically determine their own agricultural and food policies.

The term *food security* originated in international development literature in the 1960s and 1970s (Ayalew, 1997; Stringer, 2000; Ganapathy et al., 2005; Windfuhr and Jonsén, 2005) and public interest in global and domestic food security grew rapidly following the oil crisis and related food crisis of 1972-74 (Saad, 1999; Stringer, 2000; Clover, 2003), the subsequent African famine of 1984-85, and emergence of growing numbers of food banks in developed nations. *Food security* is a term with many definitions, each used differently in international, national and local contexts (Ganapathy et al., 2005). Early definitions of food security focused on aggregate food supplies at national and global levels (Clover, 2003). Over time the concept evolved and expanded to integrate a wide range of food-related issues reflecting the complexity of the role of food in human society. Much of the paradigm shift of the concepts and definitions of food security over the years can be attributed to NGO and civil societies' movements in the early 1990s that led to the birth of the concept food sovereignty.

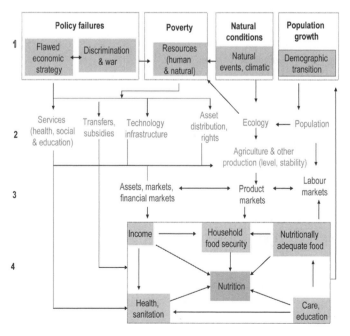

Figure 2-8. *Determinants of nutrition security: Basic causes and links.* Source: FAO, 1996a.

Notes: (i). Basic causes; (ii). Structural/institutional conditions, areas of public action; (iii). Market conditions; (iv) Micro-level conditions (household, intra-household, gender).

agricultural markets, further increasing specialization and prompting the emergence of contractual farming and vertical integration for supply and quality control, and development of special-use, high-value commodities (Barrett et al., 1999) particularly of farmed fish, livestock and specialty crop operations (Knudsen et al., 2005). Concerns have been raised regarding the impact of these structural changes on the rural poor (Lindstrom and Kingamkono, 1991; Welch and Graham, 1999; Grivetti and Olge, 2000) and marginalized urban populations.

Food safety. The right of everyone to have access to safe and nutritious food is reaffirmed by the Rome Declaration on World Food Security. Yet food-borne poisonings and illnesses represent a major daily health threat and results in significant economic costs in both developed and developing countries in spite of significant progress in the regulation of food standards, medicine, food science and technology (Box 2-11; FAO, 1999b).

The globalized food system, although it is subject to high controls and standards, can still threaten food safety, particularly for marginalized populations in industrialized and developing countries (Welch and Graham, 1999; Mol and Bulkeley, 2002). High-profile risks such as those associated with bovine spongiform encephalopathy (BSE); Belgian dioxin chickens; vegetables contaminated with Chernobyl nuclear fallout or with dioxins from waste-burning plants; and GMOs have been profiled in recent decades. Other environmental and health threats, less reported in the media, are also contributing to widespread concern about the GFS. As food passes over extended periods of time through the food production, processing, storage and distribution chain, control has become difficult, increasing the risks of exposing food to intentional, undetected or involuntary contamination or adulteration. The use of pesticides and fertilizers, the use of hormones in meat production, large-scale livestock farming, and the use of various additives by food processing industries are among the food safety concerns that are associated with the GFS. In developing countries, GFS safety concerns are compounded by rampant poverty negatively influencing policy compliance and poor infrastructure for enforcement of food control systems. Other threats to food safety in developing countries are offered by inadequate social services and service structures (potable water; health, education, transportation); population growth; high incidences of communicable diseases including Acquired Immunodeficiency Syndrome (AIDS); competitive markets and trade pressures that may encourage shortcuts that compromise food safety (CSPI, 2005).

Access to good quality food has been humankind's main endeavor from the earliest days of human existence (FAO, 1999b) with governing authorities codifying rules to protect consumers from dishonest practices in the sale of food. The first general food laws in modern times were adopted during the second half of the nineteenth century; subsequently basic food control systems were established to monitor compliance.

Efforts to deal with hazardous agents (pesticides and food additives) began in the 1940s and 50s when toxicologists derived limits on exposure for protection of human health (Rodricks, 2001). A major step in advancing a science-based food safety system was the development and implementation of Hazard Analysis and Critical Control Point (HACCP) procedures in the food industry in the 1960s. In parallel, the development of "farm to fork" strategies by the industry extended the notion of quality management along the entire supply chain (Hanak et al., 2002).

Food contamination creates a social and economic burden on communities and their health systems. The market costs of contaminated commodities cause significant export losses (Box 2-11), while sampling and testing costs and costs to food processors and consumers can be high.

The incidence of food-borne diseases may be 300 to 350 times higher than the number of reported cases worldwide. Sources of food contamination may be either microbiological or chemical and may occur throughout the food chain, from the farm to the table. Risk, particularly in developing countries, is in part due to difficulties in ensuring that appropriate procedures are followed.

Microbiological contaminants, the most reported cause of food-borne illnesses, include bacteria, fungi, viruses or parasites (Box 2-12) and usually result in acute symptoms. Over the past few decades, the incidence of reported illnesses caused by pathogenic microorganisms in food has increased significantly.

Food-borne illnesses caused by chemicals are sometimes difficult to link to a particular food, as the onset of effects may be slow and hence may go unnoticed until permanent or chronic damage occurs. Contamination by pesticides, heavy metals or other residues intentionally or unintentionally introduced into the food supply, or introduced through poor post-harvest techniques leading to mycotoxins, are included in this category (Box 2-13). On the other hand, food poisonings can also be acute with immediate adverse effects including death, such as those caused by organophosphate pesticides (Box 2-13) (Kishi, 2005).

Food irradiation is another controversial food safety issue. Although useful in reducing the risk of microbial food-borne illness, the technology also destroys vitamins (OCA, 2006); affects taste and smell; poses dangers to workers and the environment; may create toxic byproducts; and has the potential for cellular or genetic damage. The European Commission heavily regulates irradiated foods and food ingredients (EC, 1999).

Recent trends in global food production, processing, distribution, and preparation are creating a growing demand by consumers for effective, coordinated, and proactive national food safety systems. Although governments play critical roles in protecting the food supply, many countries are poorly equipped to respond to the growing dominance of the food industry and to existing and emerging food safety problems. Fraudulent practices such as adulteration and mislabeling persist and can be particularly devastating in developing countries where 70% of individual income may be spent on food (Malik, 1981). The effectiveness of HACCP is limited to large scale firms (Unnevehr and Jensen, 1999; Farina and Reardon, 2000). Export safety standards are often higher than those applied to domestic products markets particularly in developing countries. In some cases, governments have shifted the burden of monitoring product safety to the private sector, and in so doing, have become at most an auditor of the industry's programs.

Box 2-11. Food-borne illnesses: Trends and costs.

- Contaminated food contributes to 1.5 billion cases of diarrhea in children each year, resulting in more than three million premature deaths (WHO, 1999), in both developed and developing nations. One person in three in industrialized countries may be affected by food-borne illness each year. In the US food-borne diseases cause approximately 76 million illnesses annually among the country's 294 million residents resulting in 325,000 hospitalizations and 5,000 deaths (Mead et. al., 1999). Between 1993 and 2002, 21 Latin American and Caribbean countries reported 10,400 outbreaks of food- and waterborne illness causing nearly 400,000 illnesses and 500 deaths (CSPI, 2005).

- In 1995, the US experienced between 3.3-12 million cases of food-borne illness caused by seven pathogens costing approximately US $6.5-35 billion in medical care and lost productivity (WHO, 2002a).

- In the European Union, the annual costs incurred by the health care system as a consequence of Salmonella infections alone are estimated to be around EUR 3 billion (BRF, 2004).

- In the UK, care and treatment of people with the new variant of Creutzfeldt-Jakob disease (vCJD) are estimated to cost about £45,000 per case from diagnosis and a further £220,000 may be paid to each family as part of the government's no-fault compensation scheme (DHC, 2001). The range of economic impacts to the UK is from £2.5 to £8 billion, (Mathews, 2001).

- Analysis of the economic impact of a Staphylococcus aureus outbreak in India (Sudhakar, et. al., 1988) showed that 41% of the total cost of the outbreak was borne by the affected persons, including loss of wages or productivity and other expenses.

- Because of an outbreak of Cyclospora in Guatemalan raspberries in 1996 and 1997 the number of Guatemalan raspberry growers has decreased dramatically from 85 in 1996 to three in 2002.

- Realization of existence of BSE in cattle population in the US and Canada resulted in losses of $2.6 billion and $5 billion in beef exports in 2004 in the USA and Canada respectively.

- Meanwhile a new category of risks has emerged, of which BSE, genetically modified organisms (GMOs), and zootic diseases such as avian flu, are among the most prominent. The routes through which these risks may affect nature and society are more complex, less "visible" and less detectable than "conventional" risks, and are often highly dissociated over space and time (Mol and Bulkley, 2002).

Box 2-12. Common microbiological contaminants in food.

In Latin America, the most frequent bacterial agents involved were Salmonella spp. (20% of the reported outbreaks) (FAO/WHO, 2004), Staphylococcus aureus, and Clostridium perfringens (CSPI, 2005). Another pathogen, Escherichia coli O157:H7, has increased dramatically in Central and South America. Argentina has one of the highest incidences of HUS—a serious complication of E. coli infection—especially in the pediatric age group (CSPI, 2005).

Food items most commonly associated with the reported outbreaks were fish/seafood (22%), water (20%) and red meats (14%) (CSPI, 2005). Examples include a major E. coli O157:H7 outbreak in Japan linked to sprouts involving more than 9,000 cases in 1996, and several recent Cyclospora outbreaks associated with raspberries in North America and Canada, and lettuce in Germany (Bern et al., 1999; Hodeshi et al., 1999; Döller et al., 2002). In 1994, an outbreak of salmonellosis due to contaminated ice cream occurred in the USA affecting an estimated 224,000 persons. In 1988, an outbreak of hepatitis A, resulting from the consumption of contaminated clams, affected some 300,000 individuals in China (Halliday et al., 1991). In 2005 in Finland, the most common cause of food and water-borne food poisonings was noro-virus (EVIRA, 2006). A 1998 outbreak of Nipah virus typically associated with pigs and pork (WHO, 2004) killed 105 people in Malaysia. The parasitic disease trichinellosis is increasingly reported in the Balkan region among the non-Muslim population, owing in part to the consumption of pork products processed at home without adherence to mandatory veterinary controls.

Box 2-13. Chemical contamination of food: A few examples.

- Mercury: As many as 630,000 children are born each year exposed to mercury in the womb (Ahmed, 1991).
- Non-persistent organic compounds: In Spain in 1981-1982, contaminated rapeseed oil denatured with aniline killed more than 2,000 people and caused disabling injuries to another 20,000—many permanently (CDCP, 1982).
- Pesticide residues: The latest European monitoring of pesticide residues in food found 4.7% of all samples exceeding the legal threshold of pesticide residues in food and almost half of all samples had detectable levels of pesticide residues (EC, 2006); Viet Nam reports a high burden of disease associated with pesticide residues (Nguyên and Dao, 2001).
- Accidental pesticide poisonings: In India, in July 1997, 60 men were poisoned by eating pesticide-contaminated food at a communal lunch (Chaudry et al., 1998); in Tauccamarca, Peru, 24 children died in October 1999, after consuming a powdered milk substitute contaminated by the organophosphate pesticide methyl parathion, and 18 others suffered neurological damage (Rosenthal, 2003); in the Philippines, carbamate poisoning killed 28 schoolchildren and caused vomiting and diarrhea spells in 77 others in March, 2005 (Neri, 2005).
- Deliberate poisoning: In China, in 2002, more than 200 school children sickened and 38 died when rat poison was used to intentionally contaminate bakery products. (CNN, 2002).
- Naturally-occurring toxins: The chronic incidence of aflatoxin in diets is evident from the presence of aflatoxin M1 in human breast milk in Ghana, Nigeria, Sierra Leone, and Sudan and in umbilical cord blood samples in Ghana, Kenya, Nigeria, and Sierra Leone. Together with the hepatitis B virus, aflatoxins contribute to the high incidence of primary liver cancer in tropical Africa. Moreover, children exposed to aflatoxins may experience stunted growth or be chronically underweight and thus be more susceptible to infectious diseases in childhood and later life. (CSPI, 2005).
- Growth hormone: The EU banned the use of growth hormones in livestock in 1988 but the practice still continues in the US, Canada and in Australia.
- Dioxin: Exposure to dioxin causes serious adverse health effects, and remains a major public health concern in Europe, the United States and elsewhere (Schecter et al., 2001; NAS, 2003).

Major institutional arrangements. Codex Alimentarius Commission was created in 1963 by FAO and WHO to guide and coordinate world food standards for protection of consumer health and to ensure fair food trade (Heggun, 2001). Bodies that operate at regional levels include the European Food Safety Authority (EFSA); and US Food and Drug Administration (FDA). Codex food standards are considered vital in food control systems even in smaller and less developed countries. However, 96% of low-income countries and 87% of middle-income countries do not participate in the Codex actively and hence their priorities are not always reflected in the standards developed by Codex (http://www.codexalimentarius.net/web/evaluation_en.jsp). Recent findings on possible effects from low dose, chronic exposure to contaminants and development of the risk assessment procedures has led to ongoing revisions of international and national safety maximum residue levels of agrichemicals in the US, EU and Codex.

Food sovereignty. Whereas food security focuses on access to food, the concept of food sovereignty encompasses the right of peoples and sovereign states to democratically determine their own agricultural and food policies. Many definitions have emerged since the 1990s (FOEI, 2003;

Chopra, 2004; Forum for Food Sovereignty, 2007). There is currently no universally agreed public policy and regulatory framework definition for the term food sovereignty (Windfuhr and Jonsén, 2005). However, most definitions share a common reference point, starting from the perspective of those actually facing hunger and rural poverty and developing a rights-based framework that links the right to food with democratic control over local and national food production practices and policies. The concept often focuses on the key role played by small-scale farmers, particularly women, in defining their own agricultural, labor, fishing, food and land policies and practices, in ways that are environmentally sustainable, and ecologically, socially, economically and culturally appropriate to their unique circumstances (http://www.foodsovereignty .org/new/). Proponents also contend that decentralized, diverse, and locally adapted food and farming systems, based upon democratic and participatory decision-making, can ultimately be more environmentally sustainable and equitable than a globalized food system lacking such features (Cohn et al., 2006).

Via Campesina, a global farmers' movement developed the concept in the early 1990s, with the objective of encouraging NGOs and CSOs to discuss and promote alternatives to neoliberal policies for achieving food security (Windfuhr and Jonsén, 2005). The concept was publicized as a result of the International Conference of Via Campesina in Tlaxcala, Mexico, in April 1996. At the World Food Summit in 1996, Via Campesina launched a set of principles (Box 2-14) that offered an alternative to the world trade policies to realize the human right to food (Menezes, 2001; Windfuhr and Jonsén, 2005). In August the same year, reacting to the Mexican government's decision to increase maize imports from North America in accordance with the Free Trade Agreement (NAFTA), a large number of Mexican entities organized the *Foro Nacional por la Soberania Alimentaria*, underscoring the need to preserve the nation's autonomy in terms of defining its food policy (Menezes, 2001). Since then, a number of NGOs, CSOs and social movements have further developed the concept and its institutional implications (Menezes 2001; Windfuhr and Jonsén, 2005).

The concept of food sovereignty introduced into debates on food security and international trade regulation the right of each nation to maintain and develop its own capacity (particularly of small-scale farmers) to produce food to fulfill its own needs while respecting agroecosystem and cultural diversity (Menezes, 2001) and ensuring sustainable access and availability of food in order to enable people to lead quality lives and exercise democratic freedoms (Rosset et al., 2006; Riches, 1997). Market-oriented globalization of economic activity is an important driver of change in the evolution of agricultural trade and food systems. The development of the right to food based on normative qualities is another driver but with markedly different characteristics. The efforts made over the last fifty years to express in international and national laws a series of universal rights, including the right to food, has been an explicitly moral enterprise that stands in contrast to the economic processes of market-driven globalization. The right to food was included in the Universal Declaration of Human Rights adopted by the United Nations in 1948, following Franklin

D. Roosevelt's speech in 1941 that captured the world by proclaiming freedom from want and fear; freedom of speech and faith (Oshaug et al., 1994). The UN Declaration on the Right to Development Act 2 (UN, 1986; General Assembly Resolution 41/128, New York) states that ". . . the human being, being central subject to development, should be the active participant and beneficiary of the right to development." The various human rights instruments brought into force have created expectations and obligations for the behavior of individuals, social groups, and States (Oshaug and Edie, 2003). People are expected to be responsible for satisfying their needs, using their own resources individually or in association with others. States are expected to respect and protect the freedom of the people to make these efforts and the sovereignty over the natural resources around them, and are obliged to meet every individual's right to food and nutritional security.

Successive efforts have been made to build such rights, expectations, and obligations into national laws and the governance of food systems. Norway has formulated food security and the right to food as the basis of its agricultural policy, strongly driven by consumer concerns. Brazil has extended the concept of cultural heritage under Article 215 of its Constitution to include food cultures. Both these efforts have had an explicit normative quality.

The concepts of economic, social and environmental sustainability as applied to food systems have been developed in processes of negotiation and intensive discussions that reflect contrasting political priorities and ideologies (Oshaug, 2005). The food sovereignty movement is increasingly challenged to actively develop more autonomous and participatory ways of knowing to produce knowledge that is ecologically literate, socially just and relevant to context. This implies a radical shift from the existing hierarchical and increasingly corporate-controlled research system to an approach that devolves more responsibility and decision-making power to farmers, indigenous peoples, food workers, consumers and citizens for the production of social and ecological knowledge (Pimbert, 2006).

Organic agriculture. The term organic agriculture (OA) has evolved from various initiatives, including biodynamics, regenerative agriculture, nature farming, and permaculture movements, which developed in different countries worldwide from as early as 1924.[4] Since the early 1990s, OA has been defined in various ways. The most widely accepted definitions are those developed by IFOAM and the FAO/WHO Codex Alimentarius (Box 2-15). In response to the incipient marginalization of foods of local origin by supermarket chain developments; those dissatisfied with a globalizing food trade, desiring health foods or foods associated with cultural landscapes opened the way during the late 1950s and early 1960s for expansion of initiatives such as pick-

---

[4] Pioneered by German philosopher Rudolf Steiner, who theorized that a human being as part of a cosmic equilibrium has to live in harmony with nature and the environment (Stoll, 2002). Certification of biodynamic farms and processing facilities began in Europe during the 1930s under the auspices of the DEMETER Bund, a trademark chosen in 1927 to protect biodynamic agriculture.

Box 2-14. Via Campesina's food sovereignty principles.

1. *Food: A Basic Human Right:* Everyone must have access to safe, nutritious and culturally appropriate food in sufficient quantity and quality to sustain a healthy life with full human dignity. Each nation should declare that access to food is a constitutional right and guarantee the development of the primary sector to ensure the concrete realization of this fundamental right

2. *Agrarian Reform:* A genuine agrarian reform is necessary which gives landless and farming people—especially women—ownership and control of the land they work and returns territories to indigenous peoples. The right to land must be free of discrimination on the basis of gender, religion, race, social class or ideology; the land belongs to those who work it.

3. *Protecting Natural Resources:* Food Sovereignty entails the sustainable care and use of natural resources, especially land, water, and seeds and livestock breeds. The people who work the land must have the right to practice sustainable management of natural resources and to conserve biodiversity free of restrictive intellectual property rights. This can only be done from a sound economic basis with security of tenure, healthy soils and reduced use of agrochemicals.

4. *Reorganizing Food Trade:* Food is first and foremost a source of nutrition and only secondarily an item of trade. Food imports must not displace local production nor depress prices.

5. *Ending the Globalization of Hunger:* The growing influence of multinational corporations over agricultural policies has been facilitated by the economic policies of multilateral organizations such as the WTO, World Bank and the IMF. Regulation and taxation of speculative capital and a strictly enforced Code of Conduct for Trans-National-Corporations is therefore needed.

6. *Social Peace:* Everyone has the right to be free from violence. Food must not be used as a weapon. Increasing levels of poverty and marginalization in the countryside, along with the growing oppression of ethnic minorities and indigenous populations, aggravate situations of injustice and hopelessness. The ongoing displacement, forced urbanization, repression and increasing incidence of racism of smallholder farmers cannot be tolerated.

7. *Democratic control:* Small-scale farmers must have direct input into formulating agricultural policies at all levels. The United Nations and related organizations will have to undergo a process of democratization to enable this to become a reality. Everyone has the right to honest, accurate information and open and democratic decision-making. These rights form the basis of good governance, accountability and equal participation in economic, political and social life, free from all forms of discrimination. Rural women, in particular, must be granted direct and active decision making on food and rural issues.

Source: Windfuhr and Jonsén, 2005.

Box 2-15. Definitions of organic agriculture.

- *IFOAM:* Organic agriculture includes all agricultural systems that promote the environmentally socially and economically sound production of food and fibers. These systems take local soil fertility as a key to successful production. By respecting the natural capacity of plants, animals and the landscape, it aims to optimize quality in all aspects of agriculture and the environment. Organic agriculture dramatically reduces external inputs by refraining from the use of chemo-synthetic fertilizers, pesticides and pharmaceuticals. Instead it allows the powerful laws of nature to increase both agricultural yields and disease resistance.

- *FAO/WHO Codex Alimentarius Commission:* Organic agriculture is a holistic production management system that promotes and enhances agro-ecosystem health, including biodiversity cycles and soil biological activity. It emphasizes the use of management practices in preference to the use of off-farm inputs. This is accomplished by using where possible, agronomic, biological, and mechanical methods as opposed to using synthetic materials to fulfill any specific function within the system.

your-own operations and farm stands that supported a slow growth in alternative marketing channels for farm goods on which organically certified food capitalized (Roth, 1999). Consumer demand for "healthy" foods has begun to encourage large distributors and retailers also to integrate local and regional products into their offerings (Tracy, 1993; LaBelle, 2005).

Emerging evidence (Bavec and Bavec, 2006) indicates that organic farmers are able to sustain their livelihoods and increase employment in local processing and marketing, thereby increasing community economic activity and incomes (FAO, 1999b; Parrot and Marsden, 2002; Halberg et al., 2007; Kilcher, 2007; Scialabba, 2007). OA systems rely on biological processes to improve soil fertility and manage pests and are often high in crop biodiversity (Roth, 1999). The resulting increased food variety and overall per-area productivity has led to diversified and increased nutrient intake and improved food safety and food security, particularly for indigenous and resource-poor people (Roth, 1999; Scialabba, 2007; Sligh and Christman, 2007). Some studies, however, suggest that crop yields in organic farming are too low to sustain farmers' livelihoods and to produce quantities sufficient to meet growing and rapidly diversifying market needs (LaBelle, 2005) leading to concerns that more land would be needed if OA were to become widespread (Crosson and Anderson, 2002). These claims have been challenged by recent findings (Halweil, 2006; Badgley et al., 2007).

Technical challenges facing certified OA revolve around sourcing organically produced seed and fodder; consistent product quantity and quality; traceability; liability insurance of growers and processors; appropriate product attributes and pack size (LaBelle, 2005). More research is needed concerning the labor requirements of different OA systems. Labor demands in organic farming could deter younger generations from farming, but unemployment could be alleviated, since the labor is more evenly spread over a growing season (Pimentel, 1993; Sorby, 2002; Granatstein, 2003; Pimentel et al., 2005). Commercial challenges include narrowing profit margins; regulatory overload; increased competition; and the need for constant innovations to stay ahead of consumer trends (Roth, 1999), as well as uncertain implications of large-scale corporate entry into the market. These questions have prompted FAO to propose a framework for socioeconomic analysis focusing on ecological, economic and social performance as an instrument for farmers and decision makers to understand the problems, tradeoffs and outcomes in alternative scenarios for a range of OA systems (Scialabba, 2000).

Agriculture and human health. The interrelations between agriculture and human health are complex (Figure 2-9). The two are mutually and directly dependent on each others' status and performance. Agriculture contributes to good health through provision of food, fuel, fiber, fodder, materials for shelter and medicines. On the other hand, agricultural activities contribute to poor health through produce with nutritional deficiency; food-borne diseases; food poisoning; chemical pesticide residues; and a range of occupational hazards (including, for instance, induced hazards such as schistosomiasis and malaria that may be induced by irrigation developments). Similarly, human health also affects agriculture either positively or negatively. It requires a healthy individual and society to generate a productive agricultural performance. Hence individuals or societies with poor health are unable to provide the necessary quality human input in agricultural activities, leading to poor agricultural productivity (quantitatively and qualitatively) and low incomes that in turn perpetuates poor health—a vicious circle.

The interrelationship between agriculture and human health is mediated by the natural environment, human culture and technological inputs. How to achieve equitable food production delivering optimum nutrition for health requires a better understanding of the interplay between agriculture and environment, culture, and technical capacity, and how this interplay changes over time (Lang, 2006; Snowden, 2006) (Table 2-6).

2.4 Lessons from the Past: Implications for the Future

AKST encompasses different kinds of knowledge produced by numerous agencies and actors, notably but not only farmers. The complexity of the diverse and often unpredictable ways in which knowledge is generated justifies a systemic view of the processes involved in AKST. Well-evidenced but divergent and often conflicting interpretations exist of the contributions of AKST to such societal goals as increased productivity, environmental and social sustainability and equity as well as to societal knowledge about the damaging effects of agricultural technologies in different conditions of use. The resulting multiple narratives of past AKST processes and arrangements are not equally heard or recognized. Political power and economic influence has privileged some types of AKST processes, actors and institutional arrangements over others. Dominant institutional arrangements established the privileged interpretations of the day and set the agenda for searching for and implementing solutions.

The choice of historical narrative used to explain past events and the AKST options brought into farm practice has important implications for setting future priorities and projecting the future design of AKST. Special effort has been made here to render an account from differing perspectives of past and often yet unresolved controversies regarding AKST in order to present as comprehensive as possible an assessment of the effectiveness of different AKST systems in promoting innovations associated with a range of policy goals Three main lessons regarding the effectiveness of AKST in relation to the combined goals of sustainability and development are drawn: the critical importance of partnerships, the crucial role of educating farmers in their vocation and the role of public policies and regulations.

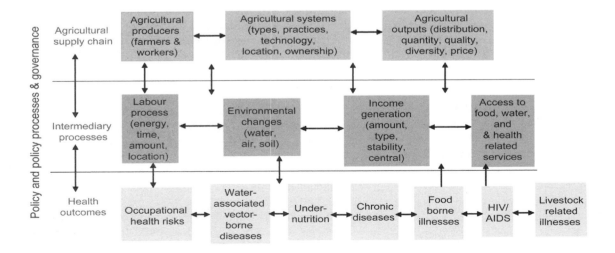

Figure 2-9. *Linkages between agriculture and health.* Source: Hawkes and Ruel, 2006.

Table 2-6. Health implications of agricultural and food revolutions.

Era/Revolution	Date	Changes in farming	Implications for food-related health
Settled agriculture	From 8500 BCE on	Decline of hunter-gathering greater control over food supply but new skills needed	Risk of crop failures dependent on local conditions and cultivation and storage skills; diet entirely local and subject to self-reliance; food safety subject to herbal skills
Iron Age	5000-6000 BCE	Tougher implements (plows, saws)	New techniques for preparing food for domestic consumption (pots and pans); food still overwhelmingly local, but trade in some preservable foods (e.g., oil, spices)
Feudal and peasant agriculture in some regions	Variable, by region/ continent	Common land parceled up by private landowners; use of animals as motive power; marginalization of nomadism	Food insecurity subject to climate, wars, location; peasant uprisings against oppression and hunger
Industrial and agricultural revolution in Europe and U.S.	Mid-18th century	Land enclosure; rotation systems; rural labor leaves for towns; emergence of mechanization	Transport and energy revolutions dramatically raise output and spread foods; improved range of foods available to more people; emergence of commodity trading on significant scale; emergence of industrial working-class diets
Chemical revolution	From 19th century on	Fertilizers; pesticides; emergence of fortified foods	Significant increases in food production; beginning of modern nutrition; identification of importance of protein; beginnings of modern food legislation affecting trade; opportunities for systematic adulteration grow; scandals over food safety result
Mendelian genetics	1860s; applied in early 20th century	Plant breeding gives new varieties with "hybrid vigor"	Plant availability extends beyond original "Vavilov" area; increased potential for variety in the diet increases chances of diet providing all essential nutrients for a healthy life.
The oil era	Mid-20th century	Animal traction replaced by tractors; spread of intensive farming techniques; emergence of large-scale food processors and supermarkets	Less land used to grow feed for animals as motive power; excess calorie intakes lead to diet-related chronic diseases; discovery of vitamins stresses importance of micronutrients; increase in food trade gives wider food choice
Green Revolution in developing countries	1960s and after	Plant breeding programs on key regional crops to raise yields; more commercialized agriculture	Transition from underproduction to global surplus with continued unequal distribution; overconsumption continues to rise
Modern livestock revolution	1980s and after	Growth of meat consumption creates "pull" in agriculture; increased use of cereals to produce meat	Rise in meat consumption; global evidence of simultaneous under-, over-, and malconsumption
Biotechnology	End of 20th century	New generation of industrial crops; emergence of "biological era": crop protection, genetic modification	Uncertain as yet; debates about safety and human health impacts and whether biotechnology will deliver food security gains to whole populations; investment in technical solutions to degenerative diseases (e.g., nutrigenomics)

Source: Hawkes and Ruel, 2006.

2.4.1 Multiple AKST actors and partnerships

In the prevailing AKST arrangements of the past, key actors often have been excluded or marginalized. Preference has been given to short-term considerations over longer-term agroecosystem sustainability and social equity and to powerful voices over the unorganized and voiceless. Strong evidence shows that development of appropriate forms of partnerships can help bring in the excluded and marginalized and open AKST to a larger set of policy goals. A large number of effective participatory approaches exist that

facilitate the establishment and operation of such partnerships. Targeted public support can help promote the use of these approaches and thereby address the biases in the hitherto dominant arrangements.

The Transfer of Technology (ToT) model, a supply-push approach, has dominated operational arrangements and policy thinking. Where the ToT model has been applied appropriately under the conditions of use necessary for achieving wide impact, it has been successful in driving yield and production gains. These conditions include properly functioning producer and service organizations, the social and biophysical suitability of technologies transferred in specific environments and proper management of those technologies at plot, farm and landscape levels. The implementation of the ToT model increased production at a faster pace than population growth in most developing countries, an achievement which did not appear likely thirty or forty years ago when the specter of famine and food crises loomed very large.

But AKST arrangements shaped by the ToT model have not been effective in meeting a broader range of goals associated with the multiple functions and roles of farm enterprises and diverse agroecosystems. Recognition of these limitations led to a growing awareness or rediscovery—documented by robust evidence—that innovation is a multisource process of demand-pull that always and necessarily involves a mix of stakeholders, organizations and types of knowledge systems. Effective innovation for combined sustainability and development goals has been led by farmers in association with a range of local institutional actors and has occurred in both OECD and tropical settings. Multiorganizational partnerships for AKST that embraces both advanced scientific understanding and local knowledge and experimental capacities have led to the development and wider adoption of sustainable practices such as participatory plant breeding, integrated pest management, precision farming and multiyear nutrient management.

Agricultural and social science research and education offer examples of diverse partnerships with potential to advance public interest science and increase its relevance to equitable and sustainable development goals. A range of knowledge, science and technology partnerships among corporate actors in the agricultural and food industries, consumer organizations, NGOs, social movements and farmer organizations have pioneered ecologically and socially sustainable approaches to food and agriculture. Experience suggests that effective and enforceable codes of conduct can strengthen multi-organizational partnerships, preserve public institutions' capacity to perform public-good research and mobilize private commercial capacity to serve sustainability and development goals.

2.4.2 AKST and education
The ability of farmers and other actors to collaborate effectively in demand-pull partnership arrangements for the generation and implementation of AKST critically depends on the quality of the formal and informal education available to them. Basic and occupational education also empowers individuals and communities to drive the evolution of farming and build agroenterprises, adapt to new job opportunities and be better prepared for migration if necessary. Over the past decades various education and extension programs have enhanced farmers' education through the integration of formal and informal AKST. Generally the most effective have built on local and indigenous knowledge and innovation systems, typically through participatory and experiential learning processes and multi-organizational partnerships. Proven options include but are not limited to experiential learning groups, 4-H clubs, farmer field schools, farmer research circles, participatory plant breeding, social forestry and related community-based forest landscape

management, study clubs and community interaction with school-based curriculum development. Their gains at local levels often are undermined by higher level interests and by economic drivers.

Measures that remove or mitigate race, ethnic and gender biases that hamper the participation in educational opportunity of marginalized community members, diverse ethnic groups and women have been essential for local progress toward social equity but have not been widely adopted. Investment in the education and training of government policymakers and public agency personnel, particularly in decentralized participatory planning and decision-making and in understanding how to work effectively with rural communities and other stakeholders has also proven effective in promoting progress toward combined sustainability and development goals; broader issues of governance remain a concern.

More generally, experience shows that investment in science-informed, farmer-centered learning and in other rural actors' educational needs develops grassroots capacity to critically assess, define and engage in positive locally-directed development and the sustainable management of their environment. Modern ICTs are beginning to open up new and potentially powerful opportunities for extending the reach and scope of educational and interactive learning opportunities. Extension and advisory services complement but do not substitute for rural and occupational education.

2.4.3 Public policy and regulatory frameworks
International agreements informed by scientific evidence and public participation have enabled decisive and effective global transitions toward more sustainable practices (for example, the Montreal Protocols, the Rotterdam and Stockholm Conventions, the FAO Code of Conduct, the EU Thematic on Sustainable Agriculture). However, new national, regional and international agreements will be needed to support further shifts towards ethical, equitable and sustainable food and agriculture systems in response to the urgent challenges such as those posed by the declining availability of clean water and competing claims of water, loss of biodiversity, deforestation, climate change, exploitative labor conditions.

Awareness of the importance of ensuring full and meaningful participation of multiple stakeholders in international and public sector AKST policy formation has increased over the period. For example, in some countries, pesticide policies today are developed by diverse group of actors including civil society and private sector actors, informed by science and empirical evidence and inclusive of public interest concerns. These policies—exemplified by the 2007 European thematic on IPM—focus on the multifunctionality of agriculture.

Three thematic narratives on the management of germplasm, pests and food systems illustrate the role of public policy and regulatory frameworks as key drivers of AKST.

- The number and diversity of actors engaged in the **management of germplasm** has declined over time, driven in large part by advancements in science, privatization of seed supply and more widespread recourse to various intellectual property regimes. This trend reduces the options available for responding to uncertainties in the future. It increases asymmetries in access to germplasm and increases the vulnerabilities of the poor. Participatory plant breeding provides strong evidence that diverse actors can engage in an effective practice for achieving and sustaining the broader goals of sustainability and development by bringing together the skills and techniques of advanced and conventional breeding and farmers' preferences and germplasm management capacities and skills, including seed production for sale. Further development and expansion would require adjustment of varietal release protocols and appropriate policy recognition under UPOV 1991.

- The debates surrounding the use of synthetic pesticides have led to new arrangements that have increased awareness, availability and effectiveness of the range of options for **pest management.** Institutional responses to evidence of harm caused by certain synthetic chemicals in actual conditions of use include the strengthening of regulatory controls over synthetic chemical pesticides at global and national levels, growing consumer and retail markets for pesticide-free and organic products, removal of highly toxic products from sale, development of less acutely toxic products and more precise means of delivery and education of users in safe and sustainable practices. What constitutes safe and sustainable practice has been defined in widely varying ways by different actors reflecting different conditions of use as well as different assessments of acceptable tradeoffs, between crop security, productivity and economic gain on the one hand and health and environmental protections on the other.

IPM exemplifies a flexible and wide-reaching arrangement of actors, institutions and AKST practices that better address the needs of diverse farmers and a more broadly acceptable balance of interests. Although definitions, interpretations and outcomes of IPM programs vary widely among actors, IPM typically incorporates KST from a broad range of sciences, including social sciences, and the experience and knowledge of a diverse set of actors. IPM has become standard practice in a number of high value production systems and has been adopted also by an increasing number of important commercial actors in food processing and retailing. Successful approaches to introducing IPM to small-scale producers in the tropics include farmer field schools, push-pull approaches, advisory services provided under contractual arrangements for supply to central processing facilities and creative use of communication tools such as short farmer-to-farmer videos and focused-message information campaigns. A combination of such approaches, backed by strong policy reform to restrict the sale of old-fashioned and highly toxic synthetic controls, will be needed to meet future development and sustainability goals. Further experimentation and operational fine-tuning of the institutional arrangements for IPM in the field in different settings is also needed. These can be evaluated by comparative assessment using a combination of social, environmental and economic measures that include both positive and negative externalities.

- **Food systems** have changed fundamentally over the last decades. Local food systems, known to sustain livelihoods at micro level, are currently challenged by globalized food systems that are evolving to meet urban demands. This trend brings opportunities but also threatens livelihoods and sovereignties of marginalized communities and indigenous peoples. Evidence based research has shown that social, ethical and cultural values in some countries can be integrated in the commercial mechanisms driving the evolution of food systems. Fair trade, territorial identities and ethnic labeling are among the options that can be considered by decision makers who wish to promote effective measures to protect the interests of the marginalized and revitalize rural livelihoods and food cultures. The promotion of geographic indicators can open development opportunities based on local resources and knowledge. They also offer opportunities for new agroenterprises such as tourism and specialty product development, as well as for collaboration with utilities such as water companies. Substantial evidence shows that production systems dominated by export markets can be weakened by erratic changes and price instability on international markets. Export-oriented food systems have sparked growing concern about the sustainability of long-distance food shipping and about the ecological footprint and social impacts of international trade in food products and agricultural commodities. Local consumption and domestic outlets for farmers' products, often enhanced by the desire to sustain cultural identities associated with the consumption of products identified with their territorial origin, can alleviate the risks for food security and food sovereignty inherent in international trade.

References

Abidin, P.E., F.A. van Eeuwijk, P. Stam, P.C. Struik, M. Malosetti, R.O.M. Mwanga et al. 2005. Adaptation and stability analysis of sweet potato varieties for low-input systems in Uganda. Plant Breed. 124 (5):491-497.

Adipala, E., P. Nampala, J. Karungi, and P. Isubikalu. 2000. A review on options for management of cowpea pests: Experiences from Uganda. Integr. Pest Manage. Rev. 5:185-196.

Agarwal, B.D. 1979. Maize on-farm research project. 1979 Report. G.B. Pant Univ. Agric. Tech., Uttar Pradesh, India.

Agrow World Crop Protection News. 2005. No. 474. June 17. Available at http:// .agrow.co.uk.

Agunga, R. 1995. What Ohio extension agents say about sustainable agriculture. J. Sustain. Agric. 5(3):169-187.

Ahmed, F.E. (ed) 1991. Seafood safety. Inst. Med., Nat. Academy Press, Washington DC.

Ahmed, I., and V.W. Ruttan (ed) 1988.

Generation and diffusion of agricultural innovations: The role of institutional factors. Gower Publ., Aldershot.

AJEA. 2000. Special issue: Improving agricultural practices and decisions. Aus. J. Exp. Agric. 40(4):493-642.

Akca, H., M. Sayili, and A. Kurunc. 2005. Trade-off between multifunctional agriculture, externality and environment. J. Appl. Sci. Res. 1:298-301.

Al-Babili, S., and P. Beyer. 2005. Golden rice — five years on the road, five years to go? Trends Plant Sci. 12:565-573.

Albrecht, H., H. Bergmann, G. Diederich, E. Grosser, V. Hoffmann, P. Keller et al. 1989. Agricultural extension: Basic concepts and methods. Vol. 1. GTZ, Eschborn.

Albrecht, H., H. Bergman, G. Diederich, E. Grober, V. Hofman, P. Keller et al. 1990. Agricultural extension: Examples and background material. Vol. 2. BMZ/GTZ/CTA, Eschborn.

Alene, A.D., P. Neuenschwander, V. Manyong, O. Coulibaly, and R. Hanna. 2005. The impact of IITA-led biological control of major pests in sub-Saharan African agriculture: A synthesis of milestones and empirical results. IITA, Ibadan.

Alex, G. and D. Byerlee. 2001. Monitoring and evaluation for AKIS Projects: Framework and options. Available at http://siteresources. worldbank.NTARD/825826-11114006361 62/20431913/monitoringandeval.pdf. World Bank, Washington DC.

Allaire, G. 1996. Transformation des systèmes d'innovation: Réflexions à partir des Nouvelles fonctions de l'Agriculture p.1-29. In G. Allaire, B. Hubert and A. Langlet (ed) Nouvelles fonctions de l'Agriculture et de l'Espace Rural; Enjeux et défis identifiés par la recherche. INRA Ed., Toulouse.

Allegri, M. 2002. Partnership of producer and government financing to reform agricultural research in Uruguay. In D. Byerlee and R.G. Echeverria (ed) Agricultural research policy in an era of privatization. CABI Publ., Wallingford.

Almekinders, C., and N. Louwaars. 1999. Farmers' seed production. In New approaches and practices. Intermediate Tech. Publ., London.

Almekinders, C., and A. Elings. 2001. Collaboration of farmers and breeders: Participatory crop improvement in perspective. Euphytica 122:525-438.

Almekinders, C., and J. Hardon (ed) 2006. Bringing farmers back into breeding. Experiences with participatory plant breeding and challenges for institutionalisation. AgroSpecial No. 5. Agromisa, Wageningen.

Alston, J.M., and P.G. Pardey.1996. Making science pay: The economics of agricultural R&D policy. Am. Enterprise Inst., Washington DC.

Altieri, M., 1985. Developing pest management strategies for small farmers based on traditional knowledge. Bull. Inst. Dev. Anthropol. 3:13-18.

Altieri, M. 1987. Agroecology: The scientific basis of alternative agriculture. Westview Press, Boulder.

Altieri, M. 1993. Ethnoscience and biodiversity: key elements in the design of sustainable pest management systems for small farmers in developing countries. Agric. Ecosyst. Environ. 46:257-272.

Altieri, M. 1999. Applying agroecology to enhance the productivity of peasant farming systems in Latin America. Environ. Dev. Sustain. 1:197-217.

Altieri, M. and D.K. Letourneau. 1982. Vegetation management and biological control in agroecosystems. Crop Prot. 1:405-430.

Altieri, M., and O. Masera. 1993. Sustainable rural development in Latin America: Building from the bottom up. Ecol. Econ. 7:93-121.

Altieri, M., P.M. Rosset and C.I. Nicholls. 1997. Biological control and agricultural modernization: Towards resolution of some contradictions. Agric. Human Values 14:303-310.

Altieri, M. and C. Nicholls. 2004. Biodiversity and pest management in agroecosystems. Food Products Press, NY.

Amalu, U.C. 2002. Food security: Sustainable food production in sub-Saharan Africa. Outlook Agric. 31:177-185.

Amanor, K.K., W. de Boef, and A. Bebbington. 1993. Genetic diversity, farmer experimentation and crop research. p. 1-13. In W. de Boef et al. (ed) Cultivating knowledge. Intermediate Tech. Publ., London.

Amilien, V. 2005. Preface: About local food. Anthropol. Food 4.

Amin, M. and Mabe, M. 2000. Impact factors: Use and abuse. Perspect. Publ. 1:1–6.

Anderson, J.R., G. Feder, and S. Ganguly. 2006. The rise and fall of training and visit (T&V) extension: An Asian mini-drama with an African epilogue. p.149-174. In A. Van den Ban and R.K. Samanta (ed) 2006. Changing roles of agricultural extension in Asian nations. B.R. Publ., Delhi.

Anderson, M. 1990. Farming with reduced synthetic chemicals in North Carolina. Am. J. Alternative Agric. 5(2):60-68.

Anderson, R.A., E. Levy, and B.M. Morrison.1991. Rice science and development politics. Research strategies and IRRI's technologies confront Asian diversity (1950-1980). Clarendon Press, Oxford.

Annerstedt, J. 1994. Measuring science, technology and innovation. In J.-J. Salomon, et al. (ed) The uncertain quest: Science, technology and development. UN Univ. Press, Tokyo.

Antle, J.M., D.C. Cole, and C.C. Crissman. 1998. Further evidence on pesticides, productivity and farmer health: Potato production in Ecuador. Agric. Econ. 2(18):199-208.

Appleton, H. (ed) l995. Do it herself: Women and technical innovation. Intermediate Tech. Publ., London.

Aranjo, A.C.P. and Telles, D.L. 1999. Endosulfan residues in Brazilian tomatoes and their impact on public health and the environment. Bull. Environ. Contamin. Toxicol. 62:671-676.

Araya, A. 2000. Masipag farmers bring nature back to farming. Available at http://www. cyberdyaryo.com/features/f2000_0620_01 .htm. CyberDyaryo.

Arnold, E. and M. Bell. 2001. Some new ideas about research for development. p. 279-316. In Partnership at the leading edge: A Danish vision for knowledge, research and development. Danish Min. Foreign Affairs, Copenhagen.

Asante, S.K., and M. Tamo. 2001. Integrated management of cowpea insect pests using elite cultivars, date of planting and minimum insecticide application. Afr. Crop Sci. J. 9(4):655-665.

Ascroft, J., N. Röling, J. Kariuki, and F. Wa Chege. 1973. Extension and the forgotten farmer. Bull. van de Afdelingen Soc. Wetenschappen. No. 37. Wageningen Landbouwhoseschool, Wageningen.

Ashby, J. 1986. Methodology for the participation of small farmers in the design of on-farm trials. Agric. Admin. Extension 22(1):1-19.

Ashby, J. 2003. Introduction: Uniting science and participation in the process of innovation - research for development. p. 1-19. In B. Pound et al. (ed) Managing natural resources for sustainable livelihoods. Uniting science and participation. Available at http://www.idrc.ca/ en/ev-43431-201-1-DO_TOPIC.html. IDRC, Earthscan, UK.

Ashby, J.A., and N. Lilja. 2004. Participatory research: Does it work? Evidence from participatory breeding. In New directions for a diverse planet. Proc. 4th Int. Crop Sci. Cong., 26 Sep - 1 Oct 2004, Brisbane. [CDROM]. Web site www.cropscience.org.au.

Aslam, A. 1996. Environment-food: Groups blast World Bank pesticides policy. [Online]. Available at www.link.no/ips/eng/intro.html. Inter Press Service, Washington DC.

Atanassov, A. 2004. To reach the poor. Next harvest study on genetically modified crops, public research and policy implications. EPTD Disc. Pap. 116. IFPRI, Washington DC.

Atchoarena, D., and L. Gasperini. 2002. Education for rural development: Towards new policy responses. FAO/IIEP-UNESCO, Paris.

Atkinson, R.C., R.N. Beachy, G. Conway, F.A. Cordova, M.A. Fox, K.A. Holbrook et al. 2003. Intellectual property rights: Public sector collaboration for IP management. Science 301:174-175.

Atlin, G.N., M. Cooper, and Å. Bjørnstad. 2001. A comparison of formal and participatory breeding approaches using selection theory. Euphytica 122:463-475.

Atteh, O.D. 1984. Nigerian farmers´ perception of pests and pesticides. Insect Sci. Appl. 5:213-220.

Ayalew, M., 1997. What is food security and famine and hunger. Internet J. Afr. Studies. Issue No. 2, Mar. Available at www.brad. ac.uk/research/ijasno2/ayalew.html.

BAAS (British Assoc. Advancement of Science). 1999. Evidence to the House of Lords Select

Committee on Science and Technology inquiry into society's relationship with science. Available at http://www.publications. parliament.uk/pa/ld199900/ldselect/ ldsctech/38/3801.htm. UK Parliament, London.

Badgley, C., J. Moghtader, E. Quintero, E. Zakem, J. Chappell, K. Avilés-Vázquez et al. 2007. Organic agriculture and the global food supply. Renewable Agric. Food Syst. 22:86-108.

Baerselman, F. 1992. The Dutch multi-year crop protection plan (MJP-G): A contribution towards sustainable agriculture. p. 141-147. In J.C. van Lenteren et al. (ed) Biological control and integrated crop protection: Towards environmentally safer agriculture. Proc. Int. Conf. organized by IOBC/WPRS. Veldhoven, Netherlands. 8-13 Sept. 1991. PUDOC, Wageningen.

Baker, B.P., C.M. Benbrook, E. Groth, and K. Lutz. 2002. Pesticide residues in conventional, integrated pest management (IPM)-grown and organic foods: Insights form three US data sets. Food Additives Contamin. 19:427-446.

Ball, J., J. Carle, and A. Del Lungo. 2005. Contribution of poplars and willows to sustainable forestry and rural development. Unasylva 56:3-9.

Balasubramanian, A.V., and T.D. Nirmala Devi (ed) 2006. Traditional knowledge systems of India and Sri Lanka. Center Indian Knowledge Syst., Chennai.

Barbier, M., and B. Lémery. 2000. Learning through processes of change in agriculture: A methodological framework. p. 381-393. In M. Cerf et al. (ed) Cow up a tree: Knowing and learning for change in agriculture, case studies from industrialised countries. INRA Ed., Paris.

Barbottin M., M. Le Bail, and M.H. Jeuffroy, 2006. Evaluation of the Azodyn crop model for cultivar decision support. Agron. Sustain. Dev. 26:107-115.

Barnett, A. 2004. From 'research' to poverty reducing 'innovation'. Sussex Res. Assoc., Brighton.

Barrett, B., P. Marenya, J. McPeak, B. Minten, F. Murithi, W. Oluoch-Kosura et al. 2006. Welfare dynamics in rural Kenya and Madagascar. J. Dev. Studies 42(2)248-277.

Barsky, O. 1984. Acumulación campesina en el Ecuador: Los productores de papa del Carchi. Facultad Latinoamericana de Ciencias Sociales, Quito.

Barzman, M., and S. Desilles. 2002. Diversifying rice-based systems and empowering farmers in Bangladesh using the farmer field school approach. Chapter 16. In N. Uphpoff (ed) Agroecological innovations: Increasing food production with participatory development. Earthscan, London.

Baum, W. 1986. Partners against hunger: CGIAR, World Bank, Washington DC.

Bavec, F., and M. Bavec, 2006. Organic production and use of alternative crops. CRC Press, Atlanta.

Bawden, R. 1995. On the systems dimension of FSR. J. Farming Syst. Res. Extension 5(2):1-19.

Bawden, R.J. and, R. Packam. 1993. Systemic praxis in the education of the agricultural systems practitioner. Systemic Practice Actions Res. 6:7-19.

Beal, G.M., W. Dissanayake, and S. Konoshima (ed) 1986. Knowledge generation, exchange and utilization. Westview Press, Boulder.

Bebbington, A., and G. Thiele (ed) 1993. Non-governmental organizations and the State in Latin America: Rethinking roles in sustainable agriculture development. Routledge, London.

Bebbington, A., D. Lewis, S. Batterbury, E. Olson, and M.S. Siddiqi. 2007. Of texts and practices: Empowerment and organisational cultures in World Bank-funded rural development programmes. J. Dev. Studies 43(4):597-621.

Beck, U., A. Giddens, and S. Lash. 1994. Reflexive modernization. Politics, tradition and aesthetics in the modern social order. Polity Press, Cambridge.

Beintema, N., and G-J. Stads. 2006. Agricultural R&D investments in sub-Saharan Africa: An era of stagnation. IFPRI, Washington DC.

Bell, R.W. 2004. Importance of micronutrients in crop nutrition. IFA Int. Symp. on Micronutrients New Delhi. 23-25 Feb 2004. Int. Fertilizer Industry Assoc.

Bellon, M.R. 1996. The dynamics of crop infraspecific diversity: A conceptual framework at the farmer level. Econ. Bot. 50:26-39.

Bellon, M.R., J.L. Pham, and M.T. Jackson. 1997. Genetic conservation: a role for rice farmers. p. 263-289. In N. Maxted et al. (ed) Plant conservation: The in situ approach. Chapman and Hall, London.

Bellows, T.S., and T.W. Fisher (ed) 1999. Handbook of biological control. Academic Press, San Diego.

Benbrook, C. 2004. Genetically engineered crops and pesticide use in the United States: The first nine years. BioTech Infonet Tech. Pap. 7. Available at http://www.biotech-info.net/ Full_version_first_nine.pdf.

Benoit-Cattin, M. 1991. Ideological and institutional antecedents of the new agricultural policy. p. 59-68. In C. Delgado, S. Jammeh (ed) The political economy of Senegal under structural adjustment. Praeger, NY.

Benor, D., J.Q. Harrison, and M. Baxter. 1984. Agricultural extension: The training and visit system. The World Bank, Washington DC.

Bentley, J.W. 1994. Fact, fantasies and failures of farmers participatory research. Agric. Human Values 11(2-3):140-150.

Bérard, L., and P. Marchenay. 1995. Tradition, regulation and intellectual property: local agricultural products and foodstuffs in France. p. 230-243. In S.B. Brush, D. Stabinsky (ed) Valuing local knowledge. Indigenous people and intellectual property rights. Island Press, Washington DC.

Berdahl, R. 2000. The privatization of public universities. Lecture presented at Erfurt Univ. Erfurt, Germany. 23 May 2000. Available at http://cio.chance.berkeley.edu/chancellor/sp/ privatization.htm. Univ. California Regents, Berkeley.

Berger, P.L., and T. Luckmann. 1966. The social construction of reality: A treatise in the sociology of knowledge. Doubleday, NY.

Berkes, F., and C. Folke (ed) 1998. Linking social and ecological systems. Management practices and social mechanisms for building resilience. Cambridge Univ. Press, UK.

Bern, C., B. Hernandez, M.B. Lopez, M.J. Arrowood, M. Alvarez de Mejia, A.M. De Merida et al. 1999. Epidemiologic studies of Cyclospora cayetanensis in Guatemala. Emerging Infect. Dis. 5:766-74.

Bettles, F. 1980. Women's access to agricultural extension services in Botswana. Min. Agric., Dept. of Land Services, Gaberone.

Bhowmik Prasanta, C. 1999. Herbicides in relation to food security and environment: A global perspective. Ind. J. Weed Sci. 31(3,4):111.

Biggs, M. 1990. The impact of peer review on intellectual freedom. Library Trends 39(1-2):145-167.

Biggs, S.D. 1978. Planning rural technologies in the context of social structures and reward systems. J. Agric. Econ. 24(3):257-274.

Biggs, S.D. 1980. On-farm research in an integrated agricultural technology development system: A case study of triticale for the Himalayan Hills. Agric. Admin. 7(2):133-145.

Biggs, S.D. 1982. Generating agricultural technology: Triticale for the Himalayan Hills. Food Policy 7(1):69-82.

Biggs, S.D. 1983. Monitoring and control in agricultural research systems: Maize in Northern India. Res. Policy 12:37-59.

Biggs, S.D. 1989. Resource-poor farmers participation in research: A synthesis of experiences from nine national agricultural research systems. OFCOR Comparative Study Pap. 3. ISNAR, The Hague.

Biggs, S.D. 1995. Participatory technology development: A critique of the new orthodoxy. AVOCADO Series 06/95. Olive Org. Dev. Training, Durban.

Biggs, S. 2006. Learning from the positive to reduce rural poverty: Institutional innovations in agricultural and natural resources research and development. Impact Assessment Workshop. 19-21 Oct 2005. CIMMYT, Mexico.

Biggs, S., and J. Farrington. 1991. Agricultural research and the rural poor. A review of social science analysis. IDRC. Ottawa.

Bigler F., D. Babendreier, and U. Kuhlmann. (ed) 2006. Environmental impact of invertebrates for biological control of Arthropods: Methods and risk assessment. CABI, Wallingford.

Bills, P.S., D. Mota-Sanchez, and M. Whalon. 2003. Background to the resistance database.

Available at www.cips.msu.edu/resistance. Michigan State Univ., East Lansing MI.

Bindraban, P., and R. Rabbinge. 2005. Development perspectives for agriculture in Africa: Technology on the shelf is inadequate. North-South Disc. Pap. WUR, Wageningen.

Bingen, J. 2003. Pesticides, politics and pest management: Toward a political ecology of cotton in Sub-Saharan Africa. p. 111-126. *In* W.G. Moseley and B.I. Logan (ed) African environment and development: Rhetoric, programs, realities. Ashgate Publ., Aldershot, UK.

Bingen, R.J., and D.W. Brinkerhoff. 2000. Agricultural research in Africa and the Sustainable Financing Initiative. Tech. Pap. 112. Office Sustain. Dev. Bur. Africa. USAID, Washington DC.

Blackburn, D.J. (ed) 1994. Extension handbook: Processes and practices for change professionals. Univ. Guelph, Ontario.

Blackmore, C.P., R. Ison and J. Jiggins (ed) 2007. Social learning: an alternative policy instrument for managing in the context of Europe's water. Special Issue. Environ. Sci. Policy 10(6):493-586.

Blandford, D., and R.N. Boisvert. 2002. Multifunctional agriculture and domestic/international policy choice. Estey Centre J. Int. Law Trade Pol. 3:106-118.

Blobaum, R. 1983. Barriers to conversion to organic farming practices in the Midwestern United States. p. 263-278. *In* W. Lockeretz (ed) Environmentally sound agriculture. Praeger Publ., NY.

Blobaum, R. 1997. Publicly funded models supporting sustainable agriculture: Four model profiles. Chapter 12. p. 193-203. *In* J. Madden, S. Chaplowe (ed) For all generations: Making world agriculture more sustainable. OM Publ., Glendale CA.

Blokker, K., S. Bruin, J. Bryden, I. Houseman, C. Okkerse, C. Van Der Meer et al. 1990. Agricultural policy and strategic investment in information technology. Knowled. Tech. Policy 3(3):76-83.

BOA (Board on Agriculture). 1995. Colleges of agriculture in the land grant universities: A profile. Nat. Academy Press, Washington DC.

Boettner, G., J. Elkinton, and C. Boet. 2000. Effects of a biological control introduction on three non-target native species of saturnid moths. Conserv. Biol. 14:1998-1806.

Bok, D. 2003. Universities in the marketplace. The commercialization of higher education. Princeton Univ. Press, Princeton NJ.

Bokonon-Ganta, A.H., H. de Groote, and P. Neuenschwander. 2002. Socio-economic impact of biological control of mango mealybug in Benin. Agric. Ecosyst. Environ. 93:367-378. Available at http://www2.hawaii.edu/~messing/Publications/Impact%20mango%20mealybug%20biocontrol.pdf

Bonneuil, C., and E. Demeulenaere. 2007. Vers une génétique de pair à pair ? L'émergence de la sélection participative. p. 122-147. *In* F. Charvolin et al. (ed) Des sciences

citoyennes? La question de l'amateur dans les sciences naturalistes. Available at http://www.environnement.ens.fr/perso/demeu/pdfs/BONNEUIL_DEMEULENAERE2007_in_Charvolin.pdf. Ed. de l'Aube, La Tour d'Aigues.

Bookchin, M. 1990. Remaking society: Pathways to a green future. South End Press, Boston.

Bordons, M., and I. Gómez. 2002. Advantages and limitations in the use of impact factor measures for the assessment of research perfomance in a peripherical country. Scientometrics 53:195-206.

Borrini-Feyerband, G., M. Pimbert, M.T. Farvar, A. Kothari, and Y. Renard. 2004. Sharing power. Learning by doing in co-management of natural resources throughout the world. IIED and IUCN/CEESP/CMWG, Paris.

Bourdieu, P. 1988. Homo academicus. Stanford Univ. Press, CA.

Borlaug, N.E., and C.R. Dowswell. 1995. Mobilising science and technology to get agriculture moving in Africa. Dev. Policy Rev. 13(2):115-29.

Boven, K. and J. Mordhashi (ed) 2002. Best practices using indigenous knowledge. Available at http://www.unesco.org/most/Bpikpub2.pdf. NUFFIC, The Hague and UNESCO/MOST, Paris.

Boyce, J.K., P. Rosset, and E.A. Stanton. 2005. Land reform and sustainable development. Working Pap. Ser. 98. Available at http://www.peri.umass.edu/fileadmin/pdf/working_papers/working_papers_51-100/WP98.pdf. Political Econ. Res. Inst., Univ. Massachusetts, Amherst.

Braun, A. 2006. Inventory of farmer study group initiatives in Africa. A synthesis report. ENDELEA (SIDA), Wageningen.

Braun, A.R., J. Jiggins, N. Roling, H. van den Berg, P. Snijders. 2006. A global survey and review of farmer field school experiences. Available at http://www.infobridge.org/asp/documents/1880.pdf. ILRI (ENDELEA), Wageningen.

Brennan, M., C. Pray, A. Naseem, and J.F.Oehmke. 2005. An innovation market approach to analyzing impacts of mergers and acquisitions in the plant biotechnology industry. AgBioForum 8:89-99.

Breusers, M. 1998. On the move. Mobility, land use and livelihood practices on the Central Plateau in Burkina Faso. Publ. PhD thesis. Agric. Univ., Wageningen.

Brewer, M.J., R.J. Hoard, J.N. Landis, and L.E. Elworth. 2004. The case and opportunity for public-supported financial incentives to implement integrated pest management. J. Econ. Entomol. 97(6):1784.

Brewster, A.L., A.R. Chapman, and S.A. Hansen. 2005. Facilitating humanitarian access to pharmaceutical and agricultural revolution. Innovation Strategy Today 1(3):203-216.

BRF (Federal Institute for Risk Assessment of Germany). 2004. The return of the germs. Available at http://www.bgvv.de/cms5w/sixcms/detail.php/4217.

Bridges, D. (ed) 1992. Crop losses due to weeds in Canada and the United States. Weed Sci. Soc. Amer., Champaign IL.

Brimblecombe, P. and C. Pfister. 1993. The silent countdown: Essays in European environmental history. Springer-Verlag, Berlin.

Briolini, G.1992. IPM in Italy with particular reference to the Emilia-Romagna region. p. 181-186. *In:* J.C. van Lenteren et al. (ed) Biological control and integrated crop protection: Towards environmentally safer agriculture. Proc. Int. Conf. Veldhoven, Netherlands. 8-13 Sept. 1991. PUDOC, Wageningen.

Broad, R. 2006. Research, knowledge, and the art of paradigm maintenance: The World Bank's Development Economics Vice-Presidency (DEC). Rev. Int. Politic. Econ. 13(3):387-419.

Brouwer, F. (ed) 2004. Sustainable agriculture and the rural environment: Governance, policy and multifunctionality. Edward Elgar Publ., Northampton MA.

Brouwer, J., L. Fussell and L. Hermann. 1993. Soil and crop growth micro-variability in the West African semi-arid tropics: A possible risk-reducing factor for subsistence farmers. Agric. Ecosyst. Environ. 45:229-238.

Brouwers, J.H.A.M. 1993. Rural people's response to soil fertility decline: The Adja case (Benin). Publ. PhD thesis. WAU Pap. 93/94. Wageningen Univ., Wageningen.

Brown, A.H.D. 2000. The genetic structure of crop landraces and the challenge to conserve them in situ on farms. p. 29-48. *In* S.B. Brush (ed.) Genes in the field: On-farm conservation of crop diversity. Int. Dev. Res. Center Plant Genetic Resour. Inst. Lewis Publ., US.

Brown, B.J., and G.G. Marten. 1986. The ecology of traditional pest management in southeast Asia. Working Pap., East-West Center, Honolulu.

Bruinsma, J. (ed) 2003. World agriculture: Towards 2015/2030 — An FAO Perspective. FAO, Rome.

Brush S., 1991. A farmer-based approach to conserving crop germplasm. Econ. Bot. 45:153-165.

Brush, S.B. 2003. The demise of 'common heritage' and protection for traditional agricultural knowledge. Available at http://law.wustl.edu/centeris/Confpapers/PDFWrdDoc/StLouis1.pdf. Washington Univ., St. Louis.

Brush, S.B. 2005. Farmer's rights and protection of traditional agricultural knowledge. CAPRi Working Pap. 36. Available at http://www.capri.cgiar.org. CGIAR, Washington DC.

Brush, S.B, M.B. Corrales, and E. Schmidt. 1988. Agricultural development and maize diversity in Mexico. Human Ecol. 16(3):307-328.

Buck, L. 1990. Planning agro-forestry extension projects: The CARE international approach in Kenya. p. 101-131. *In* W.W. Budd et al. (ed) Planning for agroforestry. Elsevier, Amsterdam.

Buck, L.E., J.P. Lassoie, E.C.M. Fernandes (ed) 1998. Agroforestry in sustainable agricultural systems. Lewis Publ., CRC Press, Boca Raton.

Buck, L.E., C.G. Geisler, J.W. Schelhas, and E. Wollenberg (ed) 2001. Biological diversity: Balancing interests though adaptive collaborative Management. CRC Press, Boca Raton.

Budd, W.W., I. Duchart, L. Hardesty, and F. Steiner. 1990. Planning for agroforestry. Elsevier, Amsterdam.

Bunch, R. 1982. Two ears of corn: A guide to people-centered agricultural improvement. World Neighbors, Oklahoma City.

Bunch, R., and G.V. Lopez. 1994. Soil recuperation in Central America: Measuring the impact four and forty years after intervention. IIED, London.

Bunyatta, D.K., J.G. Mureithi, C.A. Onyango, and F.U. Ngesa. 2005. Farmer field school as an effective methodology for disseminating agricultural technologies: Up-scaling of soil amange,ent technoogies in Trans-Nzoia District, Kenya. p. 515-526. In Proc. 21st Ann. Conf. Assoc. Int. Agric. Ext. Educ. San Antonio TX.

Busch, L., R. Allison, C. Harris, A. Rudy, B.T. Shaw, T.T en Eyck et al. 2004. External review of the collaborative research agreement between Novartis Agricultural Discovery Institute, Inc. and The Regents of the Univ. California. Inst. Food Agric. Standards, Michigan State Univ., East Lansing.

Buvinic, M., and N. Youssef. 1978. Women headed households: The ignored factor in development planning. Rep. Off. Women Dev., USAID, Washington DC.

Bye, R.A. 1981. Quelities: ethnoecology of edible greens — past, present and future. J. Ethnobiol. 1:109-114.

Byerlee, D. 1992. Technical change, productivity and sustainability in irrigated cropping systems of South-Asia: Emerging issues in the post-green revolution era. J. Int. Dev. 4(5):477-496.

Byerlee, D. 1998. The search for a new paradigm for the development of national agricultural research systems. World Dev. 26(6):1049-1055.

Byerlee, D., and G.E. Alex. 1998. Strengthening national agricultural research systems, policy issues and good practice. World Bank, Washington DC.

Byerlee, D., and R. Echeverria (ed) 2002. Agricultural research policy in an era of privatization. CABI Publ., UK.

Cahill, C. 2001. The multifunctionality of agriculture: What does it mean? EuroChoices 1(1):36-41.

Calvora, R.G. 1988. Science in its confrontation with society. Impact of Sci. Society 38(151):231-238.

Campbell, A. 1992. Taking the long view in tough times: Landcare in Australia. Third Annual Rep. Nat. Landcare Facilitator. Nat. Soil Conserv. Prog., Canberra.

Campbell, A. 1994. Landcare. Communities shaping the land and the future. Allan and Unwin, St. Leonards, Australia.

CAP (Center for American Progress)/OMB Watch. 2004. Special interest takeover: The Bush administration and the dismantling of public safeguards [Online]. Available at http://www.sensiblesafeguards.org/pdfs/finalreport.pdf. CAP and OMB Watch, Washington DC.

Carloni, A. 1983. Integration of women in agricultural projects. Case studies of ten FAO-assisted field projects. FAO, Rome.

Carney, J. (ed) 1993. Low input sustainable agriculture in Cuba. Special Issue. Agric. Human Values 10:3.

Carr, S.J. 1989. Technology for small sclae farmers in sub-Saharan Africa. Tech. pap. 109. The World Bank, Washington DC.

Carson, R. 1962. Silent Spring. Houghton Mifflin, Boston.

Castells, M. 1993. The university system: Engine of development in the new world economy. In A. Ransom et al. (ed) Improving higher education in developing countries. Econ. Dev. Inst. Seminar Ser. World Bank, Washington DC.

Castleman, B. and R. Lemen. 1998. The manipulation of international scientific organizations. Int. J. Occup. Environ. Health 4:53-55.

Cate, J.R., and M.K. Hinkle. 1994. Integrated Pest Management: The path of a paradigm. Nat. Audubon Soc., Washington DC.

Cattaneo, M.G., C. Yafuso, C.Schmidt, C. Huang, M.Rahman, C. Olson et al. 2006. Farm-scale evaluation of the impacts of transgenic cotton on biodiversity, pesticide use and yield. PNAS 103:7571-7576.

CBD. 1994. Convention on Biological Diversity, UNEP/CBD/94/1. Secretariat CBD, Montreal.

CEC (Comm. Environ.Cooperation). 2004. Maize and biodiversity: The effects of transgenic maize in Mexico. Available at http://www.cec.org/maize/.

Ceccarelli, S., S. Grando, R. Tutwiler, J. Baha, A.M. Martini, H. Salahieh et al. 2000. A methodological study on participatory barley breeding. I. Selection phase. Euphytica 111:91-104.

Ceccarelli, S., S. Grando, A. Amri, T.A. Asad, A. Benbelkacem, M. Ilarrabi et al. 2001a. Decentralized and participatory plant breeding for marginal environments. p.115-136. In D. Cooper et al. (ed) Broadening the genetic base of crop production. CABI, Wallingford.

Ceccarelli, S., S. Grando, E. Bailey, A. Amri, M. El-Felah, F. Nassif et al. 2001b. Farmer participation in barley breeding in Syria, Morocco and Tunisia. Euphytica 122:521-536.

Ceccarelli, S., D.A. Cleveland, and D. Soleri. 2002. Farmers, scientists and plant breeding. Integrating knowledge and practice. Univ. California Press, Santa Barbara.

Ceccarelli, S., S. Grando, M. Singh, M. Michael, A. Shikho, M. Al Issa et al. 2003. A methodological study on participatory barley breeding II. Response to selection. Euphytica 133:185-200.

Ceccarelli, S., and S. Grando. 2007. Decentralized-participatory plant breeding: An example of demand driven research. Euphytica 155:349-360.

Cernea, M.M., and A. Kassam (ed) 2006. Researching the culture of agri-culture: Social science research for international development. CABI, UK.

Cernea, M. 1994. Sociology, anthropology and development: An annotated bibliography of World Bank Publications 1975-1993. World Bank, ESD Mono. Ser. 3. Washington DC.

CGIAR. 2000. Equity, well-being and ecosystem health. Participatory research for natural resource management. CGIAR Prog. Participatory Res. Gender Analysis. CIAT, Cali.

CGIAR. 2001. Designing and implementing challenge programmes. Rep. CGIAR Interim Exec. Council. 30 Aug 2001. CGIAR, Washington DC.

CGIAR TAC. 2000. An evaluation of the impact of integrated pest management research at international agricultural research centres. FAO. Rome.

Chabi-Olaye, A., C. Nolte, F. Schulthess, and C. Borgemeister. 2005. Relationships of intercropped maize, stem borer damage to maize yield and land-use efficiency in the humid forest of Cameroon. Bull. Entomol. Res. 95:417-427.

Chacon, J.C., and S.R. Gliessman. 1982. Use of the non-weed concept in traditional tropical agroecosystems of south-eastern Mexico. Agroecosystems 8:1-11.

Chambers, R. 1974. Managing rural development. Scand. Inst. Afr. Studies, Uppsala.

Chambers, R. 1981. Rapid rural appraisal: Rationale and repertoire. Public Admin. Dev. 1:95-106.

Chambers, R. 1983. Rural development: Putting the last first. Longman, London.

Chambers, R., and M. Howes. 1979. Rural development: Whose knowledge counts? IDS Bull. 10(2).

Chambers, R., and B.P. Ghildyal. 1985. Agricultural research for resource-poor farmers: the farmer-first-and-last model. Agric. Admin. Ext. 20:1-30.

Chambers, R., and J. Jiggins. 1986. Agricultural research for resource-poor farmers: A parsimonious paradigm. IDS Disc. Pap. 220. Inst. Dev. Studies, Brighton (Sussex).

Chapagain, A.K., and A.Y. Hoekstra. 2003. Virtual water flows between nations in relation to trade in livestock ad livestock products. Value Water Res. Rep. Ser. 13. UNESCO-IHE, Delft.

Chaplowe, S.G. 1997a. CLADES: Latin American consortium on agroecology and sustainable development. p. 275-283. In S. Chaplowe, and J. Madden (ed) For all generations: Making world agriculture more sustainable. OM Publ., Glendale CA.

Chaplowe, S.G. 1997b. Non-governmental organizations (NGOs) in sustainable agriculture. p. 205-232. *In* J. Madden, S. Chaplowe (ed) For all generations: Making world agriculture more sustainable. OM Publ., Glendale CA.

Chaturvedi, S. 2008. Agricultural biotechnology and trends in IPR regime: Emerging challenges before developing countries. *In* D. Casle (ed) The role of intellectual property rights in biotechnology innovation. Center for Intellectual Protection Policy, Edward Elgar, Toronto.

Chaudhry, R., S.B. Lall, B. Mishra, and B. Dhawan. 1998. A foodborne outbreak of organophosphate poisoning: Indiscriminate use of organophosphates without public education on safety increases the potential threat of foodborne outbreaks of poisoning. Brit. Med. J. 317:268-269.

Chema, S., E. Gilbert, and J. Roseboom. 2003. A review of key issues and recent experiences in reforming agricultural research in Africa. Res. Rep. 24. ISNAR, The Hague.

Chiffoleau, Y., and F. Dreyfus. 2004. Cognitive styles and networks patterns: a combined approach of learning processes in sustainable agriculture. Proc. 6th IFSA Eur. Symp. Farming and rural systems research and extension. European farming and society in search of a new social contract learning to manage change. Int. Farming System Assoc. (IFSA), Vila Real, Portugal.

Chiwona-Karltun, L. 2001. A reason to be bitter. Cassava classification for the farmers' perspective. Publ. PhD thesis. Karolinska Institutet, Stockholm.

Chopra, M. 2004. Food security, rural development and health equity in Southern Africa. Equinet Disc. Pap. 22. Available at http://www.sarpn.org.za/documents/d0001067/index.php. Equinet Africa.

Christopolos, I., J. Farrington, and A.D. Kidd. 2000. Extension, poverty and vulnerability: Inception report of the study for the Neuchatel Initiative. Working Pap. 144. ODI, London.

Christou, P., and R. Twyman. 2004. The potential of genetically enhanced plants to address food insecurity. Nutr. Res. Rev. 17:23-42.

Christy, R., and L. Williamson (ed) 1992. A century of service: Land-grant colleges and their universities, 1890-1990. Transaction Publ., New Brunswick and London.

Chunyanuwat, P. 2005. Thailand Country Report. *In* Proc. Asia Workshop on the Implementation, Monitoring and Observance of the Int. Code of Conduct on the Distribution and Use of Pesticides. Available at: http://www.fao.org/docrep/008/af340e/af340e0l.htm#bm21. FAO, Rome.

CIAT. 2001. An exchange of experiences from South and South East Asia. Proc. Int. Symp. Participatory Plant Breeding and Participatory Plant Genetic Resource Enhancement. Pokhara, Nepal 1-5 May 2000. CIAT, Cali.

CIMMYT. 2002. New maize varieties offers economic lifeline for poor farmers. CIMMYT, Mexico.

CIPP. 2004. Agricultural biotechnology and intellectual property: A new framework. Centre for Intellectual Property Policy, Canada.

CIPR. 2002. Integrating intellectual property rights and development policy. Commiss. Intellectual Property Rights (CIPR), London.

City and County of San Francisco. 2007. City and County of San Francisco Municipal Environ. Code Ordinance 159-07, File Number 070411, Approved July 3, 2007. Municipal Code Corporation, FL. Available at http://www.municode.com/Resources/gateway.asp?pid=14134&sid=5

Clapp, J. 1997. Adjustment and agriculture in Africa: Farmers, the state and the World Bank in Guinea. St Martin's Press, NY.

Clark, B.R. 1983. The higher education system. Academic organization in cross-national perspective. Univ. California Press, Berkeley.

Clark, B.R. 1995. Places of inquiry: Research and advanced education in modern universities. Univ. California Press, Berkeley.

Cleveland, D.A., D. Soleri, and S.E. Smith. 1994. Do folk crop varieties have a role in sustainable agriculture? Incorporating folk varieties into the development of locally based agriculture may be the best approach. Bioscience 44:740-751.

Cleveland, D.A., and D. Soleri. 2002. Farmers, scientists and plant breeding: Integrating knowledge and practice. CABI Publ., UK.

Clover, J. 2003. Food security in Sub-Saharan Africa. Afr. Security Rev. 12(1):5-15.

CNN. 2002. Death sentence over Chinese poisonings. 30 Sept. Available at http://www.cnn.com/2002/WORLD/asiapcf/east/09/30/china.poison.

Cochrane, W.W. 1958. Farm prices, myth and reality. Univ. of Minn. Press, Minneapolis.

Cohen, J. 2005. Poorer nations turn to publicly developed GM crops. Nature Biotech 23:27-33.

Cohn, A., J. Cook, M. Fernandez, R. Reider, and C. Steward (ed) 2006. Agroecology and the struggle for food sovereignty in the Americas. Available at http://www.iied.org/pubs/pdf/full/14506IIED.pdf. Yale Sch. Forest. Environ. Studies, New Haven and IIED, London.

Colchester, M. 1994. Salvaging nature, indigenous peoples, protected areas and biodiversity conservation. DP 55. UNRISD, Geneva.

Cole, D.C., F. Carpio, and N. Leon. 2000. Economic burden of illness from pesticide poisonings in highland Ecuador. Pan Am. Rev. Public Health 8:196-201.

Cole, D.C., S. Sherwood, C. Crissman, V. Barrera, and P. Espinosa. 2002. Pesticides and health in highland Ecuadorian potato production: Assessing impacts and developing responses. Int. J. Occup. Environ. Health 8(3):182-190.

Cole, D.C., S. Sherwood, M. Paredes, L.H. Sanin, C. Crissman, P. Espinosa et al. 2007. Reducing pesticide exposure and associated neurotoxic burden in an Ecuadorian small

farm population. Int. J. Occup Environ. Health 13:281-289.

Collinson, M. (ed) 2000. A history of farming systems research. FAO, Rome and CABI Publ., Wallingford.

Combs, G.F., R.M. Welch, J.M. Duxbury, N.T. Uphoff, and M.C. Mesheim (ed) 1996. Food-based approaches to prevent micronutrient malnutrition: An international research agenda. CIIFAD, Ithaca NY.

Compton, J.L. (ed) 1989. The transformation of international agricultural research and development. Lynne Riener Publ., Boulder.

Conner, A.J., T.R. Glare, and J.P. Nap. 2003. The release of genetically modified crops into the environment. Part II. Overview of ecological risk assessment. Plant J. 33:19-46.

Conway, G. 1994. Sustainability in agricultural development: Trade-offs between productivity, stability and equitability. J. Farming Syst. Res. Extension 4(2):1-14.

Cook, S.D.N., and J.S. Brown. 1999. Bridging epistemologies: The generative dance between organizational knowledge and organizational knowing. Org. Sci. 10(4):381-400.

Cooke, B., and U. Kothari (ed) 2001. Participation: The new tyranny. Zed Books.

Costanza, R., R. d'Arge, R. de Groot, S. Farber, S. Grasso, B. Hannon et al. 1997. The value of the world's ecosystem services and natural capital. Nature 387:253-260.

Coulombe, S., J.F. Tremblay, and S. Marchand. 2004. Literacy scores, human capital and growth across fourteen OECD countries. Int. Adult Literacy Survey Mono. Ser. Statistics Canada, Ottawa.

Coutts, J.R., K. Roberts, F. Frost, and A. Coutts. 2005. The role of extension in building capacity - what works and why. Rep. r05/094. Available at http://www.rirdc.gov.au/reports/HCC/05-094.pdf. Rur. Ind. Res. Dev. Corp., Canberra.

Cox, G.W., and M.D. Atkins. 1979. Agricultural ecology. W.H. Freeman, San Francisco.

CropLife. 2003. Integrated Pest Management - the way forward for the plant science industry. Available at http://www.croplife.org/library/attachments/2312df21-ae73-4921-a54e-df06778bf4f5/5/IPM-The-Way-Forward-(Dec-2004).pdf. Croplife Int., Brussels.

CropLife. 2005a. The value of herbicides in US crop production: 2005 Update. Available at http://croplifefoundation.org/cpri_benefits_herbicides.htm. Croplife Int., Brussels.

CropLife. 2005b. The value of fungicides in US crop production: 2005 Update. Available at http://croplifefoundation.org/cpri_benefits_fungicides.htm. Croplife Int., Brussels.

CropLife. 2006. Creating opportunities for sustainable development. CropLife Int., Brussels, Belgium.

Crosson, P., and J.R. Anderson. 2002. Technologies for meeting future global demands for food. Disc. Pap. 02-02 Resources for the Future, Washington DC.

Crow, J.F. 1998. 90 years ago: The beginning of hybrid maize. Genetics 148:923-928.

Crozier, M., and E. Friedberg. 1980. Actors and systems, the politics of collective action. Chicago Univ. Press, IL.

CSPI. 2005. Food safety around the world. Available at http://www.elika.net/pub _articulos_i.asp?tipo=&articulo=137#abajo. CSPI, Washington DC.

Dalrymple, D. 1986. Development and spread of high-yielding varieties of wheat and rice in the less-developed nations. For. Agric. Econ. Rep. 95. USDA, Washington DC.

Dalrymple, D. 2006. Impure public goods and agricultural research: Toward a blend of theory and practice. Q. J. Int. Agric. 45:75-81.

Danaher, K. (ed) 1994. Fifty years is enough: The case against the World Bank and the IMF. Southend Press, Boston.

Darré, J.P. 1985. La parole et la technique: L'univers de pensée des éleveurs du Ternois. Ed. L'Harmattan, Paris.

Darré, J.P. 1995. L'invention des pratiques. Karthala, Paris.

Darré, J.P. 1999. La Production de connaissance pour l'action. INRA. Editions de la Maison des sciences de l'homme, Paris.

Dauber. R., M. Cain (ed) 1981. Women and technological change in developing countries. Westview Press, Denver.

Dayal, I., K. Mathur, M. Battacharaya. 1976. District Administration. Macmillan, New Delhi.

De Boef, W. 2000. Tales of the unpredictable. Learning about institutional frameworks that support farmer management of agro-biodiversity. Publ. PhD thesis. Wageningen Univ., Wageningen.

De Boef, W., K. Amanor, and K. Welland, with A. Bebbington. 1993. Cultivating knowledge. Genetic diversity, farmer experimentation and crop research. IT Publ., London.

De Grassi, A. 2003. Genetically modified crops and sustainable poverty alleviation in Sub-Saharan Africa: An assessment of current evidence. Available at http://www.twn.org. Third World Network Africa.

De Groote, H., O. Ajuonua, S. Attignona, R. Djessoub, and P. Neuenschwander. 2003. Economic impact of biological control of water hyacinth in Southern Benin. Ecol. Econ. 45(1):105-117.

De Haan, H., and N. Long (ed) 1997. Images and realities of rural life. Van Gorcum, The Netherlands.

de Janvry, A., and J.-J. Dethier. 1985. Technological innovation in agriculture - The Political economy of its rate and bias. CGIAR Study Pap. No. 1. Washington DC.

DeBach, P. 1964. Biological control of insect pests and weeds. Chapman and Hall, London.

DeBach, P. 1974. Biological control by natural enemies. Cambridge Univ. Press, UK.

DeBach, P. and D. Rosen. 1991. Biological control by natural enemies. 2nd ed. Cambridge Univ. Press, UK.

Deere, C.D., and M.L. de Leal. 1982. Women in Andean agriculture. ILO, Geneva.

DeGregori, T. 2001. Agriculture and modern technology: A defense. Iowa State Press and Blackwell Sci., Ames.

Delanty, G. 2001. Challenging knowledge. The university in the knowledge society. The Society for Res. into Higher Educ. Open Univ. Press, Buckingham.

Dent, D. 2005. Overview of agrobiologicals and alternatives to synthetic pesticides. In J. Pretty (ed) The pesticide detox: Towards a more sustainable agriculture. Earthscan, London.

Dent, J.B., and M.J. McGregor (ed) 1994. Rural and farming systems analysis. CABI, UK.

De Vries, B. 2000. Multifunctional agriculture in the international context: A review. Available at http://www.landstewardshipproject.org/ mba/MFAReview.pdf. The Land Stewardship Project.

Desneux, N., A. Decourtye, and J.M. Delpuech. 2007. The sublethal effects of pesticides on beneficial arthropods. Ann. Rev. Entomol. 52:81-106.

DFID. 2002. Better livelihoods for poor people: The role of agriculture. Available at http://www.dfid.gov.uk/Pubs/files/ agricultureconsult.pdf. DFID, London.

DFID. 2004. Concentration in food and retail chains. Available at http://dfid-agriculture-consultation.nri.org/summaries/wp13.pdf. DFID and IIED, London.

DFID. 2005. Growth and poverty reduction: The role of agriculture. Available at http://www .dfid.gov.uk/pubs/files/growth-poverty -agriculture.pdf. DFID, London.

DHC (UK Department of Health Compensation). 2001. Scheme for variant CJD victims announced. London, Oct. 1. Press release 2001/0457. DHC, London.

Dilts, R. 1999. Facilitating the emergence of local institutions: Reflections from the experience of the Community IPM Program in Indonesia. p. 50-65. Report of the APO Study meeting on the role of institutions in rural community development. Colombo. 21-29 Sept. 1998.

Dimitri, C., A. Effland, and N. Conklin. 2005. The 20th century transformation of U.S. agriculture and farm policy. Electronic Inform. Bull. No. 3, June 2005.

Dinham, B. 1993.The Pesticide Hazard: A Global Health and Environmental Audit. Zed Books, London.

Dinham, B. 1995. 10 years of FAO Code fails to reduce hazard problems. Available at http:// www.pan-uk.org/pestnews/pn28/pn28p6b .htm. Pesticide News No. 28.

Dinham, B. 2003. Growing vegetables in developing countries for local urban populations and export markets: problems confronting small-scale producers. Pest Manage. Sci. 59:575-582.

Dinham, B. 2005. Corporations and pesticides, p. 55-69. In J. Pretty (ed) Pesticide detox: Towards a more sustainable agriculture. Earthscan, London.

Dinham, B. 2007. Pesticide users at risk: Survey of availability of personal protective clothing when purchasing paraquat in China, Indonesia and Pakistan and failures to meet the standards of the Code of Conduct. Berne Declaration, Switzerland. Available at http:// www.evb.ch/cm_data/Paraquat-Code_ Survey_FINAL_rev1.pdf

Dixon, J., A. Gulliver, D. Gibbon. 2001. Farming systems and poverty: Improving farmers' livelihoods in a changing world. FAO, Rome.

Dolberg, F., and P.H. Petersen (ed) 1997. Maximising the influence of the user: Alternatives to the training and visit system. Proc. Workshop. 7-11 April Tune Landboskole. Denmark. DSR Forlag, Copenhagen.

Dollacker, A. 2000. Implementing sustainable agriculture: Practical approaches to IPM. Pflanzenschutz-Nachrichten Bayer 53(2-3):198.

Dollacker, A. and C. Rhodes. 2007. Integrated crop productivity and biodiversity conservation pilot initiatives developed by Bayer CropScience. Crop Prot. 26:408-416.

Döller, P.C, K. Dietrich, N. Fillip, S. Brockmann, C. Dreweck, R. Vontheim et al. 2002. Cyclosporiasis outbreak in Germany associated with the consumption of salad. Emerging Infect. Dis. 8(9):992-994.

Donald, P.F. 2004. Biodiversity impacts of some agricultural commodity production systems. Conserv. Biol. 18:17-37

Dörner, D. 1996. The Logic of failure: Recognising and avoiding error in complex situations. Addison-Wesley, Reading MA.

Dorward, A., J. Kydd. and C. Poulton. 1998. Smallholder cash crop production under market liberalisation: A new institutional economics perspective. CABI, Wallingford.

Doss, C., and A. McDonald. 1999. Gender issues and the adoption of maize technology in Africa. An annotated bibliography of Twenty-five Years on Research on Women Farmers in Africa: Lessons and implications for agricultural research Institutions. Econ. Prog. Pap. 99-02. CIMMYT, Mexico.

Doutt, R.L., and J. Nakata. 1973. The Rubus leafhopper and its egg paraistoid: An endemic biotic system useful in grape pest management. Environ. Entomol. 2:381-386.

Dreyer, H., G. Fleischer, W. Gassert, H. Stoetzer, J.P. Deguine, N. La Porta et al. 2005. IPM Europe: The European network for integrated pest management in development cooperation. Proc. EFARD 2005 Conference, Zurich.

Dufour, R. 2001. Biointensive integrated pest management IPM: Fundamentals of sustainable agriculture. July 2001. Available at http://attra.ncat.org/attra-pub/PDF/ipm.pdf.

Dupré, G. 1991. Savoirs paysans et développement. ORSTOM. Ed. Karthala. Paris.

Eastman, C., and J. Grieshop. 1989. Technology development and diffusion: Potatoes in Peru. p. 33-58. In J.L. Compton (ed) The transformation of international agricultural research and development. Westview, Boulder.

EC (European Commission). 1999. Directive 1999/2/EC of the European Parliament and of the Council of 22 February 1999 on the approximation of the laws of the Member States concerning foods and food ingredients treated with ionizing radiation. Available at http://www.fsai.ie/legislation/food/eu_docs/Manufacturing_Processing_Methods/Irradiated/Dir1999_2.pdf. Off. J. European Communities L 066, 13 March 1999.

EC (European Commission). 2000. Communication from the Commission on the Precautionary Principle, Brussels. Available at http://ec.europa.eu/dgs/health_consumer/library/pub/pub07_en.pdf.

EC (European Commission). 2001. Europeans, science and technology. Standard Eurobarometer. Available at http://europa.eu.int/comm/public_opinion/archives/eb/ebs_154_en.pdf.

EC (European Commission). 2005. Special Eurobarometer. Social values, Science and technology. Available on: http://europa.eu.int/comm/public_opinion/archives/ebs/ebs_225_report_en.pdf

EC (European Commission). 2006. Monitoting of pesticide residues in products of plant origin in the EU, Norway, Iceland and Lichtenstein. Available at http://ec.europa.eu/food/fvo/specialreports/pesticides_index_en.htm.

ECA (Economic Commission for Africa). 1992. Study on the role of the informal sector in African economies. E/ECA/PSD.7/13. 7th Session Joint Conf. African Planners, Statisticians and Demographers. Addis Ababa, 2-7 March.

Echeverria, R.G., E. Trigo, and D. Byerlee. 1996. Institutional change and effective financing of agricultural research in Latin America. Tech. Pap. 330. World Bank, Washington DC.

Edgerton, D. 2007. The shock of the old: Technology and global history since 1900. Oxford Univ. Press, UK.

Edquist, C. (ed) 1997. Systems of innovation: Technologies, institutions and organizations. Pinter, London.

EEA (European Environmental Agency). 2001. Late lessons from early warnings: The precautionary principle 1896-2000. Available at http://reports.eea.europa.eu/environmental_issue_report_2001_22/en/Issue_Report_No_22.pdf. EEA, Copenhagen.

Ehler, L. 2006. Integrated pest management (IPM): Definition, historical development and implementation and the other IPM. Pest Manage. Sci. 62:787-790.

Ehler, L., and D. Bottrell. 2000. The illusion of integrated pest management. Issues Sci Tech. 16(3):61-64.

Eicher, C., and J. Staatz. 1998. International agricultural development. Johns Hopkins Univ. Press, Baltimore.

Eisemon, T.O. 1986. Foreign training and foreign assistance for university development in Kenya: Too much of a good thing? Int. J. Educ. Dev. 6:1-13.

Eisemon, T.O., and C.H. Davis. 1992.

Universities and scientific research capacity. J. Asian Afr. Studies 27:68-93.

EJF (Environmental Justice Foundation). 2002. Death in small doses: Cambodia's pesticide problems and solutions. EKF, London.

Ekboir, J.M., G. Datrénit, V. Martinez, A. Torres-Vargas, and A. Vera-Cruz. 2006. Las fundaciones produce a diez años de su creaciòn: Pensando en el futuro. ISNAR Disc. Pap. 10. IFPRI, Washington DC.

Ekesi, S. 1999. Insecticide resistance in field populations of the legume pod-borer Maruca vitrata Fabricius (Lepidoptera: Pyralidae) on cowpea Vigna unguiculata (L.), Walp in Nigeria. Int. J. Pest Manage. 45(1):57-59.

Ekwamu, A., and M. Brown. 2005. Four years of NAADS implementation: Programme outcomes and impact. p. 25-39. In Proc. Mid-Term Review Nat. Agric. Advisory Serv. 31 May - 2 June 2005. Min. Agric. Anim. Industry Fish., Kampala.

Ellen, R., P. Parkes, and A. Bicker (ed) 2000. Indigenous environmental knowledge and its transformations. Harwood Acad. Publ., Amsterdam.

Ellis, W. 2000. A new paradigm for industry. Available at http://www.croplifeasia.org/ref_library/ipm/ipmParadigm.pdf. Asia-Pacific Crop Prot. Assoc., Bangkok.

Ellis, F. and H.A. Freeman. 2006. Rural households and poverty reduction strategies in four African countries. J. Dev. Studies 40(4):1-30.

Engel, P. 1990. Knowledge management in agriculture: Building upon diversity. Special Issue on the European Seminar on Knowledge Management and Information Technology. Wageningen. November 1989. Knowledge Society 3(3)28-35.

Engel, P.G.H., and M. Salomon. 1997. Facilitating innovation for development. A RAAKS resource box/the social organization of innovation — a focus on stakeholder interaction. Royal Trop. Inst., Amsterdam.

Enkegaard, A., and H.F. Brodsgaard. 2006. Biocontrol in protected crops: Is lack of biodiversity a limiting factor? p. 91-122. In J. Eilenberg, and H. Hokkanen (ed) An ecological and societal approach to biological control. Springer, Dordrecht.

Ericksen, P.J. 2006a. Conceptualization of food systems for global environmental change (GEC) research. GECAF Working Pap. 2. ECI/OUCE, Oxford Univ., Wallingford.

Ericksen, P.J. 2006b. Assessing the vulnerability of food systems to global environmental changes. ECI/OUCE, Oxford Univ., Wallingford.

Esman, M.J., and N. Uphoff. 1984. Local organisations - Intermediaries in rural development Cornell Univ. Press, London.

ESRC. 1999. The politics of GM food: Risk, science and public trust. Available at http://www.sussex.ac.uk/spru/documents/gecp_the_politics_of_gm_food_briefing.pdf. Spec. Briefing 5. Econ. Social Res. Council, Global Environ. Change Prog., Lancaster Univ.

ETC. 2005. Oligopoly, Inc.: Concentration in corporate power. Communique Issues #90-91.

EU. 2002. European Union consolidated versions of the treaty on European Union and of the treaty establishing the European community. Off. J. Eur. Union, C325, 24 Dec 2002, Title XIX, article 174, para. 2 and 3.

EurepGAP. 2007. The global partnership for safe and sustainable agriculture. Good agricultural practice standard. 3rd Vers. Berlin. Available at http://www.eurepgap.org/documents/infoletter/Press-conference_080207-KM2.pdf.

Evans, J. 1981. Report on extension-oriented research with women in Phalombe. Farming Systems Analysis Sec., Dep. Agric. Res., Min. Agriculture, Malawi.

Eveleens, K. 2004. The history of IPM in Asia. FAO, Rome.

Evenson, R.E. 1986. The economics of extension. p. 65-90. In G.E. Jones (ed) Investing in rural extension. Elsevier, London.

Evenson, R.E., and D. Gollin. 2003a. Assessing the impact of the Green Revolution 1960 to 2000. Science 300:758-762.

Evenson, R. E. and D. Gollin (ed) 2003b. Crop variety improvement and its effects on productivity: The impact of international agricultural research. CABI, Wallingford.

Evenson, R.E., P.E. Waggoner, and V.W. Ruttan. 1979. Economic benefits from research: An example from agriculture. Science 205:1101-1107.

EVIRA (Finnish Food Safety Authority). 2006. Number of food poisoning epidemics still on increase in 2005. Available http://www.evira.fi/portal/en/food/current_issues/archive/?id=318.

Ewen, S.W.B., and A. Pusztai. 1999. Health risks of genetically modified foods. Lancet 354:684.

Fabra, A. 2002. The intersection of human rights and environmental issues: A review of institutional developments at the international level. Background Pap. 3; Expert Seminar on Human Rights and the Environment, Geneva, 14-16 Jan. Available at http://www.unhchr.ch/environment/bp3.html.

Fairhead, J., and M. Leach. 1996. Misreading the African Landscape. Society and ecology in a forest-savanna mosaic. Cambridge Univ. Press, UK.

Falcon, W., and R. Naylor. 2005. Rethinking food security for the 21st century. Am. J. Agric. Econ. 87:113-1127.

Falcon, W.P., and C. Fowler. 2002. Carving up the commons: emergence of a new international regime for germplasm development and transfer. Food Policy 27:197-222.

Fan, S., P. Hazell, and S. Thorat. 1998. Government spending, growth and poverty: an Analysis of interlinkages in rural India. EPTD Disc. Pap. 33. Available at http://www.ifpri.org/divs/eptd/dp/papers/eptdp33.pdf. IFPRI, Washington DC.

FAO. 1995a. Prevention and disposal of obsolete pesticides. Available at http://www.fao.org/ag/AGP/AGPP/Pesticid/Disposal/index_en.htm. FAO, Rome.

FAO. 1995b. Guidelines on good labelling practice for pesticides. FAO, Rome.

FAO. 1996a. Food, agriculture and food security: developments since the World Food Conference and prospects. World Food Summit Tech. Background Doc. 1. Vol. 1. FAO, Rome.

FAO. 1999a. Committee on agriculture: Biotechnology. 25-29 January. COAG, FAO 1999. CGIAR in Latin America and the Caribbean: Interactions, achievements and prospects. FAO, Rome.

FAO. 1999b. The importance of food quality and safety for developing countries. Committee on World Food Security 25th Session. Rome, 31 May-3 June.

FAO. 2001a. Mid-term review of the Global IPM Facility. April-June. FAO, Rome.

FAO. 2001b. The state of food insecurity in the world 2001. Rome.

FAO, 2002a. Bovine Spongiform Encephalopathy (BSE) and foot and mouth disease risk assessment: Implications for the Near East. Twenty-Sixth FAO Reg. Conf. for the Near East. Tehran. 9-13 Mar 2002. NERC/02/INF/5. Available at http://www.fao.org/DOCREP/MEETING/004/Y3064E.HTM. FAO Rome.

FAO. 2002b. International code of conduct on the distribution and use of pesticides. Adopted by the 123rd Session of the FAO Council in November 2002. FAO, Rome.

FAO. 2003a. Biosecurity in food and agriculture. Committee on Agric., 17th Session. Available at http://www.fao.org/DOCREP/MEETING/006/Y8453e.HTM. FAO, Rome.

FAO, 2003b. Trade reforms and food security. Chapter 9 In The role of transnational corporations, conceptualizing the linkages. Commodities Trade Div., FAO, Rome.

FAO. 2004. The state of food insecurity in the World. Monitoring progress towards the World Food Summit and Millennium Development Goals. FAO. Rome.

FAO. 2005a. Asia regional workshop on the implementation, monitoring and observance of the international code of conduct on the distribution and use of pesticides. Bangkok. 26-28 July. Available at http://www.fao.org/docrep/008/af340e/af340e00.htm. FAO Reg. Off. Asia Pacific, Bangkok.

FAO. 2005b. International code of conduct on the distribution and use of pesticides — revised version. FAO, Rome.

FAO. 2005c. The state of food and agriculture: Agricultural trade and poverty: Can trade work for the poor? FAO, Rome.

FAO. 2005d. UN Agency says scale of toxic pesticide in Latin America 'frightening', funding for cleaning up needed. Available at http://www.un.org/News/Press/docs/2005/sag369.doc.htm. FAO, Rome.

FAO. 2005e. Understanding the Codex Alimentarius. FAO, Rome. http://www.fao.org/documents/show_cdr.asp?url_file=/docrep/008/y7867e/y7867e00.htm.

FAO. 2006a. FAO encourages early withdrawal of highly toxic pesticides; assurances given by Danish company. Available at http://www.fao.org/newsroom/en/news/2006/1000471/index.html. FAO, Rome.

FAO. 2006b. The implementation of the globally harmonized system of classification and labelling of chemicals — FAO's past and present activities. Available at http://www.unitar.org/cwg/publications/cbl/ghs/Documents_2ed/F_Guidance_Awareness_Raising_and_Training_Materials/498_FAO_past-present-activities.pdf. FAO, Rome.

FAO. 2006c. A global plant breeding initiative. FAO Agriculture 21. Available at http://www.fao.org/ag/magazine/0606sp1.htm. FAO, Rome.

FAO. 2007. New initiative for pesticide risk reduction. Committee on Agric., 20th Session 25-28 April 2007, RomeCOAG/2007/lnf.14. Available at ftp://ftp.fao.org/docrep/fao/meeting/011/j9387e.pdf. FAO, Rome.

FAO. 2008. Roles of agriculture project. Available at http://www.fao.org/es/esa/roa/index_en.asp. FAO, Rome.

FAO/WHO. 2001. Amount of poor-quality pesticides sold in developing countries alarmingly high. FAO/WHO press release, 1 Feb.

FAO/WHO. 2004. Cambodia country report on food safety. FAO/WHO Regional Conference on Food Safety for Asia and Pacific, Seremban, Malaysia, 24-27 May 2004.

Farina, E., and T. Reardon. 2000. Agrifood grades and standards in the extended Mercosur: Their role in the changing agrifood system. Am. J. Agric. Econ. 82:1170-1176.

Farrington, J., and A. Martin. 1987. Farmer-participatory research: A review of concepts and recent fieldwork. Agric. Admin. Extens. 29:247-264.

Farrington, J., A. Bebbington, K. Wellard, D. Lewis. 1993. Reluctant partners? Non-governmental organizations, the state and sustainable agricultural developments. Routledge, NY.

Farrington, J., and D.L. Lewis. 1993. Non-governmental organizations and the State in Asia: Rethinking roles in sustainable agricultural development. Routledge, London.

Farrington, J., and J.R. Witcombe. 1998. Regulatory frameworks: Why do we need them. p. xvii-xxii. In J. Witcombe et al. (ed) Seeds of choice: Making the most of new varieties for small farmers. IT Publ., London.

FAWG. 2001. Roots of change: Agriculture, ecology and health in California. Available at http://www.foodfunders.org/pdfs/roots_of_change.pdf. FAWG, San Francisco.

Feder, E. 1976. McNamara's little green revolution: World Bank scheme for self-liquidation of Third World peasantry. Econ. Pol. Weekly 10:532-541.

Feder, G., R. Murgai and J. Quizon. 2004a. The acquisition and diffusion of knowledge: The case of pest management training in farmer field schools. Indonesia. J. Agric. Econ. 55(2):221-243.

Feder, G., R. Murgai, J. Quizon. 2004b. Sending farmers back to school: The impact of farmer field schools in Indonesia. Rev. Agric. Econ. 26:45-62.

Federación Agraria Argentina (FAA). 2005. Informe Semanal Técnico NO. 167. Buenos Aires, May 20. Available at http://www.faa.com.ar/.

Ferguson, J. 1990. The anti-politics machine: Develoment, depoliticization and bureaucratic power in Lesotho. Cambridge Univ. Press, UK.

Ferleger, L., and W. Lazonick. 1994. Higher education for an innovative economy: Land-grant colleges and the managerial revolution in America. Business Econ. History 23(1):116-128.

Fernandez-Cornejo, J., and D. Schimmelpfennig. 2004. Have seed industry changes affected research effort? Available at http://www.ers.usda.gov/AmberWaves/February04/Features/HaveSeed.htm. USDA Econ. Res. Serv., Washington DC

Ferrara, J. 1998. Revolving doors: Monsanto and the regulators. The Ecologist Sept/Oct. Available at http://www.monitor.net/monitor/9904b/monsantofda.html

Fields, S. 2004. The fat of the land: Do agricultural subsidies foster poor health? Environ. Health Perspect. 112(14):A820-A823.

Fischer, K., and D. Byerlee. 2002. Managing intellectual property and income generation in public research organisations. p. 227-244. In D. Byerlee and R. Echeverría (ed) Agricultural research policy in an era of privatization. CABI, Oxon.

Fishbein, M., and I. Ajzen. 1975. Belief, attitude, intention, and behavior: An introduction to theory and research. Addison-Wesley, Reading MA.

Fisher, E.C. 2007. Opening Pandora's box: Contextualising the precautionary principle in the European Union. Oxford Legal Studies Res. Pap. 2/2007. In Ellen Vos et al. (ed) Uncertain risks regulated: National, EU and international regulatory models compared. Available at http://ssrn.com/abstract=956952. UCL Press, Cavendish Publ.

Fitzgerald, D. 2003. Every farm a factory: The industrial ideal in American agriculture. Yale Univ. Press.

FOE. 2005. Tackling GMO Contamination: making segregation and identity preservation a reality. Available at: http://www.foei.org/publications/index.html. FOE, Montreal.

FOE. 2006. Who benefits from GM crops? Available at http://www.foei.org/publications/index.html. FOE, Nigeria.

FOE/Halifax Initiative. 2002. Marketing the Earth: the World Bank and sustainable development. Available at http://www.

panna.org/campaigns/docsWorldBank/
MarketingEarth.pdf.

FOEI. 2001. Sale of the country? Peoples' food sovereignty. Friends of Earth Int., http://www.foei.org.

FOEI. 2003. Trade and people's food sovereignty. Friends of Earth Int., http://www.foei.org.

Fogel, R.W. 2004. The escape from hunger and premature death 1700-2100: Europe, America and Third World, Cambridge Univ. Press, UK.

Foko, E. 1999. Arabica coffee in the agricultural production system of the western highlands of Cameroon. Cahiers Agric. 8:197-202.

Follet, P.A., and J.J. Duan (ed) 2000. Biological control for a small planet: Nontarget effects of biological control introductions. Kluwer, Dordrecht.

Forni, F., G. Neiman, A. Bacalini, and L. Roldán. 1998. Haciendo escuela. Ciccus, Buenos Aires.

Fortmann, L. 1979. Women and agricultural development. In K.S. Kim et al. (ed) Papers on the political economy. Tanzania. Heinemann, Nairobi.

Forum for Food Sovereignty. 2007. Declaration of Nyéléni. Sélingué, Mali, 27 Feb 2007. Available at http://www.nyeleni2007.org/.

Frankel, F.R. 1971. India's green revolution: Economic gains and political costs. Princeton Univ. Press, Princeton.

Freebairn, D.K. 1995. Did the green revolution concentrate incomes? A quantitative study of research reports. World Dev. 23(2):265-279.

Fresco, L. 2002. The future of agriculture: Challenges for environment, health and safety regulation of pesticides. Available at www.fao.org/ag/magazine/oecd.pdf. FAO, Rome.

Friesen, F., A.G. Nelson, and R.C. Van Acker. 2003. Evidence of contamination of pedigreed canola (Brassica napus) seedlots in western Canada with genetically engineered herbicide resistance traits. Amer. Soc. Agron. J. 95:1342-1347.

Fukuda, W., and N. Saad. 2001. Participatory research on cassava breeding with farmers in Northeastern Brazil. Empresa Brasileira de Pesquisa Agropeccuária. EMBRAPA, Brasil.

Fukuda-Parr, S. (ed) 2007. The gene revolution. GM crops and unequal development. Earthscan, London.

Fukuoka, M. 1978. The one-straw revolution: An introduction to natural farming. Rodale Press, Emmaus.

Funtowicz, S.O., and J.R. Ravetz.1993. Science for the post-normal age. Futures 25(7):739-755.

Funtowicz, S.O., J. Ravetz, and M. O'Connor. 1998. Challenges in the use of science for sustainable development. Int. J. Sustain. Dev. 1(1):99-108.

Furuno, T. 2001. The power of duck. Integrated rice and duck farming. Tagari Publ., Tylagum, Australia.

Gaillard, J. and R. Waast. 1992. The uphill emergence of scientific communities in Africa. J. Asian Afr. Studies 27(12):41-67.

Gallagher, K. 1988. Effects of host resistance on the micro-evolution of the rice brown planthopper, Nilaparvata lugens (Stal.). PhD. Thesis.: Univ. California, Berkeley.

Gallagher, K. 1999. Frequently asked questions on HIV/AIDS facts: A Global IPM Facility Field School Guide. Global IPM Facility, Rome.

Gallin, R.S., and A. Spring (ed) 1985. Women creating wealth: Transforming economic development. Selected Papers and Speeches Assoc. Women in Dev. Conf., 25-27 April Washington DC.

Gallin, R.S., M. Aronoff, and A. Ferguson (ed) 1989. The women and international development annual. Vol. I. Westview Press, Boulder.

Gallin, R.S., and A. Ferguson (ed) 1991. The Women and International Development Annual. Vol. II. (Boulder, Colorado: Westview Press);

Gamser, M. 1988. Innovation, technical assistance and development: the importance of technology users. World Dev. 16(6):711-721.

Ganapathy, S., S.B. Duffy, C. Getz. 2005. A framework for understanding food insecurity: An anti-hunger approach, a food system approach. Centre Weight Health College Nat. Resour., Univ. California, Berkeley. Available at http://www.cnr.berkeley.edu/cwh/PDFs/Framework_Food_Insecurity_3.05.pdf.

GAO. 2001. Agricultural pesticides: Management improvements needed to further promote integrated pest management. Available at http://www.gao.gov/new.items/d01815.pdf. GAO, Washington DC.

Gao, Q., and X. Li. 2006. Agricultural information infrastructure development and information services in China. Chapter 16. p. 369-387. In A. Van den Ban, and R.K. Samanta (ed) Changing roles of agricultural extension in Asian nations. B.R. Publ., Delhi.

García-Marirrodriga, R., and P. Puig Calvó. 2007. Formación en alternancia y desarrollo local. Aidefa, Rosario.

Garett, J. 1997. Challenges to the 2020 vision for Latin America: Food and agriculture since 1970. Food Agric. Environ. Disc. Pap. 21. IFPRI, Washington DC.

Garfield, E. 1998. From citation indexes to informetrics: Is the tail now wagging the dog? Libri 48(2):67-80.

Garry, V.F., D. Schreinemachers, M.E. Harkins and J. Griffith. 1996. Pesticide appliers, biocides and birth defects in Rural Minnesota. Environ. Health Perspect. 104(4):394-399.

GECAFS. 2008. Food systems research: Why has GECAFS taken a "food systems" approach? Available at http://www.gecafs.org/research/food_system.html. Environmental Change Institute, Oxford, UK.

Gelernter, W.D. 2005. Biological control products in a changing landscape. p. 293-300. In Crop science and technology. Brit. Crop Prot. Council, Hampshire, UK.

Georghiou, G.P. 1986. The magnitude of the problem. In Pesticide resistance, strategies and tactics. Nat. Acad. Press, Washington DC.

Gentil, D. 1989. A few questions on the training and visit method. p. 25-31 In N. Roberts (ed) Agricultural extension in Africa. Available at http://www.eric.ed.gov/ERICDocs/data/ericdocs2sql/content_storage_01/0000019b/80/20/83/7c.pdf. World Bank, Washington DC.

Gentili, P. 2001. The permanent crisis of the public university. North American Congress on Latin America (NACLA) Rep. Americas 33(4):12-19.

Gepts, P. 2004. Who owns biodiversity and how should the owners be compensated? Plant Physiol. 134:295-1307.

Gerken, A., J.V. Suglo and M. Braun. 2000. Crop protection policy in Ghana. An economic analysis of current practice and factors influencing pesticide use. Univ. Hannover and GTZ, Accra.

Getz, C., and K.D. Warner. 2006. Integrated farming systems and pollution prevention initiatives stimulate co-learning extension strategies. J. Extension 44(5), Article # 5FEA4. Available at http://www.joe.org/joe/2006october/a4.shtml.

Gibbons, M., C. Limoges, H. Nowotny, S. Schwartzman, P. Scott, and M. Trow. 1994. The new production of knowledge: The dynamics of science and research in contemporary societies. Sage Publ., Thousand Oaks CA.

Gieryn, T.F. 1995. Boundaries of science. p. 393-443. In S. Jasanoff et al. (ed) Handbook of science and technology studies. Sage Publ., Thousand Oaks CA.

Gilbert, E. 1995. The meaning of the maize revolution in Sub-Saharan Africa: Seeking guidance from past impacts. Agric. Admin. Network Pap. 55, Overseas Dev. Inst., London.

Gill, G.J. 1995. Major natural resource management concerns in South Asia. 2020 Vision for Food Agric. Environ. Disc. Pap. 8, IFPRI, Washington DC.

Gill, G.J., and D. Carney. 1999. Competitive agricultural technology funds in developing countries. Overseas Dev. Inst., London.

Giller, K.E., 2002. Targeting management of organic resources and mineral fertilizers: Can we match scientists' fantasies with farmers' realities? p. 155-171. In B. Vanlauwe et al. (ed) Balanced nutrient management systems for the moist savanna and humid forest zones of Africa. CABI, Wallingford.

Girard, N., and M. Navarrette. 2005. Quelles synergies entre connaissances scientifiques et empiriques ? L'exemple des cultures du safran et de la truffe. Natures Sci. Sociétés 13:33-44.

Glasbergen, P., F. Biermann and A.P.J. Mol (ed) 2007. Partnerships, governance and sustainable development. Reflections on theory and practice. Edward Elgar, Cheltenham, UK.

Gómez, I. and M. Bordons. 1996. Limitaciones en el uso de los indicadores bibliométricos para la evaluación científica. Política Científica 46:21-26.

Gonzales, T.A. The cultures of the seed in the Peruvian Andes. Chapter 8. *In* S. Brush (ed) 1999. Genes in the field. On farm conservation of crop diversity. IPGRI, Rome and IDRC, Ottawa.

Gottlieb, R. 1993. Forcing the spring: The transformation of the American environmental movement. Island Press, Washington DC.

Graham, J.D., and J.B. Wiener. 1995. Risk vs. risk. Tradeoffs in protecting health and the environment. Harvard Univ. Press, Cambridge MA.

Granatstein, D. 2003. Tree fruit production with organic farming methods. Available at http://organic.tfrec.wsu.edu/OrganicIFP/OrganicFruitProduction/OrganicMgt.PDF. Wenatchee Center Sustain. Agric. Nat. Resour., Washington State Univ.

Granereau, A. 1969. Le livre de Lauzun. Les Éditions Gerbert, Aurillac.

Gray, M.E. and K.L. Steffey, 2007. Maximizing . crop production inputs does not equate to integrated pest management. Univ. Illinois Extension, The Bull. 8, Article 2, 18 May 2007.

Gray, R., S. Malla, and P. Phillips. 1999. The public and non-for-profit sectors in a biotechnology-based, privatising world: The canola case. p. 501-522. *In* Proc. Conf. Transitions in Agbiotech: Economics of strategy and policy. Available at http://agecon.lib.umn.edu/cgi-bin/pdf_view.pl?paperid=2192&ftype=.pdf.

Griliches, Z. 1992. The search for R&D spillovers. Scand. J. Econ. 94 (Suppl.):29-47.

Greathead, D.J. 1971. A review of biological control in the Ethiopian region. Tech. Comm. No. 5. CAB, Farnham Royal.

Greathead, D.J. 1976. A review of biological control in Western and Southern Europe. CAB, Farnham Royal.

Greathead, D.J. 2003. Historical overview of biological control in Africa. p. 1-26. *In* P. Neuenschwander et al. (ed) Biological control in IPM systems in Africa. CABI Publ., Wallingford.

Griffin, K. 1979. Underdevelopment in history. *In* C.K. Wilber (ed) 1979. The political economy of development and underdevelopment. Random House, London.

Griffin, K., and A.R. Khan. 1998. Poverty in the world: Ugly facts and fancy models. World Dev. 6(3):295-304.

Grimaldi, L. 1998. Disminuyen accidentes pot use de plaguicidas. Prensa Libre, 18-19.

Grimble, R. and K. Wellard. 1996. Stakeholder methodologies in natural resource management: a review of principles, contexts, experiences and opportunities. Agric. Syst. 65(2):173-193.

Grivetti, L.E., and B. Olge. 2000. Value of traditional foods in meeting macro- and micronutrient needs: The wild plant connection. Nutr. Res. Rev. 13:31-46.

Guerinot, M.L. 2000. Enhanced: The green revolution strikes gold. Science 287:241-243.

Guillon, M. 2004. Current world situation on acceptance and marketing of biological control agents BCAs. Available at http://www.ibma.ch/pdf/20041028%20Presentation%20BCAs%20Thailand%20%20&%20Indonesia%20Cuba.pdf. Int. Biocontrol Manufact. Assoc. Pau, France.

Guijt, I., and M.K. Shah. 1998. The myth of community. IT Publ., London.

Gurr, G., and S. Wratten (ed) 2000. Measures of success in biological control. Kluwer, Dordrecht.

Gurr, G.M., S.D. Wratten, P. Kehrli, and S. Scarratt. 2006. Cultural manipulations to enhance biological control in Australia and New Zealand: Progress and prospects. p. 154-166. *In* Proc. Int. Symp. on Biological Control of Arthropods. Davos, Switzerland. 12-16 Sept. 2005. USDA Forest Service, FHTET, Fort Collins, CO.

Gurung, B., and H. Menter. 2004. Mainstreaming gender-sensitive participatory approaches: The CIAT case study. *In* D. Pachico (ed) Scaling up and out: Achieving widespread impact through agricultural research. CIAT, Cali.

Gutierrez, A.P., P. Neuenschwander, F. Schulthess, H.R. Herren, J.U. Baumgärtner, B. Wermelinger et al.1988. Analysis of biological control of cassava pests in Africa: II. Cassava mealy bug Phenacoccus manihoti. J. Appl. Ecol. 25:921-940.

GTZ. 2004. Combating world hunger through sustainable agriculture. Available at http://www2.gtz.de/dokumente/bib/04-5888.pdf. GTZ, Germany.

Halberg, N., H.F. Alroe, M.T. Knudsen, and E.S. Kristensen. 2007. Global development of organic agriculture: Challenges and prospects. CABI Publ., UK.

Hall, A.J., B. Yoganand, R.V. Sulaiman, and N.G. Clark (ed) 2003. Post-harvest innovations in innovation: Reflections on partnership and learning. Crop Post-Harvest Programme South Asia, Aylesford.

Hall, A.J., W. Janssen, E. Pehu, and R. Rajalahti. 2006. Enhancing agricultural innovation: How to go beyond strengthening research systems. World Bank. Washington DC.

Halliday, M.L., L.Y. Kang, T.K. Zhou, M.D. Hu, Q.C. Pan, T.Y. Fu et al. 1991. An epidemic of hepatitis A attributable to the ingestion of raw clams in Shanghai, China/ J. Infect. Dis. 164(5):852-859.

Halse, M. (ed) 1966. Studies in block development and co-operative organisation. Indian Inst. Manage., Ahmedabad.

Halweil, B. 2006. Can organic farming feed us all? World Watch 19(3):19-24.

Hamburger, J., and M. Ishii-Eiteman. 2003. The struggle to reduce reliance on pesticides in World Bank projects: Can community-based monitoring lead to improved policy compliance? PANNA, San Francisco.

Hammerschlag, K. 2007. More IPM please: How USDA could deliver greater environmental benefits from Farm Bill conservation programs. Issue Pap. NRDC, San Francisco.

Hammond, R., and L. McGowan. 1992. The other side of the story: The real impacts of World Bank structural adjustment programs. The Development Gap, Washington DC.

Hanak, E., E. Boutrif, P. Fabre, M. Pineiro (ed) 2002. Food safety management in developing countries. Proc. Int. Workshop, CIRAD-FAO. Montpellier. 11-13 Dec. 2000. CIRAD-FAO. CIRAD, Montpellier.

Hanger, J., J. Moris. 1973. Women and the household economy. *In* R. Chambers, J. Moris (eds). Mwea: An irrigated rice settlement in Kenya. Weltforum Verlag Munich.

Hardell, L., M.J. Walker, B. Walhjalt, L.S. Friedman, and E.D. Richter. 2007. Secret ties to industry and conflicting interests in cancer research. Am. J. Indust. Med. 50:227-233.

Hardon, J. 2004. Plant patents beyond control: Biotechnology, farmer seed systems and Intellectual Property Rights. Agromisa Foundation, Wageningen.

Harilal, K.N., N. Kanji, J. Jeyaranjan, M. Eapen, and P. Swaminathan. 2006. Power in global value chains. Implications for employment and livelihoods in the cashew nut industry in India. Summary Rep. IIED, London.

Harris, J. 1977. The limitations of HYV technology in North Arcot District: The views from a village. *In* B.H. Farmer (ed) Green revolution? Technology and change in rice growing areas of Tamil Nadu and Sri Lanka. Westview Press, Boulder.

Harris, J. 2001. Depoliticizing development: The World Bank and social capital. LeftWord Books, Glasgow.

Harrison, J. 2004. Invisible people, invisible places: Connecting air pollution and pesticide drift in California. *In* E.M. Dupuis (ed) Smoke and mirrors: The politics and culture of air pollution. NYU Press, NY.

Harriss, B. 1978. Rice processing projects in Bangladesh: An appraisal of a decade of proposals. Bangladesh J. Agric. Econ. 1(2):24-52.

Hatchuel, A. 2000. Intervention research and the production of knowledge. p. 55-68. *In* M. Cerf et al. (ed) Cow up a tree. Knowing and learning for change in agriculture. Case studies from industrialised countries. INRA Ed., Paris.

Havelock, R.G. 1969. Planning for innovation through dissemination and utilization of knowledge. Center for Research on Utilization of Scientific Knowledge, Ann Arbor, Michigan.

Havelock, R.G. 1986. Modelling the knowledge system. p. 77-105. *In* G.M. Beal et al. (ed) Knowledge, generation, exchange and utilisation. Westview, Boulder.

Haverkort, B., J. Van der Kamp, and A. Waters-Bayer (ed) 1991. Joining farmers' experiments: Experiences in participatory technology development. Intermediate Tech., London.

Hayami, Y., and V. Ruttan. 1985. Agricultural development: An international perspective. The Johns Hopkins Univ. Press, Baltimore.

Hayenga, M.L. 1998. Structural changes in the biotech seed and chemical industrial complex. AgBioForum 1:43-55.

Hayes, T.B. 2004. There is no denying this: Defusing the confusion about atrazine. BioScience. 54:1138-1149.

Haygood, R., A.R. Ives, and D.A. Andow. 2003. Consequences of recurrent gene flow from crops to wild relatives. Proc. R. Soc. London B (Biol. Sci.) 270(1527):1879-1886.

Hawkes, C., and M.T. Ruel. 2006. Understanding the links between agriculture and health. Focus 13. May. Available at http://www.ifpri.org. IFPRI, Washington DC.

Hazell, P., and C. Ramaswamy. 1991. The green revolution reconsidered: The impact of high yielding rice varieties in South India. John Hopkins Univ. Press, Baltimore.

Hediger, W., and B. Lehmann. 2003. Multifunctional agriculture and the preservation of environmental benefits. 25th Int. Conf. Agric. Econ., Durban. 16-22 Aug. Available at http://www.iaae-agecon.org/conf/durban_papers/papers/096.pdf.

Heemskerk, W., and B. Wennink, 2005. Stakeholder-driven funding mechanisms for agricultural innovation. Case studies from Sub-Saharan Africa. Bull. 373. Dev. Policy Practice, KIT, Amsterdam.

Heggun, C. 2001. Trends in hygiene management: the dairy sector example. Food Control 12:241-246.

Heisey, P.W, C.S. Srinivasan, and C.Thirtle. 2001. Public sector plant breeding in a privatized world. ERS AIB 722. USDA, Washington DC.

Heller, M.A., and R.S. Eisenberg. 1998. Can patents deter innovation? The anticommons in biomedical research. Science 280:698-701.

Hellinger, D. 1999. Statement at the First Public Hearing of the International Financial Institutions Advisory commission on Civil Society Perspectives on the IMF and World Bank. Available at http://www.developmentgap.org/ifi_testimony.html. The Development Gap, Washington DC.

Hendrickson, M.K., and W. Heffernan. 2005. Can consolidated food systems achieve food security? p. 65-68. In Proc. New Perspectives on Food Security. 12-14 Nov. Glynwood Center, Cold Spring NY.

Heong, K.L., and M.M. Escalada. 1998. Changing rice farmers' pest management practices through participation in a small-scale experiment. Int. J. Pest Manage. 44:191-197.

Herdt, R.W., and J.W. Mellor. 1964. The contrasting response of rice to nitrogen: India and the United States. J. Farm Econ. 46:150-160.

Herdt, R.W. 2006. Establishing priorities for plant science research and developing world food security. Eur. J. Plant Pathol.115:75-93.

Herren, H.R. 1990. Biological control as the primary option in sustainable pest management: The cassava pest project. Bull. Soc. Ent. Suisse 63:405-413.

Herren, H.R., and P. Neuenschwander. 1991. Biological control of cassava pests in Africa. Ann. Rev. Entomol. 36:257-283.

Herren, H., F. Schulthness and M. Knapp. 2005. Towards zero pesticide use in tropical agroecosystems. p. 135-146 In J. Pretty (ed) The pesticide detox: Towards a more sustainable agriculture. Earthscan, London.

Higley, L.G., and L.P. Pedigo. 1993. Economic injury level concepts and their use in sustaining environmental quality. Agric. Ecosyst. Environ. 46:233-244.

Hill, P. 1982. Dry grain farming families. Hausaland (Nigeria) and Karnataka (India) compared. Cambridge Univ. Press, UK.

Hilhorst, T., and F. Muchena (ed) 2000. Nutrients on the move. Soil fertility dynamics in African farming systems. IIED, London.

Hoard, R.J., and M.J. Brewer. 2006. Adoption of pest, nutrient and conservation vegetation management using financial incentives provided by a U.S. Department of Agriculture conservation program. Hort Tech. 16:306-311.

Hobart, M. 1994. Introduction: The growth of ignorance? Chapter 1. p. 1-30. In M. Hobart (ed) An anthropological critique of development: The growth of ignorance. Routledge, London.

Hobden, A. 1995. Paradigms and politics: The cultural construction of environmental policy in Ethiopia. World Dev. 23(6):1007-1022.

Hocdé, H., J.I. Vasquez, E. Holt, and A.R. Braun. 2000. Towards a social movement of farmer innovation: Campesino a Campesino. ILEIA Newsl. 16(2):26-27

Hocdé, H., J. Lançon, and G. Trouche (ed) 2002. La sélection participative: Impliquer les utilisateurs dans l'amélioration des plantes. Actes de l'atelier. Available at http://publications.cirad.fr/une_notice.php?dk=476017. Cirad, Montpellier.

Hodeshi, M., A. Kazuhiro, M. Shunsaku, T. Satoshi, S. Nobumichi, M. Motonobu et al. 1999. Massive outbreak of Eschrichia O157:H7 infection in schoolchildren in Sakai City, Japan, associated with consumption of white radish sprouts. Am. J. Epidemiol. 150:787-796.

Hoekstra, A.Y., and A.K. Chapagtain. 2007. Water footprints of nations: Water use by people as a function of their consumption patterns. p. 35-48 In E. Craswell et al. (ed) Integrated assessment of water resources and global change. Springer Verlag, Netherlands.

Hogg, D. 2000. Technological change in agriculture: Locking into genetic uniformity. Macmillan Press, UK and St. Martin's Press, USA.

Hokkanen, H.M.T., and A.E. Hajek (ed) 2003. Environmental impact of microbial insecticides: Need and methods for risk assessment. Kluwer Academic Publ., Dordrecht.

Holl, K., G. Daily, P. Ehrlich. 1990. Integrated pest management in Latin America. Environ. Conserv. 17:341-350.

Holland, J.M. 2004. The environmental consequences of adopting conservation tillage in Europe: Reviewing the evidence. Agric. Ecosyst. Environ. 103:1-25.

Holland, J., and J. Blackburn (ed) 1998. Whose voice? Participatory research and policy change. Intermediate Tech. Publ., London.

Holmgren, D. 2002. Permaculture: Principles and pathways beyond sustainability. Holmgren Design Serv., Hepburn, Victoria, Australia.

Hopper, K.R. 2001. Research needs concerning non-target impacts of biological control introductions. p. 39-56. In E. Wajnberg et al. (ed) 2001. Evaluating indirect effects of biological control. CABI Publ., Wallingford.

Hounkonnou, D. 2001. Listen to the cradle. Building from local dynamics for African renaissance. Case studies in rural areas in Benin, Burkina Faso and Ghana. Publ. PhD thesis. Wageningen Univ., Wageningen.

Howard, P.L. (ed) 2003. Women and plants. Gender relations in biodiversity management and conservation. ZED Books, London.

Howarth, F. 1990. Environmental impacts of classical biological control. Ann. Rev. Entomol. 36:485-509.

Howell, J. (ed) 1980. Borrowers and lenders; Rural financial markets and institutions in developing countries. Overseas Dev. Inst., London.

Howell, J. 1982. Managing agricultural extension: The T&V system in practice. Agric. Admin. 11:273-284.

Howell, J. (ed) 1988. Training and visit extension in practice. Occas. Pap. 8. Overseas Dev. Inst., London.

Huang, J., Q. Wang, Y. Zhang, and J. Falck-Zapeda. 2001. Agricultural bio-technology development and research capacity in China. Working Pap. Centre for Chinese Agric. Policy, Chinese Acad. Sci., Beijing.

Huang, J., C. Pray, and S. Rozelle. 2002a. Enhancing the crops to feed the poor. Nature 418:678-684.

Huang, J., R.Hu, S.Rozelle, F.Qiao, and C.E. Pray. 2002b. Transgenic varieties and productivity of smallholder cotton farmers in China. Austral. J. Agric. Res. Econ. 46: 367-387.

Huang, J., R. Hu, C.E. Pray, F. Qiao, and S. Rozelle. 2003. Biotechnology as an alternative to chemical pesticides: a case study of Bt cotton in China. Agric. Econ. 29:55-67.

Hubert, B., R.L. Ison, and N. Röling. 2000. The „Problematique" with respect to industrialised-country agricultures. p. 14-29. In Cow up a tree: Knowing and learning for change in agriculture, case studies from industrialised countries. M. Cerf et al. (ed) INRA Ed., Paris.

Hunter, G. 1969. Modernising peasant societies. Oxford Univ. Press, London.

Hunter, G. 1970. The administration of agricultural development: Lessons from India. Oxford Univ. Press, London.

Hunter, G., A.H. Bunting, and A.F. Bottrall (ed) 1976. Policy and practice in rural development. ODI-Croom Helm, London.

Huus-Bruun, T. 1992. Field vegetables in Denmark: An example of the role of the Danish Advisory Service in reducing insecticide use. p. 201-207. In J.C. van Lenteren et al. (ed) Biological control and integrated crop protection: Towards environmentally safer agriculture. Proc. Int. Conf. organized by IOBC/WPRS. Veldhoven, Netherlands, 8-13 Sept. 1991. PUDOC, Wageningen, Netherlands.

Huylenbroeck, G., and G. Durand (ed) 2003. Multifunctional agriuclture: A new paradigm for european agriuclture and rural development. Ashgate, Burlington VT.

Hyman, E.L. 1992, Local agro-processing with sustainable technology: Sunflower seed oil in Tanzania. Gatekeeper Ser. 33. IIED, London.

IAC (InterAcademy Council) 2004. Realizing the promise and potential of African agriculture. Science and technology strategies for improving agricultural productivity and food security in Africa. IAC, Amsterdam.

IATP. 1998. Marketing sustainable agriculture: Case studies and analysis from Europe. IATP, Minneapolis.

ICIPE. 2006. Activities and impacts of major projects, 1996-2006. ICIPE, Afr. Insect Sci. Food Health, Nairobi.

Idris, A.B., and E. Grafius. 1995. Wildflowers as nectar sources for Diadegma insulare (Hymneoptera: Ichneumonidae), a parasitoid of diamondback moth (Lepidoptera: Yponomeutidae). Environ. Entomol. 24(6):1726-1735.

IDS. 2006. Science and citizens: Global and local voices. IDS Policy Briefing Issue 30. Univ. Sussex, Brighton.

IFAD. 2003. Gender programme for IFAD's strategy equitable developmenyt for women and men in the Near East and North Africa. IFAD, Rome.

IFOAM. 2003. Organic trade a growing reality. An overview and facts on worldwide organic agriculture. IFOAM, Germany.

IIED. 2006. Rapport General Espace Citoyen d'Interpellation Démocratique (ECID) sur les OGM et l'avenir de l'agriculture au Mali. Sikasso, Mali. 25-29 Jan. IIED, London.

IFPRI. 2002. Green Revolution — curse or blessing? http://www.ifpri.org/pubs/ib/ib11.pdf. IFPRI, Washington DC.

IFPRI. 2005. Proceedings of an international dialogue on pro-poor public-private partnerships for food and agriculture. Available at http://www.ifpri.org/events/conferences/2005/ppp/PPPProc.pdf. IFPRI, Washington DC.

IIRR. 1996. Recording and using indigenous knowledge: A manual. IIRR, Cavite, Philippines.

IIRR. 2005. Linking people to policy. From participation to deliberation in the context of Philippine community forestry. IIRR, Cavite, Philippines.

Imbroglini, G. 1992. Limited use of pesticides: the Italian commitment. p. 149-157. In J.C. van Lenteren et al. (ed) Biological control and integrated crop protection:

towards environmentally safer agriculture. Proc. Int. Conf. organized by IOBC/WPRS. Veldhoven, Netherlands. 8-13 Sept. 1991. PUDOC, Wageningen.

Iowa State University. 2007. Impact bioeconomy. Stories in agriculture and life sciences. College Agric. Life Sci., Iowa State Univ. Ames.

IPPC. 2005. Revision of ISPM No. 3: Guidelines for the export, shipment, import and release of biological control agents and beneficial organisms. Draft for country consultation -2004. Available at https://www.ippc.int/IPP/En/default.jsp. Int. Plant Prot. Conv. (IPPC).

IRRI. 1985. Women in rice farming. Gower Publ., UK.

Irwin, A., and H. Rothstein. 2003. Regulatory science in a global regime. p. 77-86 In F. den Hond, et al. (ed) Pesticides: problems, improvements, alternatives. Blackwell, Oxford.

Ishii-Eiteman, M. and N. Ardhianie. 2002. Community monitoring of integrated pest management versus conventional pesticide use in a World Bank project in Indonesia. Int. J. Occup. Environ. Health 8:220-231.

Ismael, Y. 2001. Smallholder adoption and economic impacts of Bt cotton in the Makhathini Flats, Republic of South Africa. Rep. DFID Nat. Resour. Policy Res. Prog., Project R7946. DFID, London.

ISNAR. 2002. Gender and agriculture in the information society. Briefing Pap. 55. Sept. Available at http://www.isnar.cgiar.org/publications/briefing/bp55.htm. ISNAR, IFPRI, Washington DC.

ITU (International Telecommunications Union). 2006. World telecommunications/ICT development report 2006: Measuring ICT for social and economic development. ITU, Geneva.

Izuno, T. 1979. Maize in Asian cropping systems: Historical review and considerations for the future. Asian Rep. No. 9. CIMMYT, New Delhi.

Jacobs, M., and B. Dinham (ed) 2003. Silent invaders: Pesticides, livelihoods and women's health. Zed Books. NY.

Jain, V. 1992. Disposing of pesticides in the third world. Environ. Sci. Tech. 26:226.

Jakobson, K.M., and A.K. Dragun (ed) 1996. Contingent valuation and endangered species. methodological issues and applications. Edward Elgar, Cheltenham.

James, C. 2004. Global status of commercialized biotech/GM crops: 2004. Brief No. 32. ISAAA, Ithaca.

Jaradat, A.A., M. Shahid, and A.Y. Al Maskri. 2004. Genetic diversity in the Batini barley landrace from Oman: I. Spike and seed quantitative and qualitative traits. Crop Sci. 44:304-315.

Jennings, B. 1997. The killing fields: Science and politics at Berkeley, California, USA. Agric. Human Values 14:259-271.

Jensen, R. 2000. Agricultural volatility and investments in children. Am. Econ. Rev. 90(2):399-404.

Jensen, J.E., and P.H. Petersen. 2001. The Danish pesticide action plan II: Obstacles and opportunities to meet the goals. p. 449-454. In The BCPC Conf. Weeds 2001, Brit. Crop Prot. Council, Farnham UK.

Jervis, M.S., M.A. Kidd, M.D. Fitton, T. Huddleson, and H. Dawah. 1993. Flower visiting by hymnopteran parasitoids. J. Nat. Hist. 27:287-294.

Jiggins, J. 1984. Rhetoric or reality: Where do women in development projects stand today? Agric. Admin. Part I, 15(3):157-175; Part II, 15(4):223-237.

Jiggins, J. 1986. Gender-related impacts and the work of the international agricultural research centers. CGIAR Impact Study Pap. No. 17. World Bank, Washington DC.

Jiggins, J. (ed) 1989 International seminar on rural extension policies. Proc. IAC, Wageningen.

Jiggins, J., and G. Poulter. 2007. A strategic vision for European ARD in 2015 and beyond. Rep. ERA-Net project, European research area - Agricultural research for development. June. Available at www.era-ard.org.

Johnson, D.G. 1996. China's rural and agricultural reforms: Successes and failure. Chinese Econ. Res. Unit (CERU), Univ. Adelaide.

Johnson, D.G. 2000. Facts, trends and issues of open markets and food security. Pap. No. 00-01. Available at www.aic.ucdavis.edu/research/Aaas.pdf. Off. Agric. Econ. Res., Univ. Chicago.

Jones, G. (ed) 1986. Investing in rural extension: Strategies and goals. Elsevier, London.

Jones, G.E., and M.J. Rolls. 1982. Progress in rural extensiona and community development. John Wiley, NY.

Jones. M. 2004. The capacity of Africa's agricultural sector to contribute to achieving the UN Millennium Development Goals. Special Pap. Ann. Rep. 2004. CTA, Wageningen.

Joshi, K.D., B.R. Sthapit, and J.R. Witcombe. 2001. How narrowly adapted are the products of decentralized breeding? The spread of rice varieties from a participatory plant breeding programme in Nepal. Euphytica 122:589-597.

Joshi, K.D., A.M. Musa, C. Johansen, S. Gyawali, D. Harris, and J.R. Witcombe. 2007. Highly client-oriented breeding, using local preferences and selection, produces widely adapted rice varieties. Field Crops Res. 100:107-116.

Jungbluth, F. 1996. Crop protection policy in Thailand-economic and political factors influencing pesticide use. Pesticide Policy Project Publ. Ser. No. 5, Institut fur Gartenbauokonomie, Univ. Hannover, Germany.

Jungbluth, F. 2000. Economic analysis of crop protection in citrus production in central Thailand. Pesticide Policy Proj. Publ. Ser. Special Issue 4, Inst. Gartenbauokonomie, Univ. Hannover, Germany.

Jutsum, A.R., S.P. Heaney, B.M. Perrin, and P.J. Wege. 1998. Pesticide resistance: Assessment of risk and development and implementation of effective management strategies. Pest. Sci. 54:435-446.

Kairo, M.T.K. 2005. Hunger poverty and protection of biodiversity: Opportunities and challenges for biological control. p. 228-236. In Proc. Int. Symp. Biological Control of Arthropods. Davos, Switzerland. 12-16 Sept. 2005. USDA Forest Service, FHTET, Fort Collins, CO.

Kalkoven, J.T.R. 1993. Survival of populations and the scale of the fragmented agricultural landscape. p. 83-90. In R.G.H. Bunce et al. (ed) Landscape ecology and agroecosystems. Lewis Publ., Boca Raton.

Kampen, J. 1997. Financing agricultural research: Lessons learned with agricultural research funds. World Bank, Washington DC.

Kamuanga, M. 2001. Farmers' perceptions of the impacts of tsetse and trypanosomosis control on livestock production: evidence from southern Burkina Faso. Trop. Anim. Health Prod. 33:(2)141.

Karel, B. 2004. The persistence of pesticide dependence: A review of World Bank projects and their compliance with the World Bank's pest management policy, 1999-2003. PANNA, San Francisco.

Kardan, N. 1993. Development approaches and the role of policy advocacy: The case of the World Bank. World Dev. 21(11):1173-1186.

Kassa, B., and H. Beyene. 2001. Efficacy and economics of fungicide spray in the control of late blight of potato in Ethiopia. Afr. Crop Sci. J. 9(1):245.

Kataki, P.K., S.P. Srivastava, M. Saifuzzaman, and H.K. Upreti. 2001. Sterility in wheat and response of field crops to applied boron in the Indo-Gangestic plains. J. Crop Prod. 4:133-165.

Keese P., O. Galman-Omitogun, and E.B. Sonaiya. 2002. Seeds of promise: Developing a sustainable agricultural biotechnology industry in sub-Saharan Africa. Nat. Resour. Forum 26:234-244.

Kelly, L. 2005. Addressing challenges of globalization: An independent evaluation of the World Bank's approach to global programs. Operations Evaluation Dep., World Bank, Washington DC.

Kenmore, P.E., F.O Carino., C.A. Perez., V.A. Dyck, and A.P. Guttierez. 1984. Population regulation of the brown planthopper within rice fields in the Philippines. J. Plant Prot. Trop. 1(1):19-37.

Kennedy, D. 2001. Enclosing the research commons: The privatization of scientific research. Science 294:2249.

Kerr, J., and S. Kolavalli. 1999. Impact of agricultural research in poverty alleviation; Conceptual framework with illustrations from literature. EPTD Disc. Pap. IFPRI, Washington DC.

Kerr, N.A. 1987. The legacy: A centennial history of the state agricultural experiment stations: 1887-1987. Univ., Missouri, Columbia.

Khan, Z.R., K. Ampong-Nyarko, P. Chiliswa, A. Hassanali, S. Kimani, W. Lwande et al. 1997. Intercropping increases parasitism of pests. Nature 388:631-632.

Khush, G.S. 2001. Challenges of meeting the global food and nutrient needs in the new millennium. Proc. Nutrition Society, 60:15-26.

Kilcher, L. 2007. How organic agriculture contributes to sustainable development. J. Agric. Res. Trop. Subtrop. (Suppl) 89:31-49.

King, J.L. 2001. Concentration and technology in agricultural input industries. Agric. Inform. Bull., 763. USDA, Washington DC.

Kipkoech, K.A., F. Schulthess, W.K. Yabban, H.K. Maritim, and D. Mithoefer. 2006. Economics of biological control of cereal stem borers in Kenya. Ann. Soc. Entomol. France 42:519-528.

Kishi, M. 2005. The health impacts of pesticides: What do we know now? Chapter 2. In J. Pretty (ed) The pesticide detox: Towards a more sustainable agriculture. Earthscan, London.

Kishi, M., N. Hirschhorn, M. Qjajadisastra, L. Satterlee, S. Strowman, and R. Dilts. 1995. Relationship of pesticide spraying to signs and symptoms in Indonesian farmers. Scand. J. Work Environ. Health 21:124-33.

Kleinman, D.L., and S.P. Vallas. 2001. Science, capitalism and the rise of the "knowledge worker": The changing structure of knowledge production in the United States. Theory Society 30:451-492.

Kline, S., and N. Rosenberg. 1986. An overview of innovation. p. 275-396. In R. Landau, and N. Rosenberg (ed) The positive sum strategy. Harnessing technology for economic growth. Nat. Acad. Press, Washington DC.

Knight, R. 2007. Interim report 2006 on investigating environmental benefits. The voluntary initiative indicator farms project. Crop Prot. Assoc. UK and Farming and Wildlife Advisory Group (FWAG). Environ. Agency, UK.

Knudsen, M.T., N. Halberg, J.E. Olesen, J. Byrne, V. Iyer, and N. Toly. 2005. Global trends in agriculture and food systems. In N. Halberg et al. (ed) Global development of organic agriculture: Challenges and promises. CABI, UK.

Koc, M., R. MacRae, L.J.A. Mougeot, and J. Welsh (ed) 1999. For hunger-poor cities: Sustainable urban food systems. IDRC, Canada.

Korten, A. 1995. A bitter pill: Structural adjustment in Costa Rica. Inst. Food Dev. Policy, San Francisco.

Korten, D.C., and F.B. Alfonso. 1983 Bureaucracy and the poor: Closing the gap. Kumarian, Westport.

Kremer, M. 2003. Encouraging private sector for tropical agriculture. 7th Int. Conf. Public Goods and Public Policy for Agric. Biotech. Ravello, Italy.

Krimsky, S. 1999. The profit of scientific discovery and its normative implications. Kent Law Rev. 75:15-39.

Krimsky, S., and D. Golding (ed) 1992. Social theories of risk. Praeger, Westport.

Kroma, M.M., and C.B. Flora. 2003. Greening pesticides: A historical analysis of the social construction of farm chemical advertisements. Agric. Human Values 20:21-35.

Kuhn, T.S. 1970. the structure of scientific revolutions. Univ. Chicago Press.

Kumar, N., and N.S. Sidharthan. 1997. Technology, market structure and internationalisation: Issues and policies for developing countries. UNU/INTECH Routledge, London.

Kumari, B., and R. Kumar. 2003 Magnitude of pesticidal contamination in winter vegetables from Hisar, Haryana. Environ. Monitor. Assess. 87(3):311-318.

LaBelle, J. 2005. How has the food system became vulnerable? Executive summary. p.1-12. In Proc. New Perspectives on Food Security. 12-14 Nov. Available at http://www.glynwood.org/programs/foodsec/Executive%20Summary.pdf. Glynwood Center, Cold Spring NY.

Ladejinsky, L.J. (ed) 1977. Agrarian reform as unfinished business: The selected papers of Wolf Ladejinsky. Oxford Univ. Press, Oxford.

Laird, S.A. (ed) 2002. Biodiversity and traditional knowledge. Equitable partnerships in practice. Earthscan, London.

Lal, R. 1989. Conservation tillage for sustainable agriculture: Tropics versus temperate environments. Adv. Agron. 42:85-197.

Lal, R., N. Uphoff, B.A. Stewart, and D.O. Hansen. 2005. Soil carbon depletion and the impending food crisis. CRC Press, Baton Rouge FL.

Lang, T. 2006. Agriculture, food and health: Perspectives on a long relationship. In C. Hawkes and M.T. Ruel (ed) Understanding the links between agriculture and health. Focus 13, Brief 2, May. Available at http://www.ifpri.org/2020/focus/focus13/focus13_02.pdf. IPPRI, Washington DC.

Latorse, M.P., and K.H. Kuck. 2006. Phytophtora infestans: Baseline sensitivity and resistance management for fluopicolide. Pflanzenschutz-Nachrichten Bayer 59 (2-3):317-322.

Latouche, K., P. Rainelli, and D. Vermersch. 1998. Food safety issues and the BSE scare: Some lessons from the French case. Food Policy 23(5):347-356.

Latour, B. 2004. Politics of nature. How to bring the sciences into democracy. Harvard Univ. Press, Cambridge MA.

Leach, M., I. Scoones and B. Wynne (ed) 2005. Science and citizens: Globalization and the challenge of engagement. Zed Books, London.

Leach, M., and I. Scoones. 2006. The slow race. Making technology work for the poor. Demos, London.

Leeuwis, C., and A. van den Ban. 2004. Communication for rural innovation. Rethinking agricultural extension. 3rd ed. Blackwell Science, Oxford.

Lele, U. 1975. The design of rural development:

Lessons from Africa. Johns Hopkins Univ. Press, Baltimore.

Lele, U. and A.A. Goldsmith. 1989. The development of national agricultural research capacity: India's experience with the Rockefeller Foundation and its significance for Africa. Econ. Dev. Cult. Change 37(2):305-343.

Lenne, J.M., and J.B. Smithson. 1994. Varietal mixtures: a viable strategy for sustainable productivity in subsistence agriculture. Aspects Appl. Biol. 39:161-170.

Leonard, D.K. 1977. Reaching the peasant farmer. Organization, theory and practice in Kenya. Univ. Chicago Press.

Lesser, W., and M. Mutschler. 2002. Lessons from the patenting of plants. p. 103-118. *In* M Rotschild, S. Newman (ed) Intellectual property rights in animal breeding and genetics. CABI, Wallingford.

Lewis, W.J., J.C. van Lenteren, S.C. Phatak, and J.H. Tumlinson. 1997. A total system approach to sustainable pest management. PNAS 94(23):12243-12248.

Liebenthal, A. 2002. Promoting environmental sustainability in development: An evaluation of the World Bank's performance. Oper. Eval. Dep., World Bank, Washington DC.

Lightfoot, C., C. Alders and F. Dolberg (ed) 2002. Linking local learners: Negotiating new relationships between village, district and nation. ISG/ARDAF/Agroforum, Greve, Denmark.

Lighthall, D. 1995. Farm structure and chemical use in the corn belt. Rural Soc. 60(3):505-520.

Lindstrom, J., and R. Kingamkono. 1991. Foods from forests, fields and fallows. Working Pap. 184. Swedish Univ. Agric. Sci., Uppsala.

Lio, M. and Meng-Chun Liu. 2006. ICT and agricultural productivity: Evidence from cross-country survey data. Agric. Econ. 34(3):221-228.

Lionberger, H. 1960. Adoption of new ideas and practices. Iowa State Univ. Press, Ames.

Lionberger, H. 1986. Towards an idealised systems model for generating and utilising information in modernising societies. p. 105-135. In G.M. Beal et al. (ed) Knowledge generation, exchange and utilization. Westview Press, Boulder.

Lionberger, H., and P. Gwin. 1991. Technology transfer from research to Users. Univ. Missouri Press, Colombia.

Lipton, M. 2001. Challenges to meet: Food and nutrition security in new millennium. Proc. Nutr. Society 60:203-214.

Lipton, M. 2005. The family farm in a globalizing world. The role of crop science in alleviating poverty. 2020 Disc.Pap. 40. IFPRI, Washington DC.

Lipton, M. and R. Longhurst. 1989. New seeds and poor people. Unwin Hyman, London.

Litsinger, J.A. 1989. Second generation insect pest problems on high yielding rices. Trop. Pest Manage. 35(3):235-242.

Lockie, S., and F. Vanclay (ed) 1997. Critical landcare. Centre Rural Social Res. Charles Sturt Univ., Wagga Wagga.

Loevinsohn, M.E. 1987. Insecticide use and increased mortality in rural Central Luzon, Philippines. The Lancet 8546:1350-1362.

Löhr, B., R. Gathu, C. Kariuki, J. Obiero, and G. Gichini. 2007. Impact of an exotic parasitoid on Plutella xylostella (Lepidoptera: Plutellidae) population dynamics, damage and indigenous natural enemies in Lenya. Bull. Entomol. Res. 97:337-350.

London, L., A. Ngowi, M. Perry, H.A. Rother, E. Cairncross, A. Solomon et al. 2005. Health and economic consequences of pesticide use: The experience of the HHED programme on pesticides in Southern Africa. Epidemiol. 16:S29.

Loomis, C.P., and J.A. Beagle. 1950. Rural social systems. Prentice Hall, NY.

Losey, J.E. and M. Vaughan 2006. The economic value of ecological services provided by insects. Bioscience 56:310-323.

Louwaars, N.P., R. Tripp, D. Eaton, V. Henson-Apollonio, R. Hu, M. Mendoza et al. 2005. Impacts of strengthened intellectual property rights regimes on the plant breeding industry in developing countries: A synthesis of five case studies. Centre for Genetic Resources, Wageningen.

Lubchenco, J. 1998. Entering the century of the environment: A new social contract for science. Science 279:491-497.

Lumbreras, L.G. 1992. Cultura, tecnologica y modelos alternatives de desarrollo. Comercio Exterior 42:199-205.

Luther, G., C. Harris, S. Sherwood, K. Gallagher, J. Mangan, and K. Touré Gamby. 2005. Developments and innovations in farmer field schools and training of trainers. p. 159-190. *In* G.W. Norton et al. (ed) Globalizing integrated pest management: A participatory research process. Blackwell Science, Oxford.

Lyson, T.A. 2005. Systems perspectives on food security. p. 65-68. *In* Proc. New Perspectives on Food Security. 12-14 Nov. Glynwood Center, Cold Spring NY.

Maathai, W. 2003. Green Belt Movement: Sharing the approach and the experience. Lantern Books, NY.

Macha, M., A. Rwazo, and H. Mkalanga. 2001. Retail sales of pesticides in Tanzania: occupational human health and safety considerations. Afr. Newsl. Occup. Health Safety 11(2):40-42.

Macharia, I., B. Löhr and H. De Groote. 2005. Assessing the potential impact of biological control of Plutella xylostella (diamondback month) in cabbage production in Kenya. Crop Prot. 24: 981-989.

Macharia, I., D. Mithoefer, and B. Löhr. 2007. Update of the ex-ante impact assessment of the biological control of Plutella xylostella (Diamondback moth) in Kenya. *In* A.M. Shelton et al. (ed) 2006. Proc. 5th Int. Workshop on Diamondback Moth and other Crucifer Insect Pests. Beijing. 24-27 Oct. 2006.

Malik, R.K. 1981. Food a priority for consumer protection in Asia and the Pacific region. Food Nutr. 7:2.

Mancini. F. 2006. An evaluation of farmer field schools in the cotton areas of southern India: Impacts on farmers' health, the environment, labour use, farming systems and livelihoods. Publ. PhD. Wageningen Univ., Wageningen.

Mancini, F., A. van Bruggen, J. Jiggins, 2007. Evaluating cotton integrated pest management (IPM) farmer field school outcomes using a sustainable livelihoods approach in India. Exp. Agric. 43:97-112.

Mangan, J., and M. Mangan. 1998. A comparison of two IPM training strategies in China: the Importance of concepts of the rice ecosystem for sustainable insect pest management. Agric. Human Values 15:209-221.

Mangione, D., S. Senni, M. Puccione, S. Gando, S. Ceccarelli. 2006. The cost of partivciaptory barley breeding. Euphytica 150:289-306.

Maredia, M.K., and D.A. Raitzer. 2006. CGIAR and NARS partner research in sub-Saharan Africa: evidence of impact to date [Online]. Available at http://www.sciencecouncil.cgiar.org/activities/spia/pubs/CGIAR_WP_AI-proof3.pdf. CGIAR Sci. Council, Washington DC.

Martin, G.J. 1995. Ethnobotany, a methods manual. WWF International, Chapman and Hall. London.

Martinez-Torres, M. 2006. Organic coffee: Sustainable development by Mayan farmers. Ohio Univ. Press.

Mason, J., A. Bailes, M. Beda-Andourou, N. Copeland, T. Curtis, M. Deitchler et al. 2005. Recent trends in malnutrition in developing regions: Vitamin A deficiency, anaemia, iodine deficiency, and child undernutrition. Food Nutr. Bull. 26(1):59-107.

Master, W.A. 2003. Research prices: A mechanism for innovation in African agriculture. 7th ICABR Int. Conf. Public Goods and Public Policy for Agricultural Biotechnology. Ravello, Italy.

Mathews, K. Jr., and J. Buzby. 2001. Dissecting the challenges of mad cow and foot-and-mouth disease. ERS/USDA, Agricultural Outlook/August, Washington DC.

Matson, P.A., W.J. Parton, A.G Power, and M.J. Swift. 1997. Agricultural intensification and ecosystem properties. Science 227:504-509.

Mattera, P. 2004. USDA Inc.: How Agribusiness has hijacked regulatory policy at the U.S. Department of Agriculture. Corporate Res. Project, Good Jobs First, Washington DC.

Matteson, P.C., M.A. Altieri and W.C. Gagne. 1984. Modification of small farmer practices for better pest management. Ann. Rev. Entomol. 24:383-402.

Maturana and Varela. 1992. The tree of knowledge: The biological roots of understanding. Shambhala, Boston.

Maunder, A.H. 1972. Agricultural extension. A reference manual. FAO, Rome.

Mayberry, B.D. 1991. A century of agriculture in the 1890 land-grant institutions and Tuskegee University—1890-1990. Vantage Press, NY.

Mazoyer M. 2005. Développement agricole inégal et sous-alimentation paysanne *In* L. Roudard and M. Mazoyer (ed) La fracture agricole et alimentaire mondiale, nourrir l'humanité aujourd'hui et demain. Encycl. Universalis, Paris.

Mazoyer, M., and L. Roudard. 2005. Brève histoire des agricultures du monde *In* M. Mazoyer and L. Roudard (ed) La fracture agricole et alimentaire mondiale, nourrir l'humanité aujourd'hui et demain. Encycl. Universalis, Paris.

Mboob, S.S. 1994. Integrated pest management: A review of the consutlants' report and its relevance to West Africa. p. 9-12. *In* IPM Working Group, Nat. Resources Inst, (ed) IPM Implementation Workshop for West Africa. Workshop Proc. Accra. NRI, Chatham.

McConnell, R., and A.J. Hruska. 1993. An epidemic of pesticide poisoning in Nicaragua: Implication for prevention in developing countries. Am. J. Public Health 83(11):1559-1562.

McCown, R.L., G.L. Hammer, J.N.G. Hargreaves, D. Holzworth, and D.M. Freebairn. 1996. APSIM: A novel software system for model development, model testing and simulation in agricultural systems research. Agric. Syst. 50:255-271.

McCown, R.L., Z. Hochman, and P.S. Carberry (ed) 2002. Learning from 25 Years of agricultural DSS: Case histories of efforts to harness simulation models for farm work. Agic. Syst. 74:1-220.

McFadyen, R.C. 1998. Biological control of weeds. Ann. Rev. Entomol. 43:369-393.

McGowan, L. 1997. Democracy undermined, economic justice denied: Structural adjustment and the aid juggernaut in Haiti. The Development Gap, Washington DC.

McMichael, P. 2001. The impact of globalization, free trade and technology on food and nutrition in the new millennium. Proc. Nutr. Society. 60(2):195-201.

Mead, P.S., L. Slutsker, V. Dietz, L.F. McCaig, J.S. Bresee, C. Shapiro et al. 1999. Food-related illness and death in The United States. Emerging Infect. Dis. 5:607-625.

Meinzen-Dick, R., M. Adato, L. Haddad, and P. Hazell. 2004. Science and poverty, an interdisciplinary assessment of the impact of agricultural research. IFPRI, Washington DC.

Meir, C., and S. Williamson. 2005. Farmer decision-making for ecological pest management. p. 82-96 in J. Pretty. (ed) The pesticide detox: Towards a more sustainable agriculture. Earthscan, London.

Mellon, M., and J. Rissler. 2004. Gone to seed: Transgenic contaminants in the traditional seed supply. Available at http://www.ucsusa.org/food_and_environment/biotechnology/page.cfm?pageID=13-15. Union Concerned Scientists, Cambridge MA.

Mellor, J.W., 1966. The economics of agricultural development. Cornell Univ. Press, Ithaca.

Mellor, J.W. 1990. Global food balances and food security. *In* C.K. Eicher and J.M. Staatz (ed) Agricultural development in the third world. 2nd ed. Johns Hopkins Univ. Press, Baltimore.

Mendelsohn, R. 2000. The market value of farmers' rights. In V. Santaniello et al. (ed) Agriculture and intellectual property rights: Economic, institutional and implementation issues in biotechnology, CABI Publ., UK.

Menezes, F. 2001. Food sovereignty: A vital requirement for food security in the context of globalization. Thematic Section: Putting people centre stage. Society Int. Dev. SAGE Publ., London.

Merton, R.K. 1957. Social theory and social structure. The Free Press, Glencoe.

Mitchell, J.P., P.B. Goodell, R. Krebill-Prather, T. Prather, K. Hembree, D. Munk et al. 2001. Innovative agricultural extension partnerships in California's central San Joaquin valley. J. Extension 39(6). Available at http://www.joe.org/joe/2001december/rb7.html.

Mkumbira, J. 2002. Cassava development for small farmers. Approaches to breeding in Malawi. PhD thesis. Agraria 365. Swedish Univ. Agric. Sci., Uppsala.

Mkumbira, J., L. Chiwona-Karltun, U. Lagercrantz, N.M. Mahungu, J. Saka, M. Hone et al. 2003. Classification of cassava into 'bitter' and 'cool' in Malawi: From farmers' perception to characterisation by molecular markers. Euphytica 132:7-22.

Mol, A.P.J., and H. Bulkley. 2002. Food risks and the environment: Changing perspectives in a changing social order. J. Environ. Policy Plan. 4:185-195.

Mollison, B. 1988. Permaculture: A designer's manual. Tagari Publications, Tylagum, Australia.

Moock, P.R. 1976. The efficiency of women as farm managers: Kenya. Am. J. Agric. Econ. 58:831-835.

Mook, B. 1974. Value and action in Indian bureaucracy. Disc. Pap. no. 65. IDS, Sussex.

Moore, D. 2004. Biological control of *Rastrococcus invadens*. Biocontrol News Inform. 25(1):17N-27N. Available at http://www.pestscience.com/PDF/BNIra67.PDF. CABI, UK.

Morales, J.A. 1991. Structural adjustment and peasant agriculture in Bolvia. Food Policy 16:58-66.

Morgan, P, and H. Baser. 1993. Making technical cooperation more effective. New approaches in international development. CIDA, Tech. Coop. Div., Hull.

Moris, J. 1981. Managing induced rural development. Int. Dev. Inst., Bloomington.

Moris, J. 1989. Extension under East African field conditions. p. 73-85. *In* N. Roberts (ed) 1989. Agricultural extension in Africa. World Bank Symp. Available at http://www.eric.ed.gov/ERICDocs/data/ericdocs2sql/content_storage_01/0000019b/80/20/83/7c.pdf. World Bank, Washington DC.

Morris, M. 1998. Maize seed industries in developing countries. Lynne Rienner Publ., Boulder CO.

Morris, M., and B. Ekasingh. 2001. Plant breeding research in developing countries: what roles for the public and private sectors. *In* D. Byerlee, R. Echeverria (ed) Agricultural research policy in an era of privatization: Experiences from the developing world. CABI, UK.

Morss, E.R., J.K. Hatch, D.R. McElewait, and C.F. Sweet. 1976. Strategies for small farmer development. An empirical study of rural development projects in Gambia, Ghana, Kenya, Lesotho, Nigeria, Bolivia, Colombia, México, Paraguay and Peru. Westview Press, Boulder CO.

Moschini, G., and H. Lapan. 1997. Intellectual property rights and the welfare effects of agricultural R&D. Am. J. Agric. Econ. 79:1229-1242.

Mosher, A.T. 1966. Getting agriculture moving: Essentials for development and modernization. Parts I & II. Agric. Dev. Council, NY.

Mosse, D. 2004a. Social analysis as product development: Anthropologists at work in the World Bank. p. 77-87. *In* A.K. Giri et al. (ed). The development of religion/The religion of development, Eburon, Delft.

Mosse, D. 2004b. Is good policy unimplementable? Reflections on the ethnography of aid policy and practice. Dev. Change 35(4):639-671.

Mowery, D.C., and B.N. Sampat. 2004. Universities in national innovation systems. p. 209-239. *In* The Oxford handbook of innovation. J. Fagerberg et al. (ed) Available at http://www.globelicsacademy.net/pdf/DavidMowery_1.pdf. Oxford Univ. Press, UK.

MRC. 2007. Report from the International Conference on the Mekong River Commission (MRC). MRC, Vietnam.

MSU. 2000. The database of arthropods resistant to pesticides. Michigan State Univ. Available at www.cips.msu.eud/resistance. MSU, East Lansing MI.

Mudimu, G.D., S. Chigume, and M. Chikanda.1995. Pesticide use and policies in Zimbabwe. Pesticide policy project Publ. Ser. no. 2, Institut für Gartenbauökonomie, Univ. Hanover, Germany.

Mukherjee, S., and V. Kathuria. 2006. Is economic growth sustainable? Environmental quality of Indian States after 1991. Int. J. Sustain. Dev. 9:1.38-60.

Mulatu, E., and H. Zelleke. 2002. Farmers' highland maize (Zea mays L.) selection criteria: Implication for maize breeding for the Hararghe highlands of eastern Ethiopia. Euphytica 127:11-30.

Mungate, D. 1983. Women: The silent farm managers in the small-scale commercial areas of Zimbabwe. Zimbabwe Agric. J. 80(6):245-249.

Muntemba, S., and R. Chimedza. 1995. Women spearhead food security. *In* Missing links: Gender equity in science and technology for development. Available at www.idrc.ca/en/ev-29519-201-1-DO_TOPIC.html. IDRC-ITDG Publ., Canada.

Murphy, B.C., J.A. Rosenheim, J. Granett, C.H. Pickett, and R.V. Dowell. 1998. Measuring the impact of a natural enemy refuge: The prune tree/vineyard example. p. 297-309. *In* C.H. Pickett and R.L. Bugg (ed) Enhancing biological control: Habitat management to promote natural enemies of agricultural pests. Univ. California Press, Berkeley.

Murray, D.L. 1994. Cultivating crisis: The human cost of pesticides in Latin America. Univ. Texas Press, Austin.

Murray, D.L., and P.L. Taylor. 2000. Claim no easy victories: evaluating the pesticide industry's global safe use campaign. World Dev. 28(10):1735-1749.

Muzale, P., with D.K. Leonard. 1985. Kenya's experience with women's groups in agricultural extension: Strategies for accelerating improvements in food production and nutrition awareness in Africa. Agricultural Admin. 19(1):13-28.

Nash, R. 1989.The rights of nature: A history of environmental ethics. Univ. Wisconsin Press, Madison.

Nash, J., and H. Safa. 1985. Women and change in Latin America. New directions in study of sex and class. Bergen and Garvey, NY.

Nathaniels, N.Q.R., M.E.R. Sijaona, J.A.E. Shoo, and N. Katinila. 2003. IPM for control of cashew powdery mildew in Tanzania: Farmers' crop protection practices, perceptions and sources of information. Int. J. Pest Manage. 49(1):25-36.

Nature Biotechnology. 2004. Drugs in crops — the unpalatable truth. Nature Biotech. 22:133.

Naylor, R., and R. Ehrlich. 1997. Natural pest control services and agriculture. p. 151-174 *In:* G.C. Daily (ed) Nature's services societal dependence on natural ecosystems. Island Press, Washington, DC.

Naylor, R., W.P. Falcon, R.M. Goodman., M.M. Jahn, T. Sengooba, H. Tefera et al. 2004. Biotechnology in the developing world: a case for increased investments in orphan crops. Food Policy 29:15-44.

Nelkin, D. 1975. Changing images of science: New pressures on old stereotypes. Newsl. Progr. Public Concept. Sci. 14:21-31.

NEPAD (New Economic Program for African Development). 2002. Comprehensive Africa agriculture development programme (CAADP). FAO/NEPAD, Rome.

Neri, M.S. 2005. Pesticide poisoned Bohol schoolchildren: Health department. SunStar News, Philippines, Tuesday, Mar. 15. Available at http://www.sunstar.com.ph/static/net/2005/03/15/pesticide.poisoned.bohol.schoolchildren.health.dep.t.html.

Neuenschwander, P. 1993. Human interactions in classical biological control. *In* M. Altieri (ed) Crop protection strategies for subsistence farmers. Westview Press, Boulder and London.

Neuenschwander, P. 2001. Biological control of the cassava mealybug in Africa: A review. Biol. Control 21:214-229.

Neuenschwander, P. 2004. Harnessing nature in Africa: Biological pest control can benefit the pocket, health and the environment. Nature 432:801-802.

Neuenschwander, P., C. Borgemeister and J. Langewald (ed) 2003. Biological control in IPM systems in Africa. CABI Publ., UK

Newell, P. and J. Wheeler (ed) 2006. Rights, resources and the politics of accountability. ZED Books, London.

Nguyên H.H., and T.A. Dao. 2001. Vietnam promotes solutions to pesticide risks. Pesticide News 53:6-7.

Nidumolu, U.B., H. van Keulen, M. Lubbers, and H. Mapfumo. 2007. Combining interactive multiple goal linear programming with an inter-stakeholder communication matrix to generate land use options. Environ. Modelling Software 22:73-83.

Norgaard, R.B. 1988. The biological control of cassava mealybug in Africa. Am. J. Agric. Econ. 70:366-371.

Northwest Area Foundation. 1994. A better row to hoe. The economic, environmental and social impact of sustainable agriculture. Northwest Area Foundation, St. Paul.

Northwest Area Foundation. 1995. Planting the future: Developing an agriculture that sustains land and community. Iowa Univ. Press, Ames.

Norton, G.W., J. Tjornhorn, D. Bosch, J. Ogrodowczyk, C. Edwards, T. Yamagiwa et al. 2005. Pesticide and IPM policy analysis. p. 191-210. *In* G.W. Norton et al. (ed) Globalizing integrated pest management: A participatory process. Blackwell Publ., Ames.

NRC (National Research Council). 1989. Alternative agriculture. Nat. Acad. Press, Washington DC.

NRC (National Research Council). 1993. Pesticides in the diet of infants and children. Nat. Acad. Press, Washington DC.

NRC (National Research Council). 1996. Understanding risk. Informing decisions in a democratic society. Nat. Acad. Press, Washington DC.

NRC (National Research Council). 2004. Biological confinement of genetically engineered crops. Nat. Acad. Press, Washington, DC.

Nurminene, T., R. Kaarina, K. Kurppa, and P.C. Holmberg. 1995. Agricultural work during pregnancy and selected structural malformations in Finland. Epidemiol. 6(1):23-30.

Nwanze, K.F., and R.A.E. Mueller. 1989. Management options for sorghum stem borers for farmers in the semi-arid tropics. *In* Int. Workshop on Sorghum Stem Borers. 17-20 Nov. 1987. ICRISAT, Patancheru, India.

Obryki, J.J., J.R. Ruberson, J.E. Losey. 2002. Interactions between natural enemies and transgenic insecticidal crops. p. 83-206. *In* L.E. Ehler et al. (ed) Genetics, evolution and biological control. CABI Publ., UK.

OCA. 2006. Global week of action against food irradiation. Statement of Food and Water Watch. Exec. Dir. Wenonah Hauter, Food and Water Watch, Nov. 28.

O'Connor, M. 2006. The AICCAN, the geGDP and the Monetisation Frontier: a typology of 'environmentally adjusted' national sustainability indicators. Int. J. Sustain. Dev. 9:61-99.

ODI. 1994. The CGIAR: What future for international agricultural research? Briefing Pap. Sept. Overseas Dev. Inst., London

OECD. 1995. Guidelines for Aid Agencies on Pest and Pesticide Management. OECD, Paris.

OECD. 2001. Multifunctionality: Towards an analytical framework. OECD, Paris.

OECD. 2006a. Promoting pro-poor growth: Agriculture. OECD, France.

OECD. 2006b. Report of the OECD pesticide risk reduction steering group: The second risk reduction survey. OECD Ser. Pesticides. No. 30. OCED, Paris.

Oesterheld, M., M. Semmartin, and A.J. Hall. 2002. Análisis bibliográfico de la investigación agronómica en la Argentina. Ciencia Hoy 12:52-62.

Ojiem, J.O., N. de Ridder, B. Vanlauwe, and K.E. Giller. 2006. Socio-ecological niche: A conceptual framework for integration of legumes in smallholder farming systems. Int. J. Sust. Agric. 4:79-93

Okoth, J., G. Khisa, and J. Thomas. 2003. Towards self-financed farmer field schools. LEISA 19(1):28-29.

Omwega, C.O., E. Muchugu, W.A Overholt, and F. Schulthess. 2006. Release and establishment of Cotesia flavipes (Hym., Braconidae) an exotic parasitoid of Chilo partellus (Lep., Cambridae) in East and Southern Africa. Ann. Soc. Entomol. France 42:511-517.

Ooi, P. 1998. Beyond the farmer field school: IPM and empowerment in Indonesia. Gatekeeper Ser. No. 78. IIED, London.

Oshaug, A. 2005. Developing voluntary guidelines for implementing the right to adequate food anatomy of a intergovernmental process. Chapter 11. p. 259-282. *In* W.B. Edie and U. Kracht (ed) Food and human rights in development. Vol. I. Legal and institutional dimensions and selected topics. Intersentia N.V., Antwerp.

Oshaug, A., W.B. Eide, and A. Eide. 1994. Human rights: A normative basis for food and nutrition-relevant policies. Food Policy 19:6, 491-516.

Oshaug, A. and W.B. Eide. 2003. The long process of giving consent to an economic, social and cultural right: Twenty-five years with the case of the right to adequate food. Chapter XIII. p. 325-369. *In* M. Bergsmo (ed) Human rights and criminal justice for the downtrodden. Essays in honour of Asbjørn Eide. Marinus Nijhoof Publ., Leiden and Boston.

OTA. 1979. Pest management strategies in crop protection. Vol. 1. US Gov. Printing Off., Washington DC.

Oxfam America. 1995. The impact of structural adjustment on community life: Undoing development. Oxfam, Boston.

Paarlberg, R.L. 2002. Governance and food security in an age of globalization. Food Agric. Environ. Disc. Pap. 36. IFPRI, Washington DC.

Paine, M.S., C.R. Burke, G.A. Verkerk, and P.J. Jolly. 2000. learning together about dairy cow fertility technologies in relation to farming systems in New Zealand. p. 163-175. In M. Cerf (ed) Cow up a tree. Knowing and learning for change in agriculture. Case studies from industrialised Countries. INRA Editions, Paris.

Paredes, A., and J. Moncado. 2000. Produce foundations in Mexico: An innovative participatory approach and demand-driven technology innovation model. GFAR meeting, Dresden, Germany.

Pardey, P.G., and N.M. Beintema. 2001. Slow magic: Agricultural R&D a century after Mendel. ASTI Initiative. IFPRI, Washington DC.

Pardey, P.G., N.M. Beintema, S. Dehmer, and S. Wood. 2006a. Agricultural research: A growing global divide? Food Policy Rep. IFPRI, Washington DC.

Pardey, P G., J.M. Alston, and R.R. Piggott (ed) 2006b. Agricultural R&D in the developing world too little, too late? IFPRI, Washington DC.

Pardey, P.G., J.M. Alston, and R.R. Piggott (ed) 2006c. Shifting ground: Agricultural R&D worldwide. Issue Brief No. 46. IFPRI, Washington DC.

Parker, P., R. Latcher, A. Jakeman, M.B. Beck, G. Harris, R.M. Argent et al. 2002. Progress in integrated assessment and modelling. Environ. Modelling Software 17:209-217.

Parrado-Rosselli, A. 2006. A participatory research approach for studying fruit availability and seed dispersal. Chapter 7. In Fruit availability and seed dispersal in terra firme rain forests of Colombian Amazon. Tropenbos PhD Ser. 2. Tropenbos Int., Wageningen.

Parrott, N., and T. Marsden. 2002. The real green revolution: Organic and agroecolocical farming in the South. Greenpeace Environ. Trust, London.

Paulson, D.D. 1995. Minnesota extension agents' knowledge and views of alternative agriculture. Am. J. Alternat. Agric. 10(3):122-128.

Paveley, N.D., D.J. Royle, R.J. Cook, U.A. Schoefl, D.B. Morris, M.J. Hims et al. 1994. Decision support to rationalise wheat fungicide use. p. 679-686. In The BCPC Conference Pests and Diseases 1994. Brit. Crop Prot. Council, Farnham, UK.

Pawar, C.S. 2002. IPM and plant science industries in India [Online]. Available at http://www.croplifeasia.org/ref_library/ipm/IPM_PSI.PDF. Agrolinks, Croplife Asia, Bangkok.

Paxson, H. 2007. Reverse engineering terroir — crafting nature-culture in American artisanal cheese production. Available at http:/web.mit.edu/newsoffice/2007/cheese.html. MIT TechTalk 11 Dec 2006.

Pearson, C.J., and R. Ison. 1997. Agronomy of grasslands systems. 2nd ed.Cambridge Univ. Press, UK. Pedigo, L.P. 1992. Integrating preventive and therapeutic tactics in soybean insect management. p. 10-19. In L.G. Coping et al. (ed) Pest management in soybeans. Elsevier, London.

Pedigo, L.P. 1989. Entomology and pest management. Macmillan, N.Y.

Pedigo, L.P. 1992. Integrating preventive and therapeutic tactics in soybean insect management. p. 10-19 In L.G. Coping et al. (ed) Pest management in soybeans. Elsevier, London.

Peeters, J.P., and J.T. Williams. 1984. Towards better use of genebanks with special reference to information. Plant Genetic Resour. Newsl. 60:22-32.

Pemsl, D., H. Waibel, and A.O. Gutierrez. 2005. Why do some Bt-cotton farmers in China continue to use high levels of pesticides? Int. J. Sustain. 3:44-56

Pengue, W.A. 2005. Transgenic crops in Argentina: The ecological and social debt. Bull. Sci. Tech. Soc. 25:314-322

Pennisi, E. 2003. Modernizing the tree of life. Science 300:1692-1697.

Penot, E. 1994. La riziculture de mangrove de la société balant dans la région de Tombali en Guinée-Bissau. p. 209-221. In Dynamique et usages de la mangrove dans les pays des Rivières du Sud (Du Sénégal à la Sierra Leone), M.C. Cormier-Salem (ed) Colloques et Séminaires, Orstom, Paris:

Pennycook, F.R., E.M. Diamand, A. Watterson, and C.V. Howard. 2004. Modeling the dietary pesticides exposures of young children. Int. J. Occup. Environ. Health 10:304-309.

Perales, Hugo R., S.B. Brush, and C.O. Qualset. 2003. Dynamic management of maize landraces in Central Mexico. Econ. Bot. 57:21-34.

Perkins, J.H. 1982. Insects, experts, and the insecticide crisis: The quest for new pest management strategies. Plenum Press, NY.

Peterson, W., C. Sands, and B. Swanson. 1989 Technology development and transfer systems in agriculture. INTERPAKS report. Off. Int. Agric. University of Illinois, Urbana.

Petit, M.J., G.E. Alex, H. Blackburn, W. Collins, J.J. Doyle, R.D. Freed et al. 1996. The emergence of a global agricultural research system. The role of the agricultural research and extension group (ESDAR). World Bank, Washington DC.

Petrick, M., and P. Weingarten. 2004. The role of agriculture in Central and Eastern European development: Engine of change or social buffer? Available at http://www.iamo.de/dok/sr_vol25.pdf. IAMO, Salle.

Pimbert, M.P., and T. Wakeford. 2002. Prajateerpu: A citizen's jury/scenario workshop on food and farming futures for Andhra Pradesh, India. London, IIED and Sussex, IDS.

Pimbert, M.P. 2006. farmers' views on the future of food and small-scale producers. IIED, London.

Pimentel, D. 1993. Economics and energetics of organic and conventional farming. J. Agric. Environ. Ethics 6:53-60.

Pimentel, D. 1997. Pest management in agriculture. p. 1-12. In D. Pimentel (ed) Techniques for reducing pesticide use: Environmental and economic benefits. John Wiley, Chichester UK.

Pimentel, D. 2005. Environmental and economic costs of the application of pesticides primarily in the United States. Environ. Dev. Sustain. 7:229-252.

Pimentel, D., U. Stachow, D.A. Takacs, H.W. Brubaker, A.R. Dumas, J.J. Meaney et al. 1992. Conserving biological diversity in agricultural/forestry systems. BioScience 42: 354-362.

Pimentel, D., H. Acquay, M. Biltonen, P. Rice, M. Silva, J. Nelson et al. 1993. Assessment of environmental and economic impacts of pesticide use. p. 47-84. In D. Pimentel, H. Lehman (ed.) The pesticide question: Environment, economics and ethics. Chapman and Hall, NY.

Pimentel, D., C. Wilson, C. McCullum, R.Huang, P.D. Wen, J. Flack et al. 1997. Economic and environmental benefits of biodiversity. Bioscience. 47:747-757.

Pimentel, D., P. Hepperly, J. Hanson, D. Douds, and R. Seidel. 2005. Environmental, energetic, and economic comparisons of organic and conventional farming systems. BioScience 55:573-582.

Pimbert, M. 2007. Transforming knowledge and ways of knowing for food sovereignty. IIED, London.

Pimbert, M.P., and T. Wakeford. 2002. Prajateerpu: A citizens jury / Scenario workshop on food and farming futures for Andhra Pradesh, India. Available at http://www.prajateerpu.org. IIED, London.

Pimbert, M.P., K. Tran-Thanh, E. Deléage, M. Reinert, C. Trehet, and E. Bennett (ed) 2006. Farmers' views on the future of food and smallscale producers. IIED, London.

Pingali, P.L., and G. Traxler. 2002. Changing locus of agricultural research: Will the poor benefit from biotechnology and privatization trends? Food Policy 27:223-228.

Plencovich, M., and A. Costantini. 2003. Dimensiones productivas, sociopolíticas y educativas de la inserción de Escuelas Agrotécnicas de la Región Pampeana. UBACYT. 2004-2007. Univ. Buenos Aires, Buenos Aires.

Plencovich, M.C., and A. Costantini. 2006. El itinerario de las escuelas agropecuarias: la educación silenciosa. XX Jornadas de Educación Agropecuaria, 2006, Paraná. FEDIAP. Available at www.fediap.com.ar/material/Lic.%20María%20Cristina%20Plencovich.ppt.

Pontius, J., R. Dilts, and A. Bartlett. 2002. From farmer field schools to community IPM: Ten years of ipm training in Asia. FAO, Reg. Off. Asia Pacific, Bangkok.

Pontius, J. 2003. Picturing impact: Participatory evaluation of community IPM in three West Java villages. p. 227-260. In CIP-UPWARD. Farmer field schools: Emerging issues and challenges. Users' perspectives with agricultural research and development. CIP, Los Baños.

Pray, C.E., J.F. Oehmke, and A. Naseem. 2005. Innovation and dynamic efficiency in plant biotechnology: An introduction to the researchable issues. AgBioForum 8:52-63.

Prescott, V., P.M. Campbell, A. Moore, J. Mattes, M.E. Rothenberg, P.S. Foster et al. 2005. Transgenic expression of bean alpha-amylase inhibitor in peas results in altered structure and immunogenicity. J. Agric. Food Chem. 53(23):9023-9030.

Pretty, J.N. 1994. Alternative systems of inquiry for sustainable agriculture. IDS Bull. 25(2):37-48.

Pretty, J.N. 1995. Regenerating agriculture: Policies and practice for sustainability and self-reliance. Earthscan, London and Nat. Acad. Press, Washington DC.

Pretty, J.N. 1999. Can sustainable agriculture feed Africa? New evidence on progress, processes and impacts. Environ. Dev. Sustain. 1:253-274.

Pretty, J.N. 2002. Agri-Culture: Reconnecting people, land and nature. Earthscan, London.

Pretty, J.N. (ed) 2005a. The earthscan reader in sustainable agriculture. Earthscan, London.

Pretty, J.N. (ed) 2005b. The pesticide detox. Towards a more sustainable agriculture. Earthscan, London.

Pretty, J., and R. Hine. 2005. Pesticide use and the environment. In J. Pretty (ed) The pesticide detox: Towards a more sustainable agriculture. Earthscan, London.

Pretty, J., and H. Waibel. 2005. Paying the price: The full cost of pesticides. p. 39-54. In J. Pretty (ed) The pesticide detox. Earthscan, London.

Pretty, J., J. Morison, R. Hine. 2003. Reducing food poverty by increasing agricultural sustainability in developing countries. Agric. Ecosyst. Environ. 95:217-234.

Prigogine, I., and I. Stengers. 1979. La nouvelle alliance. Métamorphose de la science. Ed. Gallimard, Paris.

Prokopy, R.J. 2003. Two decades of bottom-up, ecologically based pest management in a small commercial apple orchard in Massachusetts. Agric. Ecosyst. Environ. 94: 299-309.

Qaim, M., and G. Traxler. 2002. Roundup ready soybeans in Argentina: Farm level and aggregate welfare effects. Agric. Econ. 32:73-86.

Qayum, A., and K. Sakkhari. 2005. Bt Cotton in Andhra Pradesh. A tree-year assessment. Deccan Dev. Soc.Village Pastapur. Zakeerabad, Medak District, AP, India.

Quiros, C.F., R. Ortega, L .Van Raamsdonk, M. Herrera-Montoya, P. Cisneros, and E. Schmidt. 1992. Increase of potato genetic resources in their center of diversity: The role of natural outcrossing and selection by the Andean farmer. Genetic Resour. Crop Evol. 39:107-113.

Quizon J., G. Feder, and R. Murgai. 2000. A note on the sustainability of the farmer field school approach to agricultural extension. Dev. Econ. Group, World Bank, Washington DC.

Raedeke, A.H., and J.S. Rikoon.1997. Temporal and spatial dimensions of knowledge: Implications for sustainable agriculture. Agric. Human Values 14:145-158.

Ramaswami, B. and C.E. Pray. 2007. India: Confronting the challenge — the potential of genetically modified crops for the poor. p. 156-174. Chapter 9 In S. Fukuda-Parr (ed) The gene revolution. GM crops and unequal development. Earthscan, London.

Readings, B. 1996. The university in ruins. Harvard Univ. Press, Cambridge MA.

Reardon, T., C.P. Timmer, C.B. Barrett, and J. Berdegué. 2003. The rise of supermarkets in Africa, Asia and Latin America. Am. J. Agric. Econ. 85(5):1140-1146.

Reeves, M., A. Katten, and M. Guzman. 2002. Pesticide safety laws fail to protect farmworkers. In Field of poison. Available at www.panna.org. PANNA, San Francisco.

Reeves, M., and K.S. Schafer. 2003. Greater risks, fewer rights: US farmworkers and pesticides. Int. J. Occup. Environ. Health 9:30-39.

Reeves, T.G and K. Cassaday. 2002. History and past achievements of plant breeding. Aust. J. Agric. Res. 53:851-863.

Reganold, J.P., J. Glover, P. Andrews, and H. Hinman. 2001. Sustainability of three apple production systems. Nature 410:926-930.

Reij, C., I. Scoones, and C. Toulmin (ed) 1996. Sustaining the soil. Indigenous soil and water conservation in Africa. Earthscan. London.

Reij, C., and A. Waters-Bayer. 2001. Farmer innovation in Africa: a source of inspiration for agricultural development. Earthscan, London.

Reitz, S.R., G.S. Kund, W.G. Carson, P.A. Phillips, and J. Trumble. 1999. Economics of reducing insecticide use on celery through low-input pest management strategies. Agric. Ecosyst. Environ. 73:185-197.

Relyea, R.A. 2005. The impact of insecticides and herbicides on the biodiversity and productivity of aquatic communities. Ecol. Applic. 15:618-627.

Repetto, R. 1985. Paying the price: Pesticides subsidies in developing countries. Res. Rep. No. 2. World Resour. Inst., Washington DC.

Repetto, R. 1986. Skimming the water: Rent-seeking the performance of public irrigation sytems. Res. Rep. 4. World Resour. Inst., Washington DC.

Repetto, R. 1990. Deforestation in the tropics. Sci. Am. 262(4):18-24.

Repetto, R. 1992. Accounting for environmental assets. Sci. Am. 266:94-100.

Repetto, R. 1994. The "second India" revisited: Population, poverty and environmental stress over two decades. WRI, Washington DC.

Repetto, R., W. Magrath, M. Wells, C. Beer, and F. Rossini. 1989. Wasting assets: Natural resources in the national income accounts. World Resour. Inst., Washington DC.

Reus, J.A.W.A., and P.C. Leendertse. 2000. The environmental yardstick for pesticides: A practical indicator used in the Netherlands. Crop Prot. 19:637-641.

Rhoades, R.E. 2006. Seeking half our brains: Reflections on the social context of interdisciplinary research and development. p. 403-420. In M.M. Cernea, A.H. Kasam (ed) Researching the culture in agriculture: Social research for international development. CABI Publ., Oxford.

Rhoades, R.E. 1982. The art of the informal agricultural survey. Social Science Dep. Training Doc. 1982-2, CIP, Lima.

Rhoades, R.E., and R.H. Booth. 1982. Farmer-back-farmer: a model for generating acceptable agricultural technology. Agric. Admin. 11:127-137.

Richards, P. 1985. Indigenous agricultural revolution. Westview Press, Boulder.

Richards, P. 1995. Participatory rural appraisal: A quick and dirty critique. p. 13-16 In PLA Notes, No. 24. IIED, London.

Richards, P. 2002. Fighting for the rain forest: War, youths and resources in Sierra Leone. African Issues Ser. James Currey, Oxford.

Richards, P. (ed) 2005. No peace no war. An anthropology of contemporary armed conflicts. Ohio Univ. Press, Athens and James Curry, Oxford.

Riches, G. (ed) 1997.First world hunger. Food security and welfare politics Macmillan, London and St. Martins, NY.

Rickert, K. 2004. Emerging challenges for farming systems. Lessons from Australian and Dutch agriculture. Available at http://www.rirdc.gov.au/reports/Ras/03-053.pdf. Rural Industries Res. Dev. Corp., Gov. Australia.

Ricketts, T. 2001. The matrix matters: effective isolation in fragmented landscapes. The Am. Naturalist 158(1):87-99.

Riddle, J. 2004. Crop trust to conserve plant diversity. FAO newsroom. Available at http://www.fao.org/newsroom/en/news/2004/51211/index.html.

Rivera, W.M. 1996. Agricultural extension in transition worldwide: Structural, financial and managerial reform strategies. Public Admin. Dev. 16:151-161.

Rivera, W.M., and S.G. Schram. 1987. Agricultural extension worldwide: Issues, practices and emerging priorities. Green Helm, NY.

Rivera, W.M., and D.J. Gustafson (ed) 1991. Agricultural extension: Worldwide institutional evolution and forces for change. Elsevier, Amsterdam.

Rivera, W.M., and W. Zijp. 2002. Contracting for agricultural extension. International case studies and emerging practices. CABI Publ., UK.

Roberts, N. (ed) 1989. Agricultural extension in Africa. World Bank Symposium. Available at http://www.eric.ed.gov/ERICDocs/data/ericdocs2sql/content_storage_01/0000019b/80/20/83/7c.pdf. World Bank, Washington DC.

Robinson, M. 2002. Sustainable development and environmental protection. Speech by the UN High Commissioner for Human Rights delivered at the Civil Society Workshop on Human Rights, World Summit Sustain. Dev., Johannesburg, 1 Sept 2002. Available at http://www.unhchr.ch/huricane/huricane.nsf/0/A551686D4B5905D0C1256C28002BF3D6?opendocument.

Robinson, R.A., and W.J. Sutherland. 2002. Post-war changes in arable farming and biodiversity in Great Britain. J. Appl. Ecol. 39:157-176.

Rodríguez, L.C., and H.M. Niemeyer. 2005. Integrated pest management, semiochemicals and microbial antagonists in Latin American agriculture. Crop Prot. 24(7):615-623.

Rodricks, J.V. 2001. Using science-based risk assessment to develop food safety policy. p. 25-32 In Food safety policy, science and risk assessment: Strengthening the connection. Nat. Acad. Press, Washington DC.

Rogers, E.M. 1962. The diffusion of innovations. Free Press, Glencoe NY.

Rogers, E.M. 1976. Communication and development: The passing of a dominant paradigm. Comm. Res. 3:213-240.

Rogers, E.M. 1983. Diffusion of innovations. 3rd ed. Free Press, NY.

Rogers, E.M. 1995. Diffusion of innovations. 4th ed. Free Press, NY.

Rogers, E.M. 2003. Diffusion of innovations. 5th ed. Free Press, NY.

Rogers, E.M., and D.L. Kincaid. 1981. Communication networks: Towards a new paradigm for research. Free Press, NY.

Rola, A., S. Jamias, J. Quizon. 2002. Do farmer field school graduates retain and share what they learn? An investigation in Iloilo, Philippines. J. Int. Agric. Extens. Educ. 9:65-76.

Röling, N. 1988. Extension science: Information systems in agricultural development. Cambridge Univ. Press, UK.

Röling, N. 2004. Communication for development in research, extension and education. Available at http://www.fao.org/sd/dim_kn1/docs/kn1_040701a3b_en.pdf. Wageningen Univ., the Netherlands.

Röling, N., J. Ascroft, and F. Wa Chege. 1976. Diffusion of innovations and the issue of equity in rural development. Comm. Res. 3:155-71.

Röling, N., and P.G.H. Engel. 1991. The development of the concept of agricultural knowledge and information systems (AKIS): Implications for extension. p. 125-139 In W.M. Rivera, D.J. Gustafson (ed) Agricultural extension: Worldwide institu-

tional evolution and forces for change. Elsevier, Amsterdam.

Röling, N., and M.A.E. Wagemakers (ed) 1998. Facilitating sustainable agriculture. Participatory learning and adaptive management in times of environmental uncertainty. Cambridge Univ. Press, UK.

Rosenthal, E. 2003. The tragedy of Tauccamarca: A human rights perspective on the pesticide poisoning deaths of 24 children in the Peruvian Andes. Int. J. Occup. Environ. Health 9:53-58.

Rosenthal, E. 2005. Who's afraid of national laws? Pesticide corporations use trade negotiations to avoid bans and undercut public health protections in Central America. Int. J. Occup. Environ. Health 11:437-443.

Rosset, P. 2000. Lessons from the green revolution- do we need new technology to end hunger? Available at http://www.foodfirst.org/media/opeds/2000/4-greenrev.html. Inst. Food Dev. Policy, Oakland.

Rosset, P., and M. Benjamin. 1994. The greening of the revolution: Cuba's experiments with organic agriculture. Ocean Press, Melbourne.

Rosset, P., R. Patel, and M. Courville. 2006. Promised land: Competing visions of agrarian reform. Inst. Food Dev. Policy, Oakland, CA.

Roth, M. 1999. Overview of farm direct marketing industry trends. Agric. Outlook Forum. USDA, Washington DC.

Ruane, J., and A. Soninno. 2006. Results from the FAO biotechnology forum. Res. Tech Pap. No. 11. Available at ftp://ftp.fao.org/docrep/fao/009/a0744e/a0744e00.pdf. FAO, Rome.

Rukuni, M., M.J. Blackie, and C.K. Eicher. 1998. Crafting smallholder-driven agricultural research systems in Southern Africa. World Dev. 26(6):1073-1087.

Russell, E. 2001. War and nature: Fighting humans and insects with chemicals from world War I to Silent Spring. Cambridge Univ. Press, UK.

Ryan, B., and N. Gross. 1943. The diffusion of hybrid seed corn in two Iowa communities. Rur. Soc. 8:15-24.

Saad, M.B. 1999. Food Security for the Food-Insecure: new challenges and renewed commitments. Centre for Development Studies, Dublin.

Sachs, C. 1996. Gendered fields. Rural women, agriculture and the environment. Rur. Studies Ser. Westview Press, Boulder CO.

Safilios-Rothschild, C. 1982. The persistence of women's invisibility in agriculture. Theoretical and policy lessons from Leostho and Sierra Leone. Centre for Policy Studies, Working Pap. 88, The Population Council, NY.

Safrin, S. 2004. Hyperownership in a time of biotechnological promise: The international conflict to control the building blocks of life. The Am. J. Int. Law 98:641-684.

Sagenmuller, A. (ed) 1999. AgroEvo´s contribution to integrated crop management.

Hoechst Schering AgroEvo GmbH, Frankfurt.

Sakamoto, K., Y. Choi, L.L. Burmesiter. 2007. Farming multifunctionality: Agricultural policy paradigm change in South Korea and Japan. Int. J. Soc. Food Agric. 15(11):24-45.

Salazar, R. 1992. MASIPAG: Alternative community rice breeding in the Philippines. Approp. Tech. 18:20-21.

Salazar, R., N.P. Louwaars, and B. Visser, 2007. Protecting farmers' new varieties: New approaches to rights on collective innovations in plant genetic resources. World Dev. 35(9):1515-1528.

Sale, K. 1993. The green revolution: The American environmental movement, 1962-1999. Hill & Wang Sale, NY

Samanta, R.K. (ed) 1995. Women in agriculture. Perspectives, issues and experiences. M.D. Publ. New Delhi.

Samanta, R.K, and S.K. Aurora. 1997. Management of agricultural extension in global perspectives. BR Publ., New Delhi.

Samways, M. 1997. Classical biological control and biodiversity conservation: What risks are we prepared to accept? Biodivers. Conserv. 6:1309-1316.

Santaniello, V. 2003 Agricultural biotechnology: Implications for food security. Proc. 25th Int. Conf. Agric. Econ. Durban. 18-22 Aug. Available at http://www.blackwell-synergy.com/doi/pdf/10.1111/j.0169-5150.2004.00023.x

SAPRIN (Structural Adjustment Participatory Rev. Int. Network). 2002. The policy roots of economic crisis and poverty: A multi-country participatory assessment of structural adjustment. SAPRIN, Washington DC.

Scanlon, T.M. 1998. What we owe each other. The Belknap Press, Cambridge MA.

Schecter, A.J., L.C. Dai, O. Päpke, J. Prange, J.D. Constable, M. Matsuda et al. 2001. Recent dioxin contamination from Agent Orange in residents of a southern Vietnam city. J. Occup. Environ. Med. 43(5): 435-443.

Schecter, A., P. Cramer, K. Boggess, J. Stanley, O. Päpke, J. Olson et al. 2001. Intake of dioxins and related compounds from food in the U.S. population. J. Toxicol. Environ. Health (Part A) 63:1-18.

Schettler, T., G. Solomon, P. Burns, and M. Valenti. 1996. Generations at risk: How environmental toxins may affect reproductive health in Massachusetts. GBPSR and MASSPIRG, Cambridge MA.

Schillhorn van Veen, T.W., D.A. Forno, S. Joffe, D.L. Umali Deiniger, and S. Cooke. 1997. Integrated pest management, strategies and polices for effective implementation. Env. Sustain. Dev. Studies Mono. Ser. No. 13, World Bank, Washington DC.

Schulthess, F., A. Chabi-Olaye, and S. Gounou. 2004. Multi-trophic level interactions in a cassava-maize relay cropping system in the humid tropics of West Africa. Bull. Ent. Res. 94:261-272.

Schultz, T.W. 1964. Transforming traditional agriculture, Yale Univ. Press, New Haven.

Schütz, A. 1964. Collected papers II: Studies in social theory. A. Brodersen (ed) Martinus Nijhoff, The Hague.

Scialabba, N. 2000. Opportunities and constraints of organic agriculture: A socio-ecological analysis. Universitá degli Studi della Tuscia, Viterbo, 17-28 July.

Scialabba, N. 2007. Organic agriculture and food security. In Int. Conf. Organic Agric. Food Security. Rome. May 3-5. [Online]. Available at ftp://ftp.fao.org/paia/organicag/ofs/nadia.pdf. FAO, Rome.

Scoones, I., M. Melnyk, and J.N. Pretty. 1992. The hidden harvest. Wild foods and agricultural systems. A literature review and annotated bibliography. IIED, London.

Scoones, I., and J. Thompson (ed) 1994. Beyond farmer first. Rural people's knowledge, agricultural research and extension practice. Intermediate Tech., London.

Scoones, I., C. Chibudu, S. Chikura, P. Jeranyama, D. Machaka, W. Machanja et al. 1996. Hazards and opportunities. Farming livelihoods in dryland Africa: Lessons from Zimbabwe. Zed Books, London.

Seligman, M.E.P., and J.L. Hagar. 1972. Biological boundaries of learning. Meredith, NY.

Sen, A. 1981. Poverty and famines. An essay on entitlement and deprevation. Clarendon, Oxford.

Settle, W.H., H. Ariawan, E. Tri Astuti, W. Cahyana, A.L. Hakim, D. Hindayana et al. 1996. Managing tropical rice pests through conservation of generalist natural enemies and alternative prey. Ecology 77(7):1975-1988.

Seur, H. 1992. Sowing the good seed. The interweaving of agricultural change. Gender relations and religion in Serenje District, Zambia. Publ. PhD thesis. Agricultural Univ., Wageningen.

SGRP. 2006. Developing Access and Benefit Sharing Regimes: plant genetic resources for food and Agriculture, System-wide Genetic Resources Program of the CGIAR.

Shapiro, C. 2001. Navigating the patent thicket: Cross licenses, patent pools and standard setting, in innovation policy and the economy. In A. Jaffe et al. (ed) Innovation policy and the economy. vol.1. Nat. Bureau Econ. Res., Cambridge MA.

Sharma, R., and T.H. Poleman. 1993. The new economics of India's green revolution. Income and employment diffusion in Uttar Pradesh. Cornell Univ. Press, Ithaca.

Shennan, C., T. Pisani Gareau, and J.R. Sirrine. 2005. Agroecological approaches to pest management in the US. p. 147-164. In J. Pretty. (ed) The pesticide detox: Towards a more sustainable agriculture. Earthscan, London.

Sherwood, S., D. Cole, C. Crissman, and M. Paredes. 2005 From pesticides to people: Improving ecosystem health in the northern Andes. p. 147-164.In J. Pretty (ed) The pesticide detox. Towards a more sustainable agriculture. Earthscan, London.

Shiva, V. 2000. Stolen harvest: The hijacking of the global food supply. South End Press, Cambridge MA.

Shulman, S. 2006. Undermining science: Suppression and distortion in the Bush Administration. Univ. California Press, Berkeley.

Sibanda, T., H.M. Dobson, J.F. Cooper, W. Manyangarirwa, and W. Chiimba. 2000. Pest management challenges for smallholder vegetable farmers in Zimbabwe. Crop Prot. 19:807-815.

SIDA.1999. Sustainable agriculture: A summary of SIDA's experiences and priorities. Available at http://www.sida.se/sida/jsp/sida.jsp?d=118&a=3320&language=en_US&searchWords=sustainable%20agriculture. SIDA, Stockholm.

Sigsgaard, L. 2006. Biological control of arthropod pests in outdoor crops — the new challenge. Proc. Int. Workshop Implementation of Biocontrol in Practice in Temperate Regions — Present and Near Future. Flakkebjerg, Denmark. 1-3 Nov. 2005. DIAS Rep. No. 119:153-167.

Simberloff, D., and P. Stiling. 1996. Risks of species introduced for biological control. Biol. Conserv. 78:185-192.

Simmonds, N.W. 1979. Principles of crop improvement. Longman, London.

Singh, N.C. 1983. The economics of cattle tick control in dry tropical Australia. Aust. Vet. J. 60(2):37.

Sinzogan, A.A.C., J. Jiggins, S. Vodohué, D. Kossou, E. Totin, and A. van Huis. 2007. An analysis of the organizational linkages in the cotton industry in Benin. Special Issue Convergence of Sciences Research in West Africa. Int. J. Agric. Sustain. 5(2-3):213-231.

Sissoko, M. 1994. Lutte Integree contre les ennemis du mil dans la zone de Banamba au Mali. Sahel PV Info. 60:15-19.

Slaybaugh, Douglas. 1996. William I. Myers and the modernization of american agriculture. Henry A. Wallace Series on Agricultural History and Rural Life. Iowa State Univ. Press, Ames.

Sligh, M., and C. Christman. 2007. Organic agriculture and access to food. In Int. Conf. Organic Agric. Food Security. Rome. May 3-5. Available at http://www.fao.org/organicag/ofs/presentations_en.htm. FAO, Rome.

Smith, R.F., and R. van den Bosch. 1967. Integrated control. p. 295-340. In W.W. Kilglore and R.L. Doutt (ed) Pest control: Biological, physical and selected chemical methods. Academic Press, NY.

Smith, D.L., and L. Sparks. 1993. The transformation of physical distribution in retailing: The example of Tesco plc. Int. Rev. Retail Distrib. Consum. Res. 3(1):35-64.

Smith, M., G. Castillo, and F. Gomez. 2001. Participatory plant breeding with maize in Mexico and Honduras. Euphytica 122:551-565.

Snow, A., D. Andow, P. Gepts, E. Hallerman, A. Power, J. Tiedje, and L. Wolfenbarger. 2005. Genetically engineered organisms and the environment: Current status and recommendations. Ecol. Applic. 15:377-404.

Snowden, S.J. 2005. Complex acts of knowing: Paradox and descriptive self-awareness. Special Issue. J. Knowledge Manage. 6(2):100-111.

Snowden, F.M. 2006. The conquest of malaria. Italy, 1900-1962. Yale Univ. Press, New Haven.

Solow, R.M. 1957. Technical change and the aggregate production function. Rev. Econ. Statist. 39:312-320.

Song, Y. 1998. 'New' seed in 'old' China. Impact of CIMMYT collaborative programme maize breeding in South-western China. Available at http://www.cimmyt.org/Research/Economics/map/impact_studies/pdf/NewSeed.pdf. Wageningen Univ., Netherlands.

Sorby, K. 2002. What is sustainable coffee? Background Pap. Agric. Tech. Note 30. World Bank, Washington DC.

Sorby, K., G. Fleischer, and E. Pehu. 2003. Integrated pest management in development: Review of trends and implementation strategies. Agric. Rural Dev. Dep., Working Pap. No. 5. World Bank, Washington DC.

Sow, M. 1992. La riziculture traditionnelle sur les terres de mangrove en basse Guinée. IRAG, Conakry.

Spedding, C.R.W. 1975. The biology of agricultural systems. Academic Press, London.

Sperling L., M. Loevinsohn, and B. Ntabomvura. 1993. Rethinking farmers' role in plant breeding: Local bean experts and on-station selection in Rawanda. Exp. Agric. 29:509-519.

Spielman, D.J., and K. von Grebmer. 2004. Public-private partnerships in agricultural research: An analysis of challenges facing industry and the CGIAR. Environ. Prod. Tech. Div., IFPRI, Washington DC.

Srinivasan, C.S. 2004. Plant variety protection in developing countries: A view from the private seed sector in India. J. New Seeds 6(1):67-89

Stagl, S., M. Getzner, and C. Spash. 2004. Alternatives for environmental valuation. Routledge, London.

Staudt, K. 1985. Women, foreign assistance and advocacy administration. Praeger, NY.

Staudt, K., and J.M. Col. 1991. Diversity in East Africa: Cultural pluralism, public policy and the state. Women Int. Dev. Ann. 2:241-264.

Steffen, W., A. Sanderson, P.D. Tyson, J. Jäger, P.A. Matson, B. Moore et al. 2004. Global change and the Earth system: A planet under pressure. Springer-Verlag, Berlin.

Steinfeld, H., P. Gerber, T. Wassenaar, V. Castel, M. Rosales, and C. de Hahn. 2006. Livestock's long shadow: Environmental issues and options. Available at http://www.virtualcentre.org/en/library/key_pub/longshad/AOTO12EOO.pdf. FAO, Rome.

Stern, V.M., R.F. Smith, R. van den Bosch, and K.S. Hagen. 1959. The integration of chemical and biological control of the spotted alfalfa aphid. 1. The integrated control concept. Higardia 29:81-101.

Stern, N.L. 2006. Review report on the economics of climate change. Cambridge Univ. Press, UK

Sthapit, B.R., K.D Joshi, and J.R. Witcombe. 1996. Participatory rice breeding in Nepal. Crop Improve. 23:179-188.

Stiglitz, J.E. 2006. Making globalization work. W.W. Norton, NY.

Stoll, G. 2002. Asia and the international context. Organic agriculture and rural poverty alleviation: Potential and best practices in Asia. UN Econ. Social Comm. for Asia and the Pacific, Bangkok.

Stone, G.D. 2007. Agricultural deskilling and the spread of genetically modified cotton in Warangal. Curr. Anthropol. 48:67-103.

Stone, D., and C. Wright. 2007. The World Bank and governance: A decade of reform and reaction. Routledge, Oxford.

Stoop, W. 2002. A study and comprehensive analysis of the causes for low adoption rates of agricultural research results in West and Central Africa: Possible solutions leading to greater future impacts: The Mali and Ghana case studies. Interim Science Council. CGIAR, Washington DC. and FAO, Rome.

Stringer, R. 2000. Food security in developing countries. Working Pap. No. 11. Available at http://ssrn.com/abstract=231211. CIES, Univ. Adelaide.

Strong, D.R., Jr. 1997. Fear no weevil? Science 277:1058-1059.

Strong, D.R. and R.W. Pemberton. 2001. Food webs, risks of alien enemies and reform of biological control. p. 57-80. In E. Wajnberg et al. (ed) Evaluating indirect effects of biological control. CABI Publ., UK.

Stroosnijder, L.and T. van Rheenen (ed) 2001. Agro-silvo-pastoral land use in Sahelian villages. Adv. Geoecol. 33. Catena Verlag, Reiskirchen.

Sudhakar P., R.N. Rao, R. Bhat, and C.P Gupta. 1988. The economic impact of a foodborne disease outbreak due to Staphylococcus aureus. J. Food Prot. 51(11):898-900.

Swanson, B. (ed) 1984. Agricultural extension: Reference manual. 2nd ed. FAO, Rome.

Swanson, B., and W. Peterson. 1989. A field manual for analysing technology development and transfer systems. INTERPAKS report. Off. Int. Agric. Univ. Illinois, Urbana.

Swinnen, J.F.M., and L. Vranken. 2006. Patterns of land market development in transition. World Bank, Washington DC.

Symes, D., and A.J. Jansen (ed) 1994. Agricultural restructuring and rural change in Europe. Agricultural Univ., Wageningen.

Syngenta. 2003. The responsible stewardship of our crop protection products. Available at http://www.syngenta.com/en/ar2003/ social_responsibility/intro_stewardship.aspx. Syngenta, Switzerland.

Syngenta. 2006. Syngenta's commitment to IPM. Available at http://www.syngentacropprotection-us.com/enviro/ipm/index.asp?nav=ipm_syngenta. Syngenta Crop Protection US.

Tansey, G., and T. Rajotte (ed) 2008. The future control of food. Earthscan, London.

Tallontire, A., and B. Vorley. 2005. Achieving fairness in trading between supermarkets and their agrifood supply chains. UK Food Briefing Group, London.

Tee, E.S. 1999. Nutrition of Malaysians: Where are we heading? Malaysian J. Nutr. 5:87-109.

Ter Weel, P., and H. van der Wulp. 1999. Participatory integrated pest management. Policy Best Practice Doc. 3. Netherlands Min. For. Affairs, Dev. Coop. The Hague, The Netherlands.

Thapinta, A., and T. Hudak. 2000. Pesticide use and residue occurrence in Thailand. Environ. Monitor. Assess. 60:103-114.

Thayer, A.M. 2001. Owning biotech. Chem. Engineer. News 78:25-32.

Thomas, J.W. 1975. The choice of technology for irrigation tubewells in East Pakistan: An analysis of a development policy decision. In P.C.Timmer et al. (ed) The choice of technology in developing countries: Some cautionary tales. Center Int. Affairs. Harvard Univ. Press, Boston.

Thrupp, L.A. 1990. Inappropriate incentives for pesticide use: Credit requirements for agrochemicals in developing countries. Agric. Human Values 7:62-69.

Thrupp, L.A. 1996. New partnerships for sustainable agriculture. World Resour. Inst., Washington DC.

Tiffen, M., M. Mortimore, and F. Gichuki. 1994. More people, less erosion: Environmental recovery in Kenya. Wiley, Chichester.

Tittonell, P., B. Vanlauwe, P.A. Leffelar, E. Rowe, and K.E. Giller. 2005. Exploring diversity in soil fertility management of smallholder farms in western Kenya. I: Heterogeneity at regional and farm scales. Agric. Ecosyst. Environ. 110:149-165.

Tittonell, P., B. Vanlauwe, N. de Ridder, K.E. Giller. 2007. Heterogeneity of crop productivity and resource use efficiency within smallholder Kenyan farms: Soil fertility gradients or management intensity gradients? Agric. Syst. 94:376-390.

Tobin, R.J. 1996. Pest management, the environment and Japanese foreign assistance. Food Policy 21:211-228.

Ton, P. 2003. Organic cotton in sub-Saharan Africa. Available at http://www.pan-uk.org/pestnews/pn62p4.htm. PAN-UK, London.

Touni, E., B. Nyambo, and A. Youdeowei. 2007. Potential NGO roles in prevention: Components of ASP projects. African Stockpiles Prog. Prevention Res. Project. Available at http://www.pan-uk.org/Projects/Obsolete/PAN%20UK%20Arusha%20WORKSHOP%20REPORT.pdf. PAN UK, London.

Tozun, N. 2001. New policy, old patterns: A survey of IPM in World Bank projects. Global pesticide campaigner. Available at http://www.panna.org/resources/gpc/gpc_200104.11.1.pdf. PANNA, San Francisco.

Tracy, M. 1993. Food and agriculture in a market economy: An introduction to theory, practice and policy. APS, Belgium.

Traxler, G., S. Godoy-Avila, and J. Falck-Zepeda, and J. Espinoza-Arellano. 2001. Transgenic cotton in Mexico: Economic and environment impacts. Dep. Agric. Working Pap. Auburn Univ., AL.

Tribe, G.D. 2003. Biological control of defoliating and phloem- or wood-feeding insects in commercial forestry in Southern Africa. p. 113-130. In P. Neuenschwander et al. (ed) Biological control in IPM systems in Africa. CABI Publ., UK.

Trinh, L.N., J.W. Watson, Hue, N.N. De, N.V. Minh, P. Chu et al. 2003. Agrobiodiversity conservation and development in Vietnamese home gardens. Agric. Ecosyst. Environ. 97:317-344.

Tripp, R.1997. The structure of the national seed systems. p. 14-42. In R. Tripp (ed) New seeds and old laws. Regulatory reform and the diversification of national seed systems. IT Publ., London.

Tripp, R. 2000. Strategies for seed system development in sub-Saharan Africa. Working Pap. ICRISAT, Patancheru, India.

Tripp, R. 2005. The performance of low external input technology in agricultural development: A summary of three case studies. Int. J. Agric. Sustain. 3(3):143-153.

Tripp, R. 2006a. Is low external input technology contributing to sustainable agricultural development? Nat. Resour. Perspect. 102. Overseas Dev. Inst., London.

Tripp, R. 2006b. Self-sufficient agriculture: Labor and knowledge in small-scale farming. Earthscan, London.

Tripp, R., and D. Byerlee. 2000. Public plant breeding in an era of privatisation. Natural resources perspectives. No. 57. Available at http://www.odi.org.uk/NRP/57.pdf. ODI, London.

Tripp, R., M. Wijeratne, and V.H. Piyadasa. 2005. What should we expect from farmer field schools? A Sri Lankan case study. World Dev. 33:1705-1720.

Tsegaye, B. 1997. The significance of biodiversity for sustaining agricultural production and role of women in the traditional sector: The Ethiopian experience. Agric. Ecosyst. Environ. 62:215-227.

UCS (Union of Concerned Scientists). 2004. Scientific integrity in policymaking: An investigation into the Bush Administration's misuse of science. UCS, Cambridge.

UN Habitat 2007. State of the world's cities. UN Human Settlements Program, Nairobi.

UN Millennium Project. 2005. Halving hunger: It can be done. Rep. Task Force on Hunger. The Earth Institute at Columbia Univ., NY.

UNCED. 1993. Rio Declaration on Environment and Development. Principle 15. Rio de Janeiro, Brazil. Available at http://habitat .igc.org/agenda21/rio-dec.htm. UN Conf. Environ. Dev., NY.

UNCTAD. 2006. Tracking the trend towards market concentration: The case of the agricultural input industry. Available at http://www.unctad.org/en/docs/ditccom200516_en.pdf. UNCTAD report UNCTAD/DITC/COM/2005/16.

UNDP. 1995. Agroecology: Creating the synergism for a sustainable agriculture. UNDP Guidebook Ser. Sustain. Energy Environ. Div., UNDP, NY.

UNEP. 2005. One planet, many people: Atlas of our changing environment. UNEP, Nairobi.

UNESCO. 1993. World science report, 1993. UNESCO, Paris.

UNESCO. 2005. Global education digest. Comparing education statistics across the world. Available at http://www.uis.unesco .org/template/pdf/ged/2005/ged2005_en.pdf. UNESCO, Paris.

UNHCR. 2007. The state of the world's refugees. UN High Commission for Refugees, Geneva.

UNRISD. 1975. Rural institutions and planned change. UN Res. Inst. Social Dev., Geneva.

UNRISD, 2006. Land tenure reform and gender equality. Res. Policy Brief 4. UN Res. Inst. Social Dev., Geneva.

Unilever. 2005. Unilever's Colworth farm project. Putting sustainable agriuclture to the test. Unilever R&D, Colworth.

Unnevehr, L., and H.H. Jensen. 1999. The economic implications of using HACCP as a food safety regulatory standard. Food Policy 24(6):625-635.

Unver, A. 2005. The road less chosen. A personal account of rural development in Turkey (1965-2005). TKV, Ankara.

Uphoff, N. 1993. Foreword. In J. Farrington et al. (ed) Reluctant partners? Non-government organisations, the state and sustainable agricultural development. Routledge, London.

UPWARD. 2002. Papers for international farmer field school workshop. Yogyakarta, Indonesia. 21-25 Oct 2002. Available at http://www.eseap.cipotato.org/UPWARD/Events/FFS-Workshop-Yogya2002/Draft-Workshop-Papers.htm.

USAID. 2004. USAID agricultural strategy: Linking producers to markets. July 2004. Available at http://www.usaid.gov/our_work/agriculture/ag_strategy_9_04_508.pdf. USAID, Washington DC.

USAID. 2007. Publications, presentations and other products of the IPM CRSP. Cumulative compilation updated March 2007. Available at http://www.oired.vt.edu/ipmcrsp/communications/annrepts/NEW%20PubList%2006-updated.pdf. USAID, Washington DC.

USDA. 1993. The practice of integrated pest management (IPM). The PAMS approach. Available at http://www.ipmcenters.org/Docs/PAMS.pdf. USDA, Washington DC.

USDA. 2000. Genetically engineered crops: Has adoption reduced pesticide use? ERS AO #273. USDA, Washington DC.

USDA/NRCS 2001. Pest management policy. General Manual Title 190, Amend. 6. USDA, Mat. Resources Conserv. Serv., Washington DC.

USGS. 2006. The quality of our nation's waters, pesticides in the nation's streams and ground water, 1992-2001. Available at http://ca.water.usgs.gov/pnsp/pubs/circ1291/show_description.php?chapter=6&figure=5. USGS, Washington DC.

Van den Ban, A. 1963. Boer en Landbouwvoorlicting. De Communciatie ven Nieuwe Landbouwmethoden. Van Gorcum, Netherlands.

Van den Ban, A., and R.K. Samanta (ed) 2006. Changing roles of agricultural extension in Asian nations. B.R. Publ., Delhi.

Van den Berg, H., and J. Jiggins. 2007. Investing in farmers: The impacts of farmer field schools in integrated pest management. World Dev. 35:663-686.

Van Driesche, R.G., and M.S. Hoddle. 2000. Classical arthropod biological control: measuring success, step by step. p. 39-75. In G. Gurr, and S. Wratten (ed) Measures of success in biological control. Kluwer, Dordrecht.

Van Duuren, B. 2003. Integrated pest management farmer training project, Cambodia. Assessment of the impact of the IPM program at field level. DANIDA, Phnom Penh.

Van Emden, H. 1974. Pest control and its ecology. Inst. Biol., Studies in Biology No. 50. Edward Arnold, London.

Van Huis, A., and F. Meerman. 1997. Can we make IPM work for resource-poor farmers in subSaharan Africa? Int. J. Pest Manage. 43(4):313-320.

Van Ittersum, M.K., R.P. Roetter, H. Van Keulen, N. De Ridder, C.T. Honah, A.G. Laborte et al. 2004. A systems network (SysNet) approach for interactively evaluating strategic land use options at a national scale in South and South-east Asia. Land Use Policy 21:101-113.

Van Lenteren, J.C. (ed) 2006. IOBC internet book of biological control. Available at http://www.unipa.it/iobc/downlaod/IOBC%20InternetBookBiCoVersion4October2006.pdf. IOBC, Wageningen.

Van Lenteren, J.C., A.K. Minks and O.M.B. de Ponti (ed) 1992. Biological control and integrated crop protection: Towards environmentally safer agriculture. Pudoc, Wageningen.

Van Lenteren, J., and V. Bueno. 2003. Augmentative biological control of arthropods in Latin America. BioControl 48:123-139.

Van Leur, J.A.G., and H. Gebre. 2003. Diversity between some Ethiopian farmer's varieties of barley and within these varieties among seed sources. Genetic Resour. Crop Evol. 50:351-357.

Van Mele, P., and A. Salahuddin. 2005. Innovations in rural extension. CABI, UK.

Van Schoubroeck, F. 1999. Learning to fight a fly. Developing citrus IPM in Bhutan. Publ. PhD thesis. Wageningen Univ., The Netherlands.

Van Veldhuizen, L., A. Waters-Bayer, R. Ramirez, D.A. Johnson, and J. Thompson (ed) 1997. Farmers' research in practice. Lessons from the field. Intermediate Tech. Publ., London.

Vandeman, A.M. 1995. Management in a bottle: Pesticides and the deskilling of agriculture. Rev. Radical Pol. Econ. 27(3):49-55.

Vanlauwe, B., P. Tittonell, and J. Mukulama. 2006. Within-farm soil fertility gradients affect response of maize to fertiliser application in western Kenya. Nut. Cycl. Agroecosyst. 76:171-182.

Varela, F., E. Thompson, and E. Rosch. 1993. The embodied mind: Cognitive science and human experience. MIT Press, Cambridge MA.

Verhaegen, E., V. Campens, H. Wustenberghs, L. Lauwers, and R. Nutelet. 2002. Environmental aspects of agricultural accounts. Execution Rep., Centre for Agric. Econ., Brussels.

Vernooy, R., and Y.C. Song. 2004. New approaches to supporting the agricultural biodiversity important for sustainable rural livelihoods. Int. J. Agric. Sustain. 2:55-66.

Vilella, F., C. Banchero, A. Fraschina, and M.C. Plencovich. 2002. Torres de acero, torres de marfil: El reto de identificar indicadores de calidad en la relación universidad-industria del futuro. In Educación y Desarrollo para el Futuro del Mundo. Ed. Fundación para la Educación Superior I., A.C., México.

Vilella, F., M.C. Plencovich, A. Ayala Torales, and C. Bogosian. 2003. La Facultad de Agronomía de la UBA ante el proceso de autoevaluación. Editorial EFA, Buenos Aires.

Virk, D.S., M. Chakraborty, J. Ghosh, S.C. Prasad, and J.R. Witcombe. 2005. Increasing the client orientation of maize breeding using farmer participation in Eastern India. Exp. Agric. 41:413-426.

Visser, B., D. Eaton, N. Louwaars, and J. Engles. 2000. Transaction costs of germplasm exchange under bilateral agreements. Doc. No. GFAR/00/17-04-04. GFAR. Available at www.egfar.org/documents/conference/GFAR_2000/gfar1702.PDF. FAO, Rome.

Visser, B., R. Pistorius, R. van Raalte, D. Eaton, and N. Louwaars. 2005. Options for non monetary benefit sharing, an inventory. Background Study Pap. No. 30. Comm. on Genetic Resour. Food Agric. Available at ftp://ext-ftp.fao.org/ag/cgrfa/BSP/bsp30e.pdf. FAO, Rome.

Virk, D.S., D.N. Singh, S.C Prasad, J.S Gangwar, and J.R Witcombe. 2003. Collaborative and consultative participatory plant breeding of rice for the rainfed uplands of eastern India. Euphytica 132:95-108.

Virk, D.S., M. Chakraborty, J. Ghosh, S.C. Prasad, and J.R. Witcombe. 2005.

Increasing the client orientation of maize breeding using farmer participation in eastern India. Exp. Agric. 41:413-426

Vorley, B. 2003. Food, Inc: Corporate concentration from farm to consumer. UK Food Group, London.

Waage, J. 1996. Agendas, aliens and agriculture: global biocontrol in the post-UNCED era. Cornell Community Conf. Biol. Control. Apr 11-13. Cornell University, Ithaca.

Waage, J.K. 1997. Biopesticides at the cross-roads: IPM products or chemical clones? p. 11-19. In Microbial Insecticides: Novelty or necessity? BCPC Proc. No. 68. Warwick, UK, 16-18 April 1996. BCPC Publ., Bracknell, UK.

Wade, R.H. 1996. Japan, the World Bank, and the art of paradigm maintenance: The East Asian miracle in political perspective. New Left Rev. 217:3-36.

Wade, R.H. 1997. Greening the Bank: The struggle over the environment 1970-95. p. 611-734. In D. Kapur et al. (ed) The World Bank: Its first half century. Vol. 2. Brookings Inst. Press.

Wade, R.H. 2001. Making the World Development Report 2000: Attacking poverty. World Dev. 29(6):1435-42.

Wade, R.H. 2004. The World Bank and the environment. p. 72-93. In M. Boas and D. McNeill (ed) Global institutions and development: Framing the world? Routledge, London and NY.

Wafula, J.S., and C.C. Ndirutu. 1996. Capacity building needs for assessment andmanagement of risks posed by living modified organisms, perspectives of a developing country. p. 63-70. In K.J. Mulongoy et al. (ed) Transboundary movement of living modified organisms resulting from modern biotechnology. Int. Acad. Environ., Geneva.

Wajnberg, E., J.K. Scott, and P.C. Quimby (ed) 2001. Evaluating indirect effects of biological control. CABI Publ., UK.

Wale, M., F. Schulthess, E.W. Kairu, and C.O. Omwega. 2006. Cereal yield losses due to lepidopterous stemborers at different nitrogen fertilizer rates in Ethiopia. J. Appl. Entomol. 130:220-229.

Walker, D.H. 2002. Decision support, learning and rural resource management. Agric. Syst. 73:113-127.

Walker, T. 2007. Participatory varietal selection, participatory plant breeding and varietal change. Background paper for World Development Report 2008, World Bank, Washington DC.

Wargo, J. 1996. Our children's toxic legacy: How science and law fail to protect us from pesticides. Yale Univ. Press, New Haven CT.

Warner, K.D. 2006a. Agroecology in action: Social networks extending alternative agriculture. MIT Press, Cambridge.

Warner, K.D. 2006b. Extending agroecology: Grower participation in partnerships is key to social learning. Renewable Food Agric. Syst. 21(2):84-94.

Warren, D.M., and B. Rajasekaran. 1993. Putting local knowledge to good use. Int. Agric. Dev. 13:8-10.

Washburn, J. 2005. University, Inc.: The corporate corruption of american higher education. Basic Books.

Watkins, K. 1996. Free trade and farm fallacies. From the Uruguay Round to the World Food Summit. Ecologist 26:244-255

Watterson, A. 2001 Pesticide health and safety and the work and impact of international agencies: Partial successes and major failures. Int. J. Occup. Environ. Health 7(4):339-347.

Webb, P., and B. Rogers. 2003. Addressing the "in" in food insecurity. Available at http://www.fantaproject.org/publications/ffpOP1.shtml. USAID Off. Food for Peace Occas. Pap. No.1, Washington DC.

Weick, K. 1976. Educational organizations as loosely coupled systems. Admin. Sci. Q. 21(1):1-19.

Weismann, D. 2006. Global hunger index: Measurement concept, ranking of countries, and trends. FCND Discussion Paper 212. IFPRI, Washington DC.

Welch, R.M., and R.D. Graham. 1999. A new paradigm for world agriculture: meeting human needs — productive, sustainable, nutritious. Field Crops Research 60:1-10.

Werge, R. 1978. Social science training for regional agricultural development. CIMMYT Asian Rep. No. 5. CIMMYT Asian Reg. Off., New Delhi.

Wesseling C., L. Castillo, and C.G. Elinder. 1993. Pesticide poisonings in Costa Rica. Scand. J. Work Environ. Health 19:227-235.

Wesseling, C., R. McConnell, T. Partanen, and C. Hogstedt. 1997. Pesticides in developing countries: A review of occupational health impact and research needs. Int. J. Health Serv. 27:273-308.

Wesseling, C., B. van Wendel de Joode, C. Ruepert, C. Leon, P. Monge, H. Hermasillo et al. 2001. Paraquat in developing countries. Int. J. Occup. Environ. Health 7:275-286.

Wesseling, C., M. Kiefer, A. Ahlbom, R. McConnell, J. Moon, L. Rosenstock et al. 2002. Long term neurobehavioral effects of mild poisoning with organophosphate and n-methyl carbamate pesticides among banana workers. Int. J. Occup. Environ. Health 8(1):27-34.

Wesseling, C., M. Corriols, and V. Bravo. 2005. Acute pesticide poisoning and pesticide registration in Central America. Toxicol. Appl. Pharmacol. 207:S697-S705.

WHO. 1999. Public health impact of pesticides used in agriculture. WHO, Geneva.

WHO. 2002a. Global strategy for food safety: Safer food for better health. Available at http://www.who.int/foodsafety/publications/general/en/strategy_en.pdf. WHO, Geneva.

WHO. 2004. Avian influenza A (H5N1) in humans and in poultry in Asia: Food safety considerations. 24 Jan. 2004. Available at http://www.who.int/foodsafety/micro/avian1/en/. WHO, Geneva.

Whorton, D., R.M. Krauss, S. Marshall, and T.H. Milby. 1977. Infertility in male pesticide workers. The Lancet 310:1259-1261.

Williamson, S. 2001. Natural enemies and farmer decision-making. Biocont. News Inform. 22(4):91-94.

Williamson, S. 2005. Breaking the barriers to IPM in Africa: Evidence from Benin, Ethiopia, Ghana and Senegal. p. 165-180. In J. Pretty (ed) The pesticide detox: Towards a more sustainable agriculture. Earthscan, London.

Williamson, S., and D. Buffin. 2005. Towards safe pest management in industrialized agricultural systems. p. 212-225. In J. Pretty (ed) The pesticide detox: Towards a more sustainable agriculture. Earthscan, London.

Wilson, C.G., and R.C. McFadyen. 2000. Biological control in the developing world: Safety and legal issues. p. 505-511. In N.R. Spencer (ed) Proc. Xth Int. Symp. Biol. Control of Weeds. 4-14 July 1999. Montana State Univ., Bozeman.

Wilson, D. 2005.The context of food security. Proc. Conf. New Perspectives on Food Security 12-14 Nov. Available at www.glynwood.org/programs/foodsec/FSC%20Full%20Proceedings.pdf. Glywood Center, Cold Spring NY.

Windfuhr, M., and J. Jonsén. 2005. Food sovereignty: Towards democracy in localized food systems. Schumacher Centre Tech. Dev., ITDG Publ., UK.

Windsor, R.S. 1992. Economic benefits of controlling internal and external parasites in South American camelids. Annals NY Acad. Sci. 653:398.

Winter, M. 1996. Rural politics. Policies for agriculture, forestry and the environment. Routledge, London.

Witcombe, J.R., K.D. Joshi, R.B Rana, and D.S Virk. 2001. Increasing genetic diversity by participatory varietal selection in high potential production systems in Nepal and India. Euphytica 122:575-588.

Witcombe, J.R., A. Joshi, and S.N. Goyal. 2003. Participatory plant breeding in maize: A case study from Gujarat, India. Euphytica 120:413-422.

Witcombe, J.R., K.D. Joshi, S. Gyawali, A.M. Musa, C. Johansen, D.S. Virk et al. 2005. Participatory plant breeding is better described as highly client-oriented plant breeding. I. Four indicators of client-orientation in plant breeding. Exp. Agric. 41:299-319.

Wodageneh, A. 1989. Constraints confronting national biological control programs. p. 166-172. In J.S. Yaninek, and H.R. Herren (ed) Biological control: A sustainable solution to crop pest problems in Africa. Proc. Inaugural Conf. Workshop of the IITA Biological Control Program Center Africa. Cotonou. 5-9 Dec. 1988. IITA, Ibadan.

Wolfe, M. 1992. Barley diseases: Maintaining the value of our varieties. Barley Genetics 6:1055-1067.

Wolfenbarger, L., and P. Phifer. 2000. The ecological risks and benefits of genetically modified plants. Science 290:2088-2093.

Woodhill, J. 1999. Sustaining rural Australia: A political economic critique of natural resource management. PhD. Dissertation. The Australian Nat. Univ., Canberra.

Wopereis, M.C.S., A. Tamélokpo, G. Ezui, G. Gnakpénou, B. Fofana et al. 2006. Mineral fertilizer management strategies for maize on farmer fields differing in organic input history in the African Sudan savanna zone. Field Crop. Res. 96:355-362.

World Bank. 1995. Bangladesh, India and Pakistan - Seed projects. Ind. Eval. Group Rep. No. 14761. World Bank, Washington DC.

World Bank. 1998a. Operational Policy 4.09 on pest management. World Bank, Washington DC. Available at http://wbIn0018.worldbank .org/Institutional/Manuals/OpManual.nsf. World Bank, Washington DC.

World Bank. 2003. The CGIAR at 31: An independent meta-evaluation of the consultative group on international agricultural research. Vol. 3. Oper. Eval. Dep. World Bank, Washington DC.

World Bank. 2005a. Education: Building on indigenous knowledge. IK Notes No. 87. Dec. Available at www.worldbank.org/afr/ik. World Bank, Washington DC.

World Bank. 2005b. Food safety and agricultural health standards: Challenges and opportunities for developing country exports. Available at http://www.oecd.org/ dataoecd/5/24/34378759.pdf.

World Bank. 2005c. Agricultural growth for the poor. An agenda for development. Available at http://siteresources.worldbank.org/INTARD/ Resources/Agricultural_Growth_for_the_Poor. pdf. World Bank, Washington DC.

World Bank. 2005d. Sustainable pest management: Achievements and challenges.

Rep. No. 32714-GLB. Available at http://info.worldbank.org/etools/docs/ library/201658/783%5FReport.pdf. World Bank, Washington, DC.

World Bank. 2006. Cultural heritage and collective intellectual property rights. IK Notes No. 95. World Bank, Washington DC.

World Bank. 2008. Agriculture for development in the 21st Century. World Development Report. World Bank, Washington DC.

WRI/UNEP/WCSD. 2002. Tomorrow's markets. Global trends and their implications for business. WRI, Washington DC, UNEP, Paris, WBCSD (World Business Council for Sustainable Dev.), Geneva.

Wright, J. 2005. Falta Petroleo! Perspectives on the mergence of a more ecological farming and food system in post-crisis Cuba. Publ. PhD. Wageningen University, The Netherlands.

Wustenberghs, H., E. Verhaegen, L. Lauwers, and E. Mathijs. 2004. Monitoring agriculture's multifunctionality by means of integrated nation-wide accounting. 90th EAAE Seminar on Multifunctional Agriculture, Policies and Markets: Understanding the critical linkage, Rennes, France. 27-29 Oct. Available at http://merlin. lusignan.inra.fr:8080/eaae/website/pdf/45_ Wustenberghs.

Wustenberghs, H., E. Verhaegen, and L. Lauwers. 2005. Evaluation of landscape amenity benefits in multifunctional agriocultural accounts. Int. Conf. Multifunctionality of Landscapes: Analysis, Evaluation and Decision Support. Giesen, 18-20 May.

Young, E.M. 2004. Globalization and food security: Novel questions in a novel context? Progress Dev. Studies 4(1):1-21.

Yudelmann, M., A. Ratta, and D. Nygaard. 1998. Pest management and food production.

Food, Agric. Environ. Disc. Pap. 25. IFPRI, Washington DC.

Yunus, M. 1977. Story of a deep tubewell with a difference: A report on Osmania School purba beel tubewell. p. 84. In Locally Sponsored Dev. Prog. Ser. Rep. No. 28, Rural Studies Project, Dep. Econ., Chittagong University, Bangladesh.

Yunus, M. 1982. Annual Report 1981. Grameen Bank, Dhaka.

Yunus, M., and M. Islam. 1975. A report on Shahjalaler Shyamal Sylhet. In Locally Sponsored Dev. Prog. Ser. Rep. No. 1, Rural Studies Project, Dep. Econ., Chittagong University, Bangladesh.

Zahn, J. 2001. Social and environmental responsibility of large enterprises in the north/south. Globalization as opportunity: The example of Daimler Chrysler in Latin America. POEMA Pap. No.1, Belém.

Zeddies, J., R.P. Schaab, P. Neuenschwander, and H.R. Herren. 2001. Economics of biological control of cassava mealybug in Africa. Agric. Econ. 24:209-219.

Zingore, S., H.K. Murwira, R.J. Delve, and K.E. Giller. 2007. Influence of nutrient management strategies on variability of soil fertility, crop yields and nutrient balances on smallholder farms in Zimbabwe. Agric. Ecosyst. Environ. 119:112-126.

Zoundi, S.J., M.H. Collion., and H. Hocdé. 2001. Partnerships between producer organizations and research and extension institutions. p. 38-45. In P. Rondot and M.H. Collion (ed) Agricultural producer organizations: Their contribution to rural capacity building and poverty reduction. Available at http://publications.cirad.fr/ une_notice.php?dk=507572. World Bank, Washington DC.

3

Impacts of AKST on Development and Sustainability Goals

Coordinating Lead Authors
Roger Leakey (Australia) and Gordana Kranjac-Berisavljevic (Ghana).

Lead Authors
Patrick Caron (France), Peter Craufurd (UK), Adrienne Martin (UK), Andy McDonald (USA), Walter Abedini (Argentina), Suraya Afiff (Indonesia), Ndey Bakurin (Gambia), Steve Bass (UK), Angelika Hilbeck (Switzerland), Tony Jansen (Australia), Saadia Lhaloui (Morocco), Karen Lock (UK), James Newman (USA), Odo Primavesi (Brazil), Teresa Sengooba (Uganda)

Contributing Authors
Mahfuz Ahmed (Bangladesh), Elizabeth Ainsworth (USA), Mubarik Ali (Pakistan), Martine Antona (France), Patrick Avato (Germany/Italy), Debi Barker (USA), Didier Bazile (France), Pierre-Marie Bosc (France), Nicolas Bricas (France), Perrine Burnod (France), Joel Cohen (USA), Emilie Coudel (France), Michel Dulcire (France), Patrick Dugué (France), Nicholas Faysse (France), Stefano Farolfi (France), Guy Faure (France), Thierry Goli (France), David Grzywacz (UK), Henri Hocdé (France), Jacques Imbernon (France), Marcia Ishii-Eiteman (USA), Andrew Leakey (USA), Chris Leakey (UK), Andy Lowe (UK), Ana Marr (UK), Nigel Maxted (UK), Andrew Mears (Botswana), David Molden (USA), Jean-Pierre Muller (France), Jonathan Padgham (USA), Sylvain Perret (France), Frank Place (USA),

Anne Lucie Raoult-Wack (France), Robin Reid (USA), Charlie Riches (UK), Sara Scherr (USA), Nicole Sibelet (France), Geoff Simm (UK), Ludovic Temple (France), Jean-Philippe Tonneau (France), Guy Trebuil (France), Steve Twomlow (UK), Tancrède Voituriez (France)

Review Editors
Tsedeke Abate (Ethiopia) and Lorna Michael Butler (USA)

Key Messages

Key Messages

1. Agriculture is multifunctional and goes far beyond food production. Other important functions for sustainable development include provision of nonfood products; provision of ecological services and environmental protection; advancement of livelihoods; economic development; creation of employment opportunities; food safety and nutritional quality; social stability; maintenance of culture and tradition and identity. However, the promotion and achievement of multifunctionality is hindered by a lack of systematic quantitative and other data that allow a complete assessment of the impacts of wider functions. Nevertheless, enhanced recognition of the wider functions of agriculture has prompted efforts towards developing integrated land use systems that deliver a diverse set of social, economic and environmental functions, and address the tradeoffs between them.

2. Advances in AKST have enabled substantial gains in crop and livestock production, which have reduced levels of hunger and malnutrition. World cereal production has more than doubled since 1961, with average yields per hectare also increasing around 150% in many high- and low-income countries, with the notable exception of most nations in sub-Saharan Africa. Substantial gains in crop and livestock production are due to advances in many types of AKST, including biotechnology (e.g., genetic gain, stress resistance), physical (e.g., fertilizer, irrigation, mechanization), policy (e.g., intellectual property rights, variety release processes), microfinance (e.g., credit, provision of inputs), education and communication (e.g., farmer-field schools), and market and trade (e.g., demand, incentives). More recently, modern biotechnology is starting to have an impact on production. Advances have also been made in fish breeding, tree improvement and in crop and livestock husbandry. All of these advances in agricultural production have contributed to the improvement of many farmers' livelihoods and to economic growth in developed countries, although large deficiencies remain. In real terms food has become cheaper and calorie and protein consumption have increased, resulting in lower levels of hunger. On a global scale, the proportion of people living in countries with an average *per capita* caloric availability of less than 2200 kcal per day dropped from 57% in the mid-1960s to 10% by the late 1990s.

3. AKST has made some substantial positive contributions to different dimensions of livelihoods. These include:
- increased incomes, reduced hunger and malnutrition, improved health and cognitive development, improved levels of education and increased employment opportunities, reducing vulnerability to drought, pest and disease outbreaks.
- increased access to water for domestic and productive uses with positive impacts on health, food and nonfood production and environmental sustainability.
- improved relevance of AKST for different producer and consumer groups, through participatory approaches to research, extension and market assessment.
- improved support and integration of social and environmental sustainability (e.g., watershed management, community forestry management, integrated pest management (IPM) and strengthening of local seed systems) through participatory and community-based approaches to NRM at different scales.
- improved integration of gender and diversity concerns within AKST institutions, which has contributed to gender sensitive planning and awareness in AKST processes.

4. Despite much progress in agricultural development, persistent challenges remain. These include:
- *Uneven distribution of livelihood impacts*: The benefits from AKST have not been evenly distributed, varying between regions and agroecological zones, as well as between social groups. Industrialized regions have gained the most from innovations in AKST, while agroecological zones with severe biophysical constraints and marginalized social groups have benefited least. Levels of poverty, hunger, malnutrition and food insecurity still affect millions of people, particularly in SSA as well as parts of Asia, Latin America and Melanesia. Three billion people earn less than the purchasing power equivalent of US$2 per day. In some circumstances, especially in Africa, many of the poor have become ensnared in "poverty traps" without sufficient financial resources to improve or sustain their food security or livelihoods. The distributional impact of AKST has been affected by rights and access to assets—land, water, energy resources, markets, inputs and finance, training, information and communications. Despite advances in gender awareness, access to AKST products and participation in AKST processes remain limited for women and for other marginalized groups. Only limited attention has been paid to issues of vulnerability and social exclusion, or to the interaction of AKST related opportunities with social protection policies.
- *Health and human nutrition*: Globally, over 800 million people are underweight and malnourished, while changes in diet, the environment and lifestyle worldwide have resulted in 1.6 billion overweight adults; this trend is associated with increasing rates of diet-related diseases such as diabetes and heart disease. Another cause of acute and long-term human health risks arises from the misuse of toxic agrichemicals.
- *Environmental sustainability*: Agricultural use of natural resources (soils, freshwater, air, carbon-derived energy) has, in some cases, caused significant and widespread degradation of land, freshwater, ocean and atmospheric resources. Estimates suggest that resource impairment negatively influences 2.6 billion people. In many poor countries (and in marginalized communities within countries), many farmers lack access to the appropriate management interventions required to restore and sustain productivity. In addition to forest clearance and burning, the growing reliance on fossil fuels in agriculture has increased emissions of "greenhouse gases."

5. In many instances, AKST has begun to address sustainability challenges with strategies that recognize the production, livelihoods, and ecosystem service

functions required for achieving sustainable agricultural systems that span biophysical, socioeconomic and cultural diversity. The consequences of population growth and economic expansion have been a reduced resource base for future agriculture; now there are pressing needs for new agricultural land and water resources. In recent decades the development of integrated pest/water/nutrient management practices, crop/livestock systems, and crop/legume mixtures has contributed greatly to increased agricultural sustainability, but further progress is needed, especially to combat declining soil fertility. While fertilizer amendments restore fertility efficiently, many poor farmers are without the means to buy fertilizers. Consequently they suffer from a "yield gap" (the difference between crop yield potential and yield achieved). Agroforestry offers them a partial solution: biological nitrogen-fixation by leguminous trees/shrubs and crops can substantially increase crop yields. The integration of trees into field systems and by replanting watersheds, riparian and contour strips, also diversifies and rehabilitates the farming system, restoring soil organic matter, sequestering carbon in the biomass, improving water percolation and microclimate, reducing radiation losses to the atmosphere, and promoting biodiversity through the development of an agroecological succession. There are many indigenous tree species that have the potential to play these important ecological roles and also produce marketable food, fodder, and nonfood products. In this way, the ecological services traditionally obtained by long periods of unproductive fallow are provided by productive agroforests yielding a wide range of food and nonfood products. Some of these tree species are currently the subject of participatory domestication programs using local knowledge. Domestication is aimed at promoting food sovereignty, generating income and employment and enhancing nutritional benefits. Consequently, this approach brings together AKST with traditional knowledge as an integrated package capable of helping to meet development and sustainability goals.

6. Sustainable agriculture is more complex and knowledge intensive than ever before, covering sociocultural, ecological and economic dimensions. To be effective at using AKST to meet development and sustainability goals requires a wide range of actors and partnerships, and arrangements that realize the synergies between different forms of agriculture; between agriculture and other sectors; between different disciplines and between local and global organizations. Examples of measures that have contributed to realizing synergies include:

- the development of international regulatory frameworks on IPR, trade, and the environment.
- processes.
- linking multiple sources of knowledge created through the engagement of multiple stakeholders in AKST processes, including farmer organizations, civil society groups, the private sector and policy makers, as well as public sector organizations.

There is a growing recognition that the institutional, policy, financial, infrastructural and market conditions required for AKST to help meet development and sustainability goals are an intrinsic part of innovation processes. This has encouraged a growing emphasis on forging partnerships and linkages, which is beginning to have positive results. Much remains to be learned about the effective development and functioning of these partnerships to create an effective combination of different disciplines and knowledge traditions; overcome the separation of formal organizations involved in AKST and to institutionalize broader consultation processes among stakeholders with diverse interests, professional and organizational cultures, funding arrangements and capacity.

7. Since the mid-20th Century, there have been two relatively independent pathways to agricultural development: globalization and localization. Globalization, which initiated in developed countries, has dominated formal AKST and has been driven by public sector agricultural research, international trade and marketing policy. Localization has come from civil society and has involved locally based innovations, including value-addition, that meet the needs of local people and communities. Localization addresses the integration of social and environmental issues with agricultural production, but has lacked a range of market and policy linkages in support of new products and opportunities. Some current initiatives are drawing the two pathways together through public/private partnerships (e.g., fair-trade tea/coffee, forestry out-growers) involving global companies and local communities in the implementation of new regulatory frameworks and agreements that offering new paradigms for economic growth and development. Mobilizing and scaling up locally appropriate AKST in ways that integrate agricultural production with economic, social and environmental sustainability, permits localization and globalization to play complementary roles.

3.1 Methodology

The goals of this Assessment reflect an evolution of the concept of agriculture from a strong technology-oriented approach at the start of the Green Revolution to today's more human and environment-oriented paradigm. Assessing the biophysical impacts of AKST is simpler than assessing the social impacts, because of differences in complexity, and the greater emphasis on agronomic research, much of which has been on-station, rather than on-farm. This evolution of agriculture is reflected in the expansion of the Consultative Group on International Agricultural Research (CGIAR, including centers with a greater focus on natural resources systems, and more recently, on holistic and integrated approaches, including the livelihoods of poor farmers. This integration of technological advances with socially and environmentally sensitive approaches has not occurred uniformly across all sectors of AKST.

The preparation of this Chapter started with a review of the international literature (journals, conference proceedings, the reports of many and various organizations from international and nongovernmental development agencies, international conventions and development projects, and the internet). The information from this literature was then used to develop statements about the impacts and sustainability of AKST in the context of development and sustainability goals (see Chapter 1).

The main criteria used to assess the positive and negative impacts (including risks associated with technologies) of AKST were:

- Social sustainability—effects on livelihoods, nutrition and health, empowerment, equity (beneficiaries—including landless and labor), gender, access.
- Environmental sustainability—effects on natural capital, agroecosystem function, climate change.
- Economic sustainability—poverty, trade and markets, national and international development.

Levels of certainty were attributed to impact and sustainability statements based on evidence found in the international literature and the expert judgment of the authors. This certainty was associated with the range of impacts reported and to the appropriate measures of scale and specificity (Table 3-1).

3.2 Assessment and Analysis of AKST Impacts

In this subchapter we present Impact Statements (in bold), analyzed and quantified as explained above (Table 3-1).

3.2.1 Agriculture productivity, production factors and consumption

Since the mid-20th Century, there have been two relatively independent pathways to agricultural development. The first, which has dominated formal AKST, was initiated globally and has involved public-sector agricultural research coordinated by the International Agricultural Research Centers (IARCs) of the CGIAR.

3.2.1.1 Food production, consumption, and human welfare

The improvement of farm productivity was the major outcome of the Green Revolution, especially in the early years, Large benefits from resulted from the application of AKST in crop and livestock breeding, improved husbandry, increased use of fertilizers, pesticides and mechanization. However, these benefits were accompanied by some environmental issues.

Modern agricultural science and technology has positively affected a large number of people worldwide.

Goals	Certainty	Range of Impacts	Scale	Specificity
N, D	A	0 to +5	G	Especially in industrial and transitional countries

Despite large increases in population (see Chapter 1), agricultural systems have provided sufficient food resources to reduce undernourishment rates by about 50% in Asia/Pacific and Latin American/Caribbean since 1970. Large increases in agricultural production of vegetables, roots and tubers, cereals, fruits and pulses, have been made possible through genetic improvement, soil fertility management, irrigation, pesticides and mechanization (Salokhe et al., 2002; Figure 3-1). On a global scale, AKST has increased *per capita* production of calories, fats/oils, proteins and micronutrients (Evenson and Gollin, 2003ab). For example, available caloric availability increased from 2360 kcal/person/day in the mid-1960s to 2803 kcal person^{-1} day^{-1} in the 1997-1999 (Bruinsma, 2003). At present, 61% of the world's population consume >2700 kcal per day. Prices for staple foods have also declined (Bruinsma, 2003), benefiting many poor since they spend a large portion of their income on food. However, AKST benefits have been unevenly realized among and within regions and some estimates suggest that around a third of humanity has not been affected by modern agricultural science.

Agricultural science and technology has had positive impacts on the productivity (yield per unit area) of staple food crops, but these gains have not been universally realized.

Goals	Certainty	Range of Impacts	Scale	Specificity
N, D	A	+1 to +5	G	Especially in industrial and transitional countries

The cereal staples maize, rice, and wheat contribute around 60% of the caloric energy for humans on the global scale (Cassman et al., 2003). Among industrialized countries and in the developing regions of Asia and Latin and Central America (LAC), average cereal yields have sustained annual rates of increase (43 to 62 kg ha^{-1} yr^{-1}), and have more than

Table 3-1. Criteria used in the analysis of data

Goals	Certainty	Range of impacts	Scale	Specificity
Enhancement of: N = Nutrition (reduced hunger) H = Human health L = Rural livelihoods E = Environmental sustainability S = Social sustainability D = Sustainable economic development	A = Well established B = Established C = Competing explanations D = Expected, but to be confirmed E = Long-term impacts not yet available F = Speculative	-5 to +5	G = Global R = Regional N = National M-L = Multi-locational L = Local E = Experimental/ pilots	Examples: • Wide applicability, • Applicable in dry areas, • Occurs throughout tropics, • Especially in Africa, • Mainly in subsistence agriculture, • Negative in poor and positive in rich countries.

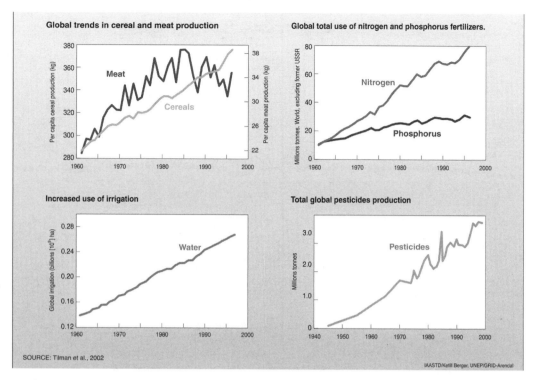

Figure 3-1. *Global trends in cereal and meat production; nitrogen and phosphorus fertilizer use; irrigation, and pesticide production.*

doubled in absolute terms since the 1960s (Figure 3-2). In contrast, in developing countries in Africa the average cereal yields have increased at a rate of 10 kg ha^{-1} yr^{-1} and productivity levels are about one-half of those achieved in industrialized countries in the early 1960s. In sub-Saharan Africa (SSA) approximately 66% of the crop production increase since 1961 is linked to area expansion. These broader trends mask significant differences among the grain staples. For example, in industrialized countries, maize productivity has grown at average rate of 122 kg ha^{-1} yr^{-1}, increasing from a base of 3 tonnes ha^{-1} in 1961 to nearly 8 tonnes ha^{-1} in 2005. In 1961, maize productivity was approximately 1 tonne ha^{-1} in developing countries. Since then, maize yields have steadily increased in developing regions of Asia (72 kg ha^{-1} yr^{-1}), demonstrated intermediate growth in Central America (37 kg ha^{-1} yr^{-1}), but achieved only slow growth among developing countries in Africa (12 kg ha^{-1} yr^{-1}). A major reason for this, especially in Africa, has been the lack of investment in public and private sector plant breeding programs (Morris, 2002). Similar trends are evident in rice and for other major commodities such as vegetables, roots, pulses and tubers (Figure 3-2).

Recently horticulture, including fruit production, has been the fastest growing food sector worldwide

Goals	Certainty	Range of Impacts	Scale	Specificity
N, D	B	+1 to +3	G	Especially China

Horticulture production has increased from 495 million tonnes in 1970 to 1379 million tonnes in 2004 (178%) (FAOSTAT, 2007). The vegetable subsector grew at an an- nual average rate of 3.6% during 1970-2004 from 255 million tonnes in 1970 to 876 million tonnes in 2004 (Ali, 2006). Most of this increased production came from area expansion with productivity per unit area increasing at less than 1% from 1970-2004. The slow improvement in the yield of horticulture crops suggests comparatively low investments in horticultural research. During 1970-2004, 52% of the increase in horticulture production came from China, 40% from all other developing countries, and remaining 8% from developed countries (Ali, 2006). This increase is having significant positive effects on income, employment, micronutrient availability and health of people in poor countries. Moreover, the share of horticulture products in trade, especially from developing countries, has increased (Ali, 2006).

Global production and consumption of livestock products have been growing dramatically over the last few decades.

Goals	Certainty	Range of Impacts	Scale	Specificity
N, H, D	A	0 to +3	G	Wide applicability

From 1979 to 2003, global meat production nearly doubled to 260 million tonnes (FAOSTAT, 2007). Among developing countries, those with large populations and rapidly growing economies (e.g., China, Brazil and India) accounted for over 50% of meat and milk production in 2005. Consumption of livestock products has also increased sharply, in part due to rising incomes and increasing urbanization in several parts of the developing world. Between 1962 and 2003 *per capita* meat consumption grew by a factor of 2.9, and milk by 1.7 in developing countries (Steinfeld et al., 2006; FAO, 2006a).

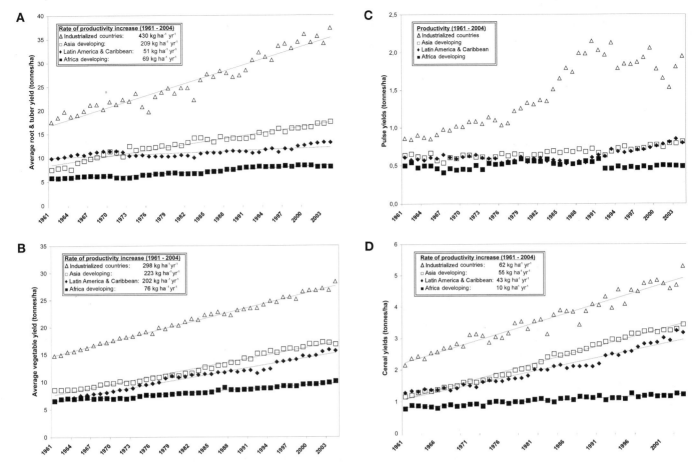

Figure 3-2. *Yield of (a) roots and tubers, (b) vegetables, (c) pulses, and (d) cereals between 1961 and 2004.* Source: FAOSTAT, 2007.

Global fish production (wild harvest and aquaculture) has increased by about 230% between 1961 and 2001

Goals	Certainty	Range of Impacts	Scale	Specificity
N, H	B	0 to +4	G	Worldwide

Between 1961 and 2001, global fish production (wild harvest and aquaculture) for all uses increased by about 230% from 39.2 million to nearly 130 million tonnes. Developing countries supply 75% of the volume and 50% of the value of the global fish trade (Kurien, 2004). Together the developing countries of Asia form the largest fish producer, with production reaching 71.2 million tonnes in 2001 (FAOSTAT, 2005). Aquaculture currently provides approximately 40% of the world's total food fish supply (Delgado et al., 2003ab; Kurien, 2004). Technological breakthroughs in aquaculture, triggered by private sector growth, increased demand for high-value fish in the world market and simultaneous changes in international laws, treaties and institutions, contributed to the rapid growth in fish supply (Ahmed and Lorica, 2002).

3.2.1.1.1 Trends in resource use (land, water, genetic resources, fertilizer, pesticides and mechanization)

Globally, land reserves have been severely depleted by cultivation

Goals	Certainty	Range of Impacts	Scale	Specificity
N, E, D	A	-1 to +2	G	Worldwide

Africa and Latin American countries do have significant tracts of undeveloped land that could be cultivated, but estimates suggest that only a small fraction these areas (7% Africa, 12% LAC) are free from the types of severe soil constraints that limit profitable and sustainable production (Wood et al., 2000). Moreover, many of the remaining undeveloped areas are of regional and global importance for biodiversity and ecosystem services (Bruinsma, 2003). The need to preserve natural areas and to avoid production on marginal lands (e.g., highly erodible hill slopes) provides strong incentives for advancing agricultural production through yield intensification (i.e., production per unit area) rather than area expansion.

The breeding and dissemination of Modern Varieties (MV) has had a major impact on food production.

Goals	Certainty	Range of Impacts	Scale	Specificity
N, L, D	A	-2 to +5	G	Widespread applicability

The breeding and dissemination of Modern Varieties with greater yield potential, better pest and disease resistance and improved organoleptic quality have, in conjunction with irrigation, fertilizer, pesticides and mechanization, had a major impact on food production (Figure 3-1). Modern Varieties, especially of cereals but also of root, protein and horticultural crops, have been widely adopted; Asia grows modern cereal varieties on 60-80% of the cultivated area (Evenson and Gollin, 2003a). Modern Varieties are also widely grown in Latin America but there has been less impact in sub-Saharan Africa and CWANA. Other than in CWANA there has been little impact of Modern Varieties on protein crops (mostly annual legumes).

Evidence relating farm size to productivity and efficiency is weak.

Goals	Certainty	Range of Impacts	Scale	Specificity
N, H, L, E, S, D	C	-4 to +4	G	Variable

Farms operated by small-scale producers are typically more efficient the smaller they are (Feder et al., 1988; Place and Hazell, 1993; Deininger and Castagnini, 2006). However, in large-scale mechanized farming economies of scale are important. For example, some regionally specific research has concluded that productivity and efficiency are positively related to farm size (Yee et al., 2004; Hazarika and Alwang, 2003), although there is also evidence that some large-scale mechanized farms are less efficient than smaller family farms (Van Zyl, 1996). The lack of clarity about the relationship between farm size and productivity and efficiency (Sender and Johnston, 2004) suggests confounding factors, such as land quality, and access to labor, markets, sources of credit and government farm policies (Van Zyl, 1996; Chen, 2004; Gorton and Davidova, 2004). For example, land per capita has been found to be a major determinant of overall household income (Jayne et al., 2003). Good management, on large- and small-scale farms, may be the most important factor affecting production efficiency. Typically, large-scale farmers with financial resources intensify agrichemical inputs and seek economies of scale, while resource-poor small-scale farmers reduce inputs, diversify, and seek risk aversion (Leakey, 2005a). Interestingly, it is often among the latter group that some of the best examples of sustainable agriculture are found, especially in the tropics (Palm et al., 2005b).

Globally there has been an extensive increase in irrigated areas, but investment trends are changing.

Goals	Certainty	Range of Impacts	Scale	Specificity
N, E	B	-1 to +5	G	Globally except SSA

Since 1961, the area of irrigated land has doubled to 277 million ha (in 2000)—about 18% of farmed land, funded initially by investments by international development banks, donor agencies, and national governments but later increasingly by small-scale private investments. Irrigation was essential to achieving the gains from high-yielding fertilizer-responsive crop varieties. Approximately 70% of the world's irrigated land is in Asia (Brown, 2005), where it accounts for almost 35% of cultivated land (Molden et al., 2007a). Forty percent of the world cereal production is from irrigated land and as much as 80% of China's grain harvest comes from irrigated land. By contrast, there is very little irrigation in sub-Saharan Africa. Trends have changed from the 1970s and early 1980s when donor spending on agricultural water reached a peak of more than US$1 billion a year. Funding fell to less than half that level by the late 1980s; benefit-cost ratios deteriorated; and as falling cereal prices and rising construction costs highlighted the poor performance of large-scale irrigation systems, opposition mounted to the environmental degradation and social dislocation sometimes caused by large dams. Today, there appears to be consensus that the appropriate scale of infrastructure should be determined by the specific environmental, social, and economic conditions and goals with the participation of all stakeholders (Molden et al., 2007a).

Increased fertilizer use is closely associated with crop productivity gains in regions that have been most successful at reducing undernourishment.

Goals	Certainty	Range of Impacts	Scale	Specificity
N, H, L, E, S	A	+2 to +5	G	Especially in Asia

On a global scale, total fertilizer consumption has increased from approximately 31 million in 1961 to 142 million tonnes in 2002 (FAOSTAT, 2007). From almost no use in the early 1960s, total fertilizer consumption rates in the developing countries of Asia (140 kg ha^{-1} yr^{-1}) now exceed those in industrialized nations (FAOSTAT, 2006) and have been a principal driver of improved crop productivity. In sub-Saharan Africa where cereal productivity has increased only modestly since the 1960s, average fertilizer consumption remains exceptionally low—under 20 kg ha^{-1} yr^{-1} (FAOSTAT, 2006). For cereal crops, approximately 50% of the yield increases observed after the introduction of modern crop

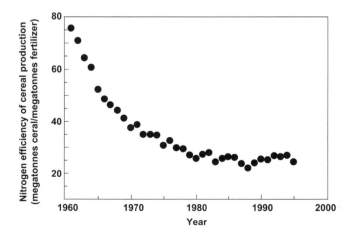

Figure 3-3. *Trend in nitrogen fertilizer efficiency of crop production calculated as annual global cereal production/annual global application of N.* Source: Tilman et al., 2002.

varieties in countries such as India can be attributed to increased fertilizer use (Bruinsma, 2003). However, there is also evidence of declining efficiency of nitrogen applications in cropping systems (Figure 3-3).

Tractors and other sources of mechanization are increasingly important to agriculture in developing countries, but many systems remain dependent on traditional forms of human and animal power.

Goals	Certainty	Range of Impacts	Scale	Specificity
N, H, L, E, S, D	B	-1 to +3	G	Developing countries

In developing countries, human, draft animal, and tractor power are used in approximately equivalent proportions in terms of total land under cultivation. There are, however, significant differences between and within countries and between regions and different types of agricultural systems. In SSA, about two-thirds of all agricultural land is cultivated by hand, whereas in LAC approximately 50% of the land is mechanically cultivated (Bruinsma, 2003). Although it is difficult to directly establish cause and effect relationships between single classes of assets and human welfare, it is generally recognized that households with animal or mechanical power tend to have better crop yields, more opportunities to pursue off-farm employment, and greater food security (Bishop-Sambrook, 2004).

Pesticide use is increasing on a global scale, but increases are not universally observed; several of the most hazardous materials are being phased out in well-regulated markets.

Goals	Certainty	Range of Impacts	Scale	Specificity
N, H, L, E, S, D	C	-5 to +4	G	Developed and developing countries

In constant dollars, global expenditures on agricultural pesticide imports has increased more than 1000% since 1960 (Tilman et al., 2001) with some estimates placing recent growth rates for pesticide use at between 4.0 and 5.4% per annum (Yudelman et al., 1998). There are exceptions to these trends, particularly in OECD countries. For example, in the US, agricultural pesticide use declined significantly after peaking in the late 1970s and has remained relatively constant since the 1990s (Aspelin, 2003). Moreover, regulatory and technological advances have, in some cases, resulted in the phase-out of particularly toxic organic compounds and the introduction of pesticides with lower non-target toxicity, which are less persistent in the environment and can be applied at lower rates (Aspelin, 2003; MA, 2005).

Total factor productivity has increased worldwide, with some regional variation.

Goals	Certainty	Range of Impacts	Scale	Specificity
D	C	-1 to +3	G	Especially in intensive systems

Total Factor Productivity (TFP), i.e., the efficiency with which all the factors of agricultural production (land, water, fertilizer, labor, etc.) are utilized, has improved over the last fifty years (Coelli and Rao, 2003). The index of TFP for world agriculture has increased from 100 in 1980 to 180 in 2000. The average increase in TFP was 2.1% per year, with

efficiency change contributing 0.9% and technical change 1.2% (Coelli and Rao, 2003). The highest growth was observed in Asia (e.g., China 6%) and North America and the lowest in South America followed by Europe and Africa. However, a positive trend does not necessarily imply a sustainable system since rapid productivity gains from new technologies may mask the effects of serious resource degradation caused by technology-led intensification, at least in the short to medium-term (Ali and Byerlee, 2002).

3.2.1.1.2 Agriculture has impacts on natural capital and resource quality

In regions with the highest rates of rural poverty and undernourishment, depletion of soil nutrients is a pervasive and serious constraint to sustaining agricultural productivity.

Goals	Certainty	Range of Impacts	Scale	Specificity
N, H, L, E, S, D	A	-1 to -5	R	SSA, ESAP

To sustain long-term agricultural production, nutrients exported from the agroecosystem by harvest and through environmental pathways (e.g., leaching, erosion) must be sufficiently balanced by nutrient inputs (e.g., fertilizer, compost, atmospheric deposition, *in situ* biological nitrogen fixation). In the tropical countries where shifting agriculture is the traditional approach to regenerating soil fertility, increasing population pressure has resulted in shorter periods of fallow and often severe reductions in soil stocks of organic carbon and nutrients (Palm et al., 2005a). Nutrient depletion is particularly acute in many of the continuous cereal production systems on the Indian sub-Continent, Southeast Asia, and sub-Saharan Africa, especially since many of the soils in these regions have low native fertility (Cassman et al., 2005). With reduced land availability for fallows, low use of fertilizer amendments, and (in some circumstances) high rates of erosion, many soils in sub-Saharan Africa are highly degraded with respect to nutrient supply capacity (Lal, 2006; Vanlauwe and Giller, 2006). It has been estimated that 85% of the arable land in Africa (ca. 185 million ha) has net depletion rates of nitrogen, phosphorous and potassium (NPK) that exceed 30 kg ha^{-1} yr^{-1} (Henao and Baanante, 2006) with 21 countries having NPK depletion rates in excess of 60 kg ha^{-1} yr^{-1}.

In high-yielding agriculture, the application of modern production technologies is often associated with environmental damage. In some cases, this damage is most attributed to inappropriate policies and management practices rather than to the technologies per se.

Goals	Certainty	Range of Impacts	Scale	Specificity
E	B	0 to -5	G	Widespread

The adoption of MVs and yield enhancing technologies like inorganic fertilizer use and irrigation have been linked to a loss of biodiversity, reduced soil fertility, increased vulnerability to pests/diseases, declining water tables and increased salinity, increased water pollution, and damage to fragile lands through expansion of cropping into unsuitable areas. A detailed assessment of the environmental impacts associated with productivity enhancing technologies concluded that empirical evidence for these associations only exists for three scenarios—salinity, lower soil fertility, and pesticides and health (Maredia and Pingali, 2001). Furthermore, many of the best documented environmental costs from agriculture are related to the misapplication of technologies or over-use of resources rather than to the direct impacts of technology *per se*. Examples of this include the subsidy-driven exploitation of groundwater for irrigation (Pimentel et al., 1997) and a lack of a complementary investment in drainage to reduce salinity problems in irrigated areas with poorly-drained soils (NAS, 1989). Some authors highlight the need for a counterfactual argument, i.e., what would have happened in the absence of yield enhancing technologies (e.g., Maredia and Pingali, 2001). For example, how much extra land would be required if yield levels had not been enhanced? Estimates suggest that at 1961 yield levels, an extra 1.4 billion ha of cultivated land would be required to match current levels of food production (MEA, 2005).

Resource-conserving technologies may reduce or eliminate some of the environmental costs associated with agricultural production with mixed results in terms of yield and overall water use.

Goals	Certainty	Range of Impacts	Scale	Specificity
H, L, E, D	A, B	-2 to +5	G	Widespread

Resource-conserving technologies (RCT) such as reduced tillage and conservation agriculture systems have been widely adopted by farmers in the last 25 years. For example, no-till systems now occupy about 95 million ha, mostly in North and South America (Derpsch, 2005), with current expansion in the Ingo-Gangetic Plain of South Asia (Hobbs et al., 2006; Ahmad et al., 2007). In general, no-till systems are associated with greatly reduced rates of soil erosion from wind and water (Schuller et al., 2007), higher rates of water infiltration (Wuest et al., 2006), groundwater recharge, and enhanced conservation of soil organic matter (West and Post, 2002). Yields can be increased with these practices, but while the physical structure of the surface soil regenerates, there can be significant interactions with crop type (Halvorson and Reule, 2006), disease interactions (Schroeder and Paulitz, 2006), surface residue retention rates (Govaerts et al., 2005), and time since conversion from conventional tillage. Other resource conserving technologies such as contour farming and ridging are also useful for increasing water infiltration, and reducing surface runoff and erosion (Reij et al., 1988; Habitu and Mahoo 1999; Cassman et al., 2005). Evidence from Pakistan (Ahmad et al., 2007) suggests that while RCT results in reduced water applications at the field scale, this does not necessarily translate into reduced overall water use as RCT serves to recharge the groundwater and then be reused by farmers through pumping. The increased profitability of RCTs also results in the expansion of the area cropped.

Modern agriculture has had negative impacts on biodiversity.

Goals	Certainty	Range of Impacts	Scale	Specificity
E	B	0 to -5	G	Widespread

The promotion and widespread adoption of modern agricultural technologies, such as modern crop and livestock varieties and management practices, has led to a reduction in biodiversity, though this is contested for some crops (Maredia and Pingali, 2001; Smale et al., 2002; Dreisigacker et al., 2003). Although biodiversity may have been temporally reduced, genetic diversity is now increasing in major cereal crops. The CGIAR and other research centers hold in trust large numbers of crop plant accessions representing diversity.

Land degradation is a threat to food security and rural livelihoods through its effects on agricultural production and the environment.

Goals	Certainty	Range of Impacts	Scale	Specificity
N, H, L, E, S, D	A	-1 to -5	G	Especially severe in the tropics

Land degradation typically refers to a decline in land function due to anthropogenic factors such as overgrazing, deforestation, and poor agricultural management (FAO/UNEP, 1996; www.unep.org/GEO/geo3). Degradation affects 1.9 billion ha and 2.6 billion people and with varying degrees of severity, and potential for recovery, encompasses a third of all arable land with adverse effects on agricultural productivity and environmental quality (Eswaran, 1993; UNEP, 1999; Esawaran et al., 2001, 2006). Inadequate replenishment of soil nutrients, erosion, and salinization are among the most common causes of degradation (Guerny, 1995; Nair et al., 1999). The GEO Report foresees that by 2030 developing countries will need 120 million additional hectares for agriculture and that this will need to be met by commercial intensification and extensification, using lands under tropical forest and with high biodiversity value (Ash et al., 2007). The restoration of degraded agricultural land is a much more acceptable option. Restoration techniques are available, but their use is inadequately supported by policy. The recovery potential of degraded land is a function of the severity, and form of degradation, resource availability and economic factors. Soil nutrient depletion can be remedied by moderate application of inorganic fertilizer or organic soil amendments, which can dramatically improve grain yields in the near-term, although responses are sensitive to factors such as soil characteristics (Zingore et al., 2007). Low-input farming systems, which are characterized by diversification at the plot and landscape scale can reverse many of the processes of land degradation, especially nutrient depletion (Cooper *et al.*, 1996; Sanchez and Leakey, 1997; Leakey *et al.*, 2005a).

Global livestock production is associated with a range of environmental problems and also some environmental benefits.

Goals	Certainty	Range of Impacts	Scale	Specificity
N, E, D	A	-3 to 0	G	Widespread applicability

The environmental problems associated with livestock production include direct contributions to greenhouse gas emissions from ruminants and indirect contributions to environmental degradation due to deforestation for pastures, land degradation due to overstocking, and loss of wildlife habitats and biodiversity (FAO, 2006d). Additionally, livestock require regular access to water resources, which they deplete and contaminate. On the other hand, extensive pastoral systems like game ranching, are more compatible with biodiversity conservation than most other forms of agriculture (Homewood and Brockington, 1999).

Intensive agricultural systems can damage agroecosystem health.

Goals	Certainty	Range of Impacts	Scale	Specificity
N, L, D	A	-1 to -5	G	Most agricultural systems

Agroecosystem health is important for nutrient, water and carbon cycling, climate regulation, pollination, pest and disease control and for the maintenance of biodiversity (Altieri, 1994; Gliessman, 1998; Collins and Qualset, 1999). Intensive production systems, such as the rice-wheat system in the Punjab, have led to deterioration in agroecosystem health, as measured by soil and water quality (Ali and Byerlee, 2002). This deterioration has been attributed to unsustainable use of fertilizer and irrigation, though whether this is due to intensification *per se* or to mismanagement is unclear. For example, in China, grain yield would have increased by 5% during 1976-89 given less erosion and less soil degradation (e.g., increased salinity) (Huang and Rozelle, 1995). More evidence is needed about the relationships between total factor productivity and long-term agroecosystem health. In some cases, intensified production on prime agricultural land may reduce negative impacts on ecosystem health by reducing the incentive to extend production onto marginal lands or into natural areas (e.g., highly erodible hillslopes).

Poor irrigation management causes land degradation with negative impact on livelihoods.

Goals	Certainty	Range of Impacts	Scale	Specificity
N, L, E, S	B	-1 to -3	R	Especially in the dry tropics

Irrigation increases crop productivity in dry areas, but can result in land degradation. Poor drainage and irrigation practices have led to waterlogging and salinization of roughly 20% of the world's irrigated lands, with consequent losses in productivity (Wood et al., 2000). While livelihoods have improved through increased production and employment, demands for irrigation water have degraded wetland biodiversity (Huber-Lee and Kemp-Benedict, 2003 quoted in Jinendradasa, 2003). Poorly conceived and implemented water management interventions have incurred high environmental and social costs, including inequity in benefit allocation and loss of livelihood opportunities. Common property resources such as rivers and wetlands, important for poor fishers and resource gatherers, have been appropriated for other uses, resulting in a loss of livelihood opportunities. Communities have been displaced, especially in areas behind dams, without adequate compensation. A large proportion of irrigation's negative environmental effects arise from the diversion of water away from natural aquatic ecosystems (rivers, lakes, oases, and other groundwater dependent wetlands). Direct and indirect negative impacts have been well documented, including salinization, channel erosion, declines in biodiversity, introduction of invasive

alien species, reduction of water quality, genetic isolation through habitat fragmentation, and reduced production of floodplain and other inland and coastal fisheries.

In some river basins, water scarcity due to irrigation has become a key constraint to food production.

Goals	Certainty	Range of Impacts	Scale	Specificity
N, H, L, S, D	A	-1 to -5	R	Especially severe in the dry tropics

Fifty years ago water withdrawal from rivers was one third of what it is today, with 70% of freshwater withdrawals (2,700 km^3 or 2.45% of rainfall) attributable to irrigated agriculture (CA, 2007). About 1.6 billion people live in water-scarce basins. Water availability is a worldwide problem (Figure 3-4) despite a decline in water withdrawal for agriculture over the past 20 years (FAO AQUASTAT, 2007) in developed (58 to 39%) and developing countries (76 to 71%), a decline of 69 to 61% globally (FAOSTAT, 2006). In both irrigated and rainfed areas, a decline in water available for irrigation, without compensating investments and improvements in water management and water use efficiency, has been found to reduce production with a consequent increase in international cereal prices and negative impacts on low-income developing countries (Rosegrant and Cai, 2001). Global investment in water distribution systems for agriculture has declined relative to other sectors during recent decades.

Agriculture contributes to degradation and pollution of water resources.

Goals	Certainty	Range of Impacts	Scale	Specificity
E, S	A	-1 to -5	G	Most agricultural systems

Traces of the herbicide "Atrazine" and other pesticides are routinely documented in shallow ground and surface waters in industrialized countries. Recent surveys in the U.S. suggest that pesticides concentrations exceed human health and wildlife safety standards in approximately 10% of streams and 1% of groundwater wells (USGS, 2006). In intensive agricultural regions, streamwater nitrogen concentrations have been found to be nearly nine times higher than downstream from forested areas (Omernik, 1977). Increasing concentrations of nitrate nitrogen in the Mississippi River have also been linked to hypoxic conditions in the Gulf of Mexico (Rabalais et al., 1996).

3.2.1.1.3 Impacts on diet and health

Patterns of food consumption are becoming more similar throughout the world,

Goals	Certainty	Range of Impacts	Scale	Specificity
N, H, L, S	B	-2 to +2	R	Widespread in the tropics

The Green Revolution did not focus on nutrient-rich foods like fruits, vegetables, legumes and seafood. The focus on cereals led to an increased *per capita* consumption of cereals, while in most developing countries, consumption of vegetables remained far below the minimum requirement level of 73 kg per person (Ali and Abedullah, 2002). Likewise, *per capita* consumption of pulses in south Asia fell

Proportion of water withdrawal for agriculture, 2001

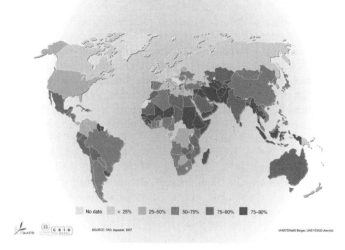

No data   < 25%   25–50%   50–75%   75–90%   75–90%

SOURCE: FAO, Aquastat, 2007.

Figure 3-4. *Agricultural water withdrawal as percentage of total water withdrawal for agricultural, domestic and industrial purposes worldwide.* Source: FAO AQUASTAT, 2007.

from 17 kg in 1971 to 12 kg in 2003 (Ali et al., 2005). Recently, however, vegetable production has increased in developing countries, through public-private collaboration in the introduction of modern varieties and technologies. The replacement of traditional plant based diets with increased consumption of more energy-dense, nutrient-poor foods with high levels of sugar and saturated fats in all world regions (Popkin, 2003) has been driven by increased incomes and other factors such as changes in food availability, and retail and marketing activities. Increased protein consumption (e.g., meat and dairy products) is occurring in developing countries, but high costs limit consumption primarily to the urban elite.

The application of modern AKST has led to a decline in the availability and consumption of traditional foods.

Goals	Certainty	Range of Impacts	Scale	Specificity
N, H, L, D	B	-2 to -4	R	Widespread in the tropics

In the past, many traditional foods were gathered from forests and woodlands, which provided rural households with food and nutritional security. With the loss of habitat through deforestation, population growth, increased urbanization and poverty and an emphasis on staple food cultivation, this wild resource has diminished. In addition, improved access to other food crops and purchased foods (Arnold and Ruiz Pérez, 1998) have contributed to the trend towards diet simplification, reduced fresh food supply, and disappearance of nutrient rich indigenous food. This simplification has had negative impacts on food diversity and security, nutritional balance, and health. Indigenous fruits and vegetables have been given low priority by policy makers, although they are still an important component of diets, especially in Africa.

Supplies of nutritious traditional food are in decline, but reversible.

Goals	Certainty	Range of Impacts	Scale	Specificity
N, H, L, S	B	-2 to -4	R	Widespread in the tropics

Deforestation and increasing pressures from urban infrastructure have reduced the fresh sources of food supply from forests and urban gardens (Ali et al., 2006). Projects to reverse this trend promote traditional foods as new crop plants (Leakey, 1999a; Leakey et al., 2005a) and encourage their consumption. For example in Zambia, the FAO Integrated Support to Sustainable Development and Food Security Program (IP-Zambia) is promoting the consumption of traditional foods (www.fao.org/sd/ip).

3.2.1.2 Biotechnology: conventional breeding and tissue culture

The modification of plants and animals through domestication and conventional plant breeding (i.e., excluding use of nucleic acid technologies and genetic engineering) has made a huge contribution to food production globally: the Green Revolution for plants, the Blue Revolution for fish and the Livestock Revolution.

3.2.1.2.1 Impact of modern varieties of crops (including trees) and improved livestock breeds
The impact of domestication and conventional breeding, especially in annual crop plants, has been well documented. Modern varieties and breeds have had positive impacts on yield and production, especially where environments have been favorable and management has been good. However, there have also been some negative effects on the environment and on biodiversity. There is also some concern that on-station and on-farm yields are stagnating.

Agriculture is dependent on very few species of animals and plants.

Goals	Certainty	Range of Impacts	Scale	Specificity
N, H, L, E	A, B	-2 to +5	G	Wide applicability

Agriculture began with the domestication of wild animals and plants. About 1000 plant species have been domesticated resulting in over 100 food and 30 non-food crops (fiber, fodder, oil, latex, etc., excluding timber). Approximately 0.3% of the species in the plant kingdom have been domesticated for agricultural purposes (Simmonds, 1976) and 4.1% for garden plants (Bricknell, 1996). These proportions rise to 0.5 and 6.5% respectively if limited to the higher plants (angiosperms, gymnosperms and pteridophytes) of which there are some 250,000 species (Wilson, 1992), but are small when compared with the 20,000 edible species used by hunter-gatherers (Kunin and Lawton, 1996). A similar pattern has occurred in animals and fish, with only a small proportion of the species traditionally consumed domesticated through AKST. Over the last 50-60 years plant and animal breeding was a major component of the Green Revolution.

Overall, the impacts of the Green Revolution have been mixed.

Goals N, L, D	Certainty C	Range of Impacts 0 to -3	Scale G	Specificity Mainly small-scale agriculture

Positive impacts on yield have been achieved in Latin America with an increase of 132% (36% from improved varieties and 64% from other inputs) on 32% less land (Evenson and Gollin, 2003a). Negative effects on yield occurred in sub-Saharan Africa even though overall yield increased 11% (130% coming from improved varieties and -30% from other inputs), since 88% more land was used. In SSA and CWANA, MVs were released but not adopted throughout the 1960s and 70s (Evenson, 2003). In some cases, MVs lacked desired organolepic qualities or were not as well adapted as Traditional Varieties (TVs). However, in many cases the lack of adoption resulted from inadequate delivery of seeds to farmers (Witcombe et al., 1988). Poor seed delivery systems remain a major constraint in many parts of Africa (Tripp, 2001).

Plants

Domestication, intensive selection and conventional breeding have had major impacts on yield and production of staple food crops, horticultural crops and timber trees.

Goals N	Certainty A	Range of Impacts +2 to +5	Scale G	Specificity Widespread applicability

Yield per unit area of the world's staple food crops, especially cereals (rice, wheat and maize) have increased over the last 50 years (Figure 3-2), as a result of publicly and privately funded research on genetic selection and conventional breeding (Simmonds, 1976; Snape, 2004; Swaminathan, 2006). Increased wheat and barley yield in the UK (Silvey, 1986, 1994), and maize yield in the USA (Duvick and Cassman, 1999; Tollenaur and Wu, 1999), e.g., is attributed equally to advances in breeding and to improved crop and soil management. Gains in productivity between 1965 and 1995 were about 2% per annum for maize, wheat and rice (Pingali and Heisey, 1999; Evenson and Gollin, 2003a), though rates have declined in the last decade. Similarly, productivity measured as total factor productivity (TFP) also increased in rice, wheat and maize (Pingali and Heisey, 1999; Evenson, 2003a). The impact of crop improvement on non-cereals has been less well documented as these crops are often far more diverse, occupy smaller areas globally and are not traded as commodities. For example, in total legumes occupy 70.1 m ha globally, but there a greater diversity of legume species is used with clear regional preferences and adaptation (e.g., cowpeas, *Vigna unguiculata*, in West Africa; pigeon pea, *Cajanus cajan*, and mung bean, *Vigna radiata*, in India). Nonetheless, plant breeding has increased yields in many protein crops (Evenson and Gollin, 2003b).

Much of the increase in crop yield and productivity can be attributed to breeding and dissemination of Modern Varieties (MV) allied to improved crop management.

Goals N, L, D	Certainty A	Range of Impacts -2 to +5	Scale G	Specificity Widespread applicability

A number of studies (Pingali and Heisey, 1999; Heisey et al., 2002; Evenson and Gollin, 2003ab; Hossain et al., 2003; Raitzer, 2003; Lantican et al., 2005) have quantified the large impact (particularly in industrialized countries and Asia) of crop genetic improvement on productivity (Figure 3-2). Much of this impact can be attributed to IARC genetic research programs, both direct (i.e., finished varieties) and indirect (i.e., parents of NARS varieties, germplasm conservation). Benefit-cost ratios for genetic research are substantial: between 2 (significantly demonstrated and empirically attributed) and 17 (plausible, extrapolated to 2011) (Raitzer, 2003). Two innovations—rice and wheat MVs rice (47% and 31% of benefits, respectively) account for most of the impact. Benefits can also be demonstrated for many other crops. For example, an analysis of the CIAT bean (*Phaseolus vulgaris*) breeding program (Johnson et al., 2003) showed that 49% of the area under beans could be attributed to the CIAT breeding program, raising yield by 210 kg ha^{-1} on average and resulting in added production value of US$177 m. For Africa, where the breeding program started later, about 15% of the area is under cvs that can be attributed to CIAT, with an added value of US$26 million. The estimated internal rate of return was between 18 and 33%, with more rapid positive returns in Africa, which built upon earlier work in LAC.

Although the adoption of MVs is widespread, many MVs may be old and farmers are therefore not benefiting from the latest MV with pest/disease resistant and superior yield.

Goals N, L, D	Certainty C	Range of Impacts -1 to -3	Scale G	Specificity High and low potential systems

Although new and potentially better MVs have been released in many countries, these have not been grown by farmers, more often than not due to the inefficiency of the varietal release and seed multiplication system (Witcombe et al., 1988) rather than poor suitability. For example, in high potential areas of the Punjab the most commonly grown wheat and rice MVs were 8-12 and 11-15 years old (Witcombe, 1999; Witcombe et al., 2001). The age of an MV in use may also vary with environment, with lower rates of turnover in more marginal areas where suitable MVs have not been released (Smale et al., 1998; Witcombe et al., 2001). Assuming that genetic gains in potential yield achieved each year are on the order of 1 to 2% (e.g., Figure 3-5), then farmers may be losing 16 to 30% of potential yield; these losses will be even higher where MVs have superior disease or pest resistance.

Gains in productivity from MVs have been greatest in high potential areas, particularly irrigated rice and wheat, but benefits have also occurred is less favorable areas.

Goals N, L, D	Certainty B	Range of Impacts +1 to +2	Scale G	Specificity Low potential environments

Yield gains of wheat on farmers' fields in more marginal environments were between 2-3% between 1979 and 1996 (Byerlee and Moya, 1993; Lantican et al., 2005), compared with increases with irrigation of about 1% per annum between 1965 and 1995 (Lantican et al., 2005). These more recent gains stem from breeding efforts based on greater

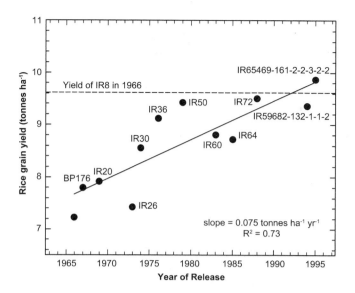

Figure 3-5. *Current yield potential of rice modern varieties (MVs) as a function of year of release.* Source: Cassman et al., 2003, derived from Peng et al., 1999.

Note: Dashed line indicates the yield potential of IR8 when it was released in 1966. Graphic illustrates the importance of "maintenance breeding" and of stagnating yield potential.

understanding of marginal environments, such as those with acid soils or heat/drought stress (Reynolds and Borlaug, 2006). In maize, about 50% of the increase in yield attributed to genetic gain is due to improvements in stress tolerance (Tollenaur and Wu, 1999), which has contributed to maize expansion in more marginal environments.

Crop improvement has reduced genetic diversity, but current breeding strategies are tackling this problem.

Goals	Certainty	Range of Impacts	Scale	Specificity
E	C	-2 to +2	G	Widespread

In Asia, MVs account for >75% area for wheat and rice and village level studies in Nepal have shown incidences of a single wheat MV, CH45, occupying 96% of the area (Evenson and Gollin, 2003b; Witcombe et al., 2001). Elsewhere, notably in Africa and CWANA, MVs occupy smaller proportions and many more TVs can be found (Evenson and Gollin, 2003b). The loss of genetic diversity due to the widespread adoption of MVs has resulted in negative environmental impacts (Evenson and Gollin, 2003ab): reducing the availability of genes for future crop improvement, creating the possibility for inbreeding depression (with negative impacts on production), reducing species ability to adapt to change (eg. climate change) and evolving resistance to new pest and disease outbreaks. However, this is disputed (Maredia and Pingali, 2001). Genetic diversity can vary both temporally and spatially, and both have to be taken into account in assessing impacts on diversity. The rapid replacement of old varieties with newer ones has increased the temporal diversity in Mexico and Pakistan, especially when current breeding programs increasingly use more genetically diverse traditional varieties in their parentage (Smale, 1997;

Smale *et al.*, 1998; Hartell *et al.*, 1998). This has been confirmed by a recent molecular study of genetic diversity in wheat (Reif *et al.*, 2005). However, molecular analysis of MVs by ages, areas and genealogies, has shown clearly that diversity in spring wheat in developing countries has not decreased since 1965 (Smale et al., 2002).

Genetic yield potential is not increasing.

Goals	Certainty	Range of Impacts	Scale	Specificity
N	C	-3 to +1	G	Widespread

Plant breeding in developed and less developed countries has to date been successful at delivering new, higher yielding varieties, largely through better adaptation, greater partitioning of biomass to seed (i.e., harvest index; Austin et al., 1980; Sayre et al., 1997) and disease resistance. However, under conditions where pests are efficiently controlled and there are no limitations to the supply of water and nutrients, there is evidence (Figure 3-5) that the yield potential of the most productive rice, wheat, and maize cultivars has not markedly increased since the Green Revolution (Duvick and Cassman, 1999; Peng et al., 1999; Sayre et al., 2006). Even in the UK, where the benefits of plant breeding have been well documented (Silvey, 1986, 1994), national wheat yields are only increasing slowly (Sylester-Bradley et al., 2005); although in any given year yields of the best varieties in National Recommended List trials show average gains >2% per year above the most recently released varieties (Austin, 1999; http://www.hgca.com/content.template/23/0/Varieties/Varieties/Varieties%20Home%20Page.mspx) It is clear that when harvest indices in some annual grasses and legumes approach their theoretical maximum, selection for increased total crop biomass and/or the exploitation of hybrid vigor will be important. Hybrid rice, which yields about 15% more than conventionally bred rice, is already grown on some 15 million ha in China (about half the total area in rice) (Longping, 2004), and hybrid sorghum shows similar promise.

Gains in yield per unit area per year are expected to remain lower than historical yields.

Goals	Certainty	Range of Impacts	Scale	Specificity
N	A	-2 to 0	G	Widespread

Conceptually, crop improvement goes through stages of domestication to produce Traditional Varieties (TVs), and then TVs are replaced by a succession of MVs (Otsuka and Yamano, 2005). In wheat, rice and maize gains were initially much higher (35-65%) when MV replaced traditional varieties (Otsuka and Yamano, 2005) Subsequent gains when MV2 replace MV1 have been lower (10-30%). This reduction in gain is to be expected, as many TVs were not necessarily well adapted, especially to changing climates, and yield may have been constrained by susceptibility to major pest and diseases, or non-biotic constraints such as lodging. Furthermore, once major constraints are tackled, most breeding efforts go into maintaining resistance and enhancing quality, and not simply increasing yield potential (Legg, 2005; Baenziger et al., 2006). Constraints due to soil fertility and structure, and diseases and pests from continuous cultivation limit increases in yield potential (see below;

Cassman et al., 2003). Nonetheless, further small gains are expected, through continued genetic gain and a better understanding and breeding for specific target environments (Reynolds and Borlaug, 2006). In developing countries and low yield potential environments the benefits of breeding for specific environments will be further enhanced with the adoption of more localized and/or participatory breeding, i.e., with the exploitation of G × E or local adaptation.

In several intensive production environments, cereal yields are not increasing.

Goals	Certainty	Range of Impacts	Scale	Specificity
N, L, E, D	A	-2 to -4	G	Intensive production systems

In several of the most important regions for irrigated rice production (e.g., areas of China, Japan, Korea) there is strong evidence of persistent yield stagnation at approximately 80% of the theoretical productivity levels predicted by simulation models (Cassman et al., 2003). This type of stalled exploitation of potential production is primarily caused by economic factors since the rigorous management practices required for yield maximization are not cost effective (Pingali and Heisey, 1999; Cassman et al., 2003). Rice yield stagnation has also been observed in areas like Central Java and the Indian Punjab at levels significantly below 80% of the theoretical productivity. In long-term cropping system experiments (LTE) with the highest-yielding rice varieties under optimal pest and nutrient management, rice yield potential declined at several locations. Subsequent evidence from a larger set of LTEs suggested that this phenomenon was not widespread, but that rice yield potential was essentially stagnant in most regions despite putative innovations in management and plant genetic resources (Dawe et al., 2000). For irrigated production systems in the maize belt of the United States, yields achieved by the most productive farmers have not increased since the mid-1980s (Duvick and Cassman, 1999). For spring wheat producers in Mexico's Yaqui Valley, only nominal increases in yield have been observed since the late 1970s.

In many regions the production potential for the staple cereal crops has not been exhausted.

Goals	Certainty	Range of Impacts	Scale	Specificity
N L, E, D	B	-2 to +2	R	Not clear

In contrast to concerns about limited future opportunities for yield improvement in cereals, there are some examples of yield increases. For example, coordinated efforts to improve management practices and profitability of Australian rice systems increased productivity from 6.8 tonnes ha[-1] in the late 1980s to 8.4 tonnes ha[-1] by the late 1990s (Ferrero and Nguyen, 2004). Farm-level maize yields in the United States are typically less than half of the climate-adjusted potential yield (Dobermann and Cassman, 2002). At the state level in India, an analysis (Bruinsma, 2003) suggests that rice productivity could be increased by 1.5 tonnes ha[-1] (ca. 50%) without exceeding the 80% criteria commonly used to establish the economically-exploitable component of the biophysical yield potential (Bruinsma, 2003).

In developing countries the productivity of many small-scale farming systems is often constrained by limited access to inputs and modern varieties (MVs) and poor management practices.

Goals	Certainty	Range of Impacts	Scale	Specificity
N, L, E, D	A	-2 to -5	G	Small-scale farms in developing countries

In upland rice systems in Laos, the importance of the adoption of improved varieties and N fertilization has been demonstrated (Saito et al., 2006). By substituting MVs for traditional landraces, rice yields doubled to 3.1 tonnes ha[-1] with a moderate dose of nitrogen fertilizer further improving yield by 1 tonne ha[-1]. Among farmers in Nepal, modern crop management practices (e.g., timely establishment, precision planting, two weedings) together with site-specific nutrient management boost rice productivity by 2 tonnes ha[-1] over typical farmer practices (Regmi and Ladha, 2005). In West Africa, rural surveys show that most farmers have limited knowledge of soil fertility management and of optimal establishment practices for rice (Wopereis et al., 1999). In these areas, nitrogen deficiency, inadequate weeding, and late planting are commonly associated with low cereal productivity (Becker and Johnson, 1999). Poor knowledge of efficient practices for maintaining soil fertility has also been identified as an important component of the low yields achieved by Bangladeshi rice farmers (Gaunt et al., 2003).

Barriers to clonal forestry and agroforestry have been overcome by the development of robust vegetative propagation techniques, which are applicable to a wide range of tree species.

Goals	Certainty	Range of Impacts	Scale	Specificity
L, E, S	B	+3 to +5	G	Widespread applicability

Techniques of vegetative propagation have existed for thousands of years (Hartmann et al., 1997), but the factors affecting rooting capacity seem to vary between species and even clones (Leakey, 1985; Mudge and Brennan, 1999). However, detailed studies of the many morphological and physiological factors affecting five stages of the rooting process in stem cuttings (Leakey, 2004) have resulted in some principles, which have wide applicability (Dick and Dewar, 1992) and explain some of the apparently contradictory published information (Leakey, 2004). Robust low-technology vegetative propagation techniques are now being implemented within participatory village-level development of cultivars of indigenous fruit/nut tree species to diversify cocoa farming systems in West Africa (Leakey et al., 2003).

Participatory domestication techniques are using low-tech approaches to cloning to develop cultivars of new tree crops for agroforestry.

Goals	Certainty	Range of Impacts	Scale	Specificity
N, H, L, E, S	A	+1 to +3	M-L	Wide applicability

Simple, inexpensive and low-tech methods for the rooting of stem cuttings have been developed for use by resource poor farmers in remote village nurseries (Leakey et al., 1990). These robust and appropriate techniques are based on a greatly increased understanding of the factors affecting successful vegetative propagation (Leakey, 2004). The identification of selection criteria is being based on the quantitative characterization of many fruit and nut traits (Atangana et al., 2001, 2002; Anegbeh et al., 2003, 2004; Waruhiu et al., 2004; Leakey, 2005b; Leakey et al., 2005bc). Using participatory approaches (Leakey et al., 2003), the implementation of these techniques is being successfully achieved by small-scale farmers from 40 communities (Tchoundjeu et al., 2006).

Clonal approaches to the genetic improvement of timber tree species result in large improvements in yield and quality traits.

Goals	Certainty	Range of Impacts	Scale	Specificity
L, E, S	A	+2 to +5	G	Widespread applicability

For example in timber species, clones of *E. urophylla* x *E. grandis* hybrid in Congo were planted in monoclonal blocks of 20-50 ha at a density of 800 stems ha^{-1} and resulted in mean annual increments averaging 35 m^3 ha^{-1}, compared with 20-25 m^3 ha^{-1} from selected provenances, and about 12 m^3 ha^{-1} from unselected seedlots (Delwaulle, 1983). In Brazil, mean annual increments between 45-75 m^3 ha^{-1} and up to 90 m^3 ha^{-1} have been recorded (Campinhos, 1999). Clonal approaches require (Leakey, 1987; Ahuja and Libby 1993ab) genetic diversity (Leakey, 1991), wise deployment (Foster and Bertolucci, 1994) and appropriate silviculture (Lawson, 1994; Evans and Turnbull, 2004) to maximize gains, minimize pest and pathogen risks and maintain species diversity in the soil microflora (Mason and Wilson, 1994), soil invertebrates (Bignell et al., 2005) and insect populations (Watt et al., 1997, 2002; Stork et al., 2003).

Increased private sector involvement in timber plantations has recently been more inclusive of social and environmental goals.

Goals	Certainty	Range of Impacts	Scale	Specificity
E	C	-1 to +3	G	Wide applicability

In the past, the cultivation of planted timber trees has mostly been implemented by national forestry agencies, often with inadequate attention to establishment techniques. In the last 20-30 years there has been increasing private sector investment, much of which has been multinational, and often in partnership with local companies or government agencies (Garforth and Mayers, 2005). These companies have focused on a few fast-growing species, especially for pulp and paper industries, often grown as exotic species outside their natural range. In these plantations genetic improvement has typically been achieved by provenance selection and clonal technologies. Increasingly, such plantations are being designed as "mosaic" estates with a view to greater synergies with both local agricultural conditions and areas protected for biodiversity (IIED, 1996) and as joint ventures with communities to provide non-fiber needs in addition to wood (Mayers and Vermeulen, 2002).

Livestock and fish

Domestication and the use of conventional livestock breeding techniques have had a major impact on the yield and composition of livestock products.

Goals	Certainty	Range of Impacts	Scale	Specificity
N, H, L	A	0 to +4	G	Widespread but, mostly in developed countries

There has been widespread use of breed substitution in industrialized countries and some developing countries, often leading to the predominance of a few very specialized breeds, and often pursuing quite narrow selection goals. Organized within-breed selection has been practiced much less widely in many developing countries, partly because of the lack of infrastructure, such as national or regional performance recording and genetic evaluation schemes. Genetic improvement—breed substitution, crossbreeding and within-breed selection—has made an important contribution to meeting the growing global demand for livestock products. Selection among breeds or crosses is a one off, non-recurrent process: the best breed or breed cross can be chosen, but further improvement can be made only by selection within the populations (Simm *et al.*, 2004). Crossbreeding is widespread in commercial production, exploiting complementarity of different breeds or strains, and heterosis or hybrid vigor (Simm, 1998). Trait selection within breeds of farm livestock typically produces annual genetic changes in the range 1-3% of the mean (Smith, 1984). Higher rates of change occur for traits with greater genetic variability, in traits that are not age- or sex-limited, and in species with a high reproductive rate, like pigs and poultry (McKay *et al.*, 2000; Merks, 2000), fish and even dairy cattle (Simm, 1998). These rates of gain have been achieved in practice partly because of the existence of breeding companies in these sectors. Typically, rates of genetic change achieved in national beef cattle and sheep populations have been substantially lower than those theoretically possible, though they have been achieved in individual breeding schemes. The dispersed nature of ruminant breeding in most countries has made sector-wide improvement more challenging.

In most species, rates of change achieved in practice through breeding have increased over the last few decades in developed countries.

Goals	Certainty	Range of Impacts	Scale	Specificity
N, H, L	A	0 to +4	G	Developed countries

The greatest gains in productivity as a result of genetic improvement have been made in poultry, pigs and, to a lesser extent, dairy cattle. Greater success through breeding programs in developed countries has been the result of better statistical methods for estimating the genetic merit (breeding value) of animals, especially best linear unbiased prediction methods; the wider use of reproductive technologies, especially artificial insemination; improved techniques for measuring performance (e.g., ultrasonic scanning to assess carcass composition *in vivo*); and more focused selection on objective rather than subjective traits, such as milk yield rather than type. Developments in the statistical, reproductive and molecular genetic technologies available have the potential to increase rates of change further (Simm et al., 2004). In recent years there has been a growing trend in developed countries for breeding programs to focus more on product quality or other attributes, rather than yield alone. There is also growing interest in breeding goals that meet wider public needs, such as increasing animal welfare or reducing environmental impact.

Gains in productivity have been variable if breeds are not matched to the environment

Goals N, H, D	Certainty B	Range of Impacts 0 to +3	Scale G	Specificity Developing countries

The gains in productivity per animal have been greatest in developed countries, and in the more "industrialized" production systems in some developing or "transition" countries. The enormous opportunities to increase productivity through wider adoption of appropriate techniques and breeding goals in developing countries are not always achieved. Breed substitution and crossing have both given rapid improvements, but it is essential that new breeds or crosses are appropriate for the environment and resources available over the entire production life cycle. Failure to do this has resulted in herds that have succumbed to diseases or to nutritional deprivation to which local breeds were tolerant, e.g., the introduction of high performing European dairy breeds into the tropics that had lower survival than pure Zebu animals and their crosses. The reproductive rate of the pure European breeds is often too low to maintain herd sizes (de Vaccaro, 1990). It is also important that valuable indigenous Farm Animal Genetic Resources are protected.

Large scale livestock production can lead to environmental problems.

Goals N, L, S	Certainty B	Range of Impacts -2 to +3	Scale G	Specificity Urban centers in developing countries

Recently, livestock production has increased rapidly, particularly in developing countries where most of the increased production comes from industrial farms clustered around major urban centers (FAO, 2005c). Such large concentration of animals and animal wastes close to dense human population often causes considerable pollution problems with possible negative effects on human health. Large industrial farms produce more waste than can be recycled as fertilizer and absorbed on nearby land. When intensive livestock operations are crowded together, pollution can threaten the quality of the soil, water, air, biodiversity, and ultimately public health (FAO, 2005c). In less intensive mixed farming systems, animal wastes are recycled as fertilizer by farmers who have direct knowledge and control of their value and environmental impact. However in industrial production, there is a longer cycle in which large quantities of wastes accumulate.

Livestock production is a major contributor of emissions of polluting gases.

Goals N, L, S	Certainty B	Range of Impacts -2 to +3	Scale G	Specificity All livestock

Livestock production is a major contributor of emissions of polluting gases, including nitrous oxide, a greenhouse gas whose warming potential is 296 times that of carbon dioxide. Livestock contributes 18% of the total global warming effect, larger even than the transportation worldwide (Steinfeld et al., 2006). The share of livestock production in human-induced emissions of gases is 37% of total methane, 65% of nitrous oxide, 9% of total carbon dioxide emissions and 68% of ammonia emissions (Steinfeld et al., 2006). This atmospheric pollution is in addition to the water pollution caused by large-scale industrial livestock systems.

Aquaculture has made an important contribution to poverty alleviation and food security in many developing countries.

Goals N, H, L, S, D	Certainty B	Range of Impacts +1 to +3	Scale G	Specificity Developing countries

Aquaculture, including culture-based fisheries, has been the world's fastest growing food-producing sector for nearly 20 years (FAO, 2002c; Delgado et al., 2003a; Bene and Heck, 2005a; World Bank, 2007b). In 1999, 42.8 million tonnes of aquatic products (including plants) valued at US$53.5 billion were produced, and more than 300 species of aquatic organisms are today farmed globally. Approximately 90% of the total aquaculture production is produced in developing countries, with a high proportion of this produced by small-scale producers, particularly in low income food deficit countries (Zeller et al., 2007). While export-oriented, industrial and commercial aquaculture practices bring in needed foreign exchange, revenue and employment, more extensive and integrated forms of aquaculture make a significant grassroots contribution to improving livelihoods among the poorer sectors of society and also promote efficient resource use and environmental conservation (FAO, 2002c). The potential of aquaculture has not yet been fully realized in all countries (Bene and Heck, 2005ab; World Bank, 2007b).

Globally, per capita fish consumption increased by 43% from 11 kg to 16kg between 1970 and 2000.

Goals N, H	Certainty B	Range of Impacts 0 to +2	Scale G	Specificity Asia particularly

In developing countries, fish have played an important role in doubling animal protein consumption per capita over the last 30 years—from 6.3 to 13.8 kg between 1970 and 2000. In the developed world, fish consumption increased by less than one-half during the same period. Urbanization, income and population growth are the most significant factors increasing fish consumption in developing countries, particularly in Asia (Dey et al., 2004).

The recent increase in aquaculture production is primarily due to advances in induced breeding or artificial propagation techniques (hypophysation).

Goals N, L, S	Certainty B	Range of Impacts -2 to +3	Scale G	Specificity Freshwater carp farming

Induced breeding and hypophysation have particularly occurred in the carp polycultures and in freshwater fish farming in rice fields, seasonal ditches, canals and perennial ponds.

However, in Bangladesh, hatchery-produced stock (mainly carps) have shown adverse effects such as reduced growth and reproductive performance, increased morphological deformities, and disease and mortalities. These effects are probably due to genetic deterioration in the hatchery stocks resulting from poor fish brood stock management, inbreeding depression, and poor hatchery operation (Hussain and Mazid, 2004).

Aquaculture has had positive and negative effects on the environment.

Goals	Certainty	Range of Impacts	Scale	Specificity
N, L, S	B	-2 to +3	G	Coastal ecosystems

There have been negative and positive impacts of aquaculture on the environment, depending on the intensification of the production systems. An incremental farmer participatory approach to the development of sustainable aquaculture in integrated farming systems in Malawi (Brummett, 1999) found that integrated farming systems are more efficient at converting feed into fish and produce fewer negative environmental impacts. The widespread adoption of integrated aquaculture could potentially improve local environments by reducing soil erosion and increasing tree cover (Lightfoot and Noble, 1993; Lightfoot and Pullin, 1995; Brummett, 1999). Negative environmental effects resulting from the aquaculture industry include threats to wild fish stocks (Naylor et al., 2000); destruction of mangrove forests and coastal wetlands for construction of aquaculture facilities; use of wild-caught rather than hatchery-reared finfish or shellfish fry to stock captive operations (often leading to high numbers of discarded by-catch of other species); heavy fishing pressure on small ocean fish for use as fish meal (depleting food for wild fish); transport of fish diseases into new waters; and non-native fish that may hybridize or compete with native wild fish. Improvements in management can help to reduce the environmental damage (Lebel et al., 2002), but only to a minor extent. However, economic impacts are site-specific. Intensive aquaculture has also had important effects on the landscape, e.g., in Thailand 50-65% of the mangroves have been replaced by shrimp ponds (Barbier and Cox, 2002).

3.2.1.2.2 Breeding for abiotic and biotic stress tolerance
Crops and plants, especially in marginal environments, are subjected to a wide and complex range of biotic (pests, weeds) and abiotic (extremes of both soil moisture and air/soil temperature, poor soils) stresses. Abiotic stresses, especially drought stress (water and heat) have proved more intractable.

Progress in breeding for marginal environments has been slow.

Goals	Certainty	Range of Impacts	Scale	Specificity
N	B	0 to +1	R	Widespread aplicability

Progress in breeding for environments prone to abiotic stresses has been slow, often because the growing environment was not characterized or understood (Reynolds and Borlaug, 2006), too many putative stress tolerant traits proved worthless (Richards, 2006), and because the complex nature of environment-by-gene interactions was not recognized and yield under stress has a low heritability (Baenziger et al., 2006). Drought, for example, is not easily quantifiable (or repeatable) in physical terms and is the result of a complex interaction between plant roots and shoots, and soil and aerial environments (Passioura, 1986). Furthermore, much effort was expended on traits that contributed to survival rather than productivity.

Although yield and drought tolerance are complex traits with low heritability, it has been possible to make progress through conventional breeding and testing methods.

Goals	Certainty	Range of Impacts	Scale	Specificity
N	D	0 to +1	R	CWANA, SSA

Breeding for marginal and stressed environments has not been easy, especially where wide-adaptation was also important. However, breeding programs that make full use of locally-adapted germplasm and TVs (Ceccarelli et al., 1987), and select in the target environments (Ceccarelli and Grando, 1991; Banziger et al., 2006) have been successful. For example, in Zimbabwe, where soil fertility is low and drought stress common, the careful selection of test environments (phenotyping) and selection indices can increase maize yields across the country and regionally (Banziger et al., 2006). Equal weight to three selection environments (irrigated, drought stress, N-stress), the use of moderately severe stress environments, and the use of secondary traits with higher heritabilities improved selection under stress. In multilocation trials, lines selected using this method outyielded other varieties at all yield levels, but more so in more marginal environments. This would seem to be a successful blue print for conventional breeding for stress environments.

Although drought tolerance is a complex trait, progress has been made with other aspects of abiotic stress tolerance.

Goals	Certainty	Range of Impacts	Scale	Specificity
N	B	0 to +2	G	Many crops

Yield is the integration of many processes over the life of a crop, and as such it is unsurprising that heritabilities are low and progress slow. In contrast, the effects of some abiotic stresses are associated with very specific stages of the life cycle (particularly flowering and seed-set) or are associated with very specific mechanisms, and these appear to be more amenable to selection. Progress has been made in breeding for tolerance to a number of stresses, including extremes of temperature (hot and cold), salt and flooding/submergence, and nutrient deficiency. For example, tolerance to extremes of temperature, which are important constraints in many crop species at and during reproductive development (i.e., in the flowering period), have been identified (Hall, 1992; Craufurd et al., 2003; Prasad et al., 2006) and in some cases genes identified and heat tolerant varieties bred (Hall, 1992). These particular responses will be increasingly valuable as climate changes.

Biological control has been successfully adopted in pest control programs to minimize the use of pesticides and reduce environmental and human health risks.

Goals	Certainty	Range of Impacts	Scale	Specificity
N, E	B	+1 to +3	M-L	Wide applicability

Ten percent of the world's cropped area involves classical biological control. The three major approaches to biological control are importation, augmentation and conservation of natural enemies (DeBach, 1964). Biological control through importation can be used in all cropping systems in developing and industrialized countries (Gurr and Wratten, 2000; van Lenteren, 2006) and has been applied most successfully against exotic invaders. Successful control is most often totally compatible with crop breeding (DeBach, 1964; Thomas and Waage, 1996), and provides economic returns to African farmers of the same magnitude as breeding programs (Raitzer, 2003; Neuenschwander, 2004). In augmentation forms of biological control, natural enemies (predators, parasitoids and pathogens) are mass produced and then released in the field, e.g., the parasitic wasp *Trichogramma* is used on more than 15 million ha of agricultural crops and forests in many countries (Li, 1994; van Lenteren and Bueno, 2003), as well as in protected cropping (Parrella et al., 1999; van Lenteren, 2000). A wide range of microbial insect pathogens are now in production and in use in OECD and developing countries (Moscardi, 1999; Copping, 2004). For example, the fungus *Metarhizium anisopliae* var *acridum* "Green Muscle"® is used to control Desert Locust (*Schistocerca gregaria*) in Africa (Lomer et al., 2001). Since agents vary in advantages and disadvantages, they must be carefully selected for compatibility with different cropping systems. However, agents are playing an increasing role in IPM (Copping and Menn, 2000). In conservation biological control, the effectiveness of natural enemies is increased through cultural practices (DeBach and Rosen, 1991; Landis et al., 2000) that enhance the efficiency of the exotic or indigenous natural enemies (predators, parasitoids, pathogens).

The economic benefits of biological control can be substantial.

Goals	Certainty	Range of Impacts	Scale	Specificity
N, L, E	A	+5	G	Wide applicability

Cultures of the predatory mite, *Metaseiulus occidentalis*, used in California almond orchards saved growers $59 to $109 ha^{-1} yr^{-1} in reduced pesticide use and yield loss (Hoy, 1992). The fight against the cassava mealy bug in Africa has had even greater economic benefits (Neuenschwander, 2004). IITA and CIAT found a natural enemy of the mealy bug in Brazil in the area of origin of the cassava crop. Subsequently, dissemination of this natural enemy in Africa saved million of tonnes of cassava per year and brought total benefits of US$ billions (Zeddies et al., 2001; Raitzer, 2003). Similar benefits for small-scale farmers have accrued from other programs on different crops and against different invaders across Africa (Neuenschwander, 2004).

Weed competition is a significant barrier to yield and profitability in most agroecosystems.

Goals	Certainty	Range of Impacts	Scale	Specificity
N, L, D	A	-2 to -5	G	Widespread

In many developing countries, hand weeding remains the prevailing practice for weed control. On small-scale farms, more than 50% of preharvest labor is is devoted to weed management, including land preparation and in-crop weed control (Ellis-Jones et al., 1993; Akobundu, 1996). Despite these labor investments crop losses to weed competition are nearly universally identified as major production constraints, typically causing yield reductions of 25% in small-scale agriculture (Parker and Fryer, 1975). Delayed weeding is a common problem caused by labor shortages, and reduced labor productivity resulting from diseases such malaria and HIV/AIDS. Hence, cost-effective low-labor control methods have become increasingly important. In Bangladesh with current methods, one-third of the farmers lose at least 0.5 tonne ha^{-1} grain to weeds in each of the three lowland rice seasons (Ahmed et al., 2001; Mazid et al., 2001). Even in areas that employ herbicides, yield losses are substantial; in the early 1990s annual losses of US$4 billion were caused by weed competition in the US. For staple cereal and legume crops like maize, sorghum, pearl millet, upland rice in semiarid areas of Africa, the parasitic witchweeds (*Striga* species) can cause yield losses ranging from 15 to 100% (Boukar et al., 2004). *Striga* infestation is associated with continuous cultivation and limited returns of plant nutrients to the soil, i.e., conditions typical of small-scale resource poor farms (Riches et al., 2005).

Intensive herbicide use has contributed to improved weed management but there are concerns about sustainable use and environmental quality.

Goals	Certainty	Range of Impacts	Scale	Specificity
N, H, L, E, D	A, B	-2 to +5	G	Widespread

Globally, approximately 1 billion kg of herbicide active ingredients are applied annually in agricultural systems (Aspelin and Grube, 1999). The benefits of judicious herbicide use are broadly recognized. In addition to tillage, prophylactic application of herbicide is the method of choice for managing weeds in industrialized countries and is also widely employed in highly productive agricultural regions in developing countries like Punjab and Haryana States in India. Herbicide use is also becoming more common in small-scale rice/wheat systems in Eastern India and in rice in countries such as Vietnam and Bangladesh where the price of labor is rising faster than crop values (Auld and Menz, 1997; Riches et al., 2005). Substitution of labor by herbicides in Bangladesh reduces weeding costs by 40-50% (Ahmed et al., 2001). Herbicides sold in small quantities are accessible to poor farmers who realize their value; rice herbicide sales have been increasing at 40-50% per year since 2002 (Riches et al., 2005). However, herbicide resistance (currently documented in 313 weed biotypes: www.WeedScience.org) and environmental contamination are growing problems. Traces of Atrazine and other potential carcinogens are routinely documented in ground and surface water resources in industrialized countries (USGS, 1999), and on a global scale the quantity of active ingredient applied as herbicide and the energy required for manufacturing and field application is larger than all other pesticides combined (FAO, 2000a). In the developing country context, acute poisoning of agricultural workers from improper handling of herbicides also poses a significant public health risk that is linked to factors such as insufficient access to high-quality protective gear, poor product labeling, and low worker literacy rates (Repetto and Baliga, 1996). However, many of the newer classes of herbicide chemistry entering the market have much more favorable environmental profiles than commonly used insecticides and can be used at very low doses. Registration of new classes of herbicides has slowed (Appleby, 2005), which places a heightened imperative on maintaining the long-term efficacy of existing herbicides. There are also concerns for the sustainable use of compounds like glyphosate that are applied in conjunction with herbicide resistant crops (HRCs). Farmers using HRCs tend to extensively rely on a single herbicide at the expense of all other weed control measures, thereby decreasing long-term efficacy by increasing the odds of evolved herbicide resistance. However these worries are less of an issue in smaller-scale systems where HRCs have not been previously used and seed systems make their widespread use less likely in the near future. Herbicides also have potential for reducing the cost of management of some important perennial and parasitic weed problems. Glyphosate is showing promise with farmers in Nigeria to reduce competition from the perennial grass *Imperata cylindrica* (Chikoye et al., 2002) and can reduce tillage inputs for management of other intractable perennial species, while in East Africa imazapyr herbicide tolerant maize has been introduced to combat *Striga* (Kanampiu et al., 2003).

Non-chemical control strategies can limit crop damage from weed competition.

Goals	Certainty	Range of Impacts	Scale	Specificity
N, H, L, E, D	B, D	+1 to +3	G	Widespread

Weed management attempts to reduce densities of emerging weeds, limit crop yield losses from established weeds, and promote the dominance of comparatively less damaging and difficult to control species. The first line of defense against weeds is a vigorous crop; basic crop management and cultural practices are important to maximize crop competitiveness and thereby reduce weed competition. Cultivars that are bred for competitive ability (Gibson *et al.*, 2003), diverse crop rotations that provide a variety of selection and mortality factors (Westerman et al., 2005), and simple management changes such as higher seeding rates, spatially-uniform crop establishment (Olsen et al., 2005), and banded fertilizer placement (Blackshaw et al., 2004) can reduce crop losses from uncontrolled weeds and, in some cases, reduce herbicide dependence. In conventional production settings, few of these options have been explicitly adopted by farmers. Cultural practice innovations for weed control work best if they are compatible and efficient complements to existing agronomic practices; hence, it is important to note the needs and constraints of farmers when developing new options for weed management (Norris, 1992). Hence participatory approaches are commonly used to ensure that practices are appropriate to farmer needs (Riches et al., 2005; Franke et al., 2006).

Parasitic weeds are major constraints to several crops but a combination of host-plant resistance and management can control them.

Goals	Certainty	Range of Impacts	Scale	Specificity
N	B	+2 to +5	G	Farmers in Africa, Asia and Mediterranean

Parasitic weeds such as *Striga* spp. and *Alectra vogelii* are major production constraints to several important crops, especially maize, sorghum and cowpea in SSA. Sources of resistance to *S. gesneroides* and *A. vogellii* were identified by traditional methods and the genes conferring resistance to and *A. vogellii* were subsequently identified using Amplified Fragment Length Polymorphism markers (Boukar et al., 2004) and successfully deployed in cowpea across W. Africa (Singh et al., 2006). Host-plant resistance to *S. asiatica* and *S. hermonthica* is now being deployed widely in new sorghum cultivars in East Africa but has been harder to find in maize. Inbred maize lines carrying tolerance to *Striga* have been developed and tolerance is quantitatively inherited (Gethi and Smith, 2004). However, the most successful strategy for controlling *Striga* in maize in West Africa is the use of tolerant cultivars used in rotation, and trap-cropping, using legumes, especially soybean, to germinate *Striga* seeds to reduce the seedbank (Franke et al., 2006). As *Striga* infestation is closely associated with low soil fertility, nutrient management, especially addition of nitrogen, can greatly increase yields of susceptible crops on infested fields. Farmers are now adopting green manures in legume/cereal rotations in Tanzania as a low-cost approach to reversing the yield decline of maize and upland rice (Riches et al., 2005). The interplanting of maize with *Desmodium* spp. within the "push-pull" system (Gatsby Charitable Foundation, 2005; Khan et al., 2006) is a promising approach to *Striga* suppression in East Africa. The broomrapes, *Orobanche* spp. are a major problem on sunflower, faba bean, pea, tomato and other vegetable crops in the Mediterranean basin, central and eastern Europe and the Middle East. Sources of resistance to broomrapes (*Orobanche* species) in a number of crops and the associated genes have been identified and mapped (Rubiales et al., 2006).

The increasing rate of naturalization and spread (i.e., invasions) of alien species introduced both deliberately and accidentally poses an increasing global threat to native biodiversity and to production.

Goals	Certainty	Range of Impacts	Scale	Specificity
E	A	-1 to -5	R	Widespread occurrence

Alien species are introduced deliberately either as new crops/livestock or as biocontrol agents; or by mistake as contamination of seed supplies or exported goods. Natural dispersal mechanisms account for only a small proportion of newly introduced species. This environmental problem has been ranked second only to habitat loss (Vitousek et al., 1996) and has totally changed the ecology of some areas (e.g., Hawaii). Negative economic and environmental impacts include crop failures, altered functioning of natural and manmade ecosystems, and species extinctions (Ewel et al., 1999). For example, in just one year the impact of the introduced golden apple snail (*Pomacea canaliculata*) on rice production cost the Philippine economy an estimated US$28-45 million, or approximately 40% of the Philippines' annual expenditure on rice imports (Naylor, 1996).

The late 20th century saw the emergence of highly virulent forms of wheat stem rust and cassava mosaic disease that are serious threats to food security.

Goals N, H ,L, S	Certainty A	Range of Impacts -5 to -4	Scale G	Specificity Most agricultural systems

The Ug99 race of *Puccinia graminis*, first discovered in East Africa, is virulent on most major resistance genes in wheat, which have provided effective worldwide protection against epidemic losses from wheat rust over the past 40 years (CIMMYT, 2005; Pretorius et al., 2002; Wanyera et al., 2006). Yield loss from Ug99 typically ranges from 40 to 80%, with some instances of complete crop failure (CIMMYT, 2005). The capacity for long-range wind dissemination of viable spores on the jet stream, the ubiquity of susceptible host germplasm, and the epidemic nature of wheat stem rust pose a significant threat to wheat producing regions of Africa and Asia, and possibly beyond. The Ug99 race recently crossed the Red Sea to Yemen, and is projected to follow a similar trajectory as the Yr-9-virulent wheat stripe rust, making its arrival in Central and South Asia possible within the next five or more years (CIMMYT, 2005; Marris, 2007).

Cassava mosaic virus (CMV) is a threat to a staple crop vital for food security.

Goals N, H, L, S	Certainty A	Range of Impacts -4 to -3	Scale R, G	Specificity Especially in Africa

In the late 1980s, CMV underwent recombinant hybridization of two less virulent virus types resulting in a severe and rapidly spreading form of cassava mosaic disease (Legg and Fauquet, 2004). CMD has expanded, via whitefly transmission and movement of infected planting stock, throughout East and Central Africa causing regional crop failure and famine (Mansoor et al., 2003; Anderson et al., 2004; Legg and Fauquet, 2004). CMD represents the first instance of a synergy between viruses belonging to the same family, which could confront agriculture with the future emergence of new and highly virulent geminivirus diseases (Legg and Fauquet, 2004). Cassava is important to future food security in Africa since it is hardy under normally low disease-pressure conditions, and has minimal crop management requirements. These qualities make it an emergency crop in conflict zones (Gomes et al., 2004), and a potentially important component of agricultural diversification strategies for adaptation to climate change.

Cereal cultivars resistant to insect pests have reduced yield losses.

Goals N	Certainty B	Range of Impacts +1 to +4	Scale G	Specificity USA,CWANA

Aphids, sun pest and Hessian fly are among the most serious pests of cereals worldwide (Miller et al., 1989; Ratcliffe and Hatchett, 1997; Mornhinweg et al., 2006). Hessian fly attacks result in yield losses of up to 30% in USA and Morocco, with estimated damage exceeding US$20 m per annum (Lafever et al., 1980; Azzam et al., 1997; Lhaloui et al., 2005). The most effective means of combating this pest has been found to be the development of cultivars with genes H1 to H31 for host plant resistance (antibiosis, antixenosis and tolerance) (Ratcliffe and Hatchett, 1997; Williams et al., 2003; Ohm et al., 2004). The development of wheat varieties resistant to the Hessian fly has been estimated to generate an internal rate of return of 39% (Azzam et al., 1997). A similar resistance approach has been taken with Russian wheat aphid, *Diuraphis noxia*) in wheat and barley in the US (Mornhinweg et al., 2006), and with soybean aphid (*Aphis glycines*). Storage pests, such as weevils, lower the quality of stored grain and seeds, and damage leads to secondary infection by pathogens, causing major economic losses. Host plant resistance has been identified against weevils, such as the maize weevil, *Sitophilus zeamais* and *Callosobruchus* spp., which also affect legumes e.g., cowpea (Dhliwayo et al., 2005).

Ethnoveterinary medicine for livestock could be a key veterinary resource.

Goals N	Certainty B	Range of Impacts +1 to +4	Scale G	Specificity USA, CWANA

Ethnoveterinary medicine (EVM) differs from the paternal approach by considering traditional practices as legitimate and seeking to validate them (Köhler-Rollefson and Bräunig, 1998). Systematic studies on EVM can be justified for three main reasons (Tabuti et al., 2003), they can generate useful information needed to develop livestock healing practices and methods that are suited to the local environment, can potentially add useful new drugs to the pharmacopoeia, and can contribute to biodiversity conservation.

3.2.1.2.3 Improving quality and postharvest techniques
Traditionally, breeding was concerned primarily with yield, adaptation and disease/pest resistance rather than quality and postharvest processing traits. In recent years, more emphasis has been given to quality, especially user-defined quality (i.e., consumer acceptance), industrial processing and bioenhancement. In particular, more breeding programs are now focusing on fodder and forage quality, and not just grain quality.

Breeding for improved and enhanced quality is increasingly important.

Goals H	Certainty C	Range of Impacts 0 to +1	Scale G	Specificity Maize, rice

Bioenhancement or biofortification is not a new concept, e.g., CIMMYT has worked on quality protein maize (QPM) for more than two decades, but concerns over micronutrient deficiencies (Bouis et al., 2000; Graham et al., 1999; www.harvestplus.org) in modern diets are driving renewed interest. Vitamin A deficiency affects 25% of all children under 5 in developing countries (i.e., 125,000 children), while anemia (iron deficiency) affects 37% of the world's population (www.harvestplus.org). Using genetic manipulation, genes for higher vitamin A have been inserted into rice (Golden Rice) (Guerinot, 2000), and efforts are underway to produce micronutrient-dense iron and zinc varieties in rice.

Breeding for fodder and forage quality and yield is becoming more important.

Goals	Certainty	Range of Impacts	Scale	Specificity
H	E	0 to +2	R	India

The recognition that most small-scale farmers use crops for multiple purposes and the rapid expansion in livestock production has resulted in breeding programs that target fodder and forage quality and yield. For example, Quantitative Trait Loci (QTLs) for stover quality traits that can be used in marker assisted breeding (MAB) have been identified in millet (Nepolean et al., 2006); ICRISAT now tests sorghum, millet and groundnut breeding lines for fodder quality and production.

A large number of postharvest technologies have been developed to improve the shelf life of agricultural produce.

Goals	Certainty	Range of Impacts	Scale	Specificity
N, H	A	+1to +3	G	Developed countries

Postharvest technologies include canning, bottling, freezing, freeze drying, various forms of processing (FFTC, 2006), and other methods particularly appropriate for large commercial enterprises. Studies on the effects of storage atmosphere, gaseous composition during storage, postharvest ethylene application and ultraviolet (UV) irradiation, and effect of plant stage on the availability of various micronutrients in different foods are being examined to provide increased understanding of the sensitivity of micronutrient availability to the ways in which foods are handled, stored and cooked (Welch and Graham, 2000; Brovelli, 2005).

3.2.1.3 Recent biotechnologies: MAS, MAB and Genetic Engineering

Nucleic acid technologies (Table 3-2) and their application in genomics is beginning to have an impact on plant (Baenziger et al., 2006; Swamininathan, 2006) and animal breeding, both through increased knowledge of model and crop species genomes, and through the use of marker assisted selection (MAS) or breeding (MAB).

Plants

The tools and techniques developed by applied modern biotechnology are beginning to have an impact on plant breeding and productivity.

Goals	Certainty	Range of Impacts	Scale	Specificity
N	B	0 to +3	G	Many crops

The use of genomic-based breeding approaches are already widespread (e.g., Generation Challenge Program: http://www.generationcp.org/index.php), particularly MAS or MAB. CIMMYT, for example, routinely uses five markers and performs about 7000 marker assays per year (Reynolds and Borlaug, 2006). These markers include two for cereal

Table 3-2. **Techniques being used to elucidate the genetic structure of populations for conservation or utilization within crop/livestock breeding programs.**

Haploid/conservative single gene markers
Polymerase Chain Reaction - Restriction fragment length polymorphism (PCR-RFLP)
Single Strand Conformation Polymorphism (SSCP)
PCR sequencing

Codominant single locus markers
Allozymes/isozymes
Microsatellites or simple sequence repeats (SSRs)
Single nucleotide polymorphism (SNP)

Dominant multilocus markers
Random amplified polymorphic DNA (RAPD)
Inter/anchored SSRs (iSSRs)
Amplified fragment length polymorphism (AFLP)

cyst nematode, one for barley yellow dwarf, one to facilitate wide crossing and one for transferring disease resistance from different genomes. Likewise, ICRISAT routinely uses MAS to incorporate genes for downy mildew resistance in pearl millet (ICRISAT, 2006). MAS can shorten the breeding cycle substantially and hence, the economic benefits are substantial (Pandey and Rajatasereekul, 1999). Using MAS, it took just over three years to introduce downy mildew resistance compared to nearly nine years by conventional breeding (ICRISAT, 2006). QTLs identified for submergence tolerance in rice have also been fine-mapped and gene-specific markers identified (Xu et al., 2006), shortening the breeding cycle with MAB to 2 years. At present, as in the examples above, most MAS is with major genes or qualitative traits and MAS is likely to be most useful in the near future to transfer donor genes, pyramid resistance genes and finger print MVs (Koebner and Summers, 2003; Baenziger et al., 2006). To date, MAS has been less successful with more complex, quantitative traits, particularly drought tolerance (Snape, 2004; Steele et al., 2006).

Knowledge of gene pathways and regulatory networks in model species is starting to have impacts on plant breeding.

Goals	Certainty	Range of Impacts	Scale	Specificity
N	B	0 to +2	G	Widespread applicability

The genome of the model plant species *Arabidopsis* and its function have been studied in great detail. One of the most important traits in crop plants is the timing of flowering and crop duration, which determines adaptation. Genes that control the circadian rhythm and the timing of flowering have been extensively studied in *Arabidopsis* (Hayama and Coupland, 2004; Bernier and Perilleux, 2005; Corbesier and Coupland, 2005) and modeled (Welch et al., 2003; Locke et al., 2005). Homologues of key flowering pathway genes have been identified in rice and many other crop plants, and flowering pathways and the control of flowering time better understood (Hayama and Coupland, 2004), thus providing an opportunity to manipulate this pathway. Drought resistance has also been studied in *Arabidopsis* and two genes, the DREB gene (Pellegrineschi et al., 2004) and the *erecta* gene (Masle et al., 2005); these confer some tolerance to water deficits or increase water-use efficiency. Promising constructs of the DREB gene have been produced in rice, wheat and chickpea (Bennett, 2006).

Modern biotechnology, no matter how successful at increasing yield or increasing disease and pest resistance, will not replace the need for traditional crop breeding, release and dissemination processes.

Goals	Certainty	Range of Impacts	Scale	Specificity
N	A	0 to +2	G	Widespread applicability

The products of most current biotechnology research are available to farmers through the medium of seed, and will therefore still go through current national registration, testing and release procedures. The same constraints to adoption by farmers apply for GM and non-GM organisms. There are arguments for shortening testing and release procedures in the case of existing varieties that have their resistance "updated" against new strains of disease. In India a new version of a widely grown pearl millet variety (HHB67) was approved for release that incorporates resistance to a new and emerging race of downy mildew (identified by DNA finger-printing and incorporated using MAS backcrossing) (ICRISAT, 2006). Only a few countries currently have biosafety legislation or research capacities that allow for testing GM crops and assessing and understanding the structure of wild genetic resources (see 3.2.2.2.3).

Livestock

There have been rapid developments in the use of molecular genetics in livestock over the past few decades.

Goals	Certainty	Range of Impacts	Scale	Specificity
N, E, D	C	0 to +3	G	Widespread applicability

Good progress has been made in developing complete genome maps for the major livestock species (initial versions already exist for cattle and poultry). DNA-based tests for genes or markers affecting traits that are difficult to measure currently, like meat quality and disease resistance, are being sought. However, genes of interest have differing effects in breeds/lines from different genetic backgrounds, and in different production environments. When these techniques are used, it is necessary to check that the expected benefits are achieved. Because of the cost-effectiveness of current performance recording and evaluation methods, new molecular techniques are used to augment, rather than replace, conventional selection methods with the aim of achieving, relevant, cost-effective, publicly acceptable breeding programs.

Biotechnologies in the livestock sector are projected to have a future impact on poverty reduction.

Goals	Certainty	Range of Impacts	Scale	Specificity
N, L, E, D	F	-2 to +4	G	North v South

At present, rapid advances in biotechnologies in both livestock production and health hold much promise for both poverty alleviation and environmental protection (Makkar and Viljoen, 2005). Areas of particular note include new generation vaccines and transgenic applications to enhance production (Cowan and Becker, 2006). Polymerase chain reaction (PCR) technology can be utilized to reduce the methane production of cattle (Cowan and Becker, 2006) and grain crops can now be genetically manipulated to lower nitrogen and phosphorous levels in animal waste. Such tools can also be utilized to characterize indigenous animal genetic resources to both understand key factors in disease resistance and adaptation and further protect local breeds. Nevertheless, the impact on poverty reduction and safety of many of these technologies is currently unknown (Nangju, 2001; Cowan and Becker, 2006).

3.2.1.4 Genetic engineering

Modern biotechnological discoveries include novel genetic engineering technologies such as the injection of nucleic acid into cells, nuclei or organelles; recombinant DNA techniques (cellular fusion beyond the taxonomic family and gene transfer between organisms) (CBD, 2000). The products of genetic engineering, which may consist of a number of DNA sequences assembled from a different organism, are often referred to as "transgenes" or "transgene constructs". Public research organizations in both high- and low-income countries and the private sector are routinely using biotechnology to understand the fundamentals of genetic variation and for genetic improvement of crops and livestock. Currently, most of the commercial application of genetic engineering in agriculture comes through the use of genetically modified (GM) crops. The commercial use of other GM organisms, such as mammals, fish or trees is much more limited.

Plants

Adoption of commercially available GM commodity crops has primarily occurred in chemical intensive agricultural systems in North and South America.

Goals	Certainty	Range of Impacts	Scale	Specificity
N, H, L, E, S, D	B	Not yet known	R	Controlled by government regulation

The two dominating traits in commercially available crop plants are resistance to herbicides and insects (Bt). Resistance is primarily to two broad spectrum herbicides: glyphosate and glufosinate. Resistance against insects is based on traits from *Bacillus thuringiensis* (Bt). The four primary GM crop plants in terms of global land area are soybean (57%), maize (25%), cotton (13%) and canola/oilseed rape (5%) (James, 2006) with the the US (53%), Argentina (18%), Brazil (11%) and Canada (6%) as major producers. In Asia, GM cotton production occurs in smaller scale systems in India (3.7%) and China (3.5%) (James, 2006). Sixteen other countries make up the remaining area (4.8%) of global GM crop production (James, 2006). GM crops are mostly used for extractive products (e.g., lecitines and oil from soybean, starch from maize) or for processed products such as cornflakes, chips or tortillas. Whole grain GM maize is only consumed as "food aid" sent to famine areas, while some parts of GM cotton plants are used as animal feed. A great diversity of novel traits and other crops plants (e.g., for pharmaceutical and industrial purposes) are under development and their impacts will need to be evaluated in the future. The main challenge here will be to keep GM pharma and industrial crops separate from crops for food (Ellstrand, 2003; Ledford, 2007).

Environmental impacts of GM crops are inconclusive.

Goals L, E, D	Certainty C	Range of Impacts Not yet known	Scale G,R	Specificity Complex interacting factors being identified

Both negative and benign impacts have been reported, depending on the studied system and the chosen comparator. Contradictory reports from laboratory and field studies with GM crops (Bt- and herbicide resistant) show a great diversity of impacts on non-target organisms, including arthropods and plants (Burke, 2003; O'Callaghan et al., 2005; Squire et al., 2005; Hilbeck and Schmidt, 2006; Sanvido et al., 2006; Torres and Ruberson, 2006). Some reports claim that GM crops do not adversely affect biodiversity of non-target organisms, or have only minor effects, while others report changes in the community composition of certain biocontrol taxa (Torres and Ruberson, 2006). Some reports find that the key experiments and fundamental issues related to environmental impacts are still missing (Wolfenbarger and Phifer, 2000; Snow et al., 2005). Another controversial topic surrounds claims that GM crops significantly reduce pesticide use and thus help to conserve biodiversity (Huang et al., 2002; Pray et al., 2002; Qaim and Zilberman, 2003; Bennett et al., 2004ab; Morse et al., 2004). Contradictory evidence has also been provided (e.g., Benbrook, 2003, 2004; Pemsl et al., 2004, 2005), which in part may be attributable to the dynamic condition of pest populations and their outbreaks over time. A further complication arises from the development of secondary pests which reduce the benefits of certain Bt crops (Qayum and Sakkhari, 2005; Wang et al., 2006). The effects of Bt crops on pesticide use and the conservation of biodiversity may depend on the degree of intensification already present in the agricultural system at the time of their introduction (Cattaneo et al., 2006; Marvier et al., 2007). A recent meta-analysis of 42 field studies (Marvier et al., 2007) in which scientists concluded that the benefits of Bt-crops are largely determined by the kind of farming system into which they are introduced, found that Bt-crops effectively target the main pest when introduced into chemical intensive industrial farming systems. This provides some support to the claim that Bt plants can reduce insecticide use. However, when Bt crops were introduced into less chemical intensive farming systems the benefits were lower. Furthermore when introduced into farming systems without the use of synthetic pesticides, (e.g., organic maize production systems), there were no benefits in terms of reduced insecticide use. In fact, in comparison with insecticide-free control fields, certain non-target taxa were significantly less abundant in Bt-crop fields. Most field studies were conducted in pesticide-intensive, large-scale monocultures like those in which 90% of all GM crops are currently grown (Cattaneo et al., 2006); consequently, these results have limited applicability to low-input, small-scale systems with high biodiversity and must be assessed separately. Introducing GM crops accompanied by an intensification strategy that would include access to external inputs could enhance benefits for small-scale systems (Hofs et al., 2006; Witt et al., 2006).

Currently there is little, if any, information on ecosystem biochemical cycling and bioactivity of transgene products and their metabolites, in above and below ground ecosystems.

Goals E	Certainty C	Range of Impacts Not yet known	Scale G	Specificity Widespread

There are multiple potential routes for the entry of Bt-toxins into the ecosystem, but there is little information to confirm the expected spread of Bt-toxins through food chains in the field (Harwood et al., 2005; Zwahlen and Andow, 2005; Harwood and Obrycki, 2006; Obrist et al., 2006). One expected route would be embedded in living and decaying plant material, as toxins leach and exude from roots, pollen, feces from insects and other animals. There is confirmation of the presence of Bt toxin metabolites in feces of cows fed with Bt-maize feed (Lutz et al., 2005). Several experiments have studied the impacts of Bt-crop plant material on soil organisms with variable results ranging from some effects, only transient effects, to no effects (e.g., Zwahlen et al., 2003; Blackwood and Buyer, 2004). However, to date there has not been a study of the ecosystem cycling of Bt toxins and their metabolites, or their bioactivity.

Evidence is emerging of herbicide and insecticide resistance in crop weeds and pests associated with GM crops.

Goals E	Certainty C	Range of Impacts Not yet known	Scale G	Specificity Widespread

Since 1995 there have been reports of an increase from 0 to 12 weed species developing resistance to glyphosate, the main broad spectrum herbicide used in GM crops from countries where herbicide-resistant GM crops are grown (van Gessel, 2001; Owen and Zelaya, 2005; Heap, 2007). In addition, the use of glyphosate has greatly increased since the introduction of herbicide tolerant crops. With the exception of Australia (http://www.ogtr.gov.au/rtf/ir/dir059final-rarmp1.rtf: 2006, Australian Gene Technology Act 2000) no resistance management plans are required for the production of herbicide resistant crops; management strategies are required for insect-resistant Bt-crops, in most countries where they are grown. There has been only one report of an insect pest showing resistance to one of the commonly used Bt-toxins (Gunning et al., 2005). Strategies are needed for efficient resistance management and the monitoring of the spread and impacts of GM-resistance genes in weed and pest populations.

There are reported incidents of unintentional spread (via pollen and seed flow) of GM traits and crops.

Goals H, N, L, E, D	Certainty C	Range of Impacts Not yet known	Scale G, R	Specificity Worldwide, controlled by government enforcement of regulations

The consequences from unintentional spread of GM traits and GM crops could be serious. GM traits and crops with varying levels of approval are spreading fast throughout the world; intentional spread occurs mainly through human transport and trade. However, a number of unapproved varieties have also spread unintentionally, creating potential genetic contamination problems that countries must be increasingly prepared to tackle (www.gmcontaminationregister.org/ or link through CBD Cartagena Protocol Biosafety Clearinghouse). In 2006, unapproved GM traits which originated in rice field trials in the US and China were found in commercial rice sold in European supermarkets; consequently farmers suffered serious economic losses due to subsequent bans on imports. Later there were additional costs in both countries for certification of freedom from unapproved GM traits. Similar controversy followed the discovery of transgenes in landraces of maize in Mexico (Quist and Chapela, 2001; Kaplinski, 2002; Kaplinski et al., 2002; Metz and Fütterer, 2002ab; Quist and Chapela, 2002; Suarez, 2002; Worthy et al., 2002). There is also evidence of increased invasiveness/weediness as a result of the gene flow of GM traits, such as herbicide and insect resistance, into cultivated or wild and weedy relatives (e.g., Snow et al., 2003; Squire et al., 2005), making them more difficult to control (Cerdeira and Duke, 2006; Thomas et al., 2007). In Canada, double and triple herbicide resistant oilseed rape volunteers occur in other crops, including other resistant soybeans and maize requiring the use of herbicides other than glyphosate or glufosinate (e.g., Hall et al., 2000; Beckie et al., 2004). The same is true for herbicide-resistant crop volunteers in the US (e.g., Thomas et al., 2007). In Canada, organic oilseed rape production in the prairies was largely abandoned because of widespread genetic contamination with transgenes or transgenic oilseed rape (Friesen et al., 2003; Wong, 2004; McLeod-Kilmurray, 2007).

Current risk assessment concepts and testing programs for regulatory approval are incomplete and still under development.

Goals	Certainty	Range of Impacts	Scale	Specificity
E	C	Not yet known	G	Wide applicability

Risk assessment concepts for genetically modified (GM) plants exist in regulations, guidelines and discussion documents in some countries, e.g., USA (Rose, 2007), Canada (Canadian Food Inspection Agency, 2004), the European Union (EC, 2002; EFSA, 2004, 2007) and internationally (OCED, 1986, 1993; Codex Alimentarius, 2003). Some groups have expressed the view that premarket testing for environmental risks of GM crops to nontarget organisms needs to follow protocols for chemicals, such as pesticides (Andow and Hilbeck, 2004), and have called for alternative approaches. A number of concepts are currently being developed and discussed (Hilbeck and Andow, 2004; Andow et al., 2006; Garcia-Alonso et al., 2006; Hilbeck et al., 2006; Romeis et al., 2006).

The development of regulatory and scientific capacity for risk assessment as well as training for farmers on proper technology use is needed to enable developing countries to benefit from biotechnology.

Goals	Certainty	Range of Impacts	Scale	Specificity
H, N, L, E, S, D	B	Not yet known	G,R	Mainly in developing countries

Realization of the benefits of GM technology in the countries will be closely linked to the understanding of the technology and the involved biosafety issues at all levels (e.g., policy, regulation, science, legal, socioeconomic, farm) and with the countries' capabilities to implement the Cartagena Protocol on Biosafety (www.cbd.int/biosafety/default.shtml). All signatory countries are currently working on the implementation of the Protocol within national contexts. However, developing countries lack national capacities on almost all involved fields, particularly biosafety. A number of capacity development projects for the implementation of the Cartagena Protocol on Biosafety are currently on-going (www.gmo-guidelines.info; www.biosafetrain.dk/, www.ribios.ch; www.unep.ch/biosafety/) but need to be complemented by efforts to develop academic educational programs for biosafety degrees (www.cbd.int/doc/newsletters/bpn/bpn-issue02.pdf).

Livestock/fish

Production of transgenic livestock for food production is technically feasible, but at an earlier stage of development than the equivalent technologies in plants.

Goals	Certainty	Range of Impacts	Scale	Specificity
N, E, D	C	Not yet known	G	Widespread applicability

Progress has been made in developing transgenic technologies in animals, including fish. To date, at least 10 species of fish have been modified for enhanced growth, including common carp, crucian carp, channel catfish, loach, tilapia, pike, rainbow trout, Atlantic salmon, chinook salmon, and sockeye salmon (Dey, 2000). These, however, have yet to be approved for commercialization (Aerni, 2001 as cited in Delgado et al., 2003). In animals there is also a focus on disease resistance through transferring genes from one breed or species to another. Coupled with new dissemination methods (e.g., cloning) these techniques are expected to dramatically change livestock production. However, there are many issues that need to be addressed regarding the lack of knowledge about candidate genes for transfer, as well as ethical and animal welfare concerns and a lack of consumer acceptance in some countries. Other constraints include the lack of an appropriate industry structure to capitalize on the technologies, and the high cost of the technologies.

3.2.1.5 Advances in soil and water management

Fertilizer and irrigation AKSTs have had a significant impact on agricultural production globally. The focus is currently on increasing the efficiency of resource use in order to reduce the negative environmental effects of over use and to reduce use of a diminishing resource.

Soil management

The use of traditional natural fallows to sustainably increase the carrying capacity of the land is now uncommon.

Goals	Certainty	Range of Impacts	Scale	Specificity
N, L, E, S	A	0 to +4	R	Mainly in the tropics

Traditionally, degraded crop fields were restored by allowing native vegetation to regenerate as a natural fallow. Fallows restore biodiversity, improve soil permeability through root activity; return organic matter to the soil; protect against erosion by rain and wind, and provide protection from direct radiation and warming (Swift and Anderson, 1993; Swift et al., 1996). Natural fallows of this sort are no longer applicable in most places because population pressure is high; consequently shorter and more efficient fallows using leguminous shrubs and trees are being developed (Kwesiga et al., 1999). When soil fertility is severely depleted, some external mineral nutrients (phosphorus, calcium) or micronutrients may be needed to support plant growth and organic matter production.

In many intensive production systems, the efficiency of fertilizer nitrogen use is low and there is significant scope for improvement with better management.

Goals	Certainty	Range of Impacts	Scale	Specificity
H, L, E, D	E	-5 to -2	G to L	Widespread

The extent of soil degradation and loss of fertility is much greater in tropical than in temperate areas. Net nutrient balances (kg ha⁻¹ per 30 years) of NPK are respectively: -700, -100, -450 for Africa; and +2000, +700, +1000 for Europe and North America. Low fertilizer recovery efficiency can reduce crop yields and net profits, increase energy consumption and greenhouse gas emissions, and contribute to the degradation of ground and surface waters (Cassman et al., 2003). Among intensive rice systems of South and Southeast Asia, crop nitrogen recovery per unit applied N averages less than 0.3 kg kg⁻¹ with fewer than 20% of farmers achieving 0.5 kg kg⁻¹ (Dobermann and Cassman, 2002). At a global scale, cereal yields and fertilizer N consumption have increased in a near-linear fashion during the past 40 years and are highly correlated. However, large differences exist in historical trends of N fertilizer usage and nitrogen use efficiency (NUE) among regions, countries, and crops. Interventions to increase NUE and reduce N losses to the environment require a combination of improved technologies and carefully crafted local policies that contribute to the adoption of improved N management practices. Examples from several countries show that increases in NUE at rates of 1% yr⁻¹ or more can be achieved if adequate investments are made in research and extension (Dobermann, 2006). Worldwide NUE for cereal production is approximately 33% (Raun and Johnson, 1999). Many systems are grossly overfertilized. Irrigated rice production in China consumes around 7% of the global supply of fertilizer nitrogen. Recent on-farm studies in these systems suggest that maximum rice yields are achieved at N fertility rates of 60-120 kg N ha⁻¹, whereas farmers are fertilizing at 180-240 kg N ha⁻¹ (Peng et al., 2006).

Good soil management enhances soil productivity.

Goals	Certainty	Range of Impacts	Scale	Specificity
N, L, E	A	+1 to +3	R	Especially important in the tropics

In the tropics, the return of crop residues at a rate of 10-12 tonnes dry matter ha⁻¹ represents an input of 265 kg carbon ha⁻¹ in the upper 10 cm soil layer (Sá et al., 2001ab; Lal, 2004). Given an appropriate C:N ratio, this represents an

increased water holding capacity of 65-90 mm, potentially a 5-12% increase in maize or soybean yield, and increased income of US$40-80 ha⁻¹ (Sisti et al., 2004; Diekow et al., 2005). Soil carbon and yields can be increased on degraded soils through conservation agriculture (e.g., no-till), agroforestry, fallows with N-fixing plants and cover crops, manure and sludge application, and inoculation with specific mycorrhiza (Wilson et al., 1991; Franco et al., 1992). Organic matter can improve the fertility of soils by enhancing the cation exchange capacity and nutrient availability (Raij, 1981; Diekow et al., 2005).

Poor nutrient recovery is typically caused by inadequate correspondence between periods of maximum crop demand and the supply of labile soil nutrients

Goals	Certainty	Range of Impacts	Scale	Specificity
N, L, E	B	+1 to +3	L	Wide applicability

The disparity between periods of maximum crop demand and the supply of labile soil nutrients (Cassman et al., 2003) can be exacerbated by overfertilization (e.g., Peng et al., 2006; Russell et al., 2006). For elements like nitrogen which are subject to losses from multiple environmental pathways, 100% fertilizer recovery is not possible (Sheehy et al., 2005). Nevertheless, precision management tools like leaf chlorophyll measurements that enable real-time nitrogen management have been shown to reduce fertilizer N application by 20-30% while maintaining rice productivity (Peng et al., 1996; Balasubramanian et al., 1999, 2000; Hussain et al., 2000; Singh et al., 2002). From 1980 to 2000 in the US, maize grain produced per unit of applied N increased by more than 40%, with part of this increase attributed to practices such as split-fertilizer applications and preplant soil tests to establish site-specific fertilizer recommendations (Raun and Johnson, 1999; Dobermann and Cassman, 2002). Despite improved management practices, average N recovery in US maize remains below 0.4 kg N per kg fertilizer N (Cassman et al., 2002), indicating significant scope for continued improvement.

Precision application of low rates of fertilizer can boost productivity among resource poor farmers.

Goals	Certainty	Range of Impacts	Scale	Specificity
N, L, E, S	B	0 to +2	N, R	Small-scale farms of the semiarid tropics.

Resource constraints prevent many small-scale farmers from applying fertilizer at rates that maximize economic returns. ICRISAT has been working in SSA to encourage small-scale farmers to increase inorganic fertilizer use and to progressively increase their investments in agricultural production. This effort introduces farmers to fertilizer use thorough micro-dosing, a concept based on the insight that farmers are risk averse, but will gradually take larger risks as they learn and benefit from new technologies (Dimes et al., 2005; Rusike et al., 2006; Ncube et al., 2007). Micro-dosing involves the precision application of small quantities of fertilizer, typically phosphorus and nitrogen, close to the crop plant, enhancing fertilizer use efficiency and improving productivity (e.g., 30% increase in maize yield in Zimbabwe). Yield gains are larger when fertilizer is combined with the application of animal manures, better weed control, and improved water management. Recent innovations have focused on formulating the single-dose fertilizer capsules.

Grain legumes can provide a significant source of nitrogen fertility to subsequent non-leguminous crops.

Goals	Certainty	Range of Impacts	Scale	Specificity
N, L, E, D	A, B	+1 to +5	G	Widespread

Nitrogen fertility is the most common constraint to crop productivity in many developing countries (Cassman et al., 2003). In industrialized countries, synthetic N fertilizer accounts for around 30-50% of the fossil fuel energy consumption in intensively cropped systems (Liska et al., 2007). Biological nitrogen fixation (BNF) from leguminous crops offers benefits in both intensive and non-intensive agricultural systems. Grain legumes are particularly attractive because they can provide an independent economic return, in addition to residual soil fertility benefits for subsequent crops. These residual benefits, however, are contingent on the amount of N that remains in the field after harvest. In Zimbabwe, sorghum grain yield following legumes increased by more than 1 tonnes ha^{-1} compared to yield achieved with continuous sorghum production (e.g., 1.62 to 0.42 tonnes ha^{-1}). Other studies in Africa have also demonstrated the value of using grain legumes such as groundnuts to improve nitrogen fertility (Waddington and Karigwindi, 2001). However, degraded soils low in soil phosphorous may limit the effectiveness of BNF (Vitousek et al., 2002). In the United States, soybean provides between 65-80 kg N ha^{-1} to subsequent grain crops and hence fertilizer applications can be reduced accordingly (Varvel and Wilhelm, 2003) (See 3.2.2.1.7).

Water management

Potential per capita water availability has decreased by 45% since 1970.

Goals	Certainty	Range of Impacts	Scale	Specificity
N, H, L, E, S, D	A	-5 to -1	G	Poor people in dry areas are most affected

Due to population growth, the potential water availability decreased from 12,900 m^3 per capita per year in 1970 to less than 7,800 m^3 in 2000 (CA, 2007). Freshwater available for ecosystems and humans globally is estimated at ~200,000 km^3 (Gleick, 1993; Shiklomanov, 1999), with the freshwater available for human consumption between 12,500 and 14,000 km^3 each year (Hinrichsen et al., 1998; Jackson et al., 2001). Groundwater represents over 90% of the world's readily available freshwater resource (Boswinkel, 2000). About 1.5 billion people depend upon groundwater for their drinking water supply (WRI, UNEP, UNDP, World Bank, 1998). The amount of groundwater withdrawn annually is roughly estimated at ~600-700 km^3, representing about 20% of global water withdrawals. The volume of water stored in reservoirs worldwide is estimated at 4,286 km^3 (Groombridge and Jenkins, 1998). A large number of the world's population is currently experiencing water stress and rising water demands greatly outweigh greenhouse warming in defining the state of global water systems to 2025 (Vörösmarty et al., 2000).

Water management schemes are resulting in increased efficiency of water use.

Goals	Certainty	Range of Impacts	Scale	Specificity
N,L,E,S	B	0 to +4	G	Wide applicability

To enhance the efficiency of water management, different forms of water resources have been identified, partitioned and quantified by land use system (Falkenmark and Rockström, 2005): basin water is "blue" water and contributes to river runoff, and green water, which passes through plants (Falkenmark, 2000). Land use changes can reallocate green water and alter the blue–green balance. There are a number of different strategies to improve water productivity values (production per unit of evapotranspiration) for both blue and green water: (1) improve timing and increase the reliability of water supplies; (2) improve land preparation and fertilizer use to increase the return per unit of water; (3) reduce evaporative losses from fallow land, lakes, rivers and irrigation canals; (4) reduce transpiration losses from nonproductive vegetation; (5) reduce deep percolation and surface runoff; (6) minimize losses from salinization and pollution; (7) reallocate limited resources to higher-value users; and (8) develop storage facilities (Molden et al., 2003, 2007b). The reallocation of water can have serious legal, equity and other social considerations. A number of policy, design, management and institutional interventions may allow for an expansion of irrigated area, increased cropping intensity or increased yields within the service areas. Possible interventions are reducing delivery requirements by improved application efficiency, water pricing and improved allocation and distribution practices (Molden et al., 2003).

Small-scale, informal types of irrigation such as water harvesting and groundwater pumps can reduce risk of crop failure and increase yield.

Goals	Certainty	Range of Impacts	Scale	Specificity
N, L	B	+1 to +3	R	Applicable in dry areas

Water harvesting is a traditional water management technology with increasing importance and potential to ease water scarcity in many arid and semi-arid regions of world. The water harvesting methods applied depend on local conditions and include such widely differing practices as bunding, pitting, microcatchments, and flood and ground water harvesting (Prinz, 1996; Critchley and Siegert, 1991). On-farm water-productive techniques coupled with improved management options, better crop selection, appropriate cultural practices, improved genetic make-up, and socioeconomic interventions such as stakeholder and beneficiary involvement can help achieve increased crop yields (Oweis and Hachum, 2004), and reduce the risk of crop failure. Most of the techniques are relatively cheap and are viable options when irrigation water from other sources is not readily available or too costly and using harvested rainwater helps in decreasing the use of groundwater.

Soil and moisture conservation, and micro-irrigation techniques have been developed to increase crop yields by small farmers.

Goals	Certainty	Range of Impacts	Scale	Specificity
N, L, E	B	+2 to +4	N, R	Small-scale farms of the semiarid tropics.

Many soil and moisture conservation and micro-irrigation techniques have been developed to increase crop yields by small farmers. Soil and moisture conservation techniques include tillage practices, planting grasses, such as vetiver, and other living barriers, terracing, bunding and contour planting (Tripp, 2006). Micro-irrigation techniques include drip irrigation, basin planting or "zai" pits, and the introduction of treadle pumps and water harvesting (Mupangwa et al., 2006). To reduce the quantities of water and nutrients used during crop establishment, ICRISAT and several NGO partners have promoted a "conservation agriculture" package based on basin planting; small basins (approx. 3375 cm³) are prepared during the dry season when labor demands are relatively low. Basin planting utilizes limited resources more efficiently by concentrating nutrients and water applications. For small-scale systems in dry areas of southern and western Zimbabwe, maize yields were 15-72% (mean = 36%) greater from basin planting than from conventional plowing and whole-field cultivation.

In many urban areas across the world, sewerage is used as source of water and nutrients in urban and peri-urban agriculture.

Goals	Certainty	Range of Impacts	Scale	Specificity
H	B	-3 to -1	L	Especially around large cities in developing countries

Global assessments show that in developing countries only a minor part of the generated wastewater is treated while the large majority enters natural water bodies used for various purposes including irrigation. Recent studies suggest that at least 2 to 4 million ha of land are globally irrigated with untreated, treated, diluted or partially treated wastewater (Furedy, 1990; Drechsel et al., 2006). Generally, it is estimated that about 25-100% of food demand in an urban environment is met through production of food in the same setting (Birley and Lock, 1999), while about 10% of wastewater generated in towns has further use in urban agriculture. These estimates take account urban horticulture, aquaculture and livestock; 25-80% of urban households engage in some form of agriculture. In many developing countries in Asia, Africa and Latin America, sewage sludge has been used for some time (Furedy, 1990; Strauss, 2000). The risks associated with downstream recycling wastewaters are especially great in countries within arid and seasonally arid zones (Strauss, 2000). New WHO Guidelines for the Safe Use of Wastewater, Excreta and Greywater (WHO, 2006) recognize the health issues concerning wastewater use in agriculture, but water pollution and its management will be an issue of concern for populations around the world for some time (Furedy, 1990; Dey et al., 2004).

Many river basins can no longer sustainably supply water for agriculture and cities.

Goals	Certainty	Range of Impacts	Scale	Specificity
N, L, E, S	A	-1 to -3	R	Especially in the dry tropics

Unsustainable use of water resources for irrigation means that extraction exceeds recharge. For example, large-scale irrigation since the 1960s has had devastating impacts on water resources and soil productivity in Central Asia. The water level of the Aral Sea has dropped by 17 m, resulting in a 50% reduction in its surface area and a 75% reduction in its volume. The resulting economic and health impacts to the Aral Sea coastal communities have also been serious (http://www.fao.org/ag/agl /aglw/aquastat/regions/fussr/index8.stm).

3.2.1.6 Advances in information and communications technologies (ICT)

Innovations in information technology have been essential for progress in biotechnology.

Goals	Certainty	Range of Impacts	Scale	Specificity
N, H	A	0 to +4	R	Mainly in developed countries

Genomics, proteomics and metabolomics generate large quantities of data that require powerful computers and large database storage capacities for effective use; advances in ICT have been fundamental to their success. The growth of the worldwide web has allowed data to be widely accessed and shared, increasing impact. The complexity and size of tasks such as describing the genome of model plants has led to global collaboration and data-sharing.

Climate and crop modeling is positively affecting crop production.

Goals	Certainty	Range of Impacts	Scale	Specificity
N,H	B	0 to +2	G	Widespread

The increasing availability of climate data and the use of simulation models, globally, regionally and locally, are having a positive impact on agricultural production. Field-scale crop growth and yield simulation models can help define breeding traits and growing environments, and analyze G x E interactions (Muchow et al., 1994; van Oosterom et al., 1996; Sinclair et al., 2005). At a larger scale, global and regional climate models (GCMs and RCMs) are producing more accurate forecasts and there is collaboration between meteorologists and crop scientists on seasonal weather forecasts (Slingo et al., 2005; Sivakumar, 2006) ranging from months to weeks; these forecasts have proved of practical and financial benefit in countries such as Australia (Stone and Meinke, 2005). More attention needs to be given to providing forecasts to farmers as climate change increases in importance.

Remote sensing and site-specific management benefit from ICT.

Goals	Certainty	Range of Impacts	Scale	Specificity
N	B	0 to +2	G	Widespread

Site-specific management and precision agriculture benefits from ICT (Dobermann and Cassman, 2002; Dobermann et al., 2002), such as global positioning systems. Remote sensing and Geographic Information Systems enable detailed monitoring, evaluation, and prediction of land use changes (see 3.2.2.1.1).

3.2.2 Impacts of AKST on sustainability, through integrated technologies and the delivery of ecosystem services and public goods

The second pathway to agricultural development has come from the grassroots of civil society and involved locally-based innovations that meet the needs of local people and communities. This pathway has its foundations in traditional farming systems and addresses the integration of social and environmental issues with agricultural production. With the realization that the globalized pathway was not leading to sustainable land use systems, numerous different types of organizations initiated efforts to bring about a change; however, the agriculture "Establishment" has in general marginalized these efforts, and they have not been mainstreamed in policy, or in agribusiness. Nevertheless, public-funded research has increasingly become involved, as illustrated by the creation of NRM programs in CGIAR Centers and and other research centers with natural resource management mandates. These and other initiatives have now given credibility to Integrated Natural Resources Management (INRM), in various forms (e.g., agroforestry and ecoagriculture) and recognized the importance of, and need for, new scientific research agendas (INRM Committee of CGIAR).

3.2.2.1 Integrated natural resource management systems

Sustainable rural development research has taken different approaches to the integration of management technologies in the search for a more holistic agricultural system (e.g., Integrated Pest Management, Integrated Water Resources Management, Integrated Soil and Nutrient Management and Integrated Crop and Livestock Management). These concepts are not foreign to developing country farmers, who traditionally have implemented various mixed farming systems appropriate to the local ecology. Research has also examined many of the ways that farmers approached integrated farm management, through various forms of mixed cropping. Over the last 25 years, agroforestry research has recognized that for millennia trees have played a role in food production both as tree crops and as providers of ecological services. Organic farming has especially focused on organic approaches to pest control, soil health and fertility rather than the use of inorganic inputs. There is a growing recognition of the importance of maintaining a functional agroecosystem capable of providing ecological services, biodiversity conservation (Cassman et al., 2005; MA, 2005c), and public goods such as water resources, watershed management, carbon sequestration and the mitigation of climate change.

Integrated Natural Resources Management (INRM) has provided opportunities for sustainable development and the achievement of development and sustainability goals.

Goals	Certainty	Range of Impacts	Scale	Specificity
N, H, L, E, S	B	+1 to +5	L, R	Wide applicability

There are good localized examples of INRM enhancing agricultural sustainability (e.g., Palm et al., 2005b). INRM, like Farming Systems Research (www.fao.org/farming systems/ifsa_mandate), aims at simultaneously improving livelihoods, agroecosystem resilience, agricultural productivity and the provision of environmental services by augmenting social, physical, human, natural and financial capital (Thomas, 2003). It focuses on resolving complex problems affecting natural resources management in agroecosystems by improving the capacity of agroecological systems to continuously supply a flow of products and services on which poor people depend. It does this by improving the adaptive capacity of systems (Douthwaite et al., 2004). INRM innovations help to restore biological processes in farming systems, greatly enhancing soil fertility, water holding capacity, improving water quality and management, and increasing micronutrient availability to farming communities (Sayer and Campbell, 2004), through such processes as the diversification of farming systems and local economies; the inclusion of local culture, traditional knowledge and the use of local species; use of participatory approaches with poor farmers to simultaneously address the issues of poverty, hunger, health/malnutrition, inequity and the degradation of both the environment and natural resources (Campbell and Sayer, 2003). INRM reduces vulnerability to risk and shocks (Izac and Sanchez, 2001) by combining concepts of natural capital and ecosystem hierarchy.

Resource-conserving technologies have been demonstrated to benefit poor farmers.

Goals	Certainty	Range of Impacts	Scale	Specificity
N, H, L, E, S, D	B, E	+1 to +3	M-L	Wide applicability

A study of projects involving IPM, integrated nutrient management (INM), conservation tillage, agroforestry with multifunctional trees in farming systems, aquaculture within farming systems, water harvesting and integrated livestock systems (Pretty et al., 2006) has examined to what extent farmers can increase food production using low-cost and available technologies and inputs, and their impacts on environmental goods and services. The multilocational study, covering 3% of cultivated land in 57 developing countries, identified very considerable benefits in productivity, which were often associated with reduced pesticide use, enhanced carbon sequestration and increased water use efficiency (WUE) in rainfed agriculture (Pretty et al., 2006). The study concluded that the critical challenge is to find policy and institutional reforms in support of environmental goods and services from resource conserving technologies that also benefit food security and income growth at national and household levels.

3.2.2.1.1 Techniques and concepts
A number of new research and monitoring techniques and tools have been developed for this relatively new area of INRM research and land management (see also 3.2.3.3).

Remote sensing and geographical information systems have provided tools for the monitoring, evaluation and better management of land use systems.

Goals	Certainty	Range of Impacts	Scale	Specificity
E	A	0 to +4	L, R	Tools with wide applicability

Monitoring land use and land use change is an integral component of sustainable development projects (Janhari, 2003; Panigrahy, 2003; Verma, 2003). Remote sensing and GIS can cost-effectively assess short- and long-term impacts of natural resource conservation and development programs (Goel, 2003). They also have useful applications in studies of (Millington et al., 2001) urbanization, deforestation, desertification, and the opening of new agricultural frontiers. For example, these technologies have been used to study the spread of deforestation, the consequences of agricultural development in biological corridors, the impact of refugee populations on the environment and the NRM impacts of public agricultural policies (Imbernon et al., 2005). Modeling can extrapolate research findings and develop simulations using data obtained through remote sensing and GIS (Chapter 4).

3.2.2.1.2 Integrated Pest Management (IPM)
IPM is an approach to managing pests and disease that simultaneously integrates a number of different approaches to pest management and can result in a healthy crop and the maintainence of ecosystem balance (Abate et al., 2000). IPM approaches may include genetic resistance, biological control and cultivation measures for the promotion of natural enemies, and the judicious use of pesticides (e.g., Lewis et al., 1997).

The success of IPM is based on effective management, rather than complete elimination, of pests.

Goals	Certainty	Range of Impacts	Scale	Specificity
N, L, E	B	+1 to +4	N, L	Wide applicability

Success is evaluated on the combination of pest population levels and the probability of plant injury. For example, when climatic conditions are conducive for disease, fungicide has been found to be ineffective in controlling *Ascochyta* blight of chickpeas (ICARDA, 1986), but when combined with host resistance, crop rotation and modified cultural practices, fewer fungicide treatments can be both more effective and more economical. As an alternative to pesticides, IPM is most beneficial in high-value crops because of additional labor costs, but when the labor costs are low or IPM is part of a wider strategy to improve yields, IPM can also be of value economically (Orr, 2003). IPM can result in reductions of pesticide use up to 99% (e.g., van Lenteren, 2000). When compared to unilateral use of pesticides, IPM provides a strategy for enhanced sustainability and improved environmental quality. This approach typically enhances the diversity and abundance of naturally-occurring pest enemies and also reduces the risk of pest or disease organisms developing pesticide resistance by lowering the single-dimension selection pressure associated with intensive pesticide use.

IPM produces positive economic, social and environmental effects.

Goals	Certainty	Range of Impacts	Scale	Specificity
N, L, E	B	+1 to +4	M-L	Wide applicability

The past 20 years have witnessed IPM programs in many developing countries, some of which have been highly successful (e.g., mealy bug in cassava, Waibel and Pemsi, 1999). Positive economic, social and environmental impacts of IPM are a result of lower pest control costs, reduced environmental pollution; higher levels of production and income and fewer health problems among pesticide applicators (Figure 3-6). IPM programs can positively affect food safety, water quality and the long-term sustainability of agricultural system (Norton et al., 2005). Agroforestry contributes to IPM through farm diversification and enhanced agroecological function (Altieri and Nicholls, 1999; Krauss, 2004). However, the adoption of IPM is constrained by technical, institutional, socioeconomic, and policy issues (Norton et al., 2005).

Within IPM, integrated weed management reduces herbicide dependence by applying multiple control methods to reduce weed populations and decrease damage caused by noxious weeds.

Goals	Certainty	Range of Impacts	Scale	Specificity
N, E	B	0 to +3	L	Wide applicability

In contrast to conventional approaches to weed management that are typically prophylactic and uni-modal (e.g., herbicide or tillage only), Integrated Weed Management (IWM) integrates multiple control methods to adaptively manage the population levels and crop damage caused by noxious weeds, thereby increasing the efficacy, efficiency, and sustainability of weed management (Swanton and Weise,

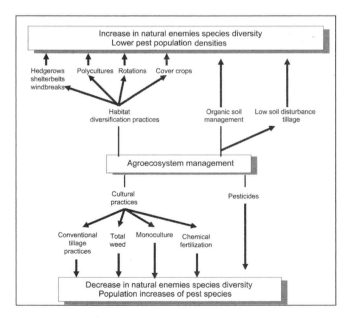

Figure 3-6. *The effects of agroecosystem management and associated cultural practices on the biodiversity of natural enemies and the abundance of insect pests.* Source: Altieri and Nicholls, 1999.

1991). IWM systems are typically knowledge intensive and make use of ecological principals. Examples of IWM elements include cultivars that are bred for competitive ability (Gibson et al., 2003), diverse crop rotations that provide a variety of selection and mortality factors (Westerman et al., 2005), and simple management changes like higher seeding rates, spatially-uniform crop establishment (Olsen et al., 2005), banded fertilizer placement (Blackshaw et al., 2004), and biological control, particularly when the weed is an exotic invader (Zimmermann and Olckers, 2003). The serious parasitic weed of cereal crops (*Striga* spp.) in Africa can be regulated in sorghum by varietal resistance (Tesso et al., 2006), and by bait crops, like *Sesbania sesban*, *Desmodium* spp. that trigger suicidal germination of *Striga* seed (Gatsby Charitable Foundation, 2005; Khan et al., 2007). Herbicide use in agriculture has not been markedly reduced by integrated weed management, as weed science has lagged behind pest and disease management initiatives in terms of developing the basic biological and ecological insights typically required for integrated management (Mortensen et al., 2000; Nazarko et al., 2005).

3.2.2.1.3 Integrated water resources management (IWRM)
IWRM acknowledges water resource management conflicts by using participatory approaches to water use and management; resource development and environmental protection (van Hofwegen and Jaspers, 1999). It recognizes that water use in agriculture, especially irrigation water, meets the needs of fisheries, livestock, small-scale industry and the domestic needs of people, while supporting ecosystem services (Bakker et al., 1999; CA, 2007).

IWRM helps to resolve the numerous conflicts associated with water use and management; resource development and environmental protection.

Goals	Certainty	Range of Impacts	Scale	Specificity
L, E, S	B	0 to +3	R	Wide applicability

Examples of IWRM at the field scale include alternate tillage practices to conserve water and low-cost technologies such as treadle pumps (Shah et al., 2000), and water-harvesting structures. IWRM recognizes the need to integrate water management at the basin level and to promote the linkages between different water uses at this level. It supports river basin management to ensure optimal (and efficient) allocation of water between different sectors and users. Through these approaches, IWRM has achieved a better balance between protecting the water resources, meeting the social needs of users and promoting economic development (Visscher et al., 1999).

Natural Sequence Farming is restoring the hydrological balance of dryland farms in Australia.

Goals	Certainty	Range of Impacts	Scale	Specificity
L, E, S	D	0 to +3	L	Wide applicability in dry areas

Many agricultural landscapes in Australia are facing a land degradation crisis as a result of increasing salinity, soil acidification and erosion, coupled with severe drought, costing the economy 2.4 billion year^{-1} (CRC Soil and Land Management 1999; Boulton, 1999, 2003). Much of this degradation has been caused by land clearance, clearance of waterways, and inappropriate European farming methods (Erskine, 1999; Erskine and Webb, 2003). Natural Sequence Farming is based on an understanding of how water functions in and hydrates the floodplain and involves techniques to slow down the drainage of water from the landscape and reinstate more natural hydrological processes (Andrews, 2005). The reported impacts (www.naturalsequence farming.com) of this have included increased surface and subsurface water storage, reduced dependence on borehole water from aquifers, significantly reduced salinity, improved productive land capacity, recharged aquifers, increased water use efficiency, increased farm productivity with lower water inputs, reduced runoff during peak inflows, and reduced use of pesticides (85%), fertilizers (20%) and herbicides (30%).

Forestry has a role in regulating water supplies for agriculture and urban areas.

Goals	Certainty	Range of Impacts	Scale	Specificity
L, E, S	B	0 to +2	R	Wide applicability

The deforestation of watersheds has led to flooding; landslides; downstream siltation of waterways, wetlands and reefs and water shortages. However, the role of forests in regulating the availability of water resources involves a complex set of relationships involving site-specific functions of slope, soil type and surface cover, associated infrastructure and drainage, groundwater regimes, and rainfall frequency and intensity (Calder, 2005). Water quality from forest catchments is well recognized as better than that from most alternative land uses (Hamilton and King, 1983; Calder, 2005). In spite of the lack of clarity of land use-hydrological relations, payment systems or markets for watershed services are becoming popular in urban areas. For example, New York City has been assisting farmers to change land use, and in doing so has avoided the cost of constructing a large water purification plant.

3.2.2.1.4 Integrated soil and nutrient management (ISNM)
There are multiple pathways for loss of soil nutrients from agroecosystems, including crop harvest, erosion, and leaching. Soil nutrient depletion is one of the greatest challenges affecting the sustainability and productivity of small-scale farms, especially in sub-Saharan Africa. Globally, N, P and K deficits per hectare per year have been estimated at an average rate of 18.7, 5.1, and 38.8 kg, respectively (Lal et al., 2005). In 2000, NPK deficits occurred respectively on 59%, 85%, and 90% of harvested area. Total annual nutrient deficit (in millions of tonnes) was 5.5 N, 2.3 P, and 12.2 K; this was associated with a total potential global production loss of 1,136 million tonnes yr^{-1} (Lal et al., 2005). Methods for restoring soil fertility range from increased fertilizer use to application of organic amendments like compost or manure. Applied in sufficient and balanced quantities, soil amendments may also directly and indirectly increase soil organic matter (see also 3.2.1.5). In addition to providing a source of plant nutrition, soil organic matter can improve the environment for plant growth by improving soil structure. A well-structured soil typically improves gas exchange,

water-holding capacity, and the physical environment for root development.

Agriculture has accelerated and modified the spatial patterns of nutrient use and cycling, especially the nitrogen cycle.

Goals	Certainty	Range of Impacts	Scale	Specificity
N, L, E	A	-3 to +3	G	Wide applicability

Nitrogen fertilizer has been a major contributor to improvements in crop production. In 2000, 85 million tonnes of N were used to enhance soil fertility (Figure 3-1). The use of N fertilizers affects the natural N cycle in the following ways:

1. increases the rate of N input into the terrestrial nitrogen cycle;
2. increases concentrations of the potent greenhouse gas N_2O globally, and increases concentrations of other N oxides that drive the formation of photochemical smog over large regions of Earth;
3. causes losses of soil nutrients, such as calcium and potassium, that are essential for the long-term maintenance of soil fertility;
4. contributes substantially to the acidification of soils, streams, and lakes; and
5. greatly increases the transfer of N through rivers to estuaries and coastal oceans.

In addition, human alterations of the N cycle have increased the quantity of organic carbon stored within terrestrial ecosystems; accelerated losses of biological diversity, especially the loss of plants adapted to efficient N use, and the loss of the animals and microorganisms that depend on these plants; and caused changes in the composition and functioning of estuarine and near-shore ecosystems, contributing to long-term declines in coastal marine fisheries (Vitousek et al., 1997).

Innovative soil and crop management strategies can increase soil organic matter content, hence maintaining or enhancing crop performance.

Goals	Certainty	Range of Impacts	Scale	Specificity
N, L, E	A	+1 to +5	G	Especially important in the tropics

The organic matter content of the world's agricultural soils is typically 50-65% of pre-cultivation levels (Lal, 2004). Strategies to increase soil organic matter (carbon) include the integration of crop and livestock production in small-scale mixed systems (Tarawali et al., 2001, 2004); no-till farming; cover crops, manuring and sludge application; improved grazing; water conservation and harvesting; efficient irrigation; and agroforestry. An increase of 1 tonnes in soil carbon on degraded cropland soils may increase crop yield by 20 to 40 kg ha^{-1} for wheat, 10 to 20 kg ha^{-1} for maize, and 0.5 to 1 kg ha^{-1} for cowpeas. The benefits of fertilizers for building soil organic matter through enhanced vegetation growth only accrue when deficiencies of other soil nutrients are not a constraint.

No-tillage and other types of resource-conserving crop production practices can reduce production costs and improve soil quality while enhancing ecosystem services by diminishing soil erosion, increasing soil carbon storage, and facilitating groundwater recharge.

Goals	Certainty	Range of Impacts	Scale	Specificity
N, L, E	B	0 to +3	R	Mostly applied in dry areas temperate/ sub-trop zone

Low-External Input Sustainable Agriculture (LEISA) is a global initiative aimed at the promotion of more sustainable farming systems (www.leisa.info). In the US, more than 40% of the cultivated cropland uses reduced or minimum tillage. At the global scale, no-till is employed on 5% of all cultivated land (Lal, 2004), reportedly covering between 60 million ha (Harington and Erenstein, 2005; Dumanski et al., 2006; Hobbs, 2006) and 95 million ha (Derpsch, 2005). Minimum tillage is a low-cost system and this drives adoption in many regions. No-till can reduce production costs by 15-20% by eliminating 4-8 tillage operations, with fuel reductions of up to 75% (Landers et al., 2001; McGarry, 2005). Conservation agriculture, which combines no-till with residue retention and crop rotation, has been shown to increase maize and wheat yields in Mexico by 25-30% (Govaerts et al., 2005). In the USA, the adoption of no-till increases soil organic carbon by about 450 kg C ha^{-1} yr^{-1}, but the maximum rates of sequestration peak 5-10 yrs after adoption and slow markedly within two decades (West and Post, 2002). In the tropics soil carbon can increase at even greater rates (Lovato et al., 2004; Landers et al., 2005) and in the Brazilian Amazon integrated zero-till/ crop-livestock-forest management are being developed for grain, meat, milk and fiber production (Embrapa, 2006). On the down-side, no-till systems often have a requirement for increased applications of herbicide and can be vulnerable to pest and disease build-up (e.g., wheat in America in late 1990s).

Short-term improved fallows with nitrogen-fixing trees allow small-scale farmers to restore depleted soil fertility and improve crop yields without buying fertilizers.

Goals	Certainty	Range of Impacts	Scale	Specificity
N, L, E, S	A	+2 to +4	R	Especially important in Africa

Especially in Africa, short-rotation (2-3 years), improved agroforestry fallows with nitrogen-fixing trees/shrubs (e.g., *Sesbania sesban* and *Tephrosia vogelii*) can increase maize yield 3-4 fold on severely degraded soils (Cooper et al., 1996; Kwesiga et al., 1999). Unlike hedgerow inter-cropping, which as a high labor demand, these fallows are well adopted (Jama *et al.*, 2006). Similar results can be achieved with legume trees and rice production in marginal, non-irrigated, low yield, conditions. The use of these improved fallows to free small-scale maize farmers from the need to purchase N fertilizers is perhaps one of the greatest benefits derived from agroforestry (Buresh and Cooper, 1999; Sanchez, 2002) and is a component of the Hunger Task Force (Sanchez et al., 2005) and the Millennium Development Project (Sachs, 2005). By substantially increasing maize yields in Africa, these easily-adopted fallows can reduce the gap between potential and achieved yields in maize.

Deeply-rooted, perennial woody plants have greater and very different positive impacts on soil properties, compared with shallow-rooted annual crops.

Goals	Certainty	Range of Impacts	Scale	Specificity
N, L, E	A	+2 to +4	G	Wide applicability: important in the tropics

The perennial habit of trees, shrubs and vines reduces soil erosion by providing cover from heavy rain and reducing wind speed. Their integration into farming systems also creates a cool, shady microclimate, with increased humidity and lower soil temperatures (Ong and Huxley, 1996; Ong et al., 1996; van Noordwijk et al., 2004). The deep and widespread roots both provide permanent physical support to the soil, and aid in deep nutrient pumping, decreasing nutrient losses from leaching and erosion (Young, 1997; Huxley, 1999). Trees also improve soils by nutrient recycling, increasing organic matter inputs from leaf litter and the rapid turnover of fine roots. This improves soil structure and creates ecological niches in the soil for beneficial soil microflora and symbionts (Lapeyrie and Högberg, 1994; Mason and Wilson, 1994; Sprent, 1994). Additionally, leguminous trees improve nutrient inputs through symbiotic nitrogen fixation. These tree attributes have been a dominant focus of agroforestry systems (Young, 1997). Most of the benefits from trees come at the expense of competition for light, water and nutrients (Ong et al., 1996). Consequently a net benefit only occurs when the tradeoffs (ecological, social and economic) are positive.

Harnessing the symbiotic associations between almost all plants and the soil fungi (mycorrhizas) on their roots is beneficial to crop growth and soil nutrient management.

Goals	Certainty	Range of Impacts	Scale	Specificity
N, L, E	A	+2 to +4	G	Wide applicability: important in the tropics

Many agricultural practices (land clearance, cultivation, fertilizer and fungicide application) have negative impacts on mycorrhizal populations, affecting the species diversity, inoculum potential, and the fungal succession. Techniques to harness the appropriate fungi, ectomycorrhizas on gymnosperms and some legumes (Mason and Wilson, 1994), and endomycorrhizas on most other plants (Lapeyrie and Högberg, 1994), include the conservation of natural soil inoculum and the inoculation of nursery stock prior to planting (Mason and Wilson, 1994). These techniques are critical for sustainable production as mycorrhizal associations are essential to plant establishment and survival, especially in degraded environments. It is now recognized that the soil inoculum of these fungal species is an important component of the soil biodiversity that enhances the sustainable function of natural ecosystems and agroecosystems (Waliyar et al., 2003).

Extensive herding, the most widespread land use on earth, is more sustainable than commonly portrayed.

Goals	Certainty	Range of Impacts	Scale	Specificity
N, L, E, S	B	+1 to +3	L	Especially important in dry Africa

Pastoralism is a widespread, ancient and sustainable form of land use. Mobile and extensive herding is highly compat-

ible with plant and animal diversity (Maestas et al., 2003). When returns to livestock are sufficient, herding can compete well economically with other forms of farming, allowing land to remain open and lightly used (Norton-Griffiths et al., 2007). Land degradation by overgrazing has been overstated with livestock playing a much smaller negative role than climate in constraining productivity in drier rangelands (Ellis and Swift, 1988; Oba et al., 2000), particularly in Africa. However, in wetter rangelands, feedbacks between livestock and vegetation can be strong and sometimes negative (Vetter, 2005). Degradation most commonly occurs when crop farming extends into marginal lands, displacing herders (Geist and Lambin, 2004) (See 3.2.2.1.9).

3.2.2.1.5 Integrated crop and livestock systems
Worldwide, livestock have traditionally been part of farming systems for millennia. Integrated systems provide synergy between crops and livestock, with animals producing manure for use as fertilizer and improvement of soil structure (as well as a source of fuel), while crop by-products are a useful source of animal and fish food. In addition, fodder strips of grasses or fodder shrubs/trees grown on contours protect soil from erosion. The production of meat, milk, eggs and fish within small-scale farms generates income and enriches the diet with consequent benefits for health. On small farms, a few livestock can be stall-fed, hence reducing the negative impacts of grazing and soil compaction.

Integrated crop and livestock production is an ancient and common production system.

Goals	Certainty	Range of Impacts	Scale	Specificity
N, L, E, S	A	+1 to +3	G	Worldwide applicability

Close linking of crops and livestock in integrated systems can create a win-win with greater productivity and increased soil fertility (McIntire et al., 1992; Tarawali et al., 2001). Without this linkage, soil fertility can fall in cereal-based systems and surplus livestock manure is wasted (Liang et al., 2005). Linking crops and livestock forms a "closed" nutrient system that is highly efficient. Crop-livestock systems are usually horizontally and vertically diverse, providing small habitat patches for wild plants and animals (Altieri, 1999) and greater environmental sustainability than crop monocultures (Russelle et al., 2007).

In small-scale crop—livestock systems, fodder is often a limiting resource, which can be supplemented by tree/shrub fodder banks.

Goals	Certainty	Range of Impacts	Scale	Specificity
N, L, E, S	A	+1 to +3	R	Worldwide applicability

In Kenya, tree-fodder from *Calliandra calothyrsus* grown in hedgerows and neglected niches has overcome the constraint of inadequate and low-quality feed resources and improved milk production and increasing income of around 1000 farmers by US\$98-124 per year (Franzel et al., 2003). Three kg of *C. calothyrsus* fodder equals 1 kg of concentrate giving a yield of >10kg milk d^{-1} with a buttermilk content of 4.5%. Likewise, in the Sahel *Pterocarpus erinaceus* and *Gliricidia sepium* are grown in fodder banks as a dry season resource for cattle and goats and this fodder is also traded in local markets (ICRAF, 1996; 1997). In western Australia, *Chamaecytisus proliferus* hedges grown on a large scale are browsed by cattle (Wiley and Seymour, 2000) and have the added advantage of lowering the water tables and thereby reducing risks of salinization.

Integrated crop and livestock production can reduce social conflict between nomadic herdsmen and sedentary farmers.

Goals N, L, E, S	Certainty B	Range of Impacts +1 to +3	Scale L	Specificity Especially important in dry Africa

Small-scale livestock producers, especially nomadic herdsmen, follow broad production objectives that are driven more by immediate needs than by the demands of a market (Ayalew et al., 2001). Conflicts between nomadic herdsmen and sedentary farmers have occurred for thousands of years. Nomadic herdsmen in the Sahel have the right during the dry season to allow their herds to graze in areas where sedentary farmers grow crops in the wet season. This leads to the loss of woody vegetation with consequent land degradation, reduced opportunities for gathering natural products (including dry season fodder), and to lowering of the sustainability of traditional farming practices. The development of living fences/hedges to protect valuable food crops and regenerating trees has the potential to enhance production for the sedentary farmers, but unless the nomads need for continued access to wells, watering holes and dry season fodder is also planned at a regional scale, may lead to worsened conflict (Leakey et al., 1999; Leakey, 2003) In this situation, effective integration of crop and livestock systems has to make provision for alternative sources of dry season fodder (e.g., fodder banks), and corridors to watering holes and grazing lands. Participatory approaches to decision making can avoid such conflicts between sedentary and nomadic herdsmen (Steppler and Nair, 1987; Bruce, 1998; UN CCD, 1998; Blay et al., 2004).

3.2.2.1.6 Agroforestry and mixed cropping

Agroforestry practices are numerous and diverse and used by 1.2 billion people (World Bank, 2004a), while tree products are important for the livelihoods of about 1.5 billion people in developing countries (Leakey and Sanchez, 1997) with many of the benefits arising from local marketing (Shackleton et al., 2007). The area under agroforestry worldwide has not been determined, but is known for a few countries (Table 3-3). In Africa trees are typically dominant in agriculture in the areas where they are a major component of the natural vegetation (Fauvet, 1996). Agroforestry practices include many forms of traditional agriculture common prior to colonization; complex multistrata agroforests developed by indigenous peoples in the last one hundred years, scattered trees in pastoral systems, cash crops such as cocoa/tea/coffee under shade, intercropping, improved fallows, and many more (Nair, 1989). As a consequence, while the number of trees in forests is declining, the number of trees on farm is increasing (FAO, 2005e). Agroforestry is the integration of trees within farming systems and landscapes that diversifies and sustains production with social, economic and environmental benefits (ICRAF, 1997). Agroforestry is therefore a practical means of implementing

Table 3-3. **Examples of land areas under agroforestry.**

Country	Area (million hectares)	Specific information
Indonesia[1]	2.80	Jungle rubber agroforests
Indonesia[2]	3.50	All multistrata agroforests
India[3]	7.40	National estimate
Niger[4]	5 to 6	Recently planted
Mali[5]	5.10	90% of agricultural land
C. America[6]	9.20	Silvopastural systems
C. America[6]	0.77	Coffee agroforests
Spain/Portugal[7]	6.00	Dehasa agroforestry
Worldwide[8]	7.80	Cocoa agroforests

[1]80% of Indonesian rubber (approximately 24% of world production); Wibawa et al., 2006.

[2]Including jungle rubber (above), durian, benzoin, cinnamon, dammar and others; M. van Noordwijk, World Agroforestry Centre, Bogor.

[3]Robert Zomer, International Center for Integrated Mountain Development (ICIMOD), Kathmandu.

[4]Gray Tappan, Science Applications International Corp. SAIC, USGS Center for Earth Resources Observation and Science, Sioux Falls.

[5]Cissé, 1995; Boffa, 1999.

[6]Costa Rica, Nicaragua, Honduras, El Salvador and Guatemala; Beer et al. 2000.

[7]Gaspar et al., 2007

[8]5.9 million ha in West and Central Africa, 1.2 million ha in Asia and 0.7 million ha in South and Central America; P. van Grinsven, Masterfoods BV, Veghel, The Netherlands.

many forms of integrated land management, especially for small-scale producers, which builds on local traditions and practices.

Increased population pressure has resulted in sustainable shifting cultivation systems being replaced by less sustainable approaches to farming.

Goals E, S	Certainty A	Range of Impacts -5 to +1	Scale G	Specificity Small-scale agriculture

Throughout the tropics, shifting (swidden) agriculture was the traditional approach to farming with a long forest fallow, representing a form of sequential agroforestry. It was sustainable until increasing population pressure resulted in the adoption of slash-and-burn systems with increasingly shorter periods of fallow. These have depleted carbon stocks in soils and in biomass, and lower soil fertility (Palm et al., 2005b), resulting in a decline in crop productivity. In the worst-case scenario, the forest is replaced by farmland that becomes so infertile that staple food crops fail. Farmers in these areas become locked in a "poverty trap" unable to afford the fertilizer and other inputs to restore soil fertility (Sanchez, 2002).

Small-scale farmers in the tropics often protect trees producing traditionally important products (food, medicines, etc.) on their farms when land is cleared for agriculture.

Goals	Certainty	Range of Impacts	Scale	Specificity
N, H, L, S	A	+1 to +3	G	Mainly small-scale agriculture

Throughout the tropics, reduced cycles of shifting cultivation with shorter periods of fallow deplete soil fertility resulting in unsustainable use of the land, loss of forest and other adverse environmental impacts. However, trees of traditionally important species have often been saved within new field systems. These trees are sometimes sacred trees, but many are protected or planted as a source of products that were originally gathered from the wild to meet the needs of local people. Now, despite the often total loss of forest in agricultural areas, these same species are commonly found in field systems, often in about a 50:50 mix with introduced species from other parts of the world (Schreckenberg et al., 2002, 2006; Kindt et al., 2004; Akinnifesi et al., 2006). A recent study in three continents has identified a number of more sedentary and sustainable alternative farming systems (Palm et al., 2005b; Tomich et al., 2005; Vosti et al., 2005). These take two forms: one practiced at the forest margin is an enrichment of the natural fallow with commercial valuable species that create an "agroforest" (Michon and de Foresta, 1999), while the second is the integration of trees into mixed cropping on formerly cleared land (Holmgren et al., 1994). It has long been recognized that deforestation of primary forest is a typical response to human population growth, but now it is additionally recognized (Shepherd and Brown, 1998) that after the removal of natural forest, there is an increase in tree populations as farmers integrate trees into their farming systems (Shepherd and Brown, 1998; Michon and de Foresta, 1999; Place and Otsuka, 2000; Schreckenberg et al., 2002; Kindt et al, 2004;) to create new agroforests. This counter intuitive relationship, found in east and west Africa (Holmgren et al., 1994; Kindt et al., 2004), the Sahel (Polgreen, 2007), and southeast Asia (Michon and de Foresta, 1999), seems to be partly a response to labor availability, partly domestic demand for traditional forest products or for marketable cash crops and partly risk aversion (Shepherd and Brown, 1998). Typically these trees are more common in small farms, e.g., in Cameroon, tree density was inversely related to area in farms ranging from 0.7-6.0 ha (Degrande et al., 2006). Accumulation curves of species diversity have revealed that a given area of land had a greater abundance and diversity of trees when it was composed of a greater number of small farms (Kindt et al., 2004). Interestingly, tree density can also be greater in urban areas than in the surrounding countryside (Last et al., 1976).

The increase in tree planting is partly due to the uptake of cash crops by small-scale farmers as large-scale commercial plantations decline.

Goals	Certainty	Range of Impacts	Scale	Specificity
N, L, E, S	B	+2 to +4	R	Mainly small-scale agriculture

The dynamics of cash-cropping is changing, with small-scale farmers increasingly becoming more commercialized and growing cash crops formerly grown exclusively by estates in mixed systems. This gives them opportunities to re-duce their risks by commercializing their cropping systems and income, and expand their income generation, making their farms more lucrative (Vosti et al., 2005). In Indonesia, many small-scale farmers now grow "jungle rubber", producing 25% of world rubber. These farmers can be classified as falling between the two extremes of being completely dependent on wage labor, and completely self-sufficient (Vosti et al., 2005).

The search for alternatives to slash-and-burn led to the identification of sites where farmers have independently developed complex agroforests.

Goals	Certainty	Range of Impacts	Scale	Specificity
N, H, L, E, S	A	0 to +5	R	Small-scale agriculture

In Indonesia, when the food crops are abandoned after 2-3 years, a commercial agroforest develops which provides a continuous stream of marketable tree products (e.g., dammar resin, rubber, cinnamon, fruit, medicines, etc). There are about 3 million ha of these agroforests in Indonesia (Palm *et al.*, 2005ab), which have been developed by farmers since the beginning of the last century (Michon and de Foresta, 1996) to replace unproductive forest fallows. These highly productive agroforests are biologically diverse, provide a good source of income, sequester carbon and methane, protect soils, maintain soil fertility and generate social benefits from the land (Palm et al., 2005ab), as well as providing other environmental services. Similar processes are occurring in many places around the world (e.g., the cocoa agroforests of Cameroon, the Highlands of Kenya, the uplands of the Philippines, and Amazonia). In the case of Cameroon, indigenous fruit and nut trees are commonly grown to provide marketable products in addition to the environmental service of shade for the cocoa (Leakey and Tchoundjeu, 2001). Interestingly, in parallel with these developments, farmers have also initiated their own processes of domesticating the indigenous fruits and nuts of traditional importance (Leakey et al., 2004). From the above examples, it is clear that traditional land use has often been effective in combining forest and cropping benefits. In many places, farmers have independently applied their own knowledge to their changing circumstances—situations which arose from such factors as deforestation, the intensification of agriculture, declining availability of land, and changes in land ownership.

There are many wild species in natural ecosystems that have traditionally been collected and gathered from natural ecosystems to meet the day-to-day needs of people.

Goals	Certainty	Range of Impacts	Scale	Specificity
N, H, L, E, S	A	+1 to +4	G	All but the harshest environments

For millennia, people throughout the tropics, as hunter-gatherers, relied on the forest as a source of non-timber forest products (NTFPs) for all their needs, such as food, medicines, building materials, artifacts (Abbiw, 1990; Falconer, 1990; de Beer and McDermott, 1996; Villachica, 1996; Cunningham, 2001). NTFPs are still of great importance to communities worldwide (Kusters and Belcher, 2004; Sunderland and Ndoye, 2004; Alexiades and Shanley, 2005).

With enhanced marketing they have the potential to support forest community livelihoods and increase the commercial value of natural forests, thus strengthening initiatives to promote the conservation of forests and woodlands, especially in the tropics. NTFPs can be a rich in major nutrients, minor nutrients, vitamins and minerals (Leakey, 1999a) and have the potential to provide future products for the benefit of humankind. However, future innovations based on NFTPs must recognize Traditional Knowledge, community practice/law/regulations and be subject to Access and Benefit Sharing Agreements, in accordance with the Convention on Biological Diversity (Marshall et al., 2006).

Non-timber forest products (NTFP) formerly gathered as extractive resources from natural forests are increasingly being grown in small-scale farming systems, and have become recognized as farm produce (Agroforestry Tree Products—AFTPs).

Goals N, H, L, S	Certainty A	Range of Impacts +1 to +3	Scale R	Specificity Relevant worldwide

Small-scale farming systems commonly include both exotic and native tree species (Schreckenberg et al., 2002, 2006; Shackleton et al., 2002; Kindt et al., 2004; Degrande et al., 2006) producing a wide range of different wood and non-wood products. Such products include traditional foods and medicines, gums, fibers, resins, extractives like rubber, and timber, which are increasingly being marketed in local, regional and international markets (Ndoye et al., 1997; Awono et al., 2002). These recent developments are generating livelihoods benefits for local communities (Degrande et al., 2006) in ways that require little investment of cash and have low labor demands. The term AFTP distinguishes these from extractive NTFP resources so that their role in food and nutritional security and in the enhancement of the livelihoods of poor farmers can be recognized in agricultural statistics (Simons and Leakey, 2004).

In the last 10 years there has been increasing investment in agroforestry programs to domesticate species producing AFTPs as new cash crops for income generation by small-scale farmers.

Goals N, H, L, E, S	Certainty A	Range of Impacts Early adoption phase	Scale M-L	Specificity Especially relevant to wet/dry tropics

Socially- and commercially-important herbaceous and woody species are now being domesticated as new crops to meet the needs of local people for traditional foods, medicines, other products (Okafor, 1980; Smartt and Haq, 1997; Guarino, 1997; Schippers and Budd, 1997; Sunderland et al., 1999; Schippers, 2000), and for expanded trade (Ndoye et al., 1997). Participatory domestication of AFTPs is in the early phases of adoption, especially in Africa (Tchoundjeu et al., 2006), small-scale farmers recognize the importance of producing and trading these traditional food species for domestic and wider use and the enhancement of food sovereignty. These programs are improving livelihoods at the household level (Schreckenberg et al., 2002; Degrande et al., 2006), and increasing food and nutritional security. Many of these new crops are important as sources of feed

for livestock (Bonkoungou et al., 1998), potential new markets, e.g., vegetable oils (Kapseu et al., 2002) and pharmaceuticals or nutriceuticals (Mander et al., 1996; Mander, 1998), for helping farmers meet specific income needs, e.g., school fees and uniforms (Schreckenberg et al., 2002), and for buffering the effects of price fluctuations in cocoa and other commodity crops (Gockowski and Dury, 1999). This emerging market orientation needs to be developed carefully as it potentially conflicts with community-oriented values and traditions. A series of "Winners and Losers" projects on the commercialization of NTFPs (now Agroforestry Tree Products—AFTPs) have examined these options (e.g., Leakey et al., 2005a; Marshall et al., 2006). These systems target the restoration of natural capital, the wellbeing of the resource-poor farmer and combine ecological benefits with cash generation (Leakey et al., 2005a), making them a component of a "Localization" strategy. The integration of domesticated indigenous fruit and nut trees into cocoa agroforests would further improve a land use system that is already one of the most profitable and biologically diverse systems (Figure 3-7).

Domesticated agroforestry trees are producing products that meet many of the needs of small-scale farmers and have the capacity to produce new agricultural commodities and generate new industries.

Goals N, L	Certainty B	Range of Impacts +2 to +3	Scale L	Specificity Mainly small-scale agriculture

Participatory rural appraisal approaches to priority setting species selected for domestication found that indigenous fruits and nuts were the species most commonly identified by rural communities (Franzel et al., 1996). Many of these fruits and nuts are important traditional foods with

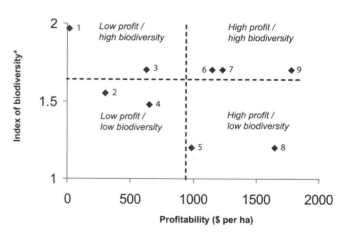

Figure 3-7. *Landuse systems in the humid zone of Cameroon in terms of profitability and plant species diversity.* Source: Izac and Sanchez, 2001.

* Based on field assessments

Note: (1= Community forest; 2 = Long fallow farming; 3 = Extensive cocoa farm; 4 = Short fallow farming; 5 = Short fallow oil palm; 6 = Extensive cocoa with fruits; 7 = Intensive cocoa; 8 = Forest oil palm; 9 = Intensive cocoa with fruits)

market potential. However, some are also sources of edible oils which are needed for cooking and livestock feed but are deficient in many tropical countries (FAO, 2003b). In West Africa, edible oils are extracted from the fruits/kernels of *Allanblackia* spp. (Tchoundjeu et al., 2006), *Irvingia gabonensis* (Leakey, 1999a), *Dacryodes edulis* (Kapseu et al., 2002), *Vitellaria paradoxa* (Boffa et al., 1996) and many other agroforestry species (Leakey, 1999a). Unilever is investing in a new edible oil industry in West Africa, using *Allanblackia* kernel oil (Attipoe et al., 2006). Many agroforestry trees are also good sources of animal fodder, especially in the dry season when pasture is unavailable, and can be grown as hedges, which can be regularly harvested or even grazed by livestock. Opportunities for cattle cake exist from by-products of species producing edible fruits and nuts (e.g., *Dacryodes edulis*, *Canarium indicum*, *Barringtonia procera*, etc.). The nuts of *Croton megalocarpus* are good poultry feed (Thijssen, 2006). In Brazil, new agricultural commodities from agroforestry systems are being used in the manufacture of innovative products for the automobile industry (Panik, 1998).

Twenty-five years of agroforestry research have developed techniques and strategies to assist farmers to reverse soil nitrogen depletion without the application of fertilizers.

Goals	Certainty	Range of Impacts	Scale	Specificity
N, E	A	+2 to +4	M-L	Mainly small-scale agriculture

Leguminous trees fix atmospheric nitrogen through symbiotic associations with soil microorganisms in root nodules (Sprent and Sprent, 1990; Sprent, 2001). The soil improving benefits of this process can be captured in ways that both improve crop yield and are easily adopted by resource-poor farmers (Buresh and Cooper, 1999), conferring major food security benefits to these farming households. Some techniques, such as alley-cropping/hedgerow intercropping are of limited adoptability because of the labor demands, while others such as short-term improved fallows are both effective and adoptable (Franzel, 1999; Kwesiga et al., 1999). Short-term improved fallows in Africa involving species such as *Sesbania sesban*, *Gliricidia sepium*, and *Tephrosia vogelii*, accumulate 100 to 200 kg N ha^{-1} in 6-24 months and to raise maize yields from about 0.5 to 4-6 tonnes ha^{-1} (Cooper et al., 1996). An external source of phosphorus is needed for active N fixation in many P-deficient tropical soils.

Tree/crop interactions are complex but can be managed for positive outcomes.

Goals	Certainty	Range of Impacts	Scale	Specificity
N, L, E	A	-2 to +3	M-L	Many situations

There are many different types of competitive interactions between trees and crops in mixed farming systems, which can be evaluated on the basis of the Land Equivalent Ratio. After 25 years of intensive study the complex physiological and ecological impacts of tree/crop interactions are now well understood (Ong and Huxley, 1996; Huxley, 1999; van Noordwijk et al., 2004); there is much evidence of the overall productivity (biomass) of agroforestry systems being greater than annual cropping systems, due to the capture of more light and water, and improved soil fertility (Ong and Huxley, 1996). Ultimately, however, it is the economic and social outcomes of beneficial interactions that usually determine the adoption of agroforestry systems (Franzel and Scherr, 2002). The numerous examples of agroforestry adoption indicate that farmers, especially small-scale farmers, recognize that the benefits are real.

Vegetated riparian buffer strips are planted for bioremediation of herbicide and nitrate pollution.

Goals	Certainty	Range of Impacts	Scale	Specificity
H, E	B	+2 to +3	L	Temperate and tropical agriculture

Vegetated buffer strips have been shown to retain >50% of sediment within the first few meters (Young et al., 1980; Dillaha et al., 1989; Magette et al., 1989; Mickelson et al., 2003). The planting of trees in strategically important parts of the catchment to maximize water capture and minimize runoff is one of the generally recognized ways of conserving water resources (Schultz et al., 1995, 2000; Louette, 2000; Lin et al., 2003, 2005). In the corn belt of the US, agroforestry strips (trees planted in grass strips) on the contour in a corn/soybean rotation had decreased loss of total P by 17% and loss of nitrate N by 37% after three years (Udawatta et al., 2004). This minimization of nutrient loss is one of the most important environmental services performed by agroforestry trees (van Noordwijk et al., 2004). Among several possible management practices, a tree-shrub-grass buffer placed either in upland fields (Louette, 2000) or in riparian areas (Schultz et al., 1995, 2000) is recognized as a cost effective approach to alleviating non-point sources of agricultural pollutants transported from cropland. Herbicide retention by buffers can also be substantial (Lowrance et al., 1997; Arora et al., 2003).

Enhanced agroecological function is promoted by agroforestry.

Goals	Certainty	Range of Impacts	Scale	Specificity
E	B	+1 to +3	M-L	Especially in the tropics

Agroecological function is dependent on the maintenance of biological diversity above and below ground, especially the keystone species at each trophic level. The ways in which biodiversity stimulates the mechanisms and ecological processes associated with enhanced agroecological function are poorly understood in any crop (Collins and Qualset, 1999); nevertheless, based on numerous studies, the principles are well recognized (Altieri, 1989; Gliessman, 1998) and are based on those of natural ecosystems (Ewel, 1999). Through the integration of trees in farming systems, agroforestry encourages and hastens the development of an agroecological succession (Leakey, 1996; Schroth et al., 2004), which creates niches for colonization by a wide range of other organisms, above and below-ground, in field systems (Ewel, 1999; Leakey, 1999b; Schroth et al., 2004; Schroth and Harvey, 2007). Integrating trees encourages and enhances agroecological function, providing enhanced sustainability as a result of active life cycles, food chains, nutrient cycling, pollination, etc., at all trophic levels, and helping to control pests, diseases, and weeds (Collins and Qualset, 1999) in about two thirds of the agroforests tested (Schroth et al., 2000). Agroforestry is thus capable of rehabilitating degraded farmland. Agroforestry systems support biodiversity conservation in human-dominated landscapes in the tropics (Schroth and Harvey, 2007), through reducing the conversion of primary habitat and providing protective ecological synergies; providing secondary habitat; and by offering a more benign matrix for "islands" of primary habitat in the agricultural landscape, especially by buffering forest edges and creating biological corridors which provide maintenance of meta-population structure (Perfecto and Armbrecht, 2003). Scaling up successful agroforestry approaches requires both improving livelihood and biodiversity impacts at the plot scale, and strategic placement within a landscape mosaic to provide ecosystem services (e.g., watershed protection, wildlife habitat connectivity).

Agroforestry strategies and techniques have been developed for the rehabilitation of degraded agroecosystems and the reduction of poverty particularly in Africa.

Goals	Certainty	Range of Impacts	Scale	Specificity
N, H, L, E, S, D	A	+1 to +4	M-L	Wide applicability, especially in tropics

Agroforestry has evolved from an agronomic practice for the provision of environmental services, especially soil fertility amelioration, to a means of enhancing agroecological function through the development of an agroecological succession involving indigenous trees producing marketable products (Leakey, 1996). In this way it now integrates environmental and social services with improved economic outputs (Leakey, 2001ab). At the community level, agroforestry can positively affect food security andthe livelihoods of small-scale farmers. It can also reverse environmental degradation by providing simple biological approaches to soil fertility management (Young, 1997; Sanchez, 2002); generating income from tree crops (Degrande et al., 2006); minimizing risk by diversifying farming systems (Leakey, 1999b) and; restoring agroecosystem services (Sanchez and Leakey, 1997). Consequently, agroforestry has been recognized as an especially appropriate alternative development strategy for Africa (Leakey, 2001 ab), where the Green Revolution has had only modest success (Evenson and Gollin, 2003).

Agroforestry can mitigate anthropogenic trace gas emissions through better soil fertility and land management, and through carbon sequestration.

Goals	Certainty	Range of Impacts	Scale	Specificity
E	B	+1 to +2	L	Small number of studies in the tropics

The integration of trees in cropping systems can improve soil organic matter, nutrient cycling and the efficient use of water, reduce erosion and store carbon due to improved plant growth. Early assessments of national and global terrestrial CO sinks reveal two primary benefits of agroforestry systems: direct near-term C storage (decades to centuries) in trees and soils, and, potential to offset immediate greenhouse gas emissions associated with deforestation and shifting agriculture. Within the tropical latitudes, it is estimated that one hectare of sustainable agroforestry can potentially offset 5 to 20 ha of deforestation. On a global scale, agroforestry systems could potentially be established on 585 to 1275×10^6 ha of technically suitable land, and these systems could store 12 to 228 (median 95) tonnes C ha^{-1} under current climate and soil conditions (Dixon, 1995). Landscape-scale management holds significant potential for reducing off-site consequences of agriculture (Tilman et al., 2002), leading to integrated natural resources management (Sayer and Campbell, 2001) (see 3.2.2.2.4).

Mixed farming systems, such as those involving cereal/legume mixtures can increase productivity and sustainability of intensive systems.

Goals	Certainty	Range of Impacts	Scale	Specificity
N, L, E, S	B	+1 to +3	R	Especially important in Asia

African savanna has a short growing season (4-5 months) with annual precipitation of 300-1300 mm. In these areas farmers typically grow maize, millet, sorghum, soybean, groundnut, and cowpea, often integrated with livestock production. Traditionally, the sustainability of intensive cereal-based systems in Asia was due to the presence of green manuring practices for soil fertility management and retention of below-ground biodiversity. However, increasing land prices and wage rates had made this option economically unviable at least in the short term and the use of green manures has declined substantially (Ali, 1998). Now short-duration grain legume varieties are available that can be incorporated in the cereal-based intensive systems (Ali et al., 1997). These grain legumes have enhanced farmers' income in the short term and improved cropping system productivity and sustainability in the long-term (Ali and Narciso, 1996). Mixed cropping also has the benefit of reducing pest infestations and diseases.

3.2.2.1.8 Watershed management

Watersheds are often mosaics that integrate many different land uses; when denuded they are very vulnerable to degradation, with severe downstream consequences in terms of flooding, landslides, siltation and reduced water quality (CA, 2007). Additionally, surface water tends to pass through deforested watersheds more quickly leaving towns and villages more susceptible to water shortages. Water storage schemes to supply urban populations and industrial complexes, or for irrigation schemes, can be wasteful and create conflicts between different water users.

Environmental sustainability of water resources is greatest when people work with natural systems and processes, rather than against them.

Goals	Certainty	Range of Impacts	Scale	Specificity
N, L, E, S	A	0 to +4	R	Wide applicability

The most successful watershed management schemes involve participation of local communities. For example, there are traditional user-managed, water catchment and management projects in many parts of the world (e.g., in southern India, the mountainous regions of the Andes, Nepal, and upland South East Asia), which are more sustainable than those imposed by hierarchical water authorities. Schemes involving local communities tend to use water more sustainably (Ruf, 2001; Molle, 2003) than modern schemes. For example, by 2001 the Syr and Amu Dar'ya rivers had decreased to less than half their size in 1957 due to intensive irrigation of cotton and rice in the former Soviet Union (UNEP, 2002).

The Lake Victoria Basin project is an integrated watershed approach to assessing the biophysical and socioeconomic effects of environmental degradation.

Goals	Certainty	Range of Impacts	Scale	Specificity
N, L, E, S	B	0 to +2	N	Widespread applicability

Lake Victoria, the world's second largest lake (68,000 km²), is located in an agricultural area with high population density (28 million people on 116,000 km² of farm land). It displays multiple water degradation problems associated with high river sediment loads from erodible soil, and unsustainable farming practices such as intense cultivation and nutrient depletion. The local communities have serious and wide-scale socioeconomic problems as a result of low crop productivity. The Lake Victoria Basin project has used an integrated watershed approach involving participatory monitoring and evaluation, coupled with spectral reflectance and remote sensing, to characterize the problems and develop agroforestry interventions and livestock exclusion trials to promote more environmentally sustainable farming practices (Swallow *et al.*, 2002).

3.2.2.1.9 Organic systems and biointensive agriculture
Organic agriculture includes both certified and uncertified production systems that encompass practices that promote environmental quality and ecosystem functionality. Organic agriculture is based on minimizing the use of synthetic inputs for soil fertility and pest management. From a consumer viewpoint, this is valuable for avoiding the perceived health risks posed by pesticide residues, growth-stimulating substances, genetically modified organisms and livestock diseases. There are also environmental benefits associated with organic production practices that arise from lower levels of pesticide and nutrient pollution in waterways and groundwater (FAO/WHO, 1999).

Organic agriculture is a small industry (1-2% of global food sales) but it has a high market share in certain products and is a fast growing global food sector.

Goals	Certainty	Range of Impacts	Scale	Specificity
D	A	+1 to +2	G	Niche marketing worldwide

Although global food sales are minimal (1-2%), there are some products with a substantial market; in Germany organic milk products have >10% market share and organic ingredients in baby food comprise 80 to 90% of market share. In the USA, organic coffee accounts for 5% of the market although it is only 0.2% worldwide (Vieira, 2001). The total market value of organic products worldwide, reached US$27.8 billion in 2004. There has been annual market growth of 20-30% (growth in the overall food production sector is 4-5% per year) (ftp://ftp.fao.org/paia/organicag/2005_12_doc04.pdf).

Food labeled as organic or certified organic is governed by a set of rules and limits, usually enforced by inspection and certification mechanisms known as guarantee systems.

Goals	Certainty	Range of Impacts	Scale	Specificity
H, E, S, D	A	+1 to +3	G	Wide applicability

There has been a steady rise in the area under organic agriculture. With very few exceptions, synthetic pesticides, mineral fertilizers, synthetic preservatives, pharmaceuticals, sewage sludge, GMOs, and irradiation are prohibited in organic standards. Sixty industrialized countries currently have national organic standards; there are hundreds of private organic standards worldwide (FAO/ITC/CTA, 2001; IFOAM, 2003, 2006). Regulatory systems for organics usually consist of producers, inspection bodies, an accreditation body for approval and system supervision and a labeling body. There are numerous informal regulatory systems outside of formal organic certification and marketing systems (peer or participatory models) that do not involve third-party inspection and often focus on local markets. The harmonization of organic standards is an issue in international trade. Harmonization has been facilitated by the organic agriculture global umbrella body, the International Federation of Organic Agriculture Movements (IFOAM) and through Codex guidelines. The Codex guidelines concern the production process and provide consumer and producer protection from misleading claims and guide governments in setting standards (FAO/WHO, 1999; El-Hage Scialabba, 2005). The extent of non-certified systems is difficult to estimate, particularly in developing countries.

Worldwide, more than 31 million ha of farmland were under certified organic management in 2006.

Goals	Certainty	Range of Impacts	Scale	Specificity
N, H, E, S	A	+1 to +3	G	Worldwide applicability

Globally organic production covers 31 million ha on more than 600,000 farms in approximately 120 countries. Organic production is rapidly expanding with an aggregate increase of 5 million hectares from 2005 to 2006. Australia has the largest area of land under organic certification systems (12.2 million ha), but Latin America has the greatest total number of organic farms (Willer and Yussefi, 2006). By region, most of the world's certified organic land is in Australia/Oceania (39%), Europe (21%), Latin America (20%), and Asia (13%). In Switzerland, more than 10% of all agricultural land is managed organically. Large areas, particularly in developing countries and some former Soviet States, are organic by default (i.e., noncertified), as farmers cannot afford to purchase fertilizers and pesticides (Willer and Yussefi, 2006). The extent of such nonmarket organic agriculture is difficult to quantify, but >33% of West African agricultural production comes from noncertified organic systems (Anobah, 2000). In Cuba which has made substantial investments in research and extension, organic systems produce 65% of the rice, 46% of fresh vegetables, 38% of non-citrus fruit, 13% of roots, tubers and plantains and 6% of the eggs (Murphy, 2000).

Yields in organic agriculture are typically 10-30% lower than those with conventional management, but in many cases organic systems are economically competitive.

Goals	Certainty	Range of Impacts	Scale	Specificity
N, H, L, E, S	B	-1 to +3	R	Widespread applicability

Yield reductions are commonly associated with adoption of organic practices in intensive production systems (Mäder et al., 2002; Badgley et al., 2007). While yields may be 10-30% lower, profits are, on average, comparable to those on conventional farms. Pest and fertility problems are particularly common during transitions to organic production. As with all production systems, the yield penalty associated with organic agriculture depends on farmer expertise with organic production methods and with factors such as inherent soil fertility (Bruinsma, 2003). In contrast to the reduced productivity responses observed in many high-yielding systems, traditional systems converted to organic agriculture, yields typically do not fall and may increase (ETC/KIOF, 1998).

Organic agriculture greatly reduces or eliminates the use of synthetic agents for pest control.

Goals	Certainty	Range of Impacts	Scale	Specificity
H, E	A	-2 to +3	G	Widespread

The use of synthetic agrochemicals, the foundation of modern agriculture, has been linked to negative impacts such as ground and surface water contamination (Barbash et al., 1999; USGS, 2006), harm to wildlife (Hayes et al., 2002), and acute poisoning of agricultural workers, particularly in the developing world where protection standards and safety equipment are often inadequate (Repetto and Baliga, 1996). Organic systems greatly reduce or eliminate synthetic pesticide use (Mäder et al., 2002), thereby diminishing these concerns. However, a small minority of the pest control substances allowed under organic standards (e.g., copper for downy mildew control in viticulture) also pose human and environmental health risks. Also, the lower efficacy of some organic pest control methods contributes to the yield penalty associated with organic systems. In the longer term, increased biodiversity and an increase in predator species can contribute to a more balanced agroecosystem.

Enhanced use of organic fertility sources can improve soil quality and sustain production, but in some situations supplies of these sources can be inadequate for sustaining high-yielding organic production.

Goals	Certainty	Range of Impacts	Scale	Specificity
H, E	A	-2 to +3	G	Widespread

Adequate soil organic matter are vital for maintaining soil quality; it is a source of macro and micronutrients for plant nutrition, enhances cation exchange capacity and nutrient retention, and facilitates aggregation and good soil structure. However, shortages of organic soil amendments are common in many developing regions (e.g., Mowo et al., 2006; Vanlauwe and Giller, 2006), especially where high population density and cropping intensity preclude rotations with N-fixing legumes or improved fallows and there are competing uses for animal manures (e.g., for cooking fuel). When population pressure is high or environments are degraded, some of the most common organic resources available to farmers (e.g., cereal stovers) are of poor quality, with low nutrient concentrations and macronutrient ratios not commensurate with plant needs. Modern best practice guidelines for conventional production systems advise the full use of all indigenous fertility sources (composts, crop residues, and animal manures), with mineral fertilizers employed to bridge deficits between crop needs and indigenous supplies (e.g., http://www.knowledgebank.irri.org/ssnm/)

Some facets of organic agriculture have clear benefits for environmental sustainability; evidence for others is mixed, neutral, or inconclusive.

Goals	Certainty	Range of Impacts	Scale	Specificity
E	A, C	-2 to +4	G	Wide applicability

Since organic agriculture is more clearly defined by what it prohibits (e.g., synthetics) than what it requires, the environmental benefits that accrue from organic production are difficult to generalize. Some evidence suggests that above and below-ground biodiversity is higher in organic systems (Bengtsson et al., 2005; Mäder et al., 2006), but neutral outcomes are also reported from long-term experiments (e.g., Franke-Snyder et al., 2001); species richness sometimes increases among a few organisms groups while others are unaffected (Bengtsson et al., 2005). Biodiversity impacts from organic agriculture are influenced by factors such as crop rotation and tillage practices, quantity and quality of organic amendments applied to the soil, and the characteristics of the surrounding landscape. Although some studies demonstrate reduced environmental losses of nitrate N in organic systems (e.g., Kramer et al., 2006), most evidence suggests that nitrate losses are not reduced in high-yielding organic systems when contrasted to conventional production system (Kirchmann and Bergstroem, 2001; DeNeve et al., 2003; Torstensson et al., 2006). While fossil energy consumption can be substantially reduced in organic systems, energy savings must be balanced against productivity reductions (Dalgaard et al., 2001). For organic systems with substantially lower yields than conventional alternatives, total enterprise energy efficiency (energy output per unit energy input) can be lower than the efficiency of conventional systems (Loges et al., 2006).

Organic markets are mostly in industrialized countries but organic markets, with a comparative advantage are emerging in developing countries.

Goals	Certainty	Range of Impacts	Scale	Specificity
D	B	+1 to +2	G	Worldwide applicability

Although the highest market growth for organic produce is in North America, the highest reported domestic market growth (approx. 30%) is in China; organic is also increasing in Indonesia. The range of marketing approaches is diverse and includes organic bazaars, small retail shops, supermarkets, multilevel direct selling schemes, community supported agriculture and internet marketing (FAO/ITC/CTA, 2001; IFOAM, 2006; Willer and Yussefi, 2006). The low external input production systems found in many developing countries are more easily converted to certified organic systems than to high external intensive production systems. Organic tropical and subtropical products such as coffee, tea, cocoa, spices, sugar cane and tropical fruits transition more easily to organic since they are generally low external input systems. The higher labor requirements of organic farming provide a comparative advantage to developing countries with relatively low labor costs (de Haen, 1999).

There are significant constraints for developing countries to the profitable production, processing and marketing of organic products for export.

Goals	Certainty	Range of Impacts	Scale	Specificity
D	B	-1 to -3	R	Developing economies

Organic markets require high quality produce and the added costs and complexities of certification. This is a constraint for developing countries where market access may be difficult due to limited and unreliable infrastructure and a lack of skilled labor. Evidence suggests that the current price premium for organic produce will decline in the long term as supply rises to meet demand and as larger corporate producers and retailers enter the market. A lower price premium may make organic agriculture uneconomical for many small-scale producers in developing countries with poor rural infrastructure and services (de Haen, 1999). However, these constraints provide an opportunity for industrialized countries to assist developing countries to expand value-adding skills and infrastructure.

Organic demand is increasingly driven by big retailers with brands that dictate standards.

Goals	Certainty	Range of Impacts	Scale	Specificity
L, D	A	-3 to +3	G	Negative in poor and positive in rich countries

Large and vertically coordinated supermarket chains now account for a major share of the retail markets for fresh and processed organic foods. Supermarket sales of organic produce range from 40% in Germany, 49% in USA, to 80% in Argentina and the UK, and 85% in Denmark. Most large food companies have acquired organic brands and small firms, initiated partnerships with organic companies, or have their own organic lines. Mergers and acquisitions of organic brands and companies affect production, processing, certification and distribution pathways, e.g., in California, 2% of organic growers represent 50% of organic sales. The world's largest organic food distributor has sales of US$3.5 billion. Increasing domination of the organic market by big companies may control market access, and lead to price regulation that reduces returns to farmers. This trend could potentially undermine one of the central principles of organic agriculture: providing a better return to farmers to support the costs of sustainable production. Industry concentration is leading to pressure to erode organic standards (El-Hage Scialabba, 2005). There may however be other benefits to some producers such as ease and scale of marketing and more standardized production.

The localization of marketing has some benefits for small-scale organic producers.

Goals	Certainty	Range of Impacts	Scale	Specificity
L, D	A	+1 to +3	M-L	Small-scale producers and traders

Some initiatives (organic farmers markets, box home delivery and community supported agriculture) have successfully empowered small-scale farmers and promoted localized food systems by supporting community-based, short food supply chains in domestic markets Generally these initiatives are small in scale but seen in total and as a global trend in industrialized and developing countries their impact is significant. One example of larger scale success is a farm in Denmark that delivered 22, 000 organic boxes per week (annual sales of Euros 20 million) in 2005. Other innovations to promote the localization of organic production are the facilitation of dialogue between different government Ministries (e.g., agriculture, trade, environment, rural development, education, health, tourism) and civil society operators (e.g., farmer associations, inspectors, accreditors, traders, retailers, consumers) and location-specific research and knowledge sharing through Organic Farmers-Field-Schools to promote location-specific research and knowledge sharing (El-Hage Scialabba, 2005).

3.2.2.2 Managing agricultural land for ecosystem services and public goods

Agroecosystems are increasingly recognized as potential providers of ecosystem services, yet typically cultivated land has lower biodiversity than natural ecosystems, and is frequently associated with reduced ecosystem services (Cassman et al., 2005), consequently necessitating tradeoffs between production and ecosystem services.

3.2.2.2.1 Water quality and quantity

The available global freshwater resource has been estimated at 200000 km^3 (Gleick, 1993; Shiklomanov, 1999), of which over 90% is groundwater (Boswinkel, 2000). Population growth has reduced annual *per capita* water availability from 12,900 m^3 in 1970 to less than 7,800 m^3 in 2000 (CA, 2007). With water a scarce resource, the role of agriculture in wise water resource management is increasing in importance (CA, 2007). Currently, 7,200 km^3 of water are used in crop production annually and this is predicted to double by 2050 (IWMI, 2006). There are two major trends in water management—government intervention on large scale projects (Molden et al., 2007b), and private and community investments in small scale projects (e.g., 26 million private small scale irrigation pumps owners in India). Large dams, reservoirs and irrigation systems are typically built by government agencies, which often continued to operate them for economic development (including agriculture, urbanization, power generation), without adequate consideration of farmer needs.

Present trends in irrigation water management within public and private sector have significant positive and negative effects on environment.

Goals	Certainty	Range of Impacts	Scale	Specificity
N, H, L, E, S, D	A	-2 to +3	G	Worldwide

Rainfall contributes about 110,000 km^3 of water per year worldwide, 40% enters rivers and groundwater (43,500 km^3) (Molden et al., 2007a). The rapid increase of irrigation in the last 50 years (Figure 3.1c) has led to dramatic modifications of hydrological systems around the world with the diversion of water from natural aquatic ecosystems (2700 km³) for irrigation having well documented negative environmental effects (Richter et al., 1997; Revenga et al., 2000; WCD, 2000; MA, 2005ab; Falkenmark et al., 2007). These include salinization (20-30 million ha—Tanji and Kielen, 2004), river channel erosion, loss of biodiversity, introduction of invasive alien species, reduction of water quality, genetic isolation through habitat fragmentation, and reduced production of floodplain and other inland/coastal fisheries. Conversely, water management practices have also contributed to environmental sustainability, with the development of irrigation reducing the amount of land required for agriculture. In recent years irrigation and water storage have also been found to create new habitats for water birds in Asia, leading to population increase (Galbraith et al., 2005). Thus the coexistence of wetlands and agriculture for 10,000 years has influenced many ecological modifications (Bambaradiniya and Amerasinghe, 2004), but now the balance tends to be negative.

Improved water management can lead to more equitable water use, but this is not common.

Goals N, H, L, E, S	Certainty A	Range of Impacts -2 to +3	Scale G	Specificity Wide applicability

Access to water is critical for poverty reduction with large positive impacts on agricultural productivity when combined with equitable distribution (Merrey et al., 2007). Targeted investments in water management in both rainfed and irrigated areas can effectively reduce inequity by providing more opportunities for the poor (Castillo et al., 2007). In China equity tends to increase with agricultural water management, because crops grown on irrigated land have the highest effect on lowering inequality (Huang et al., 2005). Equity in irrigation and agricultural water management are increased by equitable land distribution, secure ownership or tenancy rights, efficient input, credit, and product markets; access to information; and nondiscriminatory policies for small-scale producers and landless laborers (Smith, 2004; Hussain, 2005), but these conditions are rarely met and inequity occurs if wealthy and powerful people gain preferential access to water (Cernea, 2003). Interventions often exacerbate the existing imbalance between men and women's water ownership rights, division of labor and incomes (Ahlers, 2000; Chancellor, 2000; Boelens and Zwarteveen, 2002). The poorest farmers are often those at the end of irrigation systems because they receive less water and have the lowest certainty about the timing and amount delivered.

Improved water management can lead to efficient water use.

Goals N, H, L, E, S	Certainty A	Range of Impacts 0 to +3	Scale G	Specificity Wide applicability

Better water management can result in gains in water productivity, better management of rainfed agriculture, improvements in stakeholder management of schemes and reduced evaporation. In low-yielding rainfed areas and in poorly performing irrigation systems, improved water productivity

can be achieved by more reliable and precise application of irrigation water, improved soil fertility and improved soil conservation practices. Improving water productivity—gaining more yield per unit of water—is an effective means of intensifying agricultural production and reducing environmental degradation (Molden et al., 2007b). Increased agricultural productivity can also occur when women's land and water rights are strengthened and there is gender sensitivity in the targeting of credit and input provision, training, and market linkages, especially in areas where women are the farm decision makers (Quisumbing, 1995; van Koppen, 2002). However, gains in water productivity are often overstated as much of the potential has already been met in highly productive systems; a water productivity gain by one user can be a loss to another, e.g., upstream gains in agriculture may be offset by a loss in downstream fisheries, either through increased extraction or agrochemical pollution.

Water user groups are emerging as the key social tool to meet the needs of different communities.

Goals N, H, L, E, S	Certainty B	Range of Impacts +1 to +3	Scale R	Specificity Wide applicability

Access to water is critical for poverty reduction (Molden et al., 2007b). However, poor farmers often have poor access to water, as their traditional systems of water rights are overlooked by water management agencies. Smaller-scale community investments in water projects can allow better access to adequate and better quality water. One way of managing water delivery is the establishment of Water User Associations (Abernethy, 2003), but communities of water users face numerous challenges in gaining equitable and sustainable access to, and allocations of, water (Bruns and Meinzen-Dick, 2000; Meinzen-Dick and Pradhan, 2002). Social reforms to improve the equity of water allocation include providing secure water rights for users and reducing or eliminating water subsidies. Acknowledging customary laws and informal institutions can facilitate and encourage local management of water and other natural resources (CA, 2007). Clarifying water rights can ensure secure access to water for agriculture for poor women and men and other disadvantaged groups, such as the disabled (IFAD, 2006; CA, 2007) and ensure better operations and maintenance. The management of water resources can be further improved through training and capacity development. The benefits of farmer-managed irrigation schemes were confirmed in a worldwide study of 40 irrigation schemes (Tang, 1992), and a study of over 100 irrigation systems in Nepal (Lam, 1998). Management of water at the local level has to be part of an integrated process: basin, regional, national and sometimes trans-boundary (CA, 2007).

Structurally complex land use systems can enhance hydrological processes and provide some relief from water scarcity.

Goals N, L, E, S	Certainty B, E	Range of Impacts +1 to +2	Scale R	Specificity Large land masses

On a regional scale, the capacity of vegetation to trap moisture and to return it to the atmosphere by surface evaporation and transpiration affects hydrological processes and hence the distribution of rainfall (Salati and Vose, 1984). Regional-scale advection of atmospheric moisture is adversely affected by removal of woody vegetation (natural and crops), because of greater water losses to surface runoff, groundwater and a reduction of evaporation and transpiration from the canopy (Salati and Vose, 1984; Rowntree, 1988; Shuttleworth, 1988). Thus the maintenance of perennial vegetation has positive effects on rainfall patterns that enhance hydrological processes (Meher-Homji, 1988) affecting the amount of moisture that can be advected downwind to fall as rain somewhere else (Salati and Vose, 1984). Mixed perennial agricultural systems can probably mimic these hydrological functions of natural forests (Leakey, 1996).

Estuarine habitats are the interface between terrestrial freshwater and marine environments. They are important nursery grounds for the production of commercially important marine fishes, but are subject to detrimental agricultural, urban and industrial developments.

Goals	Certainty	Range of Impacts	Scale	Specificity
N, L, E	B	-1 to +3	R	Worldwide applicability

Qualitative evidence of the use of estuarine habitats by juvenile marine fishes is plentiful (Pihl et al., 2002), but recent quantitative research, including stable isotope analysis and otolith chemistry (Hobson, 1999; Gillanders et al., 2003), has confirmed and emphasized the importance of river estuaries in the connectivity between freshwater and marine habitats (Gillanders, 2005; Herzka, 2005; Leakey, 2006). While few marine fish species are considered to be dependent on estuaries, substantial energetic subsidies to fish populations are derived from their juvenile years living and feeding in estuaries (Leakey, 2006). Given the continued vulnerability of estuaries to the loss of water quality from degradation, pollution and other detrimental human impacts, information about the behavior and resource use of juvenile fishes is crucial for future fisheries management and conservation (Leakey, 2006). In the tropics, mangrove swamps are particularly important (Mumby et al., 2004).

3.2.2.2.2 Conserving biodiversity (in situ, ex situ) and ecoagriculture

Biodiversity is the total variation found within living organisms and the ecological complexes they inhabit (Wilson, 1992) and is recognized as a critical component of farming systems above and below ground (Cassman et al., 2005; MA, 2005c). It is important because there are many undomesticated species that are currently either underutilized, or not yet recognized as having value in production systems. Secondly, terrestrial and aquatic ecosystems contain many species crucial to the effective functioning of foodchains and lifecycles, and which consequently confer ecological sustainability or resilience (e.g., regulation of population size, nutrient-cycling, pest and disease control). The conservation of genetic diversity is important because evolutionary processes are necessary to allow species to survive by adapting to changing environments. Crop domestication, like this evolution requires a full set of genes and thus is grounded in intraspecific genetic diversity (Harlan, 1975; Waliyar et al., 2003).

Biological diversity plays a key role in the provision of agroecological function.

Goals	Certainty	Range of Impacts	Scale	Specificity
E	A	-3 to +4	G	General principle

Ecological processes affected by agroecosystem biodiversity include pollination, seed dispersal, pest and disease management, carbon sequestration and climate regulation (Diaz et al., 2005; MA, 2005c). Wild pollinators are essential to the reproduction of many crops, especially fruits and vegetables (Gemmill-Herren et al., 2007). To maintain a full suite of pollinators and increase agricultural productivity requires the protection of the habitats for pollinators (forests, hedgerows, etc.) within the agricultural landscape. A number of emerging management approaches to diversified agriculture (ecoagriculture, agroforestry, organic agriculture, conservation agriculture, etc.) seek to preserve and promote biodiversity (described above in 3.2.2.1).

The conservation of biological diversity is important because it benefits humanity.

Goals	Certainty	Range of Impacts	Scale	Specificity
N, H, L, E, S, D	A	-4 to +4	G	Worldwide

Humans have exploited plant diversity to meet their everyday needs for food, medicine, etc., for millennia. Agrobiodiversity is increasingly recognized as a tangible, economic resource directly equivalent to a country's mineral wealth. These genetic resources (communities, species and genes) are used by breeders for the development of domesticated crops and livestock (IPGRI, 1993). Species and ecosystems can be conserved for their intrinsic qualities (McNeely and Guruswamy, 1998), but biodiversity conservation is increasingly recognized for its importance in combating malnutrition, ill health, poverty and environmental degradation. Collecting and conserving the world's germplasm in gene banks has been estimated at US$5.3 billion (Hawkes et al., 2000), but the cost is greatly outweighed by the value of plant genetic resources to the pharmaceutical, botanical medicine, major crop, horticultural, crop protection, biotechnology, cosmetics and personal care products industries (US$500-800 billion per year) (ten Kate and Laird, 1999).

Agrobiodiversity is threatened.

Goals	Certainty	Range of Impacts	Scale	Specificity
N, H, L, E, S, D	A	-4 to 0	G	Worldwide problem

Agrobiodiversity is rapidly declining due to the destruction and fragmentation of natural ecosystems, overexploitation, introduction of exotic species, human socioeconomic changes, human-instigated and natural calamities, and especially changes in agricultural practices and land use, notably the replacement of traditional crop varieties with modern, more uniform varieties. Nearly 34,000 species (12.5% of the world's flora) are currently threatened with extinction (Walters and Gillett, 1998), while 75% of the genetic diversity of agricultural crops has been lost since the beginning of the last century (FAO, 1998a). On 98% of the cultivated area of the Philippines, thousands of rice landraces have been replaced by two modern varieties, while in Mexico and

Guatemala, *Zea mexicana* (teosinte), the closest relative of maize has disappeared. The loss of endangered food crop relatives has been valued at about US$10 billion annually (Phillips and Meilleur, 1998).

There are two major conservation strategies: ex situ and in situ.

Goals N, H, L, E, S, D	Certainty A	Range of Impacts 0 to +5	Scale G	Specificity Widely applicable

The Convention on Biological Diversity (CBD, 1992) defines *ex situ conservation* as the conservation of components of biological diversity outside their natural habitats and *in situ conservation* as the conservation of ecosystems and natural habitats and the maintenance and recovery of viable populations of species in their natural surroundings. In an ideal world It would be preferable to conserve all diversity naturally (*in situ*), rather than move it into an artificial environment (*ex situ*). However, *ex situ* conservation techniques are necessary where *in situ* conservation cannot guarantee long-term security for a particular crop or wild species. In both cases, conservation aims to maintain the full diversity of living organisms; *in situ* conservation also protects the habitats and the interrelationships between organisms and their environment (Spellerberg and Hardes, 1992). In the agrobiodiversity context, the explicit focus is on conserving the full range of genetic variation within taxa (Maxted et al., 1997). The two conservation strategies are composed of a range of techniques (Table 3-4) that are complementary (Maxted et al., 1997).

Ecoagriculture is an approach to agricultural landscape management that seeks to simultaneously achieve production, livelihoods and wildlife/ecosystem conservation.

Goals N, L, E, S	Certainty B	Range of Impacts 0 to +4	Scale G	Specificity Worldwide applicability

The Ecoagriculture Initiative secures land as protected areas for wildlife habitat in recognition that these areas may need to be cleared for future agriculture (McNeely and Scherr, 2003; Buck et al., 2004). A set of six production approaches have been proposed: (1) creating biodiversity reserves that benefit local farming communities, (2) developing habitat networks in non-farmed areas, (3) reducing land conversion to agriculture by increasing farm productivity, (4) minimizing agricultural pollution, (5) modifying management of soil, water and vegetation resources, (6) modifying farm systems to mimic natural ecosystems (McNeely and Scherr, 2003). A review of the feasibility of integrating production and conservation concluded that there are many cases of biodiversity-friendly agriculture (Buck et al., 2004, 2007), both for crop and livestock production (Neely and Hatfield, 2007). Nevertheless, economic considerations involving issues of valuation and payment for ecosystems services, as well as building a bridge between agriculturalists and conservation scientists remain a major challenge.

Modern molecular techniques for assessing and understanding the structure of wild genetic resources have greatly enhanced crop and animal breeding programs.

Goals N, E	Certainty B	Range of Impacts +1 to +4	Scale G	Specificity Relevant worldwide

Over the last 20 years, a range of molecular marker techniques (Table 3-2) have informed plant genetic resource management activities (Newton et al., 1999; Lowe et al., 2004). These techniques have revolutionized genetics by allowing the quantification of variations in the genetic code of nuclear and organellar genomes, in ways which give high quality information, are reproducible, easily scored, easily automated, and include bioinformatics handling steps. These techniques involve universal primers that can be used across a range of plant, animal and microbial taxonomic groups, avoiding the need for individual development. They also provide unequivocal measures of allele frequencies;

Table 3-4. Conservation strategies and techniques.

Methods of Conservation		
Strategies	**Techniques**	**Definition**
Ex situ conservation	Seed storage	Dried seed samples in a gene bank kept at subzero temperatures.
	In vitro storage	Explants (tissue samples) in a sterile, or cryopreserved/frozen state.
	Field gene bank	Large numbers of living material accessions transferred and planted at a second site.
	Botanic garden/ arboretum	Small numbers of living material accessions in a garden or arboretum.
	DNA/pollen storage	DNA or pollen stored in appropriate, usually refrigerated, conditions.
In situ conservation	Genetic reserve	The management of genetic diversity in designated natural wild populations.
	On-farm	Sustainably managed genetic diversity of traditional crop varieties and associated species within agricultural, horticultural or other cultivation systems.
	Home garden conservation	Sustainably managed genetic diversity of traditional crop varieties within a household's backyard.

distinguish homozygotes and heterozygotes and allow rapid identifications of gene fragments using different DNA sequences (Lowe et al., 2004).

Molecular techniques are contributing to different approaches of surveying and assessing genetic variation for management and conservation purposes.

Goals	Certainty	Range of Impacts	Scale	Specificity
N, E	A	+1 to +4	G	Relevant worldwide

Assessments of population genetic structure using molecular techniques (Table 3-2) have involved the following approaches: (1) surveys of a species to identify genetic hot spots (e.g., Lowe et al., 2000), genetic discontinuities (Moritz, 1994), genetically isolated and unique populations (Cavers et al., 2003) or populations under different geopolitical management that need to be uniformly managed for the conservation of the species (Karl and Bowen, 1999; Cavers et al., 2003); (2) identification of the genetic history of domesticated species to construct a history of introduction and likely sources of origin (Zerega et al., 2004, 2005), and weed invasions including the search for biological control agents from a relevant source region (McCauley et al., 2003); (3) examination of remnant populations of an exploited or depleted species to assess future population viability and develop appropriate management actions and determine processes and ecological factors affecting gene flow dynamics, and (4) development of genetic resource management strategies for plants in the early stages of domestication by comparisons of exploited and nonexploited populations or between domesticated and natural populations.

Domestication can lead to reduced genetic diversity.

Goals	Certainty	Range of Impacts	Scale	Specificity
N, E	A	-4 to +1	G	Relevant worldwide

The loss of genetic diversity can arise from processes associated with domestication: (1) competition for land resources resulting from the widespread planting of domesticated varieties may lead to the elimination of natural populations, (2) pollen or seed flow from cultivars in production areas can overwhelm those of remnant wild populations, causing genetic erosion of the natural populations, (3) a genetic bottleneck is formed when selective breeding of one or a few superior lines (e.g., *Inga edulis*—Hollingsworth et al., 2005; Dawson et al., 2008) results in increased inbreeding or increased genetic differentiation relative to source populations. Consequently domesticated lines often contain only a subset of the genetic variation of natural populations. Conversely, however, the breeding process can also be used to fix extreme traits or introduce additional variation in selected phenotypic characters. Agricultural diversity depends on wild sources of genes from neglected and underutilized species in order to maintain the productivity and adaptability of domesticated species. The optimization of livelihood benefits during environment change requires a stronger integration between initiatives to conserve agricultural biodiversity and wild biodiversity (Thompson et al., 2007).

Domesticated populations can have conservation value.

Goals	Certainty	Range of Impacts	Scale	Specificity
N, E	B	0 to +3	R	Relevant worldwide

Recent studies using molecular techniques have found that when domestication occurs in ways that do not lead to the loss of wild populations, genetic erosion or genetic bottlenecks, the domesticated population can itself provide a valuable contribution to genetic resource management and conservation. In Latin America, *Inga edulis*, which has been utilized by local people for several thousands of years (Dawson et al., 2008), has remained genetically diverse in five sites in the Peruvian Amazon relative to natural stands (Hollingsworth et al., 2005). In this example, genetic differentiation estimates indicated that the domesticated stands were introduced from remote sources rather than from proximate natural stands (Dawson et al., 2008). Despite maintaining high levels of diversity, this suggests that domesticated stands can also have negative impacts on long term performance through source mixing.

Village-level domestication strategies have conservation advantages in the context of global genetic resource management.

Goals	Certainty	Range of Impacts	Scale	Specificity
N, E	D	0 to +3	R	Relevant worldwide

Village-level domestication has been promoted for the development of new tree crops in developing countries (Weber et al., 2001; Leakey et al., 2003), rather than the centralized distribution of a single line or a few selected genotypes. This practice involves individual communities or villages developing superior lines of new crops from local populations or landraces that are specific to the participating communities, using established domestication practices. This strategy has the inherent advantage of harnessing adaptive variation for a range of local environmental factors, while sourcing from multiple villages ensures that a broad range of genetic variation is preserved across the species range. This strategy provides long-term benefit for genetic diversity conservation where native habitats are increasingly being lost to development. The success of this strategy lies partly in developing an appreciation for a diversity of forms within the new crop, such as has occurred in the wine industry, where customers have been educated to appreciate the diversity of flavors offered by different grape varieties.

Biodiversity and genetic diversity have been "protected" by international policies.

Goals	Certainty	Range of Impacts	Scale	Specificity
N, H, L, E, S, D	B	Expected to be positive	G	Worldwide

The Convention on Biological Diversity (CBD, 1992) was ratified in 1993 to address the broad issues of biodiversity conservation, sustainable use of its components and the equitable sharing of the benefits arising from the use of biodiversity. Its Global Strategy of Plant Conservation (GSPC) included global targets for 2010, such as *"70% of the genetic diversity of crops and other major socioeconomically valuable plant species conserved."* The International Treaty on Plant Genetic Resources for Food and Agriculture (FAO, 2002a) specifically focuses on agrobiodiversity conservation and sustainable use. The imperative to address current threats to genetic diversity was recognized by the Conference of the Parties (www.cbd.int/2010-target) to the CBD 2010 Biodiversity Target, which committed the parties "*to achieve by 2010 a significant reduction of the current rate of biodiversity loss at the global, regional and national level as a contribution to poverty alleviation and to the benefit of all life on earth.*" Thus it is recognized that the international, regional and national level conservation and sustainable use of agrobiodiversity is fundamental for future wealth creation and food security.

3.2.2.2.3 Global warming potential, carbon sequestration and the impacts of climate change

The combustion of fossil fuels, land use change, and agricultural activities constitute the dominant sources of radiatively-active gas emissions (i.e., greenhouse gases—GHG) since the advent of the industrial revolution. Expressed in CO_2 equivalents (i.e., global warming potential—GWP), agriculture now accounts for approximately 10-12% of net GWP emissions to the atmosphere from anthropogenic sources (IPCC, 2007; Smith et al., 2007), excluding emissions from the manufacture of agrochemicals and fuel use for farm practices. The IPCC also reports that nearly equal amounts of CO_2 are assimilated and released by agricultural systems, resulting in an annual flux that is roughly in balance on a global basis. In contrast, agriculture is a significant net source of the important greenhouse gases methane (CH_4) and nitrous oxide (N_2O), contributing approximately 58% and 47% of all emissions, respectively (Smith et al., 2007).

Agriculture affects the radiative forcing potential of the atmosphere (Global Warming Potential—GWP) in various ways, including: (1) heat emission from burnings of forests, crop residues and pastureland (Fearnside, 2000); (2) carbon dioxide emissions from the energy-intensive processes required to produce agricultural amendments like nitrogen fertilizers, pesticides, etc. (USEPA, 2006); (3) greater sensible heat fluxes from bare soils (Foley et al., 2003); (4) infrared radiation from bare soil (Schmetz et al., 2005) and reduced evapotranspiration from soils without vegetative cover; (5) decreased surface albedo (i.e., sunlight reflectance) when plant residues are burned (Randerson et al., 2006); (6) soil organic matter oxidation promoted by tillage (Reicosky, 1997); (7) methane emissions from ruminant livestock (Johnson and Johnson, 1995) and wetland rice cultivation (Minami and Neue, 1994); and (8) nitrous oxide emissions (Smith et al., 1997) from poorly drained soils, especially under conditions where N fertilizers are misused. In aggregate, agriculture is responsible for approximately 15% of anthropogenic CO_2 emissions, 58% of methane (CH_4) emissions and 47% of N_2O (Smith et al., 2007).

Agroecosystems can also be net sinks for atmospheric GWP. Best agricultural practices help to minimize emissions of greenhouse gases.

Goals	Certainty	Range of Impacts	Scale	Specificity
N, L, E, S	A	-3 to +3	G	Especially important in the tropics

In addition to being a source of greenhouse gas emissions, certain agricultural practices found to increase the "sink" value of agroecosystems include (1) maintaining good aeration and drainage of soils to reduce CH_4 and N_2O emissions, (2) maximizing the efficiency of N fertilizer use to limit N_2O emissions (Dixon, 1995) and to reduce the amount of CO_2 released in the energy-intensive process of its manufacture, (3) minimizing residue burning to reduce CO_2 and O_3 emissions, and (4) improving forage quality to reduce CH_4 and N_2O emissions from ruminant digestion (Nicholson et al., 2001), (5) maximizing woody biomass and (6) avoiding burning that promotes ozone formation which is photochemically active with OH radicals; OH radicals remove atmospheric CH_4 (Crutzen and Zimmerman, 1991; Chatfield, 2004).

Recent studies on wheat, soybean and rice in Free-Air Concentration Enrichment (FACE) field experiments suggest that yield increases due to enhanced CO_2 are approximately half that previously predicted.

Goals	Certainty	Range of Impacts	Scale	Specificity
N, E, S	B	-2 to +2	R	Wide applicability

Free-Air Concentration Enrichment (FACE) experiments fumigate plants with enhanced CO_2 concentrations in open air field conditions (Ainsworth and Long, 2005). Yield stimulation of major C_3 crops in elevated $[CO_2]$ is approximately half of what was predicted by early experiments in enclosed chambers (Kimball et al., 1983; Long et al., 2006), casting doubt on the current assumption that elevated carbon dioxide concentration ($[CO_2]$) will offset the negative effects of rising temperature and drought, and sustain global food supply (Gitay et al., 2001). Notably the temperate FACE experiments indicate that: (1) the CO_2 fertilization effect may be small without additions of N fertilizers (Ainsworth and Long, 2005), and (2) harvest index is lower at elevated $[CO_2]$ in soybean (Morgan et al., 2005) and rice (Kim et al., 2003).

Crop responses to elevated to CO_2 vary depending on the photosynthetic pathway the species uses.

Goals	Certainty	Range of Impacts	Scale	Specificity
N, E, S	B	-3 to +3	R	Variation between crop species

Wheat, rice and soybean are crops in which photosynthesis is directly stimulated by elevated CO_2 (Long et al., 2004). When grown at 550 ppm CO_2 (the concentration projected for 2050), yields increased by 13, 9 and 19% for wheat, rice and soybean, respectively (Long et al., 2006). In contrast, photosynthetic pathways in maize and sorghum are not directly stimulated by elevated CO_2; these crops do not show an increase in yield when grown with adequate water supply in the field at elevated CO_2 (Ottman et al., 2001; Wall et al., 2001; Leakey et al., 2004, 2006). At elevated CO_2 there is an amelioration of drought stress due to reduced water use, hence yields of maize, sorghum and similar crops might benefit from elevated CO_2 under drought stress.

Soil-based carbon sequestration (CS) can provide a significant, but finite sink for atmospheric CO_2.

Goals	Certainty	Range of Impacts	Scale	Specificity
E	B	+2 to +4	G	Worldwide

In recognition that social and economic factors ultimately govern the sustained adoption of land-based CS, strategies have been sought that sequester carbon while providing tangible production benefits to farmers (Ponce-Hernandez et al., 2004). For arable systems, no-till cultivation has been promoted as a "win-win" strategy for achieving net Global Warming Potential (GWP) reductions. Tillage disrupts soil aggregates, making organic matter pools that had been physically protected from microbial degradation more vulnerable to decomposition (Duxbury, 2005). Higher levels of soil organic matter are associated with attributes, such as crop tilth, water holding capacity and fertility that are favorable to crop growth (e.g., Lal, 1997). Although concerns have been raised about the methodologies used to assess soil carbon stocks (Baker et al., 2007), recent synthesis of data from many sites across the United States suggests that adoption of no-till (West and Post, 2002) or conversion of cropland into perennial pastures (Post and Kwon, 2000) generates soil organic carbon increases on the order of 450 kg C ha^{-1} yr^{-1}. Depending on factors such as soil texture and land use history, maximum rates of C sequestration tend to peak 5-10 yrs after adoption of CS practices and slow markedly within two decades. Hence increasing the organic matter content of soils is as an interim measure for sequestering atmospheric CO_2. Estimates from the United States suggest that if all US cropland was converted to no-till, enhanced CS rates would compensate for slightly less than 4% of the annual CO_2 emissions from fossil fuels in the U.S. (Jackson and Schlesinger, 2004). On a global scale, carbon sequestration in soils has the potential to offset from 5 to 15% of the total annual CO_2 emissions from fossil fuel combustion in the near-term (Lal, 2004).

Improved management of the vast land area in rangelands has led to significant carbon sequestration, but the benefits of carbon credit payments are not currently accessible, particularly in common property systems.

Goals	Certainty	Range of Impacts	Scale	Specificity
L, E, S	B	0 to +3	R	Wide applicability

Grazing lands cover 32 million km^2 and sequester large quantities of carbon (UNDP/UNEP/WB/WRI, 2000). Processes that reduce carbon sinks in grazing lands include overgrazing, soil degradation, soil and wind erosion, biomass burning, land conversion to cropland; carbon can be improved by shifting species mixes, grazing and degradation management, fire management, fertilization, tree planting (agroforestry), and irrigation (Ojima et al., 1993; Fisher et al., 1994; Paustian et al., 1998). But where land is held in common, mitigation is particularly complex. Mitigation activities are most successful when they build on traditional pastoral institutions and knowledge (excellent communication, strong understanding of ecosystem goods and services) and provide pastoral people with food security benefits at the same time (Reid et al., 2004).

Agroecosystems involving tree-based carbon sequestration can offset greenhouse gas emissions.

Goals	Certainty	Range of Impacts	Scale	Specificity
E	B	0 to +4	G	Wide applicability

Early assessments of national and global terrestrial CO_2 sinks reveal two primary benefits of agroforestry systems: direct near-term C storage (decades to centuries) in trees and soils, and, potential to offset immediate greenhouse gas emissions associated with deforestation and shifting agriculture. On a global scale, agroforestry systems could potentially be established on 585-1275 10^6 ha, and these systems could store 12-228 (median 95) tonnes C ha^{-1} under current climate and soil conditions (Dixon, 1995). In the tropics, within 20-25 years the rehabilitation of degraded farming systems through the development of tree-based farming systems could result in above-ground carbon sequestration from 5 tonnes C ha^{-1} for coffee to 60 tonnes C ha^{-1} for complex agroforestry systems (Palm et al., 2005a). Below-ground carbon sequestration is generally lower, with an upper limit of about 1.3 tonnes C ha^{-1} yr^{-1} (Palm et al., 2005a). Agroforestry systems with nitrogen-fixing tree species, which are of particular importance in degraded landscapes, may be associated with elevated N_2O emissions (Dick et al., 2006). The benefits of tree-based carbon sequestration can have an environmental cost in terms of some soil modification (Jackson et al., 2005) (see 3.2.2.1.7).

The value of increased carbon sequestration in agroecosystems (e.g., from no-till) must be judged against the full lifecycle impact of CS practices on net greenhouse warming potential (GWP).

Goals	Certainty	Range of Impacts	Scale	Specificity
N, L, E, S	B	-2 to +2	R	Temperate zone

Increased carbon sequestration is not the only GWP-related change induced by adoption of agronomic practices like no-till. No-till maize systems can be associated with comparatively large emissions of N_2O (Smith and Conen, 2004; Duxbury, 2005). Over a 100-yr timeframe, N_2O is 310 times more potent in terms of GWP than CO_2 (Majumdar, 2003) and higher N_2O emissions from no-till systems may negate the GWP benefits derived from increased rates of carbon sequestration. On the other hand, soil structural regeneration and improved drainage may eventually result in a fewer N_2O emissions in no-till systems. Nitrogen fertilization is often the surest method for increasing organic matter stocks in degraded agroecosystems, but the benefits of building organic matter with N fertilizer use must be discounted against the substantial CO_2 emissions generated in the production of the N fertilizer. By calculating the full lifecycle cost of nitrogen fertilizer, many of the gains in carbon sequestration resulting from N fertilization are negated by CO_2 released in the production, distribution, and application of the fertilizer (Schlesinger, 1999; Follett, 2001; West and Marland, 2002).

Climate change is affecting crop-pest relations.

Goals	Certainty	Range of Impacts	Scale	Specificity
N, L, E, S	B	0 to -3	G	Worldwide

Climate change results in new pest introductions and hence changes in pest-predator-parasite population dynamics as habitat changes (Warren et al., 2001; McLaughlin et al., 2002; Menendez et al., 2006; Prior and Halstead, 2006; UCSUSA, 2007). These changes result from changes in growth and developmental rates, the number of generations per year, the severity and density of populations, the pest virulence to a host plant, or the susceptibility of the host to the pest and affect the ecology of pests, their evolution and virulence. Similarly, population dynamics of insect vectors of disease, and the ability of parasitoids to regulate pest populations, can change (FAO, 2005a), as found in a study across a broad climate gradient from southern Canada to Brazil (Stireman et al., 2004). Changing weather patterns also increase crop vulnerability to pests and weeds, thus decreasing yields and increasing pesticide applications (Rosenzweig, 2001; FAO, 2005a). Modeling can predict some of these changes (Oberhauser and Peterson, 2003) as well as consequences hence aiding in the development of improved plant protection measures, such as early warning and rapid response to potential quarantine pests. Better information exchange mitigates the negative effects of global warming. However, the impacts of climate change are not unidirectional; there can be benefits.

There is evidence that changes in climate and climate variability are affecting pest and disease distribution and prevalence.

Goals	Certainty	Range of Impacts	Scale	Specificity
N, L, E, S	B	0 to -3	R	Worldwide

Pests and diseases are strongly influenced by seasonal weather patterns and changes in climate, as are crops and biological control agents of pests and diseases (Stireman et al., 2004; FAO, 2005a). Established pests may become more prevalent due to favorable growing conditions such as include higher winter temperatures and increased rainfall. In the UK the last decade has been warmer than average and species have become established that were seen rarely before, such as the vine weevil and red mites ` with potentially damaging economic consequences (Prior and Halstead, 2006). Temperature increase may influence crop pathogen interactions and plant diseases by speeding up pathogen growth rates (FAO, 2005a). Climate change may also have negative effects on pests.

Livestock holdings are sensitive to climate change, especially drought.

Goals	Certainty	Range of Impacts	Scale	Specificity
N, L, E, S	B	-1 to -3	R	Especially in dry tropics

Climate fluctuation is expected to threaten livestock holders in numerous ways (Fafchamps et al., 1996; Rasmussen, 2003). Animals are very sensitive to heat stress, requiring a reliable resource of drinking water, and pasture is sensitive to drought. In addition, climate change can affect the distribution and range of insect vectors of human and livestock diseases, including species like mosquitoes (malaria, encephalitis, dengue), ticks (tick typhus, lyme disease), and tsetse fly (sleeping sickness). These infectious and vector-borne animal diseases have increased worldwide and disease emergencies are occurring with increasing frequency (FAO, 2005a; Jenkins et al., 2006; Oden et al., 2006). These problems are thought to be further exacerbated by climate change because hunger, thirst and heat-stress increase susceptibility to diseases. Small-scale farmers do not have the resources to take appropriate action to minimize these risks.

The Kyoto Protocol has recognized that Land Use, Land Use Change and Forestry (LULUCF) activities can play a substantial role in meeting the ultimate policy objective of the UN Framework Convention on Climate Change.

Goals	Certainty	Range of Impacts	Scale	Specificity
E	C	0 to +3	G	Wide applicability

LULUCF activities are "carbon sinks" as they capture and store carbon from the atmosphere through photosynthesis, conservation of existing carbon pools (e.g., avoiding deforestation), substitution of fossil fuel energy by use of modem biomass, and sequestration by increasing the size of carbon pools (e.g., afforestation and reforestation or an increased wood products pool). The most significant sink activities of UNFCCC (www.unfccc.int) are the reduction of deforestation, and the promotion of tree planting, as well as forest, agricultural, and rangeland management.

3.2.2.2.4 Energy to and from agricultural systems—
bioenergy
Bioenergy has recently received considerable public attention. Rising costs of fossil fuels, concerns about energy security, increased awareness of global warming, domestic agricultural interests and potentially positive effects for economic development contribute to its appeal to policy makers and private investors. However, the costs and benefits of bioenergy depend critically on local circumstances and are not always well understood (see also Chapters 4, 6, 7).

Biomass resources are one of the world's largest sources of potentially sustainable energy, comprising about 220 billion dry tonnes of annual primary production.

Goals	Certainty	Range of Impacts	Scale	Specificity
E	B	0 to +2	G	Wide applicability

World biomass resources correspond to approximately 4,500 EJ (Exajoules) per year of which, however, only a small part can be exploited commercially. In total, bioenergy provides about 44 EJ (11%) of the world's primary energy consumption (World Bank, 2003). The use of bioenergy is especially high (30% of primary energy consumption) in low-income countries and the share is highest (57%) in sub-Saharan Africa, where some of the poorest countries derive more than 90% of their total energy from traditional biomass. Also within developing countries the use of bioenergy is heavily skewed towards the lowest income groups and rural areas. In contrast, modern bioenergy, such as the efficient use of solid, liquid or gaseous biomass for the production of heat, electricity or transport fuels, which is characterized by high versatility, efficiency and relatively low levels of pollution, accounts for 2.3% of the world's primary share of energy (FAO, 2000b; IEA, 2002; Bailis et al., 2005; Kartha, et al., 2005).

Traditional bioenergy is associated with considerable social, environmental and economic costs.

Goals L, E, S	Certainty A	Range of Impacts -3 to +2	Scale G	Specificity Especially in the tropics

The energy efficiency of traditional biomass fuels (e.g., woodfuels) is low, putting considerable strain on environmental biomass resources, which are also important sources of fodder and green manure for soil fertility restoration as well as other ecosystem services. Inefficient biomass combustion is also a key contributor to air pollution in the homestead leading to 1.5 million premature deaths per year (WHO, 2006). Collecting fuelwood is time-consuming, reducing the time that people can devote to productive activities each day e.g., farming and education (UNDP, 2000; IEA, 2002; Goldemberg and Coelho, 2004; Karekezi et al., 2004; World Bank, 2004b; Bailis et al., 2005).

Production of modern liquid biofuels for transportation, predominantly from agricultural crops, has grown rapidly (25% per year) in recent years, spurred by concerns about fossil energy security and global warming and pressures from agricultural interest groups.

Goals E	Certainty B	Range of Impacts 0 to +3	Scale G	Specificity Wide applicability

Modern liquid biofuels, such as bioethanol and biodiesel contributed only about 1% of the total road transport fuel demand worldwide in 2005 (IEA, 2006c). The main first generation products are ethanol and biodiesel. Ethanol is produced from plant-derived starch (e.g., sugar cane, sugar beet, maize, cassava, sweet sorghum), primarily in Brazil (16,500 million liters) and the US (16,230 million liters). In 2005, world production was over 40,100 million liters (Renewable Fuels Association, 2005). Sugar cane derived ethanol meets about 22% of Brazil's gasoline demand (Worldwatch Institute, 2006), much of it used in flexfuel vehicles, which can operate under different gasoline-ethanol blends (e.g., 10% ethanol: 90% gasoline). In terms of vehicle fuel economy, one liter of ethanol is equivalent to about 0.8 liters of gasoline—accounting for its lower energy content but higher octane value (Kojima and Johnson, 2005). Biodiesel is typically produced chemically from vegetable oils (e.g., rapeseed, soybeans, palm oil, *Jatropha* seeds) by trans-esterification to form methyl esters. Germany was the world's biggest producer (1,920 million liters) in 2005, followed by other European countries and the USA. Biodiesel production has been growing rapidly (80% in 2005) but overall production levels are an order of magnitude smaller than ethanol (REN 21, 2006). Biodiesel contains only about 91% as much energy as conventional diesel, and can be used in conventional diesel engines, either pure or blended with diesel oil (EPA, 2002). Other biofuels such as methanol and butanol only play a marginal role in markets today but may become more important in the future.

The production of liquid biofuels for transport is rarely economically sustainable.

Goals E	Certainty C, E	Range of Impacts Not yet known	Scale G	Specificity Mainly in developed countries

The economic competitiveness of biofuels is widely debated and depends critically on local market conditions and production methods. The main factors determining biofuels competitiveness are (1) the cost of feedstock, which typically contributes about 60-80% of total production costs (Berg, 2004; Kojima and Johnson, 2005), (2) the value of byproducts (e.g., glycerin for biodiesel and high fructose maize syrup for maize ethanol), (3) the technology that determines the scale of the production facility, the type of feedstock and conversion efficiency, and (4) the delivered price of gasoline or diesel. Brazil is widely recognized to be the world's most competitive ethanol producer from sugar cane, with 2004-2005 production costs of US$0.22-0.41 per liter of gasoline equivalent (*vs* US$0.45-0.85 per liter in USA and Europe), but the world price of sugar and the exchange rate of the Brazilian currency determine price competitiveness. Brazilian ethanol production can be competitive with oil prices at about US$40-50 per barrel (*versus* about US$65 per barrel in Europe and USA, if one takes agricultural subsidies into account). It is estimated that oil prices in the range of US$66-115 per barrel would be needed for biodiesel to be competitive on a large scale. In remote regions and landlocked countries, where exceptionally high transport costs add to the delivered price of gasoline and diesel, the economics may be more favorable but more research is needed to assess this potential (IEA, 2004ab; Australian Government Biofuels Task Force, 2005; European Commission, 2005; Henke et al., 2005; Kojima and Johnson, 2005; Henninges and Zeddies, 2006; Hill et al., 2006; IEA, 2006c; OECD, 2006a; Worldwatch Institute, 2006; Kojima et al., 2007). In order to promote production despite these high costs biofuels are most often subsidized (see Chapter 6).

Bioelectricity and bioheat are produced mostly from biomass wastes and residues.

Goals E	Certainty B	Range of Impacts 0 to +2	Scale G	Specificity Wide applicability

Both small-scale biomass digesters and larger-scale industrial applications have expanded in recent decades. The major biomass conversion technologies are thermo-chemical and biological. The thermo-chemical technologies include direct combustion of biomass (either alone or co-fired with fossil fuels) as well as thermo-chemical gasification (to producer gas). Combined heat and power generation (cogeneration) is more energy efficient and has been expanding in many countries, especially from sugarcane bagasse (Martinot et al., 2002; FAO, 2004b; REN 21, 2005; IEA, 2006a; DTI, 2006). The biological technologies include anaerobic digestion of biomass to yield biogas (a mixture primarily of methane and carbon dioxide). Household-scale biomass digesters that operate with local organic wastes like animal manure can generate energy for cooking, heating and lighting in rural homes and are widespread in China, India and Nepal. However their operation can sometimes pose technical as well as resource challenges. Industrial-scale units are less prone to technical problems and increasingly widespread in some developing countries, especially in China. Similar technologies are also employed in industrial countries, mostly to capture environmentally problematic methane emissions (e.g., at landfills and livestock holdings) and produce energy (Balce, et al., 2003; Ghosh, et al., 2006; IEA, 2006b). Despite the fact that production costs can be competitive in various settings, in the past many attempts to promote wider distribution of modernized bioenergy applications have failed. Common problems included technical difficulties and the failure to take into account the needs and priorities of consumers, as well as their technical capabilities, when designing promotion programs (Ezzati and Kammen, 2002; Kartha, et al., 2005; Ghosh et al., 2006).

Bioelectricity and bioheat production can be competitive with other sources of energy under certain conditions, especially the combination of heat and power generation within industries producing waste biomass.

Goals	Certainty	Range of Impacts	Scale	Specificity
E	B	-2 to +3	G	Wide applicability

The competitiveness of bioelectricity and bioheat depends on (1) local availability and cost of feedstocks—many of which are traded on market with strong prices variations both regionally and seasonally; (2) capital costs and generation capacity; (3) cost of alternative energy sources; and (4) local capacity to operate and maintain generators. Generally, bioelectricity production is not competitive with grid electricity but generation costs can compete with off-grid option such as diesel generators in various settings. Key to competitiveness is a high capacity utilization to compensate for relatively high capital costs and exploit cheap feedstock costs. High capacity factors can best be reached when proven technologies (e.g., thermochemical combustion) are employed on site or near industries that produce biomass wastes and residues and have their own steady demand for electricity, e.g., sugar, rice and paper mills. Estimates for power generation costs in such facilities range from US$0.06-0.12/kWh (WADE, 2004; REN 21, 2005; World Bank, 2005a; IEA, 2006b). In combined heat and power mode, when capital investments can be shared between electricity and heat generation, electricity generation costs can decrease to US$0.05-0.07/kWh, depending on the value of the heat (REN 21, 2005; IEA, 2006b). Thermochemical gasification can have higher generation costs and low capacity utilization due to weak electricity demand, and technical failures caused by improper handling and maintenance can lead to even higher production costs (Larson, 1993; World Bank, 2005a; Banerjee, 2006; Ghosh et al., 2006; Nouni et al., 2007). Data on electricity production costs with anaerobic digesters are not widely available, because most digesters are not installed commercially but through government programs to provide (1) energy access for rural households and villages, often solely for the provision of cooking fuel or heating or (2) methane capture on environmental grounds (e.g., in several industrialized countries). Overall, the economics of biomass power and heat can be improved through carbon credits.

3.2.3. Impacts of AKST on livelihoods, capacity strengthening and empowerment

3.2.3.1 Methodologies and approaches for assessing impact

Assessing the evidence for the contribution of AKST to improving livelihoods and empowerment is complex. While there is evidence of contribution to increasing productivity of agriculture and sustainability of natural resource use, the extent to which this is relevant to specific groups of people and translates into improved livelihoods, is more complicated, involving differential impacts between and within populations. The difficulty of attribution applies similarly to negative outcomes. The paths of causality are complex and highly contingent on specific conditions (Adato and Meinzen-Dick, 2007) involving interactions between AKST and the policies and institutional contexts in which AKST products are promoted and adopted. Hence the assessments of impacts are sometimes contradictory or controversial. The methodological challenges of impact assessment are considerable; especially when going beyond economic measures of impact or individual case studies. Thus it is difficult to make broader statements on the poverty and livelihood impacts of AKST investments and products across different geographical regions and client groups. Impact assessments rely on comparison—before and after a specific intervention or change, or a "with" and "without" situation (the counterfactual either being empirically measured, or theoretically constructed assuming the best available alternatives are pursued). This approach has been helpful in establishing the economic returns from agricultural research and the contribution of increased productivity, but is more difficult to construct for the livelihood dimensions.

3.2.3.1.1 Assessment of the economic impacts of AKST
Past assessments of impacts of specific AKSTs have documented adoption, productivity increases and financial returns and consequences for national food security (Hazell and Ramasamy, 1991; Evenson and Gollin, 2003a). There is evidence that agricultural productivity growth has a substantial impact on poverty reduction, although this is conditional on contextual and socioeconomic conditions, e.g., equitable land distribution (Kerr and Kolavalli, 1999; Hazell and Haddad, 2001; Jayne et al., 2003; Mathur et al., 2003; Thirtle et al., 2003). Economists have developed techniques to quantify the total economic value of the multitude of products and services (social/environmental and local/global) from agricultural programs, such as agroforestry (Pearce and Mourato, 2004).

Impact assessments of investment in agricultural research have shown that it has been highly cost effective.

Goals	Certainty	Range of Impacts	Scale	Specificity
L, E, D	B	+2 to +4	G, R, N	Wide applicability

Investment in research has resulted in substantial economic gains from increased productivity. For example, in the case of the CGIAR system benefit-cost ratios for research have been between 1.94 (significantly demonstrated and empirically attributed) and 17.26 (plausible, extrapolated to 2011) (Raitzer, 2003). Three innovations—MVs of rice (47% of benefits), MVs of wheat (31% of benefits) and cassava mealy bug biocontrol (15% of benefits) account for most of the impact using the most stringent criteria, and are worth an estimated US$30 billion [at 1990 values] (Heisey et al., 2002; Evenson and Gollin, 2003b; Hossain et al., 2003; Raitzer, 2003; Lantican et al., 2005). While focused on a very narrow range of species, as a measure of this success, the CGIAR has estimated that 30 years of agricultural research on seven major crops and three livestock products has improved yield gains so much that, had this gain not occurred, an additional 170-340 million ha of forests and grasslands would have been needed for production (Nelson and Maredia, 1999; FAO, 2003a). Other estimates of forestalled conversion of habitat to agricultural use are as high as 970 million ha (Golkany, 1999). A cost/benefit analysis by ACIAR (Raitzer and Lindner, 2005) found that research projects involving forestry/agroforestry had the greatest benefits (42.9%). Increases in total factor productivity, which contribute to increased output, are always associated with investment in research (Pingali and Heisey, 1999; McNeely and Scherr, 2003). These studies pay less attention to the social and institutional distribution of impacts or to noneconomic benefits.

3.2.3.1.2 Assessment of livelihood impacts of AKST
Systematic and detailed impact assessments of AKST's contribution to livelihood improvement and the sustainability of livelihoods over time are generally lacking. A livelihood is said to be sustainable "when it can cope with and recover from shocks and maintain or enhance its capabilities and assets both now and in the future, while not undermining the natural resource base" (Carney, 1998). Indirect impacts of AKST in relation to ownership of assets, employment on and off farm, vulnerability, gender roles, labor requirements, food prices, nutrition and capacity for collective action have been less thoroughly researched than the financial and economic impacts (Hazell and Haddad, 2001; Meinzen-Dick et al., 2004), although recent impact assessments of Participatory Methods have more comprehensively addressed these issues. Comparative case studies of livelihood change incorporating qualitative dimensions and complementing other methods have begun to document the noneconomic impacts of AKST. *(www.prgaprogram.org/ modules.php?op=modload&name=Web_Links&file=index &req=viewlink&cid=133&min=0&orderby=titleA& show=10).*

Livelihoods approaches have usefully contributed to conceptual and methodological innovations.

Goals	Certainty	Range of Impacts	Scale	Specificity
N, H, L, E, S, D	B	0 to +3	R, N, L	Wide applicability

The concept of "sustainable livelihoods" is both an AKST product and a tool, which facilitates the analysis of livelihood status and changes and the understanding of *ex ante* and *ex post* impacts. The livelihoods framework considers livelihoods as comprising the capabilities, assets and activities required for a means of living. This is a broader and more holistic view than just equating "livelihood" with income or employment (Booth et al., 1998). It links the notion of sustaining the means of living with the principle of environmental sustainability (Carney, 1998).

The elements of the livelihoods framework include the assets that people use and combine to make a living, the factors which cause vulnerability; the policies, institutions and processes which affect the environment for livelihoods; the livelihood strategies followed and the outcomes. The livelihoods framework has been used to assist situational analysis for research and development planning and to assess specific institutional, policy and technology and rural development options prior to intervention (Ashley and Carney, 1999; Shackleton et al., 2003; OECD, 2006b). More recently it has been used to assist evaluation of outcomes and impacts (Ashley and Hussein, 2000; Adato and Meinzen-Dick, 2003; Meinzen-Dick et al., 2004; Adato and Meinzen-Dick, 2007) and has complemented more macrolevel economic impact assessments. Livelihoods analysis has been further assisted by the development and refinement of participatory tools for poverty and situational analysis, especially in the context of improving client orientation and gender relevance of agricultural research and development (World Bank, 1998). Recently, the framework has helped to identify principles and processes critical to achieving sustainable livelihoods, and to understand the complexities associated with partnerships to promote local empowerment, resiliency and diversification (Butler and Mazur, 2007). Its limitations include the absence of integration of dimensions of power, the unspecified nature of "institutions and processes" which require further elaboration of knowledge, culture and innovation and the need for further tools to understand the dynamics of livelihood changes.

3.2.3.2 The contribution of AKST to livelihoods improvement
The improvement of livelihoods depends on the accessibility of the products of AKST. This depends on the factors influencing uptake, the distribution of benefits of specific technologies and their impacts. Particular attention is paid to impacts on overall levels of poverty and economic status, human health; natural and physical assets, social relationships, and vulnerability.

3.2.3.2.1 AKST and poverty

Some gains have been made in the reduction of poverty, but the contribution of AKST to increasing agricultural production and agriculture based incomes has been very different in different regions, agroecologies and for different groups of people.

Goals	Certainty	Range of Impacts	Scale	Specificity
L, D	A	-2 to +3	R, N, L	Incidence of poverty remains high in some African countries

AKST and agricultural transformation have had an important influence on the economic and social situation of many countries. Poverty is a serious global problem with 3 billion people (2.1 billion are rural poor) earning less than the purchasing power equivalent of US$2/day. The impacts of AKST are location specific and depend on complex interacting factors. Between 1990 and 2002, the proportion of people living in extreme poverty fell more rapidly in much of Asia compared with Africa, Latin America and the Caribbean (UN, 2006a); while in central and eastern Europe, the poverty rates increased. In sub-Saharan Africa, although there was a small decline in the rate of poverty, the number of people living in extreme poverty increased by 140 million. Poor countries (especially in SSA) have gained proportionately less than some richer countries (USA and Europe). Similarly, major benefits have escaped marginal agroecological regions (rain-fed dryland areas) and marginalized people (small-scale farmers, landless people, seasonally mobile populations, women and the poorest) (Fan et al., 2000; Hazell and Haddad, 2001; Sayer and Campbell, 2001). While the Green Revolution yielded large production gains in some commodity crops, basic grains and livestock, it was often at the expense of environmental degradation (Pingali and Rosegrant, 1994). Elsewhere, for example, in Uttar Pradesh and Tamil Nadu in India, it benefited the poor, including some landless laborers, reducing inequality and improving economic opportunities (Hazell and Ramasamy, 1991; Sharma and Poleman, 1993). Intensive agricultural development, particularly in Europe, led to oversupply, sanitary problems affecting livestock production and ecological issues, while the concentration of production caused economic and social decline in marginal areas (Hervieu and Viard, 1996).

Farmers have not always benefited from crop breeding.

Goals N, H, L, S	Certainty B	Range of Impacts 0 to +3	Scale G, R, N, L	Specificity Widespread

The initial success of the Green Revolution was a result of its focus on more favorable irrigated rice and wheat systems (Huang et al., 2002), but crop varieties bred for responsiveness to such conditions were less successful when the focus shifted to more marginal and variable environments (Smale et al., 1998; Witcombe et al., 2001). Although the adoption of "modern" varieties has been widespread (up to 70% in some crops) (Evenson and Gollin, 2003ab), farmers in more marginal areas have not always benefited from the latest research on pest/disease resistance and yield (Witcombe, 1999; Witcombe et al., 2001). Varieties bred on research stations have not always been well adapted to local conditions and preferences; nor for acceptable quality, utility for multipurpose uses; or acceptable postharvest characteristics (e.g., easy to thresh/process, good taste, good storability). Consequently, comparatively few of the hundreds of rice varieties released in India are grown by farmers (Witcombe et al., 1998) while some traditional varieties, e.g., a peanut variety grown in southern India, remain popular (Bantilan et al., 2003). Some new and potentially better modern varieties have failed to reach farmers due to the inefficiency of the varietal release and seed multiplication system (Witcombe et al., 1988). Participatory approaches can help overcome this inefficiency (Uphoff, 2002).

Livestock are important for rural livelihoods, but livestock technologies have made only a limited contribution to improving rural livelihoods.

Goals N, H, L, E, S, D	Certainty C	Range of Impacts +1 to +3	Scale R, N, L	Specificity New AKST more positive in industrialized countries.

Livestock are of greater importance to poor people and the landless than those with higher incomes (Delgado et al., 1999). Livestock management in difficult environments is knowledge-intensive and integrated into complex social and natural resource management systems. In general, small-scale farmers have largely relied on traditional and local knowledge to sustain their livestock production systems (Falvey and Chantalakhana, 1999). Of an estimated

600 million livestock keepers globally, most of whom are in mixed rainfed systems, 430 million are resource poor and concentrated in SSA and south Asia (Heffernan et al., 2005). The important developments in livestock technologies (feed technologies in intensive livestock production systems; artificial insemination; embryo transfer, etc.) are more widely used in the industrialized world, as there are constraints to applying these technologies in developing countries (Madan, 2005). Thus, the rapid growth in consumption of livestock products in developing countries has been due to increased numbers, rather than increased productivity (Delgado et al., 1999). Vaccination against major animal diseases has been successful in developing countries, e.g., rinderpest in Africa and Newcastle disease in Asia. In Africa, net annual economic benefit attributed to the elimination of rinderpest has been valued at US$1 billion (http://www-naweb.iaea.org/nafa/aph/stories/2005-rinderpest-eradication.html). Likewise, heat stable vaccination against Newcastle disease has led to improved village poultry production in Indonesia and Malaysia, with returns equivalent of US$1.3 million and $2.15 million respectively. The latter success was associated with understanding of the social implications and situation at village level, well developed extension packages, government leadership, and training workshops for senior policy administrators, laboratory staff and livestock officers (http://www.fao.org/docs/eims/upload/207692/7_1_1_cases.PDF). Tsetse fly eradication projects have had some success, especially where farmer-based and demand-driven approaches to control are adopted and where cohesive groups can function as the basis for collective action (Dransfield et al., 2001). Positive impacts of livestock research for poor producers have occurred through the introduction of new institutional forms, such as dairy cooperatives in India and with a supportive national policy and legislative environment. Nevertheless, many livestock projects have not had satisfactory long-term effects on the livelihoods of the poor (LID, 1999). In general, the uptake and impact of livestock technologies in developing countries is often constrained by the lack of a poverty reduction focus, their higher financial and labor demands, an overly narrow technical focus, inappropriate technologies, failure to take into account the social context of production, patterns of ownership and local knowledge and weak private sector development (Livestock in Development, 1999), or because wealthier farmers or herders captured the benefits (Heffernan et al., 2005).

Social and economic impacts of GMOs depend on the socioeconomic and institutional circumstances of the country of introduction.

Goals L, E, S, D	Certainty C E	Range of Impacts -3 to +2	Scale N	Specificity Mainly in large scale farms in industrialized countries

There have been positive farm level economic benefits from GMOs for large scale producers, but less evidence of positive impact for small producers in developing countries. The adoption of the commercially available GM commodity crops (over 90% of global area planted) has mostly occurred in large scale industrial, chemical intensive agricultural systems in North and South America (95.2% of production), with small areas in India and China (James, 2006), and the rest is shared among 16 other countries worldwide. There is little consensus among the findings from the assessments of economic and environmental impacts of GMOs. An analysis of the global impact of biotech crops from 1996 to 2006 showed substantial net economic benefits at the farm level; reduced pesticide spraying, decreased environmental impact associated with pesticide use and reduced release of greenhouse gas emission (Brookes and Barfoot, 2006). A different study of the economic impact of transgenic crops in developing countries found positive, but highly variable economic returns to adoption (Raney, 2006). In this case, institutional factors such as the national agriculture research capacity, environmental and food safety regulations, IPRs and agriculture input markets determined the level of benefits, as much as the technology itself (Raney, 2006). Adoption of GM cotton in South Africa is symptomatic, not of farmer endorsement of GM technology, but of the profound lack of farmers' choice and a failure to generate sufficient income in agroecosystems without a high level of intensification (Witt et al., 2006). Other studies have concluded that GM technologies have contributed very little to increased food production, nutrition, or the income of farmers in less-developed countries (Herdt, 2006), or even led to deskilling of farmers (Stone, 2007). In Argentina, many large scale farmers have greatly benefited from the use of herbicide resistant soybeans (Trigo and Cap, 2003; Qaim and Traxler, 2004). However significant socioeconomic and environmental problems have arisen from the increased area of soybeans linked to the introduction of GM soybean for small-scale or landless farmers, which enabled them to produce at significantly lower costs, with expansion on marginal lands (Trigo and Cap, 2003; Benbrook, 2005; Joensen et al., 2005; Pengue, 2005). In India, claims regarding benefits or damages are highly controversial with reports presenting opposing data and conclusions (e.g., Qayum and Sakkhari, 2005 *vs.* Morse et al., 2005).

3.2.3.2.2 Health and nutrition

Rates of hunger have been decreasing but hunger is still common despite the advances of AKST and the Green Revolution.

Goals	Certainty	Range of Impacts	Scale	Specificity
N, H, L, S, D	A	-3 to +4	G	Mostly in developing countries

Although the Green Revolution and other AKST have had significant impacts on increased food supply, the reduction of hunger and malnutrition has been unevenly distributed across the world. Currently, the number of people defined as hungry in 2006 was 854 million people, of whom 820 million lived in developing countries (FAO, 2006e). In parallel, food consumption per person has risen from 2358 to 2803 kcal per day between the mid 1960s and late 1990s. Now, only 10% of the global population lives in countries with food consumption below 2200 kcal, while 61% live in countries consuming over 2700 kcal (FAO, 2005c). However the incidence of hunger has not declined in many countries of sub-Saharan Africa (FAO, 2005c), where population growth (3%) outstrips increases in food production (2%). In 2005, it was estimated that 13% of the world population (850 million people) are energy-undernourished, of whom

780 million were in developing countries (FAO, 2005c). Hunger is not explained by a simple relationship between food supply and population, as adverse agricultural conditions, poverty, political instability, alone or in combination, are contributing factors (Sen, 1981).

Rates of malnutrition are decreasing, but undernutrition is still a leading cause of health loss worldwide despite AKST advances.

Goals	Certainty	Range of Impacts	Scale	Specificity
N, H, L, S, D	A	-4 to +2	G	Mostly in developing countries

AKST has been important in reducing malnutrition, especially in mothers and children. Although the world food system provides protein and energy to over 85% people, only two-thirds have access to sufficient dietary micronutrients for good health (Black, 2003). Child stunting malnutrition reduced in developing countries from 47% in 1980 to 33% in 2000, but is still a major public health problem with 182 million stunted preschool children in developing countries (70% in Asia and 26% in Africa) (de Onis, 2000). Factors implicated include low national *per capita* food availability, lack of essential nutrients due to poor diet diversity, poor child breast feeding patterns, high rates of infectious disease, poor access to safe drinking water, poor maternal education, slow economic growth and political instability (de Onis, 2000). Under nutrition remains the single leading cause of health loss worldwide (Ezzati et al., 2003), and being underweight causes 9.5% of the total disease burden worldwide. In developing countries this is linked with nearly 50% of malaria, respiratory diseases and diarrhea. Selected dietary micronutrient deficiencies (iron, vitamin A and zinc deficiency) were responsible for 6.1% of world disease burden (Ezzati et al., 2003).

A focus on increased production and food security rather than diet quality has contributed to a rise in obesity worldwide and the double burden of under- and overnutrition in developing countries.

Goals	Certainty	Range of Impacts	Scale	Specificity
N, H, L, S, D	A	-2 to +2	G	Worldwide

A focus on energy needs, rather than improved nutrition and access to a balanced and healthy diet, has been one factor in increasing overweight and obesity worldwide (Black, 2003; Hawkes, 2006). Increased food production and *per capita* availability together with a decline in world prices since the 1960s has created food energy abundance for more than 60% of the world (FAO, 2005c). Dietary and nutritional transitions have occurred worldwide, with actual patterns of diet change and hence health impacts varying (Popkin, 1998; Caballero, 2005). Socioeconomic, demographic and environmental changes have occurred that affect food availability, food choices, activity and life patterns (e.g., urbanization, work practices, transport, markets and trade) (Hawkes, 2006). Diet trends have resulted in widespread decreasing intake of fruits and vegetables and increasing intake of meat, sugar, salt and energy-dense processed foods (Popkin, 1998; WHO/FAO, 2003). Dietary fat now accounts for up to 26-30% of caloric intake, and there has been marked increases in both meat and fish intake (see 3.2.1). These dietary changes have contributed to rapidly rising obesity and its related chronic diseases such as "type 2" diabetes, hypertension, heart disease and cancers globally (WHO/FAO, 2003). In 2005 more people were overweight (1.6 billion adults [age 15+]) than underweight worldwide and 400 million adults were obese (WHO, 2005a). This problem is now increasing in low- and middle-income countries (below 5% in China, Japan and certain African nations, to 40% in Colombia, Brazil, Peru (www.iaso.org), and over 75% in the Pacific), particularly in urban settings—almost 20% in some Chinese cities (WHO, 2003). In Africa, Latin America, Asia and the Pacific, there is now the double diet-related disease burden of undernutrition and obesity (Filozof et al., 2001; Monteiro et al., 2002; Rivera et al., 2002; Caballero, 2005).

Dietary diversity is a key element of a healthy diet.

Goals	Certainty	Range of Impacts	Scale	Specificity
N, H, L, S, D	A	+2 to +4	G	Worldwide

With the increased focus on staple starch crops, and global food trends, dietary diversity has declined over recent decades (Hatloy et al., 2000; Marshall et al., 2001; Hoddinott and Yohannes, 2002). However, many studies have recognized the need for a diverse and balanced diet for optimum health (Randall et al., 1985; Krebs-Smith et al., 1987; Hatloy et al., 2000; Marshall et al., 2001). Healthy diets include fruits and vegetables, animal source proteins, and sources of fiber to (1) minimize the risks of cancer (Tuyns et al., 1987), vascular (Wahlquist et al., 1989) and cardiovascular diseases (Cox et al., 2000; Veer et al., 2000); (2) optimize birth weight of children (Rao et al., 2001), maintain overall health (Ruel, 2002), and prolong life expectancy (Kant et al., 1993), and (3) maximize earning capacity from manual labor (Ali and Farooq, 2004; Ali et al., 2006). Various measures and standards have been developed for food quality, which include Diet Quality Index (Patterson et al., 1994), Analysis of Core Foods (Kristal et al., 1990), and Healthy Eating Index (Kennedy et al., 1995). In addition, Dietary Diversity Scores are being devised to measure diet quality (Kant et al., 1993, 1995; Hatloy et al., 1998; Marshall et al., 2001; Ali and Farooq, 2004). A methodology has been developed to prioritize food commodities based on their total nutritive values (Ali and Tsou, 2000). Unlike food safety standards, measures of food quality or diet diversity have not been implemented nationally or internationally.

Food based approaches to tackle micronutrient deficiencies have long term benefits on health, educational ability and productivity.

Goals	Certainty	Range of Impacts	Scale	Specificity
N, H, L, E, S, D	B	+2 to +4	R	Mainly in developing countries

Although the potential of food based dietary diversification to reduce micronutrient deficiency disease has not been fully explored or exploited (Ruel and Levin, 2000), new approaches to overcoming micronutrient deficiencies are focusing on diet diversification and food fortification. Food fortification has to date mostly been applied in industrialized countries, as technical, sociocultural, economic and

other challenges have constrained their use in less developed countries (WHO, 2005c). Food fortification is potentially more cost effective and sustainable than treating people with food supplements and is compatible with giving greater attention to diversified production of fruits, vegetables, oilcrops and grain legumes, as well as diverse animal source proteins including fish, poultry and dairy products (FAO, 1997). It is likely that a combination of strategies, including greater emphasis on traditional foods (Leakey, 1999a), is required to tackle micronutrient malnutrition (Johns and Eyzaguirre, 2007).

Animal source protein is one component of a healthy diet but rapid increases in livestock production and red meat consumption pose health risks by directly contributing to certain chronic diseases.

Goals	Certainty	Range of Impacts	Scale	Specificity
N, H, L, E, S, D	A/B	+3 to -3	G	Worldwide

Animal source protein can be an important component of a healthy diet, but moderate consumption of meat and fish is desirable. A rapid rise in meat consumption in high, middle and some low income countries is linked to increased rates of ischaemic heart disease (particularly related to saturated fat), obesity and colorectal cancer (Law, 2000; Delgado, 2003; Popkin and Du, 2003; Larsson and Wolk, 2006). In the lowest income countries, especially Africa, consumption of animal source foods is often low, leading to malnutrition (Bwibo and Neumann, 2003). Moderate fish consumption has health benefits, e.g., reducing rates of coronary heart disease deaths (Mozaffarian and Rimm, 2006). Replacing ruminant red meat by monogastric animals or vegetarian farmed fish would create sources of animal source protein which would reduce rates of chronic diseases. A positive environmental side-effect could be reduced methane gas emissions (McMichael et al., 2007).

AKST has not solved food security problems for the rural poor.

Goals	Certainty	Range of Impacts	Scale	Specificity
N, H, L, E, S, D	B	-4 to -2	R	Rural poor in developing countries

The rural poor (who comprise 80% of those hungry worldwide) are dependent on environmental resources and services, are highly vulnerable to environmental degradation and climate change, and have poor access to markets, health care, infrastructure, fresh water, communications, and education. Wild and indigenous plants and animals are important to the dietary diversity and food security of an estimated 1 billion people (FAO, 2005b). Increased population pressures on forests and woodlands has led to a decline in gathered natural foods (Johns et al., 2006), which are often rich in nutrients, vitamins and minerals (Leakey, 1999a). The expansion of urban areas has also reduced the sources of fresh food from home gardens (Ali et al., 2006), as has the focus on large-scale, industrial production of crops and livestock at the expense of smaller mixed farming systems employed by the poor.

AKST has led to improvements in food safety although microbiological and chemical hazards continue to cause a significant health problem.

Goals	Certainty	Range of Impacts	Scale	Specificity
N, H, E, S	A	-3 to +4	G	Worldwide

The emphasis of current food safety is on reducing the transmission of food- and water-borne infectious disease related to production, processing, packaging and storage, and chemical and other non-infectious food contamination. The latter include environmental contaminants such as mercury in fish and mycotoxins, as well as food additives, agrochemicals and veterinary drugs, such as antibiotics and hormones (Brackett, 1999; Kitinoja and Gorny, 1999). To improve food safety and quality there has been increased attention to traceability, risk assessment, the provision of controls (Hazard Analysis Critical Control Point [HACCP]) and the implementation of food safety standards, such as GAP, GMP like ISO 9000, EUREP GAP and HACCP. In addition, AKST has developed both simple and high-technology solutions to extend shelf life and make stored foods safer. Techniques include low-cost, simple technology treatment of wastewater for irrigation; cost-effective methods for reducing microbial load on intact and fresh-cut fruit and vegetables; improved efficacy of water purification, such as chlorination/ozonizations (Kader, 2003); refrigeration and deep freezing; food irradiation; modified atmosphere packaging, laboratory and production-line surveillance, and genetic engineering. However public concern about the potential risks associated with new technologies has led to calls for rigorous risk assessments based on international standards (WHO, 2002). These technologies, linked to better transport have increased year-round access to healthy, safe food for many, but these public health benefits are unequally distributed and favor high-income consumers.

Emerging human and animal infectious diseases are linked to poor or limited application of AKST.

Goals	Certainty	Range of Impacts	Scale	Specificity
N, H, L, S, D	A	-2 to +2	G	Worldwide

Of 204 infectious diseases currently emerging in both high and low income countries, 75% are zoonotic (transmitted between animals and humans) (Taylor et al., 2001). They pose direct threats to human health and indirect socio-economic impacts affecting rural livelihoods due to trade restrictions. Recent high-profile examples of these animal diseases infecting humans through the food chain include Bovine Spongiform Encephalopathy (BSE) in cows and avian influenza (H5N1) in poultry. In both cases transmission has been linked to low standards in the animal feed industry and the increase of antimicrobial resistance arising from the use of antibiotics in industrialized farming systems. As this resistance will limit prevention and treatment of these diseases, the World Health Organization recommended the elimination of subtherapeutic medical antibiotic use in livestock production in 1997, and called for strict regulation and phasing out of other subtherapeutic treatments, such as growth promotants (http://europa.eu/rapid/pressReleasesAction.do?reference=IP/03/1058&format=HTML&aged=0&language=EN&guiLanguage=en). Adequate surveillance and control programs have not been introduced in many countries.

The health focus of industrial food processing and marketing has mainly been on adding value and increasing shelf-life, and not on improving nutrition.

Goals	Certainty	Range of Impacts	Scale	Specificity
N, H, L, E, S, D	B	-2 to +2	G	Worldwide

AKST has focused on adding value to basic foodstuffs (e.g., using potatoes to produce a wide range of snack foods). This has led to the development of cheap, processed food products with long shelf life but reduced nutritive value (Shewfelt and Bruckner, 2000). Postharvest treatments to extend shelf life of fruit and vegetables degrade provitamin A, such as -carotene, and reduce the bioavailability of nutrients (AVRDC, 1987; Zong et al., 1998). The benefits of this food processing technology tend to be unequally distributed between producer and retailer, with increasingly lower percentages of the final cost of processed food reaching the rural producers. In developed countries this has led to concerns that retailers may abuse their market power *vis-à-vis* other producers and consumers. The emphasis on "adding value" has also has also lowered the incentive to promote healthy fresh produce such as fruits and vegetables. Recent initiatives to develop processed "health foods" are predominantly aimed at rich consumers (Hasler, 2000). Food labeling and health claims on packaged foods are a major source of nutritional information for consumers (EHN, 2001), but voluntary labeling approaches (such as guideline daily amounts) are difficult for consumers to understand, reducing their ability to make informed choice about the nutritional value of the foods. As mentioned earlier, processed energy-dense foods (high in fat, salt and sugar) are contributing to increasing rates of obesity and associated chronic diseases (Nestle, 2003).

Agricultural production and trade policies have influenced negative trends in global nutrition and health.

Goals	Certainty	Range of Impacts	Scale	Specificity
N, H, L, S, D	A	-3 to -1	G	Worldwide

Despite the clear links between diet, disease and health, agricultural policy has been dominated by production rather than diet objectives (Lang and Heasman, 2004). There is international agreement on the requirements of a healthy diet (WHO/FAO, 2003), and the ability of diets rich in fruits and vegetables to reduce diseases like heart disease, stroke, and many cancers (Ness and Powles, 1997; WCRF/AICR, 1997; Bazzano et al., 2001; Lock et al., 2005). Saturated fatty acids (naturally present in animal fats) lead to increased serum cholesterol levels and a higher risk of coronary heart disease. Trans-fatty acids, caused by industrial hydrogenation of vegetable or marine oils by the food industry, cause higher risks of heart disease (Mozaffarian et al., 2006; Willet et al., 2006). Agricultural policies and production methods influence what farmers grow, and what people consume, through their influence on food availability and price (Hawkes, 2007). The liberalization of agricultural markets and the rise of a global, industrialized food system have had major effects on consumption patterns, resulting in high public health costs and externalities (Lang and Heasman, 2004). This has resulted in a convergence of consumption habits worldwide, with lower income groups increasingly exposed to energy dense foods, while high-income groups benefit from the global market (Hawkes, 2006).

Agrochemical use can have both positive and negative impacts on health.

Goals	Certainty	Range of Impacts	Scale	Specificity
H, L, S	A	-3 to 0	R	Mainly in developing countries

Agrochemicals have been responsible for increasing food production and as part of the control of some important human diseases such as malaria. However, they can also cause a wide range of acute and chronic health problems (O'Malley, 1997; Kishi, 2005). Chronic health effects include reproductive, neurological, developmental/learning disabilities, endocrine-disruption, and some cancers. WHO has estimated that there are at least 3 million cases each year of pesticide poisoning worldwide, one million of which are thought to be unintentional poisoning and two million suicide attempts, leading to about 220,000 deaths annually (WHO, 1986). The majority of these cases occur in developing countries where knowledge of health risks and safe use is limited, and harmful pesticides, whose use may be banned in developed regions, are easily accessible (Smit, 2002). In developing countries, acute poisoning of agricultural workers can result from poor training and lack of proper safety equipment (Repetto and Baliga, 1996), as well as an inability to read and understand health warnings. Small-scale farmers may be too poor to purchase the necessary protective equipment (if available), and may not have access to washing facilities in the fields or at home. Studies of farm workers and children living in agricultural areas in the USA and in developing countries indicate that adverse health impacts are also experienced by children playing around pesticide treated fields, and people drinking pesticide contaminated water supplies (Curl et al., 2002; Fenske et al., 2002). Pesticide related illness results in economic losses (Cole et al., 2000).

Poor health has negative impacts on agricultural productivity and the application of AKST.

Goals	Certainty	Range of Impacts	Scale	Specificity
N, H, L, E, S,	A	-4 to -2	R	Mainly developing countries

Agricultural production can be negatively affected by the poor health of agricultural workers, resulting from malnutrition, chronic noncommunicable diseases and infectious diseases (Croppenstedt and Muller, 2000; Jayne et al., 2004). Poor health also affects farmers' ability to innovate and develop farming systems (Jayne et al., 2004). Many studies show that communities with high disease prevalence experience financial and labor shortages. They respond by changing crops and reducing the area of land under cultivation, consequently decreasing productivity (Fox, 2004; Jayne et al., 2004). Ill health among families of producers can further reduce household income or other outputs of farm work as the able-bodied absent themselves from work in order to care for their sick family members (Jayne et al.,

2004). In developing countries these issues are most clearly illustrated by the impact of HIV/AIDS (Fox, 2004; Jayne et al., 2004), which, due to reductions in life expectancy, also results in loss of local agricultural knowledge and reduced capacity to apply AKST.

Agriculture has one of the worst occupational health and safety records.

Goals	Certainty	Range of Impacts	Scale	Specificity
N, H, L, E, S	A	-4 to 0	R, G	Worldwide

Irrespective of age, agriculture is one of the three most dangerous occupations (with mining and construction) in terms of deaths, accidents and occupational-related ill-health (ILO, 2000). Half of all fatal accidents worldwide occur in agriculture. Many agricultural practices are potentially hazardous to the health of agricultural workers, including use of agrochemicals and increasing mechanization. Agriculture is traditionally an underregulated sector in many countries and enforcement of safety regulations is often difficult due to dispersed nature of agricultural activity and lack of awareness of the nature and extent of the hazards. It is estimated that some 132 million children under 15 years of age work on farms and plantations worldwide due to lack of policies to prevent agricultural child labor (ILO, 2006). This work exposes them to a number of health hazards, as well as removing them from education. AKST has not addressed the tradeoffs of policies and technologies to minimize harm and maximize the health and livelihoods benefits.

The limited availability of supplies of fresh potable water is a health issue, especially in dry areas with diminishing water resources and where there are threats from nitrate pollution of water bodies and aquifers.

Goals	Certainty	Range of Impacts	Scale	Specificity
N, H, L, E, S	A	-2 to 0	R	Developing countries mainly

The lack of access to clean drinking water is estimated to be responsible for nearly 90% of diarrheal disease in developing countries (Ezzati et al., 2003). Reducing this health hazard and improving the access to clean drinking water is one of the Millennium Development Goals; currently Africa is not on track to meet these targets. In some areas of the Sahel, aquifers are becoming seriously polluted by N pulses reaching water tables (Edmunds et al., 1992; Edmunds and Gaye, 1997). This N is probably of natural origin, since N-fixing plants used dominate in natural vegetation and, in the absence of land clearance, the N was probably recycled in the upper soil profile through leaf litter deposition and decomposition. However, following deforestation, the nutrient recycling process is lost and N is slowly leached down the profile. High N contamination has serious implications for the future potability of groundwater for the human population and their livestock.

The safety of GMO foods and feed is controversial due to limited available data, particularly for long-term nutritional consumption and chronic exposure.

Goals	Certainty	Range of Impacts	Scale	Specificity
N, H, L, E	C, E	-3 to 0	N, R	Mainly in industrialized countries

Food safety is a major issue in the GMO debate. Potential concerns include alteration in nutritional quality of foods, toxicity, antibiotic resistance, and allergenicity from consuming GM foods. The concepts and techniques used for evaluating food and feed safety have been outlined (WHO, 2005b), but the approval process of GM crops is considered inadequate (Spök et al., 2004). Under current practice, data are provided by the companies owning the genetic materials, making independent verification difficult or impossible. Recently, the data for regulatory approval of a new Bt-maize variety (Mon863) was challenged. Significant effects have been found on a number of measured parameters and a call has been made for more research to establish their safety (Seralini et al., 2007). For example, the systemic broad spectrum herbicide glyphosate is increasingly used on herbicide resistant soybean, resulting in the presence of measurable concentrations of residues and metabolites of glyphosate in soybean products (Arregui et al., 2004). In 1996, EPA reestablished pesticide thresholds for glyphosate in various soybean products setting standards for the presence of such residues in herbicide resistant crop plants (EPA, 1996ab). However, no data on long-term consumption of low doses of glyphosate metabolites have been collected.

3.2.3.2.3 Access to assets

Increased returns from agriculture result in improvements in the educational status of children.

Goals	Certainty	Range of Impacts	Scale	Specificity
L, S, D	B	0 to +4	G	Wide applicability

The successful application of AKST results in improvements in the access of children to education. Enrollment in primary education has increased in developing countries (86% overall). This is highest in southern Asia (89%) but lower in some countries of Africa, western Asia and Oceania (UN, 2006a). Numbers of children out of school are much greater in poor rural areas (30%) than in urban areas (18%); 20% of girls and 17% of boys do not attend primary school. A key factor linking agriculture and education is that women are more likely to invest their assets in children's food and education when they have control of the assets and the benefits from increased productivity (Quisumbing and Maluccio, 1999) (see 3.2.3.4).

Access and rights to natural assets (agricultural, grazing, forest land and water) and the conditions and security of that access, critically affect the livelihoods of many of the world's poorest households.

Goals	Certainty	Range of Impacts	Scale	Specificity
L, E, S	A	-4 to +4	G	Wide applicability

Land tenure systems are dynamic and subject to change; e.g., in situations of population expansion, competition for land for new investment opportunities, urban expansion and road development (Platteau, 1996; Barbier, 1997; Toulmin and Quan, 2000; Chauveau et al., 2006). Differences in access to land resources relate to status and power with migrants, women and people of lower social status being the most vulnerable to expropriation (Blarel, 1994; Jayne et al., 2003). Disputes over land are common in much of Africa

(Bruce and Migot-Adholla, 1994; Place, 1995; Deininger and Castagnini, 2006). Households with land are generally better placed to make productive use of their own resources (especially labor), as well as to access capital for investment (Deininger, 2003). Conversely, land concentration and increasing landlessness may give rise to conflicts and threaten social stability, unless alternative investments and opportunities are available (Gutierrez and Borras, 2004; Mushara and Huggins, 2004; Cotula et al., 2006). In many countries, particularly in sub-Saharan Africa, there are a number of coexisting systems of authority related to land. The main contrast is between customary and statutory law, although these categories mask multiple secondary rights. Security of land tenure is seen as a precondition for intensifying agricultural production and as a prerequisite for better natural resources management and sustainable development and therefore a factor for poverty alleviation (Maxwell and Wiebe, 1998; Mzumara, 2003). Secure tenure is also important to facilitate access to credit and input markets; however, conclusions drawn about the effects of land tenure systems on investment and productivity vary considerably. Policies and programs establishing individual rights in land through land titling have not produced clear evidence showing tenure has led to greater agricultural growth (Quan, 2000), or to improved efficiency (Place and Hazell, 1993). In contrast, without supportive policies, it is difficult for poor small-scale farmers, particularly women, to enter emerging land markets (Toulmin and Quan, 2000; Quan et al., 2005). Despite women's key role in agricultural production, in many countries women's rights over land are less than those of men (Place, 1995; Lastarria-Cornhiel, 1997; Meinzen-Dick et al., 1997; Jackson, 2003). Formal rights to land for women can have an impact on intra-household decision making, income pooling, and women's overall role in the household economy as well as empowering their participation in community decision making (World Bank, 2005b). Government land registration processes have sometimes further entrenched women's disadvantage over land by excluding their rights and interests (Lastarria-Cornhiel, 1997). In some countries, land policy strategies have explored alternatives that limit open access while avoiding the rigidity of individual private ownership and titles; for example management by user groups (Ostrom, 1994) and more open participatory and decentralized policies and institutions for land and land rights management. Regarding water resources, poor communities are often adversely affected by limited access to water for drinking, domestic use, agriculture and other productive purposes. Water access has been improved by institutional and policy innovations in water management and water rights (see 3.2.4.1).

Large scale applications of modern AKST in the water sector have resulted in winners and losers among rural communities.

Goals	Certainty	Range of Impacts	Scale	Specificity
L, E, S, D	B	-3 to + 2	G	Wide applicability

Large scale irrigation schemes have had important impacts on livelihoods. However, while building the value of assets for some, the displacement of populations is one of the notable negative consequences of irrigation schemes, especially where large scale infrastructure has been built. Dams have fragmented and transformed the world's rivers, displacing 40-80 million people in different parts of the world (WCD, 2000). Criteria for land allocation do not necessarily guarantee a place in the irrigated schemes for those who have lost their land and resettlement can result in impoverishment (Cernea, 1999).

Access to energy provides important livelihood benefits and improves opportunities to benefit from AKST.

Goals	Certainty	Range of Impacts	Scale	Specificity
H, L, E, S, D	A	0 to +4	R, N L	Wide applicability

Energy is an essential resource for economic development (DFID, 2002), but more than 1.5 billion people are without access to electricity. In developing countries, approximately 44% of rural and 15% of urban households do not have access to electricity, while in sub-Sahara Africa, these figures increase to 92% and 42% respectively (IEA, 2006c). There is a direct correlation between a country's *per capita* energy consumption (and access) and its industrial progress, economic growth and Human Development Index (UNDP, 2006a). Estimates of the financial benefits arising from access to electricity for rural households in the Philippines were between $81 and $150 per month, largely due to the improved returns on education and opportunity costs from time saved, lower cost of lighting, and improved productivity (UNDP/ESMAP, 2002a). Affordable and reliable rural energy is important in stimulating agricultural related enterprises (Fitzgerald et al., 1990). However, rapid electrification, without the necessary support structures to ensure effectiveness and sustainability, does not bring benefits. Decentralized approaches to electricity provision delivered by private sector, NGOs or community based organizations are presenting viable alternatives that can improve access for rural households.

Improved utilization of biomass energy sources and alternative clean fuels for cooking can benefit livelihoods, especially for women and children.

Goals	Certainty	Range of Impacts	Scale	Specificity
H, L, E, S, D	B	0 to +4	N, L	Mainly developing countries

More than 2.5 billion people use biomass such as fuel wood, charcoal, crop waste and animal dung as their main source of energy for cooking. Biomass accounts for 90% of household energy consumption in many developing countries (IEA, 2006c). Smoke produced from the burning of biomass using simple cooking stoves without adequate ventilation, can lead to serious environmental health problems (Ezzati and Kammen, 2002; Smith, 2006), particularly for women and children (Dasgupta et al., 2004). Women and children are most often responsible for fuel collection, an activity with competes significantly with time for other activities, including agriculture (e.g., 37 hours per household per month in one study in rural India) (UNDP/ESMAP, 2002b). Simple interventions such as improved stoves can reduce biomass consumption by more than 50% and can reduce the effects of indoor smoke (Baris et al., 2006).

The successful achievement of development goals is greatest when social and local organizational development is a key component of technology development and dissemination and when resource poor farmers are involved in problem-solving.

Goals	Certainty	Range of Impacts	Scale	Specificity
L, S,D	B	0 to +3	G	Widespread in developing countries

The social and cultural components of natural resource use and agricultural decision making are fundamental influences on the outcomes from AKST. They operate both at the level of individual actors and decision makers, and at group or community level. Community based approaches have had important results in promoting social cohesion; enhancing governance by building consensus among multiple stakeholders for action around problem issues; and facilitating community groups to influence policy makers (Sanginga et al., 2007). Community based, collective resource management groups that build trust and social capital increasingly common (Scoones and Thompson, 1994; Agrawal and Gibson, 1999; Pretty, 2003). Since the early 1990s, about 0.4-0.5 million local resource management groups have been established. In the US, hundreds of grassroots rural ecosystem place-based management groups have been described as a new environmental movement (Campbell, 1994), enhancing the governance of "the commons" and investment confidence (Pretty, 2003). They have been effective in improving the management of watersheds, forests, irrigation, pests, wildlife conservation, fisheries, micro-finance and farmer's research. In conservation programs, however, there are sometimes negative impacts from social capital; the social exclusion of certain groups or categories or the manipulation of associations by individuals with self-interest (Olivier de Sardan, 1995; Pretty, 2003). When promoting community participation and decision making, it is important to set in place mechanisms to ensure the participation of the most vulnerable or socially excluded groups such as women, the poorest, or those living in remote areas, to ensure their voices are heard and their rights protected (see 3.2.3.3).

Initiatives to enhance social sustainability are strengthened if accompanied by policies that ensure the poorest can participate.

Goals	Certainty	Range of Impacts	Scale	Specificity
L, S	B	0 to +4	G	Widespread in the tropics

Poor people in the community are empowered by programs that build or transfer assets and develop human capital (health care, literacy and employment—particularly in off-farm enterprises) (IDS, 2006; UNDP, 2006b). The alternative and more costly scenario is the mitigation of livelihood and natural resource failure in poor rural areas, through long-term welfare support and emergency relief (Dorward et al., 2004).

3.2.3.2.4 Vulnerability and risk

Although AKST has had many positive impacts, it is now clear that in some circumstances it has also been a strong negative driver/factor for exclusion/marginalization processes.

Goals	Certainty	Range of Impacts	Scale	Specificity
L, S	B	-3 to 0	G	Wide applicability

Although AKST has often had positive benefits on peoples' livelihoods, there have also been negative impacts. Exclusion and marginalization processes such as poverty, hunger or rural migration, have often occurred because of differences in people's capacity to make use of knowledge and technology and to access resources (Mazoyer and Roudart, 1997). These differences are usually the result of discriminatory or exclusionary practices due to gender, class, age or other social variables. The implementation of new technology has implications for social differentiation, sometimes excluding farmers and their families from production and marketing.

Target-oriented programs have responded to this problem by building in awareness of access issues relating to AKST into project design; by monitoring poverty related indicators throughout implementation and through accompanying institutional arrangements.

Impacts of AKST have been more widely evident where they respond to, or are consistent with, the priority that the poor place on managing risk and vulnerability.

Goals	Certainty	Range of Impacts	Scale	Specificity
L, S	B	0 to +3	R, L	Widespread in developing countries

Established cultural traditions define the values and influence practices of small-scale communities. These typically emphasize low-input and risk-averse strategies which are at variance with the maximized production orientation of modern AKST. Small-scale producers make rational decisions to optimize overall benefits from limited resources (Ørskov and Viglizzo, 1994). Thus, risk management, reduction of dependence on agricultural inputs, avoidance of long-term depletion of productive potential and more careful control of environmental externalities are important to them (Conway, 1997). Local knowledge and innovation respond to these priorities; an important assessment criterion of AKST is the extent to which it has helped to reduce both short-term local risk and vulnerability to external factors (e.g., economic changes, climate variability etc). Farmers' own assessment of risk is fundamental in influencing patterns of change in farming practices. High levels of risk are likely to negatively affect adoption (Meinzen-Dick et al., 2004). Perceptions of risk and the priorities of men and women vary in relation to their asset base; especially land and labor.

The risks and costs associated with agriculture and rural development have recently been addressed by innovative microfinance initiatives.

Goals	Certainty	Range of Impacts	Scale	Specificity
L, S, D	D	0 to +3	N,L	Developing countries and poor urban areas of developed countries

Based on successful experiences in various developing countries, a model, termed agricultural microfinance, is emerging (CGAP, 2005). The model combines the most promising features of traditional microfinance, traditional agricultural finance, leasing, insurance, and contracts with processors, traders, and agribusinesses. The original features of the model include innovative savings mechanisms, highly diversified portfolio risk, and loan terms and conditions that are adjusted to accommodate cyclical cash flows and bulky investments. Perhaps two of the most innovative products contributing towards greater rural development are those related to savings and remittances (Nagarajan and Meyer, 2005). Deposits are made to mobile deposit collectors at the savers' doorstep, so reducing the transaction costs of rural farmers and households. Electronic innovations, such as the use of simple mobile phones, ATMs and remittance services, may also help drive down the costs of handling many small transactions in dispersed rural areas, and bring positive benefits to rural communities reliant on migrant labor. Successful remittance services are designed with clients to provide appropriate products and choose strategic partners at both ends of the remittance flow. Despite recent innovations, reaching the remote and vulnerable rural poor still remains a major challenge.

3.2.3.2.5 Livelihood strategies—diversification, specialization and migration

The ways in which rural people combine and use their assets to make a living varies considerably between regions, individuals, households and different social groups. Choice of livelihood strategies is affected by economic, social and cultural considerations (e.g., what is appropriate according to gender, age, status). The range of livelihood choices is generally more restricted for the "asset" poor.

Opportunities for diversification of rural income help to reduce vulnerability of the poor.

Goals	Certainty	Range of Impacts	Scale	Specificity
L, S, D	B	0 to +4	G	Widespread applicability

Where agriculture and natural resources are the basis of livelihoods, small-scale farmers often spread their risks by diversification, as for example in mixed cropping systems (Dixon et al., 2001). Diversification affects agricultural productivity in different ways, in some cases positively (Ellis and Mdoe, 2003). Diversification is a response to an environment which lacks the conditions needed to reap the benefits of agricultural specialization: enterprises with efficient market integration, input and credit supply systems, knowledge access, relatively stable commodity pricing structures and supportive policies (Townsend, 1999). However, diversification is at variance with the emphasis of much agricultural policy in developing countries, which promotes more specialization in the production of high value products for national, regional and export markets. The larger, but lower value, markets for staple food crops are perceived as less risky than higher value markets, and less dependent on technical support services and inputs. Diversification and risk reduction strategies for rural households can include non-farm income; however, this is more difficult for the extreme poor, including female-headed households (Block and Webb, 2001). While there have been advances in rural nonagricultural employment opportunities, women's share in this did not greatly increase between 1990 and 2004 (UN, 2006a). In the general context of rising youth unemployment, young rural women in particular, have difficulty in entering the labor market. Some have argued that the increasing proportion of rural income from non-agricultural sources in Africa is indicative of the failure of agriculture to sustain the livelihoods of the rural poor (Reardon, 1997; Bryceson, 1999; Ellis and Freeman, 2004). There is evidence that the larger the proportion of non-farm to farm income, the larger the overall income.

Where farm size or productivity can no longer sustain the needs of the household, alternative strategies of migration or investment are likely.

Goals	Certainty	Range of Impacts	Scale	Specificity
L, S, D	B	-1 to +3	G R	Particularly in rainfed areas in developing countries

Factors which increase vulnerability constitute severe challenges to the sustainability of livelihoods, e.g., population pressure, land and water shortages, declining productivity due to climate change, collapse of soil fertility, unstable and declining market prices. In these circumstances, some family members, often the young men, migrate to urban centers within or outside their country, in search of employment. These decisions are affected by generational and gender relationships (Chant, 1992; Tacoli, 1998; Bryceson, 1999), and contribute to the "feminization" of agriculture (Song, 1999; Abdelali-Martini et al., 2003), and the increasing dependence of poor rural households on remittances for their survival. Increasingly the migrants include young women, leaving the old and the very young on the farm. In some cases, this has negatively affected agricultural production, food security, and service provision. Labor constraints have encouraged investment in technologies and options which are less demanding in labor, e.g., the establishment of tree crops which are profitable with lower labor inputs (Schreckenberg et al., 2002; Kindt et al., 2004; Degrande et al., 2006). Off-farm remittances have in some cases also encouraged broader investments, e.g., in Andean rural communities, remittances are used for small-scale agriculture, living expenses, and construction and home improvements aimed at the agro-tourism industry (Tamagno, 2003). There is also some evidence for other aspects of more sustainable farming at very high population densities and dependence on migrant community members (see 3.2.2.1.6), combining intensification of production and erosion control (Tiffen et al., 1994; Leach and Mearns, 1996).

3.2.3.3 Participation and local knowledge systems
There is a growing body of work that systematically seeks to assess the impacts of participatory and gender sensitive approaches in agricultural research and development, and the role of local knowledge—for example the Systemwide Program on Participatory Research and Gender Analysis Program of the CGIAR (Lilja et al., 2001, 2004).

3.2.3.3.1 Participatory research approaches

Participatory approaches have developed in response to the lack of economically useful, socially appropriate and environmentally desirable applications from AKST generated by agricultural research and development organizations.

Goals	Certainty	Range of Impacts	Scale	Specificity
L, E, S, D	C	-3 to +2	G	Wide applicability

There is much evidence that the technological advances of the Green Revolution have sometimes led to environmental degradation and social injustice (Conway, 1997). This has stimulated interest in new participatory approaches, methods and techniques to meet sustainability criteria (Engel et al., 1988) and to contribute to a new development paradigm (Jamieson, 1987) targeting development goals (Garrity, 2004) (see Chapter 2). It has required major advances in the analysis of the behavior of the complex social systems found in rural communities. The growing interest in participatory approaches from the 1980s onwards, was in part a response to the contrast in the successes of Green Revolution technology in some contexts and its lack of, or negative, impact in others, particularly those characterized by high diversity, inaccessibility and weak institutions and infrastructure (Haverkort et al., 1991; Okali et al., 1994; Scoones and Thompson, 1994; Röling and Wagemakers, 1998; Cerf et al., 2000). Participatory approaches, in which development agencies and technical specialists participate, use existing local skills and knowledge as the starting point (Croxton, 1999). They are built around a process that enables farmers to control and direct research and development to meet their own needs and to ensure a sense of ownership in decisions and actions (Engel et al., 1988). The main advantages of participatory approaches have been their responsiveness to local ecological and socioeconomic conditions, needs and preferences; building on local institutions, knowledge and initiatives and fostering local organizational capacity. Criticisms have focused on their resource requirements, the difficulties of scaling-up successes from small focus areas (Cooke and Kothari, 2001), the lack of radical change in institutional relationships and knowledge sharing, and the limited engagement with market and policy actors.

Participatory approaches to genetic improvement of crops and animals results in better identification of farmer's requirements and preferences, leading to higher levels of adoption and benefit.

Goals	Certainty	Range of Impacts	Scale	Specificity
N, L, S	B	0 to +3	G, N, L	Wide applicability

In cereals and legumes, participatory approaches have been promoted in response to perceived weaknesses in conventional variety testing and formal release procedures which have not delivered suitable varieties to farmers in marginal environments, especially, but not exclusively, small-scale farmers (Witcombe et al., 1998). Formal release systems are often centralized, use a research station or other atypically favorable environment for testing, and select for average performance. Farmers or consumers are also rarely involved in this process. Consequently, varieties from these conventional release systems are often poorly adapted to small-scale farmer conditions and environments. Similarly, they have not always met the farmers' requirements for multipurpose uses (e.g., fodder and seed), or have not had acceptable postharvest characteristics (e.g., easy to thresh/process, good taste, good storability). Participatory crop development allows for the better identification of farmer preferences and the requirements of their systems of production as well as optimizing local adaptation through the capture of Genotype X Environment interactions. Genetic diversity can also benefit from participatory approaches as farmers usually select and introduce cultivars that are unrelated to the modern varieties already grown (Witcombe et al., 2001). Other benefits of the participatory approach include a shortened breeding cycle in which new varieties are grown by farmers prior to the 12-15 year period of formal multilocational testing and release. This considerably increases the cost-benefit ratios, net present value and net social benefit (Pandey and Rajatasereekul, 1999). Another benefit of participatory breeding is enhanced compatibility with local or informal seed systems, which is especially important in times of extreme climatic and other stresses. Participatory approaches in livestock research have responded to criticisms that technologies were developed but seldom delivered, or if delivered, did not benefit poor farmers/herders (Hefferman et al., 2005) and have demonstrated the importance of understanding the particular needs and circumstances of resource poor farmers, building on local knowledge. These approaches have been more appropriate to farmer circumstances and are more likely to be adopted (Catley et al., 2001; Conroy, 2005); however, the benefits for crop and livestock sectors are largely experienced at local or regional levels, and the problem of scaling-up remains.

Participatory approaches have been successfully developed for the domestication of indigenous trees for integration into agroforestry systems.

Goals	Certainty	Range of Impacts	Scale	Specificity
N, H, L, E, S,D	D, E	0 to +2	R	Especially relevant to the tropics

Throughout the tropics local tree species provide traditional foods and medicines (Abbiw, 1990; Villachica, 1996; Leakey, 1999a; Walter and Sam, 1999; Elevitch, 2006) many of which are marketed locally (Shackleton et al., 2007). Some of these species are being domesticated using a participatory approach to cultivar production (Leakey et al., 2003; Tchoundjeu et al., 2006), using simple and appropriate vegetative propagation methods (Leakey et al., 1990) so that local communities are empowered to create their own opportunities to enter the cash economy (Leakey et al., 2005a) (see 3.2.1.2.1 and 3.2.2.1.6). The use of participatory approaches ensures that the benefits of domestication accrue to the farmers. In this respect, these techniques are in accordance with the Convention on Biological Diversity (Articles 8 and 15) and provide a politically and socially acceptable form of biodiscovery. It is clear that this approach is also encouraging the rapid adoption of both the techniques and the improved cultivars (Tchoundjeu et al., 2006).

Participatory approaches are important in addressing knowledge-intensive, complex natural resource management problems.

Goals	Certainty	Range of Impacts	Scale	Specificity
L, S	D, E	0 to +2	G	Widespread applicability

In an impact assessment of participatory approaches to development of cassava based cropping systems in Vietnam and Thailand (Dalton et al., 2005), participating farmers gained additional yield benefits, compared with those who merely adopted the new planting material. The integration of management practices into the participatory learning activities resulted in a better understanding of the interrelationships between system components and led to efficiency gains.

Community entry and participatory approaches have higher initial costs, but improved efficiency in technology development, capacity strengthening and learning.

Goals	Certainty	Range of Impacts	Scale	Specificity
L, E, S, D	B, E	+2	N, L	Subsistence households of the semiarid tropics.

Crop management research increasingly involves farmers in the participatory evaluation of new technologies, identifying adoption constraints and opportunities for improving farm performance to produce more sustainable impact. Between 1999 and 2001, ICRISAT and its partners in Malawi and Zimbabwe evaluated the impact of participatory research in connection with a range of "best bet" soil fertility and water management technologies. The main findings were that community entry and participatory approaches that engage farmers in decision making throughout the research-development-diffusion-innovation process improved efficiency and impact, both through the development of relevant technology and in building farmers' capacity for experimentation and collective learning, but that these benefits had higher initial costs than traditional approaches (Rusike et al., 2006, 2007). The study recommended that public and NGO investments be targeted to build wider-scale district and village-level innovation clusters to make the projects more sustainable over a larger area. Similarly, in Colombia, participatory approaches with local agricultural research committees showed significant social and human capital benefits for members (http://www.prgaprogram.org/index.php?module=htmlpages&func=display&pid=12). However, in Honduras, where educational levels were lower and poverty higher, it was found that the process took longer; because of the need for more intensive assisted learning and social development to support the participatory technology component (Humphries et al., 2000).

3.2.3.3.2 Indigenous knowledge and innovation systems

The complex and dynamic interactions between culture, society and nature and its resources are central to social and environmental sustainability.

Goals	Certainty	Range of Impacts	Scale	Specificity
N, L, E, S	B	0 to +4	G	Worldwide

Culture and tradition are important components of social sustainability. Traditional and local knowledge are part of culture and belief systems and codified in oral forms and in cultural and religious norms. These cultural meanings are embedded in local people's understanding of the environment, the management of natural resources and agricultural practice (Warren et al., 1995; Posey, 1999). Yams are a staple crop of economic and cultural significance for the people in West Africa. For example, yams (*Dioscorea* spp.) play a vital role in society in the Dagomba ethnic group in north Ghana. About 75% of farmers in the northern region cultivate yam, as part of the African "yam zone" (Cameroon to Côte d'Ivoire) that produces 90% or 33.7 million tonnes of the world's yams each year. During the celebration of the yam festival boiled yams are smeared on the surface of stones to secure the goodwill and patronage of deities. The Dagomba invoke their gods during the communal labor through which they exchange yam germplasm. Seed yam obtained through communal labor enjoys the blessing of the gods and produces high yields according to tradition. For the Dagomba, the yam has transcended agriculture to become part of the society's culture (Kranjac-Berisavljevic and Gandaa, 2004). Failure to recognize this would result in (1) the breakdown of traditional social structure; and (2) the loss of valuable yam germplasm in many cases.

The knowledge of many indigenous communities has provided almost all their basic food, fibre, health and shelter needs as well as some products for cash income.

Goals N, H, L, E, S	Certainty A	Range of Impacts +2 to +5	Scale G	Specificity Worldwide

Typically, traditional and local KST has been developed through observation and experimentation, over many cycles, to achieve efficient and low-risk human welfare outcomes (Warren et al., 1995). A wide range of local institutions are significant in developing, disseminating and protecting this knowledge as it differs greatly from the specialized knowledge used by research and extension institutions working with agricultural science (Warren et al., 1989). The traditional actors harbor distrust for mainstream organizations and are comparatively marginalized by them. Consequently, identifying an appropriate and acceptable means of making use of traditional knowledge and protecting the valuable rights of indigenous communities to their traditional knowledge is a priority if this knowledge is not to be lost, and if the communities are to benefit (ten Kate and Laird, 1999). A good example is the patent protecting the rights of women in Botswana to traditional knowledge associated with Marula kernel oil (www.phytotradeafrica.com/awards/criteria.htm).

The important role of livestock for poor people's livelihoods has been sustained primarily through the effectiveness of indigenous knowledge.

Goals N, L, H, E, S, D	Certainty C	Range of Impacts 0 to + 4	Scale L, N	Specificity Especially in the tropics

Livestock are an important asset of many poor people, particularly in sub-Saharan Africa and south Asia (Thornton et al., 2002, 2004), providing a source of food, cash income,

manure and draft power and strengthening their capacity to cope with income shocks (Ashley et al., 1999; Heffernan and Misturelli, 2001). In India, for example, livestock holdings are more equitably distributed than land holdings (Taneja and Birthal, 2003). Livestock ownership directly and indirectly affects the nutritional status of children in developing countries (Tangka et al., 2000). In Africa, the livestock sector, particularly in arid and semiarid areas, depends to a large extent on traditional and local knowledge for animal management and animal breeding (Ayantude et al., 2007, but receives little investment in international and national research. The depth of local knowledge has advantages when developing localized initiatives, for example, in animal feeding and forage production. Productivity in animal agriculture systems can be increased under dry conditions without great external inputs (Lhoste, 2005). Participatory methods for diagnosis of animal diseases have also shown promise, both in characterization of diseases and the linkages between local knowledge and modern veterinary knowledge (Catley et al., 2001). Such participatory local analysis has been used to develop control programs adapted to local conditions and knowledge (Catley et al., 2002).

3.2.3.3.3 Linking scientific and indigenous knowledge and management capability

Significant gains have been made when farmer innovation (particularly in small-scale agriculture) is appropriately linked to formal AKST.

Goals L, S	Certainty B	Range of Impacts 0 to +4	Scale G	Specificity Especially in the tropics

Formal research and extension organizations have often not recognized the contribution of farmers' knowledge and strategies (Richards, 1985; Sibelet, 1995). However, there are good examples in plant breeding where farmers have communicated their local knowledge to researchers, and worked together in experimentation and decision making (Hocdé, 1997), researchers and stakeholders jointly designing experimentation, sharing and validating results (Liu, 1997; Gonzalves et al., 2005; Liu and Crezé, 2006). Agroforestry researchers working with farmers have investigated progressively more complex issues together, integrating biophysical and socioeconomic disciplines to resolve the sustainability problems in areas where poverty and environmental degradation coexist. This has required a unique mixture of new science (Sanchez, 1995) with local understanding of the day-to-day concerns of resource-poor farmers; the approach enhances the adoption of new ideas and technologies (Franzel and Scherr, 2002). Innovations like these evolve as a result of collective learning as well as from the pressure to constantly adapt to the changing economic environment.

The influence of social institutions on land management, based on local knowledge and norms, may be undermined by policies based on the different perspectives of professionals.

Goals L, E, S	Certainty C	Range of Impacts -3 to 0	Scale L	Specificity Widespread applicability

Local knowledge, and the local institutions associated with it, have been regarded as an important foundation for community-based natural resource management and biodiversity conservation. However, this has been challenged as a romantic view, dependent on conditions of low population density, lack of modern technology and limited consumer demand (Attwell and Cotterill, 2000). The overexploitation of natural capital has been widely attributed to a number of factors, including the loss of social institutions at the community level. In some cases this arises from changes in local systems of administration and governance. In India, the breakdown of regulations on livestock resulted in unregulated grazing (Pretty and Ward, 2001), while water resource degradation followed the replacement of collective irrigation systems by private ownership. Similarly, the failure of many formal attempts to halt rotational shifting cultivation in Thailand, Laos and Vietnam was, at its most fundamental level, associated with differing perspectives. That is, "policy makers believed that shifting cultivation was the main cause of environmental problems such as floods and landslips" (Bass and Morrison, 1994) whereas others recognized the dynamic and diverse types of shifting cultivation in which farmers engaged, and the associated economic, social, cultural and environmental values.

Institutions are crucial for sustainable development; the innovation systems approach offers more insights than previous paradigms into the complex relationships of technology development and diffusion.

Goals	Certainty	Range of Impacts	Scale	Specificity
L, S	B	0 to +3	G	Widespread applicability

The linear model of research and extension in which innovations are transferred as products from researchers to farmers via intermediaries in extension, has been challenged by experience showing that the pathways for technical changes are more diverse. In the last 15 years, the importance of knowledge in innovation processes has been more clearly recognized (Engel and Röling, 1989; Röling, 1992). Knowledge is considered as a factor of production; considered by some to be more important than land, capital and labor. More recent approaches view innovation as a complex social process (Luecke and Katz, 2003) which takes multiple forms and involves the participation and interaction of a diversity of key actors and organizations (Sibelet, 1995; Spielman, 2005). These relationships or networks, "the innovation system", operate within specific institutional and cultural contexts. Similarly, evaluation approaches have shifted from focusing on impacts of research to tracking the institutional changes and effective operation of the innovation systems (Hall et al., 2003). The innovation systems approach emphasizes continuous learning and knowledge flows, interaction of multiple actors and institutional change. Innovation Systems thinking has encouraged greater awareness of the complexity of these relationships, the processes of institutional learning and change, market and non-market institutions, public policy, poverty reduction, and socioeconomic development (Hall et al., 2003; Ferris et al., 2006). However, the approach does not explicitly engage with poverty and development agendas by examining the relationship between innovation systems, economic growth and the

distributional effects on poverty reduction and policy options which would support this (Spielman, 2005).

Devolution of resource management to local institutions has been successful where targeted support and enabling conditions were in place.

Goals	Certainty	Range of Impacts	Scale	Specificity
L, E, S, D	B	0 to +3	G	Widespread applicability

Local institutions have the capacity to manage local resources and avert possible "tragedies of the commons" (Ostrom, 1992). Rules can be created to accommodate the heterogeneity found within communities (Agrawal and Gibson, 1999; Ostrom, 2005) and there are opportunities for co-management with government (Balland and Platteau, 1996). In conservation programs, the participation of the range of stakeholders in consensus building and consideration of benefit distribution reduces the risk of conflict and the costs of implementation and control, and increases the chances of sustainability (Borrini-Feyerabend, 1997; Guerin, 2007). In some cases, (e.g., the transfer of irrigation management to communities) the drive to establish local management has led to rigid, hierarchical user associations with functional and democratic shortcomings (Agrawal and Gupta, 2005). However, research in the irrigation sector has identified that a supportive legal policy framework, secure water rights, local management capacity development and favorable cost/benefit relationships, are conditions favoring the successful transfer of management to communities (Shah et al., 2002). These characteristics encourage farmers' contributions and create a strong sense of ownership, which together lead to better subsequent operations and maintenance (Bruns and Ambler, 1992). Finally, research has shown the diversity and complexity of water rights in many developing countries and the importance of recognizing both formal legal rights and customary or indigenous rights in a "pluralistic" approach (Bruns, 2007).

Local or informal seed systems provide most seed used by farmers and are increasingly being used to deliver new varieties to farmers.

Goals	Certainty	Range of Impacts	Scale	Specificity
1	A	+1 to +5	R	Developing countries

Nearly all developing country farmers depend on their own seed, or seed obtained locally from relatives or markets, for planting (Almekinders and Louwaars, 1999; Tripp, 2001). In contrast, most new varieties released in developing countries originate from public sector organizations, Hybrid maize is the exception; it originates from the private sector and seeds are delivered through commercial networks (Morris, 2002), although these are not tailored to specific local situations. Local seed systems are therefore very important. Typically they support the local economy and are very robust and effective. Studies in India have shown that seed can move many kilometers through these informal systems, and that local entrepreneurs quickly act to meet a demand for seed (Witcombe et al., 1999). Consequently, a number of initiatives have built on informal seed systems to distribute seed. For example, relief agencies promote these systems by using seed vouchers in times of drought or civil unrest (Sperling et al., 2004). The Program for Africa's Seed Systems (funded by the Bill and Melinda Gates Foundation) is promoting the distribution of improved crops varieties through private and public channels, including community seed systems.

Scaling up the adoption of new technologies requires new approaches to partnerships and information sharing.

Goals	Certainty	Range of Impacts	Scale	Specificity
L, S	C	-2 to +2	G	Widespread applicability

Adoption and impact of new agricultural technologies have been negatively affected by overlooking the human/cultural issues, ignoring local knowledge systems, and reducing the solution of agricultural problems to merely technology (Feder et al., 1985). The factors affecting adoption of technological innovations are numerous and complex. The interaction of technologies with the economic, social, cultural and institutional context influences the scale of adoption (Feder et al., 1985). Factors shown to affect adoption include complementarity with existing systems and practices, the relative "profitability" and benefits of alternative technologies; and the incentives of the policy environment. Partnership networks and information sharing are needed for scaling up (Lilja et al., 2004); this is particularly important in non-seed based knowledge intensive technologies.

3.2.3.4 Learning and capacity strengthening
A key factor for widescale adoption of new AKST is the dissemination of information to the farmers by extension, farmer training and information management. Recent advances in ICT provide important new tools.

3.2.3.4.1 Extension and training

Education and training contribute to national economic wellbeing and growth.

Goals	Certainty	Range of Impacts	Scale	Specificity
L, S, D	B	0 to +4	G	Widespread applicability

Countries with higher levels of income generally have higher levels of education. Human capital, which includes both formal education and informal on-the-job training, is a major factor in explaining differences in productivity and income between countries (Hicks, 1987). Agricultural education plays a critical role in the transfer of technology and agricultural extension makes an important economic contribution to rural development (Evenson, 1997). Agricultural centers of excellence are yielding new technologies, and agricultural education is assisting with technology transfer activities by being part of interdisciplinary research programs. Informal mechanisms for information sharing, such as farmer-to-farmer models of agricultural development, are increasing in importance (Eveleens et al., 1996).

A better understanding of the complex dynamic interactions between society and nature is strengthening capacity for sustainable development.

Goals	Certainty	Range of Impacts	Scale	Specificity
L, E, S	B	0 to +3	G	Widespread applicability

Formal capacity development in developing countries goes beyond disciplinary expertise. It produces broad-based professionals that recognize the "systems" nature of innovation and change, and its relationships with society (Pretty, 2002; FAO, 2005c). This is needed because of the interlinking of sociological, cultural, agricultural and environmental issues and the differing and often conflicting land use needs and strategies of a multiplicity of stakeholders. Innovative methods and tools can effectively improve coordination, mediation and negotiation processes aimed at more decentralized and better integrated natural resources management (D'Aquino et al., 2003). The combined use of modeling and role-playing games helps professionals and stakeholders to understand the dynamics of these interactions (Antona et al., 2003).

Lack of appropriate education/extension and learning opportunities are a constraint to technology transfer, trade and marketing, and business development.

Goals	Certainty	Range of Impacts	Scale	Specificity
N, H, L, E, S, D	A	-5 to 0	G	Worldwide

Many developing countries have large numbers of illiterate people. This is a constraint to economic and social development, as well as agriculture (Ludwig, 1999). Some important goals include the rehabilitation of university infrastructures, particularly information and communication facilities; organizational designs that link institutions of higher education to hospitals, communities, research stations, and the private sector; and curricula and pedagogy that encourage creativity, enquiry, entrepreneurship and experiential learning (Juma, 2006).

Gender imbalances in agricultural extension, education and research systems limit women's access to information, trainers and skills.

Goals	Certainty	Range of Impacts	Scale	Specificity
L, S	A	-2 to +4	G	Worldwide

There is a severe gender imbalance in agricultural extension services (Swanson et al., 1990; FAO, 1995; FAO, 2004a). Women constitute only 12.3% of extension workers in Africa (UN, 1995). Sensitivity to gender issues and vulnerable populations (disabled, HIV/AIDS affected, youth etc.) can determine the success or failure of training/extension activities. The number of women seeking higher education in agriculture is increasing in some developing countries, although female enrolment rates remain considerably lower than males (FAO, 1995). More women are now employed in national agricultural institutions than in the 1980s, but men still comprise the overwhelming majority of those employed, especially occupying in managerial and decision making positions (FAO, 1995).

In Africa, expenditures related to agriculture and extension have been reduced in quantity and quality, thereby affecting productivity.

Goals	Certainty	Range of Impacts	Scale	Specificity
N, L, E, S, D	B	-3 to 0	R	Africa

There has been a decline in government funding to agricultural extension services in many developing countries (Alex et al., 2002; Rusike and Dimes, 2004). In the past, extension services financed by public sector (Axinn and Thorat, 1972; Lees, 1990; Swanson et al., 1997) were a key component of the Green Revolution. Today, two out of three farmers in Africa, particularly small-scale farmers and women farmers (FAO, 1990), have no contact with extension services, and worldwide publicly funded extension services are in decline. Critics of public extension claim that its services need to be reoriented, redirected and revitalized (Rivera and Cary, 1997) as the poor efficiency of traditional extension systems has undermined interest in them (Anderson et al., 2004).

Both public and private delivery services can provide agricultural extension for modern farming.

Goals	Certainty	Range of Impacts	Scale	Specificity
N, H, L, E, S, D	B	0 to +5	G	Worldwide

There is a trend towards the privatization of extension organizations, often as parastatal or quasi-governmental agencies, with farmers asked to pay for services previously received free (FAO, 1995; FAO, 2000a; Rivera et al., 2000). This trend is stronger in the North than the South (Jones, 1990; FAO, 1995). Inclusion of the private sector can ensure competition and increase the efficiency of agricultural service delivery, especially with regard to agricultural input-supply firms (Davidson et al., 2001). However, problems exist in terms of incentives and stakeholder roles. In Southern Africa, private sector led development showed that private firms have significant potential to improve small-scale crop management practices and productivity by supplying farmers with new cultivars, nutrients, farm equipment, information, capital, and other services. However, market, institutional, government, and policy failures currently limit expanded private sector participation (Rusike and Dimes, 2004).

The participation of a broad range of information providers on agricultural technologies, policies and markets, has been shown to play an important role in sustainable agricultural development.

Goals	Certainty	Range of Impacts	Scale	Specificity
N, L, E, S, D	B	0 to +3	G	Wide applicability

Currently, countries in Africa are searching for participatory, pluralistic, decentralized approaches to service provision for small-scale farmers. The private sector, civil society organizations and national and international NGOs are increasingly active in agricultural research and development (Rivera and Alex, 2004), supporting local systems that enhance the capacity to innovate and apply knowledge. In the poorest regions, NGOs have strengthened their extension activities with poor farmers by using participatory approaches and developing initiatives to empower farmer organizations (Faure and Kleene, 2004).

Community based participatory learning processes and farmer field schools (FFS) have been effective in enhancing skills and bringing about changes in practice.

Goals	Certainty	Range of Impacts	Scale	Specificity
L, S	B	0 to +3	G	Wide applicability

Agricultural extension and learning practitioners are increasingly interested in informal and community based participatory learning for change (Kilpatrick et al., 1999; Gautam, 2000; Feder et al., 2003). Group learning and interaction play an important role in changing farmer attitudes and increases the probability of a change in practice (AGRITEX, 1998) by recognizing that farming is a social activity, which does not take place in a social or cultural vacuum (Dunn, 2000). In Kamuli district in Uganda, a program to strengthen farmers' capacity to learn from each other, using participatory methods and a livelihoods approach, found that farmer group households increased their production and variety of foods, reduced food insecurity and the number of food insecure months and improved nutritional status (Mazur et al., 2006; Sseguya and Masinde, 2006).

Farmer field schools (FFS) have been an important methodological advance to facilitate learning and technology dissemination (Braun et al., 2000; Thiele et al., 2001; van den Berg, 2003). Developed in response to overuse of insecticides in Asia rice farming systems, they have become widely promoted elsewhere (Asiabaka, 2002). In FFS, groups of farmers explore a specific locally relevant topic through practical field-based learning and experimentation over a cropping season. Assessments of the impacts of farmer field schools have generally been positive, depending on the assumptions driving the assessment. FFS have significantly reduced pesticide use in rice in Indonesia, Vietnam, Bangladesh, Thailand, and Sri Lanka (where FFS farmers used 81% fewer insecticide applications), and in cotton in Asia (a 31% increase in income the year after training, from better yields and lower pesticide expenditure) (Van den Berg et al., 2002; Tripp et al., 2005). Opinions on positive impacts are not unanimous (Feder et al., 2004). Farmer field schools have been criticized for their limited coverage and difficulty in scaling up; the lack of wider sharing of learning, their cost in relation to impact (Feder et al., 2004), the lack of financial sustainability (Quizon et al., 2001; Okoth et al., 2003), the demands on farmers' time and the failure to develop enduring farmer organizations (Thirtle et al., 2003; Tripp et al., 2005; Van Mele et al., 2005). However, there are few alternative models for advancing farmers' understanding and ability to apply complex knowledge intensive technologies. There is potential for FFS to self-finance in some cases (Okoth et al., 2003). FFS can stimulate further group formation (Simpson, 2001), but sharing local knowledge and sustaining relationships with different stakeholder groups post-FFS has often not been given sufficient attention (Braun et al., 2000).

International organizations are training community workers and promoting important participatory approaches to rural development.

Goals	Certainty	Range of Impacts	Scale	Specificity
N, H, L, E, S,	B	0 to +2	G	Widespread in developing countries

The World Agroforestry Centre is an example of one international institution which is providing training to farmers, through mentorship programs with Farmer Training Schools, scholarships for women's education, support of young professionals in partner countries and the development of Networks for Agroforestry Education, e.g., ANAFE (124 institutions in 34 African countries) and SEANAFE (70 institutions in 5 South East Asian countries) (Temu et al., 2001). Similarly, agencies such as the International Foundation for Science (www.ifs.se), and the Australian Centre for International Agricultural Research (www.aciar.gov .au), provide funds to allow graduates trained overseas to reestablish at home. At IITA in West Africa, the Sustainable Tree Crops Program is training groups of Master Trainers, who then train "Trainers of Trainers", and eventually groups of farmers in the skills needed to grow cocoa sustainably (STCP Newsletter, 2003). The results of this initiative are promising (Bartlett, 2004; Berg, 2004), but there still remain crucial problems related to (1) the need for strong farmers' governance to monitor and assess extension activities, (2) sustainable funding with fair cost sharing between the stakeholders including the State, private sector, farmer organizations, and farmers, and (3) the need for Farmer field training to evolve into community-based organizations, to enable the community to continue benefiting on a sustained basis from the momentum created (Mancini, 2006).

Environmental and sustainable development issues are being included in extension programs.

Goals	Certainty	Range of Impacts	Scale	Specificity
N, H, L, E, S	B	-2 to +4	G	Wide applicability

Extension services are now including a larger number of stakeholders that are not farmers in their target groups. Increasingly environmental and sustainable development issues are being incorporated into agricultural education and extension programs (FAO, 1995; van Crowder, 1996; Garforth and Lawrence, 1997).

3.2.3.3.2 Information management
ICTs are increasingly being used to disseminate agricultural information, but new techniques require new forms of support.

Proper information management is frequently a key limiting factor to agricultural development.

Goals	Certainty	Range of Impacts	Scale	Specificity
N, H, L, E, S, D	B	-4 to +4	G	Worldwide

Information access is limited in low-income countries, but farmers have an array of informal and formal sources (extension leaflets, television, mobile films, etc.) from which they obtain information (Nwachukwu and Akinbode, 1989; Olowu and Igodan, 1989; Ogunwale and Laogun, 1997). In addition, village leaders, NGO agents and farmer resource centers are used as information hubs so that information and knowledge about new technologies and markets diffuse through social networks of friends, relatives and acquaintances (Collier, 1998; Conley and Udry, 2001; Fafchamps and Minten, 2001; Barr, 2002). Inevitably, issues of equitable access and dissemination arise as marginalized populations tend to be bypassed (Salokhe et al., 2002). The challenge is how to improve accessibility of science and technology information to contribute to agricultural development and food security. This challenge is multidimensional, covering language issues as well as those of intellectual property and physical accessibility (World Bank, 2002; Harris, 2004).

ICTs are propelling change in agricultural knowledge and information systems, allowing the dissemination of information on new technologies, and providing the means to improve collaboration among partners.

Goals	Certainty	Range of Impacts	Scale	Specificity
N, H, L, E, S, D	B	0 to +4	G	Worldwide

Information and communication technologies (ICT) are revolutionizing agricultural information dissemination (Richardson, 2006). Since the advent of the internet in the 1990s, communications technologies now deliver a richer array of information of value to farmers and rural households (Leeuwis, 1993; Zijp, 1994; FAO, 2000c); extension services deliver information services interactively between farmers and information providers (FAO, 2000c) via rural telecenters, cellular phones, and computer software packages. Important ICT issues in rural extension systems include private service delivery, cost recovery, and the "wholesaling" of information provided to intermediaries (NGOs, private sector, press, and others) (Ameur, 1994). In rural areas, ICTs are now used to provide relevant technical information, market prices, and weather reports. The Livestock Guru™ software program was created as a multimedia learning tool which enables farmers to obtain information on animal health and production and has had greater impact than more conventional media, illustrating the potential of these tools to help meet global agricultural and poverty alleviation objectives (Heffernan et al., 2005; Nielsen and Heffernan, 2006). ICTs help farmers to improve labor productivity, increase yields, and realize a better price for their produce (www.digitaldividend.org/pubs/pubs_01_overview.htm). A market information service in Uganda has successfully used a mix of conventional media, Internet, and mobile phones to enable farmers, traders, and consumers to obtain accurate market information resulting in farmer control of farm gate prices (http://www.comminit.com/strategicthinking/st2004/thinking-579.html). Similar services exist in India, Burkino Faso, Jamaica, Philippines and Bangladesh (www.digitaldividend.org/pubs/pubs_01_overview.htm). ICT also provides the opportunity to create decision support systems such as e-consultation or advisory systems to help farmers make better decisions. ICT facilitates smooth implementation of both administrative and development undertakings. However with these ICT advances comes the task of managing and disseminating information in an increasingly complex digital environment.

Advances in information technology are providing more tools for agricultural information management.

Goals	Certainty	Range of Impacts	Scale	Specificity
N, H, L, E, S, D	B	-2 to +3	G	Wide applicability

Due to advances in ICT, international organizations such as FAO have been able to respond to the need for improved information management by providing technical assistance in the form of information management tools and applications, normally in association with advice and training (http://www.fao.org/waicent). Agricultural thesauri like AGROVOC are playing a substantial role in helping information managers and information users in document indexing and information retrieval tasks.

ICTs have widened the "digital divide" between industrialized and developing countries, as well as between rural and urban communities.

Goals	Certainty	Range of Impacts	Scale	Specificity
N, H, L, E, S, D	B	-2 to +3	G	Wide applicability

Although ICT improve information flow, not all people have equal access to digital information and knowledge of the technology creating a "digital divide", a gap between the technology-empowered and the technology-excluded communities (http://www.itu.int/wsis/basic/faqs.asp; Torero and von Braun, 2006). Digital information is concentrated in regions where information infrastructure is most developed, to the detriment of areas without these technologies (http://www.unrisd.org). This, together with the ability of people to use the technology, has had an impact on the spread of digital information (Herselman and Britton, 2002). The main positive impacts on poverty from ICTs have been from radio and from telephone access and use, with less clear impacts evident for the internet (Kenny, 2002).

3.2.3.4 Gender

Farming practices are done by both men and women, but the role of women has typically been overlooked in the past. Resolving this inequity has been a major concern in recent years. For social and economic sustainability, it is important that technologies are appropriate to different resource levels, including those of women and do not encourage others to dispossess women of land or commandeer their labor or control their income (FAO, 1995; Buhlmann and Jager, 2001; Watkins, 2004).

Women play a substantial role in food production worldwide.

Goals	Certainty	Range of Impacts	Scale	Specificity
N, H, L, E, S, D	B	-3 to +3	G	Worldwide

In Asia and Africa women produce over 60% and 70% of the food respectively, but because of inadequate methodological tools, their work is underestimated and does not normally appear as part of the Gross National Product (GNP) (Kaul and Ali, 1992; Grellier, 1995; FAO, 2002b; CED, 2003; Quisumbing et al., 2005; Diarra and Monimart, 2006). Similarly, women are not well integrated in agricultural education, training or extension services, making them "invisible" partners in development. Consequently, women's contribution to agriculture is poorly understood and their specific needs are frequently ignored in development planning. This extends to matters as basic as the design of farm

tools. The key importance of the empowerment of women to raising levels of nutrition, improving the production and distribution of food and agricultural products and enhancing the living conditions of rural populations has been acknowledged by the UN (FAO, www.fao.org/gender).

Mainstreaming gender analysis in project design, implementation, monitoring and policy interventions is an essential part of implementing an integrated approach in agricultural development.

Goals	Certainty	Range of Impacts	Scale	Specificity
N, H, L, E, S, D	B	0 to +3	G	Wide applicability

The substantial roles of resource poor farmers such as women and other marginalized groups are often undervalued in agricultural analyses and policies. Agricultural programs designed to increase women's income and household nutrition have more impact if they take account of the cultural context and spatial restrictions on women's work as well as patterns of intra-household food distribution. The latter often favors males and can give rise to micronutrient deficiencies in women and children. The deficiencies impair cognitive development of young children, retard physical growth, increase child mortality and contribute to the problem of maternal death during childbirth (Tabassum Naved, 2000). Income-generating programs targeting women as individuals must also provide alternative sources of social support in order to achieve their objectives. In Bangladesh, an agricultural program aimed at improving women's household income generated more benefits from a group approach for fish production than from an individual approach to homestead vegetable production. The group approach enabled women members to overcome the gender restrictions on workspace, to increase their income and control over their income and to improve their status. In many countries of Asia, Africa, and Latin America, privatization of land has accelerated the loss of women's land rights. Titles are reallocated to men as the assumed heads of households even when women are the acknowledged household heads. Women's knowledge, which is critical to S&T and food security, becomes irreparably disrupted or irrelevant as a result of the erosion or denial of their rights (Muntemba, 1988; FAO, 2005d).

The feminization of agriculture places a burden on women who have few rights and assets.

Goals	Certainty	Range of Impacts	Scale	Specificity
L, S	B	-3 to 0	G	Especially in the tropics

Progress on the advancement of the status of rural women has not been sufficiently systematic to reverse the processes leading to the feminization of poverty and agriculture, to food insecurity and to reducing the burden women shoulder from environmental degradation (FAO, 1995). The rapid feminization of agriculture in many areas has highlighted the issue of land rights for women. Women's limited access to resources and their insufficient purchasing power are products of a series of interrelated social, economic and cultural factors that force them into a subordinate role to the detriment of their own development and that of society as a whole (FAO, 1996). The contribution of women to food security is growing as men migrate to the city, or neighboring rural areas, in search of paid jobs leaving the women to do the farming and to provide food for the family (FAO, 1998b; Song, 1999).

At the institutional and national levels, policies that discriminate against women and marginalized people affect them in terms of access to and control over land, technology, credit, markets, and agricultural productivity.

Goals L, S	Certainty B	Range of Impacts -2 to +3	Scale G	Specificity Common occurrence

Women's contribution to food security is not well reflected in ownership and access to services (Bullock, 1993; FAO, 2005c: FAO, 2006c). Fewer than 10% of women farmers in India, Nepal and Thailand own land; while women farmers in five African countries received less than 10% of the credit provided to their male counterparts. The poor availability of credit for women limits their ability to purchase seeds, fertilizers and other inputs needed to adopt new farming techniques. Although this is slowly being redressed by special programs and funds created to address women's particular needs, women's access to land continues to pose problems in most countries. In Africa, women tend to be unpaid laborers on their husbands' land and to cultivate separate plots in their own right at the same time. However, while women may work their own plots, they may not necessarily have ownership and thus their rights may not survive the death of their spouse (Bullock, 1993). In the case of male migration and *de facto* women heads of households, conflicts may arise as prevailing land rights rarely endow women with stable property or user rights (IFAD, 2004). Traditionally, irrigation agencies have tended to exclude women and other marginalized groups from access to water —for example, by requiring land titles to obtain access to irrigation water (Van Koppen, 2002). Explicitly targeting women farmers in water development schemes and giving them a voice in water management is essential for the success of poverty alleviation programs. There are insufficient labor-saving technologies to enable women's work to be more effective in crop and livestock production. Armed conflict, migration of men in search of paid employment and rising mortality rates attributed to HIV/AIDS, have led to a rise in the number of female-headed households and an additional burden on women. Women remain severely disadvantaged in terms of their access to commercial activities (Dixon et al., 2001). In the short-term, making more material resources available to women, such as land, credit and technology at the micro level is mostly a question of putting existing policies into practice. Changes at the macro-level, however, will depend on a more favorable gender balance at all levels of the power structure. In Africa, the creation of national women's institutions has been a critically important step in ensuring that women's needs and constraints are put on the national policy agenda (FAO, 1990). The introduction of conventions, agreements, new legislation, policies and programs has helped to increase women's access to, and control over, productive resources. However, rural people are frequently unaware of women's legal rights and have little legal recourse if rights are violated (FAO, 1995).

Given women's role in food production and provision, any set of strategies for sustainable food security must address women's limited access to productive resources. Ensuring equity in women's rights to land, property, capital assets, wages and livelihood opportunities would undoubtedly impact positively on the issue.

Historically, women and other marginalized groups have had less access to formal information and communication systems associated with agricultural research and extension.

Goals L, S	Certainty B	Range of Impacts -3 to 0	Scale G	Specificity Wide applicability

Worldwide, there are relatively few professional women in agriculture (Das, 1995; FAO, 2004a). In Africa, men continue to dominate the agricultural disciplines in secondary schools, constitute the majority of the extension department personnel, and are the primary recipients of extension services. Men's enrolment in agricultural disciplines at the university level is higher than women's and is also increasing (FAO, 1990). Only 15% of the world's agricultural extension agents are women (FAO, 2004a). Only one-tenth of the scientists working in the CGIAR system are women (Rathgeber, 2002) and women rarely select agricultural courses in universities.

3.2.4 Relationships between AKST, coordination and regulatory processes among multiple stakeholders

The interactions between AKST and coordination processes among stakeholders are critically important for sustainability. Technical changes in the form of inventions, strengthened innovation systems and adoption of indigenous production systems in AKST are dependent on the effectiveness of coordination among stakeholders involved in natural resources management, production, consumption and marketing, e.g., farmers, extension, research, traders (Moustier et al., 2006; Temple et al., 2006). Failure to recognize this leads to poor adoption potential of the research outputs (Röling, 1988; World Bank, 2007c). Scaling-up requires articulation between stakeholders acting at multiple levels of organizational from the farmer to international organizations and markets (Caron et al., 1996; Lele, 2004). AKST can contribute by identifying the coordination processes involved in scaling-up, but this is now recognized to involve more than the typical micro-macro analysis of academic disciplines. AKST also contributes to understanding coordination mechanisms supporting change, adaptation and technological innovation, through approaches that connect experimental/non-experimental disciplines, basic/applied research, and especially, technical, organizational, and economic variables (Griffon, 1994; Cerf et al., 2000).

3.2.4.1 Coordination and partnership toward greater collective interest

AKST affects sustainability through collective action and partnership with new stakeholders (e.g., agroforestry sector) that strengthen farmer organizations and their ability to liaise with policy-makers, and support the design of new organizations (e.g., water users associations).

Major social, economic and political changes in agricultural and rural development have emerged in the last two decades through the involvement of new civil society actors.

Goals	Certainty	Range of Impacts	Scale	Specificity
S	B	0 to +2	G	Wide applicability

Since the 1980s, civil society actors (NGO, farmer and rural organizations, etc.) have become increasingly active in national and international policy negotiations (Pesche, 2004). The emergence of new rural organizations and civil society intermediaries coincides with the trend towards decentralization (Mercoiret et al., 1997ab). More recently, federated regional civil society organizations have emerged (Touzard and Drapieri, 2003). In 2000, ROPPA (Réseau des organisations paysannes et des producteurs d'Afrique de l'Ouest) was created in West Africa, under the umbrella of UEMOA (Union Economique et Monétaire Ouest-Africaine). Similarly, in South America, Coprofam (Coordenadora de Organizaciones de Productores Familiares del Mercosur) was created at the time of the implementation of the Mercosur mechanisms, in order to defend family agriculture.

Farmer organizations representing a large number of poor agricultural producers have had great impact on rural livelihoods through the provision of services.

Goals	Certainty	Range of Impacts	Scale	Specificity
N, L, E, S, D	C	0 to +3	G	Wide applicability

Farmer organizations have enlarged their activities from enhanced production to many other support functions, and not all are for profit (Bosc et al., 2002). The support includes coordination, political representation and defense of interests, literacy and other training, and cultivation methods for sustainability of production systems and social services. In some cases, these farmer organizations have taken direct responsibility for research and dissemination (as in the Coffee Producer Federation of Colombia).

Access to water resources has been improved by water user associations and organizations ensuring access to water rights through user-based, agency and market allocations.

Goals	Certainty	Range of Impacts	Scale	Specificity
E, S	B	0 to +3	G	Mainly in tropical countries

Dissatisfaction with performance of government managed irrigation has led to the promotion of participatory irrigation management over the past twenty years. However, problems remain with efficiency of operations, maintenance, sustainability and financial capacity. The involvement of private sector investors and managers is gaining credibility as a way to enhance management skills, and relieve the government of fiscal and administrative burdens (World Bank, 2007a). Water User Association (WUA) schemes in several states in India (Rajasthan, Andhra Pradesh, Karnataka, West Bengal, Uttar Pradesh) have improved access to water resources and increased production through increased irrigation. Likewise, in Mexico, Turkey and Nepal, transferring irrigation management to farmers has resulted in improved operation, better maintenance of infrastructure, reduced government expenditure, and increased production (World Bank, 1999). In many countries, this evolution has also raised new ques-

tions regarding sustainability and social justice (Hammani et al., 2005; Richard-Ferroudji et al., 2006).

3.2.4.2 Markets, entrepreneurship, value addition and regulation

The outcomes and efficiency of market rules and organizations directly affect sustainability. Efficient trading involves (1) farmers acting within an active chain of agricultural production and marketing; (2) dynamic links to social, economic and environmental activities in the region; (3) development plans appropriate to heterogeneity of agriculture among countries; and (4) recognition of the differences in farming methods and cultural background. Many farmers have a good understanding of the nature of the demand in terms of its implications for varieties, timing, packaging and permitted chemicals. As a result of knowledge-based approaches, they progressively modify their production practices and their portfolio of products in response to changing patterns of demand. The implementation of new norms regarding the use of AKST modifies market rules and organizations and differentially affects rural livelihoods, depending on local conditions.

Both locally and internationally the food sector is processing a wider range of tropical products.

Goals	Certainty	Range of Impacts	Scale	Specificity
N, D	B	0 to +3	G	Wide applicability

Many different products can be processed from a single crop, e.g., maize in Benin is processed into forty different products, in large part explaining the limited penetration of imported rice and wheat into Benin. The branding of products by area of origin is becoming an important marketing tool affecting the competitiveness of local products in the tropical food sector (Daviron and Ponte, 2005; van de Kop et al., 2006). Competitiveness in the international market involves the promotion of distinctive properties of tropical foodstuffs (e.g., color, flavor) in products such as roots and tubers.

In aquaculture, there is increased coordination of private sector-led production and processing chains.

Goals	Certainty	Range of Impacts	Scale	Specificity
E, S	B	0 to +2	G	Wide applicability

Formal and informal links between small-scale producers and large processing companies are contributing to more efficient and competitive aquaculture (shrimp, Vietnamese catfish, African catfish and tilapia), resulting in better quality for consumers, and secured margins for producers (Kumaran et al., 2003; Li, 2003). Export certification schemes are further streamlining production, processing, distribution and retail chains (Ponte, 2006).

Seasonal fluctuation in fruit and vegetable supplies is a major problem in the marketing of perishable products.

Goals	Certainty	Range of Impacts	Scale	Specificity
N	B	-3 to +1	G	Wide applicability

Various approaches have been developed to reduce the impacts of seasonality. For example, market-based risk management instruments have been instituted, such as the promotion of the cold-storage, insurance against weather-induced damage and encouragement of over-the-counter forward contracts (Byerlee et al., 2006). Initiatives like these are enhanced by the development of varieties and production technologies that expand the productive season and overcome the biotic and abiotic stresses, which occur during the off-season (Tchoundjeu et al., 2006).

Consumers' concerns about food safety are affecting international trade regulations.

Goals	Certainty	Range of Impacts	Scale	Specificity
N, D	C	0 to +3	R	Wide applicability

The effects of the implementation of food safety standards on global trade is valued at billions of US dollars (Otsuki et al., 2001; Wilson and Otsuki, 2001). However, the regulatory environment for food safety can be seen as an opportunity to gain secure and stable access to affluent and remunerative new markets, and generate large value addition activities in developing countries (World Bank, 2005b).

Food standards are increasingly important and have implications for consumer organizations and private firms.

Goals	Certainty	Range of Impacts	Scale	Specificity
N	B	0 to +3	G	Wide applicability

New instruments of protection and competitiveness have emerged as "standards" and new forms of coordination between actors in the food chain have been developed in response to consumer and citizen concerns. Actors in the food chain work together to specify acceptable production conditions and impose them on suppliers (Gereffi and Kaplinsky, 2001; Daviron and Gibbon, 2002). Initially limited to some companies, standards are becoming accepted globally (e.g., Global Food Standard, International Food Standard [IFS], GFSI [Global Food Safety Initiative], FLO [Fair Trade Labeling Organization]) (JRC, 2007). The multiplication of these standards, which are supposed to improve food safety, preserve the environment, and reduce social disparities, etc., raises questions about international regulation, coordination, and evaluation (in the case of forests, Gueneau, 2006).

Food labeled as "organic" or "certified organic" is governed by a set of rules and limits, usually enforced by inspection and certification mechanisms known as "guarantee systems".

Goals	Certainty	Range of Impacts	Scale	Specificity
H, E, S, D	A	+1 to +3	G	Wide applicability

With very few exceptions, synthetic pesticides, mineral fertilizers, synthetic preservatives, pharmaceuticals, sewage sludge, genetically modified organisms and irradiation are prohibited in all organic standards. Sixty mostly industrialized countries currently have national organic standards as well as hundreds of private organic standards worldwide (FAO/ITC/CTA, 2001; IFOAM, 2003, 2006). Regulatory systems for organics usually consist of producers, inspection bodies, an accreditation body for approval and system supervision and a labeling body to inform the consumer (UN, 2006b). There are numerous informal organic regulation systems outside of the formal organic certification and marketing systems. These are often called "peer" or "participatory" models. They do not involve third-party inspection and often focus on local markets (UN, 2006b). The IFOAM and Codex guidelines provide consumer and producer protection from misleading claims and guide governments in setting organic standards in organic agriculture (see 3.2.2.1.9). The cultivation of GMO crops near organic crops can threaten organic certification due to the risk of cross-pollination and genetic drift.

Some food standards are now imposing minimum conditions of employment.

Goals	Certainty	Range of Impacts	Scale	Specificity
L, D	C	-3 to +2	G	Wide applicability

To face the inequalities that accrue from benefits to large-scale producers, standards have been developed to encourage small-scale producers. The most prominent example is the Fair Trade Movement (www.fairtrade.org.uk), which aims to ensure that poor farmers are adequately rewarded for the crops they produce. In 2002 the global fair trade market was conservatively estimated at US$500 million (Moore, 2004). This support has helped small organizations to market their produce directly by working similarly to that of forest certification. Where foreign buyers impose labor standards, the terms and conditions of employment in the formal supply chains are better than in the informal sector. Enforcement of food standards furthermore improve the working environment and ensure that agricultural workers are not exposed to unhealthy production practices.

The globalization of trade in agricultural products is not an import-export food model that addresses poverty and hunger in developing countries.

Goals	Certainty	Range of Impacts	Scale	Specificity
N, D	C	-4 to 0	G	Wide applicability

Many complex factors affect the economy of a country. The following evidence suggests that international policies that promote economic growth through agriculture do not necessarily resolve the issue of poverty (Boussard et al., 2006; Chabe-Ferret et al., 2006):

- An estimated 43% of the rural population of Thailand now lives below the poverty line even though agricultural exports grew 65% between 1985 and 1995.
- In Bolivia, after a period of spectacular agricultural export growth, 95% of the rural population earned less than a dollar a day.
- The Chinese government estimates that 10 million farmers will be displaced by China's implementation of WTO rules, with the livelihoods of another 200 million small-scale farmers expected to decline as a result of further implementations of trade liberalization and agriculture industrialization.
- Kenya, which was self-sufficient in food until the 1980s, now imports 80% of its food, while 80% of its exports are agricultural.
- In the USA net farm income was 16% below average between 1990-1995, while 38,000 small farms went out of business between 1995-2000.
- In Canada, farm debt has nearly doubled since the 1989 Canada-U.S. Free Trade Agreement.

- The U.K. lost 60,000 farmers and farm workers between 98-2001 and farm income declined 71% between 1995-2001.
- To provide clearer and broader figures, the World Bank has implemented the Ruralstruc project to assess the impact of liberalization and structural adjustment strategies on rural livelihoods (Losch, 2007). These examples indicate that poverty alleviation requires more than economic policies that aim at promoting global trade.

The globalization of the food supply chain has raised consumer concerns for food safety and quality.

Goals	Certainty	Range of Impacts	Scale	Specificity
N	B	-3 to 0	G	Wide applicability

The incidence of food safety hazards such as: "mad cow disease" (bovine spongiform encephalopathy), contamination of fresh and processed foods (e.g., baby milk, hormones in veal, food colorings and ionized foodstuffs in Europe, mercury in fish in Asia, etc.) have resulted in the emergence of traceability as a key issue for policy and scientific research in food quality and safety. Over the past ten years considerable research effort has been directed towards assessing risks and providing controls (Hazard Analysis Critical Control Point—HACCP). These have included the implementation of food traceability systems complying with marketing requirements (Opara and Mazaud, 2001). Consumer concerns about the safety of conventional foods and industrial agriculture as result of the use of growth-stimulating substances, GM food, dioxin-contaminated food and livestock epidemics, such as outbreaks of foot and mouth disease, have contributed to the growth in demand for organic food. Many consumers perceive organic products as safer and of higher quality than conventional ones. These perceptions, rather than science, drive the market (http://www.fao.org/DOCREP/005/Y4252E/y4252e13.htm#P11_3).

"Enlightened Globalization" is a concept to address needs of the poor and the global environment and promote democracy.

Goals	Certainty	Range of Impacts	Scale	Specificity
E, S, D	D	Not yet known	G	Wide applicability

The concept of Enlightened Globalization has been proposed to address "the needs of the poorest of the poor, the global environment, and the spread of democracy" (Sachs, 2005). It is focused on "a globalization of democracies, multilateralism, science and technology, and a global economic system designed to meet human needs". In this initiative, international agencies and countries of the industrial North would work with partners in the South to honor their commitments to international policies and develop new processing industries focused on the needs of local people in developing countries while expanding developing economies. Enlightened Globalization also is aimed at helping poor countries to gain access to the markets of richer countries, instead of blocking trade and investment.

There is new and increasing involvement of the corporate sector in agroforestry.

Goals	Certainty	Range of Impacts	Scale	Specificity
E, S	C	0 to +2	R	Wide applicability

Typically, multinational companies have pursued large-scale, high input monocultures as their production systems. However, a small number of multinational companies are now recognizing the social, environmental, and even economic, benefits of community engagement and becoming involved in agroforestry to develop new crop plants that meet specific needs in a diversifying economy. There are now several examples of new niche products becoming new international commodities (Mitschein and Miranda, 1998; Wynberg et al., 2002; Tchoundjeu et al., 2006). In Brazil, DaimlerChrysler has promoted community agroforestry for the production of a range of raw plant materials used to make a natural product alternative to fiberglass in car manufacture (Mitschein and Miranda, 1998; Panik, 1998), while in Ghana, Unilever is developing new cash crops like *Allanblackia* sp. as shade trees for cocoa (IUCN, 2004; Attipoe et al., 2006). In South Africa, the *"Amarula"* liqueur factory of Distell Corporation buys raw *Sclerocarya birrea* fruits from local communities (Wynberg et al., 2003). New public/private partnerships such as those developed by the cocoa industry can set the standard for the integration of science, public policy and business best practices (Shapiro and Rosenquist, 2004).

3.2.4.3 Policy design and implementation
Policy instruments can be introduced at many different levels: sectorial, territorial, international science policies, and international policies, treaties and conventions.

Analyses reveal that the Green Revolution was most successful when the dissemination of AKST was accompanied by policy reforms.

Goals	Certainty	Range of Impacts	Scale	Specificity
N	B	0 to +2	R	Wide applicability

Policy reform has been shown to be particularly important for the successful adoption of Green Revolution rice production technologies in Asia. When Indonesia, implemented relevant price, input, credit, extension and irrigation policies to facilitate the dissemination of the cultivation of potentially high-yielding, dwarf varieties, physical yields increased by a factor of 4-5 per unit area, as well as achieving very significant increases in labor productivity and rural employment (Trebuil and Hossain, 2004). Likewise, in Vietnam, increased rice production in the Mekong delta in 1988 was associated with the implementation of similar policies (Le Coq and Trebuil, 2005).

Agricultural policies that in the past gave inadequate attention to the needs of small-scale farmers and the rural poor are now being replaced by a stronger focus on livelihoods.

Goals	Certainty	Range of Impacts	Scale	Specificity
L	B	-3 to +3	G	Wide applicability

Agricultural policy over the last 50 years focused on the production of agricultural commodities and meeting the immediate staple food needs to avoid starvation in the growing world population (Tribe, 1994), and rarely explicitly targeted the multiple needs of the rural population (World Bank, 2007a). This situation has changed over the last 10 years with the development of a livelihoods focus in rural development projects, but in many countries, national policies are still focused on high-input farming systems with a strong emphasis on intensive farming that differs from the small-scale, low-input, mixed cropping systems of small-scale farmers which may be hurt by untargeted policy reforms (OECD, 2005). A stronger livelihood approach is based on sustainability issues, diversification of benefits, better use of natural resources, ethical trade and a more people-centric focus. Diversified farming systems often mimic natural ecosystems as noted in best-bet alternatives to slash-and-burn (Palm et al., 2005b). These typically provide radical improvements in farmer livelihoods (Vosti et al., 2005) and environmental benefits (Tomich et al., 2005).

Organizations that support and regulate the production of agricultural crops, livestock, fisheries and forestry are often poorly interconnected at the national and international level, and are also poorly connected with those responsible for the environment and conservation.

Goals	Certainty	Range of Impacts	Scale	Specificity
E	C	-3 to 0	G	Wide applicability

The creation of synergies between increased production and development and sustainability goals are often limited by the "disconnects" between agriculture and the environment. Thus the ideal of sustainable land use is often more a subject of political rhetoric than government policy. However, there are signs that some of the INRM initiatives—in agroforestry, organic agriculture, sustainable forestry certification, etc—are starting to influence environmental land use planning and agricultural authorities (Abbott et al., 1999; Dalal-Clayton and Bass, 2002; Dalal-Cayton et al., 2003), as they are also in fisheries (Sanchirico et al., 2006).

In the agricultural and food sectors, coordination of the development of international policies created by the WTO have strongly interacted with global AKST actors.

Goals	Certainty	Range of Impacts	Scale	Specificity
D	C	-1 to +2	G	Wide applicability

Changes during the period of structural adjustment had considerable impact on the ability of developing countries to define targets and find the means to implement their public research and policy interventions. The need for more "policy space" is now widely acknowledged (Rodrik, 2007), creating a wide gap between the demand for policy and the implementation of either new policy or public/private stakeholder initiatives (Daviron et al., 2004). It is not clear whether the centralized and public AST policies of the last century can be replaced by modern decentralized public/private partnerships (such as private investment on R&D, standardization initiatives, third-party certification and farmer organization credit and saving programs) targeting the reduction of poverty and increased sustainability.

3.2.4.3.1 Sectoral policies

Many of the different sectors encompassed by agriculture have policies which specifically address a particular production system, target population, or natural resource. Likewise, specific agricultural policies concern food safety and health issues. This can create problems, as these different sectors of agriculture are often poorly integrated, or even disconnected. However, a few examples (e.g., agroforestry and forestry) are emerging which illustrate some convergence between sectors.

One of the consequences of structural adjustment policies has been the abandonment of the land by poor farmers, who can no longer afford farm inputs.

Goals	Certainty	Range of Impacts	Scale	Specificity
L	B	-4 to 0	G	Mainly small-scale agriculture

Rising input prices have resulted in high migration from the countryside to urban centers in search of jobs; often low paid manufacturing jobs. In India, for example, the numbers of landless rural farmers increased from 27.9 to over 50 million between 1951 and the 1990s, hampering economic growth. This illustrates that achieving higher aggregate economic growth is only one element of an effective strategy for poverty reduction (Datt and Ravallion, 2002) and that redressing existing inequalities in human resource development and between rural and urban areas are other important elements of success.

Although governments have expanded their role in water management, particularly in large scale irrigation schemes, sustainability requires effective institutional arrangements for the management of the resource and particularly public-private coordination.

Goals	Certainty	Range of Impacts	Scale	Specificity
E	B	-2 to +2	G	Wide applicability

Large dams, reservoirs and irrigation systems have usually been built by government agencies for economic development, including agriculture, urbanization and power generation. In most countries, agriculture has been by far the largest user of water and typically its allocation and management has been a public concern of government (de Sherbinin and Dompka, 1998). In the 1980s dissatisfaction with irrigation management and sustainability was common and the importance of empowering farmers, together with their traditional systems of water rights, was recognized as important. This led to the concept of participatory irrigation management in the 1990s. Nevertheless, communities of water users have faced numerous challenges in gaining sustainable and equitable access to water (Bruns and Meinzen-Dick, 2000; Meinzen-Dick and Pradhan, 2002). Water User Associations (WUA) have emerged as an effective way of managing water delivery (Abernethy, 2003; Schlager, 2003). This approach, as well as the rise of the private sector, has led to the redefinition of the role of governments over the past 20 years. Governments are now viewed as facilitators of investments, regulators of this sector and responsible for sustainable management at the watershed scale (Hamann and O'Riordan, 2000; Perret, 2002; ComMod Group, 2004).

Deforestation is often an outcome of poorly linked inter-sectorial policies.

Goals	Certainty	Range of Impacts	Scale	Specificity
N, L, E, S	B	+1 to +4	R	Mainly small-scale agriculture

One of the common and dominant outcomes from an international study of slash-and-burn agriculture was that small-scale farmers cut down tropical forests because current national and international policies, market conditions, and institutional arrangements either provide them with incentives for doing so, or do not provide them with alternatives (Palm et al., 2005b; Chomitz et al., 2006). This trend will continue if tangible incentives that meet the needs of local people for more sustainable alternatives to slash-and-burn farming are not introduced. Some options linked to the delivery of international public goods and services, like carbon storage, may be very expensive (Palm et al., 2005a), while others like the participatory domestication of trees providing both environmental services and marketable, traditional foods and medicines (Tchoundjeu et al., 2006), that help farmers to help themselves may be a cheaper option (see 3.2.2.1.6).

Integrating forestry with other land uses has economic, environmental and social benefits.

Goals	Certainty	Range of Impacts	Scale	Specificity
E, S, D	C	0 to +3	G	Wide applicability

Recently forest agencies have recognized that tree cover outside public forests and in farmland are important for national forest-related objectives (FAO, 2006b). In forest certification the links between civil society and market action have been a key driver in the social integration of intensive forest plantations (Forest Stewardship Council www.fsc.org and Pan-European Forest Certification www.pefc.org). Consequently, certification standards are improving the direction of both forest policy and forest KST at national and international levels (Bass et al., 2001; Gueneau and Bass, 2005). Forest certification is linking land use issues from the tree stand, to the landscape, and ultimately to global levels for the production of sustainable non-timber benefits and environmental services (Pagiola et al., 2002; Belcher, 2003). When KST and market conditions are right, the flow of financial benefits can make multipurpose forest systems economically superior to conventional timber-focused systems (Pagiola et al., 2002). Non-wood forest products produce a global value of at least $4.7 billion in 2005 (FAO, 2005b).

Public interest in food safety has increased and food standards have been developed to ensure that the necessary safety characteristics are achieved.

Goals	Certainty	Range of Impacts	Scale	Specificity
N	B	0 to +2	G	Wide applicability

Public interest in the chemical residues in fresh produce (Bracket, 1999; Kitinoja and Gorny, 1999) has been heightened by the provision of quantitative data on chemical use in agriculture (OECD, 1997; Timothy et al., 2004), especially the use of banned pesticides in developing country agriculture. Of special concern is the permitted thresholds of heavy metals (Mansour, 2004), and their status as con-

taminants, especially as food administrators in developed countries have tended to set increasingly lower levels of tolerance. Traceability has become an important criterion of food quality (Bureau et al., 2001). Internationally recognized food safety standards include GAP, GMP like ISO 9000, EUREP GAP, HACCP. Similarly, various measures and standards have been developed for food quality including Diet Quality Index (Patterson et al., 1994), Analysis of Core Foods (Kristal et al., 1990), and Healthy Eating Index (Kennedy et al., 1995). Dietary Diversity Scores are also now increasingly used to measure food quality (Kant et al., 1993, 1995; Hatloy, et al., 1998; Marshall et al., 2001; Ali and Farooq, 2004), while total nutritive values are being used to prioritize food commodities (Ali and Tsou, 2000). Although consumers benefit from the better quality and greater safety attributes of food products, the enforcement of food quality standards also may increase food prices (Padilla, 1992). In addition, the cost of applying food safety standards can be a drain on public resources or may lead to disguised protection, as in the case of "voluntary certifications" which are increasingly a prerequisite for European retailers (Bureau and Matthews, 2005).

GMOs are experiencing adoption difficulties in Europe.

Goals	Certainty	Range of Impacts	Scale	Specificity
E	B	-2 to -4	G	Wide applicability

GM crops are only grown commercially in 3-4 European countries, (primarily Spain) (James, 2006) and very few GM crops and foods have been approved for commercialization. Rejection by consumers, food companies and supermarkets is responsible for poor adoption and can taken as an indication that consumer demand for GM products is almost non-existent (Bernauer, 2003). However, it is unclear to what extent consumer demand has been the result of EU regulations or *vice versa* and debate continues about the level of appropriate regulations. Before the mid-1980s, there were no GMOs on the market in Europe, but since then the EU has adopted regulations on the approval of GM crops and foods. The strict labeling laws have resulted in very few GM foods sold on the European market. There is however more tolerance of non-food GM crops in Europe and recent reports indicate that some 75% of cotton imported into the EU today is from GM varieties, mainly from the USA and China. In other parts of the world the situation with GM foods is very different, Fifteen of 16 commercial crops in China have genetically engineered pest resistance (8/16 virus, 4/16 insect, 4/16 disease resistance) and herbicide resistance (2/16) (See 3.2.1.4).

Adoption of GMOs has had some serious negative economic impacts in Canada and USA.

Goals	Certainty	Range of Impacts	Scale	Specificity
D	B	-3 to -4	G	Wide applicability

After the adoption of GM varieties, Canadian farmers lost their market for $300 million of canola (oilseed rape) to GMO-free markets in Europe (Freese and Schubert, 2004; Shiva et al., 2004). Likewise, after leading US food allergists judged Bt-corn to be a potential health hazard (Freese, 2001), US$1 billion worth of product recalls followed the discovery of animal feed Bt-corn in products for human consumption (Shiva et al., 2004). Maize exports from USA to Europe have also declined from 3.3 million tonnes in 1995 to 25,000 tonnes in 2002 due to fears about GMOs (Shiva et al., 2004). The American Farm Bureau estimates this loss has cost US farmers $300 million per year (Center for Food Safety, 2006).

3.2.4.3.2 Territorial policies

Attention to the livelihood needs of small-scale farmers and the rural poor has been insufficient, but now many developing nations are implementing policies to enhance incomes and reduce poverty.

Goals	Certainty	Range of Impacts	Scale	Specificity
L, D	B	-2 to +3	G	Wide applicability

Improving the livelihood of small-scale farmers has typically focused on market participation, through better access to information, increased efficiency of input supply systems, provision of credit, and better market chains and infrastructure (Sautier and Bienabe, 2005). In some countries, agricultural policies and market liberalization have increased economic differentiation among communities and households (Mazoyer and Roudart, 2002; IFAD, 2003). Small-scale, low-input agriculture systems have an important role as a social safety net (Perret et al., 2003), help to maintain cultural and community integrity, promote biodiversity and landscape conservation. However, the impacts of these commercialization policies on social conflict, land ownership, kinship, and resource distribution are not usually assessed (Le Billon, 2001).

Policy responses have been developed to enhance food and nutritional security, and food safety, and to alleviate the impacts of seasonal fluctuations on the poor.

Goals	Certainty	Range of Impacts	Scale	Specificity
N	B	0 to +3	G	Wide applicability

Responses to food and nutritional insecurity have included the provision of infrastructure for health facilities and parental education (Cebu Study Team, 1992; Alderman and Garcia, 1994); programs ensuring equitable distribution of nutritious foods among family members; regulations to enforce the provision by retailers of nutritional information on food purchases (Herrman and Roeder, 1998), and the improvement of safety practices for those preparing, serving and storing food (Black et al., 1982; Stanton and Clemens, 1987; Henry et al., 1990). Other approaches to supporting marketing have included linking the domestic and international markets through involvement of the private sector, developing food aid, food-for-work programs, and price instability coping mechanisms (Boussard et al., 2005).

National conservation and development strategies have increasingly promoted more integration of sustainability goals at local and national levels.

Goals	Certainty	Range of Impacts	Scale	Specificity
E, S, D	B	0 to +3	G	Wide applicability

National conservation and development strategies have recently gained as much political profile as land use planning in the past. National poverty reduction strategies, conservation strategies, and sustainable development strategies form a pool of cross-cutting approaches that seek to link institutions. This has involved the engagement of local stakeholders in participatory processes to negotiate broad visions of the future, and to focus local, regional and national institutions on poverty reduction, environmental sustainability (Tubiana, 2000), sustainable development (Dalal-Clayton and Bass, 2002) and participatory agroenterprise development (Ferris et al., 2006).

Government ministries and international agencies responsible for agriculture, livestock, fisheries and food crops are typically disconnected and in competition for resources, and power.

Goals	Certainty	Range of Impacts	Scale	Specificity
E, S, D	C	-3 to 0	G	Wide applicability

In many countries around the world the disconnections between the various subsectors of agriculture place them in competition for resources and power. Consequently, lack of compatibility between the policies and laws of different sectors make it difficult to promote sustainable development, as the potential synergies are lost, e.g., promoting forest removal for farmers to secure agricultural land tenure and grants (Angelsen and Kaimowitz, 2001). To address this problem, cross-sectoral national forums associated with international agreements/summits, have developed strategic planning initiatives to provide an integrated framework for sustainable development and poverty reduction, with mixed results. For example, the Action Plans of the Rio Earth Summit (www.un.org/esa/sustdev/documents/agenda21) and the World Summit on Sustainable Development (2002) put a premium on national level planning as a means to integrate economic, social and environmental objectives in development (Dalal-Clayton and Bass, 2002). These Action Plans have been most successful where they have (1) involved multistakeholder fora; consulted "vertically" to grassroots as well as "horizontally" between sectors; focused on different sectors' contributions to defined development and sustainability outcomes (rather than assuming sector roles); (2) been driven by high-level and "neutral" government bodies, and (3) been linked to expenditure reviews and budgets (Dalal-Clayton and Bass, 2002; Assey et al., 2007). In most countries the importance of farming for both economic growth and social safety nets is clear in such strategies, but few have stressed the links with forestry. However, due to lack of updated information, it has been difficult to progress beyond a broad, consultative approach and to identify specific tradeoff decisions, especially concerning environmental issues (Bojo and Reddy, 2003).

3.2.4.3.3 Scientific policies

Scientific policies shape the design and the use of AKST and subsequently, its impact on development, in various ways. Examples include the organization of disciplines within academic and AKST institutions, and the implementation of specific policies on intellectual property rights.

Typically, AKST development has rationalized production according to academic discipline, constraining the development of integrated production systems.

Goals	Certainty	Range of Impacts	Scale	Specificity
E	C	-3 to +1	G	Wide applicability

In the past, crop, livestock and forest sciences have typically been implemented separately. However, agroforestry integrates trees with food crops and/or livestock in a single system, improving the relationships between food crops, livestock and tree crops for timber or other products, but this level of integration is rarely visible in international institutions, national governments and markets. For example, the World Commission on Forests and Sustainable Development (1999), and the Intergovernmental Panel on Forests do not focus on agricultural links. Likewise, the InterAcademy Council Report on African Agriculture (2004) paid scant attention to forestry, or even to agroforestry. However, this is changing and a few new forms of local organization and collective action are emerging, such as Landcare (www. landcare.org), Ecoagriculture (McNeely and Scherr, 2003); community forestry associations (Molnar et al., 2005), and biological corridor conservation projects. This change has just emerged at the policy level, with the European Union approving a measure entitled "First establishment of agroforestry systems on agricultural land" (Article 44 of Regulation No 1698/2005 and Article 32 Regulation No 1974/2006, Annex II, point 5.3.2.2.2) in 2007 to provide funds for the establishment of two agroforestry systems in mainland Greece.

IPR policies are used to protect plant genetic resources that are important for food and agriculture.

Goals	Certainty	Range of Impacts	Scale	Specificity
E, S	C	-3 to +3	G	Wide applicability

Most developed countries have a system to register Plant Breeders Rights, often supported by Trade Marks and Patents. These schemes are genuinely fostering innovation and conferring benefits to innovators, while also protecting genetic resources. They are supported by the International Treaty on Plant Genetic Resources for Food and Agriculture (TRIPS) and the Convention on Biological Diversity (UNEP, 1993) which aim to promote both the conservation and sustainable use of plant genetic resources for food and agriculture and the fair and equitable sharing of the benefits arising out of their use (FAO, 2001, 2002b). The treaty addresses the exchange of germplasm between countries and required all member countries of World Trade Organization to implement an Intellectual Property Rights (IPR) system before 2000 (Tirole et al., 2003; Trommetter, 2005) "for the protection of plant varieties by patents or by an effective *sui generis* system" (Mortureux, 1999; Célarier and Marie-Vivien 2001; Feyt, 2001). Germplasm arising from international public-funded research is protected on behalf of humankind by the FAO (Frison et al., 1998; Jarvis et al., 2000; Sauvé and Watts, 2003). Agriculture is being integrated into the program and work of the CBD, including conservation of domesticated species, genetic diversity and goals for conservation of wild flora and agricultural landscapes.

Intellectual property rights regulatory frameworks currently do not protect the innovations or rights of communities or farmers in developing countries to their indigenous genetic resources.

Goals	Certainty	Range of Impacts	Scale	Specificity
E, S	B	0 to +3	G	Wide applicability

The development of IPR frameworks at international and national scales through patents, trade marks, contracts, geographical indicators and varieties do not offer much protection for poor farmers and there are many unresolved issues. For example, in developing countries many farmers do not have the ability or income to protect their rights, and the identification of the innovator can be controversial. Consequently much international activity by NGOs and farmer organizations is focused on trying to develop effective protection mechanisms for farmers and local communities based on traceability and transparency (Bazile, 2006), as for example in the Solomon Islands (Sanderson and Sherman, 2004). This is important to prevent biopiracy and to promote legitimate biodiscovery that meets internationally approved standards.

To assess and manage potential risks from LMOs and GMOs, governments are developing National Biosafety Frameworks.

Goals	Certainty	Range of Impacts	Scale	Specificity
H, L, E, S	C	Not yet known	G	Worldwide

Countries need to have capacity and mechanisms to make informed decisions as they accept or reject products of modern biotechnology (Pinstrup-Andersen and Schioler, 2001). Currently many Governments, including eighty developing countries, have developed National Biosafety Frameworks (NBF) to support the application and use of modern biotechnology in accordance with national policies, laws and international obligations, in particular the Cartagena Protocol on Biosafety (CBD, 2000). This is the first step towards the development of improved capacity for biosafety assessment and implementation of the Cartagena Protocol under the UNEP-GEF Biosafety Project (http://www.unep .ch/biosafety/news.htm). NBFs have had some success but they have not always been adopted by governments. Many African countries still lack biosafety policies and regulations and technical enforcement capacity.

3.2.4.3.4 International policy, treaties and conventions

The globalization process has been supported by international and regional trade policy frameworks, and by the policy recommendations (structural adjustment programs) of the World Bank and International Monetary Fund.

Goals	Certainty	Range of Impacts	Scale	Specificity
D	A	-2 to +2	G	Wide applicability

There are links between global trade and economic agreements and institutions, such as the World Trade Organization (WTO) and Regional Trade Agreements (e.g., NAFTA, EPA), IMF, bilateral agreements, and domestic and regional agricultural policies, technologies, R&D and natural resource use. AKST played a role in this process, particularly neo-classical economic theory which emphasized the need to shift resources in line with comparative advantages at national level, and restore price incentives to generate income at local level. Assessment of the impact of market-oriented policies has demonstrated the need for complementary and supportive public policies to cope with some of the unsustainable impacts of globalization and to reinforce the need for greater sustainability of development and growth (Stiglitz, 2002).

Development microeconomics and agricultural economics of international markets have called for sui generis policies.

Goals	Certainty	Range of Impacts	Scale	Specificity
E, S, D	D	0 to +3	G	Wide applicability

Two approaches have been taken to development economics research and policy. Firstly, there has been a shift of focus from macro issues to micro problems; e.g., from markets to households, from products to people (Sadoulet et al., 2001; Banerjee and Duflo, 2005). In this approach, research on the impacts of risk and imperfect information at the household level provided insights on the cost of market failure for households and countries (Rothschild and Stiglitz, 1976; Newberry and Stiglitz, 1979; Binswanger, 1981; Stiglitz, 1987; Boussard et al., 2006). For example, local market and institutional conditions were found to determine the success or failure of public policy. In China and other emerging economies *sui generis* macro policies have outperformed the so-called "Washington consensus" policies (Santiso, 2006). This is increasing interest in *sui generis* development and trade policies (Stiglitz and Greenwald, 2006; Rodrik, 2007). In the second approach, agricultural economics research continues to explore the value and power distribution along international commodity market chains (Gereffi and Korzeniewicz, 1994; Daviron and Ponte, 2005; Gibbons and Ponte, 2005), to determine how new patterns of labor organization throughout the chain have impact upon its overall function—and notably how they affect farmer income.

The World Trade Organization (WTO) has greatly expanded the scope of trade and commodity agreements as set out in the General Agreements on Tariffs and Trade (GATT).

Goals	Certainty	Range of Impacts	Scale	Specificity
D	C	-3 to +2	G	Wide applicability

Agricultural economic research on the causes and consequences of market instability on people and national economies (e.g., Schultz, 1949) shaped the postwar development of developing countries policies prior to Independence. These policies led to new institutional schemes to address development issues, e.g., the creation of UNCTAD and the formulation of special arrangements under GATT in the 1970s, such as the definition of rules with regard to setting trade quotas and tariffs (Ribier and Tubianz, 1996). Other matters have remained under the purview of national governments. Although not without flaws, this system has provided tools such as trade barriers which allow countries to protect their domestic markets. The Uruguay Round of negotiations, which led to the creation of the WTO, greatly

expanded the power of international arenas over agriculture, limiting the authority of national governments to fixed policies governing their own farmers, consumers, and natural resources (Voituriez, 2005). The impacts of these WTO policies on the agricultural sector have been controversial. *Ex post* analysis indicates negative impacts on the lives of poor food producers and indigenous peoples, while *ex ante* analysis on current Doha Scenarios point to possible welfare losses in the short term for some poor countries and poor households (Hertel and Winters, 2005; Polaski, 2006). Some of the losers from trade liberalization are also among the poorest (Chabe-Ferret et al., 2006). Similarly, traditional small scale farming and fishing communities worldwide have suffered from globalization, which has systematically removed restrictions and support mechanisms protecting them from the competition of highly productive or subsidized producers. To redress these negative impacts, current AKST initiatives include the examination of (1) broader special and differential treatment for developing countries, allowing them to experiment with *ad hoc* policy within a wider policy space and (2) the resort to special "rights"— e.g., the Right to food or "Food Sovereignty" under UN auspices (Ziegler, 2003).

Regional Trade Agreements have had major impacts on food exports and agriculture systems in some countries.

Goals	Certainty	Range of Impacts	Scale	Specificity
D	C	-3 to +2	R	North and South America

The implementation of North American Free Trade Agreement (NAFTA) has had major social and economic impacts on agriculture and the trading of food. For example, while beneficial to USA, corn production in Mexico collapsed with an associated decline in the real rural wage (Hufbauer and Schott, 2005). This situation arose because as a condition for joining NAFTA, Mexico had to change its Constitution and revoke the traditional "ejido" laws of communal land and resource ownership, and dismantle its system of maintaining a guaranteed floor price for corn, which sustained more than 3 million corn producers. Within a year, production of Mexican corn and other basic grains fell by half and millions of peasant farmers lost their income and livelihoods. Many of these farmers are part of the record-high number of immigrants crossing U.S. borders.

One of the side effects of the increased food trade has been worldwide increase in the number of food and food-borne diseases.

Goals	Certainty	Range of Impacts	Scale	Specificity
N	C	-3 to 0	G	Wide applicability

The World Health Organization (WHO) has identified that the increased trade of food has contributed to increased levels of human illness worldwide. In part this may simply be due to the increased volume of food imports. The WTO's Sanitary and Phytosanitary Agreement (SPS) has set criteria for member nations to follow regarding their domestic trade. These policies affect food safety risks arising from additives, contaminants, toxins, veterinary drug and pesticide residues or other disease-causing organisms. The primary goal of the SPS is to facilitate trade by eliminating differences above and below SPS standards in food, animal, and plant regulations from country to country. Independently from the international standard (Codex Alimentarius, www .codexalimentarius.net), national standards might imply an asymmetry of trade exchanges.

Structural adjustment policies (SAPs) of the World Bank and the International Monetary Fund (IMF) have significantly reshaped national agriculture policies in developing countries.

Goals	Certainty	Range of Impacts	Scale	Specificity
D	B	-3 to +1	G	Wide applicability

The structural adjustment policies were aimed at helping countries cut down their debt. Many SAPs required developing countries to cut spending. As a result, centralized seed distribution programs, price supports for food and farm inputs, agricultural research, and certain commodities (often locally consumed foods) were eliminated or downsized (Bourguignon et al., 1991). While national support systems protecting traditional livelihoods (maintaining native crops, landraces, etc.), food security, rural communities, and local cultures suffered, private corporations were given loans to partner with developing countries to develop industrial agriculture with crops mainly for export. Such financial mechanisms controversially promoted monocultural cropping that required farm inputs such as commercial seeds, chemicals, fossil-fuel based machinery, as well as requiring an increase in water usage.

Rising environment concerns and the recognition of global environmental public goods have had impacts on trade and livelihoods.

Goals	Certainty	Range of Impacts	Scale	Specificity
E, S	C	-3 to +3	G	Wide applicability

Increased interest in tropical forest conservation and the potential role of marketing non-timber forest products has led to heightened interest in the international trade of a wide range of natural products (e.g., Kusters and Belcher, 2004; Sunderland and Ndoye, 2004). The Convention on Biological Diversity has brought attention to issues of access to, and use of genetic resources of a wide range of species not formerly considered as crops, but of significance in horticulture, biotechnology, crop protection and pharmaceutical/nutriceutical and cosmetics industries (ten Kate and Laird, 1999; Weber, 2005). The CBD also outlined the ways in which these industries should interact responsibly with traditional communities, the holders of Traditional Knowledge about products from this wide array of potentially useful species when engaging in "biodiscovery" and "bioprospecting" (Laird, 2002). In particular, it has highlighted the need to appreciate the interactions between nature conservation, sustainable use and social equity through the development of "fair and equitable benefit sharing agreements" that respect the culture and traditions of indigenous people, and that support and enhance genetic diversity (Almekinders and de Boef, 2000).

3.2.4.4 Territorial governance
Territory is a new scale, intermediate between local and national issues, allowing market and state failures to be addressed. It is a portion of space delimited by a social group that implements coordination institutions and rules and thus is useful when developing integrated approaches to rural development (Sepulveda et al., 2003; Caron, 2005). Applied to agricultural production, the concept helps to address disconnects between scales with regard to ecological processes, individual decisions, collective management and policies. As it is controlled by local stakeholders, it also strengthens participation in the design of new activities and policies to reduce or prevent marginalization.

The concept of multifunctionality in agriculture and rural areas has simultaneously opened the way to changes in policies, research and operational issues.

Goals	Certainty	Range of Impacts	Scale	Specificity
N, H, L, E, S, D	B	-5 to +5	G	Worldwide

Multifunctional agriculture became a new policy goal in Europe in 2000 (www.european-agenda.com), which encouraged the transformation of rural areas towards a "multifunctional, sustainable and competitive agriculture throughout Europe". The main idea was to encourage the production of non-commodity goods or services through the subsidy of commodity outputs (Guyomard et al., 2004). Promoting multifunctionality has sometimes been the milestone of new policies, such as the French "Territorial Management Contract" (*Contrat Territorial d'Exploitation*, CTE) implemented through the 1999 Agricultural Act. The objectives have been partially achieved (Urbano and Vollet, 2005) in areas where the supply of high quality products has been increased through contracts between government and farmers, while protecting natural resources, biodiversity and landscapes. However, it is not limited to developed countries and in some developing countries, notably Brazil, multifunctional agriculture has promoted policies for family agriculture (Losch, 2004). Multifunctionality has also been advocated as a sustainable approach to land use in Africa (Leakey, 2001ab). In Europe, the concept of multifunctionality has progressed through state-of-the-art research projects (www.multagri.net), for example through new modeling tools to understand the integration of different functions.

Multifunctional approaches of rural territories contribute to the evaluation of rural development practices in which agricultural and non-agricultural business come together.

Goals	Certainty	Range of Impacts	Scale	Specificity
N, H, L, E, S, D	B	0 to +4	G	Wide applicability

Rural development to reduce poverty and improve the rural environment is recognized as an integrated activity requiring policies that take into account the holistic nature of the task. Consequently, current approaches are maintaining a broad vision of agriculture that involves: farmers integrated into the appropriate agricultural production-trade chain with dynamic links to social, economic and environmental activities in their region. Development plans are specific to the needs of the farmer and the rural development sector and recognize the heterogeneity of agriculture and its cultural setting, within and between countries (Sebillote, 2000, 2001)

In Australia, multifunctionality has stimulated a debate about Globalized Productivism versus Land Stewardship.

Goals	Certainty	Range of Impacts	Scale	Specificity
N, H, L, E, S, D	B	Not yet achieved	L	Wide applicability

In Australia, the unsustainability of agriculture lies in the application of European type of farming systems in an environment to which they are inherently unsuited (Gray and Lawrence, 2001), and, in pursuit of market liberalism, the application of neoliberal policies targeting "competitive" or "globalized" productivism (Dibden and Cocklin, 2005). In this scenario, with the increasing influence of multinational agrifood companies, landholders are pressured to increase production and extract the greatest return from the land in a competitive marketplace in ways that do not reward environmental management (Dibden and Cocklin, 2005). To reverse the social, economic and environmental decline of Australian agriculture, the Victorian government has discussed strategies with farmers for moving towards Land Stewardship. The outcome favored voluntary and education-based tools over market-based instruments and saw command-and-control regulation as a last resort (Cocklin et al., 2006, 2007). In this debate, Land Stewardship was seen as a hybrid between the "market-based instruments policy prescription" and a newer "multifunctional approach", with the recognition that people are a vital element in the sustainability equation (Cocklin et al., 2006). Multifunctionality and Land Stewardship therefore emerge as strategies promising new income streams associated with the economic diversification of the enterprise within a more spatially-variable rural space, founded on genuine social, economic and environmental integration.

Participatory land use planning has recently reemerged highlighting its political and economic nature and an increased concern with equity rather than just productivity.

Goals	Certainty	Range of Impacts	Scale	Specificity
S, D	B	0 to +2	G	Wide applicability

The disciplines of land use and rural planning now bring together the different sectors of the rural economy, especially farming, forestry and ecosystem conservation. Comparisons of actual land use with "notional potential" derived from analysis of soils, vegetation, hydrology and climate, have been based on systems of resource survey and assessment (Dalal-Clayton et al., 2003). In the post-colonial era, these systems have tended to be technocratic tools used by centrally-planned economies and development agencies that have played key roles in both the process of conversion of forest to farming, and the improvement of farm productivity (Dalal-Clayton et al., 2003), optimally at a watershed level or regional level. This hierarchical approach was not often recognized by stakeholders, especially politicians, and was neutral to all-important market influences (Dalal-Clayton et al., 2003). Consequently, land use planning has become: (1) more decentralized, often being absorbed into district authorities; (2) more focused on processes of learning based on natural resource capabilities, rather than producing one-off master plans segregating different sectoral land uses; and (3) more based on participatory approaches to recognize the need for greater equity, to identify locally-desirable land use planning options and to improve commitment and "ownership" (Caron, 2001; Lardon et al., 2001; Dalal-Clayton et al., 2003). These approaches have led to better national conservation and development strategies but they usually have major capacity constraints, which result in blunt sector-based plans and that do not realize all the potential synergies.

Modeling water allocation at the territorial level contributes to a more efficient water management.

Goals	Certainty	Range of Impacts	Scale	Specificity
E	B	0 to +2	G	Wide applicability

Optimization economic models on water allocation among competing sectors for decision support have dominated the international literature for a long time (Salman et al., 2001; Weber, 2001; Firoozi and Merrifield, 2003). Recently, there have been an increasing number of studies adopting simulation and multi-objective frameworks. Examples include water allocation between irrigation and hydropower in North Eastern Spain (Bielsa and Duarte 2001), an economic optimization model for water resources planning in areas with irrigation systems (Reca et al., 2001), a multi-objective optimization model for water planning in the Aral Sea Basin which has uncertain water availability (McKinney and Cai, 1997), and water allocation to different user sectors from a single storage reservoir (Babel et al., 2005). Links between policy and basin hydrology for water allocation are now being used to allocate water among users based on flow and shortage rights, consumptive rights and irrigation efficiencies (Green and Hamilton, 2000), although the recent implementation of new approaches needs to be better assessed.

A territorial approach to the examination of land management has mitigated issues of land insecurity, inequitable distribution of land, and social conflict.

Goals	Certainty	Range of Impacts	Scale	Specificity
S	C	-4 to +3	G	Wide applicability

Customary land tenure issues can potentially create social tension if the rights of all farmers and herdsmen are not addressed when developing new land use practices. Understanding local land management makes it possible to assess the impact of policies and to question their relevancy (Platteau, 1996; Ensminger, 1997; DeSoto, 2005), and assess the suitability of individual land rights (LeRoy et al., 1996). Local rights and institutions are now recognized by the international authorities (Deininger and Binswanger, 2001; World Bank, 2003) and entitlement policy is no longer considered to be the only solution. Beyond the identification of the various regulation authorities (Schlager and Ostrom, 1992), the territorial approach now articulates the local level with national and international levels (Lavigne Delville, 1998; Mathieu et al., 2000), thereby taking into account the plurality of systems, local authorities and land rights.

Research has paid little attention to the serious impacts of social conflicts and disorders on agricultural production.

Goals	Certainty	Range of Impacts	Scale	Specificity
N, S	D	-5 to 0	G	Wide applicability

Wars may arise from conflicts for agricultural resources (Collier, 2003), notably for land (Chauveau, 2003), or claims on forest (Richards, 1996), resulting in agricultural stagnation (Geffray, 1990; Lacoste, 2004); declining productivity of crops and livestock and the decreasing access and availability of food (Dreze and Sen, 1990; Stewart, 1993; Macrae and Zwi, 1994); destruction of storage and transformation infrastructures; ground and water pollution; higher food prices and obstacles to the transport of agricultural inputs and products. This stagnation is reinforced by factors like civil disorders, state collapse, urbanization, declining involvement of youth in agriculture, HIV and other diseases, the decline of the agricultural workforce and the development of illegal activities. Although difficult to quantify, the agricultural losses related to wars have been increasing since the 1990s (FAO, 2000a).

Postconflict programs may alleviate difficulties. This is particularly the case with the reorganization of input delivery systems, e.g., as seen in Rwanda, which was addressed by the "Seeds of Hope Project" (Mugungu et al., 1996; www.new-agri.co.uk/01-2/focuson/focuson3.html).

3.3 Objectivity of this Analysis

To determine the balance of this assessment in terms of reporting on positive and negative impacts of AKST, the frequency distribution of reported impacts was determined for each main part of the Chapter (Figure 3-8). The result indicates that about one-third of reported impacts were negative and two thirds positive. Although there were small differences between the subchapters, the trends were similar, suggesting that the authors are in general agreement about balance of this Assessment and the overall outcomes of 50 years of AKST.

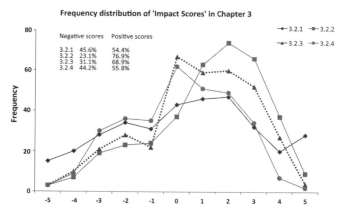

Figure 3-8. *Frequency distribution of impact scores from this Assessment*

3.4 Lessons and Challenges

The fundamental challenge for AKST in rural development is how to make agriculture both more productive and more sustainable as a source of income, food and other products and services for the benefit of all people worldwide, most of whom are living below or a little above the US$2 per day poverty line—but who also suffer many health, livelihood and environmental deprivations that are not best measured in dollars. A new approach to sustainable agriculture has to be achieved despite the growing population pressure on limited sources of all forms of natural capital (especially land, water, nutrients, stocks of living organisms and global climatic stability), many of which have already been severely degraded by former approaches to agricultural production, and which have externalized the costs of the environmental and social impacts of AKST. This Chapter has shown that the current serious situation has resulted from a culture of exploitation, coupled with a uni-dimensional approach that failed to appreciate and develop the multifunctionality of agriculture.

The overriding lesson of this chapter is that, although AKST has made great improvements in productivity, the global focus of AKST to date on production issues has been at the expense of environmental and social sustainability at the local level. Consequently, natural resources have typically been overexploited and the societies have lost some of their traditions and individuality. The sustainable implementation of AKST has been impeded by inadequate understanding, inappropriate policy interventions, socioeconomic exclusion, and a failure to address the real needs of poor people. This has been exacerbated by an overemphasis on trade with industrialized countries and a set of "disconnects" between disciplines, organizations and different levels of society that have marginalized environmental and social objectives. In developing countries, and especially in Africa, the combined effect has been that poor people's livelihoods have not benefited adequately from the Green Revolution and from globalization, due to their exclusion from the benefits of AKST. At the same time, there is a diverse body of work on improving the productivity of degraded farming systems that is based on more sustainable approaches. These are more socially-relevant, pro-poor, approaches to agriculture, with a strong reliance on both natural resources and social capital at community and landscape levels. This body of evidence, albeit disparate at present, is largely based on diversified and integrated farming systems, which are especially appropriate for the improvement of small-scale farms in the tropics. It has a stronger emphasis on environmentally and socially sustainable agriculture and offers the hope of a better future for many millions of poor and marginalized rural households. The overriding challenge is, therefore, to revitalize farming processes and rehabilitate natural capital, based on an expanded understanding of INRM within AKST. Much of this will involve the provision of appropriate information for policy-makers and farmers and the removal of the "disconnections" between different disciplines, organizations and levels of society at the heart of AKST. This will be fundamental for the integration of the different components of AKST and the scaling-up of the existing socially and environmentally sustainable agricultural practices.

This Chapter has presented an analysis of the positive and negative impacts of AKST over the last 50 years, which allows us to address the key IAASTD question: "*What are the development and sustainability challenges that can be addressed through AKST?*" We highlight ten concerns that pose the key AKST challenges to improving agriculture's sustainability, while meeting the needs of a growing population dependent on a limited and diminishing resource base:

First, the fundamental failure of the economic development policies of recent generations has been reliance on the draw-down of natural capital, rather than on production from the "interest" derived from that capital and on the management of this capital. Hence there is now the urgent *challenge* of developing and using AKST to reverse the misuse and ensure the judicious use and renewal of water bodies, soils, biodiversity, ecosystem services, fossil fuels and atmospheric quality.

Second, AKST research and development has failed to address the "yield gap" between the biological potential of Green Revolution crops and what the poor farmers in developing countries typically manage to produce in the field. The *challenge* is to find ways to close this yield gap by overcoming the constraints to innovation and improving farming systems in ways that are appropriate to the environmental, economic, social and cultural situations of resource-poor small-scale farmers. An additional requirement is for farm products to be fairly and appropriately priced so that farmers can spend money on the necessary inputs.

Third, modern public-funded AKST research and development has largely ignored traditional production systems for "wild" resources. It has failed to recognize that a large part of the livelihoods of poor small-scale farmers typically comes from indigenous plants (trees, vegetables/pulses and root crops) and animals. The *challenge* now is to acknowledge and promote the diversification of production systems through the domestication, cultivation, or integrated management of a much wider set of locally-important species for the development of a wide range of marketable natural products which can generate income for the rural and urban resource poor in the tropics—as well as provide ecosystem services such as soil/water conservation and shelter. Those food crops, which will be grown in the shade of tree crops, will need to have been bred for productivity under shade.

Fourth, AKST research and development has failed to fully address the needs of poor people, not just for calories, but for the wide range of goods and services that confer health, basic material for a good life, security, community wellbeing and freedom of choice and action. Partly as a consequence, social institutions that had sustained a broader-based agriculture at the community level have broken down and social sustainability has been lost. The *challenge* now is to meet the needs of poor and disadvantaged people—both as producers and consumers, and to reenergize some of the traditional institutions, norms and values of local society that can help to achieve this.

Fifth, malnutrition and poor human health are still widespread, despite the advances in AKST. Research on the few globally-important staple foods, especially cereals, has been at the expense of meeting the needs for micronutrients, which were rich in the wider range of foods eaten traditionally by most people. Now, wealthier consumers are also facing problems of poor diet, as urban people are choosing to eat highly processed foods that are high in calories and fat, while low in micronutrients. In addition, there are increasing concerns about food safety. The *challenge* is to enhance the nutritional quality of both raw foods produced by poor small-scale farmers, and the processed foods bought by urban rich from supermarkets. A large untapped resource of highly nutritious and health-promoting foods, produced by undomesticated and underutilized species around the world, could help to meet both these needs. Negative health impacts have also arisen from land clearance, food processing and storage, urbanization, use of pesticides, etc., creating procurement and marketing challenges for food industries and regulatory challenges for environmental and food safety organizations.

Sixth, intensive farming is frequently promoted and managed unsustainably, resulting in the destruction of environmental assets and posing risks to human health, especially in tropical and sub-tropical climates. Many practices involve land clearance, soil erosion, pollution of waterways, inefficient use of water, and are dependent on fossil fuels for the manufacture and use of agrochemicals and machinery. The key *challenge* is to reverse this by the promotion and application of more sustainable land use management. Given climate change threats in particular, we need to produce agricultural products in ways that both mitigate and adapt to climate change, that are closer to carbon-neutral, and that minimize trace gas emissions and natural capital degradation.

Seventh, agricultural governance and AKST institutions alike have focused on producing individual agricultural commodities. They routinely separate out the different production systems that comprise agriculture, such as cereals, forestry, fisheries, livestock, etc, rather than seeking synergies and optimum use of limited resources through technologies promoting Integrated Natural Resources Management. Typically, these integrating technologies have been treated as fringe initiatives. The *challenge* now is to mainstream them so that the existing set of technologies can yield greater benefits by being brought together in integrated systems. A range of biological, ecological, landscape/land use planning and sustainable development frameworks and tools can help; but these will be more effective if informed by traditional institutions at local and territorial levels. Because of the great diversity of relevant disciplines, socioeconomic strata and production/development strategies, sustainable agriculture is going to be more knowledge-intensive than ever before. This growing need for knowledge is currently associated with a decline in formal agricultural extension focused on progressive farmers and its replacement by a range of other actors who often engage in participatory activities with a wider range of farmers, but who often need greater access to knowledge. Thus part of the challenge is to reinvent education and training institutions (colleges, universities, technical schools and producer organizations), and support the good work of many NGOs by also increasing long-term investments in the upstream and downstream transfer of appropriate knowledge.

Eighth, agriculture has also been very isolated from non-agricultural production-oriented activities in the rural landscape. There are numerous organizational and conceptual "disconnects" between agriculture and the sectors dealing with (1) food processing, (2) fibre processing, (3) environmental services, and (4) trade and marketing and which therefore limit the linkages of agriculture with other drivers of development and sustainability. The *challenge* for the future is for agriculture to increasingly develop partnerships and institutional reforms to overcome these "disconnects". To achieve this it will be necessary for future agriculturalists to be better trained in "systems thinking" and entrepreneurship across ecological, business and socioeconomic disciplines.

Ninth, AKST has suffered from poor linkages among its key stakeholders and actors. For example: (1) public agricultural research is usually organizationally and philosophically isolated from forestry/fisheries/environment research; (2) agricultural stakeholders (and KST stakeholders in general) are not effectively involved in policy processes for improved health, social welfare and national development, such as Poverty Reduction Strategies; (3) poor people do not have power to influence the development of prevailing AKST or to access and use new AKST; (4) weak education programs limit AKST generation and uptake (especially for women, other disadvantaged groups in society and formal and informal organizations for poor/small farmers) and their systems of innovation are not well connected to formal AKST; (5) agricultural research increasingly involves the private sector, but the focus of such research is seldom on the needs of the poor or in public goods, (6) public research institutions have few links to powerful planning/finance authorities, and (7) research, extension and development organizations have been dominated by professionals lacking the skills base to adequately support the integration of agricultural, social and environmental activities that ensure the multifunctionality of agriculture, especially at the local level. The main *challenge* facing AKST is to recognize all the livelihood assets (human, financial, social, cultural, physical, natural, informational) available to a household and/or community that are crucial to the multifunctionality of agriculture, and to build systems and capabilities to adopt an appropriately integrated approach, bringing this to very large numbers of less educated people—and thus overcoming this and other "disconnects" mentioned earlier.

Finally, since the mid-20th Century, there have been two relatively independent pathways to agricultural development—the "Globalization" pathway and the "Localization" pathway. The "Globalization" pathway has dominated agricultural research and development, as well as international trade, at the expense of the Localization; the grassroots pathway relevant to local communities (Table 3.5). As with any form of globalization, those who are better connected (developed countries and richer farmers) tend to benefit most. The *challenge* now is to redress the balance between Globalization and Localization, so that both pathways can jointly play their optimal role. This concept, described as Third-Generation Agriculture (Buckwell and Armstrong-Brown, 2004), combines the technological efficiency of second-generation agriculture with the lower environmental impacts of first-generation agriculture. This will involve scaling up the more durable and sustainable aspects of the community-oriented "grassroots" pathway on the one hand and thereby to facilitate local initiatives through an appropriate global framework on the other hand. In this way, AKST may help to forge and develop Localization models in parallel with Globalization. This approach should increase benefit flows to poor countries, and to marginalized people everywhere. This scaling up of all the many small

Table 3-5. **Globalization and localization activities.**

Globalization	Localization
Tropical plantations for export markets	Traditional subsistence agriculture
International commodity research by CGIAR	National research by NARS
	National extension services
Green Revolution	NGOs and CBOs
Agribusiness for fertilizers/pesticides and seeds	Farmer training schools
	Participatory Rural Appraisal
Multinational companies for commodity trade	Participatory domestication and breeding
WTO trade agreements	Fair trade
Biopiracy	Water-user associations
Biotechnology	Promotion of indigenous species/germplasm
	Equity and gender initiatives
	Recognition of farmer/community
	IPR
	Agroforestry for soil fertility management

and often rather specific positive impacts of local AKST held by farmers and traders could help to rebuild natural and social capital in the poorest countries, so fulfilling the African proverb:

"If many little people, in many little places, do many little things, they will change the face of the world."

This will also require that developed country economies and multinational companies work to address the environmental and social externalities of the globalized model ("Enlightened Globalization"), by increasing investment in the poorest countries, by honoring their political commitments, and by addressing structural causes of poverty and environmental damage with locally available resources (skills, knowledge, leadership, etc). In turn, this is highly likely to require major policy reform on such issues as trade, business development, and intellectual property rights—especially in relation to the needs of poor people, notably women.

The ten lessons above have drawn very broadly on the literature. A specific lesson-learning exercise covering 286 resource-conserving agricultural interventions in 57 poor countries (Pretty et al., 2006) offers an illustration of the potential of implementing more sustainable approaches to agriculture with existing strategies and technologies. In a study covering 3% of the cultivated land in developing countries (37 million ha), increased productivity occurred on 12.6 million farms, with an average increase in crop yield of 79%. Under these interventions, all crops showed gains in water use efficiency, especially in rainfed crops and 77% of projects with pesticide data showed a 71% decline in pesticide use. Carbon sequestration amounted to 0.35 tonnes C $ha^{-1}y^{-1}$. There are grounds for cautious optimism for meeting future food needs with poor farm households benefiting the most from the adoption of resource-conserving interventions (Pretty et al., 2006). Thus great strides forward can be made by the wider adoption and upscaling of existing pro-poor technologies for sustainable development, in parallel with the development of ways to improve the productivity of these resource-conserving interventions (Leakey et al., 2005a). These can be greatly enhanced by further modification and promotion of some of the socially and environmentally appropriate AKST described in this chapter.

References

Abate, T., A. van Huis and J.K.O. Ampofo. 2000. Pest management strategies in traditional agriculture: An African perspective. Ann. Rev. Entomol. 45:631-659.

Abbiw, D.K. 1990. Useful plants of Ghana: West African uses of wild and cultivated plants. Intermediate Tech. Publ., London.

Abbott J., S. Roberts, and N. Robins. 1999. Who benefits? A social assessment of environmentally-driven trade. IIED, London.

Abdelali-Martini, M, P. Goldey, G.E. Jones and E. Bailey. 2003. Towards a feminization of agricultural labour in Northwest Syria. J. Peasant Studies 30(2):71-94.

Abernethy, L.C. 2003. Administrative and implementation concerns of water rights. In Proc. Int. Working Conf. on Water Rights: Institutional options for improved water allocation. Hanoi, Vietnam. 12-15 Feb 2003. [CD-ROM]. IFPRI, Washington DC.

Adato, M. and R. Meinzen-Dick. 2003. Assessing the impact of agricultural research on poverty using the sustainable livelihoods framework. EPTD Disc. Pap. 89/FCND Disc. Pap. 128. IFPRI, Washington DC.

Adato, M. and R. Meinzen-Dick (ed) 2007. Agricultural research livelihoods and poverty. Studies of economic and social impacts in six countries. IFPRI, Johns Hopkins Univ. Press, Baltimore.

Aerni, P. 2001. Aquatic resources and technology: Evolutionary, environmental, legal and developmental aspects. Sci. Tech. Innovation Disc. Pap. No. 13. Center Int. Dev., Cambridge MA.

Agrawal, A. and C.C. Gibson. 1999. Enchantment and disenchantment. The role of community in natural resources conservation. World Dev. 27(4):629-649.

Agrawal, A. and K. Gupta 2005. Decentralization and participation: The governance of common pool resources in Nepal's Terai. World Dev. 33(7):1101-14.

AGRITEX. 1998. Learning together through participatory extension: A guide to an approach developed in Zimbabwe. Dep. Agric., Tech. Extension Serv., Harare.

Ahlers, R. 2000. Gender issues in irrigation. In C. Tortajada (ed) Women and water management: The Latin American experience. Oxford Univ. Press, New Delhi.

Ahmad, M.D., H. Turral, I. Masih, M. Giordano, and Z. Masood. 2007. Water saving technologies: Myths and realities revealed in Pakistan's rice-wheat systems. IWMI Res. Rep. 108 IWMI, Colombo.

Ahmed, G.J.U., M.A. Hassan, A.J. Mridha, M.A. Jabbar, C.R. Riches, E.J.Z. Robinson et al. 2001. Weed management in intensified lowland rice in Bangladesh. p. 205-210. In Proc. Brighton Crop Protection Conf. — Weeds. 13-15 Nov. 2001. Farnham, UK. Brit. Crop Prot. Council.

Ahmed, M., and M. Lorica. 2002. Improving developing county food security through aquaculture development — lessons from Asia. Food Policy 27:25-141.

Ahuja, M.R., and W.J. Libby. 1993a. Clonal Forestry 1: Genetics and biotechnology. Springer-Verlag, Berlin.

Ahuja, M.R., and W.J. Libby. 1993b. Clonal Forestry 2: Conservation and application. Springer-Verlag, Berlin.

Ainsworth, E.A., and Long, S.P. 2005. What have we learned from 15 years of free-air CO2 enrichment? A meta-analytic review of the responses of photosynthesis, canopy properties and plant production in rising CO_2. New Phytol. 165:351-371.

Akinnifesi, F.K., F. Kwesiga, J. Mhango, T. Chilanga, A. Mkonda, C.A.C. Kadu et al. 2006. Towards the development of Miombo fruit trees as commercial tree crops in southern Africa. For. Trees Livelihoods 16:103-121.

Akobundu, O. 1996. Principles and prospects for integrated weed management in developing countries. p. 591-600. In Proc. Second Int. Weed Control Congress. 25-28 June 1996. Copenhagen.

Alderman, H., and M. Garcia. 1994. Food security and health security: Explaining the levels of nutritional status in Pakistan, Econ. Dev. Cult. Change 42:485-508.

Alex, G., W. Zijp, and D. Byerlee. 2002. Rural extension and advisory services: New directions. Rur. Dev. Strategy Background Pap. 9. World Bank, Washington DC.

Alexiades, M.N., and P. Shanley (ed) 2005. Forest products, livelihoods and conservation: Case studies of non-timber forest product systems. Vol. 3 - Latin America. CIFOR, Bogor, Indonesia.

Ali, M. 1998. Green manure evaluation in tropical lowland rice systems. Field Crops Res. 61:61-7.

Ali, M. 2006. Horticulture revolution for the poor: Nature, challenges and opportunities. Available at http://siteresources.worldbank.org/INTWDR2008/Resources/2795087-1191427986785/Ali_Horticulture_for_the_Poor.pdf.

Ali, M., and Abedullah. 2002. Economic and nutritional benefits from enhanced vegetable production and consumption in developing countries. J. Crop Prod. 6:145-76.

Ali, M., and D. Byerlee. 2002. Productivity growth and resource degradation in Pakistan's Punjab: a decomposition analysis. Econ. Dev. Cult. Change 50:839-864.

Ali, M., H. de Bon, and P. Moustier. 2005. Promoting the multifunctionalities of urban and peri-urban agriculture in Hanoi. Urban Agric. 15:11-13.

Ali, M., and U. Farooq. 2004. Dietary diversity to enhance rural labor productivity: Evidence from Pakistan. In Proc. Ann. Meeting AAEA. Denver. 1-4 July 2004.

Ali, M., I.A. Malik, H.M. Sabir, and B. Ahmad. 1997. Mungbean green revolution in Pakistan. Tech. Bull. No. 24. AVRDC, Shanhua, Taiwan.

Ali, M., and J.H. Narciso. 1996. Farmers perceptions and economic evaluation of green manure use in the rice-based farming system. Trop. Agric. 73(2):148-154.

Ali, M., N.T. Quan, and N.V. Nam. 2006. An analysis of food demand pattern in Hanoi: Predicting the structural and qualitative changes. Tech. Bull. 35. AVRDC, Shanhua, Taiwan.

Ali, M., and C.S. Tsou. 2000. The integrated research approach of the Asian Vegetable Res. and Dev. Center (AVRDC) to enhance micronutrient availability. Food Nutr. J. 21:472-482.

Almekinders, C. and N. Louwaards. 1999. Farmers' seed production: New approaches and practices. Intermediate Tech. Publ., London.

Almekinders, C., and W. de Boef. 2000. Encouraging diversity: The conservation and development of plant genetic resources. Intermediate Tech. Publ., London.

Altieri, M.A. 1989. Agroecology: A new research and development paradigm for world agriculture. Agric. Ecosyst. Environ. 27:37-46.

Altieri, M.A. 1994. Biodiversity and pest management in agroecosystems. Food Products Press, NY.

Altieri, M.A. 1999. The ecological role of biodiversity in agroecosystems. Agric. Ecosyst. Environ. 74:19-31.

Altieri, M.A., and C.I. Nicholls, 1999. Biodiversity, ecosystem function, and insect pest management in agricultural systems. p. 69-84. In W.W. Collins and C.O.Qualset (ed) Biodiversity in agroecosystems. CRC Press, NY.

Ameur, C. 1994. Agricultural extension: A step beyond the next step. Tech. Pap. 247. World Bank, Washington DC.

Anderson J.R., G. Feder, and S. Ganguly. 2004. The rise and fall of training and visit extension: An Asian mini-drama with an African epilogue. World Bank, Washington DC.

Anderson, P.K., A. A. Cunningham, N.G. Patel, F.J. Morales, P.R. Epstein and P. Daszak. 2004. Emerging infectious diseases of plants: pathogen pollution, climate change and agrotechnology drivers. Trends Ecol. Evol. 19(10):535-544.

Andow, D.A., and A. Hilbeck 2004. Science-based risk assessment for non-target effects of transgenic crops. BioScience 54:637-649.

Andow, D.A., G.L. Lövei and S. Arpaia 2006. Ecological risk assessment for Bt crops. Nature Biotechnology (reply) 24:749.

Andrews, P. 2005. Back from the brink. ABC Books, Australia.

Anegbeh, P.O., V. Ukafor, C. Usoro, Z. Tchoundjeu, R.B.B. Leakey, and K. Schreckenberg. 2004. Domestication of Dacryodes edulis: 1. Phenotypic variation of fruit traits from 100 trees in southeast Nigeria. New Forests 29:149-160.

Anegbeh, P.O., C. Usoro, V. Ukafor, Z. Tchoundjeu, R.R.B. Leakey, and K. Schreckenberg. 2003. Domestication of Irvingia gabonensis: 3. Phenotypic variation of fruits and kernels in a Nigerian village. Agrofor. Syst. 58:213-218.

Angelsen A., and D. Kaimowitz. 2001. Agricultural technologies and tropical deforestation. CABI, Wallingford, UK.

Anobah, R. 2000. Development of organic markets in western Africa. In T. Alföldi, W. Lockeretz and U. Nigli (ed) IFOAM 2000 -The world grows organic. Proc. 13th Int. IFOAM Scientific Conf. CH-Basel, 28-31 Aug. 2000. Hochschulverlag AG an der ETH, Zürich.

Antona, M., P. D'Aquino, S. Aubert, O. Barreteau, S. Boissau, F. Bousquet et al. (Collectif Commod). 2003. Our companion modelling approach (La modélisation comme outil d'accompagnement). J. Artificial Societies Social Simulation 6(2). Available at http://jasss.soc.surrey.ac.uk/6/2/1.html.

Appleby, A. 2005. A history of weed control in the United States and Canada - a sequel. Weed Sci. 53:762- 768.

Arnold, J.E.M., and M. Ruiz Pérez. 1998. The role of non-timber forest products in conservation and development. p. 17-41. In E. Wollenberg and A. Ingles (ed) Incomes from the forest: Methods for the development and conservation of forest products for local communities. CIFOR, Bogor, Indonesia.

Arora, K., S.K. Mickelson, and J.L. Baker. Effectiveness of vegetated buffer strips in reducing pesticide transport in simulated runoff. Trans. ASAE 46(3):635-644, 2003.

Arregui, M.C., A. Lenardon, D. Sanchez, M.I. Maitre, R. Scotta, and S. Enrique. 2004. Monitoring glyphosate residues in transgenic glyphosate-resistant soybean. Pest Manage. Sci. 60:163-166.

Ash, N., A. Fazel, Y. Assefa, J. Baillie, M. Bakarr, S. Bhattacharjya et al. 2007. Biodiversity. p. 157-192. In GEO4 Global Environmental Outlook: Environment for development. Sec. B. UNEP, Nairobi.

Ashley, C., and D. Carney. 1999. Sustainable livelihoods: Lessons from early experience. DFID Issues Seriee. DFID, London.

Ashley, C. and K. Hussein. 2000. Developing methodologies for livelihood impact assessment: Experience of the African Wildlife Foundation in East Africa. ODI Working Pap. 129. Available at http://www.odi.org.uk/publications/wp129.pdf.

Ashley, S., S. Holden, and P. Bazeley. 1999. Livestock in poverty focused development. Livestock in development. Crewkerne, UK.

Asiabaka, C. 2002. Promoting sustainable extension approaches: farmer field school FFS and its role in sustainable agricultural development in Africa. CODESRIA-IFS Sustainable Agriculture Initiative Workshop. Kampala, Uganda. 15-16 Dec. 2002. Available at http://www.codesria.org/Links/conferences/ifs/Asiabaka.pdf.

Aspelin, A. 2003. Pesticide usage in the United States: Trends during the 20th Century. CIPM Tech. Bull. #105. Center for Integrated Pest Manage., North Carolina State Univ., Raleigh NC.

Aspelin, A.L., and G.R. Grube. 1999. Pesticide industry sales and usage: 1996 and 1997 market estimates. Office of Prevention, Pesticides, and Toxic Substances, Pub # 733-R-99-001. EPA, Washington DC.

Assey, P. 2007. Environment at the heart of Tanzania's development: Lessons from Tanzania's national strategy for growth and reduction of poverty (MKUKUTA). Nat. Resour. Issues Ser, No. 6. Vice-President's Office, Dar es Salaam and IIED, London.

Atangana, A.R., Z. Tchoundjeu, J-M. Fondoun, E. Asaah, M. Ndoumbe, and R.R.B. Leakey. 2001. Domestication of Irvingia gabonensis: 1. Phenotypic variation in fruit and kernels in two populations from Cameroon. Agrofor. Syst. 53:55-64.

Atangana, A.R., V. Ukafor, P.O. Anegbeh, E. Asaah, Z. Tchoundjeu, C. Usoro et al. 2002. Domestication of Irvingia gabonensis: 2. The selection of multiple traits for potential cultivars from Cameroon and Nigeria. Agrofor. Syst. 55:221-229.

Attipoe, L., A. van Andel and S.K. Nyame. 2006. The Novella Project: Developing a sustainable supply chain for Allanblackia oil. p. 179-189 In R. Ruben, M. Slingerland, H. Nijhoff (ed) Agro-food chains and networks for development. Springer, NY. Available at http://library.wur.nl/ojs/index.php/frontis/article/view/982/553.

Attwell, C.A.M., and F.P.D. Cotterill. 2000. Postmodernism and African conservation science. Biodivers. Conserv. 9:559-577.

Auld, B., and K.M. Menz. 1997. Basic economic criteria for improved weed management in developing countries. Expert consultation on weed ecology and management. Rome. 22-24 Sept. 1997. FAO, Rome.

Austin, R.B. 1999. Yield of wheat in the United Kingdom: Recent advances and prospects. Crop Sci. 39:1604-1610.

Austin, R.B., J. Bingham, R.D. Blackwell, L.T. Evans, M.A. Ford, C.L. Morgan, and M. Taylor. 1980. Genetic improvements in winter wheat yields since 1900 and associated physiological changes. J. Agric. Sci. Camb. 94:675-689.

Australian Government Biofuels Task Force. 2005. Rep. of the biofuels task force to the Prime Minister. Commonwealth of Australia, Barton.

AVRDC (The World Vegetable Center). 1987. AVRDC 1985 Progress Rep. Shanhua, Taiwan.

Awono, A., O. Ndoye, K. Schreckenberg, H. Tabuna, F. Isseri, and L. Temple. 2002. Production and marketing of Safou (*Dacryodes edulis*) in Cameroon and internationally: Market development issues. For. Trees Livelihoods 12:125-147.

Axinn, G.H. and Thorat, S. 1972. Modernising world agriculture: A comparative study of agricultural extension education systems. Praeger, NY.

Ayalew, W., J.M. King, E. Bruns, and B. Rischkowsky. 2001. Economic evaluation of smallholder subsistence livestock production: Lessons from an Ethiopian goat development program. Milano: The Fondazione Eni Enrico Mattei Note di Lavoro Series Index: 107. Available at (http://www.feem.it/web/activ/_activ.html.

Ayantunde, A.A., M. Kango, P. Hiernaux, H.M.J. Udo, and R. Tabo. 2007 Herders' perceptions on ruminant livestock breeds and breeding management in southwestern Niger, Human Ecol. (USA) 35:139-149.

Azzam, A., S. Azzam, S. Lhaloui, A. Amri, M. El Bouhssini, and M. Moussaoui. 1997. Economic returns to research in Hessian fly resistant bread wheat varieties in Morocco. J. Econ. Entomol. 90:1-5.

Babel, M.S., A. Dasgupta, and D.K. Nayak. 2005. A model for optimal allocation of water to competing demands, Water Res. Manage. 19:693-712.

Badgley, C., J. Moghtader, E. Quintero, E. Zakem, M.J. Chappell, K. Aviles-Vazquez, A. Samulon and I. Perfecto. 2007. Organic agriculture and the global food supply. Renewable Agriculture and Food Systems 22: 86-108.

Baenziger, P., W. Russell, G. Graef, and B. Campbell. 2006. Improving lives: 50 years of crop breeding, genetics, and cytology. Crop Sci. 46:2230-2244.

Bailis, R., M. Ezzati, and D.M. Kammen. 2005. Mortality and greenhouse gas impacts of biomass and petroleum energy futures in Africa. Science 308:98-103.

Baker, J.M., T.E. Ochsner, R.T. Venterea, and T.J. Griffis. 2007. Tillage and soil carbon sequestration - What do we really know? Agric. Ecosyst. and Environ. 118:1-5.

Bakker, M., R. Barker, R. Meinzen-Dick, F. Konradsen (ed) 1999. Multiple uses of water in irrigated areas: A case study from Sri Lanka. SWIM Pap. 8. IWMI, Colombo.

Balasubramanian, V., A.C. Morales, R.T. Cruz, and S. Abdulrachman. 1999. On-farm adaptation of knowledge-intensive nitrogen management technologies for rice systems. Nutr. Cycl. Agroecosys. 53:59-69.

Balasubramanian, V., A.C. Morales, T.M. Thiyagarajan, R. Nagarajan, M. Babu, S. Abdulrachman, and L.H. Hai. 2000. Adoption of the chlorophyll meter (SPAD) technology for real-time N management in rice: A review. Int. Rice Res. Newsl. 25:4-8.

Balce, G.R., T.S. Tjaroko, and C.G. Zamora. 2003. Overview of biomass for power generation in Southeast Asia. ASEAN Center for Energy, Jakarta.

Balland, J., and J. Platteau. 1996. Halting degradation of natural resources: Is there a role for rural communities? FAO, Rome and Clarendon Press, Oxford.

Bambaradeniya, C.N.B., and F.P. Amerasinghe. 2004. Biodiversity associated with the rice field agro-ecosystem in Asian Countries: A brief review. IWMI Working Pap. 63. IWMI, Colombo.

Banerjee, A., and E. Duflo 2005. Growth theory through the lens of development economics. p. 473-552. *In* Handbook Dev. Econ., Vol. 1a. Elsevier, Amsterdam.

Banerjee, R. 2006. Comparison of options for distributed generation in India. Energy Policy 24:101-11.

Bantilan, M.C.S., U.K. Deb, and S.N. Nigam. 2003. Impacts of genetic improvement in groundnut. p. 293-313. *In* R.E. Evenson and D. Gollin (ed) Crop variety improvement and its effect on productivity. CABI, Wallingford, UK.

Banziger, M., P. Setimela, D. Hodson, and B. Vivek. 2006. Breeding for improved abiotic stress tolerance in maize adapted to southern africa. Agric. Water Manage. 80:212-224.

Barbash, J.E., G.P. Thelin, D.W. Kolpin, and R.J. Gilliom. 1999. Distribution of major herbicides in ground water of the United States. Water Resources Investigation Rep. 98-4245. USGeolgical Survery, Washington DC.

Barbier, E.B. 1997. The economic determinants of land degradation in developing countries. Philos. Trans. R. Soc. Lond. B Biol. Sci. 352(1356):891-899.

Barbier, E.B. and M. Cox. 2002. Economic and demographic factors affecting mangrove loss in the coastal provinces of Thailand, 1979-1996. Ambio 31(4):351-357.

Baris, E., S. Rivera, Z. Boehmova, and S. Constant. 2006. Indoor air pollution in cold climates: The cases of Mongolia and China. Knowledge Exchange Ser. No. 8 Dec. ESMAP, World Bank, Washington DC.

Barr, A. 2002. The functional diversity and spillover effects of social capital. J. African Econ. 11(1):90-113.

Bartlett, A. 2004. Entry points for empowerment. A report for CARE Bangladesh. Available at http://www.communityipm.org/docs/Bartlett-EntryPoints-20Jun04.pdf.

Bass, S., and E. Morrison. 1994. Shifting cultivation in Thailand, Laos and Vietnam: Regional overview and policy recommendations. Int. Inst. Environ. Dev., For. Land Use Ser. No. 2. IIED, London.

Bass, S., K. Thornber, M. Markopoulos, S. Roberts, and M. Grieg-Gran. 2001. Certification's impacts on forests, stakeholders and supply chains. IIED, London.

Bazile, D., 2006. State-farmer partnerships for seed diversity in Mali. Gatekeeper Ser., 127. IIED, London.

Bazzano, L.A., J. He, L.G. Ogden, C. Loria, S. Vupputuri, L. Myers, and P.K. Whelton. 2001. Legume consumption and risk of coronary heart disease in US men and women. Arch. Internal Med. 161:2573-2578.

Becker, M., and D.E. Johnson. 1999. Rice yield and productivity gaps in irrigated systems of the forest zone of Cote d'Ivoire. Field Crops Res. 60:201-208.

Beckie, H.J., G. Séguin-Swartz, H. Nair, S.I. Warwick, and E. Johnson. 2004. Multiple herbicide-resistant canola can be controlled by alternative herbicides. Weed Sci. 52: 152-157.

Beer, J., M. Ibrahim, and A. Schlonvoigt. 2000. Timber production in tropical agroforestry systems of Central America. p. 777-786. *In* Krishnapillay et al (ed) Forests and society: The role of research. Vol. 1 Subplenary Sessions. XXI IUFRO World Congress, Kuala Lumpur, 7-12 Aug 2000.

Belcher, B.M. 2003. What isn't an NTFP? Int. For. Rev. 5:161-168.

Benbrook, C.M. 2003. Impacts of genetically engineered crops on pesticide use in the United States: The first eight years. BioTech InfoNet Tech. Peper No 6. Available at www.biotech-info.net/technicalpaper6.html.

Benbrook, C.M. 2004. Genetically engineered crops and pesticide use in the United States: The first nine years. BioTech InfoNet Tech. Pap. 7. Available at www.biotech-info.net/Full_version_first_nine.pdf.

Benbrook, C. 2005. Rust, resistance, run down soils, and rising costs — problems facing soybean producers in Argentina. [Online] www.greenpeace.org/international_en/reports/ex-summary?item_id=715074&language_id=en.

Bene, C., and S. Heck. 2005a. Fish and food security in Africa. NAGA, WorldFish Center Quar. 28(3&4):14-18.

Bene, C., and S. Heck 2005b. Fisheries and the Millennium Development Goals: Solutions for Africa. NAGA, WorldFish Center Q. 28(3&4):19-23.

Bengtsson, J., J. Ahnstrom, and A.C. Weibull. 2005. The effects of organic agriculture on biodiversity and abundance: a meta-analysis. J. Appl. Ecol. 42(2):261-269.

Bennett, J. 2006. Evaluation and deployment of transgenic varieties [Online] http://www.generationcp.org/arm/ARM06/day_1/SP3_session/transgenic_varieties.pdf (verified Feb 07).

Bennett, R., Y. Ismael, S. Morse, and B. Shankar.

2004a. Reductions in insecticide use from adoption Bt cotton in South Africa: impacts on economic performance and toxic load to the environment. J. Agric. Sci. (Camb) 142:665-674.

Bennett, R.M., Y. Ismael, V. Kambhampati, and S. Morse. 2004b. Economic impact of genetically modified cotton in India. AgBioForum 7:96-100.

Berg, C. 2004. World fuel ethanol analysis and outlook. Available at http://www.distill.com/World-Fuel-Ethanol-A&O-2004.html.

Bernauer, T. 2003. Genes, trade and regulation: The seeds of conflict in food biotechnology. Princeton Univ. Press, Princeton, NJ.

Bernier, G., and C. Perilleux. 2005. A physiological overview of the genetics of flowering time control. Plant Biotech. J. 3:3-16.

Bielsa, J., R. Duarte. 2001. An economic model for water allocation in north eastern Spain. Water Resour. Dev. 17(3):397-410.

Bignell, D.E., J. Tondoh, L. Dibog, S.P. Huang, F. Moreira, D. Nwaga et al. 2005. Belowground biodiversity assessment: Developing a key functional group approach in best-bet alternatives to slash and burn. In C. Palm et al. (ed) Slash and burn: The search for alternatives. Columbia Univ. Press, NY.

Binswanger, H.P. 1981. Attitudes toward risk: Theoretical implications of an experiment in rural India. The Econ. J. 91(364):867-890.

Birley, M.H., and K. Lock. 1999. The health impacts of peri-urban natural resource development. Int. Centre Health Impact Assessment, Liverpool Sch. Trop. Med., Liverpool.

Bishop-Sambrook, C. 2004. Farm power and smallholder livelihoods in Sub-Saharan Africa. Available at http://www.fao.org/sd/dim_pe4/pe4_040902a1_en.htm.

Black, R., 2003. Micronutrient deficiency-an underlying cause of morbidity and mortality. Bull. WHO 2003;81:79.

Black, R.E., K.H. Brown, S. Becker, A. Arma, and M.H. Merson. 1982. Contamination of weaning foods and transmission of enterotoxigenic Escherichia coli diarrhoea in children in rural Bangladesh. Trans. R. Soc. Trop. Med. Hyg. 76:259-64.

Blackshaw, R.E., L.J. Molnar, and H.H. Janzen. 2004. Nitrogen fertilizer timing and application method affect weed growth and competition with spring wheat. Weed Sci. 52:614-622.

Blackwood, C.B., and J.S. Buyer. 2004. Soil microbial communities associated with Bt and mon-Bt corn in three soils. J. Environ. Qual. 33:832-836

Blarel, B. 1994. Tenure security and agricultural production under land scarcity: The case of Rwanda. p. 71-96. In J.W. Bruce, and S.E. Migot-Adholla (ed) Searching for land tenure security in Africa. Kendall/Hunt, Dubuque, Iowa.

Blay, D., E. Bonkoungou, S.A.O. Chamshama, and B. Chikamai. 2004. Rehabilitation of degraded lands in sub-Saharan África:

Lessons learned from selected case studies. Forestry Res. Network for Sub-Saharan Africa (Fornessa. Int. Union of Forest Res. Org.. Available at (http://www.etfrn.org/ETFRN/workshop/degradedlands/documents/synthesis_all.pdf.

Block, S., and P. Webb. 2001. The dynamics of livelihood diversification in post famine Ethiopia. Food Policy 26:333-350.

Boelens, R., and M. Zwarteveen. 2002. Gender dimensions of water control in Andean Irrigation. In R. Boelens and P. Hoogendam (ed) Water rights and empowerment. Van Gorcum, Assen, Netherlands.

Boffa, J-M. 1999. Agroforestry parklands in Sub-Saharan Africa. FAO Conserv. Guide 34. FAO, Rome.

Boffa, J-M., G. Yaméogo, P. Nikiéma, and D.M. Knudson. 1996. Shea nut (*Vitellaria paradoxa*) production and collection in agroforestry parklands of Burkina Faso. p. 110-122. In R.R.B. Leakey et al. (ed) Non-wood forest products 9. FAO, Rome.

Bojo, J. and R.C. Reddy. 2003. Status and evolution of environmental priorities in the Poverty Reduction Strategies: A review of 50 poverty reduction strategy papers. Environ. Econ. Series Pap. 93, World Bank, Washington DC.

Bonkoungou, E.G., M. Djimdé, E.T. Ayuk, I. Zoungrana, and Z. Tchoundjeu. 1998. Taking stock of agroforestry in the Sahel — Harvesting results for the future. ICRAF, Nairobi.

Booth, D., J. Holland, J. Hentschel, P. Lanjouw, and A. Herbert. 1998. Participation and combined methods in African poverty assessment: Renewing the agenda. DFID, London.

Borrini-Feyerabend, G. 1997. Beyond fences. Seeking social sustainability in conservation. IUCN, Gland, Switzerland.

Bosc P.-M., D. Eychenne, K. Hussein, B. Losch, M.-R. Mercoiret, P. Rondot et al. 2002. The role of rural producer organisations in the World Bank rural development strategy. Rural Strategy Background Pap. No 8. World Bank, Washington DC.

Boswinkel, J.A., 2000. Information Note. Int. Groundwater Resources Assessment Centre (IGRAC). Netherlands Institute of Applied Geoscience, Netherlands.

Bouis, H.E., R.D. Graham, and R.M. Welch. 2000. The Consultative Group on International Agricultural Research (CGIAR) micronutrients project: Justification and objective. Food Nutrition Bull. 21(4): 374-381. Available at http://www.ifpri.org/pubs/articles/2000/bouis00_02.pdf.

Boukar, O., L. Kong, B.B. Singh, L. Murdock, and H. W. Ohm. 2004. AFLP and AFLP-Derived SCAR markers associated with Striga gesnerioides resistance in cowpea. Crop Sci. 44:1259-1264.

Boulton, A.J. 1999. An overview of river health assessment: philosophies, practice, problems and prognosis. Freshwater Biol. 41:469-479.

Boulton, A.J. 2003. Parallels and contrasts in the effects of drought on stream macro invertebrate assemblages. Freshwater Biol. 48:1173-1185.

Bourguignon F., C. Morrisson, and J. De Melo. 1991. Poverty and income distribution during adjustment: issues and evidence from the OECD project. World Dev. 19:1485-1508.

Boussard, J.M., B. Davidron, F. Gerard, and T. Voituriez. 2005. Food security and agricultural development in Sub-Saharan Africa: Building a case for more support. Background document. Final Rep., FAO, Rome.

Boussard, J.-M., F. Gérard, M.-G. Piketty, M. Ayouz, and T. Voituriez. 2006. Endogenous risk and long run effects of liberalisation in a global analysis framework. Econ. Modelling 23:457-475.

Brackett, R.E. 1999. Incidence, contributing factors, and control of bacterial pathogens in produce. Postharvest Biol. Tech. 15:305- 311.

Braun, A., G. Thiele and M. Fernandez. 2000. Farmer Field Schools and local agricultural research committees: Complementary platforms for integrated decision making in sustainable agriculture. AgREN Network Pap. No. 105. ODI, London.

Bricknell, C. (ed) 1996. A-Z encyclopedia of garden plants. Royal Hort. Soc., Dorling Kindersley, London.

Brookes, G. and P. Barfoot. 2006. Global impact of biotech crops: Socio-economic and environmental effects in the first ten years of commercial use. AgBioForum 9:139-151.

Brovelli, E. 2005. Different paths for the fresh-market and the neutriceutical industry. Acta Hort. (ISHS) 682:859-864.

Brown, L.R. 2005. The Japan syndrome. In Outgrowing the Earth: The food security challenge in an age of falling water tables and rising temperatures. W.W. Norton, NY.

Bruce, J.W. 1998. Country profiles of land tenure: Africa, 1996. LCT Res. Pap. No.130. Univ. Wisconsin/Land Tenure Center, Madison.

Bruce, J.W. and S.E. Migot-Adholla (ed) 1994. Searching for land tenure security in Africa, Kendall/Hunt, Dubuque, Iowa.

Bruinsma, J. (ed) 2003. World agriculture: Towards 2015/2030 — An FAO perspective. FAO, Rome.

Brummett, R.E. 1999. Integrated aquaculture in Sub-Saharan Africa. Environ. Dev. Sustainability 1(3/4):315-321.

Bruns, B. 2007. Pro-poor intervention strategies. Irrigated Agric. Irrigation Drainage 56 (2-3):237-246.

Bruns, B., and J. Ambler. 1992. The use of community organizers in improving people's participation in water resources development. Promoting farmer contributions in irrigation system development. Visi, Irigasi Indonesia 6:33-41.

Bruns, B. R., and R. Meinzen-Dick. 2000. Negotiating Water Rights. Intermediate Tech. Publ., London.

Bryceson, D.F. 1999. African rural labour, Income diversification and livelihood approaches: A long term perspective. Rev. Afric. Pol. Econ. 80:171-189.

Buck, L.E., T.A. Gavin, D.R. Lee, N.T. Uphoff, D. Chandrasekharan Behr, L.E. Drinkwater et al. 2004. Ecoagriculture: A review and assessment of its scientific foundations. Cornell Univ. and Univ. Georgia, SANREM CRSP, Athens GA.

Buck, L.E., T.A. Gavin, N.T. Uphoff, and D.R. Lee. 2007. Scientific assessment of ecoagriculture. p. 20-45. In S.J. Scherr and J.A. McNeely (ed) Farming with nature: The science and practice of ecoagriculture. Island Press, NY.

Buckwell, A.E. and S. Armstrong-Brown. 2004. Changes in farming and future prospects: technology and policy. IBIS: Int. J. Avian Sci. 146 (2)14-21.

Buhlmann, H., and M. Jager (ed) 2001. Skills development in Swiss Development Cooperation: Insights and outlook. Swiss Dev. Cooperation, Berne.

Bullock, S. 1993. Women and work. Women and World Dev. Ser., Zed Publ., London.

Bureau, J.C., and A. Matthews. 2005. EU agricultural policy: What developing countries need to know. IIIS Disc. Pap. 91. Trinity College, Dublin.

Bureau, J.C., W. Jones, E. Gozlan, and S. Marette. 2001. Issues in demand for quality and trade. p. 3-33. In B. Krissof, M. Bohman, J. Caswell (ed) Global food trade and consumer demand for quality. Kluwer Academic Publ., NY.

Buresh, R.J., and P.J.M. Cooper. 1999. The science and practice of short-term fallows. Agrofor. Syst. 47:1-358.

Burke, M. 2003. Farm scale evaluations. Managing GM crops with herbicides — effects on farmland wildlife. Farmscale Evaluations Res. Consortium and Scientific Steering Committee, UK.

Butler, L.M. and R.E. Mazur. 2007. Principles and processes for enhancing sustainable rural livelihoods: Collaborative learning in Uganda. Int. J. Sustain. Dev. World Ecol. 14: 604-617.

Bwibo, N.O. and C.G. Neumann. 2003. The need for animal source foods by Kenyan children. J. Nutr. 133(Suppl.):3936S- 3940S.

Byerlee, D., and P. Moya. 1993. Impacts of international wheat breeding research in the developing world, 1966-90. CIMMYT, Mexico.

Byerlee, D., B. Myers, and T. Jayne. 2006. Managing food price risks and instability in an environment of market liberalization. ARD, World Bank, Washington DC.

CA (Comprehensive Assessment of Water Management in Agriculture). 2007. Water for food, water for life: A comprehensive assessment of water management in agriculture. Summary. Earthscan, London and IWMI, Colombo.

Caballero, B. 2005. A nutrition paradox- underweight and obesity in developing countries. New Engl. J. Med. 352:1514-1516.

Calder, I.R. 2005. Blue revolution: Integrated land and water resource management. 2nd ed. Earthscan, London.

Campbell, A. 1994. Landcare: Communities shaping the land and the future. Allen and Unwin, Australia.

Campbell, B.M., and J.A. Sayer. 2003. Integrated natural resources management: Linking productivity, the environment and development. CABI Publ., Wallingford.

Campinhos, E. 1999. Sustainable plantations of high yield Eucalyptus trees for production of fiber: The Aracruz case. New For. 17:129-143.

Canadian Food Inspection Agency. 2004. Assessment criteria for determining environmental safety of plants with novel traits. Directive 94-08. Available at http://www.inspection.gc.ca/english/plaveg/bio/dir/dir9408e.shtml.

Carney, D. 1998 Sustainable rural livelihoods: What contribution can we make? DFID, London.

Caron, P. 2001. Zonage à dires d'acteurs: des représentations spatiales pour comprendre, formaliser et décider. In p.343-357. Le cas de Juazeiro, au Brésil, Représentations spatiales et développement territorial. Hermès, Paris.

Caron, P. 2005. A quels territoires s'intéressent les agronomes? Le point de vue d'un géographe tropicaliste. EDP Sciences. Natures Sci. Sociétés 13:145-153.

Caron, P., J.-P.Tonneau, E. Sabourin. 1996. Planification locale et régionale: enjeux et limites. Le cas du Brésil Nordeste. In Globalisation, competitivness and human security: Challenge for development policy and institutional change. VIIIth Conf. European Assoc. Dev., Res. and Training Institutes. Vienne. 11-14 Sept. 1996. EADI.

Cassman, K.G., A. Dobermann, D.T. Walters. 2002. Agroecosystems, nitrogen-use efficiency, and nitrogen management. AMBIO 31:132-140.

Cassman, K.G., A. Dobermann, D.T. Walters and H. Yang. 2003. Meeting cereal demand while protecting natural resources and improving environmental quality. Ann. Rev. Environ. Resour. 28:315-358.

Cassman, K.G., S. Wood, P.S. Choo, H.D. Cooper, C. Devendra, J. Dixon et al. 2005. Cultivated systems. p. 745-794. In R. Hassan et al. (ed) Ecosystems and human well-being: current state and trends: Findings of the Conditions and Trends Working Group. The Millennium Ecosystem Assessment Ser. v. 1. Island Press, NY.

Castillo, G.E., R.E. Namara, H. Munk Ravnborg, M.A. Hanja, L. Smith and M. H. Hussein. 2007. Reversing the flow: agricultural water management pathways for poverty reduction. p. 149-191. In D. Molden (ed) Water for food, water for life: A comprehensive assessment of water management in agriculture. Earthscan, London and IWMI, Colombo.

Catley, A., P. Irungu, K. Simiyu, J. Dadye, W. Mwakio, J. Kiragu and S.O. Nyamwaro. 2002. Participatory investigations of bovine trypanosomiasis in Tana River District Kenya. Med. Vet. Entomol. 16(1):55- 66.

Catley, A., S. Okoth, J. Osman, T. Fison, Z. Njiru, J. Mwangi et al. 2001. Participatory diagnosis of a chronic wasting disease in cattle in southern Sudan. Prev. Vet. Med. 51:161-181.

Cattaneo, M.G., C. Yafuso, C. Schmidt, C-Y. Huang, M. Rahman, C. Olson et al. 2006. Farm-scale evaluation of the impacts of transgenic cotton on biodiversity, pesticide use, and yield. PNAS 103:7571-7576.

Cavers, S., C. Navarro, A.J. Lowe. 2003. A combination of molecular markers (cpDNA PCR-RFLP, AFLP) identifies evolutionarily significant units in Cedrela odorata L. (Meliaceae) in Costa Rica. Conserv. Genetics 4:571-580.

CBD. 1992. Convention on Biological Diversity: Text and Annexes. Secretariat of the Convention on Biological Diversity, Montreal.

CBD. 2000. Cartagena Protocol on Biosafety to the Convention on Biological Diversity: Text and annexes. Secretariat Convention Biol. Control, Montreal.

Cebu Study Team. 1992. A child health production function estimated from longitudinal data. J. Dev. Econ. 38:323-351.

Ceccarelli, S., and S. Grando. 1991. Environment of selection and type of germplasm in barley breeding for low-yielding conditions. Euphytica 57:207-219.

Ceccarelli, S., S. Grando, and J.A.G. Vanleur. 1987. Genetic diversity in barley landraces from Syria and Jordan. Euphytica 36:389-405.

CED. 2003. Reducing global poverty: The role of women in development. Committee Econ. Dev. NY.

Célarier, M.-F., and D. Marie-Vivien. 2001. Les droits de propriété intellectuelle. Guide pratique. Documents de la direction scientifique et de la direction des relations extérieures n°4. CIRAD, F-Montpellier.

Center for Food Safety. 2006. Genetic engineering. California food and agriculture: Report card. Genetic Engineering Policy Alliance, Morro Bay, CA, USA.

Cerdeira, A.L., and S.O. Duke. 2006. The current status and environmental impacts of glyphosate-resistant crops - A review. J. Environ. Qual. 35:1633-1658.

Cerf, M., D. Gibbon, B. Hubert, R. Ison, J. Jiggins, M. Paine et al. 2000. Cow up a tree. Knowing and learning for change in agriculture. Case studies from industrialised countries. INRA. Paris.

Cernea, M.M. 2003. For a new economics of resettlement: A sociological critique of the compensation principle. Int. Social Sci. J. 55(175):37-45.

Cernea, M.M. (ed) 1999. The economics of involuntary resettlement: Questions and challenges. World Bank, Washington DC.

CGAP. 2005. Managing risks and designing products for agricultural microfinance: Features of an emerging model. Occasional Pap. No. 11. Consultative Group to Assist the Poor, Washington DC.

Chabe-Ferret S., J. Gourdon, M.A. Marouani, and T. Voituriez. 2006. Trade-Induced changes in economic inequalities: Methodological issues and policy implications for developing countries. ABCDE World Bank Conf., Tokyo. 29-30 May 2006.

Chancellor, F. 2000. Sustainable irrigation and the gender question in Southern Africa. *In* A. Pink (ed) Sustainable development international. 3rd ed. ICG Publishing Ltd., London.

Chant, S. (ed) 1992. Gender and migration in developing countries. Belhaven Press, London.

Chatfield, R. 2004. The photochemistry of ozone. Presentation at Univ. Kwazulu-Natal. Available at http://geo.ar.c.nasa.gov/sgg/chatfield/PhotochemFreeTrop.pdf.

Chauveau, J.P. 2003. Crise foncière, crise de la ruralité et relations entre autochtones et migrants sahéliens en Côte d'Ivoire forestière. Communication au colloque OCDE — Club du Sahel et Ministère des Affaires Etrangères: Conflits dans les pays du fleuve Mano et en Côte d'Ivoire: stabiliser et reconstruire. Paris. 13-14 mai 2003.

Chauveau, J.P., J.P. Clin, J-P.Jacob, P. Lavigne Delville, and P-Y. Le Meur. 2006. Land tenure and resource access in West Africa. IIED, London.

Chen, Z. 2004. Panel econometric evidence of Chinese agricultural household behavior in the later 1990s: production efficiency, size effects and human mobility. Ph.D. Thesis, Iowa State Univ., USA.

Chikoye, D., V.M. Manyong, R.J. Carsky, G. Gbehounou, and A. Ahanchede. 2002. Response of speargrass (*Imperata cylindrica*) to cover crops integrated with hand-weeding, and chemical control in maize and cassava. Crop Protection 21:145-156.

Chomitz, K., P. Buys, G. de Luca, T.S. Thomas, and S. Wertz-Kanounnikoff. 2006. At loggerheads? Agricultural expansion, poverty reduction and the environment in the tropical forests, World Bank Policy Res. Rep., Washington DC.

CIMMYT. 2005. Sounding the alarm on global stem rust. An assessment of Race Ug99 in Kenya and Ethiopia and the potential for impact in neighboring regions and beyond. Expert Panel on the Stem Rust Outbreak in Eastern Africa, CIMMYT, Mexico.

Cisse, M.I. 1995. Les parcs agroforestiers au Mali. Etat des connaissances et perspectives pour leur amélioration. AFRENA Rep. 93. ICRAF, Nairobi.

Cocklin, C., J. Dibden, and N. Mautner. 2006. From market to multifunctionality? Land stewardship in Australia. The Geograph. J. 172:197-205.

Cocklin, C., N. Mautner, and J. Dibden, J. 2007. Public policy, private landholders: Perspectives on policy mechanisms for sustainable land management. J. Environ. Manage. 85(4):986-998.

Codex Alimentarius. 2003. Guideline for the conduct of food safety assessment of foods derived from recombinant-dna plants. CAC/GL 45-2003. Available at www.codexalimentarius.net/download/standards/10021/CXG_045e.pdf.

Coelli, T. and D.S. Prasada Rao. 2003. Total factor productivity growth in agriculture: A Malmquist Index analysis of 93 countries, 1980-2000. Working Pap. Ser. No. 02/2003. Centre for Efficiency and Productivity Analysis, School of Econ., Univ. Queensland.

Cole C., F. Carpio, and N. Leon. 2000. Economic burden of illness from pesticide poisonings in highland Ecuador. Rev. Panam. Salud Publ. 8(3):196-201. Available at http://www.scielosp.org/scielo.php?script=sci_arttext&pid=S1020-.

Collier, P. 1998. Social capital and poverty. Social Capital Initiative Working Pap. No 4. World Bank, Washington DC.

Collier, P. 2003. Breaking the conflict trap: Civil war and development policy. World Bank, Washington DC.

Collins, W.W. and C.O. Qualset. 1999. Biodiversity in agroecosystems. CRC Press, NY.

ComMod Group. 2004. ComMod: A companion modelling approach. Available at http://cormas.cirad.fr/en/reseaux/ComMod/charte.htm.

Conley, T. and C. Udry. 2001. Social learning through networks: The adoption of new agricultural technologies in Ghana. Am. J. Agric. Econ. 83 (3):668- 673.

Conroy, C. 2005. Participatory livestock research: A guide. ITDG Publ., UK.

Conway, G. 1997. The doubly green revolution: Food for all in the 21st century. Penguin Books, London.

Cooke, B. and U. Kothari (ed) 2001. Participation: The new tyranny? Zed Books, London.

Cooper, P.J., R.R.B. Leakey, M.R. Rao, and L. Reynolds. 1996. Agroforestry and the mitigation of land degradation in the humid and sub-humid tropics of Africa. Exp. Agric. 32:235-290.

Copping, L.G. 2004. The manual of biocontrol agents. Brit. Crop Prot. Council Publ., Alton, UK.

Copping, L.G., and J.J. Menn. 2000 Biopesticides a review of their action application and efficacy. Pesticide Manage. Sci. 56:651-676.

Corbesier, I., and G. Coupland. 2005. Photoperiodic floweing of arabidopsis: Integrating genetic and physiological approaches to characterization of the floral stimulus. Plant Cell Environ. 28:54-66.

Cotula, L., C. Toulmin, and J. Quan. 2006. Better land access for the rural poor. Lessons from experience and challenges ahead. IIED, FAO.

Cowan, T., and G. Becker. 2006. Biotechnology in animal agriculture: Status and current issues. CRS Rep. for Congress, March, 2006. Available at http://www.usis.it/pdf/other/RL33334.pdf.

Cox, B.D., M.J. Whichelow and A.T. Prevost. 2000. Seasonal consumption of salad vegetables and fresh fruits in relation to the development of cardiovascular disease and cancer. Public Health Nutr. 3:19-29.

Craufurd, P.Q., P.V. Vara Prasad, V.G. Kakani, T.R. Wheeler, and S.N. Nigam. 2003. Heat tolerance in groundnut. Field Crops Res. 80:63-77.

CRC Soil and Land Management. 1999. The costs of soil acidity, sodicity and salinity for Australia: Preliminary estimates, Coop. Res. Centre Soil Land Manage. Rep. CRCSLM/CTT2/6/99.

Critchley, W., and K. Siegert. 1991. Water harvesting. FAO, Rome.

Croppenstedt, A., and C. Muller. 2000. The impact of farmers' health and nutritional status on their productivity and efficiency: Evidence from Ethiopia. Econ. Dev. Cult. Change 48:475-502.

Croxton, S. 1999. Users in control: Farmer participation in technology research and development. *In* P. Starkey and P. Kaumbutho (ed) Meeting the challenges of animal traction. A resource book of the animal traction network for eastern and southern Africa (ATNESA). Harare, Zimbabawe, IT Publ., UK.

Crutzen, P.J., and P.H. Zimmerman. 1991. The changing photochemistry of the troposphere. Tellus 43AB:136-151.

Cunningham, A.B. 2001. Applied ethnobotany: People, wild plant use and conservation. Earthscan, London.

Curl, C.L., R.A. Fenske, J.C. Kissel, J.H. Shirai, T.F. Moate, W. Griffith, et al. 2002. Evaluation of take-home organophosphorus pesticide exposure among agricultural workers and their children. Environ. Health Perspect. 110(12):A787-92

D'Aquino, P., C. Le Page, F. Bousquet, and A. Bah. 2003. Using self-designed role-playing games and a multi-agent system to empower a local decision-making process for land use management: The SelfCormas experiment in Senegal. J. Artificial Soc. Social Simulation 6(3)   .

Dalal-Clayton, B., and S. Bass. 2002. Sustainable development strategies: A resource book. OECD and UNDP, Earthscan, London.

Dalal-Clayton, B., D. Dent, and O. Dubois. 2003. Rural planning in developing countries: Supporting natural resource management and sustainable livelihoods. Earthscan, London.

Dalgaard, T., N. Halberg, and J.R. Porter. 2001. A model for fossil energy use in Danish agriculture used to compare organic and conventional farming. Agric. Ecosyst. Environ. 87:51-65.

Dalton, T.J., N. Lilja, N. Johnson, and R. Howeler. 2005. Impact of participatory natural resource management research in cassava based cropping systems in Vietnam and Thailand. Working Doc. No. 23. Systemwide Program PRGA Tech. Dev. Inst. Innovation. CGIAR, Cali.

Das, M.D. 1995. Improving the relevance and effectiveness of agricultural extension activities for women farmers. Rome, FAO.

Dasgupta, S., M. Huq, M. Khaliquzzaman, C. Meisner, K. Pandey, and D. Wheeler. 2004. Monitoring indoor air pollution. World Bank, Washington DC.

Datt, G., and M. Ravallion. 2002. Is India's economic growth leaving the poor behind? Policy Res. Working Pap. 2846. World Bank, Washington DC.

Davenport, J., G. Burnell, T. Cross, S. Culloty, S. Ekaratne, B. Furness et al. 2003. Aquaculture: The ecological issues. The Brit. Ecol. Soc. Ecological Issues Ser. Blackwell, Oxford.

Davidson, A.P., M. Ahmad, and T. Ali. 2001. Dilemmas of agricultural extension in Pakistan: Food for thought. AgREN Network Pap. No. 116 July 2001. ODI, London.

Daviron B., B. Faivre-Dupaigre, V. Ribier, A. Fallot, and T. Voituriez (ed) 2004. Manuel d'élaboration des politiques agricoles. [Handbook of agricultural policies' definition and negotiation]. Editions du GRET, Paris.

Daviron, B., and P. Gibbon. 2002. Global commodity chains and African export agriculture. J. Agrarian Change 2(2):137-161.

Daviron, B., and S. Ponte. 2005. The coffee paradox. Zed Books, UK.

Dawe, D., A. Dobermann, P. Moya, S. Abdulrachman, B. Singh, P. Lal, S. Li et al. 2000. How widespread are yield declines in long-term rice experiments in Asia? Field Crops Res. 66:175-193.

Dawson, I.K., P.M. Hollingsworth, J.J. Doyle, S. Kresovich, J.C. Weber, S. Sotelo-Montes et al. 2008. Tree origins and conservation on-farm: A case study from the Peruvian Amazon. Conserv. Genetics 9:361-372.

De Haen, H. 1999. Producing and marketing quality organic products: opportunities and challenges, Sixth IFOAM Trade Conf. Florence, Oct 1999. Available at http://www .fao.org/organicag/doc/IFOAMf-e.htm.

De Beer, J.H., and M.J. McDermott. 1996. The economic value of non-timber forest products in Southeast Asia. Netherlands Committee/ IUCN, Amsterdam, The Netherlands.

De Onis, M, E.A. Frongillo, and M. Blossner. 2000. Is malnutrition declining? An analysis in levels of child malnutrition since 1980. Bull. WHO 78:1222-1333.

De Sherbinin, A., and V. Dompka (ed) 1998. Water and population dynamics: Case studies and policy implications. Introduction. AAAS, Washington, DC.

De Vaccaro, L.P. 1990. Survival of European

dairy breeds and their crosses with zebus in the tropics. Anin. Breeding Abstr. 58:475-494.

DeBach, P. 1964. The scope of biological control. In P. DeBach (ed) Biological control of insect pests and weeds. Reinhold, NY.

DeBach, P., and D. Rosen. 1991. Biological control by natural enemies. Cambridge Univ. Press, UK.

Degrande, A., K. Schreckenberg, C. Mbosso, P.O. Anegbeh, J. Okafor, and J. Kanmegne. 2006. Farmers' fruit tree growing strategies in the humid forest zone of Cameroon and Nigeria. Agrofor. Syst. 67:159-175.

Deininger, K.W. 2003. Land policies for growth and poverty reduction. Policy Res. Rep.. World Bank, Washington DC.

Deininger, K., and H. Binswanger. 2001. The evolution of the World Bank's land policy. Chapter 17 In A. de Janvry et al. (ed) Access to land: Rural poverty and public action. Oxford Univ. Press, NY.

Deininger, K.W., and R. Castagnini 2006. Incidence and impact of land conflict in Uganda. J. Econ. Behavior Org. 60(3):321-345.

Delgado, C.L. 2003. Rising consumption of meat and milk in developing countries has created a new food revolution. J. Nutr. 133:3907S-3910S.

Delgado, C.L., M. Rosegrant, H. Steinfeld, S. Ehui, and C. Courbois. 1999. Livestock to 2020. The next food revolution. Food, Agric. Environ. Disc. Pap. 28. IFPRI, Washington DC.

Delgado, C.L., N. Wada, M.W. Rosegrant, S. Meijer, and M. Ahmed. 2003a. Fish to 2020: Supply and demand in changing global markets. IFPRI, Washington DC.

Delgado, C.L., N. Wada, M.W. Rosegrant, S. Meijer, and M. Ahmed. 2003b. Outlook for fish to00202020: Meeting global demand. IFPRI, Washington, DC and WFC, Penang, Malaysia.

Delwaulle, J.C. 1983. Creation et multiplication vegetative par bouturage d'Eucalyptus hybrides en Republique Populaire du Congo. Silvicultura 8:775-778.

Deneve, S., I. Dieltjens, E. Moreels, and G. Hofman. 2003. Measured and simulated nitrate leaching on an organic and a conventional mixed farm. Biol. Agric. Hort. 21:217-229.

Derpsch, R. 2005. The extent of conservation agriculture adoption worldwide: Implications and impact. Proc. Third World Congress on conservation agriculture: Linking production, livelihoods and conservation. Nairobi, Kenya. 3-7 Oct. 2005.

DeSoto, H. 2005. Le mystère du capital. Pourquoi le capitalisme triomphe en Occident et échoue partout ailleurs? Flammarion, Paris.

Dey, M.M. 2000. The impact of genetically improved farmed Tilapia in Asia. Aquacult. Econ. Manage. 4(1-2):107-24.

Dey, M.M., M.A. Rab, K.M. Jahan, A. Nisapa,

A. Kumar, and M. Ahmed. 2004. Food safety standards and relatory measures: Implications for selected fish exporting Asian countries. Aquacult. Econ. Manage. 8 (3-4):217-236.

DFID. 2002. Energy for the poor: Underpinning the Millennium Development Goals. DFID, London.

Dhliwayo, T., K.V. Pixley, and V. Kazembe. 2005. Combining ability for resistance to maize weevil among 14 Southern African maize inbred lines. Crop Sci. 45:662-667.

Diarra, M., and M. Monimart. 2006. Landless women, hopeless women? Gender, land and decentralisation in Niger. Drylands Issue Pap. 143. Oct. 2006. IIED, London.

Diaz, S., D. Tilman, J. Fargione, F.S. Chapin III, R. Dirzo, T. Kitzberger et al. 2005. Biodiversity regulation of ecosystem services. p. 297-339. Chapter 11 In R. Hassan et al. (ed) Ecosystems and human well-being. Vol 1. Current states and trends. Millennium Ecosyst. Assessment, Island Press, Washington DC.

Dibden, J., and C. Cocklin. 2005. Sustainability and agri-environmental governance. p. 135-152. In V. Higgins and G. Lawrence (ed) Agricultural governance: Globalization and the new politics of regulation. Routledge, London.

Dick, J.M., and R.C. Dewar. 1992. A mechanistic model of carbohydrate dynamics during the adventitious root development in leafy cuttings. Annals Bot. (London) 70:371-377.

Dick, J., U. Skiba, R. Munro, and D. Deans. 2006. Effect of N-fixing and non N-fixing trees and and crops on NO and N_2O emissions from Senegalese soils. J. Biogeo. 33:416-423.

Diekow, J., J. Mielniczuk, H. Knicker, C. Bayer, D.P. Dick, and I. Koegel-Knabner. 2005. Soil C and N stocks as affected by cropping systems and nitrogen fertilization in a southern Brazil Acrisol managed under no-tillage for 1 year. Soil Till. Res. 81:87-95.

Dillaha, T.A. S.M. Renueau, and D. Lee. 1989. Vegetative filter strips for agricultural non point source pollution control. Trans. ASAE 32:.513-519.

Dimes, J., S. Twomlow, J. Rusike, B. Gerard, R. Tabo, A. Freeman, and D. Keatinge. 2005. Increasing research impacts through low-cost soil fertility management options for Africa's drought-prone areas. In G.O. Omanya, D. Paternak (ed) Sustainable agriculture systems for the drylands. Proc. Int. Symp. Sustainable Dry Land Agric. Systems. Niamey. 2-5 Dec. 2003.

Dixon, J., A. Gulliver, and D. Gibbon. 2001. Farming systems and poverty, improving farmers' livelihoods in a changing world. FAO and World Bank, Rome.

Dixon, R.K. 1995. Agroforestry systems: sources or sinks of greenhouse gases? Agrofor. Syst. 31:99-116.

Dobermann, A. 2006. Nitrogen use efficiency in cereal systems. Proc. 13th Australian Agron.

Conf. Available at http://www.regional.org.
au/au/asa/2006/plenary/soil/dobermannad
.htm

Dobermann, A., and K.G. Cassman. 2002.
Plant nutrient management for enhanced
productivity in intensive grain production
systems of the United States management for
irrigated rice in southeast China. Agron. J.
93:869-878.

Dobermann, A., C. Witt, D. Dawe, G.C. Gines,
R. Nagarajan, S. Satawathananont et al.
2002. Site-specific nutrient management for
intensive rice cropping systems in Asia. Field
Crops Res. 74:37-66.

Dorward, A., S. Fan, J. Kydd, H. Lofgren,
J. Morrison, C. Poulton et al. 2004,
Institutions and economic policies for pro
poor agricultural growth. Disc. Pap. No. 15
Dev. Strategy and Governance Div., IFPRI
and Centre for Dev. Poverty Reduction, Dep.
Agric. Sci. Imperial College, London.

Douthwaite, B., J.M. Ekboir, S.J. Twomlow,
and J.D.H. Keatinge. 2004. The concept
of integrated natural resource management
(INRM) and its implications for developing
evaluation methods. p. 321-340. In B.
Shiferaw et al. (ed) Natural resource
management in agriculture: Methods
for assessing economic impact and
environmental impacts. ICRISAT and CABI,
Wallingford.

Dransfield, B., I. Maudlin, P. Stevenson, and
A. Shaw. 2001. Reality vs rhetoric — a
survey and evolution of tsetse control in East
Africa. Agric. Human Values 18:219-233.

Drechsel, P., S. Graefe, M. Sonou, and
O.O. Cofie. 2006. Informal irrigation in
urban West Africa: An overview. Res. Rep.
102. IWMI, Colombo.

Dreisigacker, S., P. Zhang, M.L. Warburton,
M. van Ginkel, D. Hoisington, M. Bohn,
and A.E. Melchinger. 2003. SSR and
pedigree analyses of genetic diversity among
CIMMYT wheat lines targeted to different
megaenvironments. Crop Sci. 44:381-388.

Dreze, J., and A. Sen. 1990. The political
economy of hunger. Oxford, Clarendon Press.

DTI. 2006. The energy challenge. Energy Review
Rep. 2006. UK Dept, Trade and Industry
(DTI), London.

Dumanski, J., R. Peiretti, J.R. Benites,
D. McGarry and C. Pieri. 2006. The
paradigm of conservation agriculture.
p. 58-64. In Proc. World Assoc. Soil and
Water Conserv. Pap. P1-7. Available at http://
www.unapcaem.org/admin/exb/ADImage/
ConservationAgri/ParaOfCA.pdf

Dunn, A., I. Gray, and E. Phillips. 2000. From
personal barriers to community plans: A
farm and community planning approach
to the extension of sustainable agriculture.
Part 1. Extension theory. In A.D. Shulman,
and R.J. Price (ed) Case studies in increasing
the adoption of sustainable resource
management practices. Land Water Resour.
Res. Dev. Corp., Canberra.

Duvick, D.N., and K.G. Cassman. 1999. Post-
green-revolution trends in yield potential of
temperate maize in the north-central United
States. Crop Sci. 39:1622- 1630.

Duxbury, J. 2005. Reducing greenhouse warming
potential by carbon sequestration in soils:
opportunities, limits and tradeoffs. In R. Lal
et al. (ed) Climate change and global food
security. CRC Press, Boca Raton.

EC. 2002. Commission Decision of 24 July 2002
establishing guidance notes supplementing
Annex VII to Directive 2001/18/EC of the
Eur. Parliament and of the Council on the
deliberate release into the environment of
genetically modified organisms and repealing
Council Directive 90/220/EEC notified
under Doc. No. C(2002) 2715. OJ L 200,
30.7.2002. Available at http://eur-lex.europa.
eu/smartapi/cgi/sga_doc?smartapi!celexapi
!prod!CELEXnumdoc&numdoc=32002D
0623&model=guichett&lg=en. European
Commission, Brussels.

EC. 2005. Shift gear to biofuels. Results and
recommendations from the VIEWLS project
for the EC. European Commission, Brussels.

Edmunds, W.M., W.G. Darling, D.G. Kinniburgh,
S. Kotoub, and S. Mahgoub. 1992. Sources
of recharge at Abu Delaig. Sudan J. Hydrol.
131:1-24.

Edmunds, W.M., and C.B. Gaye. 1997. Naturally
high nitrate concentrations in groundwaters
from the Sahel. J. Environ. Qual. 26:1231-
1239.

EFSA. 2004. Guidance document of the Sci.
Panel GMOs for the risk assessment of
genetically modified plants and derived
food and feed. Available at http://www.efsa.
europa.eu/en/science/gmo/gmo_guidance/.
EFSA J. 99.1-94. Eur. Food Safety Authority.

EFSA. 2007. Guidance document for the risk
assessment of genetically modified plants
containing stacked transformation events by
the Sci. Panel GMOs. EFSA-Q-2003-005D.
Available at http://www.efsa.europa.eu/EFSA/
Scientific_Document/gmo_guidance_ej512_
GM_stacked_events_en.pdf. Eur. Food Safety
Authority.

EHN. 2001. Nutrition and health claims. A
European Heart Network Position Paper.
EHN, Brussels.

Elevitch, 2006. Traditional trees of Pacific
Islands: Their culture, environment and use.
Permanent Agric. Resources, Hawaii.

El-Hage Scialabba, N. 2005. Global trends in
organic agriculture markets and Countries'
demand for FAO assistance. Global Learning
Opportunity — Int. Farming Systems Assoc.
Roundtable: Organic Agriculture. 1 Nov.
2005. Available at ftp://ftp.fao.org/paia/
organicag/2005_12_doc04.pdf. FAO, Rome.

Ellis, F., and H.A. Freeman 2004. Rural
livelihoods and poverty reduction strategies
in four African countries. J. Dev. Studies
40(4):1-30.

Ellis, F., and N. Mdoe 2003. Livelihoods and
rural poverty reduction in Tanzania. World
Dev. 31(8)1367-1384.

Ellis, J., and D.M. Swift. 1988. Stability of
African pastoral ecosystems: Alternative
paradigms and implications for development.
J. Range Manage. 41:450-459.

Ellis-Jones, J., S. Twomlow, T. Willcocks,
C. Riches, H. Dhliwayo, and M. Mudhara.
1993. Conservation tillage/weed control
systems for communal farming areas in semi-
arid Zimbabwe. Brighton Crop Prot. Conf.
Weeds 3:1161-1166.

Ellstrand, N. 2003. Going to 'great lengths' to
prevent the escape of genes that produce
specialty chemicals. Plant Phys. 132:1770-
1774.

Embrapa. 2006. Sistemas de produção
sustentáveis para a amazônia legal. Available
at http://www.embrapa.br/publicacoes/
index_htm.

Engel, P., B. Haverkort, and J. Jiggins. 1988.
Concepts and processes in participative
technology development. Proc. ILEIA
Workshop on operational approaches in
participative technology development in
sustainable agriculture. E.T.C., Leusden,
Netherlands.

Engel, P., and N. Röling 1989. IKS (Indigenous
Knowledge System) and knowledge
management: Utilizing indigenous knowledge
in institutional knowledge systems. p. 101-
15. In D.M. Warren et al. (ed) Indigenous
knowledge systems: Implications for
agriculture and international development.
Iowa State Univ., Tech. Social Change Prog.,
Ames.

Ensminger, J. 1997. Changing property rights:
Reconciling formal and informal rights to
land in Africa. p. 95-119. In The frontiers of
the new institutional economics. Academic
Press, Londres.

EPA. 1996a. Pesticide tolerances for glyphosate.
Federal Register Vol. 61 No. 6: April 5.

EPA. 1996b. Pesticide thresholds for glyphosate
in various soybean media. Available at http://
www.epa.gov/EPA-PEST/1996/April/Day-05/
pr-636.html.

EPA. 2002. A comprehensive analysis of
biodiesel impacts on exhaust emissions, Draft
Technical Rep.. EPA420-P02-001. Environ.
Prot. Agency, USA. Available at http://www
.biodiesel.org/resources/reportsdatabase/
reports/gen/20021001_gen-323.pdf (accessed
22 Jan. 2007).

Erskine, W.D. 1999. Oscillatory response versus
progressive degradation of incised channels
in South Eastern Australia. p. 67-95. In
S.E. Darby and A. Simon (ed) Incised river
channels: Processes, forms, engineering, and
management. Wiley, Chichester.

Erskine, W.D. and A.A. Webb. 2003. Desnagging
to resnagging: New directions in river
rehabilitation in South Eastern Australia.
River Res. and Applications 19:233-249.

Eswaran, H. 1993. Soil resilience and sustainable
land management in the context of Agenda
21. p. 21- 32. In D.J. Greenland and
I. Szabolcs (ed) Soil resilience and sustainable
land use. CABI, Wallingford.

Eswaran, H., R. Lal and P.F. Reich. 2001. Land

degradation: an overview. *In* Bridges, E.M., I.D. Hannam et al. (ed) Responses to land degradation. Proc. 2nd. Int. Conf. Land Degradation and Desertification, Khon Kaen, Thailand. Available at http://soils.usda.gov/use/worldsoils/papers/land-degradation-overview.html. Oxford Press, New Delhi.

Eswaran, H., P. Reich, and F. Beinroth. 2006. Land degradation: an assessment of the human impact on global land resources. *In* World Congr. Soil Sci., 18th, Philadelphia. Abstract. Available at http://crops.confex.com/crops/wc2006/techprogram/P11487.HTM.

ETC/KIOF, 1998. On-farm agro-economic comparison of organic and conventional techniques in high and medium potential areas of Kenya. Netherlands, ETC and Kenya Inst. Organic Farming, Leusden/Nairobi, March.

Evans, J., and J.W. Turnbull. 2004. Plantation forestry in the tropics. 3rd ed. Oxford Univ. Press, UK.

Eveleens, K.G., R. Chisholm, E. Van de Fliert, M. Kato, P.T. Nhat, and P. Schmidt. 1996. Mid-term Review of Phase III Rep.. FAO Inter-country Prog. Dev. Appl. Integrated Pest Control in Rice in South and South-east Asia, Manila, Philippines.

Evenson, R. 1997. The economic contributions of agricultural extension to agricultural and rural development. *In* B.E. Swanson, R.P. Bentz, and A.J. Sofranko (ed) Improving agricultural extension: A reference manual. FAO, Rome.

Evenson, R., and D. Gollin. 2003a. Assessing the impact of the green revolution, 1960 to 2000. Science 300:758-762.

Evenson, R.E., and D. Gollin. 2003b. Crop genetic improvement in developing countries: overview and summary. p. 7-38. *In* R.E. Evenson and D. Gollin (ed) Crop variety improvement and its effect on productivity. CABI, Wallingford.

Ewel, J.J. 1999. Natural systems as models for the design of sustainable systems of land use. Agrofor. Syst. 45:1-21.

Ewel, J.J., D.J. O'Dowd, J. Bergelson, C.C. Daeler, C.M. D'Antonio, L. Diego Gomez et al. 1999. Deliberate introductions of species: Research needs. BioScience 49:619-630.

Ezzati, M., S.V. Hoorn, A. Rodgers, A.D. Lopez, C.D. Mathers, and C.J.L. Murray. The Comparative Risk Assessment Collobarating Group. 2003. Estimates of global and regional potential health gains from reducing multiple major risk factors Lancet 362(9380):271-80 erratum 365(9453).

Ezzati, M., and D.M. Kammen. 2002. the health impacts of exposure to indoor air pollution from solid fuels in developing countries: Knowledge, gaps, and data needs. Environ. Health Perspect. 110(11).

Fafchamps, M., and Minten, B. 2001. Social capital and trade. Am. J. Agric. Econ. 83(3):680-685.

Fafchamps, M., C. Udry, and K. Czukas. 1996. Drought and saving in West Africa: Are livestock a buffer stock? Available at SSRN http://ssrn.com/abstract=71374.

Falconer, J. 1990. The major significance of 'minor' forest products: The local use and value of forests in the West African humid forest zone. Forests Trees People. Community Forestry Note 6. FAO, Rome.

Falkenmark, M. 2000. Competing freshwater and ecological services in the river basin perspective: an expanded conceptual framework. Water Int. 25:172- 177.

Falkenmark, M., C.M. Finlayson, L.J. Gordon, E.M. Bennett, T. Matiza Chiuta, D. Coates et al. 2007. Agriculture, water and ecosystems: Avoiding the costs of going too far. p. 233-277. *In* Water for food, water for life: A comprehensive assessment of water management in agriculture. Earthscan, London and IWMI, Colombo.

Falkenmark, M., and J. Rockström. 2005. Rain: The neglected resource. Swed. Water House Policy Brief Nr. 2. SIWI.

Falvey, F., and C. Chantalakhana (ed) 1999, Smallholder dairying in the tropics. ILRI, Nairobi.

Fan, S., P. Hazell, and T. Haque. 2000. Targeting public investments by agro-ecological zone to achieve growth and poverty alleviation goals in rural India. Food Policy 25:411-428.

FAO. 1990. Women in agricultural development: Rural women and the changing socio-economic conditions in the Near East. W/U2076 R.

FAO. 1995. The Fourth Progress Report on the WCARRD Programme of Action. FAO, Rome, Italy.

FAO. 1996. Focus: Women and food security. Women and Population Div. FAO, Rome.

FAO. 1997. Preventing micronutrient malnutrition: a guide to food based approaches. Why policymakers should give priority to food based approaches. FAO, Rome.

FAO. 1998a. The state of the world's plant genetic resources for food and agriculture. FAO, Rome.

FAO. 1998b. FAO factfile: Feminization of agriculture. FAO. Rome.

FAO. 2000a. The state of food and agriculture 2000. Part II. World food and agriculture: Lessons from the last 50 years. FAO, Rome.

FAO. 2000b. Bioenergy. Background Pap. No. 2. FAO/Netherlands conf. on the multifunctional character of agriculture and land, Maastricht. 12-17 Sept.

FAO. 2000c. The role of information and communication technologies in rural development and food security. Workshop Rep. FAO, Rome.

FAO. 2001. Le traité international sur les ressources phytogénétiques pour l'alimentation et l'agriculture. FAO, Rome.

FAO. 2002a. International Treaty on Plant Genetic Resources for Food and Agriculture. FAO, Rome.

FAO. 2002b. Women, agriculture and food security. FAO fact sheet. FAO, Rome. www.fao.org/worldfoodsummit/english/fsheets/women.pdf.

FAO. 2002c. Fishery statistics: Reliability and policy implications. FAO, Rome. Available at http://www.FAOorg/fi/statist/ nature _china/30jan02.asp.

FAO. 2003a. Biological management of soil ecosystems for sustainable agriculture. Rep. of the Int. Tech. Workshop. Londrina, Brazil. 24-27 June 2002. World Soil Resour. Rep., 101. FAO, Rome.

FAO. 2003b. Food balance sheet. Available at http://www.FAOorg/.

FAO. 2004a. Gender and food security. Available at http://www.fao.org/Gender/gender.htm

FAO. 2004b. Unified bioenergy terminology - UBET. FAO, Rome.

FAO. 2005a. Special event on impact of climate change, pests and diseases on food security and poverty reduction. 31st Session of the Committee on World Food Security and Poverty Reduction. Background Document. FAO, Rome.

FAO. 2005b. Global forest resource assessment 2005. FAO, Rome.

FAO. 2005c. The state of food insecurity in the world: Eradicating world hunger. Key to achieving the Millennium Development Goals. FAO, Rome.

FAO. 2005d. Gender and land. Compendium of countries studies. FAO, Rome.

FAO. 2005e. State of the world's forests. FAO, Rome.

FAO. 2006a. Livestock report 2006. FAO, Rome.

FAO. 2006b. Responsible management of planted forests: Voluntary guidelines. FAO Forestry Working Pap., Rome.

FAO. 2006c. State of world aquaculture 2006. FAO Inland Water Resour. Aquacult. Serv., Fish. Resour. Div., Fisheries Dep. FAO Fisheries Pap. No. 500, Rome.

FAO. 2006d. Livestock's long shadow. Environmental issues and options. Available at (http://www.virtualcentre.org/en/library/key_pub/longshad/a0701e/A0701E01.pdf

FAO. 2006e. State of food insecurity in the World 2006: Eradicating world hunger — taking stock ten years after the World Food Summit. FAO, Rome.

FAO AQUASTAT. 2006. www.fao.org/ag/agl/aglw/ aquastat/main/index.htm

FAOSTAT. 2005/2006/2007. Available at http://faostat.fao.org/site/584/default.aspx.

FAO/ITC/CTA. 2001. World markets for organic fruit and vegetables. Opportunities for developing countries in the production and export of organic horticultural products. International Trade Centre, FAO, Rome. Available at http://www.fao.org/docrep/004/y1669e/y1669e00.htm#Contents.

FAO/UNEP. 1996. Our land, our future. A new approach to land use planning and management. FAO, Rome.

FAO/WHO. 1999. Guidelines for the

production, processing, labelling and marketing of organically produced foods. Codex Alimentarius Commission, CAC/GL 32-1999. FAO/WHO, Rome.

Faure G., P. Kleene. 2004. Lessons from new experiences in extension in West Africa: Management advice for family farms and farmers' governance. J. Agric. Extension Educ. 10:(1)37-49.

Fauvet, N. 1996. Tree vegetation map, CIRAD-Forêt. Available at www.state.gov/r/pa/ei/bgn/2828.htm. US Dep. State, Bur. African Affairs, Washington DC.

Feder, G., R.E. Just, and D. Zilberman. 1985. Adoption of agricultural innovations in developing countries: A survey. Econ. Dev. Cult. Change 33:255-298.

Feder, G., R. Murgai, and J.B. Quizon. 2003. Sending farmers back to school: The impact of farmer field schools in Indonesia. Policy Res. Working Pap. 3022. World Bank, Washington DC.

Feder, G., R. Murgai, and J.B. Quizon. 2004. The acquisition and diffusion of knowledge: The case of pest management training in farmer field schools, Indonesia. J. Agric. Econ. 55:221-243.

Feder, G., T. Onchan, Y. Chlamwong, and C. Hongladorom. 1988. Land policies and farm productivity in Thailand. John Hopkins Univ. Press, Baltimore.

Fenske, R.A., G. Kedan, C. Lu, J.A. Fisker-Andersen, and C.L. Curl, 2002. Assessment of organophosphorous pesticide exposures in the diets of preschool children in Washington State. J. Exposure Anal. Environ. Epidemiol, 12(1);21-28.

Ferris, S., E. Kaganzi, R. Best, C. Ostertag, M. Lundy, and T. Wandschneider. 2006. Enabling rural innovation in Africa: A market facilitator's guide to participatory agroenterprise development. CIAT Enabling Rural Inovation Guide 2. CIAT, Kampala.

Ferrero, A. and N.V. Nguyen. 2004. The sustainable development of rice-based production systems in Europe. Rice for life. Int. Rice Commission Newsl. 53. FAO, Rome.

Feyt, H., 2001. Propriété intellectuelle et sélection dans les pays du Sud. O.C.L., 8(5):546-550.

FFTC (Food and Fertilizer Technology Center). 2006. Food processing and post-harvest technology. Available at http://www.agnet.org/library/list/subcat/P.html.

Filozof, C, C. Gonzalez, M. Sereday, C. Mazza, and J. Braguinsky. 2001. Obesity prevalence and trends in Latin-American countries. Obesity Rev. 2(2):99-106.

Firoozi, F., and J. Merrifield 2003. An optimal timing model of water reallocation and reservoir construction. Eur. J. Operational Res. 145(1):165-174.

Fisher, M.J., I.M. Rao, M.A. Ayarza, C.E. Lascano, J.I. Sanz, R.J. Thomas, and R.R. Vera. 1994. Carbon storage by introduced deep-rooted grasses in the South American savannas. Nature 371:236-238.

Fitzgerald, K., D. Barnes, and G. McGranahan. 1990. Interfuel substitution and changes in the way households use energy: The case of cooking and lighting behavior in urban Java. Industry and Dev. Dep., World Bank, Washington DC.

Foley, J.A., M.H. Costa, C. Delire, N. Ramankutty, and P. Snyder. 2003. Green surprise? How terrestrial ecosystems could affect earth´s climate. Frontier Ecol. Environ. 1(1):38-44. Available at www.frontersinecology.org. Ecol. Soc. Am.

Follett, R.F. 2001. Soil management concepts and carbon sequestration in cropland soils. Soil Tillage Res. 61:77-92.

Foster, G.S., and F.L.G. Bertolucci. 1994. Clonal development and deployment: Strategies to enhancè gain while minimising risk. p. 103-111. In R.R.B. Leakey and A.C. Newton (ed) Tropical trees: The potential for domestication and the rebuilding of forest resources. HMSO, London.

Fox, M.P., S. Rosen, W.B. MacLeod, M. Wasunna, M. Bii, G. Foglia, and J.L. Simon. 2004. The impact of HIV/AIDS on labour productivity in Kenya. Trop. Med. Int. Health 9(3):318-324.

Franco, A.A., E.F. Campello, E.M.R. Silva, S.M. Faria. 1992. Revegetation of degraded soils. (In Portuguese.) Embrapa Agrobiologia. Embrapa/Centro Nacional de Pesquisa de Biologia do Solo, Comunicado Técnico, 9, Rio de Janeiro.

Franke, A.C., J. Ellis-Jones, G. Tarawali, S. Schulz, M.A. Hussaini, I. Kureh et al. 2006. Evaluating and scaling-up integrated Striga hermonthica control technologies among farmers in northern Nigeria. Crop Protection 25:868-878.

Franke-Snyder, M., D.D. Douds Jr., L. Galvez, J.G. Phillips, P. Wagoner, L. Drinkwater, and J.P. Morton. 2001. Diversity of communities of arbuscular mycorrhizal (AM) fungi present in conventional versus low-input agricultural sites in eastern Pennsylvania, USA. Appl. Soil Ecol. 16:35-48.

Franzel, S. 1999. Socio-economic factors affecting the adoption potential of improved tree fallows in Africa. Agrofor. Syst. 47:305-321.

Franzel, S., G.L. Denning, J.P.B. Lillesø and A.R. Mercado. 2003. Scaling up the impact of agroforestry: Lessons from three sites in Africa and Asia. Agrofor. Systems 61-62:329-344.

Franzel, S., H. Jaenicke, and W. Janssen. 1996. Choosing the right trees: Setting priorities for multipurpose tree improvement. p. 87. In ISNAR Res. Rep. 8. IFPRI, Washington DC.

Franzel, S., and S. Scherr. 2002. Trees on the farm: Assessing the adoption potential of agroforestry practices in Africa. CABI, Wallingford.

Freese, W., and B. Schubert. 2004. Safety testing and regulation of genetically engineered foods. p. 299-325. In S.E. Harding (ed) Biotech. Genetic Eng. Rev., Intercept, Andover.

Freese, W. 2001. The StarLink affair. Friends of the Earth, July 2001. Available at http://www.foe.org/camps/comm/safefood/gefood/starlink.pdf.

Friesen, L.F., A.G. Nelson, and R.C. van Acker. 2003. Evidence of contamination of pedigreed canola (Brassica napus) seedlots in western Canada with genetically engineered herbicide resistance traits. Agron. J. 95:1342-1347.

Frison, E., O. Smith, and M.S. Swaminathan 1998. UN Millennium Development Goals: Five years later. Agricultural biodiversity and elimination of hunger and poverty. p.12. In The Chennai Platform for Action. IPGRI, Rome.

Furedy, C. 1990. Socially aspects of human excra reuse: Implications for aquacultural projects in Asia. p. 251-66. In P. Edwards and R.S.V. Pullin (ed) Wastewater-fed aquaculture. Proc. Int. Seminar on Wastewater Reclamation and Reuse for Aquaculture. Calcutta, India. 6-9 Dec 1988. xxix. Environ. Sanitation Infor. Center, Asian Inst. Tech., Bangkok.

Galbraith, H., P. Amerasinghe, and A. Huber-Lee. 2005. The effects of agricultural irrigation on wetland ecosystems in developing countries: A literature review. CA Disc. Pap. 1. Comprehensive Assessment Secretariat, IWMI, Colombo.

Galston, W.A., and K.J. Baehler. 1995. Rural development in the United States: Connecting theory, practice, and possibilities. Island Press, Washington DC.

Garcia-Alonso, M., E. Jacobs, A. Raybould, T.E. Nickson, P. Sowig, H. Willekens et al. 2006. A tiered system for assessing the risk of genetically modified plants to non-target organisms. Environ. Biosafety Res. 5:57-65.

Garforth, C., and A. Lawrence. 1997. Supporting sustainable agriculture through extension in Asia. Nat. Resour. Perspect., No. 21, June 1997. ODI, London. Available at http://www.odi.org.uk/NRP/21.html.

Garforth M., and J. Mayers (ed) 2005. Plantations, privatisation and power: Changing ownership and management of state forests. Earthscan, London.

Garrity, D.P. 2004. World agroforestry and the achievement of the Millenium Development Goals. Agrofor. Syst. 61:5-17.

Gaspar, P., F.J. Mesías, M. Escribano, A. Rodriguez de Ledesma, and F. Pulido. 2007. Economic and management characterization of dehasa farms: implications for their sustainability. Agrofor. Syst. 71:151-162.

Gatsby Charitable Foundation. 2005. The quiet revolution: Push-pull technology and the African farmer. Occasional Pap., Gatsby Charitable Foundation, London.

Gaunt, J.L., S.K. White, J.R. Best, A.J. Sutherland, P. Norrish, E.J.Z. Robinson et al. 2003. The researcher-farmer interface in rice-wheat systems: moving from agricultural productivity to livelihoods. In Improving the productivity and sustainability of rice-wheat systems: Issues and impacts. ASA Special Publ. 65, Am. Soc. Agron., Madison, WI.

Gautam, M. 2000. Agricultural Extension: The Kenya experience: An impact evaluation. World Bank Oper. Eval. Study. Washington DC.

Geffray, C. 1990. La cause des armes au Mozambique. Anthropologie d'une guerre civile. Éditions Karthala, Paris.

Geist, H.J., and E.F. Lambin. 2004. Dynamic causal patterns of desertification. BioScience 54:817-829   .

Gemmill-Herren, B., C. Eardley, J. Mburu, W. Kinuthia, and D. Martins. 2007. Pollinators. p. 166-177. In S.J. Scherr and J.A. McNeely (ed) Farming with nature: The science and practice of ecoagriculture. Island Press, NY.

Gereffi, G., and R. Kaplinsky (ed) 2001. The value of value chains: Spreading the gains from globalisation. Special issue of the IDS Bulletin 32:3(July).

Gereffi, G., and M. Korzeniewicz (ed) 1994. Commodity chain and global capitalism. Greenwood Press and Praeger, Westport.

Gethi J.G. and M.E. Smith. 2004. Genetic responses of single crosses of maize to Striga hermonthica (Del.) Benth. and Striga asiatica (L.) Kuntze. Crop Sci. 44:2068-2077.

Ghosh, D., A.D. Sagar, and V.V.N. Kishore. 2006. Scaling up biomass gasifier use: An application-specific approach. Energy Policy 34:1566-82.

Gibbons, P., and S. Ponte. 2005. Trading down. Africa, value chains and the global economy. Temple Univ. Press, Philadelphia PA.

Gibson, K.D., A.J. Fischer, T.C. Foin, and J.E. Hill. 2003. Crop traits related to weed suppression in water-seeded rice (Oryza sativa L.). Weed Sci. 51:87-93.

Gillanders, B.M. 2005. Using elemental chemistry of fish otoliths to determine connectivity between estuarine and coastal habitats. Estuarine Coastal Shelf Sci. 64:47-57.

Gillanders, B.M., K.W. Able, J.A. Brown, D.B. Eggleston, and P.F. Sheridan. 2003. Evidence of connectivity between juvenile and adult habitats for mobile marine fauna: An important component of nurseries. Marine Ecol.-Progress Ser. 247:281-295.

Gilliom, R.J., J.E. Barbash, C.G. Crawford, P.A. Hamilton, J.D. Martin, N. Nakagaki et al. 2006. The quality of our nation's waters - Pesticides in the nation's streams and ground water, 1992-2001. Available at http://pubs.usgs.gov/circ/2005/1291/. US Geological Survey Circ. 1291, Washington DC.

Gitay, H., S. Brown, W. Easterling, and B. Jallow. 2001. Ecosystems and their goods and services, p. 237-342. In J.J. McCarthy et al. (ed) Climate change 2001: Impacts, adaptation and vulnerability. IPCC, Cambridge Univ. Press, UK.

Gleick, P.H. (ed) 1993. Water in crisis: A guide to the world's freshwater resources, Oxford Univ. Press, NY.

Gleick, P.H. 2003. Global freshwater resources: Soft-path solutions for the 21st century. Science 302:1524- 1528.

Gliessman, S.R. 1998. Agroecology: Ecological processes in sustainable agriculture. Ann Arbor Press, Chelsea.

Gockowski, J.J., and S. Dury. 1999. The economics of cocoa-fruit agroforests in southern Cameroon. p. 239-241. In F. Jiménez and J. Beer (ed) Multi-strata agroforestry systems with perennial crops. CATIE, Turrialba.

Goel, R.S. 2003. Remote sensing and GIS application in water resources development-An Overview. p. 429-436. In B. Venkateswara Rao (ed) Proc. Int. Conf. Hydrol. Watershed Manage. Vol. 2, Hyderabad.18-20 December 2002. BS Publ., Hyderabad, India.

Goldemberg, J., and S.T. Coelho. 2004. Renewable energy - traditional biomass vs. modern biomass. Energy Policy 32:711-14.

Golkany, I.M. 1999. Meeting global food needs: the environmental trade offs between increasing land conversion and land productivity. p. 256-291. In J. Morris, R. Bate (ed) Fearing food. Butterworth-Heinemann, Oxford.

Gomes, R., J. duGuerny, M.H. Glantz, and L. Hsu. 2004. Climate and HIV/AIDS: A hotspots analysis for early warning rapid response systems. UNDP, FAO, and NCAR, Bangkok.

Gonzalves, J., T. Becker, A. Braun, D. Campilan, H. De Chavez, E. Fabjer et al. (ed) 2005. Participatory research and development for sustainable agriculture and natural resource management. A sourcebook. CIP, IDRC, IFAD, UPWARD, Laguna.

Gorton, M., and S. Davidova. 2004. Farm productivity and efficiency in the CEE applicant countries: A synthesis of results. Agric. Econ. 30:1-16.

Govaerts, B., K.D. Sayre, and J. Deckers. 2005. Stable high yields with zero tillage and permanent bed planting? Field Crops Res. 94:33-42.

Graham, R.D., D. Senadhira, S.E. Beebe, C. Iglesias, and I. Ortiz-Monasterio. 1999. Breeding for micronutrient density in edible portions of staple food crops: Conventional approaches. Field Crop Res. 60:57-80.

Gray, I., and G. Lawrence, 2001. A future for regional Australia: Escaping global misfortune. Cambridge Univ. Press, UK.

Green, G.P., and J.R. Hamilton. 2000. Water allocation, transfers and conservation: Links between policy and hydrology, Water Res. Dev. 16(2):265-273.

Grellier, R. 1995. All in good time: Women's agricultural production in Sub-Saharan Africa. Nat. Resources Inst., Chatham.

Griffon, M. 1994. Analyse de filière et analyse de compétitivité. Economie des politiques agricoles dans les pays en développement. Revue Française d'Economie, Paris.

Groombridge, B., and M. Jenkins. 1998. Freshwater biodiversity: a preliminary global assessment. 2. Biodiversity in freshwaters. World Conserv. Monitoring Centre, WCMC Biodiversity Ser. No. 8. Available at: http://www.unep-wcmc.org/information_services/publications/freshwater/2.htm.

Guarino, L. 1997. Traditional African vegetables. IPGRI, Rome.

Gueneau, S. 2006. Livre blanc sur les forêts tropicales humides, La Documentation Française, Paris.

Guéneau, S., and S. Bass. 2005. Global forest governance: effectiveness, fairness and legitimacy of market-driven approaches. IDDRI, Idées pour le débat, n°13.

Guerin, K. 2007. Adaptive governance and evolving solutions to natural resource conflicts. March 2007. New Zealand Treasury Working Pap. 07/03.

Guerinot, L. 2000. Plant biology enhanced: The Green Revolution strikes gold. Science 287:241-243.

Guerny, J. 1995. Population and land degradation. Available at http://www.un.org/popin/fao/land/land.html. UN Pop. Div., Dep. Econ. Social Affairs, NY.

Gunning, R.V., H.T. Dang, F.C. Kemp, I.C. Nicholson, and G.D. Moores. 2005. New resistance mechanism in Helicoverpa armigera threatens transgenic crops expressing Bacillus thuringiensis Cry1Ac toxin. Appl. Environ. Microbiol. 71:2558-2563.

Gurr, G.M., and S.D. Wratten (ed) 2000. Biological control: Measures of success. Kluwer, Dordrecht.

Gutierrez, E.U., and S.M. Borras 2004 The Moro conflict: Landlessness and misdirected state policies. Policy Studies 8, East-West Center, Washington DC.

Guyomard, H., C. Le Mouël, and A. Gohin. 2004. Impacts of alternative agricultural income support schemes on multiple policy goals. Eur. Rev. Agric. Econ. 31:125-148.

Habitu, N., and H. Mahoo. 1999. Rainwater harvesting technologies for agricultural production: A case for Dodoma, Tanzania. In Conservation tillage with animal traction, P.G. Kambutho and T.E. Simalenga (ed) Animal Traction Network for Eastern and Southern Africa, Zimbabwe.

Hall, A., V. Rasheed Sulaiman, N. Clark, and B. Yoganand. 2003. From measuring impact to learning institutional lessons: An innovation systems perspective on improving the management of international agricultural research. Agric. Syst. 78:213-241.

Hall, A.E. 1992. Breeding for heat tolerance. p. 129-168. In J. Janick (ed) Plant breeding reviews. Vol. 10. John Wiley, NY.

Hall, L., K. Topinka, J. Huffman, L. Davis, and A. Good. 2000 Pollen flow between herbicide resistant Brassica napus (Argentine canola) is the cause of multiple resistant B. napus volunteers. Weed Sci. 48:688-694.

Halvorson, A.D., C.A. Reule. 2006. Irrigated corn and soybean response to nitrogen under no-till in Northern Colorado. Agron. J. 98:1367-1374.

Hamann, R., and T. O'Riordan. 2000. Resource management in South Africa. S. Afr. Geog. J. 82(2):23-34.

Hamilton, L.S., and P.S. King. 1983. Tropical forest watersheds: Hydrologic and soils response to major uses and conversions. Westview Press, Boulder.

Hammani, A., M. Kuper, and A. Debbarh. 2005. La modernisation de l'agriculture irriguée. Actes du séminaire euro-méditerranéen. 19-23 Avril 2004, Rabat, Maroc. Projet INCO-WADEMED, IAV Hassan II, Rabat.

Hardin, G. 1968. The tragedy of the commons. Science 162:1243-1248.

Harlan, J.R. 1975. Crops and Man. ASA/CSSA, Madison, WI.

Harrington, L., and O. Erenstein. 2005. Conservation agriculture and resource conserving technologies — a global perspective. III World Congress on Conservation Agriculture. Available at http://www.act.org.zw/postcongress/theme_02_05.asp

Harris, R. 2004. Information and communication technologies for poverty alleviation. UNDP's Asia-Pacific Dev. Infor. Prog. (UNDP-APDIP), Kuala Lumpur, Malaysia. Available at http://www.apdip.net/documents/eprimers/poverty.pdf.

Hartell, J., M. Smale, P.W. Heisey, and B. Senauer. 1998. The contribution of genetic resources and diversity to wheat production in the Punjab of Pakistan. p. 146-158. In M. Smale (ed) Farmers, gene banks and crop breeding: Economic analyses of diversity in wheat, maize, and rice. Kluwer Academic, Dordrecht.

Hartmann, H.T., D.E. Kester, F.T. Davis, and R.L. Geneve. 1997. Plant propagation: Principles and practices. 6th ed., Prentice-Hall, NJ.

Harwood, J.D., and J.J. Obrycki. 2006. The detection and decay of Cry1Ab Bt-endotoxins within non-target slugs, Deroceras reticulatum (Mollusca: Pulmonata), following consumption of transgenic corn. Biocontrol Sci. Tech. 16:77-88.

Harwood, J.D., W.G. Wallin, and J.J. Obrycki. 2005. Uptake of Bt endotoxins by nontarget herbivores and higher order arthropod predators: Molecular evidence from a transgenic corn agroecosystem. Mol. Ecol. 14:2815-23.

Hasler, C.M. 2000. The changing face of functional foods. J. Am. College Nutr. 19(90005):499S-506S.

Hatloy, A., J. Hallund, M.M. Diarra, and A. Oshaug. 2000. Food variety, socio-economic status and nutritional status in urban and rural areas in Koutiala (Mali). Public Health Nutr. 3:57-65.

Hatloy, A., L.E. Torheim, and A. Oshaug. 1998. Food variety - a good indicator of nutritional adequacy of the diet? A case study from an urban area in Mali, West Africa. Eur. J. Clinic. Nutr. 52:891-898.

Haverkort , B., J. van der Kamp, and A. Waters-Bayer (ed) 1991. Joining farmers' experiments. Inter. Tech. Publ., London.

Hawkes, C. 2006. Uneven dietary development: linking the policies and processes of globalisation with the nutrition transition, obesity and diet-related chronic diseases.

Global. Health 2:4. Available at http://www.pubmedcentral.nih.gov/picrender.fcgi?artid=1440852&blobtype=pdf.

Hawkes, C. 2007. Promoting healthy diets and tackling obesity and diet-related chronic diseases: what are the agricultural policy levers? Food Nutr. Bull. 28(2 Suppl):S312-322.

Hawkes, J.G., N. Maxted, and B.V. Ford-Lloyd. 2000. The ex situ conservation of plant genetic resources. Kluwer, Dordrecht.

Hayama, R., and G. Coupland. 2004. The molecular basis of diversity in the photoperiodic flowering responses of arabidopsis and rice. Plant Physiol. 135:6 77-684.

Hayes, T.B., A. Collins, M. Lee, M. Mendoza, N. Noreiga, A.A. Stewart, and A. Vonk. 2002. Hermaphroditic, demasculinized frogs after exposure to the herbicide atrazine at low ecologically relevant doses. PNAS 99:5476-5480.

Hazarika, G., and J. Alwang. 2003. Access to credit, plot size and cost inefficiency among smallholder tobacco cultivators in Malawi. Agric. Econ. 29:99-109.

Hazell, P., and L. Haddad, 2001. Agricultural research and poverty reduction. Food, Agric., Environ. Disc. Pap. 34. IFPRI, Washington DC.

Hazell, P.B.R.C., and C. Ramasamy. 1991. The Green Revolution reconsidered. The impact of high-yielding rice varieties in South India. IFPRI, Johns Hopkins Univ. Press, Baltimore.

Heap, I. 2007 International survey of herbicide resistant weeds. Available at www.weedscience.org/in.asp.

Heffernan, C., and F. Misturelli. 2001. Perception of poverty among poor livestock keepers in Kenya. A discourse analysis approach. J. Int. Dev. 13:863-875.

Heffernan, C., F. Misturelli, R. Fuller, S. Patnaik, N. Ali, W. Jayatilaka, and A. McLeod. 2005. The livestock and poverty assessment methodology: An overview. p. 53-69. In E. Owen et al. (ed) Livestock and wealth creation. Improving the husbandry of animals kept by resource poor people in developing countries. Nottingham Univ. Press, UK.

Heisey, P.W., M.A. Lantican, and H.J. Dubin. 2002. Impacts of international wheat breeding research in developing countries, 1966-97. CIMMYT, Mexico

Henao, J., and C. Baanante. 2006. Agricultural production and soil nutrient mining in Africa. Summary IFDC Tech. Bull. IFDC, Muscle Shoals, AL.

Henke, J.M., G. Klepper, and G. Schmitz. 2005. Tax exemption for biofuels in Germany: Is bio-ethanol really an option for climate policy? Energy 30:2617-2635.

Henninges, O. and J. Zeddies. 2006. Bioenergy in Europe: Experiences and prospects. Brief 9. In P. Hazell and R.K. Pachauri (ed) Bioenergy and agriculture: Promises and challenges. IFPRI, Washington DC.

Henry, F.J., Y. Patwary, S.R.A. Huttly, and

K.M.A. Aziz. 1990. Bacterial contamination of weaning foods and drinking water in rural Bangladesh. Epidemiol. Infect. 104:79-85.

Herdt, R.W. 2006. Biotechnology in agriculture. Annu. Rev. Environ. Resour. 31:265-295.

Herrick, J.E., and M.M. Wander. 1997. Relationships between soil organic carbon and soil quality in cropped and rangeland soils: The importance of distribution, composition, and soil biological activity. p. 405-426. In R. Lal et al. (ed) Soil processes and the carbon cycle. CRC Press, Boca Raton.

Herrmann, R., and C. Roeder. 1998. Some neglected issues in food demand analysis: retail-level demand, health information and product quality. Aust. J. Agric. Res. Econ. 42:341-367.

Herselman, M.E. and K.G. Britton. 2002. Analysing the role of ICT in bridging the digital divide amongst learners. South Afric. J. Educ. 22(4):270-274.

Hertel, T.W., and L.A. Winters (ed) 2005. Poverty and the WTO. Impacts of the Doha development agenda. World Bank, Washington DC.

Hervieu, B., and J. Viard, 1996, Au bonheur des campagnes (et des provinces), L'Aube, Marseille.

Herzka, S.Z. 2005. Assessing connectivity of estuarine fishes based on stable isotope ratio analysis. Estuarine Coastal Shelf Sci. 64: 58-69.

Hicks, N. 1987. Education and economic growth. p. 101-107. In G. Psacharopoulos (ed) Economics of education: Research and studies. Pergamon Press, Oxford.

Hilbeck, A., and D.A. Andow (ed) 2004. Environmental risk assessment of genetically modified organisms. Vol. 1. A case study of Bt maize in Kenya. CABI, Wallingford.

Hilbeck, A., D.A. Andow and E. Fontes (ed) 2006. Environmental risk assessment of genetically modified organisms. Vol. 2. Improving the scientific basis for environmental risk assessment through the case of Bt cotton in Brazil. CABI, Wallingford.

Hilbeck, A. and J.E.U. Schmidt. 2006. Another view on Bt proteins — how specific are they and what else might they do? Biopesticides Int. 2:1-50.

Hill, J., E. Nelson, D. Tilman, S. Polasky, and D. Tiffany. 2006. From the cover: Environmental, economic, and energetic costs and benefits of biodiesel and ethanol biofuels. PNAS 103:11206-11210.

Hinrichsen D., B. Robey and U.D. Upadhyay. 1998. Solutions for a water-short world. Chapter 3 In The coming era of water stress and scarcity. Popul. Rep., Ser. M No. 14. Johns Hopkins Sch. Public Health, Baltimore.

Hobbs, P.R. 2006. Conservation agriculture: What is it and why is it important for future sustainable food production? p. 15. In (ed) M.P. Reynolds et al. Challenges to international wheat improvement. Available at http://www.css.cornell.edu/faculty/hobbs/ Papers/Hobbs_final_Mexico_paper.pdf. CIMMYT, Mexico.

Hobbs, P.R., K. Sayre and R. Gupta. 2006. The role of conservation agriculture in sustainable agriculture. Phil. Trans. R. Soc. (UK). Cornell Univ. Dep. Crop Soil Sci. Available at: http://www.css.cornell.edu/faculty/hobbs/Papers/Cultivation%20Techniques%20for%20Conservation%20Agriculture%20Systems.pdf and http://www.css.cornell.edu/faculty/hobbs/publications.htm

Hobson, K.A. 1999. Tracing origins and migration of wildlife using stable isotopes: A review. Oecologia 120:314-326.

Hocdé, H. 1997. Crazy but not mad! p. 49-66. In L. van Veldhuizen (ed) Farmers' research in practice. Lessons from the field. Intermediate Tech. Publ., London.

Hoddinott, J., and Y. Yohannes. 2002. Dietary diversity as a food security indicator. FCND Disc. Pap. No. 136. IFPRI, Washington DC.

Hofs, J-L., M. Fok, and M. Vaissayre. 2006. Impact of Bt cotton adoption on pesticide use by smallholders: A 2-year survey in Makhatini Flats (South Africa). Crop Prot. 25:984-988.

Hollingsworth, P.M., I.K. Dawson, W.P. Goodall-Copestake, J.E. Richardson, J.C. Weber, C. Sotelo Montes et al. (2005). Do farmers reduce genetic diversity when they domesticate tropical trees? A case study from Amazonia. Mol. Ecol. 14:497-501.

Holmgren, P., E.J. Masakha, and H. Sjöholm. 1994. Not all African land is being degraded: A recent survey of trees on farms in Kenya reveals increasing forest resources. Ambio 23:390-395.

Homewood, K., and D. Brockington. 1999. Biodiversity, conservation and development in Mkomazi Game Reserve, Tanzania. Global Ecol. Biogeography 8:301-313.

Hossain, M., D. Gollin, V. Cabanilla, E. Cabrera, N. Johnson, G.S. Khush, and G. McLaren. 2003. International research and genetic improvement in rice: Evidence from Asia and Latin America. p. 71-108. In R.E. Evenson and D. Gollin (ed) Crop variety improvement and its effect on productivity. CABI, Wallingford.

Hoy, M.A. 1992. Biological control of arthropods: Genetic engineering and environmental risks. Biol. Control 2:166-170.

Huang, J., and S. Rozelle. 1995. Environmental stress and grain yields in China. Am. J. Agric. Econ. 77:853-864.

Huang, J., S. Rozelle, C. Pray, and Q. Wang. 2002. Plant biotechnology in China. Science 295:674-676.

Huang, J., Q. Zhang and S. Rozelle. 2005. Macroeconomic policies, trade liberalization and poverty in China. CCAP Working Pap., Chinese Acad. Sci., Beijing.

Huber-Lee, A., and E. Kemp-Benedict. 2003. Agriculture: Re-adaptation to environment. In S.S. Jinendradasa (ed) Issues of water management in agriculture: Compilation

of essays. Comprehensive Assessment Secretariat, IWMI, Colombo.

Hufbauer, G.C., and J.J. Schott. 2005. NAFTA revisited. Achievements and challenges. Inst. Int. Econ., Washington DC.

Humphries, S., J. Gonzales, J. Jimenez, and F. Sierra. 2000. Searching for sustainable land use practices in Honduras: Lessons from a programme of participatory research with hillside farmers. Agricultural Res. Extension Network Pap. No.104 July 2000, London, ODI.

Hussain, F., K.F. Bronson, S. Yadvinder, S. Bijay, and S. Peng. 2000. Use of chlorophyll meter sufficiency indices for nitrogen management of irrigated rice in Asia. Agron. J. 92: 875- 879.

Hussain, I. 2005. Pro-Poor intervention strategies in irrigated agriculture in Asia: Poverty in irrigated agriculture-Issues, lessons, options, and guidelines: Bangladesh, China, India, Pakistan, and Viet Nam. Final synthesis Rep. 1. IWMI, Colombo.

Hussain, M.G., and M.A. Mazid. 2004. Carp genetic resources in Bangladesh. p. 16-25. In D. Penman, M.V. Gupta, and M. Dey (ed) Carp genetic resources in Asia. World Fish Center, Penang.

Huxley, P. 1999. Tropical agroforestry. Blackwell Sci., London.

ICARDA. 1986. Food legume improvement program: Annual Report for 1986. ICARDA, Aleppo.

ICRAF. 1996. Annual Report 1995, ICRAF, Nairobi.

ICRAF. 1997. Annual Report 1996, ICRAF, Nairobi.

ICRISAT. 2006. Integrating MAS and conventional breeding for downy mildew resistance, p. 4-5 Nurturing the seeds of success in the semi-arid tropics. ICRISAT, Patancheru, India.

IDS. 2006. Looking at social protection through a livelihoods lens. IDS In Focus, Issue 01, May 2006.

IEA (International Energy Agency). 2002. World energy outlook 2002. IEA, Paris.

IEA (International Energy Agency). 2004a. Analysis of the impact of high oil prices on the global economy. IEA, Paris.

IEA (International Energy Agency). 2004b. Biofuels for transport. An international perspective. IEA, Paris.

IEA (International Energy Agency). 2006a. Co-utilization of biomass with fossil fuels. Summary and conclusions from the IEA Bioenergy ExCo55 Workshop. Paris.

IEA (International Energy Agency). 2006b. Energy technology perspectives 2006. In support of the G8 Plan of Action. IEA, Paris.

IEA (International Energy Agency). 2006c. World energy outlook 2006. IEA, Paris.

IFAD. 2003. Promoting market access for the rural poor in order to achieve the Millennium Development Goals. Roundtable disc. Pap. 25th Anniversary Session of IFAD's Governing Council, February 2003.

IFAD. 2004. Rural women's access to land and

property in selected countries. Progress towards achieving the aims of Articles 14, 15 and 16 of the Convention on the Elimination of All Forms of Discrimination against Women. Available at http://www.landcoalition.org/pdf/cedawrpt.pdf.

IFAD. 2006. Ghana: Upper west agricultural development project interim evaluation. IFAD, Rome.

IFOAM. 2003. The world of organic agriculture 2003 - Statistics and future prospects. Available at http://www.soel.de/inhalte/publikationen/s/s_74.pdf. FAO, Rome.

IFOAM. 2006. The world of organic agriculture 2006 - Statistics and emerging trends. 8th ed. Int. Fed. Organic Agric. Move., Bonn.

IIED. 1996. Towards a sustainable paper cycle. WBCSD, Geneva.

ILO. 2000. Safety and health in agriculture. Int. Labour Office Conf., 88th Session 2000. 30 May - 15 June. Report VI(1). ILO, Geneva.

ILO. 2006. Tackling hazardous child labour in agriculture: Guidance on policy and practice. User Guide. ILO, Geneva.

Imbernon, J., J.L. Villacorta Monzon, C.L. Zelaya, and A.A. Valle Aguirre. 2005. Fragmentación y conectividad del bosque en El Salvador. Aplicación al Corredor Biológico Mesoamericano. Bois et Forêts des Tropiques 286:15-28.

IPCC. 2007. Climate Change 2007: The physical science basis. Summary for decision makers. Contribution of the working group I to the fourth assessment report of the IPCC. Cambridge U. Press, Cambridge, UK.

IPGRI. 1993. Diversity for development. IPGRI, Rome.

IUCN. 2004. Allanblackia seeds help Ghanaian communities improve livelihoods. INCN News Release, 13 May 2004.

IWMI. 2006. Water for food, Water for life. Insights from the Comprehensive Assessment of Water Management in Agriculture. Stockholm World Water Week, 2006.

Izac, A-M., and P.A. Sanchez. 2001. Towards an NRM paradigm for international agriculture. Agric. Syst. 69:5-25.

Jackson, C. 2003. Gender analysis of land: Beyond land rights for women? J. Agrar. Change 3(4):453-480.

Jackson, R.B., E.G. Jobbàgy, R. Avissar, S.B. Roy, D.J. Barratt, C.W. Cook et al. 2005. Trading water for carbon with biological carbon sequestration. Science 310:1944-1947.

Jackson, R.B., and W.H. Schlesinger. 2004. Curbing the U.S. carbon deficit. PNAS 101:15827-15829.

Jama, B., F. Kwesiga, and A. Niang. 2006. Agroforestry innovations for soil fertility management in sub-Saharan Africa: Prospects and challenges ahead. p. 53- 60. In D. Garrity et al. World agroforestry into the future. World Agroforestry Centre, Nairobi.

James, C. 2006. Global status of commercialized biotech/GM crops. ISAAA Brief No. 35. ISAAA: Ithaca, NY.

Jamieson, N., 1987. The paradigmatic shift of rapid rural appraisal. *In* Proc. 1985 Int. Conf. Rapid Rural Appraisal. Khon Kaen Univ., Thailand.

Janhari, V.P. 2003. Summary of findings, analysis of austainability and suggestions. p. 12-22. *In* B. Venkateswara Rao (ed) Proc. Int. Conf. Hydrol. Watershed Manage. Vol. 2. Hyderabad,18-20 Dec 2002. BS Publ., Hyderabad.

Jarvis, D., B. Sthapit, and L. Sears (ed) 2000. Conserving agricultural biodiversity in situ: A scientific basis for sustainable agriculture. Proc. workshop. 5-12 July 1999. Pokhara, Nepal. IPGRI, Rome.

Jayne, T.S., P. Pingali, M. Villareal, and G. Memrich. 2004. Interactions between the agricultural sector and HIV/AIDS pandemic. Implications for agricultural policy. ESA Working Pap. No. 04-06. FAO, Rome.

Jayne, T.S., T. Yamano, M.T. Weber, D. Tschirley, R. Benfica, A. Chapoto, and B. Zulu. 2003. Smallholder income and land distribution in Africa: Implications for poverty reduction strategies. Food Policy 28(3):253-275.

Jenkins, E.J., A.M. Veitch, S.J. Kutz, E.P. Hoberg, and L. Polley. 2006. Climate change and the epidemiology of protostrongylid nematodes in northern ecosystems: *Parelaphostrongylus odocoilei* and *Protostrongylus stilesi* in Dall's sheep (Ovis d. dalli). Parasitology 132(Pt 3):387-401.

Joensen, L., S. Semino, and H. Paul. 2005. Argentina: A case study on the impact of genetically engineered soya. Available at www.econexus.info/pdf/Enx-Argentina-GE-Soya-Report-2005.pdf. The Gaia Foundation, London.

Johns, T., and P.B. Eyzaguirre. 2007. Biofortification, biodiversity and diet: a search for complementary applications against poverty and malnutrition. Food Policy 32:1-24.

Johns, T., I.F. Smith, and P.B. Eyzaguirre. 2006. Agrobiodiversity, nutrition, and health: Understanding the links between agriculture and health. 2020 Vision Briefs No. 13(12). Available at http://www.ifpri.org/2020/focus/focus13/focus13_12.pdf. IFPRI, Washington DC.

Johnson, K.A. and D.E. Johnson. 1995. Methane emissions from cattle. J. Anim. Sci. 73: 2483-2492.

Johnson, N.L., D. Pachico, and C.S. Wortmann. 2003. The impact of CIAT's genetic improvement research in beans p. 257-274. *In* R.E. Evenson and D. Gollin (ed) Crop variety improvement and its effect on productivity. CABI, Wallingford.

Jones, M. 1990. Efficiency and effectiveness in an African public administration context, Int. J. Public Sector Manage. 3:58-64.

JRC. 2007. Food quality schemes project. Joint Research Center, European Commission. Available at http://foodqualityschemes.jrc.es/en/index.html.

Juma, C. 2006. Reinventing African economies.

Technological innovation and the sustainability transition. Published keynote address for The John Pesek Colloquium on Sustainable Agriculture, Iowa State Univ., April 6-7, 2006. Available at http://www.wallacechair.iastate.edu/endeavors/pesekcolloquium/.

Kader, A.A. 2003. A perspective on postharvest horticulture (1978- 2003). Hort. Sci. 38(5):1004-1008.

Kanampiu, F.K., V. Kabambe, C. Massawe, L. Jasi, D. Friesen, J.K. Ransom et al. 2003. Multi-site, multi-season field tests demonstrate that herbicide seed-coating herbicide-resistance maize controls Striga spp. and increases yields in several African countries. Crop Prot. 22:697-706.

Kant, A.K., A. Schatzkin, and R.G. Ziegler. 1995. Dietary diversity and subsequent cause-specific mortality in the NHANES I epidemiologic follow-up study. J. Am. Coll. Nutr. 14, 233-238.

Kant, A.K., G. Block, A. Schatzkin, R.G. Ziegler, and T.B. Harris. 1993. Dietary diversity and subsequent mortality in the first national health and nutrition examination survey epidemiologic follow-up study. Am. J. Clinic. Nutr. 57:434-440.

Kaplinski, N. 2002. Conflicts around a study of Mexican crops. Nature 417:898.

Kaplinski, N., D. Braun, E. Lisch, A. Hay, S. Hake, and M. Freeling. 2002. Biodiversity (Communications arising): Maize transgene results in Mexico are artefacts. Nature 416:601-602.

Kapseu, C., E. Avouampo, and B. Djeumako. 2002. Oil extraction from Dacryodes edulis (G. Don) H.J. Lam fruit. For. Trees Livelihoods. 11:97-104.

Karekezi, S., K. Lata and S.T. Coelho. 2004. Traditional biomass energy. Improving its use and moving to modern energy use. Int. Conf. Renewable Energies, Bonn. Jan. 2004. Available at http://www.renewables2004.de/doc/DocCenter/TBP11-biomass.pdf.

Karl, S.A., and B.W. Bowen. 1999. Evolutionary significant units versus geopolitical taxonomy: molecular systematics of an endangered sea turtle (genus Chelonia). Conserv. Biol. 13:990- 999.

Kartha, S., G. Leach, and S.C. Rajan. 2005. Advancing bioenergy for sustainable development. Guidelines for policymakers and investors. ESMAP, World Bank, Washington DC.

Kaul, R.N. and A. Ali, 1992. Gender issues in African farming: A case for developing farm tools for women. J. Farming Syst. Res. Extens. 3:35-46.

Kennedy, E.T., J. Ohls, S. Carlson, and K. Fleming. 1995. The healthy eating index: Design and applications. J. Am. Dietary Assoc. 95:1103-1108.

Kenny, C. 2002. The internet and economic growth in LDCs; A case of managing expectations? UN University, World Inst. Dev. Econ. Res. Disc. Pap. No. 2002/75, Helsinki. Available at http://

papers.ssrn.com/sol3/papers.cfm?abstract_id=588483#PaperDownload.

Kerr, J. and S. Kolavalli. 1999. Impact of agricultural research on poverty alleviation: conceptual framework with illustrations from the Literature. EPTD Disc. Pap. 56. IFPRI, Washington DC.

Khan, Z.R., C.A.O. Midega, A. Hassanali, J.A. Pickett and L.J. Wadhams. 2007. Assessment of different legumes for the control of Striga hermonthica in maize and sorghum. Crop Sci. 47: 730-734.

Khan, Z.R., C.A.O. Midega, A. Hassanali, J.A. Pickett, L.J. Wadhams, and A. Wanjoya. 2006. Management of witchweed, Striga hermonthica, and stemborers in sorghum, Sorghum bicolor, through intercropping with greenleaf desmodium, Desmodium intortum. Int. J. Pest Manage. 52:297-302.

Kilpatrick, S., R. Bell, and I. Falk. 1999. The role of group learning in building social capital, J. Voc. Educ. Training 51:(1)129-144.

Kim H.Y., M. Lieffering, K. Kobayashi, M. Okada, M.W. Mitchell and M. Gumpertz. 2003. Effects of free-air CO_2 enrichment and nitrogen supply on the yield of temperate paddy rice crops. Field Crops Res. 83:261-270.

Kimball, B.A. 1983. Carbon dioxide and agricultural yield: An assemblage and analysis of 430 prior observations. Agron. J. 75:779-788.

Kindt, R., A.J. Simons, and P. van Damme. 2004. Do farm characteristics explain differences in tree species diversity among western Kenyan farms? Agrofor. Syst. 63:63-74.

Kirchmann, H., and L. Bergstrom. 2001. Do organic farming practices reduce nitrate leaching? Comm. Soil Sci. Plant Analysis 32:997-1028.

Kirk, G. 2004. The biochemistry of submerged soils. John Wiley, Chichester, UK.

Kishi, M. 2005, The health impacts of pesticides: What do we now know? p. 23-38. *In* J. Pretty (ed) The pesticide detox. Earthscan, London.

Kitinoja, L., and J.R. Gorny. 1999. Postharvest technology for small-scale produce marketers: economic opportunities, quality and food safety. Univ. CA, Davis, Postharvest Hort. Serv. 21.

Koebner, R.M.D. and R.W. Summers. 2003. 21st century wheat breeding: Plot selection or plate detection? Trends Biotech. 21:22:59-63.

Koenig, D. 2001. Toward local development and mitigating impoverishment in development-induced displacement and resettlement. Final Rep. Prepared for ESCOR R7644 and the Research Programme on Development-Induced Displacement and Resettlement. Refugee Studies Centre, Univ. Oxford, UK.

Köhler-Rollefson, I. and J. Bräunig. 1998. Anthropological veterinary medicine: The need for indigenizing the curriculum. 9th AITVM Conf. Harare. 14-18 Sept. 1998. Available at http://www.vetwork.org.uk/ilse2.htm.

Kojima, M., and T. Johnson. 2005. The potential for biofuels for transport in developing countries. Rep. 312/05. ESMAP, World Bank, Washington DC.

Kojima, M., D. Mitchell, and W. Ward. 2007. Considering biofuel trade policies. ESMAP, World Bank, Washington DC.

Kramer, S.B., J.P. Reganold, J. Glover, B.J.M. Bohannan, and H.A. Mooney. 2006. Reduced nitrate leaching and enhanced denitrifier activity and efficiency in organically fertilized soils. PNAS 103:4522-4527.

Kranjac-Berisavljevic', G., and B.Z. Gandaa. 2004. The cultural roles of yam. Geneflow. Available at http://ipgri-pa.grinfo.net/iGeneflow.php?itemid=697. IPGRI, Rome.

Krauss, U. 2004. Diseases in tropical agroforestry landscapes: The role of biodiversity. p. 397-412. In Agroforestry and biodiversity conservation in tropical landscapes, G. Schroth et al. (ed) Island Press, Washington.

Krebs-Smith, S.M., H. Smiciklas-Wright, H.A. Guthrie, and J. Krebs-Smith. 1987. The effects of variety in food choices on dietary quality. J. Am. Diet. Assoc. 87:897-903.

Kristal, A.R., A.L. Shattuck, and H.J. Henry. 1990. Patterns of dietary behavior associated with selecting diets low in fat: Reliability and validity of a behavioral approach to dietary assessment. J. Am. Dietary Assoc. 90:214-220.

Kumaran, M., N. Kalaimani, K. Ponnusamy, V. S. Chandrasekaran, and D. Deboral Vimala. 2003. A case of informal srimp farmers' association and its role in sustainable shrimp farming in Tamil Nadu. Aquacult. Asia 7(2):10-12.

Kunin, W.E. and J.H. Lawton. 1996. Does biodiversity matter? p. 283-308. In K.G. Gaston (ed) Biodiversity: A biology of numbers and difference. Blackwell, Oxford.

Kurien, J. 2004. Responsible fish trade and food security — toward understanding the relationship between international fish trade and food security. Roy. Norweg. Min. For. Affairs and FAO, Rome.

Kusters, K., and B. Belcher. 2004. Forest products, livelihoods and conservation: Case studies of non-timber forest product systems. Vol. 1. Asia. CIFOR, Jakarta.

Kwesiga, F., S. Franzel, F. Place, D. Phiri, and C.P. Simwanza. 1999. Sesbania sesban improved fallows in eastern Zambia: Their inception, development and farmer enthusiasm. Agrofor. Syst. 47:49-66.

Lacoste, Y. 2004. Géopolitique des drogues illicites. Hérodote 112. Available at http://www.herodote.org/article.php3?id_article=57.

Lafever, H.N., O. Sosa Jr., R.L. Gallun, J.E. Foster, and R.C. Kuhn.1980. Survey monitors Hessian fly population in Ohio wheat. Ohio Reporter 65:51-53.

Laird, S.A. 2002. Biodiversity and traditional knowledge: Equitable partnerships in practice. People Plants Conserv. Ser. Earthscan, London.

Lal, R. 1997. Long term tillage and maize monoculture effects on tropical Alfisol in western Nigeria. II. Soil chemical properties. Soil Till. Res. 42:161-174.

Lal, R. 2004. Soil carbon sequestration impacts on global climate change and food security. Science 304:1623-1627.

Lal, R. 2006. Managing soils for feeding a global population of 10 billion. J. Sci. Food Agric. 86:2273-2284.

Lal, R., Z.X. Tan, and K.D. Wiebe. 2005. Global soil nutrient depletion and yield reduction. J. Sustain. Agric. 26(1):123-146.

Lam, W.F. 1998. Governing irrigation systems in Nepal: Institutions, infrastructure, and collective action. Institute for Contemporary Studies. Oakland, CA.

Landers, J.N., J. Clay and J. Weiss. 2005. Integrated crop-livestock ley farming with zero tillage: Five case studies of the win-win-win strategy for sustainable farming in the tropics. p. 6. In Proc. Third World Congress on Conservation Agriculture: Linking production, livelihoods and conservation. Nairobi, Kenya. Available at www.act.org.zw/postcongress/documents/Sess4(Strat+approaches)/Landers%20et%20al.doc

Landers, J.N., H.M. Saturnino, and P.L. de Freitas. 2001. Organizational and policy considerations in zero tillage. In H.M. Saturnino and J.N. Landers (ed) The environment and zero tillage. Associação de Plantio Direto no Cerrado, Federacão Bras. Plantio Direto na Palha, Brasil.

Landis, D.A., S.D. Wratten, and G.M. Gurr. 2000. Habitat management to conserve natural enemies of arthropod pests in agriculture. Ann. Rev. Entomol. 45:175-201.

Lang, T., and M. Heasman. 2004. Food wars: The global battle for mouths, minds and markets. Earthscan, London.

Lantican, M.A., H.J. Dubin, and M.L. Morris. 2005. Impacts of international wheat breeding research in the developing world, 1988- 2002. CIMMYT, Mexico.

Lapeyrie, F., and P. Högberg, 1994. Harnessing symbiotic associations: ectomycorrhizas. p. 158-164. In R.R.B. Leakey and A.C. Newton (ed) Tropical trees: The potential for domestication and the rebuilding of forest resources. HMSO, London.

Lardon, S., P. Maurel, and V. Piveteau. 2001. Représentations spatiales et développement territorial. Hermès, Paris.

Larson, E.D. 1993. Technology for electricity and fuels from biomass. Ann. Rev. Energy Environ. 18:567-630.

Larsson S.C., A. Wolk. 2006, Meat consumption and risk of colorectal cancer: A meta-analysis of prospective studies. Int. J. Cancer 119(11):2657-2664.

Last, F.T., J.E.G. Good, R.H. Watson, and D.A. Greig. 1976. The city of Edinburgh - its stock of trees, a continuing amenity and timber resource. Scottish Forestry 30:112-126.

Lastarria-Cornhiel, S. 1997. Impact of privatization on gender and property rights in Africa. World Dev. 25:1317-1333.

Lastarria-Cornhiel, S., and J. Melmed-Sanjak. 1999. Land tenancy in Asia, Arica, and Latin America: A look at the past and a view to the future. Land Tenure Center Working Pap. 76, Univ. Wisconsin, Madison.

Lavigne Delville, P. 1998. Quelles politiques foncières en Afrique rurale? Ministère de la Coopération/Karthala, Paris.

Law, M. 2000. Dietary fat and adult diseases and the implications for childhood nutrition: an epidemiologic approach. Am. J. Clin. Nutr. 72(5 Suppl):1291S-1296S.

Lawson, G.J. 1994. Indigenous trees in West African forest plantations: The need for domestication by clonal techniques. p. 112-123. In R.R.B. Leakey and A.C. Newton (ed) Tropical trees: The potential for domestication and the rebuilding of forest resources. HMSO, London.

Le Billon, P. 2001. The political ecology of war: Natural resources and armed conflicts. Polit. Geog. 20:561-584.

Le Coq, J.F., and G. Trebuil. 2005. Impact of economic liberalization on rice intensification, agricultural diversification and rural livelihoods in the Mekong Delta, Vietnam. Southeast Asian Studies 42(4):519-547.

Le Roy, E., A. Karsenty, and A. Bertrand. 1996. La sécurisation foncière en Afrique. Pour une gestion viable des ressources renouvelables. Karthala, Paris.

Leach, M., and R. Mearns. 1996. The lie of the land: Challenging received wisdom on the Africa environment. IDS, Sussex.

Leakey, A.D.B., C.J. Bernacchi, F.G. Dohleman, D.R. Ort, and S.P. Long. 2004. Will photosynthesis of maize (Zea mays) in the US Corn Belt increase in future [CO2] rich atmospheres? An analysis of diurnal courses of CO_2 uptake under Free-Air Concentration Enrichment (FACE). Glob. Change Biol. 10:951-962.

Leakey, C.D.B. 2006. Quantifying inhabitation, feeding and connectivity between adjacent estuarine and coastal regions for three commercially important marine fishes. PhD thesis. Univ. Plymouth.

Leakey, R.R.B. 1985. The capacity for vegetative propagation in trees. p. 110-133. In M.G.R. Cannell and J.E. Jackson (ed) Attributes of trees as crop plants. Inst. Terr. Ecol., Abbots Ripton, Huntingdon, England.

Leakey, R.R.B. 1987. Clonal forestry in the tropics - A review of developments, strategies and opportunities. Commonwealth For. Rev. 66:61-75.

Leakey, R.R.B. 1991. Towards a strategy for clonal forestry: Some guidelines based on experience with tropical trees. p. 27-42. In J.E. Jackson (ed) Tree breeding and improvement, Roy. For. Soc. Engl., Wales and Northern Ireland, Tring, England.

Leakey, R.R.B. 1996. Definition of agroforestry revisited. Agrofor. Today 8:5-7.

Leakey, R.R.B. 1999a. Potential for novel food products from agroforestry trees. Food Chem. 64:1-14.

Leakey, R.R.B. 1999b. Agroforestry for biodiversity in farming systems. p. 127-145. *In* W.W. Collins and C.O. Qualset (ed) Biodiversity in agroecosystems. CRC Press, NY.

Leakey, R.R.B. 2001a. Win:Win landuse strategies for Africa: 1. Building on experience with agroforests in Asia and Latin America. Int. For. Rev. 3:1-10.

Leakey, R.R.B. 2001b. Win:win landuse strategies for Africa: 2. Capturing economic and environmental benefits with multistrata agroforests. Int. For. Rev. 3:11-18.

Leakey, R.R.B. 2003. The domestication of indigenous trees as the basis of a strategy for sustainable land use. p. 27-40. *In* J. Lemons, R. Victor, and D. Schaffer (ed) Conserving biodiversity in arid regions. Kluwer Academic Publ., Boston.

Leakey, R.R.B. 2004. Physiology of vegetative reproduction in trees. p. 1655-1668. *In* J. Burley, J. Evans and J.A. Youngquist (ed) Encycl. For. Sci. Academic Press, London.

Leakey, R.R.B. 2005a. Agricultural sustainability in FNQ: Boom or bust? A global perspective. p. 392-400. *In* Cairns 2020/2050 Business Res. Manual Project, Cairns 2020/2050 BRM Partnership, Cairns.

Leakey, R.R.B. 2005b. Domestication potential of Marula (Sclerocarya birrea subsp caffra) in South Africa and Namibia: 3. Multi-trait selection. Agrofor. Syst. 64:51-59.

Leakey, R.R.B., J.F. Mesén, Z. Tchoundjeu, K.A. Longman, J. McP. Dick., A.C. Newton et al. 1990. Low-technology techniques for the vegetative propagation of tropical tress. Commonwealth For. Rev. 69:247-257.

Leakey, R.R.B., K. Pate, and C. Lombard. 2005c. Domestication potential of Marula (Sclerocarya birrea subsp caffra) in South Africa and Namibia: 2. Phenotypic variation in nut and kernel traits. Agrofor. Syst. 64:37-49.

Leakey, R.R.B., and P.A. Sanchez. 1997. How many people use agroforestry products? Agrofor, Today 9(3):4-5.

Leakey, R.R.B., K. Schreckenberg, and Z. Tchoundjeu. 2003. The participatory domestication of West African indigenous fruits. Int. For. Rev. 5:338-347.

Leakey, R.R.B., S. Shackleton, and P. du Plessis. 2005b. Domestication potential of Marula (Sclerocarya birrea subsp caffra) in South Africa and Namibia: 1. Phenotypic variation in fruit traits. Agrofor. Syst. 64:25-35.

Leakey, R.R.B., and Z. Tchoundjeu. 2001. Diversification of tree crops: Domestication of companion crops for poverty reduction and environmental services. Exp. Agric. 37:279-296.

Leakey, R.R.B., Z. Tchoundjeu, K. Schreckenberg, S. Shackleton, and C. Shackleton. 2005a. Agroforestry Tree Products (AFTPs): Targeting poverty reduction and enhanced livelihoods. Int. J. Agric. Sustain. 3:1-23.

Leakey, R.R.B., Z. Tchoundjeu, R.I. Smith, R.C. Munro, J-M. Fondoun, J. Kengue et al. 2004. Evidence that subsistence farmers have

domesticated indigenous fruits (Dacryodes edulis and Irvingia gabonensis) in Cameroon and Nigeria. Agrofor. Syst. 60:101-111.

Leakey, R.R.B., J. Wilson, and J.D. Deans. 1999. Domestication of trees for agroforestry in drylands. Ann. Arid Zones 38:195-220.

Lebel, L., H.T. Nguyen, S. Amnuay, P. Suparb, B. Urasa and K.T. Le. 2002. Industrial transformation and shrimp aquaculture in Thailand and Vietnam: Pathways to ecological, social and economic sustainability. Ambio 31: 311-323.

Ledford, H. 2007. Out of bounds. Nature 445:132-133.

Lees, J.W. 1990. More than accountability: Evaluating agricultural extension programmes. The Rural Dev. Centre, Univ. New England, Armidale, NSW.

Leeuwis, C. 1993. Of computers, myths and modelling: The social construction of diversity, knowledge, information and communication technologies in Dutch horticulture and agricultural extension. Wageningen Studies in Sociology 36. Wageningen Agricultural Univ., Wageningen.

Legg, B. 2005. Crop improvement technologies for the 21st century. p. 31-50. *In* R. Sylvester-Bradley and J. Wiseman (ed) Yields of farmed species. Nottingham Univ. Press, UK.

Legg, J.P. and C.M. Fauquet. 2004. Cassava mosaic geminiviruses in Africa. Plant Mol. Biol. 56:585-599.

Lele, U. 2004. Technology generation, adaptation, adoption and impact: towards a framework for understanding and increasing research impact. Working Pap. 31964, World Bank, Washington DC.

Lewis, W.J., J.O., J.C. van Lanteren, S.C. Phatak, and J.H.Tumlinson, III. 1997. A total system approach to sustainable management. PNAS 94(23):12243-12248.

Lhaloui, S., M. El Bouhssini, N. Nsarellah, A. Amri, N. Nachit, J. El Haddoury, and M. Jlibene. 2005. Les cecidomyies des cereals au Maroc: Biologie, dégâts et moyens de lutte.

Lhoste, P. 2005. L'étude des systèmes d'élevage en zones tropicales concepts et méthodes. p. 97-115. *In* Manuel de zootechnie comparée Nord-Sud. INRA, Paris.

Li, L-Y. 1994. Worldwide use of Trichogramma for biological control on different crops: A survey, p. 37-54. *In* E. Wajnberg and S.A. Hassan (ed) Biological control with egg parasitoids. CABI, Wallingford.

Li, S.F. 2003. Aquaculture research and its relationship to development in China. *In* Agricultural development and opportunities for aquatic resources research in China. WorldFish Center, Penang, Malaysia.

Liang, L., T. Nagumo, and R. Hatano. 2005. Nitrogen cycling with respect to environmental load in farm systems in Southwest China. Nutr. Cycl. Agroecosyst. 73:119-134.

LID (Livestock in Development). 1999. Livestock in poverty focused development. Livestock in Development, Crewkerne.

Lightfoot, C., and R.P. Noble. 1993. A participatory experiment in sustainable agriculture. J. Farming Syst. Res. Ext. 4(1):11-34.

Lightfoot, C., and R.S.V. Pullin. 1995. Why an integrated resource management approach? *In* R.E. Brummett (ed) Aquaculture policy options for integrated resource management in SubSaharan Africa. ICLARM Conf. Proc., ICLARM, Manila.

Lilja, N., J. Ashby, and L. Sperling (ed) 2001. Assessing the impact of participatory research and gender analysis. PRGA Program, Coordination Office, CIAT, Cali.

Lilja, N., J.A. Ashby, and N. Johnson. 2004. Scaling up and out the impact of agricultural research with farmer participatory research. *In* D. Pachico (ed) Scaling up and out: Achieving widespread impact through agricultural research. CIAT, Cali.

Lin, C.H., R.N. Lerch, H.E. Garrett, and M.F. George. 2005. Incorporating forage grasses in riparian buffers for bioremediation of atrazine, isoxaflutole and nitrate in Missouri. Agrofor. Syst. 63:91-99.

Lin, C.H., R.N. Lerch, H.E. Garrett, W.G. Johnson, D. Jordan, and M.F. George. 2003. The effect of five forage species on transport and transformation of atrazine and isoxaflutole (Balance) in lysimeter leachate. J. Environ. Qual. 32:1992-2000.

Liska, A.J., H.S. Yang, V. Bremer, W.T. Walters, D. Kenney, P. Tracy et al. 2007. Biofuel energy systems simulator: LifeCycle energy and emissions analysis model for corn-ethanol biofuel (ver. 1.0, 2007). Available at http://www.bess.unl.edu. Univ. Nebraska.

Liu, M., 1997. Fondements et pratiques de la recherche-action. Logiques sociales, L'Harmattan, Paris.

Liu, M. and F. Crezé. 2006. La recherche action et les transformations sociales. Coll. Logiques sociales. L'Harmattan, Paris.

Lock, K., J. Pomerleau, L. Causer, D.R. Altmann, and M. McKee. 2005. The global burden of disease attributable to low consumption of fruit and vegetables: Implications for the global strategy on diet. Bull. WHO 83:100-108.

Loges, R., M.R. Kelm, and F. Taube. 2006. Nitrogen balances, nitrate leaching and energy efficiency of conventional and organic farming systems on fertile soils in Northern Germany. Adv. Geoecol. 39:407- 414.

Lomer C. J., R.P. Bateman, D.L. Johnson, J. Langewald, and M. Thomas. 2001. Biological control of locusts and grasshoppers. Ann. Rev. Entomol. 46:667-702.

Long, S.P., E.A. Ainsworth, A.D.B. Leakey, J. Nosberger, and D.R. Ort. 2006. Food for thought: Lower-than-expected crop yield stimulation with rising CO_2 concentrations. Science 312:1918-1921.

Long, S.P., E.A. Ainsworth, A. Rogers, and D.R. Ort. 2004. Rising atmospheric carbon dioxide: plants FACE the future. Ann. Rev. Plant Biol. 55:591-628.

Longping, Y. 2004. Hybrid rice for food security in the world. Available at http://www.fao.org/docrep/008/y5682e/y5682e06.htm#bm06. FAO, Rome.

Losch, B. 2004. Debating the multifunctionality of agriculture: From trade negotiations to development policies by the South. J. Agrarian Change 4:336-360.

Losch, B. 2007. Quel statut pour l'instabilité des prix dans les changements structurels des agricultures des Suds? p. 113-131. In La régulation des marchés agricoles internationaux: Un enjeu décisif pour le développement. L'Harmattan, Paris.

Louette, D. 2000. Traditional management of seed and genetic diversity. What is a landrace? p. 109-142. In S.B. Brush (ed) Genes in the field: On-farm conservation of crop diversity. IPGRI, Rome and IDRC, Ottawa.

Lovato, T., J. Mielniczuk, C. Bayer and F. Vezzani. 2004. Carbon and nitrogen addition related to stocks of these elements in soil and corn yield under management systems. Revista Brasileira de Ciência do Solo 28(1):175-187. (Abstract in English) Available at http://www.scielo.br/scielo.php?script=sci_arttext&pid=S0100-06832004000100017

Lowe, A.J., S.A. Harris, and P. Ashton. 2004. Ecological genetics: Design, analysis and application. Blackwell, Oxford.

Lowe, A.J., J. Wilson, A.C.M. Gillies, and I. Dawson. 2000. Conservation genetics of bush mango from central/west Africa, implications from RAPD analysis. Mol. Ecol. 9:831-841.

Ludwig, B.G. 1999, Extension professionals' perspectives on global programming. 1999 Conf. Proc. Assoc. Int. Extension Education. Port of Spain, Trinidad.

Luecke, R., and R. Katz. 2003. Managing creativity and innovation. Harvard Business School Press, Boston.

Lutz, B., S. Wiedemann, and C. Albrecht. 2005. Degradation of transgenic Cry1Ab protein from genetically modified maize in the bovine gastrointestinal tract. J. Agric. Food Chem. 53:1453-1456.

MA. 2005a. Ecosystem and human well-being: Health synthesis. Millennium Ecosystem Assessment. World Resources Inst., Washington DC.

MA. 2005b. Ecosystems and human Well-being: Wetlands and water synthesis. Millennium Ecosystem Assessment. World Resources Inst., Washington DC.

MA. 2005c. Ecosystems and human well-being. Vol 1. Current state and trends. R. Hassan et al. (ed) Millennium Ecosystem Assessment. Island Press, Washington.

Macrae, J., and A. Zwi (ed) 1994. War and hunger: Rethinking international responses in complex emergencies. Zed Books, Londres.

Madan, M.L. 2005. Animal biotechnology: Applications and economic implications in developing countries Rev. Sci. Tech. Off. Int. Epiz. 24(1):127-139.

Mäder, P., A. Fliebeta, D. Dubois, L. Gunst, P. Fried, and U. Niggli. 2002. Soil fertility and biodiversity in organic farming. Science 296:1694-1697.

Mäder P., A. Fliessbach, D. Dubois, L. Gunst, W. Jossi, F. Widmer et al. 2006. The DOK experiment (Switzerland). p. 41-58. In J. Raupp et al. (ed) ISOFAR: Long-term field experiments in organic farming. Available at http://www.isofar.org/publications/documents/raupp-etal-2006-long-term-content.pdf. ISOFAR, Verlag Dr. Köster, Berlin.

Maestas, J.D., R.L. Knight, and W.C. Gilgert. 2003. Biodiversity across a rural land-use gradient. Conserv. Biol. 17:1425-1434.

Magette, W.L., R.B. Brinsfield, R.E. Palmer, and J.D. Wood. 1989. Nutrient and sediment removal by vegetated filter strips. Trans. ASAE 32:663-667.

Majumdar, D. 2003. Methane and nitrous oxide emission from irrigated rice fields: Proposed mitigation strategies. Current Sci. 84(10):1317-1326.

Makkar, H., and G. Viljoen (ed) 2005. Applications of gene-based technologies for improving animal production and health in developing countries. Springer, The Netherlands.

Mancini, F. 2006. Impact of integrated pest management farmer field schools on health, farming systems, the environment, and livelihoods of cotton growers in Southern India. Publ. PhD, Available at http://library.wur.nl/wda/dissertations/dis3936.pdf. Wageningen Univ., The Netherlands.

Mander, M. 1998. Marketing of indigenous medicinal plants in South Africa: A case study in KwaZulu-Natal. FAO, Rome.

Mander, M., J. Mander, and C. Breen. 1996. Promoting the cultivation of indigenous plants for markets: Experiences from KwaZulu Natal, South Africa. p.104-109. In R.R.B. Leakey et al. (ed) Domestication and commercialisation of non-timber forest products. Non-Wood Forest Products No. 9. FAO, Rome.

Mansour, S.A. 2004. Pesticide exposure — Egyptian scene. Toxicology 198:91-115.

Mansoor, S.R., W. Briddon, Y. Zafar, and J. Stanley. 2003. Gemnivirus disease complexes: an emerging threat. Trends Plant Sci. 8(3):128-134.

Maredia, M., and P. Pingali. 2001. Environmental impacts of productivity-enhancing crop research: A critical review. CGIAR TAC Secretariat, FAO, Rome.

Marris, E. 2007. Wheat fungus spreads out of Africa. Available at http://www.nature.com/news/2007/070122/full/070122-3.html

Marshall, E., K. Schreckenberg, and A.C. Newton (ed) 2006. Commercialization of non-timber forest products: Factors influencing success. Lessons learned from Mexico and Bolivia and policy implications for decision makers. UNEP World Conservation Monitoring Centre, Cambridge, UK.

Marshall, T.A., P.J. Stumbo, J.J. Warren, and X.J. Xie. 2001. Inadequate nutrient intakes are common and are associated with low diet variety in rural, community-dwelling elderly. J. Nutr. 131:2192-2196.

Martinot, E., A. Chaurey, D. Lew, J.R. Moreira, and M. Wamukonya. 2002. Renewable energy markets in developing countries. Ann. Rev. Energy Env. 27:309-3048.

Marvier, M., C. McCreedy, J. Regetz, and P. Kareiva. 2007. A meta-analysis of effects of Bt cotton and maize on nontarget invertebrates. Science 316:1475-1477.

Masle, J., S.R. Gilmore and G. Farquhar. 2005. The ERECTA gene regulates plant transpiration efficiency in Arabidopsis. Nature 436:866-870.

Mason, P.A., and J. Wilson. 1994. Harnessing symbiotic associations: Vesicular-arbuscular mycorrhizas. p. 165-175. In R.R.B. Leakey and A.C. Newton (ed) Tropical trees: The potential for domestication and the rebuilding of forest resources. HMSO, London.

Mathieu, P., P. Lavigne Delville, H. Ouedrago, L. Pare and M. Zongo. 2000. Sécuriser les transactions foncières au Burkina Faso, Rapport de synthèse sur l'évolution des transactions foncières, GRET/Ministère de l'Agriculture/Ambassade de France au Burkina Faso.

Mathur, S., D. Pachico, and A.L. Jones. 2003. Agricultural research and poverty reduction, some issues and evidence. Pub. No. 335, Econ. Impact Ser. 2. CIAT, Cali.

Maxted, N., B. Ford-Lloyd, J.G. Hawkes. 1997. Plant genetic conservation: The in situ approach. Chapman and Hall, London.

Maxwell, D., and K. Wiebe. 1998. Land tenure and food security: A review of concepts, evidence and methods. Res. Pap. No. 129. Land Tenure Center, Univ. Wisconsin, Madison.

Mayers, J., and S. Vermeulen. 2002. Company-community forest partnerships: From raw deals to mutual gains? IIED, London.

Mazid, M.A., M.A. Jabber, C.R. Riches, E.J.Z. Robinson, M. Mortimer, and L.J. Wade. 2001. Weed management implications of introducing dry-seeded rice in the Barind Tract of Bangladesh. p. 211-216. In Proc. BCPC Conf. — Weeds. Brit. Crop Prot. Council, Farnham, UK

Mazoyer, M., and L. Roudart. 1997. Histoire des agricultures du monde. Du néolithique à la crise contemporaine. Seuil, Paris.

Mazoyer, M. and L. Roudart. 2002. Histoire des agricultures du monde. Du néolithique à la crise contemporaine. Points Histoire. Seuil, Paris (F).

Mazur, R., H. Sseguya, D. Masinde, J. Bbemba, and G. Babirye. 2006. Facilitating farmer-to-farmer learning and innovation for enhanced food, nutrition and income security in Kamuli District, Uganda. Innovation Africa Symposium Proceedings. Kampala. 21-23 Nov. 2006. Center for Sustain. Rural Livelihoods, Iowa State Univ..

McCauley, D.E., R.A. Smith, J.D. Linsenby, and C. Hsieh. 2003. The hierarchical spatial distribution of chloroplast DNA polymorphism across the introduced range of Silene vulgaris. Mol. Ecol. 12:3227-3235.

McGarry, D. 2005. Mitigating land degradation and improving land and environmental condition via field-practical methods of conservation agriculture. Beyond the 3th Congress on Conservation Agriculture. Nairobi. Available at http://www.act.org.zw/postcongress/theme_07_02.asp

McIntire, J., D. Bourzat, and P. Pingali. 1992. Crop-livestock interaction in sub-Saharan Africa. World Bank, Washington, D.C.

McKay, J.C., N.F. Barton, A.N.M. Koerhuis, and J. McAdam. 2000. The challenge of genetic change in the broiler chicken. In The challenge of genetic change in animal production. Brit. Soc. Animal Sci, Occas. Publ. No. 27. W.G. Hill et al (ed) Brit. Soc. Anim. Sci. Available at http://www.bsas.org.uk/publs/genchng/contents.pdf.

McKinney, D.C., and X. Cai. 1997. Multiobjective optimization model for water allocation in the Aral Sea basin. p. 44-48. In Proc. 3rd Joint USA/CIS Joint Conf. Environ. Hydrol. Hydrogeol. Am. Inst. Hydrol., St. Paul MN.

McLaughlin, J.F., J.J. Hellmann, C.L. Boggs, and P.R. Ehrlich. 2002. Climate change hastens population extinctions. PNAS 99:6070-6074.

McLeod-Kilmurray, H. 2007. Hoffmann v. Monsanto: Courts, class actions, and perceptions of the problem of GM drift. Bull. Sci. Tech. Society 27:188-201.

McMichael, A.J., J.W. Powles, C.D. Butler, and R. Uauy. 2007. Food, agriculture, energy, climate change and health. The Lancet 370:1253-1263.

McNeely, J.A., and L.D. Guruswamy. 1998. Conclusions: How to save the biodiversity of planet earth. p. 376-391. In D.G. Lakshman and J.A. McNeely (ed) Protection of global biodiversity: Converging strategies. Duke Univ. Press, Durham.

McNeely, J.A., and S.J. Scherr. 2003. Ecoagriculture: Strategies to feed the world and save wild biodiversity. Island Press, Washington DC.

Meher-Homji, M. 1988. Effects of forests on precipitation in India. In E.R.C. Reynolds, F.B. Thompson (ed) Forests, climate, and hydrology: Regional impacts. Available at http://www.unu.edu/unupress/unupbooks/80635e/80635E00.htm#Contents. The UN Univ., Tokyo.

Meinzen-Dick, R., M. Adato, L. Haddad, and P. Hazell. 2004. Science and poverty: An interdisciplinary assessment of the impact of agricultural research. Food Policy Rep. No. 16. IFPRI, Washington DC.

Meinzen-Dick, R.S., L.R. Brown, H.S. Feldstein, A.R. Quisumbing. 1997. Gender, property rights, and natural resources. FCND Disc. Pap. 29. IFPRI, Washington DC.

Meinzen-Dick, R., and R. Pradhan. 2002. Legal pluralism and dynamic property rights. CAPRi Working Pap. 22. CGIAR, Washington DC.

Menendez, R., A.G. Megias, J.K. Hill, B. Braschler, S.G Willis, Y. Collingham et al. 2006. Species richness changes lag behind climate change. Proc. R. Soc London B 273:1465-1470.

Mercoiret, M-R., J. Berthomé, P-M. Bosc, and J. Guillaume. 1997a. Les relations organisations paysannes et recherche agricole. Rapport non spécifique. CIRAD-SAR, Montpellier.

Mercoiret, M-R., J. Berthomé, D. Gentil, and P-M. Bosc. 1997b. Etats désengagés, paysans engagés; perspectives et nouveaux rôles des organisations paysannes en Afrique et en Amérique latine. Fondation Charles Léopold Meyer, Paris.

Merks, J.W.M. 2000. One century of genetic change in pigs and the future needs. In W.G. Hill (ed) The challenge of genetic change in animal production. Brit.Soc. Anim. Sci. Occas. Pub. No. 27. Brit. Soc. Animal Science.

Merrey, D.J., R. Meinzen-Dick, P.P. Mollinger, E. Karar, W. Huppert, J. Rees et al. 2007. Policy and institutional reform: The art of the possible. p. 193-231. In Water for food, water for life: A comprehensive assessment of water management in agriculture. Earthscan, London and IWMI, Colombo.

Metz, M., and J. Fütterer. 2002a. Biodiversity (Communications arising): Suspect evidence of transgenic contamination. Nature 416:600-601.

Metz, M., and J. Fütterer 2002b. Biodiversity (Communications arising): Conflicts around a study of Mexican crops. Nature 417:897.

Michon, G., and H. de Foresta. 1996. Agroforests as an alternative to pure plantations for the domestication and commercialization of NTFPs. p. 160-175. In R.R.B. Leakey et al. (ed) Domestication and commercialisation of non-timber forest products. Non-Wood Forest Products No. 9. FAO, Rome.

Michon, G., and H. de Foresta. 1999. Agroforests: Incorporating a forest vision in agroforestry. L.E. Buck, J.P. Lassoie, and E.C.M. Fernandez (ed) Agroforestry in sustainable agricultural systems. CRC Press and Lewis Publ., NY.

Mickelson S. K., J. L. Baker, and S. I. Ahmed. 2003. Vegetative filter strips for reducing atrazine and sediment runoff transport. J. Soil Water Conserv. 58:359-367.

Miller, R. H., A. Kamel, S. Lhaloui, and M. El Bouhssini. 1989. Survey of Hessian fly in Northern Tunisia. Rachis 8:27-28.

Millington, A.C., S.J. Walsh and P.E. Osborne (ed) 2001. GIS and remote sensing. Applications in biogeography and ecology. Kluwer Acad. Publ., NY.

Minami, K., and H.U. Neue. 1994. Rice paddies as a methane source. Climatic Change 27(1):13-26. Available at http://www.springerlink.com/content/t74hvj70425426w4/.

Mitschein, T.A., and P.S. Miranda. 1998. POEMA: A proposal for sustainable development in Amazonia. p. 329-366. In D.E. Leihner and T.A. Mitschein (ed) A third millennium for humanity? The search for paths of sustainable development. Peter Lang, Frankfurt am Main, Germany.

Molden, D., K. Frenken, R. Barker, C. de Fraiture, B. Mati, M. Svendsen et al. 2007a. Trends in water and agricultural development. p. 57-89. In D. Molden (ed) Water for food, water for life: A comprehensive assessment of water management in agriculture. Chapter 2. Earthscan, London and IWMI, Colombo.

Molden, D., H. Murray-Rust, R. Sakthivadivel and I. Makin. 2003. A water-productivity framework for understanding and action. In J.W. Kijne, R. Barker, D. Molden (ed) 2003. Water productivity in agriculture: Limits and opportunities for improvement. CABI and IWMI, Colombo.

Molden, D., T.Y. Oweis, P. Steduto, J.W. Kijne, M.A. Hanjra, P.S. Indraban et al. 2007b. Pathways for increasing agricultural water productivity. p. 279-310. In D. Molden (ed) Water for food. Earthscan, London and IWMI, Colombo.

Molnar, A., S. Scherr, and A. Khare. 2005. Who conserves the world's forests? For. Trends Ecoagric. Partners, Washington DC.

Molle, F. 2003. Development trajectories of river basins: A conceptual framework. Res. Rep. 72. IWMI, Colombo.

Monteiro, C.A., W.L. Conde, and B.M. Popkin. 2002. Is obesity replacing or adding to undernutriton? Evidence from different social classes in Brazil. Public Health Nutr. 5:105-112.

Moore, G. 2004. The Fair Trade movement: Parameters, issues and future research. J. Business Ethics 53:73-86.

Morgan, P.B., G.A. Bollero, R.L. Nelson, F.G. Dohleman, and S.P. Long. 2005. Smaller than predicted increase in aboveground net primary production and yield of field-grown soybean under fully open-air [CO2] elevation. Glob. Change Biol. 11:1856-1865.

Moritz, C. 1994. Defining 'evolutionarily significant units' for conservation. Trends Ecol. Evol. 9(10):373-375.

Mornhinweg, D.W, M.J. Brewer, and D.R. Porter. 2006. Effect of Russian wheat aphid on yield and yield components of field grown susceptible and resistant spring barley. Crop Sci. 46:36-42.

Morris, M.L. 2002. Impacts of international maize breeding research in developing countries: 1966- 98. CIMMYT, Mexico.

Morse, S., R. Bennett, and Y. Ismael. 2004. Why Bt cotton pays for small-scale producers in South Africa. NatureBiotech. 22:370-780.

Morse, S., R. Bennett, and Y. Ismael. 2005. Comparing the performance of official and unofficial genetically modified cotton in

India. AgBioForum 8:1-6. Available at http://www.agbioforum.org.

Mortensen, D.A., L. Bastiaans, and M. Sattin. 2000. The role of ecology in the development of weed management systems: An outlook. Weed Res. 40:49-62.

Mortureux, V. 1999. Droits de propriété intellectuelle et connaissances, innovations et pratiques des communautés autochtones et locales. Les études du BRG. Bureau des Ressources Génétiques, Paris.

Moscardi, F. 1999. Assessment of the application of baculoviruses for control of Lepidoptera. Ann. Rev. Entomol. 44:257-289.

Moustier, P., M. Figuié, N.T.T. Loc, and H.T. Son. 2006. The role of coordination in the safe and organic vegetable chains supplying Hanoi. In P.J. Batt and N. Jayamangkala (ed) Proc. First Int. Symp. on Improving the Performance of Supply Chains in Transitional Economies. Chiang Mai, Thailand. 19-23 July 2005.ISHS, Louvian.

Mowo, J.G., B.H. Janssen, O. Oenema, L.A. German, J.P. Mrema, and R.S. Shemdoe. 2006. Soil fertility evaluation and management by smallholder farmer communities in northern Tanzania. Agric. Ecosyst. Environ. 116:47-59.

Mozaffarian, D., M.B. Katan, A. Ascherio, M.J. Stampfer and W.C. Willett. 2006 Trans fatty acids and cardiovascular disease. New England J. Med. 354:1601-13.

Mozaffarian, D., and E.B. Rimm. 2006. Fish intake, contaminants, and human health: Evaluating the risks and the benefits. JAMA 296:1885-99.

Muchow, R.C., G.L. Hammer, and R.L. Vanderlip. 1994. Assessing climate risk to sorghum production in water-limited subtropical environments ii. Effects of planting dates, soil water at planting, and cultivar phenology. Field Crops Res. 36:235-246.

Mudge, K.W., and E.B. Brennan. 1999. Clonal propagation of multipurpose and fruit trees used in agroforestry, p. 157-190. In L.E. Buck, J.P. Lassoie and E.C.M. Fernandes (ed) Agroforestry in sustainable ecosystems. CRC Press/Lewis Publishers, NY.

Mugungu, C, K. Atta-Krah, A. Niang, and D. Boland. 1996. Rebuilding the Rwandan Tree Seed Centre: Seeds of hope project. For. Genetic Resour. 24. FAO, Rome.

Mumby, P.J., A.J. Edwards, J.E. Arias-Gonzalez, K.C. Lindeman, P.G. Blackwell, A. Gall et al. 2004. Mangroves enhance the biomass of coral reef fish communities in the Caribbean. Nature 427: 533-536.

Muntemba, S. 1988. Women as food producers and suppliers in the 20th century: The case of Zambia. p. 407-427. In R.E. Pahl (ed) On work: Historical, comparative, and theoretical approaches. Oxford, Basil Blackwell, UK.

Mupangwa, W., D. Love, and S.J. Twomlow. 2006. Soil-water conservation and rainwater harvesting strategies in the semi-arid Mzingwane Catchment, Limpopo Basin, Zimbabwe. Physics Chem. Earth 31:893-900.

Murphy, C. 2000. Cultivating Havana: Urban agriculture and food security in the years of crisis. Inst. Food Dev. Policy, Oakland, CA.

Mushara, H., and C. Huggins 2004. Land reform, land scarcity and post conflict reconstruction. A case study of Rwanda. Eco-conflicts. Vol. 3 No. 3. African Centre Tech. Studies, Nairobi.

Mzumara, D. 2003 Land tenure systems and sustainable development in Southern Africa. Economic Commission for Africa, Lusaka.

Nagarajan, G., and R.L. Meyer. 2005, Rural finance: Recent advances and emerging lessons, debates, and opportunities. Rural Finance Program. Ohio State Univ..

Nair, P.K. 1989. Agroforestry defined. p. 13-18. In P.K. Nair (ed) Agroforestry systems in the tropics. Kluwer Academic Publ., The Netherlands.

Nair, P.K.R., R.J. Buresh, D.N. Mugendi, and C.R. Latt. 1999. Nutrient cycling in tropical agroforestry systems: Myths and science, p. 1-31. In L.E. Buck, et al. (ed) Agroforestry in sustainable agricultural systems. CRC Press, NY and Lewis Publishers, Boca Raton.

Nangju, D. 2001. Agricultural biotechnology, poverty reduction and food security: A working paper. Asian Dev. Bank, Manila.

NAS. 1989. Irrigation induced water quality problems - What can be learned from the San Joaquin Valley experience. Nat. Academy Press, Ithaca, NY.

Naylor, R.L. 1996. Energy and resource constraints on intensive agricultural production. Ann. Rev. Energy Environ. 21:99-123.

Naylor, R.L., R.J. Goldburg, J.H. Primavera, N. Kautsky, M.C.M. Beveridge, J. Clay et al. 2000. Effect of aquaculture on world fish supplies. Nature 405:1017-1024.

Nazarko, O.M., R.C. Van Acker, and M.H. Entz. 2005. Strategies and tactics for herbicide use reduction in field crops in Canada: a review. Can. J. Plant Sci. 85:457-479.

Ncube, B., J. Dimes, S. Twomlow, W. Mupangwa, and K. Giller. 2007. Raising the productivity of smallholder farms under semi-arid conditions by use of small doses of manure and nitrogen: A case of participatory research. Nutr. Cycl. Agroecosys. 77:53-67

Ndoye, O., M. Ruiz-Perez, and A. Ayebe. 1997. The markets of non-timber forest products in the humid forest zone of Cameroon. Rural Dev. For. Network, Network Pap. 22c, ODI, London.

Neely, C.L., and R. Hatfield. 2007. Livestock systems. p. 121-142. In S.J. Scherr and J.A. McNeely (ed) Farming with nature: The science and practice of ecoagriculture. Island Press, NY.

Nelson, M., and M. Maredia. 1999. Environmental impacts of the CGIAR - An initial assessment. Tech. Advisory Comm. Secretariat, FAO, Rome.

Nepolean, T., M. Blummel, A.G. Bhasker Raj, V. Rajaram, S. Senthilvel, and C.T. Hash. 2006. QTLs controlling yield and stover quality traits in pearl millet. J. SAT Agric. Res. 2(1). http://www.icrisat.org/Journal/bioinformatics/v2i1/v2i1qtls.pdf.

Ness, A.R., and J.W. Powles. 1997. Fruit and vegetables, and cardiovascular disease: A review. Int. J. Epidemiol. 26:1-13.

Nestle, M. 2003. Food politics. How the food industry influences nutrition and health. Univ. California Press, Berkeley.

Neunschwander, P. 2004. Harnessing nature in Africa: Biological pest control can benefit the pocket, health and the environment. Nature 432:801-802.

Newberry, D.M.G., and J.E. Stiglitz. 1979. The theory of commodity price stabilisation rules: Welfare impacts and supply responses. The Econ. J. 89(356):799-817.

Newton, A.C., T. Allnutt, A.C.M. Gillies, A.J. Lowe, and R.A. Ennos. 1999. Molecular phylogeography, intraspecific variation and the conservation of tree species. Trends Ecol. Evol. 14:140-145.

Nicholson, C.F., R.W. Blake, R.S. Reid, and J. Schelhas. 2001. Environmental impacts of livestock in the developing world. Environment 43:7-17.

Nielsen, L., and C. Heffernan. 2006. New tools to connect people and places: The impact of ICTs on learning among resource poor farmers in Bolivia. J. Int. Dev. 18 (6):889-900.

Norris, R.F. 1992. Have ecological and biological studies improved weed control strategies? p. 7-33. In Proc. First Int. Weed Control Congress. Melbourne, Australia.

Norton, G.W., K. Moore, D. Quishpe, V. Barrera, T. Debass, S. Moyo, and D.B. Taylor. 2005. Evaluating socio-economic impacts of IPM. p. 225- 244. In G.W. Norton et al. (ed) Globalizing integrated pest management: A participatory process. Blackwell Publ., Ames.

Norton-Griffiths, N., M.Y. Said, S. Serneels, D.S. Kaelo, M.B. Coughenour, R. Lamprey et al. 2007. Land use economics in the Mara area of the Serengeti ecosystem. In A.R.E. Sinclair (ed) Serengeti III: Impacts of people on the Serengeti ecosystem. Chicago Univ. Press, Chicago.

Nouni, M.R., S.C. Mullick, and T.C. Kandpal. 2007. Biomass gasifier projects for decentralized power supply in India: A financial evaluation. Energy Policy 35 (2):1373-85.

Nwachukwu, I., and A. Akinbode. 1989. The use of television in disseminating agricultural technology: A case study of selected communities in Oyo State, Nigeria. Rural Dev. Nigeria 3(2):111- 115.

O'Callaghan, M., T.R. Glare, E.P.J. Burgess, and L.A. Malone 2005. Effects of plants genetically modified for insect resistance on nontarget organisms. Ann. Rev. Entomol. 50:271-292.

O'Malley, M. 1997. Clinical evaluation of pesticide exposure and poisonings. Lancet 349:1161-1166.

Oba, G., N.C. Stenseth, and W.J. Lusigi. 2000. New perspectives on sustainable grazing management in arid zones of sub-Saharan Africa. BioScience 50:35-51.

Oberhauser, K., and A. T. Peterson. 2003. Modeling current and future potential wintering distributions of eastern North American monarch butterflies. PNAS 100:14063-14068.

Obrist, L.B., A. Dutton, R. Albajes, and F. Bigler 2006. Exposure of arthropod predators to Cry1Ab toxin in Bt maize fields. Ecol. Entomol. 31:143-154.

Oden, N.H, A. Maarouf, I.K. Barker, M. Bigras-Poulin, L.R. Lindsay, M.G. Morshed et al. 2006. Climatic change and the potential for range expansion of the Lyme disease vector Ixodes scapularis in Canada. Int. J. Parasitol. 36(1):63-70.

OECD. 1986. Recombinant DNA safety considerations for industrial, agricultural and environmental applications of organisms derived by recombinant DNA techniques ('The Blue Book'). OECD, Paris.

OECD. 1993. Safety considerations for biotechnology: Scale-up of crop plants. OCED, Paris.

OECD. 1997. Impacts of fruit and vegetable production on the environmental and policy responses: European Union Community instruments relating to the environmental aspects of the fresh fruit and vegetable sector. OECD, Paris.

OECD. 2005. Review of agricultural policies — Brazil. OECD, Paris.

OECD. 2006a. Agricultural market impacts of future growth in the production of biofuels. OECD, Paris. Available at http://www.oecd .org/dataoecd/58/62/36074135.pdf.

OECD. 2006b. Promoting pro-poor growth, harmonising ex ante poverty impact assessment. OECD, Paris.

Ogunwale, A.B., and E.A. Laogun. 1997. Extension agents' avenues for information dissemination and other perception of farmers sources of information in Oyo State Agric. Dev. Program. p. 141-118. In T.A. Olowu (ed) Towards the survival of agricultural extension system in Nigeria. Proc. Third Ann. Nat. Conf. Agric. Extension Soc. Nigeria (AESON). Ilorin.

Ohm, H.W., F.L. Patterson, S.E. Ratcliffe, S.E. Cambron, and C.E. Williams 2004. Registration of Hessian fly resitant wheat germplasm line P921696. Crop Sci 44:2272-2273.

Ojima, D.S., W.J. Parton, D.S. Schimel, J.M.O. Scurlock, and T.G.F. Kittel. 1993. Modeling the effects of climate and CO2 changes on grassland storage of soil C. Water, Air, Soil Pollution 70:643-657.

Okafor, J.C. 1980. Edible indigenous woody plants in the rural economy of the Nigerian forest zone. Forest Ecol. Manage. 3:45-55.

Okali, C., J. Sumberg, and J. Farrington. 1994. Farmer participatory research. ODI, London.

Okoth, J.R., G.S. Khisa, and T. Julianus, 2003. Towards self financed farmer field schools. LEISA Mag. 19(1):28-29.

Olivier de Sardan, J.P. 1995. Anthropologie et développement. Essai en socio-anthropologie du changement social, p. 221. Hommes et sociétés. Karthala, Paris.

Olowu, T.A., and C.O. Igodan. 1989. Farmers' media use pattern in six villages of Kwara State. Rural Dev. Nigeria 3(2):98-102.

Olsen, J., L. Kristensen, J. Weiner, and H.W. Griepentrog. 2005. Increased density and spatial uniformity increase weed suppression by spring wheat. Weed Res. 45:316-321.

Omernik, J.M. 1977. Nonpoint source-stream nutrient level relationships - A nationwide study. EPA Res. Series EPA-600/3-P77-105. US EPA, Corvallis Environmental Res. Laboratory.

Ong, C.K., and P. Huxley. 1996. Tree-crop interactions: A physiological approach. CABI Wallingford.

Ong, C.K., C.R. Black, F.M. Marshall, and J.E. Corlett. 1996. Principles of resource capture and utilization of light and water. p. 73-158. In C.K. Ong and P. Huxley (ed) Tree-crop interactions: A physiological approach. CABI, Wallingford.

Opara, L.U., and F. Mazaud. 2001. Food traceability from food to plate. Outlook Agric. 30:239-247.

Orr, A. 2003. IPM for resource-poor African farmers: Is the emperor naked? World Dev. 31 (5):831-846.

Ørskov, E.R., and E.F. Viglizzo. 1994. The role of animals in spreading farmer's risks: A new paradigm for animal science. Outlook Agric. 23:81-89.

Ostrom, E. 1992. The rudiments of a theory of the origins, survival, and performance of common-property institutions. In D.W. Bromley (ed) Making the commons work: Theory, practice and policy. ICS Press, San Francisco.

Ostrom, E. 1994. Institutional analysis, design principles and threats to sustainable community governance and management of commons. p. 34-50. In R.S. Pomeroy (ed) Community management and common property of coastal fisheries in Asia and the Pacific: Concepts, methods and experiences. Workshop Reprint Ser., R94-11. ICLARM, Manila.

Ostrom, E. 2005. Understanding institutional diversity. Princeton Univ. Press, Princeton NJ.

Otsuka, K., and T. Yamano. 2005. Green revolution and regional inequality: Implications of Asian experience for Africa. p. 239-252. In The African food crisis: Lessons from the Asian green revolution. CABI, Wallingford.

Otsuki, T., J.S. Wilson, and M. Sewadeh. 2001. Saving two in a billion: Quantifying the trade effect of European food safety standards on African exports. Food Policy 29:131-146.

Ottman, M.J., B.A. Kimball, P.J. Pinter Jr, G.W. Wall, R.L. Vanderlip, S.W. Leavitt et al. 2001. Elevated CO2 on sorghum growth and yield at high and low soil water content. New Phytologist 150: 261-273.

Oweis, T., and A. Hachum. 2004. Water harvesting and supplemental irrigation for improved water productivity of dry farming systems in West Asia and North Africa. New directions for a diverse planet. Proc. 4th Int. Crop Sci. Congress. Brisbane, Australia. 26 Sept -1 Oct 2004.

Owen, M.D.K., and I.A. Zelaya. 2005. Herbicide-resistant crops and weed resistance to herbicides. Pest Manage. Sci. 61:301-311.

Padilla, M., and L. Malassis. 1992. Politiques agricoles et politiques alimentaire: Efficacite et equite. Econ. et Societes 21(6):175-192.

Pagiola, S., J. Bishop, and N. Landell-Mills. 2002. Selling forest environmental services: Market-based mechanisms for conservation and development. Earthscan, London.

Palm, C.A., M. van Noordwijk, P.L. Woomer, J.C. Alegre, L. Arévalo, C.E. Castilla et al. 2005a. Carbon losses and sequestration after land use change in the humid tropics, p. 41-63. In C.A. Palm, S.A. Vosti, P.A. Sanchez and P.J. Ericksen (ed) Columbia Univ. Press, NY.

Palm, C.A., S.A. Vosti, P.A.. Sanchez, and P. Ericksen. 2005b. Slash-and-burn agriculture: The search for alternatives. Columbia Univ. Press, NY.

Pandey, S., and S. Rajatasereekul. 1999. Economics of plant breeding: The value of shorter breeding cycles for rice in Northeast Thailand. Field Crops Res. 64:187-197.

Panigrahy, B.K. 2003. Watershed evaluation: Problems and perspectives. p. 736-739. In B. Venkateswara Rao (ed) Proc. Int. Conf. Hydrology and Watershed Management Vol. 1. Hyderabad, India.18-20 Dec 2002. BS Publ., Hyderabad, India.

Panik, F. 1998. The use of biodiversity and implications for industrial production. p. 59-73. In D.E. Leihner and T.A. Mitschein (ed) A third millennium for humanity? The search for paths of sustainable development. Peter Lang, Frankfurt am Main, Germany.

Parella, M.P., L. Stengarf Hansen, and J. van Lenteren. 1999. Glasshouse environments. p. 819-839. In T.S. Bellows and T.W. Fisher (ed) Handbook of biological control. Principles and applications of biological control. Academic Press, San Diego, London.

Parker, C., and J. Fryer. 1975. Weed control problems causing major reduction in world supplies. FAO Plant Prot. Bull. 23:83-95.

Passioura, J.B. 1986. Resistance to drought and salinity: Avenues for improvement. Aust. J. Plant Physiol. 13:191-201.

Patterson R.E., P.S., Haines, and B.M. Popkin. 1994. Diet Quality Index: capturing a multidimensional behavior. J. Am. Dietary Assoc. 94:57-64.

Paustian, K., C.V. Cole, D. Sauerbeck, and N. Sampson. 1998. CO2 mitigation by agriculture: An overview. Clim. Change 40:135-162.

Pearce, D., and S. Mourato. 2004. The economic valuation of agroforestry's environmental services. p. 67-86. *In* G. Schroth et al. (ed) Agroforestry and biodiversity conservation in tropical landscapes. Island Press, Washington.

Pellegrineschi, A., M. Reynolds, M. Pacheco, R.M. Brito, R. Almeraya, K. Yamagughi-Shinozaki, and D. Hosington. 2004. Stress-induced expression in wheat of Arabidopsis thaliana DREB1A gene delays water stress symptoms under ghreenhouse conditions. Genome 47:493-500.

Pemsl, D., H. Waibel, and A.P. Gutierrez 2005. Why do some Bt-cotton farmers in China continue to use high levels of pesticides? Int. J. Agric. Sustain. 3:44-56.

Pemsl, D., and H. Waibel, and J. Orphal. 2004. A methodology to assess the profitability of Bt-cotton: Case study results from the state of Karnataka, India. Crop Prot. 23:1249-1257.

Peng, S., R.J. Buresh, J. Huang, J. Yang, Y. Zou, X. Zhong, and G. Wang. 2006. Strategies for overcoming low agronomic nitrogen use efficiency in irrigated rice systems in China. Field Crops Res. 96:37-47.

Peng, S., K.G. Cassman, S.S. Virmani, J. Sheehy, and G.S. Khush. 1999. Yield potential trends of tropical rice since the release of IR8 and the challenge of increasing rice yield potential. Crop Sci. 39:1552-1559.

Peng, S., F.V. Garcia, R.C. Laza., A.L. Sanico, R.M. Visperas, and K.G. Cassman. 1996. Increased N-use efficiency using a chlorophyll meter on high yielding irrigated rice. Field Crops Res. 47:243- 252.

Pengue, W.A. 2005. Transgenic crops in Argentina: The ecological and social debt. Bull. Sci. Tech. Society 25:314-322.

Perfecto, I., and I. Armbrecht. 2003. The coffee agroecosystem in the neotropics: Combining ecological and economic goals. p. 157-192. *In* J. Vandermeer (ed)Tropical agroecosystems. CRC Press, Boca Raton, USA.

Perret, S. 2002. Water policies and smallholding irrigation schemes in South Africa: A history and new institutional challenges. Water Policy 4:283-300.

Perret, S., M.-R. Mercoiret. 2003. Supporting small-scale farmers and rural organisations: Learning from experiences in West Afria. A handbook for development operators and local managers. Protea Book House, Pretoria.

Pesche, D., and K.K. Nubukpo. 2004. L'afrique du coton à Cancun: Les acteurs d'une négociation. Politique Africaine 95:158-168.

Phillips, O.L., and B. Meilleur. 1998. Usefulness and economic potential of the rare plants of the United States: A status survey. Econ. Bot. 52(1):57-67.

Pihl, L., A. Cattrijsse, I. Codling, S. Mathieson, D.S. McLusky, and C. Roberts. 2002. Habitat use by fishes in estuaries and other brackish areas, p. 10-53. *In* M. Elliott and K. L. Hemingway (ed) Fishes in estuaries. Blackwell Science, UK.

Pingali, P.L., and P.W. Heisey. 1999. Cereal productivity in developing countries: past trends and future prospects. CIMMYT Economics Program Pap. 99- 03. CIMMYT, Mexico.

Pingali, P.L., and M.W. Rosegrant. 1994. Confronting the environmental consequences of the green revolution in Asia. EPTD Disc. Pap. No. 2. Environ. Prod. Tech. Div. IFPRI, Washington DC.

Pinstrup-Andersen, P., and E. Schioler, 2001. Seeds of contention: World hunger and the global controversy over GM crops. John Hopkins Univ. Press, Baltimore.

Place, F. 1995. The role of land and tree tenure on the adoption of agroforestry: Technologies in Zambia, Burundi, Uganda, and Malawi: A summary and synthesis. Land Tenure Center, Univ. Wisconsin, Madison.

Place, F. and P. Hazell. 1993. Productivity effects of indigenous land tenure systems in Sub-Saharan Africa. Am. J. Agric. Econ. 75(1):10-19.

Place, F., and K. Otsuka. 2000. Population pressure, land tenure, and tree resource management in Uganda. Land Econ. 76:233-251.

Platteau, J.P., 1996. The evolutionary theory of land rights as applied to Sub-Saharan Africa: a critical assessment. Dev. Change 27(1):29-86.

Polaski, S. 2006. Winners and losers. Impact of the Doha Round on developing countries. Carnegie Endowment Int. Peace, NY.

Polgreen, L. 2007. In Niger, trees and crops turn back the desert. p. 1. The New York Times, 11 Feb 2007.

Ponce-Hernandez, R. 2004. Assessing carbon stock and modelling win-win scenarios of carbon sequestration through land-use change. FAO, Rome.

Ponte, S. 2006. Ecolabels and fish trade: Marine stewardship council certification and South African hake industry. Tralac Working Pap. No. 9/2006. Available at http://www.tralac.org/scripts/content.php?id=5212.

Popkin, B.M. 1998. The nutrition transition and its health implications in low income countries. Public Health Nutr. 1:5-21.

Popkin, B.M., and S. Du. 2003. Dynamics of the nutrition transition toward the animal foods sector in China and its implications: A worried perspective. J. Nutr. 133(11Suppl 2):3898S-3906S

Posey, D. (ed) 1999. Cultural and spiritual values of biodiversity. Intermediate Tech. Publ., London.

Post, W.M., and K.C. Kwon 2000. Soil carbon sequestration and land-use change: Processes and Potential. Glob. Change Biol. 6:317-328.

Prasad, P.V.V., K.J. Boote, L.H. Allen Jr, J.E. Sheehy, and J.M.G. Thomas. 2006. Species, ecotype and cultivar differences in spikelet fertility and harvest index of rice in response to high temperature stress. Field Crops Res. 95:398-411.

Pray, C.E., J. Huang, R. Hu, and S. Rozelle. 2002. Five years of Bt cotton in China — the benefits continue. The Plant J. 31(4):423-430.

Pretorius, Z.A., W.H.P. Boshoff, B.D. Van Niekerk, and J.S. Komen 2002. First report of virulence in Puccinia graminis f.sp. tritici to wheat stem rust resistance genes Sr8b and Sr38 in South Africa. Plant Dis. 86(8):922.

Pretty, J. 2002. Agri-Culture: Reconnecting people, land and nature. Earthscan, London.

Pretty, J. 2003. Social capital and the collective management of resources. Science 302:1912-1914.

Pretty, J., and H. Ward. 2001. Social capital and the environment. World Dev. 29:209-227.

Pretty, J.N., A.D. Noble, D. Bossio, J. Dixon, R.E. Hine, F.W.T. Penning de Vries et al. 2006. Resource-conserving agriculture increases yields in developing countries. Environ. Sci. Tech. 40:1114-1119.

Prinz, D. 1996. Water harvesting - History, techniques, trends. Z. f. Bewaesserungswirtschaft 31:64-105.

Prior, C., and A. Halstead. 2006. Gardening in the global greenhouse. Pest and disease threats. Roy. Hort. Soc., UK.

Qaim, M., and G. Traxler 2004. Roundup ready soybeans in Argentina: Farm level, environmental and welfare effects. Agric. Econ. 32:73-86.

Qaim, M., and D. Zilberman. 2003. Yield effects of genetically modified crops in developing countries. Science 299:900-902.

Qayum, A., and K. Sakkhari 2005. Bt cotton in Andhra Pradesh: A three-year assessment. Deccan Dev. Society. Available at www.grain.org/research_files/BT_Cotton_-_A_three_year_report.pdf.

Quaak, P., H. Knoef, and H. Stassen. 1999. Energy from biomass. A review of combustion and gasification technologies. Tech. Pap. Energy Ser. No. 422. World Bank, Washington DC.

Quan, J. 2000. Land tenure, economic growth and poverty in sub-Saharan Africa. p. 31-49. *In* C. Toulmin and J. Quan (ed) Evolving land rights, policy and tenure in Africa. DFID/IIED/Univ. Greenwich Nat. Resourc. Inst., London.

Quan, J., S. Tan, and C. Toulmin (ed) 2005. Land in Africa: Market asset or secure livelihood. Proc. Land in Africa Conf. London. 8-9 Nov. 2004. IIED, NRI, and the Roy. Afric. Soc., London.

Quist, D. and I. Chapela. 2001. Transgenic DNA introgressed into traditional maize landraces in Oaxaca, Mexico. Nature 414:541-543.

Quist, D. and I. Chapela. 2002. Biodiversity (Communications arising, reply): Suspect evidence of transgenic contamination/Maize transgene results in Mexico are artefacts. Nature 416:602.

Quisumbing, A.R., L.R. Brown, H.S. Feldstein, L. Haddad, and C. Peña. 1995. Women: The key to food security. Food Policy Statement 21. IFPRI, Washington, DC.

Quisumbing, A.R., and A. Maluccio. 1999. Intrahousehold allocation and gender relations: New empirical evidence. Res. report on Gender and Development. Working Pap. Ser. No. 2. World Bank, Washington DC.

Quizon, J., G. Feder, and R. Murgain. 2001. Fiscal sustainability of agricultural extension: the case of the Farmer Field school approach. J. Int. Agric. Extension Educ. 8:13-24.

Rabalais, N.N., R.E. Turner, D. Justic, Q. Dortch, W.J. Wiseman, and B.K. Sen Gupta. 1996. Nutrient changes in the Mississippi River and system responses to the adjacent continental shelf. Estuaries 19:386-407.

Raij, B. 1981. Soil fertility evaluation. (In Portuguese). Instituto da Potassa y Fosfato, Piracicaba, SP, Brazil.

Raitzer, D.A. 2003. Benefit-cost meta-analysis of the investment in the international agricultural research centres of the CGIAR. Sci. Council Secretariat, FAO, Rome.

Raitzer, E.D.A., and R. Lindner. 2005. Review of the returns to ACIAR's bilateral R&D investments, Impact Assessment Ser. Rep. No. 35, Aug 2005.

Randall, E., M.Z. Nichaman, and C.F. Contant, Jr. 1985. Diet diversity and nutrient intake. J. Am. Diet. Assoc. 85:830-836.

Randerson, J.T., H. Liu, M.G. Flanner, S.D. Chambers, Y. Jin, P.G. Hess et al. 2006. The impact of boreal forest fire on clmate warming. Science 314:1130-1132.

Raney, T. 2006. Economic impact of transgenic crops in developing countries. Curr. Opinion Biotech. 17:174-178.

Rao, S., C.S. Yajnik, A. Kanade, C.H.D. Fall, B.M. Margetts, A.A. Jackson et al. 2001. Intake of micronutrient-rich food in rural indian mothers is associated with the size of their babies at birth. Pune maternal nutrition Study. J. Nut. 131:1217-1224.

Rasmussen, K. 2003. Effects of climate change on agriculture and environment in the semi-arid tropics, with Senegal as an example. Univ.Copenhagen/ The North/ South Priority Res. Area, Centre of African Studies, Impacts of future climate change. Available at http://www.eldis.org/static/ DOC15855.htm and http://glwww.dmi.dk/ f+u/publikation/dkc-publ/klimabog/ CCR-chap-21.pdf.

Ratcliffe, R.H., and J.H. Hatchett. 1997. Biology and genetics of the Hessian fly and resistance in wheat. p. 47-56. In K. Bondari (ed) New developments in entomology. Sci. Inform. Guild, Triv anddram, India.

Rathgeber, E.M. 2002. Female and male CGIAR scientists in comparative perspective. CGIAR. Gender and Diversity Working Pap. 37. Available at http://www.genderdiversity.cgiar. org/publications/genderdiversity_WP37.pdf. CGIAR, Nairobi.

Raun, W.R., and G.V. Johnson. 1999. Improving nitrogen use efficiency for cereal production. Agron. J. 91:357-363.

Reardon, T. 1997. Using evidence of household income diversification to inform study of the rural non-farm labour market in Africa. World Dev. 25(5):735-747.

Reca, J., J. Roldan, M. Alcaide, and E. Camacho. 2001. Optimization model for water allocation in deficit irrigation system:

I. Description of the model. Agric. Water Manage. 48(2):103-116.

Regmi, A.P., and J.K. Ladha. 2005. Enhancing productivity of rice-wheat system through integrated crop management in the Eastern-Gangetic plains of South Asia. J. Crop Improve. 15:147-170.

Reicosky, D.C. 1997. Tillage-induced CO2 emission from soil. Nutr. Cycl. Agroecosys. 49:273-285.

Reid, R.S., P.K. Thornton, G.J. McCrabb, R.L. Kruska, F. Atieno, and P.G. Jones. 2004. Is it possible to mitigate greenhouse gas emissions in pastoral ecosystems of the tropics? Environ. Dev. Sustain. 6:91-109.

Reif, J.C., P. Zhang, S. Dreisigacker, M.L. Warburton, M. van Ginkel, D. Hosington et al. 2005. Wheat genetic diversity trends during domestication and breeding. Theor. Appl. Genet. 110:859-864.

Reij, C., P. Mulder, and L. Begemann. 1988. Water harvesting for agriculture. Tech. Pap. No. 91. World Bank, Washington DC.

REN 21. 2005. Renewables 2005 global status report. Renewable Energy Policy Network, Paris and Washington DC.

REN 21. 2006. Renewables global status report 2006 update. Renewable Energy Policy Network, Paris and Washington DC.

Renewable Fuels Association. 2005. Homegrown for the homeland. Industry Outlook 2005. Washington DC.

Repetto, R., and S. Baliga. 1996. Pesticides and the immune system: The public health risks. World Resources Inst., Washington DC.

Revenga, C., J. Brunner, N. Henniger, K. Kassem, and R. Payner. 2000. Pilot analysis of global ecosystems, freshwater systems. World Resources Inst., Washington DC.

Reynolds, M.P., and N. Borlaug. 2006. Impacts of breeding on international collaborative wheat improvement. J. Agri. Sci. Camb. 144:3-17.

Ribier, V., and L. Tubiana. 1996. Globalisation, compétitivité et accords du GATT. Conséquences pour l'agriculture des pays en développement. Notes et documents CIRAD. CIRAD, Paris.

Richard-Ferroudji A., P. Caron, J-Y. Jamin, and T. Ruf. 2006. Coordinations hydrauliques et justices sociales. Programme commun systèmes irrigués. 4ème séminaire international et interdisciplinaire, Montpellier. 26 Nov 2004.

Richards, P. 1985. Indigenous agricultural revolution. Westview Press, Boulder.

Richards, P. 1996. Fighting for the rain forest: War, youth and resources in Sierra Leone. James Currey, Oxford.

Richards, R.A. 2006. Physiological traits used in the breeding of new cultivars for water-scarce environments. Agric. Water Manage. 80:197-211.

Richardson, D. 2006. ICTs — Transforming agricultural extension? CTA Working Doc. No. 8034, The ACP-EU Technical Centre for

Agricultural and Rural Cooperation (CTA), Wageningen.

Riches, C.R., A.M. Mbwaga, J. Mbapila, and G.J.U. Ahmed. 2005. Improved weed management delivers increased productivity and farm incomes for rice in Bangladesh and Tanzania. Aspects Appl. Biol. 75:127-138.

Richter, B.D., D.P. Braun, M.A. Mendelson, and L.L. Master. 1997. Threats to imperiled freshwater fauna. Conserv. Biol. 11(5):1081-1093.

Rivera, J.A., S. Barquera, F. Campirano, I. Campos, M. Safdie, and V. Tovar. 2002. Epidemiological and nutritional transition in Mexico: Rapid increase of non-communicable chronic diseases and obesity. Public Health Nutr. 5:113-122.

Rivera, W.M. and G. Alex. 2004. The continuing role of government in pluralistic extension Systems. Int. J. Agric. Educ. Extension. 11(2):41-52.

Rivera, W.M., and J.W. Cary. 1997. Privatising agricultural extension. In B.E. Swanson et al. (ed) Improving agricultural extension: A reference manual. FAO, Rome.

Rivera, W.M., W. Zijp, and G. Alex. 2000. Contracting for extension: Review of emerging practice. AKIS Good Practice Note. World Bank, Washington DC.

Rodrik, D. 2007. The cheerleaders threat to global trade. Financial Times, 27 Mar 2007.

Röling, N. 1988. Extension Science. Infor. Systems in Agric. Development. CUP, Cambridge.

Röling, N. 1992. The emergence of knowledge system thinking: Changing perception of relationships among innovation, knowledge processes and configuration. Knowledge and Policy 5(2).

Röling, N., and M.A.E. Wagermakers. 1998. Facilitating sustainable agriculture. Cambridge Univ. Press, NY.

Romeis, J., M. Meissle, and F. Bigler. 2006. Transgenic crops expressing Bacillus thuringiensis toxins and biological control. Nature Biotech. 24:63-71.

Rose, R.I. 2006. Tier-based testing for effects of proteinaceous insecticidal plant-incorporated protectants on non-target arthropods in the context of regulatory risk assessment. IOBC/ wprs Bull. 29(5):145-152.

Rosegrant, M.W., and X. Cai. 2001. Water and food production. 2020 Vision for food, agriculture, and the environment. Focus No. 9. Overcoming water scarcity and quality constraints. Brief No.2/14. Available at http://www.ifpri.org/2020/focus/focus09/ focus09_02.asp. IFPRI, Washington DC.

Rothschild, M., and J. Stiglitz. 1976. Equilibrium in competitive insurance markets: An essay on the economics of imperfect information. Q. J. Econ. 90(4):629-649.

Rowntree, P.R. 1988. Review of general circulation models as a basis for predicting the effects of vegetation change on climate. In E.R.C. Reynolds, F.B. Thompson (ed) Forests, climate, and hydrology.

Available at http://www.unu.edu/unupress/unupbooks/80635e/80635E00.htm#Contents. The UN Univ., Tokyo.

Rubiales, D., A. Perez de Luque, M. Fernandez-Aparica, J.C. Sillero, B. Roman, M. Kharrat et al. 2006. Screening techniques and sources of resistance to parasitic weeds in grain legumes. Euphytica 147:187-199.

Ruel, M.T. 2002. Is dietary diversity an indicator of food security or dietary quality? A review of measurement issues and research needs. FCND Disc. Pap. No. 140. IFPRI, Washington DC.

Ruel, M.T., and C.E. Levin. 2000. Assessing the potential for food based strategies to reduce vitamin A and iron deficiencies: A review of recent evidence. FCND Disc. Pap. No. 92. IFPRI, Washington DC.

Ruf, T. 2001. The institutional cycles of peasants' irrigation: Historical debates on ownership and control of water in south of France and in the Ecuadorian Andes. The role of water in history and development. IWHA 2nd Conf., Bergen, Norway.

Rusike, J., and J.P. Dimes. 2004. Effecting change through private sector client services for smallholder farmers in Africa. In Proc. 4th Int. Crop Sci. Congr. Brisbane. Available at http://www.cropscience.org.au/icsc2004/symposia/4/6/997_rusikej.htm .

Rusike, J., S.J. Twomlow, H.A. Freeman, and G.M. Heinrich, 2006. Does farmer participatory research matter for improved soil fertility technology development and dissemination in Southern Africa? Int. J. Agric. Sustain. 4(2):1- 17.

Rusike, J., S.J. Twomlow, and A. Varadachary. 2007. Impact of Farmer Field Schools on adoption of soil water and nutrient management technologies in dry areas of Zimbabwe. In Mapiki A and C. Nhira (ed) Land and water management for sustainable agriculture. Proc. EU/SADC Land and Water Management Applied Research and Training Programme's Inaugural Scientific Symposium, Malawi Inst. Manage., Lilongwe, Malawi. 14-16 Feb 2006.

Russell, C.A., B.W. Dunn, G.D. Batten, R.L. Williams, and J.F. Angus. 2006. Soil tests to predict optimum fertilizer nitrogen rate for rice. Field Crops Res. 97:286-301.

Russelle, M.P., M.H. Entz, and A.J. Franzluebbers. 2007. Reconsidering integrated crop-livestock systems in North America. Agron. J. 99:325-334.

Sa, J.C.M., C.C. Cerri, W.A. Dick, R. Lal, S.P. Venzke Filho, M.C. Piccolo, and B. Feigl. 2001a. Carbon sequestration in a plowed and no-tillage chronosequence in a Brazilian oxisol. p. 466-471. In D.E. Stott et al. (ed) The global farm. Selected papers from the 10th Int. Soil Conservation Org. Meeting. USDA-ARS National Soil Erosion Res. Lab. Purdue Univ., West Lafayette, IN.

Sa, J.C.M., C.C. Cerri, R. Lal, W.A. Dick, S.P. Venzke Filho, M.C. Piccolo, and B. Feigl. 2001b. Organic matter dynamics

and sequestration rates for a tillage chronosequence in a Brazilian Oxisol. Soil Sci. Soc. Am. J. 64:1486-1499.

Sachs, J. 2005. The end of poverty: How we can make it happen in our lifetime. Penguin Books, London.

Sadoulet, E., A. de Janvry, and B. Davis. 2001. Cash transfer programs with income multipliers: PROCAMPO in Mexico. World Dev. 29(6):1043-1056.

Saito, K., B. Linquist, G.N. Atlin, K. Phanthaboon, T. Shiraiwa, and T. Horie. 2006. Response of traditional and improved upland rice cultivars to N and P fertilizer in northern Laos. Field Crops Res. 96:216-223.

Salati, E., and P.B. Vose. 1984. Amazon basin: A system in equilibrium. Science 225:129-138.

Salman, A.Z., E.K. Al-Karablich, and F.M. Fisher. 2001. An inter-seasonal agricultural water allocation system (SAWAS). Agric. Syst. 68(3):233-252.

Salokhe, G., A. Pastore, B. Richards, S. Weatherley, A. Aubert, J. Keizer et al. 2002. FAO's role in information management and dissemination — challenges, innovation, success, lessons learned. Available at ftp://ftp.fao.org/docrep/fao/008/af238e/af238e00.pdf. FAO, Rome.

Sanchez, P.A. 1995. Science in agroforestry. Agrofor. Syst. 30:5-55.

Sanchez, P.A. 2002. Soil fertility and hunger in Africa. Science 295:2019-2020.

Sanchez, P.A., and R.R.B. Leakey. 1997. Land use transformation in Africa: Three determinants for balancing food security with natural resource utilization. Eur. J. Agron. 7:15-23.

Sanchez, P.A., M.S. Swaminathan, P. Dobie, and N. Yuksel. 2005. Halving hunger: It can be done. UN Millennium Project Task on Hunger. Earthscan, London.

Sanchirico, J.N., M.D. Smith, and D.W. Lipton. 2006. An approach to ecosystem-based fishery management. RFF Disc. Pap. 06-40. Resources for the Future, Washington DC.

Sanderson, J., and B. Sherman. 2004. A nut to crack a growing problem. Partners in Research and Development. Dec 2004. ACIAR, Canberra, Australia.

Sanginga, P.C., R.N. Kamugisha, and A.M. Martin. 2007. The dynamics of social capital and conflict management in multiple resource regimes: A aase of the southwestern highlands of Uganda. Ecol. Society 12(1):6.

Santiso, J. 2006. Latin America's political economy of the possible: Beyond good revolutionaries and free marketers. MIT Press, NY.

Sanvido, O., M. Stark, J. Romeis, and F. Bigler. 2006. Ecological impacts of genetically modified crops. Experience from 10 years experimental field research and commercial cultivation. Available at http://www.services.art.admin.ch/pdf/ART_SR_01_E.pdf. ART-Schriftenreihe 1, Zürich.

Sautier, D., and E. Biénabe. 2005. The role of small-scale producers' organisations in addressing market access. Beyond agriculture - making markets work for the poor p. 69-

85. In Proc. Int. Seminar. Westminster. 28 Feb- 1 March 2005. CPHP, London.

Sauvé, R., and J. Watts 2003. An analysis of IPGRI's influence on the International Treaty on Plant Genetic Resources for Food and Agriculture. Agric. Syst. 78:307-327.

Sayer, J., and B.M. Campbell. 2001. Research to integrate productivity enhancement, environmental protection, and human development. Conserv. Ecol. 5(2):32. [Online] http://www.consecol.org/vol5/iss2/art32.

Sayer, J., and B.M. Campbell. 2004. The science of sustainable development: Local livelihoods and the global environment. Cambridge Univ. Press, UK.

Sayre, K.D., A. Limon-Ortega, and R. Gupta. 2006. Raised bed planting technologies for improved efficiency, sustainability and profitability. Yield Potential Symp. March 2006. CIMMYT, Mexico.

Sayre, K.D., S. Rajaram, and R.A. Fischer. 1997. Yield potential progress in short bread wheats in northwest Mexico. Crop Sci. 37:36-42.

Schippers, R.R. 2000. African indigenous vegetables: an overview of the cultivated species. NRI/ACP-EU. Tech. Centre Agric. Rur. Coop., London.

Schippers, R.R., and L. Budd. 1997. African indigenous vegetables. IPGRI, Rome and NRI, UK.

Schlager, E., 2003. The political economy of water reform. Int. Working Conf. Water Rights: Institutional options for improved water allocation, Hanoi. 12-15 Feb 2003. IFPRI, Washington DC.

Schlager, E., and E. Ostrom. 1992. Property-rights regime and natural resources: A conceptual analysis. Land Econ. 68(3).

Schlesinger, W.H. 1999. Carbon and agriculture: Carbon sequestration in soils. Science 284:2095.

Schmetz, J., Y. Govaerts, M. Konig, H-J. Lutz, A. Ratier, and S. Tjemkes 2005. A short introduction to METEOSAT Second Generation (MSG). CIRA, Colorado State Univ., Fort Collins.

Schreckenberg, K., A. Awono, A. Degrande, C. Mbosso, O. Ndoye, and Z. Tchoundjeu. 2006. Domesticating indigenous fruit trees as a contribution to poverty reduction. For. Trees Livelihoods 16:35-51.

Schreckenberg, K., A. Degrande, C. Mbosso, Z. Boli Baboulé, C. Boyd, L. Enyong et al. 2002. The social and economic importance of Dacryodes edulis (G.Don) H.J. Lam in southern Cameroon. For. Trees Livelihoods 12:15-40.

Schroeder, K.L., and T.C. Paulitz. 2006. Root diseases of wheat and barley during the transition from conventional to direct seeding. Plant Dis. 90:1247-1253.

Schroth, G., G.A.B. Da Fonseca, C.A. Harvey, C. Gascon, H.L. Vasconcelos, and A.-M. Izac. 2004. Agroforestry and biodiversity conservation in tropical landscapes. Island Press, Washington DC.

Schroth, G., and C. Harvey (ed) 2007. Biodiversity conservation in cocoa production landscapes. Biodiversity Conserv. 16:2237-2444.

Schroth, G., U. Krauss, L. Gasparotto, J.A. Duarte-Aguilar, and K. Vohland. 2000. Pests and diseases in agroforestry systems of the humid tropics. Agrofor. Syst. 50:199-241.

Schuller, P., D.E. Walling, A. Sepulveda, A. Castillo, and I. Pino. 2007. Changes in soil erosion associated with the shift from conventional tillage to a no-tillage system, documented using 137Cs measurements. Soil Till. Res. 94:183-192.

Schultz, R.C., J.P. Colletti, T.M. Isenhart, C.O. Marquez, W.W. Simpkins, and C.J. Ball. 2000. Riparian forest buffer practices. p. 189-281. In H.E. Garrett et al. (ed) North American agroforestry: An integrated science and practice. ASA, Madison WI.

Schultz, R.C., J.P. Colletti, T.M. Isenhart, W.W. Simpkins, C.W. Mize, and M.L. Thompson. 1995. Design and placement of a multi-species riparian buffer strip system. Agrofor. Syst. 29:201-226.

Schultz, T.W. 1949. Production and welfare of agriculture. Macmillan, NY.

Scoones, I., and J. Thompson (ed) 1994. Beyond farmer first. ITDG Publ., London.

Sebillotte, M., 2000. Des recherches pour le développement local. Partenariat et transdisciplinarité, Revue d'économie Régionale et Urbaine 3:535-556.

Sebillotte, M., 2001. Des recherches en partenariat pour et sur le développement régional. Ambitions et questions. Natures Sci. Sociétés 9(3):5-7.

Sen, A. 1981. Poverty and famines: An essay on entitlement and deprivation. Oxford Univ. Press, UK.

Sepúlveda, S., A. Rodriguez, R. Echevelly, M. Portilla. 2003. El enfoque territorial del desarrollo rural. IICA, San José.

Seralini, G.E., D. Cellier, and J. Spiroux de Vendomois. 2007. New analysis of a rat feeding study with a genetically modified maize reveals signs of hepatorenal toxicity. Arch. Environ. Contamin. Toxicol. 52:596-602.

Shackleton, S.E., C.M. Shackleton, A.B. Cunningham, C. Lombard, and C. Sullivan. 2002. An overview of current knowledge on Sclerocarya birrea (A. Rich) Holst, with particular reference to its importance as a non-timber forest product in Southern Africa: A summary. Part 1. Taxonomy, ecology and role in rural livelihoods. Southern Afric. For. J. 194:27-41.

Shackleton, S.E., P. Shanley, and O. Ndoye. 2007. Invisible but viable: Recognizing local markets for non-timber forest products. Int. For. Rev. 9:697-712.

Shackleton, S.E., R.P. Wynberg, C.A. Sullivan, C.M. Shackleton, R.R.B. Leakey, M. Mander et al. 2003. Marula commercialisation for sustainable and equitable livelihoods: Synthesis of a southern African case study. In Winners and losers in forest product commercialization.

Final Technical Rep. to DFID, FRP R7795, App. 3.5 CEH, Wallingford.

Shah, T., I. Makin, R. Sakthivadivel, and M. Samad. 2002. Limits to leapfrogging: Issues in transposing successful river basin management institutions in the developing world. In C.L. Abernethy (ed) Intersectoral management of river basins. IWMI, Colombo.

Shah, T., M. Alam, M.D. Kumar, R.K. Nagar, and M. Singh. 2000. Pedaling out of poverty: Social impact of a manual irrigation technology in South Asia. IWMI Res. Rep. 45. IWMI, Colombo.

Shapiro, H-Y., and E.M. Rosenquist, 2004. Public/private partnerships in agroforestry: The example of working together to improve cocoa sustainability. Agrofor. Syst. 61:453-462.

Sharma, R., and T. Poleman. 1993. The new economics of India's Green Revolution. Cornell Univ. Press, Ithaca.

Sheehy, J.E., P.L. Mitchell, G.J.D. Kirk, and A.B. Ferrer. 2005. Can smarter nitrogen fertilizers be designed? Matching nitrogen supply to crop requirements at high yields using a simple model. Field Crops Res. 94:54-66.

Shepherd, G., and D. Brown. 1998. Linking international priority-setting to local institutional management. p. 77-87. In S. Doolan (ed) African rainforests and the conservation of biodiversity. Earthwatch, Oxford.

Shewfelt, R.L., and B. Bruckner. 2000. Fruits and vegetable quality: An integrated view. Technomic Publ., Pennsylvania.

Shiklomanov, I. A. 1999. World water resources: Modern assessment and outlook for the 21st century. Summary of world water resources at the beginning of the 21st Century, prepared in the framework of the IHP UNESCO. Federal Service of Russia for Hydrometeorology & Environment Monitoring, State Hydrological Inst., St. Petersburg.

Shuttleworth, W.J. 1988. Evaporation from Amazonian rainforest. Proc. Royal Soc., London.

Sibelet, N. 1995. L'innovation en milieu paysan ou la capacité des acteurs locaux à innover en présence d'intervenants extérieurs. Nouvelles pratiques de fertilisation et mise en bocage dans le Niumakélé (Anjouan, Comores), INA-PG, Paris.

Silvey, V. 1986. The contribution of new varieties to cereal yields in England and Wales between 1947 and 1983. J. Nat. Inst. Agric. Bot. 17:155-168.

Silvey, V. 1994. Plant breeding in improving crop yield and qulaity in recent decades. Acta Hort. (ISHS) 355:19-34. Available at http://www.actahort.org/books/355/355_2.htm

Simm, G. 1998. Genetic improvement of cattle and sheep. CABI Publishing, Wallingford, UK.

Simm, G., L. Bünger, B. Villanueva, and W.G. Hill. 2004. Limits to yield of farm species: Genetic improvement of livestock. p. 123-141. In R. Sylvester-Bradley and J. Wiseman (ed) Yields of farmed species.

Constraints and opportunities in the 21st century. Nottingham Univ. Press.

Simmonds, N.W. (ed) 1976. Evolution of crop plants. Longmans, London and NY.

Simons, A.J., and R.R.B. Leakey. 2004. Tree domestication in tropical agroforestry. Agrofor. Syst. 61:167-181.

Simpson, B.M. 2001. IPPM farmer field schools and local institutional development: Case studies of Ghana and Mali. Annex B10 in FAO 2001. Midterm review of the Global IPM Facility. FAO Rome.

Sinclair, T.R., G.L. Hammer, and E.J. Van Oosterom. 2005. Potential yield and water-use efficiency benefits in sorghum from limited maximum transpiration rate. Functional Plant Biol. 32:945-952.

Singh, B.B., O.O. Olufajo, and M.E. Ishiyaku. 2006. Registration of six improved germplasm lines of cowpea with combined resistance to Striga gesnerioides and Alectra vogelii. Crop Sci. 46:2332-2333.

Singh, B., Y. Singh, J.K. Ladha, K.F. Bronson, V. Balasubramanian, J. Singh, and C.S. Khind. 2002. Chlorophyll meter and leaf color chart based nitrogen management for rice and wheat in northwestern India. Agron. J. 94:821-829.

Sisti, C.P.J., H.P. Santos, R. Kohhann, B.J.R. Alves, S. Urquiaga, and R.M. Boddey. 2004. Change in carbon and nitrogen stocks in soil under 13 years of conventional or zero tillage in southern Brazil. Soil Till. Res. 76:39-58.

Sivakumar, M.V.K. 2006. Climate prediction and agriculture: Current status and future challenges. Clim. Res. 33:3-17.

Slingo, J.M., A.J. Challinor, B.J. Hoskins, and T.R. Wheeler. 2005. Food crops in a changing climate. Phil. Trans. R. Soc. B 360(1463):1983-1989.

Smale, M. 1997. The green revolution and wheat genetic diversity: Some unfounded assumptions. World Dev. 25:1257-1269.

Smale, M., J. Hartell, P.W. Heisey, and B. Senauer. 1998. The contribution of genetic resources and diversity to wheat production in the Punjab of Pakistan. Am. J. Agric. Econ. 80:482-493.

Smale, M., M.P. Reynolds, M. Warburton, B. Skovmand, R. Trethowan, R. Singh et al. 2002. Dimensions of diversity in modern spring bread wheat in developing countries. Crop Sci. 42:1766-1779.

Smartt, J., and N. Haq. 1997. Domestication, production and utilization of new crops. Int. Centre for Underutilized Crops, Southampton, UK.

Smit, L.A.M. 2002. Pesticides: Health impacts and alternatives. Working Pap. 45. IWMI, Colombo.

Smith, C. 1984. Rates of genetic change in farm livestock. Res. Dev. Agric. 1:79-85.

Smith, K. 2006. Rural air pollution: A major but often ignored development concern. Commission on Sustainable Development Thematic Session on Integrated Approaches to Addressing Air Pollution and Atmospheric Problems. United Nations, NY.

Smith, K.A., and F. Conen. 2004. Impacts of land management on fluxes of trace greenhouse gases. Soil Use Manage. 20:255-263.

Smith, K.A., P. McTaggarti, and H. Tsuruta. 1997. Emission of N2O and NO associated with nitrogen fertilization in intensive agriculture, and the potential for mitigation. Soil Use Manage. 13:296-304.

Smith, L.E.D. 2004. Assessment of the contribution of irrigation to poverty reduction and sustainable livelihoods. Water Resour. Dev. 20(2):243-257.

Smith, P., D. Martino, Z. Cai, D. Gwary, H. Janzen, P. Kumar et al. 2007. Agriculture. In Climate change 2007: Mitigation. Contribution of Working Group III to the Fourth Assessment Report of the Intergovernmental Panel on Climate Change. B. Metz et al. (ed) Cambridge Univ. Press, Cambridge and NY.

Snape, J.W. 2004. Challenges of integrating conventional breeding and biotechnology: A personal view! 4th Int. Crop Sci. Congress, Brisbane, Australia. Available at http://www.cropscience.org.au/icsc2004/plenary/3/1394_snapejw.htm.

Snow, A.A., D.A. Andow, P. Gepts, E.M. Hallerman, A. Power, J.M. Tiedje et al. 2005. Genetically engineered organisms and the environment: current status and recommendations. Ecol. Appl. 15:377-404. [Online] www.esa.org/pao/esaPositions/Papers/geo_position.htm

Snow, A.A., D. Pilson, L.H. Rieseberg, M.J. Paulsen, N. Pleskac, M.R. Reagon et al. 2003. A Bt trasngen reduces herbivory and enhances fecundity in wild sunflower. Ecol. Appl. 13(2):279-286.

Song, Y. 1999. Feminisation of maize agricultural production in Southwest China. Biotech. Dev. Monitor 37:6-9.

Spellerberg, I.F., and S.R. Hardes. 1992. Biological conservation. Cambridge Univ. Press, UK.

Sperling, L., T. Remington, J.M Haugen, and S. Nagoda (ed) 2004. Addressing seed security in disaster response: Linking relief with development. Available at http://www.ciat.cgiar.org/africa/seeds.htm. CIAT, Cali.

Spielman, D.J. 2005. Innovation systems perspectives on developing-country agriculture: A critical review. ISNAR Disc. Pap. 2. Sept 2005. IFPRI, ISNAR Division, Washington DC.

Spök, A., H. Hofer, P. Lehner, R. Valenta, S. Stirn, and H. Gaugitsch 2004. Risk assessment of GMO products in the European Union: Toxicity assessment, allergenicity assessment and substantial equivalence in practice and proposals for improvement and standardisation. Umweltbundesamt, Vienna.

Sprent, J.I. 1994. Harnessing symbiotic associations: the potentials of nitrogen fixing trees. p. 176-182. In R.R.B. Leakey and A.C. Newton (ed) Tropical trees: The potential for domestication and the rebuilding of forest resources. HMSO, London.

Sprent, J.I. 2001. Nodulation in legumes. Royal Botanic Gardens, Kew, London.

Sprent, J.I., and P. Sprent. 1990. Nitrogen fixing organisms. Clapham and Hall, London.

Squire, G.R., C. Hawes, D.A. Bohan, D.R. Brooks, G.T. Champion, and L.G. Firbank. 2005. Biodiversity effects of the management associated with GM cropping systems in the UK. Final Rep. to DEFRA Project EPG 1/5/198. Scottish Crop Res. Inst., Dundee, UK.

Sseguya, H., and D. Masinde. 2006. Annual Monitoring and Evaluation Results 2006. CSRL-VEDCO Kamuli Program. Center for Sustainable Rural Livelihoods. Iowa State Univ., Ames, IA.

Stanton B.F., and J.D. Clemens. 1987. A randomized trial to assess the impact of the intervention on hygiene behaviors and rates of diarrhea. Am. J. Epidemiol. 125:292-301.

STCP. 2003. Sustainable Tree Crops Program Newsl., Issue No. 3 June 2003, IITA, Ibadan.

Steele, K., A.H. Price, H. Shashidhar, and J.R. Witcombe. 2006. Marker-assisted selection to introgress rice QTLs controlling root traits into an Indian upland rice variety. Theor. Appl. Genetics 112:208-221.

Steinfeld, H., P. Gerber, T. Wassenaar, V. Castel, M. Rosales, and C. de Haan. 2006. Livestock's long shadow: Environmental issues and options. FAO, Rome.

Steppler, H., and P.K. Nair (ed) 1987. Agroforestry, a decade of development. Available at http://www.worldagroforestry.org/units/library/books/PDFs/07_Agroforestry_a_decade_of_development.pdf. ICRAF, Nairobi.

Stewart, F. 1993. War and underdevelopment: Can economic analysis help reduce the costs? J. Int. Dev. 5(4):357-380.

Stiglitz, J. 1987. Markets, market failures, and development. Am. Econ. Rev. 79(2):197-203.

Stiglitz, J. 2002. Globalization and its discontents. Penguin Press, NY.

Stiglitz, J., and B. Greenwald. 2006. Helping infant economies grow: Foundations of trade policies for developing countries. Am. Econ. Rev. 96(2):141-146.

Stireman, J.O. III., L.A. Dyer, D.H. Janzen, M.S. Singer, J.T. Lill, R.J. Marquis et al. 2005. Climatic unpredictability and parasitism of caterpillars: Implications of global warming. PNAS 102(48):17384-17387.

Stone, G.D. 2007. Agricultural deskilling and the spread of genetically modified cotton in Warangal. Curr. Anthropol. 48:67-103.

Stone, R.C., and H. Meinke. 2005 Operational seasonal forecasting of crop performance. Phil. Trans. R. Soc. B Biol. Sci. 360(1463):2109-2124.

Stork, N.E., A.D. Watt, E. Srivastava, and T. Larsen. 2003. Butterfly diversity and silvicultural practice in lowland rainforest of southern Cameroon. Biodivers. Conserv. 203:387-410.

Strauss, M. 2000. Human waste (excreta and wastewater) reuse. Contribution to ETC/SIDA Bibliography on Urban Agriculture. EAWAG/SANDEC, Duebendorf, Switzerland. Available at http://www.eawag.ch/organisation/abteilungen/sandec/schwerpunkte/wra/documents/human_waste_reuse.

Suarez, A. (plus 13 co-signatories) 2002. Letters: Conflicts around a study of Mexican crops. Nature 417:897.

Sunderland, T.C.H., L.E. Clark, and P. Vantomme. 1999. Non-wood forest products of central Africa: Current research issues and prospects for conservation and development. FAO, Rome.

Sunderland, T.C.H., and O. Ndoye (ed) 2004. Forest products, livelihoods and conservation: Case studies of non-timber forest product systems. Vol. 2 Africa. CIFOR, Indonesia.

Swallow, B., A. Okono, C. Ong, and F. Place. 2002. Case three. TransVic: Improved land management across the Lake Victoria Basin. Available at http://www.fao.org/WAIRDOCS/TAC/Y4953E/y4953e07.htm

Swaminathan, M. 2006. An evergreen revolution. Crop Sci. 46:2293-2203.

Swanson, B.E., R.P. Bentz and A.J. Sofranko (ed) 1997. Improving agricultural extension: A reference manual. Rome, FAO.

Swanson, B.E., B.J. Farmer, and R. Bahal. 1990. The current status of agricultural extension worldwide. In B.E. Swanson (ed) Report of the global consultation on agricultural extension. FAO, Rome.

Swanton, C.J., and S.F. Weise. 1991. Integrated weed management: The rationale and approach. Weed Tech. 5:657-663.

Swift, M.J., and J.M. Anderson. 1993. Biodiversity and ecosystem function in agricultural systems. p. 15-42. In E-D. Schulze and H.A. Mooney (ed) Biodiversity and ecosystem function. Springer Verlag, Berlin.

Swift, M.J., J. Vandermeer, P.S. Ramakrishnan, J.M. Anderson, C.K. Ong, and B.A. Hawkins 1996. Biodiversity and agroecosystem function. p. 261-298. In H.A. Mooney et al. (ed) Functional roles of biodiversity: A global perspective. John Wiley, NY.

Sylvester-Bradley, R., M. Foulkes, and M.P. Reynolds. 2005. Future wheat yields: Evidence, theory and conjecture, p. 233-260. In R. Sylvester-Bradley and J. Wiseman (ed) Yields of farmed species. Nottingham Univ. Press, Nottingham.

Tabassum Naved, R. 2000. Intrahousehold impact of the transfer of modern agricultural technology: a gender perspective. FCND Disc. Pap. 85, May 2000. IFPRI, Washington DC.

Tabuti, J.R.S., S.S. Dhillion, and L.A. Lye. 2003. Ethnoveterinary medicines for cattle (Bos indicus) in Bulamogi county, Uganda: Plant species and modes of use. J. Ethnopharmacol. 88:279-286.

Tacoli, C. 1998. Rural-urban interactions: A guide to the literature. Environ. Urban. 10:147-166.

Tamagno, C. 2003. Entre celulinos y cholulares: Los procesos de conectividad y la construcción de identidades transnacionales. Available at http://lasa.international.pitt.edu/Lasa2003/TamagnoCarla.pdf. Int. Congress Latin Amer. Studies Assoc (LASA), Dallas.

Taneja, V., and P. S. Birthal. 2003. Role of buffalo in food security in Asia. Asian Buffalo 1:4-13.

Tang, S.Y. 1992. Institutions and collective action: Self-governance in Irrigation. Inst. Contemporary Studies Press, San Francisco.

Tangka, F.K., M.A. Jabbar, and B.I. Shapiro. 2000. Gender roles and child nutrition in livestock production systems in developing countries. A critical review. Socioeconomics and policy Res. Working Pap. 27. ILRI, Nairobi.

Tanji, K.K., and N.C. Kielen. 2004. Agricultural drainage water management in arid and semi-arid areas. FAO Irrigation and Drainage Pap. 61. FAO, Rome.

Tarawali, S.A., J.D.H. Keatinge, J.M. Powell, O. Lyasse and N. Sanginga. 2004. Integrated natural resource management in West African crop-livestock systems. In T.O. Williams et al. (ed) Sustainable crop-livestock production for improved livelihoods and natural resource management in West Africa. ILRI, Nairobi and Technical Centre Agric. Rural Cooperation (CTA), Wageningen.

Tarawali, S.A., A. Larbi, S. Fernadez-Rivera, and A. Bationo, 2001. The role of livestock in the maintenance and improvement of soil fertility. p. 281-304. In G.Tian et al. (ed) Sustaining soil fertility in West Africa. SSSA Special Public. No. 58. SSSA/ASA, Madison.

Taylor, L.H., S.M. Latham, M.E. Woolhouse. 2001, Risk factors for human disease emergence. Philos. Trans. R. Soc. Lond. B Biol. Sci. 356(1411):983-989.

Tchoundjeu, Z., E. Asaah, P. Anegbeh, A. Degrande, P. Mbile, C. Facheux et al. 2006. AFTPs: Putting participatory domestication into practice in West and Central Africa. For. Trees Livelihoods 16:53-70.

Temple, L., M. Kwa, R. Fogain, and A. Mouliom Pefoura. 2006. Participatory determinants of innovation and their impact on plantain production systems in Cameroon. Int. J. Agric. Sustain. 4(3):233-243.

Temu, A., I. Mwanje, and K. Mogotsi. 2003. Improving agriculture and natural resources education in Africa. World Agroforestry Centre, Nairobi.

Temu, A.B., P. Rudebjer, and I. Zoungrana. 2001. Networking educational institutions for change: The experience of ANAFE. World Agroforestry Centre, Nairobi.

Ten Kate, K., and S. Laird. 1999. The Commercial use of biodiversity: Access to genetic resources and benefit-sharing. Earthscan, London.

Tesso, T., A. Deressa, Z. Gutema, A. Adugna, B. Beshir, T. Tadesse, and O. Oumer. 2006. Integrated Striga management provides relief from plague of witchweed. p. 141-158. In T. Abate (ed) Successes with value chain. EIAR, Addis Ababa.

Thiele, G., R. Nelson, O. Ortiz, and S. Sherwood. 2001. Participatory research and training: ten lessons from the Farmer Field Schools (FFS) in the Andes. Currents 27:4-11.

Thijssen, R. 2006. Croton megalocarpus, the poultry-feed tree: How local knowledge could help to feed the world. p. 226-234. In R.R.B. Leakey et al. (ed) Domestication and commercialisation of non-timber forest products. Non-Wood Forest Products No. 9. FAO, Rome.

Thirtle, C., L. Lin, and J. Piesse. 2003. The impact of research-led agricultural productivity growth on poverty reduction in Africa, Asia and Latin America. World Dev. 31:1959-1975.

Thomas, M., and J. Waage. 1996. Integration of biological control and host-plant resistance breeding. CTA, Wageningen.

Thomas, W.E., W.J. Everman, B. Scott, and J.W. Wilcut. 2007. Glyphosate-resistant corn interference in glyphosate-resistant cotton. Weed Tech. 21:372-377.

Thompson, J., T. Hodgkin, K. Atta-Krah, D. Jarvis, C. Hoogendoorn, and S. Padulosil. 2007. Biodiversity in agroecosystems. p. 46-60. In Farming with nature: The science and practice of ecoagriculture, S.J. Scherr and J.A. McNeely (ed) Island Press, NY.

Thornton, P.K., P.M. Kristiansen, R.L. Kreskas, and R.S. Reid. 2004. Mapping livestock and poverty. p. 37-50. In E. Owen et al. (ed) Responding to the Livestock revolution - The role of globalisation and implications for poverty alleviation. Brit. Soc. Anim. Sci. Pub. No. 33. Nottingham Univ. Press, UK.

Thornton, P., R. Kruska, N. Henninger, P. Kristjanson, R. Reid, F. Atieno et al. 2002. Mapping poverty and livestock in the developing world. ILRI, Nairobi.

Tiffin, M., M. Mortimore, and F. Gichuki. 1994. More people, less erosion: Environmental recovery in Kenya. ACTS Press, Nairobi.

Tillman, D., K.G. Cassman, P.A. Matson, R. Naylor, and S. Polasky. 2002. Agricultural sustainability and intensive production practices. Nature 418:671-677.

Tilman, D., J. Fargione, B. Wolff, C. D'Antonio, A. Dobson, R. Howarth et al. 2001. Forecasting agriculturally driven global environmental change. Science 292(5515):281-284.

Timothy, K., D. Donaldson, and A. Grube. 2004. Pesticides industry sales and usage: 2000 and 2001 market estimates. Off. Prevention, Pesticides, and Toxic Substances. US EPA, Washington DC.

Tirole, J., C. Henry, M. Trommetter, L. Tubiana, and B. Caillaud. 2003. Propriété intellectuelle. La documentation française, Paris.

Tollenaur, M., and J. Wu.1999. Yield improvement in temperate maize is attribuatble to greater stress tolerance. Crop Sci. 39:1597-1604.

Tomich, T.P., A. Cattaneo, S. Chater, H.J. Geist, J. Gockowski, D. Kaimowitz et al. 2005. Balancing agricultural development and environmental objectives: Assessing tradeoffs in the humid tropics. p. 415-440. In C.A. Palm et al. (ed) Columbia Univ. Press, NY.

Torero, M., and J. von Braun. 2006. Information and communication technologies for development and poverty reduction. Johns Hopkins Univ. Press, Baltimore.

Torres, J.B. and J.R. Ruberson. 2006. Canopy- and ground-dwelling predatory arthropods in commercial Bt and non-Bt cotton fields: Patterns and mechanisms. Environ. Entomol. 34:1242-1256.

Torstensson, G., H. Aronsson, and L. Bergström. 2006. Nutrient use efficiencies and leaching of organic and conventional cropping systems in Sweden. Agron. J. 98:603-615.

Toulmin, C., and J.F. Quan (ed) 2000. Evolving land rights, policy and tenure in Africa. DFID, IIED, NRI, London.

Touzard, J.-M., J-F. Draperi. 2003. Les coopératives entre territoires et mondialisation. Les cahiers de l'Economie sociale n°2. L'Harmattan, Paris.

Townsend, R. 1999. Agricultural incentives in Sub-Saharan Africa. Policy Challenges. Tech. Pap. 444. World Bank, Washington DC.

Trebuil, G., and M. Hossain. 2004. Le riz: Enjeux écologiques et économiques. Collection Mappemonde, Editions Belin, Paris.

Tribe, D.E. 1994. Feeding and greening the world: The role of international agricultural research. CABI, Wallingford.

Trigo, E.J., and E.J. Cap. 2003. The impact of the introduction of transgenic crops into Argentinian agriculture. AgBioForum 6:87-94.

Tripp, R. 2001. Seed provision and agricultural development. ODI with James Currey, Oxford and Heinemann, Portsmouth, London.

Tripp, R., 2006. Self-sufficient agriculture. Labour and knowledge in small-scale farming. Earthscan, London.

Tripp, R., M. Wijeratne, and V.H. Piyadasa. 2005. What should we expect from Farmer fields Schools? A Sri Lanka case study. World Dev. 33-10:1705-1720.

Trommetter, M. 2005. Biodiversity and international stakes: A question of access. Ecol. Econ. 53:573-583.

Tubiana, L. 2000. Environnement et développement : L'enjeu pour la France. Rapport au Premier ministre Documentation française, Paris.

Tuyns, A.J., E. Riboli, G. Doornbos, and G. Penquignot. 1987. Diet and esophageal cancer in Calvados (France). Nutr. Cancer 9:81-92.

UCSUSA. 2007. Confronting climate change in California. Available at http://www .climatechoices.org/. http://www.ucsusa.org/.

Udawatta, R.P., P.R. Motavalli, and H.E. Garrett. 2004. Phosphorus loss and runoff characteristics in three adjacent agricultural watersheds with claypan soils. J. Environ. Qual. 33:1709-1719.

UN. 1995. Report of the Fourth World Conf. Women. Beijing. 4-15 Sept. 1995. Available at http://www.un.org/esa/gopher-data/conf/ fwcw/off/a--20.en

UN. 2006a. Millennium development goals report. Available at www.un.org/millenniumgoals/pdf/ mdg2007.pdf

UN. 2006b. Strategy on solutions for harmonizing international regulation of organic agriculture.

Vol. 2, Background papers of the International Task Force on Harmonization and Equivalence in Organic Agriculture. UNCTAD, FAO, and IFOAM, Geneva and Bonn.

UN CCD. 1998. Rational use of rangelands and developments of fodder crops in Africa. UN Secretariat Convention to Combat Desertification. Rep. on the CCD/ILRI workshop. Addis Ababa. 4-7 Aug 1998. Available at http://www.unccd.int/regional/africa/meetings/regional/workshop98/report-eng.pdf.

UNDP. 2000. Bioenergy primer. Modernized biomass energy for sustainable development. UNDP, NY.

UNDP. 2006a. Human development report. UNDP, NY.

UNDP. 2006b. Social protection, the role of cash transfers, in poverty in focus. International Poverty Centre, UNDP, NY.

UNDP/ESMAP. 2002a. Rural electrification and development in the Philippines: Measuring the social and economic benefits. ESMAP Rep. 255/02. World Bank, Washington DC.

UNDP/ESMAP. 2002b. Energy strategies for rural India: Evidence from six states. ESMAP, World Bank, Washington DC.

UNDP/UNEP/WB/WRI. 2000. World resources 2000-2001: People and ecosystems- the fraying web of life. Available at: http://www.wri.org/publication/world-resources-2000-2001-people-and-ecosystems-fraying-web-life#. World Resources Inst., Washington, DC.

UNEP. 1993. Convention on biological diversity. Text and annexes. CBD/94/1. UNEP/CBD, Montreal.

UNEP. 1999. Global Environment Outlook - 2000. UNEP, DEIA&EW, Nairobi. Earthscan, London.

UNEP. 2002. Vital water graphics. Available at http://www.unep.org/dewa/assessments/ecosystems/water/vitalwater/. UNEP, Nairobi.

Uphoff, N. (ed) 2002. Agroecological innovations. Increasing food production with participatory development. Earthscan, London.

Urbano, G., and D. Vollet. 2005. L'évaluation du Contrat territorial d'exploitation (CTE). p. 69-110. In Notes et études économiques n°22. Ministère de l'Agric. et de la Pêche, Paris.

US EPA (Evironmental Protection Agency). 2006. Climate change, greenhouse gas emissions. Human-related sources and sinks of carbon dioxide. Available at http://www.epa.gov/climatechange/emissions/co2_human.html

US GS (Geological Survey). Water-resources investigations Rep. 00-4053. Sacramento, CA.

Van Crowder, L. 1996. Agricultural extension for sustainable development. SD Dimensions. FAO, Rome.

Van de Kop, P., D. Sautier, and A. Gerz. 2006. Origin-based products — Lessons for pro-poor market development. KIT, CIRAD, Paris.

Van den Berg, H. 2003. IPM Farmer schools. A synthesis of 25 Impact evaluations. Global IPM facility, Rome.

Van den Berg, H., H. Senerath, and L. Amarasinghe. 2002. The impact of particiaptory IPM in Sri Lanka. Int. learning workshop on farmer field schools FFS. Emerging issues and challenges. Yogyakarta, Indonesia. 21-25 Oct. 2002.

Van Gessel, M.J. 2001. Glyphosate-resistant horseweed from Delaware. Weed Sci. 49:703-705.

Van Hofwegen, P.J.M., and F.G.W. Jaspers. 1999. Analytical framework for integrated water resources management. IHE Mono. 2, Inter-Am. Dev. Bank, Balkema, Rotterdam.

Van Koppen, B. 2002. A gender performance indicator for irrigation: Concepts, tools and applications. Res. Rep. 59. IWMI, Colombo.

Van Lenteren, J.C. 2000. A greenhouse without pesticides: Fact or fantasy? Crop Prot. 19:375-84.

Van Lenteren, J.C. 2006. Benefits of biological control. REBECA Macrobials Meeting. April 2006. http://www.rebeca-net.de/downloads/Risk%20assessment%20macrobials%20van%20Lenteren.pdf. Int. Org. Biol. Control (IOBC-WPRS.org).

Van Lenteren, J.C. and V.H.P. Bueno. 2003. Augmentative biological control of arthropods in Latin America. BioControl 48:123-139.

Van Mele, P., A. Salahuddin, and N. Magor (ed) 2005. Innovations in rural extension: Case studies from Bangladesh. CABI, Wallingford.

Van Noordwijk, M., G. Cadish, and C.K. Ong (ed) 2004. Below-ground Interactions in tropical agroecosystems: concepts and models with multiple plant components. CABI, UK.

Van Oosterom, E.J., V. Mahalakshmi, F.R. Bidinger, and K.P. Rao. 1996. Effect of water availability and temperature on the genotype-by-environment interaction of pearl millet in semi-arid tropical environments. Euphytica 89:175-183.

Van Zyl, J. 1996. The farm size-efficiency relationship. In J. van Zyl et al. (ed) Agricultural land reform in South Africa: Policies, markets and mechanisms. Oxford Univ. Press, Cape Town.

Vanlauwe, B., and K.E. Giller. 2006. Popular myths around soil fertility management in sub-Saharan Africa. Agric. Ecosyst. Environ. 116:34-46.

Varvel, G.E., and W.W. Wilhelm. 2003. Soybean nitrogen contribution to corn and sorghum in western corn belt rotations. Agron. J. 95(5):1220-1225.

Veer, V.P., M.C. Jansen, M. Klerk, and F.J. Kok. 2000. Fruits and vegetables in the prevention of cancer and cardiovascular disease. Public Health Nutr. 3:103-107.

Verma, H.N. 2003. Major issues related to soil and water conservation in watersheds. p. 490-493. In B. Venkateswara Rao (ed) Proc. Int. Conf. Hydrology and Watershed Management. Vol. 1. Hyderabad, India. 18-20 Dec. 2002. BS Publ., Hyderabad.

Vetter, S. 2005. Rangelands at equilibrium and non-equilibrium: Recent developments in the debate. J. Arid Environ. 62:321-341.

Vieira, M. 2001. Brazil's organic cup not yet overflowing. World Organic News 13(11). January.

Villachica, H. 1996. Frutales y Hortalizas Promisorios de la Amazonia. Tratado de Cooperacion Amazonica, Lima.

Visscher, J.T., P. Bury, T. Gould, and P. Moriaty, 1999. Integrated water resources management in water and sanitation projects: Lessons from projects in Africa, Asia and South America. IRC Int. Water and Sanitation Centre, Delft, The Netherlands.

Vitousek, P.M., J.D. Aber, R.W. Howarth, G.E. Likens, P.A Matson, D.W. Schindler et al. 1997. Human alteration of the global nitrogen cycle: Causes and consequences. Ecol. Applic. 7:737-750.

Vitousek, P.M., K. Cassman, C. Cleveland, T. Crews, C.B. Field et al. 2002. Towards an ecological understanding of biological nitrogen fixation. Biogeochemistry 57:1-45.

Vitousek P.M., C.M. D'Antonio, L.I. Loope, and R. Westbrooks. 1996. Biological invasion as global environmental change. Am. Scientist 84:468-478.

Voituriez, T. 2005. Sustainable development and trade liberalisation: The opportunities and threat roused by the WTO. OCL. Oléagineux Corps Gras Lipides 12:129-133.

Vörösmarty, M.C.J., P. Green, J. Salisbury, and R.B. Lammers. 2000. Global water resources: Vulnerability from climate change and population growth. Science 289:284-288.

Vosti, S.A., J. Gockowski, and T.P. Tomich. 2005. Land use systems at the margins of tropical moist forest: Addressing smallholder concerns in Cameroon, Indonesia, and Brazil. p. 387-414. In C.A. Palm et al. (ed) Slash and burn: The search for alternatives. Columbia Univ. Press, NY.

Waddington, S.R., and J. Karigwindi. 2001. Productivity and profitability of maize + groundnut rotations compared with continuous maize on smallholder farms in Zimbabwe. Exp. Agric. 37:83-98.

WADE. 2004. Bagasse cogeneration — Global review and potential. World Alliance for Decentralized Energy, Edinburgh.

Wahlquist, M.L., C.S. Lo, and K.A. Myres. 1989. Food variety is associated with less macrovascular disease in those with type II diabetes and their healthy controls. J. Am. Coll. Nutr. 8:515-523.

Waibel, H., and D. Pemsi. 1999. An evaluation of the impact of integrated pest management research at international agricultural research centres. CGIAR, TAC, CGIAR Secretariat, Washington DC.

Waliyar, F., L. Collette, and P.E. Kenmore (ed) 2003. Beyond the gene horizon: Sustaining agricultural productivity and enhancing livelihoods through optimization of crop and crop-associated biodiversity with emphasis on semi-arid tropical agroecosystems. Proc. Workshop. Patancharu, India. 23-25 Sept 2002. ICRISAT, Hyderabad.

Wall, G.W., T.J. Brooks, N.R. Adam, A.B. Cousins, B.A. Kimball, P.J. Pinter Jr. et al. 2001. Leaf photosynthesis and water relations of grain sorghum grown in Free-Air CO2 Enrichment (FACE) and water stress. New Phytologist 152:231-248.

Walter, A., and C. Sam. 1999. Fruits of Oceania. IRD, Paris and ACIAR, Australia.

Walters, K.S. and H.J. Gillett. 1998. 1997 IUCN red list of threatened plants. World Conservation Monitoring Centre. IUCN, Gland, Switzerland.

Wang, S., D.R. Just and P. Pinstrup-Andersen. 2006. Damage from secondary pests and the need for refuge in China. In R.E. Just et al. (ed) Regulating agricultural biotechnology: Economics and policy. Nat. Resour. Manage. Policy 30:625-637.

Wanyera, R., M.G. Kinyua, Y. Jin, and R.P. Singh. 2006. The spread of stem rust caused by Puccinia graminis f. sp. tritici, with virulence on Sr31 in wheat in Eastern Africa. Plant Dis. 90(1):113.

Warren, D.M., L.J. Slikkerveer, and S.O. Titilola (ed) 1989. Indigenous knowledge systems: Implications for agriculture and international development. Studies in Tech. Social Change No. 11. Iowa State Univ., Ames, IA.

Warren, D. M., L.J. Slikkerveer, and D. Brokensha (ed) 1995. The cultural dimension of development: Indigenous knowledge systems. Intermediate Tech. Publ., London.

Warren, M.S., J.K. Hill, J.A. Thomas, J. Asher, R. Fox, B. Huntley et al. 2001. Rapid responses of British butterflies to opposing forces of climate and habitat change. Nature 414(6859):65-69.

Waruhiu, A.N., J. Kengue, A.R. Atangana, Z. Tchoundjeu, and R.R.B. Leakey, 2004. Domestication of Dacryodes edulis: 2. Phenotypic variation of fruit traits in 200 tress from four populations in the humid lowlands of Cameroon. Food Agric. Environ. 2:340-346.

Watkins, F. 2004. Evaluation of DFID development assistance: Gender equality and women's empowerment DFID's experience of gender mainstreaming: 1995 to 2004.. Available at http://www.oecd.org/dataoecd/47/52/35074862.pdf. Evaluation Dep., East Kilbride, Glasgow.

Watt, A.D., N.E. Stork, and B. Bolton 2002. The diversity and abundance of ants in relation to forest disturbance and plantation establishment in southern Cameroon. J. Appl. Ecol. 39:18-30.

Watt, A.D., N.E. Stork, C. McBeath, and G.R. Lawson. 1997. Impact of forest management on insect abundance and damage in a lowland tropical forest in southern Cameroon. J. Appl. Ecol. 34:985-998.

WCD. 2000. Dams and development: A new framework for decision-making. The report of the World Commission on Dams. Available at http://www.damsreport.org/wcd_overview.htm [Geo-1-033]. Earthscan, London.

WCRF/AICR. 1997. Food, nutrition and the prevention of cancer: A global perspective.

Available at http://www.wcrf.org/research/fnatpoc.lasso. World Cancer Res. Fund Int., Amer. Inst. Cancer Res., Washington DC.

Weber, J. 2005. Access to the benefits of sharing, p. 9-10 In Biodiversity and local ecological knowledge in France. INRA, Paris.

Weber, J.C., C. Sotelo Montes, H. Vidaurre, I.K. Dawson, and A.J. Simons. 2001. Participatory domestication of agroforestry trees: An example from the Peruvian Amazon. Dev. Practice 11(4):425-433.

Weber, M.L. 2001. Markets for water rights under environmental constraints. J. Environ. Econ. Manage. 42(1):53-64.

Welch, R.M., and R.D. Graham. 2000. A new paradigm for world agriculture: Productive, sustainable, nutritious, healthful food systems. Food Nutr. Bull. 21(4):361-366.

Welch, S., J. Roe, and Z. Dong. 2003. A genetic neural network of flowering time control in arabidopsis thaliana. Agron. J. 95:71-81.

West, T.O., and G. Marland. 2002. A synthesis of carbon sequestration, carbon emissions, and net carbon flux in agriculture: Comparing tillage practices in the United States. Agric. Ecosyst. Environ. 91:217-232.

West, T.O., and W.M. Post. 2002. Soil organic carbon sequestration rates by tillage and crop rotation: A global data analysis. J. Soil Sci. Soc. America 66:1930-1946.

Westerman, P.R., M. Liebman, F.D. Menalled, A.H. Heggenstaller, R.G. Hartzler, and P.M. Dixon. 2005. Are many little hammers effective? Velvetleaf (Abutilon theophrasti) population dynamics in two- and four-year crop rotation systems. Weed Sci. 53:382-392.

WHO. 1986. Informal consultation on planning for the prevention of pesticide poisonings. Doc. VBC/86.926. WHO, Geneva.

WHO. 2002. Global strategy for food safety: safer food for better health. WHO, Geneva.

WHO. 2003. Obesity and overweight. Fact sheet. WHO Global strategy on Diet, Physical activity and Health. Available at http://www.who.int/dietphysicalactivity/media/en/gsfs_obesity.pdf

WHO. 2005a. Global database on Body Mass Index. Available at http://www.who.int/ncd_surveillance/infobase/web/InfoBaseCommon/

WHO. 2005b. Modern food biotechnology, human health and development: An evidence based study. Available at http://www.who.int/foodsafety/publications/biotech/biotech_en.pdf

WHO. 2005c. Guidelines on food fortification with micronutrients for control of micronutrient malnutrition. WHO, Geneva.

WHO. 2006. Fuel for life: Household energy and health. WHO, Geneva.

WHO/FAO, 2003. Diet, nutrition and the prevention of chronic diseases. Rep. of the joint WHO/FAO expert consultation. WHO Technical Rep. Series, No. 916 (TRS 916).

Wibawa, G., L. Joshi, M. van Noordwijk, and E. Penot. 2006. Rubber-based agroforestry systems (RAS) as alternatives for rubber monoculture system. IRRDB Conf.,

Vietnam, 2006. Int. Rubber Res. Dev. Board, Malaysia.

Wiley, T., and M. Seymour. 2000. Systems for browsing sheep on the perennial fodder shrub Tagasaste. Agriculture Western Australia FarmNote 49/2000, Dep. Agric., Western Australia.

Willer, H., and M. Yussefi (ed) 2006. The world of organic agriculture — statistics and emerging trends. IFOAM, Bonn, Germany and FiBL, Frick, Swit.

Willett, W. 2006. Health effects of trans fatty acids intake. Testimony 30 Oct 2006 before NYC Board of Health. Available at http://www.hsph.harvard.edu/nutritionsource/Printer%20Friendly/Health_Effects_Trans_Fat.pdf.

Williams, C.E., N. Collier, C.C. Sardesai, H.W. Ohm, and S.E. Cambron, 2003. Phenotypic assessment and mapped markers for H31, a new wheat gene conferring resistance to Hessian fly (Diptera: Cecidomyiidae). Theor. Appl. Genetics 107:1516-23.

Wilson, E.O. 1992. The diversity of life. Penguin, London.

Wilson, J., R.C. Munro, K. Ingleby, P.A. Mason, J. Jefwa, P.N. Muthoka et al. 1991. Tree establishment in semi-arid lands of Kenya: Role of mycorrhizal inoculation and water-retaining polymer. For. Ecol. Manage. 45:153-163.

Wilson, J.S., and T. Otsuki. 2001. Global trade and food safety: Winners and losers in a fragmented system. DECRG, World Bank, Washington DC.

Witcombe, J.R. 1999. Do farmer-participatory methods apply more to high potential areas than to marginal ones? Outlook Agric. 28:43-49.

Witcombe, J.R., R. Petre, S. Jones, and A. Joshi. 1999. Farmer participatory crop improvement. IV. The spread and impact of a rice variety identified by participatory varietal selection. Exp. Agric. 35:471-487.

Witcombe, J.R., K. Joshi, R. Rana, and D.S. Virk. 2001. Increasing genetic diversity by participatory varietal selection in high potential production systems in Nepal and India. Euphytica 122:575-588.

Witcombe, J.R., A.J. Packwood, A.G.B. Raj, and D.S. Virk, 1988. The extent and rate of adoption of modern cultivars in India. p. 53-68. In J.R. Witcombe et al. (ed) Seeds of choice. IDTG Publ., London.

Witcombe, J.R., D. Virk, and J. Farrington (ed) 1998. Seeds of choice. ITDG Publ., London.

Witt, H., R. Patel and M. Schnurr. 2006. Can the poor help GM crops? Technology, representation and cotton in the Makhatini Flats, South Africa. Rev. Afric. Pol. Econ. 109:497-513.

Wolfenbarger, L.L., and P.R. Phifer 2000. The ecological risks and benefits of genetically engineered plants. Science 290:2088-2093.

Wong, C. 2004. Saskatchewan organic farmers file lawsuit against Monsanto and Aventis. The Canadian Press, 10 Jan 2004.

Wood, S., K. Sebastian, and S.J. Scherr. 2000. Pilot analysis of global ecosystems. IFPRI/ World Resources Inst., Washington DC.

Wopereis, M.C.S., C. Donovan, B. Nebíe, D. Guindo, and M.K. Diaye. 1999. Soil fertility management in irrigated rice systems in the Sahel and Savanna regions of West Africa. Field Crops Res. 61:125-145.

World Bank. 1998. Participation and social assessment: Tools and techniques. World Bankl, Washington DC.

World Bank. 1999. The irrigation sector. World Bank, South Asia Region and Gov. India, Min. Water Resources. Allied Publ., New Delhi.

World Bank. 2002. Extension and rural development: Converging views for institutional approaches. Workshop Summary. World Bank, Washington DC.

World Bank. 2003. Land policies for growth and poverty reduction. World Bank Policy Res. Rep. Oxford Univ. Press, UK.

World Bank. 2004a. Sustaining forest: A development strategy. World Bank, Washington DC.

World Bank. 2004b. Renewable energy for development. The role of the World Bank Group. World Bank, Washington DC.

World Bank. 2005a. Technical and economic assessment: Off grid, mini-grid and grid electrification technologies. Final Rep. Energy and Water Dep., Washington DC.

World Bank. 2005b. Gender issues and best practices in land administration projects: A synthesis report. PREM/ARD, World Bank, Washington DC.

World Bank. 2007a. World Dev. Rep. 2007. World Bank, Washington DC.

World Bank. 2007b. Changing the face of the waters: The promise and challenge of sustainable aquaculture. World Bank, Washington DC.

World Bank. 2007c. Enhancing agricultural innovation: How to go beyond strengthening agricultural research systems. World Bank, Washington DC.

Worldwatch Institute. 2006. Biofuels for transportation: Global potential and implications for sustainable agriculture and energy in the 21st Century. Worldwatch, Washington DC.

Worthy, K., R.C. Strohman, and P.R. Billings. 2002. Conflicts around a study of Mexican crops. Nature 417:897.

WRI, UNEP, UNDP, and World Bank, World Resources 1998. 1998-99 - A guide to the global environment. Oxford Univ. Press, NY.

Wuest, S.B., J.D. Williams, and H.T. Gollany. 2006. Tillage and perennial grass effects on ponded infiltration for seven semi-arid loess soils. J. Soil Water Conserv. 61:218-223.

Wynberg, R., J. Cribbins, M. Mander, S. Laird, C.R. Lombard, R. Leakey et al. 2002. An overview of current knowledge on Sclerocarya birrea (A. Rich) Holst, with particular reference to its importance as a non-timber forest product in Southern Africa: A summary. Part 2. Commercial use, tenure and policy, domestication, intellectual property rights and benefit-sharing. S. Afric. For. J. 196:67-77.

Wynberg, R.P., S.A. Laird, S. Shackleton, M. Mander, C. Shackleton, P. du Plessis et al. 2003. Marula policy brief. Marula commercialisation for sustainable and equitable livelihoods. For. Trees Livelihoods 13:203-215.

Xu, K., X. Xu, T. Fukao, P. Canlas, R. Maghirang-Rodriguez, S. Heuer et al. 2006. Sub1a is an ethylene-response-factor-like gene that confers submergence tolerance to rice. Nature 442: 705-708.

Yee, J., M.C. Ahearn, and W. Huffman. 2004. Links among farm productivity, off-farm work, and farm size in the Southeast. J. Agric. Appl. Econ. 36:591-603.

Young, A. 1997. Agroforestry for soil management. CABI, Wallingford.

Young, R.A., T. Huntrods, W. Anderson. 1980. Effect of vegetated buffer strips in controlling pollution from feedlot runoff. J.Environ. Qual. 9:483-487.

Yudelman, M., A. Ratta, and D. Nygaard. 1998. Pest management and food production: looking to the future. Food Agric. Environ. Disc. Pap. 25. IFPRI, Washington DC.

Zeddies, J., R.P. Schaab, P. Neuenschwander, H.R. Herren. 2001. Economics of biological control of cassava mealybug in Africa. Agric. Econ. 24:209-219.

Zeller, D., S. Booth and D. Pauly. 2007. Fisheries contributions to GDP: Understanding small-scale fisheries in the Pacific. Marine Resour. Econ. 21:355-374.

Zerega, N.J.C., D. Ragone, T.J. Motley. 2004. Complex origins of breadfruit (Artocarpus altilis, Moraceae): implications for human migrations in Oceania. Am. J. Bot. 91:760-766.

Zerega, N.J.C., D. Ragone, T.J. Motley. 2005. Systematics and species limits of breadfruit (Artocarpus, Moraceae). Syst. Bot. 30:603- 615.

Ziegler, J. 2003. Le droit à l'alimentation, Mille et Une Nuits, Paris.

Zijp, W. 1994. Improving the transfer and use of agricultural information - A guide to information technology. World Bank, Washington.

Zimmermann, H.G., and T. Olckers, 2003. Biological control of alien plant invaders in southern Africa. p. 27-44. In P. Neuenschwander et al. (ed) Biological control in IPM systems in Africa. CABI, Wallingford.

Zingore, S., H.K. Murwira, R.J. Delve, and K.E. Giller. 2007. Soil type, management history and current resource allocation: Three dimensions regulating variability in crop productivity on African smallholder farms. Field Crops Res. 101:296-305.

Zong, R.J., L.L. Morris, M.J. Ahrens, V. Rubatzky, and M.I. Cantwell. 1998. Postharvest physiology, and quality of Gail-lan (Brassica oleracea Alboglabra) and Choi-sum (Brassica rapa subsp. parachinensis). Acta Hort. 467:349-356.

Zwahlen, C., and D.A. Andow. 2005. field evidence for the exposure of ground beetles to Cry1Ab from transgenci corn. Environ. Biosafety Res. 4:113-117.

Zwahlen, C., A. Hilbeck, R. Howald, and W. Nentwig. 2003. Effects of transgenic Bt corn litter on the earthworm Lumbricus terrestris. Mol. Ecol. 12:1077-1086.

4

Outlook on Agricultural Change and Its Drivers

Coordinating Lead Authors
Detlef P. van Vuuren (the Netherlands), Washington O. Ochola
(Kenya), Susan Riha (USA)

Lead Authors
Mario Giampietro (Spain/Italy), Hector Ginzo (Argentina), Thomas
Henrichs (Germany), Sajidin Hussain (Pakistan), Kasper Kok (the
Netherlands), Moraka Makhura (South Africa), Monirul Mirza
(Canada), K.P. Palanisami (India), C.R. Ranganathan (India), Sunil
Ray (India), Claudia Ringler (Germany), Agnes Rola (Philippines),
Henk Westhoek (Netherlands), Monika Zurek (Germany)

Contributing Authors
Patrick Avato (Germany/Italy), Gustavo Best (Mexico), Regina
Birner (Germany), Kenneth Cassman (USA), Charlotte de Fraiture
(Netherlands), Bill Easterling (USA), John Idowu (Nigeria), Prabhu
Pingali (USA), Steve Rose (USA), Phil Thornton (UK), Stan Wood
(UK)

Review Editors
Sandra Brown (UK) and Suat Oksuz (Turkey)

Key Messages

1. Agriculture will need to respond to several key changes in driving forces in the next decades. Key drivers include an increasing global population, changes in dietary and in trade patterns, land competition, increases in agricultural labor productivity, climate change and demands for agriculture to provide ecosystem services. A driver is any natural or human-induced factor that directly or indirectly influences the future of agriculture. Categories of indirect drivers include changes in demographic, economic, sociopolitical, scientific and technological, cultural and religious and biogeophysical change. Important direct drivers include changes in food consumption patterns, natural resource management, land use change, climate change, energy and labor.

2. A range of recent global assessments provides information on plausible future developments regarding agricultural production systems and their driving forces; however, no assessment has explicitly focused on the future role of AKST. Global assessments provided by, among others, the Millennium Ecosystem Assessment, the UN Food and Agriculture Organization AT 2015/2030, and the Comprehensive Assessment of Water Management in Agriculture have explored plausible future developments in agriculture. These assessments have made use of different approaches to address future agricultural changes, and usually are based either on detailed projections or scenario analyses. These approaches do not aim to predict the future—rather they provide a framework to explore key interlinkages between different drivers and resulting changes. In that context, one should also realize that assessment is limited by the ranges of key scenario inputs that are considered. For instance, while crop prices have increased abruptly over the last few years (driven by among other an expansion of biofuel production and rapid increases in food demand) these have not been considered in most existing assessments yet.

3. Existing assessments expect increases in global population over the next 50 years (about 2-3 billion people), ongoing urbanization, and changing life patterns to lead to a strongly increasing demand for food and pressure on the agricultural system. Increasing income implies changes in diet (from carbohydrates to protein based, thus the livestock revolution), and in the manner of food preparation. These changes are expected to affect food consumption patterns and increase demand for non-home based preparation of food. Demand for food is also very likely to be affected by the demographic changes, e.g., the ageing population of many developed countries. Urbanization is projected to continue and to lead to a decline in the percentage of the population depending on agriculture for income, changes in food systems and additional pressures on arable land.

4. Most existing assessments project that international trade in agricultural commodities will increase and often predict developing countries as an aggregate will become net importers. There are many reasons for increasing agricultural trade, such as increasing demand for food, increasing interregional relationships and commodity specialization—possibly facilitated by trade liberalization. As a result, increases in agricultural trade are reported in the majority of existing scenarios, even while many assessments have used contrasting assumptions with respect to ongoing globalization for the scenarios that were developed. Globalization and liberalization will affect countries and groups within countries in different ways. While agricultural trade among developing countries is likely to increase, as a group they may become net importers of agricultural commodities with a possibility of further widening agricultural trade deficits. Conversely, industrialized countries tend to become net beneficiaries of trade arrangements as they are expected to face less pressure to reduce their support for agriculture.

5. Existing assessments highlight the importance of democratization, decentralization and other sociopolitical developments in shaping agricultural policy choices. While these factors are hard to quantify, assumptions underlying different scenarios are partly built around them (e.g., increasing international governance). Several scenarios expect participation of local farmer groups in agricultural policy formulation to increase. Most scenarios also assume that governance effectiveness is also expected to increase over time and this can reduce corruption, perceived to be prevalent in developing economies. But improving states' capacities in governance and effectiveness in policy implementation is a long term process, and impacts are still uncertain. Key options discussed in these assessments include building "soft" infrastructure, such as networks, organizations, cooperatives, in order to produce social capital that may reduce conflicts at all governance levels; facilitating common-pool agricultural resource management; and enhancing access of farmer groups to markets.

6. Existing assessments project a combination of intensification of agricultural production and expansion of cultivated land to meet increasing demands for food, feed, fiber and fuel. A major uncertainty in the scenarios presented in these assessments results from the degree of extensification versus intensification in agricultural production. Roughly 70-80% of the extra production is projected to stem from intensification. Major expansion of agricultural land is projected to take place in sub-Saharan Africa, Latin America and East Asia (excluding China). New developments in AKST are expected to focus on increases in efficiency in the entire food production chain.

7. Existing assessments indicate a major increase in bioenergy production; in the medium term this might lead to a tradeoff between energy security and food security, especially for the poor. Several scenarios, in particular those that emphasize climate policy and energy-security, indicate that agriculture may become an important producer of bioenergy. They, however, also highlight that bioenergy production can become a major use of land, possibly increasing, even in the long-term, food prices and decreasing biodiversity. Bioenergy production based on the conversion of cellulose to fuel ethanol or other hydrocarbons will impact food security and biodiversity less than first generation biofuels. Most assessments also expect higher energy prices. These higher prices (and possible changes in energy subsidies) are likely to encourage the use of more energy-efficient technologies in agricultural production as well as in processing and distributing food.

8. Existing assessments indicate that while agriculture is a major contributor to global environmental change—such as land degradation, nutrient pollution, biodiversity loss, decreasing surface and groundwater availability and climate change—it will also have to adapt to these changes. Assessments indicate an increased demand for water from the non-agricultural sectors, which could further exacerbate water limitations already felt by developing country farmers. Increasing rates of land degradation in many regions may limit the ability of agriculture systems to provide food security. Business-as-usual scenarios indicate a further increase in the already substantial negative contribution of agriculture in global environmental change. However, alternative scenarios highlight that there are also many opportunities for enhancing the positive role of agriculture in providing ecosystem services and minimizing its environmental impacts.

9. Existing assessments expect agriculture to increasingly be affected by global warming and changes in climate variability. For agriculture, changes in seasonal variability and extreme events are even more important than changes in mean temperature and precipitation. Recent studies, such as presented in IPCC's Fourth Assessment Report, indicate that negative impacts on agriculture tend to concentrate in low income regions. In temperate regions impacts could result in net positive yields. Developments in AKST will determine the capacity of food systems to respond to the likely climate changes. Agriculture is also a source of greenhouse gas emissions and therefore agriculture can play a significant role in mitigation policies. In order to play this role, new AKST options for reducing net emissions of carbon dioxide, methane and nitrous oxide need to be developed.

10. Trends observed over the last decade, as described in existing assessments, show that the share of employment in agriculture is declining and this emerging trend is expected to continue. The expected increase in urbanization and international labor migration as well as better working conditions in other sectors will catalyse a labor shift away from agriculture to other sectors. Agricultural labor productivity is expected to increase as a result of improved mechanisation and developments in AKST that are responsive to emerging agricultural and food systems.

11. There is a trend in many regions to reduce investment in traditional agricultural disciplines in favor of emerging research areas such as plant and microbial molecular biology, information technology and nanotechnology. Investment in AKST is increasingly less driven by the needs of agriculture per se, but is a spin-off of other research priorities such as human health and security. These investments mainly occur in industrialized countries and advanced developing countries and the products may not be easily accessible and applicable to least developing countries. To effectively apply advances in the emerging research areas to diverse agriculture systems requires knowledge generated in the traditional agricultural disciplines. The effect of the shift in investments on AKST is not fully explored.

4.1 Driving Forces of Agricultural Change

Changes in agriculture are a result of the developments of a range of underlying driving forces—both direct and indirect—and their many interactions. Previous chapters have described past agricultural changes in general, and change in agricultural knowledge, science and technology (AKST) in particular, in their political, economic, social, cultural, environmental contexts (Global Chapters 1-3). This assessment presents a conceptual framework to structure the analysis of agriculture and development towards reaching sustainability goals. This framework highlights that agriculture, although a central focus of this assessment, needs to be seen as part of a larger societal context and is dynamically linked to many other human activities. Changes in these activities can both directly and indirectly drive change in agriculture (Figure 4-1).

Driving forces or *drivers* are those factors that directly or indirectly induce changes in the agricultural system. A *direct* driver unequivocally influences agricultural production and services and can therefore be identified and measured with differing degrees of accuracy. An *indirect* driver operates more diffusely, often by altering one or more direct drivers, and its influence is established by understanding its effect on a direct driver. When assessing past developments and the prospects for future changes in agricultural systems and the role of AKST, it is crucial to understand the current trends in the driving forces that shape the agricultural system.

This chapter gives an overview of current literature on how agriculture and its drivers may change in the future—thus setting the context for looking specifically at how future agricultural development can be influenced by AKST (Global chapters 5-9). By identifying plausible assumptions on future changes of drivers of agricultural systems, an idea of the most prominent challenges that agriculture might face over the next 50 years emerges, and based on this understanding, key uncertainties can be distilled. Published outlooks, projections and scenario studies are presented here to give an overview of how some of the most important indirect drivers of agricultural changes are expected to unfold, based on current literature and recent international assessments—the main indirect drivers discussed in this chapter are:

1. *Demographic developments*, including changes in population size, age and gender structure, and spatial distribution (4.3.1);
2. *Economic and international trade developments*, including changes in national and per capita income, macroeconomic policies, international trade, and capital flows (4.3.2);
3. *Sociopolitical developments*, including changes in democratization and international dispute mechanisms (4.3.3);
4. *Scientific and technological developments*, including

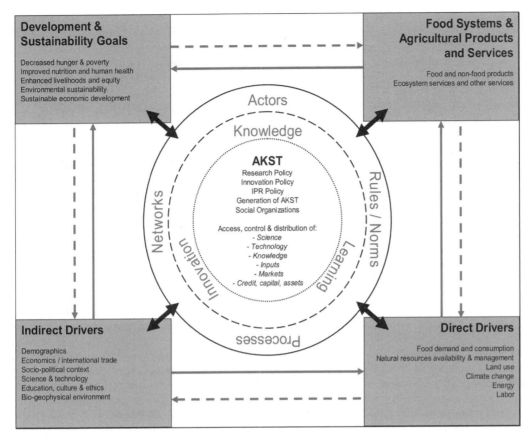

Figure 4-1. *The conceptual framework of IAASTD.*

changes in rates of investments in research and development and the rates of adoption of new technologies, including biotechnologies and information technologies (4.3.4);

5   *Education, cultural and religious developments*, including changes in choices individuals make about what and how much to consume and what they value (4.3.5);

6   *Global environmental change*, including nutrient cycling, water availability, biodiversity and soil quality—all of which are affected by global environmental change (4.3.6).

In addition, a number of direct drivers relevant to agricultural systems are discussed in this chapter and outlooks on how they might unfold over the coming decades presented, again based on published literature—the main direct drivers presented in this chapter are:

1. *Changes in food consumption patterns*, i.e., consumptions levels of crops and meat products (4.4.1);

2. *Land use change*, i.e., land availability as a constraint to agriculture (4.4.2);

3. *Natural resource management*, i.e., the impact of agriculture on natural resources and the constraints of natural resource availability and management on agriculture (4.4.3);

4. *Climate change*, i.e., the impacts of climate change on agriculture (4.4.4);

5. *Energy*, i.e., the relationship between energy and agriculture and the impact of large-scale bioenergy production (4.4.5);

6. *Labor*, i.e., the relationship between agriculture and the demand and supply of labor force (4.4.6).

Looking across the expected developments of the different individual driving forces presented in this chapter gives a first idea of how agricultural systems and their role in providing agricultural products and services might unfold over the coming decades (4.5). The future role of and options for AKST in agricultural development are explored in detail in the remainder of this report. To inform this discussion, this chapter concludes by highlighting some of the key uncertainties for future agricultural changes as well as for AKST as identified in current literature and recent international assessments (Figure 4-2).

4.2 Recent International Assessments

Recent international assessments provide a wealth of information about expected or plausible future developments—either by directly providing an outlook on expected developments in agriculture or by discussing possible developments of key driving forces and pressures that shape the future of agricultural systems. None of the existing global assessments, however, has addressed the role of and future

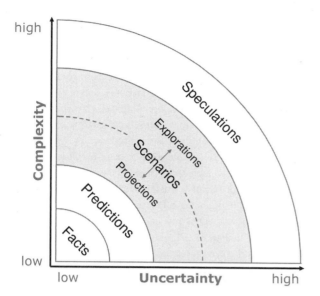

Figure 4-2. *Complexity and uncertainty associated with forward-looking assessments.* Source: Zurek and Henrichs, 2007.

prospects for agricultural science, technology and knowledge (AKST) in much detail. Nevertheless, these assessments still provide important sources of information for discussing future developments underpinning agriculture and the role of AKST. In general, assessments are helpful in exploring possible development pathways for the most important drivers and their interactions. However, assessments are limited by the ranges of key scenario inputs they consider. For example, crop production has abruptly increased in the past few years (driven by demand for biofuels and increased food demand) with prices of some crops doubling. These price increases clearly exceed the range in price projections in existing assessments. Hence, although assessments are helpful in assessing interaction forces in terms of direction of change, they need to be treated carefully as they poorly capture abrupt changes falling outside the range of modeled scenarios.

Assessments of particular relevance for the IAASTD include a wide range of exercises—some of which focus on agriculture, while others address agricultural development within the context of other issues. These assessments can be roughly grouped in two categories: Projection-based studies (which commonly set out to present a probable outlook or best estimate on expected future developments) and exploratory studies (which develop and analyze a range of alternative scenarios to address a broader set of uncertainties) (Box 4-1).

Prominent examples of projection-based assessments of direct relevance to IAASTD include the FAO's World Agriculture 2030 (Bruinsma, 2003) and IFPRI's World Food Outlook (Rosegrant et al., 2001). Important examples of recent international assessments that explore a wider range alternative scenarios include the global scenarios discussed by the Millennium Ecosystem Assessment (MA, 2005a), UNEP's Global Environment Outlooks (UNEP 2002, 2007; RIVM/UNEP 2004), IPCC's Scenarios as used in Third and

Fourth Assessment Reports (IPCC 2001, 2007abc) and the Comprehensive Assessment of Water Management in Agriculture (CA, 2007).

Most of the above mentioned assessments have either focused on informing sustainable development in general or addressed crosscutting environmental issues. Only two global scale assessments (i.e., IFPRI and FAO) have focused on projecting future agricultural production (yet even these have not addressed the full spectrum of agrarian systems and AKST from the perspective of a range of different plausible futures) (Table 4-1). It should be noted that, in addition, a range of national or regional projections of agricultural production and food security exists, but will not be discussed here.

All of the international assessments highlighted above have identified detailed assumptions about the future developments of key driving forces and reflected on a number of underlying uncertainties about how the global context may change over the coming decades. While the number of uncertainties about plausible or potential future developments is vast, a limited number of key uncertainties seem to reappear in several recent international assessments. This is well illustrated by looking at a few "archetypical" scenarios (see Raskin, 2005; van Vuuren, 2007). These scenario archetypes not only share the perspective on key uncertainties about future developments, but as a result also have similar assumptions for different driving forces (Table 4-2, 4-3):

Economic optimism/conventional markets scenarios: i.e., scenarios with a strong focus on market dynamics and economic optimism, usually associated with rapid technology development. Examples of this type of scenario include the A1 scenario (IPCC), the Markets First scenario (UNEP) or the Optimistic scenario (IFPRI).

Reformed Market scenarios: i.e., scenarios that have a similar basic philosophy as the first set, but include some additional policy assumptions aimed at correcting market failures with respect to social development, poverty alleviation or the environment. Examples of this type of scenario includes the Policy First scenario (UNEP) or the Global Orchestration scenario (MA).

Global Sustainable Development scenarios: i.e., scenarios with a strong orientation towards environmental protection and reducing inequality, based on solutions found through global cooperation, lifestyle change and more efficient technologies. Examples of this type of scenario include the B1 scenario (IPCC), the Sustainability First scenario (UNEP) or the TechnoGarden scenario (MA).

Regional Competition/Regional Markets scenarios; i.e., scenarios that assume that regions will focus more on their more immediate interests and regional identity, often assumed to result in rising tensions among regions and/or cultures. Examples of this type of scenario include the A2 scenario (IPCC), the Security First scenario (UNEP), the Pessimistic scenario (IFPRI) or the Order from Strength scenario (MA).

Regional Sustainable development scenarios: i.e., scenarios, that focus on finding regional solutions for current environmental and social problems, usually combining drastic lifestyle changes with decentralization of governance. Examples of this type of scenario include the

---

Box 4-1. Assessing the future: Projections and scenarios.

Recent international forward-looking assessments have made use of a variety of different approaches to explore key linkages between driving forces and assess resulting future developments. The type of approaches employed range from forecasts, to projections, to exploring plausible scenarios. While these approaches differ substantially, they have in common that they set out to assess possible future dynamics and understand related uncertainties and complexity in a structured manner (Figure 4-2).

Projection-based studies commonly present one (or even several) probable outlook on future developments, which is often mainly based on quantitative modelling. Commonly, such projections are based on reducing the level of uncertainty within a forward-looking assessment, either by addressing a limited time horizon or by focusing only on a subset of components of the socioeconomic and ecological system. Projections are particularly useful when they are compared against different variants to highlight expected outcomes of policy assumptions and well-defined options. Projections have also been referred to as future baselines, reference scenarios, business-as-usual scenarios, or best-guess scenarios, which usually hold many existing trends in driving forces constant.

Conversely, forward-looking assessments based on more exploratory approaches aim to widen the scope of discussion about future developments, or identify emerging issues. These types of assessments build on the analysis of alternative projections or scenarios that highlight a range of plausible future developments, based on quantitative and qualitative information. Such scenarios have been described as plausible descriptions of how the future may develop based on a coherent and internally consistent set of assumptions about key driving forces and relationships (MA, 2005a). Multiple projections or scenarios are most useful when strategic goals are discussed and reflected against a range of plausible futures, or when aiming to identify and explore emerging issues.

Determining the forward-looking approach best suited to address a specific issue depends much on the level and type of uncertainty for which one needs to account. Uncertainties have a range of sources, including the level of understanding of the underlying causal relationships (i.e., "what is known about driving forces and their impacts?"), the level of complexity of underpinning system's dynamics (i.e., "how do driving forces, impacts and their respective feedbacks determine future developments?"), the level of determinism of future developments (i.e., "to what degree do past trends and the current situation predetermine future developments?"), the level of uncertainty introduced by the time horizon (i.e., "how far into the future?"), or even surprises and unpredictable future developments (either because these factors occur randomly or because existing knowledge is not able to explore them well enough) (for a discussion of different types of uncertainties and their consequences for methods to explore the future, see Van Vuuren, 2007). As a consequence, when assessments are faced only with relatively low levels of uncertainty with regard to future developments, some approaches allow predicting—or at least—projecting plausible future developments with some degree of confidence. Conversely, where the context of high uncertainty makes predictions or projections meaningless, exploratory scenario approaches can help explore possible developments.

Whereas different approaches to developing and analyzing projections and explorative scenarios exist, some common features have emerged in past assessments (see, for example, EEA, 2002). These include:

1. *Current state,* i.e., a description of the initial situation of the respective system, including an understanding of past developments that lead to the current state;
2. *Driving forces,* i.e., an understanding of what the main actors and factors are, and how their choices influence the dynamics of their system environment;
3. *Step-wise changes,* i.e., a description of how driving forces are assumed to develop and interact, and affect the state of a system along different future time steps;
4. *Image(s) of the future,* i.e., a description of what a plausible future may look like as a consequence of assumptions on drivers, choices and their interactions;
5. *Analysis,* by looking across the scenarios to understand their implications and implied tradeoffs.

B2 scenario (IPCC) or the Adapting Mosaic scenario (MA).

Business-as-usual scenarios: i.e., scenarios that build on the assumption of a continuation of past trends. Thus these scenarios are of a somewhat different quality than the archetypes presented above, as they are not constructed around key uncertainties. Instead business-as-usual scenarios might be described as projections. Examples of this type of scenario include the Reference scenario (IFPRI) or the Agriculture Towards 2030 scenario (FAO).

4.3 Indirect Drivers of Agricultural Change

4.3.1 Demographic drivers

4.3.1.1 Driving forces behind population projections
Past and future demographic trends, such as those for fertility levels, mortality levels and migration, are influenced by varied social, economical, environmental and cultural factors. The "demographic transition" (Thompson, 1929; Notestein, 1945) has proved to be a useful concept to describe these trends in terms of several stages of transition.

Table 4-1. Overview of relevant global scenario studies.

	Main focus	Character of assessment
GSG	Sustainable development	Strong focus on storyline, supported by quantitative accounting system
IPCC-SRES	Greenhouse gas emissions	Modelling supported by simple storylines. Multiple models elaborate the same storyline to map out uncertainties.
IPCC-TAR and AR4	Climate change, causes and impacts	Assessment of available literature and some calculations on the basis of IPCC-SRES
UNEP-GEO3/ GEO4	Global environmental change	Storylines and modelling; modelling on the basis of linked models
MA	Changes in ecosystem services	Storylines and modelling; modelling on the basis of linked models
FAO-AT2020	Changes in agriculture	Single projection, mostly based on expert judgement.
IFPRI	Changes in agriculture	Model-based projections
CA	Water and agriculture	Storylines and modelling; modelling on the basis of linked models

Table 4-2. Key assumptions in different scenario "archetypes".

	Economic optimism	Reformed Markets	Global SD	Regional competition	Regional SD	Business as Usual
Economic development	very rapid	rapid	ranging from slow to rapid	slow	ranging from mid to rapid	medium (globalisation)
Population growth	low	low	low	high	medium	medium
Technology development	rapid	rapid	ranging from mid to rapid	slow	ranging from slow to rapid	medium
Main objectives	economic growth	various goals	global sustainability	security	local sustainability	not defined
Environmental protection	reactive	both reactive and proactive	proactive	reactive	proactive	both reactive and proactive
Trade	globalisation	globalisation	globalisation	trade barriers	trade barriers	weak globalisation
Policies and institutions	policies create open markets	policies reduce market failures	strong global governance	strong national governments	local steering; local actors	mixed

Note: This table summarises key assumptions in very general terms. Where differences within a set of archetypes exist, broad ranges are indicated.

Table 4-3. Recent scenario-based assessments mapped against scenario "archetypes".

	IPCC-SRES	UNEP GEO-3	GSG	MA	IFPRI	FAO
Conventional Markets	A1	Markets First	Conventional worlds		Optimistic scenario	
Reformed Markets		Policies First	*Policy reform*	*Global Orchestration*		
Global SD	*B1* (B1-450)	Sustainability First		TechnoGarden		
Regional Competition	A2	Security First	Barbarisation	Order from Strength	*Pesimistic scenario*	
Regional SD	B2		Great transitions	Adapting Mosaic		
Business as Usual	B2				Reference scenario	FAO AT2020

Note: Italics are used to indicate that scenarios are not completely consistent with the group in which it is categorised.

This notion is underlying most projections. Stage one refers to a preindustrial society where both birth and death rates are high and fluctuate rapidly. In stage two, the death rates decline rapidly due to better economic, environmental and health conditions with increase in life spans and decrease in disease attacks. This stage began in Europe during the Agricultural Revolution of the 18th century. Less developed countries entered this stage in the second half of the last century. In stage three, birth rates decline and population moves towards greater stability due to increases in urbanization, female literacy and improvements in contraceptive measures. During stage four, there are both low birth and death rates. In 43 developed countries (accounting for about 19% of the world population) fertility has dropped to well below the replacement level (two births per woman) leading to a shrinking population.

International migration is also important factor that determines the future population size and composition. However, compared to fertility and mortality, future international migration is more difficult to predict because it is often influenced by short-term changes in social, economic and political developments (see also 4.3.3). It is estimated that during 2005, about 191 million persons (representing 3% of the world population) were migrants (UN, 2005b). Of these, 60% reside in the more developed regions, while the remaining 40% reside in less developed regions. How these numbers will change is important for future regional and national demographic developments. Scenario developers have tried to capture international relationships by describing changes in these factors in the storylines (e.g., IPCC, 2000; MA, 2005a), but there has been little feedback into the demographic assumptions (the MA is an exception).

4.3.1.2 Global population: Current trends and projections
The population projections used in international assessment mostly originate from two important demographic institutions: the United Nations Population Division (UN, 2004) and the probabilistic projections from the International Institute for Applied System Analysis (IIASA) (IIASA, 2001; Lutz et al., 2001, 2004). However, population projections are also provided by the US Bureau of Census (US Census Bureau, 2003) and the World Bank (World Bank, 2004b). The range of the most commonly used projections indicates an increase of the global population from 6.5 billion today to 6.9 to 11.3 billion in 2050. The range of the latest UN scenarios spans a range from 7.7-10.6 billion (with 9.3 billion median) for 2050 (Figure 4-3). These numbers are considerably lower than demographic projections that were published in the past. The most important reason for this is that fertility trends have been revised downwards in response to recent trends. This implies that the realization of these projections is contingent upon ensuring that couples have access to family planning measures and that efforts to arrest the current spread of the HIV/AIDS epidemic are successful in reducing its growth momentum.

Different global assessments have used different population projections (Table 4-4). The Special Report on Emission Scenarios (SRES), the Global Environment Outlook and the Millennium Assessment Working Group each used scenarios that covered a wide range of possible outcomes (all within the IIASA 95% probability interval). A compari-

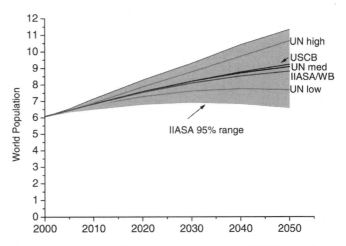

Figure 4-3. *Projected global population (present-2050) according to different scenarios.*

son of these scenarios with the most recent projections for the world shows a downward revision to the medium projections. This implies that older assessments (e.g., the IPCC SRES scenarios) tend to have higher population projections than more recent assessments (the higher population projections of the IPCC-SRES are by now less plausible but not impossible). Among the total set of demographic projections, the Millennium Ecosystem scenarios are most advanced as it used explicit storyline elements to specify trends in total population numbers and also to specify the assumptions for underlying dynamics.

All scenarios indicate that the global population is mainly driven by population increases in less developed regions. In the UN medium scenario by 2050, the population of most developed regions declines by about 1.2 million per year while, in less developed regions, there is an increase of 35 million per year and the least developed countries experience an increase of about 22 million per year. As this trend is basically repeated in all other scenarios, one concludes that least developed countries will be the primary contributors to the increase in population; this situation may aggravate poverty.

Less information is generally found on international migration. Looking at the UN medium scenario, during the 2005-2050 period, the net number of international migrants from less to more developed regions is projected to be 98 million (UN, 2005b) at the rate of 2.2 million per annum. This migration rate is likely to have substantial changes on the age structure, size and composition of the population of the receiving nations and lead to populations of mixed origin. In the MA scenarios, the migration assumptions at the global level have been coupled to storylines, with high migration rates in scenarios assuming further globalization and lower rates in scenarios assuming stronger regional emphasis.

4.3.1.2 Urbanization and ageing of world population
Most assessments do not specify the extent of urbanization; however, the underlying UN scenarios provide information. In the UN medium projection, the world's urban population reached 3.2 billion persons in 2005 and is expected

Table 4-4. Population projections in different assessments.

	Projections
IPCC-SRES	4 scenarios ranging from 8.7-11.4 billion people in 2050
MA	4 scenarios ranging from 8.1-9.6 billion people in 2050
FAO, 2001	UN medium variant
IFPRI	UN medium, high, and low variants, different UN projections
GEO4	4 scenarios, 8.0-9.5 billion people in 2050
OECD outlook	1 scenario; UN-medium (9.1 billion)

to rise to 5 billion persons by 2030 (UN, 2005a). At the same time, the rural population of the world is expected to decline slightly from 3.3 billion in 2005 to 3.2 billion in 2030. The share of urban population of the world, which was nearly 30% in 1950, is expected to reach more than 60% in 2030. The most changes are likely to happen in developing regions. The share of urban population in 2030 is expected to increase gradually to about 82% in developed countries, but increases from only 40% to 57% in less developed regions (UN, 2005b). As urbanization advances, this will have important consequences for agricultural systems; foremost will be feeding urban dwellers according to their changing diets (4.4.1) while also providing sufficient access to other resources such as clean water and other basic services. Urbanization can also increase pressure on crop areas as expansion of cities and industrial areas often occurs on good agricultural land. Finally, urbanization is likely to affect access to labor (see 4.4.6).

An important dominant demographic trend is the aging of population. During the 20th century, the proportion of older persons (60 years or above) continued to rise; this trend is expected to continue well into this century. For example, proportion of older persons was 8% in 1950, 10% in 2005 and is projected to reach about 22% by the middle of the current century (UN medium scenario). This implies that by 2050, the world is expected to have some 2 billion older persons; a tripling of the number in that age group within a span of 50 years. These trends in the UN medium scenarios are similar to those in other scenarios. The pace of population aging is much faster in the developing countries than in the developed ones. Hence, developing countries will have less time to adjust to the adverse impacts of aging. Moreover, population aging in developing countries is taking place at much lower level of socioeconomic development than that which has occurred in industrialized countries.

Trends in life expectancy are projected to continue in all scenarios. Using the UN medium scenario as an example: global average life expectancy (47 years in 1950-55) increased to to 65 years in 2000-2005 and is projected to increase to 75 years by mid-century. The UN projections assume that life expectancy may not increase beyond 85 years; some demographers believe this age represents an intrinsic limit to the human life span. Another important demo-

graphic trend is the decline in mortality in most countries, (including infant and under 5 years old mortality), because of better public hygiene and education, improved nutrition and advances in medical science (Lee, 2003). However, the mortality rate has increased in countries with deteriorating social and economic conditions and in those affected by HIV/AIDS epidemic.

4.3.1.4 Demographic change and its impact on agriculture
Although population projections have slowed from past estimates, a large absolute increase in population raises serious concerns about the capability of the agricultural production and associated natural resource base (Pimentel and Wilson, 2004). A key question is whether agriculture can feed the expanding global population in the ensuing decades. A key issue involves the effects of increased urbanization on transport of agricultural products into urban areas, with associated development of infrastructure and markets. Direct consequences include a more distant relationship between consumer and producer and an increasingly important role for actors involved in food distribution and markets (e.g., supermarkets).

4.3.2 Economics and international trade

4.3.2.1 Future trends and scenarios of economic growth and the agricultural economy
Economic change is a primary driver for future agricultural systems. The most employed indicator for economic change, GDP per capita, is used as a driver in most scenario studies.

Historically, GDP has grown substantially. Between 1950 and 2000 world GDP grew by 3.85% annually resulting in a per capita income growth rate of 2.09% (Maddison, 2003). Global GDP growth decelerated over time from 2.1% per year in the 1970s, to 1.3% per year in the 1980s, and to 1.0% annually in the 1990s (Nayyar, 2006), but this may have been a consequence of particular events such as the transition process in countries of the Former Soviet Union. Projections of future economic growth vary considerably. In many environmental assessments, GDP projections are not an outcome—but an assumption (e.g., in SRES, GEO and the MA). All studies expect economic growth to continue. The World Bank short-term outlook, for instance, uses values that are comparable to historic rates (around 2%). The four scenarios in SRES show a wide range of global annual economic growth rates from 1 to 3.1% (van Vuuren and O'Neill, 2006) (Figure 4-4 and 4-5). Many other studies use a somewhat smaller range.

All assumptions are that growth in industrialized countries will be slower than those in developing economies (Table 4-5). Among the developing regions, Asia, particularly East Asia will continue to have higher growth rates. Different outlooks exist with respect to sub-Saharan Africa. Most recent assessments assume that institutional barriers will result in slower (though positive) economic growth than in other developing regions. Other current work (OECD, 2006a), however, projects Africa to grow faster than Latin America, and slowing growth to 2030, except for MENA and sub-Saharan Africa. For the Asia region, alternative per capita income growth projections (GDP per capita) for the

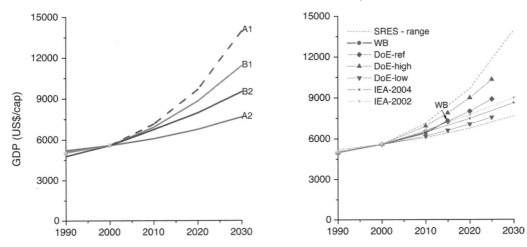

Figure 4-4. *Comparison of global GDP growth in the SRES scenarios and more recent projections.* Source: see note.

Note: SRES = Nakicenovic et al., 2000 using Scenarios A1, B1, B2, and A2; WB = WorldBank, 2004b; DoE = US.DoE, 2004; IEA = IEA, 2002, 2004.

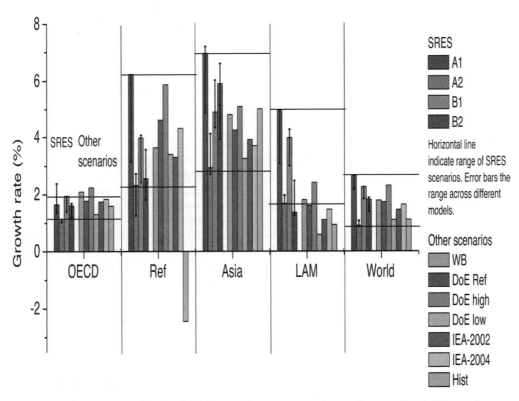

Figure 4-5. *Comparison of regional GDP annual average growth rates between 2000-2015 in the SRES scenarios and more recent studies.* Source: See note.

Note: WB = World Bank, 2004b; DoE = USDoE, 2004; IEA = IEA, 2002, 2004. Hist = Historic data from World Bank, 2003.

The horizontal lines in the figure indicate the range of growth rates set out by the SRES marker scenarios. The vertical lines showing uncertainty bars for the SRES scenarios indicate the range of different outcomes of SRES scenarios within the same family (while the bars indicate the growth rates of the Marker scenarios). The historical rate represents the 1990-2000 period.

four SRES scenarios, the four Millennium Ecosystem Assessment Scenarios, and the four GEO-3 scenarios (Westhoek et al., 2005) also show lower growth projections for the period 2025-2050 compared to 2000-2025.

4.3.2.2 Implications of income growth for agriculture

Changes in per capita income growth do affect the mix of economic activities, and this affects agriculture in a significant way, with important implications for access to labor. Along with economic development, demand for food quantity in countries initially increases and then stabilizes; food expenditures become more diverse (see also 4.4.1). At the same time, the demand for nonagricultural goods and services increases more than proportionally. A general tendency observed in the past (and assumed in most assessments) is that the economy responds to this trend by shifting resources out of agriculture; and the share of agricultural output in total economic activity declines (Figure 4-6). While high-income countries typically produce more output per hectare as a result of higher inputs, industrial and service output grows much faster so the relative contribution of agriculture declines. Technological change further replaces most of the labor force in agriculture. In this context, it is important to note that many factors discussed in elsewhere in this chapter are closely related. For instance, demographic factors (graying of the population, labor supply) (4.3.1) may have important implications for growth projections. Similarly, developments in energy supply in the coming decades (e.g., oil scarcity) could have important consequences for economic projections. High economic growth rates are likely to have an upward pressure on energy prices and, in turn, might increase the demand for alternative fuels such as bioenergy.

4.3.2.3. International trade and agriculture

Increasing trade patterns in the future. Trade is as an important distinguishing factor for scenarios in several assessments: the MA, GEO and SRES have scenarios with and without trade liberalization. Nevertheless, in terms of actual trade flows the scenarios all show an increasing trade, even in scenarios without trade liberalization (Bruinsma, 2003; MA, 2005a). An important reason for this increase is that population growth and the development of agricultural production is expected to occur unevenly. The largest increase is expected for total trade in grain and livestock products (see also Table 4-6). Some region specific patterns are expected also, e.g., the OECD region is likely to respond to increasing cereal demands in Asia and MENA. The very rapid yield and area increases projected in sub-Saharan Africa could turn the region from net cereal importer at present to net grain exporter by 2050. Net trade in meat products increases 674%. Net exports will increase in Latin America, by 23000 Gg, while the OECD region and Asia are projected to increase net imports by 15000 and 10000 Gg, respectively (MA, 2005a). Asian demand for primary commodities, such as natural rubber and soybean, is likely to remain strong, boosting the earnings of the exporters of these products. China will become the world's largest importer of agricultural commodities in terms of value by 2020, with imports increasing from US$5 billion in 1997 to US$22 billion in 2020 (UNCTAD, 2005).

At a more aggregated scale important trends can be noted (see Rosegrant et al., 2001). The overall trade surplus in agricultural commodities for developing countries may dissappear completely and by 2030 they could, as a group, become net importers of agricultural commodities, especially of temperate zone commodities (FAO, 2003). At the same time, the agricultural trade deficit of the group of least developed countries could quadruple by 2030 primarily due to the fact that industrialized countries tend to be the major beneficiaries of trade arrangements (FAO, 2006b). Many countries have faced much less pressure to reduce support for their agricultural sector primarily due to the fact that commitments to liberalize were based on historically high levels of support and protection. The future of this driver hinges on the outcomes of the Doha Round of trade talks about agricultural support and market access. Export subsi-

Table 4-5. **Per capita income growth projections, per year various assessment results.**

Region	Historic	MA		FAO		OECD	
	1971-2000	1995-2020	2020-2050	2000-2030	2030-2050	2010-2020	2020-2030
Former Soviet Union	0.4	2.24-3.5	2.64-4.91	4.5	4.3	3.7	3.4
Latin America	1.2	1.78-2.8	2.29-4.28	2.3	3.1	2.9	2.8
Middle East/ North Africa	0.7	1.51-1.96	1.75-3.42	2.4	3.1	3.6	3.9
OECD	2.1	2-2.45	1.31-1.93	2.2	2.4	2.2	2.0
Asia	5	3.22-5.06	2.43-5.28	5	4.95	4.76	4.1
Sub-Saharan Africa	-0.4	1.02-1.69	2.12-3.97	1.6	2.8	4.2	4.4
World	1.4	1.39-2.38	1.04-3	2.1	2.7	2.7	2.5

Source: MA, 2005; FAO, 2006b; OECD, 2006a.

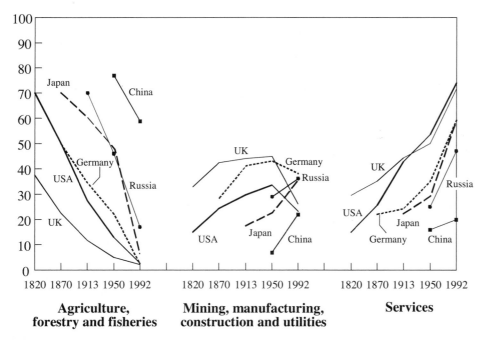

Figure 4-6. *Sectoral distribution (1820-1992) for selected countries.* Source: MA, 2005a.

dies remain substantial with EU accounting for bulk of direct export subsidies (FAO, 2006b). Removal of price support and other subsidies to 2030 could result in moderately increasing international prices, while prices would fall substantially in countries with high levels of protection (FAO, 2006b).

The impact of trade on developing countries is very controversial. Most of the conclusions that imply developing countries stand to lose in future trade arrangements are premised on the fact that developing countries will still be dependent on industrialized countries for trade. There is extensive evidence of emerging South-South trade relations (UNCTAD, 2005). These relations could imply that even if industrialized countries form a substantial component of demand for exports from developing countries, the limited outlook for growth of industrialized countries puts them behind the emerging strong demand of Asian countries, particularly China and India. Developing countries that export non-oil primary commodities benefit from increased demand and rising prices for their exports despite low commodity prices.

Terms of trade have significant impacts on affordability of food imports and food security for countries with a large share of agricultural trade. Many of the lesser developed countries have faced deteriorating terms of trade since the 1980s. Agricultural terms of trade fell by half from a peak in 1986 to a low in 2001 (Figure 4-7). Because many of these countries depend on commodity exports to finance food imports, a decline in terms of trade for agriculture threatens food security (FAO, 2004b). The region that has suffered most from declining terms of trade is sub-Saharan Africa. Since the 1970s, the deterioration of agriculture terms of trade in that region has led to a substantial reduction in the purchasing power of commodity exports. In addition to declining terms of trade, fluctuations and trends in prices negatively affected African agriculture. The declining and

fluctuating export prices and increasing import prices compound socioeconomic difficulties in the region, as well as agricultural patterns (Alemayehu, 2000). Short-term outlooks such as those from the World Bank project this situation to persist (e.g., UNCTAD, 2005).

A number of model results recently reviewed (Beierle and Diaz-Bonilla, 2003) whether trade liberalization (in the form of reduced protection and export subsidies and lowered import restrictions) would benefit small-scale farmers and others in poverty in developing countries. Several key findings from their review and other assessments are:

- Most models demonstrate negative impacts of current industrialized country (OECD) trade protection policies but positive impacts from developed country liberalization on developing country welfare, agricultural production and incomes, and food security.
- Impacts vary by country, commodity, and sector, and for regions within countries.
- OECD market access restrictions harm developing countries, although effects of production and income-support subsidies are more ambiguous.
- Developing countries tend to gain more from liberalization of their own policies than from reforms by the OECD. Consumers in developing countries benefit widely from these reforms.
- Model results differ on the basis of assumptions such as the scope of commodity coverage, mobility of resources among alternative crops and between farm and non-farm employment, availability of underutilized labor, and static versus dynamic analysis.
- Multilateral liberalization reduces the benefits derived from preferential trade agreements, but these losses are relatively small compared to the overall gains from the broader reforms.

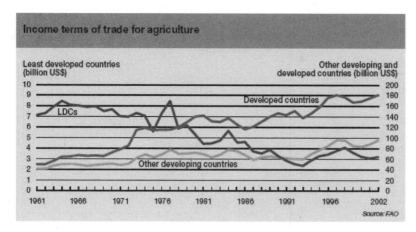

Figure 4-7. *Income terms of trade for agriculture.* Source: FAO, 2004b.

- Most models have not had sufficient resolution to analyze the impacts of reforms on small-scale and subsistence farmers, and other poor households (Gulati and Narayanan, 2002; Tokarick, 2002; Beierle and Diaz-Bonilla, 2003; Hertel and Winters, 2005).

4.3.2.4 Agricultural investments
The investment requirements to achieve projected scenarios are seldom computed. Based on an assessment of five key drivers for agricultural development (agricultural research, irrigation, rural roads, education, and clean water) investment requirements to generate modest levels of agricultural production growth have been estimated at US$579 billion during 1997-2020 (Rosegrant et al., 2001) (Table 4-7). According to this study, levels of investment required will vary from region to region, e.g., South Asia and Latin America would require the highest levels. Sub-Saharan Africa's investment requirements would total US$107 billion during 1997-2020 and would represent 19% of 1997 government spending on an annual basis. At the sector level, irrigation would account for 30% of the total investments, public agricultural research and rural roads for another 21% each, with educations' share the lowest at 13 percent.

Foreign Direct Investment (FDI) is an important source of capital flows for development. FDI for agriculture is generally lower than that of other sectors. In countries such as Vietnam, FDI in agriculture and rural areas is declining similar to other regions; IBRD/IDA commitments to the agricultural sector are also declining (Binh, 2004).

4.3.2.5 Implications for AKST
Projected income growth is likely to lead to shifts in food demand patterns, e.g., from cereals to meat consumption (see 4.4.1). It can be hypothesized that with this shift, sustainable technologies for intensive livestock production, and policy safeguards for meat safety (i.e., meat inspection services), and environmental regulation will be needed.

Higher incomes will also mean more expensive farm labor; this can be addressed through increased mechanization, and clustering of small farms, whenever applicable, for more efficient management (assuming sufficient access to capital and/or demand for labour in other economic sectors). Due to higher opportunity costs of time, supermarkets will be in high demand. In addition, food quality assurance will need to be met through certification, labeling, and appropriate packaging. These conditions are currently deficient in most developing countries.

Table 4-7. **Investment in food security under the baseline scenario, 1997-2020.**

Region/Country	Irrigation	Rural Roads	Education	Clean Water	National Agricultural Research	Total Investments
Billions of US Dollars						
Latin America	44.8	36.7	12.1	9.8	37	140.4
West Asia/North Africa	17.9	7.3	21.5	8.5	25.3	80.5
Sub-Saharan Africa	28.1	37.9	15.7	17.3	8	106.9
South Asia	61.3	27.4	14.5	27	18	148.2
India	42.5	23.5	10.5	18.4	15.6	110.5
Southeast Asia	18.6	3.9	6.8	9.4	14.1	52.6
China	3.2	6.8	2.4	14.4	14.6	41.4
Developing countries	174.6	120.3	75.9	86.5	121.7	578.9

Source: IFPRI IMPACT Projections, June, 2001.

With increased trade liberalization, developing countries would need science based regulatory frameworks for sanitary and phytosanitary issues and institutional market infrastructures to strengthen market information systems, including grades and standards. Human and organizational capacity to effectively implement the international standards will also need to be developed.

The projected decrease in FDI would create a need for more domestic investment in agriculture. Developing country decision makers will need tools to help in agricultural strategic planning and budget prioritization.

4.3.3 Sociopolitical drivers of alternative futures in agriculture and AKST

4.3.3.1 Types of sociopolitical drivers
The direct and the indirect drivers of AKST (Figure 4-1) are influenced by public policies, which are the outcome of political processes. Therefore, the sociopolitical factors that influence public policy making and the implementation of public policies are important drivers of alternative futures in AKST. These factors include (1) the political system (type of political regime, political culture, ideology, political stability); (2) public administration and its effectiveness in policy implementation; (3) the structure of society (social stratification, social values, ethnicity; social conflicts); and (4) extent of regional and global collaboration. For an assessment of alternative scenarios for the future of agriculture and AKST, it is necessary to assess major trends of change in political and administrative systems and in society, and to evaluate how these changes will influence the choice of public policies and sociopolitical events. For obvious reasons, the possibilities to project sociopolitical change are more limited than the possibilities to project trends in other drivers such as, for example, demography.

4.3.3.2 Change in political systems and public policy choices
Major political trends after World War II were an increase in authoritarian regimes (autocracies) until the early 1970s, followed by a rapid decline (see Figure 4-8). Accordingly, the number of democracies has increased rapidly since the early 1970s. The number of "anocratic" states, intermediate states where elites maintain power despite the existence of democratic procedures, has increased, too. A conventional thought is that these trends continue into the future (this underlies a large number of long-term scenarios including, for instance, most reference scenarios—but also 3 out of 4 scenarios in the Millennium Ecosystem assessment (Global Orchestration, Technogarden and Adapting Mosaic)). There has also been a trend towards increasing citizen participation in the formulation of development strategies. Sectoral policy documents such as agricultural sector strategies are increasingly developed with broad stakeholder consultation. Another important trend in political systems throughout the developing world is democratic decentralization, i.e., the transfer of political authority to lower levels of government.

How do these trends of democratization, decentralization and participatory policy making influence the choice of public policies? The relationship between democracy and economic development is complex (Bardhan, 1999), but

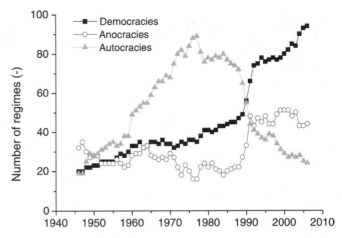

Figure 4-8. *Global regimes by type (1946-2006)*. Source: Based on Marshal, 2006.

changes in political systems that allow citizens to participate more broadly in political decision making will reduce "urban bias" (Lipton, 1977) and increase the attention given to agriculture, because this sector employs a large part of the population in developing countries. However, there is no empirical evidence of such an effect (see Fan and Rao, 2003). In democratic political regimes, agricultural interest groups are often able to exercise political pressure to obtain subsidies and protection, which typically benefit larger-scale more than small-scale farmers, whereas it is more challenging to create political pressure for investments in public goods, such as agricultural research (cf. Lopez, 2005). The influence of political regime type on other agricultural policies, such as access to land, agroenvironmental policies, and regulation against unfair competition, is not straightforward.

While the evidence on the link between political regime type and general agricultural policies is inconclusive, evidence suggests that democratization will lead to a stronger focus on food security. None of the great famines occurred in a democracy (Sen, 1981). Moreover, famines can be avoided by fairly elementary government actions, because they are rarely caused by absolute shortages in food supply (Sen, 1981). Subsequent work showed that the freedom of press does, in fact, play an important role in avoiding food crises and famines (cf. Sen and Drèze, 1989). In this light, an increase in number of democracies appears to imply that increased accountability would lead to less food crises.

For authoritarian regimes, political ideology, development orientation and the time horizon of the regime influence the commitment to agriculture and the choice of agricultural policies. Indonesia (under Suharto) and China are examples of authoritarian regimes that invested heavily in agriculture and rural development, though with limited recognition of the need for environmental sustainability. In Africa, there is evidence that military leadership has had a negative influence on public spending for agriculture (Palaniswamy and Birner, 2006). The trend towards democratization and citizen participation in policy making, which is projected to continue, has ambiguous implications with regard to alternative future scenarios for agricultural and food systems and AKST. Hence, it will be necessary to work with different assumptions when formulating scenarios.

For Asia, the political commitment to the agricultural sector is projected to continue as indicated by a relatively high budget share to this sector. In Africa, one can also observe an increased emphasis on agriculture; e.g., one indication is the Alliance for a Green Revolution in Africa, led by the former Secretary General of the United Nations, Kofi Annan. African Heads of State, in their Maputo Declaration, made a commitment to allocate at least 10% of their national budgetary resources to agricultural development (African Union, 2003). This goal is also supported by the Comprehensive Africa Agriculture Development Program (CAADP), which is high on the agenda of the New Partnership for Africa's Development (NEPAD). However, it still remains to be seen whether these commitments will indeed translate into increased investment on agriculture in Africa. In formulating scenarios, one also has to take into account regional and global trade agreements, which limit the choices that countries can make regarding their agricultural policies (see 4.2.1).

4.3.3.3 State capacity for policy implementation
To assess the impact of public policies on the development of AKST, it is necessary to consider government effectiveness, e.g., the quality of the bureaucracy, the competence and independence of the civil service, and the credibility of the government's commitment to its policies. Control of corruption is also important for the effective implementation of agricultural policies, especially in agricultural infrastructure provision, such as irrigation. Since agricultural development depends on the ability of the state to overcome market failures, which are prevalent especially in early phases of agricultural development (Dorward et al., 2004), changes in state capacity are an important sociopolitical driver. Governance problems tend to be greater in low-income countries; they are particularly prevalent in the Central African region, in spite of some recent improvements (Kaufmann et al., 2006). The state capacity for policy implementation can be improved by governance reforms, including public sector management reforms, the use of e-government, outsourcing and public-private partnerships, all of which are important in the agricultural sector. Improving state capacity is a long-term process, however, lasting often for several decades before a real impact can be achieved (Levi, 2004). Hence, for short- and medium-term scenarios, it will be useful to take into account the current variation in state capacity, as measured by governance indicator data sets (Kaufmann et al., 2006).

4.3.3.4 Social factors that shape the future of agriculture
The social factors that shape the future of agriculture include conflicts, changes in social values and social structure (related to social stratification, gender roles and ethnicity). In view of the complexity and country specificity of social factors, it is difficult to identify general global trends that can be used to formulate scenarios. There are, however, some projections on global trends; e.g., economic development gives rise to cultural changes that make individual autonomy, gender equality and democracy increasingly likely (Inglehart and Welzel, 2005).

These findings correspond to the "End of History." In this view, liberal democracy and Western values comprise the only alternative for nations in the post-Cold War world (Fukuyama, 1993). This view forms the basis of several scenarios in global assessments such as the A1b scenario in IPCC's SRES scenarios and (to some degree) the Global Orchestration scenario (MA). This view has been challenged by the controversial "Clash of Civilizations" theory (Huntington, 1996); i.e., that people's cultural and religious identity rather than political ideologies or economic factors will be the primary source of conflict in the post-Cold War world, especially between Islamic and non-Islamic civilizations. Again, this view forms the basis of several scenarios (A2 in IPCC SRES; Order from Strength in MA). It should be noted, however, there is no evidence for an increase in the frequency of intercivilizational conflicts in the post-Cold War period (e.g., Tusicisny, 2004). With regard to agricultural development, internal conflicts and civil wars matter as much, or even more, than international conflicts. The number of wars reached a peak of 187 in the mid-1980s, but was reduced by half in 2000 (Marshall et al., 2003). Most of these wars were internal conflicts, and most of them occurred in poor countries.

Instability can be defined as the incidence of revolutionary and ethnic wars, adverse regime changes, genocides or politicides (government targeting of specific communal or political groups for destruction) (Goldstone et al., 2005). The percentage of countries experiencing periods of instability reached a peak of almost 30% in the early 1990s (Figure 4-9). A predictive model with four variables (regime type, infant mortality, a "bad neighborhood" indicator—four or more bordering states in armed civil or ethnic conflict—and the presence or absence of state-led discrimination) showed that ethnically factionalized nascent democracies, without fully open access to political office and without institutionalized political competition, are particularly prone to wars and conflicts, even with favorable economic conditions (Goldstone et al., 2005).

The implications of wars and armed conflicts for agricultural development are far-reaching: crop and livestock production are reduced or abandoned due to insecurity, lack of labor, environmental degradation and destruction of infrastructure. Wars and conflicts may affect AKST in different ways, for example, by reducing the availability of public funds for agricultural research and extension, and by a loss of local knowledge due to displacement of agricultural producers.

Another important social factor shaping the future of agriculture is the capacity of communities and societies to cooperate, also referred to as social capital (see, e.g., Putnam, 1993). In agriculture, especially in small-scale agriculture, producer organizations play an important role in addressing market failures while avoiding government failures. They provide political voice to agricultural producers, help them to hold government organizations accountable, and engage in the provision of agricultural services. Their role has been increasing in recent years due to investments in their capacity and conducive factors such as democratization (Rondot and Collion, 2001).

4.3.3.5 Regional and global collaboration
The future of AKST will also depend on the development of regional and global political organizations; e.g., regional

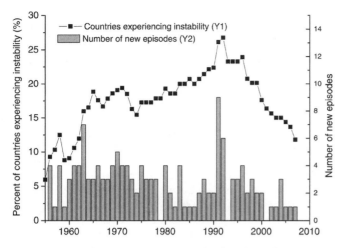

Figure 4-9. *Incidence and prevalence of political instability worldwide, 1955-2006.* Source: Update of Goldstone et al., 2005.

organizations, such as NEPAD or ECOWAS (Economic Community of West African States) play an increasing role in shaping agricultural policies. Global organizations such as the United Nations Food and Agriculture Organization (FAO), the World Organization for Animal Health, and the Consultative Group on International Agricultural Research (CGIAR) are important players in AKST, as are treaties, e.g., the International Treaty on Plant Genetic Resources, and issue-specific global networks, such as the Global Fund for Control of Highly Pathogenic Avian Influenza. Thus, the ability of the international community to cooperate to provide global public goods for agriculture is an important sociopolitical driver.

4.3.3.6 Sociopolitical factors in existing assessments
Predictive models are available and can be used for formulating future scenarios. Thus far, sociopolitical drivers have been included in assessments as scenario storylines. The IPCC SRES report used the degree of globalization versus regionalization as an important distinction between scenarios. The other axis is the extent to which a scenario focuses on social and environmental objectives versus economic objectives. This basic idea (that relates to several of the factors discussed above) has been further specified in many other scenario studies (MA, 2005a; Westhoek et al., 2005).

4.3.4 Science and technology
Scientific breakthroughs and technological innovations in the last century fueled substantial gains in agricultural productivity in many countries (see FAO statistics). These innovations not only helped meet the world's gross food and fiber needs but, along with new transport and storage technologies, transformed much of northern agriculture from subsistence to commercial market-oriented farming, thus offering more opportunities for participation in global markets. Technology is considered a core driver of future changes, affecting economic growth, social and environmental change and agriculture productivity.

4.3.4.1 Previous Assessments
The Intergovernmental Panel on Climate Change (IPCC, 2000) discusses in detail the approaches and problems encountered in predicting the impact of science and technology on global change. The IPCC identifies five commonalities in the innovation process (Box 4-2). Although the IPCC is particularly concerned with how the innovation process may affect the energy sector, these five commonalities could reasonably be applied to the agricultural sector.

There is widespread agreement that the innovation process is complex and difficult to predict. There is still no agreement on what assumptions to make regarding (1) how government and investment in industrial research and development (R&D) will impact the innovation process, (2) the motivation of producers of new technologies and (3) the role of consumers. Therefore, models and assessments mostly describe technology change in much more aggregated parameters such as exogenously assumed yield changes or learning-by-doing functions (e.g., for production costs of energy technologies). Different assumptions concerning these technology parameters have been used as important drivers to contrast the IPCC-SRES scenarios, i.e., technology change was assumed to be lower in scenarios with less globalization (resulting in lower yield improvement and less rapid economic growth). IPCC-SRES scenarios have also been built around the direction of technology change (IPCC, 2000).

The Millennium Ecosystem Assessment (MA, 2005a) recognizes science and technology as a major driver of change in ecosystems and their services. The MA identified three key concerns regarding technological trends. First, the institutions needed to foster the research and development process are not yet well established in much of the developing world. Secondly, the rate of spread of new technologies may be outpacing the time frame required to identify and address their negative consequences. Lastly, technologies can produce unexpected consequences that might lead to disruptions of ecosystems affecting large numbers of people. Like in the IPCC-SRES scenarios, the rate of technology change was used to contrast the different MA scenarios (Figure 4-10).

Box 4-2. Five commonalities in the innovation process.

1. The process is fundamentally uncertain: outcomes cannot be predicted.
2. Innovation draws on underlying scientific or other knowledge.
3. Some kind of search or experimentation process is usually involved.
4. Many innovations depend on the exploitation of "tacit knowledge" obtained through "learning by doing" or experience.
5. Technological change is a cumulative process and depends on the history of the individual or organization involved.

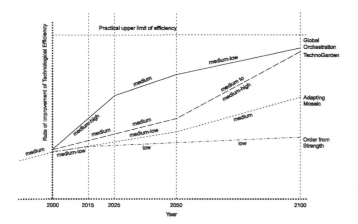

Figure 4-10. *Global trends of technological efficiencies in MA scenarios.* Source: Alcamo et al., 2005.

Note: Technological efficiency refers, for example, to the conversion efficiency of power plants, or the yield of all crops per hectare.

The Global Environment Outlook report (GEO 3) (UNEP, 2002), focused on the distribution of benefits and costs of technological developments in the future. To the extent that technological innovation is increasingly undertaken by the private sector and driven by profit, benefits are seen as primarily accruing to those who are most powerful in the marketplace. The assessment suggests that cautious government policies and empowerment of consumers may act as disincentives to technological innovation by the private sector. However, such an approach may also result in more equitable distribution of benefits. The quantitative assumptions resemble those of the IPCC-SRES scenarios.

4.3.4.2 Important trends for the future

Trends in investment. Although the innovation process is complex, investment in research and development is central. Typically, the ratio of R&D expenditures to GDP is an indicator of the intensity of R&D activities over time and in relation to other economies. OECD nations typically spend 2.26% of their GDP in overall R&D while Nicaragua can afford to spend only 0.07%.

While there is an increase in absolute R&D expenditures in agriculture, there is a concern that investment in agricultural R&D is declining in North America, Western Europe and East Asia relative to overall spending on research. While existing scenarios are not explicit on agricultural R&D trends, a plausible trend could be a further increase in absolute numbers but a decreasing ratio compared to overall GDP leading to concerns about the ability to use AKST as a response to challenges. However, several trends in R&D investment could mitigate these concerns. First, there appears to be a trend toward increasing globalization of R&D, driven by multinational corporations seeking to take advantage of knowledge of local and regional markets, technical expertise, fewer restrictions on intellectual property, and lower costs for R&D in non-OECD countries. Secondly, many countries with large public sector R&D investments continue to promote international linkages, and this emphasis is likely to become more significant as globalization continues (OECD, 2006b). Thirdly, and perhaps most importantly, China, with a very large, poor, rural population, now ranks second in R&D expenditures (Figure 4-11). It is plausible that China will shortly become the major center for agricultural research, particularly research relevant to poor rural areas. (A more detailed discussion of investment in agricultural R&D is presented in chapter 8.)

Trends in performing sector. In the last century, key innovators were national agricultural research systems, including universities, agricultural field stations, agricultural input companies, and extension services (Ruttan, 2001). Two international organizations, CIMMYT in Mexico and IRRI in the Philippines, contributed significantly to the advancement of the Green Revolution, and were mainly funded by the public sector in the first half of the last century. In contrast, in the United States, the private sector has always played a central role in the development of agricultural equipment and the performance of agricultural research and development. Private-sector research grew substantially in the last decades of the 20th century as legal rights were

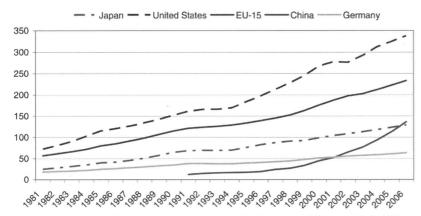

Note: (1) Figures for 2005 and 2006 are projected on the assumption that growth of R&D expenditure in 2005 and 2006 will be same as average growth over 2000-2004.
Source: OECD, Main Science and Technology Indicators, 2006-I.

Figure 4-11. *Gross domestic expenditure on R&D (billion $).* Source: OECD, 2006b.

obtained for genetic modifications (Huffman and Evenson, 1993). Given these trends, it is safe to conclude that the future of biotechnology as applied to agriculture will likely be driven by demands for specific traits to enhance production and add value. Value added output traits with consumer-oriented benefits, such as improved nutritional and other health-related characteristics, will attract the support of the private sector because these traits will turn many agricultural commodities into premium priced and quasi-specialty products (Shimoda, 1998). Again, while existing scenarios do not explicitly relate to these issues, new scenario development focusing on the agricultural sector could provide a richer assessment basis with the inclusion of these trends.

Internationally competitive biotechnology research and development systems are expected to emerge, accelerating the pace of biotechnology research. Although the investment in biotechnology is on the rise in various countries, there are scientific, political and economic uncertainties associated with it. Due to potential environmental and health risks associated with GM products, the EU has imposed stringent regulatory measures on foods containing or produced from GMOs (Meijer and Stewart, 2004). On the other hand, the production and consumption of GMOs has been widespread in other countries, such as the US and Canada. The future of agriculture will depend on how the debate on GMOs unfolds. A directly related factor, which is important for future GMO use as well, are societal choices with respect to high-input agriculture in general (Giampietro, 2007).

Another noticeable trend that could influence future agricultural development is the increase in unregulated trade in agricultural inputs and outputs in many countries (see 4.3.2). This process has created a new set of incentives for investment in private research and has altered the structure of the public/private agricultural research endeavor, particularly with respect to crop improvement (Falcon and Fowler, 2002; Pingali and Traxler, 2002).

Finally, since the World Summit on Sustainable Development (WSSD) held in Johannesburg in 2002, more research has gone into local and traditional knowledge systems. Nongovernmental organizations, research bodies, funding agencies, and the United Nations system are lending financial and technical support to locally prioritized research and development efforts that value, investigate, and protect the local and traditional knowledge systems.

Trends in adoption. The full benefits of scientific breakthroughs will not be realized without the dissemination and adoption of new technologies. There is a great deal of unused scientific knowledge and technologies "on the shelf" for immediate application, particularly for developing country agriculture. In each country, the successful local development of technologies or the transfer and adaptation of innovations from others will depend on incentives and barriers faced by investors and producers. Poor farmers can adapt new technology if small risks are associated with it; with larger risks, they may need guarantees from the state or insurance providers. Many existing on-the-shelf technologies could be adopted if the perceived risks of using them were significantly lowered or if some of the hindrances to adoption, such as missing input supply

chains, poor or nonexistent marketing channels for surplus production, and little or no access to credit or new knowledge were reduced or eliminated. Adoption of GMO material by small farmers may be limited by high costs of planting material, restrictions on the replanting of seeds, and uncertainty of market acceptability. If these concerns are not addressed, much biotechnology will likely not be adopted by poor farmers. Existing scenarios assuming high rates of technology change imply high rates of adoption.

4.3.5 Education, culture and ethics

4.3.5.1 Education
Many international organizations have addressed the issue of poverty alleviation through the diffusion and improvement of rural basic to tertiary education with global, regional or country-specific programs (see CGIAR, 2004; FAO, 2006a; UNESCO, 2006). There also are programs implemented by organizations from developed countries (e.g., Noragric[1]) to help individual developing countries identify and address problems with their rural education systems (Noragric, 2004).

Presently there are numerous, thorough studies that demonstrate education is a necessary (but not sufficient) driving force for alleviating hunger and poverty. However, there are very few assessments, scenarios or projections of plausible futures for educational policies directed toward this end. In fact, the scenarios of the major existing assessments relevant for agriculture (see 4.2) provide very little information on this issue (some attention is paid in IPFRI modeling). On historic trends, UNESCO's databases[2] show that information provided by countries on multiple educational variables seldom is complete on either a yearly or a serial basis (or both). Hence it is not surprising that few educational indicators have been projected into the future. One educational indicator that has been projected into the future is the school-age population. Two features stand out: (1) projected changes in school-age population are highly variable among countries; and (2) there may be no change in the population aged 5-14 (in some countries this age group decreased, whereas the opposite trend was predicted for other countries in this group).

One important unknown is what proportion of the population aged 15-19 and 20-29 would receive a rural (or agricultural) higher education and/or training. In poor countries with large rural populations it is likely that emphasis on rural and agricultural education will take a growing share of the total educational effort as measured in terms of GDP, but that a decreasing share of the GDP is likely in those countries in transition to a larger-scale and/or more mechanized agriculture. While this agricultural transition will require less unskilled human labor, it will require professional practitioners able to address the challenges of reduced land availability, changing climates, and increased demands for sustainable farming practices, while maintaining or increasing productivity. If sustainability is considered an important production paradigm, the curricula of rural education would

---

[1] Dep. Int. Environ. Dev. Studies, Norwegian University of Life Sciences
[2] http://www.uis.unesco.org/ev_en.php?URL_ID=3753&URL_DO=DO_TOPIC&URL_SUBCHAPTER=201

need to include management of complex production systems like agroforestry, polycultural, and silvo-pastoral systems.

There are many methods for estimating the costs and benefits of educational policies; the cost-benefit analysis (CBA) is possibly one of the most used. The results of many CBA studies for developing countries in Africa, Asia and Latin America between 1960 and 1985 have been compiled and summarized (Hough, 1993). Major conclusions from this study included the following: (1) private Rate-of-Returns[3] (ROR) are always higher (27%) than social RORs[4] (19%); (2) RORs are always highest (32%) at the lowest level of education, but vary across regions; (3) social RORs to higher education are always lower (14%) than those to secondary education (16.7%), but the converse was true with private rates (24.3% and 21.3% for higher and secondary education, respectively); (4) public subsidies are particularly high in the cases of both primary and higher education, and in general, the poorer the country, the more subsidized is its education, particularly higher education, and (5) where time series data on earnings exist, there appears to be a decline in RORs over time (Psacharopoulos and Patrinos, 2004). Finally, some types of education that exhibit higher rates-of-return are general education for women and lowest per-capita income sector, and vocational education.

4.3.5.2 Culture
Culture has had a profound influence on the creation of new agricultural systems, as well as on the continued improvement of existing ones and will continue to do so in the future. However, as with education—cultural factors are often difficult to capture in scenarios.

One factor where culture plays a role is in diets. On an aggregated level changes in diet seem to mostly follow closely changes in income, independent of cultural factors or geographical location (FAO, 2002b). However, at equivalent incomes, cultural differences become conspicuous drivers of food quality and type (FAO, 2002b) (e.g., low pork consumption rates in some regions and low beef consumption rates in others). These factors are generally taken into account in the projections of existing assessments (see also 4.4.1).

Organic agriculture has been increasing in the past, and further expansion seems likely—certainly if the actual costs of agricultural commodity and food production were reflected in both domestic and international agricultural prices. It is, however, unlikely that organic farming will become a real substitute for industrial agricultural production systems, even if organic farming yield were similar to conventional yield (see e.g., Badgley et al., 2007). Production costs would likely be higher because it is a more labor-intensive activity and it might have additional standardization costs (OECD, 2002; Cáceres, 2005). Organic farming therefore does not play a very important role in the scenarios of existing assessments though it could have an impact on poverty and hunger alleviation in, for instance, least developed countries

with large rural populations, mostly of small-scale farmers living from subsistence traditional agriculture. This trend may be explored in new scenarios.

Another factor that might be considered is that traditional and indigenous cultures may be sources of agricultural knowledge useful for devising sustainable production systems. However, the future of that knowledge is likely to be grim in a globalized world if those who retain this knowledge do not receive assistance to pursue their futures in a manner acceptable to their value system (Groenfeldt, 2003). The practical knowledge stored in traditional and indigenous agriculture could be conserved if it were the subject of interdisciplinary inquiry by research organizations and universities (Thaman, 2002; Rist and Dahdouh-Guebas, 2006) with the aim of adapting otherwise unsustainable production systems to the likely incoming environmental shocks, such as the changing climate caused by global warming (cf. Borron, 2006), natural resource depletion (e.g., irrigation water), and pollution.

4.3.5.3 Ethics
The use of biotechnology (see 4.3.4) may have considerable benefits for society, but will likely raise ethical concerns about food and environmental safety (FAO, 2002b). The adaptation of these technologies in different scenarios should therefore be related to assumptions on ethical factors. In the next decade, development of biotechnological products will be faster for issues that relate to challenges recognized by the general public (e.g., herbicide-tolerant plants) than for other areas.

4.3.6 Changes in biogeophysical environment
Over the last 50 years, the use of fertilizers, primarily N fertilizers, has increased rapidly (FAO, 2003; IFA, 2006; Figure 4-12). In the same period, the quantity of nutrients supplied in the form of manure has increased as well (Bouwman et al., 2005). Increased use made a major increase in crop production possible. However, only a portion of the supplied nutrients are taken up by crops, with the remainder lost in different forms to the environment. These losses cause progressively serious environmental problems (Galloway et al., 2002; MA, 2005a), some of which can directly affect agriculture through a reduction in water quality and through climate change, and can indirectly affect agriculture through increased pressure for agricultural systems to minimize off-site environmental impacts.

To produce more food and feed in the future, the fertilizer demand is projected to increase from 135 million tonnes in 2000 to 175 million tonnes in 2015 and to almost 190 million tonnes in 2030 (Bruinsma, 2003). These projections are based on assumed crop yield increases. In a "Constant Nitrogen Efficiency scenario" the use of nitrogen fertilizers is projected to grow from 82 million tonnes in 2000 to around 110 million tonnes in 2020 and 120-140 million tonnes in 2050 (Wood et al., 2004). In an "Improved Nutrient Use Efficiency scenario" the use increases to around 100 million tonnes in 2020 and 110-120 million tonnes in 2050. These nitrogen fertilizer projections are based on the crop yields projected by AT 2030 (FAO, 2003) and they have been used for the Millennium Ecosystem Assessment (MA, 2005a). As the number of livestock is projected to increase as well, the

---

[3] These RRs take into account the costs borne by the students and/ or their families in regard to net (post-tax) incomes.

[4] These RRs relate all the costs to society to gross (before deduction of income tax) incomes.

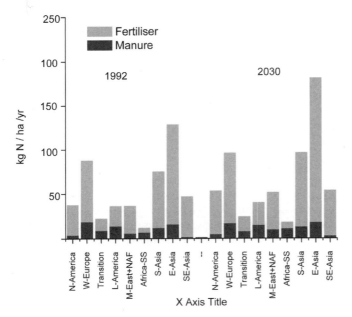

Figure 4-12. *Nitrogen application (fertilizer and manure) in 1995 and 2030.* Source: Bouwman et al., 2005.

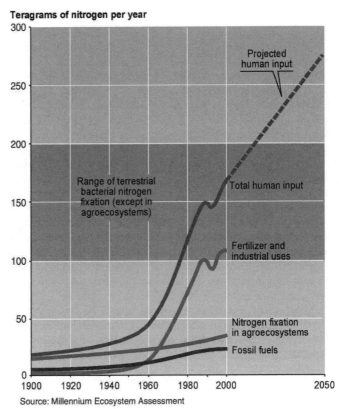

Figure 4-13. *Global trends in the creation of reactive nitrogen by anthropogenic activities.* Source: MA, 2005a.

quantity of manure is expected to increase, especially in Asia (Bouwman et al., 2005).

The increased use of fertilizer and manure will lead in many regions in the world to increased losses of reactive N and of phosphorous (P) to the environment, causing increasingly more severe environmental problems. Emissions do not only stem from agriculture; the combustion of (fossil) fuels and emissions from industries and households lead to increased levels of N and P in the environment (Figure 4-13). These emissions have already caused a range of environmental problems (MA, 2005a). The presence of excess nutrients (N, P) in water can lead to eutrophication (Bennett et al., 2001; Galloway et al., 2002) causing algal blooms, changes in resident organisms, low dissolved oxygen, and generally lower water quality. Nitrogen losses to groundwater can increase nitrate (NO_3) concentrations to levels which can have serious effects on human health. Aerial deposition of reactive N into natural terrestrial ecosystems, especially temperate grasslands, shrublands, and forests, could lead to lower plant diversity. Nitrous oxide is a powerful greenhouse gas. In 2000, nitrous oxides stemming from agriculture were responsible for more than 6% of global greenhouse gas emissions (EPA, 2006).

Nutrient loading will become an increasingly severe problem, particularly in developing countries and in East and South Asia (MA, 2005a). On the basis of projections for food production and wastewater effluents, the global river N flux to coastal marine systems may increase by 10-20% in the next 30 years. While river N flux will not change in most wealthy countries, a 20-30% increase is projected for poorer countries, which is a continuation of the trend observed in the past decades. The export is expected to reach 50 million tonnes year[1] by the year 2030 with the Pacific Ocean experiencing the greatest increase (Bouwman et al., 2005; MA, 2005a).

4.4 Assessment of Direct Drivers

4.4.1 Food consumption patterns

4.4.1.1 Expected changes in food consumption patterns and nutritional transformation

As incomes increase, direct per capita food consumption of maize and coarse grains declines as consumers shift to wheat and rice. When incomes increase further and lifestyles change with urbanization, a secondary shift from rice to wheat takes place. In general, existing assessments project a continuation of these trends. The expected income growth in developing countries (see 4.3.2) could become a strong driving force for increases in total meat consumption, in turn inducing strong growth in feed consumption of cereals. With growing urbanization, consumption is expected to shift as well towards increased consumption of fruits, vegetables, milk and milk products and to more consumer-ready, processed foods (increasingly procured in [international] supermarket chains, and fast food establishments). At the same time, growth in per capita food consumption in developed countries is expected to continue to slow as diets have reached on average reached saturation levels. These trends will lead to an increase in the importance of developing countries in global food markets (Cranfield et al., 1998; Rosegrant et al., 2001; Schmidhuber, 2003).

Several drivers of nutritional transformation are (1) gains in purchasing power of food, (2) declining food prices,

(3) shifts in demographic patterns, (4) growing urbanization, (5) changes in women's roles, (6) an enhanced understanding of the impact of diets on health, (7) government interventions towards certain foods, (8) influence exerted by the food industry, (9) growing international trade, and (10) an increasing globalization of tastes (Schmidhuber, 2003). Urbanization is generally associated with factors like higher incomes, more opportunities for women to enter the paid-work sector; and a major boost in the amount of information, goods and services. In relation to dietary habits this translates into access to a large variety of food products, exposure to different, "globalized" dietary patterns, adoption of urban lifestyles with less physically intensive activities requiring less food energy, and a preference for precooked, convenient food. Moreover, urbanization entails a physical separation of the agricultural sector from the postharvest sector and the final consumption sector (Smil, 2000; Giampietro, 2003; Schmidhuber, 2003).

Shifts in food expenditures. Decisions on food purchases will continue to be related to other household expenditure choices, such as housing, clothing, education, and health costs. With greater affluence the number of low-income countries that spend a greater portion of their budget on basic necessities, including food, will decline (Seale et al., 2003). Shifts in food expenditures for selected countries (see Table 4-8) with expected slow declines in food budget shares over time as well as slow declines in expenditures on grains are also projected (Cranfield et al., 1998).

Changes in agricultural production and retailing systems. The nutritional transformation will induce changes in agricultural production systems. Increased consumption of livestock products, e.g., will drive expansion of maize production for animal feed. Given that diets will continue to change with increasing incomes and urbanization, a doubling of cereal yields may be required. Because of the high rate of conversion of grains to meat, some analysts have argued that a reduction in meat consumption in industrialized countries, either through voluntary changes in dietary patterns, or through policies such as taxes on livestock, would shift cereal consumption from livestock to poor people in developing countries (e.g., Brown, 1995). While the long-term prospects for food supply, demand, and trade indicate a strengthening of world cereal and livestock markets, the improvement in food security in the developing world will be slow and changes in the dietary patterns in industrialized countries are not an effective route to improve food security in developing countries (Rosegrant et al., 1999).

At the same time, the agricultural production sector is catering more to globalized diets through growing industrialization and intensification of the food production process. Retailing through supermarkets is growing at 20% per annum in some countries and is expected to penetrate most developing countries over the next decades, as urban consumers demand more processed foods, shifting agricultural production systems from on-farm production toward agribusiness chains. International supermarket chains directly accelerate the nutritional transformation; e.g., the increase in the availability of yogurt and pasteurized milk has led to increases in consumption of dairy products in Brazil. Supermarkets will emerge in China and most other Asian developing countries, and more slowly in sub-Saharan Africa over the next three to five decades. The penetration of supermarkets for 42 countries based on the major drivers of change, including income, income distribution, urbanization, female participation in the labor force and openness to foreign competition through foreign direct investment, explains 90% of the variation in supermarket shares (Traill, 2006). Income growth was an important determinant for further supermarket penetration in Latin America, and further income growth and urbanization are crucial determinants for future supermarket growth in China (Traill, 2006).

The food retailing sector will increasingly serve as the primary interface between consumers and the rest of the agricultural sector (Figure 4-14). Food processing industries and supermarkets are expected improve food safety and support dietary diversification; on the other hand, they might contribute to less healthy diets through retailing of less healthy foods, such as refined white flour with reduced levels of fibers, minerals, and vitamins, or through oil hydrogenation processes.

4.4.1.2 Changing food consumption patterns in global assessments
Studies focusing on food and agriculture have seldom projected changes in food consumption patterns to 2050 at the global level; most projections in this area focus at the national level (Bhalla et al., 1999). Only two food and agriculture focused studies have done so: the FAO World Agriculture Outlook towards 2030/2050 interim report (FAO, 2006b) and IFPRI's food supply and demand projections (Von Braun et al., 2005 using the IFPRI IMPACT model) (Table 4-9).

Most studies and assessments agree that overall calorie availability continues to increase and dietary diversification continues following country and locale-specific pathways of nutritional transformation. Calorie availability levels in

Table 4-8. Projections of food budget shares and share of expenditures on grains, selected countries.

	Food budget shares		Share of expenditures on grains	
	1985	2020	1985	2020
Ethiopia	0.52	0.51	0.22	0.21
Senegal	0.41	0.37	0.13	0.11
United States	0.11	0.07	0.02	0.01

Source: Cranfield et al., 1998.

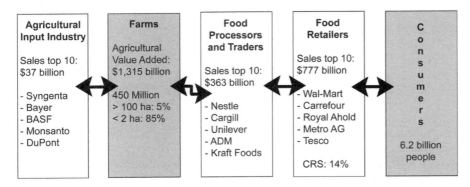

Figure 4-14. *Agricultural food business chain.* Source: Based on stock market data* and World Bank, 2005b.

Note: CR5 represents the market share of the top five companies listed in the global retail industry.

* http://www.wsj.com

these studies tend to asymptotically reach maximum availability levels of 3,500-4,000 kcal per capita (Tables 4-10, 4-11; Figures 4-15, 4-16, 4-17).
- The global consumption of meats and milk, fats, and sugars increases considerably, while consumption of roots and tubers, pulses, and cereals as food is stable or slightly declines.
- In regions with an average total daily consumption of less then 2500 kcal per capita (sub-Saharan Africa and South Asia) the situation slightly improves over time, but in 2050 the average food intake is still significantly lower then in other regions.
- In regions with low access to calories, food consumption increases in general more in more globalizing worlds (A1b, B1—IPCC SRES scenarios; Policy First—GEO-3 UNEP; GO—Millennium Ecosystem Assessment Global Orchestration scenario).
- In regions with high average total daily consumption the consumption remains stable or increases only slightly, with little or no differentiation between the scenarios.

- In middle-income regions (South East Asia, Central America, South America) food consumption slowly rises towards the level of OECD countries; with little differentiation across the scenarios.
- Differences in the consumption of animal products are much greater than in total food availability: both between regions, between scenarios and between years.
- Food demand for livestock products more or less doubles in sub-Saharan Africa and South Asia from around 200 in 2000 to around 400 kcal d[-1] by 2050; again with the highest values in globalizing scenarios. Consumption levels by 2050 can surpass 600 kcal/day in parts of Africa and South Asia.
- In most OECD countries with already high availability of kilocalories from animal products (1000 calories per capita per day or more) consumption levels are expected to barely change, while levels in South America and countries of the Former Soviet Union increase to OECD levels.

Table 4-9. **Incorporation of changing food demand patterns in global assessment studies.**

No.	Assessment Title	Publication Date	Projections timeframe	Food demand mentioned	Projections follow/adapted from
1	GEO-3 Assessment	2002	2032		FAO (2015/2030 outlook)
2	GEO-4 Assessment	2007	2000-2050	Explicitly	IFPRI IMPACT
3	IPCC 3rd Assessment	2001	Various	Not explicitly	Various, IPCC-SRES
4	IPCC 4th Assessment	2007	Various	Not explicitly	Various, IPCC-SRES
5	Millennium Ecosystem Assessment	2005	2000-2100	Explicitly	IFPRI IMPACT
6	Comprehensive Assessment of Water Management in Agriculture	2007	2000-2050	Explicitly	Watersim, based on IFPRI IMPACT
7	OECD Outlook	2006 Draft	2000-2030	Not explicitly	Partly FAO
8	World Energy Outlook	2006	2030	Not explicitly	-

Sources: UNEP, Global Environmental Outlook, 2002; IPCC, 2001, 2007; MA, 2005; de Fraiture et al., 2007; OECD, 2006; IEA, 2006.

Table 4-10. **Per capita food consumption (kcal/person/day).**

	1969/71	1979/81	1989/91	1999/01	2015	2030	2050
World	2411	2549	2704	2789	2950	3040	3130
Developing countries	2111	2308	2520	2654	2860	2960	3070
Sub-Saharan Africa	2100	2078	2106	2194	2420	2600	2830
excluding Nigeria	2073	2084	2032	2072	2285	2490	2740
Near East/North Africa	2382	2834	3011	2974	3080	3130	3190
Latin America and Caribbean	2465	2698	2689	2836	2990	3120	3200
South Asia	2066	2084	2329	2392	2660	2790	2980
East Asia	2012	2317	2625	2872	3110	3190	3230
Industrial countries	3046	3133	3292	3446	3480	3520	3540
Transition countries	3323	3389	3280	2900	3030	3150	3270

Source: FAO, 2006b.

4.4.1.4 Implications for health

Changes in food demand to 2050 are expected to contribute to increased nutrition and human health. Dietary diversification will likely increase if urbanization and income growth proceed. On the other hand, obesity rates and associated diseases are expected to increase. Obesity is increasingly becoming a public health concern as it contributes to increased mortality through noncommunicable diseases such as diabetes, hypertension, stroke, and cardiovascular diseases, among others. Factors responsible for increases in obesity include a mix of biological and ecological factors such as gene-mediated adaptation, increases in labor mechanization, urbanization, sedentary activities and lifestyle changes (Caballero, 2001). It is estimated that by 2020, 60% of the disease burden in developing countries will result from noncommunicable diseases, further exacerbated because of obesity (Caballero, 2001).

4.4.2 Natural resources

The sustainable use and management of natural resources presents a critical factor for future agriculture. The develop-

ment and adoption of appropriate AKST and management practices will be needed to ensure food security and agricultural livelihoods. One of the greatest challenges likely to continue facing global agriculture is resolving conflicts caused by growing competition for soil, water, and other natural resources on which agriculture depends (Antle and Capalbo, 2002). Conversely, the sustainable management of these natural resources will determine productivity in agriculture and food systems.

4.4.2.1 Water

Water availability for agriculture is one of the most critical factors for food security in many regions of the world, particularly in arid and semiarid regions in the world, where water scarcity has already become a severe constraint on food production (Rockstrom et al., 2003; CA, 2007). With increasing population, urbanization, changing diets and higher living standards water demand is increasing rapidly. Assuming the amount of potentially utilizable water does not increase, there will be less water available on a per capita basis. In 1989, approximately 9,000 m³ of freshwater per

Table 4-11. **Changes in the commodity composition of food by major country groups in kg/person/year.**

World	1969/71	1979/81	1989/91	1999/01	2030	2050
Cereals, food	148.7	160.1	171	165.4	165	162
Cereals, all uses	302.8	325	329.3	308.7	331	339
Roots and tubers	83.7	73.4	64.5	69.4	75	75
Sugar (raw sugar equiv.)	22.4	23.4	23.3	23.6	26	27
Pulses, dry	7.6	6.5	6.2	5.9	6	6
Vegetable oils, oilseeds and products (oil equiv.)	6.8	8.3	10.3	12	16	17
Meat (carcass weight)	26.1	29.5	33	37.4	47	52
Milk and dairy, excl. butter (fresh milk equiv.)	75.3	76.5	76.9	78.3	92	100
Other food (kcal/person/day)	216	224	241	289	325	340
Total food (kcal/person/day)	2411	2549	2704	2789	3040	3130

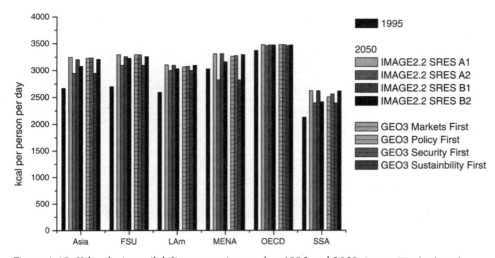

Figure 4-15. *Kilocalorie availability per capita per day, 1995 and 2050.* Source: Westhoek et al., 2005.

Notes: A1, A2, B1, and B2 are storylines used in IPCC assessments. These results are data underlying but not reported in the third IPCC Assessment Reports. In UNEP's assessment these data are not presented in the final GEO3 report.

person was available for human use. By 2000, that number had dropped to 7,800 m³ and it is expected to decline to 5,100 m³ per person by 2025, when the global population is projected to reach 8 billion. Already 1.2 billion people live in areas where water is physically scarce, and this number may double by 2050 (CA, 2007). The problem is becoming more urgent due to the growing share of food produced on irrigated land; the rapid increase of water use in industry and households; increasing water use for environmental and ecological purposes; and water quality deterioration (Rosegrant et al., 2002).

The last 50 years saw great investments in large scale, surface irrigation infrastructure as part of a successful effort to rapidly increase world staple food production and ensure food self-sufficiency. During this period, in many countries more than half of the public agricultural budget and more than half of World Bank spending was devoted to irrigation (Faures et al., 2007). Spending on irrigation reached a peak of over US$1 billion per year in the late 1970s (in constant 1980 US dollars) but fell to less than half that level by the late 1980s (Rosegrant and Svendsen, 1993). The irrigated area roughly doubled from 140 million ha in the 1960s to

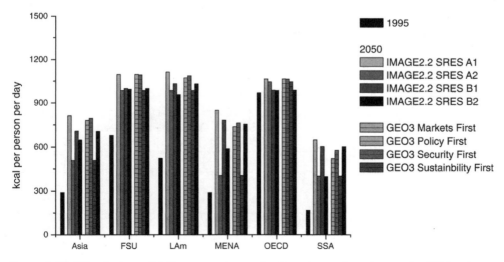

Figure 4-16. *Kilocalorie availability per capita per day from livestock products only, 1995 and 2050.* Source: Westhoek et al., 2005.

Notes: A1, A2, B1, and B2 are storylines used in IPCC assessments. These results are data underlying but not reported in the third IPCC Assessment Reports. In UNEP's assessment these data are not presented in the final GEO3 report.

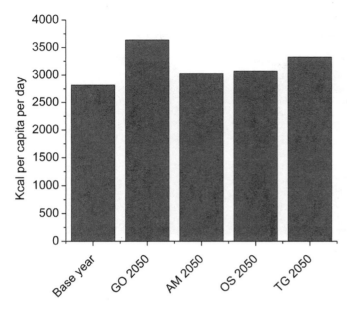

Figure 4-17. *Average global food availability, Millennium Ecosystem Assessment Scenarios.* Source: MA, 2005ab.

Notes: GO, AM, OS and TG stand for the Global Orchestration, the Adapting Mosaic, Order from Strength, and TechnoGarden Scenarios, respectively.

280 million ha in 2003, primarily in Asia (FAO, 2006c). By contrast, irrigation in sub-Saharan Africa is applied to less than 4% of the total cultivated area. The agricultural sector is expected to remain the major water user accounting for 69% of the withdrawals and 84% of the consumptive uses.

Many projections agree that water will increasingly be a key constraint in food production in many developing countries, and call for the need to improve water management and increase water use efficiency (Seckler et al., 2000; Shiklomanov, 2000; Rosegrant et al., 2002; Bruinsma, 2003; World Water Assessment Program, 2006; CA, 2007). The assessments differ in their views on the best way forward. Scenario analysis conducted as part of the Comprehensive Assessment of Water Management in Agriculture (CA, 2007) indicates that growth in global water diversions to agriculture varies anywhere between 5 and 57% by 2050 depending on assumptions regarding trade, water use efficiency, area expansion and productivity growth in rain fed and irrigated agriculture (de Fraiture et al., 2007) (Figure 4-18). Trade can help mitigate water scarcity if water-scarce countries import food from water abundant countries (Hoekstra and Hung, 2005). Cereal trade from rain fed areas in the temperate zones (USA, EU, Argentina) to arid areas (Middle East) reduces current global irrigation water demand by 11 to 13% (Oki et al., 2003, de Fraiture et al., 2004); but political and economic factors may prove stronger drivers of agricultural systems than water (de Fraiture et al., 2004).

Enhanced agricultural production from rain fed areas and higher water productivity on irrigated areas can offset the need for the development of additional water resources (Molden et al., 2000; Rosegrant et al., 2002; Rockstrom et al., 2003). However, the potential of rain fed agriculture and

the scope to improve water productivity in irrigated areas is subject to debate (Seckler et al., 2000; Rosegrant et al., 2002). Only 5% of increases in future grain production are projected to come from rain fed agriculture (Seckler et al., 2000). Over 50% of all additional grains will come from rain fed areas, particularly in developed countries, while developing countries will increasingly import grain (Rosegrant et al., 2002). Projected contribution to total global food supply from rain fed areas declines from 65% currently to 48% in 2030 (Bruinsma, 2003).

Currently, agriculture receives around 70% of total water withdrawal and accounts for 86% of consumption. Projections in growth of irrigated areas vary: 29% (Seckler et al., 2000); 24% (FAO, Bruinsma, 2003); and 12% (Rosegrant et al., 2002) (Table 4-12).The global irrigated area is expected to grow from 254 million ha in 1995 to between 280 and 350 million ha in 2025. However, towards 2050, the proportion of water used for agriculture is likely to decrease slightly, mainly at the expense of more intensive growth in other water demands such as environment, industry and public water supply. The regional water withdrawal and consumption shares for agriculture will vary as a function of stage of industrialization, climate and other management and governance factors. In many water scarce areas current per capita water consumption is unsustainable. Globally, water is sufficient to produce food for a growing and wealthier population, but continuance with today's water management practices will lead to many acute water crises in many parts of the world (CA, 2007).

While major tradeoffs will occur between all water using sectors, they will be particularly pronounced between agriculture and the environment as the two largest water demanding sectors (Rijsberman and Molden, 2001). Signs of severe environmental degradation because of water scarcity, overabstraction and water pollution are apparent in a growing number of places and the adverse impacts of irrigation on ecosystems services other than food production are well documented (Pimentel and Wilson, 2004; MA, 2005a; Khan et al., 2006; CA, 2007). Reduction in ecosystem services often has severe consequences for the poor, who depend heavily on ecosystems for their livelihoods (Falkenmark et al., 2007). Aquifer depletion and groundwater pollution threaten the livelihoods of millions of small-scale farmers in South Asia; in response, various local initiatives to recharge groundwater and stop overuse have been developed (Shah et al., 2007).

4.4.2.2 Soils and fertilizers

Sustainable management of soil is vital to agricultural productivity and food security. Among the many driving forces that will affect soils and their utility in sustaining world agriculture are population growth, land use planning and policies, land development and growth and demands for agricultural products (Blum, 2001). These driving factors operate directly and interact in different ways to produce positive (sustainability) and negative effects (degradation) on soil. Soil degradation due to improper farming practices has had more devastating effects on soil quality in many developing countries than the industrialized world. While demand for food has risen with increasing population, the present productivity from the arable land of the developing

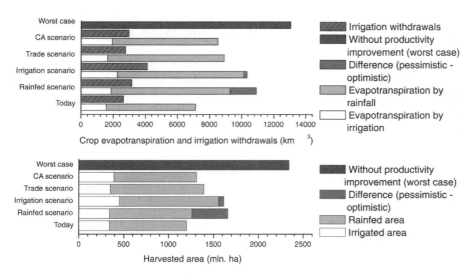

Figure 4-18. *Land and water use today and in the future under different scenarios.* Source: CA, 2007

Note: The dark gray bar represents the worst case scenario in which no productivity improvements occur in rainfed or irrigated agriculture. The 'rainfed scenario' assumes that most of future investments are targeted to upgrading rainfed agriculture. The medium-dark bar denotes the difference between optimistic and pessimistic yield assumptions and gives an indication of the risks involved in this scenario. The 'irrigation scenario' assumes a major drive in improvement of water productivity and expansion of irrigated areas. The 'trade scenario' assumes increased food trade from water abundant to water scarce areas. The 'Comprehensive Scenario' combines elements from all three scenarios depending on regional opportunities.

world has not been able to support its increasing population. The challenge is highly likely to persist towards 2050.

At the global level, out of the total ice-free land area of 13.4 x 10⁹ ha, only 4.9 x 10⁹ ha are agricultural lands. The Food and Agriculture Organization Database (WRI, 1997; FAO, 2006c) has provided individual country assessments on quantity of arable land and other indicators for national and global assessments. Out of the agricultural lands, 3.2 x 10⁹ ha are in developing countries, while 1.8 x 10⁹ ha are in industrialized countries (FAO, 2003). Some (1.3 x 10⁹ ha)

of this area has been classified as low productivity. About half of the potentially arable land is actually cultivated, while remaining lands are under permanent pastures, forests and woodland (Scherr, 1999). In the future, feeding an increasing population will remain a challenge, particularly as per capita land availability decreases and soil degradation continues.

Population pressure and improper land use practices are expected to continue giving rise to soil degradation, manifesting itself through processes such as erosion, desertification, salinization, fertility loss. Especially in regions with low

Table 4-12. **Comparison of recent global water use forecasts.**

Author	Projection period	Increase in rain-fed cereal production	Increase in irrigated yield	Increase in irrigated harvested area	Increase in cereal trade	Increase in agricultural water withdrawals
		annual growth rate	annual growth rate	annual growth rate	annual growth rate	annual growth rate
Shiklomanov 2000	1995-2025			0.74%		0.68%
Seckler et al. 2000	1995-2025	0.19%	1.13%	0.95%	0.64%	0.56%
Rosegrant et al. 2002	1995-2025	1.14%	1.14%	0.36%	2.41%	
Faures et al. 2002	1995-2030	1.10%	1.00%	0.95%	2.08%	0.43%
Alcamo et al. 2005	2000-2050			0.06%-0.18%	1.85%-2.44%	0.40%-1.22%
Fraiture et al. 2007	2000-2050	0.63%-1.03%	0.58%-1.15%	0%-0.56%	0.98%-2.01%	0.10%-0.90%

Source: adapted from CA, 2007.

(average) fertilizer use, many fields have a negative soil nutrient balance. Although the fertilizer use projections show an increased use for sub-Saharan Africa, application rates remain too low to compensate losses and crop yields will therefore remain low. Some believe that land degradation will not be a major issue in food security for the future generations (Crosson, 1994; Rosegrant and Sombilla, 1997); others argue that it will be a major constraining factor for food production in the future (Brown and Kane, 1994; Hinrichsen, 1998).

Crops are highly depended on an adequate supply of nutrients, notably N, P and potassium (K). The use of mineral fertilizer has increased significantly over the last 50 years, from 30 million tonnes in 1960, to 70 million tonnes in 1970 to 154 million tonnes in 2005 (IFA, 2006). This increased use is one of the drivers behind the increase in crop yield over the last 50 years. About two thirds of the global N fertilizer is currently used in cereal production (Cassman et al., 2003). Fertilizer use is expected to increase by 188 million tonnes by 2030 (FAO, 2004a). The projections for 2030 indicate that approximately 70% of the increase in total crop production will stem from higher yields per ha and about 30% from expansion of harvested areas. The increased (and more efficient) use of fertilizers is one of the key drivers to attain these higher crop yields.

The use of mineral and organic fertilizers is very diverse between countries and regions (Palm et al., 2004; Bouwman et al., 2005; IFA, 2006). Nitrogen input varies from virtually nil to over 500 kg N per ha; with these differences likely to continue (Daberkow et al., 2000; Bruinsma, 2003; Bouwman et al., 2005). The use of fertilizers is expected to increase further in South and East Asia; in sub-Saharan Africa the present low application rates are projected to persist and seriously hamper crop production. Low application is caused by a range of factors, which without targeted policies and interventions are likely to persist. These factors include a weak crop response to fertilizers (e.g., limited water availability, poor soil conditions), unfavorable price relations between input and output, and low net returns (Kelly, 2006).

The increase in consumption of animal products is, next to population growth, one of the major causes of the increase of global fertilizer use. World meat consumption (and production) is expected to grow by 70% in the period 2000-2030 and 120% in the period 2000-2050 (FAO, 2006b). The production and consumption of pig and poultry meat is expected to grow at a much higher speed than of bovine and ovine meat. Over the last years there has been a major expansion in large scale, vertically integrated industrial livestock systems, and this development is expected to continue over the coming decades (Bruinsma, 2003). These systems can lead to concentration of manure; although manure is a valuable source of nutrients, concentrated spreading of manure leads to significant emissions, to air, soil and water.

4.4.3 Land use and land cover change

Growing demand for food, feed, fiber and fuel, as well as increasing competition for land with other sectors (e.g., human settlement, infrastructure, conservation, and recreation), drive the need for change in the use of land already dedicated to agricultural production, and often for additional land to be brought into production. The significance of the cumulative historical growth in demand for agricultural products and services is reflected in the fact that agriculture now occupies about 40% of the global land surface. There is also clear evidence that this enormous change in land use and land cover has brought about, and continues to bring significant impacts on local, regional and global environmental conditions, as well as on economic and social welfare. In turn, such impacts spur demand for specific types of improvements in agriculture. AKST can help mitigate negative outcomes and enhance positive ones.

In this context, AKST can be seen as playing a dual role in both shaping and responding to a dynamic balance of land use and land cover conditions that deliver specific mixes of agricultural and other goods and services. As human well-being needs and preferences evolve in different societies, so too will the goals and priorities for the development of new AKST. The relative scarcities of land in Japan and labor in the USA shaped their agricultural research priorities (Hayami and Ruttan, 1985). Global experience with rampant land degradation caused by inappropriate production practices that gave rise to degradation of land cover, migration and often further expansion of the agricultural frontier has driven the search for new knowledge on sustainable farming technologies and land management practices. Land use/cover change is a complex process with multiple factors and drivers acting synergistically. In the tropics, deforestation was frequently driven by an interplay of economic, institutional, technological, cultural, and demographic factors (Geist and Lambin, 2004) (Figure 4-19). There are numerous other studies that link environmental land cover change to socioeconomic factors (e.g., Hietel et al., 2005; Xie et al., 2005).

4.4.3.1 Global land cover and land use change
Current drivers. Globally, there are a small number of recurrent drivers of land use and land use change (Figure 4-19): demographic; economic; technological; policy; and cultural. Yet, some factors play a decisive role in determining land use and thus land use change. For example, globally 78% of the increase in crop production between 1961 and 1999 was attributable to yield increases and 22% to expansion of harvested area (Bruinsma, 2003). While the pattern of yield increases outpacing increases in harvested area was true for most regions, the proportions varied. For example, 80% of total output growth was derived from yield increases in South Asia, compared to only 34% in sub-Saharan Africa. In industrial countries, where the amount of cultivated land has been stable or declining, increased output was derived predominantly through the development and adoption of AKST that served to increase yields and cropping intensities. Thus, the role of land use change and (adoption of) AKST has varied greatly between regions. Particularly in Latin America, land abundance has slowed the introduction of new technologies.

Projected global land cover and land use changes. Few global studies have produced long-term land cover and land use projections. The most comprehensive studies, in terms of land type coverage, are the Land Use and Cover Change Synthesis book (Alcamo et al., 2005), IPCC Special Report on Emissions Scenarios (SRES) (IPCC, 2000), the

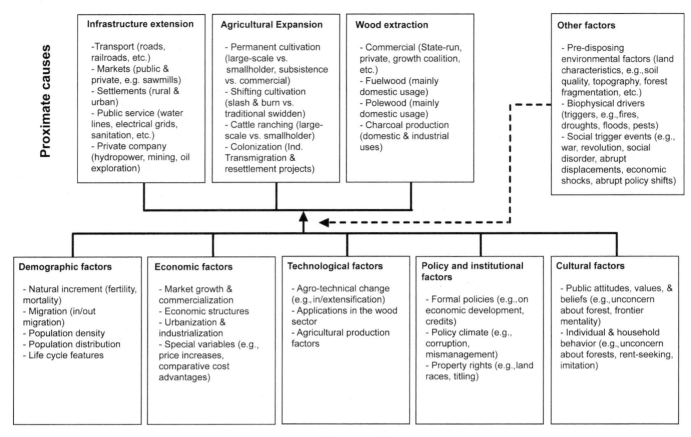

Figure 4-19. *Indirect and direct drivers of land cover change.* Source: Adapted from Geist and Lambin, 2004.

scenarios from the Global Scenarios Group (GSG) (Raskin et al., 2002), UNEP's Global Environment Outlook (UNEP, 2002), the Millennium Ecosystem Assessment (MA, 2005a) and some models from Stanford University's EMF-21 Study of the Energy Modeling Forum (e.g., Kurosawa, 2006; van Vuuren et al., 2006). Recent sector specific economic studies have also contributed global land use projections, especially for forestry (Sands and Leimbach, 2003; Sathaye et al., 2006; Sohngen and Mendelsohn, 2007; Sohngen and Sedjo, 2006) (Figure 4-20). Note that some scenario exercises are designed to span a range of diverse futures (e.g., SRES, GSG, and MA). For example, under the SRES scenarios agricultural land area could increase by 40% or decrease by 20% by 2050. Other scenarios focus on a single projected reference land-use characterization (e.g., GRAPE-EMF21, IMAGE-EMF21, GTM-2007). The more recent scenarios suggest greater agreement than under SRES or GSG, with agricultural land extent stable or growing by 10% by 2050.

In general, the recent scenarios for agricultural land use (cropland and grazing land) have projected increasing global agricultural area and smaller forest land area. The developments in forest land are, for the most part, the inverse of those for agriculture, illustrating the potential forest conversion implications of agricultural land expansion, as well as providing insights into current modeling methodologies. To date, long-term scenarios have not explicitly modeled competition between land use activities (Sands and Leimbach, 2003, is an exception). A new generation of global

modeling is forthcoming that will directly account for the endogenous opportunity costs of alternative land uses and offer new more structurally rigorous projections (e.g., van Meijl et al., 2006).

Projected changes in agricultural land are caused primarily by changes in food demand and the structure of production as defined by technology, input scarcity, and environmental condition. Scenarios with a greater extent of agricultural land result from assumptions of higher population growth rates, higher food demands, and lower rates of technological improvement that limit crop yield increases. Combined, these effects are expected to lead to a potentially sizeable expansion of agricultural land. Conversely, lower population growth and food demand, and more rapid technological change, are expected to result in lower demand for agricultural land.

There are very few published global scenarios of changes in urban areas (Kemp-Benedict et al., 2002; UNEP, 2004). All show a steep increase over the next decade, with about half estimating a stabilization of urban areas by 2025. Final total urban area is about 50% larger than in 1995. Although the total increase in area is relatively small, the implications for agriculture might be disproportionately large, since most of the urban growth is at the expense of high-value agricultural lands.

4.4.3.2 Regional and local changes
Regional and local drivers of land use change are even more complex than global drivers because a large number of di-

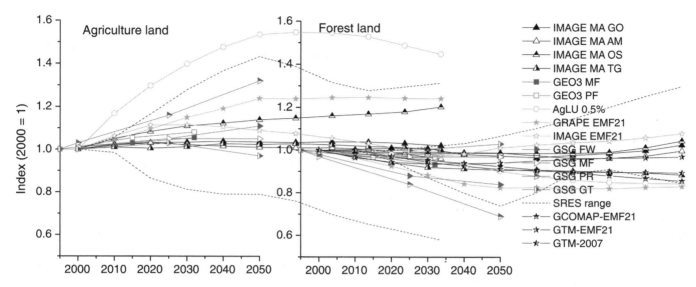

Figure 4-20. *Projected landuse changes for (a) agricultural land and (b) forest land from selected scenarios (indexed to year 2000)*

Notes: Agricultural land is an aggregate of cropland and grazing/pasture/grassland land types. The following scenarios were redrawn from Alcamo et al., 2006: GSG xx = Global Scenarios Group (Raskin et al., 2002) scenarios from the PoleStar model (MF = Market Forces, PR = Policy Reform, FW = Fortress World, GT = Great Transition); GEO-3 xx = Global Environment Outlook 3 (UNEP, 2004) scenarios using the PoleStar model (MF = Market First, PF = Policy First). The other scenarios were assembled from various sources: SRES (Special Report on Emissions Scenarios) = IPCC, 2000; IMAGE-EMF21 = van Vuuren et al., 2006 scenario from EMF-21 Study; IMAGE-MA-xx = MA, 2005 scenarios from the IMAGE model for four storylines (GO = Global Orchestration, OS = Order from Strength, AM = Adapting Mosaic, TG = TechnoGarden); AgLU-0.5% = Sands and Leimbach, 2003 scenarios with 0.5% annual growth in crop yield; GTM-EMF21 = Sohngen and Sedjo, 2006 global forest scenario from EMF-21 Study; GCOMAP-EMF21 = Sathaye et al., 2006 global forest scenario from EMF-21 Study; GTM-2007 = Sohngen and Mendelsohn, 2007 global forest scenario; GRAPE-EMF21 = Kurosawa, 2006 scenario from EMF-21 Study.

rect drivers act in addition to global (indirect) drivers. For example, in cultivated systems, cultural, socioeconomic, and educational background as well as expectations, perceptions, preferences, and attitudes toward risk of farmers and farm households can play significant roles in shaping land use choices.

Tropical deforestation depicts the connectedness of multiple drivers. In the humid tropics, deforestation is primarily the result of a combination of commercial wood extraction, permanent cultivation, livestock development, and the extension of overland transport infrastructure (e.g., Strengers et al., 2004; Verbist et al., 2005; Busch, 2006; Rounsevell et al., 2006). However, regional variations exist. Deforestation driven by swidden agriculture (see 4.5.1.2) is more widespread in upland and foothill zones of Southeast Asia than in other regions. Road construction by the state followed by colonizing migrant settlers, who in turn practice slash-and-burn agriculture, is most frequent in lowland areas of Latin America, particularly in the Amazon Basin. Pasture creation for cattle ranching is causing deforestation almost exclusively in the humid lowland regions of mainland South America. Expansion of small-scale agriculture and fuelwood extraction for domestic uses are important causes of deforestation in Africa (Geist and Lambin, 2002; FAO, 2006b) and Latin America (Echeverria et al., 2006). Recently, two new land use types that are partly related to new drivers have emerged: bioenergy production (see 4.4.5.4) and soybean expansion driven by international markets, but also by the development of GMOs has rapidly become a major threat in Latin America (see Box 4-3).

The range of combinations of factors is not infinite, although single-factor causes are rare (Reid et al., 2006). A significant share of land use changes involves lifestyle choices and shifting consumption patterns; governance; global markets and policies. Underlying causes often have a strong influence on local land use and cover changes. In the same way, land use alters in multiple ways agricultural production and AKST.

Global forces are the main determinants of land use change, as they amplify or attenuate local factors (Lambin et al., 2001). Less visible but of no lesser importance is the build-up of small impacts at lower levels of the spatial and temporal scales to generate impacts at higher levels; cumulative impacts are caused by incremental impacts at the individual level and are felt usually after some period of time at the regional or even the national level. The issue of scale is implicated in these and similar instances and makes the use of "scale-sensitive" analytical approaches imperative. Multiscale efforts bring global, regional, and local studies together (e.g., MA, 2005a).

Many recent scenarios include land cover and land use changes, and many of those include explicit information on the main land use drivers. The scenarios also acknowledge the complexity of environmental, social, and economic drivers of land use change. However, due to lack of data, a limited subset of drivers is included in the modeling efforts. The dynamics of land use (and thus of land cover) are largely governed by human (e.g., policy and socioeconomic) factors, that are well-documented as indirect drivers (see 4.2), but poorly represented as direct drivers. Important drivers

Box 4-3. Genetically modified soybeans in Latin America.

At the global scale, soybean is one of the fastest expanding crops; in the past 30 years planted area more than doubled (FAO, 2002b). Of the world's approximately 80 million ha, more than 70% are planted in the USA, Brazil and Argentina (Grau et al., 2005). Argentina's planted area increased from less than a million ha in 1970 to more than 13 million ha in 2003 (Grau et al., 2005). Soybean cultivation is seen to represent a new and powerful force among multiple threats to biodiversity in Brazil (Fearnside, 2001). Deforestation for soybean expansion has, e.g., been identified as a major environmental threat in Argentina, Brazil, Bolivia and Paraguay (Fearnside, 2001; Kaimowitz and Smith, 2001). In part, area expansion has occurred in locations previously used for other agricultural or grazing activities, but additional transformation of native vegetation plays a major role. New varieties of soybean, including glyphosate-resistant transgenic cultivars, are increasing yields and overriding the environmental constraints, making this a very profitable endeavor for some farmers (Kaimowitz and Smith, 2001). Although until recently, Brazil was a key global supplier of non-GM soya, planting of GM soy has been legalized in both Brazil and Bolivia. Soybean expansion in Brazil increased; as did research on soybean agronomy, infrastructure development, and policies aimed at risk-reduction during years of low production or profitability (Fearnside, 2001). In Brazil alone, about 100 million ha are considered to be suitable for soy production. If projected acreage in Argentina, Brazil and Paraguay are realized, an overproduction of 150 million tonnes will be reached in 2020 (AIDE, 2005).

to consider for land use dynamics are: the perceptions and values of local stakeholders land resources, its goods and services; land tenure and property rights and regulations; the development and adoption of new sources of AKST; and urban-rural connections.

4.4.4 Climate variability and climate change

Agricultural systems are already adapting to changes in climate and climate variability in many part of the world. This is in particular the case in arid areas. The IPCC concluded in its latest assessment that it is very likely that humans caused most of the warming observed during the twentieth century (IPCC, 2007a). The report also indicates that future climate change is to be expected, as a function of continuing and increasing emissions of fossil fuel combustion products, changes in land use (deforestation, change in agricultural practices), and other factors (for example, variations in solar radiation). Changes in climate will not only manifest themselves in changes in annual means (precipitation, temperature) but also in changes in variability and extremes.

4.4.4.1 Driving forces of climate change

A set of scenarios (IPCC Special Report on Emissions Scenarios—SRES) was used to depict possible emission trends under a wide range of assumptions in order to assess the potential global impact of climate change (IPCC, 2000). Subsequent calculation showed that these scenarios resulted in atmospheric concentrations of CO_2 of 540-970 parts per million in 2100 compared with around 370 parts per million in 2000. This range of projected concentrations is primarily due to differences among the emissions scenarios. Model projections of the emissions of other greenhouse gasses (primarily CH_4 and N_2O) also vary considerably by 2100 across the IPCC-SRES emissions scenarios. The IPCC scenarios are roughly consistent with current literature—with the majority of the scenarios leading to 2100 emissions of around 10-22 Gt C (Van Vuuren and O'Neill, 2006) (Figure 4-21) with projections by the IEA-2006 World Energy Outlook in the middle of this range. The IPCC-SRES scenarios do not explicitly include climate policies. Stabilization scenarios explore the type of action required to stabilize atmospheric greenhouse gas concentrations (alternative climate policy scenarios may look into the impact of a particular set of measures; or choose to peak concentrations). Ranges of stabilization scenarios giving rise to different stabilization levels are compared to development without climate policy (Figure 4-22) (IPCC, 2007c). The ranges in emission pathways result from uncertainty in land use emissions, other baseline emissions and timing in reduction rates.

4.4.4.2 Projections of climate change

IPCC calculations show that different scenarios without climate policy are expected to lead to considerable climate change: the global mean surface air temperature is expected to increase from 1990 to 2100 for the range of IPCC-SRES scenarios by 1.4 to 6.4 C° (IPCC, 2007a) (Figure 4-23). The total range given above is partly a consequence of differences in emissions, but also partly an impact of uncertainty in climate sensitivity, i.e., the relationship between greenhouse gas concentration and the increase in global mean temperature (after equilibrium is reached). Over the last few years, there has been a shift towards expressing the temperature consequences of stabilization scenarios more in terms of probabilistic expressions than single values and/or ranges. A 50% probability level for staying below 2°C corresponds approximately to 450 ppm CO_2-eq, while for 2.5°C the corresponding concentration is around 525 ppm CO_2-eq. Similarly, a scenario that would lead to 2°C warming as the most likely outcome could also lead to a 0.9 to 3.9°C warming (95% certainty). Handling uncertainty therefore represents an important aspect of future climate change policy. Costs of stabilization increase for lower concentration levels, and very low concentration levels, such as 450 ppm CO_2-eq may be difficult to reach (IPCC, 2007c).

Combining the current scenarios with climate policy and the expected temperature increase for different greenhouse gas levels shows that the former would decrease the lower bound of the expected temperature increase to about 0.5-1.0 °C above the 1990 level (i.e., based on an insensitive climate system and using a strong climate policy scenario) (Van Vuuren et al., 2006). This implies that although these values may be uncertain, climate change is very likely. High rates of temperature change are in fact most likely to occur in the first half of the century as a result of climate system inertia, the limited impacts of climate policy and lower

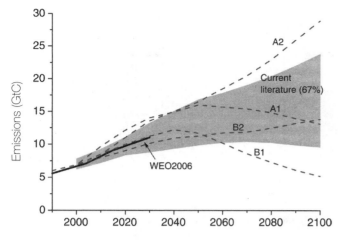

Figure 4-21. *Comparison of current CO₂ emission scenarios (scenarios since IPCC's Third Assessment Report 2001; mean + std. deviation), IPCC-SRES (A1, A2, B1, B2) and WEO2006.*

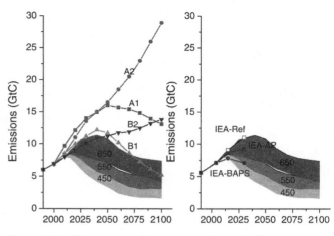

Figure 4-22. *Comparison of emission pathways leading to 650, 550 and 450 ppm CO₂-eq. and the IPCC-SRES scenarios (left) and the WEO-2006 scenarios.*

sulfur emissions. The latter are currently having a cooling effect on the atmosphere. Whereas the computation of global mean temperature is uncertain, the patterns of local temperature change are even more uncertain (Figure 4-24) (IPCC, 2007a).

For precipitation, climate models can currently provide insight into overall global and regional trends but cannot provide accurate estimates of future precipitation patterns in situations where the landscape plays an important role (e.g., mountainous or hilly areas). A typical result of climate

models is that approximately three-quarters of the land surface has increasing precipitation. However, some arid areas become even drier, including the Middle East, parts of China, southern Europe, northeast Brazil, and west of the Andes in Latin America. This will increase water stress in these areas. In other areas rainfall increases may be more than offset by increase in evaporation caused by higher temperatures.

Although climate models do not agree on the spatial patterns of changes in precipitation, they do agree that global average precipitation will increase over this century. This is

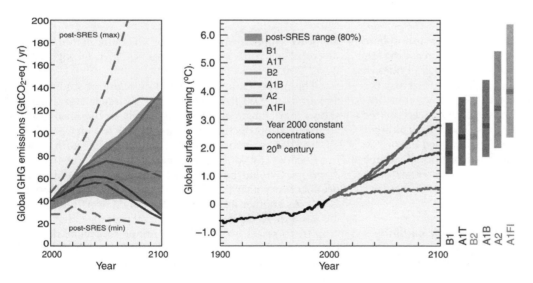

Figure 4-23. *Scenarios for GHG emissions from 2000 to 2100 (in the absence of climate policies) and projections of surface temperatures. Source, IPCC, 2007.*

Left Panel: Global GHG emissions (in GtCO2-eq) in the absence of climate policies: six illustrative marker scenarios (solid lines) and the 80th percentile range of recent scenarios published since 2000 (shaded area). Dashed lines show the full range of scenarios since 2000. Right Panel: Solid lines are multi-model global averages of surface warming for scenarios A2, A1B and B1, shown as continuations of the 20th-century simulations. The lowest line is not a scenario, but is for Atmosphere-Ocean General Circulation Model (AOGCM) simulations where atmospheric concentrations are held constant at year 2000 values. The bars at the right of the figure indicate the best estimate (solid line within each bar) and the likely range assessed for the marker scenarios at 2090-2099. All temperatures are relative to the period 1980-1999.

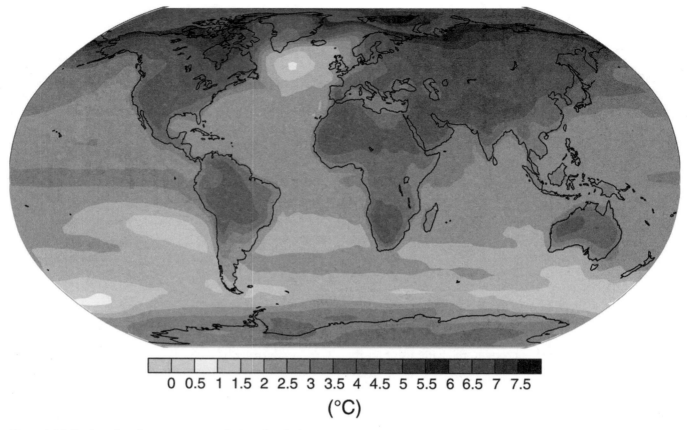

Figure 4-24. *Projected surface temperature changes for the late 21ˢᵗ century (2090-2099).* Source: IPCC, 2007. *The map shows the multi-AOGCM average projection for the A1B SRES scenario. Temperatures are relative to the period 1980-1999.*

consistent with the expectation that a warmer atmosphere will stimulate evaporation of surface water, increase the humidity of the atmosphere and lead to higher overall rates of precipitation. An important factor for agriculture is also the changes in inter-year variability, but while some report important increases in this factor, this is still very uncertain (see IPCC, 2007a). Climate change may also affect agriculture if it causes substantial melting of glaciers that feed major rivers that are used for irrigation. Without additional storage to capture increased summer runoff, more water will flow unused to the ocean, leading to water scarcity in the drier months (see IPCC, 2007a).

4.4.4.3 The potential impact of climate change on future agricultural yields

The impacts of climate change on agriculture have been assessed by IPCC (IPCC, 2007b). In fact, two combined effects have to be accounted for: the impacts of climate itself and the rising atmospheric CO_2 concentration. Increased concentrations of CO_2 can increase yields and make plants more stress-resistant against warmer temperatures and drought (although the extent to which this occurs is uncertain, and depends on crop type). Climate change can lead to both increases and decreases in yields, depending on the location of changes of temperature and precipitation (climate patterns) and the crop type (Parry et al., 1999; Alcamo et. al., 2005). A crucial factor is whether farmers are able to

adapt to temperature increases by changing planting dates or crops, or crop varieties (Droogers, 2004; Droogers and Aerts, 2005).

The preponderance of global agricultural studies (Adams et al., 1999; Parry et al., 1999, Fischer et al., 2001) shows that climate change is not likely to diminish global agricultural production by more than a few percent, if at all, by 2050; some regions may benefit (i.e., North America, Europe) and some regions may suffer (i.e., the tropics). Any losses would be on top of substantial gains in world production (which could be 55% greater than current by 2030). As another indicator of this trend, a small but growing suite of modeling studies generally predict that world crop (real) prices are likely to continue to decline through the first 2-3°C of warming before increasing with additional warming. Hence, 2-3°C of warming appears to be a crucial threshold for crop prices.

While the global situation looks manageable, there are reasons for serious concern at regional levels. The Third Assessment Report of the Intergovernmental Panel on Climate Change (IPCC), reported that a number of models simulate the potential production of temperate crops (wheat, maize, rice) to absorb 2-3°C of warming before showing signs of stress (Easterling and Apps, 2005). More recent assessment work found that agronomic adaptation extends the threshold for warming to beyond 5°C for those crops. Existing tropical crops exhibit immediate yield decline with even slight warming (Figure 4-25) because they are currently grown under conditions close to maximum temperature tolerances. Adaptation gives tropical regions a buffer of approximately 3°C of warming before yields of wheat, maize and rice decline below current levels.

Two regions that are likely to experience large negative impacts of climate change on agricultural production are Asia and Africa. Studies indicate that rice production across Asia could decline by nearly 4% over this century. In India, a 2°C increase in mean air temperature could decrease rice yield by about 0.75 tonnes ha^{-1} and cause a decline in rain fed rice in China by 5 to 12%. Sub-Saharan Africa could lose a substantial amount of cropland due to climate change-induced land degradation. Based on results from one climate model (HadCM3), as many as 40 food-insecure countries of sub-Saharan Africa may lose an average of 10 to 20% of their cereal-production potential due to climate change. Whether such declines are problematic depends on possibilities for trade and responses from agronomic research. However, some studies suggest that the impacts of climate change within a region are likely to be extremely heterogeneous, depending on local conditions. Several crop and livestock systems in sub-Saharan Africa that are highly vulnerable may experience severe climate change (Thornton et al., 2006). These include the more mar-

ginal mixed (crop-livestock) and pastoral systems in parts of the Sahel, East Africa, and southern Africa. In some areas, growing seasons may contract, and crop and forage yields may decline substantially as a result (Jones and Thornton, 2003). Vulnerable households in such places will need to adapt considerably if food security and livelihoods are to be preserved or enhanced.

While most studies still focus on changes in means, in fact changes in variability and extreme weather events may be even more important for agriculture than the changes in means. For instance, changing the frequency of dry years may seriously affect agriculture in certain areas.

Climate change will not be a major challenge to agricultural production systems in temperate regions until well into this century. In the tropics, especially Asia and Africa, however, even with adaptation, food (especially grain) production may decline with only modest amounts of climate change. Modeling studies also suggest that real food prices will reverse their long-term decline at about 3°C of warming, resulting in increasing prices thereafter.

4.4.4.4 Climate change mitigation and agriculture
According to several assessments, agriculture and forestry can play a significant role in mitigation policies (FAO, 2006d; IPCC, 2007c) as it is also a major source of green-

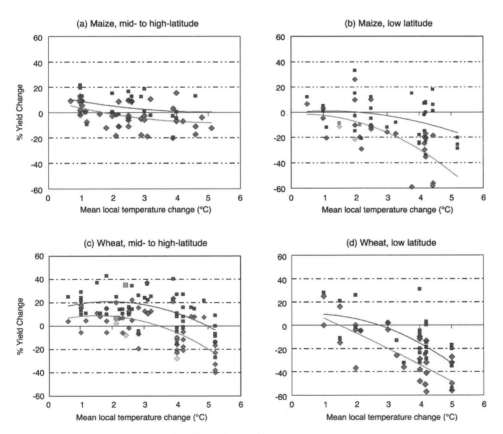

Figure 4-25. *Sensitivity of cereal yield to climate change.* Source: IPCC, 2007.

The diamond markers indicate studies reporting responses without adaptation; the block markers indicate studies reporting responses with adaptation. The lower line in each panel provides the trend line for the studies without adaptation; the upper line provides the trend line for studies with adaptation. The studies on which the figure is based span a range of precipitation ranges and CO$_2$ concentration and vary in the way how the present changes in climate variability.

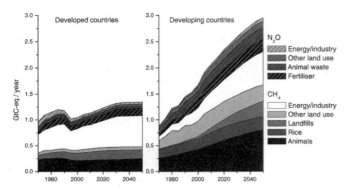

Figure 4-26. *Estimated historical and projected N_2O and CH_4 emissions from 1970-2050.* Source: Van Vuuren et al., 2007.

house gases (GHGs) (Figure 4-26). This particularly holds for methane (CH_4) and nitrous oxide (N_2O), both with higher global warming potential than CO_2. Methane emissions are primarily caused by livestock production and flooded rice fields, while N_2O emissions are related to the use of organic and inorganic N fertilizers. Finally, CO_2 emissions are also caused by land use changes and agricultural practices.

Several studies have found that reducing non-CO_2 greenhouse gases from agriculture and CO_2 emissions from land use change can effectively reduce emissions. These multigas emissions reduction scenarios are able to meet climate targets at substantially lower costs for the same targets as was found in the EMF-21 study (IPCC, 2007c; Lucas et al., 2007). A variety of options exists for mitigation of GHG emissions in agriculture (Table 4-13). Effective options are improved crop and grazing land management, restoration of drained organic (peat) soils and restoration of degraded lands (IPCC, 2007c). Lower, but still significant mitigation is possible with improved water and rice management, conservation plots, land use change and agroforestry. The relevant measures depend highly on the carbon price, i.e., the market price for reducing GHG emissions. At low carbon prices, profitable strategies are minor changes in present production systems, such as changes in tillage, fertilizer application, livestock diet formulation and manure management. Higher prices could allow the use of costly animal feed-based mitigation options, or lead to changes in land use. Effective options however, depend in general on local conditions, including climate, agricultural practices and socioeconomic circumstances; there is no universally applicable list of effective options (IPCC, 2007c).

4.4.4.5 Consequences for AKST
Based on the discussion above, challenges for AKST in the field of agriculture consist of:
- Adaptation. As climate change is likely to result in at least 2°C warming by the end of the century, and possibly >6°C, agricultural systems need to adapt to climate change. This will be even more important in developing countries than in developed countries. Effective adaptation should focus on extremes as well as changes in means. AKST will need to create less vulnerable system and AKST actors will need to provide information on options for adaptation.

- Mitigation. Agriculture is also a major source of emissions. Although some technologies already exist to reduce CH_4 and N_2O emissions from agriculture, further progress is required to reduce emissions beyond 2020.

4.4.5 Energy

4.4.5.1 Trends in world energy use
There are very important relationships between energy and agriculture. The industrial revolution led to improved access to energy services based on fossil energy (e.g., Smil, 1991) and allowed for higher production levels per unit of land or labor in the agriculture sector. In turn, this allowed for a dramatic increase in the global population, a (related) decrease in arable land per capita, and a movement of the work force away from agricultural production.

Global energy use during the last century increased by about 2.5% annually, with a clear transition in consumption of primary sources from coal to oil to natural gas (Figure 4-27). The large majority of current scenarios project a continuation of these trends. Global energy use continues to grow; in the first decades growth will primarily be based on fossil fuel consumption. Primary energy consumption projections can be found in the IPCC-SRES scenarios, World Energy Outlook (IEA, 2002; IEA, 2004) United States Department of Energy (US DoE, 2004) and the OECD environmental outlook (OECD, 2006a). The differences between the different scenarios (Figure 4-28) can be explained in terms of underlying economic growth assumptions and assumed emphasis on dematerialization. In nearly every scenario, the largest contribution to global energy increase comes from developing countries. The scenarios also share the projection that in 2030 the majority of global energy use needs come from fossil fuels. Nevertheless, clear differences in the energy mix may occur. The most important determinants are the stringency of future climate policy, differences in technology expectation, and assumed societal preferences. Studies also indicate that global energy consumption could increase by 25-100% in the next 30 years, a huge challenge to production. In the longer term, the energy mix may diversify in many different ways, ranging from almost total coal use (e.g., IPCC's A2 scenario) versus nearly total renewable energy (e.g., under stringent climate policy scenarios). The growing awareness of both the "global warming" and "peak oil" issues is finally forcing decision makers and the general public to put energy high on policy agendas.

4.4.5.2 The relationship between energy use and agriculture
The food and energy systems have historically interacted in several ways (e.g., Pimentel and Pimentel, 1979; Pimentel, 1980; Stanhill, 1984; Leach et al., 1986; Smil, 1987, 1991; Stout, 1991, 1992). As indicated above, a dominant trend during the last century was increasing energy use leading to continuous increasing productivity of agricultural land. The implication of this trend can be seen when comparing the performance of industrialized and developing countries, in the relationship between energy inputs and yields (Giampietro, 2002). Agricultural production represents only a small part of global energy consumption. However, energy consumption not only occurs in the production stage. In the EU, the food supply chain used nearly 4 EJ of primary en-

Table 4-13. Proposed measures for mitigating greenhouse gas emissions from agricultural ecosystems, their apparent effects on reducing emissions of individual gases where adopted (mitigative effect), and an estimate of scientific confidence in their reduction of overall net emissions at the site of adoption.

Measure	Examples	Mitigative effects[1]			Net mitigation[1] (confidence)	
		CO_2	CH_4	N_2O	Agreement	Evidence
Cropland management	Agronomy	+		+/-	***	**
	Nutrient management	+		+	***	**
	Tillage/residue management	+		+/-	**	**
	Water management (irrigation, drainage)	+/-		+	*	*
	Rice management	+/-	+	+/-	**	**
	Agro-forestry	+		+/-	***	*
	Set-aside, land-use change	+	+	+	***	***
Grazing land management/ pasture improvement	Grazing intensity	+/-	+/-	+/-	*	*
	Increased productivity (e.g., fertilization)	+		+/-	**	*
	Nutrient management	+		+/-	**	**
	Fire management	+	+	+/-	*	*
	Species introduction (including legumes)	+		+/-	*	**
Management of organic soils	Avoid drainage of wetlands	+	-	+/-	**	**
Restoration of degraded lands	Erosion control, organic amendments, nutrient amendments	+		+/-	***	**
Livestock management	Improved feeding practices		+	+	***	***
	Specific agents and dietary additives		+		**	***
	Longer term structural and management changes and animal breeding		+	+	**	*
Manure/biosolid management	Improved storage and handling		+	+/-	***	**
	Anaerobic digestion		+	+/-	***	*
	More efficient use as nutrient source	+		+	***	**
Bioenergy	Energy crops, solid, liquid, biogas, residues	+	+/-	+/-	***	**

Notes:

+ denotes reduced emissions or enhanced removal (positive mitigative effect)

- denotes increased emissions or suppressed removal (negative mitigative effect)

+/- denotes uncertain or variable response

[1] A qualitative estimate of the confidence in describing the proposed practice as a measure for reducing net emissions of greenhouse gases, expressed as CO_2-eq

Agreement refers to the relative degree of consensus in the literature (the more asterisks, the higher the agreement); Evidence refers to the relative amount of data in support of the proposed effect (the more asterisks, the more evidence).

Source: IPCC, 2007, adapted from Smith and Bertaglia, 2007.

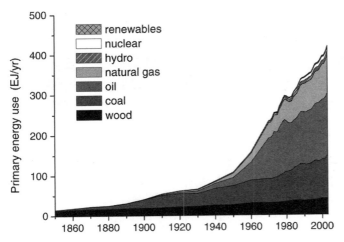

Figure 4-27. *Global energy use from 1880-2000.* Source: Van Vuuren, 2007.

ergy, or about 7% of total consumption (Ramirez-Ramirez, 2005). Within this total, nearly 45% is consumed by the food processing industries, around 25% by agriculture, around 10-15% by transport of foodstuffs and fodder and the remainder (5-10% each) for fertilizer manufacturing and transport of agricultural products.

The rapid population growth expected in developing countries is likely to have important implications for the relationship between energy use and agriculture. Increasing food production will require a strategy of intensification and consequently a further increase in the consumption of fossil energy for agricultural production. As this process is likely to coincide with structural changes in the economy, lower labor supply in agriculture will also lead to further intensification. As a result, systems in developing countries are expected to see considerable growth in energy consumption

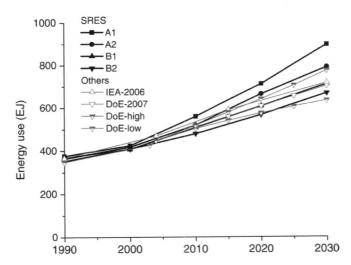

Figure 4-28. *Trends in 21st century energy use. Comparison of trends in SRES total primary energy consumption and more recent studies by US.DoE and IEA.* Source: USDoE, 2004; IEA 2004, 2006.

for food processing. Overall, this is likely to lead to further growth in agriculture energy demand, although at a lower rate than overall growth of energy consumption.

One additional factor is the role of energy prices; current high prices for oil and natural gas do have consequences, primarily on fertilizer use and transport, for agriculture. Projections for energy prices in the next decades have been revised upward in most reports (e.g., IEA, 2006), but still a considerable uncertainty remains. Higher projections are found for those projections that take into account further increased demand in Asia and restrictions (e.g., limited investments, depletion) on increases in supply.

4.4.5.3 Bioenergy

Climate change, energy security and the search for alternative income sources for agriculture have increased interest in bioenergy as an alternative fossil fuel. Many scenario studies, with and without climate policy constraints, project a strong increase in the use of bioenergy, with major implications for future agriculture (see IPCC, 2007c). However, at the same time there is a strong debate on the implications of bioenergy use; the outcome of this debate will critically influence its future use (see also Slesser and Lewis, 1979; Smil, 2003; Smeets and Faaij, 2004; Hoogwijk et al., 2005). Crucial controversies with respect to bioenergy use include whether bioenergy can provide net energy gains, reduce greenhouse gas emissions, cost-benefit ratio, environmental implications and the effects on food crop production (Box 4-4).

The potential for bioenergy production typically is based on land use projections (e.g., Smeets and Faaij, 2004). From a technical perspective, bioenergy could supply several hundred exajoules per year from 2050 onwards compared to a current global energy use of 420 EJ of which some 10% is covered by bioenergy already, predominantly in the form of traditional bioenergy. The major reason for the divergence among different estimates of bioenergy potentials is that the two most crucial parameters, land availability and yield levels in energy crop production, are very uncertain. The development of cellulosic ethanol could lead to much higher yields per hectare (see Chapter 6). Another factor concerns the availability of forest wood and agriculture and forestry residues. In particular, the use of forest wood has been identified as a potentially major source of biomass for energy (up to about 115 EJ yr^{-1} in 2050) but very low estimates are also reported.

In evaluating the information on bioenergy potential, the costs, land requirements and the environmental constraints will determine whether biomass can be transformed into a viable net energy supply to society. Hence, the drivers are (1) population growth and economic development; (2) intensity of food production systems, (3) feasibility of the use of marginal/degraded lands, (4) productivity forests and sustainable harvest levels, (5) the (increased) utilization of biomaterials, (6) limitations in land and water availability. Scenario studies evaluate bioenergy mostly in terms of competition against energy carriers and thus give an indication of demand. Bioenergy use in various energy scenarios varies widely (Figure 4-29). In these scenarios, use of bioenergy varies between 0 and 125 EJ yr^{-1} in 2030 and 25 and 250 EJ in 2050.

Box 4-4. Controversies on bioenergy use and its implications.

(a) Net energy gains and greenhouse gas emissions

There are many studies on the net energy gains of bioenergy, but results differ. These differences can often be traced to different technological assumptions, accounting mechanisms for by-products and assumed inputs (e.g., fertilizers). In some the production of ethanol from maize energy outputs has a small net gain (Farrell et al., 2006), while in others the net result is negative (Cleveland et al., 2006; Kaufmann, 2006; Hagens et al., 2006). Some other crops have a more positive energy balance, including ethanol from sugar cane, oil crops and conversion of cellulosic material (e.g., switchgrass) to second-generation biofuels. The greenhouse gas balance, a function of production patterns and agroclimatic conditions, is also important. Maize ethanol in the U.S. is believed to cut GHG emissions only by 10 to 20% compared to regular gasoline (Farrell et al., 2006) but some other crops are reported to obtain better reductions, e,g., ethanol from sugarcane—up to 90% reduction (CONCAWE, 2002; Farrell et al., 2006; Hill et al., 2006) and biodiesel up to 50-75% (CONCAWE, 2002; IEA, 2004; Bozbas, 2005; Hill et al., 2006; Rosegrant et al., 2006). More conservative analyses represent a minority, but they point to potential flaws in the mainstream lifecycle analyses, most notably with respect to assumptions about land use and nitrous oxide emissions.

(b) Costs of bioenergy

Studies on bioenergy alternatives generally find the low cost range from bioenergy to start at around $12-15 per GJ for liquid biofuels from current sugar cane to around US $15-20 per GJ for production from crops in temperate zones. In most cases, this is considerably more expensive than $6-14 per GJ for petroleum-based fuels crude oil price for oil prices from $30 per bbl to $70 per bbl. It is expected that costs of biofuels (especially the more advanced second-generation technology) will be further reduced due to technology progress, but the actual progress rate is highly uncertain. Agricultural subsidies and the economic profitability also affect the value of emission reductions under different climate policy scenarios.

(c) Impact on land use

A serious concern in the debate on biofuels is the issue of land scarcity and the potential competition between land for food pro-

duction, energy and environmental sustainability. The production of 1st generation biofuels from agricultural and energy crops is very land intensive. Land evaluation depends on (1) availability of abandoned agriculture land, (2) suitability of degraded lands for biofuel production and (3) use of natural areas. Obviously, biofuels can also compete with food production for current agricultural land and/or expansion of agricultural land into forest areas. Examples of this can already be seen where expansion of crop plantations for biofuels production have led to deforestation and draining of peatlands, e.g., in Brazil, Indonesia and Malaysia (Curran et al., 2004; FOE, 2005; FBOMS, 2006; Kojima et al., 2007).

(d) Impact on food prices

As long as biofuels are produced predominantly from agricultural crops, an expansion of production will raise agricultural prices (for food and feed). This has now become evident in the price of maize (the major feedstock in U.S. ethanol production), which increased 56% in 2006. Price rises are expected for other biofuels feedstock crops in the future (OECD, 2006; Rosegrant et al., 2006). This increase in price can be caused directly, through the increase in demand for the feedstock, or indirectly, through the increase in demand for the factors of production (e.g., land, water, etc.). More research is needed to assess these risks and their effects but it is evident that poor net buyers of food would suffer strongly under increasing prices. Some food-importing developing countries would be particularly challenged to maintain food security.

(e) Environmental implications

Whereas implications for the environment are relatively low for current small-scale production levels, high levels of biofuels feedstock production will require considerable demand for water and perhaps, nutrients. Some studies have indicated there could be tradeoffs between preventing water scarcity and biofuel production (CA, 2007). Bioenergy production on marginal lands and the use of agricultural residues could negatively affect soil organic matter content (Graham et al., 2007).

Implications for agriculture. Based on the discussion above, one possible outcome in this century is a significant switch from fossil fuel to a bioenergy-based economy, with agriculture and forestry as the leading sources of biomass (FAO, 2006f). The outcomes can be unclear. One can envision the best scenarios in which bioenergy becomes a major source of quality employment and provides a means through which energy services are made widely available in rural areas while it gives rise to environmental benefits such as carbon reductions, land restoration and watershed protection. On the other hand, one can envision worst case scenarios in which bioenergy leads to further consolidation of land hold-

ings, competition for cropland and displacement of existing livelihoods while it incurs environmental costs of decreased biodiversity and greater water stress (World Bank, 2005a).

Currently, bioenergy fuel use is rapidly expanding in response to government policies and subsidies, high energy prices and climate policy initiatives. In this context, bioenergy can also offer development opportunities for countries with significant agricultural resources, given lower barriers to trade in biofuels. Africa, with its significant sugar cane production potential, is often cited as a region that could profit from Brazil's experience and technology, though obstacles to realizing it (infrastructure, institutional, etc.)

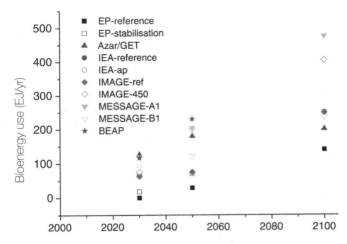

Figure 4-29. *Biomass use in the different global energy scenarios. (range of different studies).* Source: Dornburg et al., 2008.

should not be underestimated. Obviously, transporting bioenergy across the world would become a major new challenge as well.

While there is controversy on the size of the effect, major bioenergy use certainly affects environmental resources (e.g., water, land and biodiversity). Therefore, it is very important to understand and quantify the impacts and performance of bioenergy systems for determining how successful the use of biomass for energy (and materials) is, how the benefits of biomass use can be optimized and how negative impacts can be avoided.

4.4.5.5 Most important implications for AKST

Large scale use of bioenergy could transform the agricultural system into a net producer of energy. As indicated in latter parts of this assessment, the potential of bioenergy is such that it requires data and information tools for decision making based on solid technical, social and economic knowledge. The intrinsic interdisciplinary character of bioenergy means that implications for AKST will encompass areas as varied as agricultural and energy policies, natural resources and biodiversity protection and rural development. Interaction between the agricultural sector and the energy, environment and industrial sectors as well as sustainability protocols will be vital for successful bioenergy use. From the overall bioenergy chain point of view, it is important to monitor and further improve systems with respect to (1) implications for soil and water; (2) supply of agricultural inputs (fertilizer, fuel, machinery); (3) increasing overall efficiency and (4) minimizing effects on biodiversity.

4.4.6 Labor

4.4.6.1 Trends in employment of labor in agriculture

Hardly any information on labor projections is found in the currently published scenario studies. Therefore, historical trends are used here to assess future trends. Over the last 10 years, there has been a global decline in the relative share of employment in agriculture: from 46% in 1994 to 43% in 2004. However, agriculture continued to be the largest

source of employment (around 60%) in sub-Saharan Africa, South Asia and East Asia. At the same time, the share of agriculture in total employment in developed countries is small: only 4% in 2004 and likely to decline further. As the trends observed above (both in time—and across regions) are global, they are projected to continue, leading to ever lower numbers employed in agriculture. This decline underlies many of the economic projections of future scenarios. The share of agriculture in total employment will decrease dramatically in developing countries and decline more slowly in industrialized countries.

4.4.6.2 Labor productivity in agriculture

Future trends in labor productivity are expected to increase, based on the evidence over the past decade. Labor productivity in the world increased by almost 11% over the past ten years (ILO, 2005). This increase was primarily driven by the impressive growth in labor productivity in Asia and the industrialized economies. The transition economies have also contributed to the world's recent growth in productivity. The Latin America and the Caribbean realized productivity increase of just over 1% over 10 years, mainly due to the economic crisis in the beginning of the century. There were no changes in the Middle East and North Africa, while sub-Saharan Africa experienced declining productivity on average.

Similar trends in future productivity are anticipated. Based on historical data for 71 countries from 1980 to 2001, agricultural GDP per worker in sub-Saharan Africa on average grew at a rate of 1.6% per year slower than for countries in Asia, Latin America, the transition economies and the Mediterranean countries (Gardner, 2005). Agricultural GDP per laborer and national GDP were positively correlated based on data for 85 countries for 1960-2001. Within each of the regional grouping (Africa, Asia, and Latin America), the countries that grew fastest in national GDP per capita also grew fastest in agricultural GDP per worker, with a few notable exceptions, e.g., Brazil (Gardner, 2005; ILO, 2005). Levels of productivity in Latin America are the highest in the developing world, followed by the Middle East and North Africa and the transition economies. East Asia, South Asia and sub-Saharan Africa, where the majority of the poor live, have considerably lower average labor productivity (World Bank, 2004a).

An increase in agricultural labor productivity has a more significant direct effect on poverty reduction than increases in total factor productivity (ILO, 2005) and there are indirect effects on poverty from changes in food production and food prices (Dev, 1988; ILO, 2005).

Productivity gains can lead to job losses, but productivity gains also lead to employment creation, since technology also creates new products and new processes; hence, the increasing trends in productivity could lead to expanded employment in other sectors such as information and communications technologies (ILO, 2005).There will be a critical need to provide adjustment strategies (financial assistance and retraining) for displaced workers and to ensure growth in the long-term.

While in the short run, increased productivity might affect growth of employment in agriculture adversely, this outcome may not hold in the long run. Economic history shows that over the long-term, the growth of output, employment and productivity progress in the same direction (ILO, 2005). However, social costs in the short-term can be high.

4.4.6.3 Gender perspectives in agricultural labor

As agriculture and food systems evolve over the next decades, gender issues and concerns are highly likely to continue to be central to AKST development, at least in the developing countries where women have played a significant role in traditional agricultural production. Over the years improvements in agricultural technologies have seldom been targeted as recipients of improved technologies. Yet there are more women working in agriculture than men, e.g., women in rural Africa produce, process and store up to 80% of foodstuffs, while in South and South East Asia they undertake 60% of cultivation work and other food production (UNIFEM, 2000).

Employment of female vis-à-vis male workers is likely to decline in the future as women obtain more employment in other sectors. Historically, there has been a global decline in the world over the last 10 years (47% in 1994 to 43% in 2004). While this may imply further declines in the future, it may also be true that female employment in agriculture will increase as a result of changing production patterns.

The increasing participation of women in subsistence production in agriculture is highly likely to continue facilitating male out-migration to urban areas and to other sectors such as mining and commercial farming (at lower costs to society than would otherwise be possible). The greater number of men moving out of the agricultural sector is highly likely to continue. A slightly increasing feminization of the agricultural labor force in most developing countries may reflect the fact that women are entering into high value production and processing and thus less likely to abandon their agricultural ventures (Mehra and Gammage, 1999).

4.5 Existing Assessments of Future Food Systems, Agricultural Products and Services

4.5.1 Assessments relevant for changes in food systems

Existing assessments provide information how agricultural and food systems might change in response to the changes in the direct and indirect drivers discussed in the previous subchapters (note that the outcomes of these assessments may be compared to the reference scenario presented in Chapter 5). Over the past 50 years, there have been at least 30 quantitative projections of global food prospects (supply and demand balances). We have reviewed several recent global assessments (see 4.2) that provide information relevant for future agriculture and food systems, either directly (i.e., assessments with an agricultural focus) or indirectly (other assessments that include agriculture). Important organizations that provide specific agricultural outlooks at the global scale include the Food and Agriculture Organization of the United Nations (FAO), the Food and Agriculture Policy Research Institute (FAPRI), some of the research centers of the Consultative Group on International Agricultural Research (CGIAR) such as the International Food Policy Research Institute (IFPRI), the OECD, and the

United States Department of Agriculture (USDA). Other food projection exercises focus on particular regions, such as the European Union. Finally, many individual analyses and projections are implemented at the national level by agriculture departments and national level agricultural research institutions.

In subchapter 4.2, we introduced a selection of global assessments and discussed their objectives and use of scenarios. None of these—IPCC's Assessment Reports (IPCC 2001, 2007abc), UNEP's Global Environment Outlooks (UNEP 2002, 2007; RIVM/UNEP, 2004), the Millennium Ecosystem Assessment (MA, 2005a), IFPRI's World Food Outlook (Rosegrant et al., 2001), FAO's World Agriculture AT 2015/2030 (Bruinsma, 2003) and IWMI's Comprehensive Assessment of Water Management for Agriculture (CA, 2007)—address the full spectrum of the food system and AKST from the perspective of a range of different plausible futures (Table 4-14). This is not surprising given the different objectives of these assessments (see 4.2), but it does imply that an assessment that meets development objectives cannot be met solely through analyzing earlier assessments. Although the projections provided by FAO and IFPRI address agricultural production and services to some degree, the attention paid to AKST elements is relatively limited. This highlights a need for new work to integrate plausible futures with regard to the interactions between driving forces and food systems while addressing AKST in more detail. Analysis of recent scenarios exercises indicates that while some elements related to the future of food systems are touched upon, the focus of these exercises is more on production and consumption than on the distribution component of food systems. Most studies addressed qualitative and quantitative production indicators, and provide assumptions on yields for various crops, area under certain crops, input use or exchange mechanisms. Consumption as well as access to food (including affordability, allocation and preference), has often been addressed through modeling food demand in different scenarios. For example, assumptions regarding allocation of food through markets are made indirectly under different scenarios by assuming whether and how well markets and governance systems function. Food preferences are usually covered in a more qualitative manner through assumptions made about changes due to various cultural and economic factors (Zurek 2006).

The area least covered by the reviewed scenario exercises is food utilization (Zurek 2006). The IFPRI and MA exercises calculated the number of malnourished children under each scenario (a very basic indicator of hunger and whether nutritional standards are met), but nutritional outcomes under different diets and their possible changes are seldom addressed. Little, if anything, is said in any of the exercises concerning food safety issues or the social value of food, both of which can have important consequences for food preferences. The MA does quantitatively assess certain health indicators; these could be used to give a further indication on human nutritional status in different scenarios. Further in-depth research is needed on some of the specific food systems variables and their changes in the future, specifically for those related to food utilization, as well as a number related to food accessibility.

Table 4-14. **Overview of existing assessment and their relationship to agriculture.**

	IPCC/IPCC-SRES	UNEP-GEO3	MA	IFPRI 2020	FAO AT 2015/2030	CGIAR CA
Crop production levels and consequences for land	Some	Yes	Yes	Yes	Yes	Yes
Livestock production levels and consequences for land	Some	Yes	Yes	Yes	Yes	Yes
Fisheries (production and stocks)			Some	Some	Yes	Yes
Forestry		Some	Some		Yes	
Distribution		Indirect	Indirect	Indirect	Yes	Indirect
Exchange		International trade	International trade	International trade	International trade	International trade
Affordability		Some	Yes	Yes	Yes	
Allocation		Market	Market	Market	Indirect	
Preferences			Yes	Yes	Yes	
Nutritional Value			Yes	Yes		Yes
Social Value						
Food Safety			Some		Some	
Relationship with environmental variables	Climate	Yes	Yes	Some	Some	Yes
Explicit description of AKST issues			Some	Some	Some	

Source: Zurek and Henrichs, 2006.

4.5.2 Indication of projected changes

Food systems can be classified into (1) production and (2) distribution and delivery. Most assessments discussed here concentrate much more on the first. The Millennium Ecosystem Assessment, the global food projections by IFPRI and the Agriculture towards 2015/30 study by FAO provide the most relevant information in the context of the IAASTD (see Table 4-9). It should be noted the Millennium Ecosystem Assessment used four diverging scenarios (Global Orchestration, Technogarden, Adapting Mosaic and Order from Strength). Together these four scenarios cover a broad range of possible outcomes for the development of different ecological services.

4.5.2.1 Changes in production systems

Agricultural production systems can be classified in different ways. A system based on two key dimensions of cultivated systems: an agroecological dimension and an enterprise/management dimension was proposed for MA (Cassman et al., 2005). Such an approach can easily be coupled to both biogeographic factors and long-term trends in agricultural management, and hence provides a very useful structure to assess potential future changes in production systems. It pro-

vides a basis to integrate socioeconomic analysis (looking at the economic and social viability of agricultural systems) and biophysical analysis (looking at the environmental consequences). Unfortunately, however, the different existing assessments generally tend to analyze information at a much more aggregated scale, because data is lacking, particularly on agricultural management in developing countries.

General trends. In the system proposed above a useful distinction can be made along the management axis in the degree of intensification. Such a distinction would include (1) intensive (or fully colonized) agroecosystems (e.g., producing crops, often in monocultures, intensive livestock and specialized dairy farms); (2) intermediate (partially colonized) agroecosystems (e.g., pastoralism, agroforestry, slash and burn); and (3) the exploitation of uncontrolled ecosystems (e.g., fishing in the ocean or in big rivers, hunting and gathering). From a human perspective, this distinction of intensification refers to an assessment of costs and benefits. Taking out products from an exploited ecosystem requires a degree of "investment" (e.g., tilling the soil, taking care of animals, preparing fishing nets), which needs to deliver an adequate return in terms of value. This distinction is relevant also from an ecological perspective. In the case of sustainable fishing, hunting and gathering, the basic structure of the ecosystem is preserved. In partial colonization, humans manage to produce crop plants and/or livestock at a density higher than that typical of natural ecosystems. Full colonization, finally, generates agroecosystems with very little in common with the natural ecosystem that they replace. Historically, there has been a trend towards intensification of agricultural systems, although in many areas extensive systems are also still common. In crop, livestock, forestry and fishery production systems, further intensification is projected to meet increasing demand worldwide. A natural consequence of the related increase in agricultural inputs (e.g., energy, fertilizers) will also be further pressure on natural ecosystems. Without intensification increasing demands would need to be met by further expansion.

Global crop production. Worldwide, numerous cropping systems can be distinguished based on agroecological parameters, cultivation and the type of crops grown. In terms of cultivation, these categories range from irrigated systems, to high external-input rain fed and low external-input rain fed systems, shifting cultivation and mixed crop and livestock systems. In time, a noticeable trend can be observed in many countries from low-input systems to high-input systems. This shift follows from an assessment of costs and benefits, weighing the costs of inputs against the increased yield levels. The shift to high-input systems had occurred in several regions of the world by the middle of the last century, but in other areas it has occurred during the last 40 years (e.g., the Green Revolution in Asia). A basic underlying driver of this shift is increasing global food demand as a result of increasing population (see 4.3.1 and 4.4.1). Low-input systems still provide a substantial share of total agriculture, in addition to providing livelihoods for hundreds of millions of resource-poor people in developing countries. For instance, shifting cultivation is the dominant form of agriculture in tropical humid and sub-humid upland regions, and low-input rain fed systems are still important in many parts of the world (FAO, 2002b).

All assessments provide relatively little information on trends in underlying production systems for food crops; the discussion is more on an aggregated crop level with most attention focused on cereals. Worldwide, cereals represent about two-thirds of the total crop production and the total harvested area. In all assessments, the production of cereals is expected to increase (Figure 4-30). Interestingly, differences among the scenarios of these different assessments are very small. One underlying reason is that in all cases, the increase of global cereal production seems to be coupled to the increase in the global population. The increase in cereal production in the next few decades ranges from around 0.9% annually (lowest MA scenarios) to 1.3% (the IFPRI projection), which is slightly below the annual increase for the total crops production reported in these assessments. This number is, however, considerably lower than the increase in production over the past 30 years (around 2.1% and up to 3.1% annually in developing countries) (also the historic increase is nearly equal to the increase in population over the same period). These numbers are aggregated: for both the historic numbers and the projections there is a large varia-

tion at the regional and country scale, implying important trends in food trade and food security. Finally, it is important to note that in the time frame of the scenarios an increasing share of cereals will be used as animal feed to supply the very rapidly growing demand for livestock products.

There are two main sources of growth in crop production: (1) expansion of harvested land area and (2) yield increases. Globally over the last three decades yield increases for cereals have provided about 70-80% of production growth, while harvested land expansion contributed about 20-30% of growth. In the scenarios developed by these assessments, contribution of expansion of harvested land to increase in cereal production ranged from as low as 5 to around 30%. The lowest numbers are reported for the MA scenarios that assume high levels of technology change (Global Orchestration and Technogarden); all other scenarios find values that are near, or somewhat below the historic values. The lower contribution to total production from the expansion of crop area can be attributed to increasing land scarcity and possibly the lower overall rate of production increase. A decreasing quality of land brought into production, however, could imply that a greater percentage of gains in total production will be attributable to crop area expansion than has historically been the case (as indicated in the MA). Even in the two scenarios with little global expansion of harvested land, a considerable expansion of arable land still occurs in Africa, Latin America and partly in Asia, but this is compensated for by a decrease of harvested area in temperate zones. In the other scenarios, the largest expansion also occurs in these regions. The yield growth in these scenarios is about 0.6-0.9% annually at a global level. Several factors contribute to this (reasons are reported in more detail in the FAO and IFPRI assessment than in the MA), including increased irrigation and shifts from low-input to high-input agriculture. In any case, the assumed yield growth in each of the scenarios is considerably below the historic rate of change. The suggested trends in expansion of agricultural land in tropical zones are controversial, with questions about how this expansion can happen in many parts of the tropics (particularly in Africa) in any meaningful way.

For total agricultural land (all crops), similar trends are reported, although the area expansion is somewhat higher

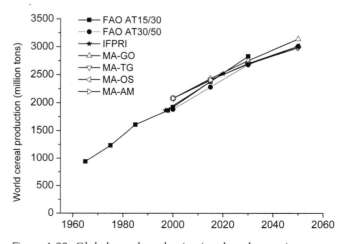

Figure 4-30. *Global cereal production in selected scenarios*

than for cereals alone. Across the assessments, the area in crop production increases from 1.5 billion ha (or 11% of the earth's land surface) to 1.60 to 1.77 billion ha. As indicated by FAO, this expansion is within the scope of total land available for crop production. The fact that the assessments considered here agree on a rather flexible continuous response of the agriculture system to demand increases is interesting, as more skeptical views have also been expressed. An important implication, however, is further loss of the area available to unmanaged ecosystems.

Global livestock production. Livestock production systems differ greatly across the world. Confined livestock production systems in industrialized countries are the source of most of the world's poultry and pig meat production, and hence of global meat supplies (FAO, 2002b). Such large-scale livestock systems are also being established in developing countries, particularly in Asia, to meet increasing demand for meat and dairy products. Livestock production also occurs in mixed crop-livestock farming systems and extensive grazing systems. Mixed crop-livestock systems, where crops and animals are integrated on the same farm, represent the backbone of small-scale agriculture throughout the developing world. Globally, mixed systems provide 50% of the world's meat and over 90% of its milk, and extensive pasture and grazing systems provide about 20-30% of beef and mutton production. To date, extensive grazing systems in developing countries have typically increased production by herd expansion rather than by substantial increases in productivity, but the scope for further increases in herd numbers in these systems is limited. The share of extensive grazing systems is declining relative to other systems, due both to intensification and to declining areas of rangeland. Considering all food production systems together, livestock production is the world's largest user of land (about a quarter of the world's land), either directly for grazing, or indirectly through consumption of feed and fodder.

As incomes increase, demand for animal products increases as well. This trend, which has been empirically established in all regions, is assumed to continue in the scenarios of the three assessments considered here. As a result, meat demand is projected to increase at a greater rate than the global population. Changing dietary preferences also contribute to this increased demand in the scenarios. Interestingly, future meat production varies considerably more than future cereal production among the scenarios (Figures 4-31, 4-32, 4-33). Assessments indicate similar growth rates for other animal products such as milk.

The increases in meat production will occur through a number of means, including changes that lead to intensified production systems, such as expansion of land use for livestock, and more efficient conversion of feed into animal products (Figure 4-34). Both the MA and FAO assessments indicate that most of the increases in world livestock production will occur in developing countries; second, while scenarios differ in their projections of future pasture area, compared with crop land area most scenarios expect very little increase in pasture land. For grazing systems, this means that some intensification is likely to occur particularly in the humid-subhumid zones where this is feasible. Considerable intensification is likely in the mixed systems,

with further integration of crop and livestock enterprises in many places. Strong growth is implied for confined livestock production systems; in the FAO scenario at least 75% of the total growth is in confined systems, although there are likely to be strong regional differences (e.g., less growth of these systems in Africa). This is a continuation of historic trends. The major expansion in industrial systems has been in the production of pigs and poultry, as they have short reproductive cycles and are more efficient than ruminants in converting feed concentrates (cereals) into meat. Industrial enterprises now account for 74% of the world's total poultry production, 40% of pig meat and 68% of eggs (FAO, 1996). At the same time, a trend to more confined systems for cattle has been observed, and a consequent rapid increase in demand for cereal- and soy-based animal feeds (these trends are included in the projections discussed in the previous subchapter) (see Delgado et al., 1999).

Finally, while there are good economic arguments for the concentration of large numbers of animals in confined systems, there can be significant impacts on surrounding ecosystems, something that is only recently started to be assessed in sufficient detail in agricultural assessments. The effects primarily involve N and P cycles. While some types of manure can be recycled onto local farmland, soils can quickly become saturated with both N and P since it is costly to transport manure.

Forestry. The FAO assessment pays considerable attention to forestry and the outlook for forestry, but mostly in a qualitative way. The MA also considers the future of forestry, but focuses more on the extent of natural forests than the development of forestry as a production system (although some data is available). Overall, both assessments agree on that the general trend over the last decades of a decreasing forest resource base and an increasing use of wood products will continue.

Important driving forces for forestry include demographic, sociopolitical and economic changes, changes in extend of agricultural land, and environmental policy. Both population and economic growth affect forestry directly via an increase in demand for wood and indirectly via the impact on agricultural production. There is strong evidence that with rising incomes, demand for forest products increases, especially for paper and panel products. The increasing demand for wood products is also assumed in the scenarios of the FAO and MA (Figure 4-35). The demand for industrial roundwood is expected to increase by about 20-80%. The lowest projection results from the Technogarden scenario (assumes a high efficiency of forest utilization in order to protect forests) while the highest projection results from the Global Orchestration scenario (reflecting the very high economic growth rate).

The use of wood products as a source of energy (fuelwood) is not expected to grow fast, and may even decline. The use of fuelwood is particularly important at lower incomes; wealthier consumers prefer and can afford other forms of energy. As a result, fuelwood consumption is a function of population growth (increasing fuelwood demand) and increased income (decreasing demand), with the net results being a small decline and rise over the next 30 years. The impact of environmental policies on forestry may

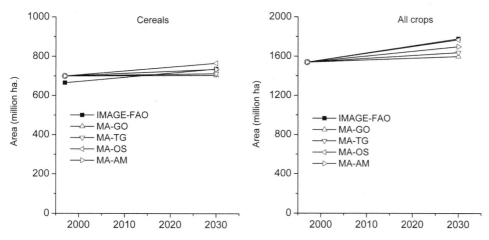

Figure 4-31. *Harvested area for cereals and all crops in selected scenarios.* Source: MNP, 2006.

Note: FAO refers to the implementation of the FAO AT2015/2030 scenario in the IMAGE model. FAO report only provides areas in developing countries.

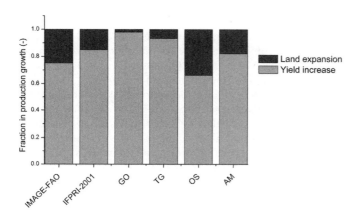

Figure 4-32. *Indication of factors underlying production growth in selected scenarios.* Source: MNP, 2006.

Note: FAO numbers refer to IMAGE implementation.

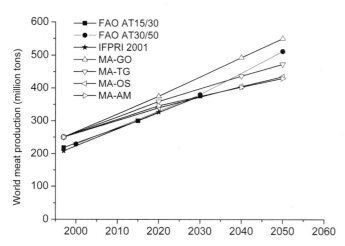

Figure 4-33. *World meat production in selected scenarios.* Source: MNP, 2006.

be important: e.g., increasing protection of forests and strategies to mitigate climate change may both result in encouraging less deforestation and reforestation initiatives to offset energy related greenhouse gas emissions.

The resulting trends in forested areas are presented in the MA for forests as a whole. The MA scenarios mostly show a further decline in forest area, but at a much slower rate than historically. In fact, the slow global deforestation trend is a result of a net reforestation in temperate zones, and a net deforestation in tropical areas. The slower deforestation trend is a direct result of the lower rate of expansion of agricultural areas coupled with greater forest conservation efforts.

Fisheries. Potential trends in world fisheries are discussed in qualitative terms in the FAO assessment, while the MA provides some projection for world fish consumption. Both assessments indicate that production of wild capture fisheries is approaching (or has passed) its sustainable limits, indicating that no real increase is expected. This implies that any growth in production will need to come from aquaculture (which is already the fastest growing component of world fisheries; especially in developing countries). It should be noted, however, that currently aquaculture mostly relies on feed that is provided by wild capture fisheries and can also cause serious pollution. Further growth, therefore, relies on finding sustainable ways to increase aquaculture. The MA reports both more conservative views (supported by ecological models) and more optimistic models (supported by agroeconomic projections). The FAO assessment presents a similar open-ended view on the future of aquaculture, indicating growth is likely to occur, but provided sustainable sources for feed are found.

4.5.2.2 Changes in food distribution and delivery
As indicated earlier, the amount of information on how other parts of the food system may change in the future is far less elaborated than the information on production systems. Based on the driving forces discussed earlier in this chapter,

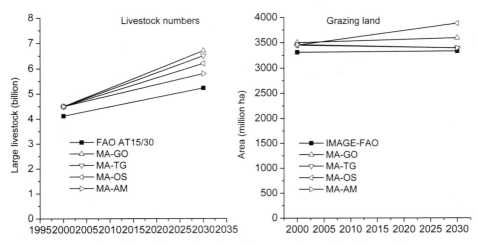

Figure 4-34. *Number of large livestock and total grazing area in selected scenarios.* Source: MNP, 2006.

Note: FAO refers to the IMAGE implementation of the FAO scenario.

and the limited information found in the 3 assessments looked at here some trends may be hypothesized:

- Assessments expect agricultural trade to increase, as indicated earlier (see 4.3.1). These increases are most pronounced in the globalization scenarios (assuming a reduction of trade barriers)—but also occur in scenarios that assume a more regional focus as a result of increasing demand for agricultural products. Obviously, this trend may have very important implications for both commercial and small-scale farmers in developing countries. Another implication may be the increasing importance of multinational companies.

- As discussed earlier, urbanization is likely to continue in all scenarios. As a consequence, food will increasingly be available to consumers via retailers and supermarkets, a trend that represents a continuation of a major trend already taking place in developing countries (Reardon et al., 2003. This trend will slowly influence

the importance of different actors in the food systems (see 4.4.1), although consequences are hard to assess. The role of farmers may, for instance, be very different in the MA's Global Orchestration scenario (with a strong market focus) than under the MA's Adaptive Mosaic scenario (in which farmers may successfully organize themselves). Important consequences of the trend towards retailers and supermarkets (and underlying urbanization) also include changes in diets (4.4.1), an increasing focus on production standards, demanding quality and safety attributes, and an increasing commercialization of upstream production processes.

- There are direct relationships between the above discussed demographic trends and agricultural production processes as well. For instance, location in relation to urban centers affects access to markets for purchased inputs and the costs of such inputs often leading to confined agricultural in periurban zones (thus reinforcing

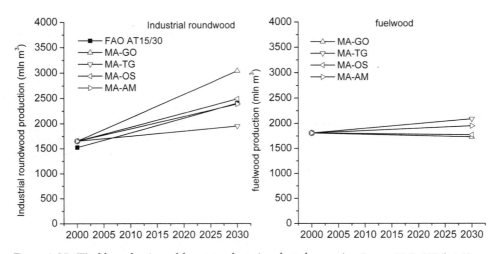

Figure 4-35. *World production of forest products in selected scenarios.* Source: FAO, 2006b; MA 2005a.

trends towards the intensified systems discussed earlier). Confined production systems facilitate the management of nutrition, breeding and health (responding to the need for production standards), but do so at the cost of increased investment demand. There are economies of scale in the provision of such processing services and the associated product marketing, and possibly in the supply of inputs (feed and feed supplements) and genetic material (e.g., day old chicks or semen). Again, this implies that under most of the scenarios discussed an increase in cooperative group activity or vertical integration of small-scale producers with large scale processing and marketing organizations.

- None of the assessment explored scenarios that completely challenge currently seen developments, such as a (1) strongly rising demand for ecologically produced food in developed countries, (2) an adoption of vegetarian diets across the world (the MA scenarios only explore slower and more rapid increase in meat demand), (3) major shifts in productivity levels as result of successful development in GMOs or other new agricultural technologies, or (4) a trend towards healthy food (vegetables, fruits) versus more high animal protein diet.

4.6 Relevance for Development and Sustainability Goals and AKST in the Future

Agriculture is a complex system that can be described by economic, biophysical, sociocultural and other parameters. However, its future is determined by an even larger set of direct and indirect drivers. Global assessments, e.g., provided by the IPCC, the MA, and FAO, and reviewed in this chapter, have addressed plausible future developments in agriculture. These assessments have made use of different approaches to address future agricultural changes, and usually employ either detailed projections accompanied by limited policy simulations or scenario analyses that consider a wide range of uncertainties in an integrated manner. Neither of these approaches aims to predict the future, but rather provide a framework to explore key interlinkages among different drivers and their resulting changes. Though these recent global assessments provide a host of information on plausible future developments regarding agricultural production systems and their driving forces, none of these assessments has explicitly focused on the future role of AKST.

4.6.1 What development and sustainability goals can to be addressed through AKST?

Some of the trends in direct and indirect drivers benefit agriculture and its role in realizing more sustainable development. Other trends, however, imply considerable challenges. Among the most important drivers identified in this chapter are:
- Land use change (balancing land claims in response to an increasing demand for agricultural products with the objective of protecting natural ecosystems)
- Changes in trade patterns (in particular consequences for smallholder farmers)
  Land degradation and water scarcity
- Climate change
- Urbanization (in particular with respect to consequences for food-supply chains)

- Demand for bioenergy
- Governance
- Breakthrough in crop and soil management, including ecological intensification, biotechnology and information technologies applied to agriculture
- Investments in AKST (both the volume and direction)

The projected increase in the global population in the next 50 years (2-3 billion people), ongoing urbanization, and changing lifestyles are likely to lead to a strongly increasing demand for agricultural products and services. Assessments indicate that this could exert pressure on the natural resource base. Historic evidence shows shifts towards more meat-intensive consumption patterns with increasing incomes, and projections are similar for the future. The demand for agricultural products will need to be met while simultaneously addressing the critical role agriculture and land use change play in global environmental problems. In this context, demand for agricultural products, land use, biodiversity and AKST are intrinsically linked. In addition to demand for food, feed and fiber, demand for bioenergy is expected to increase. A major uncertainty in the land use change scenarios presented in the literature stems from the assumed degree of extensification and intensification of agriculture. Most assessments indicate that roughly 70-80% of the extra production is projected to stem from intensification. This implies that increasing demands are also partly met by expansion of cultivated land. This is particularly the case in sub-Saharan Africa, Latin America and East Asia. AKST may help in addressing the need for productivity gains while simultaneously considering the role of agriculture and land use on local, regional and global environmental problems.

There are many reasons for increasing agricultural trade, such as increasing demand for food, increasing interregional relationships and commodity specialization, possibly facilitated by trade liberalization. Interestingly, even scenarios that assumed no further trade liberalization reported increases in agricultural trade (driven by increased demand for agricultural products). Several studies report that further globalization and liberalization will affect countries and groups within countries in different ways. While agricultural trade among developing countries is likely to increase, as a group they may become net importers of agricultural commodities with a possibility of further widening agricultural trade deficit. Conversely, industrialized countries tend to become net beneficiaries of trade arrangements as they are expected to face less pressure to reduce their support for agriculture.

Existing assessments (in particular the MA) also highlighted the role of agriculture as a major contributor to global environmental change, such as land degradation, nutrient pollution and increasing water scarcity. The rapid expansion of irrigation and associated agricultural water withdrawals for improved productivity is expected to continue to depend on availability of water resources sufficient to produce food for the growing world population while at the same time meet increasing municipal, industrial and environmental requirements. Earlier assessments indicate that water availability for agriculture is one of the most critical factors for food security, particularly in arid and semiarid regions in the world, where water scarcity has already become a severe constraint to food production. Water scarcity and increasing rates of soil degradation in many regions may limit the ability of agriculture systems to reduce food insecurity and to meet the MDG target of halving hunger by 2015. Moreover, increasing rates of land degradation in many regions may limit the ability of agriculture systems to provide food security. A final important factor is the role of agriculture in the N cycle, with effects on both local and regional scales. Decreasing these impacts may require important changes in soil fertility management. AKST must continue to address the need to develop sustainable agricultural systems in these regions. In this context, it should be noted that there are several scenarios that highlight many opportunities for enhancing the positive role of agriculture in providing ecosystem services, minimizing its environmental impacts and adapting to global environmental change.

Agriculture, a highly climate-sensitive sector, is already strongly affected by climate variability in many parts of the world, and it will be even more affected by climate change in the future. The relevant changes in climate of importance to agriculture include not only changes in mean temperature and precipitation, but even more importantly, seasonal and interannual variability and extreme events. The outcomes of the impact of climate change will vary significantly by regions. Current studies indicate that negative impacts tend to concentrate in low income regions. In some other regions, often at high latitude, there could be net positive impacts on yields. Developments in AKST will certainly influence the capacity of food systems to respond to the likely changes. Agriculture is also a source of CO_2 and non-CO_2 greenhouse gases and therefore can play a significant role in mitigation strategies. In order to play this role, new AKST options for reducing emissions of methane and NO_X from agriculture are needed.

The projected urbanization will likely coincide with a decline in the percentage of population depending directly on agriculture for their livelihood. At the same time, projected increasing income levels are likely to lead to changing diets and changing manner of food preparation. The consequences of this for the food supply chain, and in particular the role of retailers can be an important factor in future agriculture. Demand for food is also very likely to be affected by other demographic changes, e.g., the aging population in many industrialized countries. AKST will have to address the impact of changes from urbanization, consumption patterns and the agricultural labor force on agricultural production and technologies in order for food demands of the future to be met.

Energy will continue to play an increasingly important role in agriculture. Various forms of agriculture use different levels of energy; with transitions in agricultural production systems in general leading to a substitution of energy for labor. Most assessments also expect higher energy prices which could encourage the use of more energy-efficient technologies in agricultural production as well as in processing and distributing food. The most important factor with respect to energy, however, is that agriculture may become an important producer of energy in the form of bioenergy, based on both energy-security and climate change considerations. Existing assessments indicate a major increase in bioenergy production; this might lead to a tradeoff between energy security and food security, especially for the poor. In several scenarios large areas are devoted to bioenergy production. Because of potential environmental and food security impacts, bioenergy is very controversial and its value depends on assumptions about overall efficiency, tradeoffs with food production and biodiversity. Reports show that bioenergy production based on conversion of cellulose to ethanol or other hydrocarbon fuels will have less impact on food security and biodiversity than 1st generation fuels. In this context, AKST can play an important role in the development of bioenergy systems, as well as address the need to make agricultural systems more energy efficient.

While governance and other sociopolitical issues are hard to quantify in scenarios, it is known that these factor will be critically important for the future of agriculture. Scenarios primarily address these issues by building scenarios that are based on contrasting underlying assumptions concerning the role of government. Several scenarios expect governance effectiveness to increase over time (reducing the corruption that is perceived to be prevalent in developing economies). However, improving states' capacities in governance and effectiveness in policy implementation is a long term process, and effects are still uncertain. Some scenarios emphasize these uncertainties by showing consequences of failed reforms (e.g., the Order from Strength scenario of the Millennium Ecosystem Assessment). Key options discussed in existing assessments include building "soft" infrastructure, such as networks, organizations, and cooperatives, in order to produce social capital that may reduce conflicts at all governance levels. These may facilitate common-pool agricultural resource management; and enhance the access of farmer groups to markets.

4.6.2 What are the conditions needed to help AKST realize development and sustainability goals?

AKST functions within a larger system of knowledge generation, technological development and diffusion. The formal funding of this larger system will therefore affect AKST. Global spending on all research and development (R&D) is likely to increase in the future both absolutely and as a percentage of total global economic activity, though many countries outside North America, Western Europe and East Asia with small economies will probably continue to have low investments in R&D.

Public investment in AKST is increasingly less driven by the needs of agriculture per se, but is a spinoff of other research priorities such as human health and security. There is a trend in many areas to reduce investment in traditional agricultural disciplines in favor of emerging research areas such as plant and microbial molecular biology, information technology and nanotechnology. This trend is likely to be sustained and its impact on AKST is not fully explored. However, China, with a very large, poor, rural population, is now the country with the second largest total R&D expenditure. It is possible that China may make substantial investments in research relevant to poor rural areas.

Assessing potential development routes of the world agriculture system is of crucial importance if AKST is to realize development and sustainability goals. As discussed previously, there are multiple significant direct and indirect drivers of the agricultural system and many of these are likely to change significantly within the decades. Though the time horizon for research may be reduced in the future, there is now and likely to be in the future a significant lag between the recognition of development and sustainability goals and the time required for AKST to contribute to addressing those goals. Frameworks that consider important drivers of change and their interlinkages can be used to initially explore and at least partially assess the likely consequences of developing particular technologies. Additionally, the impact of these technologies can be considered when projecting future outcomes, thus giving policy makers and others the opportunity to explore and assess different approaches to AKST. However, no model provides a full description of potential changes in agriculture and AKST in the coming decades.

While a number of modelling paradigms exist, most represent agriculture primarily from a particular disciplinary perspective. Given its importance and complexity, there is a clear need for a forward looking assessment that is focused on agriculture and can consider the impact of AKST. There are two main approaches in the literature with respect to future outlooks: (1) the use of multiple scenarios and (2) the use of one central projection. The first handles uncertainties better, but is more complex and time consuming. To date, agricultural assessments use one central projection whereas most environmental assessments use multiple scenarios. The use of multiple models in assessments can help explore and understand sensitivities and uncertainties. Linking different types of models can result in a more comprehensive exploration of important issues.

References

Adams, R.M., B.A. McCarl, K. Seerson, C. Rosenzweig, K.J. Bryant, B.L. Dixon et al. 1999. The economic effects of climate change on U.S. agriculture. Chap. 2 In R. Mendelsohn and J. Neumann (ed) Impacts of climate change on the US economy. Cambridge Univ. Press, UK.

African Union. 2003. Second Summit. Mozambique.

AIDE (Advice Res. Dev. Environ.). 2005. Factsheet for soy production in South America. AIDE, Amsterdam.

Alcamo, J., D. Van Vuuren, W. Cramer, J. Alder, E. Bennett, S. Carpenter et al. 2005. Changes in ecosystem services and their drivers across the scenarios. p. 297-373 In S.R. Carpenter et al. (ed) Ecosystems and human well-being. Vol. 2. Scenarios. Millennium Ecosystem Assessment, Island Press, Washington DC.

Alemayehu, M. 2000. Industrializing Africa: Development options and challenges for the 21st century. Africa World Press, Asmara.

Antle, J.M., and S.M Capalbo. 2002. Agriculture as a managed ecosystem: Policy implications. J. Agric. Resource Econ. 27(1):1-15.

Badgley, C., J. Moghtader, E. Quintero, E. Zakem, J. Chappell, K. Avilés-Vázquez et al. 2007. Organic agriculture and the global food supply. Renew. Agric. Food Syst. 22:86-108.

Bardhan, P. 1999. Democracy and development — A complex relationship. In I. Shapiro and C. Hacker-Cordon (ed) Democracy's values. Cambridge Univ. Press, UK.

Beierle, T., and E. Diaz-Bonilla. 2003. The impact of agricultural trade liberalization on the rural poor: An overview. Workshop on agricultural trade liberalization and the poor. 3-4 Nov. RFF/IFPRI, Washington DC.

Bennett, E. M., S. R. Carpenter, and N. Caraco. 2001. Human impact on erodable phosphorus and eutrophication: A global perspective. BioScience 51:227-234.

Bhalla, G.S., P. Hazell, and J. Kerr. 1999. Prospects for India's cereal supply and demand to 2020. 2020 Vision Disc. Pap. 29. IFPRI, Washington DC.

Binh, T.N. 2004. FDI for agriculture 1988-2003 and orientations to 2010. Available at http://www.isgmard.org.vn/Information%20Service/Report/General/FDI%20in%20Agri-E.pdf. Min. Agric. Rural Dev., Hanoi.

Blum, W.E.H. 2001. Using the soil DPSIR framework — driving forces, pressures, state, impacts, and responses — for evaluating land degradation. p. 4-5. In E.M. Bridges et al. (eds) Response to land degradation. Oxford and IBH, New Delhi.

Borron, S. 2006. Building resilience for an unpredictable future: how organic agriculture can help farmers adapt to climate change. Available at ftp://ftp.fao.org/docrep/fao/009/ah617e/ah617e.pdf. FAO, Rome.

Bouwman A.F., K.W. Van Der Hoek, B. Eickhout, and I. Soenario. 2005. Exploring changes in world ruminant production systems. Agric. Syst. 84:121-153.

Bozbas, K. 2005. Biodiesel as an alternative motor fuel: Production and policies in the European Union. Renew. Sustain. Energy Rev. 9 (4).

Brown, L.R. 1995. Who will feed China? Wake-up call for a small planet. Worldwatch Environ. Alert Ser. W.W. Norton, NY.

Brown, L., and H. Kane. 1994. Full house: Reassessing the Earth's population carrying capacity. Norton, NY.

Bruinsma, J.E. (ed) 2003. World agriculture: Towards 2015/2030. An FAO perspective. Earthscan, London.

Busch, G. 2006. Future European agricultural landscapes — What can we learn from existing quantitative land use scenario studies? Agric. Ecosyst. Environ. 114:121-140.

CA (Comprehensive Assessment of Water Management in Agriculture). 2007. Water for food, water for life: A comprehensive assessment of water management in agriculture. Earthscan, London.

Caballero, B. 2001. Symposium: Obesity in developing countries: Biological and ecological factors. J. Nutr. 131(3):866.

Cáceres, D. 2005. Non-certified organic agriculture: An opportunity for resource-poor farmers? Outlook Agric. 34(3):135-140.

Cassman, K.G., A. Dobermann, D.T. Walters, and H. Yang. 2003. Meeting cereal demand while protecting natural resources and improving environmental quality. Ann. Rev. Environ. Resour. 28:315-328,

Cassman, K.G., S. Wood, P.S. Choo, H.D. Cooper, C. Devendra, J. Dixon et al. 2005. Cultivated systems. p. 745-794. Chapter 26. In R. Hassan et al. (ed) Ecosystems and human well-being: Current state and trends. Vol. 1. Conditions and trends. Millennium Ecosystem Assessment, Island Press, Washington DC.

CGIAR. 2004. The CGIAR initiative for a global open agriculture and food university (GO-FAU). Available at http://www.openaguniversity.cgiar.org. CGIAR, Washington DC.

Cleveland, C.J, C.A.S. Hall, R.A. Herendeen. 2006. Letters - energy returns on ethanol production. Science 312:1746.

CONCAWE. 2002. Energy and greenhouse gas balance of biofuels for Europe — an update. Available at www.concawe.be/download/reports/rpt_02-2.pdf. CONCAWE ad-hoc group on alternative fuels, Brussels, April 2002.

Cranfield, J.A.L., T.W. Hertel, J.S. Eales, and P. Preckel. 1998. Changes in the structure of global food demand. Am. J. Agric. Econ. 80(5):1042-1050.

Crosson P. 1994. Is U.S. agriculture sustainable? Resources 117:16-19.

Curran, L.M., S.N. Trigg, A.K. McDonald, D. Astiani, M. Hardiono, P. Siregar et al. 2004. Lowland forest loss in protected areas of Indonesian Borneo. Science 303:1000-1003.

Daberkow, S., H. Taylor, and W. Huang. 2000. Agricultural resources and environmental indicators: Nutrient use and management. USDA, Washington DC.

De Fraiture, C., X. Cai, U. Amarasinghe, M. Rosegrant, and D. Molden. 2004. Does international cereal trade save water? The impact of virtual water trade on global water use. *In* Comprehensive Assessment Res. Rep., IWMI, Sri Lanka.

De Fraiture, C., D. Wichelns, E.K. Benedict, and J. Rockstrom. 2007. Scenarios on water for food and environment. Chapter 3 In Water for food, water for life: A comprehensive assessment of water management in agriculture. Earthscan, London and IWMI, Colombo.

Delgado C., Rosegrant M., Steinfeld H., Ehui S. and Courbois C. (1999) Livestock to 2020 — the next food revolution. Food Agric. Environ. Disc. Pap. 28 IFPRI, FAO and ILRI, Rome.

Dev, S. Mahendra. 1988. Regional disparities in agricultural labour productivity and rural poverty in India. Indian Econ. Rev. 23:(2)167-205.

Dornburg, V., A. Faaij, P. Verweij, H. Langeveld, G. van de Ven, F. Wester et al. 2008. Biomass assessment: Assessment of the applicability of biomass for energy and materials. Netherlands Environmental Assessment Agency, Bilthoven the Netherlands.

Dorward, A., S. Fan, J. Kydd, H. Lofgren, J. Morrison, C. Poulton et al. 2004. Institutions and policies for pro-poor agricultural growth. Dev. Policy Rev. 22:611-622.

Droogers, P. 2004. Adaptation to climate change to enhance food security and preserve environmental quality: Example for Southern Sri Lanka. Agric. Water Manage. 66:15-33.

Droogers, P., and J. Aerts. 2005. Adaptation strategies to climate change and climate variability: a comparative study between seven contrasting river basins. Physics Chem. Earth 30:339-346.

Easterling, W.E., and M. Apps. 2005. Assessing the consequences of climate change for food and forest resources: A view from the IPCC. Climat. Change 70:165-189.

Echeverria, C., D. Coomes, J. Salas, J.M. Rey-Benayas, A. Lara, and A. Newton. 2006. Rapid deforestation and fragmentation of Chilean temperate forests. Biol. Conserv. 130:481-494.

EEA (European Environment Agency). 2002. Scenarios as a tool for international environmental assessments. Environ. Issue Rep. No. 24. EEA, Copenhagen.

EPA. 2006. Global mitigation of non-CO2 greenhouse gases. EPA Rep. 430-R-06-005. US Environ. Prot. Agency, Washington DC.

Falcon, W.P., and C. Fowler. 2002. Carving up the commons — emergence of a new international regime for germplasm development and transfer. Food Policy 27:197-222.

Falkenmark, M., C.M. Finlayson, L.J. Gordon, E.M. Bennett, T. Matiza Chiuta, D. Coates et al. 2007. Agriculture, water and ecosystems:

Avoiding the costs of going too far. p. 233-277. *In* comprehensive assessment of water management in agriculture, water for food, water for life: Earthscan, London and IWMI, Colombo.

Fan, S., and N. Rao. 2003. Public spending in developing countries — Trends, determination and impact. Environ. Prod. Tech. Div. Disc. Pap. 99. IFPRI, Washington DC.

FAO. 1996. Food for consumers - marketing, processing and distribution. FAO, Rome.

FAO. 2002a. HIV/AIDS, agriculture and food security in mainland and small countries of Africa. Proc. Twenty-First Reg. Conf. Africa, Cairo. FAO, Rome.

FAO. 2002b. The role of agriculture in the development of least developed countries and their integration into the world economy. Available at http://www.fao.org/DOCREP/007/Y3997E/Y3997E00.htm. FAO, Rome.

FAO. 2003. Compendium of agricultural-environmental indicators. Statistics Div., FAO, Rome.

FAO. 2004a. Fertilizer requirements in 2015 and 2030 revisited. FAO, Rome.

FAO. 2004b. State of agricultural commodity markets 2004. Available at ftp://ftp.fao.org/docrep/fao/007/y5419e/yy5419e00.pdf,7. FAO, Rome.

FAO. 2006a. Education for rural people (ERP). Available at http://www.fao.org/sd/erp/index_en.htm. FAO, Rome.

FAO. 2006b. World agriculture: Towards 2030/2050. Prospects for food, nutrition, agriculture and major commodity groups, Interim Rep. Global Perspect. Studies Unit, FAO, Rome.

FAO. 2006c. FAOSTAT. Available at www.fao.org.

FAO. 2006d. Livestock's long shadow. FAO, Rome.

FAO. 2006e. www.fao.org/worldfoodsummit/english/fsheets/biotech.pdf

FAO. 2006f. Presenting the international bioenergy platform. FAO, Rome.

Farrell, A.E., R.J. Plevin, B.T. Turner, A.D. Jones, M. O'Hare, and D.M. Kammen. 2006. Ethanol can contribute to energy and environmental goals. Science 311(5760):506-508.

Faures, J.M., M. Svendsen, and H. Turral. 2007. Reinventing irrigation. *In* D. Molden (ed) Water for food, water for life: A comprehensive assessment of water management in agriculture. Earthscan, London and IWMI, Colombo.

FBOMS. 2006. Impacts of monoculture expansion on bioenergy production in Brazil. Brazilian Forum of NGOs and Social Movements for the Environment and Development, Nucleo Amigos da Terra/ Brasil and Heinrich Boell Foundation, Rio de Janeiro.

Fearnside, P.M. 2001. Soybean cultivation as a threat to the environment in Brazil. Environ. Conserv. 28(01):23-38.

Fischer, G., M. Shah, H. van Velthuizen, and F. Nachtergaele. 2001. Global agro-ecological assessment for agriculture in the 21st century. IIASA, Laxenburg, Austria.

FOE. 2005. The oil for ape scandal. How palm oil is threatening orang-utan survival. Friends of the Earth, London.

Fukuyama, F. 1993. The end of history and the last man. Free Press, NY.

Galloway, J.N., E.B. Cowling, S.J. Seitzinger, and R. Socolow. 2002. Reactive nitrogen: Too much of a good thing? Ambio 31:60-63.

Gardner, B.L. 2005. Causes of rural economic development. Agric. Econ. 32(s1):21-41.

Geist, H.J., and E.F. Lambin. 2002. Proximate causes and underlying driving forces of tropical deforestation. BioScience 52(2):143-150.

Geist, H.J., and E.F. Lambin. 2004. Dynamic causal patterns of desertification. BioScience 54(9):817-829.

Giampietro, M. 2002. Energy use in agriculture. In Encyclopedia of life sciences. Available at http://www.els.net/. Nature Publ.

Giampietro, M. 2003. Multi-scale integrated analysis of ecosystems. CRC Press, Boca Raton.

Giampietro, M. 2007. The future of agriculture: GMOs and the agonizing paradigm of industrial agriculture. *In* A. Guimaraes Pereira and S. Funtowicz (ed) Science for policy: Challenges and opportunities. Oxford Univ. Press, New Delhi.

Goldstone, J.H., R.H. Bates, T.R. Gurr, M. Lustig, M.G. Marshall, J. Ulfelder et al. 2005. A global forecasting model of political instability. Ann. Meeting Am. Political Sci. Assoc. Washington, 1-4 Sept.

Graham, R.L., R. Nelson, J. Sheehan, R.D. Perlack and L.L. Wright. 2007. Current and potential U.S. corn stover supplies. Agron. J. 99:1-11.

Grau, H.R., N.I. Gasparri, and T.M. Aide. 2005. Agriculture expansion and deforestation in seasonally dry forests of north-west Argentina. Environ. Conserv. 2(2):140-148.

Groenfeldt, D. 2003. The future of indigenous values: Cultural relativism in the face of economic development. Futures 35:917-929.

Gulati, A., and S. Narayanan. 2002. Rice trade liberalization and poverty. MSSD Disc. Pap. 51. IFPRI, Washington DC.

Hagens N., R. Costanza, and K. Mulder. 2006. Letters—energy returns on ethanol production. Science 312:1746.

Hayami, Y., and V. Ruttan. 1985. Agricultural development. An international perspective. John Hopkins Univ. Press, Baltimore.

Hertel, T.W., and L.A. Winters. 2005. Estimating the poverty impacts of a prospective Doha development agenda. The World Econ. 28(8):1057-1071.

Hertel, T., H-L. Lee, S. Rose and B. Sohngen. 2006. The role of global land use in determining greenhouse gases mitigation costs. GTAP Working Pap. 36 . Global Trade and Analysis Project, Purdue Univ., West Lafayette.

Hietel, E., Waldhardt, R., Otte, A., 2005. Linking socio-economic factors, environment and land cover in the German highlands, 1945-1999. J. Environ. Manage. 75:133-143.

Hill, J., E. Nelson, D.Tilman, S.Polasky, and D. Tiffany. 2006. From the cover: Environmental, economic, and energetic costs and benefits of biodiesel and ethanol biofuels. PNAS 103:11206-11210.

Hinrichsen, D. 1998. Feeding a future world. People Planet 7(1):6-9.

Hoekstra, A.Y., and P.Q. Hung. 2005. Globalisation of water resources: International virtual water flows in relation to crop trade. Global Environ. Change 15(1):45-56.

Hoogwijk, M., A. Faaij, B. Eickhout, B. de Vries, and W. Turkenburg. 2005. Potential of biomass energy out to 2100, for four IPCC SRES land-use scenarios. Biomass Bioenergy 29:225-257.

Hough, J.R. 1993. Educational cost-benefit analysis. Educ. Res. Pap. 2. ODA, UK.

Huffman, W., and R. Evenson. 1993. Science for agriculture. Iowa State Univ. Press, Ames.

Huntington, S. 1996. The clash of civilizations and the remaking of world order. Touchstone, NY.

IEA. 2002. World energy outlook 2002. International Energy Agency, Paris.

IEA. 2004. World energy outlook 2004. International Energy Agency, Paris.

IEA. 2006. World energy outlook 2006. International Energy Agency, Paris.

IFA. 2006. Fertilizer consumption statistics. Int. Fert. Indust. Assoc., Paris.

IIASA. 2001. http://www.iiasa.ac.at/Research/POP/proj01/index.htm. Int. Inst. Applied System Analysis, Austria.

ILO. 2005. World employment report 2004-05. Available at http://www.ilo.org/public/english/employment/strat/wer2004.htm. ILO, Geneva.

Inglehart, R., and C. Welzel. 2005. Modernization, cultural change and democracy: The human development sequence. Cambridge Univ. Press, UK.

IPCC (Intergovernmental Panel on Climate Change). 2000. Special report on emission scenarios: Understanding and modeling technological change. IPCC and Oxford Univ. Press, UK.

IPCC (Intergovernmental Panel on Climate Change). 2001. Climate change 2001: Synthesis report. Cambridge Univ. Press, UK.

IPCC (Intergovernmental Panel on Climate Change). 2007a. Climate change 2007: The physical science basis: Summary for policy makers. Working group 1 Fourth Assessment Report. Cambridge Univ. Press, UK.

IPCC (Intergovernmental Panel on Climate Change). 2007b. Climate change 2007: Impacts, adaptation and vulnerability. Working group 2 Fourth Assessment Report. Cambridge Univ. Press, UK.

IPCC (Intergovernmental Panel on Climate Change). 2007c. Climate change 2007: Mitigation of climate change. Working group 3 Fourth Assessment Report. Cambridge Univ. Press, UK.

Jones, P.G., and P.K. Thornton. 2003. The potential impacts of climate change in tropical agriculture: the case of maize in Africa and Latin America in 2055. Global Environ. Change 13:51-59.

Kaimowitz, D., and J. Smith. 2001. Soybean technology and the loss of natural vegetation in Brazil and Bolivia. p. 195-211. In A. Angelsen and D. Kaimowitz (ed) Agricultural technologies and tropical deforestation. CABI Publ. UK.

Kaufmann, R.K. 2006. Letters - Energy returns on ethanol production. Science 312:1747.

Kaufmann, D., A. Kraay, and M. Mastruzzi. 2006. Governance matters: Governance indicators for 1996-2005. Policy Res. Dep. Working Pap. 4012. World Bank, Washington DC.

Kelly, V.A. 2006. Factors affecting demand for fertilizer in Sub-Saharan Africa. ARD Disc. Pap. 23. World Bank, Washington DC.

Kemp-Benedict, E., C. Heaps, and P. Raskin. 2002. Global scenario group futures: Technical notes. SEI PoleStar Ser. Rep. 9. Stockholm Environ. Inst., Boston.

Khan, S., R. Tariq, C. Yuanlai, and J. Blackwell. 2006. Can irrigation be sustainable? Agric. Water Manage. 80(1-3):87-99.

Kojima, M., D. Mitchell, and W. Ward. 2007. Considering trade policies for liquid biofuels. World Bank, Washington DC.

Kurosawa, A. 2006. Multigas mitigation: An economic analysis using GRAPE Model. Multigas mitigation and climate policy. Energy J. Spec. Issue #3.

Lambin, E.F., B.L. Turner II, H.J. Geist, S.B. Agbola, A. Angelsen, J.W. Bruce et al. 2001. The causes of land use and land-cover change: Moving beyond the myths. Global Environ. Change 11(4):261-269.

Leach G., L. Jarass, G. Obermair, and L. Hoffmann. 1986 Energy and growth. A comparison of 13 industrial and developing countries. Butterworths, London.

Lee, R. 2003. The demographic transition: Three centuries of fundamental change. J. Econ. Perspect. 17:167-190.

Levi, B. 2004. Governance and economic development in Africa: Meeting the challenge of capacity building. In B. Levi and S. Kpundeh (ed) Building state capacity in Africa. World Bank and Oxford Univ. Press, Washington DC.

Lipton, M. 1977. Why poor people stay poor: Urban bias in world development. Harvard Univ. Press, Boston.

Lopez, R. 2005. Why governments should stop non-social subsidies: Measuring their consequences for rural Latin America. Policy Res. Working Pap. No. 3609. World Bank, Washington DC.

Lucas, P.L., D.P. Van Vuuren, J.G.J. Olivier, and M.G.J. Den Elzen. 2007. Long-term reduction potential of non-CO_2 greenhouse gases. Environ. Sci. Policy 10(2):85-103.

Lutz, W., W.C. Sanderson., and S. Scherbov. 2001. The end of world population growth. Nature 412:543-545.

Lutz, W., W.C. Sanderson, and S. Scherbov. 2004. The end of world population growth in the 21th century. New challenges for human capital formation and sustainable development. Earthscan, London.

MA (Millennium Ecosystem Assessment). 2003. Ecosystems and human well-being: A framework for assessment. Island Press, Washington.

MA (Millennium Ecosystem Assessment). 2005a. Ecosystems and human well-being: Scenarios. Island Press, Washington DC.

MA (Millennium Ecosystem Assessment). 2005b. Ecosystems and human well-being: Policy responses. Island Press, Washington DC.

Maddison, A. 2003. The world economy: Historical statistics. OECD, Paris.

Marshall, M.G. 2006. Polity IV project: Political regime characteristics and transitions, 1800-2006. Available at: http://www.systemicpeace.org/polity/polity4.htm. Polity IV Project, George Mason Univ., Virginia.

Marshall, M.G., T.R. Gurr, J. Wilkenfeld, M.I. Lichbach, and D. Quinn. 2003. Peace and conflict 2003—A global survey of armed conflicts. Self-determination movements and democracy. Center Int. Dev. Conflict Manage. Univ. Maryland, College Park.

Mehra, R., and S. Gammage. 1999. Trend, countertrends and gaps in women's employment. World Dev. 27(3):533-550.

Meijer, E., and R. Stewart. 2004. The GM cold war: How developing countries can go from being dominos to being players. RECIEL (Rev. Eur. Community and Int. Environ. Law) 13(3):247-262.

MNP. 2006. Integrated modeling of global environmental change: An overview of IMAGE 2.4. Netherlands Environment Assessment Agency (MNP), Bilthoven, The Netherlands.

Molden, D., R. Sakthivadivel, and Z. Habib. 2000. Basin level use and productivity of water: Examples from South Asia. Res. Rep. 49. IWMI, Colombo.

Nakicenovic, N., J. Alcamo, G. Davis, B. de Vries, J. Fenham, S. Gaffin et al. 2000. Special report on emissions scenarios. Working Group III, IPCC. Cambridge Univ. Press,UK.

Nayyar, D. 2006. Globalization, history and development: A tale of two centuries. Cambridge J. Econ. 30:137-159.

Noragric. 2004. Poverty reduction strategies and relevant participatory learning processes in agricultural higher education. Case studies from Ethiopia, Malawi, Tanzania and Uganda. Frik Sundstøl (ed) Noragric Rep. 21A. Agric. Univ. Norway.

Notestein, F.W. 1945. Population — The long view. In T.W. Schultz (ed) Food for the world. Univ. Chicago Press, IL.

NSF. 2006. U.S. R&D continues to rebound in 2004. InfoBrief. Sci. Resourc. Stat. Nat. Sci. Foundation, Washington DC.

OECD. 2001. Environmental indicators for agriculture. Vol. 3: Methods and results. OECD, Paris.

OECD. 2002. Workshop on organic agriculture. Washington DC. Sept 2002. OECD, Paris.

OECD. 2006a. Working party on global and structural policies. Revised environmental baseline for the OECD environmental outlook to 2030. Environ. Directorate (Nov). ENV/EPOC/GSP(2006)23. OECD, Paris.

OECD. 2006b. Science, technology and industry outlook. OECD, Paris.

Oki, T., M. Sato, A. Kawamura, M. Miyake, S. Kanae, and K. Musiake. 2003. Virtual water trade to Japan and in the world. In A.Y. Hoekstra (ed) Virtual water trade: Proc. Int. Expert Meeting Virtual Water Trade. Value of Water Res. Rep. Ser. No. 12. UNESCO-IHE, Delft.

Palaniswamy, N., and R. Birner. 2006. Explaining the "political will" to support agricultural development in Sub-Saharan Africa. Paper presented at the Annual Conf. Verein fuer Socialpolitik, Res. Committee Dev. Econ. Berlin, KWF, 2-3 June 2005.

Palm, C.A., P.L.O.A. Machado, T. Mahmood, J. Melillo, S.T. Murrell, J. Nyamangara et al. 2004. Societal responses for addressing nitrogen fertilizer needs: balancing food production and environmental concerns. p. 71-92. In A.R. Mosier et al. (ed) Agriculture and the nitrogen cycle. Island Press, Washington.

Parry, M., G. Fischer, M. Livermore, C. Rosenzweig, and A. Iglesias. 1999. Climate change and world food security: A new assessment. Global Environ. Change 9:S51-S67.

Pimentel, D. (ed) 1980. Handbook of energy utilization in agriculture. CRC Press, Boca Raton.

Pimentel, D., and M. Pimentel. 1979. Food, energy and society. John Wiley, NY.

Pimentel, D., and A. Wilson. 2004. World population, agriculture, and malnutrition. World Watch Mag. 17(5).

Pingali, P.L., and G. Traxler. 2002. Changing locus of agricultural research: will the poor benefit from biotechnology and privatization trends? Food Policy 27(3):223-238.

Psacharopoulos, G., and H.A. Patrinos. 2004. Returns to investment in education: A further update. Educ. Econ. 12:111-135.

Putnam, R.D. 1993. Making democracy work: Civic traditions in modern Italy. Princeton Univ. Press, NJ.

Ramirez-Ramirez, A. 2005. Monitoring energy efficiency in the food industry. PhD thesis. Utrecht Univ., Netherlands.

Raskin, P.D. 2005. Global scenarios: Background review for the Millennium Ecosystem Assessment. Ecosystems 8:133-142.

Raskin, P., T. Banuri, G. Gallopin, P. Gutman, A. Hammond, R. Kates et al. 2002. Great transition: The promise and lure of the times ahead. Stockholm Environ. Inst., Boston.

Reardon, T., C.P. Timmer, C.B. Barrett, and J. Berdegué. 2003. The rise of supermarkets in Africa, Asia, and Latin America. Am. J. Agric. Econ. 85(5):1140-1146.

Reid, R.S., T.P. Tomich, J.C. Xu, H. Geist, and A.S. Mather. 2006. Linking land-change science and policy: Current lessons and future integration. p. 157-171. In E.F. Lambin, and H.J. Geist (ed) Land-use and land-cover change. Local processes and global impacts. Springer, Berlin.

Rijsberman, F.R., and D.J. Molden. 2001. Balancing water uses: Water for food and water for nature. p. 43-56. Thematic Background Papers, Int. Conf. Freshwater, Bonn, 3-7 Dec.

Rist, S. and F. Dahdouh-Guebas. 2006. Ethnosciences — A step towards the integration of scientific and indigenous forms of knowledge in the management of natural resources for the future. Environ. Develop. Sustain. 8:467-493.

RIVM/UNEP. 2004. The GEO-3 scenarios 2002-2032. Quantification and analysis of environmental impacts. Dutch Nat. Inst. Public Health Environ., Bilthoven.

Rockstrom, J., J. Barron, and P. Fox. 2003. Water productivity in rainfed agriculture: Challenges and opportunities for smallholder farmers in drought-prone tropical agro-ecosystems. p. 145-162. In J.W. Kijne et al. (ed) Water productivity in agriculture: Limits and opportunities for improvement. CABI Publ., Wallingford.

Rondot, P., and M. Collion. 2001. Agricultural producer organizations: Their contribution to rural capacity building and poverty reduction. World Bank, Washington DC.

Rosegrant, M., X. Cai, and S. Cline. 2002. World water and food to 2025. Dealing with scarcity. IFPRI, Washington DC.

Rosegrant, M., N. Leach, and R.V. Gerpacio. 1999. Alternative futures for world cereal and meat consumption. Proc. Nutr. Society 58(2):219-234.

Rosegrant, M., S. Msangi, T. Sulser and R. Valmonte-Santos. 2006. Biofuels and the global food balance. In: Bioenergy and Agriculture: Promises and Challenges. FOCUS 14. IFPRI, Washington, DC.

Rosegrant, M., M. Paisner, S. Meijer, and J. Witcover. 2001. Global food projections to 2020: Emerging trends and alternative futures. IFPRI, Washington DC.

Rosegrant, M., and M.A. Sombilla. 1997. Critical issues suggested by trends in food, population, and the environment to the year 2020. Am. J. Agric. Econ. 79(5):1467-1470.

Rosegrant, M., and M. Svendsen. 1993. Asian food production in the 1990s: Irrigation investment and management policy. Food Policy 18:13-32.

Rounsevell, M.D.A., I. Reginster, M.B. Araujo, T.R. Carter, N. Dendoncker, F. Ewert et al. 2006. A coherent set of future land use change scenarios for Europe. Agric. Ecosyst. Environ. 114:57-68.

Ruttan, V.W. 2001. Technology, growth, and development: An induced innovation perspective. Oxford Press, NY.

Sands, R.D., and M. Leimbach. 2003. Modeling agriculture and land use in and integrated assessment framework. Climatic Change 56(1-2):185-210.

Sathaye, J., W. Makundi, L. Dale, P. Chan, and K. Andrasko. 2006. GHG mitigation potential, costs and benefits in global forests: A dynamic partial equilibrium approach. Energy J.

Scherr, S.J. 1999. Poverty-environment interactions in agriculture: Key factors and policy implications. Paper prepared for UNDP and the European Commission expert workshop on Poverty and the Environment, Brussels, 20-21 Jan.

Schmidhuber, J. 2003. The outlook for long-term changes in food consumption patterns: Concerns and policy options. Paper prepared for the FAO Scientific Workshop on Globalization of the Food System: Impacts on food security and nutrition. 8-10 Oct. FAO, Rome.

Seale, J., A. Regmi, and J. Bernstein. 2003. International evidence on food consumption patterns. Available at http://www.ers.usda.gov/publications/tb1904/tb1904.pdf. ERS, USDA, Washington DC.

Seckler, D., D. Molden, U. Amarasinghe, and C. de Fraiture. 2000. Water issues for 2025: A research perspective. IWMI, Colombo.

Sen, A. 1981. Poverty and famines: An essay on entitlement and deprivation. Oxford Univ. Press, NY.

Sen, A., and J. Drèze. 1989. Hunger and public action. Clarendon Press, Oxford.

Shah, T., J. Burke, and K. Villholth. 2007. Groundwater: A global assessment of scale and significance. In D. Molden (ed) Water for food, water for life: A comprehensive assessment of water management in agriculture. Earthscan, London and IWMI, Colombo.

Shiklomanov, I. 2000. Appraisal and assessment of world water resources. Water Int. 25(1):11-32.

Shimoda, S.M. 1998. Agricultural biotechnology: Master of the universe? AgBioForum 1(2).

Slesser, M., and C. Lewis. 1979. Biological energy resources. Halsted, NY.

Smeets, E., and A. Faaij 2004. Biomass resource assessment on global scale for identifying biomass production and export potentials. Report prepared for NOVEM and Essent. Copernicus Inst. Sustain. Dev. Utrecht Univ., Netherlands.

Smil, V. 1987. Energy–food–environment. Realities–myths–options. Clarendon Press, Oxford.

Smil, V. 1991. General energetics. Wiley, NY.

Smil, V. 2000. Feeding the world: A challenge for the twenty first century. MIT Press, Boston.

Smil, V. 2003. Energy at the crossroads: Global perspectives and uncertainties. MIT Press, Boston.

Smith, P., and M. Bertaglia. 2007. Greenhouse gas mitigation in agriculture. In C.J. Cleveland (ed) Encyclopedia of Earth. Available at http://www.eoearth.org/article/Greenhouse_gas_mitigation_in_agriculture.

Environ. Inform. Coalition, Nat. Council Sci. Environ., Washington DC.

Sohngen, B., and R. Mendelsohn. 2007. A sensitivity analysis of carbon sequestration. *In* M. Schlesinger (ed) Human-induced climate change: An interdisciplinary assessment. Cambridge University Press, UK.

Sohngen, B., and R. Sedjo. 2006. Carbon sequestration in global forests under different carbon price regimes. Energy J. (Spec. Issue #3):109-126.

Stanhill, G. (ed) 1984. Energy and agriculture. Springer-Verlag, Berlin.

Stout, B.A. 1991. Handbook of energy for world agriculture. Elsevier, NY.

Stout, B.A. (ed) 1992. Energy in world agriculture. Elsevier, Amsterdam.

Strengers, B., R. Leemans, B. Eickhout, B. de Vries, and L. Bouwman. 2004: The land-use projections and resulting emissions in the IPCC SRES scenarios as simulated by the IMAGE 2.2 model. GeoJournal 61(4):381-393.

Thaman, K.H. 2002. Shifting sights: The cultural challenge of sustainability. Higher Educ. Policy 15:133-142.

Thompson, W.S. 1929. Population. Am. Sociol. Rev. 34(6):959-975.

Thornton, P.K., P.G. Jones, T. Owiyo, R.L. Kruska, M.P. Herrero, P. Kristjanson et al. 2006. Mapping climate vulnerability and poverty in Africa. Available at http://www .dfid.gov.uk/research/mapping-climate.pdf. ILRI, Nairobi.

Tokarick, S. 2002. Trade issues in the Doha Round: Dispelling some misconceptions. Policy Disc. Pap. IMF, Washington DC.

Traill, W.B. 2006. The rapid rise of supermarkets? Dev. Policy Rev. 24(2):163-174.

Tusicisny, A. 2004. Civilizational conflicts: More frequent, longer, and bloodier? J. Peace Res. 41(4):485-498.

UN. 2004. World population to 2300. United Nations, NY.

UN. 2005a. World population prospects: The 2004 revision highlights. United Nations, NY.

UN. 2005b. Population challenges and development goals. United Nations, NY.

UN Millennium Project. 2005. Halving hunger: It can be done. UN Taskforce on Hunger, NY.

UNCTAD. 2005. Trade and development report. Available at http://www.unctad.org/en/docs/ tdr2005_en.pdf. UN Conf. Trade Dev., NY.

UNEP. 2002. Global environment outlook 3 (GEO-3). UNEP, Nairobi and Earthscan, London.

UNEP. 2004. Global environment outlook scenario framework. Background Rep. GEO-3. UNEP, Nairobi.

UNEP. 2007. Global environment outlook 4 (GEO-4). UNEP, Nairobi and Earthscan, London.

UNESCO. 2006. http://portal.unesco.org/ education/en/ev.php-URL_ID=50558&URL_ DO=DO_TOPIC&URL_SECTION=201 .html

UNIFEM. 2000. Progress of the world's women. UN Dev. Fund Women, NY.

US Census Bureau. 2003. International data base. Updated 26 April 2005. US. Bureau of the Census, Washington DC.

US DoE. 2004. International Energy Outlook. Energy Inform. Admin., US Dep. Energy, Washington DC.

US EPA. 2006. Global anthropogenic non-CO2 greenhouse gas emissions: 19902020. Available at http://www.epa.gov/nonco2/ econ-inv/international.html. US Environ. Prot. Agency, Washington DC.

Van Meijl, T., A. Van Rheenen, A. Tabeau, and B. Eickhout. 2006. The impact of different policy environments on land use in Europe. Agric. Ecosyst. Environ. 114:21-38.

Van Vuuren, D.P. 2007. Energy systems and climate policy. Long-term scenarios for an uncertain future. PhD thesis, Utrecht Univ., Netherlands.

Van Vuuren, D.P., M.G.J Den Elzen, P.L. Lucas, B. Eickhout, B.J. Strengers, B.J. Ruijven, et al. 2007. Stabilizing greenhouse gas concentrations at low levels: An assessment of reduction strategies and costs. Climatic Change, 81:2:119-159.

Van Vuuren, D.P., B. Eickhout, P.L. Lucas, and M. Den Elzen. 2006. Long-term multi-gas scenarios to stabilise radiative forcing — Exploring costs and benefits within an integrated assessment framework. Energy J. (Spec. Iss.) 3:201-234.

Van Vuuren, D.P, and B.C. O'Neill. 2006. The consistency of IPCC's SRES Scenarios to recent literature and recent projections. Climate Change. Springer, NY.

Verbist, B., A.E.D. Putra, and S. Budidarsono. 2005. Factors driving land use change: Effects on watershed functions in a coffee agroforestry system in Lampung, Sumatra. Agric. Syst. 85:254-270.

Von Braun, J., M. Rosegrant, R. Pandya-Lorch, M.J. Cohen, S.A. Cline, M.A. Brown et al. 2005. New risks and opportunities for food security: Scenario analyses for 2015 and 2050. IFPRI, Washington, DC.

Westhoek, H., D.P. van Vuuren, and B. Eickhout. 2005. A brief comparison of four scenario studies: IPCC-SRES, GEO-3, Millennium Ecosystem Assessment, and FAO towards 2030. Netherlands Environ. Assessment Agency (MNP).

Wood, S., J. Henao, and M. Rosegrant. 2004. The role of nitrogen in sustaining food production and estimating future nitrogen fertilizer needs to meet food demand. *In* A.R. Mosier et al. (ed) Agricultural and the nitrogen cycle. Scope, Washington.

World Bank. 2003. World development indicators, 2003. World Bank, Washington, DC.

World Bank. 2004a. Agricultural investment sourcebook. Available at www.worldbank. org/agsourcebook. ARD, World Bank, Washington DC.

World Bank. 2004b. World economic prospects 2004. 1990-2000 Trends and recent projections. World Bank. Washington DC.

World Bank. 2005a. Advancing bioenergy for sustainable development. ESMAP, World Bank, Washington DC.

World Bank. 2005b. World development indicators. World Bank, Washington DC.

WRI. 1997. World resources 1996-97. World Resourc. Inst., Washington DC.

Xie, Y., Y. Mei, T. Guangjin, and X. Xuerong. 2005. Socio-economic driving forces of arable land conversion: A case study of Wuxian City, China. Global Environ. Change 15:238-252.

Zurek, M., 2006. A short review for global scenarios for food systems analysis. Working Paper 1. GECAFS, Wallingford.

Zurek, M., and T. Henrichs. 2007. Linking scenarios across geographical scales in international environmental assessments. Tech. Forecasting Soc. Change 74:1282-1295.

5

Looking into the Future for Agriculture and AKST

Coordinating Lead Authors
Mark W. Rosegrant (USA), Maria Fernandez (Peru), Anushree Sinha (India)

Lead Authors
Jackie Alder (Canada), Helal Ahammad (Australia), Charlotte de Fraiture (Netherlands), Bas Eickhout (Netherlands), Jorge Fonseca (Costa Rica), Jikun Huang (China), Osamu Koyama (Japan), Abdallah Mohammed Omezzine (Oman), Prabhu Pingali (USA), Ricardo Ramirez (Canada/Mexico), Claudia Ringler (Germany), Scott Robinson (USA), Phil Thornton (UK), Detlef van Vuuren (Netherlands), Howard Yana-Shapiro (USA)

Contributing Authors
Kristie Ebi (USA), Russ Kruska (USA), Poonam Munjal (India), Clare Narrod (USA), Sunil Ray (India), Timothy Sulser (USA), Carla Tamagno (Peru), Mark van Oorschot (Netherlands), Tingju Zhu (China)

Review Editors
R.S. Deshpande (India), Sergio Ulgiati (Italy)

Key Messages

1. Quantitative projections indicate a tightening of world food markets, with increasing resource scarcity, adversely affecting poor consumers. Real world prices of most cereals and meats are projected to increase in the coming decades, dramatically reversing trends from the past several decades. Price increases are driven by both demand and supply factors. Population growth and strengthening of economic growth in sub-Saharan Africa, together with already high growth in Asia and moderate growth in Latin America drive increased growth in demand for food. Rapid growth in meat and milk demand is projected to put pressure on prices for maize and other coarse grains and meals. Bioenergy demand is projected to compete with land and water resources. Growing scarcities of water and land are projected to increasingly constrain food production growth, causing adverse impacts on food security and human well-being goals. Higher prices can benefit surplus agricultural producers, but can reduce access to food by a larger number of poor consumers, including farmers who do not produce a net surplus for the market. As a result, progress in reducing malnutrition is projected to be slow.

2. Improved Agricultural Knowledge, Science and Technology (AKST) helps to reduce the inevitable tradeoffs between agricultural growth and environmental sustainability at the global scale. AKST can help to maximize the socioeconomic benefits of extracting natural resources from a limited resource base, through increasing water productivity and intensifying crop, livestock and fish production. Without appropriate AKST development further production increases could lead to degradation of land, water and genetic resources in both intensive and extensive systems.

3. Growing pressure on food supply and natural resources require new investments and policies for AKST. Tightening food markets indicate that a business-as-usual approach to financing and implementing AKST cannot meet the development and sustainability goals of reduction of hunger and poverty, the improvement of rural livelihoods and human health and equitable, environmentally sustainable development. Innovative AKST policies are essential to build natural, human and physical capital for social and environmental sustainability. Such policies will also require more investment in AKST. Important investments supporting increased supply of and access to food include those in agricultural research and development, irrigation, rural roads, secondary education for girls, and access to safe drinking water.

4. Continuing structural changes in the livestock sector, driven mainly by rapid growth in demand for livestock products, bring about profound changes in livestock production systems. Structural changes in the livestock sector have significant implications for social equity, the environment and public health. Projected increases in livestock numbers to 2050 vary by region and species, but substantial growth opportunities exist for livestock producers in the developing world. The availability of animal feed will however affect both the rate and extent of this growth, since competition is growing between animal and aquaculture feeds that both use fishmeal and fish oil. Livestock feeds made with fish products contribute to superior growth and survival but are increasing prices and consumption of fishmeal and fish oil in the aquaculture sector. The corresponding decrease in the use of these products in the livestock sector, especially for pigs and poultry, can affect production and increase prices. Moreover, declining resource availability could lead to degradation of land, water, and animal genetic resources in both intensive and extensive livestock systems. In grassland-based systems, grazing intensity (number of animals per ha of grazing land) is projected to increase by 50% globally, and by up to 70% in Latin America. In addition to the potential environmental impacts of more intensive livestock production systems, the sector faces major challenges in ensuring that livestock growth opportunities do not marginalize smallholder producers and other poor people who depend on livestock for their livelihoods.

Other tradeoffs are inevitably going to be required between food security, poverty, equity, environmental sustainability, and economic development. Sustained public policy action will be necessary to ensure that livestock system development can play its role as a tool for growth and poverty reduction, even as global and domestic trends and economic processes create substantial opportunities for sector growth.

5. Growing water constraints are a major driver of the future of AKST. Agriculture continues to be the largest user of freshwater resources in 2050 for all regions, although its share is expected to decline relative to industrial and domestic uses. Sectoral competition and water scarcity related problems will intensify. Reliability of agricultural water supply is projected to decline without improved water management policies. There is substantial scope to improve water management in both rainfed and irrigated agriculture. AKST and supporting interventions geared towards water conserving and productivity enhancement in rainfed and irrigated agriculture are needed to offset impacts of water scarcity on the environment and risks to farmers.

6. There is significant scope for AKST and supporting policies to contribute to more sustainable fisheries, by reducing the overfishing that has contributed to growing scarcity of resources and declining supplies of fish in the world's oceans. To date, AKST and supporting policies have not contributed to halting overfishing of the world's oceans. There are some initiatives to rebuild depleted stocks, but recovery efforts are quite variable. A common and appropriate policy response is to take an ecosystem approach to fisheries management but many governments are still struggling to translate guidelines and policies into effective intervention actions. Other policy options have included eliminating perverse subsidies, establishing certification, improving monitoring, control and surveillance, reducing destructive fishing practices such as bottom trawling bans, expanding marine protected areas and changing fishing access agreements. There are also policy responses to reduce efforts in industrial scale fishing in many areas, while also supporting small-scale fisheries through improved access to prices and market information and increasing awareness of appropriate fishing practices and post-harvest technologies.

Rapid growth in demand for aquaculture products will also be adversely affected by growing scarcity of coastal land and offshore areas and water scarcity in land-based operations. The most appropriate policy response to this problem is integrated coastal management that better utilizes these shared resources for wider benefit. Another policy option is promoting best management practices, which include looking into appropriate feeding strategies as fish oil, on which the production of high value species depends, becomes increasingly scarce.

7. Expected climate changes are likely to affect agriculture, requiring attention to harmonizing policies on climate mitigation and adaptation with others on agriculture and forest land for bioenergy and on forestry for carbon sequestration. Climate change is expected to have increasing impacts on the agriculture sector. This impact can be positive or negative. For example, CO_2 fertilization, increased precipitation and higher temperature can lengthen the growing season and improve crop yields in specific regions. Elsewhere, however, with higher temperatures and more erratic precipitation, the impact on crop yield can be negative. Under higher climate sensitivity, climate impacts are very likely to be negative for all regions. Even with small climate change projections, impacts are projected to be negative for dryland areas in Africa, Asia and the Mediterranean area. These climate impacts can be mitigated by climate policies, but very low stabilization experiments (450 ppmv CO_2-equivalents) will likely require measures such as carbon sequestration and bioenergy plantations that compete for land. Therefore, climate mitigation policy options might require reprioritizing among alternative development and sustainability goals.

8. Food safety regulations can help improve the quality of life, but need to be designed to avoid adversely affecting poor farmers' access to markets. Demand for products with high quality and safety standards is expected to grow in industrialized countries. This market will only be accessible to those developing countries with sufficient AKST capacity and knowledge to meet the higher standards, especially in post-harvest handling. Better quality standards are only likely to emerge in developing countries if consumers are educated about the benefits of consumption of perishable products, if public health regulation and liability laws are established, and if better laboratory infrastructure is built. Challenges in coming decades include ensuring safer food for consumers and raising the quality of life without reducing food availability, access and use by the poor or by creating barriers to poor countries and smaller producers by excluding their exporting produce through multinational companies. Implementation of quality and food safety control programs with intensive internal and external supervision can improve productivity without increasing costs for consumers. Government actions toward product quality standardization should consider the effect on the distribution of costs and benefits between actors.

9. Rural communities have a greater say in the future of small-scale agriculture as their access to information via information communication and technology (ICT) and to financial capital via remittance investment plans increases. The attributes of ICTs are linked directly and indirectly with the sustainability and development goals. As internet access increases in rural areas, small-scale producers will benefit from more readily available information, both traditional and local knowledge and technological and market information, if private and public institutions take up the challenge of providing climate, weather, and price data. In addition cellular phone use among national and international migrants will enhance information flows and their participation in decisions. As a result, migrant organizations in receiving countries will reinforce links with their home communities and most likely influence the choice of local development paths. Taken together, increased access to ICTs and migrant remittances will impact the land management, food security and livelihood strategies of rural communities in new ways.

10. Society benefits from involving women in all levels of processes from education to decision making and work, increasing their access and contribution to AKST. In the developing world it is expected that an increasing share of women workers would participate in rural farm activities and in agro-based industries and agro-based service sectors. Investments in health services, child care, and education are fundamental to achieving the development and sustainability goals that support women's participation in agriculture and AKST. AKST policies and investments in rural infrastructure, which improve women's status, enhance women's role, and reduce their burden through better water and energy supply, would help improve livelihoods while also supporting other AKST policies.

5.1 Scope of the Chapter and How to Use the Results

This chapter examines the potential future for agriculture and AKST using primarily quantitative methods combined with qualitative analyses of those issues that cannot easily be addressed in quantified models. For this approach a reference run is developed from 2000 to 2050, based on the assessment of drivers of agriculture and AKST explained in Chapter 4. It builds on changes in drivers used in previous assessments and uses a set of modeling tools to sketch out a plausible future based upon past trends. This reference run is used to indicate how the development and sustainability goals (see Chapter 1) might take shape out to 2050. In subchapter 5.3 the reference run is described and the results are shown. No important policy actions are assumed in the reference run to show more sharply the consequences of a noninterventionist reference case.

In a second step, in subchapter 5.4, a set of policy actions are simulated in order to assess the impact these could have on the attainment of development and sustainability goals. Here, policy experiments on investments in AKST, climate mitigation, extensive use of bioenergy, trade liberalization, changes in water productivity and in dietary changes, such as shifts to consumption of organic food or less meat, are implemented and analyzed. In this way, the quantified impact and tradeoffs of these specific policy actions can be made visible. Not all future developments, however, can be assessed with the various tools that are used in this subchapter. Subchapter 5.5 therefore describes a series of important, emerging issues related to AKST that can affect the reference world and alternative policy pathways. Subchapters 5.6 and 5.7, finally, examine synergies and tradeoffs and implications for AKST in the future, respectively.

5.2 Rationale and Description of Selected Tools

The inclusion of various tools in the assessment process enables the examination of the various relationships that transpire determined by major drivers. Also synergies and tradeoffs between specific policy interventions can be made visible through the use of modeling tools. Modeling results can be used to support policy analysis in this assessment. Clearly, models cannot provide answers for all issues. In that case, qualitative translations of the modeling results are used in this Chapter to assess the most crucial policy options that have been identified in Chapter 4.

5.2.1 Rationale for model selection
In this assessment, with its focus on agriculture and the role of AKST, the partial equilibrium agricultural sector International Model for Policy Analysis of Agricultural Commodities and Trade, or IMPACT (Rosegrant et al., 2002), plays a pivotal role. Partial equilibrium agricultural sector models are capable of providing insights into long-term changes in food demand and supply at a regional level, taking into account changes in trade patterns using macro-economic assumptions as an exogenous input. To be able to assess the environmental consequences of changes in the agricultural sector, a range of environmental models is used as well. The integrated assessment model IMAGE 2.4 (Eickhout et al., 2006) is central in this environmental assessment, while specific models like EcoOcean and GLOBIO3 (Alkemade et al., 2006) are used to provide consequences for specific issues, marine and terrestrial biodiversity, respectively. The livestock spatial location-allocation model, SLAM, (Thornton et al., 2002, 2006) and the water model WATERSIM (de Fraiture 2007) are used to give specific insights in crucial sectors for agriculture and AKST. The computable general equilibrium (CGE) model GTEM (Ahammad and Mi, 2005) is used to validate the GDP and population input data to achieve cross-sectoral consistency for the reference run. The regional models, GEN-CGE for India (Sinha and Sangeetz, 2003; Sinha et al., 2003) and the Chinese Agricultural Policy Simulation Model (CAPSiM) (Huang and Li, 2003), are used to add local flavors to the global analyses that have been performed by the other tools. India and China were chosen since future policy change in these two countries will affect global food supply, demand, prices, and food security. Moreover, China- and India-specific modeling tools are used to provide deeper insights about specific development goals such as the distributional aspects of equity and poverty which cannot be addressed by global models.

The tools used in this assessment for the reference run out to 2050 (Table 5-1). A selection of the models is also used for the policy experiments in subchapter 5.4 (Table 5-1). Short descriptions of model types are provided below

and longer descriptions, including an assessment of major uncertainties are introduced in the appendix to this chapter. Linkages among models are presented in subchapter 5.2.2.

5.2.1.1 Partial equilibrium agricultural sector models
Partial equilibrium models (PE) treat international markets for a selected set of traded goods, e.g., agricultural goods in the case of partial equilibrium agricultural sector models. These models consider the agricultural system as a closed system without linkages with the rest of the economy, apart from exogenous assumptions on the rest of the domestic and world economy. The strength of these partial equilibrium models is their great detail of the agricultural sector. The "food" side of these models generally uses a system of supply and demand elasticities incorporated into a series of linear and nonlinear equations, to approximate the underlying production and demand functions. World agricultural commodity prices are determined annually at levels that clear international markets. Demand is a function of prices, income and population growth. Biophysical information on a regional level (e.g., on land or water availability), is constraining the supply side of the model.

Food projections' models that simulate aggregations of components—regions, commodities and larger countries—tend to be more reliable (McCalla and Revoredo, 2001). PE modeling approaches require (1) consistent and clearly defined relations among all variables, (2) a transfer of the structure of interrelationships among variables, which was consistent in the past, to the future, (3) changes in complex cross-relationships among variables over time, (4) the simultaneous and managed interaction of many variables and the maintenance of consistent weights and (5) an organized and consistent treatment of massive numbers of variables and large amounts of data (McCalla and Revoredo, 2001).

Food projection models make major contributions in exploring future food outcomes based on alternative assumptions about crucial exogenous and endogenous variables. Results from alternative policy variants can be used to alert policy makers and citizens to major issues that need attention to avoid adverse food security outcomes. A test for the usefulness of these models may therefore be whether or not the analysis enriched the policy debate (McCalla and Revoredo, 2001).

While models can make important contributions at the global and regional levels, increasingly food insecurity will be concentrated in individual countries with high population growth, high economic dependence on agriculture, poor agricultural resources and few alternative development opportunities. These countries continue to be overlooked in regional and global studies, since, on aggregate, resources are sufficient to meet future food demands.

Whereas the methodology and underlying supply and demand functional forms are well established in the literature and have been validated through projections of historical trends, the driving forces and elasticities underlying the commodity and country and regional-level supply and demand functions towards the future continue to be debated in the literature. Moreover, income and population growth projections, as well as lasting external shocks contribute to the uncertainty of projection outcomes.

Table 5-1. Overview of quantitative modeling tools.

Model name	Type	Features	Output indicators	Policy experiments
IMPACT-WATER	Partial equilibrium agricultural sector model with water simulation module	Simulates food production and water based on economic, demographic, and technological change	Food supply and demand, water supply and demand, Food price and trade, number of malnourished children	Investment in AKST, Trade liberalization, Organic/change in meat demand
SLAM	Simulated Livestock Allocation Model	Simulates the allocation of land to ruminant livestock based systems using livestock numbers	Areas and density of grazing ruminants	
IMAGE	Integrated Assessment model	Simulates energy supply and demand, translates energy outcomes and food outcomes from IMPACT into environmental consequences (land use change, climate change, emissions)	Energy demand and mix, greenhouse gas emissions, land use change, temperature and precipitation change, C and N fluxes	Climate change, Bioenergy
GTEM	CGE model	Simulates the working of the global economy		Trade liberalization
WATERSIM	Partial equilibrium agricultural sector model with water simulation module		Food supply and demand, water supply and demand, Food price and trade	Water productivity
GLOBIO3	Dose-response biodiversity model	Translates environmental pressures mainly from IMAGE into indicators of biodiversity	Mean Species Abundance Index	Bioenergy
ECO-OCEAN	Marine biomass balance model	Simulates world marine capture fisheries based on the 19 FAO fishing areas	Catch, Value, Diversity, and Marine Trophic Index	
GEN-CGE	CGE model for India	Multisectoral general equilibrium model for India with gender disaggregated data	Food and nonfood supply and demand at country level, employment by worker types distinguished by gender, wages of female and male workers and income by households	Trade liberalization
CAPSiM	CAPSiM	Partial equilibrium agricultural sector model for China	Simulates food production, consumption, and farmers' income based on major driving forces of demand and supply	

Source: Compiled by authors.

5.2.1.2 Integrated assessment models
Integrated Assessment models (IAMs) are tools to address global environmental change in a consistent manner, using feedbacks from climate change, land use change and changes in atmospheric composition. They provide information on a global scale and take into account the regional interrela-

tions on many aspects like energy demand, land use change and air quality. IAMs are strong in providing insights into the consequences of specific policy options and can support policy discussions in this area.

Although the integration in most models is high from the perspective of the limited (environmental) problems they were developed for, their integration from the perspective of the IAASTD's objective is still rather low. In particular, feedbacks from ecological changes to socioeconomic drivers are limited, with some exceptions on the impacts of food production and climate policy on socioeconomic drivers.

Processes that change ecosystems and their services mostly occur at highly disaggregated levels. Models therefore require regional specificity. A tendency to increase the level of explicit geographic information in models, for instance by using a detailed grid, can be seen in the literature. Understanding interregional links, but also regional differences will be an important research issue for integrated modeling in the coming years. A nested approach to integrated assessment modeling could be a helpful way forward, in which global models provide context for detailed, regional (ecological) models.

Uncertainties are a key element in IAMs, given the high complexity and its focus on decision-making. These uncertainties include, for example, variability of parameters, inaccuracy of model specification or lack of knowledge with regard to model boundaries. Although the existence of uncertainties has been recognized early in the process of developing IAMs, uncertainty analysis is typically included only partially or not at all.

5.2.1.3 CGE models

CGE models are widely used as an analytical framework to study economic issues of national, regional and global dimension. CGE models provide a representation of national economies, next to a specification of trade relations between economies. CGE models are specifically concerned with resource allocation issues, that is, where the allocation of production factors over alternative uses is affected by certain policies or exogenous developments. International trade is typically an area where such induced effects are important consequences of policy choices. These models provide an economy-wide perspective and are very useful when:

- The numerous, and often intricate, interactions between various parts of an economy are of critical importance. As for agriculture, such interactions occur between agriculture sectors themselves (e.g., competing for limited productive resources including various types of land) as well as between agricultural sectors with other sectors/actors which either service agricultural sectors or operate in the food and fiber chain including downstream processors, traders and distributors, final consumers and governments (e.g., public policies).
- The research objective is to analyze counterfactual policy alternatives and/or plausible scenarios about how the future is likely to evolve. Examples could include the implications for agriculture of likely multilateral trade liberalization in the future, the implications for agriculture of future growth in food demand and shifts in consumer preference, or the role of bioenergy in climate change mitigation and implications for agriculture.

For analyzing such issues, the modeling of sectoral interactions is fundamental (e.g., among agriculture, energy, processing and manufacturing as well as services), trade (domestic and international), and existing policies. Given their economy-wide coverage, some variant of this type of mod-

els has become a part of the Integrated Assessment models (e.g., IMAGE; Eickhout et al., 2006).

A strength of CGE models is their ability to analyze the interaction between different sectors such as agricultural sectors, manufacturing sectors and services. In their conventional usage, CGE models are flexible price models and are used to examine the impact of relative price changes on resource allocations (of goods and factors) across a range of economic agents. Thus, in addition to providing insights into the economy-wide general equilibrium effects of policy changes, CGE models allow key interindustry linkages to be examined. However, CGEs are poor in addressing distributional issues within the regions: only average adjustments in the regional economies are simulated. Moreover, CGE models should be handled with care for long-term projections since fundamental changes in the economic structure of a region cannot be simulated by a CGE model. Therefore, CGE models are only used in this assessment for assessing the global economic consequences of trade liberalization.

5.2.1.4 Marine biomass balance models

Fisheries models, such as EcoOcean, allow managers to explore how marine systems, especially fisheries, might respond to policy changes at the scale of the ocean basin or region not addressed by most other fisheries models. This model reduces what is a highly complex and dynamic system that covers 70% of the Earth's surface to 19 regions and describes the world's fisheries for the last 50 years with reasonable accuracy (often with 10% or less variation of what is recorded by FAO between 1950 and 2003). A complete marine system is modeled that ranges from detritus to top predators including marine mammals and seabirds, and provides sufficient detail to assess changes but avoids complexity so that it is computationally possible. The predator-prey relationships between functional groups are also accounted for in the model. Because EcoOcean is based on the Ecopath suite of software and uses a trophic structure as well as predator-prey relationships, consumption rates and fishing effort, it provides a description of the ecological dynamics of the system and an indication of how the diversity of the fisheries will change over time.

The models have some weaknesses. The functional groups used in EcoOcean are broad groupings of marine organisms, which limit their ability to describe in detail how a particular species or groups of species may respond to a specific policy intervention. The model is based on biomass from published time series studies and does not necessarily include a comprehensive suite of species to provide an estimate of the biomass for each functional group. The FAO regions used in the model are broad and cannot include climate or oceanographic features. This limitation makes it difficult to accurately model the small pelagic fish group (e.g., anchoveta) which is highly influenced by changes in oceanographic conditions as seen in the offshore upwelling system in Peru. The tuna groups do not differentiate between long-lived slow-growing species such as bluefin tuna and short-lived ones such as yellowfin. This can result in overestimation of tuna landings as well as resilience. Effort, based on seven fleets, is the driver of the model and while some effort is gear-specific, such as tuna long-line and tuna purse seiners, effort for the demersal fleet is based on a range of gear including trawlers, nets, traps and hook and line that can be difficult to map to the narrative storylines. The lack of artisanal fishing information especially in Asia and several regions in Africa results in some underestimation of landings and effort. Antarctic and Arctic models are incomplete, as there is poor catch, effort and biomass data available for these areas. Consequently they are not included in this assessment.

5.2.2 Interactions of models in this assessment
The focus of the analyses in this Chapter is on the issues summarized in Figure 5-1. This figure illustrates which models address which issue.

5.2.2.1 Relations between the models
The most important inputs—population and GDP growth— are used exogenously in all modeling tools to enhance consistency of the analyses. The global CGE model is used to provide consistency among population, economic growth, and agricultural sector growth. Climate is used as an input in many of the modeling tools. Future climate change is simulated by the IMAGE 2.4 model and is then used by other models as well.

In the chain of models, IMPACT simulates food supply and demand and prices for agricultural commodities for a set of 115 countries/subregions for 32 crop and livestock commodities, including all cereals, soybeans, roots and tubers, meats, milk, eggs, oils, oilcakes and meals, sugar and sweeteners, and fruits and vegetables. The country and regional submodels are intersected with 126 river basins—to allow for a better representation of water supply and demand—generating results for 281 Food Producing Units (FPUs). Crop harvested areas and yields are calculated based on crop-wise irrigated and rainfed area and yield functions. These functions include water availability as a variable and connect the food module with the global water simulation model of IMPACT.

The SLAM model is using the livestock supply and demand from IMPACT. In the SLAM model, land is allocated to four categories: landless systems, livestock only/ rangeland-based systems (areas with minimal cropping), mixed rainfed systems (mostly rainfed cropping combined with livestock) and mixed irrigated systems (a significant proportion of cropping uses irrigation and is interspersed with livestock). The allocation is carried out on the basis of agroclimatology, land cover, and human population density (Kruska et al., 2003; Kruska, 2006). The second component of the model then allocates aggregated livestock numbers to the different systems, allowing disaggregated livestock population and density data to be derived by livestock-based system. The structure of the classification is based on thresholds associated with human population density and length of growing period, and also on land-cover information. The primary role of SLAM in this assessment is to convert the livestock outputs of the IMPACT model (number of livestock slaughtered per year per FPU) to live-animal equivalents by system, so that changes in grazing intensity by system can then be estimated. These estimates of grazing intensity are subsequently used as input data to the IMAGE model to assess the land-use change.

The crop and livestock supply and demand from IMPACT and the grazing intensity from SLAM are used as input by IMAGE 2.4, a model designed to cover the most important environmental issues. For land-use changes, the input from IMPACT and SLAM is used. For changes in the energy sector, the IMAGE energy model TIMER (van Vuuren et. al., 2007) is used. Because of the focus of the IMAGE model on land and energy, it is most suitable to also address bioenergy. The potential for bioenergy is determined by the land-use model of IMAGE 2.4 and, through price mechanisms of price supply curves, the amount of bioenergy in the total energy mix is determined by the TIMER model (Hoogwijk et al., 2005). All socioeconomic drivers are simulated for 24 regions (26 regions in the TIMER model); the land-use consequences on grid scale of 0.5 x 0.5 degrees. Through linkages of the terrestrial system to carbon and nitrogen cycle models, the atmospheric concentrations of greenhouse gases and tropospheric ozone are simulated as well. A simple climate model combined with a geographical pattern scaling procedure (Eickhout et al., 2004) translates these concentrations to local changes in temperature and precipitation.

The terrestrial changes as simulated by IMAGE are used as input by the terrestrial biodiversity model GLOBIO3 (Alkemade et al., 2006). GLOBIO3 is using dose-response relationships for each region and ecosystem type to translate environmental pressures (like climate change, nitrogen deposition, land-use change and infrastructure) to average quality values of these ecosystem types. For this analysis all ecosystems are represented by a set of representative species. The quality of the ecosystem types are therefore an approximation of the mean species abundance (MSA) present in each ecosystem type. Note that each MSA value is by definition between 0 and 1.

The fisheries EcoOcean model is used to assess the future catch, value and mean trophic index of marine systems in different oceanic parts of the world. The FAO statistical areas provide a manageable spatial resolution for dividing the world into a reasonable number of spatial units. Similarly 43 trophic groups represent the different functional groups that are found in most areas of the world's oceans. For each of the 19 regions, information from the "Sea Around Us"

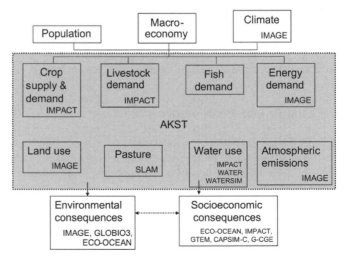

Figure 5-1. *Interaction of models in this assessment.* Source: Authors.

catch database is used for each year from 1950 to present to fit the catches. Once the model has been tuned and is deemed to perform satisfactorily, a series of future-oriented evaluations is performed. The EcoOcean model is not linked to any of the other models and is only used to add insights about the future quality of the marine systems.

IMPACT provides broad insights into socioeconomic consequences. The regional GEN-CGE and CAPSiM models provide added insights into distributional consequences and gender for China and India, respectively. The WATERSIM model is used to provide more insights into the water balance. Water demand for irrigation, domestic purposes, industrial sectors, livestock and the environment is estimated on a basin scale. Water supply for each basin is expressed as a function of climate, hydrology and infrastructure. The model iterates between basin, region and globe until the conditions of economic equilibrium and hydrologic water balance are met.

5.2.2.2 Policy experiments
IMPACT is the core modeling tool used to assess agriculture and AKST for the IAASTD. It is therefore represented in most of the policy experiments. Investments requirements in AKST and changes in diets (both organic and nonmeat) are simulated by IMPACT. Climate mitigation and bioenergy policies are implemented first in IMAGE. Trade liberalization is performed by GTEM, simulating changes in the economic structure. IMPACT then picks up changes in the economic structure and simulates the consequences for food supply and demand. The regional consequences are examined through the regional GEN-CGE and CAPSiM models for India and China. Finally, changes in water productivity are assessed by WATERSIM.

5.3 Description of Reference World, Including Quantification

5.3.1 Rationale of reference world
The reference case imagines a world developing over the next decades as it does today, without anticipating deliberate interventions requiring new or intensified policies in response to the projected developments. Current policy pathways are expected to continue out to 2050. This continuation of the "real world" is plausible. In subchapter 5.4 some of the major questions affecting the future of agriculture and AKST are simulated and results are compared to the reference world.

5.3.2 Inputs into the reference world

5.3.2.1 Population
In the reference case the global population increases from slightly more than 6.1 billion in 2000 to over 8.2 billion in 2050. Most of the growth is concentrated in middle-income and low-income countries like Brazil, India, China and Russia and the rest of the world (Table 5-2). Population growth continues to slow in high-income countries. Population growth drives changes in food demand and is an indirect driver for AKST. The data for population changes are taken from the medium variant projections of the UN (UN, 2005), based on an assessment of previous studies (see also Chapter 4).

5.3.2.2 Overall economic growth
Economic growth assumptions are loosely based on the TechnoGarden (TG) scenario of the Millennium Ecosystem Assessment (MA, 2005). Incomes are expressed as MER-based values. The economic growth assumptions of the TechnoGarden scenario are near the mid-range growth scenarios in the literature for the world as a whole and most regions. In some regions the scenario is a relatively optimistic scenario (e.g., sub-Saharan Africa). A comparison of economic growth projections in other scenarios is made in Chapter 4. Information at the regional level is provided in Table 5-3.

5.3.2.3 Agricultural productivity
Agricultural productivity values are based on the Millennium Ecosystem Assessment (MA) TechnoGarden (TG) scenario and the recent FAO interim report projections to 2030/2050 (FAO, 2006a). MA assumptions have been adjusted from the TG assumptions to allow for conformity to FAO projections of total production and per-capita consumption in meats and cereals, and to our own expert assessment. The main recent developments regarding technological change with continued slowing of growth overall have been taken into account. Growth in numbers and slaughtered carcass weight of livestock has been adjusted in a similar fashion.

5.3.2.4 Nonagricultural productivity
Growth in nonagricultural sectors is projected to be lower than in agriculture in the reference case. The nonagricultural GDP growth rates are likewise based on the MA TechnoGarden scenario but with adjustments to align with World Bank medium-term projections. While the relatively higher productivity in agriculture reflects largely the declining trends in the agricultural terms of trade, this is not translated into higher output growth in agricultural sectors relative to nonagricultural sectors.

Disparities in growth rates among countries in the developing world are projected to continue to remain high while more developed regions will see more stable growth. Developed regions will see relatively low and stable to declining growth rates between 1 and 4% per year out to 2050. Most of NAE falls into this category while several countries in ESAP (East and South Asia and the Pacific) (South Korea, Japan, New Zealand, Australia) and South Africa are quite similar in growth patterns. The LAC region will also see stable growth rates through the projection period, though slightly higher than for developed regions between 3.5 and 4.5% per year out to 2050.

East and Southeast Asia will also see stable to declining GDP growth rates through the projection period, but the rates will remain relatively high between 4 and 7% per year. In particular, China's economy will be slowing from the 10% growth in recent years to a more stable rate of 5.6% per year on average out to 2050. On the other hand, growth in South Asia will follow the strong reforms and initiatives in India focusing on macroeconomic stabilization and market reforms and should lead to projected improved income growth in that subregion of 6.5% per year out to 2050. CWANA will also see an increase in GDP growth rates through the projection period though the rates are a bit more modest and will lead to an average 4% per year out to 2050 for the region.

Table 5-2. **Population growth.**

	2000-05	2005-10	2010-15	2015-20	2020-25	2025-30	2030-35	2035-40	2040-45	2045-50
NAE	0.3%	0.3%	0.2%	0.2%	0.1%	0.1%	0.0%	0.0%	0.0%	-0.1%
CWANA	2.0%	1.9%	1.8%	1.7%	1.5%	1.3%	1.2%	1.0%	0.9%	0.8%
LAC	1.4%	1.3%	1.2%	1.0%	0.9%	0.7%	0.6%	0.5%	0.3%	0.2%
SSA	2.3%	2.2%	2.2%	2.1%	2.0%	1.9%	1.7%	1.6%	1.5%	1.4%
ESAP	1.1%	1.0%	0.9%	0.8%	0.6%	0.5%	0.4%	0.3%	0.2%	0.1%

Source: UN, 2005.

Growth in SSA has been low in the recent past, but there is much room for recovery, which will lead to strong, if modest growth. All of SSA should see an average of 3.9% growth out to 2050. Central and Western SSA will see fairly stable to slightly increasing growth with most countries experiencing annual growth in the 5-6% range. The remainder of SSA will see strong increases in GDP growth rates as recovery continues. Though many countries in East and Southern SSA will be experiencing growth less than 4% out to 2025, all of these countries are projected to see growth rates reach 6 to 9% by 2050.

5.3.2.5 Livestock

The reference run was implemented in the following way: First, global livestock systems were mapped for the baseline year (2000) and for the reference run for 2030 and 2050, using the reference populations and General Circulation Model (GCM) scenarios for these years. The latter was used to generate surfaces of length of growing period (number of days per year) to 2030 and 2050. In the absence of GCM output for diurnal temperature variation and maximum or minimum temperatures, average monthly diurnal temperature variation was estimated using a crude relationship involving average (24 hour) daily temperature and the average day-time temperature. The 0.5° latitude-longitude grid size of the GCM data was downscaled to 10 arc-minutes (0.17° latitude-longitude), and characteristic daily weather data for the monthly climate normal for the reference run in 2030 and 2050 were generated using the methods of Jones and Thornton (2003). For the second part of the analysis, the livestock numbers that were generated as output from the IMPACT model at the resolution of the FPUs were converted to live-animal equivalents using country-level ratios of live-to-slaughtered animals from FAOSTAT for 1999-2001 (the same base that was used for the IMPACT simulations). To estimate changes in grazing intensity, the extent of each system type within each FPU was estimated, and livestock numbers within each FPU were allocated to each system within the FPU on a pro-rata basis. Existing global ruminant livestock distribution maps for current conditions were used as a basis for the future variants, to derive the livestock allocation proportions appropriate to each system within each FPU.

The eleven livestock systems in the Seré and Steinfeld classification were aggregated to three: rangeland systems, mixed systems (rainfed and irrigated), and "other" systems. These "other" systems include the intensive landless systems, both monogastric (pigs and poultry) and ruminant.

5.3.2.6 Trade

Trade conditions seen today are presumed to continue out to 2050. No trade liberalization or reduction in sectoral protection is assumed for the reference world.

5.3.2.7 Water

Projections for water requirements, infrastructure capacity expansion, and water use efficiency improvement are conducted by IMPACT-WATER. These projections are combined with the simulated hydrology to estimate water use and consumption through water system simulation by IMPACT-WATER (Rosegrant et al., 2002). "Normal" priority has been given to all sectors, with irrigation being the lowest, compared with domestic, industrial and livestock uses.

The hydrology module of the IMPACT-WATER global food and water model derives effective precipitation, potential and actual evapotranspiration and runoff at these 0.5 degree pixels and scale them up to the level of FPUs, which

Table 5-3. **Per capita income growth.**

Region	2000-05	2005-10	2010-15	2015-20	2020-25	2025-30	2030-35	2035-40	2040-45	2045-50
NAE	3.3%	2.2%	2.8%	2.8%	2.7%	2.5%	2.3%	2.0%	1.8%	1.7%
CWANA	4.3%	3.6%	3.7%	3.6%	3.5%	3.8%	4.1%	4.5%	4.8%	5.0%
LAC	4.3%	1.1%	3.7%	4.6%	4.4%	4.4%	4.5%	4.6%	4.6%	4.5%
SSA	3.6%	3.4%	4.2%	4.3%	4.4%	4.6%	4.9%	5.1%	5.2%	5.2%
ESAP	3.2%	2.7%	3.7%	3.8%	3.6%	3.7%	3.7%	3.8%	3.8%	3.7%

Source: Authors (based on MEA 2005).

are also used for some of the other analyses, in the spatial operational unit of IMPACT-WATER. Projections for water requirements, infrastructure capacity expansion, and water use efficiency improvement are conducted by IMPACT-WATER. These projections are combined with the simulated hydrology to estimate water use and consumption through water system simulation by IMPACT-WATER (Rosegrant et al., 2002).

5.3.2.7 Energy use and production

As discussed in Chapter 4, the energy sector may develop in very different ways. For the reference projection, we have chosen to loosely couple future outcomes to IEA reference scenario—a scenario that lies central in the range of available energy projections. The policy variant has been developed using the IMAGE/TIMER model and incorporates the specific assumptions of the IAASTD reference projection with respect to economic growth and land use change. In terms of energy demand growth the IEA scenario is a mid-range scenario compared to full range of scenarios published in literature. For the development of the energy mix, it is a conventional development scenario assuming no major changes in existing energy policies and/or societal preferences. These assumptions are also included in the IAASTD reference projection.

5.3.2.8 Climate change

Climate change is both driving different outcomes of key variables of the reference run (like crop productivity and water availability) and is an outcome of the agricultural projections of the reference run, through land-use changes and agricultural emissions, mainly from the livestock sector (FAO, 2006b). Given the medium energy outcomes in the reference run (see 5.3.3.3), results from the B2 scenario are directly used in most of the modeling tools. From the available B2 scenario, the ensemble mean of the results of the HadCM3 model for B2 scenario was used. The pattern scaling method applied was that of the Climate Research Unit, University of East Anglia. The "SRES B2 HadCM3" climate scenario is a transient scenario depicting gradually evolving global climate from 2000 through 2100. In the IMAGE model, climate change is an output of the model. The IMAGE model uses a global climate model (MAGICC) to calculate global mean temperature change—and uses downscaling techniques to downscale this data to a 0.5 x 0.5 grid. Through this approach, different GCM results can be used to assess the consequences of the uncertainty in local climate change. For the reference run, the pattern of Hadley Centre's HadCM2 is used for the downscaling approach, which is consistent with the pattern used in the other modeling tools. For the simulations of the reference world, the medium climate sensitivity value is used of the Third Assessment Report (2.5°C), which has been adjusted slightly in the latest IPCC report. According to IPCC, the climate sensitivity is likely to be in the range of 2 to 4.5°C with a best estimate of about 3°C, and is very unlikely to be less than 1.5°C (IPCC, 2007). Climate sensitivity is not a projection but is defined as the global average surface warming following a doubling of carbon dioxide concentrations (IPCC, 2007). The uncertainties in the climate sensitivity are not assessed in the reference world. Specific sensitivity analyses will show the importance of the uncertainties in values of the climate sensitivity.

5.3.3 Description of reference world outcomes

5.3.3.1 Food sector

Food supply and demand. In the reference run, global food production increases 1.2% per year during 2000-2050. This growth is a result of rapid economic growth, slowing population growth, and increased diversification of diets. Growth of demand for cereals slows during 2000-2025 and again from 2025-2050, from 1.4% per year to 0.7% per year. Demand for meat products (beef, sheep, goat, pork, poultry) grows more rapidly, but also slows somewhat after 2025, from 1.8% per year to 0.9% annually.

Changes in cereal and meat consumption per capita vary significantly among IAASTD regions (Figures 5-2 and 5-3). Over the projection period, per capita demand for cereals as food declines in the LAC region and in the ESAP region. On the other hand demand is projected to considerably increase in the sub-Saharan Africa region and also increase in the NAE and CWANA regions. Recovery and strengthening of economic growth in sub-Saharan Africa will drive relatively fast growth in regional demand for food. In developing countries and particularly Asia, rising incomes and rapid urbanization will change the composition of cereal demand. Per capita food consumption of maize and coarse grains will decline as consumers shift to wheat and rice. As incomes rise further and lifestyles change with urbanization, there will be a secondary shift from rice to wheat. In the SSA region, growing incomes are expected to lead to a shift from roots and tubers to rice and wheat. Per capita food demand for roots and tubers in SSA is projected to decline from 171 kg to 137 kg between 2000 and 2050, while rice and wheat demand are expected to grow from 18-20 kg to 30-33 kg (Table 5-4). Under the reference run, the composition of food demand growth across commodities is expected to change considerably. Total cereal demand is projected to grow by 1,305 million tonnes, or by 70%; 50% of the increase is expected for maize; 23% for wheat; 10% for rice; and the reminder, for sorghum and other coarse grains.

Demand for meat products continues to grow rapidly across all six IAASTD regions, by 6-23 kilograms per person. The increase is fastest in the LAC and ESAP regions and slowest in the SSA and NAE regions. Rapid growth in meat and milk demand in most of the developing world will put strong demand pressure on maize and other coarse grains as feed. Globally, cereal demand as feed increases by 553 million tonnes during 2000-2050, a staggering 42% of total cereal demand increase (Figure 5-4).

Tables 5-5, 5-6, 5-7 and 5-8 present results for changes in livestock numbers for beef, sheep and goats, pigs, and poultry, respectively, for the IAASTD regions. The global population of bovines is projected to increase from some 1.5 billion animals in 2000 to 2.6 billion in 2050 in the reference run. Substantial increases are projected to occur in all regions except NAE: the number of bovines is projected to double in CWANA and ESAP, and to increase by 50% in SSA, for example. Cattle numbers are projected to peak in SSA in about 2045. Bovine populations are relatively stable in NAE to 2050 in the reference run.

Similar patterns are seen for changes in sheep and goat populations. The global population is expected to increase

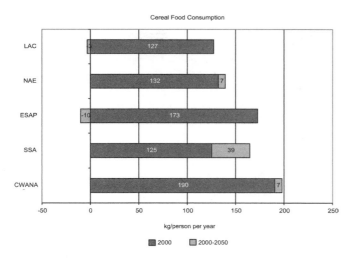

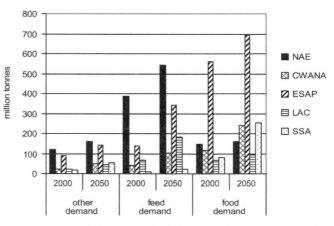

Figure 5-2. *Per capita availability of cereals as food, 2000 and 2050, reference run, by IAASTD region.* Source: IFPRI IMPACT model simulations.

Figure 5-4. *Cereal demand as feed, food & other uses, 2000 and projected 2050, reference run, by IAASTD region.* Source: IFPRI IMPACT model simulations.

from 1.7 billion in 2000 to 2.7 billion in 2050, again with substantial increases in all regions except NAE. In ESAP, sheep and goat numbers are increasing to 2050 still, but the rate of increase is declining markedly. In all other regions, numbers reach a peak sometime around 2040, and then start to decline.

Globally, pig numbers are expected to peak around 2030 and then start to decline, and numbers in no region are projected to increase between 2040 and 2050. Poultry numbers are projected to more than double by 2050. Peak numbers are reached around 2045 in NAE, with small declines thereafter, while numbers are continuing to increase somewhat in CWANA and SSA and still rapidly in LAC and ESAP. Growth in cereal and meat consumption will be much slower in developed countries. These trends are expected to lead to an extraordinary increase in the importance of developing countries in global food markets.

Sources of food production growth. How will the expanding food demand be met? For meat in developing countries, increases in the number of animals slaughtered have accounted for 80-90% of production growth during the past decade. Although there will be significant improvement in animal yields, growth in numbers will continue to be the main source of production growth. In developed countries, the contribution of yield to production growth has been greater than the contribution of numbers growth for beef and pig meat; while for poultry, numbers growth has accounted for about two-thirds of production growth. In the future, carcass weight growth will contribute an increasing share of livestock production growth in developed countries as expansion of numbers is expected to slow.

For the crops sector, water scarcity is expected to increasingly constrain production with virtually no increase in water available for agriculture due to little increase in supply and rapid shifts of water from agriculture in key water-scarce agricultural regions in China, India, and CWANA (see water resources discussion below). Climate change will increase heat and drought stress in many of the current breadbaskets in China, India, and the United States and even more so in the already stressed areas of sub-Saharan Africa. Once plants are weakened from abiotic stresses, biotic stresses tend to set in and the incidence of pest and diseases tends to increase.

With declining availability of water and land that can be profitably brought under cultivation, expansion in area is not expected to contribute significantly to future production growth. In the reference run, cereal harvested area expands from 651 million ha in 2000 to 699 million ha in 2025 before contracting to 660 million ha by 2050. The projected slow growth in crop area places the burden to meet future cereal demand on crop yield growth.

Although yield growth will vary considerably by commodity and country, in the aggregate and in most countries it will continue to slow down. The global yield growth rate for all cereals is expected to decline from 1.96% per year in 1980-2000 to 1.02% per year in 2000-2050; in the NAE

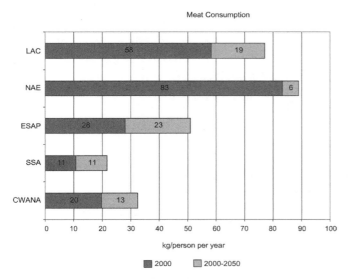

Figure 5-3. *Per capita availability of meats, 2000 and 2050, reference run, by IAASTD region.* Source: IFPRI IMPACT model simulations.

Table 5-4. **Per capita food availability, various agricultural commodities, by IAASTD region.**

	2000	2025	2050	2000	2025	2050	2000	2025	2050	2000	2025	2050	2000	2025	2050
	NAE			CWANA			ESAP			LAC			SSA		
Meat															
Beef	24	26	25	6	8	11	4	7	9	25	28	30	5	7	11
Pork	32	32	31	0	0	0	16	20	19	9	10	11	1	2	3
Lamb	2	2	2	5	7	8	1	2	3	1	1	1	2	2	2
Poultry	25	31	30	8	11	13	7	12	19	23	29	33	3	4	6
Eggs	12	12	13	4	4	5	9	11	13	8	9	9	2	2	3
Milk	102	107	113	59	64	77	22	31	42	84	86	89	17	21	31
Cereals															
Rice	6	7	8	16	17	19	94	90	82	26	26	25	18	24	30
Wheat	108	112	112	146	141	142	58	59	60	51	49	49	20	24	33
Maize	9	9	10	14	14	16	14	13	13	48	45	45	49	40	39
Sorghum	0	0	0	8	8	11	3	2	2	0	0	0	20	21	28
Millet	0	0	0	1	2	3	3	3	3	0	0	0	16	21	29
Other grain	9	7	6	5	5	5	1	1	1	1	1	1	2	2	3
Root crops & Tubers															
Potato	86	74	70	29	25	25	23	21	24	27	25	29	10	9	9
Sweet potato & yam	1	1	1	1	1	1	18	12	7	4	4	4	44	42	34
Cassava	0	0	0	2	2	2	9	8	6	25	22	19	117	111	94
Soybean	0	0	0	0	0	0	5	4	3	1	1	1	1	1	2
Vegetable	88	108	129	65	71	86	100	137	141	40	47	61	29	34	43
Sugar cane/ Beet	36	42	44	21	26	32	11	16	20	38	43	49	16	17	20
Sweetener	11	13	15	0	0	0	1	1	1	1	2	2	0	0	0
Subtropical fruit	55	66	76	59	66	85	47	65	79	90	98	122	33	38	44
Temperate fruit	24	29	37	26	28	32	10	13	14	5	6	9	0	0	0
Oils	28	31	35	13	15	20	10	16	25	14	17	26	8	10	15

Source: IFPRI IMPACT model simulations.

Table 5-5. **Bovines for the reference run, by IAASTD region (billion head).**

Region	2000	2010	2020	2030	2040	2050
CWANA	0.124	0.162	0.192	0.218	0.237	0.248
ESAP	0.578	0.745	0.911	1.055	1.165	1.209
LAC	0.349	0.430	0.507	0.566	0.610	0.627
NAE	0.268	0.288	0.306	0.311	0.304	0.282
SSA	0.179	0.219	0.253	0.273	0.278	0.270
World	1.498	1.844	2.170	2.423	2.593	2.636

Source: ILRI SLAM model simulations.

region, average crop yield growth will decline to 1.02% per year; in CWANA to 1.26% per year, and in ESAP to 0.84% annually, while cereal yield is expected to grow at a higher 1.61% and 1.68% per year in LAC and SSA, respectively.

Area expansion is significant to projected food production growth only in sub-Saharan Africa (28%) and in the LAC region (21%) in the reference run (Figure 5-5).

Table 5-9 presents regional estimates of grazing intensity in the reference world. These were calculated as the number of Tropical Livestock Units (TLU) (bovines, sheep and goats, where one bovine is equivalent to one TLU and a sheep and goat to 0.1 TLU) in the rangeland system per hectare of rangeland system occurring in each FPU. These figures were again aggregated to the five IAASTD regions. Ruminant grazing intensity in the rangelands increases in all regions in the reference run, but there are considerable regional variations. In LAC, for instance, average grazing intensities are expected to increase by about 70%, from 0.19 in 2000 to 0.32 TLU per ha for the reference run. Most of these increases will be due to higher inputs in the grazing systems in the humid and subhumid savannas. The increases are less in CWANA and SSA, and for the latter, grazing intensities are fairly stable after 2030—cattle numbers have peaked by 2040 and there are fewer in 2050 than in 2030 (see Table 5-5), small ruminant numbers by 2050 are only somewhat above those for 2030, while at the same time the model indicates some loss of grazing land in SSA to necessarily marginal mixed rainfed systems. Grazing intensities change relatively little in NAE. Again, given typical stocking rates of 10-15 ha per animal in the arid and semiarid grazing systems, these results of the reference run imply considerable intensification of livestock production in the humid and subhumid grazing systems of the world, but particularly in LAC.

It should be noted that the rate of conversion of rangeland to mixed systems will be underestimated in this analysis. The impact of infrastructural development is not taken into account, so the projected changes in grazing intensities are likely to be underestimated as a result. The analysis also makes implicit assumptions about the relative share of production that is projected to come from the rangeland versus the mixed systems in the future, in terms of relative animal numbers. Even so, given the fragility of semiarid and arid rangelands, particularly in SSA, and the uncertainties con-

Table 5-6. **Sheep and goats for the reference run, by region (billion head).**

Region	2000	2010	2020	2030	2040	2050
CWANA	0.403	0.491	0.556	0.597	0.614	0.601
ESAP	0.723	0.871	1.008	1.115	1.184	1.210
LAC	0.116	0.136	0.154	0.168	0.175	0.174
NAE	0.195	0.218	0.235	0.244	0.244	0.231
SSA	0.271	0.346	0.406	0.443	0.459	0.457
World	1.707	2.061	2.359	2.566	2.677	2.673

Source: ILRI SLAM model simulations.

Table 5-7. **Pigs for the reference run, by region (billion head).**

Region	2000	2010	2020	2030	2040	2050
CWANA	<0.001	<0.001	<0.001	<0.001	<0.001	<0.001
ESAP	0.539	0.622	0.669	0.664	0.627	0.558
LAC	0.080	0.096	0.110	0.119	0.123	0.122
NAE	0.274	0.295	0.307	0.304	0.290	0.262
SSA	0.019	0.024	0.029	0.032	0.034	0.034
World	0.912	1.038	1.115	1.121	1.076	0.978

Source: ILRI SLAM model simulations.

Table 5-8. **Poultry for the reference run, by region (billion head).**

Region	2000	2010	2020	2030	2040	2050
CWANA	1.449	1.677	1.901	2.108	2.306	2.424
ESAP	7.478	10.112	12.979	15.712	18.168	19.687
LAC	2.286	2.893	3.531	4.151	4.762	5.245
NAE	4.180	4.677	5.180	5.542	5.780	5.750
SSA	0.784	0.991	1.170	1.306	1.407	1.445
World	16.178	20.350	24.760	28.819	32.423	34.551

Source: ILRI SLAM model simulations.

Table 5-9. **Grazing intensities in rangeland systems to 2030 and 2050 for the reference run, by region (TLU per ha).**

Region	2000	2030	2050
CWANA	0.052	0.077	0.083
ESAP	0.044	0.067	0.067
LAC	0.188	0.293	0.318
NAE	0.052	0.063	0.060
SSA	0.062	0.090	0.090
World	0.064	0.094	0.098

Source: ILRI SLAM model simulations.

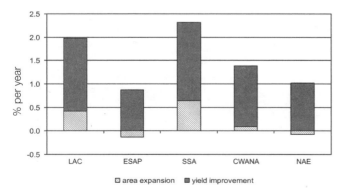

Figure 5-5. *Sources of cereal production growth, reference run, 2000-2050 by IAASTD region.* Source: IFPRI IMPACT model simulations.

cerning technological change and the institutional landscape that will affect these livestock systems in the future (Freeman et al., 2006), the shifts in production to the wetter and mixed systems that are implied are likely to have considerable potential environmental impacts in the reference run.

Food trade, prices, and food security. Real world prices of most cereals and meat are projected to increase significantly in coming decades, reversing trends from the past several decades. Maize, rice, and wheat prices are projected to increase by 21-61% in the reference world, prices for beef and pork by 40% and 30%, respectively, and for poultry and sheep and goat by 17% and 12%, respectively (Table 5-10). These substantial changes are driven by new developments in supply and demand, including much more rapid degradation on the food production side, particularly as a result of rapidly growing water scarcity, rapidly growing demands, both food and nonfood, combined with slowing yield growth unable to catch up with developments on the supply and demand side.

World trade in food will continue to increase, with trade in cereals projected to increase from 257 million tones annually in 2000 to 657 million tonnes in 2050 and trade in meat products rising from 15 million tonnes to 62 million tonnes. Expanding trade will be driven by the increasing import demand from the developing world, particularly CWANA, ESAP, and SSA, where net cereal imports will grow by 200%, 170%, and 330%, respectively (Figure 5-6). Thus, sub-Saharan Africa will face the largest increase in food import bills despite significant area and yield growth expected during the next 50 years in the reference world.

With rising prices due to the inability of much of the developing countries to increase food production rapidly enough to meet growing demand, the major exporting countries—mostly in the NAE and LAC regions—will provide an increasingly critical role in meeting food consumption needs (Figures 5-6 and 5-7). The USA, Brazil, and Argentina are a critical safety valve in providing relatively affordable food to developing countries. Maize exports are expected

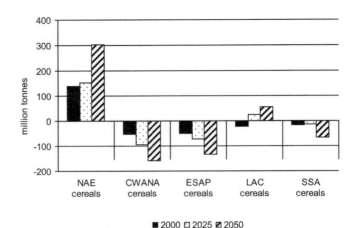

Figure 5-6. *Net trade in cereals, reference run, by IAASTD region.* Source: IFPRI IMPACT model simulations.

to double in the United States to 102 million tonnes, and to grow to 38 million tonnes in Argentina and 10-11 million tonnes in Brazil and Mexico. Wheat exports are projected to grow to 60 million tonnes in the United States, and to 37 million tonnes in Russia, 34 million tonnes in Australia, and 31 million tonnes in Argentina, respectively. Soybean exports are projected at 20 million tonnes in Argentina and 15 million tonnes in Brazil. Net meat exports are expected to increase sharply in the United States and Argentina. Europe is expected to also increase exports, mainly because of slow or no growth in demand with stable population. For rice, Myanmar is expected to join Thailand and Vietnam as particularly significant exporters.

The substantial increase in food prices will slow growth in calorie consumption, with both direct price impacts and reductions in real incomes for poor consumers who spend a large share of their income on food. As a result, there will be little improvement in food security for the poor in many regions. In sub-Saharan Africa, daily calorie availability is

Table 5-10. **Selected international food prices, 2000 and projected 2025 and 2050, reference run.**

	2000	2025	2050
Food	US$ per metric ton		
Beef	1,928	2,083	2,691
Pork	878	986	1,142
Sheep & goat	2,710	2,685	3,039
Poultry	1,193	1,192	1,399
Rice	191	223	232
Wheat	99	136	160
Maize	72	108	102
Millet	227	293	289
Soybean	186	216	216

Note: All values are three-year averages.

Source: IFPRI IMPACT model simulations.

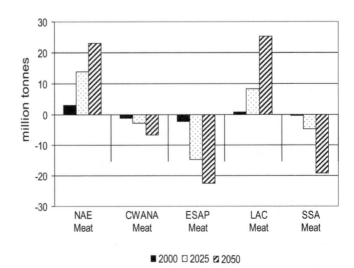

Figure 5-7. *Net trade in meat products, reference run, by IAASTD region.* Source: IFPRI IMPACT model simulations.

Box 5-1. Outcomes for China.

China's development has major impacts on both current and future food markets of the world. Key results from a disaggregated, partial agricultural equilibrium model are presented below:

Crop production

Under the baseline (or reference) run, total crop area will gradually decline. In addition, wages are predicted to rise as will the opportunity cost of land for agricultural production. Why? The main drivers of these shifts are industrialization, urbanization and the slowing of the rate of growth of population (as well as labor supply). Sown area is projected to decline by about 10%, which implies an annual rate of 0.2 over the next 50 years (Table 5-27). The decline will be largest for the cereal sector. In contrast, the sown area of crops with positive income elasticities of demand (e.g., cash crops) will expand slightly. Non-staple crop yields will grow in the reference world since the rising demand for these commodities will lead to higher prices which, in turn, will induce enhanced productivity from investment in these sectors (both in R&D and in production).

Table 5-27. **Area and yield of major agricultural commodities, China (in million hectares and tonne per hectare, respectively).**

	2004	2020	2050
Area (million ha):			
Cereal	83	75	70
Soybean + oil crops	24	21	19
Cotton	5	5	4
Sugar	2	2	2
Vegetable	18	19	20
Fruit	9	11	12
Sum of above crops	140	132	127
Yield (tonne/ha):			
Rice (in milled rice)	4.3	5.2	5.7
Wheat	4.0	4.8	5.3
Maize	5.0	6.1	6.4
Cotton	1.1	1.6	1.8
Sugar	5.6	7.8	8.7
Vegetable	19.4	25.9	27.6
Fruit	9.5	15.2	17.1

Source: CAPSIM reference run.

Implications for food security, poverty and equity

China's economic growth and trade liberalization in the reference world will facilitate many changes in the basic structure of agricultural sector. China's agriculture will be gradually shifting from crops in which its farmers have less comparative advantage (i.e., land-intensive sectors, such as grains, edible oils, sugar and cotton) to those in which farmers have more comparative advantage (labor-intensive crops, such as vegetables, fruit, pork and poultry).

Overall, China's food security will remain high. While there will be a few agricultural and food commodities that could experience a significant decline in national self-sufficiency levels (for example maize, soybeans and edible oils, sugar and ruminant meats, as shown in Table 5-28, rising imports of these few commodities will not threaten the basic food security status of either China or the world. Cereal imports will rise, but cereal self-sufficiency will remain at about 90% in 2020 and above 85% through 2050. Cereal imports rise mainly because of increasing demand for feed (especially, maize). Rising feed demand is inextricably linked to the rapidly growing livestock sector. Self-sufficiency in maize will fall from the current level of more than 100% (China actually was a net maize exporter in the 1990s and 2000s) to less than 70% after 2020. However, due to declining demand for rice and wheat (on a per capita basis) and the falling rates of population growth (with nearly no growth in the 2020s and falling population numbers thereafter), our projections suggest that China could reach near self-sufficiency in wheat and become a large exporter of rice into international markets, as long as the rest of the world liberalizes their agricultural sectors.

Table 5-28. **Self-sufficiency levels of selected major agricultural commodities in China (in percent).**

	2004	2020	2050
Cereal	102	92	86
Rice	101	107	112
Wheat	99	95	98
Maize	108	79	69
Soybean	49	41	38
Oil crops	67	63	58
Cotton	85	74	58
Sugar	91	79	65
Vegetables	101	105	106
Fruit	101	106	102
Pork	101	102	102
Beef	100	86	85
Mutton	99	94	95
Poultry	100	105	111
Milk	96	79	75

Source: CAPSIM reference run.

Outside China, a rapidly growing Chinese economy will help those countries with a comparative advantage in land-intensive products. Such countries (such as Brazil, Argentina, Brazil, the US, Canada and Australia) will expand their production and increase their exports to China. Developing countries, in particular,

continued

Box 5-1. continued

will be able to export a fairly large number of agricultural products to China. China's open trade regime and rising demand will increase the consumption of imported soybeans and other edible oils, maize, cotton, sugar, tropical and subtropical fruits, as well as some livestock products (e.g., milk, beef and mutton).

Incomes will increase across all segments of the income distribution in China. The rises will come, in part from agriculture. However, most of the growth will be based on rising nonagricultural activities, including off farm wages and self employment earnings. On average, per capita income will rise about 6% annually over the next two decades and 3-5% annually during the period from 2020-2050. Income growth from agriculture will be positive, but much lower. China's rapid economic growth and the rise in the nation's overall wealth will be accompanied by widening income inequality unless substantial efforts are undertaken to directly support the poor. Since most of the poor in China have land, improving agriculture and other activities in farming areas will positive affect the welfare of the poorest people in rural China.

As growth proceeds, China will significantly reduce its population under the poverty line. In 2001, about 11% of China's rural population was below the US$1/day poverty line (Table 5-29). With rising incomes from both the agricultural and nonagricultural sectors, the share of the poor in the total rural population is expected to be reduced to 5.4% by 2010 and to less than 1% by 2020. Moreover, under the reference run, the share of the rural population that lives in poverty would essentially be completely eliminated after 2022, a level of reduction that is faster than the targets suggested under the Millennium Development Goals of the United Nations. Specifically, the poorest 20% of China's households (based on their income levels in 2001) are expected to reduce their population share from 22.6% in 2001 to 3.9% by 2030 (Table 5-29). After about 2035, the entire rural population in the lowest income class (quintile) is expected to have graduated to the second or even third income quintiles.

Table 5-29. **Population shares by income group in rural China (in percent).**

Income group	2001	2010	2020	2030	2050
Under poverty	11.0	5.4	0.9	0.0	0.0
By household income in 2001					
1st quintile	22.6	15.8	8.9	3.9	0.0
2nd quintile	21.3	24.0	25.2	25.2	12.3
3rd quintile	20.0	18.9	17.5	16.3	19.1
4th quintile	19.0	17.4	15.2	13.0	7.6
5th quintile	17.0	24.0	33.2	41.6	61.1

Note: Households under poverty means that per capita income is less than 1$/day in PPP. Rural population with less 1$/day income accounted for 11% of total rural households in 2001. Each quintile of households accounted for 20% of total rural households in 2001, but the shares of population in lower quintiles are more than those in higher quintiles.

Source: CAPSIM reference run.

expected to only grow to 2,738 kilocalories by 2050, compared to 3,000 or more calories available, on average, in all other regions. Only the South Asia subregion has similar low gains in calorie availability—at 2,746 calories per capita per day by 2050. Calorie availability is expected to grow fastest in the ESAP region at 630 kilocalories over 2000-2050 (Figure 5-8).

In the reference run, childhood malnutrition (children of up to 60 months) will continue to decline, but cannot be eradicated by 2050 (Figure 5-9). Childhood malnutrition is projected to decline from 149 million children in 2000 to 130 million children by 2025 and 99 million children by 2050. The decline will be fastest in Latin America at 51%, followed by the CWANA and ESAP regions at 46% and 44%, respectively. Progress is slowest in sub-Saharan Africa—despite significant income growth and rapid area

and yield gains as well as substantial progress in supporting services that influence well-being outcomes, such as female secondary education, and access to clean drinking water. By 2050, an increase in child malnutrition of 11% is expected, bringing the total to 33 million children in the region.

Fisheries. The reference run is set up so that the value of landings was optimized throughout the years modeled with effort driving the model. The effort for all fleets is the same as what the effort was in the year 2003 until 2010, and after 2010 only the effort in the small pelagic fleet is allowed to vary. A second reference world was run so that after 2010, the effort in the small pelagic fleet was increased by 2% each year, which represents a modest growth in the sector, in particular carnivorous species which consume much of the small pelagic fish landed through fishmeal and fish oil. The

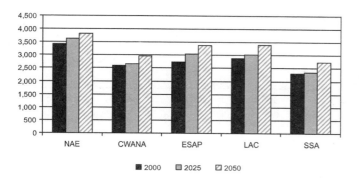

Figure 5-8. *Average daily calorie availability per capita, selected regions, reference run.* Source: IFPRI IMPACT model simulations.

2% value is based on recent FAO reports on the growth of aquaculture (FAO/WHO, 2006; see also Table 5-11).

Atlantic Ocean. Under the reference run there is an overall decline in landings (5.4%) between 2003 and 2048 but under the 2% increase in small pelagic effort variant there is a 7% increase in landings. In both alternatives the trophic level of the catch declines between 2 and 2.5%. Six FAO areas are represented in the Atlantic, and in all areas except 34 and 21 landings increase in the reference run and trophic level declines. In FAO areas where landings continue to decline, the trophic level in Area 34 is the only area where it increases while in Area 21 it continues to decline. In the 2% effort variant landings in most areas increase or remain steady except in Area 21 where it declines. The trophic level decreases in all but Area 31 where it increases slightly. The changes in biomass of the major species is seen in increases in small pelagic fish (e.g., capeline, herring) and declines in large demersal and large bentho-pelagic fish (e.g., cod, haddock).

Pacific Ocean. The baseline modeling for the Pacific results in declines in landings by 5% from 203 to 2048, while in the 2% increase in effort in the small pelagic fleet there is an overall increase (117%) in landings. The trophic level remains unchanged in the baseline but declines by 1.3% in the small pelagic effort variant. Six FAO areas are included in the Pacific and much of the change in landings in the

reference run can be attributed to Areas 77, 67 and 61 and changes in trophic levels can be attributed to Area 87. However, in the 2% effort variant the landings increase in most FAO areas but the trophic level only declines in FAO Areas 87 and 61. Much of the change in landings and trophic level are due to increasing biomass of small pelagics and declining biomass of most other groups.

Indian Ocean. In the reference run landings initially decline while stocks rebuild and then landings increase but only to 1% more than in 2003 by 2048. However, the trophic level of the catch does not decline with increased landings. Landings increase in the 2% effort variant. The growth is small, less than 5% but also the sustainability of these fisheries policy is uncertain since the trophic level of the catch continues to decline under the 2% policy from 2003 to 2008. The Indian Ocean represents two FAO Areas (51 and 57) and much of the overall increase in landings is due to increased small pelagic biomass in Area 57.

Mediterranean. Landings in the reference run increase by 7% with a corresponding decline in trophic level of 3%. In the 2% small pelagic effort variant, landings increase by 50% then level off while the trophic level declines steadily and by 3% from 2003. The sustainability of this policy is

Table 5-11. **Fisheries, reference run.**

FAO Area	Baseline	% change in trophic level 2003 to 2048	2% increase	% change in trophic level 2003 to 2048
	% change in landings 2003 to 2048		% change in landings 2003 to 2048	
Atlantic				
21	-39	-5.9	-35	-6.0
27	15	-1.5	22	-2.4
31	20	2.8	25	2.3
34	-30	1.1	3.9	-1.3
41	26	-1.2	34	-1.9
47	33	-3.1	13.9	-0.9
Pacific				
61	19	-2.3	14	-2.7
67	47	-2.8	44	-2.6
71	-15	0.5	11.4	-0.9
77	56	1.5	47	0.4
81	13	-0.1	2.8	-0.2
87	-38	3.9	13	-1.8
Indian				
51	-21	1.3	-10	-1.3
57	73	4.8	56	2.1
Med 37	71	-3.8	50	-3.1

Source: ECO-OCEAN.

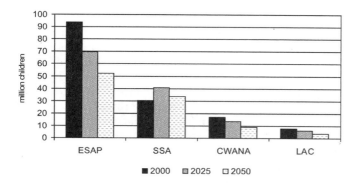

Figure 5-9. *Number of malnourished children, 2000 and projected 2025 and 2050, selected developing country regions.* Source: IFPRI IMPACT model simulations.

Table 5-12. Share of global renewable water resources and population at 2000 and 2050, reference run.

Region	IRW (Km³/year)		Share of Global IRW (%)		Share of Global Population (%)	
	2000	2050	2000	2050	2000	2050
North America and Europe (NAE)	8,677	14,802	21	32	18	13
East-South Asia and Pacific (ESAP)	12,922	15,218	31	33	54	49
Central-West Asia and North Africa (CWANA)	1,328	1,184	3	3	10	13
Latin America and Caribbean (LAC)	14,000	11,225	34	24	8	9
sub-Saharan Africa (SSA)	4,639	4,345	11	9	10	17
Developed Countries	9,946	15,424	24	33	20	14
Developing Countries	31,620	31,349	76	67	80	86
World	41,566	46,773	100	100	100	100

Note: IRW = Internal renewable water resources.

Source: IFPRI IMPACT model simulations.

uncertain since small pelagic biomass declines steadily toward the end of the modeled time period and the biomass of the large bentho-pelagic fish also declines.

5.3.3.2 Global trends in water availability and emerging challenges to water supply

Changes in human use of freshwater are driven by population growth, economic development and changes in water use efficiency. Historically, global freshwater use had increased at a rate of about 20% per decade between 1960 and 2000, with considerable regional variations due to different development pressures and efficiency changes. Because of uneven distribution of fresh water in space and time, however, today only 15% of the world population lives with relative water abundance, and the majority is left with moderate to severe water stress (Vorosmarty et al., 2005).

This global water picture may be worsened in the future under climate change and population growth. For the reference run, by 2050 internal renewable water (IRW) is estimated to increase in developed countries but is expected to decrease in the group of developing countries (Table 5-12). The disparity of changes of IRW and population by 2050 will increase the challenge to satisfy future water demands, especially for the group of developing countries.

Table 5-13 presents total water consumption, which refers to the volume of water that is permanently lost (through evapotranspiration or flow to salty sinks, etc.) and cannot be reused in the water system, typically a river basin. In the reference world, by 2050 world water consumption is expected to grow by 14%. Regionally, by 2050, SSA is projected to more than double water consumption, LAC is expected to increase water consumption by 50%, and ESAP by 13%, while in the NAE region the increase is modest, at 6%. Only CWANA reduces its water consumption—as a result of further worsened water scarcity. The IRW reduction of CWANA makes its global share of IRW decrease from 3.2% to 2.5% (Table 5-12). Combined with the increase of

population share from 10% to 13% CWANA is expected to face the largest challenge in meeting demands exerted by socioeconomic development and conservation demands to sustain ecological systems.

Irrigation is expected to continue to be the largest water user in 2050 for all regions (Table 5-14). However, it is estimated that the share of irrigation consumption in total water depletion will decrease by about 8% from 2000 to 2050, largely due to the more rapid growth of nonirrigation water demands that compete for water with irrigation (Table 5-15), and also because of projected declines in irrigated areas in some parts of the world. Actual irrigation consumption will decline significantly in CWANA due to chronically worsening water scarcity in the reference run. For individual dry river basins the IWSR could be even lower than these spatially-averaged values since abundant water in some basins mask scarcity in the dry river basins. On the other hand, significant increases are expected in the LAC and SSA regions at 45% and 77% respectively.

Constraints to water supply vary across regions. Water shortages today and in the future are not solely problems of resource scarcity, but are also closely related to stages of economic development. Three types of water scarcity constraints will become more important in the future: first is absolute resource scarcity, which will become more and more a feature of regions characterized by low and highly variable rainfall and runoff, often accompanied by high evapotranspiration potential. They include countries and subnational regions in CWANA, ESAP (for example, northwestern China), and NAE (for example, southwestern US), among others. The second type is infrastructure constraints, typically in regions where water availability is not extremely low but infrastructures to store, divert/pump, and convey water is underdeveloped. Despite rapid development of irrigation-related and other water infrastructure in the SSA region, the region will remain infrastructure constrained out to 2050. The third type, water scarcity, is caused by water

Table 5-13. Total water consumptive use, reference world.

Region	Total Water Consumption by all Economic Sectors (km³ yr⁻¹)	
	2000	2050
North America and Europe (NAE)	737	778
East-South Asia and Pacific (ESAP)	1,384	1,570
Central-West Asia and North Africa (CWANA)	519	486
Latin America and Caribbean (LAC)	252	377
sub-Saharan Africa (SSA)	61	146
Developed Countries	753	791
Developing Countries	2,199	2,567
World	2,952	3,358

Source: IFPRI IMPACT model simulations.

Table 5-14. Potential and actual consumptive water use for irrigation, and irrigation water supply reliability for 2000 and 2050.

Region	Potential Irrigation Water Consumption (km³ yr⁻¹)		Actual Irrigation Water Consumption (km³ yr⁻¹)		Irrigation Water Supply Reliability (IWSR) (%)	
	2000	2050	2000	2050	2000	2050
North America and Europe (NAE)	731	960	598	615	82	64
East-South Asia and Pacific (ESAP)	1,950	2,277	1,256	1,259	64	56
Central-West Asia and North Africa (CWANA)	758	915	489	420	65	46
Latin America and Caribbean (LAC)	268	390	224	324	83	83
sub-Saharan Africa (SSA)	50	101	50	88	99	87
Developed Countries	710	946	606	623	85	66
Developing Countries	3,047	3,697	2,010	2,085	66	56
World	3,757	4,643	2,616	2,707	70	58

Source: IFPRI IMPACT model simulations.

Table 5-15. Non-irrigation consumptive water use for 2000 and 2050 (in km³ yr⁻¹).

Region	Domestic		Industrial		Livestock		Total Non-Irrigation	
	2000	2050	2000	2050	2000	2050	2000	2050
North America and Europe (NAE)	41.0	47.8	91.2	109.7	6.0	5.3	138.2	162.9
East-South Asia and Pacific (ESAP)	64.1	153.3	48.3	133.7	16.0	23.7	128.4	310.6
Central-West Asia and North Africa (CWANA)	11.1	31.1	6.4	16.0	12.0	19.0	29.4	66.2
Latin America and Caribbean (LAC)	15.3	29.6	6.5	14.2	6.4	8.7	28.3	52.5
sub-Saharan Africa (SSA)	6.6	45.3	1.0	6.6	4.0	6.7	11.5	58.6
Developed Countries	45.4	51.4	94.4	111.4	6.7	5.9	146.5	168.7
Developing Countries	92.6	255.6	59.0	168.9	37.7	57.5	189.3	481.9
World	138.0	307.0	153.4	280.2	44.4	63.4	335.8	650.7

Source: IFPRI IMPACT model simulations.

quality constraints, which are becoming increasingly normal in regions where rivers and aquifers are contaminated by insufficiently treated or untreated industrial wastewater and nonpoint source pollution from agricultural practices. An increasing number of countries in ESAP are included in this category, for example, the Huai River Basin in China.

As a result of increased potential irrigation water consumption and reduction or moderate increase in actual irrigation consumption, irrigation water supply reliability (IWSR) is expected to decline in all regions. Globally, the IWSR decreases from 70% to 58% from 2000 to 2050. Regionally, LAC is likely to maintain a stable IWSR over the next 50 years given its abundance in water resources, although its water availability will decline by nearly 20% over this period. The IWSR of CWANA is expected to be below 50% by 2050 due to increased irrigation water demand (largely due to increased potential evapotranspiration) and reduced water availability. This would impose a significant impact on crop yield, and potentially jeopardize food security in this region.

Total harvested irrigated area is expected to increase from 433 million ha in 2000 to 478 million ha in 2025 and to then slightly decline to 473 million ha by 2050. Cereals account for more than half of all irrigated harvested area over the reference run period. Over the projections horizon, irrigated area is projected to more than double in SSA. However, by 2050, SSA is expected to still account for less than 2% of global harvested irrigated area. Increases in LAC and ESAP are projected at 41% and 12%, respectively, whereas almost no changes are projected for the CWANA and NAE regions (Figure 5-10).

Sharp increases in nonirrigation water demands are projected over the next 50 years, with increases concentrated in the group of developing countries (Table 5-15). In the reference run, globally, nonirrigation water consumption would almost double by 2050, approaching 651 km³ per year. Notably, nonirrigation consumption in developing countries is estimated to reach 482 km³, more than doubling from 2000. In comparison, total nonirrigation water consumption in developed countries only increases moderately. The most significant increase in the group of developing countries is domestic water consumption, which grows rapidly from 93 km³ to 256 km³ over 50 years. This dramatic increase is driven by both population growth and per capita demand increase due to income growth. Industrial demand would also increase significantly, with the largest relative increase in SSA (though still low by population size) and the largest absolute increase in ESAP.

5.3.3.3 Results for energy production and use
In terms of final energy demand, the reference projection shows an increase of 280 EJ in 2000 to around 500 EJ in 2030 and more than 700 EJ in 2050 (see also Figure 5-11). This is somewhat faster than the historic trend. This difference is the result of the fact that (1) historically several events have slowed down demand in energy consumption (energy crises, economic transition in FSU, Asia financial crisis), and (2) the increasing weight of developing countries with typically higher growth rates in the global total.

Most of the increase in energy demand takes place in the group of developing countries. At the same time, how-

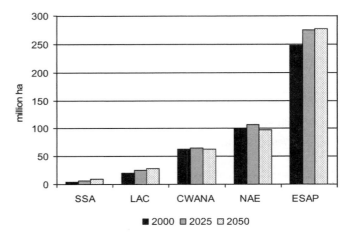

Figure 5-10. *Changes in irrigated harvested area, 2000, and projected 2025, and 2050, reference run, by IAASTD region.* Source: IFPRI IMPACT model simulations.

ever, it should be noted that per capita energy consumption remains much higher in developed countries than in developing countries. In terms of energy carriers, most of the energy consumption continues to be derived from fossil fuels—in particular oil (for transport). The growth of oil is somewhat slowed down in response to high oil prices. Modern bioenergy represents a fast growing alternative to oil—but remains small in terms of overall energy consumption. Coal use increases slightly—as high oil and gas prices imply that coal remains an attractive fuel in the industry sector. This partly offsets the trend away from coal in the buildings sector. Natural gas use increases at about the same rate as the overall growth in energy consumption. Finally, the level of electricity use increases dramatically.

In electric power, the reference run expects coal to continue to remain dominant as a primary input into power production. In fact, its share increases somewhat in response to high oil and gas prices—and as a result of high growth in electric power production in countries with high shares of coal and limited access to natural gas supplies (such as India and China). Rapidly growing alternative inputs such as solar/wind power and bioenergy gain market share, but form only about 10% of primary inputs by the end of the reference run period.

In terms of supply, it is expected that oil and natural gas production will concentrate more and more in a small number of producing countries as a result of the depletion of low-cost supply outside these countries. It is also expected that fossil fuel prices remain relatively high. In that context it should be noted that current high oil prices are mostly a result of (1) rapid increases in demand, (2) uncertainties in supply, and (3) underinvestment in production capacity. Some of these factors could continue to be important in the future—although estimates are hard to make (and strongly depend on perspectives of the future with respect to globalization and cooperation, regional tensions, etc.). In addition, depletion of low-cost resources will lead to upward pressure on prices. As a result, it is likely that fossil fuel prices remain at a relatively high level—although probably somewhat below 2005-2007 levels. Continued high price

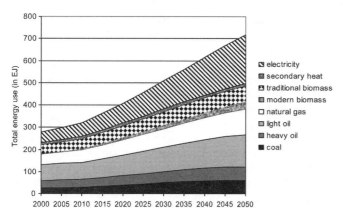

Figure 5-11. *Total energy use, reference run.* Source: IMAGE model.

levels will provide incentives to invest in alternative energy sources such as bioenergy.

5.3.3.4 Climate

Under the IAASTD reference run, the atmospheric concentration of greenhouse gas rises driven mainly by increasing emission of greenhouse gases from the energy sector (see Figure 5-12 for CO_2). The concentration of greenhouse gases reflects the balance of net fluxes between terrestrial areas, oceans and the atmosphere. By 2030 the CO_2 concentration reaches 460 ppmv, and increases further to 560 ppmv in 2050, a doubling of the preindustrial level. This trend is not stabilizing in 2050, so higher concentrations will occur on the longer term.

The effect of more greenhouse gases is a rise in global mean temperature above the preindustrial level to 1.2°C in 2030 and 1.7°C in 2050 (Figure 5-13). These values are well in line with IPCC, where best estimate values at the end of the twenty-first century range between 1.8 and 4.0°C. This range is even larger when the uncertainties in the climate sensitivity are taken into account as well. Including this range of uncertainty, the IPCC gives a temperature range between 1.1 and 6.4°C by the end of the twenty-first century. Taking uncertainties in climate sensitivity into account in the reference run, global-mean temperature increase will be around 1.0 and 2.5°C in 2050.

The calculations show that in the first few decades of the twenty-first century, the rate of temperature change is somewhat slightly slowed down compared to the current rate, in response to lower emissions, e.g., due to a slowdown in deforestation, stable methane concentration in recent decades and increasing sulfur emissions, mainly in Asia, with a cooling effect. In the following decades these trends are discontinued, driving the temperature change rate upwards again. Factors that contribute to this increase are increasing greenhouse gas concentration in combination with reduced sulfur emissions. By 2030 the temperature increases by a rate of more than 0.2°C per decade, augmenting the climatic impacts and increasing the need for adaptation (Leemans and Eickhout, 2004), especially in nature and agriculture.

Changes in global mean temperature cannot reflect the regional effects that climate change may have on crop yield, water resources and other environmental services for hu-

man development. Therefore, the regional aspects of climate change need to be addressed as well, although the extent of the regional effect is still very uncertain. Although the global mean temperature change is around 1.4°C between 1990 and 2050 (see Figure 5-13), regionally this can imply changes of more than 2.5-3°C (Figure 5-14). The IPCC concluded that many of the developing countries are most vulnerable to climate change, mainly because of their dependency on sectors, which are relying on climatic circumstances, and their low ability to adapt (IPCC, 2007). For example, agricultural production, including access to food, is projected to be severely compromised by climate variability and change in many African countries and regions (IPCC, 2007). The area suitable for agriculture, the length of growing seasons and yield potential, particularly along the margins of semiarid and arid areas, are expected to decrease.

The impact on crop growth is one of the most important direct impacts of climate change on the agricultural sector. Through CO_2 fertilization, the impact can be positive, although this effect is not larger than 20% (IPCC, 2007). However, at higher temperature increases the impacts on crop yields will be dominated by negative impacts (IPCC, 2007). In total, the impact in 2050 is still relatively small, apart from some crucial regions like South Asia (Figure 5-15). Results can become even more negative when changes in climate variability are included as well, which is not included in this analysis. These conclusions are in line with IPCC, where it was concluded that, globally, the potential for food production is projected to increase with increases in local average temperature over a range of 1-3°C, but above this it is projected to decrease (IPCC, 2007). Moreover, increases in the frequency of droughts and floods due to changes in climate variability are projected to affect local production negatively, especially in subsistence sectors at low latitudes (IPCC, 2007).

5.3.3.5 Environmental consequences—land use change

The impacts of changes in agriculture and demand for biofuels lead to changes in land use. Although expansion in pastureland is compensated by an increase in grazing intensity, and increase in crop land is partially compensated by technological improvements (see 5.3.3), total land use for

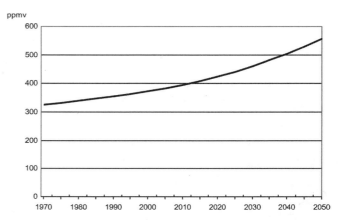

Figure 5-12. *Atmospheric CO_2 concentration out to 2050, reference world.* Source: IMAGE model.

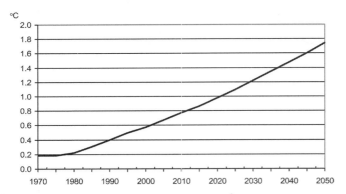

Figure 5-13. *Global surface temperature change above pre-industrial levels up to 2050, reference world.* Source: IMAGE model.

humans is still increasing until 2050 with an expansion of 4 million km². The increasing demand for bioenergy is one of the important reasons for this development (Figure 5-16). In Figure 5-17, the regional breakdown is given for bioenergy areas being used for energy purposes. Although Latin America is currently one of the most important energy crop growing regions, in the future regions where land abandonment will occur increasingly (like in Russia) are expected to

overtake this market. This is mainly due to a high increase of agriculture in Latin America which leaves little room for any additional energy crops (see also 5.3.3.3). Moreover, in this approach most of the bioenergy is grown on land that is abandoned or on land that is low productive (see Appendix for the methodology; Hoogwijk et al., 2005).

These changes in land use also affect air quality and the atmospheric composition of greenhouse gases. Land use related emissions will continue to increase, mainly due to increased animal production (CH_4 emissions) and fertilizer use (N_2O emissions) (Figure 5-18). Carbon dioxide emissions due to land use change (deforestation) are expected to stay more or less constant.

5.3.3.6 Environmental consequences—forests and terrestrial biodiversity
Forests represent valuable natural ecosystems, with high potential to provide a variety of services to mankind and rich in biodiversity measured by the number of species. Although forest can regrow with time after clear-cutting for timber production, and after abandonment of agricultural production, these areas will revert only slowly to a more natural state, if ever. As a somewhat arbitrary definition, only forests that were established before 1970 are included in the count presented here. Therefore, regrowth forests after

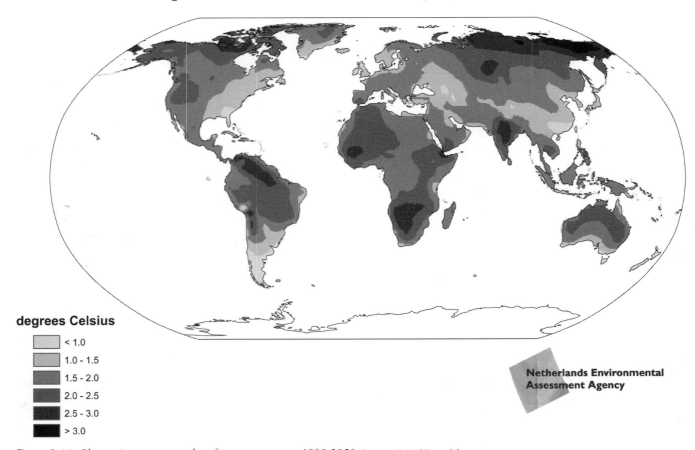

Figure 5-14. *Change in mean annual surface temperature, 1990-2050.* Source: IMAGE model.

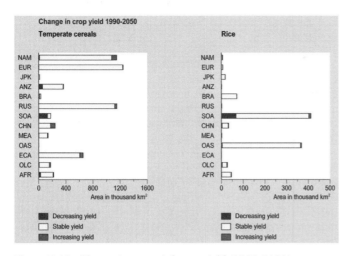

Figure 5-15. *Change in potential crop yield (1990-2050) attributed to climate change.* Source: IMAGE model.

Note: Regions are North America (NAM), Europe (EUR), Japan and Korea (JPK), Oceania (ANZ), Brazil (BRA), Russia (RUS), South Asia (mainly India; SOA), China (CHN), Middle East (MEA), Other Asia (OAS), Eastern Europe and Central Asia (ECA), Other Latin America (OLC) and Africa (AFR).

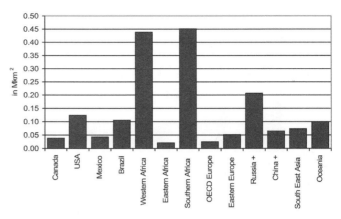

Figure 5-17. *Bioenergy area in 2050 for different countries and regions in the world, reference run.* Source: IMAGE model.

1970 are excluded from the areas shown in Figure 5-19. The only new areas included are those that change from other natural biomes to forest as a consequence of climate change and without human intervention. The figure illustrates that natural forest areas decline in all regions, but most clearly in developing regions like LAC and ESAP.

Biodiversity is expressed in terms of the indicator mean species abundance (see Appendix on GloBio3). The MSA value is affected by a range of human induced stress factors. For terrestrial biodiversity these include loss of habitat, climate change, excess nitrogen deposition, infrastructure and fragmentation. These stress factors are the direct drivers of biodiversity loss and are derived from indirect drivers such as population, GDP and energy use.

Loss of the biodiversity quality in the natural biomes started already many centuries ago, as can be seen in the his-

torical graph from 1700 to 2000 (Figure 5-20). The strongest declines occur in the temperate and tropical grasslands and forests. The remaining biodiversity is found more and more in biomes that are less suitable for human development and thus less likely to be affected, such as deserts and polar biomes. This trend continues with an anticipated and accelerating further loss of biodiversity.

At the global level, there is a substantial biodiversity loss in the reference run: the remaining MSA level drops another 10% after 2000. The rate of decrease for the period 2000-2050 is even higher than in the period 1970 to 2000 (Figure 5-21). The role of agricultural land-use change remains the largest of all pressure factors, which is clearly related to the strong increase in crop areas (see 5.3.3.5). The major contributors to the additional biodiversity loss from 2000 to 2050 are: expanding infrastructure, agriculture and climate change. The influence of nitrogen deposition and fragmentation does not increase, even though these factors share similar indirect drivers as the other factors. In fact, through expanding agriculture, less natural biomes are left where these stresses can exert their influence.

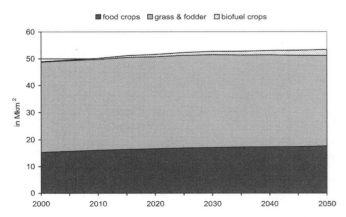

Figure 5-16. *Agricultural area (food crops, pastureland and biofuel crops) globally, 2000 to 2050, reference run.* Source: IMAGE model.

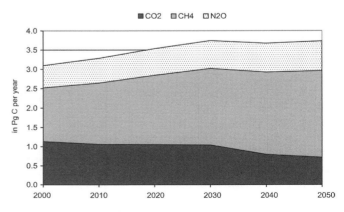

Figure 5-18. *Land-use emissions from CO_2, CH_4 and N_2O from 2000 to 2050, reference run.* Source: IMAGE model.

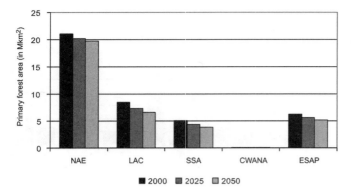

Figure 5-19. *Change in forest areas excluding regrowth, 2000, 2025, and 2050.* Source: IMAGE model.

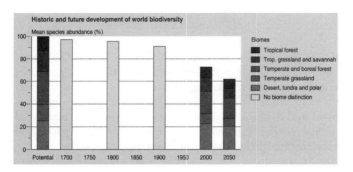

Figure 5-20. *Development of global biodiversity 1700-2050 in mean species abundance in various natural biomes.* Source: GLOBIO 3.

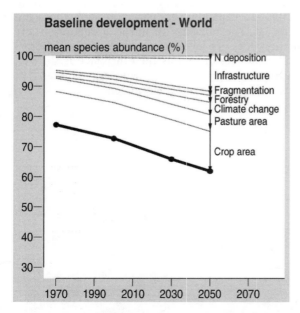

Figure 5-21. *Biodiversity development for the world, and contribution of stress factors to the decline in the reference run.* Source: GLOBIO 3.

5.4 Assessment of Selected, Major Policy Issues

5.4.1 Climate change policies and agriculture

In the previous subchapters, we have shown that the reference projection leads to a strong increase in greenhouse gas concentration—and thus considerable climate change. In contrast, the recent IPCC reports indicate that avoiding dangerous anthropogenic interference with the climate system may require stabilization of GHG concentration at relatively low levels. Current studies on emission pathways, for instance, indicate that in order to achieve the objective of EU climate policy (limiting climate change to 2°C compared to pre-industrial) with a probability of at least 50% may require stabilization below 450 ppm CO_2-eq. Stabilization at 450, 550 and 650 ppm CO_2-eq. is likely to lead to a temperature increase of respectively, about 1.2-3°C, 1.5-4°C, and 2-5°C. The Stern review recently concluded that given all evidence, limiting temperature increase to about 2°C would be an economically attractive goal (Stern, 2006). This conclusion has been debated by some authors, but also found strong support by others. For the purpose of this report, we have decided to explore the impact of stringent climate policies.

The IPCC AR4 WGIII report concludes on the basis of model-supported scenario analysis that several model studies show that it is technically possible to stabilize greenhouse gas emissions at 450 ppm CO_2-eq. after a temporary overshoot. This is also supported by the model analysis carried out for IAASTD. Emission reductions required to stabilize at 450 ppm (around 60% in 2050 and 90% in 2100, globally) can be reached through various emission reduction options (Figure 5-22). Efficiency plays an important role in the overall portfolio. Carbon capture and storage is another important technology under default assumptions—but may be substituted at limited additional costs against other zero-carbon emitting technologies in the power sector. Obviously, the concentration target forms a tradeoff between costs and climate benefits. The net present value of abatement costs (2010-2100) for the B2 baseline scenario (a medium scenario) increases from 0.2% of cumulative GDP to 1.1%, going from stabilization at 650 to 450 ppm. On the other hand, the probability of meeting the EU climate target (limiting global mean temperature increase to 2°C) increases from 0-10% to 20-70%.

One important option in the overall portfolio is also bioenergy. As discussed in Chapter 4, there is a strong debate on the advantages and disadvantages of bioenergy. The 450 ppm stabilization case is likely to lead to a strong increase in land use for bioenergy. A recent paper by van Vuuren et al. (2007) on the basis of a comparable scenario found land for bioenergy to increase from 0.4 to 1.0 Gha in 2100, while at the same time land for carbon plantation increased from 0 to 0.3 Gha.

Obviously, the climate policy variant has important benefits in reducing climate change, although some of these may only materialize in the long-term (Figure 5-23). As shown, the emission reductions are likely to reduce greenhouse gas concentration substantially in 2050. At the same time, however, the medium-term (2050) impacts on temperature increase are relatively slow. The latter is due to inertia in the

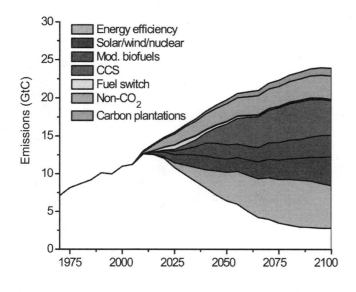

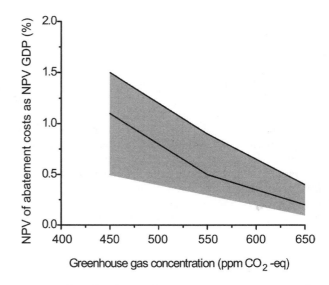

Figure 5-22. *Contribution of various options in reducing greenhouse gas emissions from baseline to the 450 ppm CO₂-eq variant and the costs associated with stabilizing greenhouse gas concentrations. Source: IMAGE-model; Van Vuuren, 2007*

Note: Net present value of abatement costs at 5%; discount rate as percentage of GDP

climate system, but also due to the fact that climate policy also reduces SO₂ emissions, reducing atmospheric aerosols that lead to a net cooling. In other words, impacts on agriculture in 2050 are similar in the stringent policy case as in the reference run (see 5.3.3.4). Uncertainty does not come from different variants—but differences in climate sensitivity. In the longer run, however, the temperature of the policy case will remain significantly below the reference case.

5.4.2 Trade policies and international market constraints

Support policies and border protection of wealthy OECD countries, valued at hundreds of billions of dollars each year, cause harm to agriculture in developing countries. Evaluating the overall effects of subsidies and trade protec-

tion policies of industrialized countries on developing countries and poverty within developing countries is challenging. The evaluation must rest on counterfactual simulation of alternative policy variants leading to a diverse set of policies and application of a range of different models.

As part of this report, GTEM was applied to two hypothetical variants representing two alternative global trade policy regimes. Variant 1 represents a global economy in which import tariffs (and tariff equivalents) on all goods are removed incrementally between 2010 and 2020 across the globe. Variant 2, on the contrary, represents a world in which trade barriers will escalate gradually between 2010 and 2020 such that by 2020 these barriers will be equivalent to twice the size of the existing tariff (and tariff equivalent) barriers across the board.

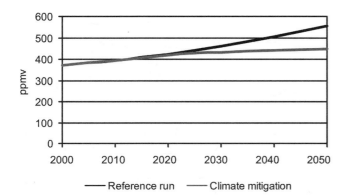

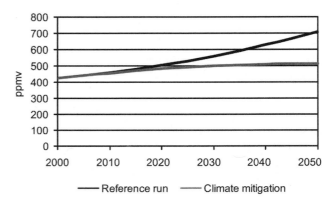

Figure 5-23. *Atmospheric CO₂ and CO₂-eq concentration between 2000 and 2050.* Source: IMAGE-model

The key impacts of the two alternative variants at 2025 are analyzed below. Unless otherwise stated, the impacts are expressed as deviations from the reference case which represents no trade liberalization or reduced sectoral protection throughout the projection period. It should be noted that the impacts of trade policy changes only represent static gains/losses associated with resource reallocation and do not encapsulate any potential dynamic gains/losses associated with any long-run productivity changes. Furthermore, except for the trade policies in question, all other policies remain unchanged as in the reference case.

Figure 5-24 shows the overall impacts of trade liberalization under variant 1 in terms of changes in gross regional product (a regional equivalent to GDP). The world economy is projected to benefit from multilateral trade liberalization. In particular, gross regional products in CWANA and SSA regions are projected to grow the most, by more than 2% relative to the reference case at 2025. However, about two-fifths of the global benefits (in today's dollars) are projected to accrue to the ESAP region. Interestingly, while the removal of trade barriers under variant 1 is projected to increase income and food consumption, the global structure of food production appears to undergo significant changes. Compared with the reference case, a significant increase in meat production is projected to occur in LAC and SSA regions with a substantial decline projected for the NAE region (Figure 5-25). The structural change in global production of nonmeat food is not as striking as in the case of meat. In nonmeat production, LAC and SSA regions are projected to register the most growth relative to the reference case at 2025 (Figure 5-26).

There is an increase globally in overall trade volume under liberalization, with a noticeably larger effect, relative to the reference case, in 2025 (soon after the liberalization is complete), compared with 2050 (Figure 5-27). Figure 5-28 presents changes in cereal production as a result of the trade liberalization variant. The removal of protection for important cereals in North Africa leads to a decline in production in the CWANA region, as well as in the NAE region. On the other hand, production in ESAP, SSA, and LAC increases. Trade liberalization leads to increased prices for cereals and meats in 2025 (Figure 5-29), but prices decline again somewhat in the later period.

While estimates of the benefits of removal of global agricultural subsidies and trade restrictions vary, other analyses have found similarly positive outcomes for Africa. One study finds that under full global agricultural trade liberalization (complete removal of trade barriers), Africa would receive annual net economic benefits of US$5.4 billion (Rosegrant et al., 2005). Another study indicated that European Union agricultural policies have reduced Africa's total potential agricultural exports by half. Without these agricultural policies, the current US$10.9 billion food-related exports annually from SSA could actually grow to nearly US$22 billion (Asideu, 2004).

Under variant 2 in which trade protection will be doubled between 2010 and 2020, all broad regional economies are projected to decline relative to the reference case (Figure 5-30). Again, CWANA and SSA regions are projected to be affected the most, declining by about 1.5% relative to the reference case at 2025.

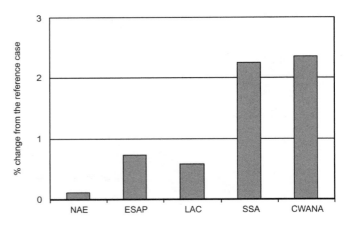

Figure 5-24. *Projected impacts on gross regional product of trade liberalisation under variant 1 at 2025.* Source: GTEM.

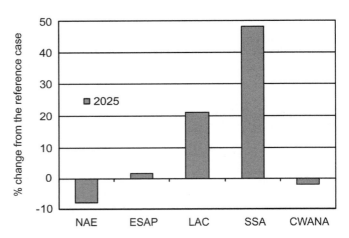

Figure 5-25. *Projected impacts on meat production under variant 1 at 2025.* Source: GTEM.

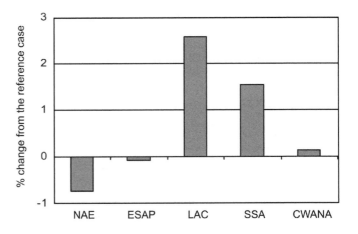

Figure 5-26. *Projected impacts on nonmeat food production under variant 1 at 2025.* Source: GTEM.

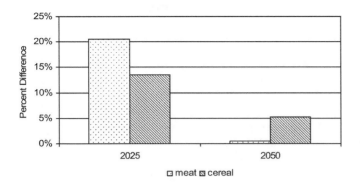

Figure 5-27. *Projected impacts on global traded volumes of meats and cereals of decreased trade protection at 2025 and 2050.* Source: IFPRI IMPACT model simulations.

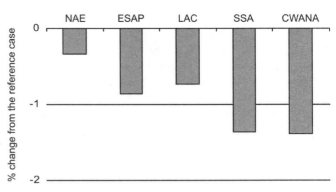

Figure 5-30. *Projected impacts on gross regional product of increased trade protection under variant 2 at 2025.* Source: GTEM.

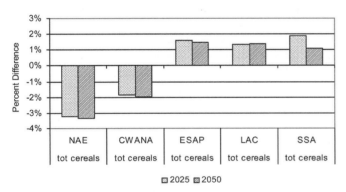

Figure 5-28. *Projected impacts on regional cereals production of decreased trade protection at 2025 and 2050.* Source: IFPRI IMPACT model simulations.

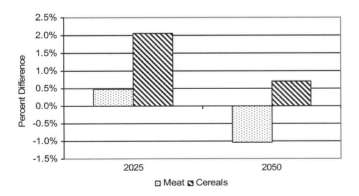

Figure 5-29. *Projected impacts on world meat and cereal prices of decreased trade protection at 2025 and 2050.* Source: IFPRI IMPACT model simulations.

5.4.3 Investment in AKST

As has been described above, the reference run describes slowly declining rates of growth in agricultural research (and extension). In the following, two alternative variations are analyzed using two sets of changing parameters. The first set of variations looks at different levels of investments in agriculture during 2005-2050. Different levels of investments can result in either higher (AKST_high) or lower crop yield and livestock numbers growth (AKST_low). The second set of variations analyzes the implications of even more aggressive or reduced growth in agricultural R&D together with advances in other, complementary sectors (AKST_high_pos and AKST_low_neg with "pos" for higher investments in complementary infrastructure and social services and "neg" for decelerating growth in these services). Such other sectors include investments in irrigation infrastructure (represented by accelerated or slowing growth in irrigated area and efficiency of irrigation water use and by accelerated or reduced growth in access to drinking water, and changes in investments of secondary education for females, an important indicator for human well-being (Tables 5-16 and 5-17).

Results of the four alternative variations are presented in Figures 5-37 to 5-45 and Table 5-18 and 5-19. The AKST high variant, which presumes increased investment in AKST, results in higher food production growth and reduced food prices and makes food more affordable for the poor when compared to the reference world. As a result, demand for cereals is projected to increase both as food and as feed, by 339 million tonnes or 13% (Figure 5-31 and Table 5-17). The combination of even more aggressive investment in AKST with sharp increases in expenditures for supporting social services results in even higher demand for cereals as both food and feed, 633 million tonnes or 24%. On the other hand, if levels of investment in AKST drop somewhat faster than in recent decades and if investments in key supporting services are not strengthened, food prices would rise, and demand would be depressed.

Despite these strong changes in AKST behavior, yield growth will continue to contribute most to future cereal production growth under both the AKST_low and AKST_high variants (Figures 5-38 and 5-39). However, under AKST_low, area expansion would contribute 38% to production

Box 5-2. Trade policy and gender, case of India

In the reference world the overall growth in agriculture would be slightly lower than the current long-term trend in Indian agricultural growth (i.e., 3%) in 2025 and would be slightly higher in 2050. Overall growth in manufacturing sector in the reference period is 10% through the first 25 years and by about 8% in the next 25 years till 2050. With such growth rates projections taken from the IMPACT model and trend growth (for non-agricultural sectors) from Indian macro economic data sets, we find that the growth in resultant investment is healthy (see Table 5-30) and decelerating inflation that reaches the 1.4% by 2050. In brief, for India the macro picture is of robust with stable growth in the economy in the reference world. However, the rural-urban divide continues while urban households continue to improve their real income. In the longer run this gap somewhat declines. Moreover, the wage gap between men and women workers in the first 25 years declines. In the reference world the consumption of the lower deciles of the population improves continuously.

The impacts of trade policies on agriculture and AKST are studied as a variant to the reference run based on the GEN-CGE model for 2025 and 2050 for the case of India. The alternative run assumes that the peak tariff rate as an average of both agricultural

Hence wage rates of male workers rise less than wages of female workers, i.e., the low intensive factors of production.

The reference world out to 2050 and related sensitivity exercises are less accurate compared to 2025. This is because various structural factors that undergo changes cannot be captured very well in the economic analysis for a 50-year projection. By 2050 wage rates generally fall as there is a contraction of domestic production in manufacturing, mainly because of the way the economy has been driven together with lower protection. Only real wage rates of the male casual workforce would witness a marginal positive growth in 2050 and in 2050-1. Driving growth only through AKST without balanced growth in non-agriculture would lead to skewed growth and adversely affect real wages in general.

The findings show that the present trend of real wage growth of the female workforce may continue until 2025 narrowing the gender gap. The wage growth of both male and female workforce would then experience a downturn. The AKST measure is sustainable till 2025 as regards improvement of wages. In the next 25 years, i.e., by 2050, AKST needs enhanced market penetration to lead to real wage growth. Otherwise low-end manufacturing may be the only expanding sector demanding casual male labor, which would explain their wage gains.

Table 5-30. **Some key economic variables for India in the reference world.**

	2000	2025	2025-1	2050	2050-1
	Level			Annual Growth	
			(%)		
CPI (Index)	100	2.44	2.2	1.42	1.4
Total Investment (Constant Prices) (Rs. 10 Million)	429,741	5.36	5.77	7.56	7.53
GDP Real (Rs. 10 Million)	1,962,996	5.23	5.23	4.87	4.87

Source: GEN-CGE model simulations.

and nonagricultural goods would fall by 88% in the first 25 years with the backdrop of WTO bindings. This alternative simulation for 2025 is noted as 2025-1. In 2050, the tariff would further fall by close to another 7%. Under this simulation, the tariff in 2050 would be around 2%.

By 2025, there would be positive growth of both casual and regular male and female workers' average real wage rate (Table 5-31). However, the rise would be higher for the female workforce, indicating a greater demand for female workers in 2025 compared to male workers. In India, the underlying production process reflected by the 2002 structure informs that female workers are less intensive in all sectors except in agro-based sectors. With AKST, there is improvement in agricultural growth, creating a higher growth for the interlinked agro-based sectors. Further, with tariff reduction, the manufacturing sector faces higher competition and experiences lower growth. Therefore, demand for more intensive factors of production in manufacturing experiences comparatively lower growth compared to agricultural and agro-based sectors.

Per capita private income increases more in urban areas (at constant prices) than in rural areas during 2025-2050 (Table 5-32). Interestingly, income in the informal sector is growing faster than wages, causing a declining share of wages in total income. Moreover, as tariff rates are rationalized the situations of both rural and urban households improve relative to a situation with a more restrictive tariff regime. Any divergence occurs only in the case of the households earning from informal activities like petty trade and low-end manufacturing both in the rural and urban areas. Moreover, rural households gain gradually through the next 25 years and significantly in the following 25 years to 2050. So by the year 2050, the extent of inequality may not be as wide as one finds today, with further improvement with reduced protection.

Table 5-33 presents population deciles and per capita consumption expenditures. For the bottom 30% urban and rural population the per capita consumption level is similar. Moreover, per capita consumption of the lowest 30% of the population

continued

Box 5-2. continued

Table 5-31. **Average real wage rate by skill for India in the reference world and under trade liberalization.**

	Base = 2000	2025	2025-1	2050	2050-1
	Unit Rs.	Annual Growth Rate (%)			
Labor casual female	1,476.32	3.00	2.77	-0.21	-0.22
Labor regular female	8,443.14	3.23	3.32	-3.43	-3.96
Total Female	**2,137.04**	**3.09**	**2.99**	**-0.93**	**-0.99**
Labor casual male	3,183.97	2.70	2.52	0.40	0.41
Labor regular male	8,865.69	0.87	0.80	-0.89	-0.89
Total Male	**4,453.40**	**1.98**	**1.84**	**-0.08**	**-0.06**
Grand Total	3,697.08	2.21	2.08	-0.21	-0.21

Note:

2025-1 = Peak tariff rate is reduced by 88 percent over 2000

2050-1 = Peak tariff rate is reduced by 98 percent over 2000

Source: GEN-CGE model simulations.

Table 5-32. **Per capita private gross income (growth rate in %). Constant prices, India.**

	Base = 2000	2025	2025-1	2050	2050-1
	Unit = Rs.	Annual Growth Rate (%)			
Rural Poor Formal	23,633	0.88	1.12	0.64	0.66
Rural Non-Poor Formal	30,433	0.88	1.12	0.64	0.66
Rural Poor Informal	19,346	2.40	2.63	2.21	2.23
Rural Non Poor Informal	17,554	1.79	1.99	3.31	3.33
Total Rural	**18,359**	**2.01**	**2.23**	**2.92**	**2.94**
Urban Poor Formal	25,952	2.77	3.01	2.13	2.16
Urban Non Poor Formal	31,763	2.77	3.01	2.13	2.16
Urban Poor Informal	18,274	4.13	4.38	3.94	3.96
Urban Non Poor Informal	23,836	3.74	3.97	4.33	4.35
Total Urban	**25,619**	**3.33**	**3.57**	**3.38**	**3.40**
Grand Total	20,283	2.20	2.43	2.93	2.95

Note: 1 USD = Rs. 43.3 in 2000.

Source: GEN-CGE model simulations.

continued

Box 5-2. continued

improves throughout 2025 to 2050 and more so in liberalized regimes; hence both rural and urban households improve their consumption. The marginally better performance in consumption of rural poor households under AKST reassures that a more agriculture oriented growth process lead to decline of the rural-urban consumption gap in the long run.

livestock" is expected to grow with annual growth rate of 23%. The availability of non-agricultural goods in the domestic market is also expected to grow ranging from 2-5% per annum. Overall, total domestic supply is expected to grow by 4-5% every year out to 2050. The availability of goods for the domestic market

Table 5-33. **Population deciles with per capita consumption expenditure changes over reference run India (in ascending order).**

	Population Deciles	Per capita consumption				
		2000	2025	2025-1	2050	2050-1
		(Rupees)				
Rural	1st Decile (poorest 10%)	1,245	1,874	2,018	5,349	5,408
	2nd Decile	1,606	2,417	2,603	6,901	6,976
	3rd Decile	1,854	2,790	3,005	7,965	8,053
	Poorest 30%	1,571	2,364	2,545	6,748	6,822
	4th Decile	2,082	3,134	3,375	8,946	9,044
	5th Decile	2,310	3,476	3,743	9,922	10,031
	6th Decile	2,575	3,874	4,172	11,060	11,182
	7th Decile	2,879	4,333	4,666	12,368	12,504
	8th Decile	3,291	4,952	5,333	14,137	14,292
	9th Decile	3,954	5,949	6,407	16,984	17,170
	10th Decile (richest 10%)	6,281	9,452	10,179	26,983	27,279
	All Rural	2,806	4,222	4,547	12,054	12,186
Urban	1st Decile (poorest 10%)	1,260	1,604	2,059	4,956	5,017
	2nd Decile	1,691	2,152	2,659	6,651	6,732
	3rd Decile	2,010	2,559	3,145	7,907	8,004
	Poorest 30%	1,653	2,105	2,621	6,504	6,583
	4th Decile	2,323	2,957	3,466	9,137	9,248
	5th Decile	2,678	3,409	3,866	10,534	10,663
	6th Decile	3,092	3,936	4,286	12,162	12,311
	7th Decile	3,604	4,588	4,811	14,177	14,351
	8th Decile	4,337	5,522	5,435	17,063	17,272
	9th Decile	5,512	7,017	6,595	21,682	21,948
	10th Decile (richest 10%)	10,226	13,019	10,437	40,227	40,719
	All Urban	3,672	4,675	4,675	14,445	14,622

Note: 1 USD = Rs. 43.3 in 2000.

Source: GEN-CGE model simulations.

Table 5-34 shows an improvement of per capita availability of different agricultural crops through 2025 and further till 2050. The domestic supply in agriculture is projected to grow by 2.87% annually to 2050, and by 4.72% to 2050. The only sector showing a decline is the "meat" sector. However, apart from "meat", "other

indicates that domestic production along with imports remains healthy even after fulfilling demand for exports. Domestic supply of goods grows more significantly for the nonagricultural sectors and then again for the later years from 2025 through 2050.

continued

Box 5-2. continued

Table 5-34. **Total domestic supply of goods and services, India, reference run and trade liberalization variant.**

	Base = 2000	2025	2025-1	2050	2050-1
	Unit Rs. 10 million	Annual Growth (%)			
Rice	170,095	1.7	0.62	2.91	2.79
Wheat	50,853.5	4.62	4.1	4.96	4.86
Maize	5,556.32	4.48	4.32	4.6	4.48
Other coarse grains	8,833.8	4.53	4.16	5.6	6.36
Pulses	21,635.1	4.59	4.28	5.03	4.93
Potatoes	7,036.53	4.59	4.27	5.12	5.28
Other crops	230,682	1.83	4.66	4.22	4.36
Oilseeds and edible oils	133,039	1.14	1.14	2.44	2.46
Meat	39,045.7	4.59	4.2	2.27	2.08
Fishing	21,015	4.6	4.08	1.54	1.9
Other livestock	115,019	4.63	4.2	5.19	5.34
Total Agriculture	802,810	2.87	2.3	3.79	3.88
Fertilizers	34,902.5	2.49	3.26	1.13	0.81
Other manufacturing	1,458,410	2.59	2.71	1.58	1.58
Other services	1,248,214	2.7	2.89	1.4	0.89
Total Nonagriculture	2,741,526	2.64	2.8	1.5	1.35
Grand Total	3,544,336	2.69	2.69	2.28	2.24

Note: 1 USD = Rs. 43.3 in 2000.

Source: GEN-CGE model simulations.

growth in SSA and LAC, and 25% in CWANA, compared to 27, 21, and 7% under the reference world. This could lead to further forest conversion into agricultural use. At the same time, rapid expansion of the livestock population under AKST_high requires expansion of grazing areas in SSA and elsewhere, which could also contribute to accelerated deforestation.

What are the implications of more aggressive production growth on food trade and food security? Under AKST_high, SSA cannot meet the rapid increases in food demand through domestic production alone. As a result, imports of both cereals and meats increase compared to the reference run, by 137% and 75%, respectively (Figure 5-34 and 5-41). Under AKST_high, ESAP would also increase its net import position for meats and cereals, while NAE would

strengthen its net export position for these commodities. Under AKST_low_neg, on the other hand, high food prices lead to depressed global food markets and reduced global trade in agricultural commodities.

Water scarcity is expected to increase considerably in the AKST_low_neg variant as a result of a sharp degradation of irrigation efficiency. The irrigation water supply reliability index therefore drops sharply (Table 5-19).

Sharp increases in international food prices as a result of the AKST_low and combined variants (Table 5-18) depress demand for food and reduce availability of calories (Figure 5-36). In the most adverse, AKST_low_neg variant, average daily kilocalorie availability per capita declines by 1,100 calories in sub-Saharan Africa, pushing the region below the generally accepted minimum level of 2,000 calories

Table 5-16. Assumptions for high/low agricultural investment variants.

Parameter changes for growth rates	2050 REFERENCE RUN	2050 High AKST variant (#1)	2050 Low AKST variant (#2)
GDP growth	3.06 % per year	3.31 % per year	2.86 % per year
Livestock numbers and yield growth	Base model output numbers growth 2000-2050 Livestock: 0.74%/yr Milk: 0.29%/yr	Increase in numbers growth of animals slaughtered by 20% Increase in animal yield by 20%	Reduction in numbers growth of animals slaughtered by 20% Reduction in animal yield by 20%
Food crop yield growth	Base model output yield growth rates 2000-2050: Cereals: %/yr: 1.02 R&T: %/yr: 0.35 Soybean: %/yr 0.36 Vegetables: %/yr 0.80 Sup-tropical/tropical fruits: 0.82%/yr	Increase yield growth by 40% for cereals, R&T, soybean, vegetables, ST fruits & sugarcane, dryland crops, cotton Increase production growth of oils, meals by 40%	Reduce yield growth by 40% for cereals, R&T, soybean, vegetables, fruits & sugarcane, dryland crops, cotton Reduce production growth of oils, meals by 40%

Source: Authors.

Table 5-17. Assumptions for high/low agricultural investment combined with high/low Investment in other AKST-related factors (irrigation, clean water, water management, rural roads, and education).

Parameter changes for growth rates	2050 BASE	2050 High AKST combined with other services (#3)	2050 Low AKST combined with other services Low (#4)
GDP growth	3.06 % per year	3.31 % per year	2.86 % per year
Livestock numbers growth	Base model output numbers growth 2000-2050 Livestock: 0.74%/yr Milk: 0.29%/yr	Increase in numbers growth of animals slaughtered by 30% Increase in animal yield by 30%	Reduction in numbers growth of animals slaughtered by 30% Reduction in animal yield by 30%
Food crop yield growth	Base model output yield growth rates 2000-2050: Cereals: %/yr: 1.02 R&T: %/yr: 0.35 Soybean: %/yr 0.36 Vegetables: %/yr 0.80 Sup-tropical/tropical fruits: 0.82%/yr	Increase yield growth by 60% for cereals, R&T, soybean, vegetables, ST fruits & sugarcane, dryland crops, cotton Increase production growth of oils, meals by 60%	Reduce yield growth by 60% for cereals, R&T, soybean, vegetables, fruits & sugarcane, dryland crops, cotton Increase production growth of oils, meals by 60%
Irrigated area growth (apply to all crops)	0.06	Increase by 25%	Reduction by 25%
Rain-fed area growth (apply to all crops)	0.18	Decrease by 15%	Increase by 15%
Basin efficiency		Increase by 0.15 by 2050, constant rate of improvement over time	Reduce by 0.15 by 2050, constant rate of decline over time
Access to water		Increase annual rate of improvement by 50% relative to baseline level, (subject to 100 % maximum)	Decrease annual rate of improvement by 50% relative to baseline level, constant rate of change over time
Female secondary education		Increase overall improvement by 50% relative to 2050 baseline level, constant rate of change over time unless baseline implies greater (subject to 100 % maximum)	Decrease overall improvement by 50% relative to 2050 baseline level, constant rate of change over time unless baseline implies less

Source: Authors.

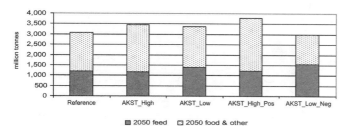

Figure 5-31. *Cereal feed, food and other demand projections, 2050, alternative AKST variants.* Source: IFPRI IMPACT model simulations.

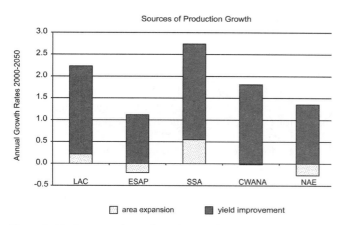

Figure 5-32. *Sources of cereal production growth, High_AKST variant, by IAASTD region.* Source: IFPRI IMPACT model simulations.

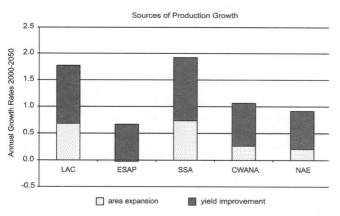

Figure 5-33. *Sources of cereal production growth, Low_AKST variant, by IAASTD region.* Source: IFPRI IMPACT model simulations.

and thus also below the levels of the base year 2000. Calorie availability together with changes in complementary service sectors can help explain changes in childhood malnutrition levels (see Rosegrant et al., 2002). Under the AKST_high and AKST_high_pos variants, the share of malnourished children in developing countries is expected to decline to 14% and 8%, respectively, from 18% in the reference world and 27% in 2000 (Figure 5-37). This translates into absolute declines of 25 million children and 55 million children,

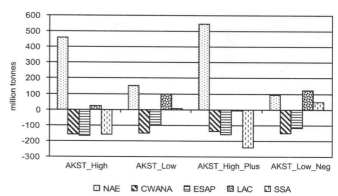

Figure 5-34. *Cereal trade in 2050, alternative AKST variants, IAASTD regions.* Source: IFPRI IMPACT model simulations.

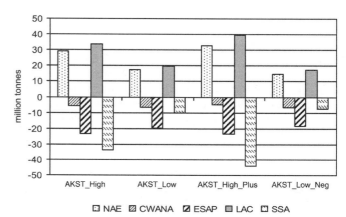

Figure 5-35. *Meat trade 2050, alternative AKST variants, IAASTD regions.* Source: IFPRI IMPACT model simulations.

respectively, under the more aggressive AKST and supporting service variations (Figure 5-38). On the other hand, if investments slow down more rapidly, and supporting services degrade rapidly then absolute childhood malnutrition levels could return or surpass 2000 malnutrition levels with 189 million children in 2050 under the AKST_low_neg variation and 126 million children under the AKST_low variation.

What are the implications for investment under these alternative policy variants? Investment needs for the group of developing countries for the alternative AKST variants have been calculated following the methodology described in Rosegrant et al. (2001) (Figure 5-39). Investment requirements for the reference run for key investment sectors, including public agricultural research, irrigation, rural roads, education and access to clean water are calculated at US$1,310 billion (see also Tables 5-16 and 5-17 for changes in parameters used). As the figure shows, the much better outcomes in developing-country food security achieved under the AKST_high and AKST_High_pos variants do not require large additional investments. Instead they can be achieved at estimated investment increases in the five key investment sectors of US$263 billion and US$636 billion, respectively.

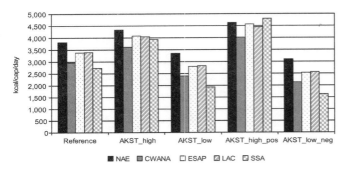

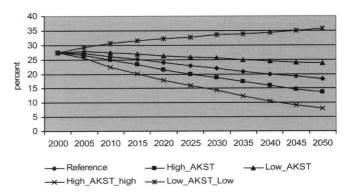

Figure 5-36. *Average daily calorie availability per capita, projected 2050, selected regions, AKST variants.* Source: IFPRI IMPACT model simulations.

Figure 5-37. *Malnourished children under alternative AKST variants in developing countries.* Source: IFPRI IMPACT model simulations.

5.4.4 Focus on bioenergy

Among the renewable sources, bioenergy deserves special attention (energy from crops, lingo-cellulosic products and timber byproducts). Currently, bioenergy is the only alternative to fossil fuels that is available for the transport sector. Studies of the potential confirm that the production of liquid fuels from biomass could meet the demand in the global transport sector. Bioenergy can also be used to produce electricity and heat. Large-scale application of biomass as an energy source will mean that in the short term bioenergy will primarily be derived from specific crops that are cultivated for energy production (sugar cane, maize, oil crops). The eventual contribution from biomass greatly depends on the expectations of extracting energy from lingo-cellulosic products (both woody and non-woody products, like poplar and grass). The large-scale cultivation of biomass for energy applications can mean a considerable change in future land use, and could compete with the use of this land for food production. Other aspects of sustainability, such as maintaining biodiversity and clean production methods, also play a role here (see Chapters 3, 4 and 6). Under scenarios in which agricultural land could become available as a result of rapid yield improvement and slow population growth, bioenergy potential is considerably higher than in land-scarce scenarios. Results for bioenergy can become more positive when the second generation bioenergy (the

lingo-cellulosic bioenergy sources) becomes available, since these sources offer more CO_2 reductions and use less land per unit of energy. However, this second generation bioenergy is not expected to become available within the coming 10 to 15 years (UN-Energy, 2007).

To explore the bioenergy potential under the IAASTD reference case, the procedure of De Vries et al. (2007) is followed in which the potential for bioenergy is defined as the amount of bioenergy that could be produced from (1) abandoned agricultural land and (2) 40% of the natural grass areas (see Appendix). Under these assumptions, the technical potential in 2050 is around 180 EJ in the absence of residues mainly from USA, Africa, Russia and Central Asia, South East Asia and Oceania. Obviously this number is very uncertain—and depends, among other factors, on (1) agricultural yields for food production, (2) yields and conversion rates for bioenergy, (3) restrictions in supply of bioenergy (to reduce biodiversity damage), and (4) uncertainty in water supply. The potential supply from residues is also very uncertain and estimates range from very low numbers to around 100 EJ. In the reference projection, a potential supply of 80 EJ is assumed. Until 2050, in this scenario the overall impact of bioenergy on biodiversity is negative, given the direct loss of land for nature versus the long-term gain of avoided climate change (SCBD/MNP, 2007).

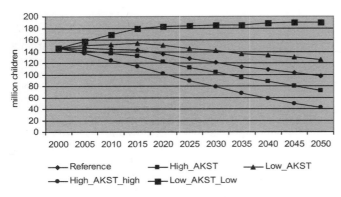

Figure 5-38. *Malnourished children under alternative AKST variants in developing countries.* Source: IFPRI IMPACT model simulations.

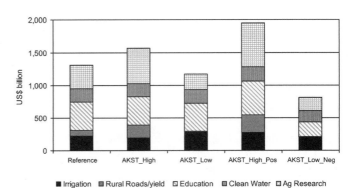

Figure 5-39. *Investment requirements, alternative AKST variants, developing countries.* Source: IFPRI IMPACT model simulations.

Table 5-18. Selected international food prices, projected to 2050, reference run and AKST variations.

Food	Reference run	AKST-high	AKST_low	AKST_high_pos	AKST_low_neg
	US$ per metric ton				
Beef	2,756	-23%	36%	-31%	63%
Pork	1,164	-29%	48%	-40%	84%
Sheep & goat	3,079	-24%	36%	-34%	60%
Poultry	1,434	-34%	62%	-46%	114%
Rice	245	-46%	105%	-62%	232%
Wheat	173	-53%	173%	-68%	454%
Maize	114	-67%	311%	-81%	882%
Millet	312	-59%	204%	-72%	459%
Sorghum	169	-57%	200%	-70%	487%
Other coarse grains	104	-74%	545%	-86%	1952%
Soybean	225	-31%	56%	-43%	106%

Source: IFPRI IMPACT model simulations.

5.4.5 The scope of improving water productivity

The reference run foresees a substantial increase in water consumption in agriculture, and particularly in non-agricultural sectors. This may be reason for concern. First, already more than a billion people live in river basins characterized by physical water scarcity (CA, 2007). In these areas water availability is a major constraint to agriculture. With increased demand for water, existing scarcity will deepen while more areas will face seasonal or permanent shortages. Second, competition for water between sectors will intensify. With urbanization, demand for water in domestic and industrial sectors will increase between 2000 and 2050. In most countries water for cities receives priority over water for agriculture by law or de facto (Molle and Berkoff, 2006), leaving less water for agriculture, particularly near large cities in water-short areas, such as MENA, Central Asia, India, Pakistan, Mexico, and northern China. Water for energy, i.e., hydropower and crop production for biofuels, will further add to the pressure on water resources. Third, signs of severe environmental degradation because of water scarcity, overabstraction and water pollution are apparent in a growing number of places (Pimentel et al., 2004; MA, 2005; Khan et al., 2006; CA, 2007) with often severe consequences for the poor who depend heavily on ecosystems for their livelihoods (Falkenmark et al., 2007). Lastly, climate change may exacerbate water problems particularly in semiarid areas in Africa were the absolute amount of rain is expected to decline, while seasonal and interannual variation increases (Wescoat, 1991; Rees and Collins, 2004; Alcamo et al., 2005; Barnett et al., 2005; Kurukulasuriya et al., 2006).

Table 5-19. Irrigation water supply reliability, projected to 2050, reference run and AKST variations.

Region	Reference	AKST_high_pos	AKST_low_neg
	Percent		
North America and Europe (NAE)	64	72	60
East-South Asia and Pacific (ESAP)	56	66	51
Central-West Asia and North Africa (CWANA)	46	52	39
Latin America and Caribbean (LAC)	83	86	75
sub-Saharan Africa (SSA)	87	92	85
Developed Countries	66	74	62
Developing Countries	56	65	51
World	58	67	53

Source: IFPRI IMPACT model simulations.

Fortunately, there is ample scope to improve water productivity and basin efficiency, to minimize additional water needs (Molden et al., 2007). AKST plays an important role in achieving these improvements. Three broad avenues to increase agricultural production while minimizing water use are to (CA, 2007) (1) improve productivity in rainfed settings, (2) increase productivity in irrigated areas, and (3) expand international agricultural trade. The scope and relevant policy measures differ considerably by region (Table 5-20).

5.4.5.1 SSA
Considering the ample physical potential and the willingness of donors to invest in African agriculture, the scope of irrigated area expansion is large. But the contribution of irrigation to food supply will likely remain limited (less than 11% of total food production), even after doubling its area. The investment cost of doubling the irrigated area is high and to make this investment economically viable massive investments in marketing infrastructure are needed

(roads, storage, communication) (Rosegrant et al., 2005). On the other hand, investments in irrigated area expansion for high-value crops (vegetables, cotton, fruits) can be an important vehicle for rural growth, and poverty alleviation particularly when geared to smallholders. Without substantial improvements in the productivity of rainfed agriculture, food production in SSA will fall short of demand. From a biophysical point of view, water harvesting techniques have proven successful in boosting yields, often up to a two or threefold increase (Röckstrom, 2003, 2007). But, low adoption rates of water harvesting techniques indicate that upscaling local successes pose major challenges for AKST.

5.4.5.2 South Asia
In South-Asia 95% of the areas suitable for agriculture are in use, of which more than half are irrigated. The biggest scope for improvement lies in the irrigated sector where yields are low compared to the obtainable level. Under a high productivity variant, all additional water and land for

Table 5-20. **Regional variation in scope for productivity improvements and area expansion.**

Region	Scope for improved productivity in rain-fed areas	Scope for improved productivity in irrigated areas	Scope for irrigated area expansion	Need for imports	High potential options in agricultural water management
sub-Saharan Africa	+++	+	+++		• water harvesting and supplemental irrigation; resource-conserving agricultural practices to mitigate land degradation • small-scale irrigation geared to smallholders • multiple use water systems to alleviate poverty • adopt development approaches that combine access to markets, soil fertility and irrigation infrastructure
MENA	+	+	-	+++	• use of low quality water • coping with increased sectoral competition and water pollution • integrating livestock with irrigation
C. Asia, E. Europe	+	++	+		• institutional reforms in irrigated areas • restore ecosystems services • modernize large-scale irrigation systems
South Asia	+++	+++	+	+	• institutional reforms in irrigated areas • integrating livestock and fisheries • water harvesting and supplemental irrigation; resource-conserving practices to mitigate land degradation
East Asia	++	+++	+	++	• water productivity in rice • reducing groundwater overdraft
Latin America	++	+	+		• land expansion and sustainable land use • support and regulation of private irrigation
OECD	+	+	+		• coping with increased sectoral competition

Key: +++ high, ++ medium, - low, - very limited

Source: Derived from CA scenario analysis (CA, 2007)

food can be met by improving land and water productivity in irrigated areas (de Fraiture et al., 2007). The scope for productivity improvement in rainfed areas is equally promising. In the high yield variant all additional land and water for food can be met by improving water productivity (de Fraiture et al., 2007). But there is considerable risk associated with this strategy. Yield improvements in rainfed areas are more uncertain than in irrigated areas because of high risk for individual farmers. If yield improvement targets are not achieved (i.e., adoption of water harvesting techniques is low or fluctuations in production due to climate variability are too high), the shortfall has to be met mainly by imports, because the scope of area expansion is limited. The scope for irrigated area expansion is limited, though groundwater expansion by private well owners will continue.

5.4.5.3 MENA
In the MENA region the scope to expand irrigated areas is very limited due to severe water shortages. Rainfed agriculture is risky due to unreliable rainfall. With climate change, variation in rainfall within the year and between years will further increase, particularly in semiarid areas. Trade will play an increasingly important role in food supply.

Table 5-21 shows the outcomes of a variant in which all high potential options are successfully implemented. The results from the WATERSIM model show that a major part of additional water use to meet future food demand can be met by increasing the output per unit of water, through appropriate investments in both irrigated and rainfed agriculture, thus relieving pressure on water resources. The output per unit water in rainfed areas increases by 31%. The potential in sub-Saharan Africa is highest (75%), while in OECD countries where productivity already is high the output per unit water increases by 20%. Overall the scope for enhancing water productivity in irrigated areas is higher than in rainfed areas (48% and 31%, respectively). In South Asia the output per unit of water can be improved by 62%. When multiple uses of water are encouraged and fisheries and livestock production are integrated, the output per unit of water in value terms may be even higher. Improvement of water productivity is often associated with higher fertilizer use, which may result in increased polluted return flows from agricultural areas. A challenge for AKST is to develop ways in which the tradeoff between enhanced water productivity and polluted return flows is minimized.

While a major part of additional water demand in agriculture can be met by improvements in water productivity on existing areas, further development of water resources is essential, particularly in sub-Saharan Africa where infrastructure is scarce. In total, irrigated areas expand by 50 million hectares (16%). In sub-Saharan Africa the expansion is largest (78%), in the MENA region the expansion is negligible because of severe water constraints. Agricultural water diversions will increase by 15% globally. A major challenge is to manage this water with minimal adverse impacts on environmental services, while providing the necessary gains in food production and poverty alleviation.

In the realization of optimistic water productivity AKST plays an essential role. Challenges for AKST are listed in Table 5-22.

5.4.6 Changing preferences for meat and certified organic products
Consumer preferences are evolving for both meat-focused diets and foods that are produced using integrated nutrient management. These two trends could (both individually and collectively) lead to several important differences from the reference case presented here. Rising interest in the health and environmental impacts, among other concerns, of conventional agriculture has pointed many consumers towards changing dietary habits away from meat and towards products that are certified in their use of better nutrient management practices (Knudsen et al., 2006, Steinfeld et al., 2006). As a result of the slowdown in meat demand, there is the potential for a shift in consumer preferences that would decrease the share of meat products in the typical person's diet and emphasize nonmeat foods. The main consequence of growing consumer demand for certified products that come from integrated nutrient management, which includes both meat and nonmeat commodities, will be the shift in production toward certified practices that would impact productivity. The impact on productivity depends on the region in which it is practiced, however. In industrial country regions, which already practice high-input intensity, conventional agriculture, the adoption of integrated nutrient management techniques would likely lower productivity and cause higher unit costs of production, while still providing greater satisfaction to those consumers who value such products. In regions like sub-Saharan Africa, on the other hand—where fairly low-intensity agriculture is still widely practiced—the adoption of integrated nutrient management techniques will likely cause an increase in yields, over and above the reference levels.

The IMPACT modeling framework, which was described earlier, was used to simulate these trends for comparing and contrasting with the reference case. Though the shift toward a less meat-intensive diet has the potential to be a global phenomenon, introduction of production techniques that practice integrated nutrient management is more practical in industrial country regions due to infrastructure and institutional requirements that are more readily available and applicable (see Halberg et al., 2006 for further discussion).

5.4.6.1 Specification of the low growth in meat demand policy issue
The global slowdown in the growth of meat demand is implemented via adjustments to the income demand elasticities for meat and vegetarian foods. Income demand elasticities for meat products (beef, pork, poultry, and sheep/goat) decline at a faster pace than in the reference case. At the same time income demand elasticities decline at a slower pace for vegetarian foods (fruits and vegetables, legumes, roots, tubers, and cereal grains). Elasticities for animal products such as dairy and eggs are left the same as in the reference case. This happens globally using a differentiated set of multipliers for developed versus developing regions, and assumes that the slowdown in meat demand is stronger in the industrialized regions, compared to that in developing regions. Regional average income demand elasticities for meat and nonmeat foods for IAASTD regions are presented in Table 5-23. The effect, in general, is that the meat income

Table 5-21. **Scenarios (policy experiment outcomes).**

Region	Irrigated area		Rain-fed area		Rain-fed cereal yield		Irrigated cereal yield		Rain-fed water productivity	
	m ha	% change	m ha	% change	t/ha	% change	t/ha	% change	kg/m3	% change
SSA	11.3	78%	174.2	10%	2.34	98%	4.37	99%	0.28	75%
MENA	21.5	5%	16.1	-12%	1.19	59%	5.58	58%	0.25	47%
C Asia, E Europe	34.7	6%	120.7	-5%	3	47%	6.06	78%	0.69	47%
South Asia	122.7	18%	83.9	-12%	2.54	91%	4.84	89%	0.46	82%
East Asia	135.6	16%	182.2	17%	3.96	51%	5.97	49%	0.57	36%
Latin America	19.5	18%	147.9	46%	3.9	58%	6.77	68%	0.63	50%
OECD	47.3	4%	179	4%	6.35	33%	8.03	22%	1.3	25%
World	394	16%	920	10%	3.88	58%	5.74	55%	0.64	31%

Region	Irrigated water productivity		Crop water depletion		Irrigation water diversions		Trade	
	Kg/m3	% change	km3	% change	km3	% change	M ton	% of consumption
SSA	0.5	58%	1,379	29%	100	46%	-25	-12%
MENA	0.82	41%	272	7%	228	8%	-127	-61%
C Asia, E Europe	1.22	51%	773	0%	271	11%	66	22%
South Asia	0.79	62%	1,700	15%	1195	9%	2	0%
East Asia	1.16	45%	1,990	19%	601	16%	-97	-12%
Latin America	0.91	52%	1,361	52%	196	12%	18	6%
OECD	1.6	20%	1,021	4%	238	2%	151	26%
World	1.01	48%	8,515	20%	2,975	14%	490	*15%*

Source: Watersim simulations (CA, 2007).

demand elasticities in industrialized countries and regions decline by 150% for meat products and 50% for nonmeat commodities compared to the reference run. In developing regions, the rates of decline are taken to be 110% and 90% of the baseline rates, for the meat and nonmeat commodities, respectively.

5.4.6.2 Specification of the adoption of integrated nutrient management.

The rise of industrialized country agricultural practices that use integrated nutrient management follows on the specification of the organic agriculture scenario in Halberg et al. (2006). This variant is specified purely as a supply-side adjustment in the industrialized world to yields of crops and livestock most easily converted to integrated nutrient management production techniques. Crops include maize, wheat, soybeans, other grains, and potatoes. Beef, dairy, and sheep/goat are the focus for livestock. The variant adjusts the yield growth rates from 2005 to 2015 such that the agronomic yields for the specified commodities achieve

the differences from the baseline specified in Halberg et al. (2006) as laid out in Tables 5-24 and 5-25. The principal change from Halberg et al. (2006) in this implementation of widespread adoption of integrated nutrient management in agriculture is that the apex of the spread is achieved in 2015, which would cover roughly half of the area harvested or managed animal herds. This year marks a turnaround in the decline of average yields for these crops, and baseline yield growths from 2015 to 2050 are achievable due to technology investments and farming system adaptations. This specification is meant to be illustrative of the potential impacts of such developments but it is an optimistic representation of such a large-scale shift to organic production.

The commodity price impacts of these two alternative outcomes compared to the reference case is fairly straightforward. In a future of increased vegetarianism, the income demand elasticities are much lower for meats and much higher for nonmeat foods than in the reference case. Prices will directly follow the changes in income demand elasticities with meat prices falling and nonmeat food prices in-

Table 5-22. **Challenges for AKST.**

Region	Challenges for AKST
sub-Saharan Africa	• Development of affordable irrigation infrastructure, suitable for smallholders, including supporting roads, and markets • Development of suitable water harvesting techniques and small supplemental irrigation methods to upgrade rain-fed areas • Creating the right institutional and economic environment for widespread adoption of these methods
MENA	• Development of environmentally sound ways to reuse return flows, often of low quality • Design of appropriate policies addressing sectoral competition and water pollution • Reduce adverse impacts of groundwater overexploitation
C. Asia, E. Europe	• Design of politically feasible institutional reforms in irrigated areas • Measures to restore ecosystems services • Adapting yesterday's large-scale irrigation systems to tomorrow's needs
South Asia	• Design of politically feasible institutional reforms in irrigated areas • Water conserving and yield boosting technologies to increase the output per unit of water in irrigated areas • Water harvesting and supplemental irrigation; resource-conserving practices to mitigate land and water degradation and the creation of enabling environment for the adoption of available techniques
East Asia	• Techniques to enhance the water productivity, particularly in rice areas (such as alternative wet-dry) • Reduce adverse impacts of groundwater overexploitation
Latin America	• Land expansion and sustainable land use • Support and regulation of private irrigation
OECD	• Policies addressing increased sectoral competition • Restoring ecosystem services

Source: Based on CA, 2007, pp 131-136.

creasing. Figure 5-40 shows the resulting differences from the reference case for the two types of foods: a 13% decrease in meat prices and a 10% increase in the price of nonmeat foods. As the rise of integrated nutrient management in agriculture results in a decline in average yields, commodity prices increase between 11-13% for major meat commodities and 3-21% for major crops like maize and soybean (Figures 5-47 and 5-48).

Per capita food consumption also shifts in these alternatives to the reference baseline. With the rise in prices in the case of increasing use of integrated nutrient management in agriculture, per capita consumption of all foods leads to decreases of up to 17%, but varies across regions, according to dietary patterns. On the other hand, the slowdown in meat demand growth shifts food preferences away from meat and toward nonmeat foods, which is commensurate with the price shifts discussed earlier, and the consumption shifts shown in Table 5-26, with a few exceptions. In sub-Saharan Africa the countervailing force of the price shifts actually leads to increased consumption of meat in addition to nonmeat foods. The price shifts in North America/Europe actually leads a slight inversion of the expected outcome, but this is due to the changes being implemented on the already low elasticities in this region not having as much effect as in other regions.

The calculation of the malnutrition indicators in the IMPACT framework (malnourished children by weight under five years old) has per capita kilocalorie consumption as an important factor and this follows the food consumption changes noted above. Nonmeat foods are denser in calories on a per kilogram basis, so a decrease in meat demand would lead to a decline in malnourishment. Figure 5-43 shows the impact on this malnutrition indicator aggregated to the developing world. Ultimately, a reduction in the growth of meat consumption with relatively more consumption of nonmeat foods sees a 0.5% decline in malnourished children while a certified organic world would see a 3% increase.

The potential evolution of consumer preferences for more use of integrated nutrient management practices in agriculture and nonmeat foods is uncertain. While the reference case presented previously already includes a certain amount of these shifting preferences, the purpose of this analysis is to highlight the potential impacts if these trends strengthen in the future. If meat demand were to decrease at a global level, the primary challenge will be to augment productivity investments on the crops that will maintain a balanced diet for consumers, particularly for crops that will constitute balanced proteins to replace meats. Increasing demands and prices for nonmeat foods will be the main challenge for agricultural production. Meanwhile, an increase in the use of integrated nutrient management practices in agriculture would raise a different set of challenges. In particular, maintaining productivity levels and controlling costs will be the most important issues to address. Alternative organic inputs for large-scale production that will maintain

Table 5-23. Changes to average income demand elasticities for meat and vegetarian foods by IAASTD region under low growth in meat demand variant.

			2000	2010	2020	2030	2040	2050
Meat	Baseline	CWANA	0.7223	0.6673	0.6095	0.5576	0.5147	0.4806
		ESAP	0.5538	0.5145	0.4809	0.4507	0.4288	0.4169
		LAC	0.5679	0.5129	0.4582	0.4023	0.3468	0.2914
		NAE	0.2761	0.2402	0.2054	0.1732	0.1438	0.1161
		SSA	0.8121	0.7966	0.7808	0.7634	0.7443	0.7221
	Low Meat	CWANA	0.7223	0.6554	0.5867	0.5253	0.4755	0.4375
	Demand	ESAP	0.5538	0.4953	0.4460	0.4064	0.3853	0.3844
		LAC	0.5679	0.5046	0.4416	0.3781	0.3164	0.2562
		NAE	0.2761	0.2178	0.1672	0.1227	0.0858	0.0533
		SSA	0.8121	0.7931	0.7736	0.7529	0.7305	0.7044
Vegetarian Foods	Baseline	CWANA	0.2486	0.2299	0.2156	0.2063	0.2021	0.2025
		ESAP	0.2243	0.2003	0.1847	0.1660	0.1438	0.1222
		LAC	0.1579	0.1421	0.1343	0.1322	0.1311	0.1324
		NAE	0.2733	0.2547	0.2387	0.2235	0.2079	0.1930
		SSA	0.3359	0.2775	0.2364	0.2027	0.1790	0.1751
	Low Meat	CWANA	0.2486	0.2337	0.2223	0.2149	0.2120	0.2134
	Demand	ESAP	0.2243	0.2138	0.2098	0.2046	0.1954	0.1848
		LAC	0.1579	0.1436	0.1367	0.1345	0.1330	0.1337
		NAE	0.2733	0.2687	0.2644	0.2599	0.2539	0.2477
		SSA	0.3359	0.2834	0.2473	0.2164	0.1941	0.1887

Source: IFPRI IMPACT model simulations.

soil nutrients and improve labor efficiencies will be rather important.

5.5. Emerging Issues that Influence the Future

5.5.1 Interface of human, animal, and plant health

5.5.1.1 Future trends
Human, animal, and plant diseases associated with AKST will continue to be of importance to future populations, including more urbanized populations in low-income countries. Two trends will be of particular importance—continued emergence and reemergence of infectious diseases and the growing human health burdens of noncommunicable diseases.[1]

Currently, 204 infectious diseases are considered to be emerging; 29 in livestock and 175 in humans (Taylor et al.,

2001). Of these, 75% are zoonotic (diseases transmitted between animals and humans). The number of emerging plant, animal, and human diseases will increase in the future, with pathogens that infect more than one host species more likely to emerge than single-host species (Taylor et al., 2001). Factors driving disease emergence include intensification of crop and livestock systems, economic factors (e.g., expansion of international trade), social factors (changing diets and lifestyles) demographic factors (e.g., population growth), environmental factors (e.g., land use change and global climate change), and microbial evolution. Diseases will continue to emerge and reemerge; even as control activities successfully control one disease, another will appear. Most of the factors that contributed to disease emergence will continue, if not intensify, in the twenty-first century (Institute of Medicine, 1992). The increase in disease emergence will impact both high- and low-income countries, with serious socioeconomic impacts when diseases spread widely within human or animal populations, or when they spill over from animal reservoirs to human hosts (Cleaveland et al., 2001).

Emerging infectious diseases of crop plants pose a significant threat to agricultural productivity and, in cases of

[1] Diseases and disabilities can be categorized into communicable diseases, maternal, and perinatal conditions, and nutritional deficiencies; noncommunicable diseases (primarily chronic diseases); and injuries.

able 5-24. **Change in average crop yields under integrated nutrient management variant.**

Region	Crop	Irrigated	Rain-fed
USA	Maize	-14	-14
European Union (15)	Maize	-14	-14
Other Developed	Maize	0	-14
Eastern Europe	Maize	0	0
USA	Wheat	-14	-14
European Union (15)	Wheat	-14	-14
Other Developed	Wheat	-14	-14
Eastern Europe	Wheat	0	0
USA	Soybean	-14	-14
European Union (15)	Soybean	-14	-14
Other Developed	Soybean	-14	-14
Eastern Europe	Soybean	-10	-10
USA	Other grains	-14	-14
European Union (15)	Other grains	-14	-14
Other Developed	Other grains	-14	-14
Eastern Europe	Other grains	0	0
USA	Potato	-20	-20
European Union (15)	Potato	-20	-20
Other Developed	Potato	-20	-20
Eastern Europe	Potato	-12.5	-12.5

Source: IFPRI IMPACT model simulations.

Table 5-25. **Change in average livestock carcass weight under integrated nutrient management variant.**

Region	Meat	Livestock
USA	Beef	-12.5
European Union (15)	Beef	-7.5
Other Developed	Beef	-12.5
Eastern Europe	Beef	-10
USA	Sheep & goat	-5
European Union (15)	Sheep & goat	-7.5
Other Developed	Sheep & goat	-5
Eastern Europe	Sheep & goat	-10
USA	Dairy	-10
European Union (15)	Dairy	-7.5
Other Developed	Dairy	-5
Eastern Europe	Dairy	0

Source: IFPRI IMPACT model simulations.

globally important staple crops, food security. The emergence of new plant diseases has largely resulted from the accidental introduction of pathogens in infected seed and in contaminated machinery and globally traded agricultural products. Furthermore, increased intensification of agricultural systems both facilitates the establishment and spread of these new pathogens, and imposes selection pressure for greater pathogen virulence (Anderson et al., 2004). Climate also plays an important role in disease emergence: winds disperse fungal and bacterial spores, nematodes and insect vectors of plant viruses; crop-canopy microclimatic conditions influence pathogen colonization of leaf surfaces; and seasonal climatic extremes mediate the extent of yield loss from plant diseases. The negative impact that increased climate variability and change will exert on host-pathogen dynamics could accelerate the process of pathogen migration into new agroecosystems, and provide conditions that elevate disease organisms from minor to major status (Coakley et al., 1999).

A second trend of importance is that noncommunicable diseases, such as heart disease, diabetes, stroke and cancer, account for nearly half of the global burden of disease (at all ages) and the burden is growing fastest in low- and middle-income countries (Mascie-Taylor and Karim, 2003). Chronic diseases are expected to rapidly increase as a result of more sedentary, urbanized lifestyles. In addition, the overall large increase in calorie availability in developing countries is ex-

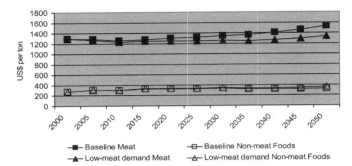

Figure 5-40. *Average world prices for meats and other foods under reference run and low growth in meat demand variant.* Source: IFPRI IMPACT model simulations.

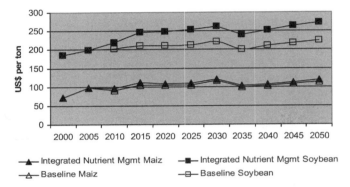

Figure 5-41. *World prices for maize and soybean under reference run and increasing use of integrated nutrient management variant.* Source: IFPRI IMPACT model simulations.

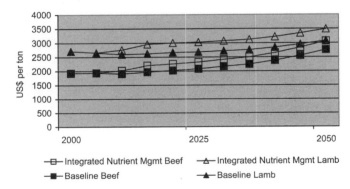

Figure 5-42. *World prices for beef and sheep/goat under reference run and increasing use of integrated nutrient management variant.* Source: IFPRI IMPACT model simulations.

Table 5-26. **Change in per capita food consumption of meats and cereals under low meat demand variant. Source: IFPRI IMPACT model simulations.**

Crop	Region	2025	2050
Cereals	NAE	1.6%	3.1%
	CWANA	0.2%	0.9%
	ESAP	0.7%	1.8%
	LAC	0.3%	1.1%
	SSA	0.4%	1.0%
Meat	NAE	-1.2%	-0.6%
	CWANA	0.5%	-1.3%
	ESAP	-4.0%	-9.8%
	LAC	1.0%	-0.1%
	SSA	2.3%	4.6%

Source: IFPRI IMPACT model simulations.

pected to lead to rapidly raising levels of obesity and associated noncommunicable diseases. Weight gain, hypertension, high blood cholesterol, and a lack of vegetable and fruit intake result in significant health burdens in both high and low-income countries (Ezzati et al., 2002). The greater supply of and demand for energy-dense, nutrient-poor foods is leading to obesity and related diseases in countries that have yet to overcome childhood undernutrition (Hawkes and Ruel, 2006).

Further, approximately 840 million people do not receive enough energy from their diets (Kennedy et al., 2003) and over three billion people are micronutrient deficient, most of them women, infants, and children in resource-poor families in low-income countries (Welch and Graham, 2005). Micronutrient deficiencies increase morbidity and mortality, decrease worker productivity, and cause permanent impairment of cognitive development in infants and children.

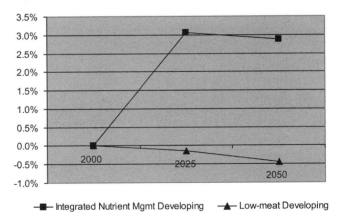

Figure 5-43. *Change in number of malnourished children in the developing world under integrated nutrient management and low growth in meat demand variants compared to reference run.* Source: IFPRI IMPACT model simulations.

5.5.1.2 Impacts of development

Development is expected to reduce some of the risks of current human, plant, and animal diseases, with new communicable diseases arising in their place. Communicable diseases are the primary cause for variations in life expectancy across countries, so reducing the burden of communicable diseases will increase life expectancy in low-income countries. Significant challenges will continue for several decades in building capacity to reduce emerging infectious diseases in low-income countries. Developing countries need laboratories and research centers (and the human and financial resources to staff and maintain them), along with resources in primary health care systems to identify, control, and treat disease outbreaks. High-income countries need to commit to additional resources for research and development on communicable diseases. Although progress is being made for some diseases, history suggests that diseases will continue to emerge faster than they are identified and controlled—infectious disease control is an ongoing process with long-term improvement, but without the possibility of eradicating all infectious diseases.

While agricultural and income growth are contributing to rapid reductions in the overall number of underweight children, the global decline masks differences across regions that will continue to adversely affect development over the coming decades. Unless more attention is paid to the problems of micronutrient deficiencies, the human health consequences will reduce the ability of nations to achieve development and sustainability goals (Welch and Graham, 2005).

Key forces that will affect development over the coming decades include demographic change; rate and degree of increase in climate variability; trends in ecosystem services; impact of climate change on freshwater resources, agricultural systems, livestock, wildlife, forests, and marine systems; economic growth and its distribution; rate of technology development; trends in governance; degree of investment in public health and other infrastructure.

A trend expected to continue is the highly inequitable distribution of health workers (WHO, 2006). The level of health expenditure is an indication of the resources for public health. Regions with the lowest relative need have the highest numbers of health workers, while those with the greatest health burden have a much smaller health workforce. Africa suffers more than 24% of the global disease burden but has access to only 3% of health workers and less than 1% of the world's financial resources, even when loans and grants are included. The Americas (Canada and the U.S.) experience 10% of the global disease burden, have 37% of the world's health workers, and spend more than 50% of the world's financial health resources.

5.5.1.3 Policies to facilitate achievement of development and sustainability goals

Reducing the threat of emerging infectious diseases requires enhancing disease surveillance and control programs through (1) strengthening existing research and monitoring facilities and establishing new laboratories and research centers for disease identification and control, (2) improving primary health care systems, and disease surveillance and control at local and global levels, and (3) developing the capacity to understand the interactions of factors that drive disease emergence. Developing these programs requires additional resources. Multidisciplinary collaboration, particularly across health and agricultural sectors, will facilitate identification of policies and measures to reduce the burden of communicable and noncommunicable diseases. One approach for reducing the burden of human and animal epidemics is the development of national networks for emergency response, with the human and financial resources to interpret forecasts, detect signs of emerging plant, animal, or human diseases, and environmental crises, and develop and implement effective responses.

Reducing undernutrition requires greater attention to food security, not just to crop yields. Although there is promising research arising on modifying crops and soil fertility to improve micronutrient content, considerable additional research is needed on new cultivars and approaches to improve the lives of billions of people worldwide. Additional resources are needed to be able to effectively deploy current and emerging technologies and cultivars. An issue likely to continue to be important is how low-income countries can afford the costs of new seeds and inputs. Development of effective policy options (including enforcement) can reduce current food safety issues.

5.5.1.4 Regional differences

As noted previously, there are large regional differences in the burden of infectious diseases and undernutrition, with the largest burdens in Africa and South and Southeast Asia. The burden of chronic diseases is now similar across most countries.

5.5.2 Information and communication technology and traditional and local knowledge

5.5.2.1 The promise of ICTs

ICT is increasing in importance for agriculture particularly for those producers who have access to markets. It is possible to attract investments when natural resource management activities are linked to the outside world (e.g., remittance workers sending funds that are invested in farm inputs) or across sectors (e.g., municipalities aggregating their health, education and local government needs for bandwidth). It follows then that the broad promise of ICTs tends to be described at the macro level:

- There is a positive link between telecommunications infrastructure and GDP, suggesting that a 1% increase in telecommunications infrastructure penetration might lead to a 0.03% increase in GDP (Torero and von Braun, 2006).
- The welfare effect of rural households is most closely associated with rural telephony which brings about immediate savings to the users (Kenny, 2002; Torero and von Braun, 2006) which is referred to as consumer surplus (Kayani and Dymond, 1997) and has been reported to represent a savings ranging from 4-9 times the costs of a single phone call (Bayes et al., 1999; Richardson et al., 2000).
- The promise of ICTs is most directly related with those MDGs that relate to health and education (Torero and von Braun, 2006).

5.5.2.2 The barriers associated with ICTs

Among the major barriers associated with ICT, we include uneven access, human resource development and local content (Torero and von Braun, 2006). Uneven access is most dramatic between urban and rural areas, and unfortunately most ICT indicators—when available—are nation-wide and therefore mask these fundamental differences. Other barriers include:

- The macroeconomic models referred to above work best for middle and high income countries; and for lower income countries the economic development impact from ICTs is expected to be modest (Torero and von Braun, 2006).

- The investments required are not simply about infrastructure. The valuation of the benefits of ICTs goes beyond the essential "access perspective", to one of "effective use". Effective use brings together several prerequisites: reliable access to infrastructure and user equipment, relevant content, cost-saving or meaning-making services, capacity development and financial sustainability (Gurstein, 2003).

- The human and organizational development aspects of ICTs have in the past been eclipsed by a fascination with the technological dimension. Capacity development refers to the training of individuals across the wide range of technologies, services, applications and content areas, and to the capacities by small and medium size enterprises to make use of ICTs.

- There is evidence that, while those with higher income and levels of education derive most benefits from ICTs, the poor spend roughly 3.7% of their monthly income on telecommunications services (Kayani and Dymond, 1997; Song and Bertolini, 2002). To date, the poverty alleviation impact of ICTs has been confirmed for radio and telephony, whereas the evidence for the Internet is less consistent (Kenny, 2002). In other words, the poor benefit from communication services significantly more than from information services.

5.5.2.3 ICTs and traditional and local knowledge

The development and spread of traditional and local knowledge can benefit directly from ICTs when holders of this knowledge have access to it and control over its utilization (Srinivasan, 2006). The integration of traditional and local knowledge and western scientific knowledge will need an interface that allows each to express its wisdom and forms, without sacrificing its cultural relevance. However, increased globalization and integration of markets presents both an opportunity and a threat to traditional and local knowledge. While knowledge will be transferred more easily across regions and countries, traditional and local knowledge might well disappear if adequate support systems are not put in place.

Challenges in terms of power and control will increase under the baseline: the consortia that own and operate ICT infrastructure work under a market logic that has little currency for respecting local and traditional knowledge. The importance of mediating organizations thus becomes evident if we are to minimize the potential abuse that such power differentials create. Moreover, mediating organizations would be necessary to coordinate the coming together of all interested parties involved. The challenge turns to the unresolved barriers of providing access to connectivity across rural and remote areas with weak demand, uneven market access and competing public investment requirements.

5.5.2.4 Policy implications

At present, Universal Access policies have failed to reach the most marginal; they have in fact given the elite groups a renewed relative advantage. This skewed impact is reminiscent of the initially uneven benefits from the Green Revolution. Under the reference conditions a divide continues to grow regarding access to ICTs between very poor and richer farmers. Hence, a more strategic targeting of policies, investment and incentive plans, and methodological innovation is necessary; the following are possible policy scenarios:

- There is an emerging understanding in ICT circles that no single approach to service delivery will satisfy the needs of all users (Ramírez and Lee, 2005). Increasing access to ICTs cannot be carried out by market forces alone. The liberalization and privatization of telecommunications created effective competition only in high density markets (in industrialized and in developing economies). Government participation in subsidizing capital infrastructure—often though competitive grants to the private sector—remains a central policy instrument, yet it needs to be adapted to the conditions of each country.

- Access is not enough. The notion of *effective use* calls for attention in tandem to a wide range of *readiness requirements*; training requirements, service development, local content development, to name but three.

- Local or "mediating" organizations that work as mediators between community needs, technology, markets and government programs can be strategic in this aspect. Mediating organizations can aggregate demand from health, education, and the business community to help attract infrastructure and service investments.

- For local content and traditional knowledge to be respected and harnessed, attention is first needed on issues of power and control over the infrastructure. The importance of policies that nurture local organizations is once again of paramount importance in the content area.

The era of seeing ICTs as magic bullets is past. ICTs are not a panacea for the poor in terms of the agriculture or natural resource management options; in contrast they do give an edge to the better off who already link with markets. Indeed for the poor the short term promise of ICTs is more evident in enhancing health and education services and especially in reducing their transaction costs (communication). On the other hand its information potential is only achievable if it is integrated with a comprehensive rural community development strategy.

5.5.3 Food safety and food security

5.5.3.1 The reference case

Trust in agricultural product quality has become one of the most important issues for consumers, since food represents security, comfort and the ability to provide basic needs to those who rely on others for protection and support (Bruemmer, 2003). Food safety will continue receiving attention in both industrialized and in less developed countries (Unnevehr, 2003) because (1) The "demand" for safe food rises as income increases. Consumers become willing to pay more for food with lower risk of microbial contamination, pesticides, and other disease-causing substances; (2) As technology improves, it is easier to measure contaminants in food and document their impact on human health; (3) Trade liberalization has increased opportunities for agricultural exports, and food safety regulations have become the binding constraint on food trade in many cases; (4) International food scares, such as BSE and avian flu, have made consumers, producers, and legislators more aware of the risks associated with agricultural food safety problems (Unnevehr, 2003; Narrod et al., 2005).

Appropriate food safety regulation is considered fundamental to expand product export from developing countries (Babu and Reidhead, 2000; Pinstrup-Andersen, 2000). However, the increased food safety standards are particularly worrisome in terms of food security and the livelihood of the poor as multinational retail companies that dominate the market often exclude small growers that find it difficult to meet foreign as well as domestic standards (Narrod et al., 2005). Labeling will be likely used to demonstrate that the food is safe to eat, however, a highly stringent label regulation, including description of origin of the ingredients of processed agricultural products, could inevitably create an unnecessary obstacle to future trade agreements (Matten, 2002).

Producers face four distinct problems: (1) How to produce safe food; (2) How to be recognized as producing safe food; (3) How to identify cost-effective technologies for reducing risk; (4) How to be competitive with larger producers who have the advantage of economies of scale in compliance for food safety requirements (Narrod et al., 2005). For some developing countries several constraints hamper the progress in implementing food safety regulations, including lack of human capacity, importance of food safety in the political agenda and inadequate postharvest and laboratory infrastructure and organization (Babu and Reidhead, 2000).

The continuous increase in urban development embedded within agricultural production areas raises concerns since it may affect both the quality of living and the safeness of final crop products. The risks posed by agricultural production systems in urban areas to health and environment may be associated with the inappropriate use of agricultural chemicals (pesticides, nitrogen, phosphorus) which may contaminate drinking water sources; cause air pollution (carbon dioxide and methane from organic matter, ammonia, nitrous oxide and nitrogen oxide from nitrates); and create odor nuisance (Carvalho, 2006). Further, close proximity of great number of people and domestic animals to production areas, accompanied with high population of wildlife creates an ideal scenario for potential disease sources (e.g., warm-blooded animals, humans) and vectors (e.g., animals, insects) to increase the risk for contamination of irrigation water and crop plants. Optimal management of urban resources and well established good agricultural practices are needed for sustainable agricultural production (Fonseca, 2006).

5.5.3.2 Identified routes

Food safety hazards can be encountered anywhere, from the farm to the table. Therefore efficient control programs are needed throughout the whole supply chain (Todd et al., 2006) especially because no intervention mechanism, other than irradiation, is currently available to completely decontaminate agricultural commodities eaten raw (Fonseca, 2006). The demand for products with high standards of overall quality and safety will continue to grow in industrialized countries. Meanwhile in developing countries, better quality standards will only occur if consumers are educated about the benefits of consumption of perishable products, if public health regulation and liability laws are established, and if better surveillance and analysis capacity is built (Berdegué et al., 2005). The development of a national food safety "culture" in the future will be influenced by who will be the regulatory/audit agents and to what extend the different actors will be involved (Schlundt, 2002; Reardon et al., 2003; Codron et al., 2005; Carvalho, 2006)

In one scenario, private rather than public standards will continue to be the predominant drivers of agri-food systems (Henson and Reardon, 2005). In developing countries where institutional capacity often limits the enforcement of mandatory public standards, firms will continue relying on private standards (Loader and Hobbs, 1999). The private sector will need to develop better training in certain agricultural practices. Implementation of quality and food safety control programs with intensive internal and external supervision can improve productivity rather than increase costs for consumers. In some developing countries large produce suppliers with dedicated and specialized perishable wholesalers will be able to save significant amount of sale-related costs as a result of production cost reduction with technical assistance, quality assurance systems and selection of preferred growers, in a semi vertical-integrated business (Berdegué et al., 2005).

In a second scenario, governmental policies will have great influence on food safety issues. In the event of food safety crises, governments will react by creating state food safety agencies, certifying and monitoring the implementation of standards and record keeping, increasing the rigor of minimum quality standards and establishing new rules for product traceability (Codron et al., 2005). In some developing countries the role of the government will still be amply intensive even in a private sector-leading scenario. This is because the infrastructure for examining water and product samples might not be economically feasible for private agencies, which in turn will force the government to provide the service (Berdegué et al., 2005). Moreover, in some developed and developing countries, with a leading private sector in the food safety area, governments will also play an important role through the establishment of job benefits. An important amount of workers and produce handlers are willing to work when they are ill because they can't afford to stay at home without pay (Fonseca and Nolte, 2007).

For some developing countries several constraints hamper progress in implementing food safety regulations, including lack of human capacity, lack of importance of food safety in the political agenda and inadequate postharvest and laboratory infrastructure and organization (Babu and

Reidhead, 2000). One way small scale producers could meet increased food safety requirements in the future is by pursuing the direction of fewer and less persistent pesticides. However in the short term the cost of these new environmentally friendly pesticides seems higher than older pesticides. Many cases aren't affordable to producers in the least developed world, where low-cost labor often compensates for the multiple applications, needed with some of the old pesticides, which might be of low quality or adulterated (Dinham, 2003; Carvalho, 2006). Countries in the tropical belt are challenged by environmental conditions and by not enough AKST developed to overcome their intrinsic productivity limitations, while complying with acceptable food safety guidelines.

5.5.3.3 Challenges as affected by policies
The main challenges for the next decades will be first to ensure safer food to consumers and raise the quality of life without creating a barrier to poor countries/producers for opportunities of success. Food security is a concern as food safety may only be "purchased" by some consumers, a situation that could be particularly notorious with products sourced from long-distance areas (Schillhorn van Veen, 2005). Secondly, in our search for mechanisms to improve food security in the world we are challenged to develop a system that will not cause the emergence of currently unknown health problems. The free trade market movement and the need to reduce internal hunger will likely result in more governments imposing their own rules or mandating the established international regulations. This will certainly create a major challenge as there are concerns about the possibility of mishandled information to affect the perception of the international consumer (Schlundt, 2002). Thus, governments will act influenced by how the actions will affect the distribution of benefits across the entire population (Codron et al., 2005).

It will become particularly difficult to control factors that compromise outbreak risks without collaborative international effort (Burlingame and Pineiro, 2007). A new approach will be necessary, one that incorporates food safety issues into the development of trade negotiations. Enhancing communication among policy makers from countries with common interests will allow the transfer of successful schemes to those in more need (Babu, 2004). The two future paths on food safety regulatory mechanisms described above (private sector providing education, auditing and analyses or government enforcement and monitoring) will be affected by the type of market to which products are directed. Moreover, even with national enforcements, some countries might continue to have regulation that differs substantially from those required in the export market and other local markets. The pressure over natural resources will determine some "natural" differences among countries (Hamilton, 2005). Some countries with known overdependence on pesticides but with the potential capacity to develop a more systematic approach will have the opportunity to improve internal standards and increase presence in the export market (Gupta, 2004). In this regard, the narrower the gap between the traditional and urban market, the more likely a country will find its way to comply with food safety expectations in the international arena (Kurien, 2005), how-

ever very little change will occur if no major effort is made in educating local consumers and developing AKST to produce for the export market.

5.5.4 Biotechnology and biodiversity

5.5.4.1 The reference case
A number of challenges—scientific, regulatory, social and economic—will fundamentally influence the degree to which genetic engineering is used in crop and livestock improvement research over the coming decades. Greater or lesser use of genetic engineering will, in turn, shape the evolution of the agricultural sector and biodiversity. Conventional breeding and genetic engineering are complements; thus the reference case development pathway includes a combination of a strong traditional plant breeding capacity together with the use of transgenic traits when useful, cost-efficient, pro-poor, and environmentally sustainable. A wide range of new traits are at various stages of development, some of which are likely to lead to varieties that are drought-resistant, exhibit improved nutritional content of feed and feedstuffs, and offer enhanced shelf-life (Graff et al., 2005). It is likely that a combination of transgenic and conventional breeding approaches will be necessary to meet the crop improvement requirements of the next 50 years. Factors shaping future adoption of new technologies include improved profits, decreased risk, increased health and well-being, and reduced effort, compared to earlier technologies used. They also include institutional and physical constraints affecting farming, like availability and terms of credit, information and product support provided by extension and technology providers, tenure conditions, and land ownership. Furthermore, the availability and quality of technology are dependent on policy and institutional variables, such as national agricultural research capacity, environmental and food safety regulations, intellectual property rights protection, and the existence of efficient agricultural input and output markets, which matter at least as much as the technology itself in determining the level and distribution of economic benefits (Raney, 2006). Potential constraints include property rights constraints, as well as evolving biosafety and food safety regulations around the world. Based on the literature assessing both constraints and benefits of crop technologies, the status quo pathway will involve continued rapid adoption of insect resistant and herbicide tolerant maize, soybean, cotton and canola varieties in the developing countries where they are already approved. More developing countries are likely to approve these crops—especially Bt cotton and yellow maize—under status quo conditions. Adoption of Bt maize in Europe will continue to expand slowly due to consumer resistance, despite growing tension between consumers and farmers (who see their competitiveness eroding in the face of competition from countries where adoption is proceeding).

5.5.4.2 Alternative pathways—more biotechnology
In spite of the limited growth in the development of transgenics, it is possible that these technologies will reemerge as a major contributor to agricultural growth and productivity. The continued safe introduction and use of the current generation of genetically engineered crops and the emergence of transgenic innovations of direct benefit to consumers or

the environment could lead to greater public acceptance of transgenic approaches and ultimately to a rationalization of regulatory regimes across countries, traits and crops. This in turn could mean that the costs (monetary and temporal) of transgenic research, development and deployment could fall significantly, leading to the rapid growth in the number of transgenic events and their pace of adoption. New biotechnological discoveries and their successful application in a country like China, where experimentation and investment in crop biotechnology continue, may lead other countries to follow suit. Finally, the concern about climate change and increasing energy prices could lead to significant investment in the development of biofuels, which, in turn, would increase rapidly growing resource scarcity and possibly higher food prices. Higher food scarcity, in turn, would increase the value of improved agricultural productivity and may lead countries to reassess the regulations restricting the growth of biotechnology. Furthermore, the development of biofuel crops may also rely on transgenic varieties and lead to enhancement of agricultural biotechnology and increase their acceptance.

Under such an alternative pathway, oilseeds with improved lipid profiles and staple grains with vitamin and mineral fortification could be introduced, and three major transgenic food crops that are already on the brink of approval (Bt rice, herbicide tolerant wheat and nutrient reinforced rice) could expand in developing countries and industrial countries alike. Furthermore, such a pathway could see traits for the remediation of polluted or degraded land or adaptation to heat and drought, which could assist in dealing with current agroenvironmental challenges and in the adaptation to rapid climate change.

5.5.4.3 Alternative pathways—less biotechnology
If society determines that the risks associated with transgenesis in agriculture exceed the benefits, the tool might be abandoned over the next several decades. Agricultural improvement research would continue, however, as it must to meet current and future challenges. Other research tools would be used more intensively, including conventional and mutagenic breeding. Non-transgenic molecular tools would also be used, such as marker assisted breeding. Under this alternative pathway, it is likely that a wider range of genetic variation would be sought within crops and wild relatives, and molecular tools would facilitate this search.

In industrial countries, more than half of all agricultural research expenditures are currently made by the private sector. Much of that research is aimed at developing patentable genetic constructs for use in crop and livestock improvement through transgenesis. The overall level of agricultural research expenditures in industrial countries could be reduced substantially if transgenic tools were abandoned, unless firms could assert binding intellectual property rights over discovered traits. At the same time, the costs associated with the regulation of transgenic crops would also be avoided. Overall, it is likely that the elimination of a powerful tool like transgenesis would slow but not stop the pace of agricultural research and improvement. As a result, humanity would likely be more vulnerable to climatic and other shocks and to increased natural resource scarcity under this alternative pathway.

5.5.4.4 Implications for the agricultural sector
For new technologies to be pro-poor, they need to relate to crops consumed by subsistence and small-scale farmers, allow for small-scale cultivation practices, and they need to be adapted to the human, physical, financial and social capital of the rural poor. Economic impacts tend to be more pro-poor where significant market competition exists in the supply of new technologies. The increased supply, as well as enhanced quality of improved technologies could contribute to reduced food prices, providing extra benefits to the urban poor. Improved food productivity can also be an important force to counter increased energy prices that are likely to contribute to increased food prices, which have a disproportionate negative effect on the poor.

5.5.4.5 Implications for biodiversity
The impacts of a rapid expansion of transgenic crops on natural and agricultural biodiversity over the next 50 years could be significant and will depend in part on how regulatory regimes evolve. Natural biodiversity could be affected through crop yields and the implications of transgenic crops for land use, potential outcrossing of transgenic material to related crop and wild species, and direct and indirect effects on non-target species. Agricultural biodiversity could be affected indirectly, much as it was by the spread of modern green revolution varieties, as well as directly through the use of the technology.

The most direct way transgenic crops could affect natural biodiversity is through their effect on crop yields and associated pressures influencing land use. To the extent that transgenic innovations support yield growth (or reduce crop losses to pests and diseases), they could alleviate pressure to expand crop production into currently uncultivated areas, endangering the natural biodiversity that exists there.

The potential for outcrossing to wild or agricultural relatives varies by crop. Transgenic varieties of crops that have a high propensity to outcross typically have not been approved for cultivation in areas where wild relatives are endemic. Most crop species, whether transgenic or not, are unlikely to be able to reproduce and persist in the wild, and management strategies can be used to minimize the risk (FAO, 2004). The potential for transgenic crop varieties to cross with conventional varieties clearly exists, although transgenic traits that do not confer a competitive advantage are unlikely to persist in farmers' fields unless they are specifically selected. Outcrossing to wild or cultivated relatives could be prevented by the use of genetic use restriction technologies, but this approach is controversial and has not been developed commercially. Whether the existence of an otherwise benign transgenic trait in an agricultural crop constitutes a meaningful loss of biodiversity is a matter of debate, particularly if it is a trait farmers have selected for (Raney and Pingali, 2005).

Whether modern variety adoption necessarily reduces agricultural biodiversity is a matter of debate. Agricultural biodiversity is important because it influences the resilience of crop ecosystems and maintains a "library" of genetic resources for current and future breeding activities. The domestication of wild plants into landraces narrowed the genetic base for these crops as farmers selected among the full range of plant types for those that produced more desirable results (Smale, 1997). Although more genetically uniform than their early relatives, landraces are characterized by a high degree of genetic diversity within a particular field. Modern varieties, on the other hand, tend to exhibit little diversity within a particular field, but each plant contains genetic material from a wide variety of progenitors and is adapted to perform well across a wide range of agroclimatic conditions. A simple count of the varieties in a particular area or measures of genetic distance among varieties thus may not tell us much about the resilience of crop ecosystems or the availability of crop genetic resources for breeding programs (Raney and Pingali, 2005).

Transgenic techniques can directly affect agricultural genetic diversity. Transgenesis permits the introduction of genetic materials from sexually incompatible organisms, greatly expanding the range of genetic variation that can be used in breeding programs. Transgenesis allows the targeted transfer of the genes responsible for a particular trait, without otherwise changing the genetic makeup of the host plant. This means that a single transgenic event can be incorporated into many varieties of a crop, including perhaps even landraces. Compared with conventional breeding in which an innovation comes bundled within a new variety that typically displaces older varieties, transgenesis could allow an innovation to be disseminated through many varieties, preserving desirable qualities from existing varieties and maintaining or potentially increasing crop genetic diversity (Raney and Pingali, 2005).

On the other hand, the widespread incorporation of a single innovation, such as the Bt genes, into many crops may constitute a kind of genetic narrowing for that particular trait. Furthermore, transgenic crops that confer a distinct advantage over landraces may accelerate the pace at which these traditional crops are abandoned or augmented with the transgenic trait (Raney and Pingali, 2005). Regulatory regimes are concerned with the potentially harmful consequences of gene flow from transgenic crops to conventional varieties or landraces. In this context, it is important to recognize that gene flow from conventional varieties to landraces frequently occurs (especially for open-pollinated crops such as maize) and is often consciously exploited by farmers. It is likely that, in the same way, farmers would consciously select for transgenic traits that confer an advantage (de Groote et al., 2005).

Regulatory decisions influence the implications of transgenic approaches for biodiversity, often in unexpected ways. For example, when biosafety procedures require the separate approval of each plant variety containing a transgenic event, it slows the development of new varieties and narrows the range of genetic diversity available to farmers. Similarly, when new transgenic approaches to address a given production constraint (such as herbicide tolerance) are delayed, the approved technology may be overused with negative consequences for biodiversity and other environmental indicators.

Finally, genetic engineering allows scientists to take advantage of biodiversity. Increased documentation of genomes and understanding of functional genomics provides information that is needed to develop new traits and new varieties that are of high value. Thus, the availability of tools for biotechnology and their development enhance the value of biodiversity, and to some extent, biotechnology and biodiversity are complementary. Furthermore, biotechnology provides tools to restore local varieties after slight modification allowing them to withstand disease or other pressures. The development of precision farming technologies that allow for the modification of application of inputs, including seeds, in response to changes in ecological conditions will provide impetus to increase crop diversity to take advantage of these new possibilities.

5.6 Implications of Policy Simulations and Emerging Policy Issues: Synergies and Tradeoffs

5.6.1 Poverty and equity
Chapter 5 examined projected changes in agriculture and AKST out to 2050 based on existing assessments and methodologies. At this point there are no established methodologies to adequately describe changes in poverty and equity out to 2050. This can only be inferred based on the state of literature and the analyses presented here. Increased agricultural productivity has been a key driver for economic and income growth in most countries at some stage of economic development and will continue to be key to growth in many agriculture-dependent developing countries out to 2050. However, although agricultural and economic growth are critical drivers for poverty reduction and explain a significant share of the historical decline in poverty in most regions of the world, policies and investments in the fields of education, health, and infrastructure are also essential for sustained poverty reduction. Lipton and Sinha (1998) argue that, while globalization is changing the outlook for the rural poor by raising average incomes, it also tends to increase income variability both across regions (leaving some regions and countries behind) and across time, thus increasing the vulnerability of those who can least afford it. Moreover, while changes in macroeconomic and trade policy tend to produce large gains for both rural and urban areas, poor farmers and (landless) agricultural laborers, who often lack the skills, health, information, or assets needed to seize new opportunities (Sinha and Adam, 2006, 2007), tend to be left out of the general economic growth process, as they may be concentrated in remote rural areas or geographic regions ill-equipped to gain from globalization/liberalization.

To redress potentially adverse impacts on equity, investments in human capital are crucial for the poor. Moreover, given emerging health and food safety issues, investments in health and nutrition are similarly important. Even with rapid economic growth and active investment in social services, some of the poor will be reached slowly if at all. And even among those who do benefit to some extent, many will remain vulnerable to adverse events. These groups will need to be reached through income transfers, or through safety nets that help them through short-term stresses or disasters.

5.6.2 Hunger, health and food security
The *reference* run has shown that a substantial increase in food prices will cause relatively slow growth in calorie consumption, with both direct price impacts and reductions in real incomes for poor consumers who spend a large share of their income on food. This in turn contributes to slow im-

provement in food security for the poor and in food sovereignty for many regions. Progress is slowest in sub-Saharan Africa—despite rapid income growth and significant area and yield growth as well as substantial progress in supporting services that influence well-being outcomes, such as female secondary education, and access to clean drinking water. By 2050, there will be a reduction by only 7% in the number of malnourished children in sub-Saharan Africa.

Alternative policy experiments show that with higher investments in AKST, the share of malnourished children in the group of developing countries is projected to decline from a baseline of 99 million by 2050 to only 74 million. If these higher investments in AKST are combined with improvements in complementary service sectors, such as health and education, the projections show that an even greater reduction, to 43 million, could be achieved. By contrast, either flatlined or slowed rates of investments in AKST will negatively affect regional food security and exacerbate childhood malnutrition, with levels that could easily surpass current malnutrition levels.

Moreover, uncertainties regarding a whole range of emerging issues, ranging from public health and food safety to policies in the areas of climate change and bioenergy could worsen (or improve) projected quantitatively modeled outcomes.

5.6.3 Natural resources and environmental sustainability

Regarding resources, scarcity is expected to become a prominent challenge for policy makers. In particular, growing water and land scarcity are projected to increasingly constrain food production growth with adverse impacts on food security and human well-being goals. Growth of population combined with shifts towards high land/fodder-intensive meat diets is resulting in additional demands for land. Although crop productivity is expected to increase (as described in the reference run) the uncertainty as to whether this productivity increase can actually be met is also increasing. The increased production of livestock is expected to come from the same or a declining resource base, and without appropriate action there are prospects that this could lead to degradation of land, water, and animal genetic resources in both intensive and extensive livestock systems. In addition, new demands on land for products such as biofuels (stimulated by concerns about climate change and energy self-sufficiency) are very likely to grow exponentially in the coming decades. This will not only impact food prices but will also lead to greater competition for land. The combination of demand factors will lead to rather grim impacts on biodiversity. The target of the Convention on Biological Diversity (CBD) to reduce the rate of loss of biodiversity significantly by 2010 seems impossible to reach. Moreover, some policy options to reduce pressures on the natural system (e.g., climate mitigation strategies as described in this chapter) have a negative impact on biodiversity through additional land-use change required for biofuels.

Water demand is projected to grow rapidly, particularly in developing countries. Irrigation remains the single largest water user over the 50-year projection period, but the increase in demand is much faster for domestic and industrial uses than for agriculture. Given significantly faster growth

in water demand in all sectors and declining water availability resulting from climate change in this baseline, developing countries are substantially more negatively affected by declining water supply reliability for irrigation and other uses than developed countries. This is especially so for developing countries with arid climates, poor infrastructure development, and rapidly increasing populations. Overall, to satisfy future water demand and secure food supply, investments in maintenance, new technology and policy reform in water and irrigation management are all necessary to maintain water supply reliability and to reduce water supply vulnerability for irrigation, especially in developing countries. Besides water supply augmentation, demand management is also of high importance in balancing future water demand and supply. Other research indicates that more investment in basin efficiency improvement would potentially bring similar effects in securing irrigation water supply as more investment is made in water infrastructures. Likewise, water saving technology and conservation measures in the industrial, rural and urban domestic sector would result in more reliable water supply in nonirrigation sectors and relieve the increasingly intensified intersectoral competition for water.

On the fisheries front, although small pelagic species are robust, the behavior of the small pelagic fish towards the end of the modeled period (2048) indicate that policies of mining small pelagic fisheries to support a growing aquaculture industry may not be sustainable in the long-term except in a limited part of the world's oceans. Caution needs to be taken even with this interpretation since small pelagic fish are extremely sensitive to oceanographic changes and if the predictions for changes in sea temperature come about, the species dynamics within this group will change significantly with potential reverberating effects up through higher trophic levels since most animals, especially marine mammals and seabirds, rely on this group of fish for much of their food. Therefore, a policy of increasing landings would need to be carefully considered in the light of climate change.

The tradeoffs between increased income for small farmers via crop production for food and fuel, livestock production, conservation and marketing of native varieties and species, and soil and water management for sustainability, will require a balancing act over the next 50 years. However, synergies do exist; biofuel crops, biotechnology, ICT, food safety standards, and globalization and trade liberalization can offer new opportunities to smallholders—if supporting policies and investments are implemented—and large agricultural producers alike making an ever more diverse range of products available to consumers.

5.7 Implications for AKST in the Future

As the reference world in 2050 and the various policy discussions show, agriculture will have to face a number of new and difficult challenges. Food security and food sovereignty are likely to still be problems 50 years from now. Agricultural production is likely to be increasingly constrained by competition for land and water and by climate change. Strategies for adapting to new regulations for food safety, and the development of biotechnology and bioenergy pose significant challenges and opportunities.

Food prices will most likely rise as a result of these opportunities and constraints. In addition, regional and national income growth, urbanization and growing global interconnectedness are expected to increase diet diversification and homogenization. Trade liberalization and greater integration of global food markets can support more reliable food supplies and lowered food prices in real terms. But as the reference run shows this is unlikely to be achieved in the coming decades.

With declining availability of water and land that can be profitably brought under cultivation, expansion in area will contribute very little to future production growth. The projected slow growth in crop area places the burden to meet future cereal demand on crop yield growth. The key to improving yields under increasingly constrained conditions lies in technology to improve agricultural productivity in order to regenerate productivity growth. Biotechnology could play an important role here. To adapt to and mitigate the various effects from climate change requires the development of new cultivars. Likewise, CO_2 emissions can be reduced through new crop management practices supported by appropriate technologies. To achieve such breakthroughs, existing global and regional research-for-development networks for agricultural production technologies and knowledge need to work closely together so that technology and knowledge can flow to allow farmers to face the risks associated with future harvests. Information and communication technologies and traditional and local knowledge could play key roles in the regeneration of future productivity growth. As the alternative policy experiments in this chapter have shown, higher, judiciously placed investments in technology, development can significantly improve outcomes for food availability and food security.

APPENDIX
Model descriptions

A.5.1 The International Model for Policy Analysis of Agricultural Commodities and Trade (IMPACT)

A.5.1.1 Introduction
IMPACT was developed in the early 1990s as a response to concerns about a lack of vision and consensus regarding the actions required to feed the world in the future, reduce poverty, and protect the natural resource base (Rosegrant et al., 1995). In 2002, the model was expanded through inclusion of a Water Simulation Model (WSM) as water was perceived as one of the major constraints to future food production and human well-being (Rosegrant et al., 2002).

A.5.1.2 Model structure and data
The current IMPACT model combines an extension of the original model with a WSM that is based on state-of-the-art global water databases (Rosegrant et al., 2002). The water module projects the evolution of availability and demand with a base year of 2000 (average of 1999-2001), taking into account the availability and variability in water resources, the water supply infrastructure, and irrigation and nonagricultural water demands, as well as the impact of alternative water policies and investments. Water demands are simulated as functions of year-to-year hydrologic fluctuations, irrigation development, growth of industrial and domestic water uses, and environmental and other flow requirements (committed flow). Off-stream water supply for the domestic, industrial, livestock, and irrigation sectors is determined based on water allocation priorities, treating irrigation water as a residual. Environmental flows are included as constraints.

The food module is specified as a set of 115 country or regional sub-models. Within each sub-model, supply, demand and prices for agricultural commodities are determined for 32 crop, livestock, fish commodities and fishmeal, sugar and sweeteners, fruits and vegetables, and low value and high value fish. These country and regional sub-models are intersected with 126 river basins—to allow for a better representation of water supply and demand—generating results for 281 Food Producing Units (FPUs). The "food" side of IMPACT uses a system of food supply and demand elasticities incorporated into a series of linear and nonlinear equations, to approximate the underlying production and demand functions. World agricultural commodity prices are determined annually at levels that clear international markets. Demand is a function of prices, income and population growth. Growth in crop production in each country is determined by crop prices and the rate of productivity growth. Future productivity growth is estimated by its component sources, including crop management research, conventional plant breeding, wide-crossing and hybridization breeding, and biotechnology and transgenic breeding. Other sources of growth considered include private sector agricultural research and development, agricultural extension and education, markets, infrastructure and irrigation. IMPACT projects the share and number of malnourished preschool children in developing countries as a function of average per capita calorie availability, the share of females with secondary schooling, the ratio of female to male life expectancy at birth, and the percentage of the population with access to safe water (see also Rosegrant et al., 2001; Smith and Haddad, 2000). The model incorporates data from FAOSTAT (FAO, 2003); commodity, income, and population data and projections from the World Bank (2000), the Millennium Ecosystem Assessment (MA, 2005), and the UN (2000) and USDA (2000); a system of supply and demand elasticities from literature reviews and expert estimates (Rosegrant et al., 2001); and rates for malnutrition (UN ACC/SCN, 1996; WHO, 1997) and calorie-child malnutrition relationships developed by Smith and Haddad (2000).

A.5.1.3 Application
IMPACT has been used for analyzing the current and future roles of agricultural commodities and impacts on food security and rural livelihoods, including the future of fisheries (Delgado et al., 2003); the role of root and tuber crops (Scott et al., 2000a, 2000b); and the "livestock revolution" (Delgado et al., 1999). IMPACT has also been applied to regional analyses as well as selected country-level studies, for example, China (Huang et al., 1997), Indonesia (SEARCA/IFPRI/CRESECENT 2004), sub-Saharan Africa (Rosegrant et al., 2005a) and Central Asia (Pandya-Lorch and Rosegrant, 2000). IMPACT has also been used to analyze structural changes, including the impact of the Asian economic and financial crisis (Rosegrant and Ringler, 2000); longer-term structural changes in rural Asia (Rosegrant and Hazell, 2000); as well as global dietary changes (Rosegrant et al., 1999). The model has also been used to describe the role of agriculture and water for achieving the Millennium Development Goals (von Braun et al., 2004; Rosegrant et al., 2005b).

Model runs have been carried out for individual centers of the CGIAR, the World Bank and the Asian Development Bank. The model has also been used for agricultural scenario analysis of the Millennium Ecosystem Assessment (Alcamo et al., 2005; MA, 2005), and is currently being used for the Global Environmental Outlook (GEO-4) assessment carried out by UNEP. Other work includes investigations into regional and global scale impacts of greenhouse gas mitigation in agriculture and theoretical large-scale conversion to organic food production.

A.5.1.4 Uncertainty
In the following tables, the points related to uncertainty in the model are summarized, based on the level of agreement and amount of evidence.

A.5.2 The Integrated Model to Assess the Global Environment (IMAGE) 2.4

A.5.2.1 Introduction
The IMAGE modelling framework had been developed originally to study the causes and impacts of climate change within an integrated context. Now the IMAGE 2.4 is used to study a whole range of environmental and global change problems, in particular aspects of land use change, atmospheric pollution and climate change. The model and its submodels have been described in detail in several publications (Alcamo et al., 1998; IMAGE team 2001; Bouwman et al., 2006).

A.5.2.2 Model structure and data
The IMAGE 2.4 framework describes global environmental change in terms of its cause-response chain, and belongs to the model family of integrated assessment models. The IMAGE model consists of two major parts: the socioeconomic system, elaborating on future changes in demographics, economy, agriculture and the energy sector, and the biophysical system, comprising land cover and land use, atmospheric composition and climate change. The IMAGE model focuses on linking those two parts through emissions and land allocation. Land allocation follows inputs from the IMPACT model, allowing an assessment of the environmental consequences of changes in the agricultural sector. One of the crucial parts of the IMAGE 2.4 model is the energy model, Targets IMage Energy Regional (TIMER). The TIMER model describes the chain, going from demand for energy services (useful energy) to the supply of energy itself through different primary energy sources and related emissions. The steps are connected by demand for energy and by feedbacks, mainly in the form of energy prices. The TIMER model has three types of sub-models: (1) a model for energy demand, (2) models for energy conversion (electricity and hydrogen production) and (3) models for primary energy supply. The final energy demand (for five sectors and eight energy carriers) is modelled as a function of changes in population, economic activity and energy efficiency. The model for electricity production simulates investments in various electricity production technologies and their use in response to electricity demand and to changes in relative generation costs.

Supply of all primary energy carriers is based on the interplay between resource depletion and technology development. Technology development is introduced either as learning curves (for most fuels and renewable options) or by exogenous technology change assumptions (for thermal power plants). To model resource depletion of fossil fuels and uranium, several resource categories that are depleted in order of their costs are defined. Production costs thus rise as each subsequent category is exploited. For renewable energy options, the production costs depend on the ratio between actual production levels and the maximum production level. Climate change mitigation policies can be implemented in the TIMER model, allowing assessing changes in the energy composition due to these policies (Van Vuuren et al., 2006; Van Vuuren et al., 2007).

The TIMER model also simulates the potential importance of biomass as an energy category. The structure of the biomass sub-model is similar to that of the fossil fuel supply models but with a few important differences (see also Hoogwijk et al., 2005). First, in the bioenergy model, depletion is not governed by cumulative production but by the degree to which available land is being used for commercial energy crops. Available land is defined as abandoned agricultural land and part of the natural grasslands in divergent land use variants for the twenty-first century and is based on IMAGE alternative variant calculations. This assumption excludes any possible competition between bioenergy and food production, which is a simplification of reality. The potential available land is categorized according to productivity levels

Table A.5.1 **Overview of major uncertainties in IMPACT Source: Based on MA, 2005**

Model Component	Uncertainty
Model structure	• Based on partial equilibrium theory (equilibrium between demand and supply of all commodities and production factors) • Underlying sources of growth in area/numbers and productivity • Structure of supply and demand functions and underlying elasticities, complementary and substitution of factor inputs. • Water simulations and connection between Water and Food modules
Parameters	Input parameters: • Base year, 3-year centered moving averages for area, yield, production, numbers for 32 agricultural commodities and 115 countries and regions, and 281 Food Producing Units • Elasticities underlying the country and regional demand and supply functions • Commodity prices • Drivers Output parameters: • Annual levels of water supply and demand (withdrawals and depletion), both agricultural and nonagricultural, food supply, demand, trade, international food prices, calorie availability, and share and number of malnourished children
Driving Force	Economic and demographic drivers: • Income growth (GDP) • Population growth Technological, management, and infrastructural drivers: • Productivity growth (including management research, conventional plant breeding) for rainfed and irrigated areas • Rainfed and irrigated area growth • Livestock feed ratios • Changes in nonagricultural water demand • Supply and demand elasticity systems Policy drivers: • Commodity price policy as defined by taxes and subsidies on commodities, drivers affecting child malnutrition, food demand preferences, water infrastructure, etc.
Initial Condition	Baseline: 3-year average centered on 2000 of all input parameters and assumptions for driving forces
Model operation	Optimization in Water Simulation Model using GAMS

that are assumed to reflect the cost of producing primary biomass. The biomass model also describes the conversion of biomass (such as residues, wood crops, maize and sugar cane) to two generic secondary fuel types: biosolid fuels (BSF) and bioliquid fuels (BLF). The solid fuel is used in the industry and power sector, and the liquid fuel in other sectors, in particular, transport.

The output of TIMER is affecting the biophysical system of IMAGE through land use changes (for bioenergy) and emissions (from the energy sector). Changes in food production are taken from IMPACT. The land cover model of IMAGE simulates the change in land use and land cover

in each region driven by demands for food, including crops, feed, and grass for animal agriculture, timber and biofuels in addition to changes in climate (Bouwman et al., 2005; Eickhout et al., 2007). The model distinguishes 14 natural and forest land cover types and six land cover types created by people. A crop module based on the FAO agroecological zones approach computes the spatially explicit yields of the different crop groups and grass and the areas used for their production, as determined by climate and soil quality (Alcamo et al., 1998). In case expansion of agricultural land is required, a rule-based "suitability map" determines which grid cells are selected. Conditions that enhance the suitabil-

Table A.5.2 **Level of confidence in different types of scenario calculations from IMPACT** Source: Based on (MA, 2005)

Level of Agreement/ Assessment	High	Established but incomplete: • Projections of Rainfed Area, Yield • Projections of Irrigated Area, Yield • Projections of Livestock Numbers, Production • Number of Malnourished Children • Calorie availability • Climate variability	Well established: • Changes in Consumption Patterns and Food Demand
	Low	Speculative:	Competing explanations: • Projections of Commodity Prices • Commodity Trade • Climate change
		Low	High
		Amount of evidence (theory, observations, model outputs)	

ity of a grid cell for agricultural expansion include potential crop yield (which changes over time as a result of climate change and technology development), proximity to other agricultural areas and proximity to water bodies. The land cover model also includes a modified version of the BIOME model (Prentice et al., 1992) to compute changes in potential vegetation. The potential vegetation is the equilibrium vegetation that should eventually develop under a given climate. The shifts in vegetation zones, however, do not occur instantaneously. In IMAGE 2.4 such dynamic adaptation is modelled explicitly according to the algorithms developed by Van Minnen et al. (2000). This allows for assessing the consequences of climate change for natural vegetation (Leemans and Eickhout, 2004). The land use system is modelled on a 0.5 by 0.5 degree grid.

Both changes in energy consumption and land use patterns give rise to emissions that are used to calculate changes in the atmospheric concentration of greenhouse gases and some atmospheric pollutants such as nitrogen and sulphur oxides (Strengers et al., 2004). Changes in the concentration of greenhouse gases, ozone precursors and species involved in aerosol formation form the basis for calculating climatic change (Eickhout et al., 2004). Next, changes in climate are calculated as global mean changes which are downscaled to the 0.5 by 0.5 degree grids using patterns generated by a General Circulation Model (GCM). Through this approach, different GCM patterns can be used to downscale the global-mean temperature change, allowing for the assessment of uncertainties in regional climate change (Eickhout et al., 2004). An important aspect of IMAGE is that it accounts for crucial feedbacks within the system, such as among temperature, precipitation and atmospheric CO_2 on the selection of crop types and the migration of ecosystems. This allows for calculating changes in crop and grass yields and as a consequence the location of different types of agriculture, changes in net primary productivity and migration of natural ecosystems (Leemans et al., 2002).

A.5.2.3 Application

The IMAGE model has been applied to a variety of global studies. The specific issues and questions addressed in these studies have inspired the introduction of new model features and capabilities, and in turn, the model enhancements and extensions have broadened the range of applications that IMAGE can address. Since the publication of IMAGE 2.1 (Alcamo et al., 1998), subsequent versions and intermediate releases have been used in most of the major global assessment studies and other international analyses, like the IPCC Special Report on Emissions Scenarios (Nakicenovic et al., 2000), UNEP's Third and Fourth Global Environment Outlook (UNEP, 2002; 2007), The Millennium Ecosystem Assessment (MA, 2006), the Second Global Biodiversity Outlook (SCBD/MNP, 2007) and Global Nutrients from Watersheds (Seitzinger et al., 2005).

A.5.2.4 Uncertainty

As a global Integrated Assessment model, the focus of IMAGE is on large scale, mostly first order drivers of global environmental change. This obviously introduces some important limitations to its results, and in particular the interpretation of its accuracy and uncertainty. An important method for handling some of the uncertainties is by using a scenario approach. A large number of relationships and model drivers whose linkages and values are either currently not known or depend on human decisions are varied in these scenarios. To explore their uncertainties, see IMAGE Team, 2001. In 2001 a separate project was performed to evaluate the uncertainties in the energy model using both quantitative and qualitative techniques. With this analysis the model's most important uncertainties were seen to be linked to its assumptions for technological improvement in the energy system, and how human activities are translated into a demand for energy (including human lifestyles, economic sector change and energy efficiency, seen in Table A.5.3).

Table A.5.3 **Overview of major uncertainties in IMAGE 2.4**

Model component	Uncertainty
Model structure	• TIMER Energy model: Integration in larger economy and dynamic formulation in energy model (learning by doing) • Land model: Rule-based algorithm for allocating land use • Environmental system: Scheme for allocating carbon pools in the carbon cycle model
Parameters	• Energy: Resource assumption and learning parameters • Land: Biome model parameter setting and CO_2 fertilization • Environment: Climate sensitivity, climate change patterns and multipliers in climate model (Leemans et al., 2002)
Driving force	• Income growth (GDP) • Population growth • Assumptions on technology change in energy model • Environmental policies
Initial condition	• Emissions in base year (2000) • Historic energy use • Initial land use/land cover map • Historical land use data (from FAO) • Climate observations in initial year
Model operation	• Downscaling method in climate change model (Eickhout et al., 2004)

The carbon cycle model has also recently been used for a sensitivity analysis to assess uncertainties in carbon cycle modeling in general (Leemans et al., 2002). Finally, a main uncertainty in IMAGE's climate model has to do with (1) "climate sensitivity," i.e., the response of global temperature computed by the model to changes in atmospheric greenhouse gas concentrations, and (2) regional patterns of changed temperature and precipitation. IMAGE 2.2 has actually been set up in such a way that these variables can be easily manipulated on the basis of more scientifically-detailed models. To summarize, most of IMAGE would need to go into the category of established but incomplete knowledge (Table A.5.4).

A.5.3 The Global Trade and Environment Model (GTEM)

A.5.3.1 Introduction
GTEM has been developed by the Australian Bureau of Agricultural and Resource Economics (ABARE) specifically to address policy issues with global dimensions and issues where the interactions between sectors and between economies are significant. These include issues such as international climate change policy, international trade and investment liberalisation and trends in global energy markets.

Table A.5.4 **Level of confidence in different types of scenario calculations from IMAGE**

Level of Agreement/ Assessment	High	Established but incomplete • Climate impacts on agriculture and biomes • Carbon cycle	Well established • Energy modeling and scenarios
	Low	Speculative • Grid-level changes in driving forces • Impacts of land degradation	Competing explanations • Global climate change, including estimates of uncertainty • Local climate change • Land use change
		Low	High
		Amount of evidence (theory, observations, model outputs)	

A.5.3.2 Model structure and data

GTEM is a multiregion, multisector, dynamic, general equilibrium model of the global economy. The key structural features of GTEM include:

- A computable general equilibrium (CGE) framework with a sound theoretical foundation based on microeconomic principles that accounts for economic transactions occurring in the global economy. The theoretical structure of the model is based on the optimizing behavior of individual economic agents (e.g., firms and households), as represented by the model equation systems, the database and parameters.

- A recursively dynamic analytical framework characterized by capital and debt accumulation and endogenous population growth, which enables the model to account for transactions between sectors and trade flows between regions over time. As a dynamic model, it accounts for the impacts of changes in labor force and investment on a region's production capabilities.

- The representation of a large number of economies (up to 87 regional economies corresponding to individual countries or country groups) that are linked through trade and investment flows, allowing for detailed analysis of the direct as well as flow-on impacts of policy and exogenous changes for individual economies. The model tracks intraindustry trade flows as well as bilateral trade flows, allowing for detailed trade policy analysis.

- A high level of sectoral disaggregation (up to 67 broad sectors, with an explicit representation of 13 agricultural sectors) that helps to minimize likely biases that may arise from an undue aggregation scheme.

- A bottom-up "technology bundle" approach adopted in modeling energy intensive sectors, as well as interfuel, interfactor and factor-fuel substitution possibilities allowed in modeling the production of commodities. The detailed and explicit treatment of the energy and energy related sectors makes GTEM an ideal tool for analysing trends and policies affecting the energy sector.

- A demographic module that determines the evolution of a region's population (and hence the labor supply) as a function of fertility, migration and mortality, all distinguished by age group and/or gender.

- A detailed greenhouse gas emissions module that accounts for the major gases and sources, incorporates various climate change response policies, including international emissions trading and quota banking, and allows for technology substitution and uptake of backstop technologies.

For each regional economy, the GTEM database consists of six broad components: the input–output flows; bilateral trade flows; elasticities and parameters; population data; technology data; and anthropogenic greenhouse gas emissions data. For the input–output and bilateral trade flows data, and the key elasticities and parameters, the GTAP version 6 database (see https://www.gtap.agecon.purdue.edu/databases/v6/default.asp) has been adapted. The databases for population, energy technology and anthropogenic greenhouse gas emissions, have been assembled by ABARE according to GTEM regions using information from a range of national and international sources. The base-year for GTEM is 2001. For this exercise, the model database has been aggregated to 21 regions that correspond to the five IAASTD sub-global regions and to 36 commodities that include 12 agricultural sectors and one fisheries sector.

GTEM equations are written in log-change forms and the model is solved recursively using the GEMPACK suite of programs (http://www.monash.edu.au/policy/gempack.htm). For IAASTD modeling purposes, the GTEM projection period extends to 2050. The model simulation provides annual projections for many variables including regional gross national product, aggregate consumption, investment, exports and imports; sectoral production, employment and other input demands; final demand and trade for commodities; and greenhouse gas emissions by gas and by source.

A detailed description of the theoretical structure of GTEM can be found in Pant (2002, 2007). Pezzey and Lambie (2001) describe the key structural features of GTEM and Ahammad and Mi (2005) discuss an update on the modeling of GTEM agricultural and forestry sectors.

A.5.3.3 Application

GTEM has been applied to a wide range of medium- to long-term policy issues or special events. These include climate change response policy analysis (e.g., Ahammad et al., 2006; Ahammad et al., 2004; Fisher et al., 2003; Heyhoe, 2007; Jakeman et al., 2002; Jakeman et al., 2004; Jotzo, 2000; Matysek et al., 2005; Polidano et al., 2000; Tulpulé et al., 1999); global energy market analysis (e.g., Ball et al., 2003, Fairhead et al., 2002; Heaney et al., 2005; Mélanie et al., 2002; Stuart et al., 2000); and on agricultural trade liberalisation issues (e.g., Bull and Roberts 2001; Fairhead and Ahammad, 2005; Freeman et al., 2000; Nair et al., 2005; Nair et al., 2006; Roberts et al., 1999; Schneider et al., 2000).

A.5.3.4 Uncertainty

(See Table A.5.3.4)

A.5.4 WATERSIM

A.5.4.1 Introduction

Watersim is an integrated hydrologic and economic model, written in GAMS, developed by IWMI with input from IFPRI and the University of Illinois. It seeks to:

- Explore the key linkages between water, food security, and environment.
- Develop scenarios for exploring key questions for food water, food, and environmental security, at the global national and basin scale

A.5.4.2 Model structure and data

The general model structure consists of two integrated modules: the "food demand and supply" module, adapted from IMPACT (Rosegrant et al., 2002), and the "water supply and demand" module which uses a water balance based on the Water Accounting framework (Molden, 1997) that underlies the policy dialogue model, PODIUM combined with elements from IMPACT (Cai and Rosegrant, 2002). The model estimates food demand as a function of population, income and food prices. Crop production depends on economic variables such as crop prices, inputs and subsidies on

Table A.5.3.4 **Overview of key uncertainties in GTEM.**

Model component	Uncertainty
Model structure	• Based on general equilibrium theory. • Conforms to a competitive market equilibrium—no "supernormal" economic profit. • Structured on nested supply and demand functions representing technologies, tastes, endowments and policies. • Incorporates the Armington demand structure—a commodity produced in one region treated as an imperfect substitute for a similar good produced elsewhere. • Total demand equals total supply—for all commodities at the global level and for production factors at the regional level.
Parameters	Input parameters: • Base year input-output flows and (bilateral) trade flows for 67 commodities and 87 countries and regions. • Numerous elasticities underlying demand and supply equations. • Technical coefficients for anthropogenic greenhouse gas emissions.
Driving Force	• Regional income growth (GDP). • Population growth. • Changes in policies (taxes and subsidies). • Technological changes—productivity growth and energy technology options. The choice of the model closure, i.e., the distinction between exogenous (drivers or shocks) and endogenous (determined or projected) variables of the model, is quite flexible. The above variables, e.g., could also be determined endogenously within the model for some specific economic closure characterized by a well specified set of economic and demographic shocks.
Initial Condition	• The 2001 global economy in terms of production, consumption and trade.
Model operation	• Suite of GEMPACK programs.

one hand, and climate, crop technology, production mode (rainfed or irrigated) and water availability on the other. Irrigation water demand is a function of the food production requirement and management practices, but constrained by the amount of available water.

Water demand for irrigation, domestic purposes, industrial sectors, livestock and the environment are estimated at basin scale. Water supply for each basin is expressed as a function of climate, hydrology and infrastructure. At basin level, hydrologic components (water supply, usage and outflow) must balance. At the global level, food demand and supply are leveled out by international trade and changes in commodity stocks. The model iterates between basin, region and globe until the conditions of economic equilibrium and hydrologic water balance are met.

Different aspects of the model use different spatial units. To model hydrology adequately, the river basin is used as the basic spatial unit. For food policy analysis, administrative boundaries should be used since trade and policy making happens at national level, not at the scale of river basins. WATERSIM takes a hybrid approach to its spatial unit of modeling. First, the world is divided into 125 major river basins of various sizes with the goal of achieving accuracy with regard to the basins most important to irrigated agri-

culture. Next the world is divided into 115 economic regions composed of mostly single nations with a few regional groupings. Finally the river basins are intersected with the economic regions to produce 282 Food Producing Units (FPUs). The hydrological processes are modeled at basin scale by summing up relevant parameters and variables over the FPUs within one basin; similarly economic processes are modeled at regional scale by summing up the variables over the FPUs belonging to one region.

The model uses a temporal scale with a baseline year of 2000. Economic processes are modeled at an annual timestep, while hydrological and climate variables are modeled at a monthly time-step. Crop related variables are either determined by month (crop evapotranspiration) or by season (yield, area). The food supply and demand module runs at region level on a yearly time-step. Water supply and demand runs at FPU level at a monthly time-step. For the area and yield computations the relevant parameters and variables are summed over the months of the growing season.

A.5.4.3 Application
Watersim has been used in the following cases:
• Scenario analysis in the Comprehensive Assessment of Water Management in Agriculture (CA, 2007)

Table A.5.6 **Overview of major uncertainties in Watersim model**

Model Component	Uncertainty
Model Structure	Food module is based on IMPACT (well established). Water module borrows from Podium, IMPACT and Water accounting methodology (all well established)
Parameters	Output: projections on water demand by basin (128) and country (115), water scarcity indices, production coming from irrigated and rainfed areas, crop water use, water productivity and basin efficiency
Driving force	• population and GDP growth • crop demand • improvements in water productivity • improvements in basin efficiency
Initial condition	Parameters are calibrated to the base year (2000). Based on the best available data sources, uncertainty minimized as far as possible, but in particular water use efficiency data are sketchy in developing countries
Model operation	Runs in GAMS

Table A.5.7 **Level of confidence for scenario calculations with Watersim model**

Level of Agreement/ Assessment	High	Established but incomplete: • Areas suffering from water scarcity • Virtual water flows due to food trade	Well-established: Global estimates and projections of crop water use
	Low	Speculative: • Crop losses due to water shortages • Impacts of environmental flow policies on food production	Competing Explanations: • Projections of irrigated areas and production • Projections of rainfed areas and production • Projections of irrigation water demand • Projections of water productivity
		Low	High
		Amount of Evidence (Theory, Observations, Model Outputs)	

- Sub-Saharan Africa investment study
- ICID – India Country paper
- Scenarios at basin level for the benchmark basins in the Challenge Program on Water and Food

A.5.4.4 Uncertainty

The water and food modules are calibrated at the base year 2000 and 1995. Model outcomes aggregated at a relatively high level (globe, continent, major basins such as the Indo-Gangetic) tend to have a better agreement than outcomes at sub-basin level. This reflects the uncertainty associated with global datasets, shown in Table A.5.6.

A.5.5 CAPSiM

A.5.5.1 Introduction

China's Agricultural Policy Simulation and Projection Model (CAPSiM) was developed at the Center for Chinese Agri-cultural Policy (CCAP) in the mid-1990s as a response to the need to have a framework for analyzing policies affecting agricultural production, consumption, prices, and trade in China (Huang et al., 1999; Huang and Chen, 1999). Since then CAPSiM has been periodically updated and expanded at CCAP to cover the impacts of policy changes at regional and household levels (Huang and Li, 2003; Huang et al., 2003).

A.5.5.2 Model structure and data

CAPSiM is a partial equilibrium model for 19 crop, live-stock and fishery commodities, including all cereals (four types), sweet potato, potato, soybean, other edible oil crops, cotton, vegetable, fruits, other crops, six livestock products, and one aggregate fishery sector, which together account for more than 90% of China's agricultural output. CAPSiM is simultaneously run at the national, provincial (31) and household (by different income groups) levels. It is the first comprehensive model for examining the effects of policies on China's national and regional food economies, as well as household income and poverty.

CAPSiM includes two major modules for supply and demand balances for each of 19 agricultural commodities. Supply includes production, import, and stock changes. Demand includes food demand (specified separately for rural and urban consumers), feed demand, industrial demand, waste, and export demand. Market clearing is reached simultaneously for each agricultural commodity and all 19 commodities (or groups).

Production equations, which are decomposed by area and yield for crops and by total output for meat and other products, allow producers' own- and cross-price market responses, as well as the effects of shifts in technology stock on agriculture, irrigation stock, three environmental factors—erosion, salinization, and the breakdown of the local environment—and yield changes due to exogenous shocks of climate and other factors (Huang and Rozelle, 1998b; deBrauw et al., 2004). Demand equations, which are broken out by urban and rural consumers, allow consumers' own- and cross-price market responses, as well as the effects of shifts in income, population level, market development and other shocks (Huang and Rozelle, 1998a; Huang and Bouis, 2001; Huang and Liu, 2002).

Most of the elasticities used in CAPSiM were estimated econometrically at CCAP using state-of-the-art econometrics, including assumptions for consistency of estimated parameters with theory. Demand and supply elasticities vary over time and across income groups. Recently, CAPSiM shifted its demand system from double-log to an "Almost Ideal Demand System" (Deaton and Muellbauer, 1980).

CAPSiM generates annual projections for crop production (area, yield and production), livestock and fish production, demand (food, feed, industrial, seed, waste, etc), stock changes, prices and trade. The base year is 2001 (average of 2000-2002) and is currently being updated to 2004 The model is written in Visual C++.

A.5.5.3 Application
CAPSiM has been frequently used by CCAP and its collaborators in various policy analyses and impact assessments. Some examples include China's WTO accession and implications (Huang and Rozelle, 2003; Huang and Chen, 1999), trade liberalization, food security, and poverty (Huang et al., 2003; Huang et. al., 2005a and 2005b), R&D investment policy and impact assessments (Huang et al., 2000), land use policy change and its impact on food prices (Xu et. al., 2006), China's food demand and supply projections (Huang et. al., 1999; Rozelle et al., 1996; Rozelle and Huang, 2000), and water policy (Liao and Huang, 2004).

A.5.5.4 Uncertainty
Tables A.5.8 and A.5.9 below summarize points related to uncertainty in the model, based on the level of agreement and amount of evidence.

A.5.6 Gender (GEN)-Computable General Equilibrium (CGE)

A.5.6.1 Introduction
The GEN-CGE model developed for India is based on a Social Accounting Matrix (SAM) using the Indian fiscal year 1999-2000 as the base year (Sinha and Sangeeta, 2001). Generally SAMs are used as base data set for CGE Models where one can take into account multi-sectoral, multi-class disaggregation. In determining the results of policy simulations generated by CGE model, a base-year equilibrium data set is required, which is termed calibration. Calibration is the requirement that the entire model specification be capable of generating a base year equilibrium observation as a model solution. There is a need for construction of a data set that meets the equilibrium conditions for the general equilibrium model, viz. demand equal supplies for all commodities, nonprofits are made in all industries, all domestic agents have demands that satisfy their budget constraints and external sector is balanced. A SAM provides the most suitable disaggregated equilibrium data set for the CGE model.

The SAM under use distinguishes different sectors of production having a thrust on the agricultural sectors and different factors of production distinguished by gender. The workers are further distinguished into rural, urban, agricultural, nonagricultural and casual and regular types. The other important feature of the SAM is the distinction of various types of households and each household type being identified with information on gender worker ratios. As the model incorporates the gendered factors of production, it is enabled to carry out counterfactual analysis to see the impact of trade policy changes on different types of workers distinguished by gender, which in turn allows the study of welfare of households again distinguished by ratio of workers by gender. Households are divided into rural and urban groups, distinguished by monthly per capita expenditure (MPCE) levels. Rural households include poor agriculturalists, with MPCE less that Rs. 350; nonpoor agriculturalists (above Rs. 351); and nonagriculturalists at all levels of income. Urban households are categorized as poor, with MPCE of less than Rs. 450 and the nonpoor, with MPCE of between Rs. 451 and 750.

A.5.6.2 Model structure and data
The GEN-CGE model follows roughly the standard neoclassical specification of general equilibrium models. Markets for goods, factors, and foreign exchange are assumed to respond to changing demand and supply conditions, which in turn are affected by government policies, the external environment, and other exogenous influences. The model is Walrasian in that it determines only relative prices and other endogenous variables in the real sphere of the economy. Sectoral product prices, factor prices, and the foreign exchange rate are defined relative to a price index, which serves as the numeraire. The production technology is represented by a set of nested Cobb-Douglas and Leontief functions. Domestic output in each sector is a Leontief function of value-added and aggregate intermediate input use. Value-added is a Cobb-Douglas function of the primary factors, like capital and labor. Fixed input coefficients are specified in the intermediate input cost function. The model

Table A.5.8 **Overview of major uncertainties in CAPSiM**

Model component	Uncertainty
Model structure	• Based on partial equilibrium theory (equilibrium between demand and supply of all commodities and production factors) • One country model (international prices are exogenous)
Parameters	Input parameters: • Some household data on production and consumption may not be consistent with national and provincial demand and supply functions • Elasticities underlying the national and provincial demand and supply functions • International commodity prices • Drivers Output parameters: • Annual levels of food and agricultural production, stock changes, food and other demands, imports and exports, and domestic prices at national level • Annual levels of food and agricultural production, food and other demands at provincial and household level
Driving force	Economic and demographic drivers: • Per capita rural and urban income • Population growth and urbanization Technological drivers: • Yield response with respect to research investment, irrigation, and others • Livestock feed rations Policy drivers: • Cultivated land expansion/control • Public investment (research, irrigation, environmental conservation, etc.) • Trade policy • Others
Initial condition	Baseline: Three-year average centered on 2001 of all input parameters and assumptions for driving forces
Model operation	Visual C++ programming language

Table A.5.9 **Level of confidence in different types of scenario calculations with CAPSiM**

Level of Agreement/ Assessment	High	Established but incomplete	Well established
	High	• Projections of R&D and irrigation investment • Projections of livestock feed ratios • Impacts on farmers income	• Changes in crop area and yield • Changes in food consumption in both rural and urban areas • Food production and consumption at household level by income group
	Low	Speculative	Competing explanations • Projections of commodity prices • Commodity trade
		Low	High
		**Amount of evidence (theory, observations, model outputs)** More than 20 papers published in Chinese and international journals based on CAPSiM	

assumes imperfect substitutability, in each sector, between the domestic product and imports. All firms are assumed to be price takers for all imports. What is demanded is the composite consumption good, which is a constant elasticity of substitution (CES) aggregation of imports and domestically produced goods. Similarly, each sector is assumed to produce differentiated goods for the domestic and export markets. The composite production good is a constant elasticity of transformation (CET) aggregation of sectoral exports and domestically consumed products. Such product differentiation permits two-way trade and gives some realistic autonomy to the domestic price system. Based on the small-country assumption, domestic prices of imports and exports are expressed in terms of the exchange rate and their foreign prices, as well as the trade tax. The import tax rate represents the sum of the import tariff, surcharge, and applicable sales tax for each commodity group. The foreign exchange rate, an exogenous variable in the base model, is in real terms. The deflator is a price index of goods for domestic use; hence, this exchange rate measure represents the relative price of tradable goods vis-a-vis nontradables (in units of domestic currency per unit of foreign currency).

A.5.6.3 Application
The GEN-CGE model can be used for studying the impact of tariff changes, removal of nontariff barriers (measured as tariff equivalents), changes in world GDP, changes in world prices, and changes in agricultural technology on employment by sector, prices, household income and welfare. One version of this model has been used for studying the impact of trade reforms in India in 2003 under a project with IDRC in Canada.

A.5.6.4 Uncertainty

A.5.7 The Livestock Spatial Location-Allocation Model (SLAM)

A.5.7.1 Introduction
Seré and Steinfeld (1996) developed a global livestock production system classification scheme. In it, livestock systems fall into four categories: landless systems, livestock only/rangeland-based systems (areas with minimal cropping), mixed rainfed systems (mostly rainfed cropping combined with livestock) and mixed irrigated systems (a significant proportion of cropping uses irrigation and is interspersed with livestock). A method has been devised for mapping the classification, based on agroclimatology, land cover, and human population density (Kruska et al., 2003). The classification system can be run in response to different scenarios of climate and population change, to give very broad-brush indications of possible changes in livestock system distribution in the future.

A.5.7.2 Model structure and data
The livestock production system proposed by Seré and Steinfeld (1996) is made up of the following types:
- Landless monogastric systems, in which the value of production of the pig/poultry enterprises is higher than that of the ruminant enterprises.
- Landless ruminant systems, in which the value of production of the ruminant enterprises is higher than that of the pig/poultry enterprises.
- Grassland-based systems, in which more than 10% of the dry matter fed to animals is farm produced and in

Table A.5.10 **Overview of major uncertainties in GEN-CGE model**

Model Component	Uncertainty
Model Structure	Labor skill
Parameters	Taken from past studies, literature
Driving force	Exogenous variables to the model
Initial condition	Base level SAM
Model operation	Data based

Table A.5.11 **Level of confidence for scenario calculations**

Level of Agreement/ Assessment	High	Established but incomplete • Trade reform analysis on employment	Well-established • Trade reform analysis on the economy
	Low	Speculative • The impact on migration of workers	Competing Explanations • Tradeoff between welfare and growth
		Low	High
	Amount of Evidence (Theory, Observations, Model Outputs): Please see references for model outputs		

which annual average stocking rates are less than ten temperate livestock units per hectare of agricultural land.

- Rainfed mixed farming systems, in which more than 90% of the value of non-livestock farm production comes from rainfed land use, including the following classes.
- Irrigated mixed farming systems, in which more than 10% of the value of non-livestock farm production comes from irrigated land use.

The grassland-based and mixed systems are further categorized on the basis of climate: arid–semiarid (with a length of growing period < 180 days), humid–subhumid (Length of Growing Period or LGP > 180 days), and tropical highlands/temperate regions. This gives 11 categories in all. This system has been mapped using the methods of Kruska et al. (2003), and is now regularly updated with new datasets (Kruska, 2006). For land-use/cover, we use version 3 of the Global Land Cover (GLC) 2000 data layer (Joint Research Laboratory, 2005). For Africa, this included irrigated areas, so this is used instead of the irrigated areas database of Döll and Siebert (2000), which is used for Asia and Latin America. For human population, we use new 1-km data (GRUMP, 2005). For length of growing period, we use a layer developed from the WorldCLIM 1-km data for 2000 (Hijmans et al., 2004), together with a new "highlands" layer for the same year based on the same dataset (Jones and Thornton, 2005). Cropland and rangeland are now defined from GLC 2000, and rock and sand areas are now included as part of rangelands.

The original LGP breakdown into arid-semiarid, humid-subhumid and highland-temperate areas has now been expanded to include hyper-arid regions, defined by FAO as areas with zero growing days. This was done because livestock are often found in some of these regions in wetter years when the LGP is greater than zero. Areas in GLC 2000 defined as rangeland but having a human population density greater than or equal to 20 persons per km^2 as well as an LGP greater than 60 (which can allow cropping) are now included in the mixed system categories.

The landless systems still present a problem, and are not included in version 3 of the classification. Urban areas have been left as defined by GLC 2000. To look at possible changes in the future, we use the GRUMP population data and project human population out to 2030 and 2050 by prorata allocation of appropriate population figures (e.g., the UN medium-variant population data for each year by country, or the Millennium Ecosystem Assessment country-level population projections). LGP changes to 2030 and 2050 are projected using downscaled outputs of coarse-gridded GCM outputs, using methods outlined in Jones and Thornton (2003).

A.5.7.3 Application
The mapped Seré and Steinfeld (1996) classification was originally developed for a global livestock and poverty mapping study designed to assist in targeting research and development activities concerning livestock (Thornton et al., 2002; 2003). Estimates of the numbers of poor livestock keepers by production system and region were derived and mapped. This information was used in the study of Perry et al. (2002), which was carried out to identify priority research opportunities that can improve the livelihoods of the poor through better control of animal diseases in Africa and Asia. Possible changes in livestock systems and their implications have been assessed for West Africa (Kristjanson et al., 2004). The methods have recently been used in work to assess the spatial distribution of methane emissions from African domestic ruminants to 2030 (Herrero et al., 2008), and in a study to map climate vulnerability and poverty in sub-Saharan Africa in relation to projected climate change (Thornton et al., 2006).

A.5.7.4 Uncertainty
Uncertainties in the scheme are outlined in Table A.5.12, together with levels of confidence for scenario calculations in Table A.5.13.

A.5.8 Global Methodology for Mapping Human Impacts on the Biosphere (GLOBIO 3)

A.5.8.1 Introduction
Biodiversity as defined by the Convention on Biological Diversity (CBD) encompasses the diversity of genes, species, and ecosystems. The 2010 target agreed on by the CBD Conference of the Parties (COP) in 2002 specifies a significant reduction in the rate of loss of biodiversity.

Biodiversity loss is defined as the long-term or permanent qualitative or quantitative reduction in components of biodiversity and their potential to provide goods and services, to be measured at global, regional and national levels. A number of provisional indicators of biodiversity loss have been listed for use at a global scale and suggested for use at a regional or national scale as appropriate (UNEP, 2006). These indicators include trends in the extent of biomes/ecosystems/ habitats, trends in the abundance or range of selected species, coverage of protected areas, threats to biodiversity and trends in fragmentation or connectivity of habitats.

The GLOBIO3 model produces a response indicator on an aggregated level, called the Mean Species Abundance (MSA) relative to the original abundance of species in each natural biome. The model incorporates this indicator in scenario projections, being uniquely able to project trends in the abundance of species (SCBD/MNP, 2007). A large number of species-climate or species-habitat response models exist, which examine the response of individual species to change. GLOBIO3 differs from these models as it measures habitat integrity through the lens of remaining species-level diversity, rather than individual species abundance.

A.5.8.2 Model structure and data
The GLOBIO 3 model framework describes biodiversity by means of estimating remaining mean species abundance of original species, relative to their abundance in primary vegetation. This measure of MSA is similar to the Biodiversity Integrity Index (Majer and Beeston, 1996) and the Biodiversity Intactness Index (Scholes and Biggs, 2005) and can be considered as a proxy for CBD indicators (UNEP, 2004).

The core of GLOBIO 3 is a set of regression equations describing the impact on biodiversity of the degree of pressure using dose–response relationships. These dose–response relationships are derived from a database of observations of species response to change. The database includes separate

Table A.5.12 **Major uncertainties in the mapped Seré & Steinfeld (1996) classification**

Model component	Uncertainty
Model Structure	• Based on thresholds associated with human population density and length of growing period • Also based on land-cover information that is known to be currently weak with respect to cropland identification • The global classification is quite coarse, and no differentiation is made of the mixed systems
Parameters	Inputs: • Land cover, length of growing period, human population density, irrigated areas, urban areas • Observed or modeled livestock densities Outputs: • Areas associated with grassland-based systems and mixed crop-livestock systems (rainfed and irrigated), broken down by AEZ (which can then be combined with other national or sub-national information, such as poverty rates)
Driving force	Even at the broad-brush level, population change and climate change will not be the only drivers of land-use change in livestock-based systems, globally
Initial condition	Some validation of the systems layers has been carried out for current conditions, but more is needed
Model operation	Assembling the input data and running the classification is not an automated procedure. It requires separate sets of FORTRAN programmes for estimating changing agroclimatological conditions; and various sets of ArcInfo scripts for spatially allocating population data and rerunning the classification

measures of MSA, each in relation to different degrees of pressure exerted by various pressure factors or driving forces. The entries in the database are all derived from studies in peer-reviewed literature, reporting either on change through time in a single plot, or on response in parallel plots undergoing different pressures.

The current version of the database includes data from about 500 reports: about 140 reports on the relationship between species abundance and land cover or land use, 50 on atmospheric N deposition (Bobbink, 2004), over 300 on the impacts of infrastructure (UNEP, 2001) and several literature reports on minimal area requirements of species.

The driving forces (pressures) incorporated within the model and their sources are as follows:
• Land cover change (IMAGE)
• Land use intensity (IMAGE / GLOBIO3)
• Nitrogen deposition (IMAGE)
• Infrastructure development (IMAGE / GLOBIO2)
• Climate change (IMAGE)

Climate change is treated differently from other drivers in GLOBIO3, as the empirical evidence compiled in GLOBIO dose-response relationships so far is limited to areas that are already experiencing significant impacts of change (such as

Table A.5.13 **Level of confidence for scenario calculations**

Level of Agreement/ Assessment	High	Established but incomplete • Agricultural and land-use intensification processes	Well-established • Impacts of human population densities on agricultural land-use
	Low	Speculative • Climate change scenarios • Human population change scenarios • Impacts of changing climate on agricultural land-use	Competing Explanations • Different or expanded sets of variables as drivers of system intensification
		Low	High
		Amount of Evidence (Theory, Observations, Model Outputs)	

the Arctic and montane forests). The current implementation in the model is based on changing temperature only. Estimates from a European model of the proportion of species lost per biome (Bakkenes et al., 2002; Leemans and Eickhout, 2004; Bakkenes et al., 2006) for increasing levels of temperature are applied within the GLOBIO3 model on a global scale. This regional bias and the absence of a modeled response to changes in moisture availability are important areas for model improvement.

Some responses to change take some time to become apparent. The loss of species from a particular area may take 30 years or may be instantaneous, depending on the type and strength of the pressure. Because of these lags, the model outcome portrays the possible impact over the short to medium term (~5 to 30 years). These lags must be better characterized, and for that the underlying databases are developed further.

There is little quantitative information about the interaction between pressures. Various assumptions can therefore be included in the model, ranging from "all interact" (only the maximum response is delivered) to "no interaction" (responses to each pressure are cumulative). The GLOBIO 3 model calculates the overall MSA value by multiplying the MSA values for each driver for each IMAGE 0.5 by 0.5 degree grid cell according to:

$$MSAi = MSA_{LU}\ MSA_N\ MSA_I\ MSA_F\ MSA_{CC}$$

where i is the index for the grid-cell, MSAXi relative mean species abundance corresponding to the drivers LU (land cover/use), N (atmospheric N deposition), I (infrastructural development), F (fragmentation) and CC (climate change). MSALUi is the area-weighted mean over all land-use categories within a grid cell.

The model relates 0.5° IMAGE maps to Global Land Cover 2000 as a base map at a 1-km scale, based on a series of simple decision rules. These maps are used to estimate the response to changes in land cover and land use intensity within each 0.5° grid cell. The land-use cover maps and the maps representing other pressures are used to generate maps of the share of remaining biodiversity, which may be derived either in terms of remaining share of original species richness, or remaining share of mean original species abundance. More data is being collated for abundance than for richness—this is the favored indicator, as it is closest to those specified by CBD. Outputs are derived at a 0.5° scale and can be scaled up to IAASTD regions.

A.5.8.3 Application
GLOBIO3 has been used in global and regional assessments. GLOBIO3 analyses contributed to an integrated assessment for the Himalaya region (Nellemann, 2004); for deserts of the world and the Global Biodiversity Outlook (SCBD/MNP, 2007).

A.5.8.4 Uncertainty
GLOBIO3 reflects a relatively new model approach. The level of confidence is highly related to the data quality and quantity, a lot of which is derived from other models, in particular, the IMAGE 2.4 model, infrastructure maps (Digital Chart of the World or DCW) and other land cover maps. The biodiversity indicator generated (MSA) is designed to be compatible with the trends in abundance of species indicator as specified by CBD. Other indicators might lead to different results. However the patterns of the global analyses are in line with earlier global analyses. Table A.5.14 provides an overview of major parameters and model structure.

Table A.5.14 **Overview of major uncertainties in the GLOBIO 3 model**

Model component	Uncertainties
Model structure	• Coupling of data from different sources and resolutions, e.g., from IMAGE, Global Land Cover database 2000. • Applying and combining statistical (regression) equations on input data to derive
Parameters	Input: • Regression parameter for relationships between drivers and biodiversity output indicator (MSA) Output: • Biodiversity indicator is Mean species abundance of original species relative to their original abundance (MSA)
Driving force	• Climate (mean annual temperature) • Land use (incl. forestry) and land use pattern • Infrastructure • Nitrogen deposition
Initial condition	• Baseline for biodiversity is "original vegetation" as simulated by the BIOME model in the IMAGE 2.4 model (Prentice et al., 1992) • Baseline for input are calculated maps for 2000 from the IMAGE model
Model operation	ArcGIS maps, Access data bases, VB scripting language

Table A.5.15 **Level of confidence in different types of scenario calculations from GLOBIO3**

Level of Agreement/ Assessment	High	**Established but incomplete** • Dose-response relationships based on existing studies with a regional bias.	**Well established** • Selection of pressure factors
	Low	**Speculative** • Interaction between pressure factors	**Competing explanations** • Use of species distribution and abundance
		Low	**High**
		Amount of evidence **(theory, observations, model outputs)**	

A.5.9 EcoOcean

A.5.9.1 Introduction

EcoOcean is an ecosystem model complex that can evaluate fish supply from the world's oceans. The model is constructed based on the Ecopath with Ecosim modeling approach and software, and includes a total of 42 functional groupings. The spatial resolution in this initial version of the Eco-ocean model is based on FAO marine statistical areas, and it is run with monthly time-steps for the time period from 1950. The model is parameterized using an array of global databases, most of which are developed by or made available through the Sea Around Us project. Information about spatial fishing effort by fleet categories will be used to drive the models over time. The models for the FAO areas will be tuned to time series data of catches for the period 1950 to the present, while forward looking scenarios involving optimization routines will be used to evaluate the impact of GEO4 scenarios on harvesting of marine living resources.

A.5.9.2 Model structure and data

The Ecopath with Ecosim (EwE) modeling approach has three main components:
- Ecopath—a static, mass-balanced snapshot of the system;
- Ecosim—a time dynamic simulation module for policy exploration; and
- Ecospace—a spatial and temporal dynamic module primarily designed for exploring impact and placement of protected areas.

The initial EcoOcean model will be composed of 19 EwE models. The EwE approach, its methods, capabilities and pitfalls are described in detail by Christensen and Walters (2004).

The foundation of the EwE suite is an Ecopath model (Christensen and Pauly, 1992; Pauly et al., 2000), which creates a static mass-balanced snapshot of the resources in an ecosystem and their interactions, represented by trophically linked biomass "pools." The biomass pools consist of a single species, or species groups representing ecological guilds. Ecopath data requirements are relatively simple, and generally already available from stock assessment, ecological

studies, or the literature: biomass estimates, total mortality estimates, consumption estimates, diet compositions, and fishery catches. The parameterization of an Ecopath model is based on satisfying two "master" equations. The first equation describes how the production term for each group can be divided:

Production = catch + predation + net migration + biomass accumulation + other mortality

The second "master" equation is based on the principle of conservation of matter within a group:

Consumption = production + respiration + unassimilated food

Ecopath sets up a series of linear equations to solve for unknown values establishing mass-balance in the same operation.

Ecosim provides a dynamic simulation capability at the ecosystem level, with key initial parameters inherited from the base Ecopath model. The key computational aspects are in summary form:
- Use of mass-balance results (from Ecopath) for parameter estimation;
- Variable speed splitting enables efficient modeling of the dynamics of both "fast" (phytoplankton) and "slow" groups (whales);
- Effects of micro-scale behaviors on macro-scale rates: top-down vs. bottom-up control incorporated explicitly.
- Includes biomass and size structure dynamics for key ecosystem groups, using a mix of differential and difference equations. As part of this EwE incorporates:
  - Age structure by monthly cohorts, density- and risk-dependent growth;
  - Numbers, biomass, mean size accounting via delay-difference equations;
  - Stock-recruitment relationship as "emergent" property of competition/predation interactions of juveniles.

Ecosim uses a system of differential equations that express biomass flux rates among pools as a function of time varying biomass and harvest rates, (Walters et al., 1997, 2000).

Predator prey interactions are moderated by prey behavior to limit exposure to predation, such that biomass flux patterns can show either bottom-up or top down (trophic cascade) control (Walters et al., 2000). Conducting repeated simulations Ecosim simulations allows for the fitting of predicted biomasses to time series data, thereby providing more insights into the relative importance of ecological, fisheries and environmental factors in the observed trajectory of one or more species or functional groups.

A.5.9.3 Application

The core of this global ocean model is Ecopath with Ecosim, which has been used for a number of regional and sub-regional models throughout the world. This global ocean model will be used for this assessment and the GEO4 Assessment.

A.5.9.4 Uncertainty

Table A.5.16 **Overview of major uncertainties in EcoOcean Model**

Model component	Uncertainty
Model structure	low
Parameters	Input parameters • Most have medium to low uncertainty; a few have high uncertainty
Driving force effort: either direct or relative	Medium to high depending on the FAO area at this stage
Initial condition	low
Model operation	medium

Table A.5.17 **Level of confidence for scenario calculations with EcoOcean model**

Level of Agreement/ Assessment	High	Established but incomplete: • Catches • Value • Landing diversity	Well-established: • Marine trophic index (MTI)
	Low	Speculative Jobs	Competing Explanations
		Low	High
	Amount of Evidence (Theory, Observations, Model Outputs)		

References

Ahammad, H., R. Curtotti, and A. Gurney. 2004. A possible Japanese carbon tax: Implications for the Australian energy sector. ABARE eReport 04.13. Available at http://www.abareconomics.com/publications_html/climate/climate_04/climate_04.html. ABARE, Canberra.

Ahammad, H., A. Matysek, B.S. Fisher, R. Curtotti, A. Gurney, G. Jakeman et al. 2006. Economic impact of climate change policy: The role of technology and economic instruments. ABARE Res. Rep. 06.7. Available at http://www.abareconomics.com/publications_html/climate/climate_06/climate_06.html. ABARE, Canberra.

Ahammad, H., and R. Mi. 2005. Land use change modeling in GTEM: Accounting for forest sinks. Australian Bureau Agric. Resource Econ. (ABARE) Conf. Pap. 05.13.

Energy Modeling Forum 22: Climate Change Control Scenarios, Stanford University, 25-27 May.

Alcamo, J., R. Leemans, and G.J.J. Kreileman. 1998. Global change scenarios of the 21st century. Results from the IMAGE 2.1 model. Pergamon and Elsevier, London

Alcamo, J., D. van Vuren, C. Ringler, W. Cramer, T. Masui, J. Alder, and K. Schulze. 2005. Changes in nature's balance sheet: model-based estimates of future worldwide ecosystem services. Ecol. Society 10(2):19. Available at http://www.ecologyandsociety.org/vol10/iss2/art19/.

Alkemade, R., M. Bakkenes, R. Bobbink, L. Miles, C. Nelleman, H. Simons et al. 2006. GLOBIO 3: Framework for the assessment of global terrestrial biodiversity. In A.F. Bouwman et al. (ed) Integrated modeling of

global environmental change. An overview of IMAGE 2.4. Netherlands Environ. Assessment Agency (MNP), Bilthoven.

Anderson, P.K., A.A. Cunningham, N.G. Patel, F.J. Morales, P.R. Epstein, and P. Daszak. 2004. Emerging infectious diseases of plants: pathogen pollution, climate change and agrotechnology drivers. Trends Ecol. Evol. 19(10):535-544.

Asiedu, E. 2004. Policy reform and foreign direct investment in Africa: Absolute progress but relative decline. Dev. Policy Rev. 22(1):41-48.

Babu, S. 2004. Future of the agri-food system: Perspectives from the Americas. Food Policy 29:669-674.

Babu, S., and W. Reidhead. 2000. Poverty, food security and nutrition in Central Asia: A case study of the Kyrgyz Republic. Food Policy 25:647-660.

Bakkenes, M., J.R.M. Alkemade, F. Ihle, R. Leemans, and J.B. Latour. 2002. Assessing effects of forecasted climate change on the diversity and distribution of European higher plants for 2050. Global Change Biol. 8:390-407.

Bakkenes, M., B. Eickhout, and R. Alkemade. 2006. Impacts of different climate stabilisation scenarios on plant species in Europe. Global Environ. Change16:19-28.

Ball, A., A. Hansard, R. Curtotti, and K. Schneider. 2003. China's changing coal industry — Implications and outlook. ABARE Res. Rep. 03.3 Available at http://agsurf.abareconomics.com/publications_html/energy/energy_03/er03_coal.pdf. ABARE, Canberra.

Barnett, T.P., J.C. Adam, and D.P. Lettenmaier. 2005. Potential impacts of a warming climate on water availability in snow-dominated regions. Nature 438(17):303-309.

Bayes, A., J. von Braun, and R. Akhter. 1999. Village pay phones and poverty reduction: Insights from a Grameen phone initiative in Bangladesh. Disc. Pap. Dev. Policy No. 8. ZEF, Bonn.

Berdegué, J.A., F. Balsevich, L. Flores, and T. Reardon. 2005. Central American supermarkets' private standards of quality and safety in procurement of fresh fruits and vegetables. Food Policy 30:254-269.

Bobbink, R. 2004. Plant species richness and the exceedance of empirical nitrogen critical loads: an inventory. Internal report. Utrecht Univ., The Netherlands.

Bouwman, A.F., T. Kram, K. Klein Goldewijk (ed) 2006. Integrated modelling of global environmental change. An overview of IMAGE 2.4. MNP Rep. 500110002. Available at http://www.mnp.nl/en/publications/2006/Integrated modellingofglobalenvironmentalchange. AnoverviewofIMAGE2.4.html. Netherlands Environ. Assessment Agency (MNP), Bilthoven.

Bouwman, A.F., K.W. van der Hoek, B. Eickhout, and I. Soenario. 2005. Exploring changes in world ruminant production systems. Agric. Syst. 84(2):121-153.

Bruemmer, B. 2003. Food biosecurity. J. Amer. Diet. Assoc. 103:687-691.

Bull, T., and I. Roberts. 2001. Agricultural trade policies in Japan — The need for reform. ABARE Res. Rep. 01.5. ABARE, Canberra.

Burlingame, B., and M. Pineiro. 2007. The essential balance: Risks and benefits in food safety and quality. J. Food Comp. Anal. 20:139-146.

CA (Comprehensive Assessment of Water Management in Agriculture). 2007. Water for food, water for life: A comprehensive assessment of water management in agriculture. Earthscan and IWMI, London and Colombo.

Cai, X., and M. Rosegrant. 2002. Global water demand and supply projections. Part 1: A modeling approach. Water Int. 27(3):159-169.

Carvalho, F.P. 2006. Agriculture, pesticides, food security and food safety. Environ. Sci. Policy 9:685-692.

Christensen, V. and Pauly, D. 1992. Ecopath II — A software for balancing steady-state ecosystem models and calculating network characteristics. Ecol. Model. 61:169-185.

Christensen, V. and C.J. Walters. 2004. Ecopath with Ecosim: Methods, capabilities and limitations. Ecol. Model. 172:109-139.

Cleaveland, S., M.K. Laurenson, and L.H. Taylor. 2001. Diseases of humans and their domestic mammals: Pathogen characteristics, host range and the risk of emergence. Phil. Trans. R. Soc. Lond. B. 356:991-999.

Coakley, S.M., H. Scherm, and S. Chakraborty, 1999. Climate change and plant disease management. Ann. Rev. Phytopathol. 37:399-426.

Codron, J-M., E. Giraud-Heraud, and L-G. Soler. 2005. Minimum quality standards, premium private labels, and European meat and fresh produce retailing. Food Policy 30:270-283.

De Brauw, A., J. Huang, and S. Rozelle. 2004. The sequencing of reform policies in China's agricultural transition. The Econ. Transition 12(3):427-465.

De Fraiture, C., D. Wichelns, E. Kemp Benedict, and J. Rockstrom. 2007. Scenarios on water for food and environment. In D. Molden (ed) A comprehensive assessment of water management in agriculture. Water for food, water for life. Earthscan and IWMI, London and Colombo.

De Groote, H., S. Mugo, D. Bergvinson, and B. Odhiambo. 2005. Assessing the benefits and risks of GE crops: Evidence from the insect resistant maize for Africa project. Inform. Syst. Biotech. ISB News Rep. Feb 2005. Available at http://www.isb.vt.edu/news/2005/artspdf/feb0503.pdf.

De Vries, H.J.M., M. Hoogwijk, and D.P. van Vuuren. 2007. Potential of wind, solar and biofuels. Energy Policy 35 (4):2590-2610.

Deaton, A., and J. Muellbauer. 1980. An almost ideal demand system. Am. Econ. Rev. 70:312-329.

Delgado, C.L., M.W. Rosegrant, H. Steinfeld, S. Ehui, and C. Courbois. 1999. Livestock to 2020. The next food revolution. 2020 Vision for Food, Agric. Environ. Disc. Pap. 28. Available at http://www.ifpri.org/2020/dp/dp28.pdf. IFPRI, Washington DC.

Delgado, C.L., N. Wada, M.W. Rosegrant, S. Meijer, and A. Mahfuzuddin. 2003. Fish to 2020. Supply and demand in changing global markets. Available at http://www.ifpri.org/pubs/books/fish2020book.htm. IFPRI, Washington DC.

Dinham, B. 2003. Growing vegetables in developing countries for local urban populations and export markets: Problems confronting small-scale producers. Pest Manage. Sci. 59:575-582.

Döll, P., and S. Siebert. 2000. A digital global map of irrigated areas. ICID J. 49(2):55-66.

Eickhout, B., M.G.J. den Elzen, and G.J.J. Kreileman. 2004. The atmosphere-ocean system of IMAGE 2.2: A global model approach for atmospheric concentrations, and climate and sea level projections. RIVM Rep. No. 481508017. Nat. Inst. Public Health Environ., Netherlands.

Eickhout, B., H. van Meijl, and A. Tabeau. 2006. Modeling agricultural trade and food production under different trade policies. In A.F. Bouwman et al. (ed) Integrated modeling of global environmental change. An overview of IMAGE 2.4. Netherlands Environ. Assessment Agency (MNP), Bilthoven.

Eickhout B., H. van Meijl, A. Tabeau, and T. van Rheenen, 2007. Economic and ecological consequences of four European land use scenarios. Land Use Policy 24:562-575.

Ezzati, M., A.D. Lopez, A. Rodgers, S. Cander Hoorn, and C.J. Murray. 2002. Comparative risk assessment collaborating group. Selected major risk factors and global and regional burden of disease. Lancet 360:1347-1360.

Fairhead, L., and H. Ahammad. 2005. China's future growth: Implications for selected Australian industries. ABARE eReport 05.13 Prepared for the Australian Government Dep. Industry, Tourism and Resources. ABARE, Canberra.

Fairhead, L., J. Mélanie, L. Holmes, Y. Qiang, H. Ahammad, and K. Schneider. 2002. Deregulating energy markets in APEC: Economic and sectoral impacts. APEC#202-RE-01.3 and ABARE Res. Rep. 02.5. Available at http://www.abareconomics.com/publications_html/energy/energy_02/apec.pdf. ABARE for Asia-Pacific Econ. Coop. Energy Working Group, Canberra.

Falkenmark, M., Finlayson, and L. Gordon. 2007. Agriculture, water and ecosystems: The cost of going too far. In D. Molden (ed) Comprehensive assessment of water management in agriculture, Water for food, water for life. Earthscan and IWMI, London and Colombo.

FAO. 2003. FAOSTAT database. Available from http://faostat.fao.org/default.aspx. FAO, Rome.

FAO. 2004. The state of food and agriculture 2003-04. Agricultural biotechnology: Meeting the needs of the poor? FAO, Rome.

FAO. 2006a. World agriculture: towards 2030/2050. Interim report. FAO, Rome.

FAO. 2006b. Livestock's long shadow. Environmental issues and options. FAO, Rome.

FAO/WHO. 2006. A model for establishing upper levels of intake for nutrients and related substances. Rep. Joint FAO/WHO Technical Workshop on Nutrient Risk Assessment, WHO Headquarters, Geneva, 2-6 May 2005. Available at http://www.who.int/ipcs/methods/nra_final.pdf/. FAO, Rome.

Fisher, B.S., G. Jakeman, K. Woffenden, V. Tulpulé, and S. Hester. 2003. Dealing with climate change: Possible pathways forward. APPEA J. 611-622.

Fonseca, J.M. 2006. Post-harvest handling and processing: sources of microorganisms and impact of sanitizing procedures. p. 85-120. *In* K.R. Matthews (ed) Microbiology of fresh produce. ASM Press, Washington DC.

Freeman, A., A. McLeod, B. Day, J. Glenn, J.A. van de Steeg, and P.K. Thornton (ed) 2006. The future of livestock in developing countries to 2030. Meeting Rep., Nairobi, 13-15 Feb 2006, ILRI-FAO, Nairobi.

Freeman, F., J. Mélanie, I. Roberts, D. Vanzetti, A. Tielu, and B. Beutre. 2000. The impact of agricultural trade liberalisation on developing countries, ABARE Res. Rep. 2000.6. ABARE, Canberra.

Graff, G. D., D. Zilberman, and A.B. Bennett. 2005. Nutritional and product quality innovation in the aricultural R&D pipeline: New crop biotechnologies and their potential economic impacts. Public Intellectual Property Resource for Agriculture (PIPRA), Davis, CA.

GRUMP. 2005. Global Urban-Rural Mapping Project (GRUMP). Available at http://beta. sedac.ciesin.columbia.edu/gpw. Center for Int. Earth Science Inform. Network (CIESIN) of the Earth Inst., Columbia Univ., NY.

Gupta, P.K. 2004. Pesticide exposure — Indian scene. Toxicology 198:83-90.

Gurstein, M. 2003. Effective use: A community informatics strategy beyond the digital divide. First Monday 8(12). Available at http://www.firstmonday.dk/issues/issues_12/gurstein/index.html.

Halberg, N., T.B. Sulser, H. Høgh-Jensen, M.W. Rosegrant, and M.T. Knudsen. 2006. The impact of organic farming on food security in a regional and global perspective. *In* N. Halberg et al. (ed) Global development of organic agriculture: Challenges and promises. CAB Int., Wallingford.

Hamilton, A.J., A. Boland, D. Stevens, J. Kelly, J. Radcliffe, A. Ziehrl et al. 2005. Position of the Australian horticultural industry with respect to the use of reclaimed water. Agric. Water Manage. 71:181-209.

Hawkes, C., and M.T. Ruel. 2006. Overview. *In* C. Hawkes and M.T. Ruel (ed) Understanding the links between agriculture and health. IFPRI, Washington DC.

Heaney, A., S. Hester, A. Gurney, L. Fairhead, S. Beare, J. Melanie et al.. 2005. New energy technologies: Measuring potential impacts in APEC. APEC Energy Working Group, APEC#205-RE-01.1 and ABARE Res. Rep. 05.1, Canberra.

Henson, S., and T. Reardon. 2005. Private agri-food standards: Implications for food policy and the agri-food system. Food Policy 30:241-253.

Heyhoe, E., Y. Kim, S. Crimp, N. Flood, P. Kokic, R. Nelson et al. 2007. Adapting to climate change: Issues and challenges in the agricultural sector. Outlook 2007. ABARE, Canberra.

Hijmans, R.J., S. Cameron, and J. Parra. 2004. WorldClim climate surfaces. Available at http://biogeo.berkeley.edu/worldclim/worldclim.htm.

Hoogwijk, M., A. Faaij, B. Eickhout, B. de Vries, and W. Turkenburg. 2005. Potential of biomass energy out to 2100, for four IPCC SRES land-use scenarios. Biomass Bioenergy 29(4):225-257. Available at doi:10.1016/j.biombioe.2005.05.002.

Huang, J., and H. Bouis. 2001. Structural changes in the demand for food in Asia: Empirical evidence from Taiwan. Agric. Econ. 26:57-69.

Huang, J., and C. Chen. 1999. Effects of trade liberalization on agriculture in China: Commodity and local agricultural studies. UN ESCAP CGPRT Centre, Bogor.

Huang, J., and N. Li. 2003. China's agricultural policy simulation and projection model-CAPSiM. J. Nanjing Agric. Univ. (Social Sci. Ed.) 3(2):30-41.

Huang, J., N. Li, and C. Chen. 2000. WTO and China's agriculture: Challenge ahead. Enriching World 2:18-19.

Huang, J., N. Li, and S. Rozelle. 2003. Trade reform, household effect, and poverty in rural China, Amer. J. Agric. Econ. 85(5):1292-1298.

Huang, J. and H. Liu. 2002. Income growth and life-style changes in rural and urban China. Chinagro Project Report WP1.7. Center for Chinese Agric. Policy, Chinese Acad. Sci., Beijing.

Huang, J., and S. Rozelle. 1998a. Market development and food consumption in rural China, China Econ. Rev. 9:25-45.

Huang, J., and S. Rozelle. 1998b. China's grain economy toward the 21st century. China's Agric. Press, Beijing.

Huang, J., and S. Rozelle. 2003. Trade reform, WTO and China's food economy in the 21st century. Pacific Econ. Rev. 8(2):143-156.

Huang, J., S. Rozelle, and M.W. Rosegrant. 1997. China's food economy to the twenty-first century: Supply, demand, and trade. 2020 Vision for Food, Agric. Environ. Disc. Pap. 19. IFPRI, Washington DC.

Huang, J., S. Rozelle, and M.W. Rosegrant. 1999. China's food economy to the 21st century: Supply, demand and trade. Econ. Dev. and Cultural Change, 47:737-766.

Huang, J., Z. Xu, N. Li, and S. Rozelle, 2005a. Trade liberalization and Chinese agriculture, poverty and equality. Issues Agric. Econ. 7:9-14.

Huang, J., Z. Xu, N. Li, and S. Rozelle. 2005b. A new round of trade liberalization: China's agriculture, poverty and environment. Bull. Nat. Natural Sci. Foundation of China 19(3) (Sum 83).

IMAGE Team. 2001. The IMAGE 2.2 implementation of the SRES scenarios. A comprehensive analysis of emissions, climate change and impacts in the 21st century. RIVM CD-ROM publication 481508018. National Inst. Public Health Environ., The Netherlands.

Institute of Medicine. 1992. Emerging infections: Microbial threats to health in the United States. Nat. Academy Press, Washington DC.

IPCC (Intergovernmental Panel on Climate Change). 2007. Climate Change 2007. Contribution of Working Group 1 to IPCC's Fourth Assessment Report. Cambridge Univ. Press, UK.

Jakeman, G., K. Hanslow, M. Hinchy, B.S. Fisher, and K. Woffenden. 2004. Induced innovations and climate change policy. Energy Econ. 26(6):937-960.

Jakeman, G., E. Heyhoe, S. Hester, K. Woffenden, and B.S. Fisher. 2002. Kyoto Protocol: the first commitment period and beyond. Aust. Commodities 9(1):176-197.

Joint Research Laboratory (JRL). 2005. GLC 2000 (Global Land Cover) data layer. JRL, Ispra, Italy.

Jones, P.G., and P.K. Thornton. 2003. The potential impacts of climate change in tropical agriculture: the case of maize in Africa and Latin America in 2055. Global Environ. Change 13:51-59.

Jones, P.G., and P.K. Thornton. 2005. Global LGP dataset. ILRI, Nairobi.

Jotzo, F., C. Polidano, E. Heyhoe, G. Jakeman, V. Tulpulé, and K. Woffenden. 2000, Climate change policy and the European Union: Emission reduction strategies and international policy pptions, ABARE Res. Rep. 2000.12, Canberra.

Kayani, R., and A. Dymond. 1997. Options for rural telecommunications development. Tech. Rep. No. 359. World Bank, Washington DC.

Kennedy, G., G. Nantel, and P. Shetty. 2003. The scourge of 'hidden hunger': Global dimensions of micronutrient deficiencies. Food Nutr. Agric. 32:8-16.

Kenny, C. 2002. Information and communication technologies for direct poverty alleviation: Costs and benefits. Dev. Policy Rev. 20(2):141-157.

Khan, S., R. Tariq, C. Yuanlai, and J. Blackwell. 2006. Can irrigation be sustainable? Agric. Water Manage. 80(1-3):87-99.

Knudsen, M.T., N. Halberg, J.E. Olesen, J. Byrne, V. Iyer, and N. Toly. 2006. Global trends in agriculture and food systems. *In* N. Halberg et al. (ed) Global development of organic agriculture: Challenges and promises. CAB Int., Wallingford.

Kristjanson, P.M., P.K. Thornton, R.L. Kruska, R.S. Reid, N. Henninger, T.O. Williams et al. 2004. Mapping livestock systems and changes to 2050: Implications for West Africa. *In* T.O. Williams et al. (ed) Sustainable crop-livestock production for improved livelihoods and natural resource management in West Africa. CTA, Wageningen.

Kruska, R.L. 2006. Seré and Steinfeld version 3. Digital data set. ILRI, Nairobi.

Kruska, R.L., R.S. Reid, P.K. Thornton, N. Henninger, and P.M. Kristjanson. 2003. Mapping livestock-orientated agricultural production systems for the developing world. Agric. Syst. 77:39-63.

Kurien, J. 2005. Responsible fish trade and security. Report of the study on the impact of international trade in fishery product on food security. FAO Fisheries Tech. Pap. 456. FAO and the Roy. Norwegian Min. Foreign Affairs, Rome.

Kurukulasuriya, P., R. Mendelsohn, R. Hassan, J. Benhin, M. Diop, H.M. Eid et al. 2006. Will African agriculture survive climate change? World Bank Econ. Rev. 20(3):367-388.

Leemans, R., and B. Eickhout. 2004. Another reason for concern: Regional and global impacts on ecosystems for different levels of climate change. Global Environ. Change 14:219-228.

Leemans, R., B. Eickhout, B. Strengers, L. Bouwman, and M. Schaeffer. 2002. The consequences of uncertainties in land use, climate and vegetation responses on the terrestrial carbon. Science in China, Ser. C, 45 (Supp.), 126.

Liao, Y., and J. Huang. 2004. A projection analysis of the grain demand in the nine major Chinese river basins in the 21st Century. South-to-North Water Transfers. Water Sci. Tech. 2(1):29-32.

Lipton, M., and S. Sinha 1998. Issues Paper for Discussion. Prepared for Brainstorming Workshop on IFAD's Strategic Focus on Poverty, Rome, 20-21 Oct. Int. Fund Agric. Dev. (IFAD), Rome.

Loader, R., and J. E. Hobbs. 1999. Strategic responses to food safety legislation. Food Policy 24(6):685-706.

MA (Millennium Ecosystem Assessment). 2005. Ecosystems and human well-being. Vol. 2: Scenarios. Findings of the Scenarios Working Group. Island Press, Washington DC.

MA (Millennium Ecosystem Assessment). 2006. Ecosystems and human well-being. Synthesis report. Island Press, Washington DC.

Majer, J.D., and G. Beeston, 1996. The biodiversity integrity index: An illustration using ants in Western Australia. Conserv. Biol. 10:65-73.

Mascie-Taylor, C.G, and E. Karim. 2003. The burden of chronic disease. Science 302:1921-2.

Matten, E. 2002. Food labeling in Codex Alimentarus. Econ. Perspect. 7:26-28

Matysek, A., M. Ford, G. Jakeman, R. Curtotti, K. Schneider, H. Ahammad et al. 2005. Near zero emissions technologies. ABARE eReport 05.1[Online]. Available at http://www.abareconomics.com/publications_html/climate/climate_05/climate_05.html. ABARE, Canberra.

McCalla, A.F., and C.L. Revoredo. 2001. Prospects for global food security: A critical appraisal of past projections and predictions. 2020 Vision for Food, Agric. Environ. Disc. Pap. 35. IFPRI, Washington DC.

Mélanie, J., R. Curtotti, M. Saunders, K. Schneider, L. Fairhead, and Y. Qiang. 2002. Global coal markets: Prospects to 2010. ABARE Res. Rep. 02.2 [Online].

Available at http://www.abareconomics.com/publications_html/energy/energy_02/coal.pdf. ABARE, Canberra.

Molden, D. 1997. Accounting for water use and productivity. SWIM Paper 1. IIMI, Colombo.

Molden, D., T. Oweis, P. Steduto, J. Kijne, M.A. Hanjra, and P. Bindraban. 2007. Pathways for increasing agricultural water productivity. In A comprehensive assessment of water management in agriculture. Water for food, water for life: Earthscan and IWMI, London and Colombo.

Molle, F., and J. Berkoff. 2006. Cities versus agriculture: Revisiting intersectoral water transfers. Potential gains and conflicts. Res. Rep. 10. Comprehensive Assessment of Water Management in Agriculture. IWMI, Colombo.

Nair, R., D. McDonald, A. Jacenko, and D. Gunasekera. 2006. Multilateral trade reform — Potential trade impacts of the Doha Round. Aust. Commodities 13(1):209-219.

Nair, R., C. Chester, D. McDonald, T. Podbury, D. Gunasekera, and B.S. Fisher. 2005. Timing of the US Farm Bill and WTO negotiations: A unique opportunity. ABARE eReport 05.11. ABARE, Canberra.

Nakicenovic N., et al. 2000. Special report on emissions scenarios. IPCC Special Rep., Cambridge Univ. Press, UK.

Narrod, C., A. Gulati, N. Minot, and C. Delgado. 2007. Food safety research priorities for the CGIAR-A concept note from IFPRI for the Science Council. IFPRI, Washington DC.

Nellemann, C. et al. 2004. The fall of the water. Available at http://www.grida.no/_documents/himalreport_scr.pdf. UNEP GRID Arendal, Norway.

Pandya-Lorch, R. and M.W. Rosegrant. 2000. Prospects for food demand and supply in Central Asia. Food Policy 25(6):637-646.

Pant, H. 2002. Global Trade and Environment Model (GTEM): A computable general equilibrium model of the global economy and environment. Available at http://www.abareconomics.com/publications_html/models/models/gtem.pdf.

Pant, H. 2007. GTEM Global Trade and Environment Model. Available at http://www.abareconomics.com/interactive/GTEM/.

Pauly, D., V. Christensen, and C. Walters. 2000. Ecopath, Ecosim, and Ecospace as tools for evaluating ecosystem impact of fisheries. ICES J. Marine Sci. 57:697-706.

Perry, B.D., J.J. McDermott, T.F. Randolph, K.R. Sones, and P.K. Thornton. 2002. Investing in animal health research to alleviate poverty. ILRI, Nairobi.

Pezzey, J.C.V., and N.R. Lambie. 2001. Computable general equilibrium models for evaluating domestic greenhouse policies in Australia: A comparative analysis. Rep. to the Productivity Commission [Online]. Available at http://www.pc.gov.au/research/consultancy/cgegreenhouse. AusInfo, Canberra.

Pimentel, D., B. Berger, D. Filiberto, M. Newton, B. Wolfe, E. Karabinakis et al. 2004. Water resources: Agricultural and environmental issues. BioScience 54(10):909-918.

Pinstrup-Andersen, P. 2000. Food policy research for developing countries: emerging issues and unfinished business. Food Policy 25:125-141.

Polidano, C., F. Jotzo, E. Heyhoe, G. Jakeman, K. Woffenden, and B.S. Fisher. 2000. The Kyoto Protocol and developing countries: Impacts and implications for mechanism design. ABARE Res. Rep. 2000.4, Canberra.

Prentice, I.C., W.P. Cramer, S.P. Harrison, R. Leemans, R.A. Monserud, and A.M. Solomon. 1992. A global biome model based on plant physiology and dominance, soil properties and climate. J. Biogeo. 19:117-134.

Ramírez, R., and R.A. Lee. 2005. Service delivery systems for natural resource stakeholders: Targeting, information and communication functions and policy considerations. Paper presented at the 5th Conf. European Fed. Inform. Tech. Agric. Food Environ. 3rd World Congress on Computers in Agric. Natural Resources. Vila Real, Portugal. 25-28 July.

Raney, T. 2006. Economic impact of transgenic crops in developing countries. Curr. Opinion Biotech. 17:1-5.

Raney T., and P. Pingali. 2005. Private research and public goods: Implications of biotechnology for biodiversity. In J. Cooper (ed) Agricultural biodiversity and biotechnology in economic development. Springer, NY.

Reardon, T., C. Timmer, C. Barrett, and J. Berdegue. 2003. The rise of supermarkets in Africa, Asia, and Latin America. Am. J. Agric. Econ. 85(5):1140-1146.

Rees, G., and D. Collins. 2004. An assessment of the potential impacts of deglaciation on the water resources of the Himalayas. HR Wallingford, UK.

Richardson, D., R. Ramírez, and M. Haq. 2000. Grameen Telecom's village phone program: A mult-media case study. Available at http://www.telecommons.com/villagephone/finalreport.pdf. CIDA, Ontario.

Roberts, I., T. Podbury, F. Freeman, A. Tielu, D. Vanzetti, N. Andrews et al. 1999. Reforming world agricultural trade policies. ABARE Res. Rep. 99.12, and RIRDC Publ. 99/96, ABARE, Canberra.

Rockström, J. 2003. Water for food and nature in drought-prone tropics: Vapour shift in rain-fed agriculture. Phil. Trans. R. Soc. B 358(1440):1997-2009.

Rockström, J. 2007. Managing water in rainfed agriculture. In A comprehensive assessment of water management in agriculture. Water for food, water for life. Earthscan and IWMI, London and Colombo.

Rosegrant, M.W., M.C. Agcaoili-Sombilla, and N. Perez. 1995. Global food projections to 2020: Implications for investment. 2020 Vision Disc. Pap. 5. IFPRI, Washington DC.

Rosegrant, M.W., X. Cai, and S.A. Cline. 2002. World water and food to 2025: Dealing with scarcity. IFPRI, Washington DC.

Rosegrant, M.W., S.A. Cline, W. Li, T.B. Sulser, and R. A. Valmonte-Santos. 2005a. Looking ahead. Long-term prospects for Africa's agricultural development and food security. 2020 Disc. Pap. 41. Available at http://www.ifpri.org/2020/dp/vp41.asp. IFPRI, Washington DC.

Rosegrant, M.W., and P.B.R. Hazell. 2000. Transforming the rural Asian economy: The unfinished revolution. Oxford Univ. Press, Hong Kong.

Rosegrant, M.W., N. Leach, and R.V. Gerpacio. 1999. Alternative futures for world cereal and meat consumption. Proc. Nutr. Society 58(2):219-234.

Rosegrant M.W., M.S. Paisner, S. Meijer, and J. Witcover. 2001. Global food projections to 2020: Emerging trends and alternative futures. Available at http://www.ifpri.org/pubs/books/globalfoodprojections2020.htm. IFPRI, Washington DC.

Rosegrant, M.W., and C. Ringler. 2000. Asian economic crisis and the long-term global food situation. Food Policy 25(3):243-254.

Rosegrant, M.W., C. Ringler, T. Benson, X. Diao, D. Resnick, J. Thurlow et al. 2005b. Agriculture and achieving the Millennium Development Goals. World Bank, Washington DC.

Rozelle, S., and J. Huang. 2000. Transition, development and the supply of wheat in China. Aust. J. Agric. Resource Econ. 44:543-571.

Rozelle, S., J. Huang, and M. Rosegrant. 1996. Why China will NOT starve the world. Choices First Q:18-24.

SCBD/MNP. 2007. Cross-roads of life on earth: Exploring means to meet the 2010 biodiversity target. Solution-oriented scenarios for Global Biodiversity Oulook 2. Sec. Convention on Biological Diversity (SCBD) and Netherlands Environ. Assessment Agency (MNP). CBD Tech. Ser. No. 31/MNP Rep. 555050001. SCBD/MNP, Montreal and Bilthoven.

Schillhorn van Veen, T.W. 2005. International trade and food safety in developing countries. Food Control 16:491-496.

Schlundt, J. 2002. New directions in foodborne disease prevention. Int. J. Food. Microbiol 78:3-17.

Schneider, K., B. Graham, C. Millsteed, M. Saunders, and R. Sturt. 2000. Trade and investment liberalisation in APEC: Economic and energy sector impacts. ABARE Res. Rep. 2000.2. ABARE, Canberra.

Scholes, R.J., and R. Biggs, 2005. A biodiversity intactness index. Nature 434:45-49.

Scott, G., M.W. Rosegrant, and C. Ringler. 2000a. Global projections for root and tuber crops to the year 2020. Food Policy 25(5):561-597.

Scott, G., M.W. Rosegrant, and C. Ringler. 2000b. Roots and tubers for the 21st century:

Trends, projections, and policy options. 2020 Vision Disc. Pap. No. 31. Available at http://www.ifpri.org/2020/dp/2020dp31.pdf. IFPRI, Washington, DC.

SEARCA/IFPRI/CRESCENT. 2004. Agriculture and rural development strategy study. Final Rep. ADB TA 3843-INO. Bogor, Indonesia.

Seitzinger, S.P., J.A. Harrison, E. Dumont, A.H.W. Beusen, and A.F. Bouwman. 2005. Sources and delivery of carbon, nitrogen, and phosphorus to the coastal zone: an overview of Global NEWS models and their application. Global Biogeochem. Cycles 19. GB4S01, doi:10.1029/2005GB002606.

Seré, C., and H. Steinfeld. 1996. World livestock production systems: Current status, issues and trends. FAO Animal Production Health Pap. 127. FAO, Rome.

Sinha, A., and N. Sangeeta. 2000. Gender in a macroeconomic framework: A CGE model analysis. Presented at the Second Ann. Meeting of the Gender Planning Network, 22-24 Nov. IDRC, Canada.

Smith, L., and L. Haddad. 2000. Explaining child malnutrition in developing countries: A cross-country analysis. Res. Rep. IFPRI, Washington DC.

Song, G., and R. Bertolini. 2002. Information and communication technologies (ICTs) for rural development: An example from rural Laos. Landnutzung und Landentwicklunk 43(2):64-79.

Sinha, A. and C. Adam. 2006. Reforms and informalization: What lies behind jobless growth in India. In B. Guha-Khasnobis and R. Kanbur (ed) Informal labour markets and development. Palgrave Macmillan.

Sinha, A. and C. Adam. 2007. Modelling the informal economy in India: An analysis of trade reforms. In B. Hariss-White and A. Sinha (ed) Trade liberalization and India's informal economy. Oxford Univ. Press, India.

Sinha, A., and N. Sangeeta. 2001. Gender in a macro economic framework: a CGE model analysis. Available at http://idl-bnc.idrc.ca/dspace/handle/123456789/29824.

Sinha, A. and N. Sangeeta. 2003. Gender in macroeconomic framework. In S. Mukhopadhaya and R.M. Sudarshan (ed) Tracking gender equity under economic reforms: Continuity and change in South Asia. Kali for Women and IDRC.

Sinha A., N. Sangeeta and K.A. Siddiqui. 2003. Informal economy: Gender, poverty and households. In R. Jhabvala et al. (ed) Informal economy centre stage. Sage Publ., Delhi.

Smale, M. 1997. The Green Revolution and wheat genetic diversity: Some unfounded assumptions. World Dev. 25(8):1257-1269.

Srinivasan, R. 2006. Indigenous, ethnic and cultural articulations of new media. Int. J. Cultural Studies 9(4):497-518.

Steinfeld, H., P. Gerber, T. Wassenaar, V. Castel, M. Rosales, and C. de Haan. 2006. Livestock's long shadow: Environmental issues and options. Available at http://

www.virtualcentre.org/en/library/key_pub/longshad/A0701E00.htm. FAO, Rome.

Stern, N. 2006. Stern review: The economic of climate change. Cambridge Univ. Press, UK.

Strengers, B., R. Leemans, B. Eickhout, B. De Vries and A.F. Bouwman. 2004. The land use projections in the IPCC SRES scenarios as simulated by the IMAGE 2.2 model. Geojournal 61:381-393.

Stuart, R., K. Schneider, and M. Stubbs. 2000. Impacts of international policies on APEC coal markets: Trade liberalisation and climate change. ABARE Conf. Pap. 2000.4 presented at the Sixth APEC Coal Flow Seminar, Coal in the New Millennium. Kyongju, Korea, 14-15 Mar.

Taylor, L.H., S.M. Latham, and M.E.J. Woolhouse. 2001. Risk factors for human disease emergence. Phil. Trans. R. Soc. Lond. B. 356:983-989.

Thornton, P.K., P.G. Jones, T. Owiyo., R.L. Kruska, M. Herrero, P. Kristjanson et al. 2006. Mapping climate vulnerability and poverty in Africa. Rep. DFID. [Online] Available at http://www.dfid.gov.uk/research/mapping-climate.pdf. ILRI, Nairobi.

Thornton, P K, R.L. Kruska, N. Henninger, P.M. Kristjanson, and R.S. Reid. 2003. Livestock and poverty maps for research and development targeting in the developing world. Land Use Policy 20:311-322.

Thornton, P.K., R.L. Kruska, N. Henninger, P.M. Kristjanson, R.S. Reid, F. Atieno et al. 2002. Mapping poverty and livestock in the developing world. Available at http://www.ilri.org/InfoServ/Webpub/Fulldocs/Mappoverty/index.htm. ILRI, Nairobi.

Todd, E., and C. Narrod. 2006. Agriculture, food safety, and foodborne diseases: Understanding the links between agriculture and health. IFPRI 2020 Vision Focus Brief 13(5). Available at http://www.ifpri.org/2020/focus/focus13/focus13_05.pdf. IFPRI, Washington DC.

Torero, M., and J. von Braun. 2006. Information and communication technologies for development and poverty reduction. John Hopkins Univ. Press, Baltimore.

Tulpulé, V., S. Brown, J. Lim, C. Polidano, H. Pant, and B.S. Fisher. 1999. The Kyoto Protocol: An economic analysis using GTEM. Energy J. (Kyoto Special Issue):257-285.

UN. 1996. Update on the nutrition situation. UN Admin. Comm. Coordination, Subcommittee on Nutrition (ACC-SCN), Geneva.

UN. 2000. World population prospects: The 2000 revision. UN Population Div. Dep. Econ. Social Affairs of the UN Secretariat. UN, NY.

UN. 2005. World population prospects: The 2004 revision. UN Population Div. Dep. Econ. Social Affairs UN Secretariat. UN, NY.

UN-Energy, 2007. Sustainable bioenergy: A framework for decision makers. UN-Energy Publ. April 2007. ftp://ftp.fao.org/docrep/fao/010/a1094e/a1094e00.pdf. UN, NY.

UNEP. 2001. GLOBIO. Global methodology for mapping human impacts on the biosphere.

Rep. UNEP/DEWA/TR 25. UNEP, Nairobi.

UNEP. 2002. Global environment outlook 3 (GEO 3). UNEP. Earthscan, London.

UNEP. 2004. Decision VII/30 Strategic plan: future evaluation of progress. Seventh Conference of the Parties to the Convention on Biological Diversity, Kuala Lumpur.

UNEP. 2006. COP8 Decision VII/30 (2006). [online] Available at http://www.biodiv.org/decisions/default.aspx?dec=VII/30. UNEP, Nairobi.

UNEP. 2007. Global environment outlook 4 (GEO-4): Environment for development. UNEP. Earthprint, Stevenage, England.

Unnevehr, L. 2003. Food safety in food security and food trade: Overview. Brief 1 in 2020 Focus 10. Food safety in food security and food trade. IFPRI, Washington DC.

USDA. 2000. Data obtained from the Economic Research Service's (ERS) Foreign Agricultural Trade of the United States database [Online] Available at http://www.ers.usda.gov/Data/FATUS/. USDA, Washington DC.

Van Minnen, J.G., R. Leemans, and F. Ihle. 2000. Defining the importance of including transient ecosystem responses to simulate C-cycle dynamics in a global change model. Global Change Biol. 6:595-611.

Van Vuuren, D.P. 2007. Energy systems and climate policy. Ph.D. thesis, Utrecht University. Netherlands Environment Assessment Agency, Bilthoven.

Van Vuuren, D.P., B. Eickhout, P.L. Lucas and M.G.J. den Elzen. 2006. Long-term multi-gas scenarios to stabilise radiative forcing — Exploring costs and benefits within an integrated assessment framework. Energy J. (special issue Nov):201-233.

Van Vuuren, D., M. den Elzen, P. Lucas, B. Eickhout, B. Strengers, B. van Ruijven et al. 2007. Stabilizing greenhouse gas concentrations at low levels: An assessment of reduction strategies and costs. Climatic Change 81(2):119-159.

Von Braun, J., M.S. Swaminathan, and M.W. Rosegrant. 2004. Agriculture, food security, nutrition and the millennium development goals. 2003-2004 IFPRI Ann. Rep. Essay. Available at http://www.ifpri.org/pubs/books/ar2003/ar2003_essay.htm. IFPRI, Washington DC.

Vorosmarty, C.J., C. Leveque, and C. Revenga. 2005. Fresh water. In Ecosystem and human well-being. Vol. 1: Current state and trends. Millennium Ecosystem Assessment. Island Press, Washington DC.

Walters, C., V. Christensen, and D. Pauly. 1997. Structuring dynamic models of exploited ecosystems from trophic mass-balance assessments. Rev. Fish Biol. Fisheries 7:139-172.

Walters, C., D. Pauly, V. Christensen, and J.F. Kitchell. 2000. Representing density dependent consequences of life history strategies in aquatic ecosystems: EcoSim II. Ecosystems 3:70-83.

Welch, R.M., and R.D. Graham. 2005. Agriculture: The real nexus for enhancing bioavailable micronutrients in food crops. J. Trace Elements Med. Biol. 18:299-307.

Wescoat, J.L., Jr. 1991. Managing the Indus River basin in light of climate change: Four conceptual approaches. Global Environ. Change 1 (5):381-95.

WHO. 1997. WHO global database on child growth and malnutrition. Programme of Nutrition. WHO Doc. WHO/NUT/97.4. WHO, Geneva.

WHO. 2006. World health report 2006 — Working together for health. WHO, Geneva.

World Bank. 2000. Global commodity markets: A comprehensive review and price forecast. Dev. Prospects Group, Commodities Team. World Bank, Washington DC.

Xu, Z., J. Xu, X. Deng, J. Huang, E. Uchida, and S. Rozelle. 2006. Grain for green versus grain: Conflict between food security and conservation set-aside in China. World Dev. 34(1):130-148.

6

Options to Enhance the Impact of AKST on Development and Sustainability Goals

Coordinating Lead Authors
Ameenah Gurib-Fakim (Mauritius) and Linda Smith (UK)

Lead Authors
Nazimi Acikgoz (Turkey), Patrick Avato (Germany/Italy), Deborah Bossio (USA), Kristie Ebi (USA), André Gonçalves (Brazil), Jack A. Heinemann (New Zealand), Thora Martina Herrmann (Germany), Jonathan Padgham (USA), Johanna Pennarz (Germany), Urs Scheidegger (Switzerland), Leo Sebastian (Philippines), Miguel Taboada (Argentina), Ernesto Viglizzo (Argentina)

Contributing Authors
Felix Bachmann (Switzerland), Barbara Best (USA), Jacques Brossier (France), Cathy Farnworth (UK), Constance Gewa (Kenya), Edwin Gyasi (Ghana), Cesar Izaurralde (Argentina), Roger Leakey (UK), Jennifer Long (Canada), Shawn McGuire (Canada), Patrick Meier (USA), Ivette Perfecto (Puerto Rico), Christine Zundel (Brazil)

Review Editors
David Bouldin (USA) and Stella Williams (Nigeria)

Key Messages

Key Messages

1. Many of the challenges facing agriculture over the next 50 years will be able to be resolved by smarter and more targeted application of existing AKST. But new science and innovation will be needed to respond to both intractable and changing challenges. These challenges include climate change, land degradation, availability of water, energy use, changing patterns of pests and diseases as well as addressing the needs of the poor, filling the yield gap, access to AKST, pro-poor international cooperation and entrepreneurialism within the "localization" pathway.

2. Smarter and more targeted application of existing best practice AKST will be critical to achieving development and sustainability goals. It is essential to build on the competences and developments in a wide range of sectors to have the maximum impact. The greatest scope for improvements exists in small-scale diversified production systems.

3. The challenges are complex, so AKST must be integrated with place-based and context relevant factors to address the multiple functions of agriculture. A demand-led approach to AKST needs to integrate the expertise from a range of stakeholders, including farmers, to develop solutions that simultaneously increase productivity, protect natural resources including those on which agriculture is based, and minimize agriculture's negative impact on the environment. New knowledge and technology from sectors such as tourism, communication, energy, and health care, can enhance the capacity of agriculture to contribute to the development and sustainability goals. Given their diverse needs and resources, farmers will need a choice of options to respond to the challenges, and to address the increasing complexity of stresses under which they operate. There are opportunities to enhance local and indigenous self-sufficiency where communities can engage in the development and deployment of appropriate AKST.

4. Advances in AKST, such as biotechnology, nanotechnology, remote sensing, precision agriculture, information communication technologies, and better understanding and use of agroecological processes and synergies have the potential to transform our approaches in addressing development and sustainability goals, but will need to be inclusive of a wide variety of approaches in order to meet sustainability and development goals. The widespread application of these breakthroughs will depend on resolving concerns of access, affordability, relevance, biosafety, and the policies (investment and incentive systems) adopted by individual countries. There will be new genotypes of crops, livestock, fish, and trees to facilitate adaptation to a wider range of habitats and biotic and abiotic conditions. This will bring new yield levels, enhance nutritional quality of food, produce non-traditional products, and complement new production systems. New approaches for crop management and farming systems will develop alongside breakthroughs in science and technology. Both current and new technologies will

play a crucial role in response to the challenges of hunger, micronutrient deficiencies, productivity, and environmental protection, including optimal soil and water quality, carbon sequestration, and biodiversity. Ecological approaches to food production also have the potential to address inequities created by current industrial agriculture.

5. Transgenic approaches may continue to make significant contributions in the long term, but substantial increases in public confidence in safety assessments must be addressed. Conflicts over the free use of genetic resources must be resolved, and the complex legal environment in which transgenes are central elements of contention needs further consideration.

6. AKST can play a proactive role in responding to the challenge of climate change and mitigating and adapting to climate-related production risks. Climate change influences and is influenced by agricultural systems. The negative impacts of climate variability and projected climate change will predominately occur in low-income countries. AKST can be harnessed to mitigate GHG emissions from agriculture and to increase carbon sinks and enhance adaptation of agricultural systems to climate change impacts. Development of new AKST could reduce the reliance of agriculture and the food chain on fossil fuels for agrochemicals, machinery, transport, and distribution. Emerging research on energy efficiency and alternative energy sources for agriculture will have multiple benefits for sustainability.

7. Reconfiguration of agricultural systems, including integration of ecological concepts, and new AKST are needed to address emerging disease threats. The number of emerging plant, animal, and human diseases will increase in future. Multiple drivers, such as climate change, intensification of crop and livestock systems, and expansion of international trade will accelerate the emergence process. The increase in infectious diseases (HIV/AIDS, malaria, etc.) as well as other emerging ones will challenge sustainable development and economic growth, and it will ultimately affect both high and low-income countries.

8. Improving water use in agriculture to adapt to water scarcity, provide global food security, maintain ecosystems and provide sustainable livelihoods for the rural poor is possible through a series of integrated approaches. Opportunities exist through AKST to increase water productivity by reducing unproductive losses of water at field and basin scales, and through breeding and soil and crop management. The poor can be targeted for increased benefit from the available water through systems that are designed to support the multiple livelihood uses of water, and demand led governance arrangements that secure equitable access to water. Economic water scarcity can be alleviated through target water resources development that includes socioeconomic options ranging from large to small scale, for communities and individuals. Allocation policies can be developed with stakeholders to take into account whole basin water needs. Integration of food production with other ecosystem services in multifunctional systems helps to achieve multiple goals, for example, integrated rice/

aquaculture systems or integrated crop/livestock systems. While the greatest potential increases in yields and water productivity are in rainfed areas in developing countries, where many of the world's poorest rural people live, equally important is improved management of large dams and irrigation systems to maintain aquatic ecosystems.

9. The potential benefits and risks of bioenergy are strongly dependent on particular local circumstances.

Research is needed on better understanding these effects and improving technologies. Expansion of biofuel production from agricultural crops (1st generation) may in certain cases promote incomes and job creation, but negative effects on poverty (e.g., rising food prices, marginalization of small-scale farmers) and the environment (e.g., water depletion, deforestation) may outweigh these benefits and thus need to be carefully assessed. Small-scale biofuels and bio-oils could offer livelihood opportunities, especially in remote regions and countries where high transport costs impede agricultural trade and energy imports. There is also considerable potential for expanding the use of digesters (e.g., from livestock manure), gasifiers and direct combustion devices to generate electricity, especially in off-grid areas and in cogeneration mode on site of biomass wastes generating industries (e.g., rice, sugar and paper mills). The next generation of liquid biofuels (cellulosic ethanol and biomass-to-liquids technologies) holds promise to mitigate many of the concerns about 1st generation biofuels but it is not clear when these technologies may become commercially available. Moreover, considerable capital costs, large economies of scale, a high degree of technological sophistication and intellectual property rights issues make it unlikely that these technologies will be adopted widely in many developing countries in the next decades. Research and investments are needed to accelerate the development of these technologies and explore their potential and risks in developing countries.

6.1 Improving Productivity and Sustainability of Crop Systems

6.1.1 Small-scale, diversified farming systems

Considerable potential exists to improve livelihoods and reduce the environmental impacts of farming by applying existing AKST in smarter ways to optimize cropping and livestock systems, especially in developing countries.

Small-scale diversified farming is responsible for the lion's share of agriculture globally. While productivity increases may be achieved faster in high input, large scale, specialized farming systems, greatest scope for improving livelihood and equity exist in small-scale, diversified production systems in developing countries. This small-scale farming sector is highly dynamic, and has been responding readily to changes in natural and socioeconomic circumstances through shifts in their production portfolio, and specifically to increased demand by increasing aggregate farm output (Toumlin and Guèye, 2003).

Small-scale farmers maximize return on land, make efficient decisions, innovate continuously and cause less damage to the environment than large farms (Ashley and Maxwell, 2001). Yet they have lower labor productivity and are less efficient in procuring inputs and in marketing, es-

pecially in the face of new requirements regarding produce quality. Land productivity of small-scale farms was found to be considerably higher than in large ones in a comparison across six low-income countries (IFAD, 2001).

AKST investments in small-scale, diversified farming have the potential to address poverty and equity (especially if emphasis is put on income-generation, value-adding and participation in value chains), improve nutrition (both in terms of quantity and quality through a diversified production portfolio) and conserve agrobiodiversity. In small-scale farming, AKST can build on rich local knowledge. Understanding the agroecology of these systems will be key to optimizing them. The challenges will be to: (1) to come up with innovations that are both economically viable and ecologically sustainable (that conserve the natural resource base of agricultural and non-agricultural ecosystems); (2) develop affordable approaches that integrate local, farmer-based innovation systems with formal research; (3) respond to social changes such as the feminization of agriculture and the reduction of the agricultural work force in general by pandemics and the exodus of the young with the profound implications for decision making and labor availability. Small-scale farming is increasingly becoming a part-time activity, as households diversify into off-farm activities (Ashley and Maxwell, 2001) and AKST will be more efficient, if this is taken into account when developing technologies and strategies for this target group.

6.1.1.1 Research options for improved productivity

To solve the complex, interlinked problems of small farmers in diverse circumstances, researchers will have to make each time a conscious effort to develop a range of options. There will be hardly any "one-size-fits-all" solutions (Franzel et al., 2004; Stoop and Hart, 2006). It is questionable if AKST will have the capacity to respond to the multiple needs of small-scale diversified farming systems (Table 6-1, 6-2).

AKST options that combine short-term productivity benefits for farmers with long-term preservation of the resource base for agriculture (Douthwaite et al., 2002; Welches and Cherrett, 2002) are likely to be most successful. In small-scale, diversified farming systems, suitable technologies are typically highly site-specific (Stoop and Hart, 2006) and systems improvements need to be developed locally, in response to diverse contexts.

Integrated, multifactor innovations. In the past, a distinction was made between stepwise improvements of individual elements of farming systems and "new farming systems design". Stepwise improvement has had more impact (Mettrick, 1993), as it can easily build on local knowledge. Recently, successful innovations of a more complex nature were developed, often by farming communities or with strong involvement of farmers. Examples include success cases of Integrated Pest Management (see 6.4.3) as well alternative ways of land management such as the herbicide-based no-till systems of South America (Ekboir, 2003), the mechanized chop-and-mulch system in Brazil (Denich et al., 2004) or the Quesungual slash-and-mulch systems in Honduras (FAO, 2005).

In the future, research addressing single problems will probably become less relevant, as the respective opportuni-

Table 6-1. Key Relationships between Future Challenges and Agricultural Knowledge, Science and Tecnology (AKST) Options for Action

AKST options for action

Columns (AKST options for action), left to right:
Water management · Resource management · Conservation agriculture · Breeding and biotechnology · Public participation · Soil conservation · Pest/pathogen management · ICT and Diagnostic technologies · Local uses of bioenergy · GIS and Remote Sensing · 2nd generation biofuels

Challenges

- Maintaining yield in high productivity systems
- Adapting to climate variability and change
- Closing yield gaps in low productivity systems
- Preserving natural resources, biodiversity and ecosystems
- Enhancing health and nutrition
- Managing water scarcity
- Diversifying agricultural systems
- Sustainable use of bioenergy
- Linking knowledge systems

Legend:
- ■ Very important for addressing this challenge
- ▦ Option contributes to addressing this challenge

IAASTD/Ketil Berger, UNEP/GRID-Arendal

ties for simple, one-factor improvements have been widely exploited already. It will be more promising to develop innovations that address several factors simultaneously (as in the above examples) and which will therefore be more context and site specific and more information-intensive.

This will require a change of emphasis in research for farming system optimization. Research needs to develop decision support tools that assist extension workers and farmers in optimizing specific farm enterprises. Such tools already exist for farm economics, site-specific nutrient management, crop protection and land use planning. Integrative approaches such as RISE (Response Inducing Sustainability Evaluation) (Häni et al., 2003), which combine economic, social and ecological aspects, aim at assessing and improving sustainability at the farm level.

Two-thirds of the rural poor make their living in less favored areas (IFAD, 2001). They will continue to depend on agriculture. Returns on investment in AKST may be limited in these areas due to their inherent disadvantages (remoteness, low-fertility soils, climatic risks) and the highly diverse systems (Maxwell et al., 2001). On the other hand, the impact of innovations on poverty, equity and environmental health may be substantial. Recent examples show that improvements are possible in less favored areas, both for simple technological changes (e.g., more productive crop varieties) as well as for more complex innovations (e.g., the mucuna cover crop system or the slash-and-mulch system in Honduras).

Sustainable alternatives to shifting cultivation. Shifting cultivation was the most widespread form of land use in the tropics and subtropics, but over the past decades, a transition occurred to managed fallows or continuous cropping with crop rotation in densely populated areas. Alternatives to slash-and-burn clearing have been developed, which better conserve the organic matter accumulated during the fallow periods. Managed fallows and sound rotations may enhance soil fertility regeneration and even produce additional benefits. This allows for extending cropping periods and reducing fallow periods without compromising sustainability. The resulting "offshoots" of shifting cultivation raise a number of issues to be addressed by AKST. Firstly, it will be important to understand the transition process, its drivers and the newly emerging problems in order to assist farmers. Secondly, for targeted up-scaling of local experiences, it will be crucial to examine the potentials and limitations of different offshoots of shifting cultivation (Franzel et al., 2004).

In less favored areas, low external input agriculture is the rule, as in these circumstances the use of mineral fertilizers and pesticides is risky and only profitable in selected cases (e.g., in high value crops). Most of the successful innovations developed for these areas built strongly on local knowledge.

Due to the site specificity of these innovations, transfer to other unfavorable environments has worked only to a very limited extent (Stoop et al., 2002). The challenge for

Table 6-2. **AKST options for addressing main challenges with related AKST gaps and needs.**

AKST potential to address challenge	AKST gaps and needs: Technology and knowledge	AKST gaps and needs: Capacity building, policies, and investments	Regional applicability
Preserving and maintaining natural resources and ecosystems			
Minimize the negative impacts of agriculture expansion on ecosystem services (6.3.1.1, 6.1.1.1).	Trade-offs analysis to assess dynamic relations between the provision of ecosystem and economic services in conflicting areas. Develop biotechnologies to reduce impacts	Training of researchers, technicians, land administrators and policy makers for the application of trade-offs analytical tools, and adoption of improved crop plants.	LAC (B)
Design of multifunctional agricultural landscapes that preserve and strength a sustainable flow of ecosystem services (6.7.5.2).	Configure systems to resemble structural and functional attributes of natural ecosystems	Enhance local capacities to develop land use strategies and policies to maximize the supply of essential ecosystem services.	LAC, SSA, tropical Asia (B)
Enhance the geographical spread of multifunctional agricultural systems and landscapes (6.7.5.2.1).	Typify the ecological service supplier as a new category of rural producer.	Implementation of public recognition and payment systems for ecological service suppliers that provide demonstrable services to society.	All regions
Creation of more conservation management areas (6.3.1.1)	Research designed to optimize productivity of the small/subsistence farmer. Incentives for in situ conservation	Promote transboundary initiatives and legislation	All regions
Sustainable management of fisheries and aquaculture (6.5)	Improved knowledge of contributions of capture and cultured fisheries to food and nutrition, food security and livelihoods	Promote alternative strategy for meeting the increasing demands for fish products Promote improved fish technology	All regions
Environmental management of dams to reduce impact on aquatic ecosystems (6.6.3)	Environmentally sound management of dams		All regions
Basin water management (6.6.3.2)	Basin management tools Benefit sharing tools for negotiation	Policies for effective water allocation	All regions
Improving water management			
Improve water productivity by reducing evaporative losses (6.6.3.1)	Biotechnologies including genetics and physiology B,C		Semiarid areas (A)
Restore existing irrigation systems (6.6.3.1)	Environmentally sound management of irrigation systems	Investment in irrigation	S and SE Asia, Central Asia, China (A); SSA (B)
Increase sustainable use of groundwater (6.6.3.2)	Hydrologic process understanding for sustainable use of groundwater		S. Asia, China (A) SSA (B)
Precision irrigation, deficit irrigation (6.6.3.1)	Technologies for use of low quality water in precision irrigation	Policies for secure access to water and for effective water allocation	NAE, MENA (A) S&SEAsia, SSA (B)
Rain water harvesting, supplemental and small scale irrigation for rainfed agriculture (6.6.3)	Affordable small scale technologies for rain water harvesting and water management	Investment in water management for rain-fed systems	S.Asia, SSA (A)
Integrated soil water and soil fertility management (6.4.2.1)			S. Asia, SSA (A)

continued

Table 6-2. **continued**

AKST potential to address challenge	AKST gaps and needs: Technology and knowledge	AKST gaps and needs: Capacity building, policies, and investments	Regional applicability
Multiple water use systems, domestic and productive uses, crops/livestock/fisheries (6.4.2.2)	Institutional and design requirements for MUS systems	Policy that promotes sector integration	All regions
Basin water management (6.4.2.2)	Basin management tools Benefit sharing tools for negotiation	Policies for effective water allocation	All regions
Linking knowledge systems			
Promote local uses of biodiversity (6.1.2; 6.8.1.2)	Mobilize and promote indigenous technologies and innovation systems, and resolve intellectual property issues.	Education, training and dissemination, extension; international coordination of IPR systems.	All regions
Enhance participatory approaches for natural resource management (6.7.5.1)	Merge farmer-based and region-specific innovation systems with formal research Improved collaborative NRM for rare species (CITES) Formal and indigenous mapping tools for monitoring of fragmented biodiversity	Gender mainstreaming Scientific and digital divide Education, training and extension, equity, transboundary initiatives and collaborations	All regions
Increase participatory research that merges indigenous and Western science (farmer field schools, seed fairs) (6.6.1; 6.7.5.1)	Develop affordable technologies that integrate local, farmer-based innovation systems with formal research	Promotion of grassroot extension, transboundary collaborations	All regions
Promote underutilized crops (6.6.1)	Develop approaches that integrate local knowledge systems with formal research	IPR, biopiracy, information and dissemination	All regions
Enhancing health and nutrition			

Detection, surveillance, and response to emerging diseases (6.7.3)
 	
  Better surveillance of zoonotic diseases
 	
  Early disease warning systems
 	
  Integrated vector and pest management
 	
  Environmental management of dams to reduce vector-borne disease
	Improve understanding of disease transmission dynamics
	
	More rapid and accurate diagnostic tools
	
	Improved vaccines
	
	Develop faster genomic-based methods for diagnostics and surveillance
		Public health infrastructure and health care systems
		
		Better integration of human and veterinary health
			SSA, S. and SE Asia (B)

Biofortification of crop germplasm (6.2; 6.7.1; 6.7.2)	Cost effective and efficient screening methods for breeding and introducing multi-gene traits Incorporate multiple nutrient traits	Public sector financing and work force Biosafety protocol Public sector investment	SSA, S. and SE Asia (A, B)
Multiple water use systems, domestic and productive uses, crops/livestock/fisheries (6.6.3)	Institutional and design requirements for MUS systems, such as Rice+Fish program; rice livestock programs	Policy that promotes sector integration; Enhance incentives for breeders	All regions
Closing yield gaps in low productivity systems			
Improve practices for root health management (6.1.3)	Genomics-based diagnostic tools for understanding root disease dynamics	Bolster S&T capacity in pest management	All regions
Conventional Breeding/rDNA assisted (6.3.1.1; 6.8.1.1)	Incorporate traits that confer stable performance like weed competitiveness, resistance to pest & diseases & tolerance to abiotic stresses	IPTGR Plant Variety Protection Public sector investment	All regions (A, B)
Transgenics (GM) (6.3)	Develop biosafety testing methodologies. Incorporate genes conferring stable performance	Biosafety protocol Public sector investment	All regions (A, B)
Improve the performance of livestock in pastoral and semi-pastoral subsistence communities. (6.2)	Enhance nutrient cycling	Improve access to grazing and water-endowed areas for nomadic and semi-nomadic communities	SSA (A, B)

Table 6-2. continued

AKST potential to address challenge	AKST gaps and needs: Technology and knowledge	AKST gaps and needs: Capacity building, policies, and investments	Regional applicability
Rain water harvesting, supplemental and small scale irrigation for rainfed systems (6.8.1.2)	Affordable small scale technologies for rain water harvesting and water management	Investment in water management for rainfed systems	SAsia, SSA (A)
Integrate soil water and soil fertility management (6.6.2.2; 6.6.3.3)	Enhance crop residue return to bolster soil organic matter levels, seed treatment of fertilizer with improved rainwater capture		SAsia, SSA (A)
Multiple water use systems, domestic and productive uses, crops/livestock/fisheries (6.6.3.2)	Institutional and design requirements for MUS systems	Policy that promotes sector integration	All regions
Maintaining yields in high productivity systems			
Conventional Breeding/rDNA assisted (6.3.1.1)	Develop varieties with higher yield potential	IPTGR; Plant Variety Protection Reinvest in plant breeding professionals	All regions
Transgenics (GM) (6.3.1.2)	Incorporate yield enhancing traits Appropriateness to small holder systems	Biosafety protocol; Public sector investment IPR issues to resolve	All regions
Soil nutrient management to reduce pollution (6.6.2.1)	Wider adoption of precision agriculture technologies	Regulations and law enforcement in developing countries	All regions
Improve performance in intensive livestock systems (6.2)	Application of production methods and techniques to optimize the use of inputs.		All regions with livestock systems
Enhance livestock productivity through use of biotechnology, genomics and transgenics for breeding (6.3.2)	Enhance capacities for gene identification and mapping, gene cloning, DNA sequencing, gene expression.		All regions
Restore existing irrigation systems (6.6.3.1)	Environmentally sound management of irrigation systems	Investment in irrigation	SE Asia, S. Asia, Central Asia, China (A); SSA (B)
Increase sustainable use of groundwater (6.6.3.2)	Hydrologic process understanding for sustainable use of groundwater		S. Asia, China (A) SSA (B)
Improve sustainability of protected cultivation (6.1.1.1)	Low-cost multifunctional films Ecologically sound management for greenhouses	Internalize externalities	NAE, Mediterranean (A) LAC, SSA (B)
Precision irrigation, deficit irrigation (6.6.3.1)	Technologies for use of low quality water in precision irrigation	Policies for secure access to water and for effective water allocation	NAE, MENA (A) S. and SE Asia, SSA (B)
Adaptation to and mitigation of climate change			
Broader adoption of soil conserving practices to reduced projected increase in soil erosion with climate change (6.8.1.1)	Prioritization of soil erosion 'hotspots'	Enhance land tenure security Strengthen conservation allotment policies.	All regions, esp. in mountainous develop. countries
Conventional breeding and biotechnology to enhance abiotic stress tolerance (6.3.1.1; 6.2; 6.8.1.1) Genetic and agronomic improvement of underutilized crops (6.8.1.1)	Change crop types; agroecosystem zone matching; Identify genes needed for GM	Biosafety protocol Public sector investment	All regions

Table 6-2. **continued**

AKST potential to address challenge	AKST gaps and needs: Technology and knowledge	AKST gaps and needs: Capacity building, policies, and investments	Regional applicability
Increase water productivity to bridge dry spells (6.8.1.2) Small-scale development of drip irrigation, treadle pumps (6.6.3.3)	Broader promotion of supplemental irrigation, soil nutrient management, improved crop establishment practices.	Policies for secure access to water Investment in risk reduction strategies	SSA, S. Asia, MENA (A)
Storage: rain water harvesting, small scale, large scale (6.6.3; 6.8.1.2)	Environmentally sound construction and management of large dams Decision support for scale of storage that is environmentally and socially sound	Enhance land tenure security Water rights and access	SSA, S. Asia (A)
Reduce agricultural GHG emissions (6.8.1.1)	Aerobic rice production (CH_4 and N_2O) Site specific nutrient management (N_2O) Animal feed improvement (CH_4 and N_2O) Expand land-based C sequestration potential	Transitional costs associated with land management changes Capacity building for outreach and extension	All regions
Sustainable use of bioenergy			
Production and use bioenergy to promote rural development (6.8.2)	Promote R&D for small-scale biodiesel and unrefined bio-oils for local use to improve energy access in local communities	Capacity building, promote access to finance	SSA, S. and SE Asia, LAC
	Promote R&D to reduce costs and improve operational stability of biogas (digesters), producer gas systems and co-generation applications	Develop demonstration projects, product standards and disseminate knowledge	All regions
Improvements in the environmental and economic sustainability of liquid biofuels for transport (6.8.2.1)	Promote R&D for 2nd generation biofuels focusing on reducing costs to make them competitive. Conduct research on environmental effects of different production pathways.	Facilitate the involvement of small-scale farmers in 2nd generation biofuels/feedstock production and low-income countries, e.g., by developing smallholder schemes, improving access to information and dealing with IPR	High-income regions (B) Low-income regions (C)

Key: A = AKST exists, B = AKST emerging, C = AKST gaps

Source: Authors' elaboration.

AKST will be to find ways for combining local knowledge with innovations developed in similar other contexts to generate locally adapted new options. The question development agents will have to address is, under which circumstances they may scale up innovations *per se* and when they should focus on scaling-up innovation *processes* (Franzel et al., 2004). In the scaling-up process, it will be crucial that research and extension act in a careful, empirical and critical way (Tripp, 2006). If wide dissemination of innovations that were successful in a certain context is attempted, this may create exaggerated expectations and hence frustration, if these innovations are not adapted in many other contexts. This happened for example with alley cropping (Carter, 1995; Akyeampong and Hitimana, 1996; Swinkels and Franzel, 2000; Radersma et al., 2004) or the system of rice intensification (SRI) developed in Madagascar (Stoop et al., 2002). Agricultural research and extension still largely works with technologies that rely strongly on external inputs, even in less favored areas (Stoop, 2002).

Potential for innovation in low external input agriculture is highest if research focuses on understanding and building on local concepts of farming such as the exploitation of within-farm variation, or intercropping. However, if research and extension work with technologies that rely strongly on external inputs, farmers will seldom adopt the results (Stoop, 2002). A further challenge is the dissemination, as farmer-to-farmer diffusion is less important than commonly assumed for such innovations (Tripp, 2006).

Low External Input Sustainable Agriculture (LEISA) comprises organic farming. Organic farming and conventional (non-labeled) LEISA can mutually benefit from each other. Organic farming with its stringent rules on external input use has to be even more innovative to solve production problems, sometimes opening up new avenues. Organic farming has the additional opportunity of deriving benefits from close links between producers and consumers. The challenge, however, is to exploit this potential.

New low external input technologies have the potential to improve productivity while conserving the natural resource base, but there is no evidence that they are specifically pro-poor (Tripp, 2006). An important concern in low external input farming is soil nutrient depletion. Across

Africa, nutrient depletion is widespread, with average annual rates of 22 kg N, 2.5 kg P and 15 kg K per ha of arable land (Stoorvogel and Smaling, 1990). Low external input technologies aiming at soil fertility improvement can seldom reduce these rates (Onduru et al., 2006).

Protected cultivation systems. Protected cultivation of high value crops has expanded rapidly in the past decades (Castilla et al., 2004), especially in the Mediterranean basin (Box 6-1). At present, however, greenhouse production with limited climate control is ecologically unsustainable as it produces plastic waste and contaminates water due to intensive use of pesticides and fertilizers. Demand for innovation thus exists with regard to reducing environmental impact, as well as enhancing productivity, product quality and diversity.

Scope exists to develop affordable plastic films that improve radiation transmission quantitatively and qualitatively. Multilayer, long-life, thermal polyethylene films can combine desirable characteristics of various materials such as anti-drop and anti-dust effects. Photoselective films have the potential to influence disease and insect pest behavior by blocking certain bands of the solar radiation spectrum (Papadakis et al., 2000) or to limit solar heating without reducing light transmission (Verlodt and Vershaeren, 2000). Protected cultivation has its own, specific pest and disease populations as well as specific challenges related climate and substrate. Plant breeding for these specific conditions has the potential to reduce significantly the amount of pollutants released, while improving productivity. Grafting vegetables to resistant rootstocks is a promising option to control soil-borne pathogens (Oda, 1999; Bletsos, 2005; Edelstein and Ben-Hur, 2006) and may help to address salt and low temperature stress (Edelstein, 2004), but needs further research to improve rootstocks. Pest and disease control with the use of antagonists has developed quickly in protected cultures in Northern Europe and Spain (Van Lenteren, 2000, 2003). There are many site and crop specific possibilities for further development of non-chemical pest control for protected cultivation.

Production in low-cost greenhouses has the potential to increase productivity and income generation, to improve water use efficiency and reduce pollution of the environment. Variability in climatic and socioeconomic conditions will require the development of location-specific solutions.

Post-harvest loss. Although reduction of post-harvest losses has been an important focus of AKST and development programs in the past, in many cases the technical innovations faced sociocultural or socioeconomic problems such as low profit margins, additional workload or incompatibility with the existing production or post-production system (Bell, 1999). The divergence between technical recommendations and the realities of rural life translated in many cases into low adoption rates.

In specific cases, large shares of food produced are lost after harvest. Yet, the rationale for improvements in the post-harvest systems has been shifting from loss prevention (Kader, 2005) to opening new markets opportunities (Hellin and Higman, 2005). Making markets work for the poor (Ferrand et al., 2004) is emerging as the new rationale of development, reflecting a shift away from governmental

---

Box 6-1. Advantages of the Mediterranean glasshouse system.

The Mediterranean greenhouse agrosystem represents greenhouse production in mild winter climate areas and is characterized by low technological and energy inputs. Strong dependence of the greenhouse microclimate on external conditions (La Malfa and Leonardi, 2001) limits yield potential, product quality, and the timing of production. It keeps production costs low as compared to the Northern European greenhouse industry. The latter is based on sophisticated structures, with high technological inputs that require important investments, and produces higher yields at higher costs (Castilla et al., 2004).

---

operation of post-harvest tasks to enabling frameworks for private sector initiatives in this field (Bell et al., 1999).

Ecological agricultural systems, which are low external input systems that rely on natural and renewable processes, have the potential to improve environmental and social sustainability while maintaining or increasing levels of food production. There is now increasing evidence of the productive potential of ecological agriculture (Pretty, 1999; Pretty, 2003; Pretty et al., 2006; Badgley et al., 2007; Magdoff, 2007).

Some contemporary studies also show the potential of ecological agriculture to promote environmental services such as biodiversity enhancement, carbon sequestration, soil and water protection, and landscape preservation (Culliney and Pimentel, 1986; Altieri, 1987; Altieri, 1999; Altieri, 2002; Albrecht and Kandji, 2003).

There is now substantial scientific evidence to show that designing and managing agricultural systems based on the characteristics of the original ecosystem is not a threat to food security. A survey of more than 200 projects from Latin America, Africa, and Asia, all of which addressed the issue of sustainable land use, found a general increase in food production and agricultural sustainability (Pretty et al., 2003). Likewise, low external input crop systems, when properly managed, have demonstrated the potential to increase agricultural yield with less impact on the environment (Bunch, 1999; Tiffen and Bunch, 2002; Rasul and Thapa, 2004; Pimentel et al., 2005; Badgley et al., 2007; Scialabba, 2007). A recent investigation comparing organic with conventional farming experiences from different parts of the world indicates that sustainable agriculture can produce enough food for the present global population and, eventually an even larger population, without increasing the area spared for agriculture (Pretty et al., 2003; Badgley et al., 2007).

In spite of the advantages of ecological agriculture in combining poverty reduction, environmental enhancement and food production, few studies address the issues of how to assess the tradeoffs (Scoones, 1998). Tradeoff analysis to assess dynamic relations between the provision of ecosystem and economic services can help to harmonize land use options and prevent potential conflict regarding the access to essential ecosystem services (Viglizzo and Frank, 2006).

Methods are focused on the identification of tradeoffs and critical thresholds between the value of economic and ecological services in response to different typologies of human intervention.

In the same way, the concept of ecological agriculture needs a better understanding of the relationship among the multiple dimensions of rural development, i.e., agricultural productivity, environmental services, and livelihood. Such questions are still open for further elaboration and pose a challenge to AKST (Buck et al., 2004; Jackson et al., 2007).

6.1.1.2 Land use options for enhancing productivity

Productivity of farming systems can be enhanced by more intensive use of space or time. Intercropping (including relay intercropping and agroforestry) is a traditional form of such intensification, widespread in food production in low-income countries, especially in less favored areas. Growing several crops or intercrops in sequence within a year offers the possibility to intensify land use in time. This intensification was made possible by changes in the crops and varieties grown (day-length-neutral or short-season varieties; varieties tolerant to adverse climatic conditions at the beginning or the end of the growing season) or in land management (no-till farming, direct seeding, etc.). On the other hand, farmers quickly change to simpler cropping systems, if economic prospects are promising (Abdoellah et al., 2006).

The development of new elements (crops or crop varieties, pest and land management options), which farmers then integrate according to a multitude of criteria into their farm systems will continue to enhance productivity. Similarly, agroforestry initiatives will be most successful, where research concentrates on developing a range of options with farmers (Franzel et al., 2004).

Intercropping has the potential to increase return to land by investing (usually) more labor. The challenge for AKST will be to strike a balance between (1) understanding the interactions in highly complex intercropping and agroforestry systems (including learning from and with farmers) and (2) developing options that farmers may add to their systems. Adding new elements may offer potential for farmers to participate in value chains and enhance income generation while ensuring subsistence. There exists considerable potential for AKST to develop germplasm of agroforestry species with commercial value (Franzel et al., 2004).

AKST has contributed substantially to intensification in time, especially in high potential areas. However, double or triple cropping in rice or rice-wheat production created new challenges on the most fertile soils (Timsina and Connor, 2001). In spite of such drawbacks, there is promise for further intensifying land use in time by optimizing rotation management and developing novel varieties that can cope with adverse conditions.

Mixed farming. In many low-income countries, integration of crop and livestock has advanced substantially for the past few decades. In densely populated areas, mixed farming systems have evolved, where virtually all agricultural by-products are transformed by animals (Toumlin and Guèye, 2003). With the demand for livestock products expected to surge in most low-income countries, potential for income generation exists. A major challenge for AKST will be to understand the tradeoffs between residue use for livestock or soil fertility and to optimize nutrient cycling in mixed systems.

Improve sustainability through multifunctional agriculture and ecosystem services. Ecosystem services are the conditions and processes through which natural ecosystems sustain and fulfill human life (Daily, 1997) and can be classified in four utilitarian functional groups: (1) provisioning (e.g., food, freshwater), (2) regulating (e.g., climate and disturb regulation), (3) cultural (e.g., recreation, aesthetic) and (4) supporting (e.g., soil formation, nutrient cycling) (MA, 2005). Given that many ecosystem services are literally irreplaceable, estimations of socioeconomic benefits and costs of agriculture should incorporate the value of ecosystem services (Costanza et al., 1997). Because of the rapid expansion of agriculture on natural lands (woodlands, grasslands) and the trend to use more external inputs (Hails, 2002; Tilman et al., 2002), the negative impact of agriculture on ecosystem services supply will require increasing attention (Rounsevell et al., 2005).

The construction of multifunctional agroecosystems can preserve and strengthen a sustainable flow of ecosystem services (Vereijken, 2002). They are best modeled after the structural and functional attributes of natural ecosystems (Costanza et al., 1997). Multifunctional agroecosystems will provide food and fiber, control disturbances (e.g., flood prevention), supply freshwater (filtration and storage), protect soil (erosion control), cycle nutrients, treat inorganic and organic wastes, pollinate plants (through insects, birds and bats), control pests and diseases, provide habitat (refugium and nursery), provide aesthetic and recreational opportunities (camping, fishing, etc.) and culture (artistic and spiritual). The evaluation of ecosystem services is an evolving discipline that currently has methodological shortcomings. However, methods are improving and site-specific valuation will be possible in the coming years. The application of tradeoff analysis to support the design of multifunctional rural landscapes will demand expertise on multicriteria analysis and participatory approaches.

Frequently recommended measures (Wayne, 1987; Viglizzo and Roberto, 1998) for addressing multifunctional needs include (1) diversification of farming activities in time and space rotational schemes, (2) the incorporation of agroforestry options, (3) conservation/rehabilitation of habitat for wildlife, (4) conservation/management of local water resources, (5) the enforcement of natural nutrient flows and cycles (exploiting biological fixation and bio-fertilizers), (6) the incorporation of perennial crop species, (7) the well-balanced use of external inputs (fertilizers and pesticides), (8) the application of conservation tillage, (9) biological control of pests and diseases, (10) integrated management of pests, (11) conservation and utilization of wild and underutilized species, (12) small-scale aquaculture, (13) rainfall water harvesting.

6.1.2 Achieving sustainable pest and disease management

Agricultural pests (insect herbivores, pathogens, and weeds) will continue to reduce productivity, cause post-harvest losses and threaten the economic viability of agricultural livelihoods. New pest invasions, and the exacerbation of existing pest problems, are likely to increase with future climate change. Warmer winters will lead to an expansion of insect and pathogen over wintering ranges (Garrett et al., 2006); this process is already under way for some plant pathogens (Rosenzweig et al., 2001; Baker et al., 2004). Within existing over winter ranges, elevation of pest damage following warm winters is expected to intensify with climate change (Gan, 2004; Gutierrez et al., 2006; Yamamura et al., 2006). Increased temperatures are also likely to facilitate range expansion of highly damaging weeds, which are currently limited by cool temperatures, such as species of *Cyperus* (Terry, 2001) and *Striga* (Vasey et al., 2005).

Several current AKST strategies for managing agricultural pests could become less effective in the face of climate change, thus potentially reducing the flexibility for future pest management in the areas of host genetic resistance, biological control, cultural practices, and pesticide use (Patterson, 1999; Strand, 2000; Stacey, 2003; Bailey, 2004; Ziska and George, 2004; Garrett et al., 2006). For example, loss of durable host resistance can be triggered by deactivation of resistance genes with high temperatures, and by host exposure to a greater number of infection cycles, such as would occur with longer growing seasons under climate change (Strand, 2000; Garrett et al., 2006). Recent evidence from CO_2-enrichment studies indicates that weeds can be significantly more responsive to elevated CO_2 than crops, and that weeds allocate more growth to root and rhizome than to shoot (Ziska et al., 2004). This shift in biomass allocation strategies could dilute the future effectiveness of post-emergence herbicides (Ziska and George, 2004; Ziska and Goins, 2006). Elevated CO_2 is also projected to favor the activity of *Striga* and other parasitic plant species (Phoenix and Press, 2005), which currently cause high yield losses in African cereal systems.

In addition to range expansion from climate change, the future increase in the trans-global movement of people and traded goods is likely to accelerate the introduction of invasive alien species (IAS) into agroecosystems, forests, and aquatic bodies. The economic burden of IAS is US$300 billion per year, including secondary environmental hazards associated with their control, and loss of ecosystem services resulting from displacement of endemic species (Pimentel et al., 2000; GISP, 2004; McNeely, 2006). The costs associated with invasive species damage, in terms of agricultural GDP, can be double or triple in low-income compared with high-income countries (Perrings, 2005).

6.1.2.1 Diversification for pest resistance

To enhance the effectiveness of agroecosystem genetic diversity for pest management, some options include shifting the focus of breeding towards the development of multi- rather than single-gene resistance mechanisms. Other options include pyramiding of resistance genes where multiple minor or major genes are stacked, expanding the use of varietal mixtures, and reducing selection pressure through diversification of agroecosystems.

Multigene resistance, achieved through the deployment of several minor genes with additive effects rather than a single major gene, could become an important strategy where highly virulent races of common plant diseases emerge, as in the case of the Ug99 race of wheat stem rust for which major gene resistance has become ineffective (CIMMYT, 2005). Integration of genomic tools, such as marker-assisted selection (MAS) to identify gene(s) of interest, will be an important element of future resistance breeding. Future breeding efforts will need to include greater farmer involvement for successful uptake and dissemination, e.g., farmer-assisted breeding programs where farmers work with research and extension to develop locally acceptable new varieties (Gyawali et al., 2007; Joshi et al., 2007). Better development of seed networks will be needed to improve local access to quality seed.

Gene pyramiding (or "stacking") has the potential to become a future strategy for broadening the range of pests controlled by single transgenic lines. For example, expressing two different insect toxins simultaneously in a single plant may slow or halt the evolution of insects that are resistant, because resistance to two different toxins would have to evolve simultaneously (Gould, 1998; Bates et al., 2005). Though the probability of this is low, it still occurs in a small number of generations (Gould, 1998); the long-term effectiveness of this technology is presently not clear. The use of gene pyramiding also runs the risk of selecting for primary or secondary pest populations with resistance to multiple genes when pyramiding resistance genes to target a primary pest or pathogen (Manyangarirwa et al., 2006). Gene flow from stacked plants can accelerate any undesirable effects of gene flow from single trait transgenic plants. This could result in faster evolution of weeds or plants with negative effects on biodiversity or human health, depending on the traits (as reviewed by Heinemann, 2007). Finally, mixtures of transgenes increase the complexity of predicting unintended effects relevant to food safety and potential environmental effects (Kuiper et al., 2001; Heinemann, 2007).

Varietal mixtures, in which several varieties of the same species are grown together, is a well-established practice, particularly in small-scale risk-adverse production systems (Smithson and Lenne, 1996). While this practice generally does not maximize pest control, it can be more sustainable than many allopathic methods as it does not place high selection pressure on pests, and it provides yield stability in the face of both biotic and abiotic stresses. For example, varietal mixtures could play an important role in enhancing the durability of resistance for white-fly transmitted viruses on cassava (Thresh and Cooper, 2005). Research on varietal mixtures has been largely neglected; more research is needed to identify appropriate mixtures in terms of both pest resistance and agronomic characteristics, and to backcross sources of pest and disease resistance into local and introduced germplasm (Smithson and Lenne, 1996).

In addition to varietal mixtures, future AKST could enhance the use of cropping system diversification for pest control through supporting and expanding, where appropriate and feasible, practices such as intercropping, mixed cropping, retention of beneficial noncrop plants, crop rotation, and improved fallow, and to understand the mechanisms of pest control achieved by these practices. The underlying principal of using biodiversity for pest control is to reduce the concentration of the primary host and to create conditions that increase natural enemy populations (Altieri, 2002). The process of designing systems to achieve multiple functions is knowledge intensive and often location specific. An important challenge for AKST will be to better elucidate underlying pest suppression mechanisms in diverse systems, such as through understanding how pest community genetics influence functional diversity (Clements et al., 2004). An equally important task will be to preserve local and traditional knowledge in diverse agroecosystems.

6.1.2.2 Tools for detection, prediction, and tracking
AKST can contribute to development through the enhancement of capacity to predict and track the emergence of new pest threats. Some recent advances are discussed below.

Advances in remote sensing. Applications include linking remote sensing, pest predictive models, and GIS (Strand, 2000; Carruthers, 2003), and coupling wind dispersal and crop models to track wind-dispersed spores and insects (Kuparinen, 2006; Pan et al., 2006). Recent advances in remote sensing have increased the utility of this technology for detecting crop damage from abiotic and biotic causal factors, thus remote sensing has good prospect for future integration with GIS and pest models. The spread of these technologies to low-income countries will likely to continue to be impeded by high equipment costs and lack of training. The further development and dissemination of low-cost thermocyclers for PCR (polymerase chain reaction) techniques could help to address this need. In general, a lack of training and poor facilities throughout most of the developing countries hinders the ability to keep up with, let alone address, new pest threats.

Advances in molecular-based tools. Emerging tools such as diagnostic arrays will help to better identify the emergence of new pest problems, and to differentiate pathovars, biovars, and races and monitor their movement in the landscape (Garrett et al., 2006). Using molecular methods for pathogen identification has excellent potential in high-income countries.

Advances in modeling pest dynamics. Recent progress in developing new mathematical approaches for modeling uncertainties and nonlinear thresholds, and for integrating pest and climate models, are providing insights into potential pest-host dynamics under climate change (Bourgeois et al., 2004; Garrett et al., 2006). Increased computational power is likely to facilitate advances in modeling techniques for understanding the effects of climate change on pests. However, the predictive capacity of these models could continue, as it currently is, to be hampered by scale limitations of data generated by growth chamber and field plot experiments, inadequate information concerning pest geographical range, and poor understanding of how temperature and CO_2 interactions affect pest-host dynamics (Hoover and Newman, 2004; Scherm, 2004; Chakaborty, 2005; Zvereva and Kozlov, 2006). Greater focus on addressing these limitations is needed. Improved modeling capacity is needed for understanding how extreme climate events trigger pest and disease outbreaks (Fuhrer, 2003). Modeling pests of tropical agriculture will likely have the greatest impact on helping AKST to address food security challenges, as these regions will be most negatively affected by climate change. This will require a substantial investment in training, education, and capacity development.

Prevention of invasive alien species. The invasive alien species issue is complex in that an introduced organism can be a noxious invasive in one context yet a desirable addition (at least initially) in another (McNeely, 2006). International assistance programs (development projects, food aid for disaster relief, and military assistance) are an important means through which IAS are introduced into terrestrial and freshwater systems, as in the case of fast growing agroforestry trees, aquaculture species, and weed seed-contaminated grain shipments (Murphy and Cheesman, 2006). Addressing this problem will require much more detailed information on the extent of the problem, as well as greater understanding of vectors and pathways. Raising awareness in the international aid community, such as through toolkits developed by the Global Invasive Species Program (GISP, 2004) are an important first step, as are prerelease risk assessments for species planned for deliberate release (Murphy and Cheesman, 2006).

More rigorous risk assessment methods are needed to determine the pest potential of accidentally introduced organisms and those intentionally introduced, such as for food and timber production, biological control, or soil stabilization. Elements needed to build risk assessment capacity include broad access to scientific literature about introduced species, access to advanced modeling software and processing time, improved expertise for determining risks related to invasive characteristics, and development of public awareness campaigns to prevent introduction (GISP, 2004).

Early detection of invasive alien species. The capacity to survey for introduction of nonnative species of concern could be enhanced. Where resources for conducting surveys are limited, surveys can prioritize towards species known to be invasive and that have a high likelihood of introduction at high risk entry points, or areas with high value biodiversity (GISP, 2004). Develop contingency planning for economically important IAS.

Management of invasive alien species. Current mechanical, chemical and biological control methods are likely to continue to be important in the future. In the case of biological control, the use of plant pathogens as natural enemies is emerging as an alternative or complement to classical biological control using arthropods, and it is being piloted in tropical Asia for controlling the highly damaging weed, *Mikania micrantha* (Ellison et al., 2005). Additionally, new and emerging genomic tools could aid IAS management, particularly for preventing the conversion of crops into weeds (Al-Ahmad et al., 2006).

Basic quantitative data on the impacts and scale of the IAS problem is still lacking in many developing countries (Ellison et al., 2005). Gaining greater knowledge of the extent of the problem will require better cross-sectoral linkages, such as between institutions that serve agriculture, natural resource management, and environmental protection.

Risk assessment for entry, establishment, and spread is a newly developing area for IAS (GISP, 2004). For example, Australia recently instituted a weed risk assessment system based on a questionnaire scoring method to determine the weed inducing potential of introduced organisms. Risk assessment is only one tool of many, and will likely have limited utility given that the number of potentially invasive species far outstrips the ability to assess the risk of each one, and high-income countries are better equipped to conduct risk assessments than low-income ones. Full eradication is generally quite difficult to achieve, and requires a significant commitment of resources. Therefore prioritization of IAS management by potential impacts, such as to those that alter fundamental ecosystem processes, and to value of habitats is an important starting point.

6.1.3 Plant root health

The ability to address yield stagnation and declining factor productivity in long-term cropping systems will depend on efforts to better manage root pests and diseases primarily caused by plant-parasitic nematodes and plant-pathogenic fungi (Luc et al., 2005; McDonald and Nicol, 2005). Soil-borne pests and diseases are often difficult to control because symptoms can be hard to diagnose and management options are limited, such as with plant-parasitic nematodes. Nematodes prevent good root system establishment and function, and their damage can diminish crop tolerance to abiotic stress such as seasonal dry spells and heat waves, and competitiveness to weeds (Abawi and Chen, 1998; Nicol and Ortiz-Monasterio, 2004). With future temperature increase, crops that are grown near their upper thermal limit in areas with high nematode pressure, such as in some cereal systems of South and Central Asia (Padgham et al., 2004; McDonald and Nicol, 2005), could become increasingly susceptible to yield loss from nematodes. Approaches for managing soil-borne pests and diseases are changing due to increasing pressure (commercial and environmental) for farmers to move away from conventional broad-spectrum soil fumigants, and greater recognition of the potential to achieve biological root disease suppression through practices that improve overall soil health.

6.1.3.1 Low input options

Soil solarization, heating the surface 5-10 cm of soil by applying a tightly sealed plastic cover, can be a highly effective means of improving root health through killing or immobilizing soilborne pests, enhancing subsequent crop root colonization by plant-growth promoting bacteria, and increasing plant-available nitrogen (Chen et al., 1991). Biofumigation of soils is achieved by the generation of isothiocynate compounds, which are secondary metabolites released from the degradation of fresh *Brassica* residues in soil. They have a similar mode of action as metamsodium, a common synthetic replacement of methyl bromide, and have been used to control a range of soilborne fungal pathogens including *Rhizoctonia, Sclerotinia,* and *Verticillium* (Matthiessen and Kirkegaard, 2006). For many plant parasitic nematodes, significant control is often achieved when solarization is combined with biofumigation (Guerrero et al., 2006).

Soil solarization is an environmentally sustainable alternative to soil fumigation, though its application is limited to high value crops in hot sunny environments (Stapleton et al., 2000), Soil solarization of nursery seedbeds is an important but underutilized application of this technology, particularly

for transplanted crops in the developing world, where farmers contend with high densities of soilborne pests and have few if any control measures. Solarization of rice seedbed soil, which is commonly infested with plant parasitic nematodes, can improve rice productivity in underperforming rice-wheat rotation areas of South Asia (Banu et al., 2005; Duxbury and Lauren, 2006). This technique has potential for broader application, such as in transplanted vegetable crops in resource-poor settings. Biofumigation using isothiocynate-producing *Brassicas* has reasonably good potential for replacing synthetic soil fumigants, especially when combined with solarization. Commercial use of biofumigation is occurring on a limited scale. However, there are significant hurdles to the broad-scale adoption of *Brassica* green manures for biofumigation related to its highly variable biological activity under field conditions compared with *in vitro* tests, and to the logistical considerations involved with fitting *Brassicas* into different cropping systems and growing environments (Matthiessen and Kirkegaard, 2006). The repeated use of chemical replacements for methyl bromide and biofumigation can lead to a shift in soil microbial communities. This shift can result in enhanced microbial biodegradation of the control agent, diminishing its effectiveness (Matthiessen and Kirkegaard, 2006).

6.1.3.2 Research needs and options

Biological control. Future nematode biocontrol could be made more effective through shifting the focus from controlling the parasite in soil to one of targeting parasite life stages in the host. This could be accomplished through the use of biological enhancement of seeds and transplants with arbuscular mycorrhiza, endophytic bacteria and fungi, and plant-health promoting rhizobacteria, combined with improved delivery systems using liquid and solid-state fermentation (Sikora and Fernandez, 2005; Sikora et al., 2005). Better biocontrol potential for both nematodes and fungi could also be achieved through linking biocontrol research with molecular biology to understand how colonization by beneficial mutualists affects gene signaling pathways related to induced systemic resistance in the host (Pieterse et al., 2001).

Disease suppression. Understanding the link between cultural practices that enhance soil health (crop rotation, conservation tillage, etc.) and the phenomena of soil disease suppressiveness would aid in developing alternative approaches to chemical soil fumigation, and could enhance appreciation of local and traditional approaches to managing soilborne diseases. Soil health indicators are needed that are specifically associated with soilborne disease suppression (van Bruggen and Termorshuizen, 2003; Janvier et al., 2007). Given the complex nature of soils, this would necessitate using a holistic, systems approach to develop indicators that could be tested across different soil types and cropping systems. Advances in genomics and molecular biology could aid in developing such indicators. Advances in the application of polymerase chain reaction (PCR)-based molecular methods of soil DNA may enable greater understanding of functional diversity, and relationships between soil microbial communities and root disease suppression linked to soil properties and changes in crop management practices (Alabouvette et al., 2004).

The loss of broad-spectrum biocides, namely methyl bromide, has created opportunities for investigating new directions in managing root diseases. Synthetic substitutes, such as chloropicrin and metam sodium, are generally less effective than methyl bromide, can cause increased germination of nutsedge and others weeds (Martin, 2003), and pose substantial health risks to farm workers and adjacent communities (MMWR, 2004).

Biocontrol of soilborne pests and pathogens will likely continue to succeed on the experimental level, and yet still have only limited impact on field-based commercial applications of biocontrol until impediments to scaling up biocontrol are addressed. These include the exceedingly high costs of registration, and lack of private sector investment (Fravel, 2005). The recent success in scaling up nematode biocontrol using a nonpathogenic strain of *Fusarium oxysporum* to control the highly destructive *Radopholus similis*, causal agent of banana toppling disease (Sikora and Pokasangree, 2004), illustrate how the alignment of multiple factors—a very effective biocontrol agent, a highly visible disease problem with significant economic impact, and substantial private-sector investment—was necessary to allow for development of a potential commercial product.

Long-term and stable organic production systems generally have less severe root disease problems than conventionally managed systems; however, the specific mechanisms that lead to soilborne disease suppression remain poorly understood (van Bruggen and Termorshuizen, 2003). Given that soilborne pests and disease play a role in the productivity dip associated with the transition from conventional to organic production, greater attention towards developing indicators of root disease suppression would help to better address development and sustainability goals.

6.1.4 Value chains, market development

Although reduction of post-harvest losses has been an important focus of AKST and development programs in the past, in many cases the technical innovations faced sociocultural or socioeconomic problems such as low profit margins, additional workload or incompatibility with the existing production or postproduction system (Bell et al., 1999). The divergence between technical recommendations and the realities of rural life translated in many cases into low adoption rates.

In specific cases, large shares of food produced are lost after harvest. Yet, the rationale for improvements in the postharvest systems has been shifting from loss prevention (Kader, 2005) to opening new markets opportunities (Hellin and Higman, 2005). Making markets work for the poor (Ferrand et al., 2004) is emerging as the new rationale of development, reflecting a shift away from governmental operation of postharvest tasks to enabling frameworks for private sector initiatives in this field (Bell et al., 1999).

Research and capacity development needs. Increasing attention is being placed on value and market-chain analysis, upgrading and innovation. Processing, transport and marketing of agricultural products are increasingly seen as a vertical integration process from producers to retailers,

to reduce transaction costs and improve food quality and safety (Chowdhuri et al., 2005).

In *market-chain analysis,* some of the challenges include improving small-scale farmer competitiveness and farmers' organizations (Biénabe and Sautier, 2005); institutional capacity building (especially access to information) (Kydd, 2002); and the reinforcement of links and trust among actors in the market chain (Best et al., 2005).

Demand driven production asks for improved market literacy of producers as a prerequisite for access to supermarkets, a challenge especially for small-scale farmers (Reardon et al, 2004; Hellin et al., 2005). Building trust among the stakeholders in the market chain is a crucial component of vertical integration (Best et al., 2005; Chowdhury et al., 2005; Giuliani, 2007). It enhances transparency of the market chain and exchange of information. Typically, actors in the market chains are at first skeptical about information sharing; when they realize that all can benefit from more transparency along the market chain they more readily provide information. Maximizing added value at farm or village level is a promising option for small-scale farmers; rural agroenterprises and household level processing can increase income generation (Best et al., 2005; Giuliani, 2007).

The creation of community-based organizations or farmers groups can result in economies of scale. Collectively, small-scale farmers are able to pool their resources and market as a group, hence reducing transaction costs (Keizer et al., 2007). It can improve their access to resources such as inputs, credit, training, transport and information, increase bargaining power (Bosc et al., 2002) and facilitate certification and labeling.

Better market access is often a key concern of small-scale farmers (Bernet et al., 2005). Promising market options directly linked to rural poor small-scale producers and processors include fair-trade channels, private-public partnership, and the creation of local niche markets (eco-labeling, certification of geographical indications of origin, tourism-oriented sales outlets, etc.). Crops neglected so far by formal research and extension hold promise for upgrading value chains (Hellin and Highman, 2005; Gruère et al., 2006; Giuliani, 2007) in which small-scale farmers have a comparative advantage.

Value-chain analysis investigates the complexity of the actors involved and how they affect the production-to-consumption process. It incorporates production activities (cultivation, manufacturing and processing), non-production activities (design, finance, marketing and retailing), and governance (Bedford et al., 2001). The analysis of livelihoods of small-scale producers, processors and traders and their current and potential relation to markets is a starting point in ensuring that markets benefit the poor. Analyzing the market chain and the requirements and potentials of all its actors allows for identifying interventions along the chain likely to provide benefits to low-income households (Giuliani, 2007).

Investments in value chain research have the potential to improve equity by opening up income opportunities for small-scale farmers. The challenge will be to make small-scale farmers competitive and to identify opportunities and develop value chains which build on their potential (labor availability, high flexibility). Increasing requirements of the market regarding food quality, safety and traceability will limit small-scale farmer participation in certain value chains. Further, access to market may be limited by inadequate infrastructure, such road systems and refrigerated transport and storage.

Successes in value chain development have been achieved through an extensive consultation processes (Bernet et al., 2005) that generate group innovations based on well-led and well-structured participatory processes. These processes stimulate interest, trust and collaboration among members of the chain. The costs and benefits of such approaches will have to be carefully assessed to determine where investment is justified; e.g., investments for upgrading the market chain could be high compared with potential benefits for niche products with limited market volume.

6.2 Improve Productivity and Sustainability of Livestock Systems

On-farm options
Mixed systems. Mixed crop-livestock systems can contributes to sustainable farming (Steinfeld et al., 1997). Improving the performance of mixed crop-livestock production systems and promoting livestock production, particularly on small-scale farms can be attained by providing access to affordable inputs for small-scale livestock keepers. Along with inputs, adequate knowledge and technologies for on-farm nutrient cycling, on-farm production of feed and fodder, and the use of crop residues and crop by-products, can also provide benefits to small-scale producers.

Intensifying the livestock component in these systems increases the availability of farmyard manure, leading to increased fodder production and increased crop yields. More research is needed on the storage and application of farmyard manure, the conservation of cultivated fodder and crop residues, and the use of crop by-products as animal feed.

Livestock keeping can improve health and nutrition in many small households and generate additional income and employment (ILRI, 2006), even when households have limited resources such as land, labor and capital (PPLPI, 2001; Bachmann, 2004). Output per farm may be small, but the combined effect of many small-scale enterprises can be large, e.g., small-scale dairy in India (Kurup, 2000), piggery in Vietnam (FAO, 2006) and backyard poultry in Africa (Guye, 2000).

Extensive systems. There is little scope for extensive livestock production systems to further extend the area presently being grazed without environmentally unsustainable deforestation (Steinfeld et al., 2006). In some areas even pasture land is decreasing as it is converted into cropland, often resulting in land use conflicts (ECAPAPA, 2005). Where pasture areas with open access remain more or less stable, productivity of land and ultimately of livestock is threatened due to overstocking and overgrazing.

Livestock productivity can be increased through the improvement of pasture and rangeland resources and better animal health. Better animal health may require improved access to veterinary services, such as the establishment of systems of community based animal health workers (Leonard et al., 2003). Feeding conserved fodder and feeds (primarily crop by-products) may help overcome seasonal shortages, while planting fodder trees, more systematic rotational grazing and fencing may improve grazing areas. Tree planting may gain further importance when linked to carbon trade programs. Fencing, on the other hand, may not be socially or culturally acceptable, in particular in areas with communal grazing land (IFAD, 2002). Land use strategies that include participatory approaches are more effective at avoiding conflicts (ECAPAPA, 2005).

Biological complexity and diversity are necessary for survival in traditional pastoral communities (Ellis and Swift, 1988). Long term conservative strategies often work best in traditional systems. The introduction of new breeding techniques (e.g., sexing of sperm straw) might cause a rapid increase in the number of cattle, but may also lead to the disappearance of local breeds and a reduction in the genetic diversity of rustic breeds of cattle, which are well adapted to extreme environments.

The overall potential of pastoral grazing systems is high (Hesse and MacGregor, 2006); the primary issue is the environmental sustainability of these systems (Steinfeld et al., 2006). Hence options to improve productivity must focus more on the application of management than the technology (ILRI, 2006).

Intensive systems. Increasingly, intensive livestock production trade is associated with a fear of contamination of air and water resources (de Haan et al., 1997; FAO, 2006). Future systems will need to consider human health aspects as well as the whole livestock food value chain (fodder and animal feed production, processing and marketing of products, etc). Since cross-regional functions such as assembly, transport, processing and distribution can cause other externalities, they must be assessed as part of an integrated system. Intensive systems are prone to disease and animals can spread zoonotic diseases like tuberculosis or bird flu that can affect humans (LEAD, 2000).

Improvements in intensive livestock production systems include locating units away from highly populated areas, and using management practices and technologies that minimize water, soil and air contamination.

6.3 Breeding Options for Improved Environmental and Social Sustainability

6.3.1 Crop breeding
Climate change coupled with population growth will produce unprecedented stress on food security. Abiotic stresses such as drought and salinity may reduce yields worldwide by up to 50% (Jauhar, 2006). Increasing demand cannot always be met by increasing the land devoted to agriculture (Kumar, 2006), however, it may be possible to improve plant productivity. Traits that are the focus of abiotic stress resistance include optimized adaptation of temperature-dependent enzymes (to higher or lower temperatures), altering day-length regulation of flower and fruit development, optimization of photosynthesis including circumventing inherent limitations in C_3 and C_4 pathways in plants (Wenzel, 2006).

6.3.1.1 Options for conventional plant breeding

The following options apply to plant breeding to help meet world demand for nutrition and higher yields in low external input production systems and lower resource demands in high external input production systems. However useful these innovations might be, biotechnology *per se* cannot achieve development and sustainability goals. Therefore, it is critical for policy makers to holistically consider biotechnology impacts beyond productivity goals, and address wider societal issues of capacity building, social equity and local infrastructure.

Modern, conventional and participatory plant breeding approaches play a significant role in the development of new crop varieties (Dingkuhn et al., 2006). The exodus of a specialist workforce in plant breeding (Baenziger et al., 2006), especially from the public sector, is a worrisome trend for maintaining and increasing global capacity for crop improvement. Critical to improved plant breeding is ensuring the continuity of specialist knowledge in plant breeding. Approaches that encourage research in the field and continuity of career structure for specialists are key to the continuation of conventional plant breeding knowledge.

There is a need for new varieties of crops with high productivity in current and emerging marginal and unfavorable (e.g., water stressed) environments; resource limited farming systems; intensive land and resource use systems; areas of high weed pressure (Dingkuhn et al., 2006); and bioenergy. Ensuring access to locally produced high-quality seeds and to opportunities for farmer-to-farmer exchanges will improve productivity, decrease poverty and hunger, encourage retention of local knowledge, safeguard local intellectual property, and further exploit the biological diversity of crop wild relatives.

Plant breeding is facilitating the creation of new genotypes with higher yield potentials in a greater range of environments (Dingkuhn et al., 2006; Hajjar and Hodgkin, 2007) mainly through recruiting genes from within the gene pool of interbreeding plants and also through biotechnology assisted hybridization and tissue regeneration (Wenzel, 2006).

Crop biodiversity is maintained both through *ex situ* and *in situ* conservation in the genomes of plants from which crops were derived, and in the genomes of crop relatives (Brush and Meng, 1998). The value of traits sourced from wild relatives has been estimated at US$340 million to the US economy every year (Hajjar and Hodgkin, 2007). Traits such as pest and disease resistance are usually determined by single genes. Wild relatives have so far contributed modestly as a source of genes for introduction of multigene traits, such as abiotic stress tolerances, but there is considerable diversity still to be tapped (Hajjar and Hodgkin, 2007).

In developing countries, public plant breeding institutions are common but their continued existence is threatened by globalization and privatization (Maredia, 2001; Thomas, 2005). Plant breeding activities differ between countries; public investment in genetic improvement may benefit from research units that include local farming communities (Brush and Meng, 1998). Moreover, differences in intellectual property protection philosophies could endanger *in situ* conservation as a resource for breeding. For example, patent protection and forms of plant variety protection place a greater value on the role of breeders than that of local communities that maintain gene pools through *in situ* conservation (Srinivasan, 2003).

Options for strengthening conservation in order to preserve plant genetic diversity include:

- Integrating material on the importance of biodiversity into curricula at all educational levels;
- Channeling more resources into public awareness at CGIAR and NGO system level;
- Facilitating national programs to conduct discussions with farmers about the long-term consequences of losing agrobiodiversity;
- Studying and facilitating the scaling up of indigenous agroecosystems that feature a high degree of agrobiodiversity awareness;
- Involving farmers in a fully participatory manner in research focused on agrobiodiversity conservation;
- Undertaking surveys of farmers and genebanks to establish which communities want their landraces back, and to find out if the landrace is still maintained in a genebank;
- Developing sustainable reintroduction campaigns;
- Developing a system whereby genebanks regenerate landraces and maintained them in farmers' fields: a hybrid *in situ* and *ex situ* conservation system;
- Involving farmers in the characterization of landraces to increase exposure and possible utilization of the material at farm level;
- Promoting the development of registration facilities that recognize a given landrace as the indigenous property of a particular area or village to enhance the importance of the landrace as an entity that is a part of local heritage;
- Developing and promoting viable and sustainable multistakeholder incentive schemes for communities who maintain local material in their agroecosystem.

Provided that steps are taken to maintain local ownership and control of crop varieties, plant breeding remains a viable option for meeting development and sustainability goals. It will be important to find a balance between exclusive access secured through intellectual property (IP) mechanisms and the need for local farmers and researchers to develop locally adapted varieties (Srinivasan, 2003; Cohen, 2005). An initial approach could include facilitating NGOs to help develop the capacity of local small-scale farmers, and providing farmer organizations with advisers to guide their investments in local plant improvement.

6.3.1.2 Optimize the pace and productivity of plant breeding

Biotechnology and associated nanotechnologies provide tools that contribute toward the achievement of development and sustainability goals. Biotechnology has been described as the manipulation of living organisms to produce goods and services useful to human beings (Eicher et al., 2006; Zepeda, 2006). In this inclusive sense, biotechnology includes traditional and local knowledge (TK) and the contributions to cropping practices, selection and breeding made by individuals and societies for millennia (Adi, 2006); it would also include the application of genomic techniques and marker-assisted breeding or selection (MAB or MAS). Modern biotechnology includes what arises from the use of in vitro modified genes. Most obvious in this category is genetic engineering, to create genetically modified/engineered organisms (GMOs/GEOs) through transgenic technology by insertion or deletion of genes.

Combining plants with different and desirable traits can be slow because the genes for the traits are located in many different places in the genome and may segregate separately during breeding. Breeding augmented by molecular screening may yield rapid advances in existing varieties. This process, however, is limited by breeding barriers or viability in the case of cell fusion approaches, and there may be a limit to the range of traits available within species to existing commercial varieties and wild relatives. In any case, breeding is still the most promising approach to introducing quantitative trait loci (Wenzel, 2006). Emerging genomics approaches are showing promise for alleviating both limitations.

Genomics. Whole genome analysis coupled with molecular techniques can accelerate the breeding process. Further development of approaches such as using molecular markers through MAS will accelerate identification of individuals with the desired combinations of genes, because they can be rapidly identified among hundreds of progeny as well as improve backcross efficiencies (Baenziger et al., 2006; Reece and Haribabu, 2007). The range of contributions that MAS can make to plant breeding are being explored and are not exhausted (e.g., Kumar, 2006; Wenzel, 2006). It thus seems reasonable that MAS has the potential to contribute to development and sustainability goals in the long term, provided that researchers consistently benefit from funding and open access to markers. MAS is not expected to make a significant improvement to the rate of creating plants with new polygenic traits, but with future associated changes in genomics this expectation could change (Baenziger et al., 2006; Reece and Haribabu, 2007).

Regardless of how new varieties are created, care needs to be taken when they are released because they could become invasive or problem weeds, or the genes behind their desired agronomic traits may introgress into wild plants threatening local biodiversity (Campbell et al., 2006; Mercer et al., 2007).

MAS has other social implications because it favors centralized and large scale agricultural systems and thus may conflict with the needs and resources of poor farmers (Reece and Haribabu, 2007). However, breeding coupled to MAS for crop improvement is expected to be easily integrated into most regulatory frameworks and meet little or no market resistance, because it does not involve producing transgenic plants (Reece and Haribabu, 2007). Varieties that are developed in this fashion can be covered by many existing IP rights instruments (e.g., Baenziger et al., 2006; Heinemann, 2007) and would be relatively easy for farmers to experiment with under "farmers' privilege" provided that suitable *sui generis* systems are in place (Sechley and Schroeder, 2002; Leidwein, 2006). The critical limitation of MAS is its ultimate dependence on plant breeding specialists to capture the value of new varieties; unfortunately, current and projected numbers of these specialists is inadequate (Reece and Haribabu, 2007).

Transgenic (GM) plants. Recombinant DNA techniques allow rapid introduction of new traits determined by genes that are either outside the normal gene pool of the species or for which the large number of genes and their controls would be very difficult to combine through breeding. An emphasis on extending tolerance to both biotic (e.g., pests) and abiotic (e.g., water stress) traits using transgenes is relevant to future needs.

Assessment of transgenic (GM) crops is heavily influenced by perspective. For example, the number of years that GM crops have been in commercial production (approximately 10 years), amount of land under cultivation (estimated in 2007 at over 100 million ha) and the number of countries with some GM agriculture (estimated in 2007 at 22) (James, 2007) can be interpreted as evidence of their popularity. Another interpretation of this same data is that the highly concentrated cultivation of GM crops in a few countries (nearly three-fourths in only the US and Argentina, with 90% in the four countries including Brazil and Canada), the small number of tested traits (at this writing, mainly herbicide and pest tolerance) and the shorter-term experience with commercial GM cultivation outside of the US (as little as a year in Slovakia) (James, 2007), indicate limited uptake and confidence in the stability of transgenic traits (Nguyen and Jehle, 2007).

Whereas there is evidence of direct financial benefits for farmers in some agriculture systems, yield claims, adaptability to other ecosystems and other environmental benefits, such as reduced alternative forms of weed and pest control chemicals, are contested (Pretty, 2001; Villar et al., 2007), leaving large uncertainties as to whether this approach will make lasting productivity gains. The more we learn about what genes control important traits, the more genomics also teaches us about the influence of the environment and genetic context on controlling genes (Kroymann and Mitchell-Olds, 2005; MacMillan et al., 2006) and the complexity of achieving consistent, sustainable genetic improvements. Due to a combination of difficult to understand gene by environment interactions and experience to date with creating transgenic plants, some plant scientists are indicating that the rate at which transgenic plants will contribute to a sustained increase in future global food yields is exaggerated (Sinclair et al., 2004).

Adapting any type of plant (whether transgenic or conventionally bred) to new environments also has the potential to convert them into weeds or other threats to food and materials production (Lavergne and Molofsky, 2007; Heinemann, 2007). This problem is particularly relevant to transgenes because (1) they tend to be tightly linked packages in genomes, making for efficient transmission by breeding (unlike many traits that require combinations of chromosomes to be inherited simultaneously), and (2) the types of traits of most relevance to meeting development and sustainability goals in the future are based on genes that adapt plants to new environments (e.g., drought and salt tolerance). Through gene flow, wild relatives and other crops may become more tolerant to a broader climatic range and thus further threaten sustainable production (Mercer et al., 2007). An added complication is that these new weeds may further undermine conservation efforts. The emergence of a new agricultural or environmental weed species can occur on a decade (or longer) scale. For example, it can take hundreds of years for long-lived tree species to achieve populations large enough to reveal their invasive qualities (Wolfenbarger and Phifer, 2000). These realities increase uncertainty in long term safety predictions.

Transgene flow also creates potential liabilities (Smyth et al., 2002). The liability is realized when the flow results in traditional, economic or environmental damage (Kershen, 2004; Heinemann, 2007). Traditional damage is harm to human health or property. Economic damage could occur if a conventional or organic farmer lost certification and therefore revenue because of adventitious presence. Environmental damage could result from, for example, harm to wildlife.

There are a limited number of properly designed and independently peer-reviewed studies on human health (Domingo, 2000; Pryme and Lembcke, 2003). Among the studies that have been published, some have provided evidence for potential undesirable effects (Pryme and Lembcke, 2003; Pusztai et al., 2003). Taken together, these observations create concern about the adequacy of testing methodologies for commercial GM plants fueling public skepticism and the possibility of lawsuits. A class-action lawsuit was filed by USA consumers because they may have inadvertently consumed food not approved for human consumption (a GM variety of maize called Starlink) because of gene flow or another failure of segregation. The lawsuit ended with a settlement against the seed producer Aventis. This suggests that consumers may have grounds for compensation, at least in the USA, even if their health is not affected by the transgenic crop (Kershen, 2004).

Farmers, consumers and competitors may be the source of claims against, or the targets of claims from, seed producers (Kershen, 2004; Center for Food Safety, 2005; Eicher et al., 2006). For example, when non-GM corn varieties from Pioneer Hi-Bred were found in Switzerland to contain novel Bt genes, the crops had to be destroyed, and compensation paid to farmers (Smyth et al., 2002).

Even if liability issues could be ignored, the industry will remain motivated to track transgenes and their users because the genes are protected as IP. Transgene flow can create crops with mixed traits because of "stacking" (two transgenes from different owners in the same genome) or mixed crops (from seed mediated gene flow or volunteers), resulting in potential IP conflicts. IP protection includes particular genes and plant varieties as well as techniques for creating transgenic plants and product ideas, such as the use of Bt-sourced Cry toxins as plant-expressed insecticides. Broad IP claims are creating what some experts call "patent thickets"; the danger of thickets is that no single owner can possess all the elements in any particular transgenic plant (Thomas, 2005).

Release of insect resistant GM potatoes in South Africa illustrates the complexity that IP and liability create for transgenic crops. The potato has elements that are claimed by two different companies. One of the IP owners has been unwilling to license the IP to South Africa for fear of liability should the potatoes cross into neighboring countries (Eicher et al., 2006).

The harms associated with transgene flow might be addressed by a combination of physical and biological strategies for containment (for a comprehensive list, see NRC, 2004). However, no single method and possibly no combi-

nation of methods would be wholly adequate for preventing all flow even though for some genes and some environments, flow might be restricted to acceptable levels (Heinemann, 2007). Future strategies for containment involving sterilization (i.e., genetic use restriction technologies, GURTs) remain highly controversial because of their potential to cause both unanticipated environmental harm and threaten economic or food security in some agroeconomic systems (Shand, 2002; Heinemann, 2007).

For transgenic approaches to continue to make significant contributions in the long term, a substantial increase in public confidence in safety assessments will be needed (Eicher et al., 2006; Herrero et al., 2007; Marvier et al., 2007); conflicts over the free-use of genetic resources must be resolved; and the complex legal environment in which transgenes are central elements of contention will need further consideration.

Epigenetic modification of traits. Epigenes are defined as units of inheritance that are not strictly based on the order of nucleotides in a molecule of DNA (Strohman, 1997; Heinemann and Roughan, 2000; Gilbert, 2002; Ashe and Whitelaw, 2007; Bird, 2007). A growing number of traits are based on epigenetic inheritance, although at present most of these are associated with disease, such as Mad Cow Disease and certain forms of cancer.

In the future, it may be possible to introduce traits based on epigenes. For example, double-stranded RNA (dsRNA) is the basis of at least two commercial transgenic plants and is proposed for use in more (Ogita et al., 2003; Prins, 2003). Small dsRNA molecules appear to be the basis for the trait in "flavr savr" tomatoes—even though at the time of development the epigenetic nature of the modification was probably not known or fully understood (Sanders and Hiatt, 2005)—and the basis for viral resistance in papaya (Tennant et al., 2001). In these cases, the epigene is dependent upon a corresponding change at the DNA level, but in time it will be possible to use the epigenetic qualities of dsRNA to infectiously alter traits without also altering the DNA content of the recipient genome using rDNA techniques. Such promise has already been demonstrated using nematodes where feeding, or soaking the worm in a liquid bath of dsRNA, was sufficient for systemic genetic modification of the worm and the stable transmission of the epigene for at least two generations (Fire et al., 1998; Cogoni and Macino, 2000). The effects of dsRNA also can be transmitted throughout a conventional plant that has been grafted with a limb modified to produce dsRNA (Palauqui et al., 1997; Vaucheret et al., 2001; Yoo et al., 2004).

RNA-based techniques will accelerate research designed to identify which genes contribute to complex traits and when and where in the organisms those genes are expressed ("turned on"). Generally, dsRNA causes transient, long-term, sometimes heritable gene silencing (turns genes "off"). While silencing that occurs by the general pathways controlled by dsRNA molecules are targeted to sequence matches between the dsRNA and the silenced genes, there are often effects on nontarget genes as well. The number of genes simultaneously silenced by a single dsRNA (including the targets) can number in the hundreds (Jackson et al.,

2003; Jackson and Linsley, 2004; Jackson et al., 2006), and a variety of dsRNAs with no sequence similarity can silence the same genes (Semizarov et al., 2003).

Once established, the effects of dsRNA may persist in some kinds of organisms, being transmitted to offspring. The instigating event is the initial combination of genetic elements with similar DNA sequences, but the silencing effect may persist even in hybrids that retain a single copy of the gene.

Furthermore, not all genes that are silenced remain so, nor are all plants grafted with tissues from silenced plants capable of acquiring the silenced phenotype. The science of infectious gene silencing is still young, leaving gaps in understanding how the molecules are transmitted and maintained, and in how the phenotype is regulated or reversed. If this or other epigenetic strategies for genetic modification are in time adopted, they must benefit from fundamentally new kinds of safety assessments in both their environmental and human health context. Importantly, these assessments should be conducted by competent researchers that are independent of the developing industry.

6.3.2 Livestock breeding options

Technologies such as artificial insemination and embryo transfer, which are routine in industrialized countries have been successfully transferred and introduced in other parts of the world (Wieser et al., 2000). However, breeding technologies are not exploited to the extent possible because animals are not adapted to local conditions, logistical problems and poor support for breeding services and information management (Ahuja et al., 2000). There is scope to further develop conventional breeding technologies, in particular through North-South cooperation. To be effective at meeting development goals breeding policies, programs and plans need to be location specific (Kurup, 2003; Chacko and Schneider, 2005).

Thus far the impact of genomics in livestock agriculture is limited to the use of transgenic animals such as chickens and cattle to produce pharmaceutical or therapeutic proteins in eggs and milk (Gluck, 2000). Genomics for diagnostics and animal vaccine development, and in feed production and formulation (Machuka, 2004) may further boost the livestock industry, although the competition from alternative sources will probably be strong (Twyman et al., 2003; Chen, 2005; Ma et al., 2005). Moreover, all these new technologies create safety risks and may not always increase sustainable production. Hence, applications should be thoroughly evaluated to ensure that they do not also undermine development and sustainability goals.

There are currently no transgenic food animals in commercial production and none likely in the short term (van Eenennaam, 2006). Over the next 10-50 years there is some potential for development and introduction of transgenic animals or birds with disease resistance, increased or higher nutritional value meat or milk production, or as biofactories for pharmaceuticals (Machuka, 2004). The science and technology is available, but the barriers include regulatory requirements, market forces and IP, safety concerns and consumer acceptance, i.e., the same range of issues as described for crops (Powell, 2003; van Eenennaam, 2006; van Eenennaam and Olin, 2006).

Responding to the increased demand for livestock products without additional threats to the environment is a major challenge for agriculture and for AKST. One option for satisfying the additional demand for animal protein is to use meat from monogastric animals (pigs and poultry) and eggs. Feed conversion rates and growth for monogastric animals are better than for ruminants, which is one reason why the increasing demand for meat tends to be met with chicken and pork. This development may be positive with regard to the direct pressure on (grazing) land caused by ruminants, but has resulted in the establishment of large pig and poultry production units which are often placed in peri-urban areas. Large volumes of animal feed are produced elsewhere and transported, while disposal of waste from these large units has become an environmental issue (FAO, 2006). Although these large livestock farms may generate some employment opportunities, the capital required excludes most small-scale farmers. One approach to increase the total efficiency and sustainability of the intensive livestock production system is area-wide integration, i.e., the integration of production with cropping activities. The main objective is to link these specialized activities on a regional scale to limit their environmental damage and enhance social benefits (LEAD, 2000).

Recent outbreaks of diseases, including some that threaten human as well animal health, highlight the need to scrutinize large livestock units and their sustainability in wider terms with regard to environment and health (Steinfeld et al., 2006).

For small-scale farmers in rural areas, local markets will remain the primary outlets for their products. These local markets may also provide opportunities for processed products. However, processing of meat and livestock products into high value niche produces for distant markets might be economically attractive. Some associated risks include the required investment in marketing for a successful enterprise may decrease the "additional" product value. In addition, rural processors may not be able to meet the quality standards to compete for distant urban or export markets (ILRI, 2006).

Further extension of grazing land to produce meat from ruminants is not a sustainable way to meet the growing demand for meat and livestock products (Steinfeld et al., 2006). Therefore, pastoralists and rangelands livestock keepers will only benefit from an increased demand for livestock products if they are able to improve their present production systems by efficient use of existing resources, i.e., breed improvement (Köhler-Rollefson, 2003) improvement of animal health and disease control (Ramdas and Ghotge, 2005), of grazing regime and pasture management, including the planting of fodder trees, and if possible supplementary feeding during times of limited grazing. Where there is potential for mixed farming, policies need to facilitate the transition of grazing systems into mixed farming systems in the semiarid and subhumid tropics through integrating crops and livestock (Steinfeld et al., 1997).

6.4 Improve Forestry and Agroforestry Systems as Providers of Multifunctionality

6.4.1 On-farm options

The ecological benefits of low-input agroforestry systems are more compatible with small-scale tropical/subtropical farming systems than for large farms. However, the coincidence of land degradation and poverty is also greatest in the tropics and subtropics and there is therefore considerable relevance of agroforestry for the attainment of development and sustainability goals. Disseminating and implementing a range of agroforestry practices, tailored to particular social and environmental conditions, on a wide scale will require large-scale investment in NARS, NARES, NGOs and CBOs, with support from ICRAF and regional agroforestry centers. Rehabilitation of degraded land and improving soil fertility can be accomplished by promoting a range of ecological/environmental services such as: (1) erosion control, (2) nutrient cycling, (3) protection of biodiversity in farming systems, (4) carbon sequestration, (5) promoting natural enemies of pests, weeds and diseases, (6) improving water availability, and (7) the restoration of agroecological function.

Agroforestry practices can also improve soil fertility in the future, which is crucial for achieving food security, human welfare and preserving the environment for smallholder farms (Sanchez, 2002; Oelberman et al., 2004; Schroth et al., 2004, Jiambo, 2006; Rasul and Thapa, 2006). An integrated soil fertility management approach that combines agroforestry technologies—especially improved fallows of leguminous species and biomass transfer—with locally available and reactive phosphate rock (e.g., Minjingu of northern Tanzania) can increase crop yields severalfold (Jama et al., 2006).

Tree crops can be established within a land use mosaic to protect watersheds and reduce runoff of water and erosion restoring ecological processes as the above- and below-ground niches are filled by organisms that help to perform helpful functions such as cycle nutrients and water (Anderson and Sinclair, 1993), enrich organic matter, and sequester carbon. (Collins and Qualset, 1999; McNeely and Scherr, 2003; Schroth et al., 2004). Many of these niches can be filled by species producing useful and marketable food and nonfood products, increasing total productivity and economic value (Leakey, 2001ab; Leakey and Tchoundjeu, 2001). A healthier agroecosystem should require fewer purchased chemical inputs, while the diversity alleviates risks for small-scale farmers. On large mechanized farming systems the larger-scale ecological functions associated with a land use mosaic can be beneficial.

As the science and practice of agroforestry are complex and comprise a range of disciplines, communities and institutions, strengthening strategic partnerships and alliances (farmers, national and international research organizations, government agencies, development organizations, NGOs, ICRAF, CIFOR, The Forest Dialogue, etc.) is crucial in order to foster the role of agroforestry in tackling future challenges. Local participation could be mobilized by incorporating traditional knowledge and innovations, as well as ensuring the scaling up and long-term sustainability of agroforestry.

Rights to land and trees tend to shape women's incentives and authority to adopt agroforestry technologies more than other crop varieties because of the relatively long time horizon between investment and returns (Gladwin et al., 2002). Agroforestry systems have high potential to help AKST achieve gender equity in property rights. This is especially true in customary African land tenure systems where planting or clearing trees is a means of establishing claims, on the trees, but also on the underlying land (Gari, 2002; Villarreal et al., 2006).

Reducing land degradation through agroforestry. Land degradation is caused by deforestation, erosion and salinization of drylands, agricultural expansion and abandonment, and urban expansion (Nelson, 2005). Data on the extent of land degradation are extremely limited and paradigms of desertification are changing (Herrmann and Hutchinson, 2005). Approximately 10% of the drylands are considered degraded, with the majority of these areas in Asia and Africa.

In all regions more threatened by deforestation, like the humid tropics, Latin America, Southeast Asia, and Central Africa, deforestation is primarily the result of a combination of commercial wood extraction, permanent cultivation, livestock development, and the extension of overland transport infrastructure (Zhang et al., 2002; Vosti et al., 2003; Nelson, 2005). Decreasing current rates of deforestation could be achieved by promoting alternatives that contribute to forest conservation. Methods may include improving forest management through multiple-use policies in natural forests and plantations of economic (cash) trees within forests (Wenhua, 2004) off-farm employment (Mulley and Unruh, 2004); and implementing an industrial development model, based on high-value added products.

Sustainable timber management implies ensuring forests continue to produce timber in long-term, while maintaining the full complement of environmental services and non-timber products of the forest. Although sustainable timber management sometimes provides reasonable rates of return, additional incentives are often needed as conventional timber harvesting is generally more profitable (Pearce and Mourato, 2004). Effective use of AKST supported by sustainable policy and legal systems and sufficient capacity is needed; the Chinese government's forest management plan implemented in 1998 offers a working example (Wenhua, 2004). However, local authorities are often inefficient in monitoring and enforcing environmental laws in large regions, as in Brazilian Amazonia where the construction of highways and the promotion of agriculture and cattle ranching facilitated the spread of deforestation. Off-farm employment can contribute significantly to forest conservation in the tropics, e.g., the tea industry in western Uganda (Mulley and Unruh, 2004).

6.4.2 Market mechanisms and incentives for agroforestry

Agroforestry is a method by which income can be generated by producing tree products for marketing as well as domestic use. There are many wild tree species that produce traditionally important food and nonfood products (e.g., Abbiw, 1990). These species can be domesticated to improve their quality and yield and to improve the uniformity of marketed products (Leakey et al, 2005) and enhance farmers' livelihoods (Schreckenberg et al., 2002; Degrande et al., 2006). Domestication can thus be used as an incentive for more sustainable food production, diversification of the rural economy, and to create employment opportunities in product processing and trade. The domestication of these species previously only harvested as extractive resources, creates a new suite of cash crops for smallholder farmers (Leakey et al., 2005). Depending on the market size, some of these new cash crops may enhance the national economies, but at present the greatest benefit may come from local level trade for fruits, nuts, vegetables and other food and medicinal products for humans and animals, including wood for construction, and fuel.

This commercialization is crucial to the success of domestication, but should be done in ways that benefit local people and does not destroy their tradition and culture (Leakey et al., 2005). Many indigenous fruits, nuts and vegetables are highly nutritious (Leakey, 1999b). The consumption of some traditional foods can help to boost immune systems, making these foods beneficial against diseases, including HIV/AIDS (Barany et al., 2003; Villarreal et al., 2006). These new nonconventional crops may play a vital role in the future for conserving local and traditional knowledge systems and culture, as they have a high local knowledge base which is being promoted through participatory domestication processes (Leakey et al., 2003; World Agroforestry Centre, 2005; Garrity, 2006; Tchoundjeu et al., 2006). Together these strategies are supportive of food sovereignty and create an approach to biodiscovery that supports the rights of farmers and local communities specified in the Convention on Biological Diversity.

A participatory approach to the domestication of indigenous trees is appropriate technology for rural communities worldwide (Tchoundjeu et al., 2006), especially in the tropics and subtropics, with perhaps special emphasis on Africa (Leakey, 2001ab), where the Green Revolution has been least successful. In each area a priority setting exercise is recommended to identify the species with the greatest potential (Franzel et al., 1996). Domestication should be implemented in parallel with the development of postharvest and value-adding technologies and the identification of appropriate market opportunities and supply chains. With poverty, malnutrition and hunger still a major global problem for about half the world population, there is a need to develop and implement a range of domestication programs for locally-selected species, modeled on that developed by ICRAF and partners in Cameroon/Nigeria (Tchoundjeu et al., 2006), on a wide scale. There will also be a need for considerable investment in capacity development in the appropriate horticultural techniques (e.g., vegetative propagation and genetic selection of trees) at the community level, in NARS, NARES, NGOs and CBOs, with support from ICRAF and regional agroforestry centers.

Agroforestry can be seen as a multifunctional package for agriculture, complemented by appropriate social sciences, rural development programs and capacity development. Better land husbandry can rehabilitate degraded land. For many poor farmers this means the mitigation of soil nutrient depletion by biological nitrogen fixation and the simultaneous restoration of the agroecosystem using low-input, easily-adopted practices, such as the diversification of the farming system with tree crops that initiate an agroecological succession and produce marketable products.

Over the last 25 years agroforestry research has provided some strong indications on how to go forward by replanting watersheds, integrating trees back into the farming systems to increase total productivity, protecting riparian strips, contour planting, matching tree crops to vulnerable landscapes, soil amelioration and water harvesting. There are many tree species indigenous to different ecological zones, that have potential to play these important roles, and some of these are currently the subject of domestication programs. In this way, the ecological services traditionally obtained by long periods of unproductive fallow are provided by productive agroforests yielding a wide range of food and nonfood products. This approach also supports the multifunctionality of agriculture as these species and products are central to food sovereignty, nutritional security and to maintenance of tradition and culture. Additionally, women are often involved in the marketing and processing of these products. Consequently this approach, which brings together AST with traditional and local knowledge, provides an integrated package which could go a long way towards meeting development and sustainability goals. The challenge for the development of future AKST is to develop this "Localization" package (Chapter 3.2.4; 3.4) on a scale that will have the needed impacts.

This integrated package is appropriate for large-scale development programs, ideally involving private sector partners (building on existing models—e.g., Panik, 1998; Mitschein and Miranda, 1998; Attipoe et al., 2006). Localization is the grassroots pathway to rural development, which has been somewhat neglected in recent decades dominated by Globalization. Programs like that proposed would help to redress the balance between Globalization and Localization, so that both pathways can play their optimal role. This should increase benefit flows to poor countries, and to marginalized people. There would be a need for considerable investment in capacity development in the appropriate horticultural and agroforestry techniques (e.g., vegetative propagation, nursery development, domestication and genetic selection of trees) at the community level, in NARS, NARES, NGOs and CBOs, with support from ICRAF and regional agroforestry centers.

By providing options for producing nutritious food and managing labor, generating income, agroforestry technologies may play a vital role in the coming years in helping reduce hunger and promote food security (Thrupp, 1998; Cromwell, 1999; Albrecht and Kandji, 2003; Schroth et al., 2004; Oelberman et al., 2004; Reyes et al., 2005; Jiambo, 2006; Rasul and Thapa, 2006; Toledo and Burlingame, 2006).

Recent developments to domesticate traditionally important indigenous trees are offering new opportunities to enhance farmer livelihoods in ways which traditionally provided household needs (especially foods) as extractive resources from natural forests and woodlands (Leakey et. al., 2005; Schreckenberg et al., 2002). These new non-conventional crops may play a vital role in the future for conserving local and traditional knowledge systems, as they have a high local knowledge base which is being promoted through participatory domestication processes (Leakey et al., 2003; World Agroforestry Centre, 2005; Garrity, 2006)

6.5 Sustainable Management of Fishery and Aquaculture Systems

Globally, fisheries products are the most widely traded foods, with net exports in 2002 providing US$17.4 billion in foreign exchange earnings for developing countries, a value greater than the combined net exports of rice, coffee, sugar, and tea (FAO, 2002). In spite of the important role that fisheries play in the national and local economies of many countries, fisheries around the globe are frequently overfished and overexploited as a result of not only weak governance, but of poor management, non-selective technology, perverse subsidies, corruption, unrestricted access and destructive fishing practices (FAO, 2002; World Bank, 2004). Reforming both the governance and management of these critical natural resources is essential to stable and long term economic development, future food security, sustainable livelihoods, poverty prevention and reduction, continuation of the ecosystem goods and services provided by these natural resources, and the conservation of biodiversity (Fisheries Opportunity Assessment, 2006; Christie et al., 2007; Sanchirico and Wilen, 2007).

Governance and management options

In most cultures, wild fisheries and marine resources are considered as common property and suffer from open, unregulated access to these valuable resources. The concept of land tenure and property rights has been instrumental in reforming terrestrial agriculture and empowering small-scale farmers. Similarly, the concepts of marine tenure and access privileges are needed to address the "wild frontier" attitude generated by open access to fisheries and to promote shared responsibilities and comanagement of resources (Pomeroy and Rivera-Guieb, 2006; Sanchirico and Wilen, 2007). Several traditional management approaches, such as in the Pacific Islands, have evolved that are based upon the concept of marine tenure.

For fisheries, major goals of zoning are to (1) protect the most productive terrestrial, riparian, wetland and marine habitats which serve as fisheries nurseries and spawning aggregation sites, and (2) allocate resource use—and thus stewardship responsibility—to specific users or user groups. Appropriate zoning would allow for the most sustainable use of various habitats types for capture fisheries, aquaculture, recreation, biodiversity conservation and maintenance of ecosystem health. Future zoning for specific uses and user groups would also shift shared responsibility onto those designated users, thus increasing self-enforcement and compliance (Sanchirico and Wilen, 2007). The greatest benefit would be in those countries where government, rule of law and scientific management capacity is weak.

Improving fisheries management is critical for addressing food security and livelihoods in many developing countries, where fishing often serves as the last social safety net for poor communities and for those who have no land tenure rights. Fisheries has strong links to poverty—at least 20% of those employed in fisheries earn less than US$1 per day—and children often work in the capture and/or processing sectors, where they work long hours under dangerous conditions.

Tenure and access privileges. Large-scale social and ecological experiments are needed to implement culturally appropriate approaches to marine tenure and access privileges that can be applied to both large-scale industrialized fisheries and small-scale artisanal fisheries (Fisheries Opportunity Assessment, 2006; Pomeroy and Rivera-Guieb, 2006). Rights-based or privilege-based approaches to resource access can alter behavioral incentives and align economic incentives with conservation objectives (Sanchirico and Wilen, 2007).

Seascape "zoning". As in terrestrial systems, zoning would protect essential and critical fisheries habitats that are necessary for "growing" fisheries populations and maintaining ecosystem health. The science of large-scale planning is relatively young and further research and implementation is needed. Future zoning should allow for the most sustainable use of various marine habitat types for capture fisheries, low trophic level aquaculture, recreation, biodiversity conservation and maintenance of ecosystem health. Ultimately, integrating landscape and seascape use designs are needed to conserve and protect ecosystem goods and services, conserve soils, reduce sedimentation and pollution runoff, protect the most productive terrestrial, wetlands and marine habitats, and promote improved water resources management.

Socioeconomic and environmental scenarios could be developed that explore the potential tradeoffs and benefits from applying different management regimes to improve wild fisheries management. Scenarios can guide the application of science to management decisions for reforming fisheries governance, both large-scale and small-scale fisheries, and incorporate cultural and traditional knowledge (Fisheries Opportunity Assessment, 2006; Philippart et al., 2007). The Locally Managed Marine Areas (LMMAs) approach in the Pacific builds upon cultural practices of setting aside specific areas as off-limits to fishing for rebuilding fisheries and biodiversity (www.LMMAnetwork.org).

Ecosystem-based management approaches focus on conserving the underlying ecosystem health and functions, thus maintaining ecosystem goods and services (Pikitch et al., 2004). Developing these approaches requires an understanding of large-scale ecological processes; identifying critical fisheries nurseries, habitats and linkages between habitats, such as between mangrove forests and coral reefs; understanding freshwater inflows into coastal estuaries and maintaining the quantity, quality and timing of freshwater flows that make wetlands some of the most productive ecosystems in the world; and how human activities, such as fishing, affects ecosystem function (Bakun and Weeks, 2006; Hiddinks et al,. 2006; Lotze et al., 2006; Olsen et al., 2006; www.worldfishcenter.org). Ecosystem based fisheries management also requires protection of essential fish habitats and large-scale regional use planning.

Ecosystem based fisheries management approaches are relatively new management tools. Given the ecological complexity of ecological systems, especially the tropical systems in many developing countries, the application of Ecosystem based fisheries management needs to be further developed and assessed. Major governance and ecological challenges exist as management is scaled up in geographic area. Institutional, governance and environmental challenges will require monitoring, evaluation and adaptive management (Christie et al., 2007).

Fisheries reserves. The design and establishment of networks of fisheries reserves are necessary to improve and protect fisheries productivity, as well as improve resilience in the face of climate change and increasing variability. Well-designed and placed fisheries reserves, which restrict all extractive uses, are needed to rebuild severely depleted ecosystems and fisheries and to serve as "insurance" against future risks; however, critical science gaps will need to be addressed before fishery reserves can be effectively utilized (Gell and Roberts, 2003).

Multispecies approaches. The concept of "maximum sustainable yield" and managing by a species-by-species or population-by-population approach has not proved effective for fisheries management given the complexity of ecosystems and foodwebs. Overfishing and "fishing down the food web" has occurred, seriously threatening the future productivity of wild fisheries (Pauly et al., 2005). Non-linear, multispecies models which incorporate trophic levels, reproductive potential and "maximum economic yield" need to be developed and applied for determining more sustainable levels, types and sizes of fish extracted (Pauly and Adler, 2005).

Environmentally friendly extraction technologies. New technology is needed that selectively removes target species and size classes, thus reducing wasteful "bycatch", allowing nonreproductive individuals to reach maturity, and protecting large individuals that disproportionately contribute to the next generation (Hsieh et al., 2006). Some advocate that destructive fishing practices—such as bottom-trawling and blast fishing—are illegal in some countries and should be prohibited and replaced with nondestructive methods (Bavinck et al., 2005; Dew and McConnaughey, 2005).

About 30% of capture fisheries are currently used to create "fish meal" destined for aquaculture and other livestock, and this percentage is expected to increase as aquaculture expands and more high-trophic level fish (such as salmon, grouper and tuna) are cultured and farmed. Ill-placed and designed aquaculture facilities have also reduced the productivity of wild fisheries and degraded environments through loss of critical habitats, especially mangrove forests and coral reefs; introduction of invasive species, pests and diseases; and use of pesticides and antibiotics.

Environmentally friendly and sustainable aquaculture. While aquaculture is one of the fastest growing food sectors in terms of productivity, this achievement has been at great cost and risk to the health and well-being of the environment, as well as the well-being of small-scale fishers and farmers. The future of aquaculture is truly at a crossroads: the future direction of aquaculture will affect the health and productivity of wild fisheries, the survival of many livelihoods, and global food security (World Bank 2006).

The future contribution of aquaculture to global food security and livelihoods will depend on the promotion of more environmentally sustainable and less polluting culture techniques; the use of low-trophic level species, especially filter-feeding species; the use of native species; appropriate siting and management approaches; and inclusion and empowerment of small-scale producers (World Bank, 2006). The culture of local, native species should be promoted to decrease the displacement of native species by escaped exotics, such as tilapia. Proper siting of aquaculture facilities is crucial to reduce environmental impact and ensure long-term sustainability and profitability; improperly sited aquaculture facilities, especially for shrimp farms, have led to the destruction of wetlands and mangrove forest that are vital to capture fisheries and the protection of coastal communities from storms, tsunamis and other coastal hazards. Enclosed, recirculating tanks that are properly sited show great promise in meeting some of these objectives and in decreasing the pollution of wild gene pools through escapes of species used in aquaculture. A more balanced approach to aquaculture is needed that incorporates environmental sustainability, integrated water resources management and equitable resources use and access to benefits (www.ec.europa.eu; www.icsf.net; www.worldfishcenter.org).

Greater emphasis is needed to develop sound fisheries "growth" practices and approaches—such ecosystem based fisheries management, networks of reserves, new quota models and new extraction technology—which will restore ecosystem productivity and resiliency. It is estimated that with proper fishing practices, capture fisheries production could be increased significantly, reversing present declines.

6.6 Improve Natural Resource Management and Habitat Preservation

6.6.1 The landscape management challenge
Losing habitats is the greatest threat to biodiversity; over the past 50 years people have destroyed or fragmented ecosystems faster and more extensively than in any period in human history (MA, 2005). Rapidly growing demands for food, freshwater, timber, and fuel driving this change have put enormous pressure on biodiversity. The creation of more conservation management areas, promotion of local biodiversity, increased participatory approaches to natural resource management (e.g., GELOSE project, Madagascar) and a close collaboration between all relevant stakeholders in biodiversity management initiatives (Mayers and Bass, 2004) will be vital to addressing further loss of existing habitats.

Restoration of fragile habitats is a way of improving degraded ecosystems or creating new areas to compensate for loss of habitat elsewhere. Enhancing transboundary initiatives (e.g., Agenda Transandina for mountain biodiversity in the Andes) has multiple benefits to conserve and restore fragile habitats. The appropriate use of technology, such as remote sensing or GIS can improve monitoring of ecosystem fragmentation (e.g., INBio Costa Rica) and can help in the protection of large areas of native vegetation within regions to serve as sources of species, individuals and genes. Landscape management can also help maintain or reestablish connectivity between native habitats at multiple scales with large contiguous areas of native vegetation for as wide a group of plant and animal species as possible. Remaining areas of native habitat within the agricultural landscape (giving priority to patches that are large, intact and ecologically important) can be conserved while further destruction, fragmentation or degradation prevented.

Active management of landscapes and land uses will be required to maintain heterogeneity at both patch and landscape levels, making agricultural systems more compatible with biodiversity conservation. Threats to native habitats and biodiversity can be identified and specific conservation strategies applied for species or communities that are of particular conservation concern. Areas of native habitat in degraded portions of the agricultural landscape can be restored and marginal lands taken out of production and allowed to revert to native vegetation.

For freshwaters, some management options include:
- Maintain or restore native vegetation buffers;
- Protect wetlands and maintain critical function zone in natural vegetation;
- Reestablish hydrological connectivity and natural patterns of aquatic ecosystems (including flooding);
- Protect watersheds with spatial configuration of perennial natural, planted vegetation and maintain continuous year-round soil cover to enhance rainfall infiltration

Nonnative, exotic species. Species that become invasive are often introduced deliberately, and many of these introductions are related to agriculture, including plants and trees introduced for agricultural and forestry purposes and species used for biological control of pests (Wittenberg and Cock, 2001; Matthews and Brandt, 2006). Policy for control of invasive species is essential, but AKST must also develop a better understanding of when and how species become invasive and how to best monitor and control them. Improved prediction and early detection of pest invasions, appears to rely heavily on the scale and frequency of introductions (not particular phenotypic characteristics of the invader) (Lavergne and Molofsky, 2007; Novak, 2007). Since the scale of introduction is a critical factor, commercial trade in all living organisms, including seeds, plants, invertebrates and all types of animals has the greatest potential to augment the invasion potential of exotic species. The most promising mechanism for targeting this critical phase in invasion is an increase in the capacity of exporting and importing nations to monitor the content of agricultural goods. This cannot be done effectively by individual countries; collective action is needed, through UN or other international bodies with appropriate global capacity development, e.g., UN Biodiversity Convention and the Cartagena Protocol.

6.6.2 Address poor land and soil management to deliver sustainable increases in productivity
The approach to addressing increased productivity will be distinctly different for fertile and low fertile lands (Hartemink, 2002).

6.6.2.1 Options for fertile lands
On-farm, low input options. The adoption of zero tillage prevents further water erosion losses, increases water use efficiency, soil organic carbon sequestration, and maintains good structure in topsoil (Díaz-Zorita et al., 2002; Bolliger et al., 2006; Steinbach and Alvarez, 2006; Lal et al., 2007).

About 95 million ha are under zero tillage management worldwide (Lal et al., 2007) in countries with industrialized agriculture, but the land area may increase in response to fuel prices and soil degradation. Zero tillage has well known positive effects upon soil properties; one negative effect is increased greenhouse gas emissions (N_2O, CH_4) due to higher denitrification rates (Baggs et al., 2003; Dalal et al., 2003; Passianoto et al., 2003; Six et al., 2004; Steinbach and Alvarez, 2006; Omonode et al., 2007). Tradeoffs between higher C sequestration and higher GHG emissions will need more assessment (Dalal et al., 2003; Six et al., 2004; Lal et al., 2007). Zero tillage can promote shallow compaction in fine textured topsoils (Taboada et al., 1998; Díaz-Zorita et al., 2002; Sasal et al., 2006) and no-till farming can reduce yield in poorly drained, clayey soils. Soil-specific research is needed to enhance applicability of no-till farming by alleviating biophysical, economic, social and cultural constraints (Lal et al., 2007).

Excessive soil compaction is of critical concern in industrial agriculture due to the use of heavier agricultural machines. A typical hazard is when high yielding crops (e.g., maize) are harvested during rainy seasons. Compaction recovery is not easy in zero tilled soils (Taboada et al., 1998; Díaz-Zorita et al., 2002; Sasal et al., 2006), which depend on soil biological mechanisms to reach a loosened condition. The alleviation and control of deep reaching soil compaction can be attained by adopting management strategies that control field traffic (Spoor et al., 2003; Pagliai et al., 2004; Hamza and Anderson, 2005; Spoor, 2006) and use mechanical (e.g., plowing) or biological (cover crop root channels) compaction recovery technology (Robson et al., 2002; Spoor et al., 2003).

A better understanding of biological mechanisms are needed, with particular focus on the role played by plant roots, soil microorganisms and meso- and macrofauna in the recovery of soil structure (Six et al., 2004; Taboada et al., 2004; Hamza and Anderson, 2005).

Increased botanical nitrogen-fixation can occur when legumes crops are rotated with cereals (Robson et al., 2002); green manure crops improve the N supply for succeeding crops (Thorup-Kristensen et al., 2003). In farms near animal production facilities (feed lots, poultry, pigs, dairy, etc.), organic animal manures may be a cheap source of essential plant nutrients and organic carbon (Edwards and Someshwar, 2000; Robson et al., 2002). The use of organic manures can be limited by problems associated with storage, handling, and transport (Edwards and Someshwar, 2000). In livestock grazing production systems, grazing intervals can be restricted and seasonal grazing intensity altered to reduce soil physical damage (Taboada et al., 1998; Menneer et al., 2004; Sims et al., 2005).

Continuous crop removal may eventually deplete native soil supplies of one or more nutrients. Some predict depletion of easily accessible P by 2025 at present annual exploitation rates of 138 million tonnes (Vance et al., 2003) while others estimate far less. Soil microbiology could potentially improve access to P, for example, through the use of P-solubilizing bacteria (Yadav and Tarafdar, 2001; Taradfar and

Claassen, 2005) and arbuscular mycorrhiza (Harrier and Watson, 2004). However, the use of microbes in P delivery to plants is complex. A better understanding of root growth is the optimal balance among plant, soil and microorganisms (Vance et al., 2003).

More field research is required to optimize the selection and production of crop varieties/species that enrich the diet with such elements as Ca, Zn, Cu, and Fe. Given the usually substantial residual effects of most of fertilizer nutrients (except N), they should be considered as investments in the future rather than annual costs. Replenishment of nutrients such as P, K, Ca, Mg, Zn through the use of agricultural by-products and biosolids and substitution and recycling of phosphorus (P) sources has been recommended (Kashmanian et al., 2000).

Soil conservation practices can reduce soil losses by wind and water erosion. Strategies for controlling sediment loss include (1) planting windbreaks and special crops to alter wind flow; (2) retaining plant residue after harvesting; (3) creating aggregates that resist entrainment, (4) increasing surface roughness; (5) improving farm equipment and (6) stabilizing soil surfaces using water or commercial products (Nordstrom and Hotta, 2004).

Improved management practices to prevent sediment loss may be effective (Nordstrom and Hotta, 2004). Many management techniques do not require sophisticated technology or great costs to implement, but they may require farmer willingness to change practices. Barriers to adoption of conservation measures include start-up or transition costs associated with new methods or equipment, inadequate education, reliance on past traditions, or a history of failed field experiments (Uri, 1999). Reluctance to implement soil conservation policies and practices can be overcome when severe erosion events associated with periods of drought remind society of the advantages of compatible methods of farming (Todhunter and Cihacek, 1999).

Shifting cultivation leads to deforestation and degradation, (Zhang et al., 2002). Most technical options to prevent agricultural expansion and abandonment are similar to those for preventing deforestation. They are also based on the promotion of off-farm employment (Mulley and Unruh, 2004), or the production of high-added value products combined with air transport. In order to increase farmers' natural capital and thereby increase long term flows of farm outputs, modifying the management of soil, water and vegetation resources, based on agroecology, conservation agriculture, agroforestry and sustainable rangeland and forest management, as well as wildlife biology and ecology has been supported (Buck et al., 2004).

Cultivation of new lands in some biomes would neither compensate nor justify the loss of irreplaceable ecological services. Other biomes are less sensitive and would not be similarly affected. The functional complementation of biomes is an effective land use option to explore on a broad scale (Viglizzo and Frank, 2006). For example, agricultural expansion in South America (Argentina, Bolivia, Brazil, Colombia) was based on the replacement of natural forests by cattle ranching and soybean cropping (Cardille and Foley, 2003; Vosti et al., 2003; Etter et al., 2006). There are potential benefits to conservation management that arise from agricultural land abandonment or

extensification. In China conversion of cultivated land has not always decreased national food security, since many converted lands had low productivity (Deng et al., 2006). Abandonment of agricultural land does increase the vulnerability of farmers. Positive outcomes in one sector can have adverse effects elsewhere (Rounsevell et al., 2006). Modern biomass energy will gain a share in the future energy market and abandoned agricultural land is expected to be the largest contributor for energy crops; the geographical potential of abandoned land for 2050 ranges from about 130 to 410 EJ yr^{-1} and for 2100, from 240 to 850 EJ yr^{-1}. At a regional level, significant potentials are found in the former USSR, East Asia and South America (Hoogwijk et al., 2005).

Large scale, high input options. Large scale approaches to soil management are available and based on the replenishment of soil nutrients, site specific nutrient management and zero tillage. These approaches include: adoption of crop models to synchronize N supply with crop demand (Fageria and Baligar, 2005; Francis, 2005); adoption of precision agriculture and variable rate technologies for inputs such as nutrients, pesticides and seeds (Adrian et al., 2005); and improvement of N fertility for non-legumes by legume fixation, fertilizers, manures and composts.

Nitrogen use efficiency is currently less than 50% worldwide, thus increasing N efficiency may reduce the use of N fertilizers (Sommer et al., 2004; Fageria and Baligar, 2005; Ladha et al., 2005). Deep rooting crops could potentially serve to redistribute N for crops in areas with nitrate polluted groundwater (Berntsen et al., 2006).

Crop models assess tradeoffs among yield, resource-use efficiency and environmental outcomes (Timsina and Humphreys, 2006), but their effective adoption requires local calibration and validation, improved farmer knowledge, cost-effective and user friendly techniques (Ladha et al., 2005). The adoption of precision and variable rate technologies by farmers is significantly affected by their perception of usefulness and net benefit (Adrian et al., 2005). To be of more benefit to farmers, crop models need to more effectively couple the spatial variability of crop yields and soil properties obtained by remote sensing and variable rate machinery needs improvement. Motivations for widespread uptake adoption of these technologies may come from environmental legislation and public concern over agrochemical use (Zhang et al., 2002).

Efficient use of N fertilizer requires that the amount and timing of the fertilizer application be synchronized with the needs of the crop (Ladha et al., 2005). The availability of the soil to supply N to the crop is closely linked with soil organic matter; maintenance of soil organic matter is a key factor in maintaining N fertility (Robson et al., 2002). Legumes are grown in rotations both for the contribution to the residual N and for the value of the crop itself (i.e., forage or food). To encourage the adoption of modern agricultural technologies governments and others will need to ensure farmers have access to technical advice, economic incentives and public education programs.

Whereas N efficiency and uptake is key for some regions, in others soil erosion control practices, such as contour cropping and terracing in soils of better quality (Popp et al., 2002), are more viable options. Soil erosion control can be costly and hence difficult to implement in developing countries (Wheaton and Monke, 2001). Governments can help by providing technical advice, economic incentives and public education programs (Warkentin, 2001). Land care schemes have been successfully adopted in several countries, and are effective in promoting "land literacy" and good agricultural practices, including leys and crop rotations and growing cover crops (Lal, 2001).

6.6.2.2 Options for low fertility lands

Agroforestry. In tropical areas, low fertility is often found in deforested areas, where critical topsoil has washed away. The replacement of traditional slash and burn cultivation by more diversified production systems based on forest products, orchard products, and forages and food products (Barrett et al., 2001; Ponsioen et al., 2006; Smaling and Dixon, 2006) and applying agroecological principles creatively (Altieri, 2002; Dalgaard et al., 2003) can improve soil fertility.

The adoption of agroforestry can maintain land productivity, decrease land degradation and improve rural people's livelihood (Albrecht and Kandji, 2003; Oelberman et al., 2004; Schroth et al., 2004; Reyes et al., 2005; Jiambo, 2006; Rasul and Thapa, 2006). At the landscape scale, the spatial organization of tree and forest landscape elements can provide filters for overland flow of water and sediments as well as corridors for forest biota, connecting areas with more specific conservation functions. At plot and regional scales, the relationship is more variable because watershed functions not only depend on plot-level land use but also on the spatial organization of trees in a landscape, infiltration, dry-season flow, and other factors (Van Noordwijk et al., 2007).

Consecutive nutrient exports may lead to extremely low K and P levels (Alfaia et al., 2004), e.g., decreased N and P availability with alley cropping (Radersma et al., 2004). Some crops, e.g., sugarcane (*Saccharum officinarum*) seem to be unsuitable for agroforestry (Pinto et al., 2005). Ecological agriculture could become an alternative if market distortions created by subsidies were removed, financial benefits were provided to resource-conserving farmers, and extension, credit, research were available (Rasul and Thapa, 2003). The adoption of integrated soil fertility management strategies at the farm and landscape scale requires consensus building activities (Barrios et al., 2006). However, promoting and supporting participatory technologies have limited impact when they are not grounded in participatory policy development and implementation (Desbiez et al., 2004; De Jager, 2005). Labor-inteseive ecoagriculture will not succeed unless farmers and the agricultural sector have higher total factor productivity including total labor productivity (Buck et al., 2004).

Soil water conservation and storage. The adoption of conservation agriculture is key to increasing water storage in marginal lands, and in most places suitable equipment is available (hand, animal-drawn, or tractor-drawn) for resource-poor farmers (Bolliger et al., 2006). Adoption of conservation agriculture also reduces soil erosion losses, (den Biggelaar et al., 2003) decreases siltation and pollution of water bodies, and has benefits for human health and biodiversity. Efforts to promote soil water conservation and storage will need to address site-specific conditions (Knowler and Bradshaw, 2007). Widespread implementation will require integration into institutions, incentive structures, and education (Molden et al., 2007) and extension outreach.

Methods to be considered include (1) conservation agriculture, including the use of water-efficient crops; (2) supplemental irrigation in rainfed areas; and (3) water harvesting in drier environments (Goel and Kumar, 2005; Hatibu et al., 2006; Oweis and Hachum, 2006).

Soil amendments. Municipal waste materials, composted or uncomposted (such as leaves and grass clippings, sludges, etc.), can be valuable soil amendments for farms near cities or towns and are inexpensive if transport costs are low (Smith 1996; Kashmanian et al., 2000). Municipal sludges can be also applied to cropland provided they possess the qualities needed by their potential users and do not possess toxins or heavy metals, such as nickel or cadmium (Smith, 1996). Other developments such as N-fixation by non-legume crops (e.g., *Azospyrilllum*), P solubilizing bacteria, and mycorrhizal associations in tropical cropping systems are expected to result from future biotechnology investigations (Cardoso and Kuyper, 2006).

The high risk of crop failure from insufficient soil moisture hinders investments in soil fertility and tilth, which in turn diminishes the potential of soils to capture and retain water, therefore increasing the vulnerability to drought. A challenge for AKST will therefore be how to couple incremental improvements in crop water relations with low-cost investments to replenish soil fertility in order to break this cycle (Rockström, 2004; Sanchez, 2005). More widespread use of practices like green manuring, composting, farmyard manure management, and use of agricultural by-products and residues can guide decision-making.

6.6.3 Sustainable use of water resources to meet on-farm food and fiber demands

A major challenge over the next 50 years will be to meet food and fiber demand with minimal increases in the amount of water diverted to agriculture. Aquatic ecosystems and people whose livelihoods depend on them are likely to be the biggest losers as more and more fresh water is diverted to agriculture on a global scale.

AKST can provide options for improving water management in agriculture that can address the growing problem of water scarcity, ecosystem sustainability and poverty alleviation. Chapters 4 and 5 present projections concerning the land and water required at the global level to produce enough food to feed the world in 2050. These include reliance on various options including intensification and expansion of rainfed and irrigated agriculture and trade as entry points to reduce the need to expand water and land diverted to agricultural production. In an optimistic rainfed scenario, reaching 80% maximum obtainable yields, while relying on minimal increases in irrigated production, the total cropped area would have to increase by 7%, and the total increase in water use would be 30%, with direct water withdrawals increasing by only 19%. In contrast, focusing on irrigation first could contribute 55% of the total value of food supply by 2050. But that expansion of irrigation would require 40%

more withdrawals of water for agriculture, surely a threat to aquatic ecosystems and capture fisheries in many areas.

The factors that contribute to optimistic and pessimistic estimates of total water needs are primarily differences in water productivity. Without gains in water productivity, water resources devoted to agricultural production will likely increase by 70-90%. On top of this is the amount of water needed to produce fiber and biomass for energy. The real world is more complex than the scenarios. Improvements will need to be made in water management across all agricultural systems, rainfed, irrigated, and combinations in between. It will be necessary to look beyond increasing water productivity to target poor people and ecosystems to benefit from these improvements. AKST will be needed that targets both physical (not enough water to meet all demands) and economic (not enough investment in water) water scarcity. Climate change and bioenergy increase the scale of the challenge, by increasing pressures on resources, and by increasing climate variability, but do not alter the nature of the challenge.

6.6.3.1 Managing evapotranspiration

Optimistic scenarios for mitigating increased water demand in agricultural systems require that water productivity be increased. This can be achieved with existing AKST, e.g., at the plot level in rainfed systems where evaporation can be very high and soil constraints are still significant, and at a system and basin level by reducing unproductive losses in landscapes. Crop breeding to gain increased benefit from water used and as yet unexplored opportunities to use precision water management to raise biomass/transpiration ratios are promising for intensive systems.

There is significant scope to reduce evapotranspiration (ET) per unit of yield by reducing evaporation and improving soil quality (Figure 6-1) (Molden et al., 2007). In many parts of the world, reducing evaporation and removing soil constraints are still important options for increasing water productivity. In very productive agricultural areas of the world, which produce most of the world's food, the historic sources of growth in water productivity—increased harvest index, soil nutrients—are being rapidly exhausted (Keller and Seckler, 2004). In contrast, currently areas with the greatest potential to increase water productivity in terms of ET are low production regions, especially sub-Saharan Africa and South Asia (Figure 6-2). These are also areas with high rates of poverty and high dependence of the poor on agriculture. Focus on these areas will both help reduce poverty, and also reduce the amount of additional water needed in agriculture.

Evaporation varies from 4-25% in irrigated systems (Burt et al., 2001), and from 30-40% and more in rainfed systems (Rockström et al., 2003) and depends on application method, climate and how much of the soil is shaded by leaves by the crop canopy; it can be very high in rainfed systems with low plant densities. Practices increasing water productivity such as mulching, plowing or breeding for early vigor of leaf expansion in order to shade the ground as rapidly as possible or longer superficial roots can reduce evaporation and increase productive transpiration.

Improvement of soil fertility can significantly improve transpiration efficiency and improving soil physical prop-

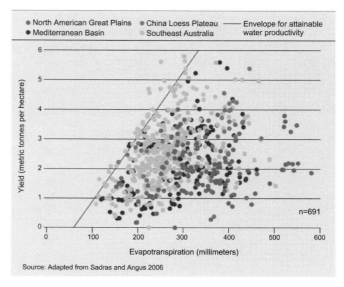

Figure 6-1. *Water productivity 'gap.'* Source: Sadras and Angus, 2006.

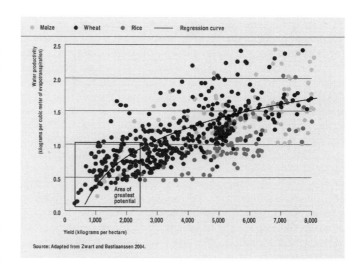

Figure 6-2. *Water productivity and yield.* Source: Adapted from Zwart and Bastiaanssen, 2004.

erties including infiltration and water storage capacity can reduce evaporation. Together these methods can result in 100% or larger increases in crop water productivity (Bossio et al., 2008). Recent examples of water productivity improvement potential through resource-conserving agricultural practices demonstrate this (Table 6-3). Only moderate effects on crop water productivity should be expected from plant genetic improvements over the next 15 to 20 years, because these gains have already been realized through breeding for increased harvest index in major grain crops. However harvest index gains through breeding strategies that target crops like millet and sorghum that have not received as much attention as the "green revolution" grains may be possible. An opportunity for improving value per unit of water also lies in enhancing nutritional quality of staple foods. Here perhaps biotechnology may offer significant potential over time (Molden et al., 2007). New precision ap-

Table 6-3. **Changes in water productivity (WP) by crop with adoption of sustainable agricultural technologies and practices in 144 projects.**

Crops	WP before intervention	WP after intervention	WP gain	Increase in WP
	-----------------kg food m^{-3} water ET-----------------			%
Irrigated				
Rice (n = 18)	1.03 (±0.52)	1.19 (±0.49)	0.16 (±0.16)	15.5
Cotton (n = 8)	0.17 (±0.10)	0.22 (±0.13)	0.05 (±0.05)	29.4
Rain-fed				
Cereals (n = 80)	0.47 (±0.51)	0.80 (±0.81)	0.33 (±0.45)	70.2
Legumes (n = 19)	0.43 (±0.29)	0.87 (±0.68)	0.44 (±0.47)	102.3
Roots and tubers (n = 14)	2.79 (±2.72)	5.79 (±4.04)	3.00 (±2.43)	107.5

Source: Pretty et al., 2006.

proaches to water management, such as irrigation of partial root systems may hold promise for increasing production per unit of water transpired in specialized production systems (Davies et al., 2002).

Besides crop and field practices, there is significant scope for reducing evaporation at the basin and landscape scales (Molden et al., 2007). High evaporation rates from high water tables and waterlogged areas can be reduced by drainage, or reducing water applications, after ensuring that these are not wetland areas supporting other ecosystem services. In degraded arid environments, up to 90% of rainfall evaporates back into the atmosphere with only 10% available for transpiration. Water harvesting in dry areas is an effective method of making available the non-beneficial evaporation of rainwater for crop transpiration (Oweis, 1999). Micro and macro-catchment techniques capture runoff and make it available for plants and livestock before evaporation, increasing the availability of beneficial rainwater, nearly halving evaporation and quadrupling increase in transpiration.

Another option is to increase the use of marginal quality water for agricultural production. While marginal-quality waters, (wastewater, saline or sodic water), potentially represent a valuable source of water for agricultural production, long term environmental and health risks are significant and must be mitigated. The prevalence of and opportunities for increasing, the use of marginal quality water in agricultural production was recently assessed (Qadir et al., 2007). Public agencies in several countries already implement policies on marginal-quality water. Egypt plans to increase its official reuse of marginal-quality water from 10% in 2000 to about 17% by 2017 (Egypt MWRI, 2004). In Tunisia in 2003 about 43% of wastewater was used after treatment. Wastewater use will increase in India, as the proportion of freshwater in agricultural deliveries declines from 85% today to 77% by 2025, reflecting rising demand for freshwater in cities (India CWC, 2002).

Worldwide, marginal-quality water will become an increasingly important component of agricultural water supplies, particularly in water-scarce countries (Abdel-Dayem, 1999). Water supply and water quality degradation are global concerns that will intensify with increasing water demand, the unexpected impacts of extreme events, and

climate change in resource-poor countries (Watson et al., 1998). State of the art systems to maximize use of saline drainage waters are currently under development in California and Australia (Figure 6-3) (Qadir et al., 2007). AKST development for sustainable use of marginal quality water is an urgent need for the future.

6.6.3.2 Multiple use livelihoods approach
Poverty reduction strategies entail elements primarily related to policy and institutional interventions to improve access for the poor to reliable, safe and affordable water. AKST contributes to increase the effectiveness agricultural water utilization by the poor. To secure water use rights now and in the future and to avoid or control the risks of unsustainable water management, it is important to understand water as a larger "bundle of rights" (water access and withdrawal rights, operational rights, decision making rights) (Cremers et al., 2005; Castillo et al., 2007). Policy and institutional interventions are described in later chapters; here the focus is on AKST options that can contribute to poverty alleviation in the future, namely, multiple use system design, small scale water management technologies, and sustainable development of groundwater resources, primarily aimed at small scale farming systems in tropical countries.

While most water use analysis focuses on crop production (particularly in irrigated systems), it is possible to increase the productivity of other components of mixed systems to provide greater overall benefit for the rural poor (Molden et al., 2007), improve health for the local population and increase biodiversity. The design, development and management of water resources infrastructure from a multiple use livelihoods perspective, can maximize the benefits per unit of water, and improve health. The integration of various water use sectors including crop, livestock, fisheries and biodiversity in infrastructure planning can result in increased overall productivity at the same level of water use, and can be compatible with improving health and maintaining biodiversity.

Livestock. Although there are few examples of research and assessments that attempt to understand the total water needs of livestock and how animal production affects water

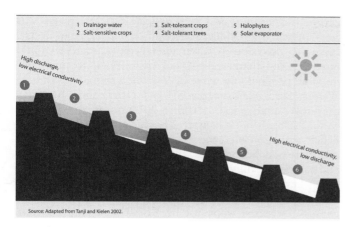

Figure 6-3. *Sequential reuse of drainage water on drainage affected lands.* Source: Qadir et al., 2007.

Note: As proposed in the San Joaquin Valley drainage Implementation Program, California.

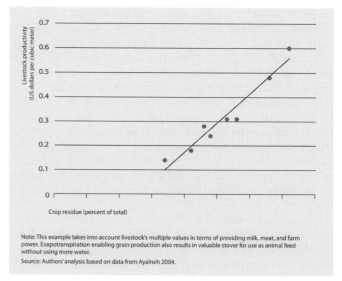

Note: This example takes into account livestock's multiple values in terms of providing milk, meat, and farm power. Evapotranspiration enabling grain production also results in valuable stover for use as animal feed without using more water.
Source: Authors' analysis based on data from Ayalneh 2004.

Figure 6-4. *Livestock water productivity relative to dietary crop residues and by-products in Ethiopia's Awash River Valley.* Source: Peden et al., 2007.

resources, a recent assessment (Peden et al., 2007) describes four entry points to maximize investment returns in water and livestock in mixed systems:

- Improving the source of feeds; e.g., in low productivity mixed systems in Ethiopia, livestock water productivity increases as the share of animal diets composed of crop residues increases (Figure 6-4) (Peden et al., 2007);
- Enhancing animal productivity through traditional animal science interventions in nutrition, genetics, veterinary health, marketing and animal husbandry;
- Conserving water resources critically need for grazing management; and
- Providing sufficient drinking water; water deprivation reduces feed intake and lowers production. For lactating cows water deprivation can greatly lower milk production (Staal et al., 2001).

While more research and site specific knowledge is needed, it is clear that securing improved outcomes in the development of agricultural water in the future will benefit from effective integration and consideration of animal use and their effect on water resources (Peden et al., 2007).

Fisheries. Fisheries can be enhanced in many existing and planned water management structures such as small dams, reservoirs, and impounded floodplains through stocking with appropriate species, greatly increasing productivity. Stocking technologies have produced high yields in lakes (Welcomme and Barley, 1998); in dams and reservoirs in Thailand, Indonesia, the Philippines and Malaysia (Fernando, 1977), in China (De Silva, 2003), and India (Sugunan and Katiha, 2004); and in floodplains in Hungary (Pinter, 1983), Bangladesh (Ahmed, 1998), and India (Sugunan and Sinha, 2001). Species introductions, and other enhancement technologies, such as fish holes, drain-in ponds, dugouts and finger ponds also effectively increase production (Dugan et al., 2007). Improved stocking management can increase production in integrated agriculture-aquaculture systems; a widespread type is integration of fish into rice paddies. While typically rice paddies produce 120-300 kg

ha^{-1} yr^{-1} of mixed fish which contribute directly to household diets, managed fish stocking and harvest can increase rice yields (due to weed control and the aeration of soils) by some 10% while producing up to 1,500 kg ha^{-1} fish (de la Cruz, 1994; Halwart and Gupta, 2004).

Health and water management systems. Under conditions that allow control of water levels, such as irrigated areas, dry season irrigation in monsoon areas and on relatively free draining soils, water management techniques can bridge the gap between agricultural and health departments (Bakker et al., 1999). These techniques include alternate wet and dry irrigation; water saving irrigation technologies; modernization of infrastructure to minimize standing water and reduce sites for disease vector breeding; and organizational initiatives such as Water Users Associations and improved extension services. Banning the use of the most toxic pesticides and promoting integrated pest management (IPM) is a high priority for preventing poisoning via water (Eddleston et al., 2002). In this case, human health and environmental interests (reducing pesticide loads) are complimentary. In addition, operation of existing dams can be re-optimized to improve health and environmental performance, such as to restore floodplain ecosystems, and new irrigation schemes can be planned and designed to minimize environmental impacts (Faurés et al., 2007).

Biodiversity. Water resources infrastructure and agricultural landscapes can be managed to maintain biodiversity and other ecosystem services beyond production of food and fiber. Water resources infrastructure can be planned and implemented in ways that minimize the impact on the native biodiversity. Biodiversity concerns need to be addressed from the earliest stages of project planning; e.g., situating infrastructure in such a way as to avoid harming critical habitats (Ledec and Quintero, 2003). At the landscape scale, the spatial organization of tree and forest landscape elements can provide filters for overland flow of water and sediments and corridors for forest biota, connecting areas with more specific conservation functions (Van Noordwijk et al., 2007). At plot and regional scales, the relationship is more variable because watershed functions not only depend on plot-level land use but also on the spatial organization of trees in a landscape, infiltration, dry-season flow, and other factors. Natural disturbance has a role in maintaining landscape biodiversity. Options for conserving biodiversity in irrigated agricultural systems include increasing water productivity and many water management designs and practices that support diverse landscapes, crops and connectivity for plant and animal movement (Molden and Tharme, 2004).

Traditional irrigation infrastructure development is one avenue for poverty alleviation; significant benefits have been demonstrated through a variety of primary and secondary effects of irrigation system development (Hussain, 2005; Castillo et al., 2007) and management strategies can improve equity in irrigation systems and can be complimentary to productivity enhancement (Hussain, 2005). As an example, land distribution that results in larger numbers of smaller holding can improve benefit sharing. Appropriate irrigation service charges can ensure adequate spending on operations and maintenance; this supports the poor, who tend to suffer the most when system level maintenance is inadequate.

6.6.3.3 Management and financing options
In order to maintain aquatic ecosystems, managers are increasingly pressed to maintain agricultural returns with reduced water delivery to irrigation systems. Reducing water delivered to irrigation requires two actions—a change in agricultural practice combined with a change in water allocation (Molden et al., 2007). Increasing blue water productivity by reducing water deliveries to agriculture, yet maintaining output, is an important strategy to retain water in aquatic ecosystems, to reallocate supplies, and to help in more precise water management, giving water managers more flexibility to deliver water to where it is needed, when it is needed. Excessive deliveries generate excessive drainage that are hard to control, require energy for pumping, reduce the quality of water and water bodies can provide breeding ground for disease vectors. Moreover, there are high ecological benefits in keeping water in rivers.

There are significant opportunities to improve irrigation water productivity through a combination of field and system management practices, and policy incentives that raise water productivity, manage salinity and increase yields (e.g., Van Dam et al., 2006). For example, there is substantial scope to reduce water deliveries to irrigation, especially to rice (Bouman et al., 2007). In addition to producing more food, there are ample opportunities in irrigation to generate more value and incur less social and environmental costs.

Supplemental irrigation, the addition of small amounts of water optimally timed to supplement rain, is probably the best way to increase water productivity of supplies. In Burkina Faso and Kenya, yields were increased from 0.5 to 1.5-2.0 tonnes ha^{-1} with supplemental irrigation and soil fertility management (Rockström et al., 2003). Yields

can be further increased with deficit irrigation, where water supplied is less than crop requirements (Zhang, 2003). Increased precision in water management is more capital intensive and therefore particularly relevant to maintaining high productivity while decreasing water diversions. In Western Syria, yields increased from 2 to 5 tonnes ha^{-1} with the timely application of 100 to 200 mm of water (Oweis et al., 2003). It must be noted, however, that precision and deficit irrigation increase risk, and therefore are most appropriate under conditions where access to water is assured, and can be carefully managed.

A key point however, is that increasing productivity of water does not necessarily drive water savings; it may encourage increased water use because it is more productive (Ahmed et al., 2007). Thus changing allocation policies is also essential to realize reduced diversions of water.

Reducing deliveries also does not necessarily save water and can have unintended detrimental side effects that can be understood by considering what happens to drainage flows. A common misperception is that because irrigation is typically 40 to 50% efficient at converting irrigation water into evapotranspiration, the focus should be on increasing efficiency and therefore reducing drainage flows (Seckler et al., 2003). Increasing efficiency can be a valuable objective for reducing uptake of water in the system and thus diminishing energy costs of pumping and operation and maintenance. However, drainage water plays an important role. Because so much drainage flow is reused downstream, there is actually much less scope in saving water in irrigation than commonly perceived. In fact, in irrigated regions in dry areas it is common to document ratios of evapotranspiration to irrigation plus rain greater than 60% reaching to over 100% when aquifers are mined. These areas include the Gediz basin in Turkey (Droogers and Kite, 1999), Egypt's Nile (Keller and Keller, 1995), Chistian sub-division in Pakistan and the Bhakra irrigation system (Molden et al., 2000), the Liu Yuan Ku irrigation system (Khan et al, 2006), the Tunuyuan irrigated area in Argentina, the Fayoum in Egypt, and Nilo Coelho in Brazil (Bos, 2004). The perennial vegetation at Kirindi Oya has been shown to evapotranspire about the same volume of water as rice and generate valuable ecosystem services; giving a different picture (65% of inflows beneficially depleted) than if paddy rice were considered alone (22% of inflows depleted by rice) (Renaud et al., 2001). In these cases, the problem is not wastage, but that high withdrawals and ET rate reduce drainage and tend to dry up rivers and wetlands, and leave little to downstream use. It is important to consider each case from a basin perspective, i.e., considering the quality and quantity of water and how drainage flows are used downstream.

Technologies such as treadle pumps, small diesel pumps, low-cost drip, and low-cost water storage can increase productivity and incomes for poor farmers (Sauder, 1992; Shah et. al., 2000; Keller et al., 2001; Polak et al., 2004). These approaches provide water at lower unit costs than large scale hydraulic infrastructure, and can be available immediately, without the long delay times of larger scale projects. Innovative development and marketing approaches that focus on increasing local private enterprise capacities and market promotion have been credited with successful dissemination to the poor (Shah et al., 2000). Credit schemes focusing on women also can have a positive effect on poverty alleviation (Van Koppen and Mahmud, 1996). By improving the precision of water delivery, these technologies can also help to increase water use efficiency, under the right conditions. There are different niches where these technologies are useful. In general treadle pumps are most suitable when water tables are within 2-4 m of soil surface. This situation is common in monsoon Asia, and exists when treadle pumps are linked to rainwater harvesting structures, but is relatively rare outside wetland or direct pumping from lakes and water bodies in Africa.

Groundwater resources. Groundwater can provide flexible, on-demand irrigation to support diversified agriculture in all climate zones. Sustainable management requires that aquifer depletion be minimized and water quality be preserved. Overwhelming evidence from Asia suggests that groundwater irrigation promotes greater gender, class, and spatial equity than do large irrigation projects. Evidence from Africa, Asia, and Latin America also suggests that groundwater is important for poor farmers to improve their livelihoods through small scale farming based on shallow groundwater (Shah et al., 2007). Small scale technologies (see above) can improve access of the poor to groundwater resources. In all parts of the developing world key common priorities for AKST are to improve the data base, upgrade the understanding of groundwater supply and demand conditions, and create effective programs for public education in the sustainable use of groundwater resources (Shah et al., 2007). Participatory approaches to sustainable groundwater management will need to combine supply-side AKST such as artificial recharge, aquifer recovery, inter-basin transfer of water, with demand-side AKST such as groundwater pricing, legal and regulatory control, water rights and withdrawal permits (see chapter 7), and promotion of water-saving crops and technologies.

Decreasing land degradation. Water use efficiency, which is often as low as only 40%, in irrigated areas (Deng et al., 2006), can be increased. This is key to reducing recharge to naturally saline areas and water tables. Where soil salinity is high, leaching fractions must be applied to remove salt from the root zone, without adding it to groundwater or mobilizing it to the river system; this is difficult and requires well thought out, innovative drainage solutions. Recognized options for management of salinity risk, or to reduce existing areas of saline soil, are revegetation with alternative species, pumping to lower the water table and construction of ditch drains for control of surface water and shallow groundwater (Peck and Hatton, 2003).

Management of salinity is complex and requires integrated solutions at catchment and basin scale with the key being to minimize mobilization of salt and reduce the amount for disposal—disposal through the stream system is undesirable and environmentally costly. All options for management of salinity risk are constrained by the economics of dry land farming and pumping or drainage is further constrained by possible environmental impacts of disposal of saline water. In Australia, the bulk of effort has been directed at "living with saline land and water," with immense public and private investment in tree planting and the search for new low recharge farming systems (Peck and Hatton, 2003). Practices to improve water use efficiency include biological mechanisms of water-saving agriculture and irrigation technologies, including low pressure irrigation, furrow irrigation, plastic mulches, drip irrigation under plastic, rainfall harvesting and terracing (Deng et al., 2006).

6.7 Using AKST to improve Health and Nutrition

AKST can improve human health and nutrition through reductions in (1) malnutrition and micronutrient deficiencies; (2) food contaminants; and (3) the emergence and reemergence of human and animal diseases, including HIV/AIDS. Key driving forces over the coming decades for these challenges include not just AKST, but also demographic change; changes in ecosystem services; global environmental change; reductions in freshwater resources; economic growth and its distribution; trade and travel; rate of technology development; governance; degree of investment in public health and health care systems; and others.

In addition, some food systems are not providing the range of nutrients needed to ensure adequate nutritional status. Approaches to improve dietary quality are needed to ensure adequate availability, accessibility, and utilization of foods with nutrients appropriate to the needs of the population.

6.7.1 On-farm options for reducing malnutrition and micronutrient deficiencies

Integrated farm systems, based on a variety of foods, can help meet the challenge of micronutrient malnutrition (Tontisirin et al., 2002). Improving crop diversity is an important part of improving dietary diversity, and thereby dietary quality. The diversity of wild and cultivated traditional plant varieties in rural areas of low-income countries provides many opportunities to identify high quality, but underutilized, nutritious foods. Increased research on locally adapted traditional varieties could lead to the development of improved varieties that are higher yielding or more resistant to pests and abiotic stresses such as drought. Household processing of wild foods collected by subsistence farmers as part of a traditional diet would increase storage life and make additional foods available during periods when food is inadequate. For example, solar drying techniques have been used to preserve foods such as mangoes, bananas and sweet potatoes.

Possible improvement of these varieties through breeding is currently limited because private and public sector breeding programs rarely focus on minor crops. Identifying and exploiting the potential of these varieties will require increased research in both high- and low-income countries. In Kenya, when farmers produced underutilized leafy green vegetable varieties, consumption was increased among farmers, and the producers found a market among middle and high income consumers who began to purchase these novel varieties (Frison et al., 2006). Once researchers identify health promoting compounds in indigenous and underutilized plants, plant breeders can develop varieties of these foods that can be produced and consumed by small-scale farmers as well as sold in high value niche markets. Beyond increasing the availability of diverse foods, preservation methods must be improved to reduce the loss of micronutrients (Ndawula et al., 2004).

In addition to increasing the range of plant foods in the diet, animal source foods, such as meat, milk, and insects from wild and domesticated sources can provide critical nutrients that may be completely unavailable in plant-based diets, such as vitamin B_{12} (Neumann et al., 2002; for Kenyan example see Siekmann et al., 2003). An effective strategy to increase the intake of animal source foods could include the improved small-scale livestock production through the use of appropriate breeds, disease prevention and control, and affordable high quality animal feeds (Brown, 2003).

Improving soil management practices, such as increasing the organic matter in the soil and mineral fertilizers (Sheldrick and Lingard, 2004), can improve food security and enable farmers to produce sufficient yields and allow for more crop diversification. These practices can optimize plant nutritional quality. For example, crops grown on zinc deficient soils often produce grains with low zinc concentrations and these seeds may produce plants with lower grain yields and poorer seed quality (Rengel, 2001). Soil management solutions have the advantage of providing a wide range of nutrients, while other approaches, such as fortification and supplements are limited to specific nutrients.

6.7.2 Research needs for reducing malnutrition and micronutrient deficiencies

Biofortified crops developed through plant breeding can improve human nutrition. Biofortification has shown promise in feeding studies in the Philippines where iron biofortified rice consumption improved iron status in the study participants (Murray-Kolb et al., 2004). While conventional processed food fortification can work well to improve the availability of critical nutrients in the diet, rural subsistence producers may not have access to fortified foods. Thus, where food processing facilities are unavailable, biofortification can improve the availability of target nutrients. In addition, where government regulation and enforcement of food fortification is still in the nascent stages of development, biofortified crops can serve as a cost-effective source of micronutrients. Dietary quality can be improved by selection of crop varieties that are more nutritionally dense when these are substituted for less nutritious alternatives. Consumption of carotenoid-rich red palm oil in lieu of other vegetable oils has improved vitamin A status in Burkina Faso (Zagre et al., 2003), while lysine and tryptophan-rich maize may offer improved growth potential for undernourished children consuming diets with low protein quality (Graham et al., 1990).

While plant breeding efforts to biofortify staple crops are underway, plant-breeding programs can also target health-related qualities such as antioxidants in fruits or vegetables (HarvestPlus, 2006). For example, plant breeders can select for high lutein content, an antioxidant with beneficial effects on eye health (Seddon, 2007) in carrots

(Nicolle et al., 2004). Plant breeding can include traditional techniques and approaches using advances in biotechnology, such as rDNA. Conventional plant breeding methods have been used to develop biofortified crops and rDNA approaches have increased carotenoid content in rice (Beyer et al., 2002). While approaches using rDNA and similar techniques have the potential to contribute to developing nutritionally improved crop varieties, research, monitoring, and evaluation are needed to ensure there are no adverse unintended consequences to human and environmental health.

Reducing food contaminants. When present in food systems, heavy metals and other contaminants, veterinary drug residues, pesticide residues, pathogens, and the toxins produced by pathogens such as mycotoxins can cause a range of short- and longer-term adverse human health consequences.

Good agricultural practices (GAPs) can lead to safer use of pesticides and veterinary drugs. GAPs can also enable the management of risks associated with pathogen contamination of foods such as fruits and vegetables. FAO has developed guidance for governments and the private sector on conducting risk assessments and to implementing risk management options throughout food systems, including on-farm practices and in food processing facilities (FAO/WHO, 2006). Hazard analysis critical control point principles can be used to target issues of biosecurity, disease monitoring and reporting, safety of inputs (including agricultural and veterinary chemicals), control of potential foodborne pathogens, and traceability (Olson and Slack, 2006). The development and adoption of GAPs for specific production systems and food safety/quality issues can be facilitated by approaches that involve broad participation. Plants can become susceptible to infection with the fungus that produces aflatoxins when they are exposed to water stress or insect damage (Dowd, 2003). There are readily available approaches management approaches (preharvest, harvest, and postharvest) to reduce aflatoxin (Mishra and Das, 2003); e.g., in tree nuts, peanuts, and cereals such as maize.

In addition, dietary approaches are being developed to counteract the effects of mycotoxins (Galvano et al., 2001). Additional research is needed to verify the detoxification ability of the proposed food components, their long-term efficacy and safety, and their economic and technical feasibility. To manage risks associated with pathogens such as *Escherichia coli* O157:H7 in fruit and vegetable production, sanitation systems throughout the food production chain are integral to GAPs guidance for preventing the presence of these organisms (Fairbrother and Nadeau, 2006). Additional strategies are being developed to reduce foodborne pathogens, e.g., chlorate as a food supplement to prevent colonization of food-producing animals by *E. coli* and other pathogens (Anderson et al., 2005).

Achieving fuller deployment of GAPs to improve food safety and public health requires establishing effective national regulatory standards and liability laws that are consistent with international best practice, along with the necessary infrastructure to ensure compliance, including sanitary and phytosanitary surveillance programs for animal and human health, laboratory analysis and research capabilities, and training and auditing programs. Challenges include harmonization of regulations establishing upper levels of intake of nutrients and other substances (Bennett and Klich, 2003), and improving food safety without creating barriers for poor producers and consumers.

Heavy metal contamination in soils affects the quality and safety of foods. For example, rice grains can accumulate cadmium (Cd) from Cd-contaminated soils, thereby exposing consumers to serious health consequences from consumption of locally produced rice (Chaney et al., 2004). Undernourished populations are particularly at risk, as iron and zinc deficiencies can cause increases in Cd absorption from the food supply (Anderson et al., 2004). While increased soil pH or maintaining soil flooding until grain maturation can reduce Cd levels in rice grains, yields can be affected (Chaney et al., 2004). Bioremediation with selected ecotypes of *Thlaspi caerulescens,* a hyperaccumulator of Cd, could effectively reduce levels in contaminated soil (Chaney et al., 2000). However, these wild ecotypes of *T. caerulescens* need to be improved for commercialization before practical applications of this technology would be available (Chaney et al., 2004).

6.7.3 Reduce factors that facilitate the emergence and reemergence of human and animal diseases

Communicable diseases are the primary cause for variations in life expectancy across countries (Pitcher et al., 2008). AKST is important for three broad categories of infectious diseases: diseases whose incidence is affected by agricultural systems and practices (e.g., malaria and bovine spongiform encephalopathy), foodborne zoonotic diseases, and epidemic zoonotic disease (e.g., avian influenza). For example, the expansion of irrigated agriculture, as a result of the need to further intensify food production and to better control water supplies under increased climate variability and change, is expected to contribute to an increased incidence of malaria in some areas and the rapidly increasing demand for livestock products could increase the likelihood of BSE to spread more widely.

The geographic range and incidence of many human and animal diseases are influenced by the drivers of AKST. Currently, 204 infectious diseases are considered to be emerging; 29 in livestock and 175 in humans (Taylor et al., 2001). Of these, 75% are zoonotic (diseases transmitted between animals and humans). The number of emerging plant, animal, and human diseases will increase in the future, with pathogens that infect more than one host species more likely to emerge than single-host species (Taylor et al., 2001). Factors driving disease emergence include intensification of crop and livestock systems, economic factors (e.g., expansion of international trade), social factors (changing diets and lifestyles) demographic factors (e.g., population growth), environmental factors (e.g., land use change and global climate change), and microbial evolution. Most of the factors that contributed to disease emergence will continue, if not intensify, this century (IOM, 1992). The increase in disease emergence will affect both high- and low-income countries.

Serious socioeconomic impacts can occur when diseases spread widely within human or animal populations, or when they spill over from animal reservoirs to human hosts (Cleaveland et al., 2001). Animal diseases not only affect animal and human health and welfare, they also influence perceptions of food safety, result in trade restrictions, adversely affect rural incomes and livelihoods, adversely affect non-livestock rural industries, have detrimental environmental effects, and adversely affect national economies for countries heavily dependent on agriculture Even small-scale animal disease outbreaks can have major economic impacts in pastoral communities (Rweyemamu et al., 2006).

6.7.3.1 On-farm options

The adoption integrated vector and pest management at the farm level, have been tested for reducing the persistence of human and animal diseases. These include environmental modification, such as filling and draining small water bodies, environmental manipulation, such as alternative wetting and drying of rice fields, and reducing contacts between vectors and humans, such as using cattle in some regions to divert malaria mosquitoes from people (Mutero et al., 2004; Mutero et al., 2006).

Specific farming practices can facilitate infectious disease emergence and reduce the incidence of certain diseases, such as malaria, in endemic regions (van der Hoek, 2004). However, the relationships between agriculture and infectious disease are not always straightforward. For example, whereas rice irrigation increases breeding grounds for the mosquito that carries malaria, in some regions the prevalence of malaria in irrigated villages is lower than in surrounding villages because better socioeconomic conditions allow greater use of antimalarials and bed nets (Ijumba et al., 2002) and/or because the mosquito vector tends to preferentially feed on cattle (Mutero et al., 2004). However, in other regions, intensification of irrigated rice reduces the capacity of women to manage malaria episodes among children, leading to a higher prevalence of malaria (De Plaen et al., 2004). Therefore, greater understanding is needed of the ecosystem and socioeconomic consequences of changes in agricultural systems and practices, and how these factors interact to alter disease risk.

In areas affected by high rates of HIV/AIDS, labor-saving agricultural technologies and systems are needed to support sustainable livelihoods. Ensuring access to diverse diets can also reduce the adverse impacts of disease on livelihoods and health. Agroforestry interventions, in particular, can improve communities' long-term resilience against HIV/AIDS and other external shocks in ways that agricultural interventions alone cannot (Gari, 2002).

In addition, improved agricultural information and knowledge exchange between experienced farmers and youth and widows is needed (Peter et. al., 2002). Agroforestry technology can respond to the cash, labor and shortages confronted by AIDS-affected communities, both in the short term and in the long term. Medicinal plants and trees often provide the only source of symptomatic relief available to the poor. Future agroforestry programs and forest policies in general should be reviewed to assess their effects on key determinants of HIV vulnerability (Villarreal et al., 2006). Using less labor intensive crops that need fewer inputs can help households allocate labor more efficiently in food producing activities (Ngwira et al., 2001). While diversifying food crop production to reduce labor demands can be helpful, the nutritional quality of the total diet must be considered.

6.7.3.2 Research and technological options beyond the farm

Resource poor farmers have limited resources to mitigate the spread of diseases. Controlling emerging infectious diseases requires early detection, through surveillance at national, regional, and international levels, and rapid intervention. For animal diseases, traceability, animal identification, and labeling also are needed. The main control methods for human and animal diseases include diagnostic tools, disease investigation facilities, and safe and effective treatments and/or vaccines. AKST under development can facilitate rapid detection of infectious pathogens, e.g., genetic tools were used in recent HPAI outbreaks to identify the viruses involved and to inform development of appropriate control programs (FAO/OIE/WHO, 2005). Syndromic surveillance of farm animals coupled with notification using internet-accessible devices is being used in some high-income countries to detect emerging diseases (Vourc'h et al., 2006).

The increasing importance of zoonotic diseases requires better integration of human and veterinary public health approaches for their detection, identification, monitoring, and control. Decreased funding in recent decades has eroded the required infrastructure and training underlying veterinary services and surveillance activities (Vallat and Mallet, 2006). Incentives to report cases of disease at the local and national levels and pay for culling of animals when appropriate could facilitate early identification of outbreaks. There is an urgent need to replenish basic capacity in many high-income countries and to increase capacity in middle- and low-income countries. Linkage of regional and international organizations and agencies is critical. Improved understanding is needed of disease transmission dynamics in order to develop more effective and efficient diagnostic systems and interventions. Diagnostic systems should be designed to process large numbers of samples and identify multiple infectious agents.

Although vaccines are a cornerstone of primary prevention, vaccine effectiveness is severely limited in remote rural areas with high infectious disease burdens, particularly Africa, South America, and Asia, due to the lack of vaccines, the lack of resources to afford vaccines, or the logistical problems of trying to use temperature-sensitive vaccines. Marker vaccines are needed so that vaccinated/treated animals can be distinguished from subclinically infected or convalescent animals in real time during epidemics (Laddomada, 2003).

The emergence and dissemination of bacteria resistant to antimicrobial agents is the result of complex interactions among antimicrobial agents (e.g., antibiotics), microorganisms, disease transmission dynamics, and the environment (Heinemann, 1999; Heinemann et al., 2000). The increasing incidence of antimicrobial resistant bacterial pathogens will limit future options for prevention and treatment of infectious diseases in animals and humans (McDermott et al., 2002). The World Health Organization has called for human and veterinary antimicrobial agents to be sold only under prescription, and for the rapid phaseout of antimi-

crobial agents used as growth promotants (WHO 2003). They also recommend that all countries establish monitoring programs for tracking antimicrobial use and resistance. Research on the use of other treatments, such as probiotics and vaccines, holds promise (Gilchrist et al., 2007). The ongoing costs of research and development, and challenges to delivery will prevent acute drug treatments from ever becoming a stand-alone solution.

6.7.4 Tackling persistent chemicals to protect human health and the environment

Persistent chemicals include potentially toxic elements like heavy metals and organic pollutants that are normally present at relatively low concentrations in soils, plants, or natural waters, and which may or may not be essential for the growth and development of plants, animals, or humans (Pierzynski et al., 2000).

6.7.4.1 On-farm options

More effective and less costly *in situ* management strategies are available to mitigate the effects of persistent chemicals and to restore soil quality. The load of persistent chemicals such as fertilizer and pesticide residues, to ground and surface waters can be significantly reduced by available technologies, such as precision agriculture. Restorative technologies like bioremediation and phytoremediation (plant based remediation) are costly and still in development. Basic research is needed on the factors affecting biotransformation processes (Adriano et al., 1999; Khan, 2005).

Intrinsic remediation using indigenous organisms can degrade industrial solvents (e.g., PCBs) and many pesticides on affected sites (Sadowski and Turco, 1999). *In situ* bioremediation can potentially treat organic and inorganic pollutants, clean soil without excavation and it is more cost effective than excavating and treating the soil on site bioremediation techniques. Such treatments remove the mobile and easily available fractions but cannot complete removal of all the contaminants (Doelman and Breedveld, 1999).

Phytoremediation refers to the extraction of contaminants via root uptake to shoot biomass and has wide application in the remediation of surface-polluted soils. Further analysis and discovery of genes for phytoremediation may benefit from recent developments in biotechnology (Krämer, 2005). Phytoremediation has potential risks, such as those associated to the use of transgenic techniques, release of nonindigenous species (potential weed) and transfer of toxic compounds to the other environmental compartments (Wenzel et al., 1999; Alkorta and Garbisu, 2001).

6.7.4.2 Off-farm technology

More effective and sensitive technologies for identifying early effects of pollution on ecosystems can also be developed. Damage could be prevented if the source of the pollution and the presence of the pollutants could be identified at minimal concentrations. Preventing or limiting the flow of chemical pollutants into the environment should be more effective than limiting damage by remediation.

New technologies that significantly increase awareness of biological impacts include biosensors and chemical approaches (Water Science and Technology Board, 2001; Heinemann et al., 2006). These approaches can also use indigenous organisms, e.g., ecotoxicological assessments of soils polluted with chromium and pentacholorophenol. The portal DATEST (http://projects.cba.muni.cz/datest) is a web-based engine that complements and stores information about a wide range of ecotoxicological tests and bioindication methods used in Ecological Risk Assessment (Smid et al., 2006).

6.7.5 Information and knowledge systems

6.7.5.1 Traditional, local knowledge options
Traditionally, many innovations for improving AKST occurred at the community level, and were diffused through community institutions (Gyasi et al., 2004). Traditional communities have domesticated dozens of plant species, have bred and conserved thousands of crop varieties and animals, and have developed farming (cropping and animal) systems and practices adapted to specific conditions (Kaihura and Stocking, 2003). Tapping on those resources and capacities and giving them recognition as well as legitimacy is a key development goal. A focus on agroecology can enrich the production and deployment of new farming practices and technologies that are environmentally, socially and culturally sustainable (Koontz et al., 2004).

Options for enhancing agricultural knowledge and innovation in local and indigenous societies include:

- Enhance local and traditional knowledge systems and grassroots innovation capacities;
- Empower communities to access knowledge and to participate in innovation processes so they have more options to respond to future changes and to biodiversity and livelihood challenges (Colfer, 2005);
- Develop a new agenda that builds on agricultural knowledge and innovation in local and indigenous societies: increase projects of international agricultural research institutions such as Bioversity International (formerly IPGRI);
- Foster participatory agricultural and environmental research projects that bring together traditional and western science (Brookfield et al., 2003; Colfer, 2004), journals such as *Etnoecologica*, and academic courses that include traditional and local knowledge.

Farmer field schools (see Chapter 2) could play a vital part as a community-based initiative for participatory research, enabling farmers to define and analyze problems, and experiment with options. Seed fairs can facilitate the selection of varieties better adapted to local conditions (Orindi and Ochieng, 2005) and adaptation to climate change. The establishment of "lead farmers" and the implementation of various grassroots extension mechanisms could reinforce the role of communities in the production and diffusion of knowledge.

6.7.5.2 Science and technology options
Advances in nanotechnology, remote sensing (RS), geographic information systems (GIS), global positioning systems (GPS) and information communication technology (ICT) can enhance progress in the application of precision and site-specific agriculture (PA).

A concern in precision agriculture is the accessibility and affordability of the technology for small farming systems. This is not surprising considering that the general trend is that farmers with large farmlands of more than 300 ha, tend to be the first to invest in the new technology, whereas small farmers are more reluctant to invest in GPS equipment (Pedersen et al., 2004). A nationwide survey in the USA concluded that adoption of PA technologies was related to farm size and large farmers are the first to adopt (Daberkow and McBride, 2001). Adoption rate is also faster in regions with larger farm sizes and more specialized in certain cash crops (Blackmore, 2000; Fountas et al., 2005). Adoption is likely to continue in countries where labor is scarce, and vast tracts of land exist, with rates of adoption accelerating when commodity prices are high and interest rates low (Swinton and Lowenberg-DeBoer, 2001).

Particularly for developing countries, the use of yield monitors, sensors, GIS and GPS, supported by advanced tools such as computer, digital camera, image processing technique, laser technology, and network system appear too complex for small-scale farmers, particularly for those whose field operations are not mechanized. Nevertheless, since precision farming being a management approach not a technology, it can be applied to developing countries industrialized countries, but the implementation may be different (Griepentrog and Blackmore, 2004).

Precision agriculture practices that can easily be adapted in developing countries include site specific nutrient management (SSNM) and simple integrated crop management (ICM) version like rice check (Lacy et al., 1999; Fairhurst et al., 2007; PhilRice, 2007). Thus, while the ownership of precision farming technologies is still an emerging option for small-scale agriculture, the adoption strategy can be adapted. Custom services can be used to help build precision farming databases while small-scale farmers gain experience with the spatial variability of their fields (Lowenberg-DeBoer, 1996).

6.7.5.2.1 Remote sensing technology
Remote sensing (RS) has a broad range of applications (urban and transportation planning, applied geosciences, land use, environmental change, etc.) in many countries, especially Europe and the United States where it is widely used, and can enhance agricultural planning for low productivity areas in developing countries.

For agriculture, RS techniques play an important role in crop identification, crop area inventory, crop yield forecasting, crop damage detection, soil and water resources inventory, and assessment of flood damage (Syam and Jusoff, 1999; Van Neil and McVicar, 2001; Patil et al., 2002). It also provides required inputs for land and water resources development plans, wasteland mapping and reclamation, irrigation development, crop-yield and crop-weather models, integrated pest management, integrated nutrient management, watershed management, agrometeorological services, and more recently, precision farming (Patil et al., 2002). Remote sensing contributes to the information needs of precision agriculture (PA) in the assessment of soil and crop conditions using multispectral imagery (Barnes and Floor, 1996).

Remote sensing is currently not widely applied in most developing countries because of timeliness, limited accessibility and cost of satellite data, and financial constraints in gathering ground data that can be correlated to the remote sensing data. It has, however, potential in improving agricultural planning in developing countries particularly in addressing food security, poverty alleviation, and sustainable development issues.

If combined with other sources of data (e.g., traditional method agrometeorological data collection) remote sensing can improve accuracy and effectiveness of various agricultural planning in developing countries. For example, RS estimates of crop yields and production of staple foods based on preharvest crop acreage and production can serve as input to a number of policy level decisions on buffer food stock (Van Neil and McVicar, 2001).

Remote sensing data can provide a sampling frame construction for agricultural statistics, crop acreage estimation, and cropland data layer or map (Allen, Hanuschak, and Craig, 2002; Saha and Jonna, 1994; Rao, 2005). Mapping soils can reveal soil properties across production fields (Dalal and Henry, 1986; Shonk et al., 1991; Mzuku et al., 2005). Remote sensing information also aids analysis of soil degradation and risk of soil erosion in agricultural lands (Thine, 2004).

By combining RS with GIS techniques, and hydrologic modeling, irrigation management can be improved for more complex water management tasks such as irrigation system performance evaluation, snowmelt runoff forecasts, reservoir sedimentation and storage loss assessments, prioritization of watersheds and their treatment, environmental impact assessment of developmental projects, prospecting of under ground water, locale specific water harvesting and recharge, interlinking of rivers and monitoring of spatial and temporal distribution of rainfall (Thiruvengadachari and Sakthivadivel, 1996). Given more time and resources, applications of RS in agricultural planning can be greatly enhanced in developing countries.

Remote sensing can also be applied to global agroenvironmental health and resources monitoring and assessment. Remote sensing can be used to assess biodiversity through (1) direct mapping of individual plants or associations of single species in relatively large, spatially contiguous units; (2) habitat mapping and predictions of species distribution based on habitat requirements; and (3) establishment of direct relationships between spectral radiance values recorded from remote sensors and species distribution patterns recorded from field observations (Nagendra, 2001; Zutta, 2003; Rao, 2005).

Satellite RS is increasingly becoming an important source of agrometeorological data (humidity, rainfall, temperature, wind, global radiation) as it can complement traditional methods of agrometeorological data collection (Sivakumar and Hinsman, 2004). Indian satellite systems, for example, operationally support disaster management by providing emergency communication links, cyclone warnings, flood forecasting data, rainfall monitoring and crop condition assessments (Rao, 2005).

Remote sensing can be used to globally monitor and assess natural resources and ecosystem for sustainable development, providing more accurate and timely information on the condition and health of agroenvironmental resources.

There are, however, some technical issues and limitations of current remote sensing technologies use (Table 6-4).

6.7.5.2.2 Information and communications technology (ICT)
ICT models can be mainstreamed and upscaled to enhance delivery of services and access to market.

Market information. In Uganda, ICT is providing farmers with reliable price data for better farm gate prices. A market information service network reaching over 7 million people each week uses conventional media, Internet, and mobile phones to enable farmers, traders, and consumers to obtain accurate market information. Over the past four years the number of markets dominated by farmers' associations has increased from 4 to 8 (Ferris, 2004).

Weather forecasting. In Africa, ICT is enabling more rapid dissemination of locally analyzed weather data. The European Meteosat Second Generation (MSG) satellite is providing detailed data and high-resolution spectral and spatial images that are expected to revolutionize the process of forecasting short-term extreme weather events, such as thunderstorms, fog and small but intense depressions that can lead to devastating storms, as well as other applications, e.g., agrometeorology, climate monitoring, and natural resource management (Taube, 2006).

Web-based marketing systems. New business models are rapidly evolving that can suit the needs of small farmers, e.g., the www.B2Bpricenow.com a free agriculture e-marketplace that provides updates via SMS messaging to farmers in the Philippines (www.digitaldividend.org/pubs/pubs_01_overview.htm). In India, e-Choupal kiosks of the agriexporter ITC Limited and "Parry's Corners" of EID Parry agricultural company provide farmers with valuable information, and allow them to sell their produce directly to these companies eliminating the middleman. E-commerce platform can also allow small farmers and farmer cooperatives to expand distribution channels for their produce (Ninomiya, 2004).

E-consultation, advisory system and training. ICT can provide farmers with electronic forums and e-consultations by email, or permit the participation of a wider electronic community in location-based seminars (Painting, 2006). Farmers can also access tools for both diagnosing field problems and making crop management decisions (e.g., TropRice [124.81.86.181/rkb/knowledgeBank/troprice/default.htm#Introduction_to_TropRice.htm] and Rice Knowledge Bank [www.knowledgebank.irri.org]). The so called "virtual academy for farmers" in the Philippines and India uses ICT through a virtual network that provides information on-demand, online learning and content development of information based on farmers' needs. Trained farmers and extension workers serve as resource persons in cyber communities thereby making ICTs accessible and user-friendly.

E-governance. India is enhancing rural development programs and improving the delivery of public services with the use of government computerization schemes, satellite com-

Table 6-4. **Current remote sensing technologies for global agroenvironmental health and resources monitoring and assessment for sustainable development.**

Types of remote sensing	Sensor description	Example of imaging sensors	Resolution	Limitations	Application in agriculture	Other applications
1. Optical Imaging a. Panchromatic	Single channel detector sensitive to broad wavelength range produce black and white imagery	IKONOS Pan	Spatial: 1 m Spectral: 1 band Temporal: 1-3 days	• Unlike microwave remote sensing, acquisition of cloud free image using optical bands is impossible because of its short wavelength that cannot penetrate clouds and rain	• Precision farming • Property damage control and verification of crop damage, e.g., drought and hail. • Farm planning, precision farming	• Highly detailed land use discrimination, urban mapping, natural resources and natural disasters mapping, environmental planning, land registration, public health, biodiversity conservation, coastal monitoring, homeland security. • Urban planning, feature and asset mapping, land use mapping
		SPOT Pan	Spatial: 10 m Spectral: 1 band Temporal: 1-26 days			
b. Multispectral	Multichannel detector with a few spectral bands. Sensitive to radiation with narrow wavelength band. The image contains brightness and color information of the targets.	Landsat MSS	Spatial: 50-80 m Spectral: 5 bands Temporal: 18 days	• Resolution tradeoff: High spatial resolution associated with low spectral resolution.	• General vegetation inventories and classification	• Environmental monitoring, land use mapping and planning, forest mapping, statistical land-use survey global-change, urban area mapping, detection of silt-water flowing and landscape analysis.
		Landast TM	Spatial: 25 m Spectral: 7 bands Temporal: 16 days		• Discrimination of vegetation types and vigor, plant and soil moisture measurement, Cropping pattern mapping, chlorophyll absorption, biomass survey, plant heat stress	• Water penetration, differentiation of snow and ice landscape analysis, lineament detection, lithological classification, urban environment assessment, delineation of water bodies, hydrothermal mapping.
1. Optical Imaging c. Superspectral	Imaging sensor has many more spectral channels (typically >10) than a multispectral sensor. The bands have narrower bandwidths that capture finer spectral characteristics of the targets.	SPOT HRV-XS	Spatial: 20 m Spectral: 3 bands Temporal: 1-26 days	• Multi, super and hyper spectral have resolution trade off: Sensors with high multispectral resolution can only offer low spatial resolution.	• Vegetation mapping and monitoring, soil erosion, agricultural boundary detection,	• Urban mapping, forestry mapping and planning, land use and land cover discrimination, maritime and coastal management, resource stewardship monitoring, habitat supply planning, wildfire mapping, landslide and mudflow detection, and rapid urban change.
		KOOS MS	Spectral: 4 bands Temporal: 1-3 days		• Precision farming, vegetation mapping, disease detection	• Environmental analysis, land management, urban growth mapping and updating, disaster mitigation and monitoring Highly detailed land use discrimination
		MODIS	Spatial: 250,500,1000 m Spectral: 36 bands Temporal: 1-2 days		• Drought detection, vegetation monitoring and forecasting	• Atmospheric temperature measurement, ozone/cloud/ atmospheric properties, ocean color, phytoplankton, biogeochemistry, land cover mapping, land use planning land cover characterization and change detection.

Table 6-4. Continued.

Types of remote sensing	Sensor description	Example of imaging sensors	Resolution	Limitations	Application in agriculture	Other applications
1. Optical Imaging (continued) d. Hyperspectral	It acquires images in about a 100 or more contiguous spectral bands. The precise spectral information enables better characterization and identification of targets.	ENVISAT MERIS	Spatial: 300, 1200 m Spectral: 15 bands Temporal: 3 days		• Inventory and yield estimation. • Crop type mapping • Monitoring of seasonal land cover changes. • Global vegetation monitoring	• Evaluation of tropohospheric aerosol properties, hazard monitoring
2. Microwave Imaging	Encompasses both active and passive remote sensing. It covers long wavelengths from 1cm-1m, which can penetrate through cloud cover, haze, dust, and all but the heaviest rainfall all day and all weather imaging.	Hyperion	Spatial: 30 m Spectral: 220 bands Temporal: 16 days		• Precision farming, crop type mapping, monitoring of crop health, moisture and maturity.	• Measures sea surface temperature, color and surface roughness • Coastal management (monitoring of phyto-planktons, pollution and bathymetry changes)
(widely used bands) a. C Band	8,000-4,000 MHz; (3.8-7.5 cm)	RADARSAT-SAR (5.6 cm)	Spatial: 8,25,30,50, 100m Spectral: C band Temporal: 24 days	• Image distortions. Extensive shadowing of areas characterized with relief. • Coarse resolution, especially for passive applications. • Radar images are rather difficult to deal with. The few commercial software packages that exist to deal with radar imagery offer a limited amount of functions. • Results are better when combined with optical images as they have been proven complimentary	• Crop monitoring and forecasting, crop mapping	• Flood detection, for disaster management, risk assessment, pollution control (oil spill), coastline mapping.
b. L Band	2,000-1,000 MHz; (15.0-30.0 cm)	ALOS-PALSAR (1270 MHZ)	Spatial: 10-100 m Spectral: L band Temporal: 46 days		• Agricultural monitoring	• Distinction of forest from grassland, land cover classification volcanic activity monitoring, flood monitoring, landslide and earthquake detection, detection of oil slick, forest biomass estimation.
3. Light Detection and Ranging (LIDAR)	An active sensor that transmits laser pulses to the targets and records the time the pulse returned to the sensor receiver. Laser is able to provide light beam with high intensity, high collimation, high coherence, high spectral purity, and high polarization purity.	LIDAR (airborne)	Spatial: 0.75 m Spectral: 1.045-1.065 μm Temporal: dependent on flight schedule	• Disadvantages are low coverage area and high cost per unit area of ground coverage. It is not cost-effective to map a large area using an airborne remote sensing system.	• Crop monitoring, plant species detection, can be used for agricultural planning and crop estimation	• Forestry management, shoreline and beach volume changes lines, flood risk analysis, habitat mapping, subsidence issues, emergency response, urban development, and monitoring of environmental changes.

Source: Authors' elaboration.

munications, and distance education and training via the Internet. Some of these projects have been quite successful suggesting that the potential impact of IT on development can be enormous, particularly in terms of improved health, hygiene, nutrition, and education (Pigato, 2001

ICT can complement conventional methods to meet the growing demand of stakeholders in accessing improved technologies and timely information and support services, improving productivity and livelihoods in poor rural communities. Although ICT allows greater and faster flow of information, due to the technical and knowledge requirements, not all people have the same level of access. ICT can further widen the "digital divide" between developed and developing countries, as well as between rural and urban communities within a country (Herselman and Britton, 2002).

6.7.5.2.3 Nanotechnology
Nanotechnology (see Glossary) may improve agriculture and resource management, particularly soil fertility, crop/animal production, pest management, veterinary medicine, product safety and quality, and farm waste management. Applications of nanotechnology in agriculture are rapidly expanding and developing (Binnig and Rohrer, 1985; Mills et al., 1997; Huang et al., 2001; Dutta, and Hofmann, 2004; Hossain et al., 2005; Graham-Rowe, 2006). Investment on nanotechnology R&D from both public and private sectors has been increasing (Kuzma and VerHage, 2006). The potential of nanotechnologies in terms of environmental impacts, including those with agriculture applications (waste management, water purification, environmental sensors, and agricultural pollution reduction) has been assessed (Defra, 2007).

Biosensors developed into nanosensors expedite rapid testing and analysis of soil, plants, and water making nutrient and water management in the farm more efficient and less laborious (Birrel and Hummel, 2001; Alocilja and Radke, 2003). Nanoporous materials such as zeolites can help release the right dosage of fertilizer at the right time owing to well-controlled stable suspensions with absorbed or adsorbed substances. Nanoelectrocatalytic systems could optimize purification of highly contaminated and salinated water for drinking and irrigation; and nanostructured materials may offer clean energy solutions through the use of solar cells, fuel cells, and novel hydrogen storage (Court et al., 2005).

Nanomaterials can provide environmental filters or as direct sensors of pollutants (Dionysiou, 2004). Nanoparticles have been used in photocatalysis that enhance degradation process in solid, farm or wastewater treatment (Blake, 1997; Herrmann, 1999). Air pollution could also be reduced (Peral et al., 1997) through on the use of photocatalysis for purification, decontamination, and deodorization of air.

The integration of nanotechnology, biotechnology, and information and communications technology could revolutionize agriculture this century (Opara, 2004). These technologies could contribute to reducing hunger and improving nutrition by optimizing plant health and eliminating pathogens or other organisms that might contaminate food.

Despite the rapidly expanding products and market of nanotechnology (nanotechnology food market in 2006 was about US$7 billion in 2006 and may reach a total of $20.4 billion by 2010 (HKC, 2006), there are some biosafety and IPR concerns. Their application in agriculture will directly introduce them into ground and surface water catchments where they may accumulate in concentrations that may undermine the goals of food safety and environmental sustainability (NSTC, 2000; ETC Group, 2005). Nanomaterials are built from nanoparticles that may be too diverse for stereotypical risk assessments (Colvin, 2003). However, since nanoscale particles have minute dimensions in common, these can direct research to likely exposure routes. For example, their small size but large-scale release may lead to their accumulation in groundwater because even particles that are not soluble in water can form colloidal species that can be carried in water (Colvin, 2003).

As with biotechnology, nanotechnologies are not evenly distributed: wealthier industrial nations produce and own the technologies. A single nanoscale innovation can be relevant for widely divergent applications across many industry sectors and companies, and patent owners could potentially put up tolls on entire industries. IP will play a major role in deciding who will capture nanotech's market, who will gain access to nanoscale technologies, and at what price (ETC Group, 2005).

6.7.5.3 Participatory approaches to AKST
Efforts to preserve natural resources and guarantee the provisioning of essential ecosystem services are frequently characterized by social, political and legal conflicts (Wittmer et al., 2006). Broad-scale approaches are necessary to face problems that extend beyond a local site and a short time span.

The asymmetric administration of shared lands and natural resources is a potential source of conflict in many trans-boundary eco-regions of the world (Viglizzo, 2001). The cross-border externalization of negative environmental impacts due to asymmetries in land conversion and intensity of farming represents a challenge to neighboring countries. The problem may become critical in shared basins with interconnected rivers and streams where downstream countries often have to pay the cost of negative impacts that have not been properly internalized upstream.

AKST can be employed to prevent or mitigate consequences of conflict over environmental resources, particularly through the use of participatory approaches supported to enhance the commitment of stakeholders to the decision-making process and to share the responsibility of managing common resources. Strategies include (1) developing stakeholder appreciation for importance of trans-boundary basin management (2) jointly designed land-use strategies to prevent potential conflicts due to negative externalities from neighboring areas, (3) environmental impact assessment for ex-ante evaluation of potentially conflicting projects, and (4) acceptance of third party independent arbitration to face current or potential conflicts when necessary.

Agricultural and environmental conflicts are characterized by the interaction of both ecological and societal complexity (Funtowicz and Ravetz, 1994). Participatory approaches (De Marchi et al., 2000) and multicriteria analysis (Paruccini, 1994) can help resolve agroenvironmental conflicts. Multiple criteria analysis uses different approaches (normative, substantive and instrumental) to deal with different types and levels of conflict resolution; it can be a powerful analytical tool in cases where a single decision-making criterion fails and where impacts (social, ecological or environmental) cannot be assigned monetary values.

Currently, most agricultural technology aims at resolving environmental problems that occur at the small spatial scale (e.g., the plot and farm level), but broad-scale technologies (Stoorvogel and Antle, 2001) are necessary to reveal impacts that are not perceived with site-specific studies. The importance of information technology increases as we scale-up to undertake problems that occur at broader geographical scales. The integration of maps, remote-sensing images, and data bases into geographic information systems (GIS) is needed to assess, monitor and account critical resources and large-scale agroenvironmental processes. This information base, coupled to models and expert systems (De Koning et al., 1999), can help support the application of participatory approaches and multicriteria analysis to resolve present or potential conflicts. Likewise, these tools become tools to support decision-making on large-scale land-use policies and managerial schemes.

The impact of climate change may exacerbate risks of conflict over resources and further increase inequity, particularly in developing countries where significant resource constraints already exist. An estimated 25 million people per year already flee from weather-related disasters and global warming is projected to increase this number to some 200 million before 2050 (Myers 2002); semiarid ecosystems are expected to be the most vulnerable to impacts from climate change refugees (Myers, 2002). This situation creates a very serious potential for future conflict, and possible violent clashes over habitable land and natural resources such as freshwater (Brauch, 2002), which would seriously impede AKST efforts to address food security and poverty reduction.

6.8 Adaptation to Climate Change, Mitigation of Greenhouse Gases

The effectiveness of adaptation efforts is likely to vary significantly between and within regions, depending on geographic location, vulnerability to current climate extremes, level of economic diversification and wealth, and institutional capacity (Burton and Lim, 2005). Industrialized agriculture, generally situated at high latitudes and possessing economies of scale, good access to information, technology and insurance programs, as well as favorable terms of global trade, is positioned relatively well to adapt to climate change. By contrast, small-scale rainfed production systems in semi-arid and subhumid zones presently contend with substantial risk from seasonal and interannual climate variability. Agricultural communities in these regions generally have poor adaptive capacity to climate change due to the marginal nature of the production environment and the constraining effects of poverty and land degradation (Parry et al., 1999).

AKST will be confronted with the challenge of needing to significantly increase agriculture output—to feed two to three billion more people and accommodate a growing urban demand for food—while slowing the rate of new GHG emissions from agriculture, and simultaneously adapting to the negative impacts of climate change on food production.

Agriculture will have to become much more efficient in its production if it is to accomplish this without significantly increasing its climate forcing potential. All of this will have to be achieved in a future where agricultural crops may be in direct competition with crops grown for energy purposes as well as without significant extensification and loss of biodiversity.

6.8.1 AKST innovations

6.8.1.1 Technological (high-input) options
Modeling. Climate simulation models indicate the intensification of the hydrologic cycle, climatic conditions which will significantly challenge efforts to control soil erosion and rehabilitate degraded lands even in well-endowed production environments (Nearing, 2004). Tropical soils with low organic matter are expected to experience the greatest impact of erosion on crop productivity because of the poor resilience of these soils to erosive forces, and the high sensitivity of yields to cumulative soil loss (Stocking, 2003; Nearing, 2004). Evidence of significant soil erosion can often be difficult to detect, and its impact on crop productivity can be masked by use of inorganic fertilizer (Knowler, 2004; Boardman, 2006). Extreme events, which significantly contribute to total erosion, are very likely to increase with climate change (Boardman, 2006), as will climate-induced changes in land use that leave soils vulnerable to erosion (Rounsevell et al., 1999).

The improvement of soil erosion modeling capacity can address the role of extreme events in soil erosion and encompass the influence of socioeconomic factors on land use change (Michael et al., 2005; Boardman, 2006). One new technique estimates the impact of more frequent extreme events under different climate scenarios by using meteorological time series projections (Michael et al., 2005). The effects of extreme events on erosion can be more simply modeled with two-dimensional hill slope approaches (Boardman, 2006); GIS can be used to develop landslide hazard maps (Perotto-Baldiviezo et al., 2004).

Recent developments in modeling techniques show potential for estimating the future impact of extreme events, through downscaling from General Circulation Models. Global climate models, however, will continue to be limited by uncertainties (Zhang, 2005). The lack of quantitative data and the technological complexity of many contemporary models are likely to limit the applicability of soil erosion modeling in less developed steep land regions (Morgan et al., 2002; Boardman, 2006). Better field-level assessments of current erosion under different crops and management practices, and, where possible, through integrating GIS into land-use planning could help developing countries assess the impacts of climate change.

Agroecological zone (AEZ) tools used by FAO (FAO, 2000) to determine crop suitability for the world's major ecosystems and climates has potential to enhance efforts to develop crop diversification strategies. The AEZ methodology, which combines crop modeling with environmental matching, allow assessment of the suitability of particular crop combinations given future climate scenarios. However, the data sets that underlie AEZ need to be improved in order to realize the full potential of these tools for crop diversification. For example the current scale of the FAO world soil maps at 1:5,000,000 needs finer resolution (FAO, 2000).

Early warning, forecasting systems. Timely forecasts, including the starting date of the rainy season, average weather conditions over the coming season, conditions within the season that are critical to staple crops and animals, and appropriate responses can increase the economic, environmental, and social stability of agricultural systems and associated communities. Advances in atmospheric and ocean sciences, a better understanding of global climate, and investments in monitoring of the tropical oceans have increased forecasting skill at seasonal to interannual timescales. Early warning systems using seasonal forecasts (such as the FAO Global Information and Early Warning System) and monitoring of local commodity markets, are increasingly used to predict likely food shortfalls with enough advance warning for effective responses by marketing systems and downstream users.

Traditional coping mechanisms depend on the ability to anticipate hazard patterns, which are increasingly erratic with the advent of climate change. One option for improving early detection and warning would be to broaden the use of GIS-based methodologies such as those employed by the Conflict Early Warning and Response Network (CEWARN), the Global Public Health Information Network (G-PHIN).

Early warning systems are important because they help to untangle the multiple but interdependent crises that characterize complex emergencies, particularly in response to climate change. In other words, continuous information gathering serves to identify the socioecological ingredients of complex crises before they escalate into widespread violence. This means technological systems are also needed. To this end, the added value of technological early warning systems should therefore be judged on their empowerment of local people-centered systems that build on the capacity of disaster-affected communities to recover with little external assistance following a disaster. Further applied research is needed on local human adaptability in decentralized settings as well as self-adaptation in dynamic disaster environments.

Linking early warning to more effective response requires a people-centered approach to climate change (UN, 2006). The quest for early warning must be more than just an "exercise in understanding how what is happening over there comes be known by us over here" (Adelman, 1998). Instead, the international community should focus on the real stakeholders and add to their capacity for social resilience. On the policy front, the lack of institutionalized early warning systems that survey the localized impact of climate change on ecological and political crises inhibits the formulation of evidence-based interventions (Levy and Meier, 2004). Regrettably, little collaboration currently exists between the disaster management and conflict prevention communities despite obvious parallels in risk assessments, monitoring and warning, dissemination and communication, response capability and impact evaluation (Meier, 2007).

Bringing climate prediction to bear on the needs of agriculture requires increasing observational networks in the most vulnerable regions, further improvements in forecast accuracy, integrating seasonal prediction with information at shorter and longer time scales, embedding crop models within climate models, enhanced use of remote sensing,

quantitative evidence of the utility of forecasts for agricultural risk management, enhanced stakeholder participation, and commodity trade and storage applications (Giles 2005; Hansen, 2005; Hansen et al., 2006; Doblas-Reyes et al., 2006; Sivakumar, 2006). For seasonal climate forecasts to be an effective adaptation tool, advances in forecasting skills need to be matched with better pathways for dissemination and application, such as by linking forecasts to broader livelihood and development priorities, and by training organizations, such as extension agencies, to facilitate the end users' ability to make effective decisions in response to forecasts (Ziervogel 2004; Garbrecht et al., 2005; Hansen 2005; Vogel and O'Brien, 2006). Substantial investments by national and international agricultural and meteorological services are needed.

Improve crop breeding potential for drought, salinity and heat tolerance. Abiotic stress of agricultural crops is expected to increase in most regions due to warmer temperatures, experienced both as episodic heat waves and mean temperature elevation, prolonged dry spells and drought, excess soil moisture, and salinity linked to higher evapotranspiration rates and salt intrusion. Expected temperature increases of 2-3°C by mid-century could significantly impair productivity of important staple crops of the developing world, such as wheat, and in truly marginal areas, millet. One-third of irrigated agricultural lands worldwide are affected by high salinity, and the area of salt-affected soils is expected to increase at a rate of 10% per year (Foolad, 2004). The magnitude of these impacts could test our capacity to achieve breakthroughs in germplasm improvement equivalent to the challenge at hand.

Advances in plant genomics, linked to the Arabidopsis model system, and the integration of genomics with physiology and conventional plant breeding could lead to the development of new varieties with enhanced tolerance to drought, heat, and salinity. Emerging genomic tools with future potential include whole-genome microarrays, marker-assisted selection using quantitative trait loci, bioinformatics, and microRNAs (Edmeades et al., 2004; Foolad, 2004; Ishitani et al., 2004; White et al., 2004; Denby and Gehring, 2005). Phenological adaptation, e.g., matching crop duration to available season length, is central to successful breeding efforts; thus conventional breeding, augmented with genomic tools, is a likely configuration of future plant breeding programs. An example of this would be the integration of phenotyping (differences in crop germplasm performance under different stress environments) with functional genomic approaches for identifying genes and mechanisms (Edmeades et al., 2004; Ishitani et al., 2004). Improvement in seasonal forecasting and in the use of remote sensing and other observational tools could also be used to further support breeding programs, through better characterization of cropping environments.

Future breakthroughs in understanding how crop plants respond to abiotic stress are very likely, given the scientific resources dedicated to investigating the *Arabidopsis thaliana*, a model system used for plant genetics and genomics studies with a small, completely sequenced genome and a short life cycle. For example, progress in genomics related to salt tolerance in Arabidopsis mutants has enhanced understanding of gene function, which could provide opportunities to exploit these mechanisms in crop species (Foolad, 2004; Denby and Gehring, 2005). However, direct extrapolation of single gene responses, gained through Arabidopsis studies, to functional abiotic tolerance of cultivated crop species could continue to be limited by differences in gene sequence between Arabidopsis and crop species (Edmeades et al., 2004; White et al., 2004). Moreover, gene expression in Arabidopsis changes when exposed to field conditions (Miyazaki et al., 2004, as reviewed by White et al., 2004), as would be expected given the influence of genotype by environment interactions. Genes for heat tolerance have been identified in a number of species, including rice, cowpea, and groundnut, which is likely to provide future opportunities for heat-tolerance breeding.

Attaining more effective use of genomics for abiotic stress-tolerance breeding will depend on closer integration of this discipline with physiology, which could lead to better understanding of how genes confer changes in whole-plant biological function and agronomic performance (genotype-to-phenotype relationships) (Edmeades et al., 2004; White et al., 2004). However, the current imbalance between genomic research and field-based physiological studies, in favor of the former, could undermine future AKST progress towards developing new stress-tolerant germplasm. Lastly, the scope of abiotic stress research needs to be extended to include more investigations of stress caused by mineral deficiencies and toxicities (Ishitani et al., 2004), as these factors strongly influence root development with implications for tolerance to climatic extremes (Lynch and St. Clair, 2004). For example, many tropical agricultural soils have high levels of exchangeable Al which stunt root system development. Bringing mineral stress tolerance more closely into the realm of abiotic stress research, while increasing the complexity of the breeding challenge, could possibly avoid short-circuiting progress on drought, heat and salinity breeding efforts when scaling up to actual field conditions where multiple and complex stresses occur.

Technological breakthroughs in breeding for abiotic stress tolerance could ultimately be limited by a potential loss of crop wild relatives to climate change. In the next 50 years, 16 to 22% of species that are wild relatives of peanut, potato, and cowpea could become extinct as a result of temperature increases and shifts in rainfall distribution, and most of the remaining species could lose over 50% of their range size (Jarvis et al., 2008). These three crops are important for food security in low-income countries, and their wild relatives are a vital genetic resource for developing future drought and pest resistant crop varieties, as well as varieties with enhanced nutritional value. Greater efforts to collect seed for gene banks (*ex situ* conservation) and to target *in situ* conservation, such as through addressing habitat fragmentation, could help to mitigate these potential losses. Strengthening links between conservation, breeding, and farmers' groups is an important component of this effort. However, diversity for its own sake is not useful, as farmers retain varieties for specific traits, not for the sake of conservation (Box 6-2).

Agronomic and genetic improvement of underutilized (or "lost") crops could provide a good opportunity to enhance agricultural diversification, particularly in Africa

where approximately 2,000 underutilized food species are consumed (NRC, 1996). Crops such as the legume Bambara groundnut (*Vigna subterranean*) and the cereal fonio (*Digitaria exilis* and *Digitaria iburua*) still figure prominently in the African diet. Fonio has very good prospects for semiarid and upland areas because it is widely consumed, tolerates poor soil and drought conditions, matures very quickly (6-8 weeks), and has an amino acid profile superior to today's major cereals (NRC, 1996). Unlocking the genetic potential of this cereal through conventional breeding and biotechnology to address low yields, small seeds, and seed shattering could help meet development and sustainability goals (Kuta et al., 2003; NRC, 1996). Similar potential exists for Bambara groundnut (Azam-Ali, 2006; Azam-Ali et al., 2001), which is still cultivated from landraces. Research needs for underutilized crops include germplasm collection, marker assisted breeding, assessments of agronomic characteristics and nutritional content, development of improved processing technologies, and market analyses. While these crops cannot replace the major cereals, their improvement could significantly enhance food security options for rural communities confronted with climate change.

Diversification of agriculture systems is likely to become an important strategy for enhancing the adaptive capacity of agriculture to climate change. Diversification strategies in the near term will need to be flexible, given that the disruptive impacts of climate change are projected to be experienced more in terms of increased variability, than as mean changes in climate. Therefore, improved skill in predicting how short-term climate phenomena, such as the El Niño Southern Oscillation and the North Atlantic Oscillation, affect seasonal and interannual variability, and the timely dissemination of forecasts will be essential for farmer decisions about whether to grow high or low water-consumptive crops and use of drought-tolerant varieties (Adams et al., 2003; Stige et al., 2006).

6.8.1.2 On-farm (low input) options

The knowledge and tools currently available could be better deployed to reduce the vulnerability of rainfed agriculture to seasonal climate variability. For example, poor crop establishment is a significant but solvable constraint in semiarid farming environments (Harris, 2006). Similarly, seasonal dry spells can be bridged using improved rainfall catchment and incremental amounts of fertilizer (Rockström, 2004). By focusing on the "manageable part of climatic variability" (Rockström, 2004), AKST could have a significant positive impact on improving the adaptive capacity of rainfed agriculture to climate change. It is also important to recognize that risk aversion practices are themselves an adaptation to climate variability, and to understand the functional linkages between existing coping strategies and future climate change adaptation.

The greatest period of risk in rainfed agriculture is the uncertainty around the timing of sufficient rainfall for crop sowing. High rainfall variability and poor quality seed leads to slow germination and emergence, causing patchy stands, and multiple and delayed replanting, making poor crop establishment a significant contributor to the productivity gap in semiarid agriculture (Harris, 2006). Emphasis can be put on targeting technologies and practices that reduce the ex-

Box 6-2. The importance of crop varietal diversification as a coping strategy to manage risk.

A study of traditional practices of conserving varieties of yam, *Dioscorea* sp., and of rice, *Oryza glaberrima*, was carried out in Ghana in 2003-2004 under an IPGRI-GEF-UNEP project on crop landraces in selected sub-Saharan African countries (Gyasi et al., 2004). It identified 50 varieties of yam and 33 varieties of rice that are managed by a wide diversity of locally adapted traditional practices in the study sites located in the semiarid savanna zone in the northern sector. The case study findings underscore the importance of crop varietal diversification as security against unpredictable rainfall, pest attack, fluctuating market and other such variable environmental and socioeconomic conditions, not to mention its importance for modern plant breeding and wider use of farm resources, notably labor and the diversity of on-farm ecological niches.

posure of sensitive crop growth stages to seasonal climate variability.

Options for addressing this challenge include improving farmer access to quality seed, adoption of improved crop establishment practices, and the use of healthy seedlings in transplant systems. Seed priming—soaking seeds in water for several hours but short of triggering germination—is an example of a simple but effective technology for improving crop establishment. Priming of some seeds results in more even and fuller stand establishment, accelerates seedling emergence and improves early growth, often leading to earlier flowering and maturity, avoidance of late-season drought and improved yields (Harris et al., 2001; Harris, 2006). Experimental crop transplanting methods in millet-sorghum areas of Africa can also reduce planting risk; e.g., staggered transplanting from seedling nurseries to allow for variable onset of the rainy season (Young and Mottram, 2001; Mottram, 2003; CAZS, 2006). This method, though more labor intensive, results in faster crop establishment with fewer gaps, and a harvest 2-3 weeks earlier than conventional seeding methods, leading to higher grain and stover yields.

By reducing crop establishment risk and decreasing the time to maturity, these technologies provide a small measure of flexibility to farmers in high-risk environments. Technologically simple approaches to improve crop establishment and seedling vigor generally have minimal downside risks, immediate and tangible benefits, and can be easily tailored to producer needs; thus they are appropriate options for small-scale rainfed systems. Seed priming, which has been tested in a wide array of dryland cereals and pulses, consistently results in average 30% increases in yield with minimal farmer investment (Harris, 2006). Similar mean yield increases have been observed with seedbed solarization of rice nurseries, though with somewhat greater farmer investment in material and time. While these are simple technologies, they do require some local testing and training to ensure that proper techniques are followed. Millet transplanting systems show good potential, though labor shortages could be an issue in some regions. An analysis of the tradeoff between labor for transplanting versus the labor and extra seed required for multiple resowing of millet fields would help to clarify the issue of labor expenditure.

Soils. Improved adoption of soil conserving practices can also mitigate the damaging effects of climate variability. Methods include the use of cover crops, surface retention of crop residues, conservation tillage, green manures, agroforestry, and improved fallow (Sanchez, 2000; Benites and Ashburner, 2003; Lal, 2005). Although these are very sound practices for soil protection, achieving broad-scale and long-term adoption of them will be a significant challenge given the current and likely future, disincentives to investment as described in the previous subchapter (Stocking, 2003; Knowler, 2004; Cherr et al., 2006; Patto et al., 2006). The resilience of conservation farming systems in the Central American highlands to recent El Niño drought (Cherrett, 1999), and to the catastrophic soil losses from Hurricane Mitch (Holt-Gimenez, 2001) provide strong evidence of conservation agriculture's potential as an adaptation response to increased rainfall variability and storm intensity with climate change.

Long-term investment in rehabilitating degraded lands is another option for addressing the negative feedback between high rainfall risks and declining soil fertility. Recent evidence of revegetation and agricultural intensification in the Sahel, catalyzed by a crisis of diminished rainfall and declining yields (Herrmann et al., 2005; Reij et al., 2005; Tappan and McGahuey, 2007; USAID, 2006), could inform future AKST efforts at integrating soil and water conservation and land reclamation into adaptation planning. Technologies and practices deployed in these areas to reclaim declining or abandoned land include rock lines, rock "Vs", and manure-amended planting pits. These techniques were used to break soil crusts, enhance water capture and retention, and regenerate N-fixing trees to improve soil fertility. Soil reclamation using these methods encompassed several hundred thousand ha in Burkina Faso and Mali, and well over a million ha in Niger (Reij et al., 2005; Tappan and McGahuey, 2007; USAID, 2006).

Important elements gleaned from these studies include:
- Legal code reforms that provided farmer, rather than government, ownership of trees was an essential precondition; the former sometimes taking the lead and the latter following;
- By improving land and claiming ownership, women were one of the main beneficiaries, and improved household food security one of the most tangible outcomes;
- Investment in fertilizer occurred after farmers invested in measures to conserve soil moisture and increase soil organic matter.

AKST could play an important role in documenting the effectiveness of these practices for seasonal climate risk management, e.g., investigating how these soil improvement practices affect soil fertility, soil moisture retention, and crop yields over a range of variable rainfall years, as well as conducting detailed socioeconomic analyses of how the benefits are distributed in local communities. Local control of the resource base is necessary for creating the enabling conditions that spur local action towards natural resource improvements, and an understanding of this dynamic is needed to effectively support local initiatives. Stabilizing and improving the natural resource base of agriculture are essential preconditions for investing in technologies for long-term adaptation to climate change (Stocking, 2003; Sanchez, 2005).

Reduction of greenhouse gas emission for agriculture. Reduction of N_2O emissions from agriculture could be achieved by better matching fertilizer application with plant demand through the use of site-specific nutrient management that only uses fertilizer N to meet the increment not supplied by indigenous nutrient sources; split fertilizer applications; use of slow-release fertilizer N; and nitrification inhibitors (DeAngelo et al, 2005; Pampolino et al., 2007). Another option to address N_2O emissions would be the use of biological means to inhibit or control nitrification in soils. Gene transfer from species exhibiting biological nitrification inhibition to cultivated species could offer another way to reduce N_2O emissions to the atmosphere and nitrate pollution of water bodies (Fillery, 2007; Subbarao et al., 2007).

Improved management of agriculture and rangelands targeted at soil conservation, agroforestry, conservation tillage (especially no-till), agricultural intensification, and rehabilitation of degraded land can yield C sequestration benefits (IPCC, 2000; Izaurralde et al., 2001; Lal, 2004). Carbon sequestration potential in soils is greatest on degraded soils (Lal, 2004), especially those with relatively high clay content (Duxbury, 2005; Lal, 2004).

Another promising approach would be to use plant material to produce biochar and store it in soil (Lehman, 2007a). Heating plant biomass without oxygen (a process known as low-temperature pyrolysis) converts plant material (trees, grasses or crop residues) into bioenergy, and in the process creates biochar as a coproduct. Biochar is a very stable compound with a high carbon content, surface area, and charge density; it has high stability against decay, and superior nutrient retention capacity relative to other forms of soil organic matter (Lehmann et al., 2006). The potential environmental benefits of pyrolysis combined with biochar application to soil include a net withdrawal of atmospheric CO_2, enhancement of soil fertility, and reduced pollution of waterways through retention of fertilizer N and P to biochar surfaces (Lehmann, 2007b). Future research is needed to more fully understand the effect of pyrolysis conditions, feedstock type, and soil properties on the longevity and nutrient retention capacity of biochar.

The robustness of soil carbon sequestration as a permanent climate change mitigation strategy has been questioned because soil carbon, like any other biological reservoir, may be reverted back to the atmosphere as CO_2 if the carbon sequestering practice (e.g., no till practice) were to be abandoned or practiced less intensively. Increasing soil organic matter through carbon sequestering practices contributes directly to the long-term productivity of soil, water, and food resources (IPCC, 2000; Lal, 2004). Thus it would seem unlikely that farmers would suddenly abandon systems of production that bring so many economic and environmental benefits. Other reports suggest that certain soil carbon sequestering practices, such as no till, may increase N_2O

emissions (Ball et al., 1999; Duxbury, 2005). This outcome, however, may be location specific (e.g., humid climatic conditions) as revealed by a comprehensive review of Canadian agroecosystem studies (Helgason et al., 2005).

Globally, farmers continue to adopt no-till as their conventional production system. As of 2001, no-till agriculture had been adopted across more than 70 million ha worldwide with major expansion in South America (e.g., Argentina, Brazil, and Paraguay) (Izaurralde and Rice, 2006). With an area under cropland estimated globally at 1.5 billion ha, there exists a significant potential to increase the adoption of no-till as well as other improved agricultural practices, which would have other environmental benefits such as improved soil quality and fertility, reduced soil erosion, and improved habitat for wildlife. Much work remains to be done, however, in order to adapt no-till agriculture to the great variety of topographic, climatic, edaphic, land tenure, land size, economic, and cultural conditions that exist in agricultural regions of the world.

In developing strategies all potential GHG emissions need to be considered for example, efforts to reduce CH_4 emissions in rice can lead to greater N_2O emissions through changes in soil nitrogen dynamics (Wassmann et al., 2004; DeAngelo et al., 2005; Yue et al., 2005; Li et al., 2006). Similarly, conservation tillage for soil C sequestration can result in elevated N_2O emissions through increased fertilizer use and accelerated denitrification in soils (Ball et al., 1999; Duxbury, 2005). However, one of the most comprehensive long-term studies of GHG emissions across several land use practices in Michigan (Robertson et al., 2000) revealed that no-till agricultural methods had the lowest Global Warming Potential when compared to conventional and organic agricultural methods.

From a GHG mitigation standpoint, strategies that emphasize the avoidance of N_2O and CH_4 emissions have a permanent effect as long as avoided emissions are tied to higher productivity, such as through increased energy efficiency and better factor productivity (Smith et al., 2007). Indeed, many of the practices that avoid GHG emissions and increase C sequestration also improve agricultural efficiency and the economics of production. For example, improving water and fertilizer use efficiency to reduce CH_4 and N_2O emissions also leads to gains in factor productivity (Gupta and Seth, 2006; Hobbs et al., 2003) while practices that promote soil C sequestration can greatly enhance soil quality (Lal, 2005). Improved water management in rice production can have multiple benefits including saving water while maintaining yields, reducing CH_4 emissions, and reducing disease such as malaria and Japanese encephalitis (van der Hoek et al., 2007). There is significant scale for achieving this "win-win" approach, with the approach largely determined by the size and input intensity of the production system, e.g., N-fixing legumes in smallholder systems and precision agriculture in large systems (Gregory et al., 2000).

There is potential for achieving significant future reductions in CH_4 emissions from rice through improved water management. For example, CH_4 emissions from China's rice paddies have declined by an average of 40% over the last two decades, with an additional 20 to 60% reduction possible by 2020 through combining the current practice of mid-season drainage with the adoption of shallow flooding, and by changing from urea to ammonium sulfate fertilizer, which impedes CH_4 production (DeAngelo et al., 2005; Li et al., 2006). There is also potential to achieve CH_4 reduction through integrating new insights of how the rice plant regulates CH_4 production and transport into rice breeding programs (Wassmann and Aulakh, 2000; Kerchoechuen, 2005).

Emerging technologies that could provide future options for reducing CH_4 and N_2O emissions from livestock include: adding probiotics, yeasts, nitrification inhibitors, and edible oils to animal feed that reduce enteric CH_4 and N_2O emissions from livestock systems (Smith et al., 2007) and controlling methanogenic archae, microorganisms that live in the rumen and generate CH_4 during their metabolism More extensive use of the antibiotic Rumensin® (monensin sodium), currently used to improve feed efficiency and prevent *Coccidiosis*, a parasitic intestinal infection, would improve energy utilization of feedstuffs through increased production of proprionic acid by rumen microorganisms and reduce the production of CH_4. However, because Rumensin is also toxic to methanogenic bacteria, it should not be fed to cattle whose waste is to be used for CH_4 generation.

Seeds. A viable option for small-scale production systems would be to refine and more widely disseminate the practice of adding small quantities of fertilizer to seed, such as through seed coating (Rebafka et al., 1993) or soaking/priming (Harris, 2006) methods. Addition of fertilizer P and micronutrients to seed, rather than soil, is an inexpensive but highly effective means for improving plant nutrition and increasing yield (> 30% average yield increase reported) on drought-prone, acidic, low fertility soils. Seed priming with dilute fertilizer has average benefit/cost ratios 20 to 40 times greater than that achieved with fertilizer addition to the soil.

This is could be an effective strategy for small-scale systems, though there are several impediments such as low availability of quality fertilizer in local markets, lack of extension services for conveying technical information, and inability of farmer to pay for fertilizer-treated seed. Imbedding these technologies within larger efforts to overhaul the seed sector, which could include credit for purchasing improved seed and information about improved crop establishment practices could facilitate farmer adoption of these technologies. These technologies also could be disseminated into local communities by targeting farmers that have made prior land improvements to increase soil water retention, and may therefore be less risk adverse.

Water resources and fisheries. While the broad implications of climate change on marine systems are known—including rising sea levels, sea surface temperatures, and acidification—the degree and rate of change is not known, nor are the effects of these physical changes on ecosystem function and productivity (Behrenfeld et al., 2006). To adjust and cope with future climatic changes, a better understanding of how to predict the extent of change, apply adaptive management, and assign risk for management decisions is needed (Schneider, 2006).

To ensure the survival of many communities, their livelihoods and global food security, new approaches to monitoring, predicting, and adaptively responding to changes in marine and terrestrial ecosystems need to be developed. Ecosystem resilience can be built into fisheries and essential fish habitats (including wetlands and estuaries) and approaches developed that reduce risk and ensure continuation of ecosystem goods and services (Philippart et al., 2007). Rising sea levels will alter coastal habitats and their future productivity, threatening some of the most productive fishing areas in the world. Changes in ocean temperatures will alter ocean currents and the distribution and ranges of marine animals, including fish populations (di Prisco and Verde, 2006; Lunde et al., 2006; Sabates et al., 2006; Clarke et al., 2007). Rising sea surface temperatures will result in additional coral reef bleaching and mortality (Donner et al., 2005). Rising atmospheric CO_2 will lead to acidification of ocean waters and disrupt the ability of animals (such as corals, mollusks, plankton) to secrete calcareous skeletons, thus reducing their role in critical ecosystems and food webs (Royal Society, 2005).

Precautionary approaches to management of fish and freshwater resources are needed to reduce the impacts from climate change, including conserving riparian and coastal wetlands that can buffer changes in sea level rise and freshwater flows. Human-induced pressures on fish populations from overfishing must be reduced so that fish populations have a chance of withstanding the additional pressures from warming seas and changes in seasonal current patterns. Human demand for increasing freshwater supplies needs to be addressed through water conservation and water reuse, thus allowing environmental flows to maintain riparian and wetland ecosystems.

Small-scale fishers, who lack mobility and livelihood alternatives and are often the most dependent on specific fisheries, will suffer disproportionately from such large-scale climatic changes. In Asia, 1 billion people are estimated to be dependent upon coral reef fisheries as a major source of protein, yet coral reef ecosystems are among the most threatened by global climate change. The combined effects of sea surface temperature rise and oceanic acidification could mean that corals will begin to disappear from tropical reefs in just 50 years; poor, rural coastal communities in developing countries are at the greatest risk and will suffer the greatest consequences (Donner and Potere, 2007; www.icsf.net). Climate change is a major threat to critical coastal ecosystems such as the Nile, the Niger and other low-lying deltas, as well as oceanic islands which may be inundated by rising sea levels. The environmental and socioeconomic costs, especially to fisheries communities in developing countries, could be enormous.

Water related risk can be reduced through adaptation and adoption of strategies to improve water productivity in rainfed farming systems. These strategies entail shifting from passive to active water management in rainfed farming systems and include water harvesting systems for supplemental irrigation, small scale off-season irrigation combined with improved cropping system management, including use of water harvesting, minimum tillage and mulch systems, improved crop varieties, improved cropping patterns (Molden et al., 2007), and particularly mitigation of soil degradation (Bossio et al., 2007). These existing technologies allow active management of rainfall (green water), rather than only managing river flows (blue water) (Rockstrom et al., 2007).

The scope for improvement is tremendous (Molden et al., 2007): rainfed farming covers most of the world croplands (80%), and produces most of the world's food (60-70%). Poverty is particularly concentrated in tropical developing countries in rural areas where rainfed farming is practiced (Castillo et al., 2007). Half of the currently malnourished are concentrated in the arid, semiarid and dry subhumid areas where agriculture is very risky due to extreme variability of rainfall, long dry seasons, and recurrent droughts, floods and dry spells (Rockstrom et al., 2007). Current productivity is generally very low (yields generally less than half of irrigated systems and in temperate regions where water risks are much lower). Even in these regions, there is generally enough water to double or often quadruple yields in rainfed farming systems. In these areas the challenge is to reduce water related risks rather than coping with absolute scarcity of water. With small investments large relative improvements in agricultural and water productivity can be achieved in rainfed agriculture. Small investments providing 1000 m^3 ha^{-1} (100 mm ha^{-1}) of extra water for supplemental irrigation can unlock the potential and more than double water and agricultural productivity in small-scale rainfed agriculture, which is a very small investment compared to the 10000-15000 m^3 ha^{-1} storage infrastructure required to enable full surface irrigation (Rockstrom et al., 2007). Provided that there are sufficient other factor inputs (e.g., N), the major hurdle for rain water harvesting and supplemental irrigation systems is cost effectiveness. Investment in R&D for low cost small scale technologies is therefore important to realize gains. This approach can address seasonal variability in rainfall (expected to increase with climate change) but have little impact in conditions of more severe interannual variability (very low rainfall), which can only be addressed by systems with storage (dams and groundwater) or buffering (lag in hydrologic response to that river flows are substantially maintained through drought periods).

Climate change will require a new look at water storage, to mitigate the impact of more extreme weather, cope with changes in total amounts of precipitation, and cope with changing distribution of precipitation, including shifts in ratios between snowfall and rainfall. Developing more storage (reservoirs and groundwater storage) and hydraulic infrastructure provides stakeholders with more influence in determining the precise allocation to desired activities including agriculture and hydropower production.

In the process of adapting to climate change multiple interests at the basin scale can be incorporated and managed, and tradeoffs with other livelihood and environmental interests included in the planning (Faurés et al., 2007). Storage will itself be more vulnerable to climatic extremes resulting from climate change, and therefore be less reliable. Furthermore, it will have proportionately greater impacts on wetland and riverine ecosystems, which are already under stress. The arguments on the relative merits of further storage will become sharper and more pressing (Molden et al., 2007). The role of groundwater as a strategic reserve will increase (Shah et al., 2007) How to plan appropriate and sustainable storage systems that address climate change is a pressing need for future AKST development.

6.8.2 Sustainable use of bioenergy

6.8.2.1 Liquid biofuels for transport
Current trends indicate that a large-scale expansion of production of 1st generation biofuels for transport will create huge demands on agricultural land and water—causing potentially large negative social and environmental effects, e.g., rising food prices, deforestation, depletion of water resources (see Chapter 4) that may outweigh positive effects. The following options are currently being discussed as means to alleviating these problems.

Reducing land and water requirements through increasing yields of agricultural feedstocks. Efforts are currently focused on increasing biofuel yields per hectare while reducing agricultural input requirements by optimizing cropping methods or breeding higher yielding crops. For example, Brazil has been able to increase yields and reduce crop vulnerability to drought and pests by developing more than 550 different varieties of sugar cane, each adapted to different local climates, rainfall patterns and diseases (GTZ, 2005). Both conventional breeding and genetic engineering are being employed to further enhance crop characteristics such as starch or oil content to increase their value as energy crops. There is a great variety of crops in developing countries that are believed to hold large yield potential but more research is needed to develop this potential (Cassman et al., 2006; Ortiz et al., 2006; Woods, 2006). However, even if yields can successfully be increased, several problems will persist for the production of liquid biofuels on a large scale.

Total land area under cultivation will still need to expand considerably in order to meet large-scale demand for biofuels and food production (Table 6-5).

Land availability and quality as well as social and environmental value and vulnerability of this land differ widely by country and region and needs to be carefully assessed at the local level (FAO, 2000; WBGU, 2003; European Environment Agency, 2006). Moreover, various studies predict that water will be a considerable limiting factor for which feedstock production and other land uses (e.g., food production, ecosystems) would increasingly compete (Giampietro et al., 1997; Berndes, 2002; De Fraiture et al., 2007). In addition to these environmental problems, special care must be taken to avoid displacement and marginalization of poor people who often have weakly enforceable or informal property and land-use rights and are thus particularly vulnerable (Fritsche et al., 2005; FBOMS, 2006; The Guardian, 2007).

Economic competitiveness will continue to be an issue. Even in Brazil, the world leader in efficient ethanol production, biofuels are competitive only under particularly favorable market conditions. To increase total land area under production, less productive areas would have to be brought into production, either for bioenergy feedstocks directly or for other agricultural crops which may be displaced on the most productive lands. This depends on economic incentives for farmers and investments in productivity enhancements and could have strong effects on agricultural systems and further accentuate food price effects.

Environmental concerns, associated with issues such as high-input feedstock production, the conversion of pristine land for agricultural production, the employment of trans-

Table 6-5. Land area requirements for biofuels production.

Percentage of total 2005 global crude oil consumption to be replaced by bioenergy	Energy yield			
	1st generation biofuels		Next generation biofuels	
	40 GJ/ha	60 GJ/ha	250 GJ/ha	700 GJ/ha
5% ~ 1500 million barrels/year	230 million ha	153 million ha	37 million ha	13 million ha
10% ~ 3010 million barrels/year	460 million ha	307 million ha	74 million ha	26 million ha
20% ~ 6020 million barrels/year	921 million ha	614 million ha	147 million ha	53 million ha

Conversion factors: 1 GJ=0.948 million BTU; 1 barrel of oil ~ 5.8 million BTU

Source: Avato, 2006.

genic crops, the depletion of water resources as well as the problematic resemblance of some biofuels feedstocks with invasive species (Raghu et al., 2006) need to be carefully assessed with special emphasis on the local context.

Producing biofuels from inedible feedstock and on marginal lands. It is often argued that using inedible energy crops for the production of biofuels would reduce pressures on food prices. Moreover, many of these crops, e.g., *Jatropha*, poplar and switchgrass, could be grown productively on marginal land, without irrigation and potentially even contributing to environmental goals such as soil restoration and preservation (GEF, 2006; IEA, 2004; Worldwatch Institute, 2006).

Inedible feedstocks. Food price increases can be caused directly, through the increase in demand for the biofuel feedstock, or indirectly, through the increase in demand for the factors of production (e.g., land and water). For example, land prices have risen considerably in the US "corn belt" over the past years—an effect that is largely attributed to the increased demand for ethanol feedstocks (Cornhusker Economics, 2007; Winsor, 2007). Such factor price increases lead to increasing production costs of all goods for which they are used as inputs. Thus, using nonedible plants as energy feedstocks but growing them on agricultural lands may only have a limited mitigating effect on food prices.

Marginal lands. Cultivating energy crops on degraded land or other land not currently under agricultural production is often mentioned as an option but it is not yet well understood. Several key issues deserve further attention: (1) The production of energy crops on remote or less productive land would increase biofuels production costs (due to lower yields, inefficient infrastructure, etc.), leading to low economic incentives to produce on these lands. In fact, while estimates of available marginal land are large, especially in Africa and Latin America (FAO, 2000; Worldwatch Institute, 2006), much of this land is remotely located or not currently suitable for crop production and may require large investments in irrigation and other infrastructure. (2) Environmental effects of bringing new stretches of land into production are problematic and need to be carefully analyzed, especially with regards to soil erosion, water resources and biodiversity.

Development of next generation biofuels. Significant potential is believed to lie with the development of new energy

conversion technologies—next generation biofuels. Several different technologies are being pursued, which allow the conversion into usable energy not only of the glucose and oils retrievable today but also of cellulose, hemicellulose and even lignin, the main building blocks of most biomass. Thereby, cheaper and more abundant feedstocks such as residues, stems and leaves of crops, straw, urban wastes, weeds and fast growing trees could be converted into biofuels (IEA, 2006; Ortiz et al., 2006; Worldwatch Institute, 2006; DOE, 2007). This could significantly reduce land requirements, mitigating social and environmental pressures from large-scale production of 1st generation biofuels (Table 6-5). Moreover, lifecycle GHG emissions could be further reduced, with estimates for potential reductions ranging from 51 to 92% compared to petroleum fuels (IEA, 2004; European Commission, 2005; GEF, 2005; Farrell et al., 2006). However there are also environmental concerns associated with potential overharvesting of agricultural residues (e.g., reducing their important services for soils) and the use of bioengineered crops and enzymes.

The most promising next generation technologies are cellulosic ethanol and biomass-to-liquids (BTL) fuels. Cellulosic ethanol is produced through complex biochemical processes by which the biomass is broken up to allow conversion into ethanol of the cellulose and hemicellulose. One of the most expensive production steps is the pretreatment of the biomass that allows breaking up the cellulose and removing the lignin to make it accessible for fermentation. Research is currently focused on how to facilitate this process, e.g., through genetically engineering enzymes and crops. BTL technologies are thermo-chemical processes, consisting of heating biomass, even lignin-rich residues left over from cellulosic ethanol production, under controlled conditions to produce syngas. This synthetic gas (mainly of carbon monoxide and hydrogen), is then liquefied e.g., by using the Fischer-Tropsch (FT) process to produce different fuels, including very high-quality synthetic diesel, ethanol, methanol, buthanol, hydrogen and other chemicals and materials. Research is also focusing on integrating the production of next generation biofuels with the production of chemicals, materials and electricity in biorefineries (Aden et al., 2002; IEA, 2004; GEF, 2006; Hamelinck and Faaij, 2006; IEA, 2006; Ledford, 2006; Ragauskas et al., 2006; Woods, 2006).

Next generation biofuels have to overcome several critical steps in order to become a viable and economic source of transport fuels on a large scale and be able to contribute to the development and sustainability goals. First, next generation biofuels technologies have not yet reached a stage of commercial maturity and significant technological challenges need to be overcome to reduce production costs. It is not yet clear when these breakthroughs will occur and what degree of cost reductions they will be able to achieve in practice (Sanderson, 2006; Sticklen, 2006; DOE, 2007). The U.S. Department of Energy has set the following ambitious goals for its cellulosic ethanol program: reducing the cost per liter from US$0.60 to 0.28 and capital investment costs from currently $0.80 to 0.49 by 2012 (DOE, 2007). Second, even if these breakthroughs occur, biofuels will have to compete with other energy technologies that are currently being developed in response to high oil prices. For example, with regards to transport fuels, technological progress is currently reducing costs of conventional (e.g., deep sea) and unconventional (e.g., tar sands) oil production and also of coal and gas to liquid technologies. Third, while countries like South Africa, Brazil, China and India are currently engaged in advanced domestic biofuels R&D efforts, high capital costs, large economies of scale, a high degree of technical sophistication as well as intellectual property rights issues make the production of next generation biofuels problematic in the majority of developing countries even if the technological and economic hurdles can be overcome in industrialized countries.

6.8.2.2 Bioenergy and rural development

Living conditions and health of the poor can be considerably improved when households have the opportunity to upgrade from inefficient, polluting and often hazardous traditional forms to modern forms of energy. Through their importance for the delivery of basic human needs such as potable water, food and lighting, these modern energy services are among the primary preconditions for advancements in social and economic development (Barnes and Floor, 1996; Cabraal et al., 2005; Modi et al., 2006). Moreover, bioenergy and ancillary industries may promote job creation and income generation. However, the balance of positive and negative effects of different forms of bioenergy is subject to significant debate and is highly context specific. Careful assessments of local needs, economic competitiveness as well as social and environmental effects are needed to determine under which circumstances modern bioenergy should be promoted.

The domestic production of biofuels from agricultural crops (1st generation) is often credited with positive externalities for rural development through creating new sources of income and jobs in feedstock production and energy conversion industries (e.g., Moreira and Goldemberg, 1999; von Braun and Pachauri, 2006; Worldwatch Institute, 2006). However, the actual effect of 1st generation biofuels production on rural economies is complex and has strong implications for income distribution, food security and the environment.

Economically, the major impact of biofuels production is the increase in demand for energy crops. In fact, biofuels have historically been introduced as a means to counteract weak demand or overproduction of feedstock corps, e.g., this was a principal reason for Brazil to introduce its ProAlcool Program in 1975 (Moreira and Goldemberg, 1999). On the one hand, this additional demand can increase incomes of agricultural producers, increase productivity enhancing investments and induce dynamic processes of social and economic development (FAO, 2000; Coelho and Goldemberg, 2004; DOE, 2005; Worldwatch Institute, 2006).

On the other hand, this needs to be evaluated against economic, social and environmental costs that may arise from large increases in biofuels production. First, even if biofuels can be produced competitively, at least part of the rise in agricultural incomes would represent a mere redistribution of income from consumers of agricultural products to producers. The extent of this redistribution depends on the degree to which food prices are affected. Second, in cases when biofuels are promoted despite having higher costs than petroleum fuels, an analogous redistribution from energy consumers to agricultural producers takes place. In both cases the effects on poverty are highly complex. Some rural poor may gain if they can participate in the energy crop production, biofuel conversion and ancillary sectors or otherwise benefit from increased economic activity in rural areas. This depends critically on aspects such as production methods (e.g., degree of mechanization) and institutional arrangements (e.g., structure of the agricultural sector, property rights of agricultural land and security of land tenure). Conversely, those rural and urban poor people who spend a considerable share of their incomes on energy and especially food are bound to lose if they have to pay higher prices. Food-importing developing countries would also suffer under globally rising food prices. Time lags in the response of producers to increased feedstock demand may lead price increases to be more accentuated in the short-term than in the medium to long-term.

Biofuels are considerably more labor intensive in production than other forms of energy such as fossil fuels and thus they are often proposed as a means for improving employment in the agricultural sector as well as in other downstream industries that process by-products such as cakes and glycerin (Goldemberg, 2004; Worldwatch Institute, 2006). However, estimating actual effects on employment is highly complex. First, any newly created employment needs to be weighed against jobs that are displaced in other sectors, including jobs that would have been created in the feedstock production sector even in the absence of biofuels production. These dynamics are complex and may involve very different industries, e.g., the livestock industry, food processors and other major user of agricultural crops (CIE, 2005).

Second, while bioenergy is labor intensive compared to other energy industries, it is not necessarily labor intensive compared to other forms of farming. In fact, energy crop production very often takes the form of large-scale mechanized farming. Thus, in cases where traditional farming is replaced by less labor intensive energy crop production, jobs may actually be lost. Similarly, no new jobs are created if biofuels production simply displaces other agricultural crops. It is unsure whether such job substitution is actually beneficial, especially considering that many jobs in feedstock production are temporary and seasonal (Fritsche et al., 2005; Kojima and Johnson, 2005; Worldwatch Institute, 2006).

Consequently, the overall effects on employment and incomes are highly complex and context specific and there is no consensus on magnitude or even direction of net effects. Even if in certain cases longer term dynamic effects may dominate for the economy as a whole, the considerable risks of welfare losses for certain stakeholders warrant careful consideration—especially with regard to the most vulnerable persons. More research is needed to develop and apply interdisciplinary tools that assess these issues more clearly (e.g., economic cost-benefit analysis).

Development of small-scale applications for biodiesel and unrefined bio-oils. The environmental and social costs of producing biofuels can be considerably lower in small-scale applications for local use due to more contained demands on land, water and other resources. At the same time, the benefits for social and economic development may be higher, especially in remote regions, where energy access and agricultural exports are complicated by high transport costs (Kojima and Johnson, 2005). Landlocked developing countries, small islands, and also remote regions within countries may fall into this category—if they can make available sufficient and cheap feedstock without threatening food security. Especially biodiesel offers potential in small-scale applications as it is less technology and capital intensive to produce than ethanol. Unrefined bio-oils offer similar benefits and their production for stationary uses such as water pumping and power generation is being analyzed in several countries, e.g., focusing on *Jatropha* as a feedstock (Indian Planning Commission, 2003; Van Eijck and Romijn, 2006). Such schemes may offer particular potential for local communities when they are integrated in high intensity small-scale farming systems which allow an integrated production of food and energy crops. More research is needed on the costs and benefits to society of these options, taking into consideration also other energy alternatives.

Conduct R&D on electricity and heat generation technologies from biomass to improve operational reliability. Some forms of bioelectricity and bioheat can be competitive with other off-grid energy options (e.g., diesel generators) and therefore are viable options for expanding energy access in certain settings. The largest potential lies with the production of bioelectricity and heat when technically mature and reliable generators have access to secure supply of cheap feedstocks and capital costs can be spread out over high average electricity demand. This is mostly the case on site or near industries that produce biomass wastes and residues and have their own steady demand for electricity, e.g., sugar, rice and paper mills. The economics as well as environmental effects are particularly favorable when operated in combined heat and power mode. Biomass digesters and gasifiers are more prone to technical failures that direct combustion facilities, especially when operated in small-scale applications without proper maintenance. More research and development is needed to improve the operational stability of these technologies as well as the design of institutional arrangements, including potential integration with biomass processing industries, livestock holdings and mixed farming. However, modern bioenergy is only one of several options available for advancing energy access and in each case local alternatives need to be compared regarding economic costs as well as social and environmental externalities (Table 6-6).

Table 6-6. **Bioenergy: Potential and limitations.**

Technological Application	Potential Benefits	Risks and Limitations	Options for Action
1st Generation Biofuels	• Energy security • Income and employment creation • GHG emission reductions	• Limited economic competitiveness • Social concerns, (e.g., pressures on food prices) • Environmental concerns (e.g., depletion of water resource, deforestation) • GHG emission reductions strongly dependent on circumstances	• R&D on improving yields of feedstocks and fuel conversion • More research on social, environmental and economic costs and benefits • Policies/initiatives furthering social and environmental sustainability
Next Generation Biofuels	• Larger production potential and better GHG balance than 1st generation • Less competition with food production	• Unclear when technology will be commercially viable • High capital costs and IPR issues limit benefits for developing countries and small-scale farmers • Issues with over-harvesting of crop residues, GMOs	• Increase R&D to accelerate commercialization • Develop approaches to improve applicability in developing countries and for small-scale farmers
Bioelectricity and Bioheat (large-scale)	• Low GHG emissions • Favorable economics in certain off-grid applications (e.g., bagasse cogeneration)	• Issues with operational reliability and costs • Logistical challenges of feedstock availability	• Develop demonstration projects, product standards • Disseminate knowledge • Access to finance
Bioelectricity and Bioheat (small-scale)	• Potential for increasing energy access sustainably in off grid areas with low energy demand using locally available feedstocks	• Costs, operational reliability, maintenance requirements	• R&D on small-scale stationary uses of biodiesel and bio-oils • Capacity building on maintenance

References

Abawi, G.S., and J. Chen. 1998. Concomitant pathogen and pest interactions. p. 135-158. *In* Plant and nematode interactions. Mono. 36. K.R. Barker et al. (ed) ASA, Madison WI.

Abbiw, D. 1990. Useful plants of Ghana: West African uses of wild and cultivated plants. Intermediate Tech. Publ., Roy. Botanic Gardens, Kew.

Abdel-Dayem, S. 1999. A framework for sustainable use of low quality water in irrigation. Proc. 17th Int. Congress on Irrigation and Drainage Granada, Spain. Available at http://lnweb18.worldbank.org/ESSD/ardext.nsf/11ByDocName/AframeworkforSustaunableUseofLowQualityWaterinIrrigation/$FILE/Lowqualitywaterinirrigation.pdf. World Bank, Washington DC.

Abdoellah, O.S., H.Y. Hadikusumah, K. Takeuchi, S. Okubo, and Parikesit. 2006. Commercialization of homegardens in an Indonesian village: Vegetation composition and functional changes. Agroforest. Syst. 68:1-13.

Adams, R.M., L.L. Houston, B.A. McCarl, M. Tiscareño, J. Matus, and R.F. Weiher. 2003. The benefits to Mexican agriculture of an El Niño-southern oscillation (ENSO) early warning system. Agric. For. Meteorol. 115:183-194.

Adelman, H. 1998. Defining humanitarian early warning. *In* S. Schmeidl and H. Adelman (ed) Early warning and early response. Available at http://www.ciaonet.org/book/schmeidl/. Columbia Int. Affairs, NY.

Aden, A., M. Ruth, K. Ibsen, J. Jechura, K. Neeves, J. Sheehan et al. 2002. Lingocellulosic biomass to ethanol process design and economics utilizing co-current dilute acid prehydrolysis and enzymatic hydrolysis for corn stover. NREL, Golden CO.

Adi, B. 2006. Intellectual property rights in biotechnology and the fate of poor farmers' agriculture. J. World Intellect. Prop. 9:91-112.

Adrian, M.A., S.H. Norwood, and P.L. Mask. 2005. Producers's perceptions and attitudes toward precision agriculture technologies. Computers Electron. Agric. 48:256-271.

Adriano, D.C., J.M. Bollage, W.T. Frankenberger, Jr, and R.C. Sims (ed) 1999. Bioremediation of contaminated soils. Agron. Mono. 37. ASA-CSSA-SSSA, Madison WI.

Ahmed, M., H. Turral, I. Masih, M. Giordano, and Z. Masood. 2007. Water saving technologies: myths and realities revealed in Pakistan's rice-wheat systems. Res. Rep. 108. IWMI, Colombo.

Ahmed, M.N. 1998. Fingerling stocking in open waters. p. 201-208. *In* H.A.J. Middendorp et al. (ed) Sustainable inland fisheries management in Bangladesh. ICLARM Conf. Proc. 58. ICLARM, Penang.

Ahuja, V., P.S. George, S. Ray, K.E. McConnell, M.P.G. Kurup, V. Gandhi, et al. 2000. Agricultural services and the poor: Case of livestock health and breeding services in India. Indian Inst. Management, Ahmedabad.

Akyeampong, E., and L. Hitimana. 1996. Agronomic and economic appraisal of alley cropping with Leucaena diversifolia on an acid soil in the highlands of Burundi. Agrofor. Syst. 33:1-11.

Alabouvette, C., D. Backhouse, C. Steinberg, N.J. Donovan, V. Edel-Hermann, and L.W. Burgess. 2004. Microbial diversity in soil — effects on crop health. p. 121-138. *In* P. Schjoning et al. (ed) Managing soil quality: Challenges in modern agriculture. CAB Int., Wallingford.

Al-Ahmad, H., J. Dwyer, M. Moloney, and J. Gressel. 2006. Mitigation of establishment of Brassica napus transgenes in volunteers using a tandem construct containing a selectively unfit gene. Plant Biotech. J. 4:7-21.

Albrecht, A., and S. Kandji. 2003. Review. Carbon sequestration in tropical agroforestry systems. Agric. Ecosyst. Environ. 99:15-27.

Alfaia, S.S., G.A. Ribeiro, A.D. Nobre, R.C. Luizão, and F.J. Luizão. 2004. Evaluation of soil fertility in smallholder agroforestry systems and pastures in western Amazonia. Agric. Ecosyst. Environ. 102:409-414.

Alkorta, I., and G. Garbisu. 2001. Review paper. Phytoremediation of organic contaminants in soils. Bioresource Tech. 79:273-276.

Alocilja, E.C., and S. Radke. 2003. Commercializing biosensors. Biosens. Bioelectron. 18:841-846.

Altieri, M.A. 1987. Agroecology: The scientific basis of alternative agriculture. Westview Press, Boulder.

Altieri, M.A. 1999. The ecological role of biodiversity in agroecosystems. Agric. Ecosyst. Environ. 74:19-31.

Altieri, M.A. 2002. Review: Agroecology: The science of natural resource management for poor farmers in marginal environments. Agric. Ecosyst. Environ. 93:1-24.

Anderson, L.S., and F.L. Sinclair. 1993. Ecological interactions in agroforestry systems. Agroforest. Abstr. 6:57-91.

Anderson, P.K., A. A. Cunningham, N.G. Patel, F.J. Morales, P.R. Epstein, and P. Daszak. 2004. Emerging infectious diseases of plants: Pathogen pollution, climate change and agrotechnology drivers. Trends Ecol. Evol. 19:10:535-544.

Anderson, R.C., R.B. Harvey, J.A. Byrd, T.R. Callaway, K.J. Genovese, T.S. Edrington et al. 2005. Novel preharvest strategies involving the use of experimental chlorate preparations and nitro-based compounds to prevent colonization of food-producing animals by foodborne pathogens. Poultry Sci. 84:649-654.

Ashe, A., and E. Whitelaw. 2007. Another role for RNA: a messenger across generations. Trends Genet. 23:8-10.

Ashley, C., and S. Maxwell. 2001. Rethinking rural development. Dev. Policy Rev. 19:395-425.

Attipoe, L., A. van Andel, and S.K. Nyame, 2006. The Novella project. Developing a sustainable supply chain for Allanblackia oil. p. 179-189. *In* R. Ruben et al. (ed) Agro-food chains and networks for development. Available at http://library.wur.nl/frontis/agro-food_chains/15_attipoe.pdf. Springer, Netherlands.

Avato, P.A. 2006. Bioenergy: An assessment. Sustainable development network background paper. World Bank, Washington DC.

Azam-Ali, S. 2006. Can underutilized crops be used for agricultural diversification in Africa's changing climates? Trop. Agric. Assoc. Newslet. 26(2):17-20.

Azam-Ali, S.N., A. Sesay, S.K. Karikari, F.J. Massawe, J. Anguilar-Manjarrez, and M. Bannayan. 2001. Assessing the potential of an underutilized crop — a case study using Bambara groundnut. Exp. Agric. 37:433-472.

Badgley, C., J. Moghtader, E. Quintero, E. Zakem, M.J. Chappell, K. Aviles-Vazquez et al. 2007. Organic agriculture and the global food supply. Renew. Agric. Food Syst. 22(2):86-108.

Baenziger, P.S., W.K. Russell, G.L. Graef, and B.T. Campbell. 2006. Improving lives: 50 years of crop breeding, genetics, and cytology (C-1). Crop Sci 46:2230-2244.

Baggs, E.M., M. Stevenson, M. Philatie, A. Regar, H. Cook, and G. Cadiz. 2003. Nitrous oxide emissions following application of residues and fertilizer use under zero and conventional tillage. Plant Soil 254:361-370.

Bailey, S.W. 2004. Climate change and decreasing herbicide persistence. Pest Manage. Sci. 60:158-162.

Baker, K.M., W.W. Kirk, J.M. Stein, and J.A. Andersen. 2004. Climatic trends and potato late blight risk in the upper Great Lakes region. HortTech. 15(3):510-518.

Bakker, M., R. Barker, R. Meinzen-Dick, and F. Konradsen (ed) 1999. Multiple uses of water in irrigated areas: A case study from Sri Lanka. SWIM Pap. no. 8. IWMI, Colombo.

Bakun, A., and S.J. Weeks. 2006. Adverse feedback sequences in exploited marine systems: are deliberate interruptive actions warranted? Fish Fisheries 7(4):316-333.

Ball, B.C., A. Scott, and J.P. Parker. 1999. Field N_2O, CO_2 and CH_4 fluxes in relation to tillage, compaction and soil quality in Scotland. Soil Till. Res. 53:29-39.

Banu, S.P., M.A. Shaheed, A.A. Siddique, M.A. Nahar, H.U. Ahmed, and M.H. Devare. 2005. Soil biological health: A major factor in increasing the productivity of the rice-wheat cropping system. Int. Rice Res. Notes 30(1):5-11.

Barany, M., A.L. Hammett, R.R.B. Leakey, and K.M. Moore. 2003. Income generating opportunities for smallholders affected by HIV/AIDS: Linking agro-ecological change and non-timber forest product markets. J. Manage. Studies 39:26-39.

Barnes, D.F., and W.M. Floor. 1996. Rural energy in developing countries: A challenge for economic development. Ann. Rev. Energy Environ. 21:497-530.

Barrett, C.B., T. Reardon, and P. Webb. 2001. Nonfarm income diversification and household livelihood strategies in rural Africa: Concepts, dynamics, and policy implications. Food Policy 26:315-331.

Barrios, E., R.J. Delve, M. Bekunda, J. Mowo, J. Agunda, and J. Ramisch. 2006. Indicators of soil quality: A South–South development of a methodological guide for linking local and technical knowledge. Geoderma 135:248-259.

Bates, S.L., J.Z. Zhao, R.T. Roush, and A.M. Shelton. 2005. Insect resistance management in GM crops: past, present and future. Nature Biotech. 23:57-62.

Bavinck, M., R. Chuenpagdee, M. Diallo, P. van der Heijden, J. Kooiman, R. Mahon et al. 2005. Interactive governance for fisheries: A guide to better practice. Eburon Acad. Publ., Delft.

Bedford, A., M. Blowfield, D. Burnett, and P. Greenhalgh. 2001. Value chains: Lessons from the Kenya tea and Indonesia cocoa sectors. Focus 3. Resource Centre Soc. Dimens. Business Practice, London.

Behrenfeld, M.J., R.T. O'Malley, D.A. Siegel, C.R. McClain, J.L. Sarmiento, and G.C. Feldman. 2006. Climate-driven trends in contemporary ocean productivity. Nature 444(7120):752-755.

Bell, A., F. Mazaud, and O. Mueck. 1999. Guidelines for the analysis of post-production systems. FAO, Rome.

Benites, J.R., and J.E. Ashburner. 2003. FAO's role in promoting conservation agriculture. p. 139-153. In L. Garcia-Torres et al. (ed) Conservation agriculture: Environment, farmers experience, innovations, socio-economy, policy. Kluwer Acad. Publ., Netherlands.

Bennett, J. W., and M. Klich. 2003. Mycotoxins. Clin. Microbiol. Rev. 16:497-516

Berndes, G. 2002. Bioenergy and water — the implications of large-scale bioenergy production for water use and supply. Global Environ. Change 12:253-271.

Bernet, T., A. Devaux, O. Ortiz, and G. Thiele. 2005. Participatory market chain approach. Beraterinnen News 1:8-13.

Berntsen, J., J.E. Olesen, B.M. Petersen, and E.M. Hansen. 2006. Long-term fate of nitrogen uptake in catch crops. Eur. J. Agron. 25:83-390.

Best, R., S. Ferris, and A. Schiavone. 2005. Building linkages and enhancing trust between small-scale rural producers, buyers in growing markets and suppliers of critical inputs. p. 19-50 In F.R. Almond, and S.D. Hainsworth (ed) Proc. Int. Seminar Beyond agriculture: Making markets work for the poor. London, 28 Feb-1 Mar 2005. CPHP and Practical Action, UK.

Beyer, P., S. Al-Babili, X. Ye, P. Lucca, P. Schaub, R. Welsch et al. 2002. Golden rice: Introducing the beta-carotene biosynthesis pathway into rice endosperm by genetic engineering to defeat vitamin A deficiency. J. Nutr. 132:506S-510.

Biénabe, E., and D. Sautier. 2005. The role of small scale producers' organizations in addressing market access. p. 69-85. In F.R. Almond, and S.D. Hainsworth (ed) Proc. Int. Seminar: Beyond agriculture: Making markets work for the poor. London, 28 Feb-1 Mar 2005. CPHP and Practical Action, UK.

Binnig, G., and H. Rohrer. 1985. Scanning tunneling microscopy. Sci. Am. 253:40-46.

Bird, A. 2007. Perceptions of epigenetics. Nature 447:396-398.

Birrel, S.J., and J.W. Hummel. 2001. Real-time multi ISFET/FIA soil analysis system with automatic sample extraction. Computers Electron. Agric. 32(1):45-67.

Blackmore, S. 2000. Developing the principles of precision farming. p. 11-13. In Proc. ICETS 2000. China Agric. Univ., Beijing.

Blake, M. 1997. Bibliography of work on photocatalytic removal of hazardous compounds from water and air. NREL/TP-430-22197. Nat. Renew. Energy Lab., Golden CO.

Bletsos, F.A. 2005. Use of grafting and calcium cyanamide as alternatives to methyl bromide soil fumigation and their effects on growth, yield, quality and Fusarium wilt control in melon. J. Phytopathol. 153:155-161.

Boardman, J. 2006. Soil erosion science: Reflections on the limitations of current approaches. Catena. 68:73-86.

Bolliger, A., J. Magid, T.J.Carneiro Amado, F. Skóra Neto, M. de F. Dos Santos Ribeiro, and A. Calegari. 2006. Taking stock of the Brazilian "zero-till revolution": A review of landmark research and farmers' practice. Adv. Agron. 91:47-109.

Bos, M.G. 2004. Using the depleted fraction to manage the groundwater table in irrigated areas. Irrig. Drain. Syst. 18:201-209.

Bosc, P., M. Eychenne, K. Hussen, B. Losch, M.R. Mercoiret, and P. Rondot. 2002. The role of rural producer organizations in the World Bank rural development strategy. Rural Dev. Strategy Background Pap. World Bank, Washington DC.

Bossio, D., W. Critchley, K. Geheb, G. van Lynden, and B. Mati. 2007. Conserving land, protecting water. p. 551-583. In D. Molden (ed) Water for food, water for life: A comprehensive assessment of water management in agriculture. Earthscan, London and IWMI, Colombo.

Bossio, D., A. Noble, D. Molden, and V. Nangia. 2008. Land degradation and water productivity in agricultural landscapes. In D. Bossio and K. Geheb (ed) Conserving land, protecting water: 'Bright Spots' reversing trends in land and water degradation. CABI, UK.

Bouman, B., R. Barker, E. Humphreys, and T.P. Tuong. 2007. Rice: Feeding the billions. In D. Molden (ed) Water for food, water for life: A comprehensive assessment of water management in agriculture. Earthscan, London, and IWMI, Colombo.

Bourgeois, G., A. Bourque, and G. Deaudelin. 2004. Modeling the impact of climate change on disease incidence: a bioclimatic challenge. Can. J. Plant Pathol. 26:284-290.

Brauch, H.G. 2002. Climate change, environmental stress and conflict. p. 9-112 In Climate change and conflict. Fed. Ministry Environ., Nature Conserv. Nuclear Safety, Berlin.

Brookfield, H., H. Parsons, M. Brookfield (ed) 2003. Agrodiversity: Learning from farmers across the world. UNU Press, Tokyo.

Brown, D.L. 2003. Solutions exist for constraints to household production and retention of animal food products. J. Nutr. 133:4042S-4047S.

Brush, S.B., and E. Meng. 1998. Farmer's valuation and conservation of crop genetic resources. Genet. Res. Crop Eval. 45: 139-150.

Buck, L.E., T.A. Gavin, D.R. Lee, N.T. Uphoff, D. Chandrasekharan Behr, L.E. Drinkwater et al. 2004. Ecoagriculture: A review and assessment of its scientific foundations. Cornell Univ. and Univ. Georgia, SANREM CRSP, Athens GA.

Bunch, R. 1999. More productivity with fewer external inputs: Central American case studies of agroecological development and their broader implications. Environ. Develop. Sustain. 1(3-4):219-233.

Burt, C.M., D.J. Howes, and A. Mutziger. 2001. Evaporation estimates for irrigated agriculture in California. p. 103-110. In Proc. Ann. Irrig. Assoc. Meeting. Available at http://www.itrc.org/ papers/ EvaporationEstimates/ EvaporationEstimates. pdf. Irrigation Assoc, Falls Church VA.

Burton, I., and B. Lim. 2005. Achieving adequate adaptation in agriculture. Clim. Change 70:191-200.

Cabraal, R.A., D.F. Barnes, and S.G. Agarwal. 2005. Productive use of energy for rural development. Ann. Rev. Environ. Resour. 30:117-144.

Campbell, L.G., A.A. Snow, and C.E. Ridley. 2006. Weed evolution after crop gene introgression: greater survival and fecundity of hybrids in a new environment. Ecol. Lett. 9:1198-1209.

Cardille, J.A., and J.A. Foley. 2003. Agricultural land-use change in Brazilian Amazonia between 1980 and 1995. Remote Sensing Environ. 87:551-582.

Cardoso, I., and T.W. Kuyper. 2006. Mycorrhizas and tropical soil fertility. Agric. Ecosyst. Environ. 116:76-84.

Carruthers, R.I. 2003. Invasive species research in the United States Department of Agriculture. Agric. Res. Service. Pest Manage. Sci. 59:827-834.

Carter, J. 1995. Alley farming: Have resource-poor farmers benefited? ODI Nat. Resource Perspect. No. 3, Jun. ODI, London.

Cassman, K., V. Eidman, and E. Simpson. 2006. Convergence of agriculture and energy: Implications for research and policy. CAST Commentary QTA2006-3. Available at http://www.cast-science.org/websiteUploads/publicationPDFs/QTA2006-3.pdf. CAST, Washington DC.

Castilla, N., J. Hernandez, and A.F. Abou-Hadid. 2004. Strategic crop and greenhouse management in mild winter climate areas. Acta Hort. 633:183-196.

Castillo G.E., R.E. Namara, H.M. Ravnborg, M.A. Hanjra, L. Smith, and M.H. Hussein. 2007. Reversing the flow: Agricultural water management pathways for poverty reduction. In D. Molden (ed) Water for food, water for life: A comprehensive assessment of water management in agriculture. Earthscan, London and IWMI, Colombo.

CAZS (Centre for Arid Zones Studies). 2006. Transplanting sorghum and millet, a key risk management in semi-arid areas. Available at http://www.bangor.ac.uk/transplanting. CAZS, Bangor, UK.

Center for Food Safety. 2005. Monsanto vs. U.S. Farmers. CFS, Washington DC.

Chakaborty, S. 2005. Potential impact of climate change on plant-pathogen interactions. Australas. Plant Pathol. 34:443-448.

Chacko, C.T. and F. Schneider. 2005. Breeding services for small dairy farmers-sharing the Indian experience. Oxford and IBH Publ., New Delhi.

Chaney, R.L, Y.M. Li, S.L. Brown, F.A. Homer, M. Malik, J.S. Angle et al. 2000. Improving metal hyperaccumulator wild plants to develop commercial phytoextraction systems: Approaches and progress. p. 129-158. In N. Terry et al. (ed) Phytoremediation of contaminated soil and water. Lewis Publ., Boca Raton FL.

Chaney, R.L., P.G. Reeves, J.A. Ryan, R.W. Simmons, R.M. Welch, and J.S. Angle. 2004. An improved understanding of soil cd risk to humans and low cost methods to remediate soil cd risks. Biometals 17:5:549-553.

Chen, M., X.Liu, Z. Wang, J. Song, Q. Qi, and P.G. Wang. 2005. Modification of plant N-glycans processing: The future of producing therapeutic protein by transgenic plants. Med. Res. Rev. 25:343-360.

Chen, Y., A. Gamliel, J.J. Stapleton, and T. Aviad. 1991. Chemical, physical, and microbial changes related to plant growth in disinfected soils. p. 103-129. In J. Katan and J.E. DeVay (ed) Soil solarization. CRC Press, Boca Raton FL.

Cherr, C.M.J., M.S. Scholberg, and R. McSorley. 2006. Green manure approaches to crop production: A synthesis. Agron. J. 98(2):302-319.

Cherrett, I. 1999. Soil conservation by stealth. Int. Agric. Develop. 19(3):13-15.

Chowdhury, S., A. Gulati, and E. Gumbira-Sa'id. 2005. High value products, supermarkets and vertical arrangements in Indonesia. MTID Disc. Pap. 83. IFPRI, Washington DC.

Christie, P., D.L. Fluharty, A.T. White, R.L. Eisma-Osorio, and W. Jatulan. 2007. Assessing the feasibility of ecosystem-based fisheries management in tropical context. Marine Policy 31:239-250.

CIE (Centre for International Economics). 2005. Impact of ethanol policies on feedgrain users in Australia. Report prepared for MLA on behalf of the Australian Beef Industry. CIE, Canberra.

CIMMYT. 2005. Sounding the alarm on global stem rust. An assessment of race Ug99 in Kenya and Ethiopia and the potential for impact in neighboring regions and beyond. CIMMYT, Mexico.

Clarke, A., E.J. Murphy, M.P. Meredith, J.C. King, L.S. Peck, and D.K.A. Barneset. 2007. Climate change and the marine ecosystem of the western Antarctic Peninsula. Philos. Trans. R. Soc. Lond. Ser. B 362(1477):149-166.

Cleaveland, S., M.K. Laurenson, and L.H. Taylor. 2001. Diseases of humans and their domestic mammals: pathogen characteristics, host range and the risk of emergence. Phil. Trans. R. Soc. Lond. Ser. B 356:991-999.

Clements, D.R., A. DiTommaso, N. Jordan, B.D. Booth, J. Cardina, and D. Doohan. 2004. Adaptability of plants invading North American cropland. Agric. Ecosyst. Environ. 104:379-398.

Coelho, S.T., and J. Goldemberg. 2004. Alternative transportation fuels: Contemporary case studies. Encycl. Energy 1:67-80.

Cogoni, C., and G. Macino. 2000. Post-transcriptional gene silencing across kingdoms. Curr. Opin. Genet. Dev. 10:638-643.

Cohen, J.I. 2005. Poorer nations turn to publicly developed GM crops. Nature Biotech. 23:27-33.

Colfer, P.C.J. (ed) 2004. The equitable forest: diversity and community in sustainable resource management. RFF Press, Washington DC.

Colfer, P.C.J. (ed) 2005. The complex forest: communities, uncertainty, and adaptive collaborative management. RFF Press, Washington DC.

Collins, W.W., and C.O. Qualset (ed) 1999. Biodiversity in agroecosystems. CRC Press, New York.

Colvin, V.L. 2003. The potential environmental impact of engineered nanomaterials. Nature Biotech. 21:1166-1170.

Cornhusker Economics. 2007. Ethanol fueling land market advances. Available at http://www.agecon.unl.edu/Cornhuskereconomics/3-21-07.pdf. Univ. Nebraska, Lincoln.

Costanza, R., R. d'Arge, R. de Groot, S. Farber, M. Grasso, and B. Hannon. 1997. The value of the world's ecosystem services and natural capital. Nature 387:253-260.

Court, E.B., A.S. Daar, and D.L. Persad. 2005. Tiny technologies for the global good. Nanotoday. May 2005. p. 14-15.

Cremers, L., M. Ooijevaar, and R. Boelens. 2005. Institutional reform in the Andean irrigation sector: Enabling policies for strengthening local rights and water management. Nat. Res. Forum 29:37-50.

Cromwell, E. 1999. Agriculture, biodiversity and livelihoods: Issues and entry points. Overseas Dev. Inst., London.

Daberkow, S.G., and W.D. McBride. 2001. Adoption of precision agriculture technologies by U.S. farmers. In P.C. Robert et al. (ed) Precision agriculture [CD-ROM]. Proc. Int. Conf., Minneapolis, 16-19 Jul 2000. ASA-CSSA-SSSA, Madison WI.

Daily, G.C. 1997. Nature's services: Societal dependence on natural ecosystems. Island Press, Washington DC.

Dalal, R.C., and R.J. Henry. 1986. Simultaneous determination of moisture, organic carbon and total nitrogen by near infrared reflectance spectrophotomery. Soil. Sci. Soc. Am. J. 50:120-123.

Dalal, R.C., W. Wang, G.P. Robertson, and W.J. Parton. 2003. Nitrous oxide emission from Australian agricultural lands and mitigation options: a review. Aust. J. Soil Res. 41:165-195.

Dalgaard, T., N.J. Hutchings, and J.R. Porter. 2003. Review. Agroecology, scaling and interdisciplinarity. Agric. Ecosyst. Environ. 100:39-51.

Davies, W.J., S. Wilkinson, and B. Loveys. 2002. Stomatal control by chemical signalling and the exploitation of this mechanism to increase water use efficiency in agriculture. New Phytologist 153(3):449-460.

De Angelo, B., S. Rose, C. Li, W. Salas., R. Beach, and T. Sulser. 2005. Estimates of joint soil carbon, methane, and N_2O marginal mitigation costs from world agriculture. p. 609-617. In A. van Amstel (ed) Proc. Fourth Int. Symp. NCGG-4. Millpress, Rotterdam.

De Fraiture, C., M. Giordano, and Y. Liao. 2007. Biofuels: Implications for agricultural water use. Paper presented Int. Conf. Linkages between energy and water management for agriculture in developing countries. Hyderabad, 29-31 Jan 2007.

De Jager, A. 2005. Participatory technology, policy and institutional development to address soil fertility degradation in Africa. Land Use Policy 22:57-66.

De Haan, C., H. Steinfeld, and H. Blackburn. 1997. Livestock and the environment: Finding a balance. European Commission Directorate-General for Development, Brussels.

De Koning, G.H.J., P.H. Verburg, A. Veldkamp, and L.O. Fresco. 1999. Multi-scales modeling of land use change dynamics in Ecuador. Agric. Syst. 61:77-93.

De la Cruz, C.R. (ed) 1994. Role of fish in enhancing ricefield ecology and in integrated pest management. ICLARM Conf, Proc. 43, WorldFish Center, Penang.

De Marchi, S., S. Funtowicz, L. Cascio, and G. Munda. 2000. Combining participative and institutional approaches with multicriteria evaluation: An empirical study for water issues in Troina, Sicily. Ecol. Econ. 34:267-282.

De Plaen, R., M.L. Seka, and A. Koutoua. 2004. The paddy, the vector and the caregiver: lessons from an ecosystem approach to irrigation and malaria in Northern Cote d'Ivoire. Acta Trop. 89:135-46.

De Silva, S. 2003. Culture-based fisheries: an underutilized opportunity in aquaculture development. Aquaculture 221:221-243.

Defra. 2007. Environmentally beneficial nanotechnologies: Barriers and opportunities. Available at http://www.defra.gov. uk/environment/nanotech/policy/pdf/ envbeneficial-report.pdf. Defra, London.

Degrande, A., K. Schreckenberg, C. Mbosso, P.O. Anegbeh, J. Okafor, and J. Kanmegne. 2006. Farmers' fruit tree growing strategies in the humid forest zone of Cameroon and Nigeria. Agroforest. Syst. 67:159-175.

Den Biggelaar, C., R. Lal, K. Wiebe, H. Eswaran, V. Brenenman, and P. Reich. 2003. The global impact of soil erosion on productivity. II. Effects on crop yields and production over time. Adv. Agron. 81:49-95.

Denby, K., and C. Gehring. 2005 Engineering drought and salinity tolerance in plants: lessons from genome-wide expression profiling in Arabidopsis. Trends Biotech. 23(11):547-552.

Deng, X., J. Huang, S. Rozelle, and E. Uchida. 2006. Cultivated land conversion and potential agricultural productivity in China. Land Use Policy 23:373-384.

Denich, M., K. Vielhauer, M.S. de A Kato, A. Block, O.R. Kato, and T.D. de Abreu Sá. 2004. Mechanized land preparation in forest-based fallow systems: The experience from Eastern Amazonia. Agroforest. Syst. 61:91-106.

Desbiez, A., R. Matthews, B. Tripathi, and J. Ellis-Jones. 2004. Perceptions and assessment of soil fertility by farmers in the mid-hills of Nepal. Agric. Ecos. Environ. 103:191-206.

Dew, C.B., and R.A. McConnaughey. 2005. Did trawling on the brood stock contribute to the collapse of Alaska's king crab? Ecol. Applic. 15(3):919-941.

Di Prisco, G., and C. Verde. 2006. Predicting the impacts of climate change on the evolutionary adaptations of polar fish. Rev. Environ. Sci. Biotech. 5(2-3):309-321.

Díaz-Zorita, M., G. Duarte, and J. Grove. 2002. A review of no-till systems and soil management for sustainable crop production in the sub-humid and semi-arid Pampas of Argentina. Soil Till. Res. 65:1-18.

Dingkuhn, M., B.B. Singh, B. Clerget, J. Chantereau, and B. Sultan. 2006. Past, present and future criteria to breed crops for water-limited environments in West Africa. Agric. Water Manage. 80:241-261.

Dionysiou, D.D. 2004. Environmental applications and implications of nanotechnology and nanomaterials. J. Environ. Engineer. 130:723-724.

Doblas-Reyes, F.J., R. Hagedorn, and T.N. Palmer. 2006. Developments in dynamical seasonal forecasting relevant to agricultural management. Climate Res. 33:19-26.

DOE. 2005. Biofuels and the economy. Available at http://permanent.access.gpo.gov/websites/ www.ott.doe.gov/biofuels/economics.html. Dep. Energy, Washington DC.

DOE. 2007. Biomass program. Multi-year technical plan. Dep. Energy, Washington DC.

Doelman, P., and G. Breedveld. 1999. In situ versus on site practices. p. 539-558. In D.C. Adriano et al. (ed) Bioremediation of cantamianted soils. Agron. Mono. 37. ASA-CSSA-SSSA. Madison, WI.

Domingo, J.L. 2000. Heath risks of GM foods: Many opinions but few data. Science 288:1748-1749.

Donner, S.D., and D. Potere. 2007. The inequity of the global threat to coral reefs. BioScience 57(3):214-215.

Donner, S.D., W.J. Skirving, C.M. Little, M. Oppenheimer, and O. Hoegh-Guldberg, 2005. Global assessment of coral bleaching and required rates of adaptation under climate change. Global Change Biol. 11:2251-2265.

Douthwaite, B., V.M. Manyong, J.D.H. Keatinge, and J. Chianu. 2002. The adoption of alley farming and mucuna: Lessons for research, development and extension. Agrofor. Syst. 56:193-202.

Dowd, P.F. 2003. Insect management to facilitate preharvest mycotoxin management. p. 327-350. In H.K. Abbas (ed) Aflatoxin and food safety part 1. J. Toxicol. Toxins Rev. 22:2&3.

Droogers, P., and G. Kite. 1999. Water productivity from integrated basin modeling. Irrig. Drain. Syst. 13(3):275-290.

Dugan, P., V.V. Sugunan, R.L. Welcomme, C. Béné, R.E. Brummett, and M.C.M. Beveridge. 2007. Inland fisheries and aquaculture. In D. Molden (ed) Water for food, water for life: A comprehensive assessment of water management in agriculture. Earthscan, London and IWMI, Colombo.

Dutta, J., and H. Hofmann. 2004. Self-organization of colloidal nanoparticles. Encycl. Nanosci. Nanotech. 9:617-640.

Duxbury, J. M. 2005. Reducing greenhouse warming potential by carbon sequestration in soils: Opportunities, limits, and tradeoffs. p. 435-450. In R. Lal et al. (ed) Climate change and global food security. Taylor and Francis, NY.

Duxbury, J., and J. Lauren. 2006. Enhancing technology adoption for the rice-wheat cropping system of the Indo-Gangetic Plains. Ann. Rep. Soil Manage. CRSP, Honolulu

HI. Available at http://www.css.cornell. edu/FoodSystems/RW%20project-a.html productivity.

ECAPAPA. 2005. Managing conflicts over pasture and water resources in semi-arid areas: Beyond misleading myths and ethnic stereotypes. Policy Brief No. 8. Eastern and Central Africa Programme for Agricultural Policy Analysis. ECAPAPA, Entebbe.

Eddleston, M., L. Karalliedde, N. Buckley, R. Fernando, G. Hutchinson, and G. Isbister. 2002. Pesticide poisoning in the developing world: A minimum pesticides list. Lancet 12(360):1163-1167.

Edelstein, M. 2004. Grafting vegetables-crop plants: Pros and cons. Acta Hort. 659:235-238.

Edelstein, M., and M. Ben-Hur. 2006. Use of grafted vegetables to minimize toxic chemical usage and damage to plant growth and yield quality under irrigation with marginal water. Acta Hort. 699:159-168.

Edemeades, G.O., G.S. McMaster, J.W. White, and H. Campos. 2004. Genomics and the physiologist: Bridging the gap between genes and crop response. Field Crops Res. 90:5-18.

Edwards, J.H., and A.V. Someshwar. 2000. Chemical, physical, and biological characteristics of agricultural and forest by-products for land application. p. 1-62. In W.R. Bartels (ed) Land application of agricultural, industrial, and municipal by-products. Book Ser. No. 6. SSSA, Madison WI.

Egypt MWRI (Ministry of Water Resources and Irrigation). 2004. The national water resources plan. MWRI, Cairo.

Eicher, C.K., K. Maredia, and I. Sithole-Niang. 2006. Crop biotechnology and the African farmer. Food Policy 31:504-527.

Ekboir, J.M. 2003. Research and technology policies in innovation systems: Zero tillage in Brazil. Res. Policy 32:573-586.

Ellis, J.E., and D.M. Swift. 1988. Stability of African pastoral ecosystems: Alternate paradigms and implications for development. J. Range Manage. 41:450-459.

Ellison, C.A., S.T. Murphy, and R.J. Rabindra. 2005. Facilitating access for developing countries to invasive alien plant classical biocontrol technologies: The Indian experience. Aspects Appl. Biol. 75:71-80.

ETC Group. 2005. Down on the farm: The impact of nanoscale technologies on food and agriculture. Available at http://www .etcgroup.org/en/materials/publications. html?pub_id=80. ETC, Ottawa.

Etter, A., C. McAlpine, K. Wilson, S. Phinn, and H. Possingham. 2006. Regional patterns of agricultural land use and deforestation in Colombia. Agric. Ecosyst. Environ. 114: 369-386.

European Commission. 2005. Shift gear to biofuels. Results and recommendations from the VIEWLS project for the European Commission, Brussels.

European Environment Agency. 2006. How much bioenergy can Europe produce without harming the environment. Rep. No. 7/2006. EEA, Copenhagen.

Fageria, N.K., and V.C. Baligar. 2005. Enhancing nitrogen use efficiency in crop plants. Adv. Agron 88:97-185.

Fairborther, J.M. and É. Nadeau. 2006. Escherichia coli: On-farm contamination of animals. Rev. Sci.Tech.Off.Int.Epiz. 25:555-569.

Fairhurst, T.H., C. Witt, R.J. Buresh, and A. Doberman. 2007. Rice: A practical guide to nutrient management. Available at www.irri .org/irrc/ssnm. IRRI, Philippines.

FAO. 2000. Bioenergy. In Cultivating our futures. Background papers. No. 2. Available at http:// www.fao.org/docrep/x2775e/X2775E02.htm . FAO, Rome.

FAO. 2002. The state of world fisheries and aquaculture 2002. Fisheries Dep., FAO, Rome.

FAO. 2005. El sistema agroforestal Quesungual: Una opción para el manejo de suelos en zonas secas de ladera. FAO, Rome.

FAO. 2006. Pollution from industrialized livestock production. Livestock Policy Brief 02. FAO, Rome.

FAO/OIE/WHO. 2005. A global strategy for the progressive control of highly pathogenic avian influenza (HPAI). FAO, Rome, OIE, Paris, WHO, Geneva.

FAO/WHO. 2006. Food safety risk analysis: A guide for national food safety authorities. FAO, Rome.

Farrell, A.E., R.J. Plevin, and B.T. Turner. 2006. Can ethanol contribute to energy and environmental goals. Science 311:506-508.

Faurés, J.M., M. Svendsen and H. Turral. 2007. Reinventing irrigation. In D. Molden (ed) Water for food, water for life: A comprehensive assessment of water management in agriculture. Earthscan, London and IWMI, Colombo.

FBOMS. 2006. Impacts of monoculture expansion on bioenergy production in Brazil. Brazilian Forum of NGOs and Social Movements for the Environment and Development, Nucleo Amigos da Terra/ Brasil and Heinrich Boell Foundation, Rio de Janeiro.

Fernando, C.H. 1977. Reservoir fisheries of Southesast Asia: Past, present and future. Proc. IPFC 17(3):475-489.

Ferrand, D., A. Gibson, and H. Scott. 2004. Making markets work for the poor: An objective and an approach for governments and development agencies. ComMark Trust. Woodmead, South Africa.

Ferris, T. 2004. The interplay of modernism, postmodernism and systems engineering. 2004. p. 1003-1007. In M. Xie et al. (ed) IEEE Int. Engineer. Manage. Conf. Proc.: Innovation and entrepreneurship for sustainable development. Vol. 3. IEEE, NY.

Fillery, I.R.P. 2007. Plant-based manipulation of nitrification in soil: A new approach to managing N loss? Plant Soil 294:1-4.

Fire, A., S. Xu, M.K. Montgomery, S.A. Kostas, S.E. Driver, and C.C. Mello. 1998. Potent and specific genetic interference by double-

stranded RNA in Caenorhabditis elegans. Nature 391:806-811.

Fisheries Opportunities Assessment. 2006. Coastal Resources Center, Univ. Rhode Island, and Florida Int. Univ.

Foolad, M.R. 2004. Recent advances in genetics of salt tolerance in tomato. Rev. Plant Biotech. Appl. Genet. 76:101-119.

Fountas, S., S.M. Pedersen, and S. Blackmore. 2005. ICT in precision agriculture — diffusion of technology. p. 15. In E. Gelb and A. Offer (ed) (e-book) ICT in Agriculture: Perspectives of technological innovation. Available at http://departments.agri.huji.ac.il/ economics/ gelb-pedersen-5.pdf. Hebrew Univ., Jerusalem.

Francis, G. 2005. Best management practices for minimizing nitrate leaching. Grower 60:68-69.

Franzel, S., G.L. Denning, J.P.B. Lilleso, and A.R. Mercado. 2004. Scaling up the impact of agroforestry: Lessons from three sites in Africa and Asia. Agroforest. Syst. 61:329-344.

Franzel, S., H. Jaenicke, and W. Janssen. 1996. Choosing the right trees: Setting priorities for multipurpose tree improvement. Res. Rep. 8. ISNAR, Washington DC.

Fravel, D.R. 2005. Commercialization and implementation of biocontrol. Ann. Rev. Phytopath. 43:337-359.

Frison, E.A., I.F. Smith, T. Johns, J. Cherfas, and P.B. Eyzaguirre. 2006. Agricultural biodiversity, nutrition, and health: Making a difference to hunger and nutrition in the developing world. Food Nutr. Bull. 27:167-179.

Fritsche, U.R., K. Hunecke, and K. Wiegmann. 2005. Kriterien zur Bewertung des Pflanzenanbaus zur Gewinnung von Biokraftstoffen in Entwicklungslaendern unter Oekologischen, Sozialen und Wirtschaftlichen Gesichtspunkten. Available at http://www.oeko.de/oekodoc/232/2004-023-de.pdf. BMZ, Berlin.

Fuhrer, J. 2003. Agroecosystem responses to combinations of elevated CO_2, ozone, and global climate change. Agric. Ecosyst. Environ. 97:1-20.

Funtowicz, S. and J.R. Ravetz. 1994. Emergent complex systems. Futures 26:568-582.

Galvano, F., A. Piva, A. Ritieni, and G. Galvano. 2001. Dietary strategies to counteract the effects of mycotoxins: A review. J. Food Prot. 64:120-131.

Gan, J. 2004. Risk and damage of southern pine beetle outbreaks under global climate change. For. Ecol. Manage. 191:61-71.

Garbrecht, J.D., H. Meinke, M. Sivakumar, R. Motha, and M. Salinger. 2005. Seasonal climate forecasts and application by agriculture: a review and recommendations. Trans. Am. Geophys. Union 86:227-228.

Garí, J.A. 2002. Agrobiodiversity, food security and HIV/AIDS mitigation in Sub-Saharan Africa: strategic issues for agricultural policy and programme responses. In Sustainable development dimensions. Available at http:// www.fao.org/sd/2002/PE0104a_en.htm. FAO, Rome.

Garrett, K.A., S.P. Dendy, E.E. Frank, M.N. Rouse, and S.E. Travers. 2006. Climate change effects on plant disease: genomes to ecosystems. Annu. Rev. Phytopathol. 44:201-221.

Garrity, D. 2006. Science-based agroforestry and the achievement of the millennium development goals. In D. Garrity et al. (ed) World agroforestry into the future. World Agroforesty Centre, Nairobi.

GEF. 2005. Technology state-of-the-art: Review of existing and emerging technologies for the large-scale production of biofuels and identification of promising innovations in developing countries. Background Pap. No. 2. STAP, Washington DC.

GEF. 2006. Report of the GEF-STAP Workshop on Liquid Biofuels. GEF/C.30/Inf.9/Rev.1. GEF, Washington DC.

Gell, F.R., and C.M. Roberts. 2003. Benefits beyond boundaries: The fishery effects of marine reserves. Trends Ecol. Evol. 18:448-455. Available at http://assets.panda.org/ downloads/benefitsbeyondbound2003.pdf

Giampietro, M., S. Ulgiati, and D. Pimentel. 1997. Feasibility of large-scale biofuel production. Bioscience 47:587-600.

Gilbert, S.F. 2002. The genome in its ecological context: Philosophical perspectives on interspecies epigenesis. Ann. NY Acad. Sci. 981:202-218.

Gilchrist, M.J., C. Greko, D.B. Wallinga, G.W. Beran, D.G. Riley, and P.S. Thorne. 2007. The potential role of concentrated animal feeding operations in infectious disease epidemics and antibiotic resistance. Environ. Health Perspect. 115:313-316.

Giles, J. 2005. Solving Africa's climate-data problem. Nature 435:863.

GISP (Global Invasive Species Programme). 2004. Invasive alien species: A toolkit of best prevention and management practices. R. Wittenberg and M.J.W. Cock (ed) CAB Int. Wallingford, UK.

Giuliani, A. 2007. Developing markets for agrobiodiversity: Securing livelihoods in dryland areas. Earthscan, London and Bioversity Int., Rome.

Gladwin, C., J. Peterson, D. Phiri, and R. Uttaro. 2002. Agroforestry adoption decisions, structural adjustment and gender in Africa. In C.B. Barrett et al. (ed) Natural resources management in African agriculture: Understanding and improving current practices. CABI Publ., Nairobi.

Gluck, R. 2000. Current trends in transgenic animal technology. Genetic Eng. News 20:16.

Goel, A.K., and R. Kumar. 2005. Economic analysis of water harvesting in a mountainous watershed in India. Agric. Water Manage. 71:257-276.

Goldemberg, J. 2004. The case for renewable energies. Available at http://www.renewables 2004.de/pdf/tbp/TBP01-rationale.pdf. Int. Conf. Renew. Energies. Secretariat, Bonn.

Gould, F. 1998. Sustainability of transgenic insecticidal cultivars: integrating pest genetics and ecology. Ann. Rev. Entomol. 43: 701-726.

Graham, G.G., J. Lembcke, E. Morales. 1990. Quality protein maize and as the sole source of dietrary protein and fat for rapidly growing young children. Pediatrics 85:85-91.

Graham-Rowe, D. 2006. Nanoporous material gobbles up hydrogen fuel. New Scientist Tech. 7 Nov. 2006.

Gray, J.S., P. Dayton, S. Thrush, and M.J. Kaiser. 2006. On effects of trawling, benthos and sampling design. Marine Pollut. Bull. 52(8):840-843.

Gregory, P.J., J.S.I. Ingram, R. Andersson, and R.A. Betts. 2000. Environmental consequences of alternative practices for intensifying crop production. Agric. Ecosyst. Environ. 1853:1-12.

Griepentrog, H.W., and B.S. Blackmore. 2004. Rural poverty reduction through research for development. Sustainable agriculture and precision farming in developing countries. 5-7 Oct 2004. Deutscher, Tropentag.

Gruère, G., A. Giuliani, and M. Smale. 2006. Marketing underutilized plant species for the benefit of the poor: A conceptual framework. EPTD Disc. 147. IFPRI, Washington DC.

GTZ. 2005. Liquid biofuels for transportation in Rio de Janeiro, Brazil. GTZ, Germany.

Guerrero, M.M., C. Ros, M.A. Martinez, and M.C. Martinez. 2006. Biofumigation vs. biofumigation plus solarization to control Meloidogyne incognita in sweet pepper. Bull. OILB/SROP 29(4):313-318.

Gupta, R., and A. Seth. 2007. A review of resource conserving technologies for sustainable management of the rice-wheat cropping systems of the Indo-Gangetic plains (IGP). Crop Prot. 26:436-447.

Gutierrez, A.P., T. D'Oultremont, C.K. Ellis, and L. Ponti. 2006. Climatic limits of pink bollworm in Arizona and California: Effects of climate warming. Acta Oecolog. 30:353-364.

Guye, E.F. 2000. The role of family poultry in poverty alleviation, food security and the promotion of gender equality in rural Africa. Outlook Agric. 29:129-136.

Gyasi, E.A., G. Kranjac-Berisavljevic, E.T. Blay, and W. Oduro (ed) 2004. Managing agrodiversity the traditional way: Lessons from West Africa in sustainable use of biodiversity and related natural resources. UN Univ. Press, Tokyo.

Gyawali, S., S. Sunwar, M. Subedi, M. Tripathi, K.D. Joshi, and J.R. Witcombe. 2007. Collaborative breeding with farmers can be effective. Field Crops Res. 101(1):88-95.

Hails, R.S. 2002. Assessing the risks associated with new agricultural practices. Nature 418:685-688.

Hajjar, R., and T. Hodgkin. 2007. The use of wild relatives in crop improvement: A survey of developments over the last 20 years. Euphytica 156:1-13.

Halwart, M., and M.V. Gupta (ed) 2004. Culture of fish in rice fields. FAO and The WorldFish Center, Penang.

Hamelinck, C.N., and A.P.C. Faaij. 2006. Outlook for advanced biofuels. Energy Policy 34:3268-3283.

Hamza, M. A., and W. K. Anderson. 2005. Soil compaction in cropping systems. A review of the nature, causes and possible solutions. Soil Till. Res. 82:121-145.

Häni, F., F. Braga, A. Stämpfli, T. Keller, M. Fischer, and H. Porsche. 2003. RISE, a tool for holistic sustainability assessment at the farm level. Available at http://www.ifama .org/conferences/2003Conference/papers/ haeni.pdf. Int. Food Agribusiness Manage. Rev. 6:4.

Hansen, J. 2005. Integrating seasonal climate prediction and agricultural models for insights into agricultural practice. Phil. Trans. Roy. Soc. B 360:2037-2047.

Hansen, J.W., A. Challinor, A. Ines, T. Wheeler, V. Moron. 2006. Translating climate forecasts into agricultural terms: Advances and challenges. Climate Res. 33:27-41.

Harris, D. 2006. Development and testing of 'on-farm' seed priming. Adv. Agron. 90:129-178.

Harris, D., A.K. Pathan, P. Gothkar, A. Joshi, W. Chivasa, and P. Nyamudeza. 2001. On-farm seed priming: using participatory methods to revive and refine a key technology. Agric. Syst. 69:151-164.

Hartemink, A.E. 2002. Soil science in tropical and temperate regions. Adv. Agron.77:269-291.

HarvestPlus. 2006. Biofortification brochure. CIAT and IFPRI, Washington DC.

Hatibu, N., K. Mutabazi, E.M. Senkondo, and A.S.K. Msangi. 2006. Economics of rainwater harvesting for crop enterprises in semi-arid areas of East Africa. Agric. Water Manage. 80:74-86.

Heinemann, J.A. 1999. How antibiotics cause antibiotic resistance. Drug Discov. Today 4:72-79.

Heinemann, J.A. 2007. A typology of the effects of (trans)gene flow on the conservation and sustainable use of genetic resources. FAO, Rome.

Heinemann, J.A., H. Rosen, M. Savill, S. Burgos-Caraballo, and G.A. Toranzos. 2006. Environment arrays: A possible approach for predicting changes in waterborne bacterial disease potential. Environ. Sci. Tech. 40:7150-7156.

Heinemann, J.A., and P.D. Roughan. 2000. New hypotheses on the material nature of horizontally transferred genes. Ann. New York Acad. Sci. 906:169-186.

Helgason, B.L., H.H. Janzen, M.H. Chantigny, and C.F. Drury. 2005. Toward improved coefficients for predicting direct N_2O emissions from soil in Candian agroecosystems. Nutr. Cycl. Agroecosyst. 72(1):87-99.

Hellin, J., A. Griffith, and M. Albu. 2005. Mapping the market: Market-literacy for agricultural research and policy to takle rural poverty in Africa. In F.R. Almond, and S.D. Hainsworth (ed) Proc. Int. Seminar Beyond agriculture: Making markets work

for the poor. London, 28 Feb-1 Mar 2005. CPHP and Practical Action, UK.

Hellin, J., and S. Highman. 2005. Crop diversity and livelihood security in the Andes. Dev. Practice 15:165-174.

HKC (Helmut Kaiser Consultancy). 2006. Study. Nanotechnology in food and food processing industry worldwide. HKC, Switzerland.

Herrero, M., E. Ibanez, P.J. Martin-Alvarez, and A. Cifuentes. 2007. Analysis of chiral amino acids in conventional and transgenic maize. Anal. Chem. 79:5071-5077.

Herrmann, J. 1999. Heterogeneous photocatalysis: Fundamentals and applications to the removal of various types of aqueous pollutants. Catalysis Today 53:115-129.

Herrmann, S.M., A. Anyamba, and C.J. Tucker. 2005. Recent trends in vegetation dynamics in the African Sahel and their relationship to climate. Global Environ. Change 15:394-404.

Herrman, S.M., and C.F. Hutchinson. 2005. The changing contexts of the desertification debate. J. Arid Environ. 63:538-555.

Herselman, M. and K. Britton. 2002. Analysing the role of ICT in bridging the digital divide amongst learners. South Afric. J. Educ. 22(4):270-274.

Hiddinks, J.G., S. Jennings, and M.J. Kaiser. 2006. Indicators of the ecological impact of bottom-trawling disturbance on seabed communities. Ecosystems 9(7):1190-1199.

HKC. 2006. Instruments and tools for nanotechnology: 2006-2010-2015. Market, technology and developments for instruments, tools, automation, information and service. HKC Consulting, Switzerland.

Hobbs, P., R. Gupta, R.K. Malik, and S.S. Dhillon. 2003. Conservation agriculture for the rice-wheat systems of the Indo-Gangetic plains of South Asia: A case study from India. In Proc. 1st World Congr. Conserv. Agric. Madrid, 1-5 Oct 2001. Available at http://www.ecaf.org/documents/ hobbs.pdf.

Holt-Gimenez, E. 2001. Measuring farmers agroecological resistance to hurricane Mitch. LEISA Magazine p. 18-21.

Hoogwijk, M., A. Faaij, B. Eickhout, B. de Vries, and W. Turkenburg. 2005. Potential of biomass energy out to 2100, for four IPCC SRES land-use scenarios. Biomass Bioenergy 29:225-257.

Hoover, J.K., and J.A. Newman. 2004. Tritrophic interactions in the context of climate change: a model of grasses, cereals Aphids and their parasitoids. Global Change Biol. 10:1197-1208.

Hossain, M.K., S.C. Ghosh, C.T. Boontongkong, and J. Dutta. 2005. Growth of zinc oxide nanowires and nanobelts for gas sensing applications. J. Metastable Nanocrystal. Materials 23:27-30.

Hsieh, C.H., C.S. Hunter, J.R. Beddington, R.M. May, and G. Sugihara. 2006. Fishing elevates variability in the abundance of exploited species. Nature 443(7113):859-862.

Huang, H.M., Y. Wu, H. Feick, N. Tran,

E. Weber, and P. Yang. 2001. Catalytic growth of zinc oxide nanowires by vapor transport. Adv. Materials (13):113-116.

Hussain, I. 2005. Pro-poor intervention strategies in irrigation agriculture in Asia: Poverty in irrigated agriculture — Issues, lessons, options and guidelines. IWMI, Colombo.

IEA. 2004. Biofuels for transport. An international perspective. IEA, Paris.

IEA. 2006. Energy technology perspectives 2006. In support of the G8 Plan of Action. IEA, Paris.

IFAD 2001. The challenge of ending rural poverty. Rural poverty report. Oxford Univ. Press, UK.

IFAD. 2002. Republic of Namibia: Northern regions livestock development project. Interim evaluation rep. 1313-NA. IFAD, Rome.

Ijumba, J.N., F.W. Mosha, and S.W. Lindsay. 2002. Malaria transmission risk variations derived from different agricultural practices in an irrigated area of northern Tanzania. Med. Vet. Entomol. 16:28-38.

ILRI. 2006. Livestock — a pathway out of poverty. ILRI's strategy to 2010. ILRI, Nairobi.

India CWC (Central Water Commission). 2002. Water and related statistics. Min. Water Resources, Water Planning and Projects Wing, New Delhi.

Indian Planning Commission. 2003. Report of the Committee on Development of Bio-Fuel Gov. India, New Delhi.

IOM (Institute of Medicine). 1992. Emerging infections: Microbial threats to health in the United States. Nat. Acad. Press, Washington DC.

IPCC (Intergovernmental Panel on Climate Change). 2000. Land use, land-use change, and forestry. Cambridge Univ. Press, UK.

Ishitani, M., I. Rao, P. Wenzl, S. Beebe, and J. Tohme. 2004. Integration of genomics approach with traditional breeding towards improving abiotic stress adaptation: Drought and aluminum toxicity as case studies. Field Crops Res. 90:35-45.

Izaurralde, R.C., and C.W. Rice. 2006. Methods and tools for designing pilot soil carbon sequestration projects. p. 457-476. In R. Lal (ed) Carbon sequestration in Latin America. Haworth Press, NY.

Izaurralde, R.C., N.J. Rosenberg, and R. Lal. 2001. Mitigation of climatic change by soil carbon sequestration: issues of science, monitoring and degraded lands. Adv. Agron. 70:1-75.

Jama, B., F. Kwesiga, and A. Niang. 2006. Agroforestry innovations for soil fertility management in sub-Saharan Africa: Prospects and challenges ahead. p. 53- 60. In D. Garrity et al. (ed) World agroforestry into the future. World Agroforestry Centre, Nairobi.

James, C. 2007. Global status of commercialized biotech/GM crops: 2006. Available at http://www.isaaa.org/resources/publications/ briefs/35/executivesummary/default.html. ISAAA, Ithaca NY.

Janvier, C., F. Villeneuve, C. Alabouvette, V. Edel- Hermann, T. Matielle, and C. Steinberg. 2007. Soil health through soil disease suppression: Which strategy from descriptors to indicators? Soil Biol. Biochem. 39:1-23.

Jarvis, A., A. Lane, and R. Hijmans. 2008. The effect of climate change on crop wild relatives. Agric. Ecosyst. Environ.

Jauhar, P.P. 2006. Modern biotechnology as an integral supplement to conventional plant breeding: The prospects and challenges. Crop Sci. 46:1841-1859.

Jiambo, L. 2006. Energy balance and economic benefits of two agroforestry systems in northern and southern China. Agric. Ecosyst. Environ. 116:255-262.

Joshi, K.D., A.M. Musa, C. Johansen, S. Gyawali, D. Harris, and J.R. Witcombe. 2007. Highly client-oriented breeding, using local preferences and selection, produces widely adapted rice varieties. Field Crops Res. 100(1):107-116.

Kader, A.A. 2005. Increasing food availability by reducing postharvest losses of fresh produce. Acta Hort. 682:2169-2176.

Kaihura, F., and M. Stocking (ed) 2003. Agricultural biodiversity in smallholder farms of East Africa. UN Univ. Press, Tokyo.

Kashmanian, R.M., D. Kluchinski, T.L. Richard, and J.M. Walker. 2000. Quantities, characteristics, barriers, and incentives for use of organic municipal by-products. p. 127-167. In W. R. Bartels (ed) Land application of agricultural. Industrial and municipal by-products. SSSA Book Ser. No. 6. SSSA, Madison, WI.

Keizer, M., F. Kruijssen, and A. Giuliani. 2007. Collective action for biodiversity and livelihoods. LEISA Mag. 23(1).

Keller, A.A., and J. Keller. 1995. Effective efficiency: A water use efficiency concept for allocating freshwater resources. Disc. Pap. 22. Center Econ. Policy Studies. Winrock Int., Washington DC.

Keller, A., and D. Seckler. 2004. Limits to increasing the productivity of water in crop production. Winrock, Arlington VA.

Keller, J., D. Adhikari, M. Peterson, and S. Suryanwanshi. 2001. Engineering low-cost micro-irrigation for small plots. Int. Dev. Enterprises, Lakewood CO.

Kerdchoechuen, O. 2005. Methane emissions in four rice varieties as related to sugars and organic acids of roots and root exudates and biomass yield. Agric. Ecosyst. Environ. 108:155-163.

Kershen, D.L. 2004. Legal liability issues in agricultural biotechnology. Crop Sci. 44:456-463.

Khan, A. G. 2005. Role of soil microbes in the rhizospheres of plants growing on trace metal contaminated soils in phytoremediation. J. Trace Elem. Med. Biol. 18:355-364.

Khan, S., R. Tariq, C. Yuanlai, and J. Blackwell. 2006. Can irrigation be sustainable? Agric. Water Manage. 80(1-3):87-99.

Knowler, D., and B. Bradshaw. 2007. Farmers' adoption of conservation agriculture: A review and synthesis of recent research. Food Policy 32:25-48.

Knowler, D.J. 2004. The economics of soil productivity: local, national, and global perspectives. Land Degrad. Dev. 15:543-561.

Köhler-Rollefson, I. 2003. Community based management of animal genetic resources — with special reference to pastoralists. p. 13-26. In FAO. Community based management of animal genetic resources. Wkshp. Proc., Mbabane, Swaziland, 7-11 May, 2001. FAO, Rome.

Kojima, M., and T. Johnson. 2005. The potential for biofuels for transport in developing countries. ESMAP Rep. 312/05. World Bank, Washington DC.

Koontz, T.M., T.A. Steelman, J. Carmin, K.S. Korfmacher, C. Moseley, and C.W. Thomas. 2004. Collaborative environmental management: what roles for goverment? RFF Press, Washington DC.

Krämer, U. 2005. Phytoremediation: novel approaches to cleaning up polluted soils. Curr. Opinion Biotech. 16:133-141.

Kroymann, J., and T. Mitchell-Olds. 2005. Epistasis and balanced polymorphism influencing complex trait variation. Nature 435:95-98.

Kuiper, H.A., G.A. Kleter, H.P.J.M. Noteborn, and E.J. Kok. 2001. Assessment of the food safety issues related to genetically modified foods. Plant J. 27:503-528.

Kumar, S. 2006. Climate change and crop breeding objectives int he twenty first century. Curr. Sci. 90:1053-1054.

Kuparinen, A. 2006. Mechanistic models for wind dispersal. Trends Plant Sci. 11(6): 296-301.

Kurup, M.P.G. 2000. Milk production in India: Perspective 2020. Indian Dairyman 52: 1:25-37.

Kurup, M.P.G. (ed) 2003. Livestock in Orissa: The socio-economic perspective. Manohar Publ., New Delhi.

Kuta, D.D., E. Kwon-Ndung, S. Dachi, M. Ukwungwu, and E. Imolehin. 2003. Potential role of biotechnology tools for genetic improvement of 'lost crops of Africa': The case of fonio (Digitaria exilis and Digitaria iburua). Afr. J. Biotech. 2(12): 580-585.

Kuzma, J., and P. VerHage. 2006. Nanotechnologies in agriculture and food production, anticipated applications. Project on emerging nanotechnologies. Available at ww.hhh.umn.edu/img/assets/7260/cv.pdf. Woodrow Wilson Int. Center Scholars, Washington DC.

Kydd, J. 2002. Agriculture and rural livelihoods: is globalization opening or blocking paths out of rural poverty? Agric. Res. Extension Network Pap. No. 121. Overseas Dev. Inst., London.

La Malfa, G., and C. Leonardi. 2001. Crop practices and techniques: Trends and needs. Acta Hort. 559:31-42.

Lacy J., W.S. Clampett, L. Lewin. R. Reinke. G. Batten, R. Williams et al.1999. Rice Check Recommendations. A guide to objective rice crop management for improving yields, grain quality and profits. NSW Agric., Australia.

Laddomada, A. 2003. Control and eradication of OIE list A diseases — the approach of the European Union to the use of vaccines. Dev. Biol. (Basel) 114:269-80.

Ladha, J.K., H. Pathak, T.J. Krupnik, J. Six, and C. Van Kessel. 2005. Efficiency of fertilizer nitrogen in cereal production: Retrospects and prospects. Adv. Agron. 87:85-156.

Lal, R. 2001. Managing world soils for food security and environmental quality. Adv. Agron. 74:155-192.

Lal, R. 2004. Soil carbon sequestration to mitigate climate change. Geoderma 123:1-22.

Lal, R. 2005. Climate change, soil carbon dynamics, and global food security. p. 113-143. In R. Lal et al. (ed) Climate change and global food security. CRC Press, NY.

Lal, R., D.C. Reicosky, and J.D. Hanson. 2007. Editorial: Evolution of the plow over 10,000 years and the rationale for no-till farming. Soil Till. Res. 93:1-12.

Lavergne, S., and J. Molofsky. 2007. Increased genetic variation and evolutionary potential drive the success of an invasive grass. PNAS (USA) 104:3883-3888.

LEAD (Livestock, Environment and Development). 2000. Introductory Paper for the Electronic Conference on Area Wide Integration of Crop and Livestock Production. Available at http://www.virtualcentre.org/en/ele/awi_2000/1session/1paper.pdf. FAO, Rome.

Leakey, R.R.B. 1999b. Potential for novel food products from agroforestry trees. Food Chem. 64:1-14.

Leakey, R.R.B. 2001a. Win: Win landuse strategies for Africa: 1. Building on experience with agroforests in Asia and Latin America. Int. For. Rev. 3:1-10.

Leakey, R.R.B. 2001b. Win: Win landuse strategies for Africa: 2. Capturing economic and environmental benefits with multistrata agroforests. Int. For. Rev. 3:11-18.

Leakey, R.R.B., K. Schreckenberg, and Z. Tchoundjeu. 2003. The participatory domestication of West African indigenous fruits. Int. For. Rev. 5:338-347.

Leakey, R.R.B., and Z. Tchoundjeu. 2001. Diversification of tree crops: Domestication of companion crops for poverty reduction and environmental services. Exp. Agric. 37:279-296.

Leakey, R.R.B., Z. Tchoundjeu, K. Schreckenberg, S.E. Shackleton, and C.M. Shackleton. 2005. Agroforestry tree products (AFTPs): Targeting poverty reduction and enhanced livelihoods. Int. J. Agric. Sustain. 3:1-23.

Ledec, G., and J.D. Quintero. 2003. Good dams and bad dams: environmental criteria for site selection of hydroelectric projects. Sustainable Dev. Working Pap. 16. World Bank, Washington DC.

Ledford, H. 2006. Making it up as you go along. Chemists make liquid fuels from biomass — or from coal. Nature 444:677-678.

Lehmann, J. 2007a. A handful of carbon. Nature 447:143-144.

Lehmann, J. 2007b. Bio-energy in the black. Frontiers Ecol. Environ. 5:381-387.

Lehmann, J., J. Gaunt, and M. Rondon. 2006. Bio-char sequestration in terrestrial ecosystems — a review. Mitigation Adaptation Strategies Global Change 11:403-427.

Leidwein, A. 2006. Protection of traditional knowledge associated with biological and genetic resources. General legal issues and measures already taken by the European Union and its member states in the field of agriculture and food production. J. World Intell. Prop. 9:251-275.

Leonard, D. K., C. Ly, and P.S.A. Woods. 2003. Community-based animal health workers and the veterinary profession in the context of African privatization. p. 19-21. In Proc. Primary animal health care in the 21st century: Shaping the rules, policies and institutions. Community-based Animal Health and Participatory Epidemiology Unit (CAPE), OAU, Mombasa.

Levy, M., and P. Meier. 2004. Early warning and assessment of environment, conflict, and cooperation. In Understanding environment, conflict, and cooperation. UNEP, Nairobi.

Li, C., W. Salas, B. DeAngelo, and S. Rose. 2006. Assessing alternatives for mitigating net greenhouse gas emissions and increasing yields from rice production in China over the next twenty years. J. Environ. Qual. 35:1554-1565.

Lotze, H.K., S.L. Hunter, B.J. Bourque, R.H. Bradbury, R.G. Cooke, M.C. Kay et al. 2006. Depletion, degradation, and recovery potential of estuaries and coastal seas. Science 312:1806-1809.

Lowenberg-DeBoer, J. 1996. Precision farming and the new information technology: implications for farm management, policy, and research: Discussion. Am. J. Agric. Econ. 78(5):1281-1284.

Luc, M., J. Bridge, and R. A. Sikora. 2005. Reflections on nematology in subtropical and tropical agriculture. p.1-10. In M. Luc, et al. (ed) Plant parasitic nematodes in subtropical and tropical agriculture, 2nd ed. CABI Publ., Wallingford.

Lund, D.C., J. Lynch-Stieglitz, and W.B. Curry. 2006. Gulf stream density structure and transport during the past millennium. Nature 444(7119):601-604.

Lynch, J.P., and S.B. St. Clair. 2004. Mineral stress: The missing link in understanding how global climate change will affect plants in real world soils. Field Crops Res. 90: 101-115.

MA (Millennium Ecosystem Assessment). 2005. Ecosystems and human well-being. Vol 1. Current state and trends. R. Hassan et al. (ed) Millennium Ecosystem Assessment. Island Press, Washington DC.

Ma, J.K.C., E. Barros, R. Bock, P. Christou, P.J. Dale, P.J. Dix et al. 2005. Molecular farming for new drugs and vaccines. EMBO Rep. 6:593-599.

Machuka, J. Agricultural genomics and sustainable development: Perspectives and prospects for Africa. Afr. J. Biotech. 3:2:127-135.

MacMillan, K., K. Emrich, H.P. Piepho, C.E. Mullins, and A.H. Price. 2006. Assessing the importance of genotype × environment interaction for root traits in rice using a mapping population II: Conventional QTL analysis. Theor. Appl. Genet. 113:953-964.

Magdoff, F. 2007. Ecological agriculture. Renew. Agric. Food Syst. 22:2:109-117.

Manyangarirwa, W., M. Turnbull, G.S. McCutcheon, and J.P. Smith. 2006. Gene pyramiding as a Bt resistance management strategy: How sustainable is this strategy? Afr. J. Biotech. 5(10):781-785.

Maredia, M.K. 2001. Application of intellectual property rights in developing countries: Implications for public policy and agricultural research institutes. Available at http://www.wipo.int/about-ip/en/studies/pdf/study_k_maredia.pdf. World Intell. Property Org., Geneva.

Martin, F.N. 2003. Development of alternative strategies for management of soilborne pathogens currently controlled with methyl bromide. Ann. Rev. Phytopathol. 41:325-350.

Marvier, M., C. McCreedy, J. Regetz, and P. Kareiva. 2007. A meta-analysis of effects of Bt cotton and maize on nontarget invertebrates. Science 316:1475-1477.

Matthews S. & Brandt K. 2006. South America Invaded: The growing danger of invasive alien species [Online]. Available at http://www.gisp.org/casestudies/showcasestudy.asp?id=274&MyMenuItem=casestudies&worldmap=&country. Global Invasive Species Programme (GISP), Nairobi.

Matthiessen, J.N., and J.A. Kirkegaard. 2006. Biofumigation and enhanced biodegradation: opportunity and challenge in soilborne pest and disease management. Crit. Rev. Plant Sci. 25:235-265.

Maxwell, S., I. Urey, and C. Ashley. 2001. Emerging issues in rural development. An issues paper. Overseas Dev. Inst., London.

Mayers, J. and S. Bass. 2004. Policy that works for forests and people: Real prospects for governance and livelihoods. Earthscan, Sterling VA.

McDermott, P.F., S. Zhao, D.D. Wagner, S. Simjee, R.D. Walker, and D.G. White. 2002. The food safety perspective of antibiotic resistance. Anim. Biotech. 13:71-84.

McDonald, A.H. and J.M. Nicol. 2005. Nematode parasite of cereals. p. 131-191. In M. Luc et al. (ed) Plant parasitic nematodes in subtropical and tropical agriculture. 2nd ed. CABI, Wallingford.

McNeely, J.A. 2006. As the world gets smaller, the chances of invasion grow. Euphytica 148:5-15.

McNeely, J.A. and S.J. Scherr. 2003. Ecoagriculture: Strategies to feed the world and save wild biodiversity. Island Press, Washington DC.

Meier, P. 2007. From theory to action: Mainstreaming disaster and conflict early warning in response to climate change. In H.G. Brauch et al. (ed) Coping with global environmental change, disasters and security threats: Challenges, vulnerabilities and risks. Springer, Berlin.

Menneer, J.C., S. Ledgard, C. McLay, and W. Silvester. 2004. The impact of grazing animals on N_2 fixation in legume-based pastures and management options for improvement. Adv. Agron. 83:181-241.

Mercer, K.L., D.A. Andow, D.L. Wyse, and R.G. Shaw. 2007. Stress and domestication traits increase the relative fitness of crop-wild hybrids in sunflower. Ecol. Lett. 10:383-393.

Mettrick, H. 1993. Development oriented research in agriculture: An ICRA textbook. Int. Center Dev. Oriented Res. Agric. (ICRA), Wageningen.

Michael, A., J. Schmidt, W. Enke, T. Deutschländer, and G. Malitz. 2005. Impact of expected increase in precipitation intensities on soil loss — results of comparative model simulations. Catena 61(2-3):155-164.

Mills, A., L. Punte, and M. Stephen. 1997. An overview of semiconductor photocatalysis. J. Photochem. Photobiol. 108:1-35.

Mishra, H.N. and C. Das. 2003. A review on biological control and metabolism of aflatoxin. Crit. Rev. Food Sci. 43:245-264.

Mitschein, T.A., and P.S. Miranda. 1998. POEMA: A proposal for sustainable development in Amazonia. p. 329-366. In D.E. Leihner and T.A. Mitschein (ed) A third millennium for humanity? The search for paths of sustainable development. Peter Lang, Frankfurt.

Miyazaki, S., M. Fredricksen, K.C. Hollis, V. Poroyko, D. Shepley, and D.W. Galbraith. 2004. Transcript expression profiles for Arabidopsis thaliana grown under open-air elevation of CO_2 and O_3. Field Crops Res. 90:47-59.

MMWR (Morbidity and Mortality Weekly Report). 2004. Brief Report: Illness Associated with Drift of Chloropicrin Soil Fumigant into a Residential Area — Kern County, California, 2003. 53(32):740-742.

Modi, V., S. McDade, D. Lallement, and J. Saghir. 2006. Energy and the Millennium Development Goals. UNDP, UN Millennium Project and World Bank, NY.

Molden, D.J., R. Sakthivadivel and Z. Habib. 2000. Basin level use and productivity of water: Examples from South Asia. Res. Rep. 49. IWMI, Colombo.

Molden, D.J., and R. Tharme. 2004. Water,

food, livelihood and environment: Maintaining biodiversity in irrigated landscapes. Int. Ecoagriculture Conf. and Practitioners' Fair, 27 Sep-1 Oct 2004. Nairobi.

Molden, D., T. Oweis, P. Steduto, J. Kijne, M. Hanjra, and P.S. Bindraban. 2007. Pathways for increasing agricultural water productivity. In D. Molden (ed) Water for food, water for life: A comprehensive assessment of water management in agriculture. Earthscan, London and IWMI, Colombo.

Moreira, J.R., and J. Goldemberg. 1999. The alcohol program. Energy Policy 27:229-245.

Morgan, R.P.C., and M.A. Nearing. 2002. Soil erosion models: present and future. p. 187-205. In J.L. Rubio et al. (ed) Man and soil at the Third Millennium. Proc. Int. Cong. Eur. Soc. Soil Conserv., Valencia, Spain, 28 Mar.-1 Apr. 2000. Inst. Water Environ., Cranfield Univ., Bedfordshire.

Mottram, A. 2003. Transplanting sorghum and pearl millet — a key to risk management in semi-arid areas. PhD Thesis, Univ. Wales, UK.

Mulley, B.D., and J.D. Unruh. 2004. The role of off-farm employment in tropical forest conservation: labor, migration, and smallholder attitudes toward land in western Uganda. J. Environ. Manage. 71:193-205.

Murphy, S.T., and O.D. Cheesman. 2006. The aid trade — international assistance programs as pathways for the introduction of invasive alien species. Biodiversity Ser., Pap. No. 109. World Bank, Washington DC.

Murray-Kolb, L.E., A. Felix, A. del Mundo, G. Gregorio, J.D. Haas, and J.L. Beard. 2004. Biofortified rice as a source of iron for iron deficient women. In FASEB J. 18:A511.

Mutero, C.M., C. Kabutah, V. Kimani, L. Kabuage, G. Gitau, J. Ssennyonga et al. 2004. A transdisciplinary perspective on the links between malaria and agroecosystems in Kenya. Acta Trop 89:171-86.

Mutero, C.M., M. McCartney, and R. Boelee. 2006. Agriculture, malaria, and water-associated diseases. p. 1-2. In Understanding the links between agriculture and health. C. Hawkes, and M.T. Ruel (ed) IFPRI, Washington DC.

Myers, N. 2002. Environmental refugees: A growing phenomenon of the 21st Century. Phil. Trans. R. Soc. Lon. Biol. Sci. B 357: 609-613.

Mzuku, M., R. Khosla, R. Reich, D. Inman, F. Smith, and L. MacDonald. 2005. Spatial variability of measured soil properties across site-specific management zones. Soil Sci. Soc. Am. J. 69:1572-1579.

Nagendra, H. 2001. Using remote sensing to assess biodiversity. Int. J. Remote Sensing 22(12):2377-2400.

Ndawula, J., J.D. Kabasa, and Y.B. Byaryhanga. 2004. Alterations in fruit and vegetable beta-carotene and vitamin C content caused by open-sun drying, visqueen-covered and

polyethylene-covered solar-dryers. Afr. Health Sci. 4:125-30.

Nearing, M.A., F.F. Pruski, and M.R. O'Neal. 2004. Expected climate change impacts on soil erosion rates: a review. J. Soil Water Conserv. 59(1):43-50.

Nelson, G.C. 2005. Drivers of change in ecosystem condition and services, p. 175-222. In Ecosystems and human well-being: Scenarios. Vol. 2. Island Press, Washington DC.

Neumann, C., D.M. Harris, and L. Rogers. 2002. Contribution of animal source foods in improving diet quality and function in children in the developing world. Nutr. Res. 22:193-220.

Nguyen, H.T., and J.A. Jehle. 2007. Quantitative analysis of the seasonal and tissue-specific expression of Cry1Ab in transgenic maize Mon810. J. Plant Dis. Prot. 114:82-87.

Ngwira, N., S. Bota, and M. Loevinsohn. 2001. HIV/AIDS, agriculture, and food security in Malawi: Background to action. Available at www.synergyaids.com/documents/rmalawi _hivAids_agricFoodSecurity.pdf.

Nickel, J.L. 1973. Pest situations in changing agricultural systems: A review. Bull. Entomol. Soc. Am. 19:136-142.

Nicol, J.M., and I. Ortiz-Monasterio. 2004. Effect of root lesion nematode on wheat yields and plant susceptibility in Mexico. Nematology 6(4):485-493.

Nicolle, C., G. Simon, E. Rock, P. Amouroux, and C. Remesy. 2004. Genetic variability influences carotenoid, vitamin, phenolic, and mineral content in white, yellow, purple, orange, and dark-orange carrot cultivars. J. Am. Soc. Hort. Sci. 129:523-529.

Ninomiya, S. 2004. Successful information technology for agriculture and rural development. Extens. Bull. 549 (Sept). Available at http://www.agnet.org/library/ eb/549/. Food Fert. Tech. Center for Asian Pacific Reg., Taiwan.

Nordstrom, K.F. and S. Hotta. 2004. Wind erosion from cropland in the USA: A review of problems, solutions, and prospects. Geoderma 121:157-167.

Novak, S.J. 2007. The role of evolution in the invasion process. Proc. Natl. Acad. Sci. 104:3671-3672.

NRC (National Research Council). 1996. Lost crops of Africa. Vol. 1 Grains. National Acad. Press, Washington DC.

NRC (National Research Council). 2004. Biological confinement of genetically engineered organisms. Nat. Acad. Press, Washington DC.

NSTC. 2006. The national nanotechnology initiative. Resaerch and development leading to a revolution in technology and industry. Supplement to the President's fiscal year 2007 budget. Nat. Sci. Tech Council, Washington DC.

Oda, M. 1999. Grafting of vegetables to improve greenhouse production. Extens. Bull. Food Fert. Tech. Center 480:1-11. Available at http://www.agnet.org/library/abstract/eb480. html. Food Fert. Tech. Center for Asian Pacific Reg., Taiwan.

Oelberman, M., R.P. Voroney, and A.M. Gordon. 2004. Review. Carbon sequestration in tropical and temperate agroforestry systems: A review with examples from Costa Rica and southern Canada. Agric. Ecosyst. Environ. 104:359-377.

Ogita, S., H. Uefuji, Y. Yamaguchi, N. Koizumi, and H. Sano. 2003. Producing decaffeinated coffee plants. Nature 423:823.

Olsen, S.B., T.V. Padma, and B.D. Richter. 2006. Managing freshwater inflows to estuaries: a methods guide. Coastal Resources Center, Univ. Rhode Island.

Olson, K.E. and G.N. Slack. Food safety begins on the farm: The viewpoint of the producer. Rev. Sci. Tech. Off. Int. Epiz. 25:2:529-539.

Omonode, R.A., T.J. Vyn, D.R. Smith, P. Hegymegi, and A. Gál. 2007. Soil carbon dioxide and methane fluxes from long term tillage systems in continuous corn and corn soybean rotation. Soil Till. Res. 92:182-195.

Onduru, D.D., A. De Jager, and G.N. Gachini. 2006. The hidden costs of soil mining to agricultural sustainability in developing countries: a case study of Machakos District, Eastern Kenya. Int. J. Agric. Sustain. 3:167-176.

Opara, L. 2004. Emerging technological innovation triad for smart agriculture in the 21st century. Part I. Prospects and impacts of nanotechnology in agriculture. Agric. Engineer. Int. The CIGR J. Sci. Res. Dev. Vol. 6 (Jul. 2004).

Orindi, V.A., and A. Ochieng. 2005. Seed fairs as a drought recovery strategy in Kenya. Bulletin 36:4. Inst. Dev. Studies, Univ. Sussex, UK.

Ortiz, R., J.H. Crouch, M. Iwanaga, K. Sayre, M. Warburton, L.L. Araus et al. 2006. Bioenergy and agricultural research for development. Brief 8 of 12. Available at http://www.ifpri.org/2020/focus/focus14/focus14_07.pdf. IFPRI, Washington DC.

Oweis, T.Y., and A.Y. Hachum. 2003. Improving water productivity in the dry areas of West Asia and North Africa. p. 179-198. In J.W. Kijne et al. (ed) Water productivity in agriculture: Limits and opportunities for improvement. CABI, Wallingford and IWMI, Colombo.

Oweis, T.Y., and A. Hachum. 2006. Water harvesting and supplemental irrigation for improved water productivity of dry farming systems in West Asia and North Africa. Agric. Water Manage. 80:57-73.

Oweis, T., A. Hachum and J. Kijne. 1999. Water harvesting and supplemental irrigation for improved water use efficiency in the dry areas. SWIM Pap. 7. IWMI, Colombo.

Oweis T., P.N. Rodrigues, L.S. Pereira. 2003. Simulation of supplemental irrigation strategies for wheat in Near East to cope with water scarcity. p. 259-272. In: G. Rossi et al. (eds) Tools for drought mitigation in Mediterranean regions. Kluwer, Dordrecht.

Padgham, J. L., G. S. Abawi, J. M. Duxbury, and M. A. Mazid. 2004. Impact of wheat on meloidogyne graminicola populations in the rice-wheat system of Bangladesh. Nematropica 34:2:183-190.

Pagliai, M., N. Vignozzi, and S. Pellegrini. 2004. Soil structure and the effect of management practices. Soil Till. Res. 79:131-143.

Painting, K. 2006. Mainstreaming ICTs: Challenges for CTA. Available at http://ictupdate.cta.int/index.php/article/eview/316/1/59. CTA, Netherlands.

Palauqui, J.-C., T. Elmayan, J.-M. Pollien, and H. Vaucheret. 1997. Systemic acquired silencing: transgene-specific post-transcriptional silencing is transmitted by grafting from silenced stocks to non-silenced scions. EMBO J. 16:4738-4745.

Pampolino, M.F., I.J. Manguiat, S. Ramanathan, H.C. Gines, P.S. Tan, T.N. Chi et al. 2007. Environmental impact and economic benefits of site-specific nutrient management (SSNM) in irrigated rice systems. Agric. Syst. 93:1-24.

Pan, Z.X., B. Yang, S. Pivonia, L. Xue, R. Pasken, and J. Roads. 2006. Long-term prediction of soybean rust entry into the continental United States. Plant Dis. 90(7):840-846.

Panik, F. 1998. The use of biodiversity and implications for industrial production. p. 59-73. In D.E. Leihner and T.A. Mitschein (ed), A third millennium for humanity? The search for paths of sustainable development. Peter Lang, Frankfurt.

Papadakis, G., D. Briassoulis, G. Scarascia Mougnoza, G. Vox, P. Feuilloley, and J.A. Stoffer. 2000. Radiometric and thermal properties of, and testing methods for, greenhouse covering materials. J. Agric. Engineer. Res. 76:7-38.

Parry, M., C. Rosenzweig, A. Iglesias, G. Fischer, and M. Livermore. 1999. Climate change and world food security: A new assessment. Global Environ. Change 9:S51:S67.

Paruccini, M. 1994. Applying multicriteria aid for decision to environmental management. Kluwer, Dordrecht.

Passianoto, C.C., T. Ahrens, B.J. Feigi, P.A. Steudler, J.B. do Carmo, and J.M. Melillo. 2003. Emissions of CO_2, N_2O and NO in conventional and no- till management practices in Rhondonia, Brazil. Biol. Fert. Soils 38:200-208.

Patil, V., A. Maru, G.B. Shashidhara, and U.K.Shanwad. 2002. Remote sensing, GIS and precision farming in India: Opportunities and challenges. In M. Fangquan (ed) Proc. AFITA Conf. Beijing, 26-28 Oct 2002.

Patterson, D.T., J.K. Westbrook, R.J.V. Joyce, P.D. Lindgren, and J. Rogasik. 1999. Weeds, insects, and diseases. Climatic Change. 43:711-727.

Patto, M.C.V., B. Skiba, E.C.K. Pang, S.J. Ochatt, F. Lambein, and D. Rubiales. 2006. Lathyrus improvement for resistance against biotic and abiotic stresses: from classical breeding to marker assisted selection. Euphytica 147(1-2):133-147.

Pauly, D., R. Watson, and J. Adler. 2005. Global trends in world fisheries: Impacts on marine ecosystems and food security. Phil. Trans. R. Soc. B. 360:5-12.

Pearce, D., and S. Mourato. 2004. The economic valuation of agroforestry's environmental services. p. 67-86. In G. Schroth et al. (ed) Agroforestry and biodiversity conservation in tropical landscapes. Island Press, Washington.

Peck, A.J., and T. Hatton. 2003. Salinity and the discharge of salts from catchments in Australia. J. Hydrol. 272:191-202.

Peden, D., G. Tadesse and A.K. Misra. 2007. Water and livestock for human development. In D. Molden (ed) Water for food, water for life: Comprehensive assessment on water in agriculture. Earthscan, London and IWMI, Colombo.

Pedersen, S.M., S. Fountas, B.S. Blackmore, M. Gylling, and J.L. Pedersen. 2004. Adoption and perspective of precision farming in Denmark. Acta Agric. Scand. Sec. B, Soil and Plant Sci. 54 (1):S.2-6

Peral, J., X. Domenech, and D.F. Ollis. 1997. Heterogeneous photocatalysis for purification, decontamination and deodorization of air. J. Chem. Tech. Biotech. 70:117-140.

Perotto-Baldiviezo, H.L., T.L. Thurow, C.T. Smith, R. F. Fisher, and X.B. Wu. 2004. GIS-based spatial analysis and modeling for landslide hazard assessment in steeplands, southern Honduras. Agric. Ecosyst. Environ. 103:165-176.

Perrings, C. 2005. Biological invasions and poverty. Available at http://www.public.asu.edu/~cperring/Perrings,%20Poverty%20and%20Invasive%20Species%20(2005).pdf.Int. Inst. Sustain., Tempe AZ.

Peter, P., P. Pinstrup-Andersen, S. Gillespie, and L. Haddad. 2002. AIDS and food security: Annual report essay reprint. IFPRI, Washington, DC.

Phillipart, C.J.M., J.J. Beukema, G.C. Cadee, R. Dekker, P.W. Goedhart, J.M. van Iperen et al. 2007. Impacts of nutrient reduction on coastal communities. Ecosystems DOI 10.1007/s10021-006-9006-7.

PhilRice (Philippine Rice Res. Inst.). 2007. PalayCheck system for irrigated lowland rice. PhilRice Nueva Ecija and FAO, Rome.

Phoenix, G.K., and M.C. Press. 2005. Effects of climate change on parasitic plants: the root hemiparasitic orobanchaceae. Folia Geobotan. 40:205-216.

Pierzynski, G.M., J. T. Sims, and G.F. Vance. 2000. Soils and environmental quality. CRC Press, NY.

Pieterse, C.M.J., J.A. van Pelt, S.C.M. van Wees, J. Ton, and L.C. van Loon. 2001. Rhizobacteria-mediated induced systemic resistance: triggering, signaling, and expression. Eur. J. Plant Pathol. 107:51-61.

Pigato, M. 2001. Information and communication technology, poverty and development in sub-Saharan Africa and South Asia. Afr. Reg. Working Pap. Ser. 20. World Bank. Washington DC.

Pikitch, E.K., C. Santora, E.A. Babcock, A. Bakun, R. Bonfil, D.O.Conover et al. 2004. Science 305:346-347.

Pimentel, D., P. Hepperly, J. Hanson, D. Douds, and R. Seidel. 2005. Environmental, energetic, and economic comparisons of organic and conventional farming systems. BioScience 55:573-582.

Pimentel, D., L. Lach, R. Zuniga, and D. Morrison. 2000. Environmental and economic costs of non-indigenous species in the United States. Bioscience 50:53-65.

Pinter, K. 1983. Hungary. In E.E. Brown (ed) World fish farming, cultivation and economics. AVI Publ., Westport CT.

Pinto, L.F.G., M.S. Bernardes, M. van Noordwijk, A.R. Pereira, B. Lusiana, and R. Mulia. 2005. Simulation of agroforestry systems with sugarcane in Piracicaba, Brazil. Agric. Systems 86:275-292.

Pitcher, H., K.L. Ebi, and A. Brenkert. 2008. Population health model for integrated assessment models. Clim. Change. [Online] DOI:10.1007/s10584-007-9286-8.

Polak, P., J. Keller, R. Yoder, A. Sadangi, J.N. Ray, T. Pattanyak et al. 2004. A low-cost storage system for domestic and irrigation water for small farmers. Int. Dev. Enterprises, Lakewood CO.

Pomeroy, R.S. and R. Rivera-Guieb. 2006. Fisheries co-management: A practical handbook. CABI and IDRC, Ottawa.

Ponsioen. T.C., H. Hengsdijk, J. Wolf, M.K. van Ittersum, R.P. Rötter, T.T. Son et al. 2006. TechnoGIN, a tool for exploring and evaluating resource use efficiency of cropping systems in East and Southeast Asia. Agric. Systems 87:80-100.

Popp, J., D. Hoag, and J. Ascough II. 2002 Targeting soil conservation policies for sustainability; New empirical evidence. J. Soil Water Conserv. 57:66-74.

Powell, K. 2003. Barnyard biotech — lame duck or golden goose? Nature Biotech. 21:965-967.

PPLPI. 2001. Pro-poor livestock policy facility; facilitating the policy dialogue in support of equitable, safe and clean livestock farming. Project description. FAO, Rome.

Pretty, J.N. 1999. Can sustainable agriculture feed Africa? New evidence on progress, processes and impacts. Environ. Dev. Sustain. 1:253-274.

Pretty, J. 2001. The rapid emergence of genetic modification in world agriculture: Contested risks and benefits. Environ. Conserv. 28: 248-262.

Pretty, J. 2003. Social capital and the collective management of resources. Science 302: 1912-1914.

Pretty, J.N., J.I.L. Morison, and R.E. Hine. 2003. Reducing food poverty by increasing agricultural sustainability in developing countries. Agric. Ecosyst. Environ. 95(1):217-234.

Pretty, J.N., A.D. Noble, D. Bossio, J. Dixon, R.E. Hine, F.T.W. Penning de Vries et al. 2006. Resource-conserving agriculture increases yields in developing countries. Environ. Sci. Tech. 40(4):1114-1119.

Prins, M. 2003. Broad virus resistance in transgenic plants. Trends Biotech. 21:373-375.

Pryme, I.F., and R. Lembcke. 2003. In vivo studies on possible health consequences of genetically modified food and feed — with paticular regard to ingredients consisting of genetically modified plant materials. Nutr. Health 17:1-8.

Pusztai, A., S. Bardocz, and S.W.B. Ewen. 2003. Genetically modified foods: potential human health effects. p. 347-371. In J.P.F. D'Mello (ed) Food safety: Contaminants and toxins. CABI, UK.

Qadir, M., D. Wichelns, L. Raschid-Sally, P.S. Minhas, P. Drechsel, A. Bahri et al. 2007. Agricultural use of marginal-quality water — opportunities and challenges. In D. Molden (ed) Water for food, water for life: A comprehensive assessment of water management in agriculture. Earthscan, London and IWMI, Colombo.

Radersma, S., H. Otieno, A.N. Atta-Krah, and A.I. Niang. 2004. System performance analysis of an alley-cropping system in Western Kenya and its explanation by nutrient balances and uptake processes. Agric. Ecosyst. Environ. 104:631-632.

Ragauskas, A.J., C.K. Williams, B.H. Davison, G. Britovsek, J. Cairney, C.A. Eckart et al. 2006. The path forward for biofuels and biomaterials. Science 311:484-489.

Raghu, S., R.C. Anderson, C.C. Daehler, A.S. Davis, R.N. Wiedenmann, D. Simberloff et al. 2006. Adding biofuels to the invasive species fire? Science 313:1742.

Ramdas, S.R. and N.S. Ghotge. 2005. Bank on hooves: Your companion to holistic animal health care. Anthra, Pune, India.

Rao, P.P.N. 2005. Space technology applications in agriculture. Available at www.digital opportunity.org/article/view/116635.

Rasul, G., and G.B. Thapa. 2003. Sustainability analysis of ecological and conventional agricultural systems in Bangladesh. World Dev. 31:1721-1741.

Rasul, G., and G.B. Thapa. 2004. Sustainability of ecological and conventional agricultural systems in Bangladesh: an assessment based on environmental, economic and social perspectives. Agric. Syst. 79(3):327-351.

Rasul, G., and G.B. Thapa. 2006. Financial and economic suitability of agroforestry as an alternative to shifting cultivation: The case of the Chittagong Hill Tracts, Bangladesh. Agric. Syst. 91:29-50.

Reardon, T.A., J.A. Berdegué, M. Lundy, P. Schütz, F. Balsevich, R. Hernández et al. 2004. Supermarkets and rural livelihoods: A research method. Michigan State Univ., East Lansing.

Rebafka, F.P., A. Bationo, and H. Marschner. 1993. Phosphorous seed coating increases phosphorous uptake, early growth and yield of pearl millet (Pennisetum glaucum (L) grown on an acid sandy soil in Niger, West Africa. Fert. Res. 35(3):151-160.

Reece, J.D., and E. Haribabu. 2007. Genes to feed the world: the weakest link? Food Policy 32:459-479.

Reij, C., G. Tappan, and A. Belemvire. 2005. Changing land management practices and vegetation on the Central Plateau of Burkina Faso (1968-2002). J. Arid Environ. 63:642-659.

Renault, D., M. Hemakumara, and D. Molden. 2001. Importance of water consumption by perennial vegetation in irrigated areas of the humid tropics: evidence from Sri Lanka. Agric. Water Manage. 46:215-230.

Rengel, Z. 2001. Agronomic approaches to increasing zinc concentration in staple food crops. In I. Cakmak and R. Weldch (ed) Impacts of agriculture on human health and nutrition. Encycl. Life Support Systems. UNESCO, EOLSS Publ., Oxford.

Reyes, T., M. Quiroz, and S. Msikula. 2005. Socio-economic comparison between traditional and improved cultivation methods in agroforestry systems, East Usambara Mountains, Tanzania. Environ. Manage. 36:682-690.

Robertson, G.P., E.A. Paul, and R.R. Harwood. 2000. Greenhouse gases in intensive agriculture: Contributions of individual gases to the radiative forcing of the atmosphere. Science 289:1922-1925.

Robson, M.C., S.M. Fowler, N.H. Lampkin, C. Leifert, M. Leitch, D. Robinson et al. 2002. The agronomic and economic potential of break crops for ley/arable rotations in temperate organic agriculture. Adv. Agron. 77:369-427.

Rockström, J. 2004. Making the best of climatic variability: Options for upgrading rainfed farming in water scarce regions. Water Sci. Tech. 49(7):151-156.

Rockström, J., P. Fox, and J. Barron. 2003. Water productivity in rain fed agriculture: challenges and opportunities for smallholder farmers in drought-prone tropical agroecosystems. In J.W. Kijne et al. (ed) Water productivity in agriculture: Limits and opportunities for improvement. CABI, UK and IWMI, Colombo.

Rockström, J., N. Hatibu, T. Oweis, and S. Wani. 2007. Managing water in rainfed agriculture. In D. Molden (ed) Water for food, water for life: Comprehensive assessment on water in agriculture. Earthscan, London and IWMI, Colombo.

Rosenzweig, C., A. Iglesias, X.B. Yang, P.R. Epstein, and E. Chivian. 2001. Climate change and extreme weather events: implications for food production, plant diseases, and pests. Glob. Change Hum. Health 2:90-104.

Rounsevell, M.D.A., P.M. Berry, and P.A. Harrison. 2006. Future environmental change impacts on rural land use and biodiversity:

a synthesis of the ACCELERATES project. Environ. Sci. Policy 9:93-100.

Rounsevell, M.D.A., S.P. Evans, and P. Bullock. 1999. Climate change and agricultural soils: Impacts and adaptation. Clim. Change. 43:683-709.

Rounsevell, M.D.A., F. Ewert, I. Reginster, R. Leemans, and T.R. Carter. 2005. Future scenarios of European agricultural land use. II. Projecting changes in cropland and grassland. Agric. Ecosyst. Environ. 107:117-135.

Royal Society. 2005. Ocean acidification due to increasing atmospheric carbon dioxide. The Royal Soc., London.

Rweyemamu, M.M., J. Musiime, G. Thomson, D. Pfeiffer, and E. Peeler. 2006. Future control strategies for infectious animal diseases: Case study of the UK and sub-Saharan Africa. p. 1-24. In Foresight, infectious diseases, preparing for the future. Available at www.foresight.gov.uk. UK Dep. Trade Industry, London.

Sabates, A., P. Martin, J. Lloret, and V. Raya. 2006. Sea warming and fish distribution: the case of the small pelagic fish, Sardinella aurita, in the western Mediterranean. Glob. Change Biol. 12(11):2209-2219.

Sadowsky, M.J., and R.F. Turco. 1999. Enhancing indigenous microorganisms to bioremediate contaminated soils. p. 274-288. In D.C. Adriano et al. (ed) Bioremediation of cantamianted soils. Agron. Mono. 37. ASA-CSSA-SSSA, Madison WI.

Sadras, V.O. and J.F. Angus. 2006. Benchmarking water use efficiency of rainfed wheat in dry environments. Aust. J. Agric. Res. 57:8.

Saha, S.K., and S. Jonna. 1994. Paddy acreage and yield estimation and irrigated cropland inventory using satellite and agrometeorological data. Asian Pac. Rem. Sens. J. 6(2):79-87.

Sanchez, P.A. 2000. Linking climate change research with food security and poverty reduction in the tropics. Agric. Ecosyst. Environ. 82:371-383.

Sanchez, P.A. 2002. Soil fertility and hunger in Africa. Science 295:2019-2020.

Sanchez, P.A. 2005. Reducing hunger in tropical Africa while coping with climate change. p. 3-19. In R. Lal et al. (ed) Climate change and global food security. Taylor and Francis, NY.

Sanchirico, J.N., and J.E. Wilen. 2007. Global marine fisheries resources: Status and prospects. Int. J. Glob. Environ. Issues 7(2,3).

Sanders, R.A., and W. Hiatt. 2005. Tomato transgene structure and silencing. Nature Biotech. 23:287-289.

Sanderson, K. 2006. A field in ferment: To move US biofuels beyond subsidized corn will be a challenge. Nature 444:673-676.

Sasal, M.C., A. Andriulo, and M.A.Taboada. 2006. Soil porosity characteristics on water dynamics under direct drilling in Argiudolls of the Argentinean Rolling Pampas. Soil Till. Res. 87:9-18.

Sauder, A. 1992. International development enterprises evaluation of marketing appropriate technology Phase II. CIDA, Canada.

Scherm, H. 2004. Climate change: Can we predict the impacts on plant pathology and pest management? Can. J. Plant Pathol. 26:267-273.

Schneider, S.H. 2006. Climate change: Do we know enough for policy action? Sci. Engineer. Ethics 12(4):607-636.

Shonk, J.L., L.D. Gultney, D.G. Schulze and G.E. Van Scoyoc. 1991. Spectroscopic sensing of soil organic matter content. Trans. ASAE 34:1978-1984.

Schreckenberg, K., A. Degrande, C. MBosso, Z. Boli Baboulé, C. Boyd, L. Enyoung et al. 2002. The social and economic importance of Dacryodes edulis (G.Don) H.J. Lam in Southern Cameroon. For. Trees Livelihoods 12:15-40.

Schroth, G., G.A.B. da Fonseca, C.A. Harvey, C. Gaston, H.L. Vasconcelos and A-M.N. Izac (ed) 2004. Agroforestry and biodiversity conservation in tropical landscapes. Island Press, Washington DC.

Scialabba, N. 2007. Organic agriculture and food security. In Int. Conf. Organic Agric. Food Security. Rome. May 3-5. Available at ftp://ftp.fao.org/paia/organicag/ofs/nadia.pdf. FAO, Rome.

Scoones, I. 1998. Sustainable rural livelihoods: A framework for analysis. IDS Working Pap. 72. IDS, Brighton.

Sechley, K.A., and H. Schroeder. 2002. Intellectual property protection of plant biotechnology inventions. Trends Biotech. 20:456-461.

Seckler, D., D. Molden and R. Sakthivadivel. 2003. The concept of efficiency in water resources management and policy. p. 37-51. In J.W. Kijne et al. (ed) Water productivity in agriculture: Limits and opportunities for improvement. CABI, UK and IWMI, Colombo.

Seddon, J.M. 2007. Multivitamin-multimineral supplements and eye disease: Age-related macular degeneration and cataract. Am.J.Clin.Nutr. 85:1:304S-307S.

Semizarov, D., L. Frost, A. Sarthy, P. Kroeger, D.N. Halbert, and S.W. Fesik. 2003. Specificity of short interfering RNA determined through gene expression signatures. PNAS USA 100:6347-6352.

Shah, T., J. Burke and K. Vilholth. 2007. Groundwater: A global assessment of scale and significance. In D. Molden (ed) Water for food, water for life: A comprehensive assessment of water management in agriculture. Earthscan, London and IWMI, Colombo.

Shah, T., M. Alam, D. Kumar, R.K. Nagar, and M. Singh. 2000. Pedaling out of poverty: Social impact of a manual technology in South Asia. Res. Rep. 45. IWMI, Colombo.

Shand, H. 2002. Terminator no solution to gene flow. Nature Biotech. 20:775-776.

Sheldrick, W.F., and J. Lingard. 2004. The use of nutrient audits to determine nutrient balances in Africa. Food Policy 29:61-98.

Siekmann, J., L.H. Allen, N.O. Bwibo, M.W. Demment, S.P. Murphy, and C.G.Neumann. 2003. Kenyan schoolchildren have multiple micronutrient deficiencies, but increased plasma vitamin B12 is the only detectable micronutrient response to meat or milk supplementation. J. Nutr. 133:3972S-3980S.

Sikora, R.A., J. Bridge, and J.L. Starr. 2005. Management practices: An overview of integrated nematode management technologies. p. 793-826. In M. Luc et al. (ed) Plant parasitic nematodes in subtropical and tropical agriculture. CABI, UK.

Sikora, R.A. and E. Fernandez. 2005. Nematode parasites of vegetables. p. 319-392. In M. Luc et al. (ed) Plant parasitic nematodes in subtropical and tropical agriculture. CABI, UK.

Sikora, R.A., and L.E. Pocasangre. 2004. New technologies to increase root health and crop production. Infomusa. 13(2):25-29.

Sims, J.T., L. Bergstrom, B.T. Bowman, and O. Oenema. 2005. Nutrient management for intensive animal agriculture: Policies and practices for sustainability. Soil Use Manage., 21(Suppl.):141-151.

Sinclair, T.R., L.C. Purcell, and C.H. Sneller. 2004. Crop transformation and the challenge to increase yield potential. Trends Plant Sci. 9:70-75.

Sivakumar, M.V.K.. 2006. Climate prediction and agriculture: Current status and future challenges. Climate Res. 33:3-17.

Sivakumar, M.V.K. and D.E. Hinsman. 2004. Satellite remote sensing and GIS applications in agricultural meteorology and WMO satellite activities. p. 1-22. In M.V.K. Sivakumar (ed) 2004. Satellite remote sensing and GIS applications in agricultural meteorology. Proc. Training Workshop. Dehra Dun, India, 7-11 Jul 2003. WMO, Geneva.

Six, J., H. Bossuyt, S.Degryze, and K. Denef. 2004. A history of research on the link between (micro) aggregates, soil biota, and soil organic matter dynamics. Soil Till. Res. 79:7-31.

Smaling, E.M.A., and J. Dixon. 2006. Adding a soil fertility dimension to the global farming systems approach, with cases from Africa. Agric. Ecosyst. Environ. 116:15-26.

Smid, R., M. Kubasek, D. Klimes, L. Dusek, J. Jarkovsky, B. Marsalek et al. 2006. Web portal for management of bioindication methods and ecotoxicological tests in ecological risk assessment. Ecotoxicology 15:623-627.

Smith, P., D. Martino, Z. Cai, D. Gwary, H. Janzen, P. Kumar et al. 2007. Policy and technological constraints to implementation of greenhouse gas mitigation options in agriculture. Agric. Ecosyst. Environ. 118:6-28.

Smith, S.R. 1996. Agricultural recycling of sewage sludge and the environment. CABI, Wallingford, UK.

Smithson, J.B., and J.M. Lenne. 1996. Varietal mixtures: a viable strategy for sustainable productivity in subsistence agriculture. Ann. Appl. Biol. 128:127-158.

Smyth, S., G.G. Khachatourians, and P.W.B. Phillips. 2002. Liabilities and economics of transgenic crops. Nature Biotech. 20:537-541.

Sommer, S.G., J.K. Schjoerring, and O.T. Denmead. 2004. Ammonia emission from mineral fertilizers and fertilized crops. Adv. Agron. 82:557-622.

Spoor, G. 2006. Alleviation of soil compaction: Requirements: Equipment and techniques. Soil Use Manage. 22:113-122.

Spoor, G., F.G. Tijink, and P. Weisskopf. 2003. Experiences with the impact and prevention of subsoil compaction in the Euporean Union. Soil Till. Res. 73:175-182.

Srinivasan, C.S. 2003. Exploring the feasibility of farmers' rights. Dev. Pol. Rev. 21:419-447.

Staal, S., M. Owango, G. Muriuki, B. Lukuyu, F. Musembi, O. Bwana et al. 2001. Dairy systems characterization of the greater Nairobi milk-shed. SDP Res. Rep., KARI and ILRI, Nairobi.

Stacey, D.A. 2003. Climate and biological control in organic crops. Int. J. Pest Manage. 49(3):205-214.

Stapleton, J.J., C.L. Elmore, and J.E. DeVay. 2000. Solarization and biofumigation help disinfect soil. Calif. Agric. 54(6):42-45.

Steinbach, H.S., and R. Alvarez. 2006. Changes in soil organic carbon contents and nitrous oxide emissions after introduction of no-till in Pampean agroecosystem. J. Environ. Qual. 35:3-13.

Steinfeld, H., C. De Haan and H. Blackburn. 1997. Livestock–Environment Interactions: Issues and options. European Commission Directorate-General for Development, Brussels.

Steinfeld, H., P. Gerber, T. Wassenaar, V. Castel, M. Rosales, and C. de Haan. 2006. Livestock's long shadow, environmental issues and options. FAO, Rome.

Sticklen, M. 2006. Plant genetic engineering to improve biomass characteristics for biofuels. Curr. Opin. Biotech. 17:315-319.

Stige, L.C., J. Stave, K.S. Chan, L. Ciannelli, N. Pettorelli, M. Glantz et al. 2006. The effect of climate variation on agro-pastoral production in Africa. PNAS 103(9):3049-3053.

Stocking, M.A. 2003. Tropical soils and food security: The next 50 years. Science 302:1356-1359.

Stoop, W. 2002. A study and comprehensive analysis of the causes for low adoption rates of agricultural research results in West and Central Africa: Possible solutions leading to greater future impacts: The Mali and Ghana case studies. Interim Sci. Council. CGIAR, Washington DC and FAO, Rome.

Stoop, W.A., and T. Hart. 2006. Research and development towards sustainable agriculture by resource-poor farmers in Sub-Saharan Africa: Some strategic and organisational considerations in linking farmer practical

needs with policies and scientific theories Int. J. Agric. Sustain. 3:206-216.

Stoop, W., N. Uphoff, and A. Kassam. 2002. A review of agricultural research issues raised by the system of rice intensification (SRI) from Madagascar: Opportunities for improving farming systems for resource-poor farmers. Agric. Syst. 71:249-274.

Stoorvogel, J.J., and J.M. Antle. 2001. Regional land use analysis: The development of operational tools. Agric. Syst. 70:623-640.

Stoorvogel, J.J., and E.M.A. Smaling. 1990. Assessment of soil nutrient depletion in Sub-Saharan Africa: 1983-2000. Winand Staring Centre, Wageningen.

Strand, J.F. 2000. Some agrometeorological aspects of pest and disease management for the 21st century. Agric. For. Meteorol. 103:73-82.

Strohman, R.C. 1997. The coming Kuhnian revolution in biology. Nature Biotech. 15:194-200.

Subbarao, G.V., M. Rondon, O. Ito, T. Ishikawa, I.M. Rao, K. Nakahara et al. 2007. Biological nitrification inhibition (BNI) — is it a widespread phenomenon? Plant Soil 294:5-18.

Sugunan, V.V. and P.D. Katiha. 2004. Impact of stocking on yield in small reservoirs in Andhra Pradesh, India. Fisheries Manage. Ecol. 11:65-69.

Sugunan, V.V., and M. Sinha. 2001. Sustainable capture and culture-based fisheries in freshwaters of India. p. 43-70. In T.J. Pandian (ed) Sustainable Indian fisheries. Nat. Acad. Agric. Sci., New Delhi.

Swinkels, R., and S. Franzel. 2000. Adoption potential of hedgerow intercropping in maize-based cropping systems in the highlands of western Kenya 2. Economic and farmers' evaluation. Exp. Agric. 33:211-223.

Swinton, S.M., and J. Lowenberg-DeBoer. 2001. Global adoption of precision agriculture technologies: Who, when and why. p. 557-562. In third European conference on precision agriculture. G. Grenier and S. Blackmore (ed) Agro Montpellier (ENSAM), Montpellier.

Syam, T., and K. Jusoff. 1999. Remote sensing (RS) and Geographic Information System (GIS) technology for field implementation in Malaysian agriculture. Seminar on repositioning agriculture industry in the next millennium, 13-14 Jul 1999. Bilik Persidangan IDEAL, Universiti Putra Malaysia, Serdang, Selangor.

Taboada, M.A., O. A. Barbosa, M.B. Rodríguez, and D.J. Cosentino. 2004. Mechanisms of aggregation in a silty loam under different simulated management regimes. Geoderma 123:233-244.

Taboada, M.A., F.G. Micucci, D.J.Cosentino, and R.S. Lavado. 1998. Comparison of compaction induced by conventional and zero tillage in two soils of the Rolling Pampa of Argentina. Soil Till. Res. 49:57-63.

Tappan, G., and M. McGahuey. 2007. Tracking

environmental dynamics and agricultural intensification in southern Mali. Agric. Syst. 94:38-51.

Tarafdar, J.C., and N. Claassen. 2005. Preferential utilization of organic and inorganic sources of phosphorus by wheat plant. Plant Soil 275:285-293.

Taube, A. 2006. ACP-EU Update: PUMA, weather satellite data for Africa. Available at http://ictupdate.cta.int/index.php/article/view/353.

Taylor, L.H., S.M. Latham, and M.E.J. Woolhouse. 2001. Risk factors for human disease emergence. Phil. Trans. R. Soc. Lond. B 356:983-989.

Tchoundjeu, Z., E. Asaah, P. Anegbeh, A. Degrande, P. Mbile, C. Facheux et al. 2006. AFTPs: Putting participatory domestication into practice in West and Central Africa. For. Trees Livelihoods 16:53-70.

Tennant, P., G. Fermin, M.M. Fitch, R.M. Manshardt, J.L. Slightom, and D. Gonsalves. 2001. Papaya ringspot virus resistance of transgenic rainbow and sunup is affected by gene dosage, plant development, and coat protein homology. Eur. J. Pl. Pathol. 107:645-653.

Terry, P.J. 2001. The Cyperaceae — still the world's worst weeds? p. 3-18. In C.R. Riches (ed) The world's worst weeds. Brighton, UK.

The Guardian. 2007. Massacres and paramilitary land seizures behind the biofuel revolution: Colombian farmers driven out as armed groups profit Lucrative 'green' crop less risky to grow than coca. The Guardian, 5 June.

Thiruvengadachari, S. and R. Sakthivadivel. 1996. Satellite remote sensing for assessment of irrigation system performance: A case study in India. IWMI, Colombo.

Thomas, Z. 2005. Agricultural biotechnology and proprietary rights. challenges and policy options. J. World Intel. Prop. 8:711-734.

Thorup-Kristensen, K., J. and J. Magid, and L. Stoumann. 2003 Catch crops and green manures as biological tools in nitrogen management in temperate zones. p. 227-302. In Advances in Agronomy. Elsevier, NY.

Thresh, J.M., and R.J. Cooper. 2005. Strategies for controlling cassava mosaic virus disease in Africa. Plant Path. 54:587-614.

Thrupp, L.A. 1998. Cultivating diversity: Agrobiodiversity and food security. World Resources Inst., Washington DC.

Tiffen, M., and R. Bunch. 2002. Can a more agroecological agriculture feed a growing world population? p. 71-91. In N. Uphoff (ed) Agroecological innovations: increasing food production with participatory development. Earthscan, London.

Tilman, D., K.G. Cassman, P.A. Matson, R. Naylor, and S. Polasky. 2002. Agricultural sustainability and intensive production practices. Nature 418:671-677.

Timsina, J., and D.J. Connor. 2001. Productivity and management of rice-wheat cropping systems: issues and challenges. Field Crops Res. 69:93-132.

Timsina, J., and E. Humphreys. 2006. Performance of CERES-Rice and CERES-Wheat models in rice-wheat systems: A review. Agric. Systems 90:5-31.

Todhunter, P.E. and L.J. Cihacek. 1999. Historical reduction of airborn dust in the Red River Valley of the North. J. Soil Water Conserv. 54:543-551.

Toledo, A., and B. Burlingame (ed) 2006. Biodiversity and nutrition: A common path. J. Food Compos. Analy. 19(6-7).

Tontisirin, K., G. Nantel, and L. Bhattacharjee. 2002. Foodbased strategies to meet the challenges of micronutrient malnutrition in the developing world. Proc. Nutr. Society 61:243-250.

Toumlin, C., and B. Guèye. 2003. Transformation in West African agriculture and the role of family farms. Issue Pap. 123. IIED, London.

Tripp, R. 2006. The performance of low external input technology in agricultural development: a summary of three case studies. Int. J. Agric. Sustain. 3:143-153.

Twyman, R.M., E. Stoger, S. Schillberg, P. Christou, and R. Fischer. 2003. Molecular farming in plants: host systems and expression technology. Trends Biotech. 21:570-578.

United Nations. 2006. Global survey for early warning systems: An assessment of capacities, gaps and opportunities toward building a comprehensive global early warning system for all natural hazards. A Report prepared at the request of the Secretary-General. UN, NY.

Uri, N., 1999. Conservation tillage in U.S. agriculture: Environmental, economic, and policy issues. Haworth Press, Binghamton NY.

USAID (US Agency Int. Dev.). 2006. Etude de la regeneration naturelle assistee dans la region de Zinder (Niger). Available at http://www. frameweb.org/ev_en.php?ID=14310_201& ID2=DO_COMMUNITY. USAID, Washington DC.

Vallat, B. and E. Mallat. 2006. Ensuring good governance to address emerging and re-emerging animal disease threats: Supporting the veterinary services of developing countries to meet OIE international standards on quality. Rev.Sci.Tech.Off.Int. Epiz. 25:1:389-401.

Van Bruggen, A.H.C., and A.J. Termorshuizen. 2003. Integrated approaches to root disease management in organic farming systems. Australas. Plant Pathol. 32:141-156.

Van Dam, J.C., R. Singh, J.J.E. Bessembinder, P.A. Leffelaar, W.G.M. Bastiaanssen, R.K. Jhorar et al. 2006. Assessing options to increase water productivity in irrigated river basins using remote sensing and modeling tools. Water Res. Dev. 22(1):115-133.

Van der Hoek, W. 2004. How can better farming methods reduce malaria? Guest editorial. Acta Tropica 89:95-87.

Van der Hoek, W., R. Sakthivadivel, M. Renshaw, J.B. Silver, M.H. Birley, and F. Konradsen. 2007. Alternate wet/dry irrigation in rice cultivation; a practical way to save water and control malaria and Japanese encephalitis? Available at http:// www.iwmi.cgiar.org/. IWMI, Colombo.

Van Eenennaam, A.L. 2006. What is the future of animal biotechnology? Cal. Agric. 60:132-139.

Van Eenennaam, A.L., and P.G. Olin. 2006. Careful risk assessment needed to evaluate transgenic fish. Calif. Agric. 60:126-131.

Van Eijck, J., and H. Romijn. 2006. Prospects for Jatropha biofuels in developing countries. an analysis for Tanzania with strategic niche management. 4th Ann. Globelics Conf. Innovation systems for competitiveness and shared prosperity in developing countries. Thiruvananthapuram, 4-7 Oct.

Van Koppen, B., and S. Mahmud. 1996. Women and water pumps in Bangladesh. The impact of participation in irrigation groups on women's status. Intermediate Tech. Publ., London.

Van Lenteren, J.C. 2000. Measures of success in biological control of arthropods by augmentation of natural enemies, p. 77-103. In S. Wratten and G. Gurr (ed) Measures of success in biological control. Kluwer Acad. Publ., Dordrecht.

Van Lenteren, J.C. 2003. Integrated pest management in greenhouses: Experiences of European countries. p. 327-339. In K. Maredia (ed) IPM in the Global Arena. CABI Publ., Wallingford.

Van Neil, T.G., and T.R. McVicar. 2001. Remote sensing of rice-based irrigated agriculture: a review. Rice CRC Tech. Rep. Available at www.ricecrc.org.

Van Noordwijk, M., F. Agua, B. Verbist, K. Hairiah, and T. Tomich. 2007. Watershed management. p. 191- 212. In S. Scherr and J. McNeely (ed) Farming with nature: The science and practice of ecoagriculture. Island Press, Washington DC.

Vance, C.P., C.Uhde-Stone, and D. L. Allan. 2003. Tansley review: Phosphorus acquisition and use: Critical adaptations by palnts for securing a nonrenewable resource. New Phytologist 157:423-447.

Vasey, R.A., J.D. Scholes, and M.C. Press. 2005. Wheat (Triticum aestivum) is susceptible to the parasitic angiosperm Striga hermonthica, a major cereal pathogen in Africa. Phytopathol. 11:1293-1300.

Vaucheret, H., C. Beclin, and M. Fagard. 2001. Post-transcriptional gene silencing in plants. J. Cell Sci. 114:3083-3091.

Vereijken, P.H. 2002. Transition to multifunctional land use and agriculture. Netherlands J. Agric. Sci. 50:171-179.

Verlodt, I., and P. Verschaeren. 2000. New interference film for climate control. Acta Hort. 514:139-146.

Viglizzo, E.F. 2001. The impact of global change on the rural environment in ecoregions of the southern cone of South America. p. 103-122. In O.T. Solbrig et al. (ed) Impact of globalisation and information society on the rural environment. Harvard Univ. Press, Cambridge.

Viglizzo, E.F., and F.C. Frank. 2006. Land-use options for Del Plata Basin in South America: tradeoffs analysis based on ecosystem service provision. Ecol. Econ. 57:140-151.

Viglizzo, E.F., and Z.E. Roberto. 1998. On trade-offs in low-input agro-ecosystems. Agric. Syst. 56:253-264.

Villar, J.L., B. Freese, A. Bebb, N. Bassey, C. Améndola, and M. Ferreira. 2007. Who benefits from GM crops? Friends of the Earth, Amsterdam.

Villarreal, M., C.H. Anyonge, B. Swallow, and F. Kwesiga. 2006. The challenge of HIV/ AIDS: Where does agroforestry fit in? p. 181-192. In D. Garrity et al. (ed) World agroforestry into the future. World Agroforesty Centre, Nairobi.

Vogel, C., K. O'Brien. 2006. Who can eat information? Examining the effectiveness of seasonal climate forecasts and regional climate-risk management strategies. Climate Res. 33:111-122.

Von Braun, J., and R.K. Pachauri. 2006. The promises and challenges of biofuels for the poor in developing countries. IFPRI, Washington DC.

Vosti, S.E., E. Muñoz, B. Chantal, L. Carpentier, M.V.N. d'Oliveira, and J. Witcover. 2003. Rights to forest products, deforestation and smallholder income: evidence from the Western Brazilian Amazon. World Dev. 31:1889-1901.

Vourc'h, G., V.E. Bridges, J. Gibbens, B.D. DeGroot, L. McIntyre, R. Poland et al. 2006. Detecting emerging diseases in farm animals through clinical observations. Emerg. Infect. Dis. 12:204-210.

Warkentin, B.P. 2001. Diffuse pollution: Lessons from soil conservation policies. Water Sci. Tech. 44:197-202.

Wassmann, R., and M.S. Aulakh. 2000. The role of rice plants in regulating mechanisms of methane emissions. Biol. Fertil. Soils 31:20-29.

Wassmann, R., H.U. Neue, J.K. Ladha, and M.S. Aulakh. 2004. Mitigating greenhouse gas emissions from rice-wheat cropping systems in Asia. Environ. Dev. Sustain. 6:65-90.

Water Science and Technology Board. 2001. Virulence-factor activity relationships. p. 143-207. In: Water Science and Tech. Board. Classifying drinking water contaminants for regulatory consideration. Nat. Acad. Press, Washington DC.

Watson, R.T., R.H. Moss, and M.C. Zinyowera (ed) 1998. The regional impacts of climate change: an assessment of vulnerability. IPCC, Cambridge Univ. Press, UK.

Wayne, R. 1987. Crop mixtures for enhanced yield and stability. Rural Res. (CSIRO) 135:16-19.

WBGU. 2003. Welt im Wandel: Energiewende zur Nachhaltigkeit. (In German) German Adv. Council Global Change, Berlin.

Welches, L.A., and I. Cherrett. 2002. The quesungual system in Honduras: An alternative to slash-and-burn. LEISA 18:10-11.

Welcomme, R.L., and C.M. Bartley. 1998. An evaluation of present techniques for the enhancement of fisheries. J. Fisheries Ecol. Manage. 5:351-382.

Wenhua, L. 2004. Degradation and restoration of forest ecosystems in China. Forest Ecol. Manage. 201:33-41.

Wenzel, G. 2006. Molecular plant breeding: Achievements in green biotechnology and future perspectives. Appl. Microbiol. Biotech. 70:642-650.

Wenzel, W.W., D.C. Adriano, D. Salt, and R. Smith. 1999. Phytoremediation: A plant-microbe based remediation system. p. 457-508. In D.C. Adriano et al. (ed) Bioremediation of cantamiated soils. Agron. Mono. 37. ASA-CSSA-SSSA, Madison WI.

Wheaton, R.Z., and E.J. Monke. 2001. Terracing as a "best management practice" for controlling erosion and protecting water quality. AE-114. Available at www.ces. purdue.edu/extmedia/AE/AE-114.HTML. Purdue Univ., West Lafayette IN.

White, J.W., G.S. McMaster, and G.O. Edemeades. 2004. Genomics and crop response to global change: what have we learned? Field Crops Res. 90:165-169.

WHO. 2003. Impacts of antimicrobial growth promoter termination in Denmark. World Health Organization int. rev. panel's evaluation of the termination of the use of antimicrobial growth promoters in Denmark, 6-8 Nov 2002. WHO/EDS/CPE/ZFK/2003.1. Copenhagen.

Wieser, M., F. Schneider and S. Waelty. 2000. Capitalisation of experiences in livestock production and dairying (LPD) in India (CAPEX). Intercooperation, Switzerland.

Winsor, S. 2007. Ethanol fuels land prices. Corn Soybean Digest 1(Feb).

Wittenberg, R. and M.J.W. Cock. 2001. Invasive alien species: How to address one of the greatest threats to biodiversity: A toolkit of best prevention and management practices. CABI, Wallingford.

Wittmer, H., F. Rauschmayer, and B. Klauer. 2006. How to select instruments for the resolution of environmental conflicts? Land Use Policy 23(1):1-9.

Wolfenbarger, L.L., and P.R. Phifer. 2000. The ecological risks and benefits of genetically engineered plants. Science 290:2088-2093.

Woods, J. 2006. Science and technology options for harnessing bioenergy's potential. Brief 6 of 12. In P. Hazell et al. (ed) Bioenergy and agriculture: Promises and challenges. IFPRI, Washington DC.

World Agroforestry Centre. 2005. Trees of change: A vision for an agroforestry transformation in the developing world. ICRAF, Nairobi.

World Bank 2004. Saving fish and fishers: Toward sustainable and equitable governance of the global fishing sector. Rep. 29090-GLB. World Bank, Washington DC.

World Bank. 2006. Aquaculture: Changing the face of the waters — meeting the promise and challenge of sustainable aquaculture. World Bank, Washington DC.

Worldwatch Institute. 2006. Biofuels for transportation: Global potential and implications for sustainable agriculture and energy in the 21st century. Worldwatch, Washington DC.

Yadav, R.S., and J.C. Tarafdar. 2001. Influence of organic and inorganic phophorus supply on the maximum secretion of acid phosphatase by plants. Biol. Fert. Soils 34:140-143.

Yamamura, K., M. Yokozawa, M. Nishimori, Y. Ueda, and T. Yokosuka. 2006. How to analyze long-term insect population dynamics under climate change: 50-year data of three insect pests in paddy fields. Popul. Ecol. 48:31-48.

Yoo, B.-C., F. Kragler, E. Varkonyi-Gasic, V. Haywood, S. Archer-Evans, Y.M. Lee et al. 2004. A systemic small RNA signaling system in plants. Pl. Cell 16:1979-2000.

Young, E.M., and A. Mottram. 2001. Transplanting sorghum and millet as a means of increasing food security in semi-arid, low-income countries.Trop. Agric. Assoc. Newslet. 21(4):14-17.

Yue, L., Q. Xiaobo, G. Qingzhu, W. Yunfan, and X. Wei. 2005. Simulation of management practices impacts on GHG emissions from agricultural soil. p. 619-626. In A. van Amstel (ed) Proc. Fourth Int. Symp. NCGG-4. Millpress, Rotterdam.

Zagre, N.M., D. Delpeuch, P. Traissac, and H. Delisle. 2003. Red palm oil as a source of Vitamin A for mothers and children: Impact

of a pilot project in Burkina Faso. Public Health Nutr. 6:733-742.

Zepeda, J.F. 2006. Coexistence, genetically modified biotechnologies and biosafety: Implications for developing countries. Am. J. Ag. Econ. 88:1200-1208.

Zhang, H. 2003. Improving water productivity through deficit irrigation: Examples from Syria, the North China Plain and Oregon, USA. p. 301-309. In J.W. Kijne et al. (ed) Water productivity in agriculture: limits and opportunities for improvement. CABI, Wallingford and IWMI, Colombo.

Zhang, Q., C.O. Justice, and P.V. Desanker. 2002. Impacts of simulated shifting cultivation on deforestation and the carbon stocks of the forests of central Africa. Agric. Ecosyst. Environ. 90:203-209.

Zhang, X.C. 2005. Spatial downscaling of global climate model output for site-specific assessment of crop production and soil erosion. Agric. For. Meteorol. 135(1-4): 215-229.

Ziervogel, G. 2004. Targeting seasonal climate forecasts for integration into household level decisions: The case of smallholder farmers in Lesotho. The Geograph. J. 170:6-21. doi: 10.1111/j.0016-7397.2004.05002.x.

Ziska, L.H., S.S. Faulkner, and J. Lydon. 2004. Changes in biomass and root shoot ration of field-grown Canada thistle (Cirsium arvense) with elevated CO_2: Implications of control with glyphosphate [N-(phosphonomethyl) glycine] in weedy species. Weed Sci. 52:584-588.

Ziska, L.H., and K. George. 2004. Rising carbon dioxide and invasive, noxious plants: Potential threats and consequences. World Res. Rev. 16(4):427-447.

Ziska, L.H., and E.W. Goins. 2006. Elevated atmospheric carbon dioxide and weed populations in glyphosphate treated soybean. Crop Sci. 46:1354-1359.

Zutta, B. 2003. Assessing vegetation functional type and biodiversity in southern California using spectral reflectance. A thesis presented to the Faculty Dep. Biol. Microbiol. Calif. State Univ., CA.

Zvereva, E.L., and M.V. Kozlov. 2006. Consequences of simultaneous elevation of carbon dioxide and temperature for plant-herbivore interactions: A metaanalysis. Glob. Change Biol. 12:27-41.

7

Options for Enabling Policies and Regulatory Environments

Coordinating Lead Authors
Anne-Marie Izac (France), Henrik Egelyng (Denmark), Gustavo
Ferreira (Uruguay)

Lead Authors
David Duthie (UK), Bernard Hubert (France), Niels Louwaars (Nether-
lands), Erika Rosenthal (USA), Steve Suppan (USA), Martin Wierup
(Sweden), Morven McLean (Canada), Elizabeth Acheampong (Ghana)

Contributing Authors
Patrick Avato (Germany/Italy), Daniel de la Torre Ugarte (Peru),
Shaun Ferris (UK), Edwin Gyasi (Ghana), Niek Koning (Netherlands),
Douglas Murray (USA), Laura T. Raynolds (USA), Peter Robbins (UK)

Review Editor
Niels Röling (Netherlands)

Key Messages

Key Messages

1. Policy approaches to improve natural resource management and the provision of environmental services can benefit from security of, access to and tenure to resources and land and the explicit recognition of the multiple functions of agriculture. Options include increased investment in sustainable surface water delivery to stop aquifer water-mining; establishment and strengthening of agencies administrating large water systems that cross traditional administrative boundaries; systems for monitoring forest conditions and forest dwellers' welfare; more resource efficient use, more transparent allocations of use and better enforcement of regulations over forests and lands; and recognition of communal rights of local and indigenous communities.

2. Mechanisms to better inform and democratize AKST policy making are fundamental to achieving development and sustainability goals. The complexities of the globalizing world require vast amounts of knowledge for informed policy development on emerging technologies, trade, environmental and other issues to support the objectives of the IAASTD. Options include increased comparative technology assessment, strategic impact assessment, and increased trade capacity development for developing countries. Strongly improved governance is needed to respond to discontinuities arising from global environmental change and conflict. Options include adoption of enhanced governance mechanisms at all levels (i.e., to institutionalize transparency, access to information, participation, representation and accountability) will help assure that social and environmental concerns, including those of the small farm sector, are better represented in local, national and international policy making.

3. Market mechanisms to internalize environmental externalities of agricultural production and reward the provision of agroenvironmental services are effective to stimulate the adoption of sustainable agricultural practices and improve natural resources management. Market mechanisms that include payment/reward for environmental services (PES) are one approach that recognizes the multifunctionality of agriculture, and creates mechanisms to value and pay for the benefits of ecosystem services provided by sustainable agricultural practices such as low-input/low-emission production, conservation tillage, watershed management, agroforestry practices and carbon sequestration. Other approaches include taxes on carbon and pesticide use to provide incentives to reach internationally or nationally agreed use-reduction targets, support for low-input/low-emission, incentives for multiple function use of agricultural land to broaden revenue options for land managers, and carbon-footprint labeling of food. Incentive and regulatory systems structured to generate stable revenue flows that contribute to long-term sustainability of service-providing landscapes will benefit small-scale farmers and local communities.

4. Decisions around small-scale farm sustainability pose difficult policy choices. Special and differential treatment is an acknowledged principle in Doha agricultural negotiations and may be warranted for small farm sectors without a history of government support. New payment mechanisms for environmental services by public and private utilities such as catchment protection and mitigation of climate change effects are of increasing importance and open new opportunities for the small-scale farm sector.

5. Opening national agricultural markets to international competition before basic national institutions and infrastructure are in place can undermine the agricultural sector, with potential long-term negative effects for poverty alleviation, food security and the environment. Some developing countries with large export sectors have achieved aggregate gains in GDP, although their small-scale farm sectors have not necessarily benefited and in many cases have lost out. The poorest developing countries are net losers under most trade liberalization scenarios. These distributional impacts call for differentiation in policy frameworks as embraced by the Doha work plan (special and differential treatment and non-reciprocal access). Trade policy reform aimed at providing a fairer global trading platform can make a positive contribution to the alleviation of poverty and hunger. Developing countries could benefit from reduced barriers and elimination of escalating tariffs for processed commodities in developed countries; deeper preferential access to developed country markets for commodities important for rural livelihoods; increased public investment in local value addition; improved access for small-scale farmers to credit; and strengthened regional markets.

6. Intensive export oriented agriculture has increased under open markets, but has been accompanied in many cases by adverse consequences such as exportation of soil nutrients, unsustainable soil or water management, or exploitative labor conditions. AKST innovations that address sustainability and development goals would be more effective with fundamental changes in price signals, for example, internalization of environmental externalities and payment/reward for environmental services.

7. Better integration of sanitary and phytosanitary standards (SPS) and policy and regulation related to food safety, plant and animal health needs to be better integrated internationally to more effectively utilize the limited national resources that are available for issues. Strong international food safety standards are important but present major regulatory costs for developing countries; lack of resources means that these countries are often only able to implement SPS standards for the purpose of trade facilitation with little benefit to domestic consumers who are affected by a wide array of food-borne illnesses. Confining Codex, OIE and IPPC to work within their constitutional mandates may be of less relevance today given the globalization of agriculture and trade. The efficacy of working within the traditional international mandates is challenged by the emergence of alternative regulatory mechanisms that integrate food safety, animal and plant health related standards and production practices in on-farm HACCP plans.

Revising SPS-related policy and regulatory measures within a biosecurity framework may be one option for promoting cross-sectoral interventions, as is increased international support for domestic application of food safety measures in developing countries.

8. IPRs may undermine research and use of AKST to meet development and sustainability goals. Even though license agreements may promote technology transfer by clarifying roles and responsibilities in some cases, policy mechanisms are needed to protect and remunerate traditional knowledge and genetic resources used to develop industrialized products. Even though IPRs have a role in a commercial approach to innovation, in many countries it is the public sector research institutions that promote the introduction of IPRs in agriculture. This promotion may be at odds with the public tasks of contributing to poverty alleviation and household nutrition security. Reliance on IPR based revenues is likely to lead to a change in public research priorities from development to business opportunities, e.g., commercial crops like maize and oil crops at the cost of research on small grains and pulses.

9. Climate mitigation options employing the agricultural sectors are not well covered under current national and international policy instruments. A much more comprehensive agreement is needed that looks forward into the future if we want to take full advantage of the opportunities offered by agriculture and forestry sectors. Achieving this could be accomplished through, among other measures, a negotiated global long-term (30-50 years), comprehensive and equitable regulatory framework with differentiated responsibilities and intermediate targets. Within such a framework there could be a modified Clean Development Mechanism, with a comprehensive set of eligible agricultural mitigation activities, including: afforestation and reforestation; avoided deforestation, using a national sectoral approach rather than a project approach to minimize issues of leakage, thus allowing for policy interventions; and a wide range of agricultural practices including zero/reduced-till, livestock and rice paddy management. Other approaches include reducing agricultural subsidies that promote GHG emissions and mechanisms that encourage and support adaptation, particularly in vulnerable regions such as in the tropics and subtropics.

7.1 Natural Resources and Global Environmental Change

"We are moving now into new, post-industrial, third-generation agriculture (TGA). The challenge for TGA is to combine the technological efficiency of second-generation agriculture with the lower environmental impacts of first-generation agriculture. . . . Policy tools, many of which are now available, must be further developed and integrated. Through a combination of regulation against pollution and degradation, the creation of markets for public goods through the rural development regulation, and enabling and educating consumers to opt for goods produced to high environmental standards, the environmental benefits of agriculture could be delivered to a high level alongside out-

puts of food and fibre." (Buckwell and Armstrong-Brown, 2004)

7.1.1. Resources, processes of change and policies
The broad history of the relation between natural resources, i.e., the natural world, and agriculture has been one of a slow transition from small patches of agriculture in a surrounding matrix of natural habitat, to one of small patches of natural habitat embedded in a matrix of agricultural or otherwise human influenced land. This trend is likely to continue at the global level over the next 50 years.

There is an obvious, but in fact poorly quantified, two-way interaction between agricultural land and natural systems. This interaction has changed significantly as the global "footprint" of agriculture has expanded. Natural systems provide "services" to agriculture both as sources of environmental goods (provisioning services) and also as sinks (regulating services), while agriculture often acts as a driver in natural resource degradation. Natural systems provide not only environmental goods and provisioning and regulating services. In Millennium Ecosystem Assessment (MA) terms, the most critical services natural systems provide to agriculture are "supporting services," such as nutrient cycling and pollination. Over the past 50 years, agriculture has gone from being a relatively minor source of off-site environmental degradation to becoming a major contributor to natural resource depletion and degradation, acting through habitat loss and fragmentation, invasive alien species, unsustainable use (over harvesting), pollution (especially of aquatic systems) and, increasingly, climate change.

Policy responses to this trend toward natural resource degradation have occurred at international, regional and local levels. An essential component of all necessary policy reforms for mitigating agricultural impacts is to integrate environmental, natural resource, and biodiversity concerns into policy making at the highest possible level in order to achieve the necessary facilitation and leverage on lower-level policies. For example, in the European Union the revised EU Sustainable Development Strategy (EU-SDS II) includes biodiversity conservation, but still lacks an overarching commitment to reduction in drivers that other sectoral policies could then address in more detail within the stronger mandate provided by EUSDS II. Further revision of the EU-SDS could provide better integration of the EU's internal and global commitments (WSSD, Doha and Monterrey) and provide better harmonization between different European sustainable development processes (Cardiff, Lisbon, Gothenburg and Johannesburg) and instruments (Extended Impact Assessment and Indicators for Sustainable Development). High level integration can also be achieved, to some extent, via Multilateral Environmental Agreements (MEAs), for example through the agreed Programme of Work for Agricultural Biodiversity of the UN Convention on Biological Diversity (CBD).

The CBD Agricultural Biodiversity work program focuses on (1) assessing the status and trends of the world's agricultural biodiversity and of their underlying causes, as well as of local knowledge of its management, (2) identifying and promoting adaptive-management practices, technologies, policies and incentives, (3) promoting the conservation and sustainable use of genetic resources of actual/potential value for food and agriculture, (4) assessing the impact of new technologies, such as modern biotechnology in general and Genetic Use of Restriction Technologies (GURTs) in particular. The work program also has cross-cutting initiatives for conservation and sustainable use of pollinators and soil biodiversity, studies the impacts of trade liberalization on agricultural biodiversity, identifies policy to promote mainstreaming and integration of biodiversity into sectoral and cross-sectoral plans and programs. But the CBD is a framework, or umbrella agreement that requires its constituent Parties to adopt policies and enact legislation for effective implementation of its Decisions.

Even if its Decisions are adopted and implemented fully at the national level, there is a danger that the CBD, like many other policy instruments, will be continually "running behind the future," (e.g., the CBD 2010 Target) to significantly reduce the rate of biodiversity loss. Historically, the principal policy instrument has been the establishment of protected areas, although this has been ineffective where prime agricultural land and high biodiversity compete, as can be seen by the underrepresentation of lowland, fertile land in the majority of current national protected area systems (WCMC, 2006).

Broadly, natural habitats around the world can be divided into three categories, each requiring different, but overlapping or integrated sets of policies to ensure their survival in the long-term (Chomitz, 2007).The first category can be defined as wilderness: the majority of the land (or aquatic) area is natural, and anthropogenic land use has had a minor impact. With the exception of the major tropical rainforest regions of Amazonia, the Congo, Indonesia and Papua New Guinea, the majority of these areas are in temperate regions and do not harbor high levels of biodiversity, although they may provide valuable ecosystem services, especially in terms of water supply and carbon sequestration. Policies that promote establishment of protected areas in these regions are still feasible due to lack of pressure from alternative land use, but even in these areas, protected area design must consider the external threats arising from climate change (e.g., increased wildfires, and global transport of pollutants).

The second major category of land could be termed frontier: land potentially suitable for agriculture that is close to an expanding agricultural system. Effective policies for the sustainable management of natural resources in these areas are difficult to design and implement. In most countries, traditional concepts of agriculture are used to develop protection policies based on the ecosystem representation and species richness as sole criteria. However, (sensu Peres and Terborgh, 1995) the development of sustainable natural resource management policies in terms of local community support and resilience in the face of climate change will be critical in coming years. Also critical will be the acknowledgement that appropriate policies and institutional arrangements (e.g., providing positive incentives to farmers to adopt sustainable soil management practices in areas where soils are depleted) can ultimately result in improved natural resources quality through agricultural use (Izac, 1997).

Increasingly, improved methods of measuring and mapping total ecosystem value of natural land are allowing land-use planners and landholders to make informed economic decisions based on a broader range of criteria than agricultural production alone (Troy and Wilson, 2006). This is allowing policy makers to introduce land-use planning "rules" (zoning) and economic incentives to better conserve natural environments in complex agricultural land-use mosaics.

At a relatively large scale, this kind of planning is increasingly emerging in the Brazilian Amazon and Atlantic rainforests, (Campos and Nepstad, 2006; Wuethrich, 2007, respectively), where government and landholders are slowly forging agreements on establishment of a complex mosaic of protected areas, sustainable use forests and agricultural land. This represents a shift in policy away from prescriptive land use decisions made by the imposition of protected area on unwilling land-users towards the use of incentives, including payments for conservation. Auction bids for direct payments for conservation services such as native forest protection, reforestation, and restoration of riparian vegetation can further improve efficiency (Chomitz et al., 2006). Under this type of policy, eligible landowners voluntarily decide whether to apply for participation, and the resultant conservation network emerges as a consequence of many independent choices about participation. Similar incentive-based schemes may be found in the US Conservation Reserve Program (CRP), the Bush Tender program in Australia and the Costa Rica Environmental Services Payment program (see references in Chomitz et al., 2006).

In the more "crowded" landscapes of Europe and the west coast of the USA, where remaining natural land exists in an agricultural and urban matrix rather than the converse, similar trends towards land use planning based on ecosystem service valuation and "multifunctionality" are being explored (Zander et al., 2007). In California, a spatially explicit conservation planning framework to explore trade-offs and opportunities for aligning conservation goals for biodiversity with six ecosystem services (carbon storage, flood control, forage production, outdoor recreation, crop pollination, and water provision) has been used. Although there are important potential trade-offs between conservation for biodiversity and for ecosystem services, a systematic planning framework offers scope for identifying valuable synergies (Chan et al., 2006).

In Europe, agroenvironmental subsidies have been used as incentives to maintain and promote biodiversity-friendly land use on agricultural land. There has been some criticism that the schemes do not deliver all of the environmental and biodiversity benefits for which they were designed, especially as the scale of implementation becomes too small and fragmented (Whittingham, 2007). One option that avoids this situation is the adoption of regional planning approaches (e.g., the OECD environmental farm plan programs) to generate more coordinated land use patterns across larger landscapes (Manderson et al., 2007).

A recent summary (Chan et al., 2006) of the policies for sustainable development at the interface between tropical forest and agriculture shows how these can be used to promote the trends described above:

At the international level:
- Mobilize carbon finance to reduce deforestation and promote sustainable agriculture.

- Mobilize finance for conservation of globally significant biodiversity.
- Finance national and global efforts to monitor forests and evaluate the impacts of forest projects and policies—including devolution of forest control.
- Foster the development of national-level research and evaluation organizations through twinning with established foreign partners.

At the national level:

- Create systems for monitoring forest conditions and forest dwellers' welfare, make land and forest allocations and regulations more transparent, and support civil society organizations that monitor regulatory compliance by government, landholders, and forest concessionaires. The prospect of carbon finance can help motivate these efforts.
- Make forest and land use regulations more efficient, reformulating them to minimize monitoring, enforcement, and compliance costs. Economic instruments can help.

In wilderness areas:

- Avert disruptive races for property rights by equitably assigning ownership, use rights, and stewardship of these lands.
- Options for forest conservation include combinations of indigenous and community rights, protected areas, and forest concessions. Some forests may be converted to agriculture where doing so offers high, sustainable returns and does not threaten irreplaceable environmental assets.
- Plan for rational, regulated expansion of road networks—including designation of roadless areas.
- Experiment with new ways of providing services and infrastructure to low-density populations.

In frontier areas:

- Equitably assign and enforce property rights.
- Plan and control road network expansion.
- Discourage conversion in areas with hydrological hazards, or encourage community management of these watersheds.
- Use remote sensing, enhanced communication networks, and independent observers to monitor logging concessionaires and protect forest-holders against encroachers.
- Consider using carbon finance to support government and community efforts to assign and enforce property rights.
- Encourage markets for environmental services in community-owned forests.

In disputed areas:

- Where forest control is transferred to communities, build local institutions with upward and downward accountability.
- Where community rights are secure and markets are feasible, provide technical assistance for community forestry.
- Make landholder rights more secure in "forests without trees."

- When forest tenure is secure, use carbon markets to promote forest regeneration and maintenance.

Mosaic lands:

- Reform regulations to reward growing trees. Promote greener agriculture—such as integrated pest management and silvo-pastoral systems—through research and development, extension efforts, community organization, and reform of agriculture and forest regulations.
- Develop a wide range of markets for environmental services—carbon, biodiversity, water regulation, recreation, pest control—to support more productive, sustainable land management.

7.1.2 Reducing the impacts of climate change and the contribution of agriculture to climate change

Agriculture can contribute to climate change in four major ways:

- Land conversion and plowing releases large amounts of stored carbon as CO_2 from original vegetation and soils;
- Carbon dioxide (CO_2) and particulate matter is emitted from fossil fuels used to power farm machinery, irrigation pumps, and from drying grain, etc., as well as fertilizer and pesticide production;
- Nitrogen fertilizer applications and related cropping practices such as manure applications and decomposition of agricultural wastes result in emissions of nitrous oxide (N_2O); and
- Methane (CH_4) is released mostly through livestock digestive processes and rice production.

The share of the agricultural sector to total global GHG emissions is approximately 58% of CH_4 and 47% of N_2O making it a significant contributor with a good deal of potential for reduction in emissions in mitigation strategies (Smith et al., 2007). With appropriate policies, each of these well-known sources of GHG can be mitigated to some extent.

Many of these mitigation options are "win-win" as long as they are supported by policy interventions that remove entry barriers and reduce transaction costs. For example, lower rates of agricultural extensification into natural habitats and the re-use/restoration of degraded land, could be encouraged through the participation of farmers in emissions trading, or biofuel production. Farmers can benefit financially depending on the amount of credits generated through carbon storage projects under the Kyoto Protocol, as is already occurring in a number of countries. Despite some transaction costs associated with quantifying and maintaining stored carbon, farmers who implement conservation agriculture; use cover crops to reduce erosion; or reforest degraded lands with tree species that have commercial value could profit financially by selling their credits in an emissions trading market. Agricultural N_2O and CH_4 mitigation opportunities include proper application of nitrogen fertilizer, effective manure management, and use of feed that increases livestock digestive efficiency. To date, there is little policy or legislation that recognizes the ability of the agricultural sector to provide GHG reductions through mitigation of N_2O and CH_4 and that provides positive incentives for farmers to adopt more sustainable practices.

Under the Kyoto Protocol Clean Development Mechanism (CDM), deliberate land management actions that enhance the uptake of carbon dioxide (CO_2) or reduce its emissions have the potential to remove a significant amount of CO_2 from the atmosphere in the short and medium term. The quantities involved may be large enough to satisfy a portion of the Kyoto Protocol commitments for some countries (but are not large enough to stabilize atmospheric concentrations without additional major reductions in fossil fuel consumption). Carbon sequestration options or sinks that include land-use changes (LUCs) can be deployed relatively rapidly at moderate cost and could play a useful bridging role while new energy technologies are being developed. A challenge remains to find a commonly agreed and scientifically sound methodological framework and equitable ways of accounting for carbon sinks. These should encourage and reward activities that increase the amount of C stored in terrestrial ecosystems but at the same time avoid rewarding inappropriate activities or inaction. Collateral issues, such as the effects of LUC on biodiversity and on the status of land degradation, should be addressed simultaneously with the issue of carbon sequestration in order to exploit potential synergies between the goals of UNCBD, UNCCD, UNFCCC and the Kyoto Protocol. Such measures would also improve local food security and alleviate rural poverty (FAO, 2004b).

7.1.3 Managing the natural resource base of agriculture

7.1.3.1 Soils, nutrients and pests
Soils. Multifunctional agriculture recognizes the many ecosystem services of soil, including: (1) services that support the growth of plants, including nutrient regulation, water supply and water cycle; (2) storage of carbon in soil organic matter and hence regulation of GHGs; (3) regulation of the impact of pollutants through biological activities and absorption on soil particles; (4) habitat for a very large component of biodiversity (e.g., soil microorganisms and invertebrates); (5) biodiversity pool, such as habitats, species and genes; (6) physical and cultural environment for humans and human activities; (7) source of raw materials; (8) archive of geological and archeological heritage (Kibblewhite et al., 2007). The framework European Commission strategy for soil protection (CEC, 2006) is based on identification of risk of loss of function, and putting in place remediation measures to mitigate threat. Many of these remediation measures could be applied to agricultural lands, but will need to be driven by a different mix of command and control, incentive-based, or market-based trading policy measures appropriate to different situations. Policies based on payments per tonne C or market sales of C are likely to be more efficient than those based on a per hectare basis, but will require new methods and techniques to provide cost effective information about the relationship between carbon sequestration and land quality, use and management in addition to estimates of base line for effective enforcement and verification (Antle and Mooney, 2002). Policy measures that promote carbon sequestration in soils would most likely generate positive results for the other functions listed above (Swift et al., 2004).

Projected increases in certified organic agriculture raise additional sets of opportunities for AKST to contribute to maintaining productivity and soil nutrient levels while controlling costs and improving labor efficiencies (Chapter 5). Policy options for reforming institutional environments, policies and programs to be more conducive to sustainable agricultural methods (Egelyng and Høgh-Jensen, 2006) include:

- Investing in the development of organic certification in developing countries.
- Reforming tax systems to shift the conditions under which certified organic farming compete with energy intensive agricultural systems, involving a shift from taxing wages towards taxing pollution and consumption of resources. (Chapter 2)
- Increasing awareness of organic certification to domestic consumers in developing countries;
- Supporting development of methods for organic certification compliant pest (and weed) and soil nutrient management, particularly non-proprietary, methods for the public good, such as biocontrol using natural enemies, non-chemical, and cultural methods of pest management.
- Supporting AKST to further energy efficiency in organic agriculture;
- Developing certified organic seeds that are better adapted to low-input farming landscapes (Chapter 2).
- Investing in low external input technologies aimed at soil fertility improvement (Chapter 6)

Nutrients. Although in many countries policies for reductions of point source pollution have been successfully introduced, controlling non-point source pollution remains a more difficult challenge. Agriculture's contribution to non-point source pollution varies widely as a complex function of land use, cropping system, soil type, climate, topography, hydrology, animal density, and nutrient management techniques. Despite this complexity, research based nutrient management practices that are effective at reducing non-point source pollution are available. Wider implementation of currently recommended nutrient management plans is important for further gains in environmental quality.

Site-specific, nutrient management planning should guide the implementation of agricultural nutrient management practices that will be profitable and protective of the environment. Modern agricultural science innovations can increase not only efficiency of production, but also efficiency of nutrient use. Examples include (1) increased plant nutrient recovery and nutrient retention by animals; (2) improved understanding and modeling of the fate and transport of nutrients in soils; and (3) development of mitigation and bioremediation strategies such as wetlands, riparian buffers, and filter strips to limit total nutrient exports from agricultural systems.

Adoption of efficient agricultural nutrient management practices may be limited by current market processes that do not provide for positive or negative externalities and the politics of crop and animal production. Social and political pressures to prevent nutrient overloads from agriculture are increasing, but many in the sector cannot afford the high transaction costs to introduce mitigation measures and maintain profits under current agricultural business models without subsidies. In countries where these subsidies have been introduced the key policy challenge is to improve cost effectiveness through competitive bidding; environmental cost-benefit analysis; and performance-based payments for farmers to remove environmentally sensitive land from crop production.

Pests. Invasive alien species (IAS) are a threat to global biodiversity and can have devastating effects on both agricultural and natural systems at large scales after small isolated introductions. A major policy challenge from IAS is the fact that the vast majority of current and future IAS were either poorly known species, or were unknown as pests before their introduction to a new location. This is the main reason for the failure of past policies to deal with IAS, even those using the best available risk assessment methodologies (Keller et al., 2007).

Future IAS policies should be based on the following principles in order to mitigate this weakness.

- National IAS systems should be linked to regional and global databases of known IAS and their treatment.
- IAS control systems should be based on a "pathways of entry" approach where detection and control effort is focused on the most likely points of entry into a country (or region). Introductions of IAS occur through various channels or pathways, both intentionally and unintentionally. Primary pathways of intentional introduction of potential IAS include horticultural products, food products, and exotic pets, the use of nonnative organisms in aquaculture and for restocking of marine and inland water systems for commercial and recreational fisheries; scientific research; horticulture; trade in pets and aquarium species; biocontrol agents; and *ex situ* breeding projects. Pathways of unintentional introductions include ballast water and ballast sediments, ship hulls, packaging materials and cargo containers, garbage and waste, international assistance programs; tourism; military activities, and unprocessed materials, such as timber.
- Risks posed by pathways of IAS prior to introduction and establishment should be addressed and mitigated both before the IAS reach the border and at the border. Preventing introductions before they occur is the most effective and cost-efficient approach to addressing IAS issues. Removing IAS once they have become established requires significantly more financial, technical, and personnel resources than preventing their introduction; and, often, complete removal is not even possible.
- An operating principle of the system should be that it is based on a list of approved species for deliberate introduction, and that any species not on the list must pass through a risk assessment process before being approved for entry.

A number of policy initiatives have been undertaken for specific major pathways of introduction including:

- *Importation of living plants and plant material.* Many attempts are being made to address plant-related pathways of invasive species. One voluntary initiative, based on the Missouri Botanic Garden St. Louis Declaration, is developing and implementing self-governed and self-regulated codes of conduct for nursery professionals, government agencies, the gardening public (specifically garden clubs), landscape architects, and botanic gardens/arboreta, designed to stop use and distribution of invasive plant species. Working with these respective industries, the process has generally appealed to the responsible use and import of horticultural products by the private sector to minimize the introduction of IAS. There is an urgent need for the IPPC to more effectively address, perhaps through a quarantine/sterilization-based international sanitary and phytosanitary measure (ISPM) based the problem of "hitchhikers" on horticultural products, which are potential IAS, but may not be considered plant pests per se (e.g., spiders, ants).
- *FAO Code of Conduct for Responsible Fisheries.* This code includes a section encouraging the use of legal and administrative frameworks to promote responsible aquaculture, including discussions with neighboring states prior to the introduction of nonindigenous species, minimizing the impacts of nonindigenous or genetically altered fish stocks, as well as minimizing any adverse genetic or disease impacts. While the Code serves as a useful guide, it is not focused on specific prevention, management and control measures related to IAS within the field of aquaculture and fisheries.

Given the role of trade in the production and transport of goods, approaches to regulating pathways of IAS should consider relevant trade rules and agreements. The World Trade Organization's (WTO) Agreement on Sanitary and Phytosanitary Measures (SPS Agreement) defines the basic rights and obligations of WTO members regarding use of sanitary and phytosanitary measures to protect human, animal or plant life or health from the entry, establishment or spread of pests, diseases, disease carrying organisms; and prevent or limit other damage from the entry, establishment or spread of pests (see 7.3.3 for details).

7.1.3.2 *Genetic resources and agrobiodiversity*
Three major types of policy tools are available to support conservation of genetic resources (1) public investment in *in situ* and *ex situ* conservation; (2) clearer intellectual property rights, including for farmer innovations, particularly in developing countries; and (3) material transfer agreements (Rubenstein et al., 2005). Apart from ecological approaches to agriculture, connected to nature management, strategies for conservation and sustainable use of agricultural genetic resources also include "*ex situ*" and "*in situ*, on farm" approaches. *Ex situ* conservation in gene banks is well established for major crops under the auspices of the FAO by the centers of the CGIAR, and at national plant and farm animal gene banks. A Global Crop Diversity Trust has recently been initiated to generate funds for the sustainable conservation of the most important collections worldwide, on behalf of all future generations, and Norway is hosting a long-term conservation facility in the Arctic at Svalbard. Public policies converged progressively through the International Undertaking (1985) to the International Treaty on Plant Genetic Resources for Food and Agriculture (IT PGRFA – 2005) providing special rules for the conservation

of PGRFA, their sustainable use and the sharing of benefits arising from their use. It also contains mechanisms for facilitated access and benefit sharing through its Multilateral System. Many signatory countries (116 in June 2007) are yet to implement the Treaty. Policy options range from contributing to the Global Crop Diversity Trust (currently mainly OECD countries, but also countries like Colombia and Brazil), to establishing or expanding national *ex situ* collections (e.g., India), and for liberal or restrictive regimes for access to these collections, including the sharing of information on these resources. The agricultural sector generally supports liberal access regimes in order to promote availability of genetic resources for plant breeding in support of food security and rural development.

On-farm policy approaches to management of genetic resources include various types of support to farmers who maintain and further develop genetic resources, such as payments based on cultural heritage (e.g., historic cattle breeds), technical support to foundations for crop-hobbyists (e.g., old apple variety clubs in Europe) to participatory plant breeding strategies in many developing countries (Almekinders and Hardon, 2006). Such mechanisms may conflict with existing policies and laws that focus on seed system development, including seed laws, plant breeder's rights, etc. The EU recently developed the concept of "conservation varieties" for this reason. National policies to bring these objectives in harmony with each other are supported by the concept of Farmers' Rights of the IT PGRFA. However, countries may make these rights "subject to national law and as appropriate" (Art. 9 – IT PGRFA) which provides broad options for national priorities and implementation strategies. The use of a range of standard economic tools (taxes, subsidies, "cap and trade" with permits) can help maintain higher level of plant genetic diversity in seed markets recognizing that information barriers limit market efficiency (Heal et al., 2004).

Recent advances in molecular biology have provided a whole new set of tools for investigating biodiversity, including the diversity of agricultural plants and animals. While there has undoubtedly been significant loss of diversity over time of plant and animal genetic resources for agriculture at the varietal level, there is some evidence that overall losses of genetic diversity when measured at the genetic level have not been so great and that modern biotechnological breeding tools can regenerate some of this diversity, especially if the tools can be transferred to developing country agricultural research levels through support for initiatives such as African Agricultural Technology Foundation and Public Sector Intellectual Property Resource for Agriculture (PSAPRA) (USDA, 2003).

Livestock. The livestock sector is an important source of greenhouse gases and factor in the loss of biodiversity, while in developed and emerging countries it is a significant source of water pollution.[1] Current decision-making on the livestock–environment–people nexus is characterized by se-

vere underpricing of virtually all natural processes that go into livestock production process. This includes neglect of major downstream externalities. Limiting livestock's impact on the environment to more socially optimal levels requires measures to reduce land and other natural resources requirements for livestock production. This could be achieved by intensification of the most productive arable and grassland used to produce feed or pasture; and retirement of marginally used land, where this is socially acceptable and where other uses of such land, such as for environmental purposes, are in demand, and have higher value. Intensification can lead to gradual reductions of resource use and waste emissions across the sector. For example, precision feeding and use of improved genetics can greatly reduce emissions of gases (carbon dioxide, methane, etc.) and of nutrients per unit of output.

The major policy goals for addressing environmental pressure points arising from current policy and market processes in the livestock sector are:
- Controlling expansion into natural ecosystems;
- Managing rangeland in a sustainable way;
- Reducing nutrient loading in livestock concentration areas; and
- Reducing the environmental impact of intensive feed production.

Because the major stressors arising from the livestock sector differ in different parts of the world—ranging from overgrazing in Australia and sub-Saharan Africa, biodiversity loss from pasture expansion in Latin America, to pollution arising from intensive pig production in Europe and SE Asia—the mix and emphasis of the policy instruments will need to be different in different parts of the world, but could include measures to:
- Limit livestock's land requirements,
- Correct distorted prices,
- Strengthen land titles,
- Price water and water quality internalizing all externalities,
- Remove subsidies,
- Liberalize trade, and
- Support intensification and promoting research and extension of cutting edge technology.

The choice of policy instruments should take into consideration the broader goals of efficiency; effectiveness and equity (Hahn et al., 2003), given the major economic contribution and social role played by the livestock sector globally.

Aquaculture. "Traditional" aquaculture has been an integral part of one of the world's most sustainable agricultural systems—the polyculture Chinese fish-farm (FAO/IPT, 1991) for around 3000 years, but recent rapid expansion of commercial aquaculture, in the absence of appropriate policies, is generating negative environmental and social impacts that threaten to undermine the long-term sustainability of the industry. In recognition of these growing negative impacts, broad environmental management principles have been agreed for the sector (e.g., FAO Code of Conduct for Responsible Fisheries, Article 9; FAO Technical Guidelines

---

[1] (The following text draws heavily on a major review of the negative impact of livestock production on the environment by Steinfeld et al., 2006.)

for Responsible Fisheries No. 5. Development of Aquaculture). However, these have often not been well integrated into national policy and legislative frameworks.

A number of policy tools have been used to regulate unsustainable aquaculture expansion, e.g., banning of mangrove utilization for aquaculture practices, determining maximum production per area, standards for feed, rules for disease control, and the use of drugs.

Regulations focused on individual production units alone cannot guarantee sustainability at the landscape level because they do not consider the cumulative impacts of multiple farms on a particular area. In addition, existing regulatory structures for aquaculture mostly do not allow, or facilitate, a production mode or approach that is conducive to long-term sustainability. Nutrient cycling and reutilization of wastes by other forms of aquaculture (polyculture) or local fisheries are frequently prohibited or discouraged.

FAO and progressive industries are now increasingly promoting an ecosystem approach to aquaculture which will "balance diverse societal objectives, taking into account knowledge and uncertainties of biotic, abiotic and human components of ecosystems including their interactions, flows and processes and applying an integrated approach to aquaculture within ecologically and operationally meaningful boundaries" (FAO, 2006a). The purpose of an ecosystem approach should be to plan, develop and manage the sector in a manner that addresses the multiple needs and desires of societies, without jeopardizing the options for future generations to benefit from the full range of goods and services provided by aquatic ecosystems.

Policy options for improved environmental performance of the aquaculture sector could include:

- Further development of guidelines for environmentally sound and sustainable aquaculture industry and promotion of compliance with the guidelines;
- Promotion of the adoption of exclusion zones that protect wild stocks in areas considered to be essential to their continued survival in the wild and to maintain commercial wild fisheries;
- Improved integration of aquaculture development with wild fish stock management, including, where appropriate, enhancement strategies for aquatic species to help wild stock fisheries recover and to provide additional recreational opportunities;
- Promotion and enforcement of regulations that require Strategic Environmental Impact Assessment of potential aquaculture developments at the landscape level and develop land use plans that maintain total production within environmentally sustainable limits;
- Adoption of production unit design and management practices that encourage integration, recycling and reuse of effluents, provide for disposal and/or processing of wastes;
- Adoption of production unit design and management practices that minimize and, where practicable, eliminate the use of agriculture and veterinary chemicals and ensure the correct use and disposal of registered chemicals;
- Support for the development and use of diets and feeding strategies which minimize adverse environmental impacts;

- Promotion of improved monitoring and enforcement of management systems to reduce the risk deliberate and unintentional releases.
- Development of appropriate protocols regarding the safe transfer and culture of exotic species and the translocation of live product within and between states, including living modified organisms (see Myhr and Dalmo, 2005)
- Promotion of industry training and education opportunities in environmental awareness, clean production methods and best practice; and
- Promotion of an information clearinghouse and information dissemination system for environmentally sound aquaculture.

7.1.3.3 Water scarcity, water quality and the distribution of water

The broad policy recommendations which can be made for improved water management in the agricultural sector have their roots in the same fundamental paradigm shift that is required for all aspects of sustainable development—full cost accounting and recognition of the multifunctionality and interdependence of landscapes. There is a need for overall reform in the water sector which must address the following: getting technical water bureaucracies to see water management as a social and political as well as a technical issue; supporting more integrated approaches to agricultural water management; creating incentives to improve equity, efficiency, and sustainability of water use; improving the effectiveness of the state itself, particularly its regulatory role; developing effective coordination and negotiation mechanisms among various water development and management sectors; and empowering marginalized groups, including women to have a voice in water management (Merrey et al., 2007).

Improve investment in sustainable irrigation. There are four principal reasons to invest in irrigation over the next three to five decades:

1. To reduce rural poverty—in countries and regions that rely on agriculture for a large portion of their GDP (much of SSA), increased agricultural productivity is the most viable option for poverty reduction;
2. To keep up with food demand and changing food preferences;
3. To adapt to changing condition—increasing competition for water will require investments that enable farmers to grow more food with less water; increasing climate variability and extremes, due to climate change, may require further irrigation development and changes in the operation of existing schemes; and
4. To increase multiple benefits and ecosystem services from existing systems, while reducing negative impacts (Faurès et al., 2007).

Investment has traditionally meant public expenditure on new irrigation systems (capital investment). A broader definition is needed that includes public investment in irrigation and drainage development, institutional reform, improved governance, capacity building, management improvement, creation of farmer organizations, and regulatory oversight, as well as farmers' investment in joint facilities, wells, and on-farm water storage and irrigation equipment. The appropriate focus for both policy and investment will depend on both the scale and type of irrigation, and the structure of national economies (Faurès et al., 2007)

Develop locally relevant groundwater management strategies that support aquifer recharge and manage demand. A large body of evidence from Asia suggests that groundwater irrigation promotes greater interpersonal, intergender, interclass, and spatial equity than do large irrigation projects. Evidence from Africa, Asia, and Latin America also suggests that groundwater is important in settings where poor farmers find opportunities to improve their livelihoods through small-scale farming based on shallow groundwater circulation. However, pumping costs are rising, and irrigation-supporting subsidies are compromising the viability of rural energy providers. India is a prime example. Moreover, the impacts of groundwater depletion on water quality, stream flows, wetlands, and down-gradient users in certain pockets are rapidly threatening to undermine the benefits. In arid regions, where fossil groundwater is a primary source of water for all uses, intensive groundwater irrigation may threaten future water security. In addition, with anticipated shifts in precipitation patterns induced by climate change, groundwater's value as strategic reserve is set to increase worldwide (Shah et al., 2007). Because groundwater use and dependency will continue to grow in many parts of the developing world, participatory approaches to sustainable groundwater management will need to combine supply-side measures, such as artificial recharge, aquifer recovery, interbasin transfer of water and the like, with demand-side measures, such as groundwater pricing, legal and regulatory control, water rights and withdrawal permits and promotion of water-saving crops and technologies (Shah et al., 2007).

Establish and strengthen the authority of agencies administrating large water systems that cross traditional administrative boundaries. The state has historically played a leading role in water development, both in supporting large-scale irrigation, hydropower, and flood control as well as facilitating private and small-scale farmer managed irrigation (Merrey et al., 2007). There are good reasons for the state's central role in regulating and managing this vital public good resource. While the state remains the main actor to initiate reforms, these reforms are needed at all jurisdictional levels, from local to national level, and even at regional level. A recent trend has been to promote river basin organizations to manage competition for water at the basin level. There is general agreement on the long-term benefits of effective integrated management of river basins, especially with increasing competition and environmental degradation. But attempts to impose particular models of river basin organizations in developing countries, especially models derived from the experiences of rich countries, are not likely to succeed because the objectives and institutional contexts differ so greatly (Shah et al., 2005). An externally imposed one-size-fits-all strategy for managing such complexity is unlikely to be effective. Numerous models for institutional

arrangement for basin water governance exist, and their effectiveness will depend on local basin and national conditions (Molle et al., 2007).

Better integration of water use between agricultural and industrial users. Water use by agriculture could limit the amount available for other uses when water becomes scarce, however usually the opposite is true. Higher value uses (e.g., domestic purposes and industries in urban areas) have precedence; hence agriculture must adapt to reduced allocations. Uses with lower priority than agriculture are aquatic ecosystems and the environment (CA, 2007). Industrial and domestic use can also affect agriculture through the discharge of untreated wastewater from urban areas into surface-water system, decreasing the quality of water used in irrigation. Intersectoral water allocation is to a large extent a product of broader political and economic considerations, such as the political clout of urban areas and industrial interests (see Molle and Berkoff, 2005). Negotiating and crafting new types of organizational arrangements for managing irrigation, therefore, are not possible without considering broader institutional arrangements and policies in the water, agricultural, and rural sectors as well as currency, trade, and overall macroeconomic policies (Merrey et al., 2007).

Water markets to better allocate water amongst uses and users. Water markets are playing an increasingly important role in the developed world in allocating water on a regional basis. There are examples in which government has used markets or market-like arrangements to resolve vexing problems of allocation. Water pricing is one market vehicle that has received considerable attention. The difficulties of implementing water pricing in developing countries, however, are substantial. Pricing policies for full cost recovery of infrastructure development and operation and maintenance, for example, risk seriously aggravating water deprivation and poverty (Dinar, 2000; Molle and Berkoff, 2007). A sliding-scale pricing strategy is one possible solution (Schreiner and van Koppen, 2001). Another water market reform mechanism is tradable water rights, which represents the greatest degree of privatization in water management. In addition to clearly defined water rights (including transfer rights), water markets require physical infrastructure that allows water to be transferred, and institutional arrangements to protect against negative impact on third parties (Easter et al., 1998). Earlier enthusiasm for market-based water reforms was at best premature (Merrey et al., 2007). The conditions necessary for market-based reforms to contribute to sustainable water management in agriculture are extremely rare in developing countries and uncommon even in rich countries. The Chile and Valencia (Spain) water market reforms have been held up as examples, but closer inspection raises many questions (Bauer, 1997, 2005; Ingo, 2004; Trawick, 2005). As in all market and private property rights situations, questions of regulation (who sets the rules and what are the rules?) and capture of benefits (who wins and who loses in imperfect markets?) are central for assessing market-inspired reforms. A phased approach of vesting rights in existing users and currently excluded users and of clarifying regulatory mechanisms before developing detailed water market mechanisms may be more appropriate and politically more feasible than a rush to markets (see Bruns et al., 2005).

Encourage water-saving irrigation practices and technology. Farmers in most industrialized countries have only recently begun to adopt water saving practices, whereas in developing countries they have been relying on traditional water saving practices for a long time. Low levels of adoption of water-saving may be because the knowledge and incentives are not in place for farmers to benefit directly by saving water. There is an important role for the private sector in making low-cost agricultural water management technologies such as treadle pumps, small power pumps, and bucket and drip kits more widely available. Such technologies can be readily acquired and used by individual small-scale farmers, both men and women, and in many situations can substantially improve nutrition and incomes (Shah et al., 2000; Namara et al., 2005; Mangisoni, 2006; Merrey et al., 2006). Restrictive policies in some sub-Saharan African countries are retarding the wider use of these technologies, in marked contrast to South Asian countries.

Reform of irrigation management to involve local stakeholders. The establishment of Water User Associations and contracting the management of lateral canals to individuals can improve water management by providing incentives for users and managers to conserve water and improve fee collection to increase irrigation revenues. However, pilot projects to transfer management from the state to user groups on government-built schemes have rarely been scaled up effectively to cover larger areas. Many governments were reluctant, even when project documents promised to do so. Another reason was the failure to recognize the critical differences between government- and farmer-managed irrigation systems. Management transfer programs in countries as diverse as Australia, Colombia, Indonesia Mali, Mexico, New Zealand, Senegal, Sri Lanka, Turkey, and the United States have demonstrated some positive results from involving farmers and reducing government expenditures, but they have rarely shown improvements in output performance or quality of maintenance (Vermillion, 1997; Vermillion and Garcés-Restrepo, 1998; Samad and Vermillion, 1999; Vermillion et al., 2000). The few notable exceptions are middle-income developing countries such as Mexico and Turkey and high income countries such as New Zealand and the United States. Research in the 1990s on irrigation management transfer processes and outcomes produced many case studies and some useful guidelines for implementation (e.g., Vermillion and Sagardoy, 1999). There is broad agreement on the necessary conditions, but very few cases where they have been met on a large scale (Merrey et al., 2007).

Further "coping" strategies proposed for addressing water scarcity (see also Chapter 6 for more details on the options) need attention at policy levels to incorporate their potential into water management agendas to optimize the use of limited water resources:

a. *Desalinization:* Currently, the costs of desalinated water remain too high for use in irrigated agriculture, with the exception of intensive horticulture for high-value cash crops, such as vegetables and flowers (mainly in greenhouses), grown in coastal areas (where safe waste disposal is easier than in inland areas), but recent advances in membrane technology are reducing costs. At the global level the volume of desalinated water produced annually (estimated at 7.5 km^3) is currently quite low, representing about 0.2% of the water withdrawn for human use (FAO, 2006b).

b. *Urban wastewater:* Two features complicate policies pertaining to wastewater use in agriculture: most wastewater is generated outside the agricultural sector, and many individuals and organizations have policy interests pertaining to wastewater use (Qadir et al., 2007). Millions of small-scale farmers in urban and peri-urban areas of developing countries use wastewater for irrigating crops or forest trees or for aquaculture, reducing pressure on other freshwater resources. Surveys across 50 cities in Asia, Africa and Latin America have shown that wastewater irrigation is currently a common reality in three-quarters of cities (IWMI, 2006). Most domestic wastewater generated in developing countries is discharged into the environment without treatment but the dominant trend is for more wastewater treatment as countries develop national integrated water resources management plans or improved environmental policies, for example in Mexico, Brazil, Chile and Costa Rica (UNCSD, 2005). Israel currently uses 84% of its treated sewage effluent in agricultural irrigation and in a few cities, such as Windhoek in Namibia, the water is treated to a very high standard so that it can even be used as drinking water (UNIDO, 2006).

c. *Virtual water and food trade:* The import of food from water-rich countries allows water-poor countries to save water they would have used to grow food (equivalent to the import of "virtual water"), and scarce water reserves can be used for more valuable domestic, environmental and industrial purposes. Countries with limited water resources might also change their production patterns to prioritize production of agricultural commodities requiring less water and to import those requiring more water (FAO/IFAD, 2006). Whereas the strategy of importing virtual water is appealing from a water perspective, political, social and economic issues, rather than water abundance or scarcity, drive much of the current world food trade.

d. *Improving the productivity of water use in agriculture* (see Chapter 6 for detailed options). Productivity gains could improve overall water use efficiency in irrigated and rainfed agriculture. Agronomic improvements to improve overall productivity will also reduce the global "water footprint" of agriculture. This could be achieved by, for example, improving the efficiency of fertilizer use; improving soil moisture retention capacity through buildup of organic matter; preventing crop productivity losses due to insects, diseases and weeds; or reducing post-harvest losses due to insects, fungi and bacteria. Each of these is an area for research and technology development, or even for the reintroduction of older management systems, to promote water use efficiency gains which places a high demand on AKST (CA, 2007; Hsiao et al., 2007).

7.2 Trade and Markets

7.2.1 Trade and markets: The enabling policy context for AKST contributions to sustainability and development goals

Market and trade policies can limit or enhance the ability of agricultural and AKST systems to drive development, strengthen food security, maximize environmental sustainability, and help make the small-scale farm sector profitable to spearhead poverty reduction. Over seventy percent of the world's poor are rural and most are involved in farming; about 2.5 billion people, or 40% of the world population, depend on agricultural activities for their livelihoods, an increase of a billion over the past half century (FAOSTAT, 2005). In the poorest countries agriculture is the engine of the rural economy (Diouf, 2007).The steep decline in commodity prices and terms of trade for agriculture-based economies has had significant negative effects on the millions of small-scale producers. Although there has been a recent upturn in commodity prices, in part due to an increase in demand compounded by a weak US dollar, market analysts do not anticipate that will continue (FAO, 2006c). Continuing overproduction in NAE countries contributes to these depressed world commodity prices.

Supporting the rural farm sector has been and continues to be a preferred option in NAE and other countries; it offers a compelling option for reducing poverty in developing countries (Lappe et al., 1998; CA, 2007; Diouf, 2007). Agriculture provides multiple public goods, such as the conservation of ecosystem goods and services (e.g., biodiversity and watersheds), poverty reduction and food security (Inco and Nash, 2004; McCalla and Nash, 2007). How to structure trade and market policy platforms to drive development and support the multiple functions of agriculture is a highly debated issue, discussed in a large body of literature. It is generally acknowledged that analyses, projections and related policy options derived from research on trade and market policy are controversial and susceptible to different interpretations. Some studies suggest that liberalization has been associated with reduced poverty and enhanced food security, whereas others indicate the opposite (FAO, 2006c).

The uptake of AKST by farmers does not occur in an economic vacuum. It takes place in a national market environment that is, in turn, partly determined and shaped by international trade (and its effects on national and market processes).

Policy options are determined by distinct national circumstances and different states of development (Dorward et al., 2004; FAO, 2006c; Morrisey, 2007; Morrison and Sarris, 2007). Policy options will differ as a function of a country's stage of economic development and governance overall; the stage of development, composition and competitiveness of its agricultural sector; and its initial factor conditions and endowments.

The IAASTD mission statement leads to an assessment of policies that pay particular attention to poorer rural sectors and poorer countries. A "business as usual" policy will not enable these countries to address development and sustainability goals. Agricultural export trade can offer opportunities for the poor, but there are major distributional impacts among countries and within countries that in many cases have not been favorable for the small-scale farm sector and rural livelihoods. There is growing concern that developing countries have opened their agricultural sectors to international competition too extensively and too quickly, before basic institutions and infrastructure are in place, thus weakening their agricultural sectors with long-term negative effects for poverty, food security and the environment (Diouf, 2007; Morrison and Sarris, 2007). The poorest developing countries are net losers under most trade liberalization scenarios (FAO, 2006c).

The assertion that greater openness will benefit poor developing countries irrespective of their stage of agricultural development (and the trade policies and implementation practices of their trading partners) is increasingly questioned in the literature, and by developing countries and other relevant stakeholders (African Group, 2006). This literature indicates that the investments required to allow shifts of resources out of traditional agricultural activities into higher value alternative activities (either agricultural or nonagricultural) are not likely to occur where market failures are pervasive without some form of state intervention (Morrison and Sarris, 2007). For countries at earlier stages of development, trade liberalization can be damaging to food security in the short to medium term if introduced before a package of policy measures to raise productivity and maintain employment has been implemented (FAO, 2006c).

There is broad agreement across the IAASTD sub-Global reports and in the literature on the need to increase investment in human capital, land tenure (titling and expansion of land ownership by small producers and landless workers), water access, technology, infrastructure, nonagricultural rural enterprises, organizations of small scale farmers, and other forms of expansion of social capital and political participation for the poor and vulnerable (Díaz-Bonilla et al., 2002). A reinvigorated look at how these policy packages can be funded, given that developing country general revenues are often reduced when tariffs are reduced, and overseas development assistance (ODA) to the agricultural sector has been flagging. Developing countries must address significant local production, marketing and institutional constraints. There is wide agreement in the literature that a renewed donor effort is urgently needed if development and sustainability goals are to be advanced, and specifically to enable a supply response to any opportunities for the small-scale farm sector that may arise from future trade negotiations (Diaz-Bonilla et al., 2003; FAO, 2006c; Diouf, 2007).

Trade policy reform aimed at providing a fairer global trading platform can make a positive contribution to the alleviation of poverty and hunger (FAO, 2006c; Diouf, 2007; Morrison and Sarris, 2007). Approaches that are tailored to distinct national circumstances and different stages of development, and that target increasing the profitability of the small-scale farm sector, are effective options to reduce poverty in developing countries. Flexibility and differentiation in trade policy frameworks will enhance the ability of developing countries to benefit from agricultural trade; pursue food security, poverty reduction and development goals; and minimize potential dislocations associated with trade liberalization.

7.2.2 Policy challenges and tradeoffs

Whether and how ASKT systems are generated, delivered and used in ways that promote poverty reduction and environmental sustainability can be enhanced or limited by trade and market policies. The sub-Global IAASTD reports identify many policy challenges:

1. crafting trade rules that allow developing countries needed flexibility to pursue development, poverty reduction and food security agendas, and that address the distributional impacts of welfare benefits and loses from trade liberalization;
2. achieving remunerative prices for small-scale farmers;
3. increasing the value captured by small-scale producers in vertically integrated agrifood chains;
4. addressing the increased regulatory responsibilities required by trade agreements with limited tax revenues, which can be diminished by tariff reductions;
5. addressing the environmental externalities of agriculture; and
6. improving governance of agriculture sector policy making, including decisions about AKST research, development and delivery, and trade policy decision-making.

There are also important synergies and tradeoffs between policy options that merit special consideration. Potential liberalization of biofuels trade is a clear example, presenting tradeoffs between food security, greenhouse gas (GHG) emission reductions, and rural livelihoods which need to be carefully assessed for different technologies and regions. There are likely to be significant tradeoffs between for example policies to promote agricultural development, such as the reduction of agricultural subsidies and increased investment in roads to help rural farmers, and environmental and social impacts such as increased tropical deforestation and increased agricultural land concentration in some parts of the developing South. Forest protection policies in many of these countries may not be sufficiently strong to resist the increased economic pressure to expand the agricultural land and increase tropical deforestation. Note that these concerns also apply to other policy interventions that may work to increase agricultural rents including increased road building and other market access measures that tend to increase the pressure on forests (Angelsen and Kaimowitz, 1999).

7.2.3 Policy flexibility to pursue development, poverty reduction and food security agendas

There is broad acknowledgement that agricultural trade can offer opportunities for the poor under the right circumstances. Numerous studies show that the impact of the economic growth spurred by trade on poverty, hunger and the environment depends as much on the nature of the growth as on its scale. National agricultural trade policy to advance sustainability and development goals will depend upon the competitiveness and composition of the sector. Appropriate trade policy for the agriculture sector in developing countries will depend upon the extent to which the sector produces exportable products, import substituting basic foods, and non-tradable products (Morrison and Sarris, 2007). Advice to developing countries has tended to focus on promoting opportunities for increased exports to international markets (traditional and nontraditional crops) rather than enhancing competitiveness or market opportunities in domestic and regional markets; greater balance among these policy approaches may be indicated.

Market conditions and opportunities for domestically or regionally produced staples are potentially more favorable to poorer developing countries than are the opportunities for expanding exports to the global market (Diao and Hazell, 2004; Morrison and Sarris, 2007). For example, Africa imports 25% of basic grains such as maize, rice, and wheat. Domestic production could potentially replace some of these imports.

Appropriate agricultural trade policy at early stages of development, for countries with an important agricultural sector, may include moderate levels of import protection, and in countries where applied tariffs are already low, further liberalization may not be appropriate (Morrison and Sarris, 2007). A recent FAO study concluded that, for countries at earlier stages of development, trade liberalization can be damaging to food security in the short to medium term if introduced before a package of policy measures to raise productivity and maintain employment has been put in place (FAO, 2006c). Lower tariffs will imply intensified competition from imported foods for the domestic agricultural sector. Reduced tariffs also increase vulnerability of domestic production to competition arising from import surges. A number of instances have been reported in which a developing country's agricultural production has been negatively affected by such sudden, short-run increases in food imports (FAO, 2006c).

For many developing countries sustainable food security depends on local food production; ensuring policy space for these countries to maintain prices for crops that are important to food security and rural livelihoods is essential (FAO, 2006c). Trade policies designed to ensure sufficient levels of domestic production of food, not just sufficient currency reserves to import food, and to balance domestic production with food stocks and foreign exchange reserves, are reported as important components of food security and sovereignty in the ESAP, LAC and SSA IAASTD sub-Global reports. Agricultural policies in industrialized countries, including export subsidies, have reduced commodity prices and thus food import costs; however this has undermined the development of the agricultural sector in developing countries, and thus agriculture's significant potential growth multiplier for the whole economy (Diaz-Bonilla et al., 2003). Reducing industrialized countries' trade distorting policies including subsidies is a priority, particularly for commodities such as sugar, groundnuts and cotton where developing countries compete. Some observers point out that net agricultural importing countries (particularly net food importers in Africa) will suffer a balance of payments loss from the negative terms of trade effect as world commodity prices rise, but at the same time stress the importance of lifting subsidies to benefit the rural sectors of these countries, where poverty is concentrated, because of higher world agricultural prices (Panagariya, 2004).

The steep secular decline in commodity prices (Figure 7-1) and terms of trade for agriculture-based economies has had significant negative effects on the millions of small-scale producers. For example, from 1980 to 2001, the price of robusta coffee fell from 411.7 to 63.3 cents kg⁻¹; cotton fell from 261.7 to 110.3 cents kg⁻¹ and rice (Thai) dropped from $521.40 to 180.20 per tonne. (Ong'wen and Wright, 2007). Even the best performing agricultural subsector (horticultural products) saw an annual 1.35% price decline over 1961-2001 (FAO, 2005a). Although the increase in the volume of exports over the past two decades has resulted in a 30% aggregate trade revenue increase for developing countries as a whole, this volume-driven increase accrued to a small number of net exporting developing countries. Export earnings of the least developed countries (LDCs) fell by 30% during that time, with countries in SSA suffering most from the fall in prices and incomes (FAO, 2005a).

Although most aggregate agricultural production is not traded internationally, and most primary producers do not supply global commodity chains, national agricultural planning is increasingly oriented towards exports. Intergovernmental institutions are advising governments how to integrate small scale producers into these supply chains, with the goal of reducing poverty, particularly in developing countries dependent on commodity exports for the majority of their hard currency revenues (UNCTAD, 2005). This export focus has left small scale producers, the majority of the rural poor, ever more vulnerable to international market factors (Figure 7-2). For example, as a result of a supermarket price war in the United Kingdom, Costa Rican banana plantation workers wages fell from US$12-15 a day in 2000 to $7-8 in 2003 (Vorley and Fox, 2004).

The increase in absolute numbers of agriculturally dependent populations during the past two decades, together with the inability of primary producers to capture more than a small fraction of those increased trade incomes, has meant that growth in agricultural trade flows have had on aggregate a very modest effect on poverty reduction. This implies that policy options are needed to provide greater opportunities for small-scale producers to increase their profitability in such an international context. Failure to do so will result in missed opportunities to promote sustainable development.

7.2.3.1 Special products, special safeguard mechanisms and deeper preferences

There is broad agreement that the rules of the international trading system should recognize the food security and development needs and priorities of developing countries (FAO, 2006c). Flexibility and differentiation in trade policy frameworks (i.e., "special and differential treatment") can enhance developing countries' ability to benefit from agricultural trade; pursue food security, poverty reduction and development goals; and minimize negative impacts of trade liberalization. This includes the principle of nonreciprocal access, i.e., that the developed countries and wealthier developing countries should grant nonreciprocal access to less developed countries has a significant history and role to play in trade relations to foster development. Preferential market access for poorer developing countries, least developed countries and small island economies will be important.

At the household level depressed prices can mean inability to purchase AKST, the need to sell productive assets,

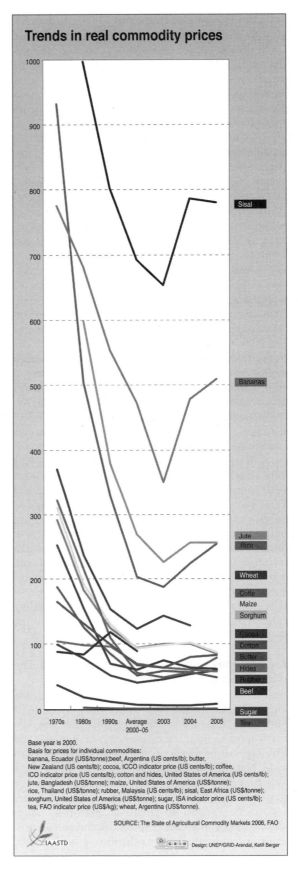

Figure 7-1. *Trends in real commodity prices.* Source: FAO, 2006c

"What are you complaining about? It's a level playing field."

Figure 7-2. *Level playing field.* Source: Barsotti, 2004.

or missed school fees. World Trade Organization country categories that better reflect the heterogeneity of food security conditions in developing countries' food security could help ensure that no food insecure country is denied use of these mechanisms. These measures aim to provide tariff options to developing countries so that they may support rural livelihoods. The formula for applying safeguards under some regional and bilateral trade agreements can limit their effectiveness (see e.g., CAFTA) and may need to be revisited if safeguards are to effectively address rural livelihood issues (Priyadarshi, 2002; Stiglitz, 2006).

Flexibility and differentiation in trade policy will thus enhance developing countries' ability to benefit from agricultural trade, and pursue food security, poverty reduction and development goals. Multilateral trade regime is currently based on the principle of "reciprocity for and among all countries" with the principle of reciprocity among equals, but differentiation between those countries in markedly different circumstances. The principle of nonreciprocal access, i.e., that the developed countries and wealthier developing countries should grant nonreciprocal access to countries less developed than themselves, has a significant history and role to play in trade relations to foster development. The European Union for example followed this approach by unilaterally opening its markets to the poorest countries of the world, and eliminating most tariffs and trade restrictions without demanding any reciprocal concessions (Stiglitz, 2006; Stiglitz and Charlton, 2005).

The WTO July Framework Agreement of 2004 acknowledges that developing countries will need to designate some products as special products based on livelihood security, food security and rural development concerns. Developing countries may require significant time periods for investments in their agriculture sectors, including targeted ASKT research, development and delivery to the small scale sector, enhanced institutional and organization capacity and governance, to make the sector competitive on the international market (Polaski, 2006).

When they signed the Agreement on Agriculture, some developing countries bound tariffs on important food se-

curity and other sensitive crops at very low levels, increasing the vulnerability of their farmers to the drop in global commodity prices. At the same time many DCs did not reserve the right to use emergency safeguard measures. The experience of the GATT round shows that following trade liberalization agricultural imports in developing countries have risen more rapidly than have exports, leading to import surges and a deterioration of net agricultural trade.

7.2.3.2 Distributional impacts of welfare benefits and losses from trade liberalization

Most of the gains from any further liberalization are likely to accrue to developed countries and the larger, wealthier developing countries (FAO, 2006c). For developing countries, the projected, or potential, welfare benefits resulting from the most likely Doha Round scenarios for the Agreement on Agriculture and non-agricultural market access outcomes are just US$6.7 billion and concentrated in just a few developing country WTO members (Anderson and Martin, 2005) (Figure 7-3, 7-4). The poorest countries including those of SSA, except South Africa, are net losers according to these estimates (Anderson et al., 2005; Jaramillo and Lederman, 2005).

There are major differences and distinct distributional impacts between regions, among countries within a region, and between different farm sectors within any particular country (Figure 7-5). One analysis of World Bank CGE projections (Anderson and Martin, 2005) for likely Doha outcomes calculated that the "benefit to the developing countries is more than $17 per person per year, or almost $.05 per person per day" whereas high income countries would realize more than ten times the per capita welfare benefits of developing countries (Ackerman, 2005).

Model-based analyses that have been used to bolster the case for further trade liberalization are often overoptimistic in their assumptions as to the ability for resources to be invested in "higher return" activities (and the assumption of full employment in developing countries); the use of their results in arguing for further trade liberalization in poorer economies could be misleading (Morrison and Sarris, 2007). The models assume that markets function competitively (ignoring vertical integration within value chains that can limit competition); assume that within highly aggregated regions producers have access to similar technologies; and assume faithful implementation of commitments by all parties (Morrison and Sarris, 2007). Additionally, many CGE models are based on assumptions, such as a net zero tax revenue impact from tariff reductions, and full employment, that be difficult or impossible to realize in the real world.

"[T]here is a general consensus that the trade agreements, reforms and policies adopted throughout Latin America and the Caribbean within the last ten to fifteen years have had uneven impacts, with many of the benefits concentrated in the hands of the elite few, while the poorest often bear the brunt of the ills wrought by greater exposure to the world market. The fact is that trade liberalization has not reduced poverty nor inequity. And clearly there are winners and losers" (IADB, 2006).

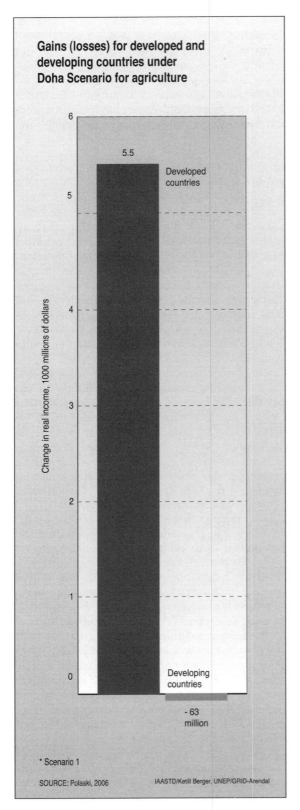

Figure 7-3. *Projected gains (losses) for developed and developing countries under Doha scenarios for agriculture.*

7.2.3.3 Meeting new regulatory costs associated with international trade

Developing countries are facing significant new regulatory costs related to international trade, e.g., meeting international SPS standards, with fewer resources due to tariff revenues losses (which represent a significant percentage of collected tax revenues in many countries). For many countries, decreased tariff revenues means decreased funds available for social and environmental programs and agriculture sector development, as other taxes (such as consumption taxes) can be politically and administratively difficult to collect. Concern that the high costs of regulatory measures to comply with sanitary and phytosanitary standards will divert resources from national food and animal safety priorities is an example. The fundamental practical question of how developing countries can advance sustainability and development objectives without significant increases in donor driven efforts is noted across the developing South.

Tariffs represent about a quarter of tax revenue in developing countries; other taxes are hard to collect in poor countries, particularly with large informal sectors (Panayatou, 2000; Bhagwhati, 2005). Tariff revenue reduction as a result of liberalization can represent a significant proportion of government revenues (Díaz-Bonilla et al., 2002; FAO, 2006c). This compounds the effects of structural adjustment programs, which weakened the institutional capacity of developing governments to carry out basic functions such as tax collection, enforcement of laws, and provision of basic health, sanitation and education services (Jaramillo and Lederman, 2005).

7.2.4 Policy options to address the downward pressure on prices for the small-scale sector

7.2.4.1 Subsidies

Price stability at remunerative prices is an important factor in determining farmer's capacity to invest and innovate rather than pursue low-return, risk-averse behavior (African Group, 2006; Murphy, 2006). Reducing or eliminating agricultural subsidies and protectionism in industrialized countries, especially for those commodities in which developing countries compete (e.g., cotton, sugar and groundnuts) is an important objective of trade reform to reduce the distortionary impact in those markets (Figure 7-6) (Díaz-Bonilla et al., 2002; Dicaprio and Gallagher, 2006; Nash and McCalla, 2007).

For developed countries agriculture is a very small share of the economy and employment, yet subsidies and other supports are highest, unfairly tilting the benefits of agricultural trade in their favor (Watkins and Von Braun, 2002). Agricultural research, farmers' support and investment in infrastructure have also been greatest in these countries (Pardey and Beintema, 2001). The result is increased concentration of agricultural production capacity in the very few countries. This leads to a relevant question: where are subsidies used, what would be the impact of their elimination on the redistribution of production capacity? Developing country income would be some 0.8% higher by 2015 if all merchandise trade barriers and agricultural subsidies were removed between 2005 and 2010, with about two-

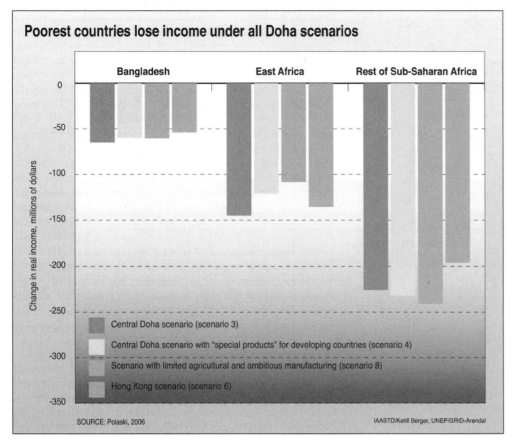

Figure 7-4. *Poorest countries lose income under all Doha scenarios.*

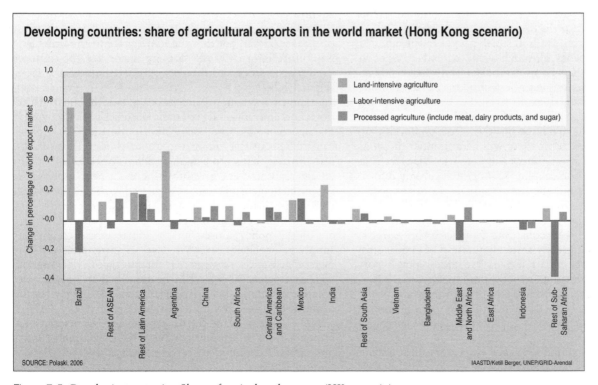

Figure 7-5. *Developing countries: Share of agricultural exports (HK scenario).*

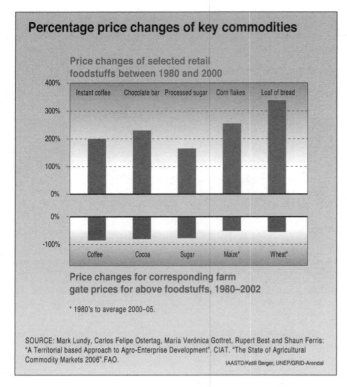

Percentage price changes of key commodities

Price changes of selected retail foodstuffs between 1980 and 2000

Price changes for corresponding farm gate prices for above foodstuffs, 1980–2002

* 1980's to average 2000–05.

SOURCE: Mark Lundy, Carlos Felipe Ostertag, María Verónica Gottret, Rupert Best and Shaun Ferris: "A Territorial based Approach to Agro-Enterprise Development". CIAT. "The State of Agricultural Commodity Markets 2006".FAO.

IAASTD/Ketill Berger, UNEP/GRID-Arendal

Figure 7-6. *Price change of key commodities.* Source: Lundy et al., 2005.

thirds of the total gain coming from agricultural trade and subsidy reform (Anderson et al., 2005).

In globally integrated markets, international prices affect domestic prices across the globe, even for small farmers who grow only for the domestic market (Stiglitz, 2006). Nevertheless reducing trade distorting export subsidies in industrialized countries, although widely agreed to be necessary, is also acknowledged as insufficient by itself to establish higher world prices for many commodities. For example, the reduction or lifting of export subsidies by the US and EU is critically important for some commodities such as cotton, but is unlikely to have a large positive effect on developing countries as a whole, e.g., there will be a number of countries that gain but also a number of countries that lose (Ng et al., 2007). The short-run impact of global subsidy reform will largely depend on whether a country is a net importer or exporter of the products concerned. Countries such as Argentina, for which products subject to export subsidies for some WTO members constitute a large share of exports, are likely to benefit greatly from elimination of export subsidies. Conversely, countries such as Bangladesh that export virtually no products that are subsidized in industrial countries but import a substantial share of such products (13% of imports) are unlikely to benefit in the short run from removal of export subsidies (Ng, Hoekman, and Olarreaga, 2007).

Econometric simulations suggest that removal of trade distorting subsidies would increase agricultural commodity prices only modestly; for example, even cotton, which is heavily subsidized, would increase an average of merely 4 to 13.7%, depending on policy scenario assumptions, defined

baseline and other factors (Baffes, 2006). It is questionable however if such a price increase from depressed agricultural commodity prices reported by FAO (2005b) would suffice to reach the "normal" price, which, according to the WTO, is the zero degree of trade distortion.

Policy tools in addition to subsidy cuts may be needed to raise agricultural prices to remunerative levels (African Group, 2006). Proposals for a plurilateral commitment from major exporting countries not to allow trade at prices below cost of production (CoP)—dumping—and for OECD member countries to publish full CoP figures annually are options that merit further study. (Full CoP would include the primary producer's production costs + government support costs [Producer Subsidy Estimates] + transportation and handling on a per unit basis.) Publication of full CoP figures, when compared to freight on board (FoB) export prices would enable calculation of the percentage of the price that is dumped on world markets (Murphy et al., 2005). Further refinements of the dumping calculation methodology have been made in the context of determining the extent to which industrialized animal production receive input subsidies from below CoP feed grains (Starmer et al., 2006).

7.2.4.2 Supply management for tropical commodities
On average, prices of tropical products (taking dollar inflation into account) are only about one seventh of what they were in 1980 (UN General Assembly). Essentially, less income is earned as more commodities are produced. At the same time, retail prices of products made from coffee (roasted and instant coffee) have increased substantially over the same period. This phenomenon also applies to many other primary commodities produced by developing countries, e.g., cocoa, sugar, cotton, maize, spices. An OECD report acknowledges that "there is concern not only that oligopolistic retailing and processing structure will lead to abuse of market power, but that the lion's share of the benefits of any future reforms in the farming sector may be captured by the processors and retailers . . ." (Lahidji et al., 1996) (Figure 7-7).

The view on supply management held by most institutions and conventional economic perspective is that supply management has been tested and is too costly and prone to problems of free-riding and quota abuse. However, it is also the case that supply management is being used in many commercial markets, given this success a new approach to supply management, that is regulated through the private sector rather than government, may be an effective and fundamental solution to a growing world problem. A variant on this policy approach is to refocus global commodity supply management on the concept of sustainable development. The option suggests that the International Commodity Agreements (ICAs) could be reformed to reduce price volatility, building on the coffee, cocoa and sugar lessons of the 1980s.

The African proposal to explore supply management mechanisms is an option to achieve the production control mechanisms common to other economic sectors (African Group, 2006).

Policy options to help meet the sustainability and development objectives include a bundle of mechanisms to stabilize and increase prices. Supply management mecha-

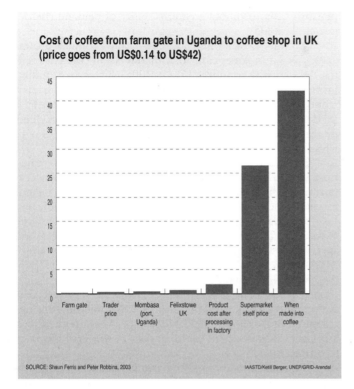

Cost of coffee from farm gate in Uganda to coffee shop in UK (price goes from US$0.14 to US$42)

SOURCE: Shaun Ferris and Peter Robbins, 2003   IAASTD/Ketill Berger, UNEP/GRID-Arendal

Figure 7-7. *Cost of coffee between Uganda production and UK retail.* Source: Ferris and Robbins, 2003

nisms should be investigated, market by market, to determine their potential to do this. One critical policy issues is whether the objective should be price stabilization or price increases (Lines, 2006). To address the continued slide in global commodity prices an increasing number of development groups and policy analysts are suggesting that supply management can provide a viable means of dealing with this chronic problem. In the OECD supply management is used to regulate the supply and demand of more than 50 goods on the world market.

7.2.4.3 Escalating tariffs
Current tariff structure is a disincentive for investment in the creation of value-added agroprocessing in the developing south, because developed countries use escalating tariffs. Escalating tariffs discourage development by placing higher tariffs on manufactured goods than on raw commodities and materials. Levying much higher tariffs on processed agricultural products than raw commodities makes it more difficult for developing countries to promote and gain from value-added local agroprocessing industries, which could provide much needed off-farm rural employment (Wise, 2004). Reducing or eliminating tariff escalation would greatly facilitate off-farm diversification in developing countries (Koning et al., 2004) and encourage value-added agroprocessing (Stiglitz, 2006).

The fisheries sector in many of the poorest countries face trade barriers to diversifying production and exports towards value-added processing products. These barriers include tariff escalation, stringent standards, and rules of origin requirements, among others (ITCSD, 2006). Ne-

gotiations on regulating fisheries subsidies have attracted considerable attention at the WTO, but other areas that are critical to the fisheries sector, including market access, non-tariff barriers, and measures taken under multilateral environmental agreements, have not been addressed. Many stakeholders in the debate, foremost among them the fishing communities whose livelihoods are at stake, have been marginalized in these discussions (ITCSD, 2006).

7.2.5 Options to increase market size, competition, and value capture in commodity chains to increase incomes for small-scale farm sector

7.2.5.1 Regional integration
Regional integration to create larger, regional markets with common external tariffs but no restrictions to internal trade, as a substitute for lack of a large domestic market, can help maintain more consistent demand and stable, higher prices, for locally produced commodities. Large domestic internal markets have often been found to be a prerequisite to agriculture based growth in Asian economies, since they facilitated the shifting of the commodity from surplus to deficit areas, helping to ensure effective demand was maintained even in times of surplus and therefore assisting in stabilizing prices (Morris and Sarris, 2007).

Supporting food production encourages local and regional market integration. In many low income countries especially in Africa, emphasis on cash crop production for export has encouraged transportation networks linking rural areas direct to ports bud neglecting internal connections such as local market feeder roads that would benefit small scale farmers producing for local and regional markets. Greater emphasis on food production for local markets reduces the need for domestic farmers to contract as suppliers to multinationals and encourages greater independence. While some producers will continue to find it profitable to link into global commodity chains, regional integration can create the opportunity for small scale producers to diversity their markets with potentially significant benefits for rural livelihoods (Morrisey, 2007).

7.2.5.2 State trading enterprises for developing country export commodities
Many observers note that that in many marginalized markets the private sector has not filled the gap left by withdrawal of the state from its significant role in providing secure outlets for small-scale producers including more remote producers or producers in higher risk environments (Morrison and Sarris, 2007). The reestablishment of state trading enterprises (STEs) for developing country commodities, if designed with improved governance mechanisms to reduce rent-seeking, may provide enhanced market access for small-scale farmers in developing countries and create competition in concentrated export markets. Export state-trading enterprises can thus offer a competitive counterweight to concentrated export markets.

STEs have real costs and it is widely acknowledged that they have been marred by corruption and cronyism in some countries. Nonetheless, properly overseen and controlled by farmers' organizations, they offer important benefits, especially in developing countries where the private sector is undercapitalized (Stiglitz, 2006). Nonetheless, STEs can potentially provide a useful counterbalance to the market power of global agribusiness thereby increasing competition. STEs may be only market for producers in remote areas of developing countries, and governments can insist that STEs provide this service, whereas they cannot demand it of private corporations (Murphy, 2006).

Current WTO rules require that governments complete questionnaires about any STEs operating in their country, but no similar requirement applies to transnational agribusiness, although they may control a significant share of global trade in a particular commodity. This information generation requirement could be expanded to include any company—private or public—with, for example, more than a given percentage of the import or export market. This information could be gathered by the WTO or under the auspices of the UN Conference on Trade and Development which has a long-standing mandate to monitor restrictive business practices (Murphy, 2006).

7.2.5.3 Microfinance and microcredit

Almost all small-scale agricultural systems benefit significantly from rural credit; this credit will not flow from commercial sources, so policy action is needed (Najam et al., 2007). Microfinance and microcredit programs and banks present a key alternative strategy for many developing countries' agricultural market infrastructure.

Because so much of the developing South's agricultural output is generated by small-scale farmers and other microentrepreneurs, microfinance (as the set of financial services whose scale matches the needs of micro and small producers) is the mechanism by which agricultural producers are able to expand their production, buy fertilizer and other inputs and technologies, smooth seasonal fluctuations in household and enterprise income, and introduce flexibility into small-farm/microproducer investment and asset building.

Newer financial services and products, such as crop or rain insurance, are critical to reducing the risk associated with adopting new technology, transitioning to sustainable agricultural practices, and innovating production and marketing methods. Credit terms tailored to agricultural production and marketing, such as loan repayment terms that track with seasonal crop production, are vitally important to enabling agricultural producers to take advantage of economic opportunities.

7.2.5.4 Alternative trade channels: Fair Trade, certified organics, and mark of origin

As a means of developing pro-poor procurement, initiatives such as Fair Trade and environmentally linked production systems, such as organic and eco-friendly production, were introduced as alternatives to the mainstream commodity markets. While these models offer small-scale producers better terms of trade, the market share for these trading systems has been slow to grow and still only occupies a small percentage of global trade. Nevertheless, the principles were proven and a new generation of business models needs to be designed that can provide windows for the less endowed producers to enter mainstream markets through trading

platforms that promote greater stability of demand (Berdegue et al., 2005).

There are now almost 600 Fair Trade producer groups across 54 Latin American, African and Asian countries selling 18 certified products (FLO, 2008). Over five million farmers, farm workers and their families currently benefit from Fair Trade with many more seeking to enter these markets (FLO, 2008). Sales of Fair Trade certified products increased 42% between 2005 and 2006 with a value of over US$2 billion in 2006 (FLO, 2007).

Fair Trade is no longer a niche market with certified products now sold by large mainstream food processing corporations (such as Proctor and Gamble and Nestle), giant retailers (such as Carrefour, Costco, and Sam's Club), and fast food chains (such as McDonald's and Dunkin' Donuts) (Krier, 2005; Raynolds et al., 2007). (For a business perspective on this growth see Kroger, 2004; Roosevelt, 2004). Market research suggests that there is a very large pool of potential Fair Trade consumers. In the UK, the ethical food market is currently valued at US$3.2 billion per year (Cooperative Bank, 2003). In the US, 68 million consumers with purchases of US$230 billion per year are identified as "Lohas" (lifestyles of health and sustainability) shoppers (Cortese, 2003).

Fair Trade is increasingly envisioned in Latin America as an avenue for bolstering small-scale production in domestic markets and for South-South trade, in addition to northern markets (Bisaillon et al., 2005). Mexico has already developed its own Fair Trade labeling organization (Comercio Justo Mexico) and certification agency (Certimex). This expanded vision of Fair Trade's future is encouraging efforts to expand Fair Trade to include basic food products, such as corn and beans.

There are a number of policy options for promoting Fair Trade as a concrete vehicle for ameliorating poverty and hunger and bolstering environmental sustainability and rural livelihoods. These policies can foster Fair Trade by strengthening the involvement of Fair Trade organizations, producers, traders, and consumers. Governments and multilateral organizations could complement existing ethical business and civil society initiatives, and thus broaden Fair Trade's benefit streams, via a suite of policy options. These include: educating consumers, producers, and businesses about Fair Trade; supporting producer cooperatives and worker organizations to ensure that they have the capacity, information, contacts, and product quality to enter Fair Trade networks; committing to source Fair Trade items; and encouraging the creation of new Fair Trade networks, for example, for basic food stuffs to promote South-South trade.

Similarly, certified organics can work as an effective policy instrument to promote broader rural development and environmental protection goals. Policy options exist to make institutional and policy environments more conducive to certified organic agriculture and less conducive to energy intense (net energy consuming) agriculture. A number of recent case studies confirm the benefits for small-scale farmers of participating in certified organic farming schemes (Eyhorn, 2007; FAO, 2007).

With organic global sales now approaching US$40 billion, certified organic agriculture (COA) offers a challenging, but attractive rural development pathway for small-scale producers and for policy makers wishing to support the production of global public goods. Organic agriculture can help develop an alternative global market that improves agricultural performance through better access to food, relevant technologies, and environmental quality and social equity (FAO, 2007).

COA is value-added agriculture, which is accessible to small farmers who cannot purchase off-farm synthetic inputs such as fertilizers and pesticides (Egelyng and Høgh-Jensen, 2006) and the knowledge intensive methods practiced in COA are particularly compatible with traditional and local knowledge capacity for innovation. COA may provide a way out of poverty for developing country farmers. Widening adoption is therefore a clear policy option; several governments now have targets for the expansion of certified or compliance assessed organic production. To this end, the FAO/IFOAM/UNCTAD International Task Force on Harmonization and Equivalence in Organic Agriculture has generated international organic guarantee tools, adoption of which would help support further development of organic market.

Policy options generated by the FAO International Conference on Organic Agriculture and Food Security (Rome, 3-5 May 2007) include increased advocacy and training on organics, investing in organic awareness in agricultural and environmental education, building organic knowledge in university and research institutions, transitional crop insurance, providing organic training to extension officers, and supporting investments that facilitate the transition of small-scale producers to organic agriculture. Some of these have been incorporated into developing countries, national policies and legal frameworks (FAO, 2007).

A number of Latin American countries have also adopted policies and legal frameworks to promote organic agriculture. For example, Costa Rica adopted a national law to develop organic agriculture which sets out a series of mechanisms and incentives to support the organic sector (Asamblea Legislativa de Costa Rica, 2006). These policy tools include incentives to promote increased professional education on organic production, organic certification options for the national and international market, crop insurance for farmers transitioning to organic production, special credit lines for small and medium scale organic producers, and tax exemptions on inputs for organic production as well as on profits from the sale of organic products.

Mark of Origin or Appellation is an approach that has been widely used in France as a means of locking in added value via protection of specific spatial areas, such as a defined geographic area known to produce a high quality brand, or an area that has traditionally developed a specific type of food processing. The classic examples of this are the wine denominations that allow buyers to purchase products based on geographic location, grape variety and year. Whereas this has proven to be very effective in areas that respect such legal definitions, the products are generally based on long-term consumer loyalty and cultural standards. As such this system is unlikely to be applied to mainstream products unless this strategy is used in combination with other standards such as air-miles and or carbon footprints.

Certification is another approach to locking in access to higher value markets is to join a certification scheme such as those offered for organic production and rainforest production. All of these movements aim to capture a premium price for producers who can provide evidence that they are meeting and have been monitored to prove their compliance with specific ethical and environmental standards. While the area of certification is gaining appeal, the system is extremely expensive and unless charges can be passed onto consumers the ability of poor producers to comply with such regulations will be doubtful.

An Agricultural Market Analysis Unit could be established and supported in developing countries. This unit would be concerned with coordinating and developing policy on the development of market-oriented strategy in agriculture and setting policy guidelines for agricultural research. The Unit would also coordinate its activities with relevant regional bodies and work closely with the private sector and, especially, with those private-sector support groups working to stimulate production for growth markets.

Many actors in the agricultural sectors in poor developing countries are still not familiar with the idea of competitive markets. A National Market Education Programme could be established targeted, primarily at farmers, traders and agricultural product processors. Such a programme needs to be linked to the Agricultural Market Analysis Unit (see above). Market Information Services (see below) and run in conjunction with other stakeholders including Ministries of Agriculture, Education and Trade, farmers' and traders' associations and other private sector actors and with extension services.

The program needs to set targets for training farmers to understand how competitive markets work, to take advantage of market information and to inform them of the difficulties and opportunities associated with market conditions. Issues addressed need to include the stimulation of collective activity to improve economies of scale, linking supply variety and quality to market needs, negotiation of sales and inputs and the use of credit and business management. The program should have a limited duration and should be administered efficiently as a separate unit within a national agricultural development reform program.

Many small- and medium-scale farmers, traders and processors in poor developing countries have limited access to information about prices and market conditions of the commodities they produce. Farmers find themselves in a weak bargaining position with traders which results in lower-than-market farm-gate prices, high transaction costs and wastage. Market Information Services need to be established at local, national, and regional levels to collect, process, and disseminate market information in the appropriate language of intended recipients. Such services need to be fully coordinated with each other and involve full participation of stakeholders.

To assist developing countries to compete successfully in the world economy research and extension institutions need to develop or acquire new skills and expertise in market analysis and market linkage. Producers need to ensure that there are viable markets for any existing or new products. They need to ensure that the quality and packaging of those products meet the requirements of customers both on the domestic and export market. Research and extension services have a vital role to play in this effort and must be prepared to reform quickly to meet the challenges of globalization.

In many respects national research programs have succeeded in their goal to achieve food security, the current emphasis should now be to develop dynamic and commercially oriented research that supports improved market analysis, market access and added value processing. Extension services should now focus on assisting producers to trade more effectively within a liberalised market. Special attention should be given to aspects such as linkage of production to markets, access to credit and collective marketing which will enable the millions of atomised, small-scale farmers to gain from economies of scale in their input and output markets.

Government research services need to work closely with the private sector which is increasingly developing its own research capacity, particularly in regard to higher value commodities and research related to issues and problems further up the value chain.

7.2.6 Market mechanisms to optimize environmental externalities

Agriculture generates environmental externalities (see 7.1). There are currently few market mechanisms that internalize these externalities.

The environmental impacts of agricultural trade stem at least in part from the globalization of market failures, as well as the lack of market mechanisms to internalizing the environmental externalities of production and account for the positive externalities (Boyce, 1999). Trade liberalization leading to the displacement of traditional jute production in Bangladesh by imported synthetic fibers is an example. Nearly the entire price advantage enjoyed by synthetics over jute would be eliminated if environmental externalities were factored into the price (Boyce, 1999). At the same time, traditional producers receive no compensation for the positive environmental externalities, e.g., biodiversity conservation, associated with many forms of traditional production. Similarly, U.S. corn production which requires significant energy and agrochemical inputs that cause significant environmental externalities is sold at below the cost of production in Mexico, displacing traditional corn production in the small and medium farmers who plant diverse traditional varieties (Nadal and Wise, 2004).

Trade agreements bring two distinct kinds of production into direct competition, with vastly different environmental impacts and with significant ramifications. In both cases the market price for the modern product fails to internalize or account for significant environmental externalities. At the same time, the positive environmental externalities that are present in many forms of traditional agriculture are not assessed.

7.2.6.1 Policy options for internalizing environmental externalities

Some OECD countries adopted economic measures, including environmental taxes on agricultural inputs as a part of a policy package to reduce the environmental impacts

of pesticides, fertilizer and manure waste. Denmark, Norway and Sweden, for example, have introduced taxes on pesticide use, as incentives to reach pesticide use reduction targets. Similarly, the Netherlands imposed an excise manure tax. The recent reforms of the European CAP may be interpreted as a move towards rewarding farmers, not only as producers of food, but as caretakers of natural resources and environmental services. European support for organic agriculture is an important aspect of this recognition (Egelyng and Høgh-Jensen, 2006).

Many critical ecosystem services are undervalued or unvalued; there are no market signals that would spur technological development of alternative supplies (Najam et al., 2007). Charges to internalize cost of transportation energy expenditure in globalized agriculture, such as "food mile" taxes are one policy approach. Food mile taxes could help internalize the social and environmental externalities of transport, including the climate impacts, pollution, and the cross-border movement of pests and livestock pathogens, among others (Jones, 2001).

Policy approaches to assist small-scale producers to articulate their carbon rating will be key, especially as an oversimplified response may be to simply ban long haul agricultural goods, and provide greater support to local food systems and season procurement policies that could end year round supply of off-season goods. In some cases though, an integrated analysis of energy costs and GHG emission from distant developing country production as compared to local northern country production will be favorable for developing country production. For example a recent analysis showed that Kenyan flower production exported long distances to the European market nonetheless generated fewer GHG emissions than hot-house flower production in the Netherlands (DFID, 2007).

7.2.6.2 Payments for agroenvironmental services

Ecosystem services remain largely unpriced by the market. These services include climate regulation, water provision, waste treatment capacity, nutrient management, watershed functions and others. Payments for environmental services (PES) reward the ecosystem services provided by sustainable agriculture practices. PES is a policy approach that recognizes the multifunctionality of agriculture and creates mechanisms to value and pay for these benefits. In principle, payments for environmental services (PES) such as watershed management, biodiversity conservation and carbon sequestration, can advance the goals of both environmental protection and poverty reduction (Alix-Garcia et al., 2005).

PES is an approach that, like economic instruments used for pollution prevention, seeks to support positive environmental externalities through the transfer of financial resources from beneficiaries of the services to those who protect or steward the environmental resources that provide the service. PES schemes often focus on environmental services provided by forest conservation, reforestation, sustainable forest extraction, and certain agroforestry and silvo-pastoral practices. Carbon sequestration services are also involved in several PES schemes, both to increase active absorption through reforestation or to avoid carbon emissions through forest conservation.

A key objective of PES schemes is to generate stable revenue flows that can help ensure long-term sustainability of the ecosystem that provides the service; and to structure the arrangement so that small farmers and communities, not just large landowners, may participate and benefit (this may involve increased transaction costs, and tends to be more effective where farmers are well organized). Examples in Latin America show that community participation and equitable rules are key; promoting rural livelihoods must be a stated objective of the PES program otherwise the lion's share of benefits will go to wealthy landowners. In one example in Costa Rica 70% of PES for carbon sequestration in one year went to a single wealthy landowner (Rosa et al., 2004).

PES revenues can be generated by user fees, taxes, subsidies, and grants by IFIs and donor organizations and NGOs. Long-standing programs, including those established by New York City and Quito, Ecuador, which levy increased fees on water users to fund watershed conservation are well known. A similar, smaller programs in the Cauca Valley of Colombia works on a similar principle; farmer associations organized a PES program which levies additional water use fees to promote the adoption of conservation measures on over one million hectares and maintain dry-season water flows (Mayrand and Paquin, 2004).

PES schemes may also include measures to assist local communities with market development and revenue diversification as part of the compensation, or payment, package for the environmental service protected and provided. For example, in Brazil, rubber tappers receive payments for forest conservation services they provide through their management of forest resources. In the US, the Conservation Reserves Program provides funding to farmers to remove sensitive lands from production, prevent land degradation and preserve biodiversity.

Other projects promote the adoption of improved silvopastoral practices in degraded pasture areas that may provide valuable local and global environmental benefits, including biodiversity conservation; payment-for-service mechanism are being employed to encourage the adoption of silvopastoral practices in three countries of Central and South America: Colombia, Costa Rica, and Nicaragua. The project has created a mechanism that pays land users for the global environmental services they are generating. Another example is the Coffee and Biodiversity project supported by the GEF and the World Bank in El Salvador, which provides marketing and technical support as a proxy for direct payments, to promote biodiversity protection and habitat creation on shade-grown coffee plantations via niche marketing of "shade-grown," song-bird friendly coffee (Pagiola and Agostini, 2002).

Supportive national policy environments are important. In 1997 Costa Rica reformed its forest law to allow land users to receive payments for specified land uses, including new plantations, sustainable logging, and forest conservation. The amended law recognizes four types of environmental services: carbon sequestration, biodiversity conservation services, hydrological services, and scenic beauty and ecotourism. The law also introduced a fuel tax to finance forest conservation and established an agency (Fonafifo) to raise funds and manage the PES scheme. Similarly, the Ecuadorian National Biodiversity Policy recommends the establishment of markets for environmental services, and the establishment of the mechanisms for water and watershed conservation, coastal protection, global climate changes services, and compensation to landowners—importantly, to both individuals and communities (Mayrand and Paquin, 2004).

Another variant of a PES scheme is the BioCarbon Fund established by the World Bank to buy certified emission reductions from land-use, land-use change, and forestry projects admissible under the Kyoto Protocol. The Fund is designed to target agricultural and forestry projects that enhance other ecosystem services, such as biodiversity and watershed protection, while improving the livelihoods of local people. Projects include conservation agriculture, such as shade-grown coffee, agroforestry to restore degraded areas, improved agricultural practices, such as shifting from subsistence farming to organic agriculture, and reforestation (Kumar, 2005).

Bioenergy and biofuels: subsidies and standards. Large direct and indirect subsidies, including tax credits for biofuels, have been used to build bioenergy production and markets. Fuel blending mandates and import restrictions, particularly tariffs on ethanol likewise have helped to build domestic markets (UN Energy, 2007). How the bioenergy value chain is structured is crucial for determining the development benefits of this sector. Policy options to support small and medium size enterprises in bioenergy should be considered because of studies showing the multiplier economic and development effects of local ownership in local economies (Morris, 2007; UN Energy, 2007).

First generation liquid biofuels: trade, subsidy and sustainability issues. When subsidies are granted to biofuels, they should be tied to objectively observable positive externalities. Biofuels policies set incentives for producers that directly affect the extent of externalities, the primary justification for granting the subsidies in the first place. In the case of current policies in most countries it is apparent that these incentives are rarely closely linked to the externalities they are allegedly supposed to provide. In fact, the majority of policies in OECD countries create incentives to maximize production of 1st generation biofuels, irrespective of quality and quantity of externalities. Consequently, many biofuels are produced with intensive use of energy inputs, leading to low energy balances and GHG emission reductions while contributing to environmental problems.

Biofuels produced from agricultural feedstocks (first generation) are rarely competitive with other forms of energy and practically all producing countries support their biofuels industries through a complex set of federal and state-level policies. The most common forms of support are reductions on excise taxes that are designed to foster consumption by reducing the cost of biofuels relative to conventional fuels. On the supply side, these policies are often complemented with direct production support, e.g., payments of Euro 45 ha^{-1} for energy crops grown on non-set-aside land in the EU and subsidized credit for producers in Brazil and the U.S. In addition, biofuels also benefit indirectly from highly distorted agricultural markets in OECD countries, e.g., the U.S. maize sector, the primary ethanol feedstock in the country, received US$37.4 billion in subsidies between 1995-2003 (UNCTAD, 2006). In many countries, subsidies are accompanied by blending mandates, e.g., the E.U. set a voluntary 5.75% biofuels target for 2007, supported by several mandatory targets at the country level. The differential treatment of ethanol and biodiesel under international trade rules (ethanol is classified as an agricultural product, biodiesel is classified as a chemical/industrial product) has important implications on international market access and also affects how the fuels would be treated under a proposed WTO category of "environmental goods and services" (IEA, 2004; IEA, 2006; Kopolow, 2006; UNCTAD, 2006; USDA, 2006; Kojima et al., 2007).

Together, these forms of policy support generate substantial economic costs—reducing funds available for other policy goals, including energy conservation and support for other alternative energy generation technologies. Current levels of subsidies are considerable. For example, total annual subsidies to liquid biofuels in the US are estimated at US$5.1-6.8 billion, corresponding to US$0.38-0.49 and US$0.45-0.57 per liter of petroleum equivalent ethanol and biodiesel, respectively (Kopolow, 2006). Moreover, taxes on fuels represent a significant source of government income in many countries and reductions are often difficult to compensate. While blending mandates are attractive to policy makers because they do not directly affect government budgets, they too create considerable economic costs. In addition, blending mandates create inefficiencies by guaranteeing a market for biofuels producers irrespective of costs and limiting competition. This reduces incentives to develop more efficient and cheaper production—an effect that is reinforced by trade barriers.

Against these costs stand potential benefits in terms of rural development, climate change mitigation and energy security as well as possible negative effects on the environment and food prices. Consequently, decision makers need to carefully assess whether the full social costs of bioenergy and associated promotion policies are worth achievable benefits.

Policy options to reduce the social and environmental externalities of 1st generation biofuels production such as sustainability standards are widely noted in the literature, but developing effective standards that balance environmental and social interests with access to export markets for developing countries is a significant challenge. Given the potentially adverse social and environmental effects of large-scale increases in biofuels production (see Chapters 4 and 6), the development of sustainability standards is being discussed in different private and government supported forums.

In the absence of universal regulations and enforcement, standards are viewed as key to limiting negative effects and improving benefits for small-scale farmers (O'Connell et al., 2005; Reijnders, 2006; WWF, 2006). In addition to disagreements on the definition of these standards, with large differences of opinion between industrialized and developing countries, uncertainty persists on how effective such standards can actually be. Given that biofuels are fungible export commodities, their effectiveness would depend on the participation of all major consumers and producers. Moreover, qualifying for standards and obtaining certification can be a considerable financial and institutional burden

for poor producing countries. It is therefore essential that developing countries are included and supported in the process of the development of sustainability standards to ensure that environmental and social considerations are balanced with the broader needs of developing countries, including considerations about the needs of small-scale farmers, farmer cooperatives and access to the markets of industrialized countries.

Liberalization of biofuels trade would shift production to developing countries. There is significant question as to whether this would benefit small-scale farmers and in the absence of effective safeguards the resulting expansion of production in these countries could magnify social and environmental costs. Growing crops for biofuels could worsen water shortages; biofuel crop production in addition to food crops will add another new stress on water use and availability (de Fraiture et al., 2007b).

Second generation biofuels. The U.S. Department of Energy calculates that if all corn now grown in the US were converted to ethanol, it would satisfy only about 15% of the country's current transportation needs (USDOE, 2006); others put that figure as low as 6% (ETC, 2007). A second approach is to produce ethanol from cellulose, which has the potential to obtain at least twice as much fuel from the same area of land as corn ethanol, because much more biomass is available per unit of land. Thus promoting research and development for second generation biofuels is an often noted policy option. Synthetic biology approaches to break down cellulose and lignin, crucial for second generation "cellolosic" biofuels production, are still years off but may be promising. Importantly, efforts are also needed to allow developing countries and small-scale farmers to profit from the resulting technologies (Diouf, 2007). While some countries have recently increased their support for research and development on 2nd generation biofuels, more public efforts international efforts are needed to focus on developing means by which 2nd generation biofuels may benefit small-scale farmers and developing countries. This includes tackling the high capital intensity of technologies, facilitating farmer cooperatives and dealing with intellectual property rights issues.

If any of the synthetic biology approaches are successful, the agricultural landscape could quickly be transformed as farmers plant more switchgrass or miscanthus as feedstock crops—not only in North America, but also across the global South. By removing biomass that might previously have been returned to the soil, fertility and soil structure would also be compromised. As presently envisioned, large-scale, export-oriented biofuel production in the global South could have significant negative impacts on soil, water, biodiversity, land tenure and the livelihoods of farmers and indigenous peoples (de Fraiture et al., 2007a).

Bioelectricity and bioheat. There is considerable potential for bioelectricity and bioheat to contribute to economic and social development (see Chapters 3 and 6) and a number of clear policy options to promote a better exploitation of this potential (Stassen, 1995; Bhattacharya, 2002; Kishore et al., 2004; Kartha et al., 2005; Ghosh et al., 2006). Promotion of R&D, development of technical standards as well as better access to information and finance are needed to better exploit the potential of bioelectricity and bioheat in developing countries.

Promoting research and development to improve the operational stability and reducing capital costs promises to improve the attractiveness of bioenergy, especially of small and medium-scale biogas digesters and thermo-chemical gasifiers, is important for the developing South. The development of product standards and dissemination of knowledge is also key. A long history of policy failures and a wide variety of locally produced generators with large differences in performance have led to considerable skepticism about bioenergy in many countries. The development of product standards as well as better knowledge dissemination can contribute to increase market transparency and improve consumer confidence.

Experience of various bioenergy promotion programs has shown that proper operation and maintenance are key to success and sustainability of low-cost and small-scale applications. Therefore, building local capacity, ensuring that local consumers are closely engaged in the development as well as the monitoring and maintenance of facilities, and increased access to finance for bioenergy are necessary. Compared to other off-grid energy solutions, bioenergy often exhibits higher initial capital costs but lower long-term feedstock costs. This cost structure often forces poor households and communities to forego investments in modern bioenergy, even when payback periods are very short. Improved access to finance can help to reduce these problems.

7.2.7 Enhancing governance of trade and technologies

Agricultural policymaking and AKST investment are affected by global governance issues that may apply in a number of economic sectors, including agriculture. This section addresses a suite of governance issues in trade and environmental decision-making, including the democratization of global trade regimes, as well as international competition policy to govern corporate power over commodity markets and promote more equitable distribution of agricultural rents that could help drive development and improve rural livelihoods. The section also reviews policy options for international instruments (agreements and intuitions) to assess the impact of proposed trade agreements and emerging technologies against the development and sustainability goals; these processes, including strategic impact assessments of proposed trade agreements and comparative technology assessments, could help educate policy makers and stakeholders, increase transparency, and assist in making decisions that would support development goals.

7.2.7.1 Governance of trade and environmental decision-making

If trade negotiation processes were made more transparent, social and environmental concerns would likely be better represented in the resulting agreements. The principles of good governance, such as representation, transparency, accountability, access to information and systematic conflict resolution should be fully internalized and implemented by international trade and environmental institutions (Stiglitz, 2006). Developing countries, which often lack personnel

and institutional capacity to deal with the complexity of trade negotiations are at a distinct disadvantage negotiating for the interests of their rural sectors in these fora, and often lack resources to analyze important and highly complex issues, to develop negotiating positions and to respond quickly and effectively to their various negotiating teams. Civil society participation is limited from negotiations through dispute resolution process, much of which takes place behind closed doors.

Policies to strengthen developing country negotiating capacity in trade talks are important. Trade capacity development, as a part of "aid for trade" packages, are one option. Consideration may also be given to establishing national and regional teams of experts with the necessary authority to analyze the interests of their stakeholder groups and to establish appropriate negotiating positions.

Another option is to develop CSO consultative committees to support negotiators, giving farmer organizations, business and NGOs the opportunity to provide valuable input and support negotiators. A number of countries, for example Kenya, the Philippines and India, have created national consultative committees to the WTO (Murphy, 2006).

Without effective global environmental governance, nation-states, subject to the pressures of globalization, may drift towards a low-level environmental policy convergence that is insensitive to local ecological conditions and does not respect the diversity of preferences and priorities across and within nations (Zarsky, 1999). The creation of a United Nations Environmental Organization, perhaps modeled on the World Health Organization, has been proposed as one policy approach to address this significant global governance deficit and promote technologies and behaviors that respect ecosystems more effectively (Esty, 1994; Friends of the UNEO, 2007).

7.2.7.2 International competition policy and antitrust: governing commodity markets to promote development goals

Vertical and horizontal concentration in global commodity markets is a primary cause of market distortion. Possible policy responses include an international review mechanism for proposed mergers and acquisitions among agribusiness companies that operate in a number of countries simultaneously (Stiglitz, 2006), the establishment of international competition policy, and the reestablishment of state trading enterprises.

One of the major anticompetitive effects of globalization has been a rapid concentration of market power away from producers into the hands of a limited number of trade and retail companies (Vorley et al., 2007). What looks like buying and trading between countries is often the redistribution of capital among subsidiaries of the same parent multinational corporation (Shand, 2005). As a result, the negotiating power within agricultural chains over the past 20 years has moved rapidly away from the producer end of the market chain. The first level of consolidation was made at the wholesale level through a series of mergers, acquisitions and takeovers that reduced the number of international traders from hundreds of family based enterprises to a handful of international trade houses that dominate particular commodities, such as Archer Daniel Midland, Unilever and Cargill.

This situation means that even when farmers organize and aggregate, produce quality goods, and sell collectively, they have insufficient volumes of sale to negotiate effectively with four to five giant corporations. There is increasing concern that lack of competition in the marketplace is having seriously negative social effects on agricultural producers; the most vulnerable are the poorly organized, resource poor farmers in developing nations (Figure 7-8).

One approach to address this imbalance in trade relationships is the establishment of international competition policy in the form of multilateral rules on restrictive business practices. A potential model for this approach is the French law (*Loi Galland*) that prohibits selling at a loss and "excessively low prices."

Another policy option that is widely noted is the reintroduction of price bands as a means of cushioning the impact of world price instability. For example Chile's Free Trade Agreements with EU and Canada allowed it to keep its agricultural price band which was designed to stabilize import costs of agricultural staples (including wheat, sugar, oil) through adjustment to tariffs on such with the objective of allowing a fair rate of return to Chilean farmers even if they were competing with heavily subsidized US farmers. In contrast the US-Chile Free Trade Agreement committed Chile to phase out its agricultural price band system. An international competition policy framework might also include creation of an independent UN agency to address some of the issues that UN Center for Transnational Corporations used to address.

7.2.7.3 Strategic impact assessment and comparative technology assessment

There is often a dearth of information on the potential social and economic benefits and risks of proposed trade agreements and emerging technologies alike. Policy tools to allow developing countries to better analyze benefits, risks and tradeoffs of proposed trade agreements and the introduction of new technologies are needed. Policy approaches to redress this issue include Strategic Impact Assessment (SIA) of trade agreements and Comparative Technology Assessment for emerging technologies. Additionally, increased research and more sensitive trade policy stimulations tailored to countries at different stages of development with different characteristics to their agricultural sectors may be helpful to inform policy choices to address development and sustainability goals (Morrison and Sarris, 2007).

SIAs have provided early warnings as well as research evidence that failing to mitigate negative environmental effects can substantially reduce net economic and welfare gains from trade. In this way, these assessments can provide critical information to governments and stakeholders allowing them to consider whether or not to reject or mitigate a trade policy proposal that is likely to worsen poverty, inequity or environmental degradation in certain sectors.

Strategic impact assessment of trade agreements that have been undertaken for regional agreements, such as NAFTA and multiple EU trade agreements, aim to give negotiators a fuller understanding of potential environmental impacts in their own countries, such that they may be taken

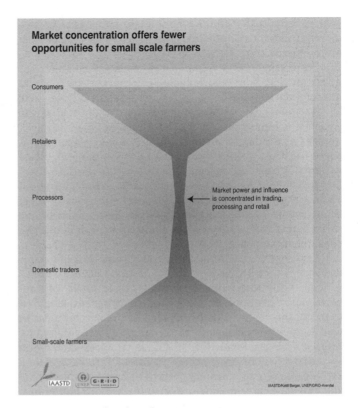

Market concentration offers fewer opportunities for small scale farmers

Consumers

Retailers

Processors

← Market power and influence is concentrated in trading, processing and retail

Domestic traders

Small-scale farmers

IAASTD UNEP G·R·I·D Arendal

IAASTD/Ketill Berger, UNEP/GRID-Arendal

Figure 7-8. *Agricultural market concentration.*

into account alongside the economic and social considerations on which trade negotiations have traditionally been based. The fuller information on environmental issues enables negotiators to make more reliable tradeoffs, in those cases where the effects do not provide win-win outcomes for national and international economic, social and environmental concerns.

The European Commission for example has defined the goal of SIA as generating information to integrate sustainability into European trade policy by assessing a proposed trade agreement's potential impacts on sustainable development. SIAs, which are public documents, inform negotiators and interested stakeholders of the possible social, environmental, and economic consequences of a trade agreement; provide analysis that will help maximize benefits of the agreement through better management of environmental, social and economic resources; and inform the design of policy options, including capacity building and international regulation, that may maximize the benefits and reduce the negative impacts of the proposed trade agreement (George and Kirkpatrick, 2003).

Another noted policy option to increase information and transparency is the establishment of an intergovernmental framework for the comparative assessment of new technologies as they evolve from initial scientific discovery through to possible commercialization. For example, observers have noted that rapid developments in nanotechnologies and nano-material production may outcompete developing countries' primary commodities in international markets in the near and mid-term (ETC Group, 2005). All stakeholders, perhaps especially including developing country governments that are negotiating market access for their agricultural commodities and raw materials in various multilateral, regional and bilateral agreements, could be provided with information on how future technology development may affect them and the markets that are essential for their economies.

The potential benefits and risks of nanotechnologies present an example of the benefits for the realization of development and sustainability goals that a technology assessment agreement or agency might afford. There has been considerable reporting and analysis of the potential benefits of nanoscale technologies for developing countries, particularly with regard to water and energy. The potential health and environmental risks of this new technology platform, as well as nanotechnology's potential impacts on commodity markets and the social and economic disruption that may cause, are less well studied. Nanotechnologies are still very new; nonetheless if a new engineered nano-material outperforms a conventional material, including for example cotton textiles, copper or rubber, that are key commodities for developing country economies, significant economic dislocation may result (ETC Group, 2005).

Emerging technologies, including nanoscale technologies, require scientific, socioeconomic and societal evaluation in order for governments to make informed decisions about heir risks and benefits. Rather than approaching technology assessment in a piecemeal, technology-by-technology fashion, governments and the international community could consider longer term strategies to address technology introduction on an ongoing basis. One option for the international community is to consider an independent body that is dedicated to assessing major new technologies and providing an early warning and early listening system. Comparative technology assessment could help policy makers and stakeholders monitor and assess the introduction of new technologies and their potential socioeconomic, health and environmental impacts.

One policy approach might be to reinvigorate the capacity of the UN System to Conduct Technology Assessment for Development. The UN Commission on Science and Technology for Development has become a subsidiary body of the Economic and Social Council, where it operates with greatly reduced staff and funding. This commission could be strengthened, or another specialized UN agency could be given the mandate to both conduct technology assessments and build capacity in developing countries to assess technologies, with the goals of promoting poverty reduction, health and environmental protection, and sustainable development (ETC Group, 2005).

Another policy option could be the establishment of a legally-binding multilateral agreement on comparative technology assessment, potentially negotiated through a specialized agency such as UNCTAD, the ILO, or ECOSOC's Commission on Sustainable Development. The objective of such a convention would be to provide an early warning and assessment framework capable of monitoring and assessing emerging technologies in transparent processes and their potential benefits as well as costs and risks for human health, the environment, and poverty reduction and development. At the same time, such an agreement might help to generate information that would help educate citizens and stakeholder groups, via participatory and transparent processes, support broader societal understanding of emerging technologies, encourage scientific innovation, and facilitate equitable benefit and risk-sharing. Alternatively, a specialized Technology Assessment Agency could be created, within the UN system to conduct comparative technology assessments of new and emerging technologies.

7.3 Food Safety, Animal and Plant Health

The management of food safety, animal and plant health issues along the farm to fork continuum requires a level of coordination and integration that often is not provided by the current international policy and regulatory framework for agriculture. Instead, these three issues are largely addressed in terms of international standards elaboration through parallel programs developed by the Codex Alimentarius Commission, World Animal Health Organization (OIE) and the International Plant Protection Convention (IPPC) for food safety, animal health and plant health respectively. These standards and related sanitary and phytosanitary (SPS) measures are implemented and enforced to a greater or lesser degree through an array of often uncoordinated national initiatives variously managed by ministries such as agriculture, health, environment, forestry, fisheries, trade, commerce and international affairs. Related to this lack of coordination, or perhaps because of it, alternative regulatory mechanisms such as third party standard and certification systems mandated by private sector retailers, in response to increased consumer demand for improved food safety and food quality, have been implemented. Much of the cost burden for meeting these private regulatory requirements is borne by primary producers.

The increasing internationally traded volume and variety of food, food ingredients, feed, animals and plants poses many challenges for private quality assurance programs and government SPS programs. SPS system failures affect both exporting and importing countries. For example, recent U.S. imports of contaminated pet food from China resulted not only in the deaths and illnesses of an unknown number of pets and the closing of 180 processing plants in China, but a U.S. Congressional proposal for reorganization of US food import inspection. Yet proposals to equip SPS authorities and private establishments with adequate personnel and technology to enforce standards sometimes encounter not only bureaucratic resistance and/or opposition from industry segments of the supply chain, but broader resistance based on credible threats of trade retaliation in nonfood and agriculture sectors (Barboza, 2007; Clayton, 2007; Weiss 2007). Whether or not food and agriculture trade expands to the extent projected by FAO for 2030 (FAO, 2006c), the cost-benefit framed tension between measures to protect human, animal and plant health and broader trade pressures is likely to remain.

7.3.1 Food safety

7.3.1.1 Surveillance challenges

The lack of reliable data or data that are comparable between countries on the prevalence and severity of foodborne disease, despite several WHO initiatives to develop global and regional surveillance and outbreak reporting systems, continues to impede the development of evidence-based food safety interventions in many WHO member countries (WHO, 2002a, 2004). For example, the EU has undertaken a fully harmonized baseline surveillance study for *Salmonella* in poultry production which is the basis of targeted interventions to manage the prevalence of this foodborne pathogen (EFSA, 2006a).

Epidemiological uncertainty about the origin, prevalence and severity of much foodborne illness makes it difficult to target resources and do comprehensive and proactive food safety control planning. More than 200 known diseases are transmitted by food, however underreporting, illnesses caused by unknown pathogens and other factors, such as water sanitation, obscure the origin of foodborne illness. The confluence of these factors impedes estimates to characterize the burden of existing foodborne illness, much less the evolution of future pathogens.

Pathogens featured in today's headlines, such as *Listeria monocytogenes* or *E. coli* O157:H7, were not identified as major causes of foodborne illness 20 years ago (Mead et al., 1999). However, for most foodborne infections effective preventive interventions can be taken despite a lack of exact epidemiological knowledge. Furthermore, the majority of foodborne infections in most countries are caused by a few pathogens e.g., in the EU, *Salmonella* and *Campylobacter* accounted for about 96% of reported zoonoses cases in 2005 (EFSA, 2006a). In developing countries, actions such as water sanitation and heat treatment of food in combination with measures for basic sanitary and hygienic routines would have significant health benefits, even without immediate support of detailed surveillance data. Diarrhea is the leading causes of illness and death in less developed countries, killing an estimated 1.9 million people annually worldwide and almost all deaths are caused by food or waterborne microbial pathogens (Schlundt et al., 2004). This incidence of morbidity and mortality is consequent to the fact that globally > 1 billon people, and in sub-Saharan Africa > 40%, lack access to clean drinkable water and 2.4 billion do not have basic sanitation (CA, 2007). In practice, this means that these people have to drink water with fecal contamination from humans and animals and their intestinally excreted pathogens.

For countries with weak surveillance and outbreak detection systems, estimating the burden of foodborne illness is even a more daunting challenge, despite the assistance provided by WHO's Global Salm-Serv, Global Outbreak and Response Network (GOARN), International Food Safety Authorities Network (INFOSAN) and the FAO/OIE/WHO Global Early Warning System and Response for zoonotic disease surveillance (Flint et al., 2005). Further complicating the future of foodborne disease surveillance is the likelihood that as a result of climate change, new pathogens will emerge, particularly in fish and shellfish raised in water whose quality is degraded or contaminated (Rose et al., 2001).

The timeliness and efficacy of preventive or prophylactic food safety interventions depend on accurate, comprehensive and timely surveillance information. The factors of uncertainty in calculating the burden of foodborne illness are compounded by weak national surveillance systems upon which the international systems depend. Many governments, particularly in least developed countries, are unable to finance the development of such surveillance systems as part of national health system planning.

Since welfare benefits from agricultural trade are expected to increase for only a few developing countries as a result of the WTO Doha Round of negotiations (Bouet et al., 2004; Anderson and Martin, 2005; Polaski, 2006), it is unlikely that non-benefiting countries will be able to pay the costs of foodborne disease surveillance systems and SPS interventions from trade revenues. Therefore, in what follows we assume that some form of public finance and donor assistance will be required for capacity building in surveillance and other food safety activities. Furthermore, public finance may be involved in helping to insure against global foodborne illnesses risks that are not and perhaps cannot be insured by private firms.

7.3.1.2 Financing a public good

The globalization of the food and feed trade enables a broader and more rapid transmission of foodborne illness, particularly from high-risk microbial pathogens of animal origin (OIE, 2006). Development of surveillance data often becomes a priority only if a food contamination incident or zoonosis threatens trade, e.g., BSE and avian influenza. Such threats to trade usually focus only on emerging diseases and less on those that are prevalent and perennially cause major problems. Yet the costs of foodborne illness far exceed those that can be recovered from inspection fees or other forms of trade related SPS financing, even when the origin of an illness can be traced back to a specific source. Whereas the costs of food safety measures can be internalized to some extent in the cost of a product, there is no adequate mechanism for financing the public health costs resulting from transborder foodborne illness. FAO and WHO recognize that "[f]ood safety is an essential public health issue for all countries," but the normative framework and technical assistance planning for food safety in developing countries is largely a function of trade policy, or more broadly of the economics of private markets. Donor interest in and exporting country demand for SPS related assistance tends to be triggered by the threat of trade disruption or to ensure that food imports are safe (World Bank, 2005).

Although food safety is characterized as a global public good, economic analysis of food safety interventions often is framed largely in cost/benefit terms of market failures, in this case, the failure to internalize such negative cross border externalities as foodborne illness. Attempts to mitigate these externalities on an *ad hoc* emergency basis "is a costly and unsustainable form of assistance" (World Bank, 2005). The role of public food safety management is defined in terms of serving the market, without, however, an adequate financing mechanism designed to enable the development of food safety as a public good and taking into consideration the full and considerable cost for foodborne infection *e.g.*, loss of labor time and cost for medical care. New proposed public finance mechanism (*e.g.*, Kaul and Conceição, 2006) could be adapted to the provision of food safety, both on a global and regional basis, as a global public good. A World Bank study has argued cogently for a more proactive and preventative supply and demand approach to providing capacity building for food safety to facilitate trade (World Bank, 2005) but such capacity building need not be limited to trade facilitation.

7.3.1.3 Food safety standards for domestic public health benefits

In theory, trade related food safety standards and control measures may also be applied readily to domestic food safety programs. In practice, according to developing country official respondents to survey input into the FAO/WHO Food Standards Programme Evaluation, developing countries adopt few international food standards into domestic legislation because they lack the resources and technical capacity to implement and enforce the standards (CAC, 2002).

The unmet challenge remains, how to apply food safety measures not only for internationally traded products but for the great share of global food production that is not traded internationally. Of particular concern is the implementation of standards and other guidance to prevent foodborne illness resulting from new foodborne pathogens in domestically produced and consumed foods or from existing pathogens whose prevalence or severity has increased. The challenge of applying standards domestically for public health benefits is even greater in countries where food safety control systems are not integrated into the public health system but are instead largely confined to export establishments and import inspection. Policy options, outlined below, to meet this challenge should take into account capacity building challenges.

Despite the proliferation of international public and private standards, compliance with which is required for market entry, there are relatively few public studies of sanitary/phytosanitary (SPS) compliance costs for developing country agricultural exports (Pay, 2005). These few quantified studies indicate that existing levels and kinds of trade related technical assistance are far from providing the necessary facilities, such as accredited laboratories for measuring pesticide residue levels, to enable SPS standards implementation and enforcement (e.g., Larcher Carvalho, 2005). However, in some developing countries, qualitative needs assessments should suffice to demonstrate the desirability of donor financing of basic SPS infrastructure and training.

Notwithstanding the technical capacity shortfall to implement the SPS requirements of trade agreements, the view that "aid for trade" should be a binding, scheduled and enforceable part of trade negotiations for least developed countries (WTO, 2004; Stiglitz and Charlton, 2006) has not received support from developed country WTO members. While "best endeavor" capacity building can be helpful, the tradeoff in depending largely on private sector SPS infrastructure investment is that WTO members not integrated into transnational corporate food supply chains likely will be unable to ensure that their agricultural exports meet SPS requirements.

In the absence of adequate funding, proliferation of unfunded negotiating mandates may result in attempts to avoid SPS rule compliance. Furthermore, domestic adoption of international standards will not be enhanced by a simple increase in current capacity building initiatives, since there is a considerable disjuncture between the sanitary-phytosanitary technical assistance requested by developing countries,

particularly for SPS infrastructure, the assistance provided by donors that is often limited to training to understand SPS rules (CAC, 2002).

7.3.2 Animal health

Internationally, policies aimed at managing infectious animal diseases, including emerging or reemerging human diseases caused by animal-borne pathogens (Taylor et al., 2001), have been directed to improving preparedness. Methods of controlling and responding to zoonoses have been proposed, through developing and strengthening surveillance systems and identifying risks, including the economic, sociological and political implications and the need for intersectoral collaboration (e.g., WHO, 2004). It is particularly challenging for developing countries to try to meet internationally defined or driven regulations and policies for the animal health sector as these are continuously shifting in response to the increasing needs and ambitions of developed countries. This is assessed below in relation to the major groups of infectious animal diseases and the current principles for their control and regulatory support (Figure 7-9).

7.3.2.1 Major epizootic diseases and impacts on trade and developing countries

The effectiveness of current policies (eradication and SPS standards for maintaining disease free status) successfully applied in developed countries to prevent outbreaks of the major epizootic diseases (Leforban and Gerbier, 2002; DG SANCO, 2007) means that many developing countries will continue to be excluded from accessing the high-value international markets. This is generally because of the conflict between free trade and the protection of health status. The eradication of important epizoootic diseases, a core principle of the OIE who determine the animal health standards within the SPS Agreement, is not considered achievable in developing countries in the next decades because it requires significant efforts and investments in surveillance and veterinary service to meet eradication and control policies (Scoones and Wolmer, 2006). The magnitude of the challenges involved is demonstrated by the fact that an estimated 200 million poultry producers in Asia are on small holdings (e.g., 97% in Thailand and 75% in Cambodia (FAO, 2004a; Gilbert et al., 2006).

International debate on this dilemma has focused on an increased implementation in developing countries of other policies such as using a risk- and commodity-based approach that allows an alternative to the total restriction in trade of animals and animal products (Brückner, 2004; Thomson et al., 2004; Perry et al., 2005; Thomson et al., 2006). The concept is that different commodities pose very different risks for the spread of pathogens. For example, deboned meat has a reduced risk in relation to whole carcasses and is applied by certain countries to facilitate import from certain foot and mouth disease infected countries (Sabirovic et al., 2005). Similarly, policies that limit import restrictions to certain export producing areas (regions) instead of restrictions on whole countries or continents are also recommended. Such regionalization is considered as a useful additional tool in maintaining trade flow by limiting import restrictions in the case of new outbreaks of animal

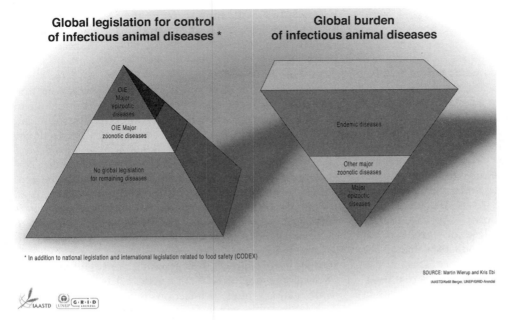

Global legislation for control of infectious animal diseases *

OIE Major epizootic diseases

OIE Major zoonotic diseases

No global legislation for remaining diseases

Global burden of infectious animal diseases

Endemic diseases

Other major zoonotic diseases

Major epizootic diseases

* In addition to national legislation and international legislation related to food safety (CODEX)

SOURCE: Martin Wierup and Kris Ebi

IAASTD/Ketill Berger, UNEP/GRID-Arendal

Figure 7-9. *Infectious animal diseases.*

diseases, and also allowing import from individual countries or regions based on their improvement of the animal health status food products (DG SANCO, 2006). However, these policies require a reliable and independent system of certification based on international standards (Thomson et al., 2006).

Instead of focusing on achieving high value exports from African countries to Europe and the U.S, bilateral agreements between developing countries that protect exporting countries and producers could be promoted (Scoones and Wolmer, 2006). A third alternative e.g., for African countries, is to initially focus on local trade and markets to supply the growing local and regional demand for meat (Kulibaba, 1997; Diao et al., 2005; Scoones and Wolmer, 2006). These alternative policies for developing countries emphasize benefits to their producers by using food safety and animal health standards needed for the local and regional market.

In addition to the introduction of advanced and new methods for improved and more cost effective disease and outbreak control (e.g., DIVA vaccines), the recent pandemic of highly pathogenic avian influenza virus demonstrated the importance of providing international support to developing countries when coordinated interventions are required to manage international emergencies, and also that sustained improvements in national disease control systems are required so that countries view such activities as investments rather than internationally imposed costs (Lokuge and Lokuge, 2005).

7.3.2.2 Zoonoses as foodborne infections—policies for integrated approach

The BSE crisis, the avian influenza pandemic and the threat of global warming with vectors and diseases moving into new areas has highlighted the importance of the animal–human link via the food chain and the need for capacity building for surveillance and control of zoonotic diseases (FAO, 2006d). In addition to these and other emerging zoonotic diseases, as highlighted above, there is also a basic need for effective policies for the prevention of the majority of the foodborne outbreaks that in most part of the world are caused by agents like *Salmonella* and *Campylobacter*.

In the US it is estimated that *Campylobacter* causes 2 million cases of foodborne infections annually and *Salmonella* is estimated to cause another 1.4 million infections, the latter at a total estimated annual cost of US$3 billion annually (Mead et al., 1999; USDA, 2007). In developing countries the situation is likely to be at least of the same magnitude. The vast majority of these infections primarily originate from animal production so the overriding aim for the animal health sector is safe food and consumer protection (Schlundt et al., 2004). A problem is that these infections usually cause no or very limited economic losses to animal production. Thus efforts are needed to implement policies with economic incentives for producers to improve hygiene in their animal production in order to decrease the input of potential pathogens to the food chain.

The need for integrated approaches is emphasized when interventions are needed along the whole food chain. Of particular interest are challenges posed by the increasing global demand for protein as animal feed, in response to the increasing global demand for meat (Morgan and Prakash, 2006). To meet that demand soybean production has increased e.g., in Brazil, which has resulted in deforestation and monoculture followed by environmental degradation from high pesticide use and significant problems with pesticide residues in the soy products produced (Klink and Machado, 2005). In addition, many countries have experienced an increased risk of *Salmonella* contamination in soy meal, which constitutes an important route for introducing *Salmonella* into animal production when used as animal feed (Hald et al., 2006; EFSA, 2006b). A pandemic spread of *Salmonella* occurred when contaminated fishmeal from South America was exported to the U.S. and Europe, causing more than one million human cases in the U.S. alone (Clark et al., 1973; Crump, Griffin and Angulo, 2002). *Salmonella* contamination has become a significant challenge to the global marketing of animal feed and food products (Plym-Forshell and Wierup, 2006).

7.3.2.3 Endemic diseases—the major challenge and potential

Categorization of livestock diseases is critical for the determination of public intervention, as highlighted in the recent assessment of the EU animal health policy (DG SANCO, 2006). International and national policy and legislation focuses on the control of the major epizootic diseases and, increasingly, on the foodborne zoonotic diseases. Economic compensation in case of outbreaks, surveillance and other measures are generally limited to these, so-called, listed diseases (Table 7-1) (WHO, 2006).

Endemic diseases comprise the majority of animal diseases and, in developed countries, continuous implementation of disease prevention measures directed against these endemic diseases is necessary for efficient production. The economic importance of endemic diseases is recognized and in many developed countries a number of the endemic diseases have been successfully eradicated or controlled (e.g., Aujeszky´s disease in pigs, infectious bovine respiratory disease and bovine virus diarrhea in cattle). Such programs have been found to be very cost effective (e.g., Valle et al., 2005). The increasing focus on reducing antibiotic use to prevent resistance and on animal welfare further emphasizes the importance of control and/or eradication of animal diseases (Wierup, 2000; Angulu et al., 2004).

Control of animal diseases and the promotion and protection of animal health are essential components of any effective animal breeding and production program (FAO, 1991). However, despite remarkable technical advances in the diagnosis, prevention and control of animal diseases, the condition of animal health through the developing world remains generally poor, causing substantial economic losses and hindering any improvement in livestock productivity (FAO, 1991, 2002). Consequently, in addition to efforts to minimize the negative effects of the major epizootic and foodborne diseases, policy could also focus on the prevention and control of endemic diseases, even though the producer is generally considered to bear the responsibility for production losses caused by this group of diseases. However, such actions could also have a direct strengthening effect on food safety and food security and, in this respect it has been emphasized that a focus on safe food in the context of strengthening export capacities of developing countries should come second to the primary objective of improving food safety for local consumption (Byrne, 2004).

The global burden of animal diseases when also including the cost for public health and loss of labor is also estimated to be dominated by the endemic diseases, in contrast to the public focus on the control of the epizootic diseases.

7.3.2.4 Animal welfare

The protection of animal welfare and the demand for a sustainable animal production system, which is increasingly being considered in animal health policies and in SPS-associated regulations, can be an additional constraint for developing countries trying to access international markets. However, sustainable extensive livestock production practices in developing countries that promote animal welfare could open niche market opportunities in developed countries. This is in contrast to intensive livestock production in many sectors of the industrialized world where in the short term the implementation of systems for improved animal welfare often are associated with increased cost of production (OIE, 2005a; Kyprianou, 2006).

The veterinary services of developing and transition countries are in urgent need of the necessary resources and capacities that will enable their countries to benefit more

Table 7-1. **Estimated global burden of infectious animal diseases.**

Classification of infectious animal diseases	Qualitative estimation of relative number	Qualitative estimation of relative cost and importance for major stakeholders			
		Public sector cost		Producers cost	
		Animal health	Public health	Developed countries	Developing countries
Major epizoootics	+	+++++	-	-*	++
Other major diseases including major zoonoses	++	+++	+++	++	++
Endemic diseases including "neglected" zoonoses	+++++	+	+++++	+++++	+++++

*Diseases eradicated or absent. Elimination policy applied in case of outbreaks when significant costs may occur.

**Losses of production and labor and costs of control and medical treatment.

Source: M. Wierup and K. Ebi.

fully from market access opportunities in trade agreements, while at the same time providing greater protection for human and animal health, animal welfare and reducing the risks linked to zoonoses (OIE, 2004; Thomson et al., 2006). It is of utmost importance that the ongoing initiatives from OIE and others to support veterinary services, in particular in developing countries, continue. OIE emphasizes the need for veterinary services to support access of animals and their products into national markets, indicating the importance of animal health control in a safe and secure food supply. A challenging factor is the limited availability of veterinarians trained in veterinary public health (WHO, 2002b), which in developing countries has opened discussions on the need of paraprofessionals such as community animal health workers (Scoones and Wolmer, 2006).

7.3.2.5 Priority setting for disease control technologies

Historically significant resources have been directed towards tools to implement eradication policies and research often focuses on the production of a vaccine that simply should be the key to success. These resources are also often directed to diseases that gain special attention in relation to international trade but that might be of less economic importance in an endemic situation in a developing country (Scoones and Wolmer 2006). However, effective vaccines are available only for a limited number of infections and therefore preventive actions need to come into focus. Many important diseases have been successfully controlled through the application of simple, preventive hygienic methods; a "bottom-up" approach to priority setting can therefore be recommended (Scoones and Wolmer, 2006). Recommending that milk be boiled prior to consumption in South Africa could more simply and cheaply limit human health risks due to *Brucellosis* than a comprehensive vaccination control program in a cattle population where the disease caused relatively limited production losses (Mokaila, 2005).

7.3.3 Plant health

Food availability depends in the first instance on the actual production of food, which is influenced by agroecological production potential as well as by available production technologies and input and output markets (FAO, 2005b). Plant pests are key constraints to achieving the true yield potential of food and fiber crops, particularly in tropical and subtropical regions where conditions necessary for the reproduction of pests may be present year-round (FAO, 2005c). In addition to their direct and deleterious effect on the yield and quality of plant products, plant pests can also pose an absolute barrier to imports when countries apply phytosanitary measures to regulate the entry or plants, plant products or others materials capable of harboring plant pests.

7.3.3.1 The challenge of international phytosanitary standards

International phytosanitary standards recognized as authoritative by the SPS Agreement can be a positive driver in developing countries. When applied to high value food products, these have played a beneficial role in stimulating improvements to existing regulatory systems and the adoption of safer and more sustainable production practices (World Bank, 2005). More commonly, however, international phytosanitary standards are considered as barriers to trade that particularly discriminate against developing country stakeholders who can neither afford to meet the high costs of compliance associated with these nor participate effectively in their development by international standard setting bodies like the International Plant Protection Convention (IPPC) (e.g., Simeon, 2006). Governments, institutions and farmers may respond to such standards in a number of ways: support or participate in programs that will address the management of the pest problem; find alternative foreign markets for nationally produced goods; focus on increasing domestic demand for trade-prohibited plants and products; or exit production, with or without compensation and/or incentives to promote diversification into other crops.

Governments generally divide resources applied to address phytosanitary considerations in two ways: (1) to meet the phytosanitary requirements of importing countries (export certification); and (2) to meet domestic phytosanitary requirements, including those applied to imported agricultural products. In both developed and developing countries these regulatory tasks are typically addressed through an array of plant protection and quarantine (PPQ) programs. Core services of traditional PPQ programs include activities such as: detection and control or management of plant pests of quarantine or economic significance; undertaking pest risk analyses; and managing import, export and/or domestic certification programs. These programs are being challenged by increases in the volume and kinds of agricultural products being traded internationally, the number of countries exporting such products, and international travel which creates more opportunities for the rapid introduction and spread of new pest species (FAO, 2003).

7.3.3.2 Opportunities through regionalism

For some countries, particularly those with limited resources applied to national PPQ programs, regional or subregional programs may be a workable alternative. Regional initiatives to harmonize standards where trade between the participating member countries for specific plant products is significant and where an international standard is not needed (i.e., a different, less restrictive or less economically punitive standard will suffice). Regional pooling of scientific resources (human and institutional) to collectively manage plant pests and implement surveillance programs can enable developing countries to meet the surveillance and pest risk assessments required for compliance with import requirements. Surveillance data is important to ensure that domestic phytosanitary measures are equivalent to those applied to imported commodities so that discrimination against imports based on pest exclusion is not supported. Efforts to collect these data for key pests that affect movement of plant material from or within a specific region may be best addressed by establishing harmonized protocols for data collection and then pooling resources to acquire the necessary information to demonstrate pest-free status. Initiatives to promote meaningful, results-based regional cooperation to address plant health issues will require incentives to promote cooperation both within and between national agricultural systems. Where regional regulatory programs may be government to government, these should also actively encourage the inclusion of other stakeholders, especially the private sector and producer groups.

7.3.3.3 Biosafety and plant protection

With the ratification of the Cartagena Protocol on Biosafety, many governments are in the process of developing or implementing national biosafety regulatory programs (GEF, 2006). With the rapid adoption and global trade of transgenic maize, soybean, cotton and canola the primary focus of these new programs is typically the regulation of transgenic crops. National, bilateral and international support for the establishment of biosafety regulatory programs has favored the creation of new regulatory entities under ministries other than agriculture. Given the shared nature of many of the regulatory functions of PPQ and biosafety programs (e.g., risk assessment, monitoring and inspection activities) and the inclusion of Living Modified Organisms in ISPM No. 11 (Pest Risk Analysis for Quarantine Pests, Including Analysis of Environmental Risks and Living Modified Organisms), there exists an opportunity to apply new resources available for biosafety regulatory capacity building to strengthen existing PPQ programs so that the objectives of both can be achieved without building redundant administrative services. This could be achieved under the umbrella of "plant biosecurity" to include plant health, plant biosafety and also invasive alien plant species. Inputs for programs related to plant biosafety or, more broadly, plant biosecurity should be actively sought from, if not led by, ministries of agriculture.

7.3.3.4 Meeting the plant health needs of small-scale farmers

Control of plant pests that are important from a trade perspective may be of little or no significance to small-scale farmers who are not exporting their plant products. Instead, their priorities are likely to be management of local pests that will have a direct impact on their harvested or postharvest yield. Policy makers could ensure that the small-scale farmer, whose fields may be an inoculum source of a trade-prohibited pest, is provided with incentives to assist in the management of such pests so that export certification of the commodity in question can still be achieved. This could come in the form of support that links breeding or pest management programs designed to address the priorities of the small farmer with activities that will also assist in the management of the prohibited plant pest. Similarly, a government could strengthen the capacity of regulators to enforce compliance with internationally relevant phytosanitary standards but couple this with direct support for the primary producer where production practices may have to be modified so that pest exclusion goals can be attained.

An alternative policy option is to realign public sector AKST funding to support research explicitly directed to improving small-scale, diversified farming practices that promote improved yields and enhanced food quality through sustainable pest management practices. These could variously include IPM, organic farming, and improved plant breeding programs, including the development of pest resistant varieties through marker assisted selection or recombinant DNA techniques. National prioritization of the needs of resource-poor farmers may be more important in the future as scientific and agricultural technology spillovers from developed countries that are adapted by developing countries may be less available (Alston et al., 2006).

7.3.3.5 The private sector and third party certification

The private sector has responded to enhanced consumer awareness and concern about food safety by developing their own phytosanitary (and sanitary) standards, enforced through third party certification (Hatanaka et al., 2005). This means that participating primary producers have to meet an array of requirements that go beyond those mandated in government regulations, such as implementing traceability programs or participating in accreditation programs that add expense and complexity to more traditional production systems. While there are examples of developing country farmers who have benefited from third party certification (Hatanaka et al., 2005), arguably these private sector standards discriminate against resource poor farmers who cannot afford the high costs of participation. In response, governments may decide to align their public sector investment to ensure that AKST is applied to assisting producers to meet only statutory phytosanitary standards, through agricultural research, extension and/or education systems. Individual farmers or commodity-specific producer associations would have to use their own resources to meet additional private-sector requirements. Alternatively, governments could strategically invest in AKST that will promote the participation of small-scale farmers in third party certification, through the provision of education programs and technical assistance. This may also provide a stimulus for the development of off-farm employment opportunities through the provision of services such as third-party accreditation of farms or production systems. Internationally, the private sector in developed countries, which is driving third party certification, should promote the harmonization of private sector standards and streamline accreditation, especially where these apply to plant products produced in developing countries (Jaffee, 2005).

7.3.3.6 Climate change and plant health

A significant consideration for policy makers tasked with addressing plant health issues is the impact that climate change will have on plant production. Climate change can affect plant health by: modifying the encounter rate between host and pest by changing the ranges of the two species; introducing new hosts, vectors and/or pests; causing social changes such as shifts in agricultural labor; and shifting land use patterns that will alter the potential for populations of plants and pests to migrate to fragmented landscapes (Garrett et al., 2006). In response to this, policy makers will be challenged to decide if investments in development and deployment of AKST will be proactive (e.g., inclusion of climate prediction in forecasting models of plant disease) or reactive (e.g., deployment of resistant varieties after the emergence of a new plant disease). Action to mitigate the impacts of climate change on crop production will require integrated strategies developed and implemented in a participatory fashion that emphasizes the need to include non-traditional players in agricultural research. Coherent policies could be developed cooperatively through multidisciplinary partnerships within government (e.g., Ministries of Agriculture, Energy, Trade, Health and Commerce) and with significant guidance from academic, agricultural, nongovernmental and private sector players.

7.3.4 Ways forward

Recognizing that food safety, animal health and plant health are global public goods, new mechanisms to support the development and, most importantly, implementation of proactive and preventative policies and programs to facilitate compliance with SPS standards could be explored. Internationally, donor support could be targeted to specifically assist those countries that cannot adequately finance SPS standard implementation nationally but attention could also be paid to ensuring that trade facilitation is not the only driver of SPS program delivery. The application of AKST to address yield and quality losses associated with pests or pathogens that are of domestic, but not international, importance may have more impact on reducing hunger and poverty, and improving nutrition and health, particularly in the least developed countries, than applying these resources exclusively to accessing international markets. For small developing countries, the possibility of regional food safety "trusts" to provide a continuous funding source for shared SPS related surveillance programs, infrastructure and personnel should be considered. An international SPS insurance mechanism that would supplement or replace current *ad hoc* funding to detect and mitigate transborder food contamination incidents, zoonoses and plant health contagion should also be considered.

Given the globalization of agriculture and trade, the institutional separation of Codex, OIE and IPPC may be of limited relevance in the future. The traditional mandates of these international organizations are already challenged by the emergence of alternative regulatory mechanisms that integrate food safety, animal and plant health related standards and production practices e.g., Good Agricultural Practices, Good Manufacturing Practices, on-farm HACCP plans and other retailer-driven certification programs. Revising SPS-related policy and regulatory measures within an explicitly coordinated biosecurity framework may be one option for promoting cross-sectoral interventions. Internationally, policy and regulation related to food safety, plant and animal health could be better integrated if the mandates of Codex, OIE and IPPC were recast to remove areas of duplication, identify sources of conflict and promote opportunities for policy and program coordination to more effectively utilize the limited resources that are applied to SPS issues.

Policy options

- For smaller and contiguous developing countries, strengthening or starting regional foodborne, animal and plant health surveillance systems may be a viable option, particularly where dietary patterns, agricultural practices, and natural resources for agriculture are similar.
- Consideration should be given to establishment of national or regional food safety trust funds invested to ensure a continuous funding mechanism to gradually build the national or regional surveillance systems upon which effective food safety interventions depend. The trusts could be financed from an increase in ODA and from an increase in agrifood corporate taxes. Alterna-

tively, governments can continue to respond *ad hoc* to food safety emergencies or SPS related threats to trade, financed by voluntary funds for each purpose.

- Governments should consider expanding current "aid for trade" commitments to include the financing of specific SPS infrastructure requested by WTO members with documented incapacity to finance that infrastructure from domestic sources. Since it is unlikely that governments will support binding and enforceable "aid for trade" commitments, governments should consider developing a model contract for expedited needs assessment that is not tied to import of SPS technology or training from any one donor.
- Considering that SPS standards are largely implemented in developing countries for the purpose of trade facilitation, often with little benefit to local consumers of domestically produced food, policies that focus on domestic food production and domestic priorities for animal and plant health, food safety and public health could receive greater attention.
- Weak national SPS surveillance systems could be strengthened to improve the timeliness and efficacy of preventative or prophylactic food safety, animal and plant health interventions. Even where there is an absence of detailed epidemiological or surveillance data, foodborne infections and animal and plant diseases could be better managed through policies that promote simple, workable SPS programs implemented at the farm or community level. Capacity building could be redirected from training to understand SPS rules to technical support needed to operationalize such programs.
- Eradication of the major epizootic animal diseases is unlikely to be achieved in the foreseeable future in many developing countries in spite of significant investment and effort to do so. An alternative, commodity based approach could instead be used as a tool to promote access to international markets which would also allow resources to be allocated for the prevention of losses caused by other animal and zoonotic diseases.
- Governments could align their public sector investment to ensure that AKST is applied to assisting producers to meet only statutory SPS standards, through agricultural research, extension and/or education systems.
- Governments could strategically invest in AKST to promote the participation of small-scale farmers in third party certification, through the provision of education programs and technical assistance.
- The ongoing initiatives from OIE and others to support veterinary services in developing countries could continue as a means to support access of animals and their products into national and international markets and to improve food safety and secure food supply. Policies that recognize and support the training of paraprofessionals such as community animal health workers could be promoted to compensate for the limited availability of veterinarians trained in veterinary public.
- Policies could support the provision of international support to developing countries when coordinated interventions are required to manage international emergencies (e.g., highly pathogenic avian influenza virus) and sustained improvements in national disease control

systems could be viewed as investments rather than internationally imposed costs.

- FAO, WHO and OIE could consider establishing a joint task force to examine what those agencies and their member governments might do to prepare their SPS surveillance and intervention systems to identify SPS risks and hazards that may result from anticipated effects of climate change on food and agriculture production and distribution.

7.4 Knowledge and Knowledge Management—Property Rights

The generation, dissemination and maintenance of AKST increasingly depend on property rights, placing AKST in private, community, and public domains. As opportunities to protect AKST increase, access to innovations, local knowledge and genetic resources become restricted through different regulatory systems. Public research may result in privately controlled knowledge, either as a result of institutional policies or of public-private partnerships. IPRs have multiple objectives, ranging from stimulating investments in R&D, facilitating technology transfer and bringing knowledge to the public domain through publication and setting time limits to any exclusive rights. However, the validity of these in a low income country (LIC) and development context of low technical capacity is contested. Nevertheless, public research institutions have to decide how to deal with these developments and how far to go in developing capacities to manage proprietary knowledge and materials (Egelyng, 2005).

Opportunities to legally protect knowledge can be analyzed in terms of the likely impact on the generation of and access to such knowledge for development purposes. This analysis refers to both the strength of intellectual property rights in general and to the policies towards the use of the protection systems by public (research) organizations. IPRs fit in a paradigm of market-led development which is essentially different from both the concept of sharing ideas that characterize most farming communities (and which is essentially different from medicinal knowledge in many communities) and from the public goods paradigm which dominated the agricultural research for development policies for over 50 years. Strong intellectual property rights may support commercial investments in research, but may not be effective in stimulating research for non-market uses to serve the need of the poor.

Changes in property rights systems impact the roles of the stakeholders in AKST. National policies, such as the Bayh-Dole Act in the USA, promoting "protection" of IP by public universities, led to new commercialization strategies for publicly developed AKST, including exclusive licensing of IP to companies in exchange for follow-up university research contracts and product commercialization. Reduced public expenditure on agricultural research in a number of countries, and the expansion of public-private partnerships in agricultural research also tend to stimulate the protection of knowledge by public research institutions in order to generate income (Louwaars et al., 2006). An analysis of the impact of the Bayh-Dole Act in the USA (Rosenberg and Nelson, 1994) indicates that, as a result of the high costs

of managing IP, very few schools make a net profit on their R&D investment.

The design of systems of rights and the forms in which these are implemented are examples of interactions between various levels of organization ranging from international conventions and commitments to local forms of interpretation with or without the filter of national policy and legal systems. If different rights systems do not specify how results produced by AKST system are used, exploited and disseminated, those knowledge products and technologies may be unused by the intended AKST beneficiary. If the public sector is stimulated to use the rights to create a flow of revenue, public sector researchers are likely to change their programming away from the needs of the poor (World Bank, 2006).

7.4.1 Public research and the generation of public goods

The status and nature of AKST as "public/private good" is critical for its value in development.

Anthropologists and sociologists (Fuller, 1993; Callon, 1994) hold that science is a public good. Innovation, a change in order to solve a constraining situation, is both a key for human development and a tool for competitiveness. The report of a Commission initiated by the British government to look at how IPR might work better for poor people and developing countries remains a most important analysis of this challenge (Commission on Intellectual Property Rights, 2002).

Both the international priorities for the production of global public goods in global schemes (e.g., in the CGIAR), and national innovation programs can to be viewed in this changing perspective when they are to deal with multiple development objectives. The introduction of private, community and national rights creates a wide range of challenges for public research.

Realization of environmental objectives are very appropriate areas of inquiry for publicly funded AKST: research towards fulfilling these objectives can frame natural resources as public goods requiring collective management, such as climate, air quality, water, landscapes. Knowledge can help private stakeholders, like farmers, forest owners, rural factories, to develop environmentally friendly practices even in those areas that are privately owned.

7.4.2 Multilateral negotiations on rights systems

Attempts to import concepts from one multilateral agreement to another to enhance their mutual compatibility have met with strong opposition.[2] For example, the proposal to make the CBD's Prior Informed Consent in the use of traditional knowledge and genetic resources a substantive

---

[2] How to protect and license the use of intellectual property (IP) and traditional knowledge (TK) continues to be fiercely debated in World Intellectual Property Organization (WIPO), the WTO negotiations to amend the Agreement on Trade Related Aspects of Intellectual Property (TRIPs) and in various civil society forums. The implementation of national sovereignty over genetic resources and arrangements for Access and Benefit Sharing (ABS) as debated within the Convention on Biological Diversity (CBD) and the International Treaty on Plant Genetic Resources for Food and Agriculture (ITPGRFA) also link with the protection of IP and TK.

requirement of patentability in TRIPS has not prospered (Visser, 2004).

WIPO and WTO. Proposed binding WIPO norms to protect traditional knowledge and genetic resources from unauthorized and unremunerated misappropriation (i.e., "biopiracy") have been rejected as a threat to WTO IP rules (e.g., paragraph 211, WIPO, 2006a). The nature of the threat is, however, not specified, but likely relate to the collective stewardship of traditional knowledge, which from a classical IP viewpoint would represent an undermining of the individual status of patent ownership (Finger, 2004). It remains an option to further explore a developing country proposal to amend TRIPs Article 29 to require disclosure in patent applications of traditional knowledge and genetic resources used in the development of patented products (WTO, 2006a).[3] Proponents of disclosure argue that disclosure would improve patent quality (Article 27.1), prevent abuse of the patent system and promote the public interest (Article 8), provide social and economic benefits to WTO members (Article 7) and make TRIPs supportive of the CBD, particularly its ABS provisions (Articles 1 and 15). Opponents of disclosure opponents contend that ABS is best implemented through contracts that offer a cash payment or other benefits in exchange for the rights to patent products developed from an agreed number of genetic resource samples (WTO, 2006b).

Intellectual property regimes alone, no matter how comprehensive, fully implemented, and mutually supportive of other multilateral treaties, are insufficient to enable development of the seed systems needed to fulfill goals, and poorly designed and implemented regimes can be detrimental to achieving these objectives (World Bank, 2006).

Genetic resources in agriculture. Challenges to bringing the private rights of IPRs in harmony with the collective rights over traditional knowledge and local genetic resources are further complicated by the rights based on national sovereignty over the physical genetic resources, as established in the CBD. Apart from conceptual and legal challenges, this complication has led in the past years to practical problems in the exchange of genetic resources, which affects the agricultural use of genetic resources in plant and animal breeding more that any other type of use. Important steps have been taken in the sharing of benefits derived from the use of these resources in a multilateral way through the IT PGRFA's conclusion of the Standard Material Transfer Agreement The IT PGRFA confirms the Farmers' Right of protection of TK, which established a link with the debate in WIPO, the right of benefit sharing linking it further to the CBD, and the right to participate in decision making at the national level on matters related to the conservation and sustainable use of plant genetic resources for food and agriculture. The IT PGRFA refers the implementation of these rights to the national level (Article 9.2C). However, it may prove difficult for national policy makers to implement these Farmers' Rights while avoiding conflicts among IPR, biodiversity and seed regulations with the right of farmers to save, use, exchange and see farm-saved seed.

Traditional knowledge (TK) and genetic resources. A review of technical papers in support of the WIPO negotiations has proposed that an "international enforcement pyramid" be constructed from existing practices to enable developing countries to control and sustainable use traditional knowledge and genetic resources (Drahos, 2006). The "enforcement pyramid" would integrate indigenous and national government practices and would be coordinated by a Global BioCollecting Society under the aegis of WIPO, FAO and the CBD. Complicating the construction of an effective enforcement mechanism for traditional knowledge and genetic resources are differences between indigenous customary law and governance, and national government jurisdiction, particularly where indigenous territories cross national boundaries (IIED, 2006).

WIPO negotiations for a Substantive Patent Law Treaty (SPLT) present a framework for IP protection and enforcement very different from an enforcement pyramid based on national and indigenous group enforcement practices for traditional knowledge and genetic resources protection. The SPLT is part of a Patent Agenda to create and enforce a "global patent" with mechanisms far more specific and powerful than the TRIPs enforcement provisions (Article 41), and reduced transaction costs (WIPO, 2001). The U.S., EU and Japan are the main SPLT advocates and cooperate in patent matters. Some IP scholars are concerned that the SPLT could negatively affect public AKST and access to publicly held genetic resources, particularly in countries where rules on plant variety protection do not yet limit farmers' rights to save or exchange seed (e.g., Tvedt, 2005). The SPLT may also limit developing countries ability to shape their patent laws to their own specific needs, taking into account the development stage that they are in. SPLT is thus seen as supporting only a trade agenda rather than supporting a Development Agenda (WIPO, 2004). The debate is ongoing.

Nevertheless, elements of the draft SPLT are being carried forward in Bilateral Investment Treaties (BITs) and so-called "TRIPs plus" provisions in bilateral Free Trade Agreements (FTAs)[4]. BITs with many of these parties define

---

[3] The information embedded in the genetics of the seed and the associated farmers' and scientific knowledge comprise a significant part of AKST. The value of TK appropriated for use in patented agricultural and medical products would represent at least $5 billion annually in royalties to developing countries, if TK were protected and licensed as patents are (McLeod, 2001). Just half of such a sum, if invested for the *in situ* conservation of agrobiodiversity and if distributed effectively to the often collective and indigenous stewards of that biodiversity, could help realize development and sustainability goals.

[4] "TRIPS plus" agreements assert TRIPS as a foundation but add some provisions that arguably conflict with TRIPS provisions. Recent FTAs require the patenting of biological resources, thus overriding the patenting exemption in TRIPs Article 27.3, and require countries to become members of the Union for the Protection of new Varieties of Plants (UPOV), thus closing the door for alternative breeder's rights protection systems, including earlier versions of the UPOV Act that are more compatible with farmers' seed systems (World Bank, 2006). These FTAs also prohibit parties from citing resource constraints as a legal defense for nonenforcement of IP obligations (Fink and Reichenmiller, 2005).

IP and genetic resources in ABS agreements as "investments" and allow a very broadly defined "investor" to sue states for nonenforcement or inadequate enforcement of investor rights, no matter how resource constrained the developing countries parties may be (Correa, 2004).

IP and TK economics. Desegregated data on agricultural IP costs[5] could help policy makers make more informed decisions about whether to assume the costs and legal obligations of specific patented agricultural products. The cost is difficult to justify in light of the aggregate 53% price drop in agricultural export commodities from 1997 to 2001 (FAO, 2005a), nor from the expected 2.8% price increase resulting from WTO Doha Round (Bouët et al., 2004). In our assessment, the terms of trade rationale for investment in patented AKST becomes weaker still when taking into account the costs of state liability for non-enforcement of IP as "investment" in BITs and individual producer liability for violating patent holder rights, e.g., of agricultural biotechnology firms.

There is no agreed methodology for estimating the economic value of traditional and local knowledge and genetic resources, for the purpose of licensing its use in specific patented products (Drahos, 2006). Agreement on such a methodology might be derived on the basis of experience with studies estimating the value of traditional knowledge and genetic resources used in specific patent products. A royalty or licensing fee system based on the value in the seed market is incorporated in national laws in some countries (e.g., the gene fund in India) and in the IT PGRFA that will be used to fund genetic resource conservation. Seed sales royalty funding therefore remains a policy option.[6]

The global costs of withholding access to genetic resources due to national access regimes are insufficiently researched. Given the interdependence of countries on genetic resources (Flores-Palacios, 1997) and the fact that exchange of agricultural genetic resources among developing countries is much more frequent than transfer from South to North (Fowler et al., 2001), such costs are likely to be borne to a large extent by developing countries. Providing financing mechanisms to facilitate genetic resource access and transfer from agrobiodiverse rich developing countries to agrobiodiverse poor developing countries is one option for remedying this situation.

Other policy options regarding intellectual property and traditional knowledge include:

1. Insofar as traditional knowledge and genetic resources may form part of the prior art of a patented AKST product, adoption of a disclosure amendment in TRIPS could serve to enhance patent quality and might be a disincentive of misappropriation. Information so disclosed could be part of the legal basis for any licensing agreement on traditional knowledge and genetic resources used in a patented product. WTO members might also wish to consider adopting a weaker interim standard of TK protection, such as a Declaration on Trade and Traditional Knowledge (Gervais, 2005).

2. Given the impasse at the WTO over disclosure, developing country members are seeking to protect traditional knowledge and genetic resources with strong norms in WIPO negotiations. However, in the absence of the capacity to enforce such norms, members could agree on the design of mechanisms for and the financing of an "enforcement pyramid" (Drahos, 2006) for any norms that are agreed, preferably under the coordination of a joint agency under CBD, FAO and WIPO. This could ensure that local and indigenous stewards of agrobiodiversity would participate in member government decisions on the licensing of traditional knowledge and genetic resources, as called for in Article 9 of the ITPGRFA.

3. Both multilateral IP negotiations and implementation discussion lack ex ante and ex post economic analysis of the cost and benefits of adopting IP commitments and patented AKST. Analysis of, not rhetorical claims about, IP benefits for developing countries, could inform bilateral agreements and national government IP policy and legislation (Park and Lippoldt, 2004). Multilateral and technical assistance could include financing for IP economic analysis.

4. A methodology for valuing the ex-post contribution of traditional knowledge and genetic resources in patented AKST option to agree, as a prerequisite for any agreement on how to license traditional knowledge and genetic resources in patented products. Decision makers may consider ex-post studies of traditional knowledge and genetic resources valuation in existing patented products to better delineate the elements for an agreed methodology.

5. Facilitating access to and sustainable use of traditional knowledge and genetic resources among developing countries, particularly for the benefit of agrobiodiverse poor developing countries may be a an issue that policy makers may wish to move higher in the priorities of the multilateral AKST agenda, Both for food security and agroenvironmental sustainability reasons, it remains an option for policy makers to consider developing guidelines for a specific facility for the benefit of agrobiodiverse poor developing countries.

7.4.3 Effects of rights on AKST at the national and institutional levels

IPRs on products and processes that are relevant to agricultural development in the widest sense create novel con-

---

[5] One estimate suggests that developing country TRIPs commitments in the Uruguay Round amounted to $60 billion annually in implementation costs, licensing fees and royalties (Finger, 2004).

[6] As part of a project to measure genetic resource erosion and to suggest how royalties be paid to source countries of genetic resources might provide incentives for conservation, it was estimated that a 1% royalty on sales of patented seeds incorporating traditional knowledge and genetic resources would return about $150 million per annum to the source countries, most of them developing countries (FAO, 1998). However, with the creation of the Global Crop Diversity Trust in 2004, FAO's erstwhile interest in using royalties to pay for protection and enhancement of crop diversity has diminished. The Trust Fund has shifted away from FAO's prior dependence on annual donor government contributions towards funding Trust activities from an endowment supplemented by foundation grants. Nevertheless, the Trust states that there are many crop diversity conservation activities that it cannot and will not finance.

ditions for the use of AKST at different levels. The trend towards privatization of AKST is particularly felt in the sphere of plant breeding and biotechnology.

Stimulating private investments in research. IPRs are meant to stimulate private investment in research, but even though evidence of such effects in various industries may be available, it is very weak in the agricultural sector, notably plant breeding (Pray, 1991; Alston and Venner, 2000). IPRs protected in LDCs with a limited research capacity are more likely to improve access to proprietary technologies from abroad (e.g., Bt cotton). There are claims of positive effects of protection of breeder's rights for a selective number of cases; however, without taking into account alternative explanations for the observed effects and without providing data for other crops in the same case study countries (UPOV, 2005). Other studies show inconclusive results of the value of IP protection for the plant breeding sector in LDCs. Based on evidence in five developing countries, IPRs may support the development of a private seed industry, but only when this sector has reached a certain level of maturity; IP protection is not a major stimulus for initial investments in the sector (World Bank, 2006).

Public-private partnerships in research. In a market system, IPRs provide a way to share benefits among the different chain partners through the transfer of technology fees (royalties). They are the basis for negotiating partnerships in research between private and public partners, notably private IPR-holders and public research institutions in accessing technologies in a certain country. However, the reliance on negotiated license agreements also introduces prospects for unequal sharing of benefits based on differences in negotiating capabilities and power of the partners.

Financial support to the public research systems through IPRs. Even though IPRs fit in a commercial approach to innovation, it is, in many countries, the public sector research institutions that promote the introduction of IPRs in agriculture. This promotion is based primarily on a perception that these institutes may obtain significant revenue when their inventions (e.g., plant varieties) may be protected. This revenue is welcomed when there is underinvestment in public research (common in many countries since the 1990s), but may be viewed differently if such benefits can only be obtained in commercial markets (e.g., seed markets). Reliance on IPR based revenues is likely to lead to a change in public research priorities, in some cases to commercial crops like maize and oil crops to the detriment of research on small grains and pulses, and to benign ecologies and market oriented farmers, to the detriment of a small-scale farmer focus (Fischer and Byerlee, 2002). Such research shifts may fit in market orientation priorities of national development strategies, but may at the same time challenge to some extent the public tasks of contributing to poverty alleviation and household nutrition security (Louwaars et al., 2006).

The most common alternative strategy for a public research institute may be to publish its innovations, i.e., place in the public domain. This strategy reduces opportunities to obtain financial revenue and may limit public-private partnerships.

Challenges to technology transfer—thickets of rights. Even though license agreements may promote technology transfer by clarifying roles and responsibilities, IPRs may also pose serious limitations to research and the use of technologies in development. Particularly in advanced research, so-called thickets of rights lead to the tragedy of the anti-commons leading to underinvestment and underutilization of technologies (Heller and Eisenberg, 1998). Property rights on research tools, processes and products create very complex situations for researchers and their institutions, potentially leading to underutilization of technologies. Research institutes have to learn how to establish and negotiate their freedom to operate on these technologies. The quality and enforceability of the claims of a patent may significantly differ between jurisdictions; negotiating access to a technology can be very difficult when unequal partners are involved; so-called humanitarian use licenses (license on a technology for R&D for development with "soft" conditions) may be granted when the use of a technology is unlikely to challenge the commercial interests of the rights holder, but the "small print" license details can create significant obligations for the recipient.

These are new policy challenges for most developing countries, the actual impact of which cannot be readily assessed yet (World Bank, 2006). The rights on enabling technologies create challenges for producing public goods, which has been the main focus of public research, and more specifically for the centers of the Consultative Group on International Agricultural Research. When more and more technologies are protected by IPRs in their target countries, producing international public goods may become more and more difficult (Fischer and Byerlee, 2002). Currently, these centers are venturing in license strategies on their protected technologies that provide a public good status for the purpose of poverty alleviation and food security in developing countries, while maintaining ownership in commercial markets both in developing and industrialized countries.

Costs of compliance. Compliance with the rights of IPR-holders requires public and private research institutions alike to invest in capacities that they had not required in the past, notably legal and commercial specialists. There are already commercial seed companies that spend far more on legal services than on research. This preponderance of legal over research expense in fighting through the patent thicket may be a "warning" to public research institutions that emulating commercial plant breeding practices to produce public goods may be a less an optimal production pathway. Legal advice is not only needed to channel the use of research results in development oriented and commercial markets through contracts that need to be negotiated and concluded and court and settling disputes. Legal considerations are also more and more influencing the research itself. Scientists may be required to use old (free or cheap) technologies instead of effective ones which may be costly or not available.

Scientists frequently feel stifled by the legal advisors who have to make sure that third party IP rights of contributive technologies are respected and that the IP produced by the scientists can be protected, by putting restrictions on scientific communication before a patent application is filed.

Another compliance cost is the need to transfer obligations derived from contracts downstream, i.e., a research institute working in plant breeding with genetic materials that have been obtained through contracts may have to require farmers involved in local testing of potential new varieties to sign contracts restricting their use of the varieties that they obtained (e.g., farmers participating in rice research in the Philippines).

Humanitarian use licenses on individual parts of AKST can reduce these transaction costs to a limited extent since the negotiations that lead to such licenses may be lengthy. One policy option is more generic approaches that limit such costs have been initiated by international consortiums of research institutions forming the "Generation Challenge Program" (Barry and Louwaars, 2005), and those collaborating in PIPRA (Public Intellectual Property Resource for Agriculture).

Application of open-source approaches to genetic technologies (www.bios.net) is a policy option for providing more sustainable solutions to the emerging patent thicket, but its impact is yet limited.

Private, community and national rights. It is not only the private rights (primarily IPRs) that affect the organization of agricultural research for development. Community rights, such as those based on traditional knowledge, and sovereign national rights (on genetic resources based on the CBD) affect research institutions in a similar way. Transaction costs are increased through the need to negotiate access and terms, the opportunities to use the best available inputs in research are reduced, and the use of the research results may be restricted (Safrin, 2004; Louwaars, 2006). Research institutions need to trace all the knowledge, technologies and genetic materials in the various research programs and may have to check at the start of every program or experiment whether third party rights may interfere with their program or experimental design. These institutions may have to consult with legal advisors regarding these rights at every step of making their new technologies available to farmers. One policy option is to expand and strengthen the International Treaty on Plant Genetic resources for Food and Agriculture, the implementation of which is likely to reduce the transaction costs at least for the use of genetic resources of the major field crops and pasture species covered by the Standard Material Transfer Agreement.

Challenges for public research and policy options. Whether or not public research organizations intend to obtain revenue through protecting their own intellectual property, they need to develop institutional policies how to deal with such rights. Such policies need to be supported at the national level of policy and regulation.

Option to strengthen awareness of the issues and professional capacity in IP-strategy and marketing (Erbisch and Fischer, 1998) can focus on three different levels: scientists, research managers and policy makers (Cohen et al., 1999), which often requires the establishment of specialized technology transfer offices (Maredia and Erbisch, 1999).

Above all, national policy makers responsible for agricultural development and the national agricultural research systems need to be aware of the challenges that new rights regimes on intellectual property, traditional knowledge and genetic resources pose in the public research institutes and their relation with an emerging private sector. Policies that reduce public expenditure, that promote the use of IPRs by public research institutions or that restrict access to genetic resources and traditional agricultural knowledge could be based on a thorough understanding of the role of public research in the arena of access, development and use of AKST in development.

Policy options at the national level to make sure that thickets of rights do not develop in technologies and materials that are important for development and sustainability goals particularly include mechanisms to exempt the use of knowledge and materials for use for these goals when these are protected by private, community and national rights.

7.4.4 Rights systems on natural resources: from simple ownership to bundle of rights

Scientific knowledge takes into account the frames through which the real world is perceived by stakeholders, such as scientists (fundamental and applied), local innovators, policy makers, businessmen, negotiators in international arenas. The knowledge on local management systems of natural resources and the theories to which this knowledge refers are the basis upon which decisions and agreements are made. Appropriate AKST can contribute to the improvement of the understanding of what is relevant at the field level and with local situations.

There are a wide variety of rights and management systems for natural resources. For example, one may own the land but not the subsoil resources, or the trees in a forest. A participant in a common property regime may have guaranteed exclusive use of a parcel she has cleared, or that parcel may be subject to reassignment by a tribal elder. An untitled farmer at the agricultural frontier may have what is commonly considered "ownership" of the "improvement" to the land, which may not be *de jure*, but sufficiently enshrined in a *de facto* sense that those improvements can be bought and sold in the market. Some common property regimes have proven to be far more sustainable than individual property regimes. Commons are open access resources, the property of which is not allocated to individuals but supposedly owned in common. Commons are not excludable and are *in se* not rivalrous. (Kaul et al., 1999; Wouters and de Meester, 2003). Strengthening the focus on new rules and international agreements that take into account more complex situations in regard of property rights and regimes is an option.[7]

---

[7] There are many examples of successful management "in common", based on a variety of rights which are used to regulate access to, usage, exploitation, ownership, alienation, exclusion, etc. of such resources. Even though land is a rival and excludable good, many traditional societies maintain nonexclusive grazing and hunting grounds. And some communities effectively manage as commons such natural resources as land, forests, water and plant and animal species (Demsetz, 1967; Bromley, 1990; Barzel, 1997), thus reconfirming that excludable resources do not necessarily have to be made private or exclusive. Doing so is a policy choice.

This issue is raised without further details in Article 10c of the CBD (Sustainable Use of Components of Biological Diversity): "protect and encourage customary use of resources in accordance with traditional cultural practices that are compatible with conservation or sustainable use requirements." This Article is now in legal tribunals by native populations experiencing difficulties with norms they feel are being imposed on them (Goldman, 2004). Such rules may lead to confusion about ownership or accountability of "resources" that have meaning and values at the local as well as at the global levels or aggravate the situation of those who are marginalized by the negotiated rules (Allier, 1997, 2002).

Studies on local management systems can contribute to designing new systems that better fit to evolving and dynamic conditions. Conceptual analyses has greatly benefited from scientific research since the late 1980s (Schlager and Ostrom, 1992; Sandberg, 1994; Le Roy, 1996; Chauveau, 1998; Lavigne Delville, 1998; Karsenty, 2003). Taking into account the different forms of knowledge involved, e.g., "explicit" and "incorporated," can lead to a more complex view of what is at stake in a range of situations (Box 7-1).

The principle of legal plurality facilitates operational understanding of two coexisting legal worlds These normative productions were defined as "*droits de la pratique*", i.e., rights based on practice, as a "plural set based on different ages and particular stakes, actors and formalisms", specifying what is commonly designated as the "law of the land".

A piece of land may be viewed as a "good" while the resources may be seen as "things" free of access or as an "having" (as defined above) open to harvesting by people other than the owner with his/her authorization. All these management practices may be subject to seasonal variation depending on the types of resources to be taken (grass, crops, berries, mushrooms, game, fish, etc.). The "right to hand over" (Chauveau, 1998) between the right of exclusion and of alienation, as hybrid forms of access to land, such as buying land for migrants, which gives them the right to pass it on to their heirs, but not the right to sell it. This traditional order may evolve with time. This system commonly falls within the more general social norms, and follows an intrinsic evolution as a result of overall change in the customary order, and of interactions with the positive law implemented by the modern state. The trend towards commoditization of land and resources will challenge the authority of these different modes and is likely to lead to individual property and ownership as understood by capitalist economy and modern law.

However when the excess capacity of common goods is limited, congestion may turn the consumption of the good as rival, i.e., when an additional unit of the good consumed by one member negatively affects other members' satisfaction of the public good. An example of this situation is the fish-stock in oceans. Overfishing depletes the world's fish-stock and threatens endangered species with extinction (Wouters and de Meester, 2003). Thus a complex set of laws and agreements have completed the UN Convention on the Law of the Sea (1982), which introduced two fundamental principles: (1) the territorial sea, providing a coastal state with the right to control a narrow band of sea as an exten-

---

Box 7-1. Clarifying a bundle of rights.

Analysis of property rights on aquatic resources in Lapland (Schlager and Ostrom, 1992; Sandberg, 1994) specified the notion of access, catch, management, exclusion, alienation according to a cumulative gradient (for instance, the right to alienate includes all other rights). Each right is associated with a category of users, the *proprietor* holding the right of exclusion and the *owner* the right to alienate, which clarifies the distinction between property and ownership. This latter term only encompasses the meaning of absolute private ownership. Thus the distinction between the authorized (who has the right to "harvest") and unauthorized user (who only has right of access) enables detection of tacit rights of free access for small catches. Following this hierarchy of rights, action is organized on three levels: the *constitutional* level in which rights are being constantly elaborated and challenged, the *operational* level on which the rights of access and catches are exercised, and an *intermediary* level at which management and exclusion decisions are taken, and which the authors define as the "collective" level (Schlager and Ostrom, 1992). This denomination throws light on the individual or collective nature of rights and decisions, according to the hierarchical rank of a right but does not, however, specify the collective level concerned for each type of right. The theory of "*maîtrises foncières/fruitières*" (Le Roy, 1996) allows increased genericity on this aspect and better characterization of the diversity of actual arrangements. It enriches the typology by the added dimension of comanagement modes of the holders of these property rights and distinguishes five forms of property rights, i.e. modes of appropriation linked to comanagement modes. These comanagement modes include:

1. "Undifferentiated property rights" with rights of access to "a thing";
2. "Priority property rights" encompassing access and extraction rights (notion of "having");
3. "Exclusive property rights" encompassing the same rights as above plus the right to exclude (notion of "functional property");
4. "Specialized property rights" encompassing rights of access, of extraction and management (notion of "ownership") and;
5. "Absolute exclusive property rights", i.e., the right to use and sell, hand over, etc., therefore to alienate what can consequently be called a "good".

These different types of property rights may be applied to public commons (belonging to all) or appropriated by "one or *n* groups" that are internal or external to a defined community, or even privately appropriated; they rely on how knowledge (on the objects, the interaction with objects and the relationship within people) is shared between the stakeholders.

sion of its sovereignty offshore; and (2) freedom of the high seas, meaning the freedoms of navigation and fishing in the high sea beyond that offshore coastal area (Joyner, 2000).

The first principle relies on the comanagement between states and coastal communities in planning, regulating, and conducting resource management (Borgese, 1999). One of the main issues is the obligation for states to maintain or restore populations of harvested fish at levels that produce a "maximum sustainable yield". "Non-exploitive users", i.e., the rest of society's citizens, also have a right of access to the Exclusive Economic Zone for other functions, which include permission to locate aquaculture installations, mineral mining, shipping access, etc. decisions on which remain with government (Caddy, 1999).

On the second principle, a UN Agreement for the Conservation and Management of Straddling Fish Stocks and Highly Migratory Fish Stocks has been adopted in 1995, mandating states to establish subregional and regional conventions and organizations to facilitate conservation and management of living resources, and an International Seabed Authority for the deep ocean floor and non living marine resources. Except for sedentary species of the sea floor, international fisheries agreements do not speak in terms of ownership of resources but of access rights. This distinction raises the fine point as to the timing of the access and even whether this right could be extended to include the progeny of the resource share in future rights. A corpus of international law has evolved around the 1982 Law of the Sea Convention for protecting and managing the world's oceans (Joyner, 2000), which will likely be extended in the future (Caddy, 1999). Assessments such as the MA, point out that these arrangements are insufficient to avoid a decline of populations of harvested fisheries and 25% of the oceans are overfished, creating problems for both the fish species and the fishermen depending on them. The setting of fishing quotas doesn't take into account the effects of the withdrawal of one species on the functioning of the whole marine ecosystem: it alters not only the targeted fish population but also the other trophic levels concerned by this species as prey or predator. In most situations there is insufficient knowledge on the functioning of marine complex ecosystems to design better management rules.

Challenges for public research and policy options. Scientific knowledge has to help to understand the complexity of such situations, in the oceans as well as on the continents, to formalize these different sets of right regimes and also to design new ways for collective action for the fair implementation of such rights, and reach optimally sustainable management of renewable natural resources. Such knowledge has to guide the design of laws, incentives, contracts, taxes, quotas, permits and licenses that take into account the diversity of situations and that avoid blueprint solutions.

Natural Resources Management Policies. Since state appropriation of NRM based on positive law may coexist with the modalities of local rights systems, which distinguish access to, usage, exploitation, ownership, alienation, exclusion, of "common" goods at a collective level, one option is to recognize that the "law of the land" may further involve land tenure

systems that cannot be reduced to individual ownership. Collective ownership and management of natural resources is protected in Article 10c of the CBD (Sustainable Use of Components of Biological Diversity). Indigenous groups have referenced this Article to help defend their collective rights and NRM practices against governments that would ignore these rights in fulfilling commitments to protect "global" resources. New instruments for collective action have to make explicit and feasible the fair implementation of collective rights and NRM practices in order to obtain the best and sustainable management of renewable natural resources. Formal institutions have to take into account this diversity of NRM knowledge and avoid conforming only to a concept of individual ownership and rights.

7.5 Pro-Poor Agricultural Innovation

7.5.1 Technology supply push and the global agricultural treadmill

The dominant policy model for promoting innovation is called the linear model (Kline and Rosenberg, 1986), or the transfer of technology model (Chambers and Jiggins, 1987; Chapter 2). Also known as "technology supply push," this approach relies on the agricultural treadmill (Cochrane, 1958) i.e., market-propelled waves of technological change that squeeze farm-gate prices, stimulate farmers to capture economies of scale, deliver high internal rates of return to investments in agricultural research (Evenson et al., 1979), but also encourage externalization of significant social and environmental costs (Lal et al., 2005; Mukherjee and Kathuria, 2006).

While the technology push model provided the basis for the positive impacts of the Green Revolution in favorable areas (Castillo, 1998) and under defined conditions that typically included high subsidies on fertilizers and pesticides (Pontius et al., 2002), it has not served nearly as well as resource-poor areas that are highly diverse, rain fed, and risk prone, and that currently hold most of the world's poor (Anderson et al., 1991; Biggs and Farrington, 1991; Vanlauwe et al., 2006).

The market-propelled diffusion of innovations called "the agricultural treadmill" (Cochrane, 1958) has been ongoing in developed market economies for 50 years or more. The literature observing the process for hybrid maize in the American Midwest goes back to 1943 (Ryan and Gross, 1943). During these 50 years, farmers in those economies have been able to capture significant economies of scale. The treadmill process in those economies has been heavily supported in terms of public funding of agricultural research, education and extension, credit subsidies, land and irrigation development, supportive legislation, access to inputs, services and markets, and the evolution of farmers' organizations and their lobbies that represent farmers' interests at state and federal or EU levels. One can now speak of a "global treadmill" that allows farmers in developed economies to export their (sometimes subsidized) products to developing countries and compete with local small-scale farmers.

Value added per agricultural worker in 2003 (constant 2000 US$) in developed market economies was 23,081 with a growth over 1992-2003 of 4.4% (FAO, 2005b). For sub-

Saharan Africa the figures are 327 and 1.4%, respectively. As long as the global treadmill is operating, even with all OECD subsidies removed, efforts to uplift rural poverty will remain severely handicapped and it will continue to be difficult to enlist the vast arable lands in developing countries that are now underperforming and degrading for purposes of global food security. In these circumstances, to continue with a technology-supply push conception of innovation seems inappropriate. The rural poor are not on the global treadmill; instead the global treadmill prevents them from development. Required are institutional framework conditions that provide realistic opportunities to subsistence farmers to become small-scale commercial farmers.

In imperfect markets the benefits are uneven and do not always reach the poor. Policy responses of proven historical efficacy to addressing unevenness in competitiveness and opportunity include institutional framework conditions within which AKST can play a more positive role, i.e., by stimulating targeted investment in creating small farmers' access to market opportunities, inputs, alternative employment and to creating value-adding enterprises and by temporary market protection to infant agro-industries. The contemporary and future challenge is to achieve positive policy outcomes in ways that internalize the environmental and social costs as well.

7.5.2 Brokered long-term contractual arrangements

Brokered long-term contractual arrangements (BLCA; a term used here to designate a suite of modern contractual arrangements) have proven effective in improving the livelihoods of poor farmers and fostering rural innovation (see Box 7-2) (Little and Watts, 1994; Key and Runsten, 1999). However, the set of conditions required for this policy option to be attractive are rather restrictive. BLCAs were initiated to use the good aspects of state trading enterprises (STEs) because STEs proved sensitive to corruption, rent seeking, gender discrimination and externalization of costs to farmers (Hobart, 1994; Dorward et al., 1998). A major challenge facing expanded use of BLCAs as a policy option is to avoid repetition of the historical record that provides ample evidence of the misuse and abuse of nationalized BLCA-like (STE) schemes.

BLCAs, under favorable social conditions with transparency and strong farmer organization, provide a policy option for public sectors to invest in the creation of opportunities for poor farmers. Synergies between long term contractual arrangements and the organic and fair trade markets increase when such types of contractual arrangements are coupled with group certification of small-scale organic producers. Policy options include retooling abolished STEs and creating legal, financial and technical support for emerging new BLCAs that are pro-poor.

7.5.3 Endogenous development and traditional knowledge

Endogenous development draws mainly on locally available resources, local knowledge, culture and leadership, with an openness that allows for integration of outside knowledge and practices (Haverkort et al., 2002; Millar, 2005).

Traditional knowledge can be effective and reliable (Brammer, 1980; Warren et al. 1991; Reij et al., 1996; Brammer, 2000; Balasubramanian and Devi, 2005) with respect

---

Box 7-2. Pineapple export in Ghana.

Ghana traditionally exported Cayenne pineapples. But since 2002, international demand has shifted to the extra sweet MD2 variety with quite dramatic consequences for Ghana's exporters and small-scale producers. Many of the latter quit production altogether, while the former faced loss of their market contracts in Europe unless they could change to MD2. That was no sinecure. An acre requires 22,000 suckers and some of the larger exporters grow hundreds of hectares. Initially tissue culture material from Latin America was imported, but this proved expensive and some mishaps occurred. Then BOMARTS Farms Ltd (about 400 ha pineapple), faced with termination of its contract, decided to set up a commercial tissue culture lab with assistance from scientists of the Department of Botany at the University of Ghana. Millions of plantlets were produced, some of which were sold to commercial producers who in turn could provide their out-growers. MD2 makes many suckers per plant, so that farmers themselves can quickly multiply the variety. At the time, most small-scale producers were not ready to spend money on buying plantlets. The Government stepped in to save Ghana's second largest export crop and contracted BOMARTS to produce over a two-year period 4.8 million plantlets at cost (3 eurocents per plantlet). Twice a week, the Ministry of Agriculture collects 44,000 plantlets and distributes them to farmers through Sea Freight Pineapple Exporters Ghana (SPEG) and Horticultural Association of Ghana (HAG) on credit at a tenth of the price. BOMARTS itself has few out-growers and largely exports its own produce.

At the other extreme are exporters who have no farm operations themselves. The typical setup is a mix with out-growers making a substantial contribution to the consignment of the exporter. Exporting companies make detailed contracts with out-growers, providing inputs on credit, specifying the times of planting, force flowering (uniformity) and harvesting, so that the company has a steady supply. Around harvesting time, the company will inspect and spray the crop and it harvests and transports the fruits. Companies exert very strict quality control (e.g., water content). The sanctions are high: costs of destruction of a rejected assignment in Europe are deducted by importers. Farmers whose crop is rejected have to sell in the local market, often below cost price. At the time of writing, Ghana's pineapple exports are getting back on track and the number of small farmers growing MD2 is rapidly expanding. For many, pineapple is their main source of monetary income.

Source: E. Acheampong.

---

to: (1) knowledge about the agroecosystem and seasonality in which the farmers operate; (2) information about what local people need, want and have capacity for in terms of resources, access to markets; (3) locally adapted technical knowledge and practices and (4) a system view based on having to live by the results.

Farmers may innovate at the system level. For example, farmers on the very densely inhabited Adja Plateau in Benin have developed an "oil palm fallow" rotation that allows them to suppress *Imperata cylindrica*, restore soil fertility for annual crops, and make money from distilling palm wine once the palms are cut down (Brouwers, 1993). But traditional knowledge may have weaknesses such as attributing plant disease to rain and thus foregoing useful management measures (Almekinder and Louwaars, 1999) or an inability to respond to rapidly changing circumstances, e.g., climate change. Experience with multiagent approaches suggests that mobilizing the intelligence of a great many actors to address a new and complex problem can be an effective and efficient way to solve such systemic complexity (Funtowicz and Ravetz, 1993; Gilbert and Troitzsch, 1999).

Policy options for promoting endogenous development include decentralization; use of rapid rural appraisals and participatory approaches; empowerment initiatives; multistakeholder processes; and strengthening farmer organizations. Decentralization as in India or Uganda, however, may strengthen and widen the base for democratic participation in agricultural research decision making, open new opportunities for collaboration in agroenterprise innovations and service delivery, address specific local development problems, and improve responsiveness to the needs of the poor (e.g., SNV and CEDELO, 2004).

Rapid rural appraisal (RRA) and participatory approaches may supply more accurate or insightful information than questionnaire surveys or more relevant or better adapted technologies than the experiments of scientists conducted in conditions and places remote from the fields (e.g., Collinson, 2000). Participation has long been dominant in pro-poor development approaches and may range from simple consultation to support for autonomous decision-making (e.g., Pretty, 1994; Biggs, 1995). RRA and participatory approaches may be poorly performed and insufficient, however, for addressing the multiple scales of policy intervention required (Biggs, 1978; Biggs, 1995; Cleaver, 2001; Cooke and Kothari, 2001). The challenge in meeting development and sustainability goals is to create complementarity that draws on best practice across the range of pro-poor approaches and policies (Biggs, 1982; Biggs, 1989; Bunders, 2001; Ceccarelli et al., 2002; Chema et al., 2003).

Participatory Technology Development (PTD) (Jiggins and De Zeeuw, 1992) is a concrete approach to the design of complementary action that is relevant for achieving development and sustainability goals but has some negatives associated with it. With very small windows of opportunity, it is not easy to reduce poverty by enhancing productivity at the farm level, even through PTD. The challenge is to *stretch* those windows through access to markets, better prices, the development of services, and the removal of extractive practices and patrimonial networks. Given opportunities, West African farmers have time and again considerably increased their production without major technical change. Technology becomes important once framework conditions begin to improve (Box 7-3).

Empowerment. The corollary of recognizing resource-poor farmers as partners in complementary and collaborative approaches to development is to accept their empowerment.

Box 7-3. The Convergence of Sciences Program (CoS) in Ghana and Benin.

To ensure that the research problems chosen were based on the needs and opportunities of resource-poor farmers, CoS pioneered a new pathway for science that used technography, diagnostic studies, and with farmer participatory experimental field research (van Huis et al., 2007). A key component was ex-ante impact assessment and pre-analytical choice making that optimized sensitivity to context and avoided cul-de-sac path dependency. Technography (Richards, 2001) was used to map the coalitions of actors, processes, client groups, framework conditions and contextual factors at a macro level, so as to identify realistic opportunities. Given the small windows of opportunity, technography identified space for change. Diagnostic studies (Nederlof et al., 2004; Röling et al., 2004) ensured that research outcomes would work in the local context, be appropriate to prevailing land tenure, labor availability and gender, and take into account farmers' opportunities, livelihood strategies, culture, and felt needs. The diagnostic studies also identified and established forums of stakeholders for learning from a concrete experimental activity, and gave farmers a say in the design of field experiments. CoS conducted 21 experiments with small farmers on themes such as soil fertility and weed management, crop agrobiodiversity and integrated pest management (IPM). The studies showed that participatory low external input *technology* development within carefully identified windows of opportunity *can* be beneficial. However, the researchers also ran into the limitations of this approach and started to include experiments with creating space for change through *institutional* innovation. Soil fertility improvement depends on land tenure (Saïdou et al., 2007). They negotiated land use rules between migrant farmers and landowners that allowed improving soil management practices. In Ghana, an organization was established to procure Neem seeds from the North as a condition for small-scale cocoa farmers to reduce their use of synthetic pesticides (Dormon et al., 2007). This in turn stimulated collective arrangements for processing Neem seeds because their use in maize mills is unacceptable due to their bitter taste.

With very small windows of opportunity, it is not easy to reduce poverty by enhancing productivity at the farm level, even through PTD. The challenge is to *stretch* those windows through access to markets, better prices, the development of services, and the removal of extractive practices and patrimonial networks. Given opportunities, West African farmers have time and again considerably increased their production without major technical change. Technology becomes important once framework conditions begin to improve.

Source: Hounkonnou et al., 2006; Van Huis et al., 2007.

It can be more efficient to increase farmers' countervailing power than to increase an agency's intervention power through investing in more vehicles, agent training or budget support. Farmer field schools (FFS) (Box 7-4) is an option that warrants further empirical research to determine the conditions under which this may be so and the kinds of policy environment that best enable empowerment strategies to be effective in meeting development and sustainability goals. (Van den Berg and Jiggins, 2007, for a review and assessment of IPM FFS literature).

Multistakeholder processes. A special participatory approach is the facilitation of multistakeholder processes (Leeuwis and Pyburn, 2002; Wals, 2007). Especially in resource dilemmas, where different categories of interdependent stakeholders make competing claims on common pool resources, sustainable solutions cannot come from regulation, technology or market interventions only. The way forward is a facilitated process of negotiation, shared (social) learning, and agreement on concerted action, based on trust, fairness and reciprocity. There is increasing evidence that humans are capable of agreeing on sustainable solutions and of creating institutional conditions that support the implementation of such solutions if drawn into appropriate knowledge processes (e.g., Ostrom et al., 1992; Blackmore et al., 2007). Multistakeholder processes increasingly are important with respect to climate change adaptation, when agreements have to be reached to avoid crisis or when loss of ecosystem services becomes a key cause of poverty.

The Chain-Linked Model. Commercial innovation studies give a central place to the entrepreneur who sees a possibility to capture an opportunity by mobilizing resources, including knowledge (Kline and Rosenberg, 1986). The driver of innovation in these situations typically is the entrepreneur spotting or creating market-related or social organizational opportunity. Policy support to innovation in these cases is provided by helping entrepreneurs to access specialized sources of knowledge, services and skills (Coehoorn, 1994; Crul, 2003). International experience of supporting innovation in small and medium enterprises in nonfarming sectors can be useful in guiding pro-poor agricultural enterprise development.

Strengthening farmer organizations. Investing in people's organizations is a policy option (Toulmin, 2005) with a long history. The experience of the USA and Europe shows that strong farmers' organizations can be a necessary condition for commercially efficient agricultural development (Bigg and Satterthwaite, 2006). An African example is provided by ROPPA in West Africa (Koning and Jongeneel, 2006; ROPPA, 2006). Organizations such as AGRITERRA in The Netherlands attempt to strengthen farmers' organizations in developing countries through training, delegating research funds to farmers' organizations, and building farmers' capacities as effective partners in the negotiation of contracts as well as in research-priority setting. Since farmers' organizations need allies in other sectors or at other levels if they are to become strong and act effectively in collaborative AKST partnerships (Wennink and Heemskerk, 2006) it is a useful policy option to invest in "platforms" (or organized

---

Box 7-4. Farmer field schools.

The invention of the Farmer field schools (FFS) by the Indonesian FAO team that introduced IPM in rice after the emergence of the Brown Planthopper was an enormous breakthrough, given the prevalence of the TandV system of extension at the time (Pontius et al., 2002). The FFS turned the linear model upside down: instead of ultimate users, farmers became experts; technology transfer was replaced by experiential learning; and instead of teaching content up front, the agent stayed in the back and facilitated the process. Evaluations of FFS programs (Van de Fliert, 1993; Van den Berg, 2003) indicate that FFS participants increase their productivity, reduce pesticide use, lower costs, and show remarkable signs of empowerment, in terms of speaking in public, organizational skills, and self-confidence. The effect is so remarkable that the most effective ways to convince politicians and senior civil servants of FFS impact is to expose them to an FFS in action. Such visitors quickly grasp what the FFS can do in terms of enlisting the elusive small-scale farmer in the national project.

It is one thing to implement an effective FFS pilot, quite another to scale it up to the national level. A certain set of practices determines FFS quality. Erosion of these practices soon leads to loss of fidelity and loss of the remarkable effects. Vulnerabilities include the curriculum (e.g., use of a field as the main tool for teaching), process facilitation (e.g., avoiding reverting to technology supply push or promoting government agendas), training facilitators in non-directive methods, timeliness (i.e., coinciding with the growing season), financing (e.g., utilizing public funds for snacks for farmers). FFS programs are vulnerable to corruption by the pesticide industry (e.g., Sherwood, 2005).

The FFS does not fit a bureaucratic, centralized, hierarchical government system. The FFS is a form of farmer education rather than a form of extension, which is not "fiscally sustainable" in the short term (Feder et al., 2004).

---

social arenas) where farmers and researchers can meet on a level playing field. The inclusion of small farmers' representatives on such platforms (as in the PRODUCE foundations in Mexico) may require special effort but may still end up favoring those with sufficient assets to seize commercial opportunity.

One of the persistent experiences in agricultural development is that, while it can be relatively easy to promote pro-poor endogenous development, collaborative AKST partnerships and the mobilization of indigenous knowledge in pilot projects, the prevailing governance conditions make it difficult to scale up and embed successful pilot experiences in routine institutional behaviors The difficulty in part lies in social realities that position power and opportunity as highly contested zero-sum contests. In 1986, when Java's rice fields were devastated by resurgent waves of brown plant hoppers (BPH) resistant to pesticides destroyed the natural enemies or predators of the BHP, it took considerable time for the government to respond. The problem was a principle called "asal bapak senang" that may be translated as "as long as father is happy" (with the sense of "to avoid upsetting your boss with negative information"). At each level in the hierarchy, the bad news about the devastation in the rice fields was watered down. It was only when the people from his own village came to the President directly to ask for help that he learned that something was seriously amiss. In our assessment, policy initiatives that aim at empowerment and endogenous development would be most accepted where democratic forms of government and a strong civil society exist; most poor people live in countries where these conditions are not present.

7.5.4 Innovation systems (IS)

Innovation is the emergent property of the interaction among organizations and people who make the complementary contributions required for innovation to take place (Röling and Engel, 1991; Bawden and Packam, 1993). The configuration of actors is not fixed (Engel and Salomon, 1997).

The empirical research of successful and innovative economies that stimulated the recent interest in innovation Systems has found that "the essential determinant of innovation appeared to be that the suppliers of new knowledge were intimately engaged with the users of that knowledge" (Barnett, 2006).

Older traditions of systems thinking and practice (e.g., Checkland, 1981; Checkland with Scholes, 1990) drew attention to linkages, relationships, interfaces, conflicts, convergence, and reciprocity in innovation processes. The application of such thinking and practice to pro-poor development in agriculture has been stimulated also by the evidence that it appears to be suited to dealing with the kind of institutional development that The New Institutional Economics (North, 2005) sees as a precursor to growth.

The "innovation systems" approach in recent years has become an ex-ante policy model (World Bank, 2007) that draws on the aforementioned traditions as well as on empirical research on the emergence of Asian economies. Such models are an increasingly important tool for stimulating innovation at the interface of agriculture, sustainable natural resource management and economic growth, for instance in the context of the EU's Water Framework Directive (e.g., Blackmore et al., 2007) and Land Care and more recently Catchment Management Authorities in Australia (Campbell, 1994). These experiences also show up the weakness of the IS approach: absent appropriate enabling policy frameworks and economic drivers at higher system scales, successful lower scale innovations can peter out or become

frustrated. The lessons may be linked to the widespread confidence that rational choice theory offers an appropriate foundation for policy designed to support innovation; the empirical evidence suggests to the contrary that, given the public good character of development and sustainability goals, policies based on an understanding of the role of collective management in innovation processes may be more appropriate (Ostrom et al., 1993; Gunderson et al., 1995).

Conditions under which the policy options may be conducive to meeting development and sustainability goals. The following concrete steps have been proposed to make an innovation systems approach work in resource-poor environments (see Tripp, 2006; McCann et al., 2006; Van Huis and Houkonnou, 2007):

- Public, private and civil society agencies identify a number of priority themes based on national plans, or poverty reduction strategies;
- For each theme, rapid appraisal of agricultural knowledge systems (RAAKS) (Engel and Salomon, 1997) or other methods are used to identify configurations of stakeholders (including researchers, farmer organizations, etc.) that constitute promising innovation systems. Such configurations include actors at the both the national and the decentralized local government level;
- Key representatives of these stakeholders are facilitated to form a "Community of Practice" (COPs) (Wenger, 1998) at decentralized (e.g., district) and national levels, where the national level has the power and ability to create conducive institutional framework conditions for the concrete activities at the decentralized level. An IS approach thus requires trained facilitators who operate within a national mandate that recognizes the importance of IS;
- For each COP, diagnostic studies identify concrete opportunities that can be realized through concerted action by the stakeholders;
- Each COP submits proposals to a national fund set up for this purpose;
- Each COP is monitored to allow national learning about the IS approach as a basis for staff training and increasing management effectiveness.

The IS approach assumes considerable political will and an understanding of processes that cannot be captured by hierarchy and market since creating windows of opportunity for small-scale producers will require new kinds of institutional innovation (Egelyng, 2000).

References

Ackerman, F. 2005. The shrinking gains from trade: A critical assessment of Doha Round projections. Global Dev. Environ. Inst. Tufts Univ., Medford MA.

African Group. 2006. Communication to the Committee on Agriculture. TN/AG/GEN/18. June 7, 2006. World Trade Organization, Geneva.

Alix-Garcia, J., A. de Janvry, E. Sadoulet and J.M. Torres. 2005. An assessment of Mexico's payment for environmental services program. Prepared for FAO, Rome. Available at http://are.berkeley.edu/~sadoulet/papers/FAOPES-aug05.pdf. Univ. California, Berkeley.

Allier, J.M. 1997. Ecological economics (energy, environment and society). Blackwell, Oxford.

Allier, J.M. 2002. The environmentalism of the poor — A study of ecological conflicts and valuation. Edward Elgar, Cheltenham.

Almekinder, C., and J. Hardon (ed) 2006. Bringing farmers back into breeding. Experiences with participatory plant breeding and challenges for institutionalisation. Agromisa Special 5, Agromisa, Wageningen.

Almekinder, C. and N. Louwaars (ed) 1999. Farmers' seed production: A new handbook. ITDG Publ., London.

Alston, J.M., P.P. Pardey, and R.R. Piggott. 2006. Synthesis of themes and policy issues. p. 361-372. In P.P. Pardey (ed) Agricultural R&D in the developing world: Too little, too late? IFPRI, Washington DC.

Alston, J.M., and R.J. Venner. 2000. The effects of the US Plant Variety Protection Act on wheat genetic improvement. EPTD Disc. Pap. 62. IFPRI, Washington DC.

Anderson, K., and W. Martin. 2005. Agricultural trade reform and the Doha development agenda. World Bank, Washington DC.

Anderson, K., W. Martin, and D. van der Mensbrughhe. 2005. Market and welfare implications of Doha reform scenarios. In K. Anderson and W. Martin (ed) Agricultural trade reform and the Doha development agenda. World Bank, Washington DC.

Anderson, R.A., E. Levy, and B.M. Morrison. 1991. Rice science and development politics. Research strategies and IRRI's technologies confront Asian diversity (1950-1980). Clarendon Press, Oxford.

Angelsen, A., and D. Kaimowitz. 1999. Rethinking the causes of deforestation: Lessons from economic models. World Bank Res. Observ. 14:73-98

Angulu, J.A , J.A. Nunnery, and H.D. Bair. 2004. Antimicrobial resistance in zoonotic enteric pathogens. Rev. Sci. Tech. 23:485-96.

Antle, J.M., and S. Mooney. 2002. Designing efficient policies for agricultural soil carbon sequestration. p. 323-336. In J. Kimble (ed) Agriculture practices and policies for carbon sequestration in soil. CRC Press, Boca Raton FL.

Asamblea Legislativa de la República de Costa Rica. 2006. Ley para el desarrollo, promoción y fomento de la actividad agropecuaria orgánica. Expediente No. 16.028. Texto Final Aprobado. Primer debate: 5 de sept. Segundo debate: 7 de sept 2006.

Baffes, J. 2006. Cotton and the developing countries: Implications for developing countries. p. 119-128. In R. Farmer (ed) Trade, Doha and development. Available at http://siteresources. worldbank.org/INTRANETTRADE/ Resources/239054-1126812419270/9. Cotton&.pdf. World Bank, Washington DC.

Balasubramanian, A.V., and T.D. Nirmala Devi (ed) 2005. Traditional knowledge systems of India and Sri Lanka. Centre for Indian Knowledge Systems, Chennai.

Barboza, D. 2007. Food safety crackdown in China. The New York Times. June 28.

Barnett, A. 2006. Innovations: Lessons from the UK funded crop post-harvest research. p. 48-50 In RAWOO knowledge makes a difference. Science and the Millennium Development Goals. Report of a Seminar.

The Hague RAWOO Sept. 2006. Publ. 30.

Barry, G., and N. Louwaars. 2005. Humanitarian licenses: Making proprietary technology work for the poor. p. 23-34. In N. Louwaars (ed) Genetic resource policies and the generation challenge programme. Available at http://www.generationcp. org/latestnews/GenResPol_and_GCP.pdf. Generation Challenge Program, CIMMYT, Mexico.

Barzel, Y. 1997. Economic analysis of property rights. 2nd ed. Cambridge Univ. Press, UK.

Barsotti, C. 2004. Level playing field. Cartoon. Nov. 22. The New Yorker, New York.

Bauer, C.J. 1997. Bringing water markets down to earth: The political economy of water rights in Chile, 1976-1995. World Dev. 25:639-656.

Bauer, C.J. 2005. Siren song: Chilean water law as a model for international reform. RFF Press, Washington DC.

Bawden, R.J., and R. Packam. 1993. Systems praxis in the education of the agricultural systems practitioner. Paper presented at the 1991 Ann. Meeting Int. Soc. Systems Sciences. Östersund, Sweden. Systems Practice 6:7-19.

Berdegué, J.A., F. Balsevich, L. Flores, and T. Reardon. 2005. Central American supermarkets private standards of quality and safety in procurement of fresh fruits and vegetables. Food Policy 30(3):254-269.

Bhagwati, J. 2005. In defense of globalization. Oxford Univ. Press, NY.

Bhattacharya, S.C. 2002. Biomass energy in Asia: A review of status, technologies and policies in Asia. Energy Sust. Dev. 6:5-10.

Bigg, T., and D. Satterthwaite (ed) 2005. How to make poverty history: The central role of local organisations in meeting the MDGs. IIED, London.

Biggs, S.D. 1978. Planning rural technologies in the context of social structures and reward systems. J. Agric. Econ. 29:257-274.

Biggs, S.D. 1982. Generating agricultural technology: Triticale for the Himalayan hills. Food Policy 7:69-82.

Biggs, S.D. 1989. Resource-poor farmers' participation in research: a synthesis of experiences from nine national agricultural research systems. OFCOR Comparative Study Pap. 3. ISNAR, The Hague.

Biggs, S.D. 1995. Participatory technology development: A critique of the new orthodoxy, p. 1-10. In Two articles focusing on participatory approaches. AVOCADO Series 06/95. Olive Organization Development and Training, Durban.

Biggs, S., and J. Farrington. 1991. Agricultural research and the rural poor. A review of social science analysis. IDRC, Ottawa.

Bisaillon, V., C. Gendron, and M.-F. Turcotte. 2006. Fair trade and the solidarity economy: The challenges ahead. Ecole des Sciences de la Gestion, Univ. Montreal, Canada.

Blackmore, C., R. Ison, and J. Jiggins (ed) 2007. Social learning: An alternative policy instrument for managing in the context of

Europe's water. Special Issue. Environ. Sci. Policy 10(6).

Borgese, E.M. 1999. Global civil society: Lessons from ocean governance. Futures 31:983-991.

Bouet, A., J.-C. Bureau, Y. Decreux, and S. Jean. 2004. Multilateral agricultural trade liberalization: The contrasting fortunes of developing countries in the Doha Round. Working Pap. No. 2004-18. Centre d'Études Prospectives et d'Informations Internationales, Paris.

Boyce, J.K. 1999. The globalization of market failure? International trade and sustainable agriculture. Political Economy Research Institute (PERI), Amherst MA.

Brammer, H. 1980. Some innovations do not wait for experts: A report on applied research by Bangladesh peasants. CERES 13:24-28.

Brammer, H. 2000. Agroecological aspects of agricultural research in Bangladesh. Univ. Press, Dhaka.

Bromley, D.W. (ed) 1990. Essay on the commons. Univ. Wisconsin Press, Madison.

Brouwers, J. 1993. Rural people's knowledge and its response to declining soil fertility. The Adja case (Benin). Wageningen Papers. Agric. Univ., Wageningen.

Brückner, G.K. 2004. Working towards compliance with international standards. Rev. Sci. Tech. 23:95-107.

Bruns, B.R., C. Ringler, and R.S. Meinzen-Dick (ed) 2005. Water rights reform: Lessons for institutional design. IFPRI, Washington DC.

Buckwell, A., and S. Armstrong-Brown. 2004. Changes in farming and future prospects. Tech. Policy 146(Supp. 2):14-21.

Bunders, J. 2001. Utilization of technological research for resource-poor farmers: the need for an interactive innovation process. p. 27-37. In Utilization of research for development cooperation: Linking knowledge production to development policy and practice. Publ. 21. Netherlands Dev. Assistance Res. Council (RAWOO), The Hague.

Byrne, D. 2004. The impact of EU sanitary and phytosanitary legislation on developing countries. Meeting with World Bank Executive Directors Washington DC. SPEECH/04/139. Available at http.//www .eurunion.org/news/speeches/2004/040318db .htm.

CA (Comprehensive Assessment on Water Management in Agriculture). 2007. D. Molden (ed) Water for food, water for life: A comprehensive assessment of water management in agriculture. Earthscan, London and IWMI, Colombo.

CAC (Codex Alimentarius Commission). 2002. Report of the evaluation of the Codex Alimentarius and other FAO and WHO Food Standards Work (ALINORM 03/25/3). CAC, Rome.

Caddy, J.F. 1999. Fisheries management in the twenty-first century: Will new paradigms apply? Rev. Fish Biology Fish. 9:1-43.

Callon, M. 1994. Is science a public good? Sci. Tech. Human Values 19(4):395-424.

Campos, M.T., and D.C. Nepstad. 2006. Smallholders, the Amazon's new conservationists. Conserv. Biol. 20(5): 1553-1556.

Campbell, A. 1994. Landcare. Communities shaping the land and the future. Allan and Unwin, St Leonards, Australia.

Castillo, G.T. 1998. A social harvest reaped from a promise of springtime: User-responsive participatory agricultural research in Asia. Chapter 11, p. 191-214. In N. Röling and M. Wagemakers (ed) Facilitating sustainable agriculture. Participatory learning and adaptive management of environmental uncertainty. CUP, Cambridge.

CEC. 2006. Proposal for a directive of European Parliament and of the Council establishing a framework for the protection of soil and amending Directive 2004/35/EC. Commission of the European Communities, Brussels.

Ceccarelli, S., D.A. Cleveland, and D. Soleri. 2002. Farmers, scientists, and plant breeding. Integrating knowledge and practice. Univ. California Press, Santa Barbara.

Chambers, R., and J. Jiggins. 1987. Agricultural research for resource-poor farmers. Part I: Transfer-of technology and farming systems research. Part II: A parsimonious paradigm. Agric. Admin. Extension 27:35-52, 109-128.

Chan, K.M.A., M.R. Shaw, D.R. Cameron, E.C. Underwood, and G.C. Daily. 2006. Conservation planning for ecosystem services. PLoS 4(11):e379.

Chauveau, J.-P. 1998. La logique des systèmes coutumiers. p. 66-75 In P. Lavigne Delville (ed) Quelles politiques foncières pour l'Afrique rurale?, Karthala — Coopération française, Paris.

Checkland, P. 1981. Systems thinking, systems practice. John Wiley, Chicester.

Checkland, P., and J. Scholes. 1990. Soft systems methodology in action. John Wiley, Chicester.

Chema, S., E. Gilbert, and J. Roseboom. 2003. A review of key issues and recent experiences in reforming agricultural research in Africa. Res. Rep. 24. ISNAR, The Hague.

Chomitz, K.M. 2007. At loggerheads? Agricultural expansion, poverty reduction, and environment in the tropical forests. World Bank, Washington DC.

Chomitz, K.M., G.A.B. Da Fonseca, K. Alger, D.M. Stoms, M. Honzák, E. Charlotte Landau et al. 2006. Viable reserve networks arise from individual landholder responses to conservation incentives. Ecol. Society 11(2):40. Available at http://www.ecologyandsociety.org/vol11/iss2/art40/.

Clark, G. M., A.F. Kaufmann, E.J. Gangarosa, and M.A. Thompson.1973. Epidemiology of an international outbreak of Salmonella agona. Lancet 2:490-3.

Clayton, C. 2007. USDA eyed for food inspection. DTN. April 26. Available at http://www.dtn.com

Cleaver, F. 2001. Institutions, Agency and the limitations of participatory approaches to

development. p. 36-55. In B. Cooke and U. Kothari (ed) Participation: The new tyranny. Zed Books, London.

Cochrane, W.W. 1958. Farm prices, myth and reality. Univ. of Minnesota Press, Minneapolis.

Coehoorn, C.A. 1994. The Dutch innovation centres: Implementation of technology policy or facilitation of small enterprises. Ph.D. thesis, RUG, Groningen.

Coelho, S.T. 2005. Biofuels — advantages and trade barriers. Working Pap. UNCTAD/DITC/2005/1. UNCTAD, Geneva.

Cohen, J.I., C. Falconi, J. Komen, S. Salazar, and M. Blakeney, 1999. Managing proprietary science and institutional inventories for agricultural biotechnology. p. 249-260 In J.I. Cohen (ed) Managing agricultural biotechnology: Addressing research programme needs and policy implications. Biotechnol. Agric. Ser. No. 23. CABI, Wallingford.

Collinson, M. (ed) 2000. History of farming systems research. FAO and CABI, Wallingford.

Commission on Intellectual Property Rights. 2002. Integrating intellectual property rights and development policy. Rep. Commission on Intellectual Property Rights. Available at www.iprcommission.org

Cooke, B., and U. Kothari (ed) 2001. Participation: The new tyranny. Zed Books, London.

Cooperative Bank. 2003. The ethical consumerism report. The Cooperative Bank, London.

Correa, C.M. 2004. Bilateral investment agreements: Agents of new global standards for the protection of intellectual property rights Available at http://www.grain.org/briefings/?id=186

Cortese, A. 2003. They care about the world (and they shop, too). New York Times, 20 July.

Crul, M. 2003. Eco-design in Central America. PhD thesis. Technical Univ., Delft.

Crump, J.A., P.M. Griffin, and F.J. Angulo. 2002. Bacterial contamination of animal feed and its relation to human foodborne illness. Clin. Infect. Dis. 35:859-65.

De Fraiture, C., X. Cai, U. Amarasinghe, M. Rosegrant, and D. Molden. 2004. Does cereal trade save water? The impact of virtual water trade on global water use. Comprehensive Assessment of Water Management in Agriculture Research Report 4. IWMI, Colombo.

De Fraiture, C., M. Giordano, and L. Yongsong. 2007a. Biofuels: Implications for agricultural water use. IWMI, Colombo.

De Fraiture, C., D. Wicheins, J. Rockstrom, and E. Kemp-Benedict. 2007b. Looking ahead to 2050: Scenarios of alternative investment approaches. In D. Molden (ed) Water for food, water for life: A comprehensive assessment of water management in agriculture. Earthscan, London and IWMI, Colombo.

Demsetz, H., 1967. Towards a theory of property rights. Am. Econ. Rev. 57(2): 347-359.

DFID. 2007. Balancing the cost of food air miles: Listening to trade and environmental concerns. September. Available at http://www.dfid.gov.uk/news/files/foodmiles.asp

DG SANCO. 2006. Evaluation of the Community Animal Health Policy (CAHP) 1995-2004 and alternatives for the future. Tender No 2004/ S 243-208899. Available at http://ec.europa.eu/food/animal/diseases/strategy/cahp_termsref_en.pdf. European Commission, Brussels.

DG SANCO. 2007. A new animal health strategy for the European Union (2007-2013) where prevention is better than cure. Available at http://ec.europa.eu/food/animal/diseases/strategy/index_en.htm. European Commission, Brussels.

Diao, X., and P. Hazell. 2004. Exploring market opportunities for African smallholders. 2020 Africa Conf. Brief No. 6. IFPRI, Washington DC.

Diao, X., M. Johnson, S. Gavian, and P. Hazell. 2005. Africa without borders: Building blocks for regional growth. IFPRI Issue Brief 38, Washington DC. Available at www.ifpri.org/pubs/ib/ib38.pdf.

Díaz-Bonilla, E., X. Diao, and S. Robinson. 2003. Thinking inside the boxes: Protection in the development and food security boxes versus investments in the green box. In E. Diaz-Bonilla et al., WTO negotiations and agricultural liberalization: The effect of developed countries' policies on developing countries. CABI Publ., UK.

Díaz-Bonilla, E., M. Thomas, and S. Robinson. 2002. On boxes, contents, and users: Food security and WTO negotiations. TMD Disc. Pap. 82. IFPRI, Washington DC.

Dicaprio, A., and K.P. Gallagher. 2006. The shrinking of development space: How big is the bite? In J. World Investment Trade 7(5).

Diouf, J. 2007. Biofuels should benefit the poor, not the rich. The Financial Times. 15 Aug.

Dinar, A. (ed) 2000. The political economy of water pricing reforms. Oxford Univ. Press, NY.

Dorward A., J. Kydd, and C. Poulton (ed) 1998. Smallholder cash crop production under market liberalisation: A new institutional economics perspective. CABI, Wallingford.

Dorward, A., J. Kydd, J. Morrison, and I. Urey. 2004. A policy agenda for pro-poor agricultural growth. World Dev. 32(1):73-89.

Drahos, P. 2006. Towards an international framework for the protection of traditional group knowledge and practice. UNCTAD-Commonwealth Sec. Workshop. 2-4 Feb 2006. UNCTAD, Geneva.

Easter, K.W., M.W. Rosegrant, and A. Dinar (ed) 1998. Markets for water: Potential and performance. Kluwer Academic Publishers, Boston.

EFSA. 2006a. Community summary report on trends and sources of zoonoses, zoonotic agents and antimicrobial resistance in the European Union in 2005. The European Food Safety Authority. The EFSA J. 94.

EFSA. 2006b. Opinion of the Scientific Panel on Biological Hazards (BIOHAZ) and of the Scientific Panel on Animal Health and Welfare (AHAW) on Review of the Community Summary Report on trends and sources of zoonoses, zoonotic agents and antimicrobial resistance in the European Union in 2004 (EFSA-Q-2006-050/051).

Egelyng, H. 2000. Managing agricultural biotechnology for sustainable development: the case of semi-arid India. Int. J. Biotech. 2(4).

Egelyng, H. 2005. Evolution of capacity for institutionalized management of intellectual property at International Agricultural Research Centers: A strategic case study. AgBioForum 8(1):7-17.

Egelyng, H., and H. Høgh-Jensen. 2006. Towards a global research programme for organic research and farming. p. 232-342 *In* N. Halberg et al. (ed) Global development of organic agriculture: Challenges and promises. CABI, Wallingford.

Engel, P.G.H., and M. Salomon. 1997. Facilitating innovation for development. A RAAKS Resource Box. KIT, Amsterdam.

Erbisch, F.H., and A.J. Fischer. 1998. Transferring intellectual properties. p. 31-48. *In* F.H. Erbisch and F.M. Maredia (ed) Intellectual property rights in agricultural biotechnology. CABI, Wallingford.

ETC Group. 2005. The potential impacts of nano-scale technologies on commodity markets: The implications for commodity dependent developing countries. November. ETC and The South Centre, Canada.

Evenson, R.E., P.E. Waggoner and V.W. Ruttan. 1979. Economic benefits from research: An example from agriculture. Science 205:1101-1107.

Eyhorn, Frank. 2007. Organic farming for sustainable livelihoods in developing countries? The case of cotton in India. VDF, Zürich.

FAO. 1991. Integrated livestock-fish production systems. *In* Proc. FAO/IPT Workshop on Integrated Livestock-Fish Production Systems. Inst. Advanced Studies, Univ. Malaya, Kuala Lumpur, 16-20 Dec. 1991.

FAO. 1998. State of the world's plant genetic resources for food and agriculture. FAO, Rome.

FAO. 2002. Improved animal health for poverty reduction and sustainable livelihoods. Animal Prod. Health Pap. 153. FAO, Rome.

FAO. 2003. Biosecurity in food and agriculture. COAG/2003/9. FAO, Rome.

FAO. 2004a. Animal health special report avian influenza - related issues: Socioeconomic costs. WATT poultry global e-news. FAO, Rome.

FAO. 2004b. Assessing carbon stocks and modeling win-win scenarios of carbon sequestration through land-use changes. FAO, Rome.

FAO. 2005a. State of agricultural commodities markets 2004. FAO, Rome.

FAO. 2005b. The state of food and agriculture. Available at www.fao.org/docrep/008/a0050e/a0050e10.htm. FAO, Rome.

FAO. 2005c. Special event on impact of climate change, pests and diseases on food security and poverty reduction: Background document. 31st Sess. Committee World Food Security. FAO, Rome.

FAO. 2006a. State of world aquaculture 2006. Fisheries Tech. Pap. No. 500. Rome. Available at http://www.fao.org/docrep/009/a0874e/a0874e00.htm. Fisheries Dep., FAO, Rome.

FAO. 2006b. Water desalination for agricultural applications. *In* J. Martínez Beltrán and S. Koo-Oshima (ed) Proc. FAO expert consultation on water desalination for agricultural applications. Rome, 26-27 April 2004. FAO, Rome.

FAO. 2006c. The state of agricultural commodities markets 2006. Rome, FAO,

FAO. 2006d. Capacity building for surveillance and control of zoonotic diseases. FAO/WHO/OIE Expert and Technical Consultation. Rome, 14-16 June 2005.

FAO. 2007. Organic agriculture and food security. Int. Conf. Rome, 3-5 May 2007. Available at ftp://ftp.fao.org/docrep/fao/meeting/012/j9918e.pdf. FAO, Rome.

FAO/IFAD. 2006. Water for food, agriculture and rural livelihoods. Chapter 7 of the 2nd UN World Water Development Report: Water, a shared responsibility. FAO and IFAD, Rome.

FAOSTAT. 2005. Available at http://faostat.fao.org/site/584/default.aspx.

Farina, E.M.M.Q. 1999. Challenges for Brazil's food industry in the context of globalization and Mercosul consolidation. Int. Food Agribusiness Manage. Rev. 2:315-330.

Faurès, J.-M., M. Svendsen, and H. Turral. 2007. Reinventing irrigation. *In* D. Molden (ed) Water for food, water for life: A comprehensive assessment of water management in agriculture. Earthscan, London and IWMI, Colombo.

Ferris, R.S.B, and P. Robbins. 2003. The coffee conundrum: Retail prices rising, farmers income falling. ACIAR Newsl. March.

Finger, J.M. 2004. Introduction and overview. p. 1-36. *In* J.M. Finger and P. Schuler (ed) Poor people's knowledge: Promoting intellectual property in developing countries. World Bank and Oxford Univ. Press, Washington DC.

Fink, C., and P. Reichenmiller. 2005. Tightening TRIPs: The intellectual property provisions of recent U.S. Free Trade Agreements. Trade Note No. 20. World Bank, Washington DC.

Fischer, K., and D. Byerlee. 2002. Managing intellectual property and income generation in public research organizations. p. 227-244. *In* D. Byerlee and R. Echeverría. Agricultural research policy in an era of privatization. CABI, Oxon.

Flint, J.A., Y.T. Van Duynhoven, F.J. Angulo,

S.M. DeLong, P. Braun, M. Kirk et al. 2005. Estimating the burden of acute gastroenteritis, foodborne disease and pathogens commonly transmitted by food: An international review. Clinical Infect. Dis. 41:698-704.

FLO. 2007. Shaping global partnerships. Annual Report 2006/7. Fairtrade Labelling Organizations Int. (FLO).

FLO. 2008. About FLO. Available at http://www.fairtrade.net/sites/aboutflo. Fairtrade Labelling Organizations Int. (FLO).

Fowler, C., M. Smale, and S. Gaiji. 2001. Unequal exchange? Recent transfers of agricultural resources. Dev. Policy Rev. 19(2):181-204.

Friends of the UNEO. 2007. Presidency overview report. Agadir, 12-13 April. Available at www.reformtheun.org/index.php?module=uploads&func=download&fileId=2277.

Fuller, S. 1993. Knowledge as product and property. p.157-190. *In* N. Stehr and R. Ericson (ed) The culture and power of knowledge. Walter de Guyter, Berlin.

Funtowicz, S.O., and J.R. Ravetz 1993. Science for the post-normal age. Futures 25(7):739-755.

Garrett, K.A., S.P. Dendy, E.E. Frank, M.N. Rouse, and S.E. Travers. 2006. Climate change effects on plant disease: Genomes to ecosystems. Ann. Rev. Phytopathol. 44: 489-509.

GEF. 2006. Evaluation of GEF support for biosafety. Global Environment Facility (GEF), Washington DC.

George, C., and C. Kirkpatrick. 2003. Sustainability impact assessment of world trade negotiations: Current practice and lessons for further development. Impact Assessment Research Center, Inst. Dev. Policy and Management (IDPM), Univ. Manchester. UK.

Gervais, Daniel. 2005. Traditional knowledge and intellectual property: A TRIPS compatible approach. Michigan State Law Rev. Spring.

Ghosh, D., A.D. Sagar, and V.V.N. Kishore. 2006. Scaling up biomass gasifier use: An application-specific approach. Energy Policy 34:1566-1582.

Gilbert, N., and K. Troitzsch. 1999. Simulation for the social scientist. Open U. P., Buckingham.

Gilbert, M., P. Chaitaweesub, T. Parakamawongsa, S. Premashthira, T. Tiensin, W. Kalpravidh et al. 2006. Free-grazing ducks and highly pathogenic avian influenza, Thailand. Emerg. Infect. Dis. 12(2):227-34.

Goldman, M. 2004. Eco-governmentality and other transnational practices of a 'green' World Bank. p. 166-192. *In* R. Peet and M. Watts (ed) Liberation Ecology. Routledge, London and New York.

Gunderson, L.H., C.S. Hollings and S. Light (ed) 1995. Barriers and bridges to the renewal of ecosystems and institutions. Colombia Press, NY.

Hahn, R.W., S.M. Olmstead, and R.N. Stavins.

2003. Environmental regulation during the 1990s: A retrospective analysis. Harvard Environ. Law Rev. 27:377-415.

Halberg, N., H. Alroe, M. Knudsen, and E. Kristensen (ed) 2006. Global development of organic agriculture: Challenges and promises. CABI, UK.

Hald, T, A. Wingstrand, T. Brondsted, and D.M. Wong. 2006. Human health impact of salmonella contamination in imported soybean products: A semiquantitative risk assessment. Foodborne Pathogens Dis. 3:422-431.

Hall, A.J., V. Rasheed Suliaman V., N.G. Clark, and B. Yoganand. 2003. From measuring impact to learning institutional lessons: An innovation systems perspective on improving the management of international agricultural research. Agric. Syst. 78:213-241.

Hatanaka, M., C. Bain, and L. Busch. 2005. Third-party certification in the global agrifood system. Food Policy 30:354-369.

Haverkort, A.W., K. Van't Hooft, and W. Hiemstra (ed) 2002. Ancient roots, new shoots, endogenous development in practice. Zed Books, London.

Heal, G.M., G. Walker, S. Levine. 2004. Genetic diversity and interdependent crop choices in agriculture. Resourc. Energy Econ. 26(2):175-184.

Heller, M., and R. Eisenberg, 1998. Can patents deter innovation? The anticommons in biomedical research. Science 280:698-701.

Herdt, R.W., and J.W. Mellor. 1964. The contrasting response of rice to nitrogen: India and the United States. J. Farm Econ. 46: 150-160.

Hobart, M. (ed) 1994. An anthropological critique of development: The growth of ignorance. Routledge, London.

Hounkonnou, D., D. Kossou, T.W. Kuyper, C. Leeuwis, P. Richards, N. Röling et al. 2006. Convergence of sciences: The management of agricultural research for small-scale farmers in Benin and Ghana. NJAS Wageningen J. Life Sci. 53(3-4): 343-367.

Hsiao, T.C., P. Steduto, and E. Fereres. 2007. A systematic and quantitative approach to improve water use efficiency in agriculture. Irrig. Sci. 25:209-231.

IADB. 2006. The poverty impact of trade integration. June 27. Available at www.iadb .org. IADB, Washington DC.

IEA. 2004. Biofuels for transport. An international perspective. IEA, Paris.

IEA. 2006. World energy outlook 2006. IEA, Paris.

IIED. 2006. Protecting community rights over traditional knowledge: Implications of customary laws and practices. Nov. 2006. Int. Inst. Environ. Dev. (IIED), London.

Inco, M., and J. Nash. 2004. Agriculture and the WTO: Creating a trading system for development. Oxford Univ. Press and World Bank, Washington DC.

Ingo, G. 2004. El derecho local a los recursos

hídricos y la gestión ambiental regional de Chile: estudios de caso. In F. Peña (ed) Los pueblos indígenas y el agua: Desfíos de siglo XXI. El Colegio de San Luis, Water Law and Indigenous Rights (WALIR), Mexican Sec. Environ. (SEMARNAT), and Mexican Inst. Water Tech. (IMTA), San Luis de Potosí, Mexico.

ITCSD. 2006. Fisheries. ITCSD Dev. Series. Int. Trade Sustain. Dev., Geneva.

IWMI. 2006. Recycling realities: Managing health risks to make wastewater an asset. Water Policy Briefing 17. IWMI, Colombo.

Izac, A-M. 1997. Developing policies for soil carbon management in tropical regions. Geoderma 79:261-276.

Jaffee, S.M. 2005. Food safety and agricultural health standards and developing country exports: Rethinking the impacts and policy agenda. Trade Note 25. World Bank, Washington DC.

Jaramillo, C.F., and D. Lederman. 2005. DR-CAFTA: Challenges and opportunities for Central America. World Bank, Washington DC.

Jones, A. 2001. Eating oil — Food in a changing climate. Sustain. Elm Farm Res. Centre, London.

Joyner, C.C. 2000. The international ocean regime at the new millennium: A survey of the contemporary legal orders. Ocean Coastal Manage. 43:163-203.

Karsenty, A. 2003. Différentes formes de droit dans l'accès et la gestion des ressources en Afrique et à Madagascar. p. 263-283. In E. Rodary et al. (ed) Conservation de la nature et développement: l'intégration impossible. Karthala, GRET, Paris.

Kartha, S., G. Leach, and S.C. Rajan. 2005. Advancing bioenergy for sustainable development. Guidelines for policymakers and investors. ESMAP, World Bank, Washington DC.

Kaul, I., and P. Conceição. 2006. The changes underway: Financing global challenges through international cooperation behind and beyond borders. In I. Kaul and P. Conceição (ed) The new public finance: Responding to global challenges. UNDP/ Oxford University Press, London.

Kaul, I., I. Grunberg, and M.A. Stern (ed) 1999. Global public goods. International Cooperation in the 21st Century. Oxford Univ. Press, Oxford.

Keller, R.P., D.M. Lodge, and D.C. Finnoff. 2007. Risk assessment for invasive species produces net bioeconomic benefits. PNAS 104:202-207.

Key, N., and D. Runsten. 1999. Contract farming, smallholders, and rural development in Latin America: The organization of agroprocessing firms and the scale of outgrower production. World Dev. 27: 381-402.

Kibblewhite, M.G., K. Ritz, and R.S. Swift. 2007. Soil health in agricultural systems Phil. Trans. R. Soc. B

Kishore, V.V.N., P.M. Bhandari, and P. Gupta. 2004. Biomass energy technologies for rural infrastructure and village power - Opportunities and challenges in the context of global climate change concerns. Energy Policy 32:801-810.

Kline, S., and N. Rosenberg. 1986. An overview of innovation. p. 275-306. In R. Landau and N. Rosenberg (ed) The positive sum strategy. Harnessing technology for economic growth. National Acad. Press, Washington DC.

Klink, C., and R. Machado.2005. Conservation of Brazilian cerrado. Conserv. Biol. 19: 707-713.

Kojima, M., D. Mitchell, and W. Ward. 2007. Considering biofuel trade policies. ESMAP, World Bank, Washington DC.

Koning, N., M. Calo, and R. Jongeneel. 2004. Fair trade in tropical crops is possible: International commodity agreements revisited. Wageningen UR North-South Centre, The Netherlands.

Koning, N., and P. Jongeneel. 2006. Food sovereignty and export crops, could ECOWAS create an OPEC for sustainable cocoa? Paper for the Forum on Food Sovereignty. Niamey, Sep 2006.

Kopolow, D. 2006. Government support for ethanol and biodiesel in the United States [Online]. Available at www.globalsubsidies. org. IISD Global Subsidies Initiative (GSI).

Krier, J.M. 2005. Fair trade in Europe 2005: Facts and figures on fair trade. In P. Kristiansen, et al. (ed) Organic agriculture: A global perspective. CABI, UK.

Kroger, C. 2004. Fair trade product elbows organic domain. The Packer, May 3, 2004.

Kulibaba, N.P. 1997. Good governance in sheep's clothing: Implementing the action plan for regional facilitation of the livestock trade in West Africa's central corridor. Implementing Policy Change Project, Case Study 3. USAID, Washington DC.

Kumar, P. 2005. Market for ecosystem services. Int. Inst. Sustain. Dev. (IISD). Winnipeg, Canada.

Kyprianou, M. 2006. Animal welfare — a part of EU food chain policy. Int. Conf. organized by the Austrian Presidency Brussels, Eur. Commissioner for Health and Consumer Prot. Ref., SPEECH/06/211.

Lahidji, R., W. Michalski, and B. Stevens. 1998. The future of food: An overview of trends and key issues. The future of food: Long-term prospects for the agrofood sector. OECD, Washington DC

Lal, R., N. Uphoff, B.A. Stewart, D.O. Hansen. 2005. Soil carbon depletion and the impending food crisis. CRC Press, Baton Rouge, Fl.

Lappe, F.M., J. Collins, and P. Rosset. 1998. World hunger: Twelve myths. Food First and Grove Press, NY.

Larcher Carvalho, A. 2005. Cost of agrifood safety and SPS compliance: Mozambique, Tanzania and Guinea tropical fruits. UN Conf. Trade Development, Geneva.

Lavigne Delville, P. 1998. L'environnement, dynamiques sociales et interventions externes, construire et gérer l'interface entre acteurs. p. 381-394. *In* G. Rossi et al. (ed) Sociétés rurales et environnement. Festion des ressources et dynamiques locales au Sud. Karthala, Paris.

Le Roy, E. 1996. La théorie des maîtrises foncières. p. 59-76. *In* E. Le Roy et al. (ed) La sécurisation foncière en Afrique. Pour une gestion viable des ressources renouvelables. Karthala, Paris.

Leeuwis, C., and R. Pyburn (ed) 2002. Wheelbarrows full of frogs. Social learning in natural resource management. Koninklijke Van Gorcum, Assen.

Leforban, Y., and G. Gerbier. 2002. Review of the status of foot and mouth disease and approach to control/eradication in Europe and Central Asia. Rev. Sci. Tech. 21:477-92.

Lines, T. 2006. Market power, price formation and primary commodities. November. South Centre, Geneva.

Little, P.D., and M. Watts. 1994. Living under contract: Contract farming and agrarian transformation in sub-Saharan Africa. Univ. Wisconsin Press, Madison.

Lokuge B., and K. Lokuge. 2005. Avian Influenza, world food trade and WTO Rules: The economics of transboundary disease control. Available at http://cgkd.anu.edu.au/menus/workingpapers.php. Regulatory Institutions Network, The Australian Nat. Univ., Canberra.

Louwaars, N. 2006. Ethics watch: Controls over plant genetic resources — a double-edged sword. Nature Rev. Genetics 7:241.

Louwaars, N., R. Tripp, and D. Eaton. 2006. Public research in plant breeding and intellectual property rights: A call for new institutional policies. World Bank, Washington DC.

Lundy, M. C.F. Ostertag, M.V. Gottret, R. Best, and S. Ferris. 2005. A territorial based approach to agro-enterprise development. CIAT, Colombia.

Manderson, A.K., A.D. Mackay, and A.P. Palmer. 2007. Environmental whole farm management plans: Their character, diversity, and use as agri-environmental indicators in New Zealand. J. Environ. Manage. 82(3):319-331.

Mangisoni, J. 2006. Impact of treadle pump irrigation technology on smallholder poverty and food security in Malawi: A case study of Blantyre and Mchinji Districts. IWMI, Pretoria.

Maredia, K.M., and F.H. Erbisch. 1999. Capacity building in intellectual property management in agricultural biotechnology. p. 49-62 In F.H. Erbisch and F.M. Maredia (ed) Intellectual property rights in agricultural biotechnology. CABI, Wallingford.

Mayrand, K. and Paquin, M. 2004. Payments for environmental services: A survey and assessment of current schemes. Unisféra

Int. Center Comm. Environ. Coop. North America, Montreal.

McCalla, A.F., and J. Nash (ed) 2007. Reforming agricultural trade for developing countries: Issues, challenges, and structure of the volume. Vol. 2. World Bank, Washington DC.

McCann, J., T. Dalton, and M. Mekuria. 2006. Breeding Africa's new smallholder maize paradigm. Int. J. Agric. Sustainability (IJAS) 4(2):99-107.

McLeod, K. 2001. Owning culture: Authorship, ownership and intellectual property law. *In* T. Miller (ed) Popular culture and everyday life series. Peter Lang Publ.

Mead, P.S., L. Slutsker, V. Dietz, L.F. McCaig, J.S. Bresee, and C. Shapiro. 1999. Food-related illness and death in the United States. Emerg. Infect. Dis. 5:607-625.

Merrey, D.J., R. Meinzen-Dick, P.P. Mollinga, and E. Karar. 2007. Policy and intuitional reform: the art of the possible. *In* D. Molden (ed) Water for food, water for life: A comprehensive assessment of water management in agriculture. Earthscan, London and IWMI, Colombo.

Merrey, D. J., R. Namara, and M. de Lange. 2006. Agricultural water management technologies for small scale farmers in Southern Africa: An inventory and assessment of experiences, good practices, and costs. IWMI, Pretoria.

Millar, D. 2005. Endogenous development: Some issues of concern. Ghana J. Dev. Studies 2 (1):92-109.

Mokaila, P. 2005. Bovine Brucellosis eradication scheme in South Africa, is the battle won or not? Understanding veterinary policy processes. Institutional and Policy Support Team (IPST)/AU-IBAR. Case study prepared for the workshop on Understanding Veterinary Policy Processes. Aberdare, Kenya. 31 Jan. - 3 Feb. 2005. Institutional and Policy Support Team (IPST)/AU-IBAR, Nairobi.

Molle, F., and J. Berkoff. 2005. Cities versus agriculture: Revisiting intersectoral water transfers, potential gains, and conflicts. Comprehensive assessment of water management in agriculture. Res. Rep. 10. IWMI, Colombo.

Molle, F., and J. Berkoff (ed) 2007. Irrigation water pricing policy in context: Exploring the gap between theory and practice. CABI, Wallingford and IWMI, Colombo.

Molle, F., P. Wester, and P. Hirsch. 2007. River basin development and management. *In* D. Molden (ed) Water for food, water for life: A comprehensive assessment of water management in agriculture. Earthscan, London and IWMI, Colombo.

Morgan, N., and A. Prakash. 2006. International livestock markets and the impact of animal disease. Rev. Sci. Technol. Off. Int. Epiz. 25:517-528.

Morris, D. 2007. Energizing rural America: Local ownership of renewable energy

production is the key. January. Center Am. Progress, Washington DC.

Morrisey, O. 2007. What types of WTO-compatible trade policies are appropriate for different stages of development? p. 58-78. *In* J. Morrison and A. Sarris (ed) WTO rules for agriculture compatible with development. FAO, Rome.

Morrison, J., and A. Sarris. 2007. Determining the appropriate level of import protection consistent with agriculture led development in the advancement of poverty reduction and improved food security. p. 14-57. *In* J. Morrison and A. Sarris (eds.) WTO rules for agriculture compatible with development. FAO, Rome.

Mukherjee, S., and V. Kathuria. 2006. Is economic growth sustainable? Environmental quality of Indian States after 1991. Int. J. Sustain. Dev. 9:38-60.

Murphy, S. 2006. Concentrated market power and agricultural trade. Ecofair Trade Dialogue Disc. Pap. Wuppertal Inst. and Heinrich Boëll Foundation, Berlin.

Myhr, A.I., and R.A. Dalmo. 2005. Introduction of genetic engineering in aquaculture: Ecological and ethical implications for science and governance. Aquaculture 250 (3-4):542-554.

Nadal, A., and T.A. Wise. 2004. The environmental costs of agricultural trade liberalization: Mexico-U.S. maize trade under NAFTA. Disc. Pap. DP04. Available at http://ase.tufts.edu/gdae/Pubs/rp/DP04NadalWiseJuly04.pdf. Working Group on Dev. Environ in Americas.

Najam, A., D. Runnalls, and M. Halle. 2007. Environment and globalization: Five propositions. Int. Inst. Sustain. Dev (IISD), Winnipeg, Canada.

Namara, R., B. Upadhyay, and R.K. Nagar. 2005. Adoption and impacts of microirrigation technologies from selected localities of Maharashtra and Gujarat States of India. IWMI Res. Rep. 93. IWMI, Colombo.

Nash, J., and A.F. McCalla. 2007. Agricultural trade reform and developing countries: Issues, challenges, and structure of the volume. *In* A.F. McCalla and John Nash (ed) Reforming agricultural trade for developing countries. The World Bank, Washington DC.

Ng, F., B. Hoekman, and M. Olarreaga. 2007. The impact of agricultural support policies on developing countries. *In* A. McCalla and J. Nash. (ed) Reforming agricultural trade for developing countries. World Bank, Washington DC.

North, D. 2005. Understanding the process of economic change. Princeton Univ. Press, NJ.

O'Connell, D., B. Keating, and M. Glover. 2005. Sustainability guide for bioenergy: A scoping study. RIRDC/L&WA/FWPRDC/MDBC Joint Venture Agrofor. Prog. RIRDC Publ. 05/190. Gov. Australia, Kingston.

OIE (World Organization for Animal Health). 2004. Veterinary institutions in the developing world: Current status and future needs. Rev. Sci. Tech. Off. Int. Epiz. 23 (1).

OIE (World Organization for Animal Health). 2005a. Animal welfare: Global issues, trends and challenges. Rev. Sci. Tech. Off. Int. Epiz. 24(2).

OIE (World Organization for Animal Health). 2006. Report of the fifth meeting of the OIE working group on animal production food safety. Appendix XXXVIII. Cooperation between the Codex Alimentarius Commission and the OIE on food safety throughout the food chain. OIE, Paris.

Ong'wen, O., and S. Wright. Small farmers and the future of sustainable agriculture. 2007. EcoFar Trade Dialogue Disc. Pap. No. 7.

Ostrom, E. 1992. Governing the commons. The evolution of institutions for collective action. Cambridge Univ. Press, UK.

Ostrom, E., L. Schroeder, and S. Wynne. 1993. Institutional incentives and sustainable development. infrastructural policies in perspective. Westview Press, Boulder.

Pagiola, S., P. Agostini, J. Gobbi, C. De Haan, M. Ibrahim, E. Murguietio et al. 2002. Paying for biodiversity conservation services in agricultural landscapes. Available at ftp://ftp.fao.org/docrep/nonfao/lead/x6154e/x6154e00.pdf.

Panayotou, T. 2000. Economic growth and the environment. Center for Int. Dev. Harvard Univ., Cambridge MA.

Panagariya, A. 2004. Subsidies and trade barriers: Alternative perspective. In B. Lomborg (ed) Global crises, global solutions. Cambridge Univ. Press, UK.

Park, W., and D.C. Lippoldt. 2004. International licensing and the strengthening of intellectual property rights in developing countries. Trade Policy Working Pap. 10, OECD, Paris.

Pay, E. 2005. Overview of the sanitary and phytosanitary measures in quad countries on tropical fruits and vegetables imported from developing countries. Trade-Related Agenda, Development and Equity (TRADE) Res. Pap. No.1:45-46. South Centre, Geneva.

Peres, C. A., and J. Terborgh. 1995. Amazonian nature reserves: An analysis of the defensibility status of existing conservation units and design criteria for the future. Conserv. Biol. 9:34-46.

Perry, B., A. Nin Pratta, K. Sones, and C. Stevens. 2005. An appropriate private level of risk: Balancing the need for safe livestock products with fair market access for the poor. Pro-Poor Livestock Policy Initiative, Working Pap. No 23. FAO, Rome.

Plym-Forshell, L., and M. Wierup. 2006. Salmonella contamination - a significant challenge to the global marketing of animal food products. Rev. Sci. Tech. Off. Int. Epiz. 25: 541-554.

Polaski, S. 2006. Winners and losers: Impact of the Doha Round on developing countries. Carnegie Endowment Rep. 2006. Carnegie Endowment, Washington DC.

Pontius, J., R. Dilts, and A. Bartlett. 2002. From farmer field schools to community IPM. Ten years of IPM training in Asia. FAO, Regional Off. Asia and the Pacific, Bangkok.

Pray, C., 1991. Plant breeder's rights legislation, enforcement and RandD: lessons for developing countries. In G.H. Peters and B.F. Stanton (ed) Sustainable agricultural development: the role of international cooperation. Proc. 21st Int. Conf. Agric. Economists. Dartmouth Publ., Aldershot UK.

Pretty, J. 1994. Alternative systems of inquiry for sustainable agriculture. IDS Bull. 25(2):37-49.

Priyadarshi, S. 2002. Reforming global trade in agriculture: A developing country perspective. Trade environment and development. Carnegie Endowment Int. Peace, Washington DC.

Qadir, M., D. Wichelns, L. Raschid-Sally, P.S. Minhas, P. Drechsel, A. Bahri, and P. McCornick. 2007. Agricultural use of marginal-quality water- opportunities and challenges. In D. Molden (ed) Water for food, water for life: A comprehensive assessment of water management in agriculture. Earthscan, London and IWMI, Colombo.

Raynolds, L.T., D. Murray, and A. Heller. 2007. Regulating sustainability in the coffee sector: A comparative analysis of environmental and social certification initiatives. Agric. Human Values 24:147-163.

Reardon, T., and C.H. Barrett. 2000. Agroindustrialization and international development: An overview of issues, patterns and determinants. Agric. Econ. 23:195-205.

Reij, C., I. Scoones, and C. Toulmin (ed) 1996. Sustaining the soil. Indigenous soil and water conservation in Africa. Earthscan. London.

Reijnders, L. 2006. Conditions for the sustainability of biomass based fuel use. Energy Policy 33:863-876.

Röling, N., and P. Engel (1991). The development of the concept of Agricultural Knowledge and Information Systems (AKIS): Implications for extension. p. 125-139. In W. Rivera and D. Gustafson (ed) Agricultural extension: Worldwide institutional evolution and forces for change. Elsevier Science Publishers, Amsterdam.

ROPPA. 2006. Appèl de Niamey pour la soveraineté alimentaire de l'Afrique de 'lOuest. ROPPA, Niamey.

Rosa, H., D. Barry, S. Kandel, and L. Dimas. 2004. Compensation for environmental services and rural communities: Lessons from the Americas. Polit. Econ. Res. Inst., Univ. Massachusetts, Amherst.

Rose, J.B., P.R. Epstein, E.K. Lipp, B.H. Sherman, S.M. Bernard, and J.A. Patz. 2001. Climate variability and change in the United States: Potential impacts on water and foodborne diseases caused by microbiologic agents. Environ. Health Perspect. 109(Suppl.2): 211-220.

Rosenberg, N., and R.R. Nelson. 1994. American universities and technical advance in industry. Res. Policy 23:323-348.

Roosevelt, M. 2004. The coffee clash: Many firms see a marketing advantage in selling politically correct beans: Will Starbucks get hurt? Time, 8 Mar 2004.

Rubenstein, K., P. Heisey, R. Shoemaker, J. Sullivan, and G. Frisvold. 2005. Crop genetic resources: An economic appraisal. Econ. Inform. Bull. No. 2. ERS, USDA, Washington DC.

Ryan, B., and N. Gross. 1943. The diffusion of hybrid seed corn in two Iowa communities. Rural Soc. 8:15-24.

Sabirovic, M., P. Grimely, and F. Landeg. 2005. Foot and mouth disease in Brazil (EU Exporting Area). DEFRA, London.

Safrin, S., 2004. Hyperownership in a time of biotechnological promise. The international conflict to control the building blocks of life. Am. J. Int. Law 98: 641-685

Samad, M., and D.L. Vermillion. 1999. Assessment of the impact of participatory irrigation management in Sri Lanka: Partial reforms, partial benefits. IWMI Res. Rep. 34. IWMI, Colombo.

Sandberg, A. 1994. Gestion des ressources naturelles et droits de propriété dans le grand nord norvégien: éléments pour une analyse comparative. Natures Sci. Soc. 2(4):323-333.

Schlager, E., and E. Ostrom. 1992. Property rights regimes and natural resources: A conceptual analysis. Land Econ. 68:249-262.

Schlundt, J., H. Toyofuku, J. Jansen, and S.A. Herbst. 2004. Emerging food-borne zoonoses. Rev. Sci. Tech. 23:513-33.

Schreiner, B., and B. van Koppen. 2001. From bucket to basin: poverty, gender, and integrated water management in South Africa. In C.L. Abernethy et al. (ed) Intersectoral management of river basins. Proc. int. workshop on integrated water management in water stressed river basins in developing countries. Strategies for poverty alleviation and agricultural growth. IWMI, German Foundation of International Development, Colombo.

Scoones, I., and W. Wolmer. 2006. Livestock, disease, trade and markets: Policy choices for the livestock sector in Africa. IDS Working Pap. 269:1-53.

Shah, T., M. Alam, D. Kumar, R.K. Nagar, and M. Singh. 2000. Pedaling out of poverty: Social impacts of a manual irrigation technology in South Asia. IWMI Res. Rep. 45. IWMI, Colombo.

Shah, T., I. Makin, and R. Sakthivadivel. 2005. Limits to Leapfrogging: Issues in transposing successful river basin management institutions in the developing world. In M. Svendsen (ed) Irrigation and river basin management: Options for governance and institutions. CABI Publishing, Wallingford.

Shah, T., J. Burke, and K. Villholth. 2007. Groundwater: A global assessment of scale and significance. In Water for Food, water for life: A comprehensive assessment of water management in agriculture. Earthscan, London and IWMI, Colombo.

Shand, H. 2005. Oligopoly, Inc. ETC Group, Ottawa.

Simeon, M. 2006. Sanitary and phytosanitary measures and food safety: Challenges and opportunities for developing countries. Rev. Sci. Tech. Off. Int. Epiz. 25:701-712.

Smith, P., D. Martino, Z. Cai, D. Gwary, H. Janzen, P. Kumar et al. 2007. Agriculture. In B. Metz et al. (ed) Climate change 2007: Mitigation. Contribution of Working Group III to the Fourth Assessment Report of the Intergovernmental Panel on Climate Change. Cambridge Univ. Press, UK.

SNV and CEDELO. 2004. La décentralisation au Mali: du discours à la practique. Bull. 358. KIT, Amsterdam.

Starmer, E., A. Witteman and T.A. Wise. 2006. Feeding the factory farm: Implicit subsidies to the broiler chicken industry. Working Pap. 06-03. Available at http://www.ase.tufts.edu/gdae/Pubs/wp/06-03BroilerGains.pdf. Global Dev. Environ. Inst., Tufts Univ., Medford MA.

Stassen, H. 1995. Small-scale biomass gasifiers for heat and power. A global review. ESMAP, World Bank, Washington DC.

Steinfeld, H., P. Gerber, T. Wassenaar, V. Castel, M. Rosale, and C. de Haan. 2006. Livestock`s long shadow: Environmental issues and options. FAO, Rome.

Stiglitz, J.E. 2006. Making globalization work. W.W. Norton, NY.

Stiglitz, J.E., and A. Charlton. 2005. Fair trade for all. Oxford Univ. Press, NY.

Swift, M.J., A.M. Izac, and M. van Noordwijk. 2004. Biodiversity and ecosystem services in agricultural landscapes: Are we asking the right questions? Agric. Ecosyst. Environ. 104(1):113-134.

Taylor L.H., S.M. Latham and M.E. Woolhouse. 2001. Risk factors for human disease emergence. Phil. Trans. R. Soc. Lond. B Biol. Sci. 356(1411):983-989.

Thomson, G.R., B.D. Perry, A. Catley, T.J. Leyland, M.L. Penrith, and A.I. Donaldson. 2006. Certification for regional and international trade in livestock commodities: The need to balance credibility and enterprise. Vet. Rec.159:53-7.

Thomson, G.R., E.N. Tambi, S.K. Hargreaves, T.J. Leyland, A.P. Catley, G.G. van't Klooster et al. 2004. International trade in livestock and livestock products: The need for a commodity-based approach. Vet. Rec.155:429-33.

Toulmin, C. 2005. Foreword In T. Bigg and D. Satterthwaite (ed) How to make poverty history: The central role of local organisations in meeting the MDGs. IIED, London.

Trawick, P. 2005. Going with the flow: The state of contemporary studies of water management in Latin America. Latin American Res. Rev. 40:443-456.

Tripp, R. 2006. Self-sufficient agriculture: Labour and knowledge in small-scale farming. Earthscan, London.

Troy, A., and M.A. Wilson. 2006. Mapping ecosystem services: Practical challenges and opportunities in linking GIS and value transfer. Ecol. Econ. 60:435-449.

UNCSD. 2005. Sanitation: Policy options and possible actions to expedite implementation. Report of the Secretary-General. 13th Sess. UN Comm. Sustainable Development. 11-22 April 2005.

UNCTAD 2005. Commodity policies for development: A new framework for the fight against poverty. UN Comm. Trade and Development. T/B/COM 1/75. Dec 8. UNCTAD, Geneva.

UNCTAD. 2006. The emerging biofuels market: Regulatory, trade and development implications. UNCTAD/DITC/TED/2006/4. UN Conf. Trade and Development, Geneva.

UN Energy. 2007. Sustainable bioenergy: A framework for decision makers. April. Available at http://esa.un.org/un-energy/pdf/susdev.Biofuels.FAO.pdf. United Nations Energy, Rome.

UNIDO. 2006. Water and industry. Chapter 8 of the 2nd UN World Water Development Report: Water, a shared responsibility. UN Industrial Dev. Org.

UPOV. 2005. Report on the impact of plant variety protection. Union for the Protection of New Varieties of Plants (UNPOV), Geneva.

USDA. 2003. 21st century agriculture: A critical role for science and technology. USDA, Washington DC.

USDA. 2006. Bio-fuels: Alternative future for agriculture 2006. GAIN Rep. CH6049. Available at www.fas.usda.gov/gainfiles/200608/146208611.pdf. Foreign Agric. Service, USDA, Washington DC.

USDA. 2007. Foodborne illness cost calculator. USDA, Washington DC. Available at http://www.ers.usda.gov/data/foodborneillness

Valle, P.S., E. Skjerve, S.W. Martin, R.B. Larssen, O. Osteras, and O. Nyberg. 2005. Ten years of bovine virus diarrhoea virus (BVDV) control in Norway: A cost-benefit analysis. Prev. Vet. Med. 72:189-207.

Van de Fliert, E. 1993. Integrated pest management. Farmer field schools generate sustainable practices: A case study in Central Java evaluating IPM Training. WU Papers 93-3. Agricultural University, Wageningen.

Van den Berg, H. 2003. IPM farmer field schools: A synthesis of impact evaluations. Global IPM Facility, FAO, Rome.

Van den Berg, H., and J. Jiggins. 2007. Investing in farmers: The impact of farmer field schools in relation to integrated pest management. World Dev. 35(4):663-686.

Van Huis, A., and D. Hounkonnou (ed) 2007. CoS-SIS: Convergence of sciences: Strengthening innovation systems in Benin, Burkina Faso, Ghana and Mali. 5th vers., June 2007. Research Proposal submitted to DGIS and Royal Netherlands Embassies. WUR, Entomology, Wageningen.

Van Huis, A., J. Jiggins, D. Kossou, C. Leeuwis, N. Röling, O. Sakyi-Dawson et al. 2007. Can convergence of agricultural sciences support innovation by resource-poor farmers in Africa? The cases of Benin and Ghana. Int. J. Agric. Sustain. 5:91-108.

Vanlauwe, B., P. Tittonell, and J. Mukalama. 2006. Within-farm soil fertility gradients affect response of maize to fertilizer applications in western Kenya. Nutr. Cycl. Agroecosyst. 76:171-182.

Vermillion, D.L. 1997. Impacts of irrigation management transfer: A review of the evidence. IWMI Res. Rep. 11. IWMI, Colombo.

Vermillion, D.L., and C. Garcés-Restrepo. 1998. Impacts of Colombia's current irrigation management transfer program. IWMI Res. Rep. 25. IWMI, Colombo.

Vermillion, D.L., M. Samad, S. Pusposutardjo, S.S. Arif, and S. Rochdyanto. 2000. An assessment of small-scale irrigation management turnover program in Indonesia. IWMI Res. Rep. 38. IWMI, Colombo.

Vermillion, D.L., and J.A. Sagardoy. 1999. Transfer of irrigation management services, guidelines. FAO Irrig. Drainage Pap. 58. FAO, Rome.

Visser, C. 2004. Making intellectual property laws work for traditional knowledge. p. 207-240. In J.M. Finger and P. Schuler (ed) Poor people's knowledge: Promoting intellectual property in developing countries. Oxford Univ. Press, UK.

Vorley, B., and T. Fox. 2004. Global food chains — constraints and opportunities for smallholders. Paper for OECD DAC POVNET. Agriculture and pro-poor growth team workshop. 22. OECD, Paris.

Vorley, B., A. Fearne and D. Ray. 2007. Regoverning markets: A place for small-scale producers in modern agri-food chains? Gower Publ., IIED, London.

Wals, A. (ed) 2007. Social learning: Towards a sustainable world. Wageningen Acad. Publ., The Netherlands.

WCMC. 2006. World database of protected areas. Online database. http://www.unep-wcmc.org/wdpa/index.htm. UNEP, Nairobi.

Warren, D.M., L.J. Slikkeveer, and D. Brokensha (ed) 1991. Indigenous knowledge systems: the cultural dimension of development. Kegan Paul Int., London.

Watkins, K., and J. Von Braun. 2002. Time to stop dumping on the world's poor. 2002-2003 Annual Report: Trade policies and food security. IFPRI, Washington, DC.

Wei, A., and J. Cacho. 2000. Establishing linkages between globalization, developing country economic growth and agroindustrialization: The case of East Asia. Int. Food Agribusiness Manage. Rev. 2(3-4).

Weiss, R. 2007. Tainted Chinese imports common. The Washington Post. May 20.

Wennink, B. and W. Heemskerk (ed) 2006. Farmers' organisations and agricultural innovation. Case studies from Benin, Rwanda and Tanzania. Bull. 374. KIT, Amsterdam.

Whittingham, M.J. 2007. Will agri-environment

schemes deliver substantial biodiversity gain, and if not why not? J. Appl. Ecol. 44(1):1-5.

Wierup, M. 2000. The control of microbial diseases in animals: Alternatives to the use of antibiotics. Int. J. Antimicrobial. Agents 14:315-319.

WHO. 2002a. WHO global strategy for food safety: Safer food for better health. WHO, Geneva.

WHO. 2002b. Future trends in veterinary public health. WHO Tech. Rep. Ser. 907. WHO, Geneva.

WHO. 2004. Report of the WHO/FAO/OIE joint consultation on emerging zoonotic diseases. Geneva, Switzerland. 3-5 May 2004. WHO/CDS/CPE/ZFK/2004.9. WHO, Geneva.

WHO. 2006. The control of neglected zoonotic diseases- A route to poverty alleviation. Report of a joint WHO/DFID-AHP meeting with the participation of FAO and OIE. Geneva. 20-21 Sept. 2005, WHO/SDE/FOS/2006.1. WHO, Geneva.

WIPO (World Intellectual Property Organization). 2001. Agenda for development of the international patent system. A/36/14 Aug 6. WIPO, Geneva.

WIPO (World Intellectual Property Organization). 2004. Proposal by Argentina and Brazil for the establishment of a development agenda for WIPO. WO/GA/31/11. 27 Aug 2004. WIPO General Assembly. WIPO, Geneva.

WIPO (World Intellectual Property Organization). 2006a. Intergovernmental Comm. Intellectual Property and Genetic Resources, Traditional Knowledge and Folklore, 9th Sess., WIPO/GRTKF/IC/9/14 Prov. 20 July 2006. Paragraph 211. WIPO, Geneva.

Wise, T.A. 2004. The paradox of agricultural subsidies: Measurement issues, agricultural dumping, and policy reform. Available at http://ase.tufts.edu/gdae/Pubs/wp/04-02AgSubsidies.pdf . Global Dev. Environ. Inst. Working Pap. 04-02. Tufts Univ., MA.

World Bank. 2005. Food safety and agricultural health standards: Challenges and opportunities for developing country exports. Rep. No. 31207. World Bank, Washington DC.

World Bank. 2006. Intellectual property rights. Designing regimes to support plant breeding in developing countries. ARD Rep. No. 35517-GLB. World Bank, Washington DC.

World Bank. 2007. Enhancing agricultural innovation: How to go beyond strengthening agricultural research systems. World Bank, Washington DC.

Wouters, J., and B. de Meester. 2003. The role of international law in protecting public goods. regional and global challenges. Leuven Interdisciplinary Research Group on International Agreements and Development (LIRGIAD), Working Pap. no.1.

WTO. 2004. The future of the WTO: Addressing institutional challenges in the new millennium. A report to the Director-General by a Consultative Board chaired by P. Sutherland World Trade Organization Secretariat, Geneva.

WTO. 2006a. Doha Work Programme — The outstanding implementation issues on the relationship between the TRIPs agreement and the Convention on Biological Diversity. General Council, WT/GC/W/564. 31 May 2006.

WTO. 2006b. Amending the trips agreement to introduce and obligation to disclose the origin of genetic resources and traditional knowledge in patent applications. WT/GC/W/566. 14 June 2006.

WTO. 2007. Communication from the African Group on Commodities. Committee on Agriculture. JOB (07)/118. July 2, 2007. World Trade Organization, Geneva.

Wuethrich, B. 2007. Reconstructing Brazil's Atlantic rainforest. Science 315:1070-1072.

WWF. 2006. Sustainability standards for bioenergy. World Wide Fund for Nature Germany, Frankfurt.

Zander, P., A. Knierim, J.C.J. Groot, and W.A.H. Rossing. 2007. Multifunctionality of agriculture: Tools and methods for impact assessment and valuation. Agric. Ecosyst. Environ. 120:1-4.

Zarsky, L. 1999. Havens, halos and spaghetti: Untangling the relationship between FDI and the environment. p. 47-73. In Foreign direct investment and the environment. OECD, Paris.

8

Agricultural Knowledge, Science and Technology: Investment and Economic Returns

Coordinating Lead Authors
Nienke Beintema (The Netherlands), Ahmet Ali Koc (Turkey)

Lead Authors
Ponniah Anandajayasekeram (Australia), Aida Isinika (Tanzania), Frances Kimmins (UK), Workneh Negatu (Ethiopia), Diane Osgood (Switzerland/USA), Carl Pray (USA), Marta Rivera-Ferre (Spain), V. Santhakumar (India), Hermann Waibel (Germany)

Contributing Authors
Jock Anderson (Australia), Steven Dehmer (USA), Veronica Gottret (Bolivia), Paul Heisey (USA), Philip Pardey (Australia)

Review Editors
Harriet Friedman (USA) and Osamu Ito (Japan)

Key Messages

Key Messages

1. On average, investments in agricultural research and development (R&D) are still growing but at a decreasing rate for the public sector during the 1990s. However, there has been an increasing diversity in investment trends among countries. Investment in public agricultural R&D in many developed countries has stalled or declined and has become a small proportion of total Science & Technology (S&T) spending. Many developing countries are also stagnating or slipping in terms of public agricultural R&D investments, except for a selected few (often the more industrialized countries). The slowing growth in agricultural R&D investment in the public sector has implications for attaining the development goals. Investments by the private sector in developed countries have been increasing, but have remained small in most developing countries. There is a knowledge gap in other areas of AKST investments such as extension, traditional knowledge, farming systems, social sciences, ecosystems services, mitigation and adaptation of climate change, and health in agriculture.

2. Funding for public agricultural R&D in developing countries is heavily reliant on government and donor contributions, but these sources have declined. Despite declining government budgets for agriculture in general, and agricultural R&D specifically, government remains the major source of funding for public agricultural R&D in most developing countries. The trend indicates that donor support for agricultural R&D has substantially declined since the mid-1980s with the majority of this smaller amount supporting global research rather than research at the country level.

3. The participation of nongovernmental agencies in agricultural R&D is increasing. AKST in the more developed world is increasingly undertaken by the private sector. Private sector research is also growing in the developing world, but is concentrated in a few countries where the private sector thinks it can make a profit. In addition, higher education agencies, NGOs, foundations, and producer groups are also increasing their participation in agricultural R&D. Still, publicly funded research in developing countries is mostly conducted by government-sponsored agencies.

4. There is evidence of underinvestment in research in agriculture. Rates of Return (ROR) in AKST across commodities, countries and regions on average are high (40-50%) and have not declined over time. They are higher than the rate at which most governments can borrow money, which suggests underinvestment in AKST. Although limited, evidence indicates that the investments in agricultural R&D perform equally well or better than the other public sector investments in the agricultural sector.

5. Public investments in AKST have significantly contributed to overall economic growth, but this has not always translated into poverty reduction. Public investments in AKST have in some countries significantly contributed to poverty reduction, but AKST's impact on poverty varies greatly depending on the policies, institutions, and access to resources of the country. Before AKST investments are made, distributional aspects should be explicitly taken into account. Additional analysis is required to understand better who has benefited from this additional growth and why it did not always translate into commensurate improvement of poverty and food security. Likewise, agricultural price policies and trade policies influence the distributional impacts of productivity-increasing technology, as do land and access patterns.

6. Rates of return alone are not sufficient to guide AKST investment decisions. AKST investment generates economic, social, environmental, health and cultural costs and benefits to society, some of which are considered as externalities (positive or negative) and spillovers. These non-economic impacts are also important to society, but often not included in conventional RoR analysis due to quantification and valuation problems. The challenge is to factor these aspects into the macro-level decision-making process. RoR analysis needs to be complemented by other approaches to estimating impact of AKST investment on poverty reduction, ecosystem services and well-being. More evidence is needed on the economic and social impact of AKST investment in sectors such as forestry and fisheries, as well as in policy-oriented social science research.

7. AKST investments could have been more effective and efficient in achieving sustainable development goals had more attention been given to governance. Governance is an important determinant of mobilization of resources for AKST. It also plays a major role in allocation of resources between different components of AKST. Increased demand for effectiveness, efficiency, responsiveness to stakeholder needs, accountability and transparency is a driving force leading to changes in AKST investment decisions. High transaction costs in knowledge generation and transfer, inefficiency in resource allocation and utilization, lack of transparency, exclusion of some stakeholders, unequal access, and fear of private monopoly over technologies developed through public AKST institutions have prompted changes in AKST systems. The ability to allocate resources more effectively will also depend on a significant improvement in the capacity in public and private sectors to forecast and respond to environmental, social, and economic changes, locally and globally. This will include the capacity to make strategic technological choices, create effective public policy and regulatory frameworks, and pursue educational and research initiatives.

8. Increasing participation of nongovernmental stakeholders and more appropriate incentive systems are required to improve the effectiveness of AKST investments. Institutional arrangements for AKST resource mobilization and allocation have, in the past, largely excluded users of research information, resulting in inefficiency and ineffectiveness in AKST investments. These arrangements have also resulted in unequal access to technologies. The demands for enhanced stakeholder participation, improved accountability and transparency are leading to institutional innovations around AKST investment governance issues. These are new and unproven arrangements, so investments will be needed to carefully monitor, evaluate and learn from the lessons in order to derive best practices. Investment is also required for strengthening capacity in order to institutionalize such practices.

9. More government funding and better targeted government investments in AKST in developing countries can make major contributions to meeting development and sustainability goals. Developing countries need to increase the intensity of AKST investments. This would involve a major increase in public sector investments, which is justified given the high rates of return to research and the evidence that AKST investments can reduce poverty. However, to do this, public investments must be targeted using evidence other than simply overall RORs, as they usually do not include environmental and human health impacts, positive or negative, or information on the distribution of costs and benefits among different groups.

10. Major public and private research and development investments will be needed in plant and animal pest and disease control. Continued intensification of agricultural production, changes in agriculture due to global warming, the development of pests and diseases that are resistant to current methods of controlling them, and changes in demand for agricultural products, will lead to new challenges for farmers and the research system. Investments in this area by the public and private sector have provided high returns in the past and are likely to provide even higher returns in the future. In addition, these investments could lead to less environmental degradation by reducing the use of older pesticides and livestock production methods; increased demand for labor, which could reduce poverty; and positively improve human health of farmers and their families by reducing their exposure to pesticides. This is an area in which public and private collaboration is essential.

11. Increasing investments in agricultural research, innovation, and diffusion of technology by for-profit firms can also make major contributions to meeting development and sustainability goals. Private firms (large and small) have been and will in the future continue to be major suppliers of inputs and innovations to commercial and subsistence farmers. They will not provide public goods or supply good and services for which there is no market, but there could be spillovers from private suppliers of technology to farmers and consumers. To make the best use of private investments in AKST, governments must provide regulations to guard against negative externalities and monopolistic behavior, and support good environmental practices, while at the same time providing firms with incentives to invest in AKST.

12. AKST investment that increases agricultural productivity and improves existing traditional agricultural and aquaculture systems in order to conserve scarce resources such as land, water and biodiversity remains as a high priority; these investments can improve livelihoods and reduce poverty and hunger. The major resource constraint on increasing agricultural production in the future will continue to be agricultural land. AKST must focus on increasing output per unit of land through technology and management practices. Water is the next most important resource constraint to agricultural production and is likely to become more of a constraint in the future. AKST resources need to be reallocated into water-saving techniques, improved policies and management techniques. Fossil fuels reserves are limited; high fuel prices and environmental concerns have recently focused attention on the need for agriculture to more efficiently use this resource. Government investment in AKST may be necessary to reduce the dependence of the agricultural sector on petroleum.

13. AKST investment to reduce greenhouse gas emissions and provide other ecosystem services is another priority investment area. AKST investments are needed to develop policies, technologies and management strategies that reduce agriculture's contribution to greenhouse gas (GHG) emissions, and consequent global warming. This requires the development of new farming systems, which use better technologies, produce less GHG, and build on local and traditional knowledge to improve current cropping systems in order to become more sustainable. Investments are also needed to underpin policies such as payments for environmental services to farmers, which could induce the development and adoption of practices that provide stronger environmental services. Some of the agricultural technologies and policies that provide these ecosystem services can be designed to use the assets of the poor, such as labor in labor-abundant economies.

8.1 Spending and funding trends in AKST

8.1.1 Trends in agricultural R&D spending

8.1.1.1 Public sector spending
Worldwide, public investments in agricultural research and development (R&D) increased, in inflation-adjusted terms, over the past two decades from an estimated $15 billion in 1981 to $23 billion in 2000 (in 2000 international dollars); an increase of about one-half (Table 8-1 and Figure 8-2).[1] [2] The share of the developing countries as a group have increased considerably over the years; during the 1990s the

---

[1] Public includes government, higher education, and nonprofit.

[2] Unless otherwise stated, financial figures in this subchapter have been expressed in inflation adjusted "international dollars" using the benchmark year 2000 and purchasing power parities (PPPs). PPPs are synthetic exchange rates used to reflect the purchasing power of currencies, typically comparing prices among a broader range of goods and services than conventional exchange rates. Using PPPs as conversion factors to denominate value aggregates in international dollars results in more realistic and directly comparable agricultural research spending amounts in countries than if market exchange rates are used. This is because the latter tends to underestimate the quantity of spending used in economies with relatively low prices while overestimating the quantity for those countries with high prices. This is particularly a problem when valuing something like expenditures on agricultural R&D, where normally about two-thirds of the resources are spent on local scientist and support staff salaries and not on capital or other goods and services that are normally traded internationally.

group invested more on public agricultural R&D than the combined total in the industrialized world. Investments by Asia and Pacific countries as a group grew relatively resulting in an increasing share of the global total; the regional share was 33% in 2000 compared to only 20% in 1981. Most of this growth took place during the late-1990s. In contrast, the corresponding share for sub-Saharan Africa continued to decline, falling from 8 to 6% of the global total between 1981 and 2000.

Public agricultural R&D has become increasingly concentrated in just a handful of countries. Among the rich countries, the United States (US) and Japan accounted for 54% of public spending in 2000; about the same as two decades earlier. Three developing countries, China, India, and Brazil, spent 47% of the developing world's public agricultural research total, an increase from 33% in 1981. Meanwhile, only 6% of the agricultural R&D investments worldwide were conducted in 80 countries that combined had a total to more than 600 million people in 2000.

Growth in inflation-adjusted spending has slowed down since the 1970s when most regions experienced high growth rates (Figure 8-2). Overall spending in the Asia and Pacific region increased with an annual growth rate of 3.9% during the 1990s; lower than the regional growth in the 1980s (Beintema and Stads, 2006). However the average growth

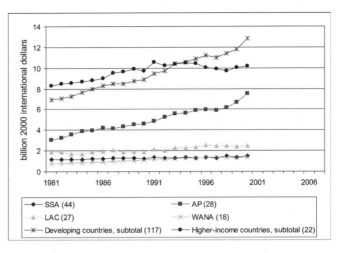

Figure 8-1. *Total public agricultural research expenditures by region, 1981-2000. Source: Pardey et al., 2006b based on Agricultural Science and Technology Indicators (ASTI) data at www.asti.cgiar.org*

Note: See Table 8-1

Table 8-1. **Total public agricultural research expenditures by region, 1981, 1991, and 2000.**

	Agricultural R&D spending			Shares in global total		
	1981	1991	2000	1981	1991	2000
	(million 2000 international dollars)			*(percent)*		
Asia & Pacific (28)	3,047	4,847	7,523	20.0	24.2	32.7
China	1,049	1,733	3,150	6.9	8.7	13.7
India	533	1,004	1,858	3.5	5.0	8.1
Latin America & Caribbean (27)	1,897	2,107	2,454	12.5	10.5	10.7
Brazil	690	1,000	1,020	4.5	5.0	4.4
sub-Saharan Africa (44)	1,196	1,365	1,461	7.9	6.8	6.3
West Asia & North Africa (18)	764	1,139	1,382	5.0	5.7	6.0
Developing countries, subtotal (117)	*6,904*	*9,459*	*12,819*	*45.4*	*47.3*	*55.7*
Japan	1,832	2,182	1,658	12.1	10.9	7.2
USA	2,533	3,216	3,828	16.7	16.1	16.6
Subtotal, higher-income countries (22)	*8,293*	*10,534*	*10,191*	*54.6*	*52.7*	*44.3*
Total (139)	**15,197**	**19,992**	**23,010**	**100.0**	**100.0**	**100.0**

Notes: The number of countries included in regional totals is shown in parentheses. These estimates exclude East Europe and former Soviet Union countries. The high-income countries total excludes a number of high income countries such as South Korea and French Polynesia (which has been grouped in the Asia and Pacific total), Bahrain, Israel, Kuwait, Qatar, and United Arab Emirates (grouped in West Asia and North Africa), and Bahamas (Latin America and Caribbean). To form these regional totals national spending estimates were scaled up for countries that represented 79% of the reported sub-Saharan African total, 89% of the Asia and Pacific total, 86% of the Latin America and Caribbean total, 57% of the West Asia and North Africa total, and 84% of the high-income total.

Source: Pardey et al., 2006b based on Agricultural Science and Technology Indicators (ASTI) data at www.asti.cgiar.org.

rate in total spending in China and India increased during the 1990s. This was in part due to an increase in total agricultural R&D spending in both countries during the second half of the 1990s, which reflects new government policies to revitalize public agricultural research and improve its commercialization prospects. Two other regions, Latin America and the Caribbean, and West Asia and North Africa, both experienced relative less growth in total spending during the 1990s (2.0 and 3.3%, respectively). In contrast, the increase in total spending in sub-Saharan Africa decreased in the 1990s from 1.3 to 0.8% compared to a decade earlier. An even more severe drop in spending is found in many sub-Saharan African countries. In about half of the 24 countries for which time series data were available, the public sector spent less on agricultural R&D in 2000 than 10 years earlier.

Noteworthy is the decline in total agricultural R&D spending among the rich countries; during the 1990s total spending declined by an annual rate of 0.6%. Specifically Japan, and to a lesser degree a few European countries, reduced their investments in agricultural research. Support for publicly performed agricultural research among rich countries has declined over a long period in time due to changes in government spending priorities and a shift toward privately performed agricultural R&D. These slowdowns in agricultural R&D spending may curtail the future spillovers of technologies from rich to poor countries (Pardey et al., 2006a) (see 8.2.7).

The allocation of resources among various lines of research is a significant policy decision and takes place at different levels and, in theory (although not always in practice), follows the priorities set across commodity and multidisciplinary research programs. More than one half of the full-time equivalent (fte) researchers in a sample of 45 developing countries conducted crops research while 15% focused on livestock and 8% on natural resources research (Table 8-2). Asia-Pacific had relatively less livestock research-

ers (13%) than sub-Saharan Africa and Latin America (18% each). Forestry, fisheries, and postharvest accounted for 4 to 6% each. The remaining 9% of the research staff in the developing world conducted research in other agriculture related sciences.

For all three regions, fruits and vegetables are among the major crops being researched. Unsurprisingly, rice is a relatively important crop in the Asia-Pacific region while maize has high importance in Latin America.

The allocation of resources above does not cover the full scope of AKST, e.g., areas of importance in the future may include bioenergy, climate change, transgenics and biodiversity. The Stern Review on the Economics of Climate Change (Stern, 2007) concludes that an annual investment of 1% of global GDP is required to mitigate the negative effects of climate change. Although economists argue whether the figures in the Stern review are right, most agree that the cost of failing to tackle climate change will so vastly outweigh the cost of succeeding that further refinement of the calculations are largely irrelevant to the political and investment choices that must be made now. Among these could be the creation of incentives for investment in low-carbon technologies.

Some limited information on the budget levels in bioenergy R&D in OECD countries is available through the International Energy Agency (IEA). Total R&D budgets for bioenergy are estimated to have increased almost three-fold since 1992 to a total of $271 million (in 2005 international dollars) in 2005. Despite the increased interest in renewable energy and energy-saving technological innovations to mitigate climate change, total budgets on energy R&D in OECD countries, in adjusted terms, have remained flat since 1992. As a result, the share of bioenergy R&D in total energy R&D investments also grew almost threefold during 1992-2005 (IEA, 2006).

In this chapter, public agricultural research includes research performed by government, higher education, and nonprofit agencies. There are substantial differences among

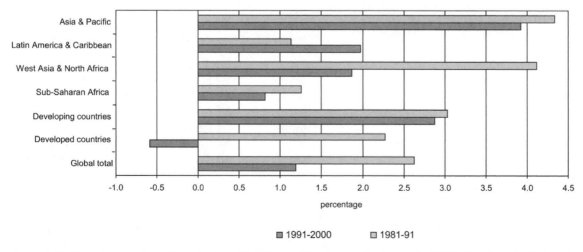

Figure 8-2. *Growth rates of public agricultural R&D spending. Source: Pardey et al., 2006b based on Agricultural Science and Technology Indicators (ASTI) data at www.asti.cgiar.org*

Notes: See Table 8-1. Annual growth rates were calculated using the least-squares regression method, which takes into account all observations in a period. This results in growth rates that reflect general trends that are not disproportionately influenced by exceptional values, especially at the end point of the period.

Table 8-2. Commodity focus by main research area, various years.

Major commodity area	Asia-Pacific (10), 2002/03	sub-Saharan Africa (26), 2000/01	Latin America (9), 1996	Total developing countries (45)
	(percent)			
Crops	52.5	48.1	53.5	52.1
Livestock	13.2	17.8	17.9	14.7
Forestry	6.5	6.1	4.8	6.2
Fisheries	5.8	4.8	4.3	5.4
Post-harvest	3.6	6.5	3.9	4.1
Natural Resources	8.6	7.1	8.8	8.4
Other	9.8	9.5	6.7	9.2
Major crops				
Wheat	6.2	4.9	4.3	5.7
Rice	18.0	7.6	6.1	14.4
Maize	5.4	8.0	13.8	7.3
Cassava	0.6	5.8	2.2	1.6
Vegetables	9.4	9.0	18.6	11.0
Fruits	11.7	11.0	17.4	12.7
Sugarcane	5.0	4.9	3.7	4.7
Coffee	0.6	3.0	6.3	2.0
Other	43.3	45.7	27.4	40.7

Note: Shares based on allocation of full-time equivalent researchers.

Source: ASTI database, 2007.

countries and between regions in the structure of the public research sector (Figure 8-3). Public research in the United States is done mainly in state agricultural experiment stations located primarily in colleges of agriculture and in federally administered, but often regionally located, laboratories. A large share of public agricultural R&D in Asia-Pacific and Latin America is conducted by government agencies (about three-quarters of the total). This is similar to the government agency share in a 27-country sub-Saharan African total. A small, but growing proportion of public agricultural research in Latin America and sub-Saharan Africa is conducted by nonprofit institutions. Nonprofit institutions are often managed by independent boards not directly under government control. Many are closely linked to producer organizations from which they receive the large majority of their funding, typically by way of taxes levied on production or exports (see 8.3.3).

8.1.1.2 Private sector spending
Agricultural R&D investments by the private sector have grown in recent years and in the industrialized world now account for more than half of the sum of the public and private research investments. Although private sector performed agricultural R&D appears to have increased in some developing countries, overall the role of the private sector is still small and will likely remain so given weak funding incentives for private research. In addition, many of the private sector R&D activities in developing countries

focus solely on the provision of input technologies or technological services for agricultural production with most of these technologies produced in the industrialized world.

Private sector share of total agricultural research investments are estimated at 37% (Table 8-3). Most of which was performed in the industrialized countries (94%) where they spent on average more on agricultural research than the

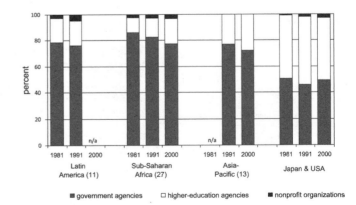

Figure 8-3. *Organizational orientation of public agricultural R&D, 1981, 1991, 2000.* Source: Pardey et al., 2006b based on ASTI data.

Note: The number of countries included in regional totals is shown in parentheses. The reported shares for Japan and the United States may understate the role of nonprofit institutions. n/a indicates not available.

Box 8-1. Plant breeding and biotechnology research.

Trends in multinational plant and biotech research

One of the most rapidly growing areas of private sector agricultural research has been the plant biotech area. This research started in the 1970s, increased very rapidly in the late 1980s and 1990s to over a billion dollars of research in response to the technological opportunities offered by the breakthroughs of cellular and molecular biology and also due to stronger intellectual property rights particularly in the US. Some of this change was due to companies shifting research resources from chemical research to biological research.

Since 1999, several of the six largest biotech firms, which dominate private biotech research worldwide, have reduced their agricultural biotechnology research, and in the aggregate agricultural biotechnology research expenditures probably stagnated. Monsanto reduced its research expenditure, which is about 85% agricultural biotechnology and plant breeding, from US$588 million in 2000 to US$510 million in 2003 before increasing back to $588 million in 2005. Syngenta's plant science R&D expenditures declined from $161 million in 2000 to $109 million in 2003 and to $100 million in 2005 (Syngenta, 2006). In contrast Bayer and BASF seem to be increasing their investments in biotech. Bayer purchased Aventis Crops Sciences, which had a major biotech research program, in 2001. Bayer has made a substantial investment in Agricultural biotech R&D since then and now spends about $80 million on seed and biotech research expenses (Garthof, 2005). BASF spent approximately $82 million in 2004 (Garthof, 2005). They recently (2006) acquired the Belgium biotech firm CropDesign and have committed themselves to spending $320 million on biotech research over the new three years (Nutra Ingredients, 2006).

Public sector investment in agricultural biotech growing rapidly in some large developing countries

Despite the controversy about transgenic crops and generally sluggish investments in biotechnology, government investments in agricultural biotechnology research and development are growing rapidly in some large developing countries. The most dramatic growth in public biotech investments is in China from under 300 million yuan in 1995 to over 1.6 billion yuan in 2003 (equivalent to US$ 200 million). This 1.3 billion yuan increase accounts for between 25 to 33% of the increase in all agricultural research in the same time period (Huang et al., 2005). In addition Chinese cities and provinces have announced major government programs to commercialize the results of public sector biotech research such as the new center in Beijing, which will invest US$160 million over the next three years to nurture 100 companies and 500 labs (Gong, 2006).

National governments in Brazil, Malaysia, and South Africa are also making major investments in agricultural biotech research and some provincial governments such as Sao Paolo in Brazil and Andhra Pradesh in India are also making substantial investments. In July 2006 the Brazilian government announced that it would invest US$3.3 billion over the next 10 years to develop biotechnology for health, industry, and agriculture (checkbiotech.org). Malaysia announced that it would invest US$3.12 billion in agriculture in the next plan period and that agricultural biotechnology would play a major role (Government of Malaysia, 2006). Indian officials said in the spring of 2006 that it will invest US$100 million and the US will add US$24 million on agricultural biotechnology in India (Jayaraman, 2006). South Africa launched Plantbio (www.plantbio.org.za) in late 2004 to support the commercialization of plant biotech products.

---

public sector. In contrast, only 8% of total spending in the developing world was conducted by private firms with the remaining 92% by public agencies. In the developing world, private sector involvement in agricultural research was relatively higher in the Asia and Pacific region with an average of 11% in 2000 (Pardey et al., 2006b).

Private sector involvement in agricultural R&D in OECD countries differs from one country to another. In 2000, more than 80% of total agricultural R&D spending in Belgium, Sweden, and Switzerland was done by the private sector. In contrast, private sector shares were below 25% in Australia, Austria, Iceland, and Portugal that same year. Private and public sectors are involved in different types of research. In 1993 only 12% of the private research in five industrialized countries (Australia, the Netherlands, New Zealand, UK, and the US) focused on farm-oriented technologies compared to 80% in the public sector. Food and other postharvest accounted for 30 to 90% of agricultural R&D investments in Australia, Japan, the Netherlands, and New Zealand. Chemical research accounted for 40 and 75% of

private research in the UK and US, but was less important in Australia, and almost negligent in New Zealand (Alston et al., 1999).

A survey of seven Asian countries during the mid-1990s showed that the share of private investments had grown in three countries (China, India, and Indonesia) even more than the increases in public sector investments (Pray and Fuglie, 2001). However, this growth was uneven across subsectors. Total investments in the agricultural chemical industry in Asia, which includes mostly pest control chemicals and, to a lesser extent, fertilizer and biotechnology, tripled during mid-1980s and mid-1990s. Private spending on livestock research also grew considerably, but growth was substantially slower in other subsectors such as plantation crops and machinery. Both locally-owned and multinational firms played similar important roles in agricultural R&D. Multinational firms accounted for an average of 45% of total private research spending in the seven Asian countries, but with substantial differences among countries. Almost all research in China by truly private firms (rather than govern-

Table 8-3. **Estimated public and private agricultural R&D investments, 2000.**

	Expenditures			Shares	
	Public	Private	Total	Public	Private
	(millions 2000 international dollars)			(percent)	
Asia & Pacific	7,523	663	8,186	91.9	8.1
Latin America & Caribbean	2,454	124	2,578	95.2	4.8
sub-Saharan Africa	1,461	26	1,486	98.3	1.7
West Asia & North Africa	1,382	50	1,432	96.5	3.5
Developing countries, subtotal	12,819	862	13,682	93.7	6.3
Higher-income countries, subtotal	10,191	12,086	22,277	45.7	54.3
Total	**23,010**	**12,948**	**35,958**	**64.0**	**36.0**

Source: Pardey et al., 2006b based on ASTI data.

ment-owned, commercial firms) was by multinational firms in the mid-1990s while in Malaysia only 10% of private sector investment from multinationals. Foreign firms were concentrated in the agricultural chemical and livestock subsectors; i.e., those with the highest growth rates (Pray and Fuglie, 2001).

In SSA, only 2% of total agricultural R&D is conducted by the private sector.[3] Almost two-thirds of the region's private research was done in South Africa. Most firms in SSA have few research staff with low total spending and they focus on crop improvement research, often export crops (Beintema and Stads, 2006).[4] Similarly as in the Asian region, multinationals and locally owned companies play a similarly important role. Given the tenuous market realities facing much of African agriculture, it is unrealistic to expect marked and rapid development of locally conducted private R&D. Yet there may be substantial potential for tapping into private agricultural R&D done elsewhere through creative public-private joint venture arrangements (Osgood, 2006).

In 2000, total investments in all sciences conducted by the public and private sectors combined were over $700 billion (in 2000 international prices) (Table 8-4). The regional shares in the global total differ substantially from the shares in agricultural R&D spending. Industrialized countries combined accounted for about 80% of total science and technology (S&T) spending while SSA's share was less than one percent. There are also considerable differences in the shares of public and private agricultural R&D spending in total S&T spending. Agricultural R&D spending in SSA accounted for more than one-third of the region's total sci-

ence spending while in the other regions in the developing world these shares were considerably lower (9 to 12%). In the industrialized world spending in agricultural R&D was only 4% of the total S&T investments.

8.1.1.3 Intensity of research
In order to place a country's agricultural R&D efforts in an internationally comparable context, measures other than absolute levels of expenditures and numbers of researchers are needed, e.g., the intensity of investments in agricultural research. The most common research intensity indicator is a measure of total public agricultural R&D spending as a percentage of agricultural output (AgGDP).[5] The industrialized countries as a group spent $2.36 on public agricultural R&D for every $100 of agricultural output in 2000, a large increase over the $1.41 they spent per $100 of output two decades earlier, but slightly down from the 1991 estimate of $2.38 (Figure 8-4). This longer-run increase in research intensity is in stark contrast to the group of developing countries; this group has seen no measurable growth in the intensity of agricultural research since 1981. In 2000, the developing world spent just 53 cents on agricultural R&D for every $100 of agricultural output. Agricultural output grew much faster in the developing countries as a group than in the industrialized countries. As a result, intensity ratios remained fairly stable for the developing regions as a group despite overall higher growth rates in agricultural R&D spending in the developing countries, and the intensity gap between rich and poor countries has widened over the years. More than half of the industrialized countries for

---

[3] The private sector does, however, play a stronger role in funding agricultural research, as opposed to performing research itself. Many private companies contract government and higher-education agencies to perform research on their behalf.

[4] Examples are cotton in Zambia and Madagascar and sugar cane in Sudan and Uganda.

[5] Some exclude for-profit private agricultural research expenditures when forming this ratio, presuming that such spending is directed toward input and postharvest activities that are not reflected in AgGDP. For reasons of consistency with these other studies, we excluded national and multinational private companies (but not nonprofit institutions) from the calculated intensity ratios.

Table 8-4 **Total S&T spending by region and shares agriculture in total, 2000.**

	S&T spending	Shares in global total S&T spending	Agricultural R&D as a share of total S&T spending
	(millions 2000 international dollars)	*(percent)*	
Asia & Pacific (26)	94,950	13.4	8.6
Latin America & Caribbean (32)	21,244	3.0	12.1
sub-Saharan Africa (44)	3,992	0.6	37.2
West Asia & North Africa (18)	14,893	2.1	9.6
Developing countries, subtotal (120)	135,079	19.1	10.1
Higher-income countries (23)	573,964	80.9	3.9
Total (143)	**709,043**	**100**	**5.1**

Note: These estimates exclude East Europe and former Soviet Union countries. The number of countries included in regional totals is shown in parentheses. Regional sample sizes are slightly different from those in Table 8-1.

Source: Pardey et al., 2006b.

which data exists have higher research intensity ratios in 2000 than they did in 1981 (and the majority of them spent in excess of $2.50 on public agricultural R&D for every $100 of AgGDP). Most countries in our Asian and Latin American sample (9 out 11 Asian countries and 8 out of 11 Latin American countries) increased their intensity ratios over the 1981-2000. Only six of the 26 countries in SSA had higher intensity in 2000 compared to two decades earlier.

The large and growing gap between developing and industrialized countries as groups is even larger in terms of total, i.e., public and private, agricultural research spending (Figure 8-5). In 2000, the intensity of total spending was nine times higher in rich countries than in poor ones; and four times higher than when only public research spending is used as the basis of the intensity calculation.

Other research intensity ratios can be calculated as well. The industrialized countries as a group spent $692 on public

agricultural research per agricultural worker in 2000, more than double the corresponding 1981 ratio (Table 8-5). The developing countries as a group spent just $10 per agricultural worker in 2000, substantially less than double the 1981 figure. These differences are not too surprising considering that a much smaller share of the workforce in industrialized countries is employed in agriculture, and the absolute number of agricultural workers declined more rapidly in these countries than it did in the developing countries.

Expressing agricultural R&D spending per capita gives a different trend than the other two intensity calculations. Spending per capita for the industrialized countries as a group increased substantially from 1981 to 1991, but has declined since then. About half of the rich countries experienced declining levels of spending per capita; Japan most severely due to the sharp decline in agricultural R&D spending in that country during the 1990s. Spending per capita

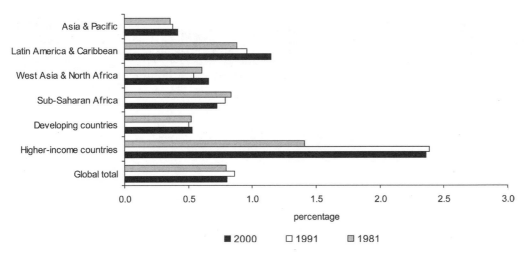

Figure 8-4. *Intensity of public agricultural R&D investments over agricultural output.* Source: Pardey et al., 2006b based on ASTI data.

Note: The intensity ratios measure total public agricultural R&D spending as a percentage of agricultural GDP.

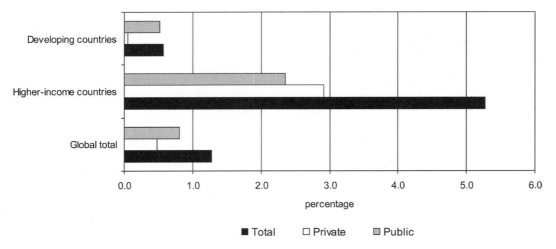

Figure 8-5. *Public, private and total agricultural research intensities, 2000.* Source: Pardey et al., 2006b based on ASTI data.

Note: The intensity ratios measure total public and private agricultural R&D spending as a percentage of agricultural GDP.

levels are much lower for the developing countries. Most countries, especially those in Africa, spent less than $3 per capita in 2000; while 59% of the industrialized countries invested more than $10 per capita in 2000. In contrast to the group of rich countries, agricultural R&D spending per capita for the developing countries as a group continued to increase from $2.12 per capita in 1981 to $2.72 in 2000. The exception is SSA where spending per capita has declined during the 1981-2000 period.

8.1.1.4 International agricultural R&D
International agricultural research efforts began in the middle of the 20th century when the Ford and Rockefeller Foundations placed agricultural staff in developing countries to collaborate with national scientists. These efforts evolved into the establishment of four international organizations

during the 1960s. In 1971 these centers formed the basis for the Consultative Group on International Agricultural Research (CGIAR or CG). Currently there exist 15 centers (see Chapter 2.2.4), with a total budget of US$415 million in 2004—US$384 million in 2000 prices. Although the CG system has played an important role in the Green Revolution, it only spends a small part of total of the global agricultural R&D investment. In 2000, the CG represented 1.6% of the US$23 billion global public sector investment in agricultural R&D (from 0.8% in 1981); 2.9% when spending by the rich countries is excluded (Pardey et al., 2006a).

After an initial expenditure of US$7 million in 1960, total spending rose to US$13 million per year in 1965, in inflation-adjusted terms. By 1970, the four founding centers (IRRI, CIMMYT, IITA, and CIAT) were allocated a total of US$15 million annually. During the next decade, the total

Table 8-5. **Other intensity ratios, 1981, 1991 and 2000.**

	Public agricultural R&D spending					
	Per capita			Per capita of economically active agricultural population		
	1981	1991	2000	1981	1991	2000
	(2000 international dollars)					
Asia & Pacific	1.31	1.73	2.35	3.84	5.23	7.57
Latin America & Caribbean	5.43	4.94	4.96	45.10	50.54	60.11
sub-Saharan Africa	3.14	2.69	2.28	9.79	9.04	8.22
West Asia & North Africa	3.24	3.63	3.66	19.15	27.30	30.24
Developing countries, subtotal	*2.09*	*2.34*	*2.72*	*6.91*	*8.14*	*10.19*
Higher-income countries, subtotal	10.91	13.04	11.92	316.52	528.30	691.63
Total	*3.75*	*4.12*	*4.13*	*14.83*	*16.92*	*18.08*

Source: Pardey et al., 2006b based on ASTI data.

number of centers increased to twelve, and the funding per center increased. This led to a tenfold increase in nominal spending to US$141 million in 1980. During the 1980s, spending continued to grow, more than doubling in nominal terms to reach US$305 million in 1990. The rate of growth had slowed but was still substantial. In the 1990s, however, although the number of centers still grew, funding did not grow enough to maintain the level of spending per center and growth rates declined. Since 2000, funding has grown in total but with a continuing trend toward earmarked support for specific projects and programs of research involving multiple centers and other research providers outside the CGiAR. In 1980, the share of the four founding centers of total CG spending was 54%, but by 2004 it had slipped to only 36% (Pardey et al., 2006a).

8.1.2 Determinants of public and private R&D investments

A conceptual model of the factors that influence these investments is needed to make a critical assessment of research investments trends.

8.1.2.1 Determinants of public research

In the absence of public intervention, private firms will under-invest in research when the output of that research has the characteristics of a public good—that is, the outputs of research are often non-rival and non-excludable. Because of the public goods nature of research, the social benefits are much higher than the private benefits, and hence the justification for public intervention. While many public investments have high social benefits, public investment will only be justified if the return is higher than other forms of public investment. A review of the RORs to research (see 8.2) shows that public investments do have high payoffs, often 40 to 50% or more. Considering that private companies and governments usually can obtain credit at interest rates below 10% and the public RORs on other types of government investments are considerably lower than 40%, these returns are very high.

Whereas studies that show high social RORs to research investments may convince economists that agricultural research is a good investment, most policy makers who actually do not appear to have been sufficiently convinced that these high social RORs warrant large investments. Rather, public investments in agricultural research respond to many of the same forces that influence the amount and direction of private research (Hayami and Ruttan, 1985).

Public agricultural R&D increases when there are advances in basic knowledge and technology in fields such as biology, chemistry, engineering, and information technology that increase the possibility of an innovation or reduce the cost of developing an innovation. This is referred to as an increase in technological opportunity. The discovery of dwarfing genes in rice and wheat created the opportunity for plant breeders around the world to produce many new types of varieties that would respond to higher doses of nutrients and water. These opportunities increased the potential return to research and led to major increases in public sector plant breeding research around the world. Likewise, the tools of biotechnology have created a major shift in the innovation possibility curve for plant and animal breeding,

pest control, and abiotic stress tolerance, and many governments have again responded by increasing their investments in research (Box 8-1).

Changes in the demand for agricultural products by farmers and consumers induced public R&D investments (Hayami and Ruttan, 1985). Historically in Asia, population and per capita income increases increased the demand for basic food grains such as rice; farmers were unable to increase production rapidly due to limited land and agricultural prices increased. Private firms did not attempt research to fulfill this demand because profits were projected to be insufficient. When farmers and consumers were sufficiently well organized and demanded a solution, Asian governments invested in agricultural research. For example, following World War I Japan had very high rice prices and consumers demanded cheaper food, but Japanese farmers had no land for expansion. The government responded by investing in research that eventually led to nutrient-responsive rice varieties; this resulted in biological technologies and inexpensive fertilizers increased yields per unit of land. In the late 1950s, national governments, nonprofit foundations, and aid donors responded in a similar manner to the food crisis and high food prices caused by rapid population growth in other Asian countries and invested in the international agricultural research centers and the national agricultural research systems.

Demand for solutions to specific problems, such as a new disease or pest or the shortage of a key input (such as the aforementioned land shortages in Asia and resulting development of land-saving technologies), also lead to public sector research investments and can direct the allocation of investments. The worldwide public sector response to Avian Influenza is a current example. There are also demands that receive insufficient investments, for example, research on diseases such as malaria or investment in appropriate agricultural technologies for poor people. Whether these factors will actually lead to more or less R&D investments by governments depends on the structure of the government, its ability to raise money, and the power of various interest groups to influence government spending decisions.

Some governments are more committed to R&D as a major tool for economically sustainable development. They will put a larger share of their budget into research of all types including agriculture (Anderson, 1994). The structure of the research system will also influence the size and direction of agricultural research (Morris and Ekasingh, 2002). Some governments have structured their R&D system to be more responsive to the demands of the agricultural sector while others are more responsive to demands of the food consumers, agricultural scientists, or foreign aid donors. The size and power of different interest groups can also have a major impact on the size and direction of agricultural R&D. Commercial farmers can push their governments for large investments in research that is likely to concentrate on reducing costs crop production or increasing demand. If the textile industry is strong, research will be focused on bringing down the cost of cotton. Strong consumer lobbies are likely to lead to research that lower food prices. In countries where private research on topics such as maize breeding or poultry breeding become strong, the companies that are doing this research will lobby for government to stop competing with them in the applied research of the development of new varieties and to move upstream to work on things like germplasm enhancement (Pray and Dina Umali-Deininger, 1998; Pray, 2002).

8.1.2.2 Determinants of private research

For private firms agricultural R&D is an investment that they hope will increase their profits. The returns to private research improve in the presence of sizable expected demand for the research products, the availability of exclusion mechanisms to appropriate part of the benefits from the new product or process, favorable market structure, and a favorable business environment that permits efficient operations (Pray and Echeverría, 1991). The profitability of private research also depends on technological opportunities (Pray et al., 2007).

Potential demand for inputs and consumer products developed through research, and thus market size, varies among regions depending on the size of the population, the purchasing power of the prospective buyers, local agroclimatic conditions, and sectoral and macroeconomic policies that influence input and output prices. In 2000, for example, the size of the global crop protection market was estimated to be US$28 billion (Syngenta, 2004), and consequently the first generation of biotechnology traits were designed to capture a portion of this market by either substituting for, or enhancing the productivity of, existing chemicals. Firms introduced these traits into crops with large markets, thereby enhancing their ability to extract rents.

Changes in the incentive environment affect the demand for research services and the speed at which countries can adopt new agricultural innovations. Macroeconomic and sectoral policies alter the relative profitability of agricultural activities which in turn affect the expected profitability of adopting different agricultural innovations, as well as the capacity of different segments of the farm community to acquire the new technologies (Anderson, 1993). The effectiveness of agricultural support services delivery (public and private), in particular agricultural extension, and rural infrastructure (roads, markets, irrigation) will also have a major influence on the types and range of technologies introduced and the speed of adoption. Bilateral and multilateral trade agreements and phytosanitary legislation reshape trading rules and influence market access and thus potential market size (Spielman and von Grebmer, 2004).

Government policies that affect the local business environment directly influence the returns to private research. Examples of such policies are government marketing of inputs that reduce the market share of private firms and licensing and investment regulations that favor smaller firms over larger firms (Pray and Ramaswami, 2001).

Appropriability is an important precondition for private for-profit firms to participate in agricultural research. If firms can not capture (appropriate) some of the social benefits of their research, they cannot make profits on their research investments and will stop investing (Byerlee and Fischer, 2002). To capture some of the benefits from the innovation, the innovating firm must be able to prevent imitators from using the innovation. The ability to do this is a function of the characteristics of the technology, the laws on intellectual property and their enforcement, the struc-

ture of the industry that is producing the technology and the industry that is using it. The legal means of protection against unauthorized use include patents, plant breeder's rights, contracts, and trademarks. They also control their use by keeping inventions or key parts of their inventions secret, which in some countries is protected by trade secrecy law. These legal means tend to give limited protection in developing countries (Pray et al., 2007).

Inventors can also protect their inventions by biological means such as putting new characteristics into hybrid cultivars or including other technical means to prevent copying. In the case of hybrids the seeds will yield 15 to 20% less. This is usually sufficient incentive for farmers to purchase new seeds each year. In the case of genetic use restriction techniques some of the proposed techniques (none are in commercial use yet) would use genetically engineered crops, which would produce sterile seed unless the seed had been treated with a specific chemical.

The degree of appropriability achieved is a function of the strength of intellectual property laws, and other factors causing farmers to prefer to purchase a technology, the degree to which government agencies can enforce the law which exist, the structure of industry that reduces the cost of enforcing IPRs, and the technical capacity of firms to balance the value they can charge farmers for their products, which ultimately depends on the farmers receiving more value than they pay for, protect their varieties through the use of hybrids (Pray et al., 2007).

Private research investments are also determined by the potential costs of the agricultural research program and the associated risks (Pray and Echeverria, 1991). The cost of research is the combination of quantity and price of research inputs, the number of years needed to develop a new technology, and available knowledge in the area of science. Such costs decrease with the supply of research inputs, the presence of a favorable business environment, the stock of existing knowledge and technology, and available human capital for conducting research activity. Research costs increase in the presence of anticompetitive markets or when firms have to meet certain regulatory requirements.

The supply of research inputs and thus their price depends on the availability and accessibility of research tools and knowledge, many of which are produced by the public sector. For example, private breeders, to add desirable traits to new private varieties, may use improved populations of crop germplasm developed by public research programs as parent material. The advances in biotechnology knowledge have led to a significant increase in private investment in agricultural research in the United States and Europe over the past two decades. Greater private sector R&D implies that the marginal cost of applied agricultural research will decline as firms take advantage of economies of scale and scope. However, the concentration of key research inputs amongst a few firms raises the possibility that the cost of conducting research for those who do not have access to such technologies will increase (Pray et al., 2007).

The domestic supply and quality of human capital, a key input to the research activity, influences the level of research investments. In the Philippines, the availability and low cost of hiring local well-trained research personnel encouraged some multinational firms to transfer their research programs to teams of Filipino scientists (Pray, 1987). The domestic supply of skilled personnel is heavily dependent on the level and composition of public and private expenditures on education.

Several aspects of the business environment affect the level and productivity of research costs. Industrial policy can influence the degree of market concentration, the intensity of competition, and the prices of research inputs and outputs. Various government incentive programs, such as government contracts for new products and processes, grants and concessional loans, technical information services, and tax incentives, reduce research costs. Indirectly, the development of capital markets makes it easier for firms to raise funds for research (for example, venture capital). Bilateral and multilateral agreements also improve trade opportunities by facilitating access to intermediate technologies.

Regulation such as product quality standards, quality testing regulations and seed certification procedures can greatly increase the costs of commercializing research output and they can delay the adoption to new technology which reduces the incentive to innovate and reduces the benefits to farmers. Regulations that have been put in place in many countries to ensure that products developed using biotechnology are environmentally benign and safe for human consumption are necessary to gain consumer acceptance, but they have greatly increased the cost of developing and releasing transgenic plant varieties. For example, one seed company spent US$1.6 to 1.8 million to obtain regulatory approval for Bt cotton in India. This is more than the annual research budgets of most Indian seed companies. As a result, only the largest companies can afford to attempt to commercialize genetically modified crops (Pray et al., 2005). Bangladeshi regulations that required irrigation pumps and diesel engines meet efficiency standards of wealthy countries delayed the commercialization of inexpensive Chinese irrigation equipment and slowed the spread of high-yielding rice varieties by 5 to 10 years (Gisselquist et al., 2002)

8.1.3 Investments in other AKST components

Investment data for other AKTS components, such as education and mainstreaming traditional knowledge, are difficult to obtain.

Due to the public good attributes of extension services, it not surprising that the great majority of official extension workers worldwide are publicly-funded and most extension is delivered by civil servants. Universities, autonomous public organizations, and NGOs deliver perhaps 10% of extension services, and the private sector may deliver another 5% (Anderson and Feder, 2003).

The structure and function of national extension systems continue to change, particularly as the level and source of funding, especially public funding, changes across different countries. In many countries, there is a continuing effort to shift the cost of extension to farmers, although these different approaches to privatizing extension or to increase cost recovery by public extension systems have met with different levels of success (Anderson, 2007), private sector involvement remains small.

Given the numbers of extension personnel and the likely costs incurred in the different country contexts, agricultural extension investment is of the same order of magnitude

(although likely lower) as the agricultural research world presented in expenditure terms (Table 8-1); so it is surprising that it has been subject to relatively little critical data collection and analysis. In contrasting differences between developing and more industrialized countries, one feature is the even more extreme differentiation between public and private entities; however, the situation is not fully clear (World Bank, 2006; Anderson, 2007).

8.1.4 Funding agricultural R&D in developing countries

Although various new funding sources and mechanisms for agricultural research have emerged in recent decades (see 8.3), the government remains the principal source of funding for many developing countries. For example, the principal agricultural research agencies in the largest countries (in terms of agricultural R&D investments) such as Brazil, China, India, Mexico, Nigeria, and South Africa are still mostly funded by the government. In contrast, the principal agencies in a number of countries have been able to diversify their sources of support through contract research (for example, Chile and Cote d'Ivoire) or a commodity tax on agricultural production or export (for example, Uruguay, Malaysia, Colombia) (ASTI, 2007).

Bilateral and multilateral funding has been an important source for agricultural R&D for many countries. Since 1970, both multilateral and bilateral assistance grew in real terms, but began to decline after the early 1990s to only US$51.2 billion by 2001. In recent years, ODA has increased again (Table 8-7). After several decades of strong support, international funding for agriculture and agricultural research began to decline around the mid-1980s. This decrease is mostly related to the significant increase in the share of ODA spent on social infrastructure and services (FAO, 2005a). Data on the sectoral orientation of aid are available for bilateral funds only. The agricultural component of bilateral assistance grew steadily and accounted for 16% in 1985, declining thereafter to 4% in 2003. Regionally the largest proportional reductions in assistance occurred in Asia. ODA to agriculture halved in SSA and decreased by 83% in South and Central Asia during the period 1980-2002 (FAO, 2005a).

Data on aggregate trends of donor funding for agriculture and agricultural research are unavailable, but information on agricultural R&D grants and loans from the World Bank and the United States Agency for International Development (USAID) is accessible. The amount of funding that USAID directed toward agricultural research conducted by national agencies in less-industrialized countries declined by 75% in inflation-adjusted terms from the mid-1980s to 2004. Again, Asian countries experienced the largest losses, but funding to Africa and LAC was also cut severely (Pardey et al., 2006b). Over the past two decades, World Bank lending to the rural sector has been erratic, but after adjusting for inflation, the general trend has been downward as well. The exception is the large amount of lending in 1998, which resulted mostly from loans with large research components approved for India, China, and Ethiopia (Pardey et al., 2006b).

There appears to be no single cause for the decline to the donor support for agriculture between1980-2003, although

Table 8-6. **Aid to agriculture, 1970-2004.**

Year	Total official development assistance (ODA)	Bilateral aid	
		Amount	Share to agriculture
	(million 2000 U.S. dollars)		*(percent)*
1970	24,719	20,886	4.91
1975	35,448	26,233	11.13
1980	49,166	31,875	16.63
1985	41,773	30,782	15.93
1990	67,071	47,540	11.39
1995	64,077	44,129	9.82
2000	53,749	36,064	6.36
2003	65,502	47,222	4.22
2004	74,483[a]	50,700[a]	n/a

Note: n/a indicates not available.

[a]Preliminary estimate

Source: Pardey et al., 2006b.

the following factors could have contributed (Morrison et al., 2004):

- Loss in donor confidence in agriculture;
- Perceived high transaction costs and complexities in agricultural investments;
- Changes in definitions in aid statistics;
- Weaker demand for assistance to agriculture from many developing country governments;
- Changes in development policy and approaches to more market led approaches
- Shifting emphasis towards the education and health sectors;
- Changes in aid modalities, such as the movement away from the green revolution technologies of the 1960s to 1980s and the integrated rural development projects of the 1980s and 1990s, to the current sector wide approaches and support to poverty reduction strategies (Eicher, 2003).

However, science and the use of new ideas have been acknowledged by many as being important in delivering the MDGs and there has been renewed interest by the donor community on the role of agriculture in promoting economic growth and poverty reduction. In addition, a number of new funding sources such as the Bill and Melinda Gates Foundation have become available.

A number of developing countries, especially in SSA, have become increasingly dependent on donor funding. Although the share of donor contributions in total funding for SSA agricultural R&D has declined slightly in the later half of the 1990s (Figure 8-6). These declines resulted in part from the termination of a large number of World Bank projects in support of agricultural R&D or the agricultural sector at large. Donor contributions (including World Bank

loans) accounted for an average of 35% of funding to principal agricultural research agencies in 2000. Five years earlier, close to half the funding of the 20 countries for which time series data were available was derived from donor contributions. These regional averages mask great variation among countries. In 2000, donor funding accounted for more than half of the agricultural R&D funding in seven of the 23 sample countries. Eritrea, in particular, was highly dependent on donor contributions. In contrast, donor funding was virtually insignificant in Botswana, Malawi, Mauritius, and Sudan (under 5%) (Beintema and Stads, 2006).

Since the International Conference on Financing for Development convened in Monterrey 2002, the share of aid to least developed countries in donor gross national income (GNI) has increased to 0.08%, and longer term commitments to reach 0.7% have been made by donors but it is still short of the target, and the level of the external assistance to agriculture has remained unchanged (FAO, 2005a). However the situation continues to change.

To improve upon past efforts to achieve food security, the New Partnership for Africa's Development (NEPAD) has developed the Comprehensive Africa Agriculture Development Programme (CAADP). In line with CAADP's goal of improving agricultural productivity with an average of 6 percent per year, is the recommendation to double the region's intensity in agricultural research by 2015 (IAC, 2004). Doubling Africa's agricultural research intensity ratio from 0.7% in 2000 to about 1.5% by 2015 would require an average annual growth rate in agricultural R&D spending of 10% (Beintema and Stads, 2006). This goal seems unlikely considering that growth in Africa's R&D spending averaged 1% per year during the 1990s as reported earlier. There is no evidence that governments and donor organizations have substantially increased their funding to agricultural research

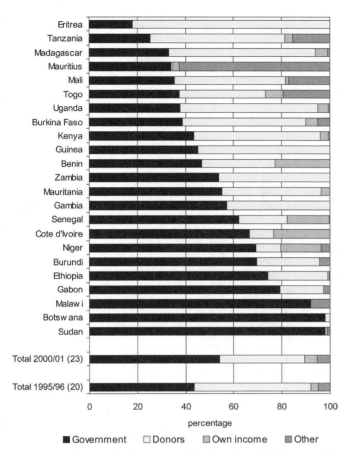

Figure 8-6. *Country-level sources of funding in sub-Saharan Africa, 1995/96 and 2000.* Source: Beintema and Stads, 2006.

Notes : Figure includes only funding data from the main agricultural research agencies in each of the respective countries. Combined, these agencies accounted for 76% of total spending for the 23-country sample in 2000. Data for West Africa, with the exception of Nigeria, are for 2001.

since the late 1990s and it is unlikely that the high level of donor support will continue indefinitely.

8.2 Impacts of AKST Investments

The purpose of undertaking and impact assessment of agricultural AKST depends on when the assessment is done in relation to the project cycle. It can be undertaken before initiating the research (ex-ante) or after completion of the research activity (ex-post). Ex-ante impact studies (proactive) can indicate the potential benefits from research and, therefore, assist managers in planning, priority setting and, consequently, in allocating scarce resources. They can also provide a framework for gathering information to carry out an effective ex-post evaluation. Ex-post studies (reactive) can demonstrate the impacts of past investments in achieving the broader social and economic benefits. Most commonly, ex-post impact assessments are carried out because decision makers and research managers usually require them as a precondition for support. They are undertaken to (1) help managers by providing better and more convincing advice on strategic decisions about future AKST investment;

(2) make scientists and researchers aware of the broader implications, if any, of their research; (3) Identify weak links in the research to affect pathways; and (4) better inform managers on the complementarities and tradeoffs between different activities within a research program (Maredia et al., 2001).

8.2.1 Conceptual framework

AKST investments generate different outputs including technologies of various types, management tools and practices, information, and improved human resources. In the literature the term impact is used in many different ways (DANIDA, 1994; Cracknell, 1996; Pingali, 2001). In this chapter we refer to impact of AKST investment as the broad long-term economic, social and environmental effects (SPIA, 2001). Impact assessment is a process of measuring whether a research program has produced its intended effects, such as increase in production and/or income, improvement in the sustainability of production systems (Anderson and Herdt, 1990) or improvements in livelihood strategies. In any comprehensive impact assessment, it is necessary to differentiate between the research results (outputs) and the contribution of research to development efforts (outcomes) and both aspects should be addressed simultaneously. A conceptual framework for assessing impacts (Figure 8-7) incorporates the multifaceted consequences of AKST investment in terms of both institutional and developmental impacts including spillover effects. This framework recognizes the multiple impacts of AKST investments and the need for multi-criteria analysis as well as RORs earned by such investments.

A comprehensive impact assessment requires multiple techniques using both qualitative and quantitative assessment. This means that not all impacts associated with AKST investment can be quantified and valued in monetary terms, although new techniques are emerging that could complement ROR measures especially in valuing social and environmental consequences (Anandajayasekeram et al., 2007). There are also concerns about exclusively reliance on ROR for decision making. The portfolio approach considers the internal rates of return across projects rather than considering them in isolation and aims to maximize the expected returns to the entire AKST investment. Despite its shortcomings the ROR to investment is the most commonly used measure to compare the relative performance of investments and a frequently used measure of research efficiency. Most literature on impact assessment of AKST investment is largely based on RORs and these studies are assessed in the following sections (Alston et al., 2000a; Anandajayasekeram et al., 2007).

8.2.2 Economic impact assessment

Economic impact measures economic benefits produced by an AKST project or program and relates these benefits with the economic costs associated with the same project or program. This information is used to compute measures like benefit-cost ratio, internal rate of return (IRR) and net present value of benefits (NPV). Economic impact evaluations are intended to measure whether a project or program actually had (or expected to have) an economic impact and compare this impact with project or program costs. They do not measure whether it was designed or managed and ex-

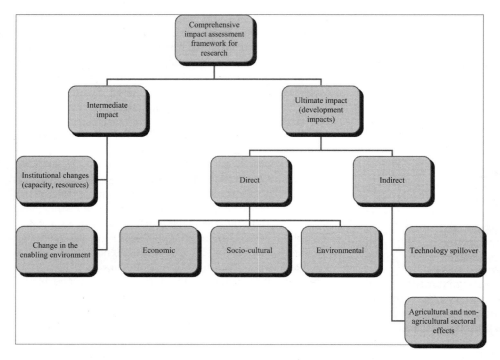

Figure 8-7. *Comprehensive impact assessment framework for R&D investment.* Source: Shrestha and Bell, 2002.

ecuted optimally (Evenson, 2001). An AKST program may have other relevant impacts, such as social (poverty reduction, enhanced nutrition, equity) and environmental effects; and benefits of the research may be distributed in different ways. Some nonmarket impacts such as environmental or health effects of AKST could potentially be given economic value and incorporated into economic analysis. Measurement in these cases is, however, usually more difficult than the measurement of economic impacts that are observable in product or input markets. These attributes should be accounted for in some way, even if economic values cannot be ascertained, when a more realistic evaluation of research impacts is required. In any meaningful empirical analysis, a multi-criteria approach is recommended to assess the impact of AKST assessment.

The literature on economic impact studies includes a wide range of levels of impact analysis. The economic basis for government involvement in agricultural AKST is the perception of market failure leading to private underinvestment (Nelson, 1959; Arrow, 1962; Alston and Pardey, 1998). The appropriate criterion for the assessment of policy aiming to correct market failure is the effect on net social benefits, and this can be expressed as a social rate of return (ROR) to public investment in agricultural AKST (Alston et al., 2000a).[6]

8.2.3 Methodological limitations of impact measurements

Although there have been significant developments in impact assessment methodologies a number of issues still need further attention. Key among these are the issues of attribution, incrementality, causality, defining counterfactual situations, and estimating economic impacts for organizational and institutional innovation and social science and policy research. The issue of counterfactual situations refers to the significant problem of determining what the pattern of productivity growth would have been in the absence of a particular research investment (Alston and Pardey, 2001). This is associated with dynamics of productivity factors even in the absence of AKST investment. AKST programs operate in environments in which ordinary or "natural" sequences of events influence outcomes. Impact assessment and ROR estimates must arrive at estimates of net intervention effects, i.e., they should measure the changes attributable to the intervention.

Causality is another issue that merits attention. In measuring the impacts of AKST investments, it is important to ensure that the impacts measured are the results of the technologies and activities undertaken within the program/project. However, as one moves from the direct product/output to broader economic, social and environmental effects, the chain of causal events is too long and complex, and the variables affecting ultimate outcomes are too numerous to permit the identification and measurement of impacts of specific interventions (Biggs, 1990; Rossi and Freeman, 1993). This is further complicated by the time lag between initial investment and reaping its return.

Attribution problems arise when one believes or is trying to claim that a program has resulted in certain outcomes,

---

[6] In the literature the terms financial, economic and social rates of returns mean different things, but in this chapter the term economic rate of return and social rates of return are used interchangeably. This is because the various meta-analyses do not explicitly make this distinction.

and there are alternative plausible explanations. Relating an impact indicator with a specific research investment is only valid in the absence of other effects on indicators, such as markets and policies (Ekboir, 2003). In addition, many ROR estimates often fail to account for the effects of work done by others in the research development continuum. Temporal aspects of the attribution problem would result when assuming a specific time lag between research results and their implementation. At times, the period over which research affects productivity may be overestimated. A number of strategies can be used to address attribution, called contribution analysis (Mayne, 1999), which may enhance the validity of the estimates, but do not eliminate the problem.

In many estimates the spill-over effects are not usually included as benefits (see 8.2.7). In others, the effects resulting from changes in rural employment, health and education policies and programs are excluded. Environmental impacts, both negative and positive, are often ignored (see 8.2.5) as well as those costs arising from institutional marketing arrangements. This issue can be addressed through estimations of the ROR for research and complementary services, as well as research. Increasingly it is likely that valuation of nonmarket impacts of agricultural AKST will be incorporated into economic analysis. Past studies did not address these issues because of measurement difficulties and the fact that research impacts directly observable in commodity and factor markets were abundantly available. Research systems are now increasingly called upon to provide positive nonmarket environmental or health benefits as well as to mitigate past negative impacts. A number of economic tools are now available to measure nonmarket environmental benefits and costs such as environmental quality, longevity, and health effects (Freeman, 1985; Feather et al., 1999; Dolan, 2000; Hurley, 2000).

There has been little empirical work in the area of assessing the ROR for social science research, and organizational and institutional innovations including capacity strengthening. This lack is associated with the difficulty of attributing any change in policy, institutions or process and the linked economic growth or poverty reduction to research information generated by social science or other factors (Alston and Pardey, 2001). Attempts to quantify the ROR to social science research have used esoteric methods, utilizing "incredible identifying assumptions" that cannot be robustly defended (Gardner, 2003; Schimmelpfennig and Norton, 2003; Schimmelpfennig et al., 2006). In addition, social and applied science research is often undertaken jointly and difficult to isolate.

8.2.4 Empirical evidence
There are many studies on the ROR of investment in agricultural development. This critique draws heavily upon a number of meta-analyses (Alston et al., 2000a; Evenson, 2001; Evenson and Rosegrant, 2003; Thirtle et al., 2003), which cover 80% of published materials (more than one thousand studies). Most of these studies are associated with production technologies, far fewer address other research outputs e.g., post harvest processing, marketing, policy, organizational or institutional innovations. We recognize the shortcomings in these data but these reviews provide useful insights into impacts of AKST investments across con-

tinents, commodities and components (research, extension, private sector research.

8.2.4.1 Rate of returns to national AKST investments
One meta-analysis estimated the economic impact of agricultural R&D investment at the national level for 48 selected developing countries in Africa, Asia, and Latin America (Thirtle et al., 2003) The analysis revealed that R&D expenditures per unit of land have an elasticity of 0.44 in terms of productivity. It was also noted that the elasticity of agricultural R&D is positive and highly significant in all cases and is slightly larger for Africa than Asia and both are over 50% greater than the Latin America's elasticity. The elasticity of value added per unit of land with respect to agricultural R&D was used to calculate ROR to agricultural R&D at the country and the continental level (Table 8-7). The estimated ROR for the sample countries in Africa ranged between -12 and 58%. In only three cases the gains were less than the expenditures. For the Asian countries, the estimated ROR ranged between -1 and 50%; and appears to be less varied and generally higher. The mean of the country RORs for Asia (26%) is better than for Africa (18%) and the weighted mean (31%) is still higher. These means are dominated by the huge agricultural sectors of China and India, both of which seem to have done well in economic terms. In the case of Latin America, only five of the thirteen countries had positive RORs. The estimated ROR ranged between -22 and 40%. The poor results for the Latin American countries are at least partly due to the limitation of data availability (Thirtle et al., 2003).

8.2.4.2 Rates of return to crop genetic improvement investments
Over the years, a significant amount of AKST resources have been devoted to genetic improvement. A second assessment of economic consequences of crop genetic improvement estimated the economic impact of 17 commodities and 35 country/regions using a "global market equilibrium" model (Evenson and Rosegrant, 2003). Benefit/cost ratio (using 6% as the external interest rate) and IRRs of crop genetic improvement programs by region have been computed for both national agricultural research systems (NARS) and international agricultural research centers (IARCs) (Table 8-8). The IRRs for the NARS ranged between 9 and 31%, which are considerably lower than the ones reported in individual studies. This is primarily because most individual studies tend to ignore the research costs to build the germplasm stock that is required to reach the stage where benefits are produced. The lowest IRR was observed for SSA (9%). The IRRs for the IARC programs are very high and ranged between 39 and 165%. The lowest IRR was observed for Latin America. These high RORs reflect the leveraging associated with the high production of IARC crosses and high volume of IARC germplasm (Evenson and Rosegrant, 2003).

8.2.4.3 Economic impacts of research and extension investments
A number of economic impact studies were assessed to evaluate the contribution of agricultural research and extension programs both public and private, using the estimated ROR on investment to index economic impacts (Evenson,

Table 8-7. Comparison of Rate of Return (ROR) for national agricultural R&D expenditureacross sub-regions.

Sub-regions	Countries	Mean ROR (%)	Weighted mean ROR (%)	Countries with negative ROR
Africa	Algeria, Botswana, Ethiopia, Cote d'Ivoire, Ghana, Guinea-Bissau, Kenya, Lesotho, Mauritania, Morocco, Rwanda, Senegal, Tanzania, Tunisia, Uganda, Zambia, Zimbabwe	18	22	Lesotho, Senegal, Tanzania
Asia	Bangladesh, China, India, Indonesia, Jordan, Malaysia, Nepal, Pakistan, Philippines, Sri-Lanka, Thailand	23	26	Sri-Lanka
Latin America	Bolivia, Brazil, Chile, Colombia, Costa-Rica, Dominican Republic, Guatemala, Honduras, Jamaica, Mexico, Panama, Peru, Venezuela	10	-6	Brazil, Dominican Republic, Jamaica, Mexico, Panama, Peru, Venezuela

Source: Thirtle et al., 2001.

2001) (Table 8-9). The benefit exceeded cost in SSA almost 15 years later than was the case for Latin America and Asia, causing the low IRR. The available evidence suggests that the economic RORs to agricultural R&D are high (Evenson, 2001). The broad scope of the evidence for high returns suggests considerable international spillovers. Economic RORs to agricultural research are likely to be above most public and private rates (Fuglie et al., 1996; Alston et al., 2000a; Evenson, 2001). Recently studies have been carried out on the impact of natural resource management research. In most cases however these were satisfactory but lower than in germplasm improvement research (Waibel and Zilberman, 2007).

Due to a variety of market failures, private returns to R&D are far smaller than economic returns as private developers cannot appropriate many of the benefits associated with their research (Evenson and Westphal, 1995; Scotchmer, 1999; Shavall and van Ypserle, 2001) (see 8.1.2). In agriculture in particular, firms often have difficulty in capturing much of the economic benefits of their investments (Huffman and Evenson, 1993); e.g., in the US seed companies retained 30 to 50% of the economic benefits from enhanced hybrid seed yields and 10% of benefits from non-hybrid seed during 1975-1990 (Fuglie et al., 1996). A key market failure that inhibits developers from recovering the

cost of R&D in agriculture is the potential for resale of seeds (Kremer and Zwane, 2005). The gap between social and private returns may be more acute in tropical agriculture, where market failures are particularly severe and the intellectual property rights (IPR) environment is weaker (Pray and Umali-Deininger 1998; Kremer and Zwane, 2005).

Very often the reported higher economic ROR is attributed to the selectivity bias. First, highly successful programs are likely to be evaluated. Second, the unsuccessful evaluations are less likely to be published than evaluations showing impact. However, one can compare the studies covering aggregate programs, which includes both successful and unsuccessful, with studies of specific commodity programs and the evidence is based on a substantial part the world's agricultural research and extension programs (Evenson, 2001). Returns to AKST investments vary across continents, commodities, types of research, methods of estimation, public versus private, and over time (Alston et al., 2000a) (Tables 8-10, 8-11, 8-12).

The distribution of ROR for crops, livestock, and multiple commodities is similar to that for the entire sample (Table 8-11). A substantial difference in the distribution of ROR is observed for resources research; these estimates mostly include forestry research, for which the research lags are relatively long, contributing to the relatively low average

Table 8-8. Costs-benefits and internal rate of return for NARS and IARC crop genetic improvement programs by region.

	NARSs		IARCs	
	Estimated benefits		Estimated	Lower range
	IRR	B/C	IRR	B/C
Latin America	31	56	39	34
Asia	33	115	115	104
West Asia-North Africa	22	54	165	147
sub-Saharan Africa	9	4	68	57

Note: The International Model for Policy Analysis of Agricultural Commodities and Trade (IMPACT) model developed by IFPRI is a partial equilibrium model covering 17 commodities and 35 country/regions.

Source: Evenson and Rosegrant, 2003.

Table 8-9. **Summary of Internal Rate of Return (IRR) estimates.**

	Number of IRRs reported	Distribution						Approx. median IRR
		0-20	21-40	41-60	61-80	81-100	100+	
	(count)	(percent)						
Extension								
Farm observation:	16	56	0	6	6	5	6	18
Aggregate observations	29	24	14	7	0	27	27	80
Combined research and extension	36	14	42	28	03	8	16	37
By region:								
OECD	19	11	31	16	0	11	16	50
Asia	21	24	19	19	14	.09	14	47
Latin America	23	13	26	34	8	.08	.09	46
Africa	10	40	30	20	10	0	0	27
All extension	81	26	23	16	3	.19	13	41
Applied research								
Project evaluation	121	25	31	14	18	6	7	40
Statistical	254	14	20	23	12	10	20	50
Aggregate programs	126	16	27	29	10	9	9	45
Commodity programs:								
Wheat	30	30	13	17	10	13	17	51
Rice	48	8	23	19	27	8	14	60
Maize	25	12	28	12	16	8	24	56
Other cereals	27	26	15	30	11	7	11	47
Fruits and vegetables	34	18	18	09	15	9	32	67
All crops	207	19	19	14	16	10	21	58
Forest products	13	23	31	68	16	0	23	37
Livestock	32	21	31	25	9	3	9	36
By region:								
OECD	146	15	35	21	10	07	11	40
Asia	120	08	18	21	15	11	26	67
Latin America	80	15	29	29	15	7	6	47
Africa	44	27	27	18	11	11	5	37
All applied research	375	18	23	20	14	8	16	49
Pre-invention science	12	0	17	33	17	17	17	60
Private sector R&D	11	18	9	45	9	18	0	50
Ex-ante research	87	32	34	21	6	1	6	42

Source: Evenson, 2001.

Table 8-10. **Ranges of rates of return.**

Sample	Number of observations	Rate of return				
		Mean	Mode	Median	Minimum	Maximum
	(count)	(percent)				
Full sample[a]						
Research only	1,144	99.6	46.0	48.0	-7.4	5,645
Extension only	80	84.6	47.0	62.9	0	636
Research and extension	628	47.6	28.0	37.0	-100.0	430
All observations	1,852	81.3	40.0	44.3	-100.0	5,645
Regression sample[b]						
Research only	598	79.6	26.0	49.0	-7.4	910
Extension only	18	80.1	91.0	58.4	1.3	350
Research and extension	512	46.6	28.0	36.0	-100.0	430
All observations	1,128	64.6	28.0	42.0	-100.0	910

[a]The original full sample included 292 publications reporting 1,886 observations. Of these, 9 publications were dropped because, rather than specific rates of return, they reported results such as >100% or <0. As a result of these exclusions, 32 observations were lost. Of the remaining 1,854, two observations were dropped as extreme (and influential) outliers. These two estimates were 724,323% and 455,290% per year.

[b]Excludes outliers and observations that could not be used in the regression owing to incomplete information on explanatory variables.

Source: Alston et al., 2000a.

rates of return. The highest ROR observed for all agriculture, field crops, livestock, tree crops, resources and forestry were 1,219; 1,720; 5,645; 1,736; 457; and 457, respectively. All studies related to livestock and trees had a positive ROR. The mean ROR for livestock R&D was around 121. These data demonstrate that the estimated RORs for livestock species are comparable to the rates estimated for the other sectors. In addition, in this study the overall estimated ROR for animal research was 18% but when this was decomposed, the ROR for animal health research and animal improvement research were found to be 15 and 27%, respectively; indicating the underestimation of ROR for the overall investment. Probably, the decomposition by species would also show different RORs associated to each of them.

Although the mean ROR estimates for industrialized countries is higher than that for developing countries (98 and 60%, respectively), the median are virtually identical (46 versus 43%) (Table 8-12). While there are not many studies from Africa assessing the returns to R&D, the existing analyses generally indicate high returns in the range of 4 to 100% for country level studies (Anandajayasekeram and Rukuni, 1999).

The key findings of the last meta-analysis were (Alston et al., 2000a):
- Research has much higher ROR than extension only or both research and extension combined;

- There is no measurable difference in estimated ROR between privately and publicly performed research;
- The RORs were 25% per year higher for research on field crops and 95% per year lower for research on natural resources than for total agriculture;
- There is no significant difference in rates of return related to whether studies reported basic or other categories of research;
- The estimate also indicates that if research took place in an industrialized country, the ROR was higher by 13% per year, but this effect was not statistically significant at the 10% level. The estimated rates of return tended to be lower in Africa and West Asia and North Africa than in Latin America and the Caribbean or Asia;
- There is no evidence that the ROR to agricultural R&D has declined over time;
- Unable to detect any effect of accounting for spillovers or market distortions on measured rates of return to research.

8.2.4.4 Agricultural research and education investments and agricultural growth
A summary of studies that have applied decomposition analysis to agricultural growth in developing countries suggests that past investments in agricultural research may have contributed anywhere from 5 to 65% of agricultural

Table 8-11. **Rates of return by commodity orientation.**

Commodity orientation	Number of observations	Rate of return				
		Mean	Mode	Median	Minimum	Maximum
	(count)			(percentage)		
Multicommodity[a]	436	80.3 (110.7)	58.0	47.1	-1.0	1,219.0
All agriculture	342	75.7 (110.9)	58.0	44.0	-1.0	1,219.0
Crops and livestock	80	106.3 (115.5)	45.0	59.0	17.0	562.0
Unspecified[b]	14	42.1 (19.8)	16.4	35.9	16.4	69.2
Field crops[c]	916	74.3 (139.4)	40.0	43.6	-100.0	1,720.0
Maize	170	134.5 (271.2)	29.0	47.3	-100.0	1,720.0
Wheat	155	50.4 (39.4)	23.0	40.0	-47.5	290.0
Rice	81	75.0 (75.8)	37.0	51.3	11.4	466.0
Livestock[d]	233	120.7 (481.1)	14.0	53.0	2.5	5,645.0
Tree crops[e]	108	87.6 (216.4)	20.0	33.3	1.4	1,736.0
Resources[f]	78	37.6 (65.0)	7.0	16.5	0.0	457.0
Forestry	60	42.1 (73.0)	7.0	13.6	0.0	457.0
All studies	1,772	81.2 (216.1)	46.0	44.0	-100.0	5,645.0

Notes: See Table 8-10. Standard deviations are given in parentheses. Sample excludes two extreme outliers and includes only returns to research only and combined research and extension, so that the maximum sample size is 1,772. In some instances further observations were lost owing to incomplete information on the specific characteristics of interests.

[a]Includes research identified as all agriculture or crops and livestock, as well as unspecified.

[b]Includes estimates that did not explicitly identify the commodity focus of the research

[c]Includes all crops, barley, beans, cassava, sugar cane, groundnuts, maize, millet, other crops, pigeon pea or chickpea, potato, rice sesame, sorghum and wheat.

[d]Includes beef, swine, poultry, sheep or goats, all livestock, dairy, other livestock, pasture, dairy and beef.

[e]Includes other tree and fruit and nuts.

[f]Includes fishery and forestry.

Source: Alston et al., 2000a.

growth, depending on the country and time period (Pingali and Heisey, 2001). Decomposition of recent measurements of African agricultural growth suggests that up to one-third of the growth in aggregate agricultural productivity is attributable to past investments in agricultural research (Oehmke et al., 1997). This roughly corresponds to a contribution of agricultural research to economic growth of ¼ of a percentage point.

A study on agriculture growth and productivity in the United States demonstrated similar results (Shane et al., 1998). During 1974-1991 annual growth rate of agriculture productivity was estimated to be 2.2% and entire economy productivity growth was 0.2% in the U.S. total factor productivity (TFP) growth rate was 2.3% during 1959-91. During 1949-91, productivity growth in agriculture can be attributed to four major factors: public investment in agricultural R&D (50%), public expenditure on infrastructure (25%), private investment in R&D, and technological advances embodied in material inputs such as fertilizers and chemicals (combined 25%).

Technological advancements depend on the quality and quantity of scientific capacity of the national institutes (Mashelkar, 2005). There is a positive relationship between science enrolment and technology achievement indices indicating that increased investment in human capacity can result in better technological advancement. This relationship has not been fully understood or analyzed in the development literature. In general, there is a lack of analysis or questioning of the role of increased capacity in explaining growth performance in developing counties (Ul Haque and Khan, 1997).

Detailed evidence on the ROR of investments in agricultural education is very limited, but a number of studies have showed that education has a positive impact on economic growth and positive benefits to health and other noneconomic benefits (UNESCO and OECD 2003; Evenson, 2004; World Bank, 2007). Economic growth increases 4% for every additional year of schooling of the adult population; particularly high attainment levels in secondary and tertiary education are relatively more important for economic growth (UNESCO/OECD 2003). The available evidence on the impacts of tertiary education on economic growth and poverty reduction in the case of SSA Africa show that an increase of one year of tertiary education will result in growth increase of 6% in the first year, and about 3% after five years (Bloom et al., 2007). Another study found that an increase in the number of degrees awarded in natural sciences and engineering in East Asia have a strong positive relationship on GDP per capital levels (Yusuf and Nabeshima, 2007). The expected rates of return to investments in education at primary, secondary and higher level for selected countries in Africa range from 24% for primary education, 18.2% for secondary to 11.4% for higher education (Psacharopoulos, 1994).

Per capita income levels are higher among counties that have invested in educating their population (Babu and Sengupta, 2006) and increased capacity will result in better implementation of programs and policies that reduce poverty and hunger (Kaufman et al., 2003). Whereas there is a positive relationship between increased capacity and better governance, there is a need for better understanding of the nature and magnitude of the capacity needed to increase governance of development programs and policies that translate the development goals into development outcomes (Babu and Sengupta, 2006).

There is ample evidence available from the literature that AKST investments have contributed significantly to organizational and institutional innovations in the form of methods, tools development, capacity strengthening, and understanding how institutes interact with each other in achieving developmental goals. However, not much work has been done on assessing the RORs on investments in agricultural training and capacity strengthening. Assessing the economic impacts of non-research products such as training, networking and advisory services policy and institutional reforms need greater emphasis. Detailed analysis of causality of agricultural education and capacity strengthening on development outcomes will require identifying indicators that reflect institutional and human capacity to contribute better development processes.

8.2.4.5 Rates of return to CGIAR investments

The CGIAR Science Council's Standing Panel on Impact Assessment (SPIA) commissioned an independent study to weigh the measurable benefits of CGIAR research against the total cost of system operation to 2001 (Raitzer, 2003). The analysis found that the value of documented benefits generated by the CGIAR surpasses total investment in the system. The analysts did not calculate a single benefit-cost ratio for all potential audiences. Instead, they offered five different versions of the benefit-cost ratio to allow for its sensitivity to different assumptions regarding the credibility of the values derived for key measures of benefit. The most restrictive assessment yields a benefit cost ratio of 1.9, i.e., returns of nearly 2 dollars for every dollar invested. The most inclusive estimate puts the benefits–cost ratio nearly nine times higher.

The analysis excluded the benefits from the vast majority of CGIAR work, which has not been subject to large-scale ex-post economic assessment. The analysis aggregated only published large-scale economic assessments that met a strict set of criteria for plausibility and demonstration of causality. As a result, only a few isolated examples of success are used to produce these substantial benefit levels, and many probable impacts that lack reliable quantification are omitted. To underscore this point, the economic value of benefits derived from just three CGIAR innovations is estimated to be greater than the entire US$7 billion (in 1990 prices) invested in the IARCs since the CGIAR was established. Under very conservative assumptions, benefits generated (through 2001) from (1) new, higher yielding rice varieties in Latin America, Asia, and West Africa; (2) higher-yielding wheat in West Asia, North Africa, South Asia, and Latin America; and (3) cassava mealybug biocontrol throughout the African continent combined almost twice the aggregate cumulative CGIAR costs. If slightly more generous assumptions were applied, the estimated benefits generated to date by these three technologies rise to more than eight times the total funds invested in CGIAR research and capacity-building programs. If impact assessment were applied to a larger proportion of the system's portfolio then these three innovations will result in much higher aggregate benefit values. Furthermore, the aggregated studies do not take into account multiplier effects that result from stimulated growth in the nonfarm economy, or nonmarket benefits. As a result, even the most generous of the values reported may be considered as conservative (Raitzer, 2003).

8.2.4.6 Rates of return to agricultural R&D investments in sub-Saharan Africa

A compilation of the available case studies on ROR for African agricultural R&D investment support findings of four meta-analyses (Table 8-13). Of the 27 RORs to past investments in agricultural technology development and dissemination (TDT), 21 show RORs in excess of 12%. Detailed investigations into the lower RORs suggest that researchers had not yet found the right mix of activities to produce cost-effective solutions in challenging agroecological environments. Examining the future potential impact of innovations released or still in the development stage, 24 of 30 forward-looking RORs show expected returns in excess of 12%. The second study reviewed the impact studies conducted in Eastern and Southern Africa (ESA) during 1978-2005 (Anandajayasekeram et al., 2007). The RORs for those studies using the noneconometric methods ranged from 0 to 109. For those studies using the econometric methods ranged between 2 to 113%. Only 10 out of the 86 observations were below 12% under the worst case scenario. These compilations confirm that returns to research in SSA are similar to those found elsewhere, showing a high payoff for a wide range of programs.

8.2.5 Environmental impacts of AKST investments

The success of modern agriculture in recent decades has often masked significant externalities that have positively and negatively affected natural resources. Externalities of agriculture include the depletion of resources such as fossil fuel, water, soil and biodiversity; pollution of the environment by the products of fuel combustion, pesticides and fertilizers; and economic and social costs to communities. In the past, the objectives of AKST investments have been largely to increase quality, quantity and to improve food security. Thus the environmental impacts of agricultural technologies were not usually considered in ROR and other decision-making tools. Their importance, however, is increasingly understood because of the positive link between ecologically sustainable development and poverty reduction (UNEP, 2004a). A good example at the national level is a study that estimates the total external environmental and health costs of "modern agriculture" in the United Kingdom at a total cost of £2343 million in 1996 (Pretty et al., 2000). This is equivalent to 89% of average net farm income and £208 per hectare of arable and permanent pasture. These estimates only include those externalities that give rise to financial costs, and so are likely to underestimate the total negative impacts to the environment.

The quality and size of environmental impacts depend on many external forces. Different agroecological zones, market conditions, and financial and social incentives as well as specific technologies play significant roles in determining impacts. In order to quantify and value the environmental impact of an agricultural R&D investment, it is important to understand the source of the impact, the nature, and the relationship between the impact and those variables that can affect producers and consumers. Agriculture globally (including livestock and land use, but excluding transport of agricultural products) has an estimated contribution to GHG emissions of 32% (Stern, 2007).

Although many economic valuation techniques have been developed and refined over the last twenty years, obtaining monetary values of environmental impacts is difficult due to two basic reasons. First, the issue of time and scale complicates the data collection and valuation. Many of the environmental impacts are accumulative by nature, and thus time is a critical constraint in estimating values. Geographic scale is also critical, for example, captive shrimp production leads to a large-scale pollution of marine environments and destruction of mangroves (Clay, 2004). In general, environmental and ecological economists consider the scale either through an ecosystem-centric lens: the plot, the farm, the watershed, and region (Izac and Swift, 1994) or a human-centric lens: the individual farmer, the local community,

Table 8-13 **Summary of results of Economic Assessment of African R&D Investments.**

Author	Type of analysis	Number of observations	Range of RORs	Range of B/C ratio	Geographical coverage
		(count)		*(percent)*	
Oehmke et al. (1997)	Ex-post	27	< 0 to 135	—	sub-Saharan Africa
	Ex-ante	19	< 0 to 271	1.35 :1 to 149 :1	
	Combined	46	< 0 to 271	1.35 :1 to 149 :1	
Anandajayasekeram et al. (2007)	Econometric methods	25	2 to 113	—	East and Southern Africa
	Non econometric methods	61	< 0 to 109	1.35:1 to 149	
	Combined	86	< 0 to 113	1.35:1 to 149	

Sources: Oehmke et al., 1997; Anandajayasekeram et al., 2007.

downstream communities, national citizens and the global population. These issues of scale are seldom incorporated into ROR calculations or other decision-making tools, and thus they exclude the critical elements of "who pays, who benefits". Second, reconciling different levels of aggregation to obtain reliable estimates is complex. For example, movement of pesticides through soil is determined by several factors such as specific soil characteristics (physical and chemical), properties of the soil, the climate, crop management practices, and so on. The problem is how to generate information that reflects the complex of physical, biological and technical factors. One of the more difficult areas to estimate is the value of impacts on and by biodiversity, because the links between biodiversity and ecosystem functions and services is less understood than many other environmental interactions. Moreover, no monetary values can be given to it and the value is also context-specific and relative to the livelihoods and uses given to biodiversity. The willingness is related to the knowledge of the impacts of biodiversity loss, including the impact of climate change (Turpie, 2003).

New development in economic science, such as ecological economics, can bring promising tools in the future to measure externalities and tackle the problems identified above (Proops, 1989; Jacobs, 1996). One example is the evaluation of one century of agricultural production in the Rolling Pampas of Argentina by analysing energy flows within systems (Ferreyra, 2006). The ecological footprint quantifies the amount of resources required by a production method or a technology related to AKST, and thus can give an idea of the environmental impact (Wackernagel and Rees, 1997) and was used to assess the resource use and development limitations in shrimp and tilapia aquaculture (Kautsky et al., 1997).

Due to the complexity of agriculture and the links with the food chain, most studies, particularly ecological economic studies, examine the impacts of food systems and not the technologies in isolation. There is a significant paucity of data and studies on environmental impacts (see below as well as Table 8-14).

8.2.5.1 Agriculture
Biodiversity loss. Reduction in the use of biodiversity in agriculture is driven by the increased pressures and demands of urban and rural populations and by the global development paradigm, which favors specialization and intensification (FAO, 2003). Most studies combine influences and impacts from crop and livestock systems. The total economic benefits of biodiversity with special attention to the services that soil biota activities provide worldwide is estimated to be US$1,542 billion per year (Pimentel et al., 1997). The estimated total damage to UK's wildlife, habitats, hedgerows and drystone walls was £125 million in 1996 (Pretty et al., 2000).

Soil erosion. Scientists estimate the global cost of soil erosion at more than US$400 billion per year. This includes the cost to farmers as well as indirect damage to waterways, infrastructure, and health (Pimentel et al., 1995). In the UK, the combined cost of soil erosion with organic carbon losses was £106 million in 1996 (Pretty et al., 2000).

Pesticides and chemical fertilizers. Agricultural runoff pollutes ground and surface waters with large amounts of nitrogen and phosphorus from fertilizers, pesticides and agricultural waste. Agriculture is the main cause of pollution in US rivers and contributes to 70% of all water quality problems identified in rivers and streams (Walker et al., 2005). In the UK the cost of contamination of drinking water with pesticides is £120 million per year (Pretty et al., 2000).

Carbon sink. Agricultural systems contribute to CO_2 emissions through several mechanisms: (1) the direct use of fossil fuels in farm operations; (2) indirect use of fossil fuels through inputs, such as fertilizers; and (3) the loss of soil organic matter. On the other hand, agricultural systems accumulate carbon when organic matter is accumulated in the soil, or when above-ground woody biomass acts either as a permanent sink or is used as an energy source that substitutes for fossil fuels (Pretty and Ball, 2001). A 23-year ongoing research project by the Rodale Institute in the US found that if 10,000 medium-sized farms in the US converted to organic production, the carbon stored in the soil would equal taking 1.2 million cars off the road, or reducing car travel by 27 billion kilometers (The Rodale Institute, 2003). Forty sustainable agriculture and renewable-resource-management projects in China and India (Pretty et al., 2002) increased carbon sinks in soil organic matter and above-ground biomass; avoided carbon emissions from farms by reducing direct and indirect energy use; and increased renewable energy production from biomass. The potential income from carbon mitigation is $324 million at $5 tonne^{-1} of carbon (Pretty et al., 2002).

Water use. Agriculture consumes about 70% of fresh water worldwide. For example, the water required for food and forage crops growing ranges from about 300 to 2,000 liter kg^{-1} dry crop yield, and for beef production 43,000 liter kg^{-1} (Pimentel et al., 2004). Virtual water refers to the water used in the production process of an agricultural product (Chapagain and Hoekstra, 2003). Using a virtual water approach, some countries are net importers of water while others are exporters. It is expected that in the future, approaches to quantify the amount of water used by different countries or regions will be extremely important.

8.2.5.2 Livestock
The livestock sector has enormous impacts on the environment: it is responsible for 18% of GHG emissions measured in CO_2 equivalents, and 9% of anthropogenic CO_2 emissions, including the combustion of fossil fuels to make the additional inputs. Globally, it accounts for about 8% of human water use, mostly for the irrigation of feed crops (Steinfeld et al., 2006). It is estimated that 1 kg of edible beef results in an overall requirement of 20 to 43 tonnes water per kg of meat (Smil, 2002; Pimentel et al., 2004). The total area occupied by grazing is equivalent to 26% of the world land; the total agricultural area dedicated to feedcrop production is 33%. In all, livestock production accounts for 70% of agricultural land and 30% of land globally (Steinfeld et al., 2006). It is probably the largest sectoral source of water pollution. In the US, livestock are responsible for

Table 8-14. **Relative size of environmental impacts of high-external input farming systems in absence of monetized or otherwise quantified assessments.**

		Individual farmer and/or household	Local community	Downstream community	Global society	Data availability for economic quantification[a]
Agriculture[b]	Biodiversity loss, off and on-farm species and plant genetic resources	--	--	-	-----	Some
	Erosion and soil quality	----	-	----	-----	Many
	Run-off of agro-chemicals (eutrophization)	0	-	-----	---	Many
	Pesticides and impact on non-target species	-	---	----	---	Many
	Water table loss	-	--	-----	-	Few
	Fossil Fuel Use: Non-renewable and climate change impact	-- (financial cost)	0	0	-----	Some
	Genetic improvement	++++++	0	0	++	Many
	Carbon sequestration	0	0	0	+++++	Some
	Land saving	+++++	++	0	++++	Many
Livestock	Biodiversity loss, off- and on-farm species and animal genetic resources	--	--	-	-----	Few
	Water contamination (surface and underground) and eutrophization	-	--	-----	-	Some
	Fossil Fuel Use	--	0	0	---	Some
	Land use and deforestation for animal nutrition	0	0	-	-----	Many
	GHG emissions	0	0	0	-----	Some
	Genetic improvement	++++++	0	0	++	Some
Aquaculture	Fisheries decline (due to fishmeal production and capture of wild gravid females and/or post larvae seeds)	-----	-----	0	--	Few
	Destruction of coast forest (e.g., mangrove for shrimp production)	-----	-----	-----	-----	Some
	Erosion and release of CO_2 into atmosphere	--	--	--	-----	Few
	Fossil Fuel Use	--	0	0	---	Few
	Soil and water salinization	----	-----	-----	-	Few
	Runoff of agro-chemicals	0	-	-----	---	Few
	Biodiversity loss (due to diseases, hybridization and competition with wild fish)	-	-	--	-----	Few

Note: + = positive impact. - = negative impact. Degree of impact: + minimal; ++ moderate; +++ high; ++++ very high, and +++++ very high likelihood of some irreversibility.

[a]Economic valuation refers both to monetary and other methods of valuation tools.

[b]Agriculture impacts include the livestock and aquaculture impact derived from the crops required for the animal and fish nutrition industry (e.g., 33% of world feedcrop land is dedicated to animal nutrition, thus, it is a livestock impact added to agriculture).

Source: Authors' elaboration.

55% of soil erosion and sediment, 37% of pesticide use, 50% of antibiotic use and a third of the loads of nitrogen and phosphorus into freshwater resources (Steinfeld et al., 2006). Data are not available to estimate these impacts from an economic perspective. "Emergy" evaluations have recently been used to evaluate the costs of grazing cattle in Argentina's Pampas (Rótolo et al., 2007); to compare soy production systems in Brazil (Ortega et al., 2003); and to compare organic and conventional production systems (Castellini et al., 2006).

The rapid spread of large-scale industrial livestock production focused on a narrow range of breeds is the biggest threat to the world's farm animal diversity (FAO, 2007). Traditional livestock losses worldwide range from one breed per week (Thrupp, 1998) to one per month (FAO, 2007). Many traditional breeds have disappeared as farmers focus on new breeds of cattle, pigs, sheep, and chickens. In the year 2000, over 6,300 breeds of domesticated livestock were identified; of these, over 1,300 are now extinct or considered to be in danger of extinction. Many others have not been formally identified and may disappear before they are recorded or widely known. When breeds without recorded population data are included, the number at risk may be as high as 2,255. Europe records the highest percentage of extinct breeds or breeds at risk (55% for mammalian and 69% for avian breeds). Approximately 80% of the value of livestock in low-input developing-country systems can be attributed to non-market roles, while only 20% is attributable to direct production outputs (FAO, 2007). By contrast, over 90% of the value of livestock in high-input industrialized-country production systems is attributable to the latter. How to measure or evaluate the importance of considering nonmarket values of livestock when planning AKST investment is lacking. Obtaining such data frequently requires the modification of economic techniques for use in conjunction with participatory and rapid rural appraisal methods (FAO, 2007).

8.2.5.3 Forestry

There are different forestry systems ranging from systems of monoculture of trees (aiming to obtain products such as cellulose, wood or other products) to systems that cultivate different tree species with other agricultural products, including livestock. Agroforestry systems (AFS) provide a mix of market and nonmarket goods and services with a high level of output per purchased investments and minimal environmental impacts (Diemont et al., 2006). An agricultural system that includes agroforestry is more profitable than a conventional system (Neupane and Thapa, 2004); agroforestry has great potential to minimize the rate of soil degradation, increase crop yields and food production, and raise farm income in a sustainable manner.

8.2.5.4 Aquaculture

Intensification of aquaculture has resulted in higher impacts into the environment. A deeply analyzed case is that of shrimp farming. In a simple cost-benefit analysis, industrial shrimp farming is usually found to be profitable; however, cost-benefit analyses that include environmental costs, can contradict these findings. For example, a study performed in India concluded that shrimp culture caused more economic harm than good. The economic damage (loss of mangroves, salinization and increasing unemployment) outweighed the benefits by 4 to 1 or 1.5 to 1, depending on the areas considered (Primavera, 1997). In Thailand, the total economic value of an intact mangrove exceeds that of shrimp farming by 70% (Castellini et al., 2006). The estimated internal benefits of developing shrimp farms are higher than the internal costs in the ratio of 1.5 to 1 (Gunawardena and Rowan, 2005). When the wider environmental impacts are more comprehensively evaluated, the external benefits are much lower than the external costs in a ratio that ranges between 1 to 6 and 1 to 11.

In Malawi, the ecological footprint approach applied to integrated aquaculture showed that when waste from each farming enterprise is recycled into other enterprises, the economic and ecological efficiencies of all are increased (Brummet, 1999).

8.2.5.5 Traditional and local knowledge

Traditional knowledge and local farming systems associated are often either ignored or sidelined by new technologies and profit-oriented interventions (Upreti and Upreti, 2002). Though there is no economic valuation estimated in monetary terms, it is well recognized that there is tremendous value in traditional knowledge for maintaining and improving farming systems, particularly with regard to agrobiodiversity management and utilization.

In recent years researchers have started to address this significant gap. For example, a new conceptual framework was developed to assess the value of pastoralism that goes beyond conventional economic criteria (Hesse and McGregor, 2006). The objective is to provide fresh insights to its contribution to poverty reduction, sustainable environmental management and the economically sustainable development of dryland areas of East Africa in the context of increasing climate uncertainty. One can associate environmental impacts of pastoralist traditional knowledge in terms of sustainable land use and risk management in disequilibrium environments, biodiversity conservation and improved agricultural returns, but these too are rarely captured in national statistics or recognized by policy makers.

8.2.6 Health impacts of agricultural R&D investments

The interactions between agriculture and human health are well recognized. Agricultural technologies through their effects on productivity, income, and food quality and security can improve the health status of producers and consumers (Table 8-15); healthier people will generally be more productive than people who suffer from sickness or who are undernourished. On the other hand, agricultural technologies can have negative effects on the health status of farmers, farm laborers, farm household members and consumers (Table 8-16).

Pesticides are an example of positive (increases in productivity), and negative (environment and human health) effects (see also 8.2.5). There are at least 1 million cases of pesticide poisoning annually, with women and children in developing countries disproportionably affected (WHO, 1990; UNEP, 2004b). The total number of unintentional fatal poisonings from all sources, including agricultural chemicals, is 350,000 per year (WHO, 2006). These global

Table 8-15. **Positive contributions of AKST for human health by sector.**

Sector	AKST product	Consequences for human health	Data availability for economic quantification
Crops	Micronutrient trait in crop varieties	Prevent human diseases	Ex-ante assessments for Biofortification
Livestock	Animal Protein	Balanced diet	Unknown
Aquaculture	Animal Protein and micronutirents	Balanced diet	Unknown
Forestry	Non-timber products and food from natural resources	Prevent food insecurity	Unknown

Source: Authors' elaboration.

figures are not collected systematically or on a regular basis; estimation of the incidence of pesticide poisoning is difficult as surveillance systems may be inadequate and tend to underreport (PAHO, 2002; London and Bailie, 2001). Hence official reports represent lower bound estimates. Farmers in the developing world experience high rates of exposure to human health risks when using pesticides (Jeyaratnam et al., 1982 and 1987; Kishi et al., 1995; Ajayi, 2000; Rola and Pingali, 1993; Antle et al., 1998; Crissman et al., 1994 and 1998). Some authors (Cuyno et al., 2001; Garming and Waibel, 2006) established that farmers reveal a willingness to pay for reducing the negative health effects from chemical pesticides.

Economic studies carried out in industrialized countries found that health costs make up about 10% of the total ex-

ternality costs of pesticides (Pimentel et al., 1993ab; Waibel et al., 1999). Pesticide use globally continues to rise and hence the concerns about the implications for human health remain (Ecobichon, 2001).

Considerable AKST investments were made by the public and the private sector to minimize the negative health effects of pesticides. These investments included two major products, safe use technology packages and IPM. Chemical companies have developed modules on safe use training, which is an example of a private sector AKST. Pilot projects were carried out in Mexico, India and Zimbabwe (Atkin and Leisinger, 2000), Guatemala, Kenya and Thailand (Hurst, 1999). Successes were very limited; Farmers often went back to their old practices shortly after the training (Atkin and Leisinger, 2000). In addition, some safe use technolo-

Table 8-16. **Negative effects of AKST for human health by sector.**

Sector	Type of effect	Consequences	Data availability for economic quantification
Crops		Acute and chronic diseases	Case studies including economic evaluations and occasional country statistics
	Water pollution with pesticides and nitrogen fertilizer	Intoxication/death from drinking water	Statistics and case studies mainly in developed countries
	Air pollution (e.g., transport, fertilizers production, deforestation)	Respiratory and allergic diseases	Unknown
Livestock	Increase of cheap meat production and consumption	Obesity (cancer, diabetes, coronary diseases)	Studies mainly for the U.S.
	Antibiotics use	Increasing resistance to antibiotics	Few studies
	Water pollution with animal wastes	Intoxication/death from drinking water	Unknown
	Air pollution (transport, GHG emissions, deforestation)	Respiratory and allergic diseases	Unknown
	Increasing animal trade	Increasing number of zoonosis	Recent studies
Aquaculture	Antibiotics use	Increasing resistance to antibiotics	Few studies
	Air pollution (transport)	Respiratory and allergic diseases	Unknown
	Residues in aquaculture feed (mercury, dioxins, polychlorinated bromides	Intoxication, neurotoxicity	Few studies

Source: Authors' elaboration.

gies (e.g., protective equipment) and management practices, (e.g., hygienic measures after spraying, compliance with re-entry intervals, safe storage of equipment and pesticides) were often found unfeasible in tropical climates and under the conditions of poor countries (Cole et al., 2000).

A good of example of public sector AKST investment to mitigate the negative impact of pesticide use is IPM; IPM technologies are site-specific in that they need to be developed for specific agroecological, socioeconomic and policy conditions. As a result a wide range of examples exist in both developing and industrialized countries. Despite the large amount of investment in IPM, global impact studies are rare. A meta-analysis for the CGIAR in 1999 showed that, although no aggregate ROR could be established, the ROR for IPM was above 30%, and this does not include the significant environmental and health benefits. In the industrialized countries several successful IPM programs have been implemented in selected crops (e.g., Norton, 2005), but successes on the aggregate level remain questionable due to a lack of enabling policy conditions (Waibel et al., 1999).

Iron, zinc and iodine deficiencies are widespread nutritional imbalances (WHO, 2002; FAO, 2004; Hotz and Brown, 2004; UN-SCN, 2004). The adverse health outcomes of micronutrient deficiencies include child and maternal mortality, impaired physical and mental activity, diarrhea, pneumonia, stunting or blindness, among others (Stein et al., 2005). Biofortification research aims to reduce malnutrition by breeding essential micronutrients into staple crops. The CGIAR HarvestPlus Challenge Program concentrates on increasing iron, zinc and beta-carotene (provitamin A) content in six staple crops species (rice, wheat, maize, cassava, sweet potatoes and beans). In addition, the program supports exploratory research in ten additional crops (Qaim et al., 2006). Most biofortified crops are in R&D phase, except for beta-carotene rich orange fleshed sweet potatoes and Golden Rice (Low et al., 1997; Goto et al., 1999; Ye et al., 2000; Lucca et al., 2001; Murray-Kolb et al., 2002; Drakakai et al., 2005, Ducreux et al., 2005).

Thus far, only ex-ante economic analyses exist for biofortified crops. An evaluation of the potential health benefits of Golden Rice in the Philippines showed that micronutrient deficiencies can lead to significant health costs, which could be reduced through biofortification (Zimmermann and Qaim, 2004). In an ex-ante impact assessment using disability adjusted life years (DALYS) approach (Qaim et al., 2006) the estimated Internal Rate of Return (IRR) was very high, ranging 31 to 66% (pessimistic scenario) and 70 to 168% (optimistic scenario). Ex-ante studies on the expected impact of biofortification research under HarvestPlus have been conducted for rice in the Philippines, beans in Brazil and Honduras, sweet potato in Uganda, maize in Kenya and cassava in Nigeria and Brazil; health-cost reductions range from 3 to 38% in the pessimistic scenario and from 11 to 64% in the optimistic scenario depending on crop and location (Meenakshi et al., 2006).

To find out if biofortified crops will be adopted by growers on a large scale requires research including ex-post studies building on observable data to verify the preliminary results. Further research is also needed on the bioavailability and micronutrient interactions in the human body. The key conclusion emerging from the available ex-ante studies is that biofortification could play an important role in achieving nutrient security in particular situations. However, its benefits will depend on the necessary institutional framework that can facilitate the effective introduction of these technologies as well as an enabling policy framework.

Other impacts of AKST on health, both positive and negative, can be shown with the development of industrial livestock. Livestock products contribute to improved nutrition globally and are linked to disease, such as cardiovascular disease, diabetes and certain types of cancer (Walker et al., 2005).

8.2.7 Spillover effects

The wide applicability of research results over a range of agricultural production conditions or environments often cutting across geographical and national boundaries are generally referred to as spillover effects. Spillover effects are a combination of four effects: price effects from the increased production caused by reduced costs which are captured in the supply and demand framework (Hesse and McGregor, 2006). Spill-over technology from country "Y" which can be adopted without any research in country "X"; spillover of technology from country "Y" which requires adaptive research before it is applicable in country "X"; and spillover of scientific knowledge which ultimately enhances future research in many areas.

Technological spillovers increase the returns to research and can be spill-ins or spill-outs. Spill-ins take place when a country is adapting a technology developed elsewhere. This reduces the national research costs and shortens the time required for developing and disseminating the finished product. The gains from spill-ins are important to all research organizations, but are higher in smaller systems. Spill-outs take place when research findings are used by other countries. Spill-outs are important when one is interested in the total benefits occurring to the country where the technology was developed as well as the country where it was adopted. This aspect is critical when performing impact assessment of a regional network (Anandajayasekeram et al., 2007).

It has been long recognized that AKST spillovers are both prevalent and important (Evenson, 1989; Griliches, 1992). A study that fails to account appropriately for spill-ins will overestimate the benefits from its own research investment.[7] Similarly if state to state or nation to nation spillovers are important—as in the case of regional research networks—and the study measures its own benefit at the national level and ignores the "spill-outs", this will underestimate the ROR. Only 12% of the 292 studies in the sample of one of the aforementioned meta analysis made any allowance for technology spillovers; even fewer allowed for international spillovers (Alston et al., 2000a). They also noted that by far the majority of research impact studies that have allowed for international agricultural technology spillovers were commodity specific studies, rather than national aggregate studies, and mostly they were studies of crop varietal improvements.

---

[7] Farmer to farmer spill in/outs are also important, not just locally but where they happen through travel, guest worker return etc, but not easy to capture.

A study covering twelve different commodities and using a multicountry trade model, found that spillover effects from regions where research is conducted to over regions with similar agroecologies and rural infrastructures ranged from 64 to 82% of total international benefits (Davis et al., 1987). An analysis of 69 national and international wheat improvement research programs found that given the magnitude of potential spill-ins from the international research system, many wheat programs could significantly increase the efficiency of resource use by reducing the size of their wheat research programs and focusing on the screening of varieties developed elsewhere (Davis et al., 1987). The impact of research conducted within individual Latin American and Caribbean countries covering edible beans, cassava, maize, potatoes, rice, sorghum, soybeans and wheat showed that when allowance was made for spillovers to other regions of the world, the resulting price impacts had important consequences for the distribution of benefits between producer and consumers and thus among countries within Latin America and the Caribbean (Alston et al., 2000b). At least for the United States, the locational range of spill-in effects for crop production is lower than for livestock production (Evenson, 1989). Crop genetic improvements in the United States had spillover effects into the rest of the world, with consumers in the rest of the world gaining but producers outside the United States losing (Frisvold et al., 2003). Overall increases in net global welfare from United States crop improvements were distributed 60% to the US, 25% to other industrialized countries, and the remainder to developing and transitional economies.

Growth in public funding for international research has slowed over the last twenty years (see 9.1). Thus, understanding the ROR of the CGIAR is very important, including the spill in and spill out impacts. Over the years, a number of studies have attempted to value the benefits to particular countries from research conducted at CG centers, in some cases comparing them against donor support provided by the countries in question (Brennan, 1986, 1989; Burnett et al., 1990; Byerlee and Moya, 1993; Bofu et al., 1996; Fonseca et al., 1996; Pardey et al., 1996; Brennan and Bantilan, 1999; Johnson and Pachico, 2000; Brennan et al., 2002; Heisey et al., 2002). For the period 1973-1984, Australia gained US$747 million in terms of cost savings to wheat producers as a United States benefit from its adoption of wheat varieties from CIMMYT and rice varieties from IRRI (Brennan, 1986, 1989). Depending on the attribution rule used, the United States' economy gained at least US$3.4 billion and up to US$14.6 billion from 1970 to 1993 from the use of improved wheat varieties developed by CIMMYT and US$30 million and up to US$1 billion through the use of rice varieties developed by IRRI.[8] These estimates did not account for the world price impact as a result of the rest of the world having adopted CIMMYT wheat varieties and thereby driving down the price of wheat.

Assessments were made of Australia's benefits from research conducted by ICRISAT and ICARDA taking explicit account of the world price impacts (Brennan and Bantilan, 1999; Brennan et al., 2002). Research on sorghum

(ICRISAT) resulted in a national benefit of US$3.6 million (producer loss of US$1.7 million and consumer gain of US$5.3 million) for Australia. Similarly, ICRISAT's research on chickpeas would have given a national benefit of US$1.2 million (producer loss of US$2.6 million and a consumer gain of US$3.8 million). The average estimated net gain to Australia as a result of the overall research effort at ICARDA in five crops (durum wheat, barley, chick pea, lentils and faba bean) is US$7.4 million per year (in 2001 dollars and exchange rates) over the period to 2002 (Brennan et al., 2002). This represents 1% of the gross value of Australia's production of the five crops. Most of those gains are achieved in the faba bean and lentil industries. Producers receive most of the welfare gains in Australia, amounting to US$6.5 million of the total.

The main findings of the various studies are (Alston, 2002):

- Intra national and international spillovers of public agricultural AKST results are very important.
- Spillovers can have profound implications for the distribution of benefits from research between consumers and producers and thus among countries, depending on their trade status and capacity to adopt the technology.
- It is not easy to measure these impacts, and the results can be sensitive to the specifics of the approach taken, but studies that ignore spillovers are likely to obtain seriously distorted estimates of ROR.
- Because spillovers are so important, research resources have been misallocated both within and among nations.

The estimation of these state, national or multinational impacts is data intensive, difficult, and adds to the measurement problems (Alston, 2002). However, there can be little doubt that agricultural AKST generates very large benefits and that a very large share of those benefits comes through spillovers. The omission or mismeasurement of spillover effects may have contributed to a tendency to overestimate ROR to agricultural AKST in some instances. Clearly, the issue of international research spillovers is an important one for the allocation of resources for research both nationally and internationally. The spillover benefits to industrialized countries from international agricultural research have positive funding implications. More work is needed in this area to develop better methods to measure spillovers and also to develop the necessary policy institutional arrangements to harness the full potential of spillover effects of AKST technologies (Alston, 2002; Anandajayasekeram et al., 2007).

Agricultural machinery and agricultural chemicals are obvious cases where industrial AKST is directed towards the improvement of agricultural inputs. Recent studies conclude that when new industrial products first come on the market, they are priced to only partially capture the real value of the improvement (most new models of equipment are better buys than the equipment that they replace) (Evenson, 2001). This produces a spill-in impact. Another type of spill-in that is recognized in few studies is the "recharge" spill-in from pre-invention science. Many of the studies summarized in the meta-analysis actually covered a wide range of research program activities including many pre-invention science activities. Some studies specifically identified pre-invention ex-

---

[8] For a discussion of the issues related to these estimate see Alston (2002) and Pardey et al. (2002).

penditures and activities as well as industrial spill-ins (Table 8-17). These studies report relatively high rates of return and are roughly equal to the social RORs to public agricultural research.

8.2.8 Impacts of public sector agricultural R&D investments on poverty

Several recent studies (Fan et al., 2000; 2004ab, 2005; Fan ·and Zhang, 2004) clearly indicate that public investment in agricultural R&D is the most efficient public sector investment (with the exception of Ethiopia). In terms of number of poor people moving out of poverty, agricultural R&D investments ranked among the top three. Although limited, this evidence indicates that the investment in agricultural R&D performs equally as well or better than the other public sector investments and contributes significantly to poverty reduction. These studies measured the effects of public spending on growth and poverty reduction in selected Asian and African countries using pooled time-series and cross-region data (Table 8-18).To assess the impact of public investment on poverty, the number of poor people who would come out of poverty for a fixed investment across different

sectors was estimated. Similar to estimate the economic benefit of the investment the benefit/cost ratios were estimated at the national level based on the increase in household income and/or productivity per unit of investment. In Asian countries, the growth effects of investments in agricultural research, roads and education are found to be large. Regional differences were observed within the countries. This demonstrates that there is an opportunity to improve the growth and poverty impacts of total public investments through better regional targeting of specific types of investment (Fan et al., 2005).

Effects of improved technology on income distribution across farms with different resource endowments have been ambiguous. Poverty reduction is largely about distribution. The benefits of agricultural research investments are large and undisputed, but their actual levels and distributional effects remain under discussion (Alston and Pardey, 2001). Measurement of distributional effects can, in principle, be made using economic surplus methods (Alston et al., 1995), although such measurement is not common. One reason why the debates continue may be that discussions of research impacts on poverty implicitly refer to only one of two separate

Table 8-17. **Economic impact studies: Private sector R&D spill-in and pre-invention science spill-in.**

Study	Country/region	Period of study	IRR	
Private sector R&D spill-in:				
Rosegrant and Evenson, 1993	India	1956-87	Domestic	50+
			Foreign	50+
Huffman and Evenson, 1993	US	1950-85	Crops	41
Ulrich et al., 1985	Canada		Malting barley	35
Gopinath and Roe, 1996	US	1991	Food processing	7.2
			Farm machinery	1.6
			Total social	46.2
Evenson, 1991	US	1950-85	Crop	45-71
			Livestock	81-89
Evenson and Avila, 1996	Brazil	1970-75-80-85		NC
Pre-invention science spill-in:				
Evenson, 1979	US	1927-50		110
		1946-71		45
Huffman and Evenson, 1993	USA	1950-85	Crops	57
			Livestock	83
			Aggregate	64
Evenson et al., 1999	India	1954-87	Domestic	
			Foreign	
Evenson and Flores, 1978	Int. (IRRI)	1966-75		74-100
Evenson, 1991	US	1950-85	Crops	40-59
			Livestock	54-83
Azam et al., 1991	Pakistan	1966-68		39

NC= Not calculated

Source: Evenson, 2001.

concepts, absolute and relative poverty. Absolute poverty is a measure of how many people lie below a certain income threshold; relative poverty measures the degree of income inequality. Studies that show positive effects of agricultural R&D on poverty alleviation may implicitly be considering absolute poverty; studies that indicate negative effects may be more likely to refer to relative poverty (Foster, 1998).

There are three sets of factors which further conflate attempts to analyze the impact of AKST on poverty reduction. First, what is the role of underlying socioeconomic conditions in determining the benefits/costs of AKST? It is easy to find cases in which poor farmers with small land holdings have benefited as much as large-scale farmers, and those in which the benefits of new technology were confined to wealthy, more commercialized farms only. Which outcome predominates depends primarily on the underlying socioeconomic conditions of a particular case rather than the characteristics of the technology per se (Kerr and Kolavalli, 1999). Second, what is the role of the underlying social and

political institutions? A review of the impacts of agricultural research on the poor (Kerr and Kolavalli, 1999) shows that it is difficult to make generalizations about the impacts of agricultural research on the poor and the distribution of benefits depends on the underlying social and political institutions rather than technology per se. Effects of improved technology on income distribution across farms with different resource endowments have been ambiguous. About 80% of a review of 324 papers on the distributional impacts of the green revolution argued that inequity worsened, but there were significant variations within the data set (Freebairn, 1995). Innovations in agricultural research will not reduce poverty in the absence of poverty-focused policy and action (Gunasena, 2003). Third, in the absence of specific data on the impacts of AKST on poverty alleviation, one cannot simply use economic growth nor yield increases as a proxy for poverty reduction. The effect of agricultural research on poverty is usually linked in the literature through its effects on agricultural productivity (Kerr and Kolavalli,

Table 8-18. **Ranking of public investment effects in selected Asian and African countries.**

	China	India	Thailand	Vietnam	Uganda	Tanzania	Ethiopia
Ranking of returns in agricultural production							
Agricultural R&D	1	1	1	1	1	1 (52.46)	3
Irrigation	5	4	5	6			
Education	2	3	3	3	3	3 (9.00)	2
Roads	3	2	4	4	2	2 (9.13)	1
Telecommunications	4			2			
Electricity	6	8	2	5			
Health		7			4		
Soil and water conservation		6					
Anti-poverty programs		5					
Ranking of returns in poverty reduction							
Agricultural R&D	2	2	2	1	1	3*	
Irrigation	7	7	5	6			
Education	1	3	4	3	3	2	
Roads	3	1	3	4	2	4	
Telecommunications	4			2		1	
Electricity	5	8	1	5			
Health		6			4		
Soil and water conservation		5					
Antipoverty programs	6	4					

Sources: Fan et al., 2000, 2004ab, 2005; Fan and Zhang, 2004; Mogues et al., 2006.

1999). However, increasing productivity is not enough to decrease poverty (Palmer-Jones and Sen, 2006). There are other factors that can affect poverty which are not affected by the increase in productivity, such as the distribution of the income, the adoption of the technology or the suitability of the technology for the rural community. In addition, increased food supply does not automatically mean increased food security for all. What is important is who produces the food, who has access to the technology and knowledge to produce it, and who has the purchasing power to acquire it (Pretty and Hine, 2001). Sub-Saharan Africa has experienced growth in the agricultural sector (FAO, 2005b); additional analysis is needed to understand who has benefited from this growth, and why this growth did not translate into increased food security.

There is no agreement in the literature as to what kind of technologies would have the biggest impact on the reduction of hunger and poverty. While some authors agree that the main problem is not the technology itself, but the access of the poor to new technologies, others would argue that the problem is that the technologies developed are not pro-poor, and benefit only the wealthier farmers. Often, poor farmers do not have access to the technologies and do not participate in the decision process on technologies. Thus the research process ignores farmers' knowledge and experience even though they may offer insights that could help identify and/or develop effective technologies for unfavorable areas. For that reason the technologies developed may increase productivity but this may not be their main objective. Such systems may perpetuate a sense of helplessness among resource-poor farmers as they wait for effective technological solutions. Recently, some explicit efforts are being made to include farmers' needs in the R&D agenda. All actors seem to agree that participatory approaches are needed; participatory plant breeding seems to be promising (Almekinders and Eling, 2001). What remains unclear is what role the industry would play in this democratization process.

The poverty reduction effect can be substantial and it is free, in the sense that R&D has already paid for itself, whereas redistribution can be counterproductive due to its negative effects on growth (Thirtle et al., 2003). A long-run view of technological change must take into account the distributional effects of agricultural research investments. These research investments go beyond technology and include institutional innovations and the structure of the innovation system catering to agriculture. The distributional impact of technological change ultimately depends on the particular context of policies, markets, and institutions and on interregional connectedness through infrastructure (von Braun, 2003) (Figure 8-8). Adoption of technologies and success depends on many factors, e.g., land ownership, access to water, and availability and efficient use of diverse plant genetic resources (von Braun, 2003). This is in line with the argument that at the farm level, prices, access to inputs and resources, credit and markets, education levels and the distribution of land, affect both the rate of uptake of improved technologies and the extent to which they benefit the poor (Hazell, 1999). Improved technologies may fail to benefit poor farmers, not because they are inherently biased against the poor, but because the distribution of land, or

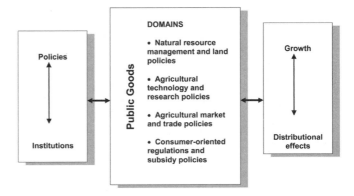

Figure 8-8. *The conditioning of agricultural growth and distributional effects.* Source: Von Braun, 2003

access to resources and markets is inequitable. When these are taken into account it becomes possible to explain why similar technologies can have very different impacts on the poor in different regions, or at different times.

Research needs to focus on commodities used by poor people, and on areas where the poor are concentrated (rain fed highlands, semiarid tropics and marginal lands) (Gunasena, 2003). Without adequate investment in infrastructure, technology and human development in these areas, conditions are likely to deteriorate further. Technologies likely to succeed in these areas include mixed farming systems—livestock and agroforestry, improved fallows, and cover crops (Gunasena, 2003). In all cases marketing institutions need to be developed to support the small-scale farmers.

Although less controversial than biotechnology, low-external-input agriculture (LEIA) is also the subject of considerable disagreement (DIFD, 2004). Debate on the relevance of these technologies is unfortunately often clouded by ideology. A dataset containing information on 208 cases from 52 developing countries show that in these projects and initiatives, about 9 million farmers have adopted sustainable agriculture practices and technologies on 29 million hectares (Pretty and Hine, 2001). This demonstrates that sustainable agriculture can reduce food poverty through (1) appropriate technology adapted by farmers' experimentation; (2) a social learning and participatory approach between projects and farmers; (3) good linkages between projects/initiatives and external agencies, together with the existence of working partnerships between agencies; (4) presence of social capital at local level. A variety of options are available to increase the returns to families from their production, either by reducing losses to pests (better storage and treatment) and inefficient processes (e.g., fuel-saving stoves); or by adding value before sale or use (conversion of primary products through processing). Adding value through direct or organized marketing may involve improvements to physical infrastructure (e.g., roads, transport); or through direct marketing and sales to consumers (thus cutting out wholesalers and "middlemen").

Other aspects that need attention are the effects of the crop technology adoption on gender, for example, in the distribution of work roles in the cropping (Von Braun and Webb, 1989) and the significant spatial dependence on growth rates of agricultural output (Palmer-Jones and Sen, 2006).

Additional important issues to be considered in terms of effects of AKST on poverty reduction include (1) how researchers are evaluated (Gunasena, 2003); (2) poverty alleviation as specific target in agricultural policy (Gunasena, 2003); (3) relationship of increased productivity to reduced food prices, and evidence that rural poverty is linked to international world prices (Minota and Daniels, 2005; Yavapolku et al., 2006). In addition, indebted countries may have economic growth but not poverty reduction as they must pay a substantial part of GDP to external debt.

The future is not just about the need for more scientific effort and technical breakthroughs generated by both more public funding and private sector interventions, but about the political economy of agriculture and food in the developing world (Scoones, 2003). Two basic components of well-being are having a secure livelihood to meet one's basic needs, and realizing and expanding one's capabilities in order to achieve fulfillment. For that reason measuring the link between poverty and agricultural growth by using the human development index or developing new indexes may be necessary.

8.3 Governance of AKST Investments: Towards a Conceptual Framework

8.3.1 Demand for improved governance

Particularly since the mid-1980s, there has been increasing demand for AKST systems to be accountable to various stakeholders. These demands have been prompted by the high transaction costs of conventional agricultural research systems in knowledge generation and transfer as well as inefficiency in resource allocation and utilization (Von Oppen et al., 2000). Other reasons for recent demands include lack of transparency, exclusion of other stakeholders from the process of setting research agendas, unequal access to technologies emanating from research and fear of private sector monopoly over technologies, particularly in biotechnology (McMahon, 1992; Reisfschneider et al., 1997; Echeverria, 1998; Von Oppen et al., 2000).

The pressure for accountability is varied across countries and regions. For example, in industrialized countries issues of efficiency and pluralism in the research process are becoming more important (Heemskerk and Wennink, 2005). In most Asian and Latin American countries, the pressure for more accountability seems to be driven by local stakeholders (Byerlee and Alex, 1998; Von Oppen et al., 2000; Hartwich and Von Oppen, 2000). In the case of sub-Saharan Africa, it is the donors, who provide more than one-half of the funding for agricultural research in some countries (see 8.1), who pressure for accountability (Herz, 1996). These demands for accountability have resulted in changes in both the sources and the mechanisms for funding AKST (8.1.4) and hence the rules and modalities which govern the mobilization and utilization of AKST investments.

8.3.2 Defining and judging governance in relation to AKST investments

The changes in governance of AKST can be viewed as part of an "induced institutional innovation" (Ruttan, 2003),

which sees changes in institutions or governance driven by factors of demand and supply. On the demand side, the contemporary economic and social realities (including developments of new technologies) are pushing for changes in governance and institutions mediating AKST investments globally, nationally and at lower levels within nations. On the supply side, advances in social science knowledge are increasingly an important source of shifts in the supply of institutional solutions (Ruttan, 2003). Thus the accumulated knowledge (both theoretical and empirical) on the functioning of institutions can be viewed as facilitating the supply of new institutional solutions.

The discussion of governance and the criteria to judge good governance can be approached in several ways. These criteria can be based on certain outcomes such as how efficient or effective is the governance in meeting predetermined objectives (Box 8-2).

Box 8-2. On the theoretical framework to analyze governance.

There are different streams of theoretical literature informing the discussion on governance. One such framework is that of New Institutional Economics (NIE), an extended framework of neoclassical economics. It takes into account demand factors such as the role of relative prices since such prices play an important role in deciding what is an appropriate institution in a given context. However NIE admits the possibility that the evolution of appropriate institutional innovation need not be an automatic process. There can be social, political, and even institutional reasons that distort or blunt the evolution of appropriate institutions. There has been significant development in institutional analysis during the last two decades highlighting the possibilities of persistence of institutional inefficiency due to reasons of path dependence, political economy and informational problems. An alternative framework is that of the national innovation system (NIS) (Freeman, 1987; Lundvall, 1992). It treats R&D as an innovation system in which both the producers and users are seen as parts of the same system and attempts to identify certain patterns in system relationships, governance, capacity-building or learning, evolving roles, and wider institutional contexts (Hall and Yoganand, 2002). However from the point of view of NIE, NIS approach lacks a coherent theoretical framework, and thus is unable to develop consistent stories or explanations of different institutional changes taking place in different socioeconomic contexts. Meanwhile, the criticism of the innovation system proponents on the NIE-based approach would be that the latter is inadequate to handle power structures and learning. However the issues of incorrect learning and information problems have become part of the agenda of NIE increasingly in the nineties (North, 1991) and the New Political Economy takes into account the role of power struggles in facilitating or blocking beneficial institutional changes.

Governance has three core functions: (1) To identify what is the "optimal" institutional structure; (2) To manage institutions, which implies monitoring, sustaining, fine-tuning, and facilitating of all these activities (3) to change the existing institutions or bring about newer ones to close the gap between the existing and the "optimal" structures. Institutions are the formal and informal rules, including norms and practices. Organizations are not institutions but actors within institutions. Institutions often include markets too. However nonmarket or meta-market institutions are also required for AKST investments because of multiple forms of market failures. Some of these market failures are observable in any R&D investment requiring institutional interventions such as patent rules. In general, market failures arise from the public good nature of some forms of AKST, implying that it is very difficult or costly to exclude people from using a technology (Norton, 1991; Stiglitz, 2000).

The framework for assessing governance of AKST systems comprises of a set of characteristics that the outcomes of institutional interventions mediated through good governance is expected to have. These are briefly discussed below to provide clarity of the analytical framework:

- Good governance should address identifiable market failure problems, in order to allocate public resources in areas where uncoordinated action of private individuals would be inadequate or inappropriate. Nonetheless, correcting market failure is only justified where the losses from the failure are higher than the cost of correcting it, including both direct and indirect costs of institutional intervention. Where there are multiple forms of market failure occurring at different points of the technology development process, each failure must be addressed accordingly.

- Where agricultural growth is not the constraining factor to poverty reduction or to attaining economic growth, prioritization of appropriate intervention among sectors or between agricultural research and other interventions within agriculture is another important outcome. There may be cases where factors other than R&D become binding constraints to agricultural competitiveness (Garelli, 1996). The appropriate level of resource allocation for agricultural R&D is also related to sectoral prioritization (Tabor et al., 1998). AKST decisions are made by many actors; each of them takes into account the expected action of others. For example, poor and small developing countries may want to depend on international technology transfer and this may be the best option for certain technologies in order to avoid unnecessary and costly duplication of efforts (Tabor et al., 1996). However, such dependence may not always be possible for tropical countries, whose commodities (crops, livestock, fisheries, and forests) are unlikely to be cultivated or raised in the industrialized world.

- Good governance should ensure that institutions and organizations (as well as individuals who work within these organizations) serve the intended purpose effectively and efficiently under all conditions. The achievement of efficiency in research investments is complex due to problems of economies (some times diseconomies) of scale and scope, which determines the required degree of specialization or diversification of specific research organizations. These considerations may also lead to contracting out or contracting in of specific activities, and also the extent of decentralization in decision-making. The role of governance is to enable the internalization of such efficiency concerns in decision-making. This requires design of institutions with: (1) The ability to shape specific objectives to suit socioeconomic realities. This implies that there should be mechanisms to discontinue research programs that are no longer acceptable to stakeholders. (Sometimes inefficient institutions may continue to persist due to path dependence and lock-ins. Meanwhile, efficient institutions have a built-in ability to adapt to changing realities through feedback.) (2) The ability to meet the objectives with reasonable assessment of risk and uncertainty. (3) The ability to assess current and potential future demands of AKST investments. (4) The ability to carry out assignments and tasks in the most efficient manner, which entails producing given output at the cheapest possible cost or achieving maximum output for a given cost. Higher institutional efficiency can be achieved by aligning the incentives of actors to be in tune with institutional objectives, which in turn should change with evolving economic environment.

- Good governance should aim at following procedures that ensure transparency and accountability for minimizing mistakes and errors of judgment to ensure that broader societal priorities are reflected in decision making without being captured by distributional struggles of narrow interest groups. For example, although local scientists may be better informed about national agricultural priorities, their decisions on funding priority may be biased towards maximizing the flow of funds to their own area of work (Tabor et al., 1998).

There can also be process-based criteria for good governance, where the concern is not only on outcomes but also on how these outcomes are produced. For example, participation of specific stakeholders can be viewed as important for efficiency or effectiveness of outcomes but also as an important element on its own, with the assumption that pursuing participation is good irrespective of its impact on efficiency or effectiveness. Thus there have also been arguments that good governance should follow certain procedural correctness which should permit

- Negotiation of diverse interests and the identification of common interests;
- Negotiation of clear rules and norms among multiple stakeholders, their effective implementation and the setting-up of control mechanisms for compliance to these rules and norms;
- Equitable access to resources (economic/financial, human, natural, social, physical) and AKST;
- Participation in strategic decision-making of all relevant stakeholders;
- Adequate equilibrium among power forces in decision-making and implementation of strategic decisions, and
- Capacity to influence policy making.

Governance can also be viewed at multiple levels; at the level of a research station, a national research system, a regional research network as well as at the global level. When we analyze the issues of governance of a research station, we take the external environment including the objectives given to the station as exogenous, and try to see how the governance of the station can be improved to meet these given objectives within the resource constraints. One can also analyze the larger question of governance at which one critically looks at whether the objectives defined by, or resources given to the station are appropriate and meet the criteria of good governance. Based on the conceptual framework as given here, one can develop a set of questions that are relevant for analyzing the governance of, and institutions involved in AKST investments (Table 8-19).

8.3.3 Analyzing the experience of governing AKST investments

8.3.3.1 Public funding/public sector research
The model of public sector research organization came to exist in many parts of the world during the second half of the nineteenth century. The founding of the public research organization was based on an assumption that nongovernmental agencies (including private firms and farmers themselves) are unable to mobilize adequate resources and skills required to generate agricultural research (see also 2.2.2). It was then assumed that farmers generally needed to be educated on the benefits of new technologies and did not have any major role in the generation of technology directly. Thus government, either national or regional, provided the resources for the establishment of these research establishments from the taxes, international aid or other assets such as state-owned land. This perception of the farmer, however, has changed in recent years.

Investment for AKST by governments has been successful on certain counts. It enhanced the capacity of a number of countries to carry out good quality research. In many poor countries, there would not have been any significant level of agricultural R&D without these institutions due to the limited capacity as well as inadequate interest of the private or not-for-profit sectors to provide agricultural R&D, which mostly falls in the public good domain. Government-funding for AKST has also played an important role in enhancing the awareness of farmers, in creating a wide pool of trained personnel and informing policy making at the national level in a number of countries.

Despite such achievements, this model has had several problems. For example, it did not perform well in assessing the needs of farmers in many parts of the world and it has been fairly slow in responding to social and economic changes. There have been innumerable cases where research investments were directed in a way that they failed to meet set objectives, even if the uncertainty inherent in R&D activities is accounted for. Public organizations were not very successful in taking into account local agroclimatic and socioeconomic features in their research programs (Santhakumar and Rajagopalan, 1995). Efficiency of public R&D organizations is also open to question, and one feature noted in many developing countries is the spending of a greater amount of financial resources to provide the salary of permanently employed staff, with little left for actual research activities, which in turn affect the research output and hence the research efficiency (Eicher, 2001). This may not be directly evident in ROR calculations of agricultural research. It is possible to have high ROR even with these levels of inefficiency. There have been inappropriate resource allocations between capital and operating expenditures in the public sector, resulting in a pool of inadequately trained and equipped personnel, research laboratories without sufficient operational and maintenance funds, or other inefficiencies.

The fiscal problems of the governments of many developing countries have led to a reduction of resources made available to public research systems, which often reduced

Table 8-19. Guiding questions for institutional assessment on governance.

Issue/Actor	Guiding Question
Governance	1. What are the appropriate intervention strategies in different sectors given the overall social objectives?
	2. What is the appropriate intervention given the objectives in the agricultural sector?
	3. What is the problem of market failure to be addressed?
	4. What is the institutional mechanism required given the problem of market failure?
	5. How to ensure that governance decisions are accountable and transparent?
Institutions	6. Is the institutional arrangement capable of meeting the objective?
	7. Is the institutional arrangement capable of internalizing the requirements or demands of its potential clients?
	8. Is the institutional arrangement leading to efficient decisions given its alternatives?
	9. Does the arrangement have flexibility to evolve in tune with the changing socioeconomic realities?
Organizations	10. What kind of feedback is likely to be generated by the organizations operating within this institutional framework?
Individuals	11. Are the incentives (monetary as well as other nonmonetary rewards) of the individual actors aligned with the stated objectives of the organizations?

the funds available for recurring and operating costs (Premchand, 1993; Eicher, 2001). The rewards of the agricultural staff tend to be misaligned leading to difficulties in keeping the best talent on the one hand, while the indexed salaries of employees without much concern for market wages tend to balloon the overall budget for this purpose. There is also the widely discussed problem of wage erosion, meaning the loss of salary purchasing power, which impacts negatively on commitment and morale of research staff.

Sometimes allocations of public resources can lead to spending being spread too thinly across commodities, regions and research themes. There can also be other inefficiencies within public organizations leading to wastage of resources, corruption and poor planning in public-funded research. Public sector scientists can continue with research on commodities (crops, livestock, and natural resources) and technologies even when farmers move out of these areas due to economic reasons. Some studies show that returns to public sector agricultural extension became low due to the multitude of "non-extension" duties, and that extension agents were not the main sources of technical information to farmers (Isinika and Mdoe, 2001).

During the 1980s, public research models were reformed to become more participatory. This was to make public research organizations more responsive to the requirements of farmers, especially those that are poor and live in resource-deprived areas (Kaimowitz, 1993). Only limited successes were achieved through such participatory research models. This could be due to the fact that the structure of public research organizations was not reformed. The channels of priority setting do not correspond to the funding channels, in other words, funding is provided from other sources than those setting research priorities (Hartwich and von Oppen, 2000). Another reason could be that the incentives for individual researchers were not always adequately oriented to participatory research. These incentives include not merely additional money but also additional facilities to carry out participatory research, but also intangible ones.

Public research organizations have also responded to the criticisms on their inefficiency by adopting impact assessment of their efforts, priority-setting exercises, and also the introduction of operation and management reforms through measures such as decentralization, accountability, transparency and cost recovery among others (Hall et al., 2000). Moreover there have been efforts to give more autonomy to research organizations, remove them from civil service regulations and to provide greater flexibility to manage their physical, financial and human resources (World Bank, 2000). One can see such examples from the industrialized world. There has also been decentralization of research and extension systems in developing countries including Uganda, Tanzania, Kenya, Zambia and Ethiopia (Anandajayasekeram and Rukuni, 1999). Similar examples of more pluralism in AKST systems were documented for various other African and Latin American countries (Shao, 1996; Byerlee, 1998; Echeverria, 1998; Heemskerk and Wennink, 2005). But the experiences in different countries are mixed. Research practices and administrative and financial procedures of national research systems have not witnessed any major changes in a number of countries. On the other hand, reforming compensation in the national agricultural system

of Chile has made the public research sector more attractive to talented agricultural researchers (Venezian and Muchnik, 1994).

The resource allocation for public sector research, though ideally driven by considerations of social welfare, is determined in reality by the political economy[9], i.e., the struggle between the interests of different societal sections (social groups, regions, growers of specific crops, gender), and also those who dominate decision-making. Evidence from different parts of the world indicates the influence of such political-economy factors in resource allocation for agricultural research. Research and extension spending is linked to the political effectiveness of farm interests (Rose-Ackerman and Evenson, 1985). A study of 37 countries show that structural changes in the economy have important effects on the political incentives to invest in public agricultural research (Swinnen et al., 2000).

Thus even when agricultural research provides higher returns or has the potential to reduce poverty; it does not get enough investments in the public allocation of resources. Sometimes ideological considerations lead to high priority being placed on certain crops, thus making investments economically inappropriate. For example, concerns about food security in certain states of India have led to excessive research investments on some crops, and farmer adoption of commercial crops unsuited for the region (Santhakumar and Rajagopalan, 1995; Santhakumar et al., 1995). Gender is an area where political economy influences research investments and outcomes. This manifests itself in certain situations through inadequate investment in research on crops cultivated by women or technologies which would reduce the drudgery of female agricultural workers. In certain other situations, new technologies produced through research lead to the displacement of women workers. An example is access women have to ICT, which may be limited because of their reduced physical access to resources and infrastructure, social and cultural norms, education and skills, and poverty and financial constraints (Hambly Odame et al., 2002).

Does democracy help in achieving the socially desirable objectives through AKST investments? There is no straightforward answer evident from the literature (e.g., Diamon and Plattner, 1995) on whether democracy vs. the responsiveness of governments has a higher degree of such achievement. Even in democratic countries the political process can be captured by narrow interest groups, whose goals do not necessarily aim at overall social welfare. Even if the role of such groups are controlled, democracy is likely to be driven by the preference of the median voter, and there are situations in which the interest of such a voter need not be in tune with the maximization of the overall welfare of the society. Thus, though democracy is valuable by itself, and provides greater opportunity for wider participation in political decision-making, there is no assurance that it will lead to decisions that enhance the welfare of the society as a whole. The lesson for AKST is that democratic governance is not sufficient to ensure effective and efficient investments aimed at achieving larger development goals.

---

[9] The term *political economy* is used within the framework of "new political economy" (e.g., Bardhan, 1997).

8.3.3.2 International donors

Broadly, international donors are motivated by three objectives for extending funding for ASKT to developing countries. These are:

- International charity or resource transfer based on altruistic considerations;
- Correction of international market failure or the provision of international public goods; and/or
- Expansion of the markets of the donor countries.

These objectives have motivated international donors to support agricultural research and extension capacity to enhance food production in many developing countries during the last 50 or more years.

Although international funding for AKST is a major source of support in the developing or poorer countries and domestic research would not have developed without this crucial support, international funding can also create distortions. The availability of international funding at times may encourage domestic players from mobilizing internal resources. This is most visible in Africa where donor support to agricultural research has increased in relation to domestic support so that nearly half of the agricultural investment in Africa is from donors including development banks (see 8.1.4). This has perpetuated donor dependence and undermined efforts to develop domestic political support for sustainable funding, especially for the smallholder sector (Rukuni et al., 1998; Eicher, 2001). The allocations of international funds between different types of expenditures, such as between capital and recurring costs, do not need to adequately reflect the domestic opportunity cost of the resources. There have been instances where external aid has compounded the inefficiencies in AKST investment decisions in developing countries. The risk of bad investment goes up when grants are easily available (Tollini, 1998).

Correcting market failures at the international level could be another force driving international donors to fund AKST systems or generation. There are at least two major forms of market failures. There can be international negative externalities, which need action at the international level, but there may also be instances where it is efficient for the international community to take action to address certain problems within the developing countries that have the potential for global impact. The recent incidence of avian flu is a good example. Even if the interest in the industrialized world is to protect itself, financing some activities in developing world on preventative measures at the source of the problem would be a more effective and efficient strategy rather than spending money only on protective activities within the industrialized world. Similar arguments apply for international public goods. Certain technologies or technology generation systems themselves can be seen as international public goods. The ideal strategy would be for the industrialized and developing world to pool their resources together, but there are problems of coordinating such efforts. The severity of lacking such public goods perceived in the industrialized world would encourage them to take proactive steps, whereas developing countries who face other more pressing problems would give low priority. How far AKST investments driven by the requirements of correcting international market failure reflect the economic variables of the world as a whole, would determine their effectiveness, efficiency and outcomes. Moreover, it is important to see that such investments made in the developing world do not create distortions in their economies.

The expansion of markets or cost-reduction of global production has also driven industrialized countries, multinational firms and multilateral agencies to make AKST investments in developing countries. These, however, raise a number of issues: (1) Trade and nontrade barriers (and associated transaction costs) might influence where such investments take place and at what cost; (2) Since the domestic institutions in many developing countries are weak, this may lead to an intensification of "market failure" problems in such countries. For example, there are apprehensions on increasing field research of new (genetically modified) seed varieties in developing countries as part of international contract research, without taking adequate safeguards against the unknown long-term impacts of such seed varieties and also for the preservation of local genetic materials.

The urge to expand the lending of multilateral funding agencies has also received criticism during the last decade. The incentives of the personnel in these agencies could be directed towards excessive lending, and this, combined with the incentive of political and administrative decision makers of developing countries to borrow excessively (more than what is warranted by the domestic economy considerations), can lead to excessive loans. Whether this incentive problem has affected the efficiency of multilateral funding for AKST in developing countries is an issue that needs to be analyzed.

8.3.3.3 Competitive funding

Block grants have been used for allocating research resources for many years. Now block grants have become less attractive as concerns have been raised about inefficiency in resource allocation, effectiveness and relevance of research as well as exclusion of other stakeholders in the research process, from priority setting to execution of research projects/programs (McMahon, 1992; Echeverria, 1998; Reisfschneider et al., 1998; Von Oppen et al., 2000). This has led to the gradual evolution of competitive funding mechanisms at the international and national levels. Competitive grants:

- Allow for a wider network of actors to participate in the research process broadening the scientific talent available (Von Oppen et al., 2000);
- Allow for a possibility to seek a diversity of funding sources (Byerlee, 1998);
- Improve research quality (Byerlee and Alex, 1998);
- Improve allocation of research resources (Alston et al., 1995).

However, competitive funds have the disadvantage of having high transaction costs (Echeverría, 1998). Competitive grants take scientists' time (funded through core funding) for preparation of research proposals, and evaluation (Huffman and Johnson, 2001). There is also significant increase in administrative costs for managing research competition. Another disadvantage of competitive grants is that they do not contribute to capacity development in terms of infrastructure and human capital development. They also tend to be of short term in nature, which may divert attention from more crucial research topics and national priorities (Echeverría, 1998). It has been noted in Africa that competitive grants (1) fail to include beneficiaries in the research process; (2) fail to prioritize and hence tend to spread resources too thinly; (3) create uncertainty as to whether the funds are truly competitive and are able to link to performance, given the limited number of researchers in the region; (4) are expensive to operate; and (5) are not sustainable without external donor support. The inherent ex-ante uncertainty in research, asymmetric information that makes monitoring of scientists by administration difficult, and the sharing of risk between funding agencies, administrators and scientists are issues that may make contract-oriented reforms in R&D complex even in industrialized countries.

8.3.3.4 Commodity boards and growers' associations
The growing role of commodity boards, producer-funded or growers' associations, in research is also a related development. Nonprofit organizations constitute a comparatively large share of agricultural research in Colombia and some Central American countries (Beintema and Pardey 2001). Colombia has twelve nonprofit institutions, which accounted for about one-quarter of the country's agricultural research investments during the mid-1990s. Many of these agencies began conducting research several decades ago and are funded largely through export or production taxes or voluntary contributions (Beintema et al., 2006). In Africa examples include agencies conducting research on tea (Kenya, Tanzania, Malawi, Zimbabwe), coffee (Uganda, Kenya, Tanzania), cotton (Zambia), and sugar (Mauritius, South Africa). There are, however, other forms of nonprofit institutions in a number of countries, including Madagascar and Togo, although these play a limited role in agricultural research (Beintema and Stads, 2006). There is no evidence that the involvement of growers' associations or private sector has added more investment for AKST or have been replacing government funding.

How far research driven by these agencies is different in terms of efficiency and effectiveness from that in state-funded organizations, especially in the developing world, is a question requiring further investigation. In tea research in India the R&D carried out under planters' association leads to the development of appropriate technology due to the greater awareness of clients' requirements, and faster or timely communication of these technologies to the users (Muliyar, 1983). If commodity boards have also a mandate for marketing and/or the provision of other support services (including subsidies), they may have a greater incentive for being effective in terms of technology generation and extension, even if these boards function under the government (Narayana, 1992). In Kenya, acceptable ratios of personnel/operations cost prevail in coffee and tea research, which is financed by a cess. But there are also cases in Kenya where growers' associations became politicized and hence being less accountable to the growers (Kangasniem, 2002).

One concern is that the producers' associations or commodity boards focus on the sole benefit to producers and thereby mostly neglecting the welfare of the consumers and the economy as a whole. It is not uncommon to see the growers' associations and commodity boards lobbying for enhanced protection of their products in domestic markets or support for exports, both of which may have a negative impact on domestic consumers. Moreover, the provision of subsidies associated with the propagation of specific technologies, as well as the bureaucratic compulsions of commodity boards may also lead to excessive inducement of farmers to adopt specific production systems, which may not be sustainable in a more market-determined situation. Finally, it is possible that producer organizations may not be the best suppliers of research services except for adaptive on-farm research (Echeverria et al., 1996). These shortcomings provide a justification for continuation of government funding for basic and strategic research even in industrialized countries. Moreover, for crops that have a large number of cultivators such as rice or wheat, the concept of growers' association becomes unmanageable and would have problems similar to those of government-owned research. Additionally, to what extent the small farmers are represented by these associations remains unclear and depends on the commodity and the countries.

8.3.3.5 Private research
In the industrialized countries and the more advanced developing countries, the inadequacies of the public research model led to the gradual emergence of private sector (or broadly market-oriented) reforms in agricultural R&D investments in the late seventies and eighties. This was facilitated by the interests and the capability that the private sector has developed in AKST investments. The structural adjustment policies implemented in many developing countries,[10] the global changes in trade regime and developments in biotechnologies, have also facilitated this transition.[11] This transition is manifested in the increase in private sector funding in public sector organizations and universities, and the increase of the research directly carried out by private sector organizations. The commercial or application-orientation of the private sector to some extent fills the gap between technology generation and extension that existed in the public research model. There has been an increasing involvement of the private sector in agricultural extension as well (Umali and Schwartz, 1994).

There are variations between countries and regions in terms of the contribution of private sector in agriculture research (see 8.1.1). Though private sector investments play an important role in OECD countries, their share in many developing countries continues to remain insignificant. Not surprisingly, there may be a linkage between national income of the countries and the role of the private sector in agricultural research (McIntire, 1998). But the lack of significant private research is also often the result of the legal and administrative environment in many countries (Ahmed and Nagy, 2001).[12] There are indications that mutually

---

[10] See Tabor (1995) for a number of articles dealing with the impact of structural adjustment policies on agricultural research system.

[11] Private sector involvement in agricultural biotechnology research started much before, and by the 1990s, private sector investment in this regard has exceeded that of the universities and government owned laboratories (Lewis, 2000).

[12] On the other hand some countries (for example, Thailand) seem to have government policies favorable to private sector research.

negative perceptions of public and private players, unresolved issues of risk and liability, high transaction and opportunity costs act as barriers against the development of public-private partnerships (Spielman, 2004; Spielman and Grebmer, 2004).

Each of these funding mechanisms has advantages and disadvantages. In developing countries where governance structures are still weak, the advantages may not be apparent during initial stages of the funding options (Box 8-3).

8.3.4 AKST governance and changes in the larger institutional environment

So far we have considered only the institutions directly governing AKST investments. However the broader institutional environment encompassing the ownership of rights over land, water, and other common property resources would also influence indirectly the governance of AKST investments. The institutions under this category can include land reform, water management, forest protection, international standards related to food products and agricultural imports, international law of the seas, global agreements on climate change and so on. These institutions that set the rules for managing natural resources locally, nationally and internationally would have a direct bearing on the effectiveness, nature and content of AKST investments. Similar is the impact of emerging organizational forms in the trade of agricultural and related commodities. For example, contract farming for export-oriented horticultural crops is expanding in many developing countries, and this will have a bearing on how AKST is generated and used, and consequently how investments are made for this purpose (Porter and Phillips-Howard, 1997; Haque, 1999). It is not only that the effectiveness of AKST investments is influenced by institutions governing natural resource management and use, but, increasingly AKST investments are also seen as solutions, albeit partially for sustaining the natural resource base. This is especially important in a context where urban and environmental interests in resources such as land and water compete with farming interests (Farrell, 2004). AKST investments and the institutions of natural resource management are in turn influenced by the wider political and economic institutions of countries and the world. The market expansion in developing countries,[13] changes in world trade regime,[14] structural adjustment policies in many countries, and others are going to influence not only natural resource management but also investments in AKST.

In addition to these institutions, the way human consumption especially that of food and agricultural commodities changes in the future would have a strong influence on the nature of AKST investments. Though economic vari-

---

[13] Poorly developed market infrastructure can influence the distribution of gains from agricultural research (Dasgupta and Stiglitz, 1980).

[14] The opening up of an economy may result in having less farmers influence prices; hence they may become less capable of being the major beneficiaries of agricultural innovations. Changes in trade regime may have a greater potential in changing the distribution of direct benefits of agricultural research in a country than other routes such as better targeting of agricultural research expenditure (Voon, 1994; Sexton and Sexton, 1996).

Box 8-3. Experience of new funding options in African countries.

Many African countries have implemented new governance enhancing strategies such as separation of policy making, funding and service provision, decentralization of public administration, deconcentration of service provision, and empowerment of communities and farmers organizations. Experience from Tanzania and Benin (Heemskerk and Wennink, 2005) have shown that local R&D funding schemes have contributed significantly to financial diversification for agricultural innovation. However, real and substantial empowerment of farmers' organizations in controlling financial research for adaptive research and pre-extension is still low. Although downward accountability has improved, real client control of funds has stagnated and farmers' representation in management teams of competitive grant schemes remains weak due to traditional top down attitudes of researchers and research managers.

Decentralization and deconcentration of local innovation development funds have been more successful in technology generation, and in fostering the competitive element, which has enhanced the quality of research and the sense of ownership. Nonetheless, other concerns such as developing more viable mechanisms for client representation, priority focus and pro-poor focus of available funds, level of co-sharing and cost sharing are all yet to be resolved. In addition, some of the competitive grants and commodity based innovation development funds are insufficiently integrated into the national financing system.

In terms of effectiveness and efficiency, there is evidence that more adaptive technologies are flowing to farmers under competitive funding, but there is no effective mechanism to systematize the information on the innovation adoption process. There has also been improvement in priority setting, planning and implementation, but not as much in monitoring and financing. Competitive grants tend to spread resources too thinly. Experience in Tanzania showed that effectiveness of competitive grants could be improved by focusing on a single theme using the value chain approach. Another disadvantage in the African context is that competition may be limited due to insufficient numbers of competent researchers. In addition, competitive funds in African have been dependent on donors, whose pledges by donors have sometimes not been forthcoming. Cofinancing from local sources has also been unpredictable. Competitive funds are also expensive to operate due to high transaction cost especially for monitoring and evaluation (Lema and Kapange, 2005ab).

---

ables such as income play an important role, social, cultural and ideological factors do have significant influence on the evolution of human food and consumption systems. There need not be a linear evolution from traditional and home-based subsistence consumption to a full reliance on globally integrated markets for commodities produced with factory-based inputs and modern technology. There are indications from India and China that economic growth and development do not lead to a decline in (if not an increase of) the demand for the so-called traditional systems of food-making or nature-dependent health care systems. This underscores the importance of visualizing different scenarios of future and their likely influence on the investments of AKST. However one probable scenario on the governance of AKST in the near future is outlined below.

8.3.5 The future roles of governance and institutional structure

In many developing countries the domestic private sector may continue to play only a small role in the near future. Even in industrialized countries, the new set of research instruments is not going to replace the conventional public research model. It is envisaged that there will be a combination of public and private investments with the latter increasing over time. The additional costs associated with competitive funding would encourage the persistence of a combination of conventional forms of funding (such as formula funding) and competitive grants in the near future. However competitive funding as a mechanism complementary to the regular budgetary support seems to be inevitable (Gage et al., 2001), or project funding and institutional grants may have to coexist (Becker, 1982).

Similarly one should not expect that the private sector is going to replace the public sector even in areas such as agricultural biotechnology in which private organizations have an upper hand. Private sector research will concentrate on areas where (a greater part of the) benefits can be privately appropriated as in export or plantation crops, hybrid seed development or in off-farm processing of agricultural products, and in the diffusion of capital goods such as agrochemicals. For example, USAID recognizes that the private sector will not deliver biotechnology applications for many crops (such as minor or food security crops), will not address all biotic and abiotic production constraints, which are important in developing countries nor will it realize the development of commercial markets in all developing countries (Lewis, 2000). Public sector research will have to fill these gaps. Moreover, some of the conventional market failures associated with agricultural R&D are still important and hence some form of societal or state intervention may continue to be necessary. Some of these market failures, which make private investments alone inadequate, are the following:

- Given the scale economies in specific research initiatives, competition and existence of multiple firms may not be economical. This would lead to monopoly powers of the existing firms, which would warrant certain regulations to remove entry barriers in order to avoid social losses;
- Given the features of positive externality or public good associated with the development of agricultural innovations and knowledge, it is very likely that there can be underinvestment (less than the socially optimal levels) by private firms in such cases. This may be particularly so in the creation of what can be called basic or pure knowledge where the appropriation or excludability problem is acute;

- Certain innovations or technologies may have negative externalities especially with regard to environmental pollution or long-term health hazard. This is an area where institutional intervention by the state or society is required to make the private firms internalize these externalities;
- There can also be a distributional issue which would prompt governments to intervene (that need not necessarily be through state-owned research organizations) to see that technologies that help poorer farmers living in less resource-endowed areas (for example drought prone) are also generated. It is argued that the disbursement of funds in public sector research through competitive grants is likely to generate regional disparities as well as less money for activities such as managing natural resources and the environment, which need not be profitable in market value terms. This too can encourage public support for research, which are not solely based on commercial considerations;
- Agricultural research has to stand on the firm foundation of higher education. In many countries, including those in the industrialized world, higher education in AKST is closely linked to research laboratories. Higher education is unlikely to thrive solely on profit-oriented investments. This would necessitate the functioning of public/private organizations involved in agricultural research based albeit partially on public funds and endowments or other nonprofit oriented investments.

However it is very likely that there is more and more rethinking on the specific roles governments (both national and local), funding organizations and public sector research organizations in AKST investments. It is quite possible that state-owned institutions devote more resources on technologies to be used by the poor, and also on environmental conservation and other related areas where due to the externalities, private firms are less likely to invest adequately. (This is based on the assumption that the distributional struggles, political economy and the overall governance, including the role of democracy, are such that poverty reduction and mitigation of externalities become priorities of the governments.) In future there will be more and more public private partnerships in agricultural research and here the experience from OECD countries seem to be successful in making research systems more responsive to the rapid transformation of economy and their innovation requirements (Guinet, 2004). There are multiple ways of enlisting private partnership in public research and here the choice of mechanisms is very important to enhance the overall benefits. Governments and public sector organizations may be more involved in regulation and quality control of products and technologies developed by the scientists from both public organizations and private firms. Scientists may have to encounter more competition in getting research funds not only from international organizations but also from their national governments. The labor market for scientists may also become more flexible with shorter-period incentive-based contracts rather than permanent jobs. Though there is evidence that participation by private partners enables publicly funded research to concentrate on areas where private incentives are weaker (Day-Rubenstein and Fuglie,

1999), care is needed to ensure that institutional changes in public sector and changing sources of funding do not undermine the research agenda of public institutions, especially the generation of knowledge, which may not seem to be profitable and viable by the private firms.

8.4 Investment options

The goal of this international assessment of AKST is to provide policy makers with investment options for meeting the development and sustainability goals. Since no single investment can meet all goals at once, a portfolio of AKST investments are needed. Countries are likely to have different weights on the importance of the different objectives and so alternative combinations of AKST investments will be presented based on whether countries place more weight on environmental goals, improving health and nutrition, reducing poverty and hunger, or maximizing economic growth.

This subchapter focuses on the research investment options of governments, international organizations, and foundations that support AKST in order to achieve development and sustainability goals. The questions that these organizations would like to be answered include:

- How much should governments invest in AKST versus other public goods?
- How should AKST resources be allocated? Which commodities? Where—for example, less favored land, small poor countries? What type of technology—for example, labor using, land saving, or water saving technologies? Which disciplines? Which components of AKST? Which institutions?
- What methods should be used to decide how much money to invest and how to set AKST priorities?

The answers to the first two questions need to incorporate multiple criteria, which should include at least public RORs to research as well as the impacts on poverty, human health, and environment (see 8.2). Societies and policy makers who place more emphasis on poverty reduction rather than economically sustainable development or environmental sustainability could place more weight on the AKST investments that reduce poverty than societies that favor improving the environment. Societies with more poor people may place more weight on research to improve the livelihood of the poor than on research to reduce greenhouse gases Countries in which agribusiness plays a big role in the economy and a large role in governance of the public research institutes, may invest more in developing productivity-increasing change.

Formal priority-setting methods, including those based on ROR studies, are in practice only occasionally used to set research priorities, and formal multi-criteria techniques for research resource allocation are used even less (Alston et al., 1995). This is because they are expensive, time consuming, and some factors are difficult, if not impossible, to quantify. The impacts of agricultural research on environment, health and poverty have been particularly difficult to measure (see 8.2.5, 8.2.6, 8.2.8). As a result, most of the studies that we were able to assess and base our policy options on are those of the ROR type. However, making mistakes when investing in AKST can also be a problem. Investing money in an AKST project that has little social or economic importance, large negative consequences, or very little chance of succeeding can be even more expensive than formal priority setting. Thus investing in formal priority setting can save money and have high payoffs. Changes in governance that incorporate users of this technology into the priority setting and evaluation processes can also be productive.

When looking forward—particularly 50 years forward—people who decide on AKST investments often simply have to look for major problems that appear to be coming and invest to fill gaps in knowledge.

8.4.1 Criteria and methods for guiding AKST investments

"Any research resource allocation system, regardless of how intuitive or how formal in its methodology, cannot avoid making judgments on two major questions. What are the possibilities of advancing knowledge or technology if resources are allocated to a particular commodity, problem or discipline? What will be the value to society of the new knowledge or the new technology if the research effort is successful?" (Ruttan, 1982).

ROR studies and broader comprehensive impact assessments can be undertaken before initiating any AKST investment (ex-ante) or after completion of the R&D activities (ex-post) depending on the purpose. The purpose of undertaking ex-ante assessments is to study the likely economic impact of the proposed investment, to formulate research priorities by examining the relative benefits of the different AKST investments, to identify the optimal portfolio of investments and to provide a framework for gathering information to carry out an effective and efficient ex-post assessment. Thus the greatest benefit of ex-ante assessment is derived from its power to assist decision makers to make informed decisions on investments i.e., in setting priorities to allocate the scarce resources.

AKST investment priorities are set at both micro and macro levels. More formal quantitative methods are used at the macro level and participatory methods are increasingly being used at the micro level. Priority setting is carried out explicitly or implicitly in all AKST investments through allocation of research resources to different commodities, regions, disciplines problems and type of technology. Since priority setting occurs at various levels of decision making, the resource allocation questions and methods employed vary depending on the level at which priorities are set. Priority setting also requires intensive consultation among and between politicians, administrators, planners, researchers as well as the beneficiaries. Formal procedures facilitate this process as they systematize the consideration of key variables and multiple objectives in the analysis and allow an interactive process to develop.

Priority setting based on ex-ante assessment employs a range of methods that can be broadly classified into supply- and demand-oriented approaches; although some combination of these approaches is often used in empirical studies. Supply-oriented approaches to priority setting and resource allocation often are conducted at the more aggregative regional and national level and use a variety of methods from informal methods based on previous allocations; discussions and consensus among research managers taking into account national agricultural goals and strategies; to formal quantitative methods such as scoring models, congruency analysis, domestic resource cost ratio, mathematical programming, and simulation techniques. The more sophisticated approaches such as programming and simulation rely on mathematical optimization of a multiple goal objective function to select the optimum portfolio of AKST investment. These are data and skill intensive and thus often quite costly to undertake. Many attempts have been made in the past to use a formal priority setting exercise to ensure that research resources are allocated in ways that are consistent with national and regional objectives and needs. Studies which have been undertaken to assess ex-ante AKST investment priorities have included those employing criteria which include equity and distributive concerns (Fishel, 1971; Pinstrup-Andersen et al., 1976; Binswanger and Ryan, 1977; Oram and Bindlish, 1983; Pinero, 1984; Von Oppen and Ryan, 1985; ASARECA, 2005) those focusing more on efficiency criteria such as congruency (Scobie, 1984); those employing the notion of comparative advantage using domestic resource cost analysis (Longmire and Winkelmann, 1985); those using economic surplus to examine research priorities (Schuh and Tollini, 1979; Norton and Davis, 1981; Ruttan, 1982; Davis et al., 1987, Omamo et al., 2006); and those using an optimization routine (Pinstrup-Andersen and Franklin, 1977; Mutangadura, 1997). One of the most comprehensive studies of research resource allocation lists methods for allocating research resources (Alston et al., 1995). These combine information from scientists, technicians and other experts on the expected output of science, their probability of success and possible timelines with information from economists and other social scientists on what the potential economic and social payoff would be if the research investment is successful. The formal methods have been extended to include environmental consequences of AKST investments (Crosson and Anderson, 1993). The overall aim is to foster consistency of research priorities with goals and objectives and to improve the efficiency of the AKST investments in meeting the needs of the producers, consumers and society at large.

In demand-oriented approaches, priorities are set based on the perspective of major stakeholders from outside the research system—especially the users. These might employ consultative and participatory methods using various forms of ranking techniques or users themselves might be empowered to make decisions on research priorities. However, it is worth keeping in mind that demand-led and supply-led approaches are not mutually exclusive. Better results can be obtained by combining formal supply-led priority setting with participatory approaches leading to better ownership of resulting priorities and greater chances that the priorities will be translated into actual resource allocation. Even the imperfect participation and empowerment of beneficiaries is likely to produce better results than conventional supply-led approaches on both efficiency and equity grounds, as they can improve the probability of broad-based adoption of technologies and knowledge generated, thereby enhancing innovation capacity. The challenge is to develop a judicious blend of bottom-up (demand-led) and top-down (supply-led) approaches to priority setting incorporating the multiple goals of AKST investments.

Formal models exist for the ex-ante evaluation of research projects, which are being used increasingly in more industrialized countries to allocate research funds but this is less common in developing countries (Pardey et al., 2006a). Few formal ex-ante models incorporate the goals of reducing poverty and hunger and the environmental consequences as explicit criteria for allocating research resources. Some progress has been made recently to incorporate these aspects in the analytical process. The two ex-ante studies reported in Eastern and Southern Africa (ASARECA, 2005; Omamo et al., 2006) consider the ex-ante benefits of all major commodities and the economic and poverty reduction potential of research investments. In addition, there are specific studies on site-specific maize research in Kenya (Mills et al., 1996) and the research priority setting under multiple objectives for Zimbabwe (Mutungadura, 1997; Mutungadura and Norton, 1999). The extent to which such results are actually used for setting the R&D agenda remains unclear. These approaches (based on expected costs and benefits) are very useful in allocating resources among applied and adaptive research programs and projects. However, they are of very little use to allocate resources between basic, strategic, applied and disciplinary research.

It is not just methods per se that are problematic; it is also the ability of would-be analysts gaining the requisite skills to use what methods are available. In the context of NARS, the task of developing the needed capacity to address aspects such as environmental and economic assessment of agricultural technology consequences on NRM (Crosson and Anderson, 1993) is still not yet adequately developed, especially in an era of profound underfunding of research, at local, national and regional levels. An important issue in developing and implementing AKST investment priorities is to explicitly incorporate the requirements of those who are expected to benefit from such investments.

Our approach in this study, which presents the empirical evidence available on the economic, health and environmental impacts of research but does not try to use a formal priority setting process to weight the importance of different criteria, reflects the discussion of well-intentioned, but often misguided attempts to deal with such multi-criteria formulations of research priorities (Alston et al., 1995). The review of methods based on scoring models suggests that there are definitely methodological challenges in such work yet to be satisfactorily dealt with. This fact shows the need of more resources to develop easier and more effective evaluation methods that can include environmental and societal (poverty, nutrition and health) impacts, both positive and negative.

8.4.2 Investment options
The ideal social planner would be able to rank research investments by their expected contribution to economically sustainable development, decreased hunger and poverty, improved nutrition and health, and environmental sustainability; and then would solicit weights from society based on the relative value society places on these expected contributions. Each country will have different weights based on the governance of the system and the countries' available re-

sources, their culture, their institutions and their technology. More investment in AKST can make important contributions to the goals of economically sustainable development, hunger and poverty reduction, environmental sustainability, and improvement of nutrition and health.

The private sector will not make major investments in the provision of public goods, poverty reduction, and the provision of environmental services and health services for which there is no market (see 8.3.5). Therefore most governments, especially in developing countries, need more public sector investments into AKST that will produce public goods and services necessary to reach development and sustainability goals. Few countries are likely to reach the 2% research intensity level of OECD countries, but they will need a major increase in investment in agricultural research intensity from the current level of 0.5% (see 8.1).

As reported in 8.2.8, studies of seven countries in Asia and Africa showed the returns to agricultural research were high relative to other investments that countries could make such as irrigation, roads, electricity, and other government programs (Fan et al., 2000, 2004ab, 2005; Fan and Zhang, 2004). Agricultural research was one of the leading investments that governments could make to reduce poverty. Research by itself will not lead to poverty reduction, but it can be an important component of a poverty reduction strategy. The other component of AKST in these studies was primary education which also made a major contribution to poverty reduction. The evidence shows that research alone cannot reduce poverty and thus funding for AKST must be accompanied by other pro-poor policies, such as access to natural resources, equity of distribution, good governance practices, and local market development.

Projections in Chapter 5 of this report show that the baseline scenario will have a limited impact on reducing child malnutrition—it would decline 15% in the reference world (see 5.3). However, with increased levels of AKST investments accompanied by other complementary investments, the share of malnourished children is expected to decline. In addition, projections suggest that returns to research will stay high. Under the business-as-usual scenario, the demand for agricultural products will continue to grow rapidly in the next 50 years; resources that are now used to produce agricultural products will be increasingly in short supply—water, land, and clean air; and basic science will move rapidly ahead creating new opportunities for applied science and technology, which will also increase returns to research.

An additional factor that will be required to keep returns high is good governance (see 8.3). Specifically, the farmers, who will be the primary users of the research, must be included in determining how public money is invested in AKST and how that funding is allocated. In addition, consumers of food and other ecosystem services from agriculture must be represented. Finally, the private sector, which provides inputs to the agricultural sector and purchases, markets and processes agricultural products, must also be represented.

Private sector investments in agricultural research, innovation, and diffusion of technology and management systems in developing countries are also essential to meeting development and sustainability goals (Pray et al., 2007).

While the private sector will not make major investment in the provision of public goods and poverty reduction for which there is no market, private agricultural input companies can be an efficient way to provide poor farmers with inputs such as improved seeds and livestock, which can help improve the incomes of the poor and other private companies can develop and supply farmers with inputs needed to increase their supply of ecosystem services (see 8.1.2 and 8.3.5). By encouraging companies to develop and supply technology and management systems to the commercial sector, the public sector can concentrate its limited resources on research to produce public goods, the development and supply of technology to the poor and the development and diffusion of environmental and health services.

It appears that the underinvestment in private research in developing countries is even greater than in public sector research. Because of the spillovers of the benefits of technology from private suppliers of technology to farmers and consumers, substantial benefits have accrued to farmers and consumers from private sector research (see 8.2.4 and 8.2.7). The median rate of return to society from research by private firms is 50% (Evenson, 2001). Aggregate studies in India (Evenson et al., 2001) and the US (Huffman and Evenson, 1993) have shown that private research and private imported technology have made major contributions to agricultural productivity growth. Case studies of specific private research programs have shown that the benefits of private research can reach farmers growing poor peoples' crops such as pearl millet and sorghum in rainfed environments such as the semiarid tropics of India (Pray et al., 1991). Despite these benefits, research investments by the private sector in developing countries lags even farther behind OECD countries than by the public sector investments, both in absolute amounts and in research intensity. Private research investment as a share of Agricultural GDP is 0.03% in developing countries and almost 3% in industrialized countries (see 8.1). To induce more private research, governments can invest in educating scientists and technicians and developing research infrastructure such as *ex situ* and *in situ* germplasm collections and basic research programs such as enhancing the diversity of plant and animal germplasm, which will generate ideas for new technology in the private sector. It also requires an enabling business environment for private investment (see 8.1). The components of an enabling business environment include a system for protecting intellectual property rights, the ability to enforce contracts, a stable regulatory environment, functioning markets for agricultural inputs and outputs, and so on. To make sure that private investments in AKST meet societies' goals, governments need to put in place incentives that will induce private firms to meet social goals. These incentives can be positive such as payments for environmental services that could induce firms to develop technology to more effectively provide those services. These incentives can also be negative, for example environmental and food safety regulations and liability laws that penalize negative externalities from introduced technologies. In addition, industrial policies that limit monopoly power will also be needed. Government alone cannot enforce regulations and industrial policies. The active involvement of NGOs and other parts of civil society is essential.

Tradeoffs occur between different development goals when different choices of AKST investments on specific commodities or types of institutions are made (Table 8-20). For example, RORs to wheat research have been high (see 8.2.4), but the research that produced high-yielding wheat varieties may also have induced more irrigation of wheat in poorly drained regions, which has led to increased salinity, destroyed land, and displaced farmers. Pesticide use on wheat is limited so there has been little negative impact of pesticides. Green revolution wheat varieties reduced prices of wheat, which increased consumption of wheat by the poor improving their health. The high yielding wheat varieties during the green revolution period in South Asia increased demand for labor and thus the incomes of the poor (Lipton, 2001). An example of research that has positive effects on economically sustainable development, but a negative impact on other development goals is research to increase the productivity of intensive livestock production. It has high RORs but major negative environmental effects through water and air pollution, and negative health impacts through *E. coli* and other public health crises (see 8.2.5 and 8.2.6). At the same time it can have positive health impacts through dramatic declines in the price of meat and poultry, which in turn facilitates access for more people to animal protein and other essential nutrients.

8.4.2.1 Options for societies aiming to give major support to environmental sustainability

For these societies investment in AKST can have three different, but complementary, alternatives: reducing the negative environmental impacts of farming systems, enhancing existing agricultural systems that have been shown to be environmentally sustainable, and developing new agricultural systems. They will have to focus on providing ecosystem services such as reduced greenhouse gas emissions, absorption of the carbon dioxide, reduced water pollution and slowing the loss of biodiversity.

We have made judgments about the most important negative impacts of agricultural technologies on the environment (see 8.2.5); unfortunately, data is not available to know which of the impacts are most important or which negative impacts could be mitigated most effectively through investments in AKST. This gap suggests that the first important need for AKST investment is for social and ecological scientists working with other scientists to develop methodologies and to quantify the externalities of high and low external input farming systems from a monetary perspective as well as from other perspectives such as the concept of energy flows used in "emergy" evaluations. Evidence on these externalities' potential implications on food security also needs to be analysed.

There are three other types of AKST investments in which countries can invest. First is research to develop management practices, technologies, and policies that reduce the ecological footprint of agriculture, such as reducing agriculture's use of fossil fuels, pesticides and fertilizers. This would include AKST investments to develop management practices such as: no-tillage systems to reduce use of fossil fuels for tillage, integrated pest management strategies to avoid overuse of inorganic pesticides, integrated soil management technologies to reduce the need for inorganic fertilizer, rotational grazing and support of mixed farming systems to improve the nutrient cycling within agriculture and livestock production. In this area, investments on sustainable and low-input farming practices would also be recommended. AKST investments can also increase agriculture's role as a carbon sink. The greatest dividends would come from conversion from grain crops to agroforestry as there is a benefit from both increased soil organic matter and the accumulation of above-ground woody biomass. Thus agroforestry can play a major role in the two key dimensions of climate change: mitigation of greenhouse gas emissions and adaptation to changing environmental conditions (Garrity, 2004). Other management strategies such as including grasslands within rotations, zero-tillage (or no-till) farming, green manures, and high amendments of straw and manures, would also lead to substantial carbon sequestration (Pretty and Ball, 2001).

A second type of AKST activity would be the development of biological substitutes for industrial chemicals or fossil fuels. These would include new biopesticides, improvements in biological nitrogen fixation, and search for alternative sources of energy that do not compete with food production and do not induce deforestation. There is some evidence that research in this area can provide a good economic ROR, and the RORs are likely to rise as more governments put policies in place that reward farmers for the provision of these services.

Third, research to support traditional knowledge on effective ways of using and conserving available resources such as soils, water, and biodiversity to improve rural livelihoods will be required. This knowledge has been neglected but research and management systems based on this knowledge have been shown to have positive ecological and economic impacts in all areas of agriculture (crops, livestock, aquaculture and agroforestry). New nonconventional crops and breeds may play a vital role in the future for conserving local and indigenous knowledge systems and culture, as they have a high local knowledge base which is being promoted through participatory domestication processes (Leakey et al., 2005; World Agroforestry Centre, 2005; Garrity, 2006; Tchoundjeu et al., 2006).

This may be an area of AKST that had lower returns to public investments research than some other types of research historically. This is due in part to the difficulty of measuring the impact of research in this area and the lack of studies of the impact of these types of research. It is also partly due to the fact that the complementary policies and institutions needed to implement solutions developed by AKST are often not in place. Considering that agriculture and land use contribute to 32% of global emissions, more research is needed to analyse the potential contribution of new and existing but ignored agricultural technologies and practices that could contribute to decreasing global warming and climate change. Another important type of research investment needed is social science research which develops recommendations for policy and institutional changes that reward farmers for reducing the negative externalities, enhancing the multiple functions of agriculture, and for the provision of ecosystem services. Investments in incentives for private sector to develop technologies that assist farmers to provide ecosystems services are also needed.

Table 8-20. **Summary of impacts of productivity increasing technology—economic returns, externalities and spillovers.**

	Median of ROR for Productivity Increases		Environmental externalities	Health externalities	Impact on poor
	Evenson (2001)	Alston et al. (2001a)			
All crops	57	44			0
Wheat	51	40	-- Irrigation with poor drainage + high yields reduce need to clear forest	0/+	+/-
Rice	60	51	-- over irrigation & high pesticide use + high yields reduce need to clear forest	- pesticides	+/-
Maize	56	47	-- over irrigation & high pesticides + high yields reduce need to clear forest	- pesticides	+/-
Other cereals	57	n/a			+/-
Fruits and vegetables	67	n/a	-- high pesticide use	-- high pesticides affect laborers & consumers + improves nutrients in diet	+ home gardens/- commercial
Livestock	36	53	-- for intensive livestock production which can lead to nitrogen and phosphorus pollution of water	- zoonotic diseases - food poisoning + increases protein & minerals in diet	+ if subsistence or milk coops - if intensive or contract production???
Forestry	n/a	14			+ if agroforestry
Forest products	37	n/a			?
Tree crops	n/a	33	- plantations that replace uncultivated land can reduce biodiversity + plantations that replace crops could be a carbon sink		- if plantations
Resource management	n/a	17	++ for more effective management which substitutes labor for chemicals	+ if reduce use of pesticides	+ if saving resources of poor or tech is labor intensive
Developing countries	37-67	43			
CGIAR	39-165	40			++
Private	50	34	-- intensive livestock and pesticide use, but management & biotech can reduce chemical pesticides	- if increases pesticide use + if it reduces pesticide use	- or 0

Note: - small negative impact; -- large negative impact; + small positive impact; and ++ large positive impact; n/a means not available.

Sources: Evenson (2001), Alston et al. (2000a) and the judgments of the authors.

For AKST for environmental services to have adequate levels of funding and for the funding to be sustainable, consumers and environmental groups must have a role in the governance of the research system.

8.4.2.2 Options for societies aiming to give major support to improving nutrition and human health

As in the case of the environmental sustainability, this area of AKST investment can adopt several different but complementary goals: to reduce the negative documented impact of agriculture on health and to develop policies and technologies aiming to improve the nutritional and health status of population. The key areas of AKST investment could be in improved quantity and nutritional quality of culturally appropriate food to the poor, safer management or reduction in use of pesticides and research to improve food safety. This is another area for which evidence is lacking and where investments are needed to obtain data on the size of the problem and the potential of AKST to solve it.

AKST has positive and negative effects on human health (see 8.2.6). Increased plant and animal productivity have reduced prices of these food products and often reduced undernutrition of the poor and led to more balanced diets. At the same time, increased productivity has led to environmental pollution (of water and air) and overuse of antibiotics and pesticides (including toxic residues in plants and animals and resistance to antibiotics) have lead to serious health impacts. Additionally, problems of growing obesity in industrialized and developing countries are also indirectly linked to AKST.

The evidence on negative impacts suggests that one area of major AKST investment needs to be on improved pesticide management and the reduction in use of dangerous pesticides and antibiotics (see 8.2.6). In particular investments in IPM and substituting less dangerous chemicals or biopesticides for dangerous pesticides appear to be important investments. Farming systems which improve productivity while using little or no pesticide, chemical fertilizer and antibiotics need to be studied and developed in order to improve their management and increase their potential to feed the local population. Organic agriculture is one type of farming system that reduces pesticide use and has a growing demand, so investments in research to increase the productivity and resilience of organic agriculture would be appropriate.

AKST investment to develop and implement schemes for food safety and quality standards to improve public health and consumer confidence is a major area in the health portfolio. In addition, investments to increase the nutritional values of crops and livestock products with the objective of improving the nutritional status of global population, such as biofortification, need more emphasis in plant breeding research. Biofortification is one of the few areas where there have been careful studies both of the size of the health problem and that AKST investments can reduce these problems (see 8.2.6).

Farming system diversification can also improve nutrition. The expansion of vegetable and fruit tree cultivation on farms can have a significant effect on the quality of child nutrition. Many indigenous fruits, nuts and vegetables are highly nutritious (Leakey, 1999). The consumption of some

traditional foods can also help to boost immune systems, making these foods beneficial against diseases, including HIV-AIDS (Barany et al., 2003; Villarreal et al., 2006). If countries neglect investments to improve the farming systems of both subsistence and commercial farmers, major health problems would increase.

Reduction of nutritional imbalances would require research on educational programs and policy mechanisms to provide appropriate incentives for facilitating the access to healthier products and healthier consumption patterns while penalizing in the market those products leading to nutritional problems, for example, through the internalization in the final price of the products the health costs calculated by means of AKST. Still, more AKST from social sciences is needed in order to find and develop the best policy strategies to avoid malnutrition. The AKST investments to continue the reduction in the numbers of the undernourished through productivity increasing research or better distributional or commercialization strategies of food are described in more detail below in the section on poverty.

8.4.2.3 Options for societies aiming to give major support to hunger and poverty reduction

These societies will need to target investments in research, policy and institutional change in organizations that provide research to produce public goods. These include public research, extension and education programs as well as the international research centers of the CGIAR.

AKST investments can increase the productivity of major subsistence crops such as rice, wheat, and other basic staples that are grown and/or consumed by the poor (see 8.4.3.1) while respecting the culture and livelihoods of those who produce the food. Investment can also be allocated to the productivity-increasing research in regions where the poor are located, such as rain-fed and marginal areas, even if these are not the areas which would increase total Agricultural GDP the most. Also, investments to preserve biodiversity and traditional systems that maintain the livelihoods of millions of people are required in order to increase the wealth of poor populations in many countries. For example, research on animal genetic resources conservation programs could be directed to increase drought resistance or disease resistance of local domestic breeds. This implies appropriate technologies that do not destroy the environment while at the same time aims to improve the existing local knowledge of traditional farming systems towards the needs of the farmers.

Research for the poor should aim to develop and maintain crop and animal production techniques that allow extending the assets controlled by the poor such as labor, management skills, or biodiversity with assets owned by the wealthy, such as land.

Investments in institutional change and policies which improve the access of the poor to food, education, land, water, seeds, markets and improved technology for producing food, better access to jobs, and more influence on the governance of research systems are a major need for reducing poverty (see 8.3.4). The investments in improved institutions might include AKST programs that support small scale agricultural and food industry innovators and public private partnership with the aim to (1) encourage adaptation and adoption of pro-poor technologies from the public research agencies; (2) adapt scientific discoveries from industrialized countries and, if needed, import technology from them that increases the productivity of poor farmers; and (3) transfer technology from neighboring countries, which may have developed technology that is more appropriate for poor farmers in developing countries than those of industrialized countries. These technologies can flow through multinational corporations, local private firms, public sector research systems and their regional networks, and farmer-to-farmer communication.

8.4.2.4 Options for societies aiming to give major support to economically sustainable development

These societies should consider investing in AKST which provides evidence of high future (ex ante) RORs. These investments will include some areas that had high ROR in the past such as yield increasing technologies and promise high returns in the future since there will be continued demand for these technologies and science. AKST investments in water management and pest and disease management, which have less history of high returns but are likely to have high returns in the future because of high demand or recent advances in science, would also be included. In addition, some AKST investments that did not have high returns in the past, such as NRM, are likely to have high ROR in the future if policies, such as carbon trading under the Kyoto protocol or subsidies for good environmental practices in agricultural policies, provide incentives to adopt these technologies. It is clear that economically sustainable development can only be achieved if the environment is at the same time preserved.

Governments must continue to invest in AKST to develop productivity-increasing technology and management systems that save on the use or reduce the misuse of scarce resources such as land, water, and in fossil fuels. The major resource constraint to increasing agricultural production in the future will continue to be agricultural land. In the future AKST must focus on increasing output per unit of land through technology and management practices. RORs to land saving research are high. There are a limited number of studies which show substantial returns to land management research. However, future AKST must avoid the negative externalities of past investments in this area.

Water is the next most important resource constraint to agricultural production and is likely to be even more of a constraint in the next 50 years. AKST resources are being reallocated into water-saving techniques, improved policies and management techniques. The expected ROR to investments in water productivity research may be lower than for germplasm research. Nevertheless, comparing the cost of science-based studies with the costs of inaction (growing poverty, malnutrition and disaster relief) indicates that the benefits of science-based actions vastly exceed the costs (Kijne and Bennet, 2004). Still, a few examples of water-saving research, which were evaluated by SPIA had high returns (Waibel and Zilberman, 2006), and some of the research on drought tolerant crops (both breeding new varieties and recovering existing ones) looks very promising. However, the development of these technologies will take time, and major changes in water pricing policies are likely to be needed to

give farmers in irrigated areas incentives to adopt such technologies.

Fossil fuels in the long run may run out. Concerns about their impact on global warming, and the high price of fossil fuel has once again focused attention on the need for agriculture to save on the use of this scarce resource and support agricultural systems that have higher outputs per unit of sustainable energy. In this context, low-external-input agriculture could bring promising opportunities. There is little evidence yet from the ROR literature of high returns, but the demand is there and agricultural research has the capacity to produce appropriate technologies (see also chapter 6). Since prices are likely to continue to fluctuate due to politics as much as to scarcity, AKST investments by governments will be necessary to develop these technologies and to inform farmers how they may best reduce agricultural use of fossil fuels.

Major public and private R&D investments will be needed in emerging issues such as plant and animal pest and disease control. Continued intensification of agricultural production, changes in agriculture due to global warming, the development of pests and diseases that are resistant to current methods of controlling them, or changes in demand for agricultural products such as the increasing demand for organic products, will lead to new challenges for farmers and the research system. Investments in this area by the public and private sector have provided high returns in the past and are likely to provide even higher returns in the future. In addition, these investments could lead to less environmental degradation by reducing the use of older pesticides and improving livestock production methods. These technologies could also use more labor, which in labor abundant countries, could reduce poverty. They would also positively impact human health. Pest and disease control is an area in which public and private collaboration is essential.

Pre-invention, strategic, and basic research can be justified in many countries and in international research centers. The studies that try to estimate the separate impacts of different components of AKST find that both applied and more basic research investments have high returns (see 8.2.4). Advances in basic biological knowledge such as genomics and proteomics, nanotechnology, ICT, and other new advances in AKST will create major new opportunities for meeting development and sustainability goals (see Chapter 6). Emerging knowledge of agroecological processes and synergies, and the application of resultant technologies, will play a crucial role in future AKST investments. Both new and existing but neglected knowledge can pay off by increasing public and private development of technologies and management practices that improve agricultural production, mitigate climate change, improve health or reduce poverty. Thus it is not inherently productivity increasing or polluting but is needed to achieve economically sustainable development. A major increase in private sector research will be needed to increase agricultural productivity growth for developing countries.

8.4.2.5 A portfolio of AKST investments to meet multiple goals

If, as has been argued earlier in this subchapter, a large infusion of public funding in AKST is needed, a coalition of interest groups will have to lobby for this increased funding. This suggests that policy makers and advocates for AKST activities that increase environmental sustainability, achieve economically sustainable development, improve nutrition and health, and reduce poverty, should attempt to put together an AKST investment portfolio that attracts groups beyond the traditional agricultural community. The investment areas listed above, which can meet multiple criteria, could be attractive to these different groups. As indicated above, many AKST investments can meet multiple goals. Other investments primarily meet one goal but still play a valuable role and should not be eliminated because they do not make major contributions to all of the goals. For example, private research to increase poultry productivity may create increased pollution, but this does not mean that governments should try to prevent private poultry research. A more appropriate approach may be to encourage the private sector to do productivity-enhancing research but at the same time prevent the potential pollution through more effective enforcement of laws against pollution, by mandating waste management plans or by public sector research to development management systems which reduce pollution and improved public health.

One strategy is to make small public investments in an enabling policy environmental that would encourage private research and shift public research into the production of public goods and meeting other social goals such as improving the environment or developing technology for resource poor farmers or into basic research. For example, many countries could reduce their public research investments on improving the productivity hybrid maize, which will be done by the private sector, and shift those resources into productivity-enhancing research on cassava or open pollinated varieties of maize grown by poor people. Or the resources could be shifted into fertilizer and pest management to reduce overuse of chemicals that create pollution and can harm human health. Shifting more public AKST investments to increase the productivity and adoption of organic agriculture for which markets are available can also reduce the use of nonorganic pesticides and chemical fertilizers.

8.4.3 Future AKST investment levels and priorities

8.4.3.1 Levels of AKST investments
More government funding and better targeted government investments in AKST in developing countries can make major contributions to meeting development goals. The evidence of returns to AKST investments shows that public investments have high payoffs, in the order of 40-50% and can reduce poverty (see 8.2.4 and 8.2.8). These returns are high compared to other public sector investments and evidence shows that AKST investments are one of the most effective ways to reduce poverty. In addition, public investments in AKST can be used to reduce agriculture's contribution to global warming and to improve public health. However, to do this public investments must be targeted using evidence other than the ROR, which usually do not include environmental and human health impacts, positive or negative, or the distribution of costs and benefits among different groups.

Increasing investments in agricultural research, innovation, and diffusion of technology by for-profit firms can also make major contributions to meeting development and sustainability goals. Private firms both large and small have been and in the future will continue to be major suppliers of inputs and innovations to both commercial and subsistence farmers. They will not provide public goods or supply goods and services for which there is no market; but evidence shows that there are spillovers from private suppliers of technology to farmers and consumers. However, private research intensity in developing countries is only one hundredth of the corresponding ratio in industrialized countries. To make the best use of private investments in AKST, governments must provide both government regulations to guard against negative externalities and monopolistic behavior and support good environmental practices providing firms with incentives to invest in AKST.

8.4.3.2 Allocation of AKST resources
Social science research to assist priority-setting, to measure the impact of past AKST investments in health and the environment, to improve AKST and complementary institutions and policies, and to link with indigenous knowledge is a high priority investment. One of the major constraints of this assessment is the lack of evidence on both the positive and negative impact of AKST on the environment, human health, and, to a lesser extent, on poverty reduction (see 8.2.5, 8.2.6, 8.2.8). Investments are needed to develop better methodologies and indicators to measure these impacts, both with monetary and nonmonetary values. In addition, investments are needed to develop better methods for measuring the contributions of indigenous knowledge, social science research on institutions and policies, the value of improving governance systems, and better priority-setting tools and methods. Finally more investments are needed in research priority setting processes in developing countries which include both social and natural scientists and input from stakeholders (see 8.4.1).

AKST investments that can increase the productivity of agriculture and improve the existing traditional systems of agriculture and aquaculture in order to conserve scarce resources such as land, water and biodiversity remains a high priority. The major resource constraint on increasing agricultural production in the future will continue to be agricultural land. AKST must focus on increasing output per unit of land through technology and management practices.

AKST investment to reduce greenhouse gas emissions and provide other ecosystem services is another priority investment area. Agriculture and land use contribute 32% of total GHG emissions (Stern, 2007). Thus, AKST investments to develop policies, technologies and management strategies that reduce agriculture's contribution could facilitate to decreasing global warming. This requires the development of new farming systems, which use fewer technologies, produces less GHG, and builds on indigenous knowledge to improve current cropping systems to be more sustainable. These systems could include practices such as no-tillage systems, integrated pest management strategies, integrated soil management technologies, rotational grazing and support of mixed farming systems to improve the nutrient cycling. A second, complementary type of AKST activity is the development of policies such as payments for environmental services from farmers, which could induce the development and adoption of practices that provide environmental services. In addition, some of the agricultural technologies and policies for provide these ecosystem services can be designed to use the assets of the poor, such as labor in labor-abundant economies which would reduce poverty.

Major public and private research and development investments will be needed in plant and animal pest and disease control. Continued intensification of agricultural production, changes in agriculture due to global warming, the development of pests and diseases that are resistant to current methods of controlling them, and changes in demand for agricultural products, will lead to new challenges for farmers and the research system. Investments in this area by the public and private sector have provided high returns in the past and are likely to provide even higher returns in the future. In addition, these investments could lead to: less environmental degradation by reducing the use of older pesticides and livestock production methods; more labor use, which could reduce poverty; and positively improve human health of farmers and their families by reducing their exposure to pesticides. This is an area in which public and private collaboration is essential.

References

Ahmed, M., and J.G. Nagy. 2001. Private investment in agriculture research: Pakistan. ERS, USDA, Washington DC.

Ajayi, O.O.C. 2000. Pesticide use practices, productivity and farmers' health: The case of cotton-rice systems in Cóte d'Ivoire, West Africa. Pesticide Policy Publ. Series. Spec. Issue 3. Univ. Hannover.

Akinnifesi, F.K., R. Leakey, A.J. Simons, Z. Thoundjeu, and P. Matakala (ed) 2007. Domestication, utilization and commercialization of indigenous fruit trees and products in the tropics. CABI, Wallingford.

Alavalapati, J.R.R., R.K. Shrestha, G.A. Stainback, and J.R. Matta. 2004. Agroforestry development: An environmental economic perspective. Agrofor. Syst. 61:299-310.

Almekinders, C.J.M., and A. Eling. 2001. Collaboration of farmers and breeders: Participatory crop improvements perspective. Euphytica 122(3):425-438.

Alston, J.M. 2002. Spillovers. Aust. J. Agric. Res. Econ. 46:315-346.

Alston, J.M., C. Chan-Kang., M.C. Marra., P.G. Pardey, and T.J. Wyatt. 2000a. A meta-analysis of rates of return to agricultural R&D: Ex pede herculem? IFPRI Res. Rep. No. 113. IFPRI, Washington DC.

Alston, J.M., G.W. Norton, and P.G. Pardey. 1995. Science under scarcity: Principles and practice for agricultural research valuation and priority setting. Cornell Univ. Press, Ithaca.

Alston, J.M., and P.G. Pardey. 2001. Reassessing research returns: Attribution and related problems. In G. Peters and P. Pingali (ed) Tomorrow's agriculture: Incentives, institutions, infrastructure, and innovations. Proc. 24th Int. Conf. Agric. Economists. Ashgate, Aldershot, UK.

Alston, J.M., P.G. Pardey, and J. Roseboom. 1998. Financing agricultural research: International investment patterns and policy perspectives. World Dev. (26)6:1057-1071.

Alston, J.M., P.G. Pardey, and V.H. Smith. 1999. Paying for agricultural productivity. Johns Hopkins Univ. Press, Baltimore.

Alston, J.M., P.G. Pardey, S. Wood, and L. You. 2000b. Strategic technology investments for LAC agriculture: A framework for evaluating the local and spillover effects of R&D. Paper presented as part of an IDB-sponsored project policies on food, agriculture and the environment, and indicators and priorities for agricultural research. IFPRI, Washington DC.

Anandajayasekeram, P. and M. Rukuni. 1999. Agricultural research and poverty alleviation: Lessons from Eastern and Southern Africa. Int. workshop on assessing the impact of agricultural research on poverty alleviation. San Jose, Costa Rica, 14-16 Sept.

Anandajayasekeram, P., M. Rukuni, S. Babu, F. Liebenberg, and C.L. Keswani. 2007. Impact of science on African agriculture and food security. CABI, Wallingford.

Anderson, J.R. 1993. The economics of new technology adaptation and adoption. Rev. Marketing Agric. Econ. 61(2):301-9.

Anderson, J.R. (ed) 1994. Agricultural technology: Policy issues for the international community, CABI, Wallingford.

Anderson, J.R. 2007. Agricultural advisory services, Background Pap. 2008 WDR Agric. Dev., World Bank, Washington DC.

Anderson, J.R. and R.W. Herdt. 1990. Reflections on impact assessment. In R.G. Echeverría (ed) Methods for diagnosing research system constraints and assessing the impact of agricultural research. Vol. 1. Diagnosing agricultural system constraints. ISNAR, The Hague.

Anderson, J.R. and G. Antony, and J.S Davis. 1990. Research priority setting in a small developing country: The case of Papua New Guinea. p. 169-78. In R.G. Echeverría (ed) Methods for diagnosing research system constraints and assessing the impact of agricultural research. Vol. 2. Assessing the impact of agricultural research. ISNAR, The Hague.

Antle, J. M., D.C. Cole, and C. Crissman. 1998. The role of pesticides in farm productivity and farmer health. Economic, environmental, and health tradeoffs in agriculture. In C.C. Crissman, J. M. Antle, S.M. Capalbo (ed) Pesticides and the sustainability of Andean potato production. Kluwer, Dordrecht.

ASARECA (Assoc. Strengthening Agricultural Research in Eastern and Central Africa). 2005. Strategic choices and programme priorities for ASARECA Animal Agricultural Network (A-AARNET). ASARECA Secretariat, Entebbe, Uganda.

ASTI (Agricultural Science and Technology Indicators). 2007. ASTI [Online]. Available at www.asti.cgiar.org

Atkin, J., and K.M. Leisinger. 2000. Safe and effective use of crop protection products in developing countries. CABI, Wallingford.

Azam, O.T., E.A. Boom, and R.E. Evenson. 1991. Agricultural research productivity in Pakistan. Pakistan Agric. Res. Council, Islamabad, Pakistan.

Babu, S.C., and D. Sengupta, 2006. Capacity development as a research domain: Frameworks, approaches, and analysis. ISNAR Disc. Pap. 9. IFPRI, Washington DC.

Barany, M., A.L. Hammett, R.R.B. Leakey, and K.M. Moore. 2003. Income generating opportunities for smallholders affected by HIV/AIDS: Linking agro-ecological change and non-timber forest product markets. J. Manage. Studies 39:26-39.

Bardhan, P.K. 1997. The role of governance in economic development: A political economy approach. OECD, Paris.

Becker Jr., W.E. 1982. Behavior and productivity implications of institutional and project funding of research: Comment. Amer. J. Agric. Econ. 64:595-598.

Beintema, N.M., and G.J. Stads. 2006. Agricultural R&D investments in sub-Saharan Africa: An era of stagnation. ASTI Background Rep., IFPRI, Washington DC.

Biggs, S.D. 1990. A multiple source of innovation model of agricultural research and technology promotion. World Dev. 18:1481-1499.

Binswanger, H.P., and J.G. Ryan. 1977. Efficiency and equity issues in *ex ante* allocation of research resources. Indian J. Agric. Econ. 32(3):217-31.

Birner, R., and J.R. Anderson. 2007. How to make agricultural extension demand-driven? The case of India's agricultural extension policy. Disc. Pap. IFPRI, Washington DC.

Bloom, D., D. Canning, and K. Chan. 2006. Higher education and economic development in Africa. Africa Region Human Dev. Dep., World Bank, Washington DC.

Bofu, S., T. Weiming, W. Jimin, W. Chunlin, Y. Zhengui, W. Shengwu et al. 1996. Economic impact of CIP-24 in China. *In* T.S. Walker, C.C. Crissman (ed) Case studies of the impact of CIP-related technologies. CIP, Lima.

Bouis, H.E., 2002. Plant breeding: A new tool for fighting micronutrient malnutrition. J. Nutri. 132:491S-494S.

Brennan, J.P. 1986. Impact of wheat varieties from CIMMYT on Australian wheat production. Agric. Econ. Bull. No. 5, NSW Dep. Agric., Sydney.

Brennan, J.P. 1989. Spillover effects of international agricultural research: CIMMYT-based semi-dwarf wheat in Australia. Agric. Econ. 3:323-332.

Brennan, J.P., A. Aw-Hassan, K.J. Quade, and T.L. Nordblum. 2002. Impact of ICARDA research on Australian agriculture. Report prepared for ACIAR. Econ. Res. Rep. No. 11, NSW Agric., Wagga Wagga, Australia.

Brennan, J.P., and M.C.S. Bantilan. 1999. Impact of ICRISAT research on Australian agriculture. Econ. Res. Rep. No. 1, ACIAR, NSW Agric., Wagga Wagga, Australia.

Brummett, R. 1999. Integrated aquaculture in sub-Saharan Africa. Environ. Dev. Sustain. 1:315-321.

Burnett, P.A., G.O. Edmeades, J.P. Brennan, H.A. Eagles, J.M. McEwan, and W.B. Griffin. 1990. CIMMYT's contributions to wheat production in New Zealand', p. 311-316. *In* Proc. Sixth Assembly of the Wheat Breeding Society of Australia.

Byerlee, D. 1998. The search for a new paradigm for the development of national agricultural research systems. World Dev. 26(6):1049-1055.

Byerlee, D., and G.E. Alex. 1998. Strengthening national agricultural research systems: Policy issues and good practices. World Bank, Washington DC.

Byerlee, D., and K. Fischer. 2002. Accessing modern science: Policy and institutional options for agricultural biotechnology in developing countries, World Dev. 30(6): 931-48.

Byerlee, D., and P. Moya. 1993. Impacts of international wheat breeding research in the developing world, 1966-90. CIMMYT, Mexico City.

Byerlee, D., G.E. Alex, and R. Echeverría. 2002. The evolution of public research systems in developing countries: Facing new challenges, *In* D. Byerlee and R. Echeverría (ed) Agricultural research policy in an era of privatization. CABI, Wallingford.

Castellini, C., S. Bastianoni, C. Granai, A. Dal Bosco, and M. Brunetti. 2006. Sustainability of poultry production using the emergy approach: Comparison of conventional and organic rearing systems. Agric. Ecosyst. Environ. 114:343-350.

Chambers, R. 1997. Whose reality counts? Intermediate Tech. Publ., London.

Clay, J. 2004. World agriculture and the environment. Island Press, Washington DC.

Cole, D.C., F. Carpio and N. León. 2000. Economic burden of illness from pesticide poisonings in highland Ecuador. Panam. J. Public Health 8(3):196-201.

Cracknell, B. 1996. Evaluating development aid: Strengths and weaknesses. Evaluation 2(1):23-34.

Crissman, C.C., J.M. Antle, and S.A. Capalbo. 1998. Economic, environmental, and health tradeoffs in agriculture: Pesticides and the sustainability of Andean potato production. Kluwer Acad. Publ., London.

Crissman, C.C., D.C. Cole, and F. Carpio. 1994. Pesticide use and farm worker health in Ecuadorian potato production. Am. J. Agric. Econ. 76 (8):593-597.

Crosson, P., and J.R. Anderson. 1993. Concerns for sustainability: Integration of natural resource and environmental issues for the research agendas of NARS. Res. Rep. 4. ISNAR, The Hague.

Cuyno, L.C.M., G.W. Norton, and A. Rola. 2001. Economic analysis of environmental benefits of integrated pest management: A Philippine case study. Agric. Econ. 25:227-233.

Diamon, L. and M.F. Plattner (ed) 1995. Economic reform and democracy. John Hopkins Univ. Press, Baltimore.

DANIDA (Danish International Development Agency). 1994. Evaluation report: Agricultural sector evaluation. Impact studies. Vol. 2. Methods and findings. DANIDA, Copenhagen.

Dasgupta, P., and J. Stiglitz. 1980. Industrial structure and innovative activity. Econ. J. 90:266-293.

Davis, J.S., P.A. Oram, and J.G. Ryan. 1987. Assessment of agricultural research priorities: An international perspective. ACIAR Mono. No. 4, ACIAR, Canberra.

Dawe, D., R. Robertson, and L. Unnevehr. 2002. Golden rice: What role could it play in alleviation of vitamin A deficiency? Food Policy 27:541-560.

Day-Rubenstein, K., and K. Fuglie. 1999. Resource allocation in joint public-private agricultural research. J. Agribusiness 17:123-134.

Desai, M., S. Fukuda-Parr, C. Johansson, and F. Sagasti. 2002. Measuring the technology achievement of nations and the capacity to participate in the network age. J. Human Dev. 3(1): 95-122.

Dey, M.M., P. Kambewa, M. Prein, D. Jamu, F.J. Paraguas, D.E. Pemsl, and R.M. Briones. 2007. Impact of the development and dissemination of integrated aquaculture-agriculture technologies in Malawi. *In* Waibel, H. and D. Zilberman (ed) International research on natural resource management: Advances in impact assessment. CABI, Wallingford.

DFID (UK Dep. Int. Dev.). 2004. What is pro-poor growth and why do we need to know? Pro-poor growth briefing note 1. DFID, London.

Diamon, L., and M.F. Plattner (ed) 1995. Economic reform and democracy. John Hopkins Univ. Press, Baltimore.

Diemont, S., J. Martin, and S. Levy-Tacher, S.I. 2006. Emergy evaluation of Lacandon Maya indigenous swidden agroforestry in Chiapas, Mexico. Agrofor. Syst. 66:23-42.

Dolan, P. 2000. The measurement of health-related quality of life for use in resource allocation decisions in health care. p.1723-1760. *In* A.J. Culyer and J.P. Newhouse (ed) Handbook of health economics. Vol. I. Elsevier Science, Amsterdam.

Drakakaki, G., S. Marcel, R.P. Glahn, E.K. Lund, S. Pariagh, R. Fischer et al. 2005. Endosperm-specific co-expression of recombinant soybean ferritin and aspergillus phytase in maize results in significant increases in the levels of bioavailable iron. Plant Mol. Biol. 59: 869-880.

Ducreux, L.J.M., W.L. Morris, P.E. Hedley, T. Shepherd, H.V. Davies, S. Millam, and M.A. Taylor. 2005. Metabolic engineering of high carotenoid potato tubers containing enhanced levels of β-carotene and lutein. J. Exp. Bot. 56:81-89.

Echeverria, R.G. 1998. Will competitive funding improve performance of agricultural research? Disc. Pap. 98-16. ISNAR, The Hague.

Echeverria, R.G., E. J. Trigo and D. Byerlee. 1996. Institutional change and effective financing of agricultural research in Latin America. Tech. Pap. 330. World Bank. Washington DC.

Ecobichon, D.J. 2001. Pesticide use in developing countries. Toxicology 160:27-33.

Eicher, C.K. 2001. Africa's unfinished business: building sustainable agricultural research systems. Dep. Agric. Econ. Staff Pap. 2001-10. Michigan State Univ., East Lansing MI.

Eicher, C.K. 2003. 50 years of donor aid to African agriculture. Successes in African Agriculture Conf., Pretoria, 1-3 Dec.

Ekboir, J. 2003. Why impact analysis should not be used for research evaluation and what the alternatives are. Agric. Systems 78(2):166-182.

Evenson, R.E. 1979. Agricultural research, extension and productivity change in U.S. agriculture: A historical decomposition analysis. Agric. Res. Extension Evaluation Symp. Moscow, Idaho. 21-24 May.

Evenson, R.E. 1989. Spillover benefits of agricultural research: Evidence from US experience. Am. J. Agric. Econ. 71(2):447-452.

Evenson, R.E. 1991. Human resource and technological development. Seminar on Overseas Education for Development. E.T.S. May 27-29.

Evenson, R.E. 1999. Economic impacts of agricultural research and extension. In B.L. Gardner and G.C. Rausser (ed) Handbook of agricultural economics. Elsevier Science, Amsterdam.

Evenson. R.E. 2001. Economic impacts of agricultural research In B. Gardner and G. Rausser (ed) Handbook of agricultural economics. North Holland/Elsevier, Amsterdam.

Evenson, R.E. 2004. Private and public values of higher education in developing countries: Guidelines for investment. J. Higher Educ. Africa 2(1):151-176.

Evenson, R.E., and A.F. Avila. 1996. Productivity change and technological transfer in the Brazilian grain sector. R. Econ. Rural 34(2):93-109.

Evenson, R.E., and P. Flores. 1978. Economic consequence of new rice technology in Asia. IRRI, Los Banos.

Evenson, R.E., C.E. Pray, and M.W. Rosegrant. 1999. Agricultural research and productivity growth in India. Res. Rep. No. 109. IFPRI, Washington DC.

Evenson, R.E., and M. Rosegrant. 2003. The economic consequences of crop genetic improvement programmes. In R. Evenson and D. Gollin (ed) Crop variety improvement and its effect on productivity. The impact of international agricultural research. CABI, Wallingford.

Evenson, R.E., and L.E. Westphal, 1995. Technological change and technology strategy. p. 2209-2299. In J. Behrman and T.N. Srinivasan (ed) Handbook of development economics. Elsevier Sci., NY.

Fan, S., P. Hazell, and S. Thorat. 2000. Government spending, agricultural growth and poverty in rural India. Am. J. Agric. Econ. 82:1038-1051.

Fan, S., P. Huong, and T. Long. 2004a. Government spending and poverty reduction in Vietnam. IFPRI, Washington DC.

Fan, S., D. Nyange, and N. Rao. 2005.

Public investment and poverty reduction in Tanzania: Evidence from household survey data. DSGD Disc. Pap. 18. IFPRI, Washington DC.

Fan, S., and X. Zhang. 2004. Investment, reforms and poverty in rural China. Econ. Dev. Cult. Change 52(2):395-422.

Fan, S., X. Zhang, and N. Rao. 2004b. Public expenditure, growth and poverty reduction in rural Uganda. DSG Disc. Pap. 4. IFPRI, Washington DC.

FAO. 2003. Why should soil biodiversity be managed and conserved? [Online] Available at http://www.fao.org/ag/agl/agll/soilbiod/consetxt.stm

FAO, 2004. The state of food insecurity in the world 2004. FAO, Rome.

FAO. 2005a. The state of the food insecurity in the world 2005. FAO, Rome.

FAO. 2005b. Food security and agricultural development in sub-Saharan Africa: Building a case for more public support. FAO, Rome.

FAO. 2007. The state of the world's animal genetic resources for food and agriculture. Comm. Genetic Resourc. Food Agric. CGRFA-11/07/Inf.6. Rome.

Farrell, K.R. 2004. Public–private relationships in agricultural research: Policy implications, presented at inaugural lecture of the Farrell distinguished public policy lectureship. Univ. Guelph, 1 Dec.

Feather, P, D. Hellerstein, and L. Hansen. 1999. Economic valuation of environmental benefits and the targeting of conservation programs: The case of the CRP. Agric. Econ. Res. Rep. No. 778. ERS, USDA, Washington DC.

Ferreyra, C. 2006. Emergy analysis of one century of agricultural production in the Rolling Pampas of Argentina. Int. J. Agric. Resourc. Govern. Ecol. 5(2-3):185-205.

Fishel, W.L. 1971. The Minnesota agricultural research resource allocation system and experiment. In W.L. Fisher (ed) Resource allocation in agricultural research. Univ. Minnesota Press, Minneapolis.

Fonseca, C., R. Labarta, A. Mendoza, J. Landeo, and T.S. Walker. 1996. Economic impact of the high-yielding, late-blight resistant variety Canchan-INIAA in Peru. p. 51-63 In T.S. Walker and C.C. Crissman (ed) Case studies of the impact of CIP-related technologies. CIP, Lima.

Foster, J.E. 1998. Absolute vs. relative poverty. Am. Econ. Rev. 88(2):335-341.

Freebairn, D.K. 1995. Did the green revolution concentrate incomes? A quantitative study of research reports. World Dev. 23(2):265-279.

Freeman, A.M. 1985. Methods for assessing the benefits of environmental programs. p.223-270. In A.V. Kneese and J.L. Sweeney (ed) Handbook of natural resource and energy economics. Vol. I. Elsevier Sci., Amsterdam.

Freeman, C. 1987. Technology policy and economic performance: Lessons from Japan. Pinter, London.

Frisvold, G., J. Sullivan, and A. Raneses. 2003.

Genetic improvements in major U.S. crops: The size and distribution of benefits. Agric. Econ. 28(2):109-119.

Fuglie, K., N. Ballenger, K. Day, C. Ollinger, M. Reilly, J. Vassavada, and J. Yee. 1996. Agricultural research and development: Public and private investments under alternative markets and institutions. Agric. Econ. Rep. No. 735. ERS, USDA, Washington DC.

Gage, J.D., C.T. Sarr, and C. Adoum. 2001. Sustainable agricultural research: institutional and financing reforms in Senegal. Sustainable Financing Country Study No. 3. Bureau for Africa, Off. Sustain. Dev., USAID, Washington DC.

Gardner, B. 2003. Global public goods from the CGIAR: Impact assessment. Thematic working paper for the CGIAR at 31: An independent meta-evaluation of the CGIAR. OED, World Bank, Washington DC.

Garelli, S. 1996. The fundamentals of world competitiveness. In IMD (ed) World Competitiveness Yearbook. IMD, Lausanne, Switzerland.

Garming, H., and H. Waibel. 2007. Pesticides and farmer health in Nicaragua - A willingness to pay approach. Annual Conf. Verein für Socialpolitik Res. Comm. Dev. Economics, Göttingen, 29-30 June 2007.

Garrity, D.P. 2004. Agroforestry and the achievement of the Millennium Development Goals. Agrofor. Syst. 61:5-17.

Garrity, D.P. 2006. Science-based agroforestry and the achievement of the Millennium Development Goals. In D.P. Garrity et al. (ed) World agroforestry into the future. World Agroforesty Centre, Nairobi.

Garthoff, B. 2005. Innovation leading future growth. Presentation given at Bayer R&D Investor Day, London, December.

Gisselquist, D., C. Pray, and J. Nash. 2002. Deregulating technology transfer in agriculture: impact on technical change, productivity, and incomes. World Bank Res. Observer 17(2):.237-265.

Gopinath, M., and T.L. Roe. 1996. Sources of growth in U.S. GDP and economy-wide linkages to the agricultural sector. J. Agric. Res. Econ. 78(4):325-340.

Goto, F., T. Yoshihara, N. Shigemoto S. Toki, and F. Takaiwa. 1999. Iron fortification of rice seed by soybean ferritin gene. Nature Biotech. 17:282-286.

Government of Malaysia. 2006. Ninth Malaysia Plan 2006-2010 Econ. Planning Unit Prime Dep., Putrajaya, Malaysia.

Griliches, Z. 1992. The search for R&D spillovers. Scandinavian J. Econ. 94 (Suppl.):29-47.

Guinet. 2004. Public-private partnerships for research and innovation: An evaluation of the Dutch experience. OECD, Paris.

Gunasena, H.P.M. 2003. Food and poverty: Technologies for poverty alleviation. South Asia conf. on technologies for poverty reduction. New Delhi, 10-11 Oct.

Gundwardana, M. 2005. Economic valuation of a mangrove ecosystem threatened by shrimp aquaculture in Sri Lanka. Environ. Manage. 36(4):535-550.

Hall, A.J., R. V. Sulaiman., N.G. Clark, M.V.S. Sivamohan, and B. Yoganand. 2000. Public and private sector partnerships in Indian agricultural research: Emerging challenges to creating an agricultural innovation system. Twenty-fourth Int. Conf. Agric. Economists. Berlin, 13 Aug.

Hall, A.J., and B. Yoganand. 2002. New institutional arrangements in agricultural R&D in Africa: Concepts and case studies. In H.A. Freeman, D.D. Rohrbach, and C. Ackello-Ogut (ed.). Targeting agricultural research for development in the semi-arid tropics of Sub-Saharan Africa. Proc. workshop, 1-3 July 2002. ICRISAT, Nairobi.

Hambly Odame, H., N. Hafkin, G. Wesseler, and I. Boto. 2002 Gender and agriculture in the information society, CTA/ISNAR Briefing Pap. 55. ISNAR, The Hague.

Haque, T. 1999. Impact of contract farming in India: A case study of Pepsico, (Punjab), Hindustan Lever (Punjab) and VST Natural Products Ltd (Andhra Pradesh). NCAP, New Delhi.

Hartwich, F., and M. Von Oppen. 2000. Knowledge brokers in agricultural research and extension. p. 445-454. In F. Graef et al. (ed) Adapted farming in West Africa: Potentials and perspectives. Verlag Ulrich E. Grauer, Stuttgart.

Hayami, Y., and V.W. Ruttan. 1985. Agricultural development: An international perspective. Rev. Ed. Johns Hopkins Univ. Press, Baltimore.

Hazell, P.B.R. 1999. The impact of agricultural research on the poor: A review of the state of knowledge. CIAT workshop on assessing the impact of agricultural research on poverty alleviation. San Jose, Costa Rica,14-16 Sept.

Heemskerk, W., and B. Wennink (ed) 2005. Stakeholder-driven funding mechanisms for agricultural innovation:Case Studies from Sub-Saharan Africa. KIT Development, Policy and Practice, Bull. 373. Roy. Trop. Inst., Amsterdam.

Heisey, P.W., M.A. Lantican, and H.J. Dubin. 2002. Impacts of international wheat breeding research in developing countries, 1966-97. CIMMYT, Mexico City.

Hesse, C., and J. MacGregor. 2006. Pastoralism: Drylands' invisible asset? Developing a framework for assessing the value of pastoralism in East Africa. Issue Pap. 142. IIED, London.

Hotz, C., and K.H. Brown (ed), 2004. Assessment of the risk of zinc deficiency in populations and options for its control. Int. Zinc Nutrition Consultative Group Tech. Doc. 1, Food Nutr. Bull. 25:S91-S204.

Huffman, W.E., and R.E. Evenson. 1993. Science for agriculture: A long-term perspective. Iowa State Univ. Press, Ames.

Huffman, W.E., and M.A. Johnson. 2001.

Research, extension, and education policy. p. 209-214. In J.L. Outlaw and E.G. Smith (ed) 2002 Farm bill: Policy options consequences. The Farm Foundation, Oak Brook IL.

Hurley, J. 2000. An overview of the normative economics of the health sector. p 55-118. In A.J. Culyer and J.P. Newhouse (ed) Handbook health econ., Vol. I. Elsevier Sci., Amsterdam.

Hurst, P. 1999. The global pesticide industry's "safe use and handling" training project in Guatemala. Int. Union Food Agric. Workers 43, Geneva.

IAC (InterAcademy Council). 2004. Inventing a better future: A strategy for building worldwide capacities for science and technology. Roy. Netherlands Acad. Arts Sci., Amsterdam.

IEA (International Energy Agency). 2006. R&D statistics: Version 2005 [online]. Available at http://www.iea.org/Textbase/stats/rd.asp.

Isinika, A.C., and N.S.Y. Mdoe, 2001. Improving farm management skills for poverty alleviation: The case of Njombe District. Research for poverty alleviation (REPOA) Rep. 01.1, Mkuki Na Nyota, Dar-es-Salaam.

Izac, A.M., and M. Swift. 1994. On agricultural sustainability and its measurements in small-scale farming in Sub-Saharan Africa. Ecol. Econ. 11(2):105-126.

Jacobs, M. 1996. What is socioecological economics? Ecol. Econ. Bull. 1(2):14-16.

Jayaraman, S.K. 2006 US-Indian agbiotech deal under scrutiny. Nature Biotech. 24(5):481.

Jeyaratnam, J., K. C. Lun, and W.O. Phoon. 1987. Survey of acute pesticide poisoning among agricultural workers in four Asian countries. Bull. WHO 65(4):521-527.

Jeyaratnam, J., R.S. de Alwis Seneviratne, and J.F. Copplestone. 1982. Survey of pesticide poisoning in Sri Lanka. Bull. WHO 60(4):615-619.

Johnson, N., and D. Pachico. 2000. Impact of past research on crop genetic improvement. In D. Pachico (ed) Impact assessment annual report. CIAT, Cali.

Kaimowitz, D. (ed) 1993. Making the link: Agricultural research and technology transfer in developing countries. West-View Press, Boulder CO.

Kangasniem, J. 2002. Financing agricultural research by producer organizations in Africa. p. 81-104. In D. Byerlee, and R.G. Echeverria (ed) Agricultural research policy in an era of privatization. CABI, Wallingford.

Kautsky, N., H. Berg, C. Folke, J. Larsson, and M. Troell. 1997. Ecological footprint for assessment of resource use and development limitations in shrimp and tilapia aquaculture. Aquaculture Res. 28:753-766.

Kaufman, D., A. Kraay, and M. Mastruzzi. 2003. Governance matters III: Governance indicators for 1996-2002. Res. Pap. 3106. World Bank, Washington DC.

Kerr, J., and S. Kolavalli. 1999. Impact of

agricultural research on poverty alleviation: Conceptual framework with illustrations from the literature. EPTD Disc. Pap. 56. IFPRI, Washington DC.

Kijne, J., and J. Bennet. 2004. Science-based crop water productivity improvement: What is it and who pays? Available at www.knowledgebank.irri.org/theme1/pdfs/sciencebasedcwp.pdf.

Kishi, M., N. Hirschhorn, M. Djajadisastra, L.N. Satterlee, S. Strowman and R. Dilts 1995. Relationship of pesticide spraying to signs and symptoms in Indonesian farmers. Scand. J. Work Environ. Health 21:124-133.

Kremer, M., and A.P. Zwane. 2005. Encouraging private sector research for tropical agriculture. World Dev. 33(1):87-105.

Law, M.T., G.J. Miller, and J.M. Tonon. 2004. Earmarked: The political economy of agricultural research appropriations. Univ. Vermont, Burlington and Washington Univ., St. Louis.

Leakey, R.R.B. 1999. Potential for novel food products from agro-forestry trees. Food Chem. 64:1-14.

Leakey, R.R.B., Z. Tchoundjeu, K. Schreckenberg, S.E. Shackleton, and C.M. Shackleton. 2005. Agro-forestry tree products (AFTPs): Targeting poverty reduction and enhanced livelihoods. Int. J. Agric. Sustain. 3:1-23.

Lema, N., and B. Kapange. 2005a. The national agricultural research fund in Tanzania. In W. Heemskerk and B. Wennink (ed) Stakeholder-driven funding mechanisms for agricultural innovation: Case studies from Sub-Saharan Africa. KIT Dev., Policy Practice Bull. 373. Roy. Trop. Inst., Amsterdam.

Lema, N., and B. Kapange. 2005b. Zonal agricultural research funds in Tanzania. In W. Heemskerk and B. Wennink (ed) Stakeholder-driven funding mechanisms for agricultural innovation: Case studies from Sub-Saharan Africa. KIT Dev., Policy Practice Bull. 373. Roy. Trop. Inst., Amsterdam.

Lewis, J. 2000. Leveraging partnerships between the public and the private sector: Experience of USAID™ agricultural biotechnology programs. p. 196-199. In G.J. Persley and M.M. Lantin (ed) Proc. Int Conf. Agricultural biotechnology and the poor. 21-22 Oct. 1999. CGIAR, Washington DC.

Lipton, M. 2001. Reviving global poverty reduction: What role of genetically modified plants? J. Int. Dev. 13:823-846.

London, L. and R. Bailie. 2001. Challenges for improving surveillance for pesticide poisoning: Policy implications for developing countries. Int. J. Epidemiol. 30:564-570.

Longmire, J., and D. Winkelmann. 1985. Research allocation and comparative advantage. Paper presented at the 19th Int. Conf. Agric. Economists, Malaga, Spain, 26 Aug-4 Sept 1985. Econ. Prog., CIMMYT, Mexico.

Low, J., P. Kinyae, S. Gichuki, M.A. Oyunga, V. Hagenimana, and J. Kabira. 1997. Combating vitamin A deficiency through the use of sweetpotato. CIP, Lima.

Lucca, P., R. Hurrell, and I. Potrykus. 2001. Genetic engineering approaches to improve the bioavailability and the level of iron in rice grains. Theor. Appl. Genetics 102:392-397.

Lundvall, B.A. (ed.) 1992. National innovation systems: Towards a theory of innovation and interactive learning. Pinter, London.

Mansfield, E., J. Rapoport, A. Romeo, S. Wagner, and G. Beardsley. 1977. Social and private rates of return from industrial innovations. Q. J. Econ. 91(2):221-240.

Maredia, M.K., and D. Byerlee. 2000. Efficiency of research investments in the presence of international spillovers: Wheat research in developing countries. Agric. Econ. 22:1-16.

Maredia, M.K., D. Byerlee, and J.R. Anderson. 2001. Ex post evaluation of economic impacts of agricultural research programs: A tour of good practice. p. 5-42. In The future of impact assessment in the CGIAR: Needs, constraints and options. Proc. workshop organized by SPIA of the Tech. Advisory Committee (TAC), 3-5 May 2000. FAO, Rome.

Mashelkar, R.A. 2005. Nation building through science and technology: a developing world perspective. 10th Zukerman Lecture, Roy. Soc. London. Innov. Strategy Today 1:16-22.

Mayne, J. 1999. Addressing attribution through contribution analysis using performance measures sensibly. Disc. Pap. Off. Auditor General of Canada, Ottawa.

McIntire, J. 1998. Coping with fiscal stress in developing country agricultural research. p. 81-96. In S. Tabor et al. (ed) Financing agricultural research: A sourcebook. ISNAR, The Hague.

McMahon, M. 1992. Getting beyond the national institute model for agricultural research in Latin America: A cross -country study of Brazil, Chile, Colombia and Mexico. Latin America and the Caribbean Tech. Dep., Reg. Studies Prog. Rep. 20. World Bank, Washington DC.

Meenakshi, J.V., N. Johnson, V.M. Manyong, H. De Groote, D. Yanggen, J. Javelosa et al. 2006. Analyzing the cost-effectiveness of biofortification: A synthesis of the evidence. Paper presented at the Ann. AAEA Meeting, Long Beach CA, 23-26 July.

Mills, B.R., R. Hassan and P.G. Pardey. 1996. Ex-ante benefits from site-specific agricultural research: Maize in Kenya. p. 133-162. In Global agricultural science policy for the Twenty First Century. Melbourne, Australia, 26-28 Aug.

Minota, N., and L. Daniels. 2005. Impact of global cotton markets on rural poverty in Benin. Agric. Econ. 33(Suppl):453-466.

Mogues, T. 2006. Shocks, livestock asset dynamics and social capital in Ethiopia. DSGD Disc. Pap. 38. IFPRI, Washington DC.

Morris, M., and B. Ekasingh. 2002. Plant breeding research and developing countries: What roles for the public and private sectors. In D. Byerlee and R. Echeverría (ed) Agricultural research policy in an era of privatization. CABI, Wallingford.

Morrison, J., D. Bezermer, and C. Arnold. 2004. Official development assistance to agriculture. DFID, London.

Moyo, S., G.W. Norton, J. Alwang, I. Rhinehart, and M. Deom. 2007. Peanut research and poverty reduction: Impacts of variety improvement to control peanut viruses in Uganda. Am. J. Agric. Econ. 89(2):448-60.

Muliyar, M.K. 1983. Transfer of technology in plantation crops. J. Plantation Crops 11(1):1-12.

Murray-Kolb, L.E., F. Takaiwa, F. Goto, T. Yoshihara, E.C. Theil, and J.L. Beard, 2002. Transgenic rice is a source of iron for iron-depleted rats. J. Nutr. 132:957-960.

Mutangaduraa, G., and G.W. Norton. 1999. Agricultural research priority setting under multiple objectives: An example from Zimbabwe. Agric. Econ. 20(3): 277-286.

Mutungadura, G. 1997. Agricultural research priority setting under multiple objectives: The case of Zimbabwe. Ph.D. thesis, Virginia Polytechnic Inst. and State Univ.

Myers, N. 1999. The next green revolution: Its environmental underpinnings. Curr. Sci. 76(4):507-513.

Nadiri, I. 1993. Innovations and technological spillovers. NBER Working Pap. 4423. Nat. Bur. Econ. Res., Boston.

Narayana, D. 1992. Interaction of Price and technology in the presence of structural specifications: An analysis of crop production in Kerala. Ph.D. dissertation, Indian Statistical Inst., Koltaka.

Nelson. R. 1959. The simple economics of basic scientific research. J. Political Econ. 67:297-306.

Neupane, R.P., and G.B. Thapa. 2001. Impact of agroforestry intervention on farm income under the subsistence farming system of the middle hills, Nepal. Agrofor. Syst. 53(1): 31-37.

North, C.D. 1991. Institutions. J. Econ. Perspect. 5(1):97-112.

Norton, G.W. and J.S. Davis. 1981. Evaluating returns to agricultural research: A review. Am. Agric. Econ. 63(4) 685-699.

Norton, G.W., E.A. Heinrichs, G.C. Luther, and M.E. Irwin (ed) 2005. Globalizing integrated pest management: A participatory research process. Blackwell, Ames IA.

Nutra Ingredients. 2006. BASF expands crop biotech capabilities. [online] Available at http://www.agbios.com/main.php?action=ShowNewsItem&id=7558.

Oehmke, J.F., P. Anandajayasekeram, and W.A. Masters. 1997. Agricultural technology development and transfer in Africa: Impacts achieved and lessons learned. USAID Tech. Pap. 77. USAID, Washington DC.

Omamo, S.W., X. Diao, S. Wood, J. Chamberlin, L. You, S. Benin et al. 2006. Strategic priorities for agricultural development in Eastern and Central Africa. Res. Rep. No. 150. IFPRI, Washington DC.

Oram, P., and V. Bindlish. 1983. Investment in agricultural research in developing countries: progress, problems, and the determination of priorities. IFPRI, Washington DC.

Ortega, E., M. Miller, M. Anami, and P.R. Beskow. 2001. From emergy analysis to public policy: soybean in Brazil. Proc. Second Biennial Emergy Res. Conf. 20-23 Sept., Univ. Florida, Gainesville, FL.

Osgood, D. 2006. Living the promise? The role of the private sector in enabling small scale farmers to benefit from agro-biotech. Int. J. Tech. Global. 2(1/2):30-45.

PAHO (Pan-American Health Organization). 2002. Epidemiological situation of acute pesticide poisoning in Central America 1992-2000. Epidemiol. Bull. Pan Am. Health Org. 23:5-9.

Palmer-Jones, R., and K. Sen. 2006. It is where you are that matters: The spatial determinants of rural poverty in India. Agric. Econ. 34:229-242.

Pardey, P.G., J.M. Alston, J.E. Christian, and S. Fan. 1996. Hidden harvest: US benefits from international research aid. Food Policy Rep., IFPRI, Washington DC.

Pardey, P.G., J.M. Alston, and R.R. Piggott (ed) 2006a. Agricultural R&D in the developing world: Too little, too late? IFPRI, Washington DC.

Pardey, P.G., and N.M. Beintema. 2001. Slow magic: Agricultural R&D a century after Mendel. IFPRI, Washington DC.

Pardey, P.G., N.M. Beintema, S. Dehmer, and S. Wood. 2006b. Agricultural research: A growing global divide? Food Policy Rep. IFPRI, Washington DC.

Pardey, P.G., C. Chan-Kang, and J.M Alston. 2002. Donor and developing-country benefits from international agricultural research: Double dividend? IFPRI, Washington DC.

Pimentel, D. 1997. Economic and environmental benefits of biodiversity. BioScience 47(11):747-758.

Pimentel, D., H. Acquay, M. Biltoneen, P. Rice, M. Silva, J. Nelson et al. 1993a. Assessment of environmental and economic impacts of pesticide use. p. 46-84. In D. Pimentel and H. Lehmann (ed) The pesticide question - environment, economics, and ethics. Chapman and Hall, London.

Pimentel, D., B. Berger, D. Filiberto, M. Newton, B. Wolfe, E. Karabinakis et al. 2004. Water resources: Agricultural and environmental issues. Bioscience 54(10):909-918.

Pimentel, D., C. Harvey, P. Resosudarmo, K. Sinclair, D. Kurz, M. McNair et al. 1995. Environmental and economic costs of soil erosion and conservation benefits. Science 267(5201):1117-1123.

Pimentel, D., L. McLaughlin, A. Zepp, B. Lakitan, T. Kraus, P. Kleinman et al. 1993b. Environmental and economic effects of reducing pesticide use in agriculture. Agric. Ecosyst. Environ. 46:273-288.

Pinero, M.E. 1984. An analysis of research priorities in the CGIAR system: A discussion paper. FAO, Rome.

Pingali. P.L. 2001. Milestones in impact assessment research in the CGIAR, 1970-1999. With an annotated bibliography of impact assessment studies conducted in the CGIAR, 1970-1999, prepared by M.P. Feldmann. SPIA of Tech. Advisory Comm. CGIAR, Mexico City.

Pingali, P.L., and P.W. Heisey. 2001. Cereal-crop productivity in developing countries: Past trends and future prospects. p. 56-82. *In* J.M. Alston et al. (ed) Agricultural science policy: Changing global agendas. Johns Hopkins Univ. Press, Baltimore.

Pinstrup-Andersen, P., N.R. de Londono, and E. Hoover. 1976. The impact of increasing food supply on human nutrition: implications for commodity priorities in agricultural research and policy. Am. J. Agric. Econ. 58(2):131-42.

Pinstrup-Andersen, P., and D. Franklin. 1977. A systems approach to agricultural research resource allocation in developing countries. p. 416-35. *In* G.M. Arndt et al. (ed) Resource allocation and productivity in national and international agricultural research. Univ. Minnesota Press, Minneapolis.

Porter, G. and K. Phillips-Howard. 1997. Comparing contracts: An evaluation of contract farming schemes in Africa. World Dev. 25(2):227-238.

Pray, C.E. 1987. Private sector agricultural research in Asia p 411-432. *In* V.W. Ruttan and C.E. Pray (ed) Policy for agricultural research. Westview, Boulder CO.

Pray, C.E. 2002. The growing role of the private sector in agricultural research. *In* D. Byerlee and R. Echeverría (ed) Agricultural research policy in an era of privatization. CABI, Wallingford.

Pray, C.E., and R.G. Echeverría. 1991. Determinants and scope of private sector agricultural research in developing countries. p. 343-364. *In* P.G. Pardey et al. (ed) Agricultural research policy: International quantitative perspectives. Cambridge Univ. Press, Cambridge.

Pray, C.E., and K.O. Fuglie. 2001. Private investments in agricultural research and international technology transfer in Asia ERS Agric. Econ. Rep. 805. ERS, USDA, Washington DC.

Pray, C.E., and B. Ramaswami. 2001. Technology, IPR, and reform options: A case study of the seed industry with implications for other input industries. Int. Food Agric. Marketing Rev. Spec. Issue 2(3/4):407-420.

Pray, C.E., and L. Umali-Deininger 1998. The private sector in agricultural research system: Will it fill the gap? World Dev. 26(6):1127-1148.

Pray, C.E., P. Bengali, and B. Ramaswami. 2005. Costs and benefits of biosafety regulation in India: A preliminary assessment. Quar. J. Int. Agric. 44(3):267-289.

Pray, C.E., D. Johnson, and K.O Fuglie. 2007. Private sector research. *In* R.E. Evenson and P. Pingali (ed) Handbook of agricultural economics 3 -Agricultural development: Farmers, farm production and farm markets. Elsevier, Amsterdam.

Pray, C.E., S. Ribeiro, R.A.E. Mueller, and P.P. Rao. 1991. Private research and public benefit: The private seed industry for sorghum and pearl millet in India. Res. Policy 20:315-324.

Premchand, A. 1993. Public expenditure management. IMF, Washington DC.

Pretty, J.N. 2000. Towards sustainable food and farming systems in industrialized countries. Int. J. Agric. Resourc. Govern. Ecol. 1(1): 77-94.

Pretty, J.N., and A.S. Ball. 2001. Agricultural influences on carbon emissions and sequestration: A review of evidence and the emerging trading options. Centre for Environ. Society Occas. Pap. 2001-03. Univ. Essex, UK.

Pretty, J.N. and R. Hine. 2001. Reducing food poverty with sustainable agriculture: A summary of new evidence. Final Report from the SAFE-World (The Potential of Sustainable Agriculture to Feed the World). Univ. Essex, UK.

Pretty, J.N., A.S. Ball, L. Xiaoyun, and N.H. Ravindranath. 2002. The role of sustainable agriculture and renewable-resource management in reducing greenhouse-gas emissions and increasing sinks in China and India. Phil. Trans. R. Soc. London A, 360:1741-1761.

Pretty, J.N., C. Brett, D. Gee, R.E. Hine, C.F. Mason, J.I.L. Morison et al. 2000. An assessment of the total external costs of UK agriculture. Agric. Syst. 65:113-136.

Primavera, J.H. 1997. Socioeconomic impacts of shrimp culture. Aquaculture Res. 28:815-827.

Proops, J.L.R. 1989. Ecological economics: Rationale and problem areas. Ecol. Econ.1:59-76.

Psacharopoulos, G. 1994. Returns to investment in education: A global update. World Bank, Washington DC.

Qaim, M., A.J. Stein, and J.V. Meenakshi. 2006. Economics of biofortification. A plenary paper prepared for the Int. Assoc. Agric. Economists (IAAE) Conf., Gold Coast, Australia, 12-16 Aug.

Raitzer, D.A. 2003. Benefit-cost meta-analysis of investment in the international agricultural research centres of the CGIAR. CGIAR Sci. Council Secretariat, Rome.

Reifschneider, F.J.B., U. Lele, and A.D. Portugal. 1997. Strategies for financing agricultural research in Brazil: Sustainable and diversification for a competitive future. Agric. Res. Extension Unit, World Bank, Washington DC.

Robleto, M.L., and W. Marcelo. 1992. La deuda ecológica. Una perspectiva sociopolítica. Area Internacional, Inst. de Ecologia Política, Santiago, Chile.

Rola, A.C., and P.L. Pingali. 1993. Pesticides, rice productivity, and farmer's health: An economic assessment. IRRI, Manila, Philippines.

Rose-Ackerman, S., and R.E. Evenson. 1985. The political economy of agricultural research and extension: Grants, votes, and reapportionment. Am. J. Agric. Econ. 67:1-14.

Rosegrant, M., and R.E. Evenson. 1993. Agricultural productivity growth in Pakistan and India: A comparative analysis. Pakistan Dev. Rev. 32(4):433-438.

Rossi, P.H., and H.E. Freeman. 1993. Evaluation: A systematic approach. Sage Publ., Beverly Hills.

Rótolo, G.C., T. Rydberg, G. Lieblein, and C. Francis. 2007. Emergy evaluation of grazing cattle in Argentina's Pampas. Agric. Ecosyst. Environ. 119:383-395.

Rukuni, M., M.J. Blackie, and C.K. Eicher. 1998. Crafting smallholder-driven agricultural research systems in Southern Africa. World Dev. 26:1049-1056.

Ruttan, V.W. 1982. Agricultural research policy. Univ. Minnesota Press, Minneapolis.

Ruttan, V.W. 2003. Social science knowledge and economic development: An institutional design perspective. Univ. Michigan Press, Ann Arbor.

Santhakumar, V., and R. Rajagopalan. 1995. Green revolution in Kerala? A discourse on technology and nature. South-Asia Bull. 15(2):109-119.

Schimmelpfennig, D., and G.W. Norton. 2003. Measuring the benefits of international agricultural economics research. Quar. J. Int. Agric. 42(2):207-222.

Schimmelpfennig, D., C. O'Donnell, and G.W. Norton. 2006. Efficiency effects of agricultural economics research in the U.S. Agric. Econ. 34(3):273-280.

Schuh, G.E. and H. Tollini. 1979. Costs and benefits of agricultural research: State of the arts. World Bank Staff Working Pap. 360. World Bank, Washington DC.

Scobie, G.M. 1984. Investment in agricultural research: Some economic principles. Working Pap., CIMMYT, El Batan, Mexico.

Scoones, I. 2003. Can agricultural biotechnology be pro-poor? Democratizing biotechnology: Genetically modified crops in developing countries. Briefing 2. Inst. Dev. Studies, Brighton, UK.

Scotchmer, S. 1999. On the optimality of the patent renewal system. RAND J. Econ. 30(2):181-196.

Sexton, R.J. and T.A. Sexton. 1996. Measuring research benefits in an imperfect market. Agric. Econ. 13:201-204.

Shane, M., T. Roe, and M. Gopinath. 1998. U.S. agricultural growth and productivity: An economy wide perspective. USDA/ERC Agric. Econ. Rep. AER-758. USDA, Washington DC.

Shao, F.M. 1996. Funding of agricultural research in Tanzania. *In* Funding agricultural research in Sub-Saharan Africa. p 113-141. *In* FAO (ed) Report of the FAO/SPAAR/KARI expert consultation on funding of agricultural research in Sub-Saharan Africa. Kenya Agric. Res. Inst. (KARI), 6 -7 July 1993. FAO, Rome.

Shavall, S., and T. Ypserle. 2001. Van rewards versus intellectual property rights. J. Law Econ. 44(2):525-547.

Shrestha, S., and M.A. Bell. 2002. Impact from research and collaboration with the national agricultural research and extension systems: A brief review. IRRI, Philippines.

SPIA (Standing Panel on Impact Assessment). 2001. The future of impact assessment in the CGIAR needs, constraints and options. Proceedings of a workshop organized by SPIA, Tech. Advisory Comm. Rome, 3-5 May. CGIAR Sci. Council, Rome.

Spielman, D.J., and K. Von Grebmer. 2004. Public-private partnerships in agricultural research: An analysis of challenges facing industry and the Consultative Group on International Agricultural Research. EPTD Disc. Pap. 113. IFPRI, Washington DC.

Spielman, D.J. 2004. Agricultural sector investment and the role of public-private partnerships. African development and poverty reduction. The Macro-Micro Linkages Forum Paper. Cornell Univ., Ithaca NY.

Stein, A.J., J.V. Meenakshi, M. Qaim, P. Nestel, H.P.S. Sachdev, and Z.A. Bhutta, 2005. Analyzing the health benefits of biofortified staple crops by means of the disability-adjusted life years approach: A handbook focusing on iron, zinc and vitamin A. Harvest Plus Tech. Mono. 4. IFPRI, Washington DC and CIAT, Cali.

Steinfeld, H., P. Gerber, T. Wassenaar, V. Castel, M. Rosales, and C. de Haan. 2006. Livestock's long shadow: Environmental issues and options. Livestock, Environment and Development Initiative (LEAD). FAO, Rome.

Stern, N. 2007. The economics of climate change: The Stern review. Cambridge Univ. Press.

Stiglitz, J.E. 2000. Economics of the public sector. WW Norton, NY.

Swanson, B.E. 2006. The changing role of agricultural extension in a global economy. J. Int. Agric. Extension Educ. 11(3):5-17.

Swinnen, J.F.M., H. de Gorter, G.C. Rausser, and A.N. Banerjee. 2000. The political economy of public research investment and commodity policies in agriculture: An empirical study. Agric. Econ. 22(2):111-122.

Syngenta. 2004. Annual Report 2003. Syngenta, Basel.

Syngenta. 2006. Annual Report 2005. Syngenta, Basel.

Tabor, S., H. Tollini, and W. Janssen. 1996. Globalization of agriculture research: Do winners take all? Disc. Pap. 96-11. ISNAR, The Hague.

Tabor, S.R. (ed) 1995. Agricultural research in an era of adjustment: Policies, institutions, and progress. EDI Seminar Series. Econ. Dev. Inst., World Bank and ISNAR, Washington DC.

Tabor, S.R., W. Janssen and H. Bruneau (ed) 1998. Financing agricultural research: A sourcebook. ISNAR, The Hague.

Tchoundjeu, Z., E. Asaah, P. Anegbeh,

A. Degrande, P. Mbile, C. Facheux et al. 2006. AFTPs: Putting participatory domestication into practice in West and Central Africa. For. Trees Livelihoods 16:53-70.

The Rodale Institute. 2003. The Rodale Institute farming systems trial. [Online]. Available at www.newfarm.org/depts/NFfield_trials/1003/carbonsequest.shtml. Rodale, PA.

Thirtle, C., L. Lin. and J. Piess. 2003. The impact of research-led agricultural productivity growth on poverty reduction in Africa, Asia and Latin America. World Dev. 31(12):1959-1975.

Thrupp, L.A. 1998. Cultivating diversity: Agrobiodiversity and food security. World Resourc. Inst., Washington, DC.

Tollini, H. 1998. Capital investment policies and agricultural research. p. 29-45. In S. Tabor et al. (ed) Financing agricultural research: A sourcebook. ISNAR, The Hague.

Turpie, J.K. 2003. The existence value of biodiversity in South Africa: how interest, experience, knowledge, income and perceived level of threat influence local willingness to pay. Ecol. Econ. 46(2):199-216.

UI Haque, N., and A.M. Khan. 1997. Institutional development: Skill transference through reversal of "human capital flight" or technical assistance. Working Pap. 97/89. IMF, Washington DC.

Ulrich, A., W.H. Furtan, and A.Schmitz 1985. Public and private returns from joint venture research in Canada. Univ. Calgary Press, Calgary.

Umali, D.L., and L. Schwartz. 1994. Public and private agricultural extension: Beyond traditional frontiers. Disc. Pap. 236. World Bank, Washington DC.

UNEP. 2004a. Exploring the links: Human well-being, poverty and ecosystem services. UNEP, Nairobi.

UNEP. 2004b. Childhood pesticide poisoning: Information for advocacy and action. Chemicals Programme of the United Nations. UNEP, Nairobi.

UNESCO and OECD. 2003. Financing education: Investments and returns: Analysis of world education indicators. 2002 ed. UNESCO and OECD, Paris.

UN-SCN (United Nations Standing Committee on Nutrition), 2004. 5th Report on the World Nutrition Situation. UN-SCN, Geneva.

Upreti, B.R., and Y.G. Upreti. 2002. Factors leading to agro-biodiversity loss in developing countries: The case of Nepal. Biodivers. Conserv.11:1607-1621.

Vasconcelos, M., K. Datta, N. Oliva, M. Khalekuzzaman, L. Torrizo, S. Krishnan et al. 2003. Enhanced iron and zinc accumulation in transgenic rice with the Ferritin gene. Plant Sci. 164:371-378.

Venezian, E., and E. Muchnik. 1994. Structural adjustment and agricultural research in Chile. Briefing Pap. 9. ISNAR, The Hague.

Villarreal, M., C. Holding Anyonge, B. Swallow, and F. Kwesiga. 2006. The challenge of HIV/

AIDS: Where does agroforestry fit in? In D. Garrity (ed) Science-based agroforestry and the achievement of the Millennium Development Goals. World Agroforestry Center, Nairobi.

Von Braun, J. 2003. Agricultural economics and distributional effects. Presidential address of the 25th Int. Conf. Agric. Economists. Durban, South Africa. 16-22 Aug.

Von Braun, J., and P.J. Webb. 1989. The impact of new cropping technology on the agricultural division of labor in a West African setting. Econ. Dev. Cult. Change 37:513-534.

Von Oppen, M., and J.G. Ryan. 1985. Research resource allocation: Determining regional priorities. Food Policy 10(3):253-264.

Von Oppen, M., S. Abele, E. van den Akker, F. Hartwich, E. Krüsken, and U. von Poschinger-Camphausen. 2000. Alternatives to public and private funding of agricultural research in developing countries: Potentials of the third sector. Paper presented at the Deutscher Tropentag 2000, Hohenheim, 11-12 Oct.

Voon, J.P. 1994. Measuring research benefits in an imperfect market. Agric. Econ. 10:89-93.

Wackernagel, M., and W.E. Rees. 1997. Perceptual and structural barriers to investing in natural capital: Economics from an ecological footprint perspective. Ecol. Econ. 20:3-24.

Waibel, H., and D. Zilberman (ed) 2007. International research on natural resource management: Advances in impact assessment. CABI, Wallingford.

Waibel, H., G. Fleischer and H. Becker. 1999. The economic benefits of pesticides: A case study from Germany. German J. Agric. Econ. 48(6):219-230.

Walker, P., P. Rhubart-Berg, S. McKenzie, K. Kelling, and R.S. Lawrence. 2005. Public health implications of meat production and consumption. Public Health Nutri. 8(4): 348-356.

WHO. 1990. Public health impact of pesticides used in agriculture. WHO, Geneva.

WHO. 2002. The World health report 2002: Reducing risks, promoting healthy life. WHO, Geneva.

WHO. 2006. Agrochemicals: Linking health and environmental management. Policy Brief. Available at http://www.who.int/heli/risks/toxic/chemicals/index.html. WHO, Geneva.

World Agroforestry Centre. 2005. Trees of change: A vision for an agroforestry transformation in the developing world. World Agroforestry Centre, Nairobi.

World Bank. 2000. Agricultural technology notes. No. 25. Rural Dev. Dep., World Bank, Washington DC.

World Bank. 2006. Investments in agricultural extension and information services. Module 3 in the AgInvestment Sourcebook. Available at http://web.worldbank.org/WBSITE/EXTERNAL/TOPICS/EXTARD/EXTAGISO U/0,,contentMDK:20930620~menuPK:275

6949~pagePK:64168445~piPK:64168309~ theSitePK:2502781,00.html #. World Bank, Washington DC.

Wright, B.D. 1983. The economics of invention incentives: Patents, prizes, and research contracts. Am. Econ. Rev. 73:691-707.

Yavapolkul, N., G. Munisamy, and A. Gulatic.

2006. Post–Uruguay round price linkages between developed and developing countries: The case of rice and wheat markets. Agric. Econ. 34:259-272.

Ye, X., S. Al-Babili, A. Klöti, J. Zhang, P. Lucca, P. Beyer, and I. Potrykus, 2000. Engineering the provitamin A (β-carotene) biosynthetic

pathway into (carotenoid-free) rice endosperm. Science 287:303-305.

Zimmermann, R., and M. Qaim. 2004. Potential health benefits of golden rice: A Philippine case study. Food Policy 29:47-168.

Authors and Review Editors

Argentina

Walter Ismael Abedini • La Plata National University

Héctor D. Ginzo • Ministerio de Relaciones Exteriores, Comercio Internacional y Culto

Maria Cristina Plencovich • Universidad de Buenos Aires

Sandra Elizabeth Sharry • Universidad Nacional de La Plata

Miguel Taboada • Universidad de Buenos Aires

Ernesto Viglizzo • INTA Centro Regional La Pampa

Australia

Helal Ahammad • Department of Agriculture, Fisheries and Forestry

Tony Jansen • TerraCircle Inc.

Roger R.B. Leakey • James Cook University

Andrew Lowe • Adelaide State Herbarium and Biosurvey

Andrew Mears • Majority World Technology

Bolivia

Manuel de la Fuente • National Centre of Competence in Research North-South

Botswana

Baone Cynthia Kwerepe • Botswana College of Agriculture

Brazil

André Gonçalves • Centro Ecológico

Odo Primavesi • Embrapa Pecuaria Sudeste (Southeast Embrapa Cattle)

Canada

Jacqueline Alder • University of British Columbia

Harriet Friedman • University of Toronto

Thora Martina Herrmann • Université de Montréal

Sophia Huyer • UN Commission on Science and Technology for Development.

JoAnn Jaffe • University of Regina

Shawn McGuire

Morven A. McLean • Agriculture and Biotechnology Strategies Inc. (AGBIOS)

M. Monirul Qader Mirza • University of Toronto, Scarborough

Ricardo Ramirez • University of Guelph

China

Jikun Huang • Chinese Academy of Sciences

Colombia

Maria Veronica Gottret • CIAT

Costa Rica

Marian Perez Gutierrez • National Centre of Competence in Research North-South, Centre Suisse de Recherche Scientifique

Côte d'Ivoire

Guéladio Cissé • National Centre of Competence in Research North-South

Denmark

Henrik Egelyng • Danish Institute for International Studies (DIIS)

Thomas Henrichs • University of Aarhus

Egypt

Mostafa A. Bedier • Agricultural Economic Research Institute

Salwa Mohamed Ali Dogheim • Agriculture Research Center

Ethiopia

P. Anandajayasekeram • International Livestock Research Institute

Berhanu Debele • National Centre of Competence in Research North-South

Workneh Negatu Sentayehu • Addis Ababa University

Gete Zeleke • Global Mountain Program

Finland

Riikka Rajalahti • Ministry of Foreign Affairs

France

Martine Antona • Centre International de Recherche en Agriculture pour le Développement

Didier Bazile • Centre International de Recherche en Agriculture pour le Développement

Pierre-Marie Bosc • Centre International de Recherche en Agriculture pour le Développement

Nicolas Bricas • Centre International de Recherche en Agriculture pour le Développement

Jacques Brossier • Institut National de la Recherche. Agronomique (INRA)

Perrine Burnod • Centre International de Recherche en Agriculture pour le Développement

Patrick Caron • Centre International de Recherche en Agriculture pour le Développement

Emilie Coudel • Centre International de Recherche en Agriculture pour le Développement

Fabrice Dreyfus • University Institute for Tropical Agrofood Industries and Rural Development

Michel Dulcire • Centre International de Recherche en Agriculture pour le Développement

Patrick Dugué • Centre International de Recherche en Agriculture pour le Développement

Stefano Farolfi • Centre International de Recherche en Agriculture pour le Développement

Guy Faure • Centre International de Recherche en Agriculture pour le Développement

Nicolas Faysse • Centre International de Recherche en Agriculture pour le Développement

Thierry Goli • Centre International de Recherche en Agriculture pour le Développement

Henri Hocdé • Centre International de Recherche en Agriculture pour le Développement

Bernard Hubert • Institut National de la Recherche Agronomique (INRA)

Jacques Imbernon • Centre International de Recherche en Agriculture pour le Développement

Jean-Pierre Müller • Centre International de Recherche en Agriculture pour le Développement

Sylvain Perret • Centre International de Recherche en Agriculture pour le Développement

Michel Petit • Institut Agronomique Mediterraneen Montpellier

Anne-Lucie Raoult-Wack • Agropolis Fondation

Nicole Sibelet • Centre International de Recherche en Agriculture pour le Développement

Ludovic Temple • Centre International de Recherche en Agriculture pour le Développement

Jean-Philippe Tonneau • Centre International de Recherche en Agriculture pour le Développement

Guy Trebuil • Centre International de Recherche en Agriculture pour le Développement

Tancrede Voituriez • Centre International de Recherche en Agriculture pour le Développement

The Gambia
Ndey Sireng Bakurin • National Environment Agency

Germany
Anita Idel • Mediatorin (MAB)
Hermann Waibel • Leibniz University of Hannover

Ghana
Elizabeth Acheampong • University of Ghana
Edwin A. Gyasi • University of Ghana
Gordana Kranjac-Berisavljevic • University for Development Studies
Carol Markwei • University of Ghana

India
Sachin Chaturvedi • Research and Information System for Developing Countries (RIS)
Purvi Mehta-Bhatt • Science Ashram
Poonam Munjal • CRISIL Ltd
K.P. Palanisami • Tamil Nadu Agricultural University
C.R. Ranganathan • Tamil Nadu Agricultural University
Sunil Ray • Institute of Development Studies
V. Santhakumar • Centre for Development Studies
Anushree Sinha • National Council for Applied Economic Research (NCAER)

Indonesia
Suraya Afiff • KARSA (Circle for Agrarian and Village Reform)

Italy
Gustavo Best • Independent
Michael Halewood • Bioversity International
Anne-Marie Izac • Alliance of the CGIAR Centres
Prabhu Pingali • FAO
Sergio Ulgiati • Parthenope University of Naples
Keith Wiebe • FAO
Monika Zurek • FAO

Jamaica
Audia Barnett • Scientific Research Council

Japan
Osamu Ito • Japan International Research Center for Agricultural Sciences (JIRCAS)
Osamu Koyama • Japan International Research Center for Agricultural Sciences (JIRCAS)

Jordan
Mahmud Duwayri • University of Jordan

Kenya
Tsedeke Abate • International Crops Research Institute for the Semi-Arid Tropics
Boniface Kiteme • Centre for Training and Integrated Research in Arid and Semi-arid Lands Development
Washington Ochola • Egerton University
Frank M. Place • World Agroforestry Centre

Kyrgyz Republic
Ulan Kasymov • Central Asian Mountain Partnership Programme

Malaysia
Khoo Gaik Hong • International Tropical Fruits Network

Mauritius
Ameenah Gurib-Fakim • University of Mauritius

Mexico
Jesus Moncada • Independent
Scott S. Robinson • Universidad Metropolitana - Iztapalapa

Morocco
Saadia Lhaloui • Institut National de la Recherche Agronomique

Netherlands
Nienke Beintema • International Food Policy Research Institute
Bas Eickhout • Netherlands Environmental Assessment Agency (MNP)
Judith Francis • Technical Centre for Agricultural and Rural Cooperation (CTA)
Janice Jiggins • Wageningen University
Toby Kiers • Vrije Universiteit
Kaspar Kok • Wageningen University
Niek Koning • Wageningen University
Niels Louwaars • Wageningen University
Niels Röling • Wageningen University

Mark van Oorschot • Netherlands Environmental Assessment Agency (MNP)

Detlef P. van Vuuren • Netherlands Environmental Assessment Agency (MNP)

Henk Westhoek • Netherlands Environmental Assessment Agency (MNP)

New Zealand
Jack A. Heinemann • University of Canterbury

Nigeria
Stella B. Willliams • Obafemi Awolowo University

Oman
Abdallah Mohamed Omezzine • University of Nizwa

Pakistan
Syed Sajidin Hussain • Ministry of Environment

Peru
Maria E. Fernandez • National Agrarian University
Carla Tamagno • Universidad San Martin de Porres

Philippines
Mahfuz Ahmed • Asian Development Bank
Dely Pascual Gapasin • Institute for International Development Partnership Foundation
Agnes Rola • University of the Philippines Los Baños
Leo Sebastian • Philippine Rice Research Institute

South Africa
Moraka Makhura • Development Bank of Southern Africa
Urmilla Bob • University of KwaZulu-Natal

Spain
Mario Giampietro • Universitat Autònoma de Barcelona
Marta Rivera-Ferre • Autonomous University of Barcelona

Sri Lanka
Deborah Bossio • International Water Management Institute
Charlotte de Fraiture • International Water Management Institute
David Molden • International Water Management Institute

Sudan
Balgis M.E. Osman-Elasha • Higher Council for Environment & Natural Resources (HCENR)

Sweden
Martin Wierup • Swedish University of Agricultural Sciences

Switzerland
Felix Bachmann • Swiss College of Agriculture
David Duthie • United Nations Environment Programme
Markus Giger • University of Bern
Ann D. Herbert • International Labour Organization
Angelika Hilbeck • Swiss Federal Institute of Technology
Udo Hoeggel • University of Bern
Hans Hurni • University of Bern
Andreas Klaey • University of Bern
Cordula Ott • University of Bern
Brigitte Portner • University of Bern

Stephan Rist • University of Bern
Urs Scheidegger • Swiss College of Agriculture
Juerg Schneider • State Secretariat for Economic Affairs
Christine Zundel • Research Institute of Organic Agriculture (FiBL)

Taiwan
Mubarik Ali • World Vegetable Center

Tanzania
Aida Cuthbert Isinika • Sokoine University of Agriculture
Rose Rita Kingamkono • Tanzania Commission for Science & Technology

Thailand
Thammarat Koottatep • Asian Institute of Technology

Turkey
Nazimi Acikgoz • Ege University
Hasan Akca • Gaziosmanpasa University
Ahmet Ali Koc • Akdeniz University
Suat Oksuz • Ege University

Uganda
Theresa Sengooba • International Food Policy Research Institute

United Kingdom
Steve Bass • International Institute for Environment and Development
Stephen Biggs • University of East Anglia
Norman Clark • The Open University
Peter Craufurd • University of Reading
Cathy Rozel Farnworth • Independent
Chris Garforth • University of Reading
David Grzywacz • University of Greenwich
Andy Hall • United Nations University – Maastricht
Frances Kimmins • NR International Ltd
Chris D.B. Leakey • University of Plymouth
Karen Lock • London School of Hygiene and Tropical Medicine
Ana Marr • University of Greenwich
Adrienne Martin • University of Greenwich
Ian Maudlin • Centre for Tropical Veterinary Medicine
Nigel Maxted • University of Birmingham
Johanna Pennarz • ITAD
Charlie Riches • University of Greenwich
Peter Robbins • Independent
Geoff Simm • Scottish Agricultural College
Linda Smith • Department for Environment, Food and Rural Affairs (end Mar 2006)
Philip Thornton • International Livestock Research Institute
Jeff Waage • London International Development Centre

United States
Emily Adams • Independent
Elizabeth A. Ainsworth • U.S. Department of Agriculture
Jock Anderson • The World Bank
Patrick Avato • The World Bank
Debbie Barker • International Forum on Globalization
Barbara Best • US Agency for International Development
Regina Birner • International Food Policy Research Policy Institute
David Bouldin • Cornell University

Sandra Brown • Winrock International
Lorna M. Butler • Iowa State University
Kenneth Cassman • University of Nebraska, Lincoln
Gina Castillo • Oxfam America
Medha Chandra • Pesticide Action Network North America
    Regional Center (PANNA)
Joel I. Cohen • Independent
Daniel de la Torre Ugarte • University of Tennessee
Steven Dehmer • University of Minnesota
William E. Easterling • Pennsylvania State University
Kristie L. Ebi • ESS, LLC
Shaun Ferris • Catholic Relief Services
Jorge M. Fonseca • University of Arizona
Constance Gewa • George Mason University
James C. Hanson • University of Maryland
Paul Heisey • U.S. Department of Agriculture
Omololu John Idowu • Cornell University
Marcia Ishii-Eiteman • Pesticide Action Network North America
    Regional Center (PANNA)
R. Cesar Izaurralde • Joint Global Change Research Institute
Moses T.K. Kairo • Florida A&M University
Russ Kruska • International Livestock Research Institute
Andrew D.B. Leakey • University of Illinois
A.J. McDonald • Cornell University
Patrick Meier • Tufts University
Douglas L. Murray • Colorado State University
Clare Narrod • International Food Policy Research Institute
James K. Newman • Iowa State University
Diane Osgood • Business for Social Responsibility
Jonathan Padgham • World Bank

Philip Pardey • University of Minnesota
Ivette Perfecto • University of Michigan
Cameron Pittelkow • Independent
Carl E. Pray • Rutgers University
Laura T. Raynolds • Colorado State University
Robin Reid • Colorado State University
Susan Riha • Cornell University
Claudia Ringler • International Food Policy Research Institute
Steven Rose • U.S. Environmental Protection Agency
Mark Rosegrant • International Food Policy Research Institute
Erika Rosenthal • Center for International Environmental Law
Sara Scherr • Ecoagriculture Partners
Jeremy Schwartzbord • Independent
Matthew Spurlock • University of Massachusetts
Timothy Sulser • International Food Policy Research Institute
Steve Suppan • Institute for Agriculture and Trade Policy
Stan Wood • International Food Policy Research Institute
Angus Wright • California State University; Sacramento
Howard Yana Shapiro • MARS, Inc.
Tingju Zhu • International Food Policy Research Institute

Uruguay
Gustavo Ferreira • Instituto Nacional de Investigación
    Agropecuaria (INIA), Tacuarembó

Zimbabwe
Stephen Twomlow • International Crops Research Institute for
    the Semi-Arid Tropics

Peer Reviewers

Argentina
Stella Navone • Buenos Aires University
Sandra Sharry • Redbio
Victor Trucco • Argentine Association of Producers in Direct Seeding

Australia
Lindsay Falvey • University of Melbourne
Simon Hearn • Australian Centre for International Agricultural Research
Stuart Hill • University of Western Sydney
Geoffrey Lawrence
Gabrielle Persley • Doyle Foundation
Jim Ryan • Science Council, CGIAR
David Vincent • Australian Centre for International Agricultural Research
Sarah Withers • Department of Foreign Affairs and Trade

Austria
Elfriede Fuhrmann • BMLFUW

Bangladesh
Mohamed Ataur Rahman • Homestead Cropping and Ecoagriculture Research Center for Sustainable Rural Development

Belgium
Kevin Akoyi • Vredeseilanden
Helen Holder • Friends of the Earth Europe
Melanie Miller • Independent
Christian Verschueren • CropLife International

Benin
Shellemiah Keya • WARDA
Peter Neuenschwander • IITA

Brazil
Government of Brasil
Francisco Reifschneider • Embrapa

Canada
Saikat Basu • University of Lethbridge
R. Thomas Beach • Agricultural Institute of Canada
Eleanor Boyle • Independent
David Coates • Convention on Biological Diversity
Donald C. Cole • University of Toronto
David Cooper • Convention on Biological Diversity
Howard Eliot • Independent

Harriet Friedman • University of Toronto
Cathy Holtslander • Saskatchewan Organic Directorate
Sophia Huyer • UN Commission on Science and Technology for Development
JoAnn Jaffe • University of Regina
Muffy Koch • Agbios
Lucio Munoz • Independent
Vaclav Smil • University of Manitoba
Iain C. MacGillivray • Canadian International Development Agency
Mary Stockdale • University of British Columbia, Okanagan

Chile
Sarah Gladstone • Independent

Colombia
Andrew Jarvis • Bioversity International/CIAT

Costa Rica
Carlos Araya Fernandez • Universidad Nacional

Denmark
Frands Dolberg • University of Aarhus
Henrik Egelyng • Danish Institute for International Studies (DIIS)
Niels Halberg • University of Aarhus
Høgh Jensen • University of Copenhagen

Egypt
Ayman Abou-Hadid • Agricultural Research Center
Malcolm Beveridge • WorldFish Center
Salah Galal • Ain Shams University

Ethiopia
David Spielman • International Food Policy Research Institute
Wilberforce Kisamba-Mugerwa • International Food Policy Research Institute

Finland
Traci Birge • University of Helsinki
Jukka Peltola • MTT Agricultural Research
Riika Rajalahti • Ministry of Foreign Affairs
Marja-Liisa Tapio-Biström • Ministry of Agriculture and Forestry

France
Louis Aumaitre • EAAP
Jacques Loyat • Ministry of Agriculture
Michèle Tixier-Boichard • Ministry of Higher Education and Research

Sophie Valleix • Centre National de Ressources en Agriculture
   Biologique
Bruno Vindel • Ministère de l'Agriculture et de la Pêche

Germany
Annik Dollacker • Bayer CropScience
Jan van Aken • Greenpeace International

India
Ramesh Chand • NCAP
C.P. Chandrasekhar • Jawaharlal Nehru University
Pradip Dey • Indian Council of Agricultural Research
Nata Duvvury • International Center for Research on Women
Jayati Ghosh • Jawaharlal Nehru University
Indian Council of Agricultural Research
Pramod Joshi • NCAP
Sudhir Kochhar • Indian Council of Agricultural Research
Sushil Kumar • NDRI
Aditya Misra • Project Directorate on Cattle
Suresh Pal • NCAP
V.N. Sharda • Central Soil & Water Conservation Research &
   Training Institute
C. Upendranadh • Institute for Human Development

Indonesia
Suraya Affif • KARSA (Circle for Agrarian and Village Reform)
Russell Dilts • Environmental Services Program
Yemi Katerere • CIFOR
David Raitzer • CIFOR

Iran
Mohammad Abdolahi-Ezzatabadi • Iran Pistachio Research
   Institute
Reza Sedaghat • Iran Pistachio Research Institute
Farhad Saeidi Naeini • Iranian Research Institute of Plant
   Protection

Ireland
Government of Ireland
Charles Merfield • University College Dublin
Sharon Murphy • Department of Agriculture, Fisheries and Food

Italy
Agriculture Department • FAO
Gustavo Anriquez • FAO
Susan Braatz • FAO
Jelle Bruinsma • FAO
Jorge Csirke • FAO
Forestry Department • FAO
Theodor Friedrich • FAO
Peter Gardiner • FAO
Mario Giampietro • National Research Institute on Food and
   Nutrition
Michael Halewood • Bioversity International
Paul Harding • Bioversity International
Toby Hodgkin • Bioversity International
Devra Jarvis • Bioversity International
Timothy Kelley • Science Council, CGIAR
Peter Kenmore • FAO
Yianna Lambrou • FAO
Annie Lane • Bioversity International

Brian Moir • FAO
Piero Morandini • University of Milan
Shivaji Pandey • FAO
Plant Protection Service • FAO
Teri Raney • FAO
Per G. Rudebjer • Bioversity International
Jan Slingenbergh • FAO
Andrea Sonnino • FAO
Clive Stannard • FAO
Henning Steinfeld • FAO
Jeff Tschirley • FAO
Harry van der Wulp • FAO

Japan
Kozo Mayumi • The University of Tokushima

Jordan
Ahmad Abu Awaad
Barakat Abu Irmaileh • University of Jordan
Amad Hijazi

Kenya
Sabrina Barker • United Nations Environment Programme
Christian Borgemeister • International Center for Insect
   Physiology and Ecology
Salif Diop • United Nations Environment Programme
Dennis Garrity • World Agroforestry Center
Mario Herrero • ILRI
Jan Laarman • World Agroforestry Center
Marcus Lee • United Nations Environment Programme
Bernhard Löhr • International Center for Insect Physiology and
   Ecology
Dagmar Mithöfer • International Center for Insect Physiology
   and Ecology
Evans Mwangi • University of Nairobi
Manitra Rakotoarisoa • ILRI
Thomas Randolph • ILRI
Robin Reid • Colorado State University
Karl Rich • ILRI
Fritz Schulthess • International Center for Insect Physiology and
   Ecology
Deborah Scott • ACORD
Nalini Sharma • United Nations Environment Programme
Gemma Shepherd • United Nations Environment Programme
Anna Stabrawa • United Nations Environment Programme

Malaysia
Stephen Hall • WorldFish Center
Froukje Kruijssen • Bioversity International
Li Ching Lim • Third World Network
Percy Sajise • Bioversity International

Madagascar
Xavier Rakotonjanahary • FOFIFA

Mexico
Bruce G. Ferguson • El Colegio de la Frontera Sur
Luis García-Barrios • El Colegio de la Frontera Sur
Rodomiro Ortiz • CIMMYT
Armando Paredes • Consejo Coordinador Empresarial

Netherlands
Huub Loffler • Wageningen University
Juan Lopez Villar • Friends of the Earth International
Rudy Rabbinge • Science Council, CGIAR
Johan C. van Lenteren • IOBC Global
Henk Westhoek • Netherlands Environmental Assessment Agency (MNP)

New Zealand
A. Neil Macgregor • Journal of Organic Systems
Simon Terry • Sustainability Council

Oman
Ahmed Al Rawahi • University of Nizwa

Pakistan
Pakistan Environmental Protection Agency (Pak-EPA)

Philippines
Teodoro Mendoza • University Los Banos
Patrick Safran • Asian Development Bank

Poland
Ursula Soltysiak • AgroBio Test

Qatar
Mohammad Ghanim Abdulla • Supreme Council for Environment and Natural Reserves

Russia
Eugenia Serova • IET, Center AFE

Sri Lanka
Sithara Atapattu • IWMI
Charlotte de Fraiture • IWMI
David Molden • IWMI
Hugh Turral • IWMI

Sweden
Ulf Herrström • Independent
Permilla Malmer • Swedish Biodiversity Center
Martin Wierup • Swedish University of Agricultural Sciences

Switzerland
Else Katrin Bueneman • Swiss Federal Institute of Technology
Adrian Dubock • Syngenta
David Duthie • United Nations Environment Programme
Annabé Louw-Gaume • Institute of Plant Sciences ETHZ
Jeffrey A. McNeely • IUCN-The World Conservation Union

Tanzania
Jamidu Katima • University of Dar es Salaam

Tunisia
Rym Ben Zid • Independent

Turkey
Suat Oksuz • Ege University

Uganda
Kevin Akoyi • Vredeseilanden
Henry Ssali • Kawanda Agricultural Research Institute

United Kingdom
Azra Awan-Hamlyn • CABI
Roy Bateman • Imperial College London
Ken Becker • CABI
Andrew Bennett • Syngenta Foundation
Claire Brown • United Nations Environment Programme
Philip Bubb • United Nations Environment Programme
Matthew Cock • CABI
Jeremy Cooper • University of Greenwich
Janet Cotter • Greenpeace International, Exeter University
Stuart Coupe • Practical Action
Peter Craufurd • Reading University
Sue D'Arcy • Masterfoods UK
UK Department of Environment, Food and Rural Affairs
UK Department for International Development
Barbara Dinham • Pesticide Action Network
Mary Dix • U.S. Department of Agriculture
Carol Ellison • CABI
Les Fairbank • Institute of Grassland & Environmental Research
Cathy Farnworth • Independent
Ioan Fazey • University of Wales
Arabella Fraser • Oxfam UK
Michael Fullen • University of Wolverhampton
John Gowing • Newcastle University
Lesley Hamill • Department for International Development
David Harris • Bangor University
Jim Harvey • Department for International Development
Geoff Hawtin • World Biodiversity Trust
Emma Hennessey • Department of Environment, Food and Rural Affairs
Mark Holderness • CABI
Janice Jiggins • University of Wageningen
Brian Johnson • English Nature
Amir Kassam • Reading University
Deborah Keith • Syngenta
Greg Masters • CABI
Graham Matthews • Imperial College London
Lera Miles • United Nations Environment Programme
Joe Morris • Cranfield University
Patrick Mulvaney • Practical Action
Rebecca Murphy • CABI
Trevor Nicholls • CABI
Clare Oxborrow • Friends of the Earth England, Wales and Northern Ireland
Sam Page • CABI
Helena Paul • EcoNexus
Michel Pimbert • International Institute for Environment and Development
Pete Riley • GM Freeze
Jo Ripley • Independent
Jonathan Robinson • Independent
Flavie Salaun • Department of Environment, Food and Rural Affairs
Peter Sanguinetti • Crop Protection Association
Tracey Scarpello • Institute of Food Research, Norwich
Francis Shaxson • Tropical Agriculture Association
Julian Smith • Central Science Laboratory, York

Janet Stewart • CABI
Margaret M. Stewart • The Springs Foundation
Steve Suppan • Institute for Agriculture and Trade Policy
Harry Swaine • Syngenta
Geoff Tansey • Independent
Dan Taylor • Find your Feet/ UK Food Group
Reyes Tirado • Greenpeace International
Jeff Waage • Imperial College London
Jim Waller • CABI
Christof Walter • Unilever
Hilary Warburton • Practical Action
Stephanie Williamson • Pesticide Action Network, UK

United States
Jonathan Agwe • The World Bank
Kassim Al Khatib • Council for Agricultural Science and
    Technology
Miguel Altieri • University of California, Berkeley
Jock Anderson • The World Bank
Molly Anderson • Winrock International
Michael Arbuckle • The World Bank
Patrick Avato • The World Bank
Catherine Badgley • University of Michigan
Donald C. Beitz • Council for Agricultural Science and
    Technology
Philip L. Bereano • 49th Parallel Biotechnology Consortium
Philip Berger • U.S. Department of Agriculture
Charles Bertsch • U.S. Department of Agriculture
Regina Birner • International Food Policy Research Institute
Jan Bojö • The World Bank
William Boyd • U.S. Department of Agriculture
John M. Bonner • Council for Agricultural Science and
    Technology
Lynn Brown • The World Bank
Michael Brewer • Michigan State University
Marilyn Buford • U.S. Forest Service
Russ Bulluck • U.S. Department of Agriculture
Jim Byrum • Michigan Agribusiness Association
Kitty Cardwell • U.S. Department of Agriculture
Glenn Carpenter • U.S. Department of Agriculture
Janet Carpenter • U.S. Department of Agriculture
Jean-Christophe Carret • The World Bank
Cheryl Christensen • U.S. Department of Agriculture
Harold Coble • U.S. Department of Agriculture
Leonard Condon • US Council for International Business
Daren Coppock • National Association of Wheat Growers
CropLife International
Jonathan Crouch • CIMMYT
Tom Crow • U.S. Forest Service
Ralph Crawford • U.S. Forest Service
Dana Dalrymple • U.S. Agency for International Development
Christopher Delgado • The World Bank
Charles Di Leva • The World Bank
Mary Dix • U.S. Forest Service
Rex Dufour • National Center for Appropriate Technology
Edgar Duskin • Southern Crop Protection Association
Denis Ebodaghe • U.S. Department of Agriculture
Svetlana Edmeades • The World Bank
Indira Ekanayake • The World Bank
Norman Ellstrand • University of California, Riverside
Gershon Feder • The World Bank

Erick Fernandes • The World Bank
Jorge Fernandez-Cornejo • U.S. Department of Agriculture
Steven Finch • U.S. Department of Agriculture
Mary-Ellen Foley • The World Bank
Lucia Fort • The World Bank
Christian Foster • U.S. Department of Agriculture
Bill Freese • Center for Food Safety
John Gowdy • Rensselaer Polytechnic Institute
James T. Greenwood • Biotechnology Industry Association
Peter Gregory • Cornell University
Doug Gurian-Sherman • Union of Concerned Scientists
Jeanette Gurung • Women Organizing for Change in Agriculture
    and Natural Resource Management (WOCAN)
Julie Guthman • University of California at Santa Cruz
Kimberly Halamar • United States Council for International
    Business
Michael Hansen • Consumers Union of US
Karen Hauda • US Patent and Trademark Office
Paul Heisey • U.S. Department of Agriculture
Brian Hill • Pesticide Action Network North America
Kenneth Hinga • U.S. Department of Agriculture
Marcia Ishii-Eiteman • Pesticide Action Network North America
Cheryl Jackson • U.S. Agency for International Development
Gregory Jaffe • Center for Science in the Public Interest
Steven Jaffee • The World Bank
Willem Janssen • The World Bank
Randy Johnson • U.S. Forest Service
Tim Josling • Stanford University
John Keeling • National Potato Council
Kieran Kelleher • The World Bank
Drew Kershen • University of Oklahoma
Nadim Khouri • The World Bank
Lisa Kitinoja • Extension Systems International
Kristie Knoll • Knoll Farms
Jack Kloppenburg • University of Wisconsin
Susan Koehler • U.S. Department of Agriculture
Masami Kojima • The World Bank
Anne Kuriakose • The World Bank
Saul Landau • California Polytechnic, Pomona
Andrew W. LaVigne • American Seed Trade Association
Josette Lewis • U.S. Agency International Development
Jennifer Long • University of Illinois, Chicago
R.S. Loomis • University of California, Davis
Douglas Luster • U.S. Department of Agriculture
Karen Luz • World Wildlife Fund
William Martin • The World Bank
Rachel Massey • Tufts University and University of
    Massachusetts, Lowell
William Masters • Purdue University
Peter Materu • The World Bank
Shaun McKinney • U.S. Department of Agriculture
Rekha Mehra • The World Bank
Stephen Mink • The World Bank
Andrew D. Moore • National Agricultural Aviation Association
Helen Murphy • University of Washington
Douglas L. Murray • Colorado State University
Farah Naim • U.S. Department of Agriculture
John Nash • The World Bank
Gerald Nelson • University of Illinois, Urbana-Champaign
World Nieh • US Forest Service
Nwanze Okidegbe • The World Bank

Craig Osteen • U.S. Department of Agriculture
Susan J. Owens • U.S. Department of Agriculture
Jon Padgham • World Bank
John Parrota • U.S. Forest Service
Mikko Paunio • The World Bank
Eija Pehu • The World Bank
Ivette Perfecto • University of Michigan
Cameron Pittelkow • Pesticide Action Network North America
C.S. Prakash • Ag Bioworld
Catherine Ragasa • The World Bank
Margaret Reeves • Pesticide Action Network North America
Adam Reinhart • U.S. Agency International Development
Peter Riggs • Forum on Democracy & Trade
Claudia Ringler • International Food Policy Research Institute
Jane Rissler • Union of Concerned Scientists
Peter M. Robinson • US Council for International Business
Naomi Roht-Arriaza • University of California Hastings College of Law
Mary Rojas • Chemonics
Jill Roland • U.S. Department of Agriculture
Matt Rooney • U.S. Department of State
Erika Rosenthal • Center for International Environmental Law
Julienne Roux • The World Bank
Phrang Roy • The Christensen Fund
E.C.A. Runge • Council for Agricultural Science and Technology
Marita Sachiko • The World Bank
Marc Safley • U.S. Department of Agriculture
William Saint • The World Bank
Michael Schechtman • U.S. Department of Agriculture

Sara Scherr • Ecoagriculture Partners
Seth Shames • Ecoagriculture Partners
Hope Shand • ETC Group
Animesh Shrivastava • The World Bank
Jimmy W. Smith • The World Bank
John Soluri • Carnegie Mellon University
Meredith Soule • US Agency International Development
Robert Spitzer • U.S. Department of Agriculture
Doreen Stabinsky • College of the Atlantic
Lorann Stallones • Colorado State University
Steven Sweeney • U.S. Department of Agriculture
Peter Tabor • U.S. Department of Agriculture
Greg Traxler • Auburn University
Gwendolyn H. Urey • California Polytechnic, Pomona
Evert Van der Sluis • South Dakota State University
John Vandermeer • University of Michigan
Bea VanHorne • US Forest Service
Jay Vroom • CropLife America
Jennifer Washburn • New America Foundation
Ross Welch • U.S. Department of Agriculture
Sybil Wellstood • U.S. Department of Agriculture
Ford West • The Fertilizer Institute
Pai-Yei Whung • U.S. Department of Agriculture
Keith Wiebe • U.S. Department of Agriculture
David Winickoff • University of California, Berkeley
Barbara Wolff • U.S. Department of Agriculture
Angus Wright • California State University, Sacramento
Stacey Young • U.S. Agency International Development
Cristóbal Zepeda • U.S. Department of Agriculture

Agriculture A linked, dynamic social-ecological system based on the extraction of biological products and services from an ecosystem, innovated and managed by people. It thus includes cropping, animal husbandry, fishing, forestry, biofuel and bioproducts industries, and the production of pharmaceuticals or tissue for transplant in crops and livestock through genetic engineering. It encompasses all stages of production, processing, distribution, marketing, retail, consumption and waste disposal.

Agricultural biodiversity Encompasses the variety and variability of animals, plants and microorganisms necessary to sustain key functions of the agroecosystem, its structure and processes for, and in support of, food production and food security.

Agricultural extension Agricultural extension deals with the creation, transmission and application of knowledge and skills designed to bring desirable behavioral changes among people so that they improve their agricultural vocations and enterprises and, therefore, realize higher incomes and better standards of living.

Agricultural innovation Agricultural innovation is a socially constructed process. Innovation is the result of the interaction of a multitude of actors, agents and stakeholders within particular institutional contexts. If agricultural research and extension are important to agricultural innovation, so are markets, systems of government, relations along entire value chains, social norms, and, in general, a host of factors that create the incentives for a farmer to decide to change the way in which he or she works, and that reward or frustrate his or her decision.

Agricultural population The agricultural population is defined as all persons depending for their livelihood on agriculture, hunting, fishing or forestry. This estimate comprises all persons actively engaged in agriculture and their non-working dependants.

Agricultural subsidies Agricultural subsidies can take many forms, but a common feature is an economic transfer, often in direct cash form, from government to farmers. These transfers may aim to reduce the costs of production in the form of an input subsidy, e.g., for inorganic fertilizers or pesticides, or to make up the difference between the actual market price for farm output and a higher guaranteed price. Subsidies shield sectors or products from international competition.

Agricultural waste Farming wastes, including runoff and leaching of pesticides and fertilizers, erosion and dust from plowing, improper disposal of animal manure and carcasses, crop residues and debris.

Agroecological Zone A geographically delimited area with similar climatic and ecological characteristics suitable for specific agricultural uses.

Agroecology The science of applying ecological concepts and principles to the design and management of sustainable agroecosystems. It includes the study of the ecological processes in farming systems and processes such as: nutrient cycling, carbon cycling/sequestration, water cycling, food chains within and between trophic groups (microbes to top predators), lifecycles, herbivore/predator/prey/host interactions, pollination, etc. Agroecological functions are generally maximized when there is high species diversity/perennial forest-like habitats.

Agroecosystem A biological and biophysical natural resource system managed by humans for the primary purpose of producing food as well as other socially valuable nonfood goods and environmental services. Agroecosystem function can be enhanced by increasing the planned biodiversity (mixed species and mosaics), which creates niches for unplanned biodiversity.

Agroforestry A dynamic, ecologically based, natural resources management system that through the integration of trees in farms and in the landscape diversifies and sustains production for increased social, economic and environmental benefits for land users at all levels. Agroforestry focuses on the wide range of work with trees grown on farms and in rural landscapes. Among these are fertilizer trees for land regeneration, soil health and food security; fruit trees for nutrition; fodder trees that improve smallholder livestock production; timber and fuelwood trees for shelter and energy; medicinal trees to combat disease; and trees that produce gums, resins or latex products. Many of these trees are multipurpose, providing a range of social, economic and environmental benefits.

AKST Agricultural Knowledge, Science and Technology (AKST) is a term encompassing the ways and means used to practice the different types of agricultural activities, and including both formal and informal knowledge and technology.

Alien Species A species occurring in an area outside of its historically known natural range as a result of intentional or accidental dispersal by human activities. Also referred to as introduced species or exotic species.

Aquaculture The farming of aquatic organisms in inland and coastal areas, involving intervention in the rearing process to enhance production and the individual or corporate ownership of the stock being cultivated. Aquaculture practiced in a marine environment is called mariculture.

Average Rate of Return Average rate of return takes the whole expenditure as given and calculates the rate of return to the global set of expenditures. It indicates whether or not the entire investment package was successful, but it does not indicate whether the allocation of resources between investment components was optimal.

Biodiversity The variability among living organisms from all sources including, inter alia, terrestrial, marine and other aquatic ecosystems and the ecological complexes of which they are part; including diversity within species and gene diversity among species, between species and of ecosystems.

Bioelectricity Electricity derived from the combustion of biomass, either directly or co-fired with fossil fuels such as coal and natural gas. Higher levels of conversion efficiency can be attained when biomass is gasified before combustion.

Bioenergy (biomass energy) Bioenergy is comprised of bioelectricity, bioheat and biofuels. Such energy carriers can be produced from energy crops (e.g. sugar cane, maize, oil palm), natural vegetation (e.g. woods, grasses) and organic wastes and residues (e.g. from forestry and agriculture). Bioenergy refers also to the direct combustion of biomass, mostly for heating and cooking purposes.

Biofuel Liquid fuels derived from biomass and predominantly used in transportation. The dominant biofuels are ethanol and biodiesel. Ethanol is produced by fermenting starch contained in plants such as sugar cane, sugar beet, maize, cassava, sweet sorghum or beetroot. Biodiesel is typically produced through a chemical process called trans-esterification, whereby oily biomass such as rapeseed, soybeans, palm oil, jatropha seeds, waste cooking oils or vegetable oils is combined with methanol to form methyl esters (sometimes called "fatty acid methyl ester" or FAME).

Bioheat Heat produced from the combustion of biomass, mostly as industrial process heat and heating for buildings.

Biological Control The use of living organisms as control agents for pests, (arthropods, nematodes, mammals, weeds and pathogens) in agriculture. There are three types of biological control:

Conservation biocontrol - the protection and encouragement of local natural enemy populations by crop and habitat management measures that enhance their survival, efficiency and growth.

Augmentative biocontrol - the release of natural enemies into crops to suppress specific populations of pests over one or a few generations, often involving the mass production and regular release of natural enemies.

Classical biocontrol - the local introduction of new species of natural enemies with the intention that they establish and build populations that suppress particular pests, often introduced alien pests to which they are specific.

Biological Resources Include genetic resources, organisms or parts thereof, populations, or any other biotic component of ecosystems with actual or potential use or value for humanity.

Biotechnology The IAASTD definition of biotechnology is based on that in the Convention on Biological Diversity and the Cartagena Protocol on Biosafety. It is a broad term embracing the manipulation of living organisms and spans the large range of activities from conventional techniques for fermentation and plant and animal breeding to recent innovations in tissue culture, irradiation, genomics and marker-assisted breeding (MAB) or marker assisted selection (MAS) to augment natural breeding. Some of the latest biotechnologies, called 'modern biotechnology', include the use of *in vitro* modified DNA or RNA and the fusion of cells from different taxonomic families, techniques that overcome natural physiological reproductive or recombination barriers.

Biosafety Referring to the avoidance of risk to human health and safety, and to the conservation of the environment, as a result of the use for research and commerce of infectious or genetically modified organisms.

Blue Water The water in rivers, lakes, reservoirs, ponds and aquifers. Dryland production only uses green water, while irrigated production uses blue water in addition to green water.

BLCAs Brookered Long-term Contractual Arrangements (BLCAs) are institutional arrangements often involving a farmer cooperative, or a private commercial, parastatal or a state trading enterprise and a package (inputs, services, credit, knowledge) that allows small-scale farmers to engage in the production of a marketable commodity, such as cocoa or other product that farmers cannot easily sell elsewhere.

Catchment An area that collects and drains rainwater.

Capacity Development Any action or process which assists individuals, groups, organizations and communities in strengthening or developing their resources.

Capture Fisheries The sum (or range) of all activities to harvest a given fish resource from the 'wild'. It may refer to the location (e.g., Morocco, Gearges Bank), the target resource (e.g., hake), the technology used (e.g., trawl or beach seine), the social characteristics (e.g., artisanal, industrial), the purpose (e.g., [commercial, subsistence, or recreational]) as well as the season (e.g., winter).

Carbon Sequestration The process that removes carbon dioxide from the atmosphere.

Cellulosic Ethanol Next generation biofuel that allows converting not only glucose but also cellulose and hemi-cellulose—the main building blocks of most biomass—into ethanol, usually using acid-based catalysis or enzyme-based reactions to break down plant fibers into sugar, which is then fermented into ethanol.

Climate Change Refers to a statistically significant variation in either the mean state of the climate or in its variability, persisting for an extended period (typically decades or longer). Climate change may be due to natural internal processes or external forcing, or to persistent anthropogenic changes in the composition of the atmosphere or in land use.

Clone A group of genetically identical cells or individuals that are all derived from one selected individual by vegetative propagation or by asexual reproduction, breeding of completely inbred organisms, or forming genetically identical organisms by nuclear transplantation.

Commercialization The process of increasing the share of income that is earned in cash (e.g., wage income, surplus production for marketing) and reducing the share that is earned in kind (e.g., growing food for consumption by the same household).

Cultivar A cultivated variety, a population of plants within a species of plant. Each cultivar or variety is genetically different.

Deforestation The action or process of changing forest land to non-forested land uses.

Degradation The result of processes that alter the ecological characteristics of terrestrial or aquatic (agro)ecosystems so that the net services that they provide are reduced. Continued degradation leads to zero or negative economic agricultural productivity.

For loss of *land* in quantitative or qualitative ways, the term *degradation* is used. For water resources rendered unavailable for agricultural and nonagricultural uses, we employ the terms *depletion* and *pollution. Soil* degradation refers to the processes that reduce the capacity of the soil to support agriculture.

Desertification Land degradation in drylands resulting from various factors, including climatic variations and human activities.

Domesticated or Cultivated Species Species in which the evolutionary process has been influenced by humans to meet their needs.

Domestication The process to accustom animals to live with people as well as to selectively cultivate plants or raise animals in order to increase their suitability and compatibility to human requirements.

Driver Any natural or human-induced factor that directly or indirectly causes a change in a system.

Driver, direct A driver that unequivocally influences ecosystem processes and can therefore be identified and measured to different degrees of accuracy.

Driver, endogenous A driver whose magnitude can be influenced by the decision-maker. The endogenous or exogenous characteristic of a driver depends on the organizational scale. Some drivers (e.g., prices) are exogenous to a decision-maker at one level (a farmer) but endogenous at other levels (the nation-state).

Driver, exogenous A driver that cannot be altered by the decision-maker.

Driver, indirect A driver that operates by altering the level or rate of change of one or more direct drivers.

Ecoagriculture A management approach that provides fair balance between production of food, feed, fuel, fiber, and biodiversity conservation or protection of the ecosystem.

Ecological Pest Management (EPM) A strategy to manage pests that focuses on strengthening the health and resilience of the entire agro-ecosystem. EPM relies on scientific advances in the ecological and entomological fields of population dynamics, community and landscape ecology, multi-trophic interactions, and plant and habitat diversity.

Economic Rate of Return The net benefits to all members of society as a percentage of cost, taking into account externalities and other market imperfections.

Ecosystem A dynamic complex of plant, animal, and microorganism communities and their nonliving environment interacting as a functional unit.

Ecosystem Approach A strategy for the integrated management of land, water, and living resources that promotes conservation and sustainable use in an equitable way.

An ecosystem approach is based on the application of appropriate scientific methodologies focused on levels of biological organization, which encompass the essential structure, processes, functions, and interactions among organisms and their environment. It recognizes that humans, with their cultural diversity, are an integral component and managers of many ecosystems.

Ecosystem Function An intrinsic ecosystem characteristic related to the set of conditions and processes whereby an ecosystem maintains its integrity (such as primary productivity, food chain biogeochemical cycles). Ecosystem functions include such processes as decomposition, production, pollination, predation, parasitism, nutrient cycling, and fluxes of nutrients and energy.

Ecosystem Management An approach to maintaining or restoring the composition, structure, function, and delivery of services of natural and modified ecosystems for the goal of achieving sustainability. It is based on an adaptive, collaboratively developed vision of desired future conditions that integrates ecological, socioeconomic, and institutional perspectives, applied within a geographic framework, and defined primarily by natural ecological boundaries.

Ecosystem Properties The size, biodiversity, stability, degree of organization, internal exchanges of material and energy among different pools, and other properties that characterize an ecosystem.

Ecosystem Services The benefits people obtain from ecosystems. These include provisioning services such as food and water; regulating services such as flood and disease control; cultural services such as spiritual, recreational, and cultural benefits; and supporting services such as nutrient cycling that maintain the conditions for life on Earth. The concept "ecosystem goods and services" is synonymous with ecosystem services.

Ecosystem Stability A description of the dynamic properties of an ecosystem. An ecosystem is considered stable if it returns to its original state shortly after a perturbation (resilience), exhibits low temporal variability (constancy), or does not change dramatically in the face of a perturbation (resistance).

Eutrophication Excessive enrichment of waters with nutrients, and the associated adverse biological effects.

Ex-ante The analysis of the effects of a policy or a project based only on information available before the policy or project is undertaken.

Ex-post The analysis of the effects of a policy or project based on information available after the policy or project has been implemented and its performance is observed.

Ex-situ Conservation The conservation of components of biological diversity outside their natural habitats.

Externalities Effects of a person's or firm's activities on others which are not compensated. Externalities can either hurt or benefit others—they can be negative or positive. One negative externality arises when a company pollutes the local environment to produce its goods and does not compensate the negatively affected local residents. Positive externalities can be produced through primary education—which benefits not only primary school students but also society at large. Governments can reduce negative externalities by regulating and taxing goods with negative externalities. Governments can increase positive externalities by subsidizing goods with positive externalities or by directly providing those goods.

Fallow Cropland left idle from harvest to planting or during the growing season.

Farmer-led Participatory Plant Breeding Researchers and/or development workers interact with farmer-controlled, managed and executed PPB activities, and build on farmers' own varietal development and seed systems.

Feminization The increase in the share of women in an activity, sector or process.

Fishery Generally, a fishery is an activity leading to harvesting of fish. It may involve capture of wild fish or the raising of fish through aquaculture.

Food Security Food security exists when all people of a given spatial unit, at all times, have physical and economic access to sufficient, safe and nutritious food to meet their dietary needs and food preferences for an active and healthy life, and that is obtained in a socially acceptable and ecologically sustainable manner.

Food Sovereignty The right of peoples and sovereign states to democratically determine their own agricultural and food policies.

Food System A food system encompasses the whole range of food production and consumption activities. The food system includes farm input supply, farm production, food processing, wholesale and retail distribution, marketing, and consumption.

Forestry The human utilization of a piece of forest for a certain purpose, such as timber or recreation.

Forest Systems Forest systems are lands dominated by trees; they are often used for timber, fuelwood, and non-wood forest products.

Gender Refers to the socially constructed roles and behaviors of, and relations between, men and women, as opposed to sex, which refers to biological differences. Societies assign specific entitlements, responsibilities and values to men and women of different social strata and subgroups.

Worldwide, systems of relation between men and women tend to disadvantage women, within the family as well as in public life. Like the hierarchical framework of a society, gender roles and relations vary according to context and are constantly subject to changes.

Genetic Engineering Modifying genotype, and hence phenotype, by transgenesis.

Genetic Material Any material of plant, animal, microbial or other origin containing functional units of heredity.

Genomics The research strategy that uses molecular characterization and cloning of whole genomes to understand the structure, function and evolution of genes and to answer fundamental biological questions.

Globalization Increasing interlinking of political, economic, institutional, social, cultural, technical, and ecological issues at the global level.

GMO (Genetically Modified Organism) An organism in which the genetic material has been altered anthropogenically by means of gene or cell technologies.

Governance The framework of social and economic systems and legal and political structures through which humanity manages itself. In general, governance comprises the traditions, institutions and processes that determine how power is exercised, how citizens are given a voice, and how decisions are made on issues of public concern.

Global Environmental Governance The global biosphere behaves as a single system, where the environmental impacts of each nation ultimately affect the whole. That makes a coordinated response from the community of nations a necessity for reversing today's environmental decline.

Global Warming Refers to an increase in the globally-averaged surface temperature in response to the increase of well-mixed greenhouse gases, particularly CO_2.

Global Warming Potential An index, describing the radiative characteristics of well-mixed greenhouse gases, that represents the combined effect of the differing times these gases remain in the atmosphere and their relative effectiveness in absorbing outgoing infrared radiation. This index approximates the time-integrated warming effect of a unit mass of a given greenhouse gas in today's atmosphere, relative to that of carbon dioxide.

Green Revolution An aggressive effort since 1950 in which agricultural researchers applied scientific principles of genetics and breeding to improve crops grown primarily in less-developed countries. The effort typically was accompanied by collateral investments to develop or strengthen the delivery of extension services, production inputs and markets and develop physical infrastructures such as roads and irrigation.

Green Water Green water refers to the water that comes from precipitation and is stored in unsaturated soil. Green water is typically taken up by plants as evapotranspiration.

Ground Water Water stored underground in rock crevices and in the pores of geologic materials that make up the Earth's crust. The upper surface of the saturate zone is called the water table.

Growth Rate The change (increase, decrease, or no change) in an indicator over a period of time, expressed as a percentage of the indicator at the start of the period. Growth rates contain several sets of information. The first is whether there is any change at all; the second is what direction the change is going in (increasing or decreasing); and the third is how rapidly that change is occurring.

Habitat Area occupied by and supporting living organisms. It is also used to mean the environmental attributes required by a particular species or its ecological niche.

Hazard A potentially damaging physical event, phenomenon and/or human activity, which my cause injury, property damage, social and economic disruption or environmental degradation.

Hazards can include latent conditions that may represent future threats and can have different origins.

Household All the persons, kin and non-kin, who live in the same or in a series of related dwellings and who share income, expenses and daily subsistence tasks. A basic unit for socio-cultural and economic analysis, a household may consist of persons (sometimes one but generally two or more) living together and jointly making provision for food or other essential elements of the livelihood.

Industrial Agriculture Form of agriculture that is capital-intensive, substituting machinery and purchased inputs for human and animal labor.

Infrastructure The facilities, structures, and associated equipment and services that facilitate the flows of goods and services between individuals, firms, and governments. It includes public utilities (electric power, telecommunications, water supply, sanitation and sewerage, and waste disposal); public works (irrigation systems, schools, housing, and hospitals); transport services (roads, railways, ports, waterways, and airports); and R&D facilities.

Innovation The use of a new idea, social process or institutional arrangement, material, or technology to change an activity, development, good, or service or the way goods and services are produced, distributed, or disposed of.

Innovation system Institutions, enterprises, and individuals that together demand and supply information and technology, and the rules and mechanisms by which these different agents interact.

In recent development discourse agricultural innovation is conceptualized as part and parcel of social and ecological organization, drawing on disciplinary evidence and understanding of how knowledge is generated and innovations occur.

In-situ Conservation The conservation of ecosystems and natural habitats and the maintenance and recovery of viable populations of species in their natural habitats and surroundings and, in the case of domesticated or cultivated species, in the surroundings where they have developed their distinctive properties and were managed by local groups of farmers, fishers or foresters.

Institutions The rules, norms and procedures that guide how people within societies live, work, and interact with each other. Formal institutions are written or codified rules, norms and procedures. Examples of formal institutions are the Constitution, the judiciary laws, the organized market, and property rights. Informal institutions are rules governed by social and behavioral norms of the society, family, or community. Cf. Organization.

Integrated Approaches Approaches that search for the best use of the functional relations among living organisms in relation to the environment without excluding the use of external inputs. Integrated approaches aim at the achievement of multiple goals (productivity increase, environmental sustainability and social welfare) using a variety of methods.

Integrated Assessment A method of analysis that combines results and models from the physical, biological, economic, and social sciences, and the interactions between these components in a consistent framework to evaluate the status and the consequences of environmental change and the policy responses to it.

Integrated Natural Resources Management (INRM) An approach that integrates research of different types of natural resources into stakeholder-driven processes of adaptive management and innovation to improve livelihoods, agroecosystem resilience, agricultural productivity and environmental services at community, eco-regional and global scales of intervention and impact. INRM thus aims to help to solve complex real-world problems affecting natural resources in agroecosystems.

Integrated Pest Management The procedure of integrating and applying practical management methods to manage insect populations so as to keep pest species from reaching damaging levels while avoiding or minimizing the potentially harmful effects of pest management measures on humans, non-target species, and the environment. IPM tends to incorporate assessment methods to guide management decisions.

Intellectual Property Rights (IPRs) Legal rights granted by governmental authorities to control and reward certain products of human intellectual effort and ingenuity.

Internal Rate of Return The discount rate that sets the net present value of the stream of the net benefits equal to zero. The internal rate of return may have multiple values when the stream of net benefits alternates from negative to positive more than once.

International Dollars Agricultural R&D investments in local currency units have been converted into international dollars by deflating the local currency amounts with each country's inflation ration (GDP deflator) of base year 2000. Next, they were converted to US dollars with a 2000 purchasing power parity (PPP) index. PPPs are synthetic exchange rates used to reflect the purchasing power of currencies.

Knowledge The way people understand the world, the way in which they interpret and apply meaning to their experiences. Knowledge is not about the discovery of some finale objective 'truth' but about the grasping of subjective culturally-conditioned products emerging from complex and ongoing processes involving selection, rejection, creation, development and transformation of information. These processes, and hence knowledge, are inextricably linked to the social, environmental and institutional context within which they are found.

Scientific knowledge: Knowledge that has been legitimized and validated by a formalized process of data gathering, analysis and documentation.

Explicit knowledge: Information about knowledge that has been or can be articulated, codified, and stored and exchanged. The most common forms of explicit knowledge are manuals, documents, procedures, cultural artifacts and stories. The information about explicit knowledge also can be audio-visual. Works of art and product design can be seen as other forms of explicit knowledge where human skills, motives and knowledge are externalized.

Empirical knowledge: Knowledge derived from and constituted in interaction with a person's environment. Modern communication and information technologies, and scientific instrumentation, can extend the 'empirical environment' in which empirical knowledge is generated.

Local knowledge: The knowledge that is constituted in a given culture or society.

Traditional (ecological) knowledge: The cumulative body of knowledge, practices, and beliefs evolved by adaptive processes and handed down through generations. It may not be indigenous or local, but it is distinguished by the way in which it is acquired and used, through the social process of learning and sharing knowledge.

Knowledge Management A systematic discipline of policies, processes, and activities for the management of all processes of knowledge generation, codification, application and sharing of information about knowledge.

Knowledge Society A society in which the production and dissemination of scientific information and knowledge function well, and in which the transmission and use of valuable experiential knowledge is optimized; a society in which the information of those with experiential knowledge is used together with that of scientific and technical experts to inform decision-making.

Land Cover The physical coverage of land, usually expressed in terms of vegetation cover or lack of it. Influenced by but non synonymous with land use.

Land Degradation The reduction in the capability of the land to produce benefits from a particular land use under a specific form of land management.

Landscape An area of land that contains a mosaic of ecosystems, including human-dominated ecosystems. The term cultural landscape is often used when referring to landscapes containing significant human populations.

Land Tenure The relationship, whether legally or customarily defined, among people, as individuals or groups, with respect to land and associated natural resources (water, trees, minerals, wildlife, and so on).

   Rules of tenure define how property rights in land are to be allocated within societies. Land tenure systems determine who can use what resources for how long, and under what conditions.

Land Use The human utilization of a piece of land for a certain purpose (such as irrigated agriculture or recreation). Land use is influenced by, but not synonymous with, land cover.

Leguminous Cultivated or spontaneous plants which fix atmospheric nitrogen.

Malnutrition Failure to achieve nutrient requirements, which can impair physical and/or mental health. It may result from consuming too little food or a shortage or imbalance of key nutrients (eg, micronutrient deficiencies or excess consumption of refined sugar and fat).

Marginal Rates of Return Calculates the returns to the last dollar invested on a certain activity. It is usually estimated through econometric estimation.

Marker Assisted Selection (MAS) The use of DNA markers to improve response to selection in a population. The markers will be closely linked to one or more target loci, which may often be quantitative trait loci.

Minimum Tillage The least amount possible of cultivation or soil disturbance done to prepare a suitable seedbed. The main purposes of minimum tillage are to reduce tillage energy consumption, to conserve moisture, and to retain plant cover to minimize erosion.

Model A simplified representation of reality used to simulate a process, understand a situation, predict an outcome or analyze a problem. A model can be viewed as a selective approximation, which by elimination of incidental detail, allows hypothesized or quantified aspects of the real world to appear manipulated or tested.

Multifunctionality In IAASTD, multifunctionality is used solely to express the inescapable interconnectedness of agriculture's different roles and functions. The concept of multifunctionality recognizes agriculture as a multi-output activity producing not only commodities (food, feed, fibers, agrofuels, medicinal products and ornamentals), but also non-commodity outputs such as environmental services, landscape amenities and cultural heritages (See Global SDM Text Box).

Natural Resources Management Includes all functions and services of nature that are directly or indirectly significant to humankind, i.e. economic functions, as well as other cultural and ecological functions or social services that are not taken into account in economic models or not entirely known.

Nanotechnology The engineering of functional systems at the atomic or molecular scale.

Net Present Value (NPV) Net present value is used to analyze the profitability of an investment or project, representing the difference between the discounted present value of benefits and the discounted present value of costs. If NPV of a prospective project is positive, then the project should be accepted. The analysis of NPV is sensitive to the reliability of future cash inflows that an investment or project will yield.

No-Till Planting without tillage. In most systems, planter-mounted coulters till a narrow seedbed assisting in the placement of fertilizer and seed. The tillage effect on weed control is replaced by herbicide use.

Obesity A chronic physical condition characterized by too much body fat, which results in higher risk for health problems such as high blood pressure, high blood cholesterol, diabetes, heart disease and stroke. Commonly it is defined as a Body Mass Index (BMI) equal to or more than 30, while overweight is equal to or more than 25. The BMI is an index of weight-for-height and is defined as the weight in kilograms divided by the square of the height in meters (kg/m^2).

Organic Agriculture An ecological production management system that promotes and enhances biological cycles and soil biological activity. It is based on minimal use of off-farm inputs and on management practices that restore, maintain and enhance ecological harmony.

Organization Organizations can be formal or informal. Examples of organizations are government agencies (e.g., police force, ministries, etc.), administrative bodies (e.g., local government), non governmental organizations, associations (e.g., farmers' associations) and private companies (firms). Cf. with Institutions.

Orphan Crops Crops such as tef, finger millet, yam, roots and tubers that tend to be regionally or locally important for income and nutrition, but which are not traded globally and receive minimal attention by research networks.

Participatory Development A process that involves people (population groups, organizations, associations, political parties) actively and significantly in all decisions affecting their lives.

Participatory Domestication The process of domestication that involves agriculturalists and other community members actively and significantly in making decisions, taking action and sharing benefits.

Participatory Plant Breeding (PPB) Involvement of a range of actors, including scientists, farmers, consumers, extension agents, vendors, processors and other industry stakeholders—as well as farmer and community-based organizations and non-government organization (NGOs) in plant breeding research and development.

Participatory Varietal Selection (PVS) A process by which farmers and other stakeholders along the food chain are involved with researchers in the selection of varieties from formal and farmer-based collections and trials, to determine which are best suited to their own agroeco-systems' needs, uses and preferences, and which should go ahead for finishing, wider release and dissemination. The information gathered may in turn be fed back into formal-led breeding programs.

Pesticide A toxic chemical or biological product that kills organisms (e.g., insecticides, fungicides, weedicides, ro-denticides).

Poverty There are many definitions of poverty.

Absolute Poverty: According to a UN declaration that resulted from the World Summit on Social Development in 1995, absolute poverty is a condition characterized by severe deprivation of basic human needs, including food, safe drinking water, sanitation facilities, health, shelter, education and information. It depends not only on income but also on access to services.

Dimensions of Poverty: The individual and social characteristics of poverty such as lack of access to health and education, powerlessness or lack of dignity. Such aspects of deprivation experienced by the individual or group are not captured by measures of income or expenditure.

Extreme Poverty: Persons who fall below the defined poverty line of US$1 income per day. The measure is converted into local currencies using purchasing power parity (PPP) exchange rates. Other definitions of this concept have identified minimum subsistence requirements, the denial of basic human rights or the experience of exclusion.

Poverty Line: A minimum requirement of welfare, usually defined in relation to income or expenditure, used to identify the poor. Individuals or households with incomes or expenditure below the poverty line are poor. Those with incomes or expenditure equal to or above the line are not poor. It is common practice to draw more than one poverty line to distinguish different categories of poor, for example, the extreme poor.

Private Rate of Return The gain in net revenue to the private firm/business divided by the cost of an investment expressed in percentage.

Processes A series of actions, motions, occurrences, a method, mode, or operation, whereby a result or effect is produced.

Production Technology All methods that farmers, market agents and consumers use to cultivate, harvest, store, process, handle, transport and prepare food crops, cash crops, livestock, etc., for consumption.

Protected Area A geographically defined area which is designated or regulated and managed to achieve specific conservation objectives as defined by society.

Public Goods A good or service in which the benefit received by any one party does not diminish the availability of the benefits to others, and/or where access to the good cannot be restricted. Public goods have the properties of non-rivalry in consumption and non-excludability.

Public R&D Investment Includes R&D investments done by government agencies, nonprofit institutions, and higher-education agencies. It excludes the private for-profit enterprises.

Research and Development (R&D) Organizational strategies and methods used by research and extension program to conduct their work including scientific procedures, organizational modes, institutional strategies, interdisciplinary team research, etc.

Scenario A plausible and often simplified description of how the future may develop based on explicit and coherent and internally consistent set of assumptions about key driving forces (e.g., rate of technology change, prices) and relationships. Scenarios are neither predictions nor projections and sometimes may be based on a "narrative storyline". Scenarios may be derived from projections but are often based on additional information from other sources.

Science, Technology and Innovation Includes all forms of useful knowledge (codified and tacit) derived from diverse branches of learning and practice, ranging from basic scientific research to engineering to local knowledge. It also includes the policies used to promote scientific advance, technology development, and the commercialization of products, as well as the associated institutional innovations. *Science* refers to both basic and applied sciences. *Technology* refers to the application of science, engineering, and other fields, such as medicine. *Innovation* includes all of the processes, including business activities that bring a technology to market.

Shifting Cultivation Found mainly in the tropics, especially in humid and subhumid regions. There are different kinds; for example, in some cases a settlement is permanent, but certain fields are fallowed and cropped alternately ('rotational agriculture'). In other cases, new land is cleared when the old is no longer productive.

Slash and Burn Agriculture A pattern of agriculture in which existing vegetation is cleared and burned to provide space and nutrients for cropping.

Social Rate of Return The gain to society of a project or investment in net revenue divided by cost of the investment, expressed by percentage.

Soil and Water Conservation (SWC) A combination of appropriate technology and successful approach. Technologies promote the sustainable use of agricultural soils by minimizing soil erosion, maintaining and/or enhancing soil properties, managing water, and controlling temperature. Approaches explain the ways and means which are used to realize SWC in a given ecological and socio-economic environment.

Soil Erosion The detachment and movement of soil from the land surface by wind and water in conditions influenced by human activities.

Soil Function Any service, role, or task that a soil performs, especially: (a) sustaining biological activity, diversity, and productivity; (b) regulating and partitioning water and solute flow; (c) filtering, buffering, degrading, and detoxifying potential pollutants; (d) storing and cycling nutrients; (e) providing support for buildings and other structures and to protect archaeological treasures.

Staple Food (Crops) Food that is eaten as daily diet.

Soil Quality The capacity of a specific kind of soil to function, within natural or managed ecosystem boundaries, to sustain plant and animal productivity, maintain or enhance water and air quality, and support human health and habitation. In short, the capacity of the soil to function.

Subsidy Transfer of resources to an entity, which either reduces the operating costs or increases the revenues of such entity for the purpose of achieving some objective.

Subsistence Agriculture Agriculture carried out for the use of the individual person or their family with few or no outputs available for sale.

Sustainable Development Development that meets the needs of the present without compromising the ability of future generations to meet their own needs.

Sustainable Land Management (SLM) A system of technologies and/or planning that aims to integrate ecological with socioeconomic and political principles in the management of land for agricultural and other purposes to achieve intra- and intergenerational equity.

Sustainable Use of Natural Resources Natural resource use is sustainable if specific types of use in a particular ecosystem are considered reasonable in the light of both the internal and the external perspective on natural resources. "Reasonable" in this context means that all actors agree that resource use fulfils productive, physical, and cultural functions in ways that will meet the long-term needs of the affected population.

Technology Transfer The broad set of deliberate and spontaneous processes that give rise to the exchange and dissemination of information and technologies among different stakeholders. As a generic concept, the term is used to encompass both diffusion of technologies and technological cooperation across and within countries.

Terms of Trade The *international terms* of trade measures a relationship between the prices of exports and the prices of imports, this being known strictly as the barter terms of trade. In this sense, deterioration in the terms of trade could have resulted if unit prices of exports had risen less than unit prices for imports. The *inter-sectoral terms of trade* refers to the terms of trade between sectors of the economy, e.g., rural and urban, agriculture and industry.

Total Factor Productivity A measure of the increase in total output which is not accounted for by increases in total inputs. The total factor productivity index is computed as the ratio of an index of aggregate output to an index of aggregate inputs.

Tradeoff Management choices that intentionally or otherwise change the type, magnitude, and relative mix of services provided by ecosystems.

Transgene An isolated gene sequence used to transform an organism. Often, but not always, the transgene has been derived from a different species than that of the recipient.

Transgenic An organism that has incorporated a functional foreign gene through recombinant DNA technology. The novel gene exists in all of its cells and is passed through to progeny.

Undernourishment Food intake that is continuously inadequate to meet dietary energy requirement.

Undernutrition The result of food intake that is insufficient to meet dietary energy requirements continuously, poor absorption, and/or poor biological use of nutrients consumed.

Urban and Peri-Urban Agriculture Agriculture occurring within and surrounding the boundaries of cities throughout the world and includes crop and livestock production, fisheries and forestry, as well as the ecological services they provide. Often multiple farming and gardening systems exist in and near a single city.

Value Chain A set of value-adding activities through which a product passes from the initial production or design stage to final delivery to the consumer.

Virtual Water The volume of water used to produce a commodity. The adjective 'virtual' refers to the fact that most of the water used to produce a product is not contained in the product. In accounting virtual water flows we keep track of which parts of these flows refer to green, blue and grey water, respectively.

The real-water content of products is generally negligible if compared to the virtual-water content.

Waste Water 'Grey' water that has been used in homes, agriculture, industries and businesses that is not for reuse unless it is treated.

Watershed The area which supplies water by surface and subsurface flow from precipitation to a given point in the drainage system.

Watershed Management Use, regulation and treatment of water and land resources of a watershed to accomplish stated objectives.

Water Productivity An efficiency term quantified as a ration of product output (goods and services) over water input.

Expressions of water productivity. Three major expressions of water productivity can be identified: (1) the amount of carbon gain per unit of water transpired by the leaf or by the canopy (photosynthetic water productivity); (2) the amount of water transpired by the crop (biomass water productivity); or (3) the yield obtained per unit amount of water transpired by the crop (yield water productivity).

Agricultural water productivity relates net benefits gained through the use of water in crop, forestry, fishery, livestock and mixed agricultural systems. In its broadest sense, it reflects the objectives of producing more food, income, livelihood and ecological benefits at less social and environmental cost per unit of water in agriculture.

Physical water productivity relates agricultural production to water use—more crop per drop. Water use is expressed either in terms of delivery to a use, or depletion by a use through evapotranspiration, pollution, or directing water to a sink where it cannot be reused. Improving physical water productivity is important to reduce future water needs in agriculture.

Economic water productivity relates the value of agricultural production to agricultural water use. A holistic assessment should account for the benefits and costs of water, including less tangible livelihood benefits, but this is rarely done. Improving economic water productivity is important for economic growth and poverty reduction.

Annex D
Acronyms, Abbreviations and Units

ABS	Access and Benefit Sharing
AEZ	agricultural zone
AIDS	Acquired immune deficiency syndrome
AFS	agroforestry systems
AFTP	Agroforestry Tree Product
AKST	Agricultural knowledge, science and technology
ARC	Agricultural Research Council
ARI	agricultural research institute
ASARECA	Association for Strengthening of Agricultural Research in Eastern and Central Africa
ASB	Alternatives to slash and burn
AST	Agricultural science and technology
ASTI	Agricultural Science and Technology Indicators
billion	one thousand million
BIT	Bilateral Investment Treaty
BLCA	brokered long-term contractual arrangement
BNF	Biological nitrogen fixation
BPH	brown plant hoppers
BSE	Bovine spongiform encephalopathy
Bt	soil bacterium *Bacillus thuringiensis* (usually refers to plants made insecticidal using a variant of various *cry* toxin genes sourced from plasmids of these bacteria)
C	carbon
Ca	calcium
CA	Comprehensive Assessment of Water Management in Agriculture
CAADP	Comprehensive Africa Agriculture Development Program
CBD	Convention on Biological Diversity
CBN	Cassava Biotechnology Net
CBO	Community-based organization
Cd	cadmium
CDM	Clean Development Mechanism
CGE	computable general equilibrium
CGIAR	Consultative Group on International Agricultural Research
CH_4	methane

CIAT	International Center for Tropical Agriculture
CIFOR	Center for International Forestry Research
CIMMYT	International Maize and Wheat Improvement Center
CIP	International Potato Center
CIPAF	Center for Research and Technological Development for small-scale family agriculture
CMD	cassava mosaic disease
CMV	cassava mosaic virus
C:N	carbon to nitrogen ratio
CO_2	carbon dioxide
CO_2-eq	carbon dioxide equivalent
$[CO_2]$	carbon dioxide concentration
COA	certified organic agriculture
Codex	Codex Alimentarius
CoP	cost of production
COP	community of practice
Coprofam	Coordenadora de Organizaciones de Productores Familiares del Mercosur
CORAF	Central African Council for Agricultural Research and Development
CoS	Convergence of Sciences program
CS	carbon sequestration
CSO	civil society organization
Cu	copper
CWANA	Central and West Asia and North Africa
Defra	UK Department of Environment, Food and Rural Affairs
DFID	UK Department of International Development
DNA	deoxyribonucleic acid
dsRNA	double-stranded ribonucleic acid
EFSA	European Food Safety Authority
EJ	Exajoules
Embrapa	Brazilian Agriculture and Livestock Research Company
EPA	US Environmental Protection Agency
ESAP	East and South Asia and the Pacific
ET	evapotranspiration
EU	European Union

EVM	Ethnoveterinary medicine		HYV	High yielding variety
FACE	Free-Air Concentration Enrichment		IA	institutional arrangement
FAO	Food and Agriculture Organization of the United Nations		IAASTD	International Assessment of Agricultural Knowledge, Science and Technology for Development
FARA	Forum for Agricultural Research in Africa		IAM	Integrated assessment model
FDA	US Food and Drug Administration		IARC	International Agricultural Research Center
FDI	Foreign Direct Investment		IAS	invasive alien species
Fe	iron		IBRD	International Bank of Rural Development
FFS	farmer field school		ICA	International Commodity Agreement
FLO	Fair Trade Labeling Organization		ICAR	Indian Council of Agricultural Research
FoB	freight on board		ICARDA	International Center for Agricultural Research in the Dry Areas
FPRE	Farmer participatory research and extension		ICRAF	World Agroforestry Center
FPU	food producing unit		ICRISAT	International Crops Research Institute for Semi-arid Tropics
FSRE	Farming systems research and extension		ICT	information and communication technologies
FTA	Free Trade Agreement		IDA	International Development Agency
g	gram (10^{-3} kg)		IEA	International Energy Agency
G × E	gene by environment		IFAD	International Fund for Agricultural Development
GBA	Global Biodiversity Assessment		IFC	International Finance Corporation
GCM	general circulation model		IFI	international financial institution
GDP	Gross domestic product		IFOAM	International Federation of Organic Agriculture Movements
GE	genetic engineering/genetically engineered		IFPRI	International Food Policy Research Institute
GEF	Global Environment Facility		IFS	International Food Standard
GFAR	Global Forum on Agricultural Research		IIASA	International Institute for Applied System Analysis
GEO	Global Environment Outlook		IITA	International Institute for Tropical Agriculture
GFS	globalized food system		IK	Indigenous knowledge
GFSI	Global Food Safety Initiative		ILO	International Labour Organisation
Gg	gigagram (10^6 kg)		ILRI	International Livestock Research Institute
Gha	gigahectare (10^9 hectare)		IMF	International Monetary Fund
GHG	greenhouse gas		INM	Integrated Nutrient Management
GHI	Global Hunger Index		INRM	Integrated Natural Resources Management
GIS	geographic information system		IP	intellectual property
GISP	Global Invasive Species Program		IPCC	Intergovernmental Panel on Climate Change
GLASOD	Global assessment of human-induced soil degradation		IPGRI	Bioversity International
GM	genetically modified/genetic modification		IPM	Integrated pest management
GMO	genetically modified organism		IPPC	International Plant Protection Convention
GNP	Gross National Product		IPR	intellectual property rights
GPS	global positioning system		IR	insect resistant
GR	Green revolution		IRR	internal rate of return
GSG	Global Scenarios Group		IRRI	International Rice Research Institute
GSPC	Global Strategy of Plant Conservation		IS	innovation systems
Gt	gigaton/gigatonne; 10^{19} tonnes		ISNM	Integrated soil and nutrient management
GURT	Genetic Use of Restriction Technologies		ISPM	International sanitary and phytosanitary measure
GWP	global warming potential		ITU	International Telecommunications Union
ha	hectare (10^4 m^2)		IWM	Integrated Weed Management
HACCP	Hazard Analysis Critical Control Point			
HI	harvest index			
HIV	Human immunodeficiency virus			
HR	herbicide resistant			
HRC	herbicide resistant crop			
HT	herbicide tolerant			

IWMI	International Water Management Institute	PA	precision agriculture
IWRM	Integrated water resources management	PCR	Polymerase chain reaction
IWSR	irrigation water supply reliability	PE	partial equilibrium
K	potassium	PES	Payments for environmental services
kcal	kilocalorie	PGRFA	Plant Genetic Resources for Food and Agriculture
kg	kilogram, 10^3 grams	PIPRA	Public-Sector Intellectual Property Resource for Agriculture
km	kilometer		
kWh	kilowatt hour	PPB	Participatory plant breeding
LAC	Latin America and the Caribbean	ppm	parts per million
LDC	least developed countries	ppmv	parts per million by volume
LEISA	Low-External Input Sustainable Agriculture	PPP	Purchasing Power Parity
LIC	low income countries	PPQ	plant protection and quarantine
LFS	local food system	PVP	plant variety protection
LLMA	Locally Managed Marine Area	QPM	quality protein maize
LTE	long-term cropping system experiments	QTL	Quantitative Trait Loci
LUC	land use change	R&D	research and development
m	10^2 cm	RCT	Resource-conserving technologies
MA	Millennium Ecosystem Assessment	RISE	Response Inducing Sustainability Evaluation
MAB/S	marker assisted breeding/selection	RNA	ribonucleic acid
MASIPAG	Farmer-Scientist Partnership for Development, Inc	ROPPA	Réseau des organisations paysannes et des producteurs d'Afrique de l'Ouest
MDG	Millennium Development Goals	ROR	rates of return
Mg	magnesium	RRA	rapid rural appraisal
mg	milligram (10^{-3} grams)	RS	remote sensing
MIGA	Multilateral Investment Agency	S&T	science and technology
MLS	multilateral system	SAP	Structural adjustment policies
MRL	maximum residue level	SEARICE	South East Asian Regional Initiatives for Community Empowerment
MSA	mean species abundance		
MV	Modern variety	SIA	Strategic Impact Assessment
N	nitrogen	SODP	Seeds of Development Program
NAE	North America and Europe	SPIA	Standing Panel on Impact Assessment
NAFTA	North American Free Trade Agreement	SPLT	Substantive Patent Law Treaty
NARI	National agricultural research institute	SPS	Sanitary and Phytosanitary
NARS	national agricultural research systems	SRES	Special Report on Emission Scenarios
NBF	National Biosafety Frameworks	SSA	Sub-Saharan Africa
NEPAD	New Partnership for Africa's Development	SSNM	site specific nutrient management
ng	nanogram (10^{-9} grams)	STE	state trading enterprise
NGO	nongovernmental organization	TFP	Total Factor Productivity
N_2O	nitrous oxide	TG	Technogarden scenario
NPK	nitrogen, phosphorus, potassium	TGA	third generation agriculture
NPV	net present value of benefits	TK	traditional knowledge
NRM	Natural resource management	tonne	10^3 kg (metric ton)
NTFP	non-timber forest product	ToT	Transfer of Technology
NUE	nitrogen use efficiency	TRIPS	Trade-Related Aspects of Intellectual Property Rights
O_3	ozone		
OA	organic agriculture	T&V	training and visit
ODA	overseas development assistance	TV	Traditional variety
OECD	Organization of Economic Cooperation and Development	UEMOA	Union Economique et Monétaire Ouest-Africaine
OH	hydroxyl	UNCBD	UN Convention on Biodiversity
OIE	World Animal Health Organization	UNCCD	UN Commission to Combat Desertification
p	phosphorus		

UNCED	UN Conference on Environment and Development	WARDA	Africa Rice Center
UNCTAD	UN Conference on Trade and Development	WHO	World Health Organization
UNDP	United Nations Development Program	WIPO	World Intellectual Property Organization
UNEP	United Nations Environment Programme	WP	water productivity
UNFCCC	United Nations Framework Convention on Climate Change	WRI	World Resources Institute
UPOV	International Union for the Protection of New Varieties of Plants	WSSD	World Summit on Sustainable Development
		WUA	Water User Association
USDA	U.S. Department of Agriculture	WUE	water use efficiency
UV	ultraviolet	WWF	World Wildlife Fund
		yr	year
		Zn	zinc

Steering Committee for Consultative Process and Advisory Bureau for Assessment

Steering Committee

The Steering Committee was established to oversee the consultative process and recommend whether an international assessment was needed, and if so, what was the goal, the scope, the expected outputs and outcomes, governance and management structure, location of the secretariat and funding strategy.

Co-chairs

Louise Fresco, Assistant Director General for Agriculture, FAO
Seyfu Ketema, Executive Secretary, Association for Strengthening Agricultural Research in East and Central Africa (ASARECA)
Claudia Martinez Zuleta, Former Deputy Minister of the Environment, Colombia
Rita Sharma, Principal Secretary and Rural Infrastructure Commissioner, Government of Uttar Pradesh, India
Robert T. Watson, Chief Scientist, The World Bank

Nongovernmental Organizations

Benny Haerlin, Advisor, Greenpeace International
Marcia Ishii-Eiteman, Senior Scientist, Pesticide Action Network North America Regional Center (PANNA)
Monica Kapiriri, Regional Program Officer for NGO Enhancement and Rural Development, Aga Khan
Raymond C. Offenheiser, President, Oxfam America
Daniel Rodriguez, International Technology Development Group (ITDG), Latin America Regional Office, Peru

UN Bodies

Ivar Baste, Chief, Environment Assessment Branch, UN Environment Programme
Wim van Eck, Senior Advisor, Sustainable Development and Healthy Environments, World Health Organization
Joke Waller-Hunter, Executive Secretary, UN Framework Convention on Climate Change
Hamdallah Zedan, Executive Secretary, UN Convention on Biological Diversity

At-large Scientists

Adrienne Clarke, Laureate Professor, School of Botany, University of Melbourne, Australia
Denis Lucey, Professor of Food Economics, Dept. of Food Business & Development, University College Cork, Ireland, and Vice-President NATURA
Vo-tong Xuan, Rector, Angiang University, Vietnam

Private Sector

Momtaz Faruki Chowdhury, Director, Agribusiness Center for Competitiveness and Enterprise Development, Bangladesh
Sam Dryden, Managing Director, Emergent Genetics
David Evans, Former Head of Research and Technology, Syngenta International
Steve Parry, Sustainable Agriculture Research and Development Program Leader, Unilever
Mumeka M. Wright, Director, Bimzi Ltd., Zambia

Consumer Groups

Michael Hansen, Consumers International
Greg Jaffe, Director, Biotechnology Project, Center for Science in the Public Interest
Samuel Ochieng, Chief Executive, Consumer Information Network

Producer Groups

Mercy Karanja, Chief Executive Officer, Kenya National Farmers' Union
Prabha Mahale, World Board, International Federation Organic Agriculture Movements (IFOAM)
Tsakani Ngomane, Director Agricultural Extension Services, Department of Agriculture, Limpopo Province, Republic of South Africa
Armando Paredes, Presidente, Consejo Nacional Agropecuario (CNA)

Scientific Organizations

Jorge Ardila Vásquez, Director Area of Technology and Innovation, Inter-American Institute for Cooperation on Agriculture (IICA)
Samuel Bruce-Oliver, NARS Senior Fellow, Global Forum for Agricultural Research Secretariat
Adel El-Beltagy, Chair, Center Directors Committee, Consultative Group on International Agricultural Research (CGIAR)
Carl Greenidge, Director, Center for Rural and Technical Cooperation, Netherlands
Mohamed Hassan, Executive Director, Third World Academy of Sciences (TWAS)
Mark Holderness, Head Crop and Pest Management, CAB International
Charlotte Johnson-Welch, Public Health and Gender Specialist and Nata Duvvury, Director Social Conflict and Transformation Team, International Center for Research on Women (ICRW)
Thomas Rosswall, Executive Director, International Council for Science (ICSU)

Judi Wakhungu, Executive Director, African Center for Technology Studies

Governments

Australia: Peter Core, Director, Australian Centre for International Agricultural Research

China: Keming Qian, Director General Inst. Agricultural Economics, Dept. of International Cooperation, Chinese Academy of Agricultural Science

Finland: Tiina Huvio, Senior Advisor, Agriculture and Rural Development, Ministry of Foreign Affairs

France: Alain Derevier, Senior Advisor, Research for Sustainable Development, Ministry of Foreign Affairs

Germany: Hans-Jochen de Haas, Head, Agricultural and Rural Development, Federal Ministry of Economic Cooperation and Development (BMZ)

Hungary: Zoltan Bedo, Director, Agricultural Research Institute, Hungarian Academy of Sciences

Ireland: Aidan O'Driscoll, Assistant Secretary General, Department of Agriculture and Food

Morocco: Hamid Narjisse, Director General, INRA

Russia: Eugenia Serova, Head, Agrarian Policy Division, Institute for Economy in Transition

Uganda: Grace Akello, Minister of State for Northern Uganda Rehabilitation

United Kingdom Paul Spray, Head of Research, DFID

United States: Rodney Brown, Deputy Under Secretary of Agriculture and Hans Klemm, Director of the Office of Agriculture, Biotechnology and Textile Trade Affairs, Department of State

Foundations and Unions

Susan Sechler, Senior Advisor on Biotechnology Policy, Rockefeller Foundation

Achim Steiner, Director General, The World Conservation Union (IUCN)

Eugene Terry, Director, African Agricultural Technology Foundation

Advisory Bureau

Non-government Representatives

Consumer Groups

Jaime Delgado • Asociación Peruana de Consumidores y Usuarios
Greg Jaffe • Center for Science in the Public Interest
Catherine Rutivi • Consumers International
Indrani Thuraisingham • Southeast Asia Council for Food
    Security and Trade
Jose Vargas Niello • Consumers International Chile

International organizations

Nata Duvvury • International Center for Research on Women
Emile Frison • CGIAR
Mark Holderness • GFAR
Mohamed Hassan • Third World Academy of Sciences
Jeffrey McNeely • World Conservation Union (IUCN)
Dennis Rangi • CAB International
John Stewart • International Council of Science (ICSU)

NGOs

Kevin Akoyi • Vredeseilanden
Hedia Baccar • Association pour la Protection de l'Environment
    de Kairouan
Benedikt Haerlin • Greenpeace International
Juan Lopez • Friends of the Earth International
Khadouja Mellouli • Women for Sustainable Development
Patrick Mulvaney • Practical Action
Romeo Quihano • Pesticide Action Network
Maryam Rahmaniam • CENESTA
Daniel Rodriguez • International Technology Development Group

Private Sector

Momtaz Chowdhury • Agrobased Technology and Industry
    Development
Giselle L. D'Almeida • Interface
Eva Maria Erisgen • BASF
Armando Paredes • Consejo Nacional Agropecuario
Steve Parry • Unilever
Harry Swaine • Syngenta (resigned)

Producer Groups

Shoaib Aziz • Sustainable Agriculture Action Group of Pakistan
Philip Kiriro • East African Farmers Federation
Kristie Knoll • Knoll Farms
Prabha Mahale • International Federation of Organic Agriculture
    Movements

Anita Morales • Apit Tako
Nizam Selim • Pioneer Hatchery

Government Representatives

Central and West Asia and North Africa

Egypt • Ahlam Al Naggar
Iran • Hossein Askari
Kyrgyz Republic • Djamin Akimaliev
Saudi Arabia • Abdu Al Assiri, Taqi Elldeen Adar, Khalid Al
    Ghamedi
Turkey • Yalcin Kaya, Mesut Keser

East and South Asia and the Pacific

Australia • Simon Hearn
China • Puyun Yang
India • PK Joshi
Japan • Ryuko Inoue
Philippines • William Medrano

Latin America and Caribbean

Brazil • Sebastiao Barbosa, Alexandre Cardoso, Paulo Roberto
    Galerani, Rubens Nodari
Dominican Republic • Rafael Perez Duvergé
Honduras • Arturo Galo, Roberto Villeda Toledo
Uruguay • Mario Allegri

North America and Europe

Austria • Hedwig Woegerbauer
Canada • Iain MacGillivray
Finland • Marja-Liisa Tapio-Bistrom
France • Michel Dodet
Ireland • Aidan O'Driscoll, Tony Smith
Russia • Eugenia Serova, Sergey Alexanian
United Kingdom • Jim Harvey, David Howlett, John Barret
United States • Christian Foster

Sub-Saharan Africa

Benin • Jean Claude Codjia
Gambia • Sulayman Trawally
Kenya • Evans Mwangi
Mozambique • Alsácia Atanásio, Júlio Mchola
Namibia • Gillian Maggs-Kölling
Senegal • Ibrahim Diouck

Secretariat and Cosponsor Focal Points

Secretariat

World Bank
Marianne Cabraal, Leonila Castillo, Jodi Horton, Betsi Isay, Pekka Jamsen, Pedro Marques, Beverly McIntyre, Wubi Mekonnen, June Remy

UNEP
Marcus Lee, Nalini Sharma, Anna Stabrawa

UNESCO
Guillen Calvo

With special thanks to the Publications team: Audrey Ringler (logo design), Pedro Marques (proofing and graphics), Ketill Berger and Eric Fuller (graphic design)

Regional Institutes

Sub-Saharan Africa – African Centre for Technology Studies (ACTS)
Ronald Ajengo, Elvin Nyukuri, Judi Wakhungu

Central and West Asia and North Africa – International Center for Agricultural Research in the Dry Areas (ICARDA)
Mustapha Guellouz, Lamis Makhoul, Caroline Msrieh-Seropian, Ahmed Sidahmed, Cathy Farnworth

Latin America and the Caribbean – Inter-American Institute for Cooperation on Agriculture (IICA)
Enrique Alarcon, Jorge Ardila Vásquez, Viviana Chacon, Johana Rodríguez, Gustavo Sain

East and South Asia and the Pacific – WorldFish Center
Karen Khoo, Siew Hua Koh, Li Ping Ng, Jamie Oliver, Prem Chandran Venugopalan

Cosponsor Focal Points

GEF	Mark Zimsky
UNDP	Philip Dobie
UNEP	Ivar Baste
UNESCO	Salvatore Arico, Walter Erdelen
WHO	Jorgen Schlundt
World Bank	Mark Cackler, Kevin Cleaver, Eija Pehu, Juergen Voegele

Annex G
Reservations on Global Report

Australia: Australia recognizes the IAASTD initiative and reports as a timely and important multistakeholder and multidisciplinary exercise designed to assess and enhance the role of AKST in meeting the global development challenges. The wide range of observations and views presented however, are such that Australia cannot agree with all assertions and options in the report. The report is therefore noted as a useful contribution which will be used for considering the future priorities and scope of AKST in securing economic growth and the alleviation of hunger and poverty.

Canada: In recognizing the important and significant work undertaken by IAASTD authors, Secretariat and stakeholders on the background Reports, the Canadian Government notes these documents as a valuable and important contribution to policy debate which needs to continue in national and international processes. While acknowledging the valuable contribution these Reports provide to our understanding on agricultural knowledge, science and technology for development, there remain numerous areas of concern in terms of balanced presentation, policy suggestions and other assertions and ambiguities. Nonetheless, the Canadian Government advocates these reports be drawn to the attention of governments for consideration in addressing the importance of AKST and its large potential to contribute to economic growth and the reduction of hunger and poverty.

United States of America: The United States joins consensus with other governments in the critical importance of AKST to meet the goals of the IAASTD. We commend the tireless efforts of the authors, editors, Co-Chairs and the Secretariat. We welcome the IAASTD for bringing together the widest array of stakeholders for the first time in an initiative of this magnitude. We respect the wide diversity of views and healthy debate that took place.

As we have specific and substantive concerns in each of the reports, the United States is unable to provide unqualified endorsement of the reports, and we have noted them.

The United States believes the Assessment has potential for stimulating further deliberation and research. Further, we acknowledge the reports are a useful contribution for consideration by governments of the role of AKST in raising sustainable economic growth and alleviating hunger and poverty.

Index

Note: Any italicized page number is for a page that has a table, box, or figure.

About Island Press

Since 1984, the nonprofit Island Press has been stimulating, shaping, and communicating the ideas that are essential for solving environmental problems worldwide. With more than 800 titles in print and some 40 new releases each year, we are the nation's leading publisher on environmental issues. We identify innovative thinkers and emerging trends in the environmental field. We work with world-renowned experts and authors to develop cross-disciplinary solutions to environmental challenges.

Island Press designs and implements coordinated book publication campaigns in order to communicate our critical messages in print, in person, and online using the latest technologies, programs, and the media. Our goal: to reach targeted audiences—scientists, policymakers, environmental advocates, the media, and concerned citizens—who can and will take action to protect the plants and animals that enrich our world, the ecosystems we need to survive, the water we drink, and the air we breathe.

Island Press gratefully acknowledges the support of its work by the Agua Fund, Inc., Annenberg Foundation, The Christensen Fund, The Nathan Cummings Foundation, The Geraldine R. Dodge Foundation, Doris Duke Charitable Foundation, The Educational Foundation of America, Betsy and Jesse Fink Foundation, The William and Flora Hewlett Foundation, The Kendeda Fund, The Andrew W. Mellon Foundation, The Curtis and Edith Munson Foundation, Oak Foundation, The Overbrook Foundation, the David and Lucile Packard Foundation, The Summit Fund of Washington, Trust for Architectural Easements, Wallace Global Fund, The Winslow Foundation, and other generous donors.

The opinions expressed in this book are those of the author(s) and do not necessarily reflect the views of our donors.